COMPREHENSIVE POLYMER SCIENCE

IN 7 VOLUMES

COMPREHENSIVE POLYMER SCIENCE

The Synthesis, Characterization, Reactions & Applications of Polymers

CHAIRMAN OF THE EDITORIAL BOARD

SIR GEOFFREY ALLEN, FRS

Unilever Research and Engineering, London, UK

DEPUTY CHAIRMAN OF THE EDITORIAL BOARD

JOHN C. BEVINGTON

University of Lancaster, UK

Volume 2

Polymer Properties

VOLUME EDITORS

COLIN BOOTH & COLIN PRICE

University of Manchester, UK

PERGAMON PRESS

**OXFORD · NEW YORK · BEIJING · FRANKFURT
SÃO PAULO · SYDNEY · TOKYO · TORONTO**

U.K.	Pergamon Press plc, Headington Hill Hall, Oxford OX3 0BW, England
U.S.A.	Pergamon Press, Inc., Maxwell House, Fairview Park, Elmsford, New York 10523, U.S.A.
PEOPLE'S REPUBLIC OF CHINA	Pergamon Press, Room 4037, Qianmen Hotel, Beijing, People's Republic of China
FEDERAL REPUBLIC OF GERMANY	Pergamon Press GmbH, Hammerweg 6, D-6242 Kronberg, Federal Republic of Germany
BRAZIL	Pergamon Editora Ltda, Rua Eça de Queiros, 346, CEP 04011, Paraiso, São Paulo, Brazil
AUSTRALIA	Pergamon Press (Australia) Pty Ltd., P.O. Box 544, Potts Point, NSW 2011, Australia
JAPAN	Pergamon Press, 5th Floor, Matsuoka Central Building, 1-7-1 Nishishinjuku, Shinjuku-ku, Tokyo 160, Japan
CANADA	Pergamon Press Canada Ltd., Suite No. 271, 253 College Street, Toronto, Ontario, Canada M5T 1R5

First edition 1989

Library of Congress Cataloging-in-Publication Data

Comprehensive polymer science.

Includes index.
Contents: v. 1. Polymer characterization/volume editors. Colin Booth & Colin Price—v. 2. Polymer properties/volume editors, Colin Booth & Colin Price—[etc.]—v. 7. Specialty polymers & polymer processing/volume editor, Sundar L. Aggarwal.
1. Polymers and polymerization. I. Allen, G. (Geoffrey), 1928– . II. Bevington, J. C.
QD381.C66 1988   547.7   88–25548

British Library Cataloguing in Publication Data

Comprehensive polymer science.

Vol. 2: Polymer Properties.

1. Polymers
I. Allen, Geoffrey   II. Bevington, John C.
III. Booth, Colin   IV. Price, Colin
547.7

ISBN 0-08-036206-0 (vol. 2)
ISBN 0-08-032515-7 (set)

Printed in Great Britain by A. Wheaton & Co. Ltd., Exeter

Contents

Preface

It is only 60 years since Staudinger's model of the molecular nature of a polymer was becoming universally accepted and the physical states of rubbers, plastics and fibres understood. Unfortunately, for some time many academic chemists continued not to appreciate the full significance of polymerization reactions and physicists tended to regard polymeric materials as inevitably being of indeterminate composition and unamenable to study by conventional physical methods.

Nevertheless, in the 1930s the foundations were laid for the understanding of the main polymerization mechanisms. An industry based on synthetic rubbers, plastics and fibres was soon established. In World War II it played a major strategic role and afterwards grew to be one of the main elements of the heavy chemicals industry. It became recognized that synthetics may be superior to natural materials in their properties and that they may be used for completely new purposes.

Alongside the production of well-defined materials there grew the ability to characterize the structure of polymer molecules and to understand the relationships between methods of preparation and subsequent treatment, structure and properties, both chemical and physical. As a result, a vast literature of polymer science and technology has been generated and four Nobel prizes awarded specifically for contributions to polymer science. Add to this the fact that many biological molecules, including polypeptides, enzymes, antibodies, carbohydrates and so on, are polymers of varying degrees of complexity, then the universality of polymers in the physical and biological sciences and technologies forms a dominant modern theme.

Comprehensive Polymer Science is a series of volumes designed to set down the structure of this vast subject in such a way that researchers and teachers of polymer science and workers in associated fields can find an authoritative and comprehensive account of the topic of immediate interest. That topic is set out in a framework of related subjects. The text is focused on synthetic polymers with little reference to biological macromolecules *per se* but the science underpins both physical and biological systems.

To ensure that the wide coverage is maintained at an authoritative level, more than 250 authors from 20 countries have been enlisted. Their contributions have been organized into a series of major themes:

Volume 1	Polymer Characterization
Volume 2	Polymer Properties
Volumes 3–5	Polymerization Mechanisms
Volume 6	Polymer Reactions
Volume 7	Specialty Polymers & Polymer Processing

Because of the wide coverage the editors were presented with a particularly difficult decision with regard to symbols and nomenclature. The latter does not follow strictly the recommendations of IUPAC nor are symbols consistent throughout the whole work. However, usage in a particular chapter is consistent with the practice in the current literature. Thus a reader will be able to frame new publications in the context of the information presented in this series of volumes.

We should like to acknowledge the way in which the staff at the publisher, particularly Dr Colin Drayton (who initially proposed the project), Dr Helen McPherson and their editorial team, have supported the editors and authors in their endeavour to produce a text that is both complete and up-to-date and that will appeal to industrial and academic researchers alike. *Comprehensive Polymer Science* is a milestone in the literature of the subject in terms of coverage, clarity and a sustained high level of presentation.

GEOFFREY ALLEN
London

JOHN C. BEVINGTON
Lancaster

Contributors to Volume 2

Professor A. Abe
Department of Polymer Chemistry, Tokyo Institute of Technology, 2-12-1 O-okayama, Meguro-ku, Tokyo 152, Japan

Professor R. C. Armstrong
Department of Chemical Engineering, Massachusetts Institute of Technology, Cambridge, MA 02139-4307, USA

Professor D. C. Bassett
Department of Physics, J. J. Thompson Physical Laboratory, University of Reading, Whiteknights, PO Box 220, Reading RG6 2AF, UK

Dr G. C. Berry
Department of Chemistry, Carnegie Mellon University, 4400 Fifth Avenue, Pittsburgh, PA 15213, USA

Professor R. B. Bird
Department of Chemical Engineering, University of Wisconsin-Madison, 1415 Johnson Drive, Madison, WI 53706-1691, USA

Professor D. Bloor
Department of Chemistry, Queen Mary College, Mile End Road, London E1 4NS, UK

Dr D. Briggs
ICI Plc, Petrochemicals and Plastics Division, PO Box 90, Wilton Centre, Middlesbrough, Cleveland TS6 8JE, UK

Dr B. J. Briscoe
Department of Chemical Engineering, Imperial College, Prince Consort Road, London SW7 2BY, UK

Dr R. A. Brown
Department of Chemistry, University of Manchester, Manchester M13 9PL, UK

Dr S. J. Candau
Laboratoire de Spectrométrie et d'Imagerie Ultrasonores, Université Louis Pasteur, Unité Associée au CNRS n° 851, 4 rue Blaise Pascal, F-67070 Strasbourg Cedex, France

Professor E. F. Casassa
Department of Chemistry, Carnegie Mellon University, 4400 Fifth Avenue, Pittsburgh, PA 15213, USA

Professor C. F. Curtiss
Department of Chemistry, University of Wisconsin-Madison, 1101 University Avenue, Madison, WI 53706-1396, USA

Professor Sir Sam Edwards
Cavendish Laboratory, University of Cambridge, Madingley Road, Cambridge CB3 OHE, UK

Dr P. Gradin
PGI (Swedish Plastics and Rubber Institute), Fabriksgatan 32, S851 71 Sundsvall, Sweden

Dr O. Hassager
Instituttet fur Kemiteknik, Danmarks tekniske Højskole, DK-2400, Lyngby, Denmark

Dr P. G. Howgate
PGI (Swedish Plastics and Rubber Institute), Fabriksgatan 32, S851 71 Sundsvall, Sweden

Dr J.-F. Joanny
Ecole Normale Supérieure de Lyon, 46 Allée d'Italie, F-69364 Lyon Cedex 07, France

Dr C. W. Lantman
Mobay Corporation, Mobay Road, Pittsburgh, PA 15205-9741, USA

Dr R. D. Lundberg
Scientific Advisor, Paramins, Technology Division, EXXON Chemical Company, PO Box 536, Linden, NJ 07036, USA

Dr G. B. McKenna
Polymers Division, Bldg. 224, Room A209, National Bureau of Standards, Gaithersburg, MD 20899, USA

Professor W. J. MacKnight
Department of Polymer Science and Engineering, University of Massachusetts, Amherst, MA 01003, USA

Dr L. Mandelkern
Department of Chemistry and Institute of Molecular Biophysics, The Florida State University, Tallahassee, FL 32306, USA

Professor J. E. Mark
Department of Chemistry and Polymer Research Centre, University of Cincinnati, Cincinnati, OH 45221, USA

Dr A. J. Masters
Department of Chemistry, University of Manchester, Manchester M13 9PL, UK

Professor M. Muthukumar
Institute for Polymer Science and Technology, University of Massachusetts, Amherst, MA 01003, USA

Dr T. de V. Naylor
BP Research Centre, Chertsey Road, Sunbury-on-Thames, Middlesex TW16 7NL, UK

Professor S. Nomura
Department of Polymer Science and Engineering, Faculty of Textile Science, Kyoto Institute of Technology, Matsugasaki, Sakyo-ku, Kyoto 606, Japan

Dr J. R. Owen
Department of Chemistry, University of Salford, Salford M5 4WT, UK

Professor R. A. Pethrick
Department of Pure and Applied Science, University of Strathclyde, Thomas Graham Building, 295 Cathedral Street, Glasgow G1 1XL, UK

Dr C. Price
Department of Chemistry, University of Manchester, Manchester M13 9PL, UK

Dr J.-P. Queslel
Michelin Manufacture Francaise des Pneumatiques Michelin, 63 Clermont-Ferrand, France

Dr D. G. Rance
ICI Plc, Petrochemicals and Plastics Division, PO Box 90, Wilton Centre, Middlesbrough, Cleveland, TS6 8JE

Dr B. W. Ready
Department of Chemistry, University of Essex, Wivenhoe Park, Colchester CO4 3SQ, UK

Dr I. D. Robb
Unilever Research Laboratory, Port Sunlight, Bebington, Wirral, Merseyside L63 3JW, UK

Dr R. Seldén
PGI (Swedish Plastics and Rubber Institute), Fabriksgatan 32, S851 71 Sundsvall, Sweden

Professor R. I. Tanner
Department of Mechanical Engineering, University of Sydney, Sydney, New South Wales 2006, Australia

Dr J. G. Tomka
Department of Textile Industries, University of Leeds, Leeds LS2 9JT, UK

Dr R. Van der Haegen
DSM Research BV, PO Box 18, 6160 MD Geleen, The Netherlands

Dr A. S. Vaughan
Central Electricity Research Laboratories, Leatherhead, Surrey KT22 7SE, UK

Dr D. J. Walsh
CR and D 356/135, Experimental Station, E. I. Du Pont de Nemours & Co., Wilmington, DE 19898, USA

Professor G. Williams
Edward Davies Chemical Laboratories, University College of Wales, Aberystwyth, Dyfed SY23 1NE, UK

Professor R. J. Young
Department of Polymer Science and Technology, University of Manchester Institute of Science and Technology, Manchester M60 1QD, UK

Dr X. F. Yuan
Department of Chemistry, University of Manchester, Manchester M13 9PL, UK

Contents of All Volumes

Volume 1 Polymer Characterization

Volume 2   Polymer Properties

Volume 3   Chain Polymerization I

Volume 4   Chain Polymerization II

1

Chain Statistics and Scaling Concepts

M. MUTHUKUMAR
University of Massachusetts, Amherst, MA, USA
and
S. F. EDWARDS
University of Cambridge, UK

1.1 INTRODUCTION

In this article, we will focus on the various conceptual ideas used to explain a variety of global properties of polymeric systems with details of conformational aspects being dealt with in Chapter 2 of this volume by Abe. What do we mean by global properties?

Consider, for example, a polymer represented by (**1**) where n can be as large as 10^5. Depending on the chemical nature of R, conformational states[1-5] with differing energies arise at a local level. For example, there are three conformational states (rotational isomers) at every C—C bond. These states and the torsional energy as a function of rotation about the middle C—C bond are represented in Figure 1.

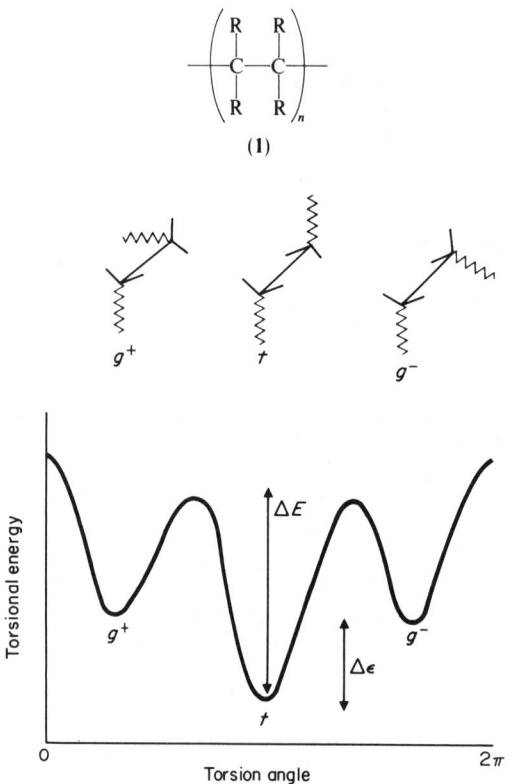

Figure 1 The three conformational states and a plot (schematic) of the torsional energy as a function of the rotational angle about the middle C–C bond; $\Delta\varepsilon$ and ΔE set the scales for local and dynamical inflexibilities, respectively

If $\Delta\varepsilon \ll k_B T$ (k_B is Boltzman's constant, T is the absolute temperature), there exists complete static flexibility. For higher values of $\Delta\varepsilon/k_B T$, there is a local preference for the *trans* conformation in comparison to g^+ and g^- states. Nevertheless, at a large scale the chain will still look like a flexible one. Depending on the value of $\Delta\varepsilon/k_B T$, we can define a length called persistence length l over which the local inflexibility persists. But, beyond this length, the conformations are totally uncorrelated. The parameter which determines the overall chain flexibility is l/L (L is the chain contour length). If $l/L \ll 1$, the chain has complete static flexibility; for $l/L > 1$, the chain is a rigid rod.

Similarly, $\Delta E/k_B T$ determines the dynamical flexibility. If $\Delta E/k_B T \ll 1$, the time $\tau \ [\sim \exp(\Delta E/k_B T)]$ required for the isomerizations $t \rightleftharpoons g^\pm$ is in picoseconds. That is, the chain is dynamically flexible. For higher values of $\Delta E/k_B T$, dynamical inflexibility arises locally. However, if we are looking at large scale motions of the chain involving times much greater than τ (or frequency, $\omega \ll \tau^{-1}$), the chain can be taken to be dynamically flexible.

We will focus in this chapter only on those properties of polymer chains which do not depend on local properties and therefore restrict ourselves to very large characteristic lengths and small frequencies. Such properties which are independent of the local static and dynamic details are the global properties. By global, we mean the universal properties which are independent of the monomer structure, nature of the solvent, *etc*. In the present discussions, the local properties will appear only through phenomenological parameters. It has been found that in the appropriate variables all macroscopic polymer properties can be plotted on universal curves so that, although the detail of polymer structure is essential to calculate the glass temperature, the friction coefficients required in viscoelastic theory and related spatial parameters, if these are taken as given parameters, the theory is then universal.

1.2 SINGLE CHAIN STATISTICS

If *l* is the distance along the chain contour, beyond which the conformations are uncorrelated, the chain configuration can be imagined to be a series of steps of length *l*. Alternatively, the chain can be taken to be a string of segments each of length *l*. (Clearly, each segment may contain many monomers.) This results in the simple Gaussian chain model. In addition, the monomers of the chain interact among themselves and the molecules of the solvent (in which the polymer is present) through short range interactions. This interaction is generically called the *excluded volume interaction* due to the fact that two monomers cannot occupy the same volume at the same time.

1.2.1 Gaussian Chain Model

Assume that a chain is composed of $n + 1$ segments joined consecutively and numbered 0, 1, 2, ... n. Let r_j be the position vector of the centroid position of the jth segment. Assuming that (i) the magnitude and direction of each bond vector is independent of those of all the preceding bond vectors, and (ii) there are no potential interactions other than the connectivity of the segments, then the probability distribution function G_0 of the end-to-end distance $\boldsymbol{R}$ of this Gaussian model chain in three dimensions is given by[6-10]

$$G_0 = \prod_{i=1}^{n} \int dr_i^3 \left(\frac{3}{2\pi l^2}\right)^{3/2} e^{-\frac{3}{2l^2}\sum_{i=1}^{n}(r_i - r_{i-1})^2} \delta^3(r_n - r_0 - \boldsymbol{R}) \tag{1}$$

The root mean square centroid separation is *l*, the Kuhn step length. $\delta^3(r)$ is the three-dimensional Dirac delta function. The distribution G_0 can be evaluated to be

$$G_0(\boldsymbol{R}, L) = \left(\frac{3}{2\pi L l}\right)^{3/2} e^{-\frac{3R^2}{2Ll}} \tag{2}$$

where $L(= nl)$ is the contour length of the chain.

1.2.1.1 Average size

The mean square end-to-end distance of the Gaussian chain follows from equation (2) as

$$\langle R^2 \rangle_0 = \int d^3 R R^2 G_0(\boldsymbol{R}, L) \bigg/ \int d^3 R G_0(\boldsymbol{R}, L) = Ll \tag{3}$$

Another quantity which measures the size of the chain is the mean square radius of gyration, R_g, defined by[1,6]

$$\langle R_g^2 \rangle_0 = \frac{1}{2n^2} \left\langle \sum_i \sum_j R_{ij}^2 \right\rangle \tag{4}$$

where the angular brackets indicate the averaging over all possible chain configurations. *i* and *j* represent the segment labels. $\boldsymbol{R}_{ij}$ is the distance vector between the *i* and *j* segments. If *i* and *j* are far apart along the contour of the chain, then the probability distribution function of $\boldsymbol{R}_{ij}$ and the mean square segment-to-segment distance are given by

$$G_0(\boldsymbol{R}_{ij}) = \left(\frac{3}{2\pi l^2 |i - j|}\right)^{3/2} e^{-\frac{3R_{ij}^2}{2l^2|i - j|}}$$

$$\langle R_{ij}^2 \rangle_0 = l^2 |i - j| \tag{5}$$

It follows from equations (4) and (5) that

$$R_g^2 \;\equiv\; \langle R_g^2 \rangle_0 \;=\; Ll^2/6 \;=\; \langle R^2 \rangle_0/6 \tag{6}$$

1.2.1.2 Static structure factor, S(k), the Debye function[6,7,11]

Figure 2   Schematic scattering experiment with either light, X-ray or neutrons: $k \simeq (4\pi/\lambda)\sin(\theta/2)$, where λ is the wave length of the incident beam

The spectral density of the scattered field in a typical experiment of Figure 2 is proportional to $S(k)$,

$$S(k) \;\equiv\; \frac{1}{n^2}\Big\langle \sum_{i,j} e^{i\boldsymbol{k}\cdot\boldsymbol{R}_{ij}} \Big\rangle \;=\; \frac{1}{n^2}\sum_{ij}\int d^3R_{ij}\,e^{i\boldsymbol{k}\cdot\boldsymbol{R}_{ij}}G_0(\boldsymbol{R}_{ij}) \;=\; \frac{2}{k^4R_g^4}(e^{-k^2R_g^2}-1 \;+\; k^2R_g^2) \tag{7}$$

where equations (5) and (6) have been used. For small k, i.e. $kR_g < 1$

$$S(k) \;=\; 1 \;-\; \frac{R_g^2}{3}k^2 \;+\; 0(k^4)$$

so that the size[12] of the chain can be obtained from the initial slope of the plot of $S(k)$ vs. k^2. For k such that $kR_g > 1$, that is, the typical distance probed in the experiment is smaller than the size of the chain, equation (7) gives

$$S(k) \;=\; 12/k^2Ll \tag{8}$$

Since the monomer pair correlation function is the Fourier transform of $S(k)$, the pair correlation function decays as r^{-1} for distances $r < R_g$.

1.2.1.3 Relation to field theory

It proves very convenient to introduce a Fourier transform of the distribution $G_0(\boldsymbol{R}, L)$ with respect to the position R and a Laplace transform with respect to the length L. We define the characteristic function[6] in space dimension d

$$\hat{G}_0(\boldsymbol{k}, L) \;=\; \int d^dR\,e^{-i\boldsymbol{k}\cdot\boldsymbol{R}}G_0(\boldsymbol{R}, L) \;=\; e^{-k^2Ll/2d} \tag{9}$$

and the propagator

$$\tilde{G}_0(\boldsymbol{k}, E_0) \;=\; \int_0^\infty dL\,e^{-E_0L}\hat{G}_0(\boldsymbol{k}, L) \;=\; 1/(E_0 \;+\; k^2l/2d) \tag{10}$$

In the free field theory,[13] the probability distribution for a real scalar field $\phi(\boldsymbol{r})$ is given by

$$P\{\phi\} \;=\; e^{-1/2\int d^dr[[\nabla\phi(r)]^2 \;+\; \mu^2\phi^2(r)]} \tag{11}$$

where μ is the mass. Such distributions also arise in magnetic systems where $\mu^2 \propto T - T_c$, where T_c is the critical temperature of a phase transition in the system. The spatial correlation function is

$$G_0(r) \;=\; \int \delta\phi\,\phi(r)\phi(o)P\{\phi\} \Big/ \int \delta\phi\,P\{\phi\} \tag{12}$$

The symbol $\delta\phi$ means integration over all fields, which reduces to simpler terms if the system is in a box when $\delta\phi = \pi d\phi\hat{k}$ where $\phi\hat{k}$ are all of the Fourier components of $\phi(r)$.

Upon Fourier transforming

$$\tilde{G}_0(k) = \int d^d r\, e^{-ik\cdot r} G_0(r) = 1/(k^2 + \mu^2) \tag{13}$$

Also, the correlation length ξ becomes

$$\xi^2 = \int d^d r\, r^2 G_0(r) \Big/ \int d^d r\, G_0(r) = \mu^{-2} \tag{14}$$

Since $\mu \propto |T - T_c|^{1/2}$ in spin systems, equation (14) can be rewritten with the use of the critical exponent v

$$\xi \sim |T - T_c|^{-v}; \quad v = 1/2 \tag{15}$$

The comparison of equations (10) and (13) suggests that μ^2 (mass squared) can be identified with the variable E_0 (Laplace variable conjugate to chain length). Since E_0 has an inverse relation to L, equation (14) can be identified with equation (3) so that we can write

$$\langle R^2 \rangle_0 \sim L^{2v}; \quad v = 1/2 \tag{16}$$

Therefore we may make the formal correspondence between the Gaussian chain and the free field theory. This correspondence is summarized in Table 1.

Table 1 The Relation between Gaussian Chain and the Free Field Theory

Gaussian chain	*Free field theory*
1. Probability distribution function for the end-to-end distance $G_0(R, L)$	Spatial correlation function $\langle \phi(r)\phi(o) \rangle$
2. Fourier–Laplace transform of $G_0(R, L)$, $(E_0 + k^2 l/2d)^{-1}$	Fourier transform of $\langle \phi(r)\phi(o) \rangle (k^2 + \mu^2)^{-1}$
3. E_0	μ^2 (mass squared or $T - T_c$)
4. Mean square end-to-end distance $\langle R^2 \rangle_0 \sim L^{2v}; v = 1/2$	Correlation length squared, ξ^2 $\xi \sim \mu^{-1} \sim |T - T_c|^{-v}; v = 1/2$

1.2.2 Excluded Volume Effect

All the potential energy arising from the interactions of the various non-bonded monomer–monomer molecules, polymer–solvent molecules and the solvent–solvent molecules can be put in terms of a two-point potential function (provided the density of polymer is not so high as to need three-point terms)

$$\text{Potential energy} = \sum_{u \le i < j \le n} \frac{u(R_{ij})}{k_B T} \tag{17}$$

It can be shown that it is valid for many purposes to assume that u is short ranged and is represented by a pseudopotential of strength w, a step justified in the section on scaling later

$$u(R_{ij})/k_B T = w\delta(R_{ij}) \tag{18}$$

The excluded volume parameter w can be interpreted[1,6,7,14] as the angular averaged binary cluster integral for a pair of segments

$$wl^2 \equiv \langle \int d^d r_{ij}[1 - e^{-u(r_{ij})/k_B T}] \rangle \tag{19}$$

Clearly w depends on the nonuniversal nature of the polymer–solvent system and the temperature. The θ temperature is defined to be the temperature where the Gaussian chain statistics are valid due to the apparent cancellations between the various attractive polymer–solvent interactions and the repulsive polymer–polymer interactions. Near θ it is possible to write the temperature dependence of w as

$$w \sim (T - \theta)/T \tag{20}$$

The consequences of the excluded volume effect are summarized below within the mean field, scaling and renormalization group theories.

1.2.2.1   *Flory's formula*[1,7,15]

If R is about the average radius of the polymer coil, the repulsive energy of the chain due to the monomer–monomer interactions over a volume of R^d is proportional to $k_B T w(T) l^2 \rho^2 R^d$, where ρ is the monomer density supposed proportional to n/R_d. Therefore

$$\text{Repulsive energy} \quad \sim \quad k_B T w l^2 n^2 / R^d \tag{21}$$

Thus the excluded volume effect tends to swell the chain which in turn will reduce the entropy according to

$$\text{Entropic energy} \quad = \quad -T k_B ln G_0(R, L) \quad = \quad \frac{3}{2} \frac{k_B T R^2}{\langle R^2 \rangle_0} \quad + \quad \text{term independent of } R \tag{22}$$

The compromise between these two opposing forces is struck by obtaining R_F which minimizes the total free energy (sum of equations 21 and 22)

$$F_{\text{total}}/k_B T \quad = \quad \frac{R^2}{\langle R^2 \rangle_0} \quad + \quad \frac{w L^2}{R^d} \quad + \quad \text{terms independent of } R$$

$$\left(\frac{\partial}{\partial R} F_{\text{total}}/k_B T \right)_{R = R_F} \quad = \quad 0$$

$$R_F \quad \sim \quad L^{3/(d + 2)}$$

$$v \quad = \quad 3/d \quad + \quad 2 \tag{23}$$

This is the Flory's value of the critical exponent v. This exponent is in good, but not perfect, agreement with the experimental results in the asymptotic region of large excluded volume (*i.e.* in good solutions). Writing the excluded volume interaction as a potential, Flory and his co-workers have obtained a mean field expression for the mean square radius of gyration $\langle R_g^2 \rangle$ for arbitrary strength of excluded volume effect, in the limit of very large L

$$\alpha^5 \quad - \quad \alpha^3 \quad = \quad a w L^{1/2}/l^{3/2} \tag{24}$$

where a is some number, $\alpha^2 = \langle R_g^2 \rangle / \langle R_g^2 \rangle_0$. This formula gives an interpolation between $\langle R^2 \rangle \sim L$ for the Gaussian limit (θ solutions) $\langle R^2 \rangle \sim L^{6/5}$ for the self-avoiding limit (good solutions) as in Figure 3.

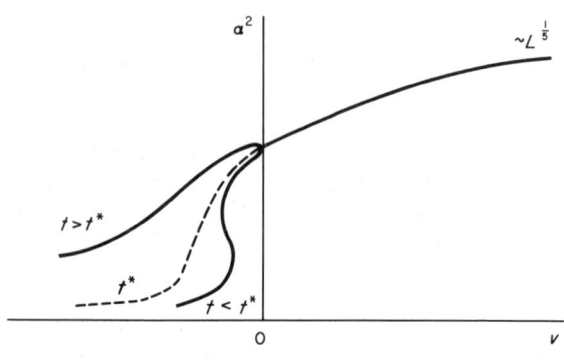

Figure 3   Schematic plot of α *vs.* the dimensionless two-body interaction

In the original formula[1,6] of Flory, $a = 2.6(3/2\pi)^{3/2}$. Stockmayer[16,17] suggested that the choice of $a = 4/3(3/2\pi)^{3/2}$ gives the exact result of first order perturbation theory[6] near θ temperature and the result of equation (23) in good solutions. With this choice of a, equation (24) is called the modified Flory formula which can be derived in a systematic way.[18] Many approximate closed expressions for α exist in the literature which are adequately reviewed.[6,18−27]

1.2.2.2 Note on coil–globule transition

The full theory[28−43] of this is complex and controversial. We will give a brief account. Introduction of three-body interactions into the analysis gives (instead of equation 24)

$$\alpha^5 \;-\; \alpha^3 \;=\; v \;+\; t/\alpha^3 \tag{25}$$

where v and t are the strengths of the two-body and three-body interactions expressed in dimensionless units. This is plotted in Figure 3. Clearly, there is a critical behavior for a set of values of v and t. The coil is in the swollen state for $v > v^*$. If $t > t^*[= 1/4(9/20)^3]$, the chain contracts smoothly as the value of v is lowered. However, the coil size shrinks abruptly at a particular value of $v > v^*$ $(= -27/50\sqrt{5})$ provided $t < t^*$. As shown in Figure 3, one actually has an unstable branch which is typical of van der Waals-like theories. The value of v at the 'coil–globule transition' is obtained by drawing the tie line as in the van der Waals theory of gas–liquid phase transition. When the variational analysis of equation (23) is applied to equation (25) for the globule (the collapsed state)

$$v \;=\; 1/d \tag{26}$$

1.2.2.3 Perturbation theory

There is a mathematical formalism ideally suited to describing the global properties of polymers known as path integration.[9,44] In this formalism, the polymer is described by a position function $\mathbf{r}(s)$ of the arc length s which is the continuum limit of the label n of the nth monomer.

By making a continuum approximation to the sums of equations (1), (17) and (18)

$$\frac{1}{l^2}\sum_{i=1}^{n} (\mathbf{r}_i \;-\; \mathbf{r}_{i-1})2 \;\rightarrow\; \frac{1}{l}\int_0^L ds \left(\frac{\partial \mathbf{r}(s)}{\partial s}\right)^2$$

$$\sum_{0 \le i < j \le n} w\delta(\mathbf{R}_{ij}) \;\rightarrow\; \frac{w}{2}\int_0^L ds \int_0^L ds'\, \delta^d[\mathbf{r}(s) \;-\; \mathbf{r}(s')] \tag{27}$$

The probability distribution function $G(\mathbf{R}, L)$ of the end-to-end distance $\mathbf{R}$ of a chain with the excluded volume interaction is given by the path integral[8,9,14]

$$G(\mathbf{R}, L) \;=\; \int_{\mathbf{r}(0)=0}^{\mathbf{r}(L)=\mathbf{R}} D[\mathbf{r}]\exp\left\{ -\frac{d}{2l}\int_0^L ds \left(\frac{\partial \mathbf{r}(s)}{\partial s}\right)^2 \;-\; \frac{w}{2}\int_0^L ds \int_0^L ds'\, \delta^d[\mathbf{r}(s) - \mathbf{r}(s')] \right\} \tag{28}$$

The expression $\int D[\mathbf{r}]$ means the continuum limit of $\pi d\mathbf{r}_n$. Another way to look at this is to realize that $\int D[\mathbf{r}]$ implies the summation over all possible paths between the ends of the chain $\mathbf{r}(0)$ which is taken to be the origin and $\mathbf{r}(L) = \mathbf{R}$.

Perturbation theory[6,45−54] is made by deriving an expansion for G in terms of w. This is obtained in a very straightforward manner by simply expanding the exponential, $\exp\left\{ -\dfrac{w}{2}\displaystyle\int_0^L ds \int_0^L ds'\, \delta^d[\mathbf{r}(s) - \mathbf{r}(s')] \right\}$. Defining

$$G \;=\; \sum_{m=0}^{\infty} (-w)^m G_m \tag{29}$$

the various averages can be obtained as perturbation series. For example,

$$\langle R^2 \rangle \;=\; \int d^d R R^2 G \Big/ \int d^d R G \tag{30}$$

This method was initiated by Fixman[49] and has recently been taken to as many as six terms.[54]

With the definition of the functional integral as in equation (28), G(R, L) is not a normalized probability distribution. In fact, because of the singular nature of the pseudopotential, equation (19) is ill defined.[55-57] One particular symptom of this malaise is that short distance divergent integrals (UV divergence) appear when one attempts to write G(R, L) as a perturbation series in w as in equation (29). This problem arises also in the standard calculations[6] of the perturbation series which do not employ functional integrals. Typically, however, we are interested in $\langle R^2 \rangle$ or similar ratios of G factors where the terms of the perturbation series can be so arranged that no divergences are apparent for the ratios. By introducing a Fourier transform of the distribution function G(R, L) with respect to the position R and a Laplace transform with respect to the length L, i.e.

$$\tilde{G}(k, E_0) = \int_0^\infty dL \, e^{-E_0 L} \hat{G}(k, L)$$

$$\hat{G}(k, L) = \int d^3 R \, e^{-ik \cdot R} G(R, L) \qquad (31)$$

The mean square end-to-end distance $\langle R^2 \rangle$ can be written[54] as

$$\langle R^2 \rangle = l \mathscr{L}^{-1}\{K/E^2\}/\mathscr{L}^{-1}\{1/E\}, \qquad (32)$$

where $\mathscr{L}^{-1}$ is the inverse Laplace transform operator $\int \dfrac{dE_0}{2\pi i} \exp(E_0 L)$ and

$$E = 1/\tilde{G}(k = 0, E_0)$$

$$K = \frac{6}{l} \frac{\partial}{\partial k^2} [1/\tilde{G}(k, E_0)] \Big|_{k = 0}$$

$$J \equiv \frac{dE_0}{dE} \qquad (33)$$

The series expansions for J and K of equation (33) are easily derived.[54,58] The series coefficients a_m and b_m in

$$J = 1 - \sum_{m=1} a_m(-\lambda/\sqrt{E})^m$$

$$K = 1 + \sum_{m=1} b_m(-\lambda/\sqrt{E})^m \qquad (34)$$

have been determined to order $m = 6$. Here $\lambda = \dfrac{w}{\pi}(3/2l)^{3/2}$. In order to obtain $\langle R^2 \rangle$ as a series in λ, one must first obtain the series for

$$E_0 - C = \int dE \, J \qquad (35)$$

The integration constant C is, strictly speaking, infinity for the two parameter model with no UV cut-off. However, since equation (32) involves only the ratio of inverse Laplace transforms, both numerator and denominator in equation (32) can be modified with a common multiplicative constant. That is, one can redefine

$$L^{-1} = \int \frac{dE_0}{2\pi i} \exp(E_0 - C)L = \int \frac{dE}{2\pi i} J \exp(L \int dE \, J) \qquad (36)$$

which now involves only finite and calculable quantities. As discussed by des Cloizeaux,[20] the prescription of equation (36) is equivalent to multiplying every G(R, L) in the problem by e^{-CL} and the combination $\exp(-CL)G(R, L)$ is finite in the limit of no UV cut-off. Note that such a factor simply contributes an additive term proportional to L to the free energy of the polymer and as such would be indistinguishable from other microscopic energy contributions proportional to the polymer molecular weight. Also, this is the only 'renormalization' or redefinition that one needs to perform and shows why the UV divergences of the two parameter model are innocuous in this formulation.

The remaining evaluation of $\mathscr{L}^{-1}\{1/E\}$ and $\mathscr{L}^{-1}\{K/E^2\}$ in powers of w is fairly straightforward and leads to

$$\langle R^2 \rangle = Ll\left(1 + \sum_i c_i z^i\right) \tag{37}$$

$$z = (3/2\pi l)^{3/2} w L^{1/2} \tag{38}$$

where the first six coefficients are known so far.[54] The modulus of the coefficients increases so fast that the series must be divergent but asymptotic, in accordance with analytic prediction.[59,60]

1.2.2.4 Numerical calculations on lattice chains

An alternative approach[51] to equation (31) is based on lattice model walks with some site potential energy to simulate chain contact repulsion. A key feature of this approach is the use of the generating functions[61]

$$P(x) = \sum_{n=0}^{\infty} p_n x^n \tag{39}$$

where p_n are probability functions describing chains of n steps. Working with $P(x)$ rather than p_n greatly reduces the complexity of the problem. By considering the limit of very large numbers of steps in this procedure, Barrett and Domb[52] could simulate the two-parameter continuum model and determine correctly c_1, c_2 and c_3 of equation (31) in three dimensions.

For the case of fully developed excluded volume limit, the exact enumeration method[62-69] consists of carrying out enumerations of the number, C_n, of all possible nonintersecting random walks of n steps on a lattice and the number, $f_n(R)\Delta R$, of those walks whose end points lie between R and $R + \Delta R$ from the origin. Since all possible configurations are *a priori* equally possible

$$\langle R^2 \rangle = C_n^{-1} \Sigma_R R^2 f_n(R) \Delta R \tag{40}$$

and a plot of $\langle R_{n+1}^2 \rangle / \langle R_n^2 \rangle$ against $1/n$ approaches a straight line of slope 2ν. Estimates of ν for lattice self-avoiding walks continue to be investigated.[70,71]

There is an alternative method, called the Monte Carlo method, to obtain ν where nonintersecting random walks are simulated on various lattices and statistical data describing the chain are accumulated. This method is restricted in accuracy by statistical fluctuations in contrast to the exact enumeration method. But long walks can be studied here. This method has been applied to understand a variety of phenomena involving polymers as reviewed in refs. 73–75.

1.2.2.5 Scaling laws

It follows from equation (28) that the partition function of the chain is given by

$$Z = \cfrac{\int D[\mathbf{r}(s)] e^{-\frac{d}{2l}\int_0^L ds \left(\frac{\partial \mathbf{r}(s)}{\partial s}\right)^2 - \frac{w}{2}\int_0^L ds \int_0^L ds' \delta^d[\mathbf{r}(s) - \mathbf{r}(s')]}}{\int D[\mathbf{r}(s)] e^{-\frac{d}{2l}\int_0^L ds \left(\frac{\partial \mathbf{r}(s)}{\partial s}\right)^2}} \tag{41}$$

In going to the continuous notation of equation (27) we have allowed every point along the contour of the chain to undergo excluded volume interaction with every other point along the chain. To justify that this simplification is valid we can introduce a cut-off length along the chain below which the monomers do not have excluded volume interaction, *i.e.*

$$\int_0^L ds \int_0^L ds' \rightarrow \int_0^L ds \int_0^L ds' |s - s'| \geq \Lambda \tag{42}$$

Λ is thus an additional parameter introduced to characterize the model chain, but we will show that it plays no significant role for L large, *i.e.* the global properties of the chain are independent of Λ in the asymptotic limit. By changing[76] the variables $s/l \to s$ and $r/l \to r$, Z becomes

$$Z = \frac{\int D[\mathbf{r}(s)]e^{-\frac{d}{2}\int_0^n ds\left(\frac{\partial \mathbf{r}(s)}{\partial s}\right)^2 - \frac{v}{2}\int_0^n ds\int_0^n ds' \delta[\mathbf{r}(s) - \mathbf{r}(s')]|s - s'|^2 \Lambda/l}}{\int D[\mathbf{r}(s)]e^{-\frac{d}{2}\int_0^n ds\left(\frac{\partial \mathbf{r}(s)}{\partial s}\right)^2}} = Z(n, v, \Lambda/l) \tag{43}$$

Thus the partition function is a function of n, v and Λ/l only (where $v \equiv wl^{2-d}$).

Let us now perform the following changes of variables

$$s_1 \equiv as, \quad a > 0$$
$$\mathbf{r}_1(s_1) \equiv a^{-1/2}\mathbf{r}(s) \tag{44}$$

and look at the consequences. We take a to be totally arbitrary and we have absolute freedom to choose any value we want for a. Under the scale transformation of equation (44), we get from equation (43)

$$Z(n, v, \Lambda/l) = Z(an, va^{(d-4)/2}, a\Lambda/l) \tag{45}$$

Thus the functional nature of the partition function is unchanged by the transformation; only the variable are regrouped. It is easy to calculate $\langle R^2 \rangle$ from Z and the result is

$$\langle R^2 \rangle = \frac{l^2}{a}f(an, va^{(d-4)/2}, a\Lambda/l) \tag{46}$$

where f is some function. We now make a choice of the arbitrary variable a in such a way that our choice reduces the number of independent variables in equation (46). Such a choice is

$$a = \frac{1}{n} \tag{47}$$

This then leads to

$$\langle R^2 \rangle = nl^2 f(1, vn^{-(d-4/2)}, \Lambda/L) \tag{48}$$

In the limit of $L \gg \Lambda$, the cut-off dependence of the function f may be ignored so that

$$\langle R^2 \rangle = nl^2 g(vn^{(4-d)/2}) \tag{49}$$

where g is the appropriate function. It is obvious from equation (49) that: (i) the only dimensionless parameter of the problem is $vn^{-(d-4)/2}$; (ii) for $v = 0$, the Gaussian result $\langle R^2 \rangle = Ll$ follows exactly; and (iii) for $d \geq 4$, g is independent of n (for $n \to \infty$) so that random walk statistics prevail, thus defining 4 as the upper critical dimension.[7, 77-80]

For $d < 4$, the dimensionless parameter can be very large. In view of Flory's calculation of v, we expect a power law behavior of g for asymptotically large values of the argument

$$\lim_{x \to \infty} g(x) \to x^? \tag{50}$$

where ? is some unknown exponent. Since we have already introduced v as the exponent to describe the n-dependence of $\langle R^2 \rangle$, the ? exponent can be related to v and the scaling behavior in the asymptote is

$$\langle R^2 \rangle_d \sim n^{2v_d} v^{2(2v_d - 1)/(4-d)} \tag{51}$$

where v_d is taken to be an unknown exponent.

Similar to equation (46), it is possible to derive the scaling form for the mean square end-to-end distance of a chain in d-dimensional space confined by two parallel plates separated by an orthogonal distance of D. The result is

$$\langle R^2 \rangle_{d,D} = \langle R^2 \rangle_d h(D/\langle R^2 \rangle_d^{1/2}) \tag{52}$$

where h is an unknown function, but one such that

$$\lim_{x \to \infty} h(x) \to 1 \tag{53}$$

Now we make an important *assumption*. Let

$$h(x) \sim x^{-y} \quad \text{for } x \ll 1 \tag{54}$$

where y is some unknown power. However, we know from physical grounds that in this limit of $x \ll 1$, the chain is forced into the space of $(d-1)$ dimension

$$\lim_{x \ll 1} \langle R^2 \rangle_{d,D} \sim \langle R^2 \rangle_{d-1} \tag{55}$$

Therefore it follows from equations (52–55) that

$$\langle R^2 \rangle_d (D/\langle R^2 \rangle_d^{1/2})^{-y} \sim \langle R^2 \rangle_{d-1} \tag{56}$$

The use of equation (51) in equation (56) yields

$$n^{2v_d}{}^{(1+y/2)}v^{2(2v_d-1)(1+y/2)/(4-d)} \sim n^{2v}{}^{(d-1)}v^{2(2v_{d-1}-1)/(5-d)} \tag{57}$$

By matching the exponents of n and v on both sides of equation (57) gives

$$(2 - v_d^{-1}) = \frac{(4-d)}{(5-d)}(2 - v_{d-1}^{-1}) \tag{58}$$

Since in $d = 1$, the chain with excluded volume effect can only line up as a rod with $v_1 = 1$, equation (50) gives

$$v_d = \frac{3}{d+2}, \quad d \le 4 \tag{59}$$

recovering the Flory formula.

Notice that in the derivation of equation (59), the only input is the assumption of equation (54) and the obvious results of equation (55) and $v_1 = 1$, and everything else is simple changes of variables. This is to be contrasted with Flory's method of optimizing the chain free energy to obtain v_d.

1.2.2.6 Renormalization group theory

The problem of a polymer chain with excluded volume interaction is related to the ϕ^4 field theory,[13] the Lagrangian of which is given by

$$\mathscr{L}[\phi(r)] = 1/2[(\nabla\phi)^2 + \mu^2\phi^2 + \lambda\phi^4] \tag{60}$$

where the field variable ϕ has n components. (In this nomenclature the Ising model is $n = 1$, the xy model $n = 2$, the Heisenberg model $n = 3$, *etc.*) The bare coupling constant λ is analogous to the excluded volume parameter w and the bare mass μ is related to $E_0^{1/2}$ as in Table 1. It can be exactly shown that the generating function of the end-to-end probability distribution for self-avoiding chains of length L is in fact identical[81,82] to the two-point correlation function for the n-vector model (equation 60) of interacting fields in the $n = 0$ limit.

The calculation of the correlation functions in the field theory is well documented in the literature.[13] For $d = 4$, the calculation of the various quantities encounter divergence problems (due to the presence of the cut-off Λ of equation (42) which are removed by the proper renormalizations of the bare mass E_0 and the bare coupling constant w. The analogs of the renormalization group equations (such as those of Gellman and Low, 'tHooft and Veltman, and Callan and Symanzik) which describe the dependence of the correlation functions (and the vertex functions) on arbitrary scale factors can be readily written for the polymer case also.[20-27,77-80] A study of the Wilson function, $\beta(g)$,[13,80] which describes the dependence of the renormalized coupling constant g on an arbitrary momentum k for a given bare excluded volume interaction

$$\beta(g) = \left(k\frac{\partial g}{\partial k}\right)_w = -\left[\frac{\mathrm{d}}{\mathrm{d}g}\ln g_{\mathrm{B}}\right]^{-1} \tag{61}$$

$$g_{\mathrm{B}} \equiv \frac{4w}{\pi}\left(\frac{3}{2l}\right)^{3/2} E^{-1/2}$$

(where E is the renormalized 'mass') shows that, at the 'fixed point' $g = g^*$, the effective interaction is independent of the length scale thus defining the criticality. This is shown in Figure 4.

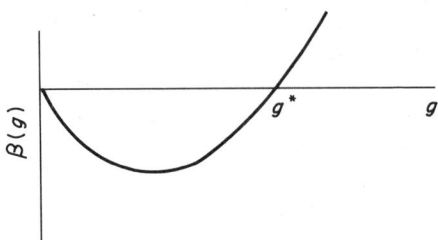

Figure 4 Shape of the Wilson function *vs.* the renormalized interaction parameter: g^* is the fixed point

From the knowledge of g^* and the tangent of $\beta(g^*)$ and the definition of the critical exponents, values of the critical exponents can be obtained.

The most common method used in the calculation of v is the $\varepsilon(=4-d)$ expansion[13,80] and the result[83,84] was applied to polymers by de Gennes[7] to give

$$v = \frac{1}{2} + \frac{\varepsilon}{16} + \frac{15}{512}\varepsilon^2 - 0.0224\varepsilon^3 + 0.0667\varepsilon^4 + 0.2430\varepsilon^5 + \ldots \tag{62}$$

Although this does not converge, it is taken to be an asymptotic series, *i.e.* one uses it as long as the terms are diminishing and in the various research works in the polymer area, the properties of interest are calculated to either first or second order in ε and attempts are made to compare[22,26] with experimental data.

1.2.2.7 *Rigorous results on crossover for* $\langle R^2 \rangle$

Since $\langle R^2 \rangle$ is given by equation (37) for very weak excluded volume interactions and by $\langle R^2 \rangle$ proportional L^{2v} in the asymptotically large z limit, it is convenient to define the effective exponent, 2σ by

$$2\sigma = \frac{\mathrm{d}\ln\langle R^2\rangle/Ll}{\mathrm{d}\ln z} \tag{63}$$

which is proportional to z for small z and approaches the constant $4v - 2$ for large z. The results of σ from the above mentioned schemes based on mean field arguments,[1,6] variational calculations,[14,18] semi-self consistent procedures,[6] renormalization group procedures combined with ε expansion approximations,[20-27] *etc.*, differ from each other depending on the severity of the intrinsic approximation of the method employed. Worse yet, because the approximations are uncontrolled there is no way of estimating the uncertainties. An accurate calculation of $\langle R^2 \rangle$ has recently been performed[85] in order to assess the extent of the deviations of the various approximate schemes.

Since $\langle R^2 \rangle$ is given by equations (32), (34) and (36), it can be determined for all $z > 0$ by performing the series analysis on J and K first to obtain these as functions valid for all values of their (complex) argument and then to perform the inverse Laplace transforms in equation (32) numerically. An alternative procedure[86] is to perform a series analysis on equation (37) directly. There are two key features which are common to both methods. The first is that the methods build in the expected scaling behavior of the unknown function J and K or $\langle R^2 \rangle$ in the limit of large argument $\lambda/\sqrt{E}$ or z respectively. The second is that both methods use, as an independent variable, a variable whose physical range is finite; in the analysis of J and K the variable is the so-called renormalized coupling constant g defined below in relation to the second virial coefficient,[1,6,12] while in the analysis of $\langle R^2 \rangle$ it is the effective exponent σ defined in equation (63).

(i) Scaling arguments

First we discuss the scaling aspects.[80] We expect, in the nonperturbative regime

$$\frac{\langle R^2 \rangle}{Ll} = B_R L^{2v-1}\{1 + b_R L^{-\Delta} + \ldots\} = B'_R z^{4v-2}\{1 + b'_R z^{-2\Delta} + \ldots\} \tag{64}$$

for large L or z. Δ is the leading correction to scaling exponent which is universal in the sense that other functions such as the second virial coefficient A_2 have the same $z^{-2\Delta}$ relative correction to scaling behavior. The scaling laws for J and K can be guessed in part from the scaling we already assume for $\langle R^2 \rangle$ and in part from the polymer–magnetic system critical point analogy.[13] As in Table 1, E_0, which is the Laplace variable conjugate to the chain contour length, is identified with $T - T_0$ in the critical phenomena where T is the temperature and T_0 is the mean field critical temperature. Writing E_{0c} to represent $T_c - T_0$, where T_c is the actual critical temperature, $T - T_c$ is identified with $E_0 - E_{0c}$. Furthermore, in the limit of small k, where terms of $0(k^4)$ can be neglected, $\tilde{G}(k, E_0)$ becomes

$$\tilde{G}(k, E_0) = (E + k^2 l K/6)^{-1} \tag{65}$$

which is of the familiar Ornstein–Zernike form with correlation length ξ given by $\xi^2 = l K/6E$. Since the correlation length exponent v is defined through $\xi \propto (T - T_c)^{-v} - v \propto (E_0 - E_{0c})^{-v}$

$$\frac{K}{E} = A_\xi (E_0 - E_{0c})^{-2v}\{1 + a_\xi (E_0 - E_{0c})^{\Delta} + \ldots\} \tag{66}$$

where Δ is the same correction to the scaling exponent as in equation (64) and we have subscripted the constants A and a by ξ as a mnemonic to remind us of the connection between K/E and the correlation length. Let us also write

$$\frac{1}{E} = A_\chi (E_0 - E_{0c})^{-\gamma}\{1 + a_\chi (E_0 - E_{0c})^{\Delta} + \ldots\} \tag{67}$$

since E as inverse propagator in the $k = 0$ limit is the polymer analog of the inverse magnetic susceptibility χ and γ is the susceptibility exponent. The correlation length exponent v is related to γ by

$$\gamma = (2 - \eta)v \tag{68}$$

a relation we may here simply use as the definition of η. Amplitudes A_i and a_i in equations (66) and (67) are special to the two-parameter model but renormalization group analysis predicts the leading correction to scaling amplitude ratio a_ξ/a_χ will be universal.

The total number of polymer configurations is $\int d^3R\, G(\mathbf{R}, L)$ which, as defined by equation (31), is $\hat{G}(0, L) = \mathcal{L}^{-1}\{\frac{1}{E}\}$. On substituting equation (67) into equation (31) we get

$$\hat{G}(0, L) = e^{(E_{0c} - C)L} A_\chi \frac{L^{\gamma-1}}{\Gamma(\gamma)}\left\{1 + a_\chi \frac{\Gamma(\gamma)}{\Gamma(\gamma - \Delta)} L^{-\Delta} + \ldots\right\} \tag{69}$$

Note that, as discussed in the context of the perturbation theory, this contains the arbitrary multiplication constant e^{-CL}. For $\langle R^2 \rangle$, which is the ratio given in equation (32), this arbitrariness disappears. On using equation (64) in equation (36) and the above result for $\hat{G}(0, L)$ we can derive equation (64) with the constants given by

$$B_R = A_\xi \frac{\Gamma(\gamma)}{\Gamma(\gamma + 2v)}, \quad b_R = (a_\xi + a_\chi)\frac{\Gamma(\gamma + 2v)}{\Gamma(\gamma + 2v - \Delta)} - a_\chi \frac{\Gamma(\gamma)}{\Gamma(\gamma - \Delta)} \tag{70}$$

Although equations (66) and (67) are obviously useful to deduce the asymptotic real space polymer scaling properties, the series expansions for J and K are in terms of the variable $\lambda/\sqrt{E}$. We therefore combine these equations to obtain

$$K = A_\xi A_\chi^{-\frac{2\nu}{\gamma}} E^{-(2-\eta)} \left\{ 1 + \left(a_\xi - \frac{2\nu}{\gamma} a_\chi \right) A_\chi^{\frac{\Delta}{\gamma}} E^{\frac{\Delta}{\gamma}} + \ldots \right\} \tag{71}$$

Similarly, $J = \dfrac{dE_0}{dE}$ is given by

$$J = \frac{1}{\gamma} A_\chi^{\frac{1}{\gamma}} E^{-1+\frac{1}{\gamma}} \left\{ 1 + a_\chi \left(\frac{\Delta+1}{\gamma} \right) A_\chi^{\frac{\Delta}{\gamma}} E^{\frac{\Delta}{\gamma}} + \ldots \right\} \tag{72}$$

(ii) Renormalized coupling constant

As mentioned above, an independent variable (the so-called renormalized coupling constant) g is used in the analysis of J and K. It is related[85] to the second virial coefficient and so its designation as an interaction constant or 'coupling constant' is not without merit. However, the designation 'renormalized' is historical and a complete misnomer in the present context. The quantity g is a well-defined and finite quantity in the two-parameter model and involves absolutely no renormalization. It is given by

$$g = -4 \sum_{n=0} \Gamma_n (-\lambda/\sqrt{E})^{n+1} / K^{3/2} \tag{73}$$

where Γ_n are the coefficients of the perturbation series of the four-point vertex function[58,85] in w.

By construction,

$$g = g^* + 0 \left(\left(\frac{\lambda}{\sqrt{E}} \right)^{-2\Delta/\gamma} \right) \tag{74a}$$

for $\lambda/\sqrt{E} \to \infty$. For $\lambda/\sqrt{E} \to 0$,

$$g = \frac{4\lambda}{\sqrt{E}} + 0 \left(\left(\frac{\lambda}{\sqrt{E}} \right)^2 \right) \tag{74b}$$

so that the series (73) can be reverted to yield $\lambda/\sqrt{E}(g)$.

The constant g^* is found to be universal and not just a property of the two-parameter model. The remaining crucial property that makes g a useful independent variable for series expansion is the monotonicity of g with respect to $\lambda/\sqrt{E}$.

(iii) Series analysis

The evaluation of $\langle R^2 \rangle$ for arbitrary values of z involves three steps.

First, all series expansions in $\lambda/\sqrt{E}$ are reevaluated in terms of g. It follows from equation (74) that

$$\frac{\lambda}{\sqrt{E}} \propto (g^* - g) \tag{75}$$

and

$$\beta(g) \equiv -\left[\frac{d}{dg} \ln \left(\frac{\lambda}{\sqrt{E}} \right) \right]^{-1} = \sum_{n=1} \beta_n g^n \tag{76}$$

has a zero at $g = g^*$ with slope

$$\left. \frac{d\beta}{dg} \right|_{g*} = \frac{2\Delta}{\gamma} \tag{77}$$

Similarly, it has been shown that

$$j(g) \equiv -\beta \frac{d}{dg} \ln J(g) = \sum_{n=1} j_n g^n \tag{78}$$

is analytic at $g = g^*$ with

$$j(g^*) = 2\left(1 - \frac{1}{\gamma}\right) \tag{79}$$

and thus can be used to determine the exponent γ. Also with

$$k(g) \equiv -\beta\frac{d}{dg}\ln K(g) = \sum_{n=1} k_n g^n \tag{80}$$

it follows from equation (71) that

$$k(g^*) = \frac{2\eta}{2 - \eta} = 2\eta v/\gamma \tag{81}$$

The second step involves the determination of $\beta(g)$, $j(g)$ and $k(g)$. The series coefficients β_n, j_n and k_n are determined from the J and K series of equations (33) and (34). These series are only asymptotically convergent so the corresponding functions $\beta(g)$, $j(g)$ and $k(g)$ must be determined by a Borel resummation procedure.[87−90] The functions E(g), J(g) and K(g) are then determined by integration of their logarithmic derivatives to be

$$E(g) = \frac{16\lambda^2}{g^2}\left(1 - \frac{g}{g^*}\right)^{\gamma/\Delta}\exp\left\{2\int_0^g dx\left[\frac{1}{\beta(x)} + \frac{1}{x} + \frac{\gamma}{2\Delta(g^* - x)}\right]\right\}$$

$$J(g) = \left(1 - \frac{g}{g^*}\right)^{(1-\gamma)/\Delta}\exp\left\{-\int_0^g dx\left[\frac{j(x)}{\beta(x)} + \frac{(\gamma - 1)}{\Delta(g^* - x)}\right]\right\}$$

$$K(g) = \left(1 - \frac{g}{g^*}\right)^{-\eta v/\Delta}\exp\left\{-\int_0^g dx\left[\frac{k(x)}{\beta(x)} + \frac{\eta v}{\Delta(g^* - x)}\right]\right\} \tag{82}$$

The final step involves the evaluation of the inverse Laplace transforms in equation (32). Expressing the differential dE_0 as

$$dE_0 = J\,dE = \frac{2JE}{\beta}dg \tag{83}$$

where equations (33) and (76) are used.
Then

$$E_0 - E_0' = 2\int_{g'}^g dx\frac{JE}{\beta} \tag{84}$$

where the integration constant C of equation (35) has been explicitly taken to be $E_0' = E_0(g')$ with g' any convenient value different from zero. The full expression for $\langle R^2\rangle$ of equation (32) now becomes

$$\langle R^2\rangle = l\int_c dg\frac{JK}{\beta E}e^{(E_0 - E_0')L}\Big/\int_c dg\frac{J}{\beta}e^{(E_0 - E_0')L} \tag{85}$$

where the contour C corresponds[85] to the original E_0 plane contour deformed into the left half plane to surround the real axis cut from $-\infty$ to $E_{0c}[=E_0(g^*)]$. Thus equations (82), (84) and (85) constitute a practical numerical solution to the determination of $\langle R^2\rangle$ for any value of z.

As mentioned in the beginning of this section, an alternative method[86] is to perform a series analysis of equation (37). Since the effective exponent 2σ is proportional to z for small z and approaches the constant $2\sigma^* = 4v - 2$ as $z^{-2\Delta}$ for large z. If $\sigma(z)$ is monotonic then the series for $\sigma(z)$ can be reverted to yield $z(\sigma)$ and this function will diverge as $\sigma \to \sigma^*$ as $(\sigma^* - \sigma)^{-1/2\Delta}$. If 'corrections' to these already asymptotic corrections are negligible, the logarithmic derivative $d\ln Z/d\theta$ will be dominated by a simple pole at $\sigma = \sigma^*$ with residue $1/2\Delta$. The inverse function $W(\sigma) = (d\ln Z/d\sigma)^{-1}$ will have a simple zero at $\sigma = \sigma^*$ and it makes sense to *directly* use the series for $W(\sigma)$ to determine it on the *entire* interval $0 \le \sigma \le \sigma^*$. des Cloizeaux *et al.*[86] actually use a very elementary extrapolation procedure to improve on the truncated series for $W(\sigma)$ but this is a minor technical point and does not qualitatively alter their results. Now once the function $W(\sigma)$ has been determined, the integrations $\ln z = \int\frac{d\sigma}{W}$ and $\langle R^2\rangle/Ll = \exp(2\int\sigma\,d\ln z)$ are elementary and lead to the desired explicit results valid for all $z > 0$.

Comparison of the results from these two methods and many other formulas for $\langle R^2 \rangle$ in the literature is made below.

(iv) Results

The values of the various critical exponents calculated from equations (77)–(85) are

$$v = 0.5886, \quad \gamma = 1.1613, \quad \Delta = 0.465 \quad \text{and} \quad \eta = 0.0270 \tag{86}$$

in complete agreement with the results of Le Guillou and Zinn-Justin.[91]
For $z \to \infty$

$$\frac{\langle R^2 \rangle}{Ll} = 1.5310\, z^{0.3544}[1 \quad + \quad 0.1204\, z^{-0.930} \quad + \ldots] \tag{87}$$

and the full expansion factor $\alpha^2 = \langle R^2 \rangle/Ll$ as given by equation (85) and the effective exponents, as defined by equation (63), are plotted against z in Figures 5 and 6. The results obtained from

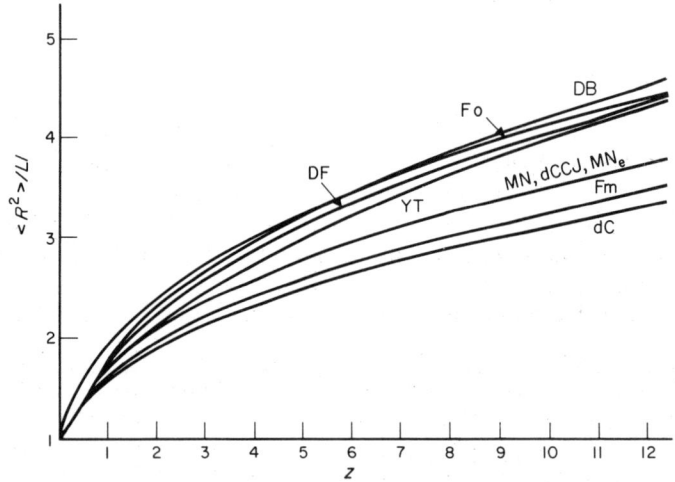

Figure 5 Theoretical values of $\langle R^2 \rangle/Ll$ calculated from various theories: curves MN, Fo, Fm, YT, DB, dC, DF, dCCJ and MN$_e$ correspond to equations 85, 89–96, respectively

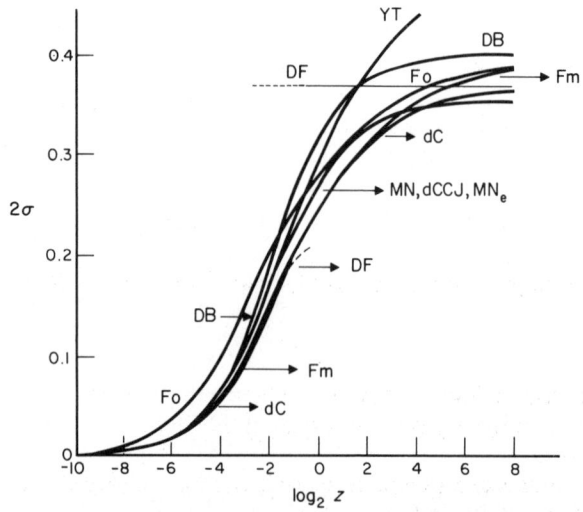

Figure 6 Theoretical values of the effective exponent 2σ plotted against $\log_2 z$ (the labels are the same as in Figure 5)

equation (85) are labelled as MN. The large z behavior of the equation of des Cloizeaux, *et al.*[86] is

$$a^2 = 1.5338 \, z^{0.3538} [1 + 0.1195 \, z^{-0.947} + \ldots] \tag{88}$$

Equations (87) and (88) are in excellent agreement. Furthermore the values of 2σ for intermediate z values obtained by Muthukumar and Nickel[85] (labelled as MN) and des Cloizeaux, *et al.*[86] (labelled as dCCJ) are indistinguishable from each other in Figures 5 and 6.

It is in part because of this remarkable agreement between two completely independent analyses of the six term series that these results represent the 'exact' solution to $\langle R^2 \rangle$ for the two-parameter model. Another reason is that the analysis of Muthukumar and Nickel,[85] although in philosophy the same as that of Le Guillou and Zinn-Justin,[91] differ in detail as does the analysis of Baker, *et al.*[92,93] Yet these analyses, and closely related $0(\varepsilon^5)$ ε-expansion analyses, all predict the same excluded volume model exponent $v \approx 0.588$, and for the Ising model, exponents in agreement with the results of high temperature lattice series analyses.[94a] In Figures 5 and 6, the plots of the following earlier crossover formulas are included for comparison with these rigorous results.

(a) Original Flory formula[1,6] (Fo)

$$\alpha^5 - \alpha^3 = 2.6z \tag{89}$$

(b) Modified Flory formula[6,16,17] (Fm)

$$\alpha^5 - \alpha^3 = 4/3z \tag{90}$$

(c) Yamakawa–Tanaka formula[50] (YT)

$$\alpha^2 = 0.572 + 0.428 (1 + 6.23z)^{1/2} \tag{91}$$

(d) Domb–Barrett empirical formula[19] (DB)

$$\alpha^2 = (1 + 20/3z + 4\pi z^2)^{1/5} \tag{92}$$

(Strictly speaking, this formula is not directly comparable. As stated very clearly by Domb and Barrett, only in a special limit does the Comb–Joyce model become the two-parameter model whereas the formula (92) is determined from series in another limit, *viz.* the self-avoiding walk limit.)

(e) des Cloizeaux ε^2 formula[21] (dC)

$$\alpha^2 = \{1 - g/0.266)^{-0.562} \exp(-1.113g + 0.676g^2) \tag{93}$$

$$z = g(1 - g/0.266)^{-1.531}$$

(f) Douglas–Freed ε^2 formula[27] (DF)

$$\alpha^2 = \left(1 + \frac{32}{3}Z\right)^{1/4} \left[1 - 0.125 \frac{32Z/3}{\left(1 + \frac{32Z}{3}\right)}\right], \quad Z \le 0.15$$

$$\alpha^2 = 1.734 z^{0.3672}, \quad z \ge 0.75 \tag{94}$$

(g) des Cloizeaux–Conte–Jannink formula[86] (dCCJ)

$$\alpha = \left(1 - \frac{\sigma}{0.1769}\right)^{-0.1868} \left(1 + \frac{\sigma}{0.2184}\right)^{-0.0057} \left[1 - \frac{(0.2150\sigma - \sigma^2)}{0.3214}\right]^{0.0963}$$

$$\exp\left[0.0199 \arctan\left(\frac{0.5567\sigma}{0.3214 - 0.1075\sigma}\right)\right] \tag{95}$$

Furthermore, in an attempt to obtain a very simple extrapolation formula which can fit α^2 for arbitrary z the curve

$$\alpha^2 = (1 + 7.524Z + 11.06Z^2)^{0.1772} \tag{96}$$

labelled as MN_e is included in Figures 5 and 6. The coefficients of z and z^2 and the exponent in equation (96) are chosen to fit the asymptotic results of the first order perturbation theory for small z and equation (87) for $z \to \infty$.

It is evident that all formulas except equations (95) and (96) deviate substantially from our results. The plots of equations (95) and (96) in Figures 5 and 6 are essentially indistinguishable from the MN curve.

It is interesting to observe that F_m formula is the closest to the rigorous result for small z. Such rigorous results for the radius of gyration, second virial coefficient, *etc.* are not yet available.

1.2.3 Branching

Most of the syntheses of linear polymers introduce chain branching as defects. In addition, certain branched polymers possess specific technologically important properties over linear chains. Consequently, extensive research has been carried out in calculating the various configurational properties corresponding to well-defined branching architecture.

Branched polymers can be classified into stars, combs, random branching, *etc.* The mean square radius of gyration, $\langle s^2 \rangle_b$, of branched polymers where each strand obeys Gaussian statistics is known exactly in the literature.[6] For example, for a star polymer with f arms emanating from the center,

$$\langle S^2 \rangle_b = g\langle S^2 \rangle_1, \qquad \langle S^2 \rangle_1 = \frac{1}{6}nl^2, \qquad g = \frac{3}{f} - \frac{2}{f^2}$$

where n is the total number of Kuhn segments in the chain and l is the Kuhn step length. $\langle S^2 \rangle_1$ is the mean square radius of gyration of a linear chain with the same number of segments. Explicit expressions of g for other molecular architectures are reviewed in ref. 6.

1.2.3.1 *Excluded volume effect*

The incorporation of the excluded volume effect in the calculation of $\langle S^2 \rangle_b$ and other related quantities is straightforward but tedious and is analogous to the calculations in the case of linear chains. For branched polymer chains, the perturbation series for the excluded volume effect may be written as

$$\langle S^2 \rangle_b = g\langle S^2 \rangle_1 \left[1 + \sum_{i=1}^{\infty} a_i^b z^i \right]$$

where z is the excluded volume parameter defined in Section 1.2.2.3. The value of a_1^b is known[6] for uniform stars and combs, based on the classical cluster expansion method.

1.2.3.2 *Scaling*

A scaling argument similar to that for a linear chain in Section 1.2.2.5 has been attempted for stars by Daoud and Cotton.[94b] They assumed that there are three regimes of monomer density as a function of the radial distance from the center of the star chain. In the innermost region near the center, the monomer density is assumed to be uniform. In the central region, the monomer concentration is large so that the excluded volume interaction is fully screened with the chain arms obeying Gaussian statistics. In the outermost region, the monomer density is assumed to be so low that the arms behave according to the self-avoiding walk statistics. The predications of this scaling description are found to be in qualitative agreement with some experimental results.

1.2.3.3 *Renormalization group*

A chain conformational renormalization group method has been developed by Miyaki and Freed[94c] for star polymers. Explicit expressions for the distribution functions for intersegment distance vectors and their moments, $\langle S^2 \rangle_b$, the osmotic second virial coefficient and the related functions, *etc.*, have been reported based on the ε-expansion approximation. A detailed comparison of these results with those of linear chains, scaling predictions and the experimental data is also reported[94d] in the literature.

1.2.3.4 *Random branching*

The calculation of $\langle S^2 \rangle_b$ for a large molecule which has undergone branching at random points during its chemical synthesis, was originally carried out by Zimm and Stockmayer[94e] using combinatorial analysis. Assuming that there are no loops and ignoring the excluded volume effect, they found that

$$\langle S^2 \rangle_b \sim n^{1/2}$$

The unphysical value of the exponent (lower than even the value of 2/3 for a compact globule) is due to the neglect of the excluded volume effect. The same result can be simply obtained by the field theoretic method discussed below as pointed out by de Gennes.[94f] If $G_0(k, E_0)$ is the propagator representing a Gaussian strand of certain contour length and with certain end-to-end distance (where E_0 and k are the Fourier and Laplace variables conjugate to the contour length and the end-to-end distance respectively), then the full propagator $G(k, E_0)$, for the randomly branched structure in Figure 7a can be diagrammatically represented by Figure 7b.

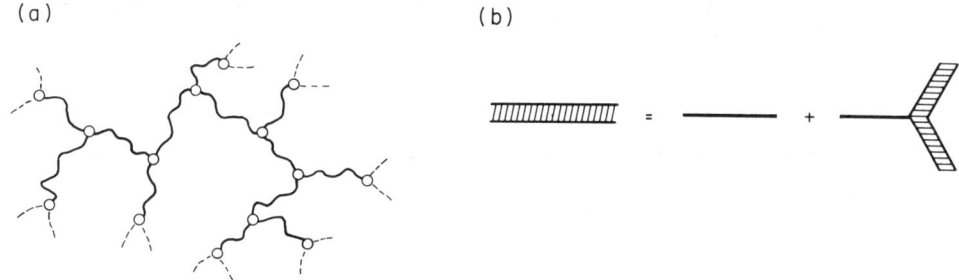

(a)   (b)

Figure 7 (a) Randomly branched structure; (b) diagrammatic representation of the propagator $G(K, E_0)$

Here, G and G_0 are denoted respectively by shaded and thin lines. The vertex point of the graph on the right hand side is given the weight μ corresponding to the branching probability. The rules to evaluate these graphs are exactly the same as in Section 1.2.2.3. Specifically, the above equation becomes

$$G(k, E_0) = G_0(k, E_0)[1 + \mu G(0, E_0) G(k, E_0)]$$

Since $G_0(k, E_0) = (E_0 + k^2 1/6)^{-1}$, G becomes

$$G(k, E_0) = \frac{1}{\dfrac{k^2 l}{6} + E_0 - V}$$

where V is self consistently given as

$$V = \frac{\mu}{E_0 - V}$$

The solution therefore follows as

$$V(E_0) = \frac{1}{2}(E_0 - \sqrt{E_0^2 - 4\mu})$$

Thus, G and V have a branch point on the real E_0 axis at $E_0 = 2\mu^{1/2}$. The partition function is given by

$$Z_n = \int_{-i\infty}^{i\infty} \frac{dE_0}{2\pi i} G(k = 0, E_0) \exp(LE_0) = \frac{1}{2\sqrt{\pi}} \left(\frac{1}{\sqrt{\mu n}}\right)^{3/2} \exp(2n\sqrt{\mu})$$

Since the mean square radius of gyration is given by

$$\langle S^2 \rangle_b = \frac{l^2 \displaystyle\sum_{i=1}^{n} Z_{n-i} Z_i (n - i)i}{n \displaystyle\sum_{i=1}^{n} Z_{n-i} Z_i}$$

it follows from the last two equations that

$$\langle S^2 \rangle_b = \frac{1}{6} n l^2 g$$

with

$$g = \frac{3\sqrt{\pi}}{2} (n\sqrt{\mu})^{-1/2}$$

which gives the same Zimm–Stockmayer exponent.

The Flory derivation of the mean field exponent for a single self-avoiding walk chain has been generalized by Lubensky and co-workers[94g] to branched polymers. Writing the total free energy of the chain with excluded volume interaction as (see Section 1.2.2.1)

$$F = \frac{R^2}{\langle R^2 \rangle_0} + \frac{wn^2}{R^d}$$

where R is the average radius of the chain and $\langle R^2 \rangle_0$ is the mean square radius of the chain in the absence of the excluded volume effect, the minimization of F with respect to R yields

$$R \approx (wn^2 \langle R^2 \rangle_0)^{1/(d+2)}$$

Using the mean field result of $\langle R^2 \rangle_0 \sim n^{1/2}$ gives

$$R \approx w^{1/(d+2)} n^{5/2(d+2)}$$

In three dimensions, the average radius of a branched polymer thus appears to correspond to the Gaussian statistics.

1.3 MANY CHAINS IN SOLUTION

Historically the first problem studied was that of the thermodynamics of very high concentration.[1,6,95-99]

1.3.1 Flory–Huggins Theory

For very high concentrations of polymer solutions where the density fluctuations are not dominant, one can expect mean field arguments to have validity. The first theory to obtain the thermodynamics of polymer solutions is due to Flory and Huggins.[97,98] In the Flory–Huggins theory each polymer chain is represented as a chain of n segments, each exactly equal in volume to a solvent molecule (which is comparable to l^3). A polymer solution may now be represented by a lattice divided into cells, each cell of which may be occupied either by a segment of the polymer molecule or by a solvent molecule. A two-dimensional representation of such a lattice is shown in Figure 8.

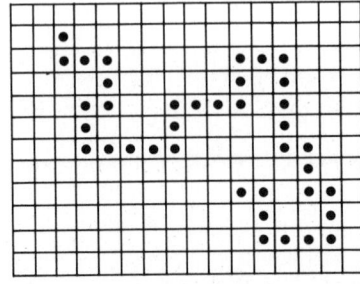

Figure 8   Schematic diagram of the lattice representation of the polymer solution in the Flory–Huggins theory

The total volume of the solution is divided into $n_0 (= n_1 + nn_2)$ cells, where n_1 and n_2 represent the numbers of solvent and polymer molecules respectively. It is possible to calculate the entropy of mixing ΔS_m of n_1 solvent molecules and n_2 polymer molecules by knowing the number of possible configurations of placing n_1 solvent molecules and n_2 polymer molecules (each with n segments) on a lattice with n_0 sites. Assuming that the chains are randomly distributed and no correlations of any sort are present, ΔS_m is given by

$$\Delta S_m / k_B = -n_1 \ln \phi_1 - n_2 \ln \phi_2 \tag{97}$$

where $\phi_1 = n_1/n_0$ is the volume fraction of the solvent and $\phi_2 = 1 - \phi_1 = nn_2/n_0$ is that of the polymer.

If ε_{11}, ε_{22} and ε_{12} are the interaction energies between the nearest neighbor solvent molecules, polymer segments and solvent–polymer segment respectively, the enthalpy of mixing ΔH_m in forming the polymer solution is

$$\Delta H_m / k_B T = \chi n_0 \phi_1 \phi_2 \tag{98}$$

where the Flory–Huggins 'χ parameter' is

$$\chi \equiv \frac{z}{2 k_B T} (\varepsilon_{11} + \varepsilon_{22} - 2\varepsilon_{12}) \tag{99}$$

with z as the coordination number of the lattice. It follows from equations (97) and (99) that the free energy of mixing per site is given by

$$\Delta F_m / k_B T = \frac{\phi}{n} \ln \phi + (1 - \phi) \ln (1 - \phi) + \chi \phi (1 - \phi) \tag{100}$$

where ϕ is the polymer concentration ϕ_2. This is the celebrated Flory–Huggins free energy formula for polymer solutions.

The osmotic pressure π of the solution is given by

$$-\pi V_1 / k_B T = [\mu_1(\phi) - \mu_1(0)/k_B T = \left[\frac{\partial}{\partial n_1} (\Delta F_m / k_B T) \right]_{n_2 T} = \ln (1 - \phi) + \left(1 - \frac{1}{n} \right) \phi + \chi \phi^2 \tag{101}$$

where μ_1 is the chemical potential of the solvent and V_1 is the volume of each cell ($\sim l^3$). Defining the monomer concentration ρ to be ϕ/V_1, equation (101) yields at very low concentrations

$$\pi / k_B T = \frac{\rho}{n} + \frac{1}{2} w l^2 \rho^2 + \cdots \tag{102}$$

with

$$w l^2 \equiv (1 - 2\chi) V_1 \tag{103}$$

Equation (103) gives the relation between the χ parameter and the binary cluster integral of equation (19). When $\chi = 1/2$, the second virial coefficient A_2 vanishes and the osmotic pressure is given by the 'ideal gas law'

$$\pi = \rho k_B T / n \tag{104}$$

The temperature at which $\chi = 1/2$ is called the ideal (θ) temperature and the Gaussian statistics prevails.

For good solvents ($\chi < 0$), the Flory–Huggins theory predicts that A_2 is proportional to w and is independent of chain molecular weight. However the experimental data show that $A_2 \sim L^{-0.2}$ Attempts to explain this molecular weight dependence of the second virial coefficient have been made by Flory and Krigbaum[100] and others.[101–103] A complete theory will not involve a lattice model and treat the averages more carefully. It is still an active subject.[104–114]

1.3.2 Screening

Flory[1,115] intuitively argued that although the solvent is good (so that chains are swollen in dilute solutions), the chains attain ideal dimensions at high concentrations. The neutron-scattering experiments[116,117] have verified this conjecture. Edwards theoretically established[118-121] this 'screening' of the excluded volume interaction as the polymer concentration is increased. For a dense solution containing Gaussian chains with interchain excluded volume interaction of the type

$$u(r_{\alpha i} - r_{\beta j})/k_B T = wl^2 \delta(r_{\alpha i} - r_{\beta j}) \tag{105}$$

the effective potential energy $\tilde{u}$ between ith segment of αth chain and the jth segment of βth chain has been derived to be

$$\tilde{u}(r_{\alpha i} - r_{\beta j})/k_B T = wl^2 \left[\delta(r_{\alpha i} - r_{\beta j}) - \frac{e^{-|r_{\alpha i} - r_{\beta j}|/\xi_E}}{4\pi \xi_E^2 |r_{\alpha i} - r_{\beta j}|} \right] \tag{106}$$

where the Edwards screening length is given by[7,118]

$$\xi_E^{-2} = 12\rho w \tag{107}$$

Thus as ρ increases, ξ_E decreases resulting in enhanced screening of the excluded volume effect. The size of the polymer in this regime is given by[118]

$$\langle R^2 \rangle = Ll \left(1 + \frac{12\omega\xi}{\pi l^2} \right) = Ll \left(1 + \frac{\omega}{\rho} \frac{12}{\pi l^2} \right) \tag{108}$$

so that the expansion parameter is $\omega/\rho l^4$. However if L is very large, ρ can be small and the chains are still overlapping. So this theory is inadequate and it was shown by de Gennes *et al.*[7] that scaling arguments could resolve the problem.

1.3.3 Scaling Laws

In this section we derive the scaling laws[7] for polymer solutions in good solvents at nonzero polymer concentrations. The polymer segment concentration $\rho(= Nn/V)$ where the polymer solution with volume V contains N chains each with n segments is the key variable to describe the thermodynamic and dynamic properties of polymer solutions under nondilute conditions. Since we are in search of a dimensionless variable in order to apply the scaling method, we look for a fundamental concentration which can make the polymer concentration dimensionless. This is the so-called overlap concentration ρ^*, at which all the polymer coils are imagined to just begin to overlap (thus essentially behaving like hard spheres). See Figure 9.

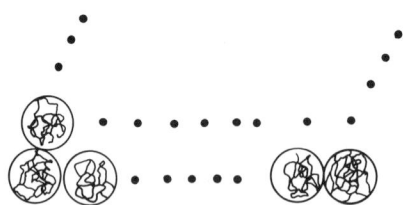

Figure 9 Crude definition of the overlap concentration

As we specialize in good solutions here, the overlap segment concentration ρ^* is given by

$$\rho^* \sim \frac{Nn}{NR_F^3} \sim n^{1-3\nu} \sim n^{-4/5} \tag{109}$$

where Flory's result for ν in $d = 3$ has been used. We take ρ/ρ^* to be the dimensionless parameter for the many chain system. This can be rigorously proved.[7] The situation where ρ/ρ^* is small, but w large is called *semidilute*, and we derive by means of scaling, expressions for the osmostic pressure, the correlation length and the mean square end-to-end distance of a labeled chain.

1.3.3.1 Osmotic pressure π

The osmotic pressure π can, in general be written as

$$\frac{\pi}{k_B T} = \frac{\rho}{n} f\left(\frac{\rho}{\rho^*}\right) \tag{110}$$

where f is some unknown function such that $f(x) \to 1$ and $x \to 0$, since the 'ideal gas law' of $\pi/k_B T = \rho/n$ must be recovered for very dilute solutions. The functional nature of f may be guessed based on the following argument. As far as the thermodynamic properties are concerned, for $\rho > \rho^*$, there should not be any difference between a polymer solution containing a single chain with a very large number of monomers, N_0, and that with many x chains each with X_0 monomers such that $N_0 = x X_0$. In other words, this means that π is independent of n for $\rho > \rho^*$. This clearly implies that $f(\rho/\rho^*)$ *must* go like a power of (ρ/ρ^*) for $\rho > \rho^*$. Let this unknown power be denoted as y. Therefore

$$\frac{\pi}{k_B T} \sim \frac{\rho}{n}\left(\frac{\rho}{\rho^*}\right)^y \sim \rho^{1+y} n^{-1+\frac{4y}{5}} \tag{111}$$

Since this must scale as n^0, we get

$$-1 + \frac{4y}{5} = 0$$

so that $y = \frac{5}{4}$. Substituting this value in equation (111), we obtain

$$\pi \sim \rho^{9/4} \tag{112}$$

which is a nonclassical result.

1.3.3.2 Correlation length ξ

The concept of screening survives in semidilute conditions. At the overlap concentration, by definition, the average distance between the chains is proportional to the Flory radius (R_F) in a good solvent. Therefore we can guess the following form for ξ

$$\xi \sim R_F\left(\frac{\rho}{\rho^*}\right)^{y_1}, \qquad \rho \geq \rho^* \tag{113}$$

where y_1 is unknown. Nevertheless, we can determine y_1 from the fact that ξ should be independent of n for $\rho > \rho^*$, i.e.

$$R_F(\rho n^{4/5})^{y_1} \sim n^0 \tag{114}$$

Since $R_F \sim n^{3/5}$, we get

$$\frac{3}{5} + \frac{4}{5}y_1 = 0 \tag{115}$$

so that $y_1 = -\frac{3}{4}$. Inserting this value of y_1 in equation (113)

$$\xi \sim \rho^{-3/4} \tag{116}$$

1.3.3.3 ⟨R²⟩ of a labeled chain

At $\rho = \rho^*$, the mean square end-to-end distance of a labeled chain should be proportional to R^2 suggesting the general scaling form

$$\langle R^2 \rangle \sim R_F^2\left(\frac{\rho}{\rho^*}\right)^{y_2}, \qquad \rho \geq \rho^* \tag{117}$$

where y_2 is not known. We can picture the chain having its configuration being determined by its own excluded volume until it meets another, and then starting a new excluded walk, *i.e.* being a series of blobs to use de Gennes' term. This suggests that on the large scale the chain will still be Gaussian and the requirement of this condition on equation (117) yields

$$R_F^2(\rho n^{4/5})^{y_2} \sim n \tag{118}$$

which then gives

$$\frac{6}{5} + \frac{4}{5}y_2 = 1 \tag{119}$$

Thus clearly

$$y_2 = -\frac{1}{4} \tag{120}$$

and

$$\langle R^2 \rangle \sim \rho^{-1/4} \tag{121}$$

To use the blob idea directly, let us imagine that the labeled chain is made up of n/g blobs each of length ξ containing g segments (see Figure 10).

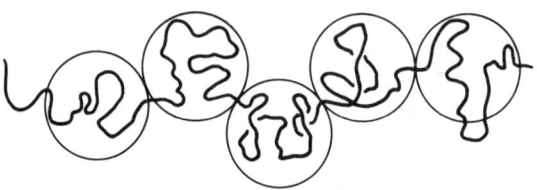

Figure 10   A chain is imagined to be made of n/g blobs each of length ξ containing g segments

Thus a blob is an effective step along the contour of the chain containing many (g) segments. Furthermore, let us assume that: (i) the segments inside the blob obey the excluded volume chain statistics so that $\xi \sim g^{3/5}$; and (ii) the n/g blobs obey the random walk statistics such that $\langle R^2 \rangle = (n/g)\xi^2$. Thus ξ is taken to be the distance up to which the native self-avoidance due to the excluded volume interaction is completely correlated and beyond which it is totally uncorrelated. Since $g \sim \xi^{5/3}$

$$\langle R^2 \rangle \sim \frac{n}{g}\xi^2 \sim d\xi^{1/3} \sim n\rho^{-1/4} \tag{122}$$

which follows from equation (116). This is the same result as in equation (121). Therefore ξ defined as the average distance between different chains in the earlier section is proportional to ξ defined here as the correlation length for the excluded volume effect beyond which the latter is screened.

1.3.4   Extrapolation Formulas

The above scaling laws are valid in semidilute solutions where density fluctuations are dominant. What is the crossover behavior between the semidilute solutions and the dense solutions where the Flory–Huggins theory is adequate? In addition, a knowledge of the various numerical prefactors is lacking in the scaling results. This problem has yet to be solved convincingly but we offer some background here.

In order to provide a general description of the polymer solution at any polymer concentration, a variational theory[120] has recently been presented. The salient features of this theory are as follows.

(i) The correlation of the density fluctuations leads to a renormalization of the bare δ function pseudopotential (excluded volume) interaction, $wl^2\delta(r)$, into an effective screened interaction $\Delta(r)$

$$\Delta(r) = wl^2[\delta(r) - (4\pi\xi^2 r)^{-1}\exp(-r/\xi)] \tag{123}$$

(ii) When small wave vectors dominate the problem, the screening length ξ is given by

$$\xi^{-2} = 6w\rho l/l_1(1 + 27w\xi/8\pi l_1^2)$$

$$l_1^3(1/l - 1/l_1) = 4.216\, w\xi \tag{124}$$

where

$$\langle R^2 \rangle \equiv Ll_1 \tag{125}$$

(iii) In the region of criticality, limit $L \to \infty$ and $\rho \to 0$, equations (124) and (125) reproduce the scaling results

$$l_1 \sim \rho^{-1/4} \quad \text{and} \quad \xi \sim \rho^{-3/4} \tag{126}$$

In the other limit of $\rho \to \infty$, l_1 and ξ are given by

$$l_1 = l + 0\left(\frac{w}{\rho}\right)^{1/2}$$

$$\xi_E = (6w\rho)^{-1/2} \tag{127}$$

thus recovering Gaussian chain dimensions for very high concentrations and the Edwards screening length.

(iv) The total free energy of mixing, ΔF, for a polymer solution of N chains in volume V is

$$\Delta F/k_B T = \left[\frac{\phi}{n}\ln\phi + (1 - \phi)\ln(1 - \phi) + \left(\frac{1}{2} - \frac{w}{l}\right)\phi(1 - \phi)\right]V - \frac{9}{16\pi}\frac{wLN}{l_1\xi} + \frac{V\xi^{-3}}{24\pi} \tag{128}$$

where ϕ is the volume fraction of the polymer, ρl^3. In the limit of $L \to \infty$ and $\rho \to 0$, the osmotic pressure π becomes

$$\frac{\Pi}{k_B T} = \frac{\rho l}{L} + \frac{40\pi}{243}\left(\frac{16\pi\alpha^3}{9}\right)^{1/4} w^{3/4}l^3\rho^{9/4} \tag{129}$$

verifying the scaling result. For very dense solutions

$$\frac{\Pi}{k_B T} = \frac{\rho l}{L} - \frac{5\sqrt{6}}{32\pi}w^{3/2} + wl^2\rho^2$$

$$\to \frac{\rho l}{L} + wl^2\rho^2 \tag{130}$$

the Flory–Huggins result.

As originally pointed out by de Gennes,[35] for temperatures θ and below and at higher concentrations, three-body interactions become important.[122,123] It has also been demonstrated that higher order interactions from four-body onwards lead to corrections[123] of $0(n^{-1/2})$ to the results of calculations of three-body terms so that these higher order terms can safely be ignored. The following formula for the free energy of mixing, ΔF, for a polymer solution has been proposed by a consideration of the three-body interactions in the theory of Muthukumar and Edwards[120]

$$\frac{\Delta F}{k_B T} = \frac{\phi}{n}\ln\phi + (1 - \phi)\ln(1 - \phi) + \chi\phi(1 - \phi) + \left(w_3 - \frac{1}{6}\right)\phi^3 + (24\pi\xi^3)^{-1}$$

$$- \frac{9}{16\pi}\frac{(1/2 - \chi + w_3\phi)\phi}{\alpha^2\xi} \tag{131}$$

$$\xi^{-2} = \frac{6(1/2 - \chi + w_3\phi)\phi}{[\alpha^2 + 27/8\pi(1/2 - \chi + w_3\phi)\xi\alpha^{-2}]} \tag{132}$$

$$\alpha^6 - \alpha^4 = 4.216(1/2 - \chi + w_3\phi)\xi \tag{133}$$

$$\alpha^2 = l_1/l$$

Here $(1/2 - \chi)$ and w_3 are the strengths of two-body and three-body interactions, respectively, in dimensionless units. ξ is expressed in units of l. These formulas recover the scaling law,[7] $\xi \sim \phi^{-1}$ at the θ temperature and at very high polymer concentrations, in addition to the various scaling laws pertinent to the semidilute regime of good solutions.

1.3.5 Phase Equilibria

1.3.5.1 *Dilute polymer solutions*

When the osmotic pressure of a polymer solution given by equation (101) is plotted against the polymer concentration ϕ, a phase separation is predicted if $\chi > \chi_c$ as shown in Figure 11.

This phase separation has been observed experimentally and this prediction is one of the major successes of the Flory–Huggins mean field theory. In general the free energy of the solution depends on ϕ as shown in Figure 12.

It follows from equation (100) that the coexistence curve is obtained as

$$\partial(\Delta F_{\mathrm{m}}/k_{\mathrm{B}}T)/\partial\phi \;=\; \frac{1}{n} \,-\, 1 \,+\, \frac{1}{n}\ln\phi \,-\, \ln(1 \,-\, \phi) \,+\, \chi(1 \,-\, 2\phi) \;=\; 0 \qquad (134)$$

and the spinodal curve is

$$\partial^2(\Delta F_{\mathrm{m}}/k_{\mathrm{B}}T)/\partial\phi^2 \;=\; \frac{1}{n\phi} \,+\, \frac{1}{(1 \,-\, \phi)} \,-\, 2\chi \;=\; 0 \qquad (135)$$

which are schematically plotted in Figure 13.

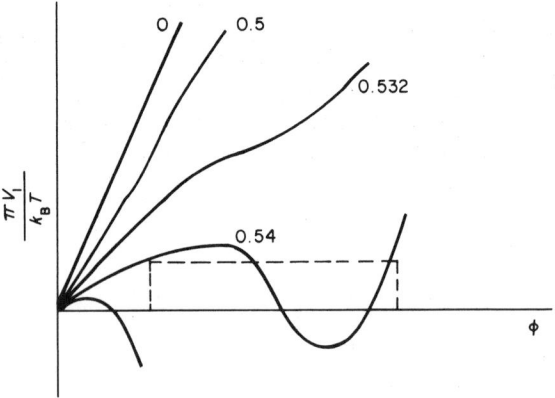

Figure 11   Plot of $\pi\, V_1/k_{\mathrm{B}}T$ *vs.* ϕ for $n = 1000$: the values of χ are indicated on the curves

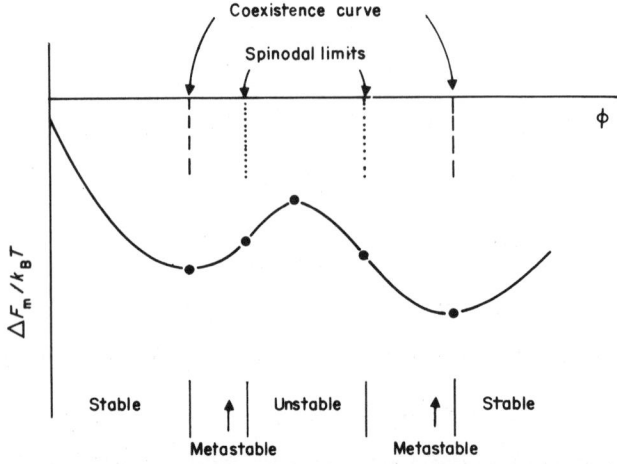

Figure 12   Typical concentration dependence of the free energy

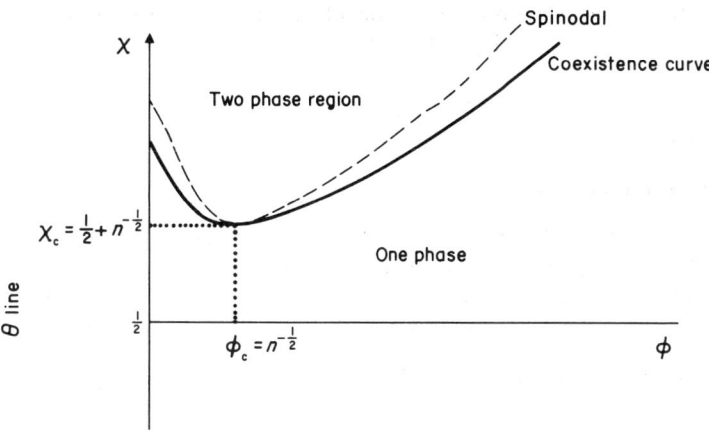

Figure 13 Phase diagram of polymer–solvent system

The critical values of χ and ϕ are obtained from $\partial^3(\Delta F_m/k_B T)/\partial\phi^3 = 0$ and combining with equation (135). The result is

$$\chi_c = (1 + n^{1/2})^2/2n \rightarrow 1/2 + n^{-1/2}, \quad n \rightarrow \infty$$

$$\phi_c = (1 + n^{1/2})^{-1} \rightarrow n^{-1/2}, \quad n \rightarrow \infty \tag{136}$$

As discussed above, the incorporation of density fluctuations leads to corrections to the Flory–Huggins free energy. For example, when three-body interactions are significant, as in the case for temperatures below θ and where phase separation occurs, it has been shown[124] that $\phi_c \sim n^{-1/3}$ and the coexistent polymer volume fraction ϕ along the spinodal curve near the critical point obeys $|\phi - \phi_c|\phi_c \sim n^{1/9}$. These exponents of 1/3 and 1/9 are to be compared with the Flory–Huggins exponents of 1/2 and 1/4 respectively. The experimental data[125–128] support the new exponents.

1.3.5.2 *Polymer blends*

For a polymer 'solution' of polymer 1 in another polymer 2, the Flory–Huggins free energy is

$$\Delta F_m/k_B T = \frac{\phi_1}{n_1}\ln\phi_1 + \frac{\phi_2}{n_2}\ln\phi_2 + \chi\phi_1\phi_2 \tag{137}$$

where n_1 is the number of segments per chain of type 1 and n_2 is that of type 2. $\phi_1 = \phi$ and $\phi_2(= 1 - \phi)$ are the volume fractions of components 1 and 2, respectively. This system also shows phase separation and the critical values are

$$\chi_c = (n_2^{1/2} + n_1^{1/2})^2/2n_1 n_2$$

$$\phi_c = n_2^{1/2}/(n_1^{1/2} + n_2^{1/2}) \tag{138}$$

For the symmetric blend, $n_1 = n_2 = n$, the phase diagram is outlined in Figure 14. Since χ_c is small, strong incompatibility is observed in polymer mixtures.

Based on the 'random phase approximation' the static structure factor $S(k)$ of the blend is given by[7,129]

$$S^{-1}(k) = \frac{1}{\phi n_1 S_1(k)} + \frac{1}{(1 - \phi)n_2 S_2(k)} - 2\chi \tag{139}$$

where $S_i(k)$ is the static structure factor of a single chain of type i (see equation 7). This satisfies the 'sum rule'

$$S^{-1}(0) = \partial^2(\Delta F/k_B T)/\partial\phi^2 \tag{140}$$

Therefore at each point of the spinodal curve, the scattered intensity [which is proportional to $S(k)$] in small-angle scattering diverges. For the symmetric blend, $n_1 = n_2 = n$, it follows from

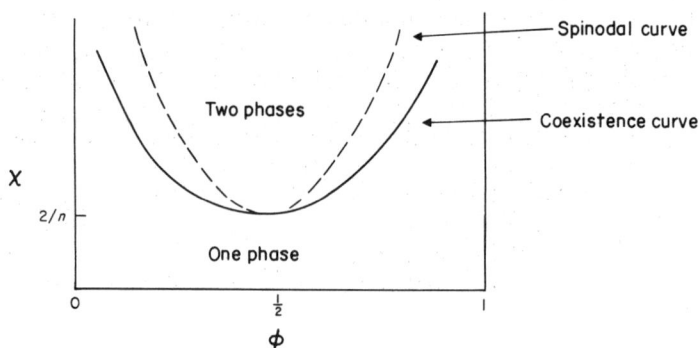

Figure 14   Phase diagram for a binary polymer blend

equations (7) and (139) at small k that

$$S^{-1}(k) = 2(\chi_s(\phi) - \chi) + \frac{k^2 l^2}{18\phi(1 - \phi)} \tag{141}$$

where $\chi_s = 1/n\phi(1 - \phi)$ is the value of χ on the spinodal. Equation (141) shows that $S(k)$ is of 'Ornstein–Zernike form'

$$S(k) \propto (1 + k^2 \xi_s^2)^{-1}$$

$$\xi_s = \frac{l}{6}[\phi(1 - \phi)(\chi_s(\phi) - \chi)]^{-1/2} \tag{142}$$

where ξ_s diverges near the spinodal.

Equation (139) is used to determine the interaction parameter χ by measuring $S^{-1}(k \to 0)$. It has been found that χ so determined is composition dependent.[130-133] The composition dependence of χ can actually be attributed[134] to the contribution of density fluctuations to the free energy of the blend which are ignored in equation (139).

1.4   DYNAMICS

1.4.1   Dynamics of Chains in Dilute Solutions

When a large rigid sphere is suspended in a solvent, its dynamical properties, such as the velocity autocorrelation function, mean square displacement, dynamic structure factor, *etc.*, are well known,[135-161] and form the basis of suspension theory from Einstein onwards.

If this Brownian particle is not a rigid sphere but made up of a linear flexible chain, then how do the various dynamical properties get modified? Clearly we do not expect any change in the dynamical laws if our probe explores very large length and time scales which are so large that only dynamics characteristic of the motion of the *whole* chain is monitored. However, if the length and time scales correspond to the dynamics inside the coil, we expect different dynamical laws. On the other hand, if the time scale is too short, such as nanoseconds, and the length scale is also short, such as a few bonds, then we expect to see conformational isomerization dynamics characteristic of monomers only. Thus we can identify three regimes for the dynamics of an isolated chain in a dilute solution: (i) the short time regime displaying the characteristics of the monomer; (ii) the very long time regime corresponding to the Brownian motion of the whole chain; and (iii) the intermediate regime where the attributes of the chain connectivity are dominant.[162-220]

Since, in the intermediate regime, the time explored is longer than the characteristic time for the various *trans–gauche* conformational transitions (which in turn depends on the chemical nature of the monomers), the dynamical behavior of the polymer chain in this regime is independent of the chemistry of the chain. We call this independence of the local details as the global features of the chain dynamics.

We need to know the characteristic time beyond which we will cease to observe the global chain dynamics but only the center of mass motion. In answering such a question we need the relaxation time for a polymer chain which has been disturbed from its equilibrium configuration, and the mean square displacement of a labeled monomer as a function of time.

The most important interaction present in a dilute suspension of a chain is the hydrodynamic interaction. In addition, the excluded volume interaction may be present depending on the nature of the solvent, polymer and temperature; this interaction vanishes at the θ temperature, so there is already an important problem when all interactions are ignored. However the interachain entanglement effects corresponding to the uncrossability of different portions of the same chain are always present. The simplest model which cannot actually be realized physically is to both ignore interactions and hydrodynamic effects and assume *ad hoc* that the solvent attributes a phenomenological friction coefficient for every monomer. This model is called the Rouse model.

1.4.1.1 *Rouse model*

Historically Rouse took the chain to be a Gaussian chain with n beads and Kuhn step length l, so that the chain contour length is $L = nl$. The equation of motion for an arbitrary bead, say ith, belonging to this model chain can be readily written from the following physical considerations. First, there is the random force $\mathbf{f}(t)$ from the solvent acting on the ith bead. Second, the ith bead experiences an elastic force from the fact that it is connected to its neighbors $i + 1$ and $i - 1$ (if i is not the end of the chain; if not, obvious modifications can be made). The force exerted on the ith bead by its neighbors is given by

$$-\frac{3k_{\mathrm{B}}T}{l^2} \sum_{j=1}^{n} A_{ij}\mathbf{R}_j \tag{143}$$

where A_{ij}, called the Rouse matrix, is defined to be

$$A_{ij} = 2\delta_{ij} - \delta_{i,j+1} - \delta_{i,j-1} \tag{144}$$

with δ_{ij} as the Kronecker delta.

The sum of these random and elastic forces must balance all the other forces acting on i which are: (i) inertial force, $m_i\ddot{\mathbf{R}}_i(t)$, where $\ddot{\mathbf{R}}_i$ is the double time derivative of $\mathbf{R}_i(t)$, and m_i is the effective mass of the bead; (ii) frictional force, $\zeta_b\dot{\mathbf{R}}_i(t)$, where ζ_b is the bead friction coefficient which is obviously related to the random force $\mathbf{f}_i$ as in the case of Brownian motion.

Thus the equation of motion for the ith bead is

$$m\ddot{\mathbf{R}}_i(t) + \zeta_b\dot{\mathbf{R}}_i(t) = \mathbf{f}_i(t) - \frac{3k_{\mathrm{R}}T}{l^2} \sum_{j=1}^{n} A_{ij}\mathbf{R}_j(t) \tag{145}$$

This is the Langevin equation[10] for the chain bead. Since the polymer chain is so large that the acceleration of the various beads is very small, the inertial term of equation (145) can be ignored. This is a very good approximation for experimental situations where only small frequencies are important. With this assumption we get

$$\zeta_b\dot{\mathbf{R}}_i(t) + \frac{3k_{\mathrm{B}}T}{l^2} \sum_{j} A_{ij}\mathbf{R}_j(t) = \mathbf{f}_i(t) \tag{146}$$

If we suppress the random force term, equation (146) is called the Rouse equation and the dynamical properties can be calculated by transforming equation (146) into an eigenvalue problem. It is important to realize that beads and springs, although hallowed by time in the polymer literature, have no reality and one can derive the Rouse result directly by noting the free energy of a Gaussian coil is $kT\int\left(\frac{\partial r}{\partial s}\right)^2 ds$ so that either by differentiating this, or by taking equation (146) in the continuous notation, we obtain

$$\zeta\dot{\mathbf{R}}(s, t) - \frac{3k_{\mathrm{B}}T}{l}\frac{\partial^2\mathbf{R}(s, t)}{\partial s^2} = \mathbf{f}(s, t) \tag{147}$$

where all the terms of equation (146) have been divided by a factor of Kuhn length (for technical convenience) so that $\zeta \equiv \frac{\zeta_b}{l}$. We now define the Fourier transforms

$$\mathbf{R}(q, t) = \int_0^L ds\, \mathbf{R}(s, t)\exp(iqs) \tag{148}$$

$$\mathbf{R}(q, \omega) = \int_{-\infty}^{\infty} dt\, \mathbf{R}(q, t) \exp(i\omega t) \tag{149}$$

The variable q, which is conjugate to the arc length position variable s, labels the normal mode. If equation (148) is written in discretized (bead variable) notation, q is actually $2\pi p/L$ where $p = 0, 1, 2, 3, \ldots, n$. The integer variable p is called the Rouse mode variable. Thus a small q corresponds to the lower Rouse modes and hence large length scales inside the chain; and *vice versa* for large q values. As shown below, $p = 0$ mode represents the motion of the center of mass of the chain. Also notice that for small values of p, q is proportional to L^{-1}. The frequency ω and time t are Fourier conjugate variables.

Multiplying each term of equation (147) by $\exp(iqs + i\omega t)$ and then integrating over s and t, we get the equation

$$\left(i\omega\zeta + \frac{3k_{\mathrm{B}}T}{l}q^2 \right)\mathbf{R}(q, \omega) = \mathbf{f}(q, \omega) \tag{150}$$

Assuming that

$$\langle \mathbf{f}(q, \omega) \cdot \mathbf{f}(q', \omega') \rangle \propto \delta(q + q')\delta(\omega + \omega') \tag{151}$$

and that the proportionality factor is so as to recover the equilibrium result for the mean square end-to-end distance

$$\langle [\mathbf{R}(0) - \mathbf{R}(L)]^2 \rangle = Ll \tag{152}$$

the time-dependent correlation function of segment-to-segment distance is given by

$$\langle [\mathbf{R}(s, t) - \mathbf{R}(s', t')]^2 \rangle = \frac{2l}{\pi} \int_0^{\infty} \frac{dq}{q^2} \left[1 - \cos q(s - s') \exp\left(-\frac{|t - t'|}{\tau_q} \right) \right] \tag{153}$$

where

$$\tau_q = \frac{l\zeta}{3k_{\mathrm{B}}Tq^2} \tag{154}$$

τ_q is the relaxation time for the wave number variable q. The result of equation (154) may be obtained rather readily by looking at the structure of equation (150). $3k_{\mathrm{B}}Tq^2/l\zeta$ is the characteristic frequency which matches with the experimental frequency ω. The reciprocal of this characteristic frequency is the relaxation time. Writing q in terms of the Rouse mode index p, $q = \dfrac{2\pi p}{L}$, we get

$$\langle [\mathbf{R}(s, t) - \mathbf{R}(s', t')]^2 \rangle = \frac{Ll}{\pi^2} \sum_{p=1}^{\infty} p^{-2} \left[1 - \cos\left(\frac{2\pi p(s - s')}{L} \right)\exp\left(-\frac{|t - t'|}{\tau_p} \right) \right] \tag{155}$$

where the pth Rouse mode relaxation time τ_p is given by

$$\tau_p = \frac{L^2 l\zeta}{12\pi^2 k_{\mathrm{B}}Tp^2} \tag{156}$$

Thus the longest relaxation time τ_1 is proportional to L^2 and hence to the square of the molecular weight. τ_1 is called the Rouse relaxation time. Various dynamical quantities of interest can now be deduced directly from equation (156).

(i) Mean square displacement of a labeled monomer

Substituting $s' = s$ in equation (153), we get the following result for the mean square displacement of the segment position s

$$\langle [\mathbf{R}(s, t) - \mathbf{R}(s, t')]^2 \rangle = \frac{2l}{\pi} \int_0^{\infty} \frac{dq}{q^2} [1 - \exp(-3k_{\mathrm{B}}Tq^2|t - t'|/l\zeta)] \tag{157}$$

For extremely short times the exponential terms appearing in the integrand of equation (157) can be expanded as a power series in $|t - t'|$ and the leading term is

$$\lim_{t \to t'} \langle [\mathbf{R}(s, t) - \mathbf{R}(s, t')]^2 \rangle = \frac{6k_B T}{\pi \zeta} \int_0^{2\pi/l} dq |t - t'| \tag{158}$$

$$= \frac{12k_B T}{\zeta_b} |t - t'| \tag{159}$$

In equation (158) the correct upper limit of $2\pi/l$ (instead of the approximate infinite limit) has been used. Thus for very short times such that the segment is not influenced by the neighbor segments, it behaves just like a Brownian particle. However, for times that are not small, equation (157) can be rewritten with the change of variable, $x = 3k_B Tq^2 |t - t'|/l\zeta$

$$\langle [\mathbf{R}(s, t) - \mathbf{R}(s, t')]^2 \rangle = \frac{2l}{\pi} \left(\frac{3k_B T|t - t'|}{l\zeta} \right)^{1/2} \int_0^\infty \frac{dx}{x^2} [1 - \exp(-x^2)] \propto |t - t'|^{1/2} \tag{160}$$

Due to the coupling to the neighbors, the motion of the segment is even slower than the Brownian motion. The fact that the segments are hinged together in the fashion of a random flight, reduces the time exponent for the mean square displacement of a segment by a factor of two. Thus one can identify three regimes if $\ln\langle [\mathbf{R}(s, t) - \mathbf{R}(s, 0)]^2 \rangle$ is plotted against $\ln t$ as in Figure 15.

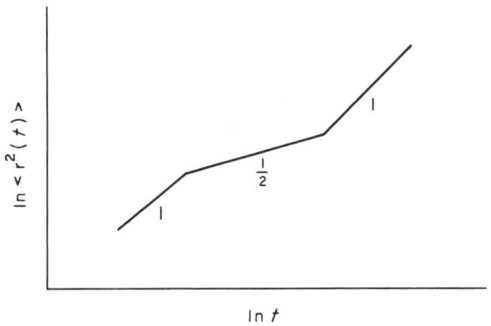

Figure 15 Double logarithmic plot of mean square monomer displacement *vs.* time: the first two regimes follow from equations (159) and (160) and the third regime corresponds to the Brownian motion

(ii) Mean square displacement of the center of mass of the chain

The center of mass of the chain, $\mathbf{R}_0(t)$ is given by

$$\mathbf{R}_0(t) = \frac{1}{L} \int_0^L ds\, \mathbf{R}(s, t) \tag{161}$$

and

$$\langle [\mathbf{R}_0(t) - \mathbf{R}_0(t')]^2 \rangle = \frac{6k_B T}{\zeta L} |t - t'| \tag{162}$$

Therefore the center of mass of the chain obeys the diffusive law, although the monomer displacement has a weaker time dependence.

The above results for the Rouse model are applicable to the experimental conditions where the hydrodynamic and excluded volume interactions and the entanglement effects can be completely ignored. We shall identify such an experimental regime later on. Now, we attempt to incorporate the hydrodynamic interaction in describing the chain dynamics in infinitely dilute solutions. The Rouse chain model incorporating the effect of hydrodynamic interaction is called the Kirkwood–Riseman model[179] or Zimm[170] model. These models differ from each other in certain subtle features and the numerical prefactors only; the predicted molecular weight dependence of the longest relaxation time, viscosity of the solution, diffusion coefficient, *etc.* are the same.

1.4.1.2 Kirkwood–Riseman–Zimm Model for hydrodynamic interaction

Consider a certain volume of a solvent to which a polymer will be introduced later on. Since, in general, we do not worry about the dynamics at a single solvent molecule level, we approximate the dynamics of the solvent as that of a continuous hydrodynamic continuum. The equation of motion for such a continuum is assumed to be adequately described by the linearized Navier–Stokes equation.[138,139] Assuming that the fluid is incompressible the velocity field $v(r)$ at r is related to the force field $\mathbf{F}(r')$ at r' as given by

$$v(r) = \int dr' \, G(\mathbf{r} - \mathbf{r}') \cdot \mathbf{F}(r') \tag{163}$$

where G, called the Oseen tensor, is

$$G(\mathbf{r} - \mathbf{r}') = \frac{1}{8\pi\eta_0|\mathbf{r} - \mathbf{r}'|}\left[1 + \frac{(\mathbf{r} - \mathbf{r}')(\mathbf{r} - \mathbf{r}')}{|\mathbf{r} - \mathbf{r}'|^2} \right] \tag{164}$$

where η_0 is the solvent viscosity. The important aspect of the hydrodynamic interaction is the long-ranged $|r - r'|^{-1}$ dependence of the coupling between the velocity and the force at two different points. Incorporating this, the Rouse equation (147) is modified to

$$\zeta\dot{\mathbf{R}}(s, t) - \frac{3k_B T}{l}\int_0^L ds' \, \mathbf{D}[\mathbf{R}(s) - \mathbf{R}(s')]\cdot\frac{\partial^2}{\partial s'^2}\mathbf{R}(s', t) = \mathbf{f}(s, t) \tag{165}$$

with

$$\mathbf{D}[\mathbf{R}(s) - \mathbf{R}(s')] = 1\delta(s - s') + \frac{\zeta}{8\pi\eta_0|\mathbf{R}(s) - \mathbf{R}(s')|}\left[1 + \frac{(\mathbf{R}(s) - \mathbf{R}(s'))^2}{|\mathbf{R}(s) - \mathbf{R}(s')|^2} \right] \tag{166}$$

The R dependence of $\mathbf{D}$ makes equation (165) nonlinear and presents technical difficulties[6] in obtaining the various dynamical properties of the chain. In order to simplify the calculation, we perform the preaveraging approximation[6] of replacing $\mathbf{D}[\mathbf{R}(s) - \mathbf{R}(s')]$ by its configuration-averaged value

$$1D(s - s') = \langle \mathbf{D}[\mathbf{R}(s) - \mathbf{R}(s')]\rangle = 1[\delta(s - s') + \frac{1}{6\pi\eta_0}\langle|\mathbf{R}(s) - \mathbf{R}(s')|^{-1}\rangle] \tag{167}$$

which depends only on $s - s'$. Therefore the preaveraged Langevin equation for the chain dynamics becomes

$$\zeta\dot{\mathbf{R}}(s, t) - \frac{3k_B T}{l}\int_0^L ds' \, D(s - s')\frac{\partial^2}{\partial s'^2}\mathbf{R}(s', t) = \mathbf{f}(s, t) \tag{168}$$

Using the Fourier transforms given by equations (148) and (149), we obtain

$$\left(i\omega\zeta + \frac{3k_B T}{l}D(q)q^2 \right)\mathbf{R}(q, \omega) = \mathbf{f}(q, \omega) \tag{169}$$

where

$$D(q) = 1 + \zeta/\eta_0(lq)^{1/2} \tag{170}$$

In the hydrodynamic limit of small q, the second term dominates so that $D(q) \sim q^{-1/2}$. This in turn modifies the longest relaxation time of the chain.

Following the same procedure as in the case of the Rouse model, the time-dependent correlation function of segment-to-segment distance is obtained from equation (169)

$$\langle[\mathbf{R}(s, t) - \mathbf{R}(s', t')]^2\rangle = \frac{2l}{\pi}\int_0^\infty \frac{dq}{q^2}\left[1 - \cos q(s - s')\exp\left(-\frac{|t - t'|}{\tau(q)} \right) \right] \tag{171}$$

where the relaxation times are given by

$$\tau(q) \quad = \quad \frac{l\zeta}{3k_{B}Tq^{2}D(q)} \tag{172}$$

$$= \quad \frac{l}{3k_{B}Tq^{2}\left[\dfrac{1}{\zeta} \quad + \quad 1/\eta_{0}(lq)^{1/2}\right]} \tag{173}$$

We have used equation (170) in obtaining equation (173). Two limits have been identified in the literature.

(i) Free draining limit. In this limit, the hydrodynamic interactions are not important so that the second term on the right of equation (170) is ignored. This is the same as the Rouse limit as discussed in the previous section.

(ii) Non-free draining limit. In this limit, the hydrodynamic interactions are dominant so that the first term on the right of equation (170) can be ignored. Now

$$\tau(q) \quad = \quad \frac{\eta_{0}l^{3/2}}{3k_{B}Tq^{3/2}} \tag{174}$$

Substitution of the relation $q = 2\pi p/L$ yields

$$\tau_{p} \quad = \quad \frac{\eta_{0}l^{3/2}L^{3/2}}{3(2\pi)^{3/2}k_{B}Tp^{3/2}} \tag{175}$$

In this Zimm limit τ_{1} is proportional to $L^{3/2}$ and τ_{p} is proportional to $p^{-3/2}$. Substitution of equation (174) into equation (171) and a change of variable gives

$$\langle[\mathbf{R}(s, t) \quad - \quad \mathbf{R}(s, t')]^{2}\rangle \quad \sim \quad |t \quad - \quad t'|^{2/3} \tag{176}$$

Thus the monomer displacement is faster and the longest relaxation time is shorter in the Zimm model than in the Rouse model.

1.4.2 Many Chains

These bead–spring models of Rouse and Kirkwood–Riseman–Zimm suffer from the artificiality of the beads and springs. The bead friction coefficient is an *ad hoc* phenomenological coefficient. This should arise naturally from the frictional forces coupling the polymer and solvent directly with the continuous version of the chain without beads and springs.

The experimental data on the viscosity of a solution of linear flexible chains show that the Zimm model provides an adequate description at very low polymer concentrations. However, at higher concentrations (but not very high) the viscosity of the solution scales with the molecular weight, a result predicted by the Rouse model which ignores the hydrodynamic interaction. This suggests that the long-ranged hydrodynamic interaction coupling any two segments of a chain is screened by the presence of other chains as the polymer concentration is progressively increased until ultimately the hydrodynamic interaction is screened out. The concentration dependent screening of r^{-1}-like hydrodynamic interaction is analogous to the familiar Debye screening of the electrostatic interaction in a solution of electrolytes. The difference is that hydrodynamic interaction is of dynamical origin while the electrostatic screening is of static origin. The idea of the screening of hydrodynamic interaction has been first treated by Freed and Edwards for a polymer solution at arbitrary concentration. Their theory has been later refined by Edwards and Freed and their collaborators.[176,177] Instead of giving all the details of these derivations, we present below only the salient features of the idea of hydrodynamic screening.

1.4.2.1 Basic equations

Consider a solvent containing N Gaussian chains where the equations of motion of the chains and the solvent are coupled by the 'no-slip' boundary condition

$$\frac{\partial}{\partial t}\mathbf{R}_{\alpha}(s_{\alpha}, t) \quad = \quad \mathbf{v}[\mathbf{R}_{\alpha}(s_{\alpha}t)] \tag{177}$$

where $\mathbf{R}_\alpha(s, t)$ is the position vector of the arc length position s_α of chain α. (It turns out that other boundary conditions can be used without major changes.)[191] Representing this constraint as a Lagrange multiplier[176] $\sigma_\alpha(s_\alpha)$, the coupled equations for the polymer solution become (in the zero-frequency limit)

$$-\eta_0 \nabla^2 v(r) \ + \ \nabla p(r) \ = \ \mathbf{F}(r) \ + \ \sum_{\alpha=1}^{N} \int_0^L ds_\alpha \delta[\mathbf{r} \ - \ \mathbf{R}_\alpha(s_\alpha)]\sigma_\alpha(s_\alpha) \tag{178}$$

$$-\frac{3k_B T}{l}\frac{\partial^2 \mathbf{R}_\alpha(s_\alpha)}{\partial s_\alpha^2} \ = \ -\sigma_\alpha(s_\alpha) \tag{179}$$

where p is the pressure. All other terms have already been defined. The objective now is to determine $\sigma_\alpha(s_\alpha)$ and then calculate the microscopic velocity field from equation (178) which upon averaging over the distributions of polymer chain configurations gives the various transport coefficients. Within the multiple scattering formalism[186] the effective equation of motion for the polymer solution becomes

$$-\eta_0 \nabla^2 \langle v(r)\rangle \ + \ \nabla\langle p(r)\rangle \ + \ \int dr' \, \xi_H^{-2}(r \ - \ r') \cdot \langle v(r)\rangle \ = \ \mathbf{F}(r) \tag{180}$$

where the angular brackets indicate the above-mentioned average and the structure of the kernal ξ_H^{-2} is given below.

Using the Fourier transform we get from equation (180)

$$[\eta_0 k^2 \mathbf{1} \ + \ \xi_H^{-2}(k)] \cdot \langle v(k)\rangle \ + \ ik'p(k)\rangle \ = \ \mathbf{F}(k) \tag{181}$$

$\xi_H^{-2}(k)$ contains all the consequences of the presence of polymer chains in the solution. Since we are interested in the hydrodynamic properties which reflect on large length scale behavior, we can expand $\xi_H^{-2}(k)$ as a Taylor series in k^2

$$\xi_H^{-2}(k) \ = \ \xi_H^{-2}(k \ = \ 0) \ + \ \xi_1 k^2 \ + \ \xi_2 k^4 \ + \dots \tag{182}$$

where ξ_1, ξ_2, *etc.* are coefficients. It is clear from the structure of equations (181) and (182) that the transverse component of ξ_1 should be the change $\delta\eta$ in the shear viscosity of the solvent due to the added polymers

$$\delta\eta \ = \ \eta \ - \ \eta_0 \ = \ \lim_{k\to 0} k^{-2}(\mathbf{1} \ - \ \hat{k}\hat{k})\cdot\xi_H^{-2}(k) \tag{183}$$

where η is the shear viscosity of the polymer solution. We expect the polymer solution to have the same fluid flow symmetry as that of the pure solvent obeying the Navier–Stokes equation. This means that $\xi_H^{-2}(k = 0)$ must identically be zero. A correct theory[176] of $\xi_H^{-2}(k)$ should possess this virtue in the hydrodynamic limit of $k \to 0$.

However, it turns out that $\xi_H^{-2}(k)$ is independent of k over a broad range of k values with interesting consequences. Writing the transverse component of the k-independent $\xi_H^{-2}(k)$ as ξ_H^{-2}, it is obvious from equation (181) that ξ_H is the hydrodynamic screening length. Clearly, the hydrodynamic interaction cannot be screened in the large length scales of the hydrodynamic limit $(k \to 0)$, as argued above. Thus the question is under what conditions does such screening occur, *i.e.* when is $\xi_H^{-2}(k)$ independent of k? Furthermore, how does ξ_H depend on the chain length, polymer concentration, solvent quality, *etc*?

1.4.2.2   Results

For arbitrary polymer concentrations, an analysis of the effective medium theory provides the following coupled integral equations for $\xi_H^{-2}(k)$ [which is the transverse component of $\xi_H^{-2}(k)$]

$$\xi_H^{-2}(k) \ = \ \frac{cN_A L}{\pi M}\int_{2\pi/L}^{\infty} dq \frac{k^2 l_1(q, c)/3}{\left\{\left[\dfrac{k^2 l_1(q, c)}{6}\right]^2 + q^2\right\}}G^{-1}(q) \tag{184a}$$

$$G(q) \ = \ \frac{1}{3\pi^2}\int_0^{\infty} dj \frac{j^2}{[\eta_0 j^2 \ + \ \xi_H^{-2}(j)]}\frac{j^2 l_1(qc)/3}{\left\{\left[\dfrac{j^2 l_1(qc)}{6}\right]^2 + q^2\right\}} \tag{184b}$$

where l_1 is given by equation (124), c is the polymer concentration, N_A is the Avogadro number and M is the molecular weight. If the excluded volume effect were completely absent, l_1 is the Kuhn length. Also at very high polymer concentrations where the excluded volume interaction is fully screened $l_1 = l$. At intermediate concentrations, l_1 is a function of c. In dilute solutions, l_1 is mode dependent as given by

$$l_1^{5/2}\left(\frac{1}{l} - \frac{1}{l_1}\right) = \frac{2\sqrt{3}}{\pi}wq^{-1/2} \tag{185}$$

We now summarize the consequences of equation (184) in the various interesting limiting cases.

(i) Infinitely dilute limit

As $\xi_H^{-2}(k)$ is directly proportional to c according to equation (139a), the $\xi_H^{-2}(j)$ of equation (139b) can be ignored in this limit. It then follows that $\xi_H^{-2}(k)$ need only be solved where k^{-1} is macroscopic, i.e. the $k^2 l_1$ in the denominator in equation (184a) can be ignored and $\xi_H^{-2} \sim k^2$. It is only if one studies the interior of the polymer that ξ^{-2} depends on k differently and we do not need this information.

When the excluded volume effect is absent, the intrinsic viscosity, $[\eta]$, is obtained from equations (183) and (184) to be

$$[\eta] = \lim_{c \to 0} \frac{\eta - \eta_0}{\eta_0 c} \tag{186}$$

$$= \left(\frac{1}{6\pi}\right)^{1/2}\frac{N_A}{M}(Ll)^{3/2}\sum_{p=1}^{\infty} p^{-3/2} \tag{187}$$

It can be rewritten as

$$[\eta] = \frac{RT}{M\eta_0}\sum_{p=1}^{\infty}\tau_p \tag{188}$$

$$\tau_p = \eta_0(Ll)^{3/2}/(6\pi)^{1/2}k_B T p^{3/2} \tag{189}$$

where R is the gas constant and τ_p is the relaxation time of the pth mode.

Thus, for dilute θ solutions, the intrinsic viscosity is proportional to $M^{1/2}$, the longest relaxation time is proportional to $M^{3/2}$ and the pth mode relaxation time is proportional to $p^{-3/2}$.

When the excluded volume effect is fully present a similar analysis shows that

$$[\eta] \propto M^{0.8}, \quad \tau_p \propto p^{-1.8}, \quad \tau_1 \propto M^{1.8} \tag{190}$$

in the asymptotic limit of $M \to \infty$.

(ii) Screening conditions

The equations (184a) and (184b) must now be solved. Here, although the macroscopic viscosity comes from

$$\xi_H^{-2} = \frac{cN_A Lk^2}{\pi M}\int_{2\pi/L}^{\infty}dq\frac{l_1}{3q^2}G^{-1}(q)$$

we need ξ_H^{-2} for all k since it appears significantly in equation (184b) and arguments can be made to obtain the asymptotic behavior. Over the important range of integration in equation (184b), $\xi_H^{-2}(k)$ is not dependent on k. This independence of ξ_H^{-2} on k implies that the hydrodynamic interaction is screened. The hydrodynamic screening length ξ_H is given by

$$\xi_H^{-1} = \pi c N_A Ll_1\eta_0^{1/2}/2M \tag{191}$$

Therefore ξ_H is proportional to $c^{-1/2}$ in θ solutions. If the solvent is good, l_1 can have concentration dependence, depending on what the polymer concentration is. (For example, l_1 is proportional to $c^{-1/4}$ under semidilute conditions of a good solution so ξ_H is proportional to $c^{-3/4}$.)

However, for small k, $\zeta_H^{-2}(k)$ is proportional to k^2 as given by

$$\zeta_H^{-2} = (1/24)(cN_AL/M)^2\eta_0 l_1^3 L k^2 \tag{192}$$

Insertion of equation (192) into equation (183) gives the viscosity of the solution η to be

$$\frac{\eta - \eta_0}{\eta_0} = (cN_AL/M)^2 l_1^3 L \tag{193}$$

$$\propto c^2 l_1^3 L \tag{194}$$

The proportionality of the specific viscosity to L is the Rouse limit.

Thus, when $\ln(\eta - \eta_0)$ is plotted against $\ln(c^2 M)$ as in Figure 6 two regimes may be identified. The Zimm regime has a slope of $1/2$ and the Rouse regime has a slope of 1. The crossover is due to the screening of hydrodynamic interaction.

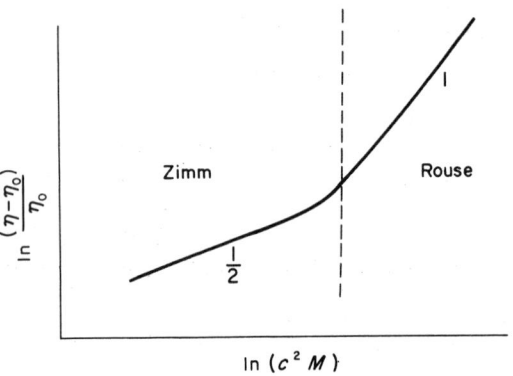

Figure 16 Zimm and Rouse regimes

(iii) Intermediate concentrations

For low, but nonzero concentrations, the concentration dependence of τ_p has been calculated to be

$$\frac{\tau_p}{\tau_p^0} = 1 + (cAp^{-k}) - 2^{1/2}(cAp^{-k})^{3/2} + 2(cAp^{-k})^2 + 0(c^{5/2}) \tag{195}$$

where τ_p^0 is the value of τ_p at infinite dilution, A is a constant in units of inverse concentration. The positive constant k is $1/2$ in θ solvents and 0.8 in good solvents. An explicit expression for A is known. The predictions of equation (195) are compared in Figure 17 with the flow–birefringence data on polystyrene in Aroclor solvent by Schrag et al.[189]

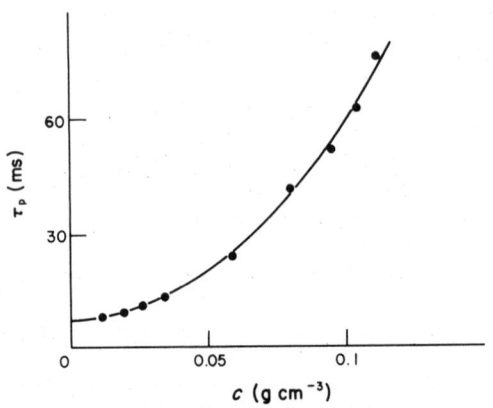

Figure 17 Concentration dependence of τ_p

1.4.3 Entanglements

All the previous results are valid only if the entanglement effects are absent.[183] However, for a polymer solution above a certain critical concentration C^{**} containing chains with molecular weight higher than a critical molecular weight M_c, new universal features emerge in the dynamical properties. Under these experimental conditions, the polymer solution is a tangled state as depicted in Figure 18.

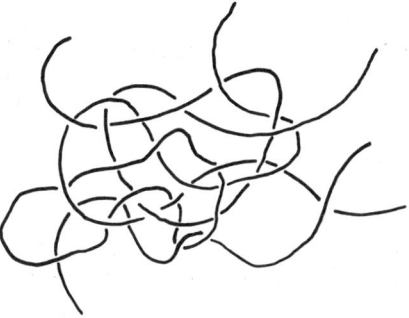

Figure 18   Entangled polymer solution

The entanglement constraints that two portions of either the same chain or different chains cannot intersect lead to very long relaxation times. The viscoelastic and mechanical properties of a system with dominant entanglement effects are distinctly different from those of a solution where Rouse dynamics are valid. For example, the zero shear rate viscosity varies as $M^{3.4}$ instead of the Rouse behavior of $\eta \sim M$ (as shown in Figure 19).

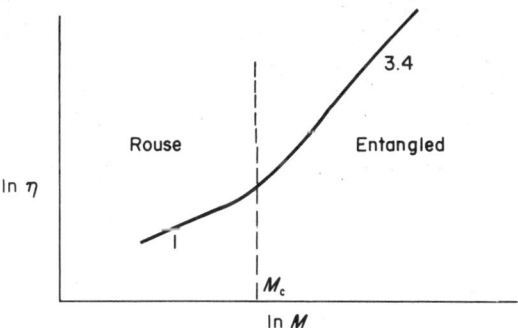

Figure 19   Rouse and entanglement regimes

How do we describe the entanglements and how do we calculate the relaxation times of a chain and other dynamical properties of the entangled system? The statistical mechanics of such a disordered system with the uncrossability constraints is very difficult to formulate. Nevertheless there have been three kinds of approaches.

1.4.3.1   *Topological constraints*

If the entanglement effect can be represented[192-207] in terms of a constraint, then the reduction in the number of configurations of chains can be calculated due to the presence of such constraints. From the change in the number of configurations, the transport properties can, in principle, be calculated by finding the local entropic force from the probability distribution function. The representation of these constraints should be such that it distinguishes various different topological states. In other words, we need an invariant to represent the invariance of a topological state under deformation. Two invariants have been attempted so far.

(i) Integral invariant[192-201]

Consider for two curves $\mathbf{r}_1(s)$ and $\mathbf{r}_2(s)$, the Gauss looping integral is

$$\frac{1}{4\pi}\oint_1\oint_2 \frac{(d\mathbf{r}_1 \times d\mathbf{r}_2)\cdot(\mathbf{r}_1 - \mathbf{r}_2)}{|\mathbf{r}_1 - \mathbf{r}_2|^3} = m \qquad (196)$$

where m is the number of times $\mathbf{r}_2$ winds around $\mathbf{r}_1$ (for example see Figure 20).

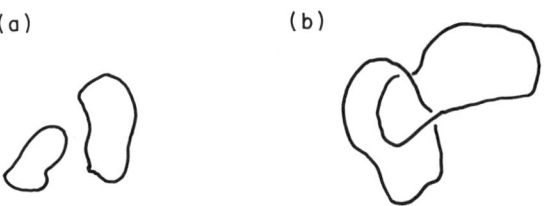

Figure 20   Illustration of the entanglement index, m, for two curves $r_1(s)$ and $r_2(s)$: (a) $m = 0$; (b) $m = 1$

Using this constraint, the configuration sum of a polymer with entanglement index m (with respect to another chain) can be calculated. This problem is the same as the Aharonov–Bohm effect encountered in quantum mechanics. Although the entropic force from the topological constraints has approximately been calculated, the dynamical consequences are still unexplored.

(ii) Algebraic invariant[202-207]

The Alexander polynomial, encountered in the algebraic topology, can be used to distinguish different topological states particularly in computer simulations. Interesting quantities like the probability that two rings are entangled as a function of distance between their centers of mass can readily be computed. However, the modification on the Rouse relaxation time is unknown.

It is to be noted that both invariants discussed above are not complete invariants and more work needs to be done.

1.4.3.2   *Mean field ideas*

(i) Primitive path

When a labeled chain in the entangled state of Figure 16 is projected on a plane, the plane will look as shown in Figure 21. The continuous curve is the projection of the labeled chain. The dots represent the various chains passing through the plane. Since the collective disappearance of all the dots surrounding the continuous curve will take a long time, Edwards assumed that the labeled chain is essentially localized in a tube-like region.[208] The contour of this tube can be approximately obtained by holding one end of the chain, pulling the other end and finding the shortest path of the continuous curve which originally contains slacks. This contour is called the primitive path which is the shortest curve with the same topology as the real chain relative to other chains. The real chain is wriggling around the primitive path. Since the primitive path is only a mathematical construct, we further assume for the sake of simplicity, that it is a random walk with N steps such that

$$Na^2 = Ll \qquad (197)$$

The step length a of the primitive path is assumed to be comparable to the tube diameter which is unknown.

(ii) Reptation

de Gennes[209] argued that the motion of the polymer chain into the walls of the 'tube' cannot be dominant in the zeroth-order picture and assumed that the chain changes its configuration only through the diffusion along the tube (primitive path). He visualized kinks along the chain as one dimensional defects which eventually exited from the tube. This one dimensional random walk of the chain is called *reptation*. Thus it is obvious from the model that in order for the chain to change its

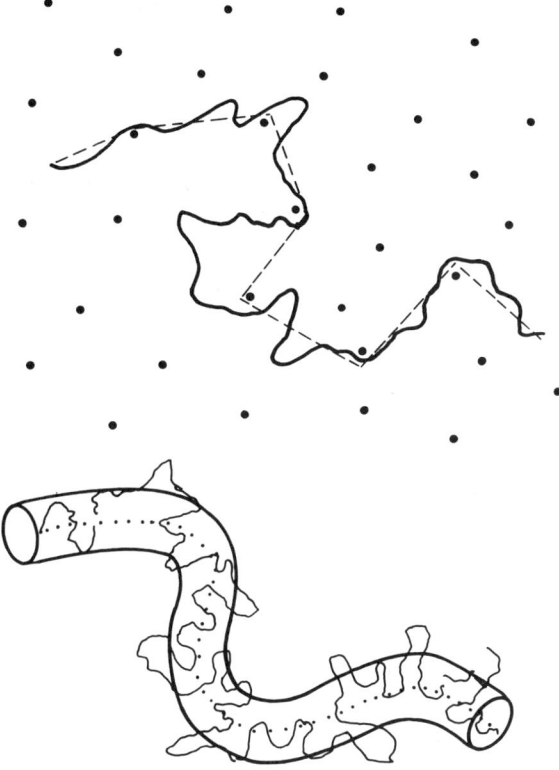

Figure 21 Definition of the primitive path and the imagination of the tube

configuration, it must 'disengage' from its original tube. The time required for this process is called the disengagement time, τd, and other related quantities can be calculated using simple scaling arguments.

(iii) Scaling laws[7]

(a) Disengagement time, τ_d. Since the chain is diffusing along the tube, the friction coefficient of the chain along the tube is proportional to the tube length, which in turn is proportional to the contour length of the chain L, according to equation (152). Therefore, the curvilinear diffusion coefficient of the chain, D is proportional to L^{-1}

$$D \quad \propto \quad D_1/L \tag{198}$$

where D_1 is the monomer diffusion coefficient in the appropriate units. For the chain to get out of its original tube, it must reptate over a length comparable to L. Since the reptation is analogous to the one-dimensional random walk (distance)^2 ~ $(D_{\text{curvilinear}})$ (time)

$$L^2 \quad \sim \quad \frac{D_1}{L}\tau_d \tag{199}$$

from which it follows that τ_d varies as L^3

$$\tau_d \quad \sim \quad L^3 \tag{200}$$

(b) Center of mass diffusion, D_g. In time τ_d, a marked segment of the chain must have undergone a random walk of n steps since Gaussian statistics is assumed for the tube. Therefore, we expect the center of mass, R_g, of the chain (which might lie on one of the segments of the chain) also to undergo a random walk of n steps in time τ_d. For the random walk of n steps, the square of the displacement of R_g scales with n

$$\langle (\Delta R_g)^2 \rangle \quad \sim \quad n$$

But for the same random walk in time τ_d,

$$\langle (\Delta R_g^2) \rangle \sim D_g \tau_d$$

where D_g is the diffusion coefficient of the center of mass of the chain. Therefore, it follows that

$$D_g \sim n/\tau_d \sim L^{-2} \qquad (201)$$

This prediction is borne out by many experiments.[7,213,214] Similarly the viscosity was shown by de Gennes to be proportional to L^{-3}. This does not agree with experiment in that $M^{3.4}$ rather than M^3 is well attested, and there is a large but inconclusive literature why one of de Gennes predictions is very accurate and the other only approximate.

(c) Mean square displacement of one monomer, $r^2(t)$. Four regimes may be identified.[210]
(i) For large times far greater than τ_d, the monomer behaves like the center of mass of the chain, so that

$$r^2(t) \sim D_g t$$

(ii) $\tau_{Rouse} < t < \tau_d$, where τ_{Rouse} is the Rouse relaxation time. In this regime, the displacement $s(t)$ of the primitive path along the tube is given by

$$s^2(t) \sim Dt$$

But the mean square displacement of a monomer is related to the displacement along the tube according to

$$r^2(t) \sim s(t)$$

so that

$$r^2(t) \sim t^{1/2}$$

(iii) $t < \tau_{Rouse}$. In this regime, the relaxation of the chain occurs only over a certain number of monomers, say $m(<n)$. Since the time under consideration is shorter than τ_{Rouse}, this relaxation corresponds to a Rouse mode excitation involving m monomers. Therefore, m^2 scales like t (according to Rouse mechanism). The displacement along the tube is given by

$$s^2(t) \sim \frac{D_1}{m}t \sim t^{1/2}$$

Since $r^2(t) \sim s(t)$

$$r^2(t) \sim t^{1/4}$$

(iv) t very small so that the relaxation does not yet feel the presence of the tube. Now the dynamics is fully Rouse-like and equation (115) is the dynamical law

$$r^2(t) \sim t^{1/2} \qquad (202)$$

These four regimes are identified in Figure 22. The quarter power law of $r^2(t)$ is taken to be the earmark of reptation dynamics.

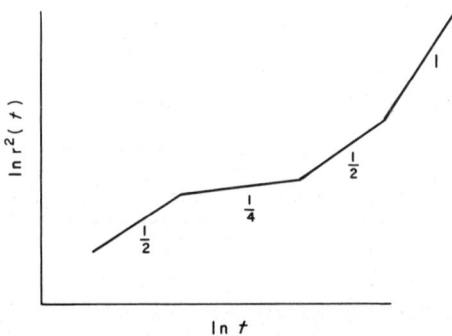

Figure 22 Logarithmic plot of mean square monomer displacement *vs.* time, t, where t is very small

(iv) Doi–Edwards equations[211–214]

These authors studied how tube models with reptation dynamics could be turned into a full theory of viscoelasticity. To do this one needs to describe the dynamics of the primitive path. Let the primitive chain make one step in time Δt. Define a random variable, $\xi(t)$ which is $+1(-1)$ if the primitive chain moves backward (forward). Let ρ be a random vector of length a which is the new position of one of the ends of the primitive chain after one move (see Figure 23). Since the primitive chain is assumed to be made of N points $\mathbf{R}_1, \dots, \mathbf{R}_N$, connected by bonds of constant length a, the Langevin equation of the primitive chain is given by

$$\mathbf{R}_i(t + \Delta t) = (1 + \xi)\mathbf{R}_{i+1}(t)/2 + (1 - \xi)\mathbf{R}_{i-1}(t)/2, \quad 2 \le i \le N - 1$$

$$\mathbf{R}_1(t + \Delta t) = (1 + \xi)\mathbf{R}_2(t)/2 + (1 - \xi)[\mathbf{R}_1(t) + \rho(t)]/2$$

$$\mathbf{R}_N(t + \Delta t) = (1 + \xi)[\mathbf{R}_N(t) + \rho(t)]/2 + (1 - \xi)/\mathbf{R}_{N-1}(t)/2 \tag{203}$$

Figure 23 Illustration of the random vector, ρ, which is the new position of one of the ends of the primitive chain after one move

Assuming that ξ and ρ are white noises and their correlations are local in time, the mean square displacement of a monomer is given by

$$\langle [\mathbf{R}(s, t) - \mathbf{R}(s, 0)]^2 \rangle = \frac{2Dat}{L} + \sum_{p=1}^{\infty} \frac{4aL}{p^2\pi^2} \cos^2\left(\frac{p\pi s}{L}\right)(1 - e^{-p^2 t/\tau_d}) \tag{204}$$

$$\tau_d = \frac{L^2}{D\pi^2} \sim l^3$$

Thus τ_d is also the longest relaxation time. It can readily be shown that the mean square displacement of the center of mass of the chain is given by

$$\mathbf{R}_G \equiv \frac{1}{N} \sum_{i=1}^{N} \mathbf{R}_i(t)$$

$$\langle [\mathbf{R}_G(t) - \mathbf{R}_G(0)]^2 \rangle = 2Dt/N \sim L^{-2}t \tag{205}$$

Similar to the results of equations (204) and (205) which are predicted by scaling arguments also, other related quantities can be calculated. For example, the fraction of the initial tube which is still occupied at time t becomes, from a calculation of the first passage time

$$F(t) = \frac{8}{\pi^2} \sum_{p=\text{odd}} p^{-2} \exp(-p^2 t/\tau_d) \tag{206}$$

The Doi–Edwards theory also predicts that the stress relaxation in the terminal region is given by

$$G(t) = G_N^0 F(t) \tag{207}$$

where G_N^0 is the plateau modulus. It follows from equations (206) and (207) that the zero shear rate viscosity is

$$\eta = \int_0^{\infty} dt\, G(t) = \frac{\pi^2}{12} G_N^0 \tau_d \sim M^3 \tag{208}$$

thus the predicted value of the molecular weight exponent for the viscosity is different from the

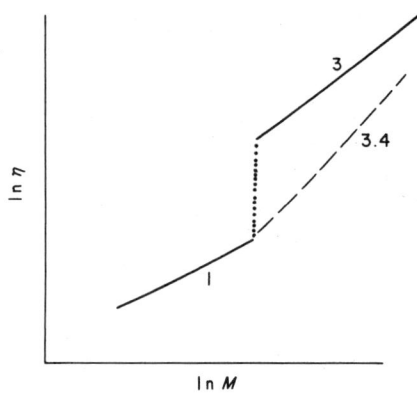

Figure 24 Solid curve suggested by Doi–Edwards theory: the broken line corresponds to the experimental data (schematic)

experimental value. A careful analysis of the various numerical prefactors suggests that the viscosity predicted by the Doi–Edwards theory is the upper bound as shown in Figure 24.

It is to be noted that the de Gennes–Doi–Edwards reptation theory assumes that the polymer system is entangled and then calculates the consequences. The existence of the entangled state is not derived *a priori*, i.e. the crossover from the Rouse to the entangled state is not yet available.

Furthermore, the Doi–Edwards theory predicts that the stress σ in the terminal region due to the strain E is given by

$$\sigma(t) \;=\; \frac{15}{4} G_N{}^0 F(t) Q(E) \tag{209}$$

where Q is a universal tensor with components

$$Q_{ij} \;=\; \frac{1}{\langle |E\cdot u| \rangle} \left\langle \frac{(E\cdot u)_i (E\cdot u)_j}{|E\cdot u|} \right\rangle \tag{210}$$

and $\langle\;\rangle$ means the average taken over all directions of the unit vector u.

1.4.3.3 Monte Carlo simulations

Many simulations of polymer dynamics have been performed[75,214–220] to test the ideas of reptation. When 30 chains each of about 100 connected hard spheres are simulated at a density of 0.7, the reptation dynamics is *not* observed. However, if one chain is mobile and every other chain is frozen in, then reptation is observed. Thus, the most recent extensive simulations cast doubt on the presence of reptation under realistic conditions. On the other hand, the chains might be too short to observe the real dynamics.

1.4.4 Spinodal Decomposition[221–231]

For a binary blend, the relaxation of the volume fraction $\phi(r, t)$ of one of the components is described by the continuity equation

$$\frac{\partial}{\partial t}\phi(r, t) \;-\; \nabla\cdot\int dr'\, \Lambda(r \;-\; r')\nabla'\mu(r')/k_B T \;=\; 0 \tag{211}$$

where $\mu(r)$ is the chemical potential

$$\mu(r) \;=\; \frac{\delta(\Delta F/k_B T)}{\delta\phi(r)} \tag{212}$$

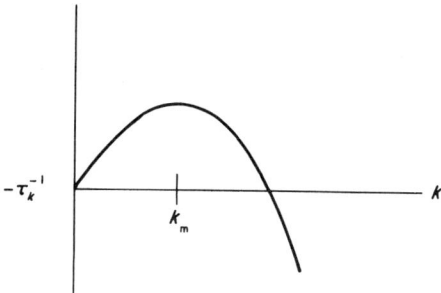

Figure 25 Plot of $-1/\tau_k$ *vs.* k

with ΔF given by the extended Flory–Huggins free energy

$$\Delta F/k_B T = \frac{\phi}{n}\ln\phi + \frac{(1-\phi)}{n}\ln(1-\phi) + \chi\phi(1-\phi) + \frac{a^2}{18\phi(1-\phi)}(\nabla\phi)^2 \tag{213}$$

Λ appearing in equation (211) is the nonlocal Onsager coefficient.[129,210,221] Writing the above equations in terms of the Fourier variable k, we obtain

$$\frac{\partial\phi}{\partial t} + k^2\Lambda(k)\mu(k)/k_B T = 0 \tag{214}$$

where $\mu(k)$ is the Fourier transform of equation (212). Combining equations (212)–(214) and retaining only linear terms in the density fluctuations $\delta\phi(k)$, one obtains

$$\frac{\delta}{\delta t}\delta\phi k(t) = -\frac{1}{\tau k}\delta\phi k(t) \tag{215}$$

with the relaxation time

$$\frac{1}{\tau k} = k^2\Lambda(k)\left[\frac{1}{n\phi(1-\phi)} - 2\chi + \frac{l^2k^2}{18\phi(1-\phi)}\right] \tag{216}$$

If the length scale is such that $k^2 Ll \gg 1$, then analysis[129] shows that

$$\Lambda(k) \sim L^{-2}k^{-2}\phi(1-\phi) \tag{217}$$

The plot of $1/\tau_k$ *vs.* k is shown in Figure 25 (schematic). Thus the linear theory predicts the wave number k_m of the composition fluctuations that grow most rapidly.[226–231]

1.5 REFERENCES

1. P. J. Flory, 'Principles of Polymer Chemistry', Cornell University Press, Ithaca, New York, 1953.
2. M. V. Volkenstein, 'Configurational Statistics of Polymer Chains', Interscience, New York, 1963.
3. T. M. Birshtein and O. B. Ptitsyn, 'Conformations of Macromolecules', Interscience, New York, 1966.
4. P. J. Flory, 'Statistical Mechanics of Chain Molecules', Interscience, New York, 1969.
5. D. Poland, 'Cooperative Equilibria in Physical Biochemistry', Clarendon Press, Oxford, 1978.
6. H. Yamakawa, 'Modern Theory of Polymer Solutions', Harper and Row, New York, 1971.
7. P. G. de Gennes, 'Scaling Concepts in Polymer Physics', Cornell University Press, Ithaca, New York, 1979.
8. S. F. Edwards, in 'Molecular Fluids', ed. R. Balian and G. Weill, Gordon and Breach, London, 1976.
9. K. F. Freed, *Adv. Chem. Phys.*, 1972, **22**, 1.
10. S. Chandrasekhar, *Rev. Mod. Phys.*, 1943, **15**, 1.
11. P. Debye, *J. Phys. Colloid Chem.*, 1947, **51**, 18.
12. C. Tanford, 'Physical Chemistry of Macromolecules', Wiley, New York, 1961.
13. D. J. Amit, 'Field Theory, the Renormalization Group, and Critical Phenomena', McGraw-Hill, New York, 1978.
14. S. F. Edwards, *Proc. Phys. Soc. (London)*, 1965, **93**, 605.
15. M. E. Fisher, *J. Phys. Soc. Jpn., Suppl.*, 1969, **26**, 44.
16. W. H. Stockmayer, *J. Polym. Sci.*, 1955, **15**, 595.
17. W. H. Stockmayer, *Makromol. Chem.*, 1960, **35**, 54.

18. S. F. Edwards and P. Singh, *J. Chem. Soc., Faraday Trans. 2*, 1979, **75**, 1001.
19. C. Domb and A. J. Barrett, *Polymer*, 1976, **17**, 179.
20. J. des Cloizeaux, in 'Phase Transitions', ed. M. Levy, J. C. Le Guillou and J. Zinn-Justin, Plenum Press, Cargese, 1980.
21. J. des Cloizeaux, *J. Phys. (Orsay. Fr.)*, 1981, **42**, 635.
22. Y. Oono, *Adv. Chem. Phys.*, 1985, **61**, 301.
23. Y. Oono, T. Ohta and K. F. Freed, *J. Chem. Phys.*, 1981, **74**, 6458.
24. Y. Oono and K. F. Freed, *J. Phys. A: Math. Gen.*, 1982, **15**, 1931.
25. A. L. Kholodenko and K. F. Freed, *J. Chem. Phys.*, 1983, **78**, 7390.
26. K. F. Freed, *Acc. Chem. Res.*, 1985, **18**, 38.
27. J. F. Douglas and K. F. Freed, *Macromolecules*, 1984, **17**, 2344; 1985, **18**, 201.
28. I. M. Lifshitz, A. Y. Grosberg and A. R. Khokhlov, *Rev. Mod. Phys.*, 1978, **50**, 683.
29. C. Williams, F. Brochard and H. L. Frisch, *Annu. Rev. Phys. Chem.*, 1981, **32**, 433.
30. P. J. Flory, *J. Chem. Phys.*, 1949, **17**, 303.
31. W. H. Stockmayer, *Br. Polym. J.*, 1977, **9**, 89.
32. S. F. Edwards, *N. B. S. Spec. Publ. (US)*, 1966, **273**.
33. I. M. Lifshitz, *Sov. Phys. JETP (Engl. Transl.)*, 1969, **28**, 1280.
34. C. Domb, *Polymer*, 1974, **15**, 259.
35. P. G. de Gennes, *J. Phys. (Orsay, Fr.)*, 1975, **36**, L55; 1978, **39**, L299.
36. J. M. Stephen, *Phys. Lett. A*, 1975, **53**, 363.
37. M. A. Moore, *J. Phys. A: Math. Gen.*, 1977, **10**, 305.
38. I. C. Sanchez, *Macromolecules*, 1979, **12**, 980.
39. C. B. Post and B. H. Zimm, *Biopolymers*, 1979, **18**, 1487.
40. A. L. Kholodenko and K. F. Freed, *J. Chem. Phys.*, 1984, **80**, 900.
41. E. A. Di Marzio, *Macromolecules*, 1984, **17**, 969.
42. M. Muthukumar, *J. Chem. Phys.*, 1984, **81**, 6272.
43. J. Dayantis, *Makromol. Chem.*, 1986, **187**, 1035.
44. R. P. Feynman and A. R. Hibbs, 'Quantum Mechanics and Path Integrals', McGraw-Hill, New York, 1965.
45. B. H. Zimm, *J. Chem. Phys.*, 1946, **14**, 164.
46. B. H. Zimm, W. H. Stockmayer and M. Fixman, *J. Chem. Phys.*, 1953, **21**, 1716.
47. E. Teramoto, *Busseiron Kenkyu*, 1951, **39**, 1.
48. T. B. Grimley, *Trans. Faraday Soc.*, 1959, **55**, 681.
49. M. Fixman, *J. Chem. Phys.*, 1955, **23**, 1656.
50. H. Yamakawa and G. Tanaka, *J. Chem. Phys.*, 1967, **47**, 3991.
51. C. Domb and G. S. Joyce, *J. Phys. C*, 1972, **5**, 956.
52. A. J. Barrett and C. Domb, *Proc. R. Soc. London, Ser. A*, 1979, **367**, 143; 1981, **376**, 361.
53. Y. Chikahisa, G. Tanaka, K. Solc and M. Takahashi, *Rep. Prog. Polym. Phys. Jpn.*, 1981, **24**, 33.
54. M. Muthukumar and B. G. Nickel, *J. Chem. Phys.*, 1984, **80**, 5839.
55. J. des Cloizeaux, *Phys. Rev. A*, 1974, **10**, 1665.
56. D. J. Burch and M. A. Moore, *J. Phys. A: Math. Gen.*, 1976, **9**, 435.
57. T. A. Witten, Jr. and L. Schäfer, *J. Chem. Phys.*, 1981, **74**, 2582.
58. T. A. Witten, Jr. and L. Schäfer, *J. Chem. Phys.*, 1981, **74**, 2582.
59. S. F. Edwards, *J. Phys. A: Math. Gen.*, 1975, **8**, 1171.
60. Y. Oono, *J. Phys. Soc. Jpn.*, 1975, **39**, 25; 1976, **41**, 787.
61. M. E. Fisher, *J. Chem. Phys.*, 1966, **44**, 616.
62. C. Dobm, *Adv. Chem. Phys.*, 1960, **15**, 229.
63. D. S. McKenzie, *Phys. Rep.*, 1976, **27C**, 35.
64. M. E. Fisher and M. F. Sykes, *Phys. Rev.*, 1959, **114**, 45
65. C. Domb, *Adv. Phys.*, 1969, **9**, 149.
66. C. Domb and M. F. Sykes, *J. Math. Phys. (N.Y.)*, 1961, **2**, 63.
67. M. F. Sykes, *J. Math. Phys. (N.Y.)*, 1961, **2**, 52.
68. M. E. Fisher and B. J. Hiley, *J. Chem. Phys.*, 1961, **34**, 1253.
69. C. Domb, *J. Chem. Phys.*, 1963, **38**, 2957.
70. I. Majid, Z. V. Djordjevic and H. E. Stanley, *Phys. Rev. Lett.*, 1983, **51**, 1282.
71. D. C. Rapaport, *J. Phys. A: Math. Gen.*, 1985, **18**, 113.
72. F. T. Wall, S. Windwer and P. J. Gans, *J. Chem. Phys.*, 1962, **37**, 1461; 1963, **38**, 2220, 2228.
73. S. Windwer, in 'Markov Chains and Monte Carlo Calculations in Polymer Science', ed. G. G. Lowry, Dekker, New York, 1970.
74. W. H. Stockmayer, in 'Fluides Moleculeires', ed. R. Balian and G. Weill, Gordon and Breach, New York, 1976.
75. A Baumgartner, in 'Applications of the Monte Carlo Method in Statistical Physics', ed. K. Binder, Springer-Verlag, Berlin, 1984.
76. M. K. Kosmas and K. F. Freed, *J. Chem. Phys.*, 1978, **69**, 3647.
77. K. G. Wilson and J. Kogut, *Phys. Rep.*, 1974, **12C**, 77.
78. S. K. Ma, *Rev. Mod. Phys.*, 1973, **45**, 589.
79. S. K. Ma, 'Modern Theory of Critical Phenomena', W. A. Benjamin, New York, 1976.
80. E. Brezin, J. C. LeGuillou and J. Zinn-Justin, in 'Phase Transitions and Critical Phenomena', ed. C. Domb and M. S. Green, Academic Press, New York, 1976, vol. 6.
81. P. G. de Gennes, *Phys. Lett. A*, 1972, **38**, 339.
82. J. des Cloizeaux, *J. Phys. (Orsay, Fr.)*, 1975, **36**, 281.
83. S. G. Gorishny, S. A. Larin and F. V. Tkachov, *Phys. Lett. A*, 1984, **101**, 120.
84. J. C. Le Guillou and J. Zinn-Justin, *J. Phys., Lett. (Orsay. Fr.)*, 1985, **46**, L137.
85. M. Muthukumar and B. G. Nickel, *J. Chem. Phys.*, 1986, **85**.
86. J. des Cloizeaux, R. Conte and G. Jannink, *J. Phys. Lett. (Orsay, Fr.)*, 1985, **46**, L595.
87. E. Brezin, J. C. LeGuillou and J. Zinn-Justin, *Phys. Rev. D*, 1977, **15**, 1544.

88. E. Brezin and G. Parisi, *J. Stat. Phys.*, 1978, **19**, 269.
89. D. I. Kazakov and D. V. Shirkov, *Fortschr. Phys.*, 1980, **28**, 465.
90. J. Westwater, *Commun. Math. Phys.*, 1982, **84**, 459.
91. J. C. Le Guillou and J. Zinn-Justin, *Phys. Rev. B: Condens. Matter*, 1980, **21**, 3976; *Phys. Rev. Lett.*, 1977, **39**, 95.
92. G. A. Baker, Jr., B. G. Nickel and D. I. Meiron, *Phys. Rev. B: Solid State*, 1978, **17**, 1365.
93. G. A. Baker, Jr., B. G. Nickel, M. S. Green and D. I. Meiron, *Phys. Rev. Lett.*, 1976, **36**, 1351.
94. (a) M. E. Fisher and J. -H. Chen, *J. Phys. (Orsay, Fr.)*, 1985, **46**, 1645; (b) M. Daoud and J. P. Cotton, *J. Phys. (Orsay, Fr.)*, 1982, **43**, 531; (c) A. Miyaki and K. F. Freed, *Macromolecules*, 1983, **16**, 1228; 1984, **17**, 678; (d) J. F. Douglas and K. F. Freed, *Macromolecules*, 1984, **17**, 2344; (e) B. H. Zimm and W. H. Stockmayer, *J. Chem. Phys.*, 1949, **17**, 1301; (f) P. G. de Gennes, *Biopolymers*, 1968, **6**, 715; (g) J. Isaacson and T. C. Lubensky, *J. Phys. (Orsay, Fr.)*, 1980, **41**, L469.
95. H. Tompa, 'Polymer Solutions', Butterworths, London, 1956.
96. T. L. Hill, 'Introduction to Statistical Thermodynamics', Addison-Wesley, Reading, MA, 1960.
97. P. J. Flory, *J. Chem. Phys.*, 1942, **10**, 51.
98. M. L. Huggins, *J. Phys. Chem.*, 1942, **46**, 151; *Ann. N. Y. Acad. Sci.*, 1942, **41**, 151; *J. Am. Chem. Soc.*, 1942, **64**, 1712.
99. D. H. Napper, 'Polymeric Stabilization of Colloidal Dispersions', Academic Press, London, 1983.
100. P. J. Flory and W. R. Krigbaum, *J. Chem. Phys.*, 1950, **18**, 1086.
101. D. J. Burch and M. Moore. *J. Phys. A: Math. Gen.*, 1976, **9**, 435.
102. L. Schafer and T. A. Witten, Jr., *J. Chem. Phys.*, 1977, **66**, 2121.
103. L. Schafer and T. A. Witten, Jr., *Proc. Stat. Phys., Annal. Israel Phys. Soc.*, 1978, **13**(2), 974.
104. I. C. Sanchez, in 'Polymer Compatibility and Incompatibility: Principles and Practises', ed. K. Solc, Harwood, Cooper Station, NY, 1982.
105. I. C. Sanchez, *Annu. Rev. Mater. Sci.*, 1983, **13**, 387.
106. I. C. Sanchez and R. H. Lacombe, *J. Phys. Chem.*, 1976, **80**, 2352, 2568.
107. R. Koningsveld, L. A. Kleintjens and A. R. Schultz, *J. Polym. Sci., Part A-2*, 1970, **8**, 1261.
108. R. Koningsveld and L. A. Kleintjens, *Macromolecules*, 1971, **4**, 637.
109. R. Koningsveld, W. H. Stockmayer, J. W. Kennedy and L. A. Kleintjens, *Macromolecules*, 1974, **7**, 73.
110. L. A. Kleintjens, R. Koningsveld and W. H. Stockmayer, *Br. Polym. J.*, 1976, **8**, 144.
111. J. Dayantis, *Macromolecules*, 1982, **15**, 1107.
112. M. Daoud and G. Jannink, *J. Phys. (Orsay, Fr.)*, 1976, **37**, 973.
113. M. Daoud, *J. Polym. Sci., Polym. Symp.*, 1977, **61**, 305.
114. I. C. Sanchez, *Macromolecules*, 1984, **17**, 967.
115. P. J. Flory, *J. Chem. Phys.*, 1949, **17**, 303.
116. M. Daoud, J. P. Cotton, B. Farnoux, G. Jannink, G. Sarma, H. Benoit, R. Duplessix, C. Picot and P. G. de Gennes, *Macromolecules*, 1975, **8**, 804.
117. R. W. Richards, A. Maconnachie and G. Allen, *Polymer*, 1978, **19**, 266.
118. S. F. Edwards, *Proc. Phys. Soc. (London)*, 1966, **88**, 265.
119. S. F. Edwards, *J. Phys. A: Math. Gen.*, 1975, **8**, 1670.
120. M. Muthukumar and S. F. Edwards, *J. Chem. Phys.*, 1982, **76**, 2720.
121. J. des Cloizeaux and G. Jannink, 'Les Polymeres en Solution: Modelisation et leur structures', Edition de Physique, Paris, 1987, chapt. 13, section 2.5.5.
122. B. Duplantier, *J. Phys. (Orsay, Fr.)*, 1980, **41**, L409; 1982, **43**, 991.
123. I. C. Sanchez, *Macromolecules*, 1979, **12**, 980.
124. M. Muthukumar, *J. Chem. Phys.*, 1986, **85**.
125. T. Dobashi, M. Nakata and M. Kaneko, *J. Chem. Phys.*, 1980, **72**, 6685, 6692; 1984, **80**, 948.
126. K. Shinozaki, T. van Tan, Y. Saito and T. Nose, *Polymer*, 1982, **23**, 728.
127. Y. Izumi and Y. Miyake, *J. Chem. Phys.*, 1984, **81**, 1501.
128. I. C. Sanchez, *J. Appl. Phys.*, 1985, **58**, 287.
129. K. Binder, *J. Chem. Phys.*, 1983, **79**, 6387.
130. C. T. Murray, J. W. Gilmer and R. S. Stein, *Macromolecules*, 1985, **18**, 996.
131. R. Koningsveld and L. A. Kleintjens, *J. Polym. Sci., Polym. Symp.*, 1977, **61**, 221.
132. R. Koningsveld, L. A. Kleintjens and H. M. Schoffeleers, *Pure Appl. Chem.*, 1974, **39**, 1.
133. F. S. Bates and G. Wignall, *J. Chem. Phys.*, in press.
134. M. Muthukumar, to be published.
135. G. G. Stokes, *Trans. Camb. Philos. Soc.*, 1851, **9**, 8.
136. A. Einstein, *Ann. Phys. (Leipzig)*, 1906, **19**, 289, 1911, **34**, 591.
137. A. Einstein, 'The Theory of Brownian Movement', Dover, New York, 1956.
138. J. Happel and H. Brenner, 'Low Reynolds Number Hydrodynamics', Prentice-Hall, New York, 1965.
139. L. D. Landau and E. M. Lifshitz, 'Fluid Mechanics', Pergamon Press, New York, 1959.
140. B. J. Berne and R. Pecora, 'Dynamic Light Scattering', Wiley, New York, 1976.
141. N. Wax, 'Noise and Stochastic Processes', Dover, New York, 1954.
142. G. K. Batchelor, 'An Introduction to Fluid Dynamics', Cambridge University Press, 1970.
143. G. K. Batchelor, *J. Fluid Mech.*, 1970, **41**, 545.
144. G. K. Batchelor and J. T. Green, *J. Fluid Mech.*, 1972, **56**, 401.
145. G. K. Batchelor, *J. Fluid Mech.*, 1976, **74**, 1.
146. A. R. Altenberger and J. M. Deutch, *J. Chem. Phys.*, 1973, **59**, 894.
147. A. R. Altenberger, *Chem. Phys.*, 1976, **15**, 269.
148. B. U. Felderhof, *Physica A (Amsterdam)*, 1977, **89**, 373; *J. Phys. A: Math. Gen.*, 1978, **11**, 929.
149. G. D. J. Phillies, *J. Chem. Phys.*, 1974, **60**, 976, 983; 1977, **67**, 4690; 1981, **74**, 2436.
150. W. Hess and R. Klein. *J. Phys. A: Math. Gen.*, 1982, **15**, 1669.
151. B. J. Ackerson, *J. Chem. Phys.*, 1976, **64**, 242; 1978, **69**, 684.
152. W. B. Russel and A. B. Glendinning, *J. Chem. Phys.*, 1981, **74**, 948.
153. D. Bedeaux, R. Kapral and P. Mazur, *Physica A (Amsterdam)*, 1977, **88**, 88.
154. J. M. Peterson and M. Fixman, *J. Chem. Phys.*, 1963, **39**, 2516.

155. F. W. Wang and B. H. Zimm, *J. Polym. Sci., Polym. Phys. Ed.*, 1974, **12**, 1619, 1639.
156. M. Muthukumar, in 'Polymer-Flow Interaction', ed. Y. Rabin, *AIP Conf. Proc.*, 1985, **137**.
157. M. Muthukumar and K. F. Freed, *J. Chem. Phys.*, 1983, **78**, 497, 511; 1982, **76**, 6195.
158. K. F. Freed and M. Muthukumar, *J. Chem. Phys.*, 1978, **68**, 2088; 1978, **69**, 2657; 1982, **76**, 6186.
159. M. Muthukumar, *J. Chem. Phys.*, 1982, **77**, 959; 1983, **78**, 2764.
160. M. Muthukumar and M. DeMeuse, *J. Chem. Phys.*, 1983, **78**, 2773.
161. D. Hatziavramidis and M. Muthukumar, *J. Chem. Phys.*, 1985, **83**, 2522.
162. P. E. Rouse, Jr., *J. Chem. Phys.*, 1953, **21**, 1272.
163. P. H. Verdier and W. H. Stockmayer, *J. Chem. Phys.*, 1962, **36**, 227; P. H. Verdier, *J. Chem. Phys.*, 1970, **52**, 5512.
164. R. A. Orwoll and W. H. Stockmayer, *Adv. Chem. Phys.*, 1969, **15**, 305.
165. K. Iwata, *J. Chem. Phys.*, 1971, **54**, 12.
166. S. F. Edwards and A. G. Goodyear, *J. Phys. A: Gen. Phys.*, 1972, **5**, 965, 1188.
167. H. J. Hilhorst and J. M. Deutch, *J. Chem. Phys.*, 1975, **63**, 5153.
168. W. Kuhm, *Kolloidn. Zh.*, 1934, **68**, 2.
169. K. Sölc and W. H. Stockmayer, *J. Chem. Phys.*, 1971, **54**, 2756.
170. B. H. Zimm, *J. Chem. Phys.*, 1956, **24**, 269.
171. B. H. Zimm, G. M. Roe and L. F. Epstein, *J. Chem. Phys.*, 1962, **24**, 279.
172. J. E. Hearst, *J. Chem. Phys.*, 1962, **37**, 2547.
173. S. F. Edwards and K. F. Freed, *J. Chem. Phys.*, 1974, **61**, 1189; K. F. Freed and S. F. Edwards, *J. Chem. Phys.*, 1974, **61**, 3626.
174. S. F. Edwards and M. Muthukumar, *Macromolecules*, 1984, **17**, 586.
175. M. Muthukumar and S. F. Edwards, *Polymer*, 1982, **23**, 345.
176. M. Muthukumar, *J. Phys. A: Math. Gen.*, 1981, **14**, 2129.
177. Y. Oono and M. Kohmoto, *J. Chem. Phys.*, 1983, **78**, 520.
178. A. Lee, P. R. Baldwin and Y. Oono, *Phys. Rev. A*, 1984, **30**, 968.
179. J. G. Kirkwood and J. Riseman, *J. Chem. Phys.*, 1948, **16**, 565.
180. P. Debye and A. M. Bueche, *J. Chem. Phys.*, 1948, **16**, 573.
181. P. J. Flory and T. G. Fox, *J. Am. Chem. Soc.*, 1951, **73**, 1904.
182. P. G. de Gennes, *Physics*, 1967, **3**, 37; E. Dubois-Violette and P. G. de Gennes, *Physics*, 1967, **3**, 181.
183. J. D. Ferry, 'Viscoelastic Properties of Polymers', Wiley, New York, 1980.
184. K. Osaki, *Adv. Polym. Sci.*, 1973, **12**, 1.
185. W. W. Graessley, *Adv. Polym. Sci.*, 1974, **16**, 1; 1982, **47**, 67.
186. K. F. Freed, in 'Progress in Liquid Physics', ed. C. A. Croxton, Wiley, London, 1978.
187. A. Perico and K. F. Freed, *J. Chem. Phys.*, 1984, **81**, 1466, 1475; K. F. Freed, *J. Chem. Phys.*, 1983, **78**, 2051.
188. K. F. Freed and A. Perico, *Macromolecules*, 1981, **14**, 1290.
189. C. J. T. Martel, T. P. Lodge, M. G. Dibbs, T. M. Stokich, R. L. Sammler, C. J. Carriere and J. L. Schrag, *Faraday Symp. Chem. Soc.*, 1983, **18**, 173.
190. M. Muthukumar, *Macromolecules*, 1984, **17**, 971.
191. M. Bixon, *Annu. Rev. Phys. Chem.*, 1976, **27**, 65.
192. M. Delbruck, in 'Mathematical Problems in the Biological Sciences', *Proc. Symp. Appl. Math., 14th*, American Mathematical Society, Providence, RI, 1962.
193. S. Prager and H. L. Frisch, *J. Chem. Phys.*, 1967, **46**, 1475.
194. S. F. Edwards, *Proc. Phys. Soc. (London)*, 1967, **91**, 513; *J. Phys. A: Gen. Phys.*, 1968, **1**, 15.
195. N. Saito and Y. Chen, *J. Chem. Phys.*, 1973, **59**, 3701.
196. F. W. Wiegel, *J. Chem. Phys.*, 1977, **67**, 469.
197. R. Alexander-Katz and S. F. Edwards, *J. Phys. A: Gen. Phys.*, 1972, **5**, 674.
198. M. G. Brereton and S. Shah, *J. Phys. A: Gen. Phys.*, 1981, **14**, L51.
199. K. Iwata and T. Kimura, *J. Chem. Phys.*, 1981, **74**, 2039.
200. K. Iwata, *J. Chem. Phys.*, 1983, **78**, 2778; 1982, **76**, 6363, 6375.
201. F. Tanaka, *Prog. Theor. Phys.*, 1982, **68**, 148, 164.
202. J. W. Alexander, *Trans. Am. Math. Soc.*, 1928, **30**, 275.
203. J. des Cloizeaux and M. L. Mehta, *J. Phys. (Orsay, Fr.)*, 1979, **40**, 665.
204. R. Ball and M. L. Mehta, *J. Phys. (Orsay, Fr.)*, 1981, **42**, 1193.
205. A. V. Vologodskii, A. V. Lukashin, M. D. Frank-Kamenetskii and V. V. Anshelevich, *Sov. Phys. JETP (Engl. Transl.)*, 1974, **39**, 1059.
206. A. V. Vologodskii, A. V. Lukashin and M. D. Frank-Kamenetskii, *Sov. Phys. JETP (Engl. Transl.)*, 1975, **40**, 932.
207. J. P. J. Michels and F. W. Wiegel, *Phys. Lett. A*, 1982, **90**, 381.
208. S. F. Edwards, *Proc. Phys. Soc. (London)*, 1967, **92**, 9.
209. P. G. de Gennes, *J. Chem. Phys.*, 1971, **55**, 572.
210. P. G. de Gennes, *J. Chem. Phys.*, 1980, **72**, 4756.
211. M. Doi and S. F. Edwards, *J. Chem. Soc., Faraday Trans. 2*, 1978, **74**, 1789, 1802, 1818; 1979, **75**, 38.
212. C. F. Curtiss and R. B. Bird, *J. Chem. Phys.*, 1981, **74**, 2016, 2026.
213. W. W. Graessley, *Faraday Symp. Chem. Soc.*, 1983, **18**, 1.
214. M. Doi and S. F. Edwards, 'The Theory of Polymer Dynamics', Oxford University Press, 1986.
215. K. E. Evans and S. F. Edwards, *J. Chem. Soc., Faraday Trans. 2*, 1981, **77**, 1891, 1913, 1929. .
216. A. Baumgärtner and K. Binder, *J. Chem. Phys.*, 1981, **75**, 2994.
217. A. Baumgärtner, K. Kremer and K. Binder, *Faraday Symp. Chem. Soc.*, 1983, **18**, 37.
218. J. Deutsch, *Phys. Rev. Lett.*, 1982, **49**, 926; 1983, **51**, 1924.
219. K. Kremer, *Phys. Rev. Lett.*, 1983, **51**, 1923.
220. K. Kremer and K. Binder, *J. Chem. Phys.*, 1984, **81**, 6381.
221. J. D. Gunton, M. San Miguel and P. S. Sahni, in 'Phase Transitions and Critical Phenomena', ed. C. Domb and M. S. Green, Academic Press, London, 1983, vol. 8.
222. P. Pincus, *J. Chem. Phys.*, 1981, **75**, 1996.

223. J. W. Cahn and J. E. Hilliard, *J. Chem. Phys.*, 1958, **28**, 258; 1959, **31**, 688.
224. J. W. Cahn, *J. Chem. Phys.*, 1965, **42**, 93.
225. H. E. Cook, *Acta Metall.*, 1970, **18**, 297.
226. T. Hashimoto, J. Kumaki and H. Kawai, *Macromolecules*, 1983, **16**, 641.
227. H. L. Snyder, P. Meakin and S. Reich, *J. Chem. Phys.*, 1983, **78**, 3334; *Macromolecules*, 1983, **16**, 757.
228. T. Nishi, T. T. Wang and T. K. Kwei, *Macromolecules*, 1975, **8**, 227.
229. T. Hashimoto, K. Sasaki and H. Kawai, *Macromolecules*, 1984, **17**, 2812.
230. K. Sasaki and T. Hashimoto, *Macromolecules*, 1984, **17**, 2818.
231. H. Yang, M. Shibayama, R. S. Stein, N. Shimizu and T. Hashimoto, *Macromolecules*, 1986, **19**, 1667.

2
Chain Conformations

AKIHIRO ABE

Tokyo Institute of Technology, Japan

2.1 INTRODUCTION

The late Professor Flory, in his Nobel Prize address,[1] described the underlying principle of the conformational analysis of high polymers as follows:

'It is noteworthy that the chemical bonds in macromolecules differ in no discernible respect from those in 'monomeric' compounds of low molecular weight. The same rules of valency apply; the lengths of the bonds, *e.g.* C–C, C–H, C–O, *etc.*, are the same as the corresponding bonds in monomeric molecules within limits of experimental measurement. This seemingly trivial observation has two important implications: first, the chemistry of macromolecules is coextensive with that of low molecular substances; second, the chemical basis for the special properties of polymers that equip them for so many applications and functions, both in nature and in the artifacts of man, is not therefore to be sought in peculiarities of chemical bonding but rather in their macromolecular constitution, specifically, in the attributes of long molecular chains.'

Recent advances in conformational analysis have confirmed this statement to be valid within the accuracy of the experiments employed in determining various average properties of chain molecules. In this chapter, the author has chosen to mention several important aspects regarding the reliability, applicability and limitations of the technique, rather than attempting to cover the field comprehensively as a review.

Properties of polymers are intimately related to the configurational characteristics of their molecular chains, and in principle, can be evaluated by averaging over all configurations of the chain.[2] For a random flight chain, the mean-square end-to-end distance r of the chain is given by the familiar expression

$$\langle r^2 \rangle_0 = nl^2 \tag{1}$$

where the angle bracket denotes the average taken over all configurations permitted in the unperturbed state, n is the number of bonds and l^2 is the mean-square bond length. Accounting for the restriction imposed by the fixed bond angle leads to the freely rotating chain model. For a limit of large n (*i.e.* $n \to \infty$),

$$C_\infty \equiv \langle r^2 \rangle_0 / n l^2 = (1 + \cos\theta)/(1 - \cos\theta) \qquad (2)$$

where θ denotes the supplement of the bond angle. For a tetrahedrally bonded chain, $\cos\theta = 1/3$ and $C_\infty = 2$.

In real chain molecules, bond rotations are usually subject to strong mutual correlations. The rotational states of a given bond are influenced by those of its immediate neighbors in the chain skeleton. Hindrances to free rotation extend the range of correlation. Since bond lengths and bond angles ordinarily may be regarded as fixed, the bond rotations are the variables over which averaging must be carried out. The rotational isomeric state (RIS) scheme, which has abundant precedent for the interpretation of conformation of small molecules,[3] has provided a simplification in treating this complex problem without sacrificing much accuracy in the results.

The validity of the RIS model was critically examined by Volkenstein in his pioneering work.[4] Adaptation of this model to a polymeric chain is straightforward in concept. In practice, however, its application is complicated by the interdependence of rotations about neighboring bonds along the chain. The mathematical treatment of the problem was attempted independently around 1960 by Gotlib,[5] Birshtein and Ptitsyn,[6] Lifson,[7] Nagai[8] and Hoeve.[9] Versatile methods involving serial multiplication of generator matrices were introduced by Flory[1,2] in 1964. Quantitative treatment of the configuration-dependent properties of polymer chains in terms of their detailed molecular structure has thus been realized with the same accuracy as is achieved in the conformational analysis of smaller molecules.

2.2 FORMULATION OF THE CONFORMATIONAL PARTITION FUNCTION: GENERAL PROCEDURE

In the classical form, the average conformation of a polymer with an end-to-end distance r can be generally defined by a partition function such as

$$Z(r)\,dr = \int \cdots \int_{dr} \exp[-(K+E)/RT] \prod_i q_i p_i \qquad (3)$$

where K and E in the Boltzmann expression represent the kinetic and potential energy terms respectively, q_i denotes the coordinate of the ith unit, and p_i designates its momentum. The integration over momentum space immediately leads to

$$Z(r)\,dr = (2\pi m R T)^{3(n+1)/2} \int \cdots \int_{dr} \exp(-E/RT) \prod_i q_i \qquad (4)$$

In this simplification, the molecule is treated as a rigid body, and it is convenient to adopt the internal coordinate system for q_i. Then, we have for the conformational part,

$$Z(r)\,dr = \int \cdots \int_{dr} \exp(-E/RT) \prod_i q_i \qquad (5)$$

In general, however, moments of inertia for overall rotation are not, in principle, independent of the coordinates of internal rotation. This problem was discussed for simple molecules by Pitzer *et al.*[10] some years ago. The couplings of the kinetic energy term with the internal rotation have been recently reinvestigated by Morita *et al.*[11] These couplings are essentially long-range effects, and are intractable within the framework of the present RIS scheme. The effect arising from this source should be quite small, however, in view of the fact that a large number of RIS calculations have been carried out successfully in treating a wide variety of conformation-dependent properties of polymeric chains.

In the RIS approximation, each molecule (or bond) is treated as occurring in one or another of several discrete rotational states. These states are ordinarily chosen to coincide with potential minima. Fluctuations about the minima are ignored. The RIS approximation should be valid for those bonds having distinct rotational minima separated by barriers substantially greater than RT. For bonds with rotational barriers less than RT, the RIS approximation is deprived of its primary physical basis. It is, however, still useful as a mathematical device, and may be used legitimately even in those cases where rotational barriers are low, or negligible, in height. According to this scheme, the bond rotational partition function, formally defined for bond i by[2]

$$z_i = \int_0^{2\pi} \exp[-E_i(\phi_i)/RT]d\phi_i \tag{6}$$

may be replaced by a sum of a discrete set of parameters. Thus,

$$z_i = \sum_j u_j \tag{7}$$

where u_j is usually defined as the Boltzmann factor for state j of conformational energy E_j

$$u_j = \exp(-E_j/RT) \tag{8}$$

The neighbor-dependent character of the bond rotation can be taken into account by adopting a matrix representation, such as

$$U_i = [u_{\delta\varepsilon}]_i \tag{9}$$

with states (δ) for bond $i-1$ indexing the rows and those (ε) for bond i the columns. It has been assumed that the bond rotational interdependence does not extend beyond first neighbors. The conformational partition function Z for a chain of n bonds may be generated by

$$Z = U_1\left[\prod_{i=2}^{n-1} U_i\right]U_n \tag{10}$$

where U_1 and U_n are defined so as to contribute the required sum of elements, by multiplication as specified by equation (10).

Expressions for the n-alkane chain, including polymethylene (PM), are illustrative of the application of the scheme outlined above.[2,12] As is well known, the internal C–C bond of an n-alkane chain exists in one of the three staggered conformations: *trans* (t) and two *gauche* forms ($g^\pm$). Corresponding to equation (9), the statistical weight matrix for $1 < i < n$ may be formally given in a 3×3 scheme,

$$U_i = \begin{bmatrix} u_{tt} & u_{tg^+} & u_{tg^-} \\ u_{g^+t} & u_{g^+g^+} & u_{g^+g^-} \\ u_{g^-t} & u_{g^-g^+} & u_{g^-g^-} \end{bmatrix} \tag{11}$$

where the rows are indexed in the order t, g^+, g^- to the state of bond $i-1$, and the columns to the state of bond i in the same manner. The suffixes appended to each statistical weight factor specify the rotational states of the two consecutive internal bonds. In consideration of the chemical constitution of the n-alkane chain, the following form is conventionally adopted

$$U_i = \begin{bmatrix} 1 & \sigma & \sigma \\ 1 & \sigma & \sigma\omega \\ 1 & \sigma\omega & \sigma \end{bmatrix} \tag{12}$$

where σ is the statistical weight factor assigned to the first-order interaction taking place between two methylene groups which are three bonds apart, and ω is introduced to account for the repulsive second-order interaction characteristic of the $g^\pm g^\mp$ conformation, all statistical weights being defined relative to the *trans* form. In the simple RIS scheme, we may choose to set $\sigma = \exp(-E_\sigma/RT)$ and $\sigma\omega = \exp[-(E_\sigma + E_\omega)/RT]$, where E_σ and E_ω respectively represent the conformational energy involved in the first- and second-order interactions in the rotational states specified by the suffixes. (For details, see the conformational energy surface shown in Section 2.3.1). Matrices U_1 and U_n are the row and column vectors

$$U_1 = [1 \quad 0 \quad 0]$$ (13)

and

$$U_n = \begin{bmatrix} 1 \\ 1 \\ 1 \end{bmatrix}$$ (14)

As may be demonstrated by a simple manipulation, equation (10) yields the required sum of the statistical weights of all possible conformations of the chain.

For a simple chain of infinite length, the treatment may be simplified by adopting the largest-eigenvalue method.[2] The eigenvalues of matrix U_i (see equation 12) are readily found to be

$$\lambda_{1,2} = (\tfrac{1}{2})\{1 + \sigma(1+\omega) \pm \sqrt{[1-\sigma(1+\omega)]^2 + 8\sigma}\}$$ (15)

and

$$\lambda_3 = \sigma(1-\omega)$$ (16)

where $\lambda_1 > \lambda_2 > \lambda_3$. For a large n, the conformational partition function Z can be approximated by

$$Z = \left(A_{11} \sum_{\delta=1}^{3} B_{1\delta} \right) \lambda_1^{n-2}$$ (17)

where the eigencolumn A_1 and eigenrow $B_1 = A_1^{-1}$ corresponding to the largest eigenvalue λ_1 are given as

$$A_1 = \begin{bmatrix} 1-\lambda_2 \\ 1 \\ 1 \end{bmatrix}$$ (18a)

and

$$B_1 = (\lambda_1 - \lambda_2)^{-1}[1 \quad (\lambda_1-1)/2 \quad (\lambda_1-1)/2].$$ (18b)

We then have

$$Z = [(1-\lambda_2)/(\lambda_1-\lambda_2)]\lambda_1^{n-1}$$ (19)

In most applications it is the logarithm of Z which is required. Hence, for sufficiently large n, it will be permissible to ignore the front factor, and let

$$Z = \lambda_1^{n-2} \quad (n \to \infty)$$ (20)

2.3 ESTIMATION OF BOND CONFORMATIONS OF THE POLYMETHYLENE (PM) CHAIN FROM THE CONFORMATIONAL ENERGY SURFACE

2.3.1 Comparison of the Simple RIS Scheme with More Elaborate Methods

Molecular force field calculations frequently provide information useful in assessing the overall conformational energy surface as a function of the skeletal rotational angles.[2,13,14] Such calculations also provide very important information regarding the high energy conformations, which generally elude direct observation by conventional techniques.

The results of calculations for *n*-alkanes, which are applicable to the PM chain as well, provide an illustrative example.[12,15,17] In Figure 1, an energy contour map calculated for two consecutive internal C–C bonds is shown at intervals of 1 kcal mol^{-1} (4.2 kJ mol^{-1}). In these calculations, an intrinsic threefold potential with barrier height of 2.8 kcal mol^{-1} (11.7 kJ mol^{-1}) was assigned to rotation about the C–C bond, and Buckingham's 6-exp energy functions were employed for the evaluation of nonbonded interatomic interactions

$$E(\phi) = \sum_i (E_0/2)(1 - \cos 3\phi_i) + \sum_{k,l} [a_{kl}\exp(-b_{kl}r_{kl}) - c_{kl}/r_{kl}^6]$$ (21)

where the first sum includes all rotatable bonds (ϕ_i) of the molecule, or of a given molecular fragment, and the second summation is taken over all atom pairs k,l, whose distance of separation r_{kl} depends on one or more of the ϕ_i. Parameters required in these expressions were adjusted to reproduce the observed energy difference between the *gauche* and *trans* states. The bond lengths and bond angles were taken to be invariant. As an examination of models reveals, steric interactions dependent upon rotations about more than three consecutive bonds are disallowed by interferences of shorter range and hence may be ignored. It suffices therefore to consider only near neighbors. From the potential energy surface thus obtained, the probability of occurrence of a bond in a given state may be estimated. In a simple RIS scheme, rotational states are defined by the minima of the corresponding potential wells, and the conformational energies are given as the differences between the minimum energies. Statistical weights for the PM chain are then expressed as

$$\sigma = \exp(-550/RT), \quad \omega = \exp(-2200/RT) \tag{22}$$

with RT given in cal mol^{-1} (1 cal = 4.2 J). These expressions have been successfully used[12] in the calculation of the unperturbed dimension $\langle r^2 \rangle_0$ of the PM chain and its temperature coefficient d ln $\langle r^2 \rangle_0/\mathrm{d}T$. In the three-state model, here employed, effects associated with the g^+g^- doublet[12,15] appearing at $(115°, -80°)$ and $(80°, -115°)$ (*cf.* Figure 1) are ignored, and we assume instead that the pair g^+g^- can be represented as occurring at normal *gauche* angles.

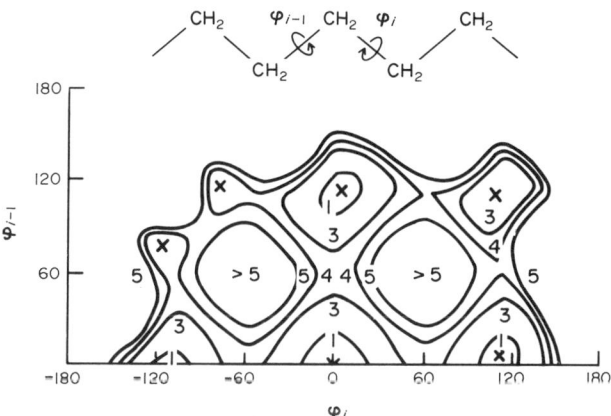

Figure 1 Energy map for the internal rotations of the PM chain. The neighboring C–C bonds are fixed in the *trans* states $(\phi_{i-2} = \phi_{i+1} = 0)$. Energy contours are given at intervals of 1 kcal mol^{-1} (4.2 kJ mol^{-1}) relative to the *tt* state $(\phi_{i-1} = \phi_i = 0)$. Positions of minima are marked by X. The diagram for the range $-180° \leq \phi_{i-1} \leq 0°$ may be generated by inversion through the origin $(0, 0)$

In a more elaborate RIS scheme, account is taken of differences in the effective shapes of potential wells. According to the procedure developed by Flory *et al.*,[2,18] bond partition functions z and average energies $\langle E \rangle$ can be evaluated for each potential energy minimum on the basis of the energy contour map (Table 1). These z and $\langle E \rangle$ values are used to determine the preexponential factors and the average conformational energies required in the generalized statistical weight expression. Thus, for *n*-alkanes, we obtain

$$\sigma = 0.9 \exp(-560/RT), \quad \omega = 1.1 \exp(-1940/RT) \tag{23}$$

Values of the preexponential factor depart from unity only by *ca.* 10%. This suggests that the effect arising from the dissimilarity of the shapes of the potential energy wells should be comparatively small in the PM system. In Table 1, values of $\langle \phi_{i-1} \rangle$ and $\langle \phi_i \rangle$ obtained by taking averages over the entire domain are also included.

For an infinitely long chain ($n \rightarrow \infty$), where the largest-eigenvalue method is applicable, bond conformation probabilities may be estimated for a consecutive pair of bonds according to the expression

$$f_{\delta\varepsilon} = \partial \ln \lambda_1/\partial \ln u_{\delta\varepsilon} \tag{24}$$

where $u_{\delta\varepsilon}$ denotes the (δ, ε) element of U_i given in equation (12). Values of f have been calculated for the *tt*, $tg^\pm$, $g^\pm g^\pm$ and $g^\pm g^\mp$ conformations by using the two sets of parameters given in equations

Table 1 Partition Functions, Average Energies and Averaged Rotation Angles for the PM Chain Deduced from the Potential Energy Contour Map (Figure 1) for a Temperature of 25 °C

Rotational state	z^a	$\langle E \rangle$ (kcal mol^{-1})^b	$\langle \phi_{i-1} \rangle, \langle \phi_i \rangle$ (°)^c
tt	1.000	0.85	0.0, 0.0
tg^+	0.430	1.42	0.9, 116.4
g^+g^+	0.201	1.97	116.4, 116.9
g^+g^-	0.018	3.93	109.0, $-$109.0

^{a} Expressed relative to a value of unity for the tt state. ^{b} Expressed relative to zero for the minimum in the tt state: 1 kcal = 4.2 kJ. ^{c} Rotation angles are measured from the *trans* state ($\phi_t = 0$).

Table 2 Comparison of Bond Conformation Probabilities of the PM Chain Estimated by Various Methods for a Temperature of 140 °C

Bond conformation probability	RIS approximation^a Simple scheme: Equation (22)	RIS approximation^a Refined scheme: Equation (23)	Integration method^b
f_{tt}	0.337	0.361	0.351
$f_{tg^\pm} + f_{g^\pm t}$	0.545	0.531	0.534
$f_{g^\pm g^\pm}$	0.110	0.098	0.101
$f_{g^\pm g^\mp}$	0.008	0.010	0.012
f_t	0.609	0.626	0.618
f_g	0.391	0.374	0.382

^{a} A. Abe, *Polym. J.*, 1982, **14**, 427. ^{b} T. Oyama and K. Shiokawa, *Polym. J.*, 1981, **13**, 1145.

(22) and (23),[17] and the results are summarized in Table 2. The probabilities for single bonds, f_t and $f_g (= 1 - f_t)$, can be deduced therefrom, and are also included in the table. Adoption of the second parameter set (equation 23) leads to a somewhat higher value of f_t (0.626) than that of the simple RIS scheme (0.609), but the difference is trivial.

Oyama *et al.*[16] have calculated bond conformation probabilities for the PM chain by a direct integration method. They adopted the formulation prescribed by Saito.[19] According to this scheme,

$$\lambda_m \psi_m(\phi_{i-1}) = \int K(\phi_{i-1}, \phi_i) \psi_m(\phi_i) \, d\phi_i \tag{25}$$

where λ_m is the mth eigenvalue, and $\psi_m(\phi)$ is its eigenfunction. The kernel K is defined by

$$K(\phi_{i-1}, \phi_i) = \exp\left\{ -\frac{1}{RT}[\tfrac{1}{2}E(\phi_{i-1}) + \tfrac{1}{2}E(\phi_i) + E(\phi_{i-1}, \phi_i)] \right\} \tag{26}$$

Since $K(\phi_{i-1}, \phi_i) = K(\phi_i, \phi_{i-1})$ (*cf.* Figure 1), it can be expanded as

$$K(\phi_{i-1}, \phi_i) = \sum_m \lambda_m \psi_m(\phi_{i-1}) \psi_m(\phi_i) \tag{27}$$

Equation (25) may then be solved numerically on the basis of the potential energy contour diagram shown in Figure 1. For an infinitely long chain, only the terms associated with the largest eigenvalue are important. The average bond conformation may be calculated from the eigenfunctions. The results obtained by Oyama *et al.*[16] are cited in the fourth column of Table 2. The values obtained by this method compare favorably with those derived from the aforementioned RIS schemes. The results of calculations summarized in Table 2 indicate that the simplest RIS scheme, in which statistical weight parameters are expressed as a simple Boltzmann factor for the minimum energy, gives a satisfactory representation of the bond conformations of the PM chain.

The same conclusion has been drawn from the treatment of the unperturbed dimension of the PM chain.[17] The characteristic ratio C_∞ and its temperature coefficient $d \ln C_\infty / dT$ were calculated using the parameter set given in equation (22). When the geometrical parameters $\angle \text{CCC} = 112°$, $\phi_t = 0$ and $\phi_{g^\pm} = \pm 116.5°$ (*cf.* Table 1) were used, the following values were derived: $C_\infty = 7.65$ and $d \ln C_\infty / dT = -1.06 \times 10^{-3} \text{ K}^{-1}$ at 140 °C. Use of the statistical weights assembled in the second parameter set (equation 23) gave $C_\infty = 7.84$ and $d \ln C_\infty/dT = -1.23 \times 10^{-3}$ K^{-1}. These results are in close agreement with those obtained by the direct integration method.[16]

Finally, it should be emphasized that, in the refined RIS scheme, due consideration can be given to effects arising from variation in the shape of the potential energy minima. With this device, the RIS scheme has gained a wider applicability in the conformational analysis of polymer chains. In fact, as demonstrated by Suter *et al.*[18] in the treatment of polypropylene (PP), this method allows a reasonable interpretation of the experimental results even when the shape of the potential energy well is highly distorted due to intramolecular steric conflicts.

2.4 DETERMINATION OF CONFORMATIONAL ENERGIES AND STATISTICAL WEIGHTS

2.4.1 Polyoxyethylene

As described in the preceding section, various conformation-dependent properties of polymeric chains can be treated within the framework of the RIS approximation provided that the neighbor-dependent character of the conformational energies is properly taken into account. In these conformational analyses, an accurate estimate of rotational states and their relative energies is essential. The required information is often obtained by direct spectroscopic measurements on small molecules having structural features similar to those of the chain molecule under investigation. Molecular force field calculations[2, 13, 14] provide a supplementary technique and allow calculation of the overall conformational energy surface. The energy expressions used in these calculations are, however, semiempirical and therefore their validity must be carefully tested against appropriate experimental data.[20] A typical example of this approach is the calculation of the energy surface for the PM chain, as illustrated in Figure 1. In that treatment, parameters for the nonbonded interactions were adjusted to reproduce the experimental energy difference E_σ of 0.5 kcal mol^{-1} (2.1 kJ mol^{-1}) between the *gauche* and *trans* states. The same parameters have been used in the estimation of barrier heights for rotations in propane, *n*-butane, isobutane and neopentane, with results which compared favorably with experiment.[12]

One of the most characteristic features of the polyoxyethylene (POE) chain, $\text{[CH}_2\text{CH}_2\text{O]}_x$, is its strong preference for the *gauche* state about the constituent OC–CO bond, which makes the chain highly flexible.[2, 20–24] In this arrangement, *gauche* O $\cdots$ O interactions take place around the skeletal C–C bond. Semiempirical energy calculations using customary expressions have failed to reproduce the observed *gauche* preference of the C–C bond.[22, 23] Such a stabilization phenomenon has been termed the *gauche* oxygen effect.[22] A similar *gauche* effect is known for a variety of disubstituted ethanes, XCH$_2$CH$_2$X, X being an electron-withdrawing group.[25] The MO investigation of these effects indicates that the intramolecular correlation is confined to a rather short range.[26, 27] From the practical point of view, the discrepancy between the observed and calculated energy differences may be amended by introducing an appropriate empirical correction term. With such revisions, semiempirical energy expressions should remain effective in the conformational analysis of chain molecules.[23]

Statistical weight factors required for the description of the POE chain have been determined from the NMR coupling constant measurements of 1,2-dimethoxyethane (DME), a monomer model compound, with supplemental use of semiempirical energy calculations. In this analysis, the torsional part of the potential function for the C–C bond was modified by inclusion of an extra stabilization term ΔE. Thus,

$$E_{tor}(\phi_{CC}) = (E_0/2)(1 - \cos 3\phi_{CC}) - (\Delta E/2)(1 + \cos 3\phi_{CC}) \quad (28)$$

where ΔE is the correction term for the *gauche* oxygen effect defined as $\Delta E = 0$ for $-60° \leq \phi_{CC} \leq 60°$ and $\Delta E = 0.95$ kcal mol^{-1} (4.0 kJ mol^{-1}) for $60° < \phi_{CC} < 300°$. The height of the torsional barrier located on both sides of the *trans* state (E_0) was taken to be 2.8 kcal mol^{-1} (11.7 kJ mol^{-1}). This expression, when used in combination with the equations introduced previously for the nonbonded van der Waals and coulombic interactions, leads to a value of $E(0°, 120°) = -0.4$ kcal mol^{-1} (-1.7 kJ mol^{-1}) for the *gauche* minimum as expressed relative to the *trans* state. The energy contour diagram obtained in this manner is shown in Figure 2.[23]

Calculations were carried out for the energies $E(\phi_{CO}, \phi_{CC})$ taken at 10° intervals: all regions for which $E \leq 3$ kcal mol^{-1} (12.6 kJ mol^{-1}) relative to the *tt* minimum were included. The conformational partition functions z, average energies $\langle E \rangle$, and average rotation angles $\langle \phi \rangle$ were estimated for the five nonequivalent states. The results are summarized in Table 3. Statistical weight parameters are given in the form

$$\delta = \delta_0 \exp(-E_\delta/RT) \quad (29)$$

Chain Conformations

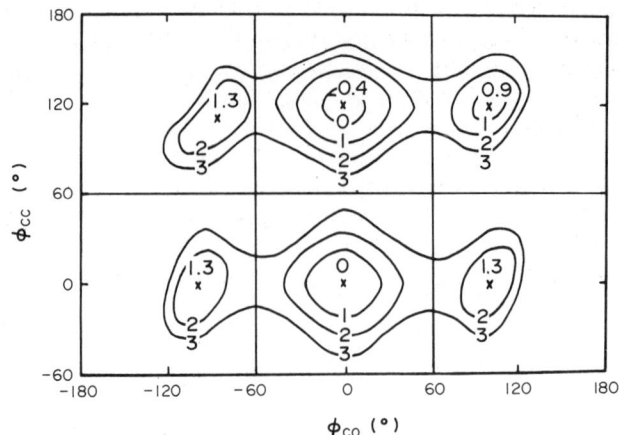

Figure 2 Energy map for the bond pair C–O, C–C of 1,2-dimethoxyethane, expressed in kcal mol^{-1} (4.2 kJ mol^{-1}) relative to the *tt* minimum ($\phi_{CO} = \phi_{CC} = 0$). The symbol × denotes local minima

where δ represents any of the statistical weights ρ, v and ω, as defined in Table 4. The preexponential factor δ_0 and the average energy E_δ can be deduced from the z and $\langle E \rangle$ values listed in Table 3, and the results are summarized in Table 4.[23] The factor ρ_0 departs significantly from unity, indicating that the corresponding potential well is sizeably skewed.

Table 3 Conformational Characteristics of DME as Estimated from the Potential Energy Contour Map (Figure 2) for a Temperature of 25 °C[a]

Rotational state	Z	$\langle E \rangle$ (kcal mol^{-1})	$\langle \phi_{OC} \rangle$, $\langle \phi_{CC} \rangle$ (°)
tt	1.00	0.61	0.0, 0.0
$tg^{\pm}$	1.88	0.23	0.0, 117.0
$g^{\pm}t$	0.09	1.72	98.1, 0.0
$g^{\pm}g^{\pm}$	0.18	1.33	98.1, 117.0
$g^{\mp}g^{\pm}$	0.06	1.87	−85.0, 110.6

[a] See the footnote to Table 1.

Table 4 Parameters for the Statistical Weight Factors of DME Deduced from Energy Calculations[a]

E_v^b (kcal mol^{-1})	E_ρ^b (kcal mol^{-1})	E_ω^b (kcal mol^{-1})	v_0	ρ_0	ω_0
−0.38	1.11	0.53	0.99	0.61	0.88

[a] Statistical weights v and ρ are used for the interactions dependent on a single bond rotation, while ω represents those associated with two consecutive bond rotations:
[b] 1 kcal = 4.2 kJ.

With the statistical weight parameters thus evaluated, the fractions of conformers may be calculated for given bonds of the chain. For DME, analytical expressions, as shown below, can be obtained by simple manipulation.

$$f_g^{CC} = 1 - f_t^{CC} = pv/(1 + pv) \tag{30}$$

where

$$p = 2[(1 + \rho + \rho\omega)/(1 + 2\rho)]^2 \tag{31}$$

and

$$f_g^{CO} = 1 - f_t^{CO} = q\rho/(1 + q\rho) \tag{32}$$

where

$$q = 2[1 + v(1 + \omega) + \rho(2 + v + 2v\omega + v\omega^2)]/[1 + 2v(1 + \rho\omega) + 2\rho(1 + v)] \tag{33}$$

Fractions were calculated to be $f_g^{CC} = 0.78$ and $f_g^{CO} = 0.12$, respectively, around the C–C and C–O bond for a temperature of 25 °C. For an infinitely long POE chain, the bond conformation probabilities can be easily calculated by the largest-eigenvalue method (*cf.* equation 24). The values of f_g^{CC} and f_g^{CO} thus obtained were found to be nearly identical with those of the monomer model, suggesting that bond conformations are predominantly determined by the local structure in this polymer system.

In NMR spectra of DME, the ^{13}CH satellite side bands provide the information for the CH$_2$–CH$_2$ bond, and similarly for the rotation about the adjoining C–O bond, the vicinal ^{13}C–^{1}H coupling constant associated with the terminal methyl group is useful. The RIS representations of these coupling constants are given in Figure 3. The observed values correspond to rotational averages[28] such as

$$^3J_{HH} = {}^3J_{AB} = {}^3J_{A'B'} = J_G f_t^{CC} + (\tfrac{1}{2})(J_T' + J_G'')f_g^{CC} \tag{34}$$

$$^3J_{HH}' = {}^3J_{AB'} = {}^3J_{A'B} = J_T f_t^{CC} + J_G' f_g^{CC} \tag{35}$$

and

$$^3J_{CH} = J_G f_t^{CO} + (\tfrac{1}{2})(J_T + J_G')f_g^{CO} \tag{36}$$

(In these expressions, notations J_T, J_G, J_G' appear in both couplings, *i.e.* ^{1}H–C–C–^{1}H and ^{13}C–O–C–^{1}H. Numerical values of these coupling constants should vary depending on the bonding system.) To elucidate conformational energies from the NMR measurements, the following prescriptions are adopted for the statistical weight parameters

$$v = 0.99 \exp(-E_v/RT)$$

$$\rho = 0.61 \exp(-E_\rho/RT) \tag{37}$$

$$\omega = 0.88 \exp(-E_\omega/RT)$$

where the numerical values of the preexponential factors are those estimated by the energy calculation, as given in Table 4. The effect arising from the difference in the shape of the potential well (*cf.* Figure 2) is thus taken into account. In this study, we treat E_v and E_ρ as adjustable parameters, instead of adopting the numerical results previously calculated. The magnitude of E_v

(a)

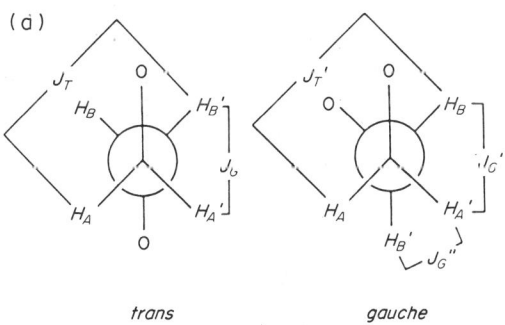

trans   *gauche*

(b)

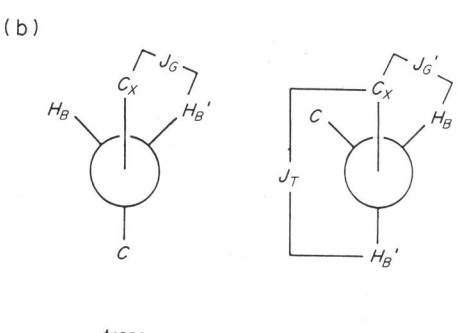

trans   *gauche*

Figure 3 (a) The preferred conformations about the C–C bond, and definition of the *trans* and *gauche* vicinal ^{1}H–^{1}H couplings. (b) Definition of the *trans* and *gauche* vicinal ^{13}C–^{1}H couplings for the C–O bond

may be estimated from the analysis of the experimental values of $^3J_{HH}$ and $^3J'_{HH}$. The probable range of E_ρ is similarly deduced from the study on $^3J_{CH}$. For the second-order interaction involved in the $g^\mp g^\pm$ arrangement of the OC–C–OC sequence, the calculated value of $E_\omega = 0.53\,\text{kcal mol}^{-1}$ (2.2 kJ mol^{-1}) is used (*cf.* Table 4).

To facilitate the analysis, we assume that

$$J_T = J'_T$$
$$J_G = J'_G = J''_G \tag{38}$$

Substitution of these relations into equations (34) and (35), followed by elimination of f_t^{CC} and f_g^{CC} yields

$$^3J_{HH} = -\tfrac{1}{2}\,^3J'_{HH} + \tfrac{1}{2}J_T + J_G \tag{39}$$

Values of J_T, J_G and E_v can be estimated by using the relations set forth in equations (34), (35) and (39). Following Gutowsky,[28] best fit values of these parameters were found so as to reproduce the experimental data of $^3J_{HH}$ and $^3J'_{HH}$ observed at different temperatures. The E_v values thus obtained are summarized in Table 5.

Table 5 Conformational Energies, E_v and E_ρ, Estimated for DME and POE from the Observed Temperature Dependence of Vicinal Coupling Constants, $^3J_{HH}$, $^3J'_{HH}$ and $^3J_{CH}$

Solvent	Temperature range (°C)	$E_v^{a,b}$ (kcal mol^{-1})	$E_\rho^{a,c}$ (kcal mol^{-1})
DME			
c-C$_6$D$_{12}$	40.0–100.0	−0.5	0.8
(CD$_2$CD$_2$O)$_2$	40.0–100.0	−0.5	0.8
CDCl$_3$	40.0–100.0	−0.9	1.1
CD$_3$OD	40.0– 80.0	−0.9	0.8
(CD$_3$)$_2$SO	24.8–117.0	−0.8	0.8
D$_2$O	24.8–102.0	−1.2	1.1
neat	20.0– 80.0	−0.6	0.9
POE			
(CD$_2$CD$_2$O)$_2$	20.0– 90.0	−0.5	1.0
D$_2$O	40.0– 92.0	−1.2	1.1

[a] 1 kcal = 4.2 kJ. [b] The *trans* and *gauche* coupling constants for the H–C–C–H moiety determined concomitantly are in the range $J_T = 11.1$ to 11.8 Hz and $J_G = 2.2$ to 2.4 Hz. [c] The *trans* and *gauche* coupling constants required in the analysis (C–O–C–H) were taken from those observed for 1,4-dioxane.

As defined in equations (32), (33) and (36), the temperature dependence of $^3J_{CH}$ arises through the fraction term $f_g^{CO}(=1-f_t^{CO})$, which is in turn closely related to E_ρ (*cf.* equation 37). Adoption of the *trans* and *gauche* couplings deduced from the observed $^3J_{CH}$ value of 1,4-dioxane leads to the E_ρ values shown in the last column of Table 5. In these treatments, the neighbor-dependent character of the bond rotation along the chain is rigorously taken into account, as given in expressions (30)–(36). In the first step, conformational energy E_v was tentatively deduced by adopting the value of E_ρ calculated in the preceding section. The value of E_v thus derived was used in the estimation of E_ρ in the following step. In principle, such procedures should be iteratively repeated until a set of mutually consistent values of E_v and E_ρ is obtained. However, E_v and E_ρ were found to vary quite insensitively with each other. Thus, in practice, the iteration was terminated after the first trial. In equations (30)–(33), the statistical weight factor ω explicitly represents the neighbor-dependent character of the bond rotation. Since ω is always accompanied by ρ (*i.e.* $\omega\rho \ll 1$) in equations (31) and (33), neither f_g^{CC} nor f_g^{CO} are significantly affected by changes in E_ω within the range $E_\omega = 0.5 \pm 0.5\,\text{kcal mol}^{-1}$ (2.1 ± 2.1 kJ mol^{-1}). Table 5 also includes the results obtained for the neat DME liquid. The analysis has been extended to the polymer, POE, for which the largest-eigenvalue method is applicable (*cf.* equation 24). Values of E_v obtained for DME are negative, ranging from −0.5 (*cyclo*-C$_6$D$_{12}$) to −1.2 kcal mol^{-1} (D$_2$O) (1 kcal = 4.2 kJ). The *gauche* form tends to be more stabilized as the dielectric constant of the medium increases.

A similar phenomenon is found for the 1,2-disubstituted ethanes.[25] An increase in the *gauche* fraction is also observed in hydrogen bond forming solvents such as CDCl$_3$ and D$_2$O. As for the C–O bond, IR and Raman studies[29] on some simple dialkyl ethers such as diethyl ether or ethyl methyl ether have yielded conformational energies in the range from 1.1 to 1.5 kcal mol^{-1} for the difference between the *gauche* and *trans* states in the neat liquid, in solution, as well as in the gas phase. The results of electron diffraction studies[30] on ethyl methyl ether have led to a value of 1.2 kcal mol^{-1} (5.0 kJ mol^{-1}). In the present study we have adopted a preexponential factor $\rho_0 = 0.61$ as suggested by the conformational energy calculations. If we tentatively employ a simple Boltzmann factor for ρ by setting $\rho_0 = 1.0$, the value of E_ρ is raised from 0.8 to 1.0 kcal mol^{-1} (*cf.* Table 5). In view of the ambiguity involved in the choice of J_G (*cf.* equation 38), the difference between simple ethers and DME may be insignificant.

It should be noted here that the E_v and E_ρ values derived for POE exhibit a good correspondence with those for DME. The conformational energies thus established have been used in the RIS calculations of various conformation-dependent properties of the POE chain.[24a] Values of the RIS parameters have been adjusted somewhat to optimize the agreement with experimental data for the unperturbed dimension $C_\infty = \langle r^2 \rangle_0 / nl^2$, dipole moment ratio* $\langle \mu^2 \rangle / nm^2$, and molar Kerr constant $\langle_m K \rangle / x$. The final set of parameters derived in this manner is as follows: $\angle CCO = \angle COC = 111.5°$; $\phi_t^{CC} = \phi_t^{CO} = 0°$, $\phi_g^{CC} = \pm 112.0°$, $\phi_g^{CO} = \pm 100.0°$; $E_v = -0.5$, $E_\rho = 0.8$, $E_\omega = 0.53$, units for energies being given in kcal mol^{-1} (1 kcal = 4.2 kJ). Results of calculations for the polymer are compared with those observed in Table 6. A somewhat larger discrepancy may be noted between the calculated and observed values of d ln C_∞ / dT as well as $\langle \mu^2 \rangle / nm^2$. In both cases, the agreement can be improved by adopting a smaller value of E_ρ: 0.4 to 0.6 kcal mol^{-1} (1.7 to 2.5 kJ mol^{-1}). A value of E_ρ less than 0.6 kcal mol^{-1} is, however, incompatible with the experimental results obtained in the oligomeric region.[23] The RIS scheme established in this manner has been found to be compatible with the X-ray scattering data obtained for α,ω-diiododerivatives of ethylene oxide oligomers.[24b]

Table 6 Comparison of the Calculated and Observed Values of Some Conformation-Dependent Properties of POE

	C_∞	[d ln C_∞ / dT] (10^{-3} K^{-1})	$\langle \mu^2 \rangle / nm^2$	[d ln$\langle \mu^2 \rangle / dT$] (10^{-3} K^{-1})	$\langle_m K \rangle / x$ $(10^{-12} \text{ cm}^5 \text{ statvolt}^{-2} \text{ mol}^{-1})$
Calculated	5.2	0.74	0.35	3.0	-2.9
Observed	5.2[a]	0.23[b]	0.53[d]	2.6[d]	-2.0[e]
		0.2(± 0.2)[c]	0.48[e]		

[a] D. R. Beech and C. Booth, *J. Polym. Sci., Part A-2*, 1969, **7**, 575.
[b] P. J. Flory, 'Statistical Mechanics of Chain Molecules', Interscience, New York, 1969.
[c] S. Bluestone, J. E. Mark and P. J. Flory, *Macromolecules*, 1974, **7**, 325.
[d] K. Bak, G. Elefante and J. E. Mark, *J. Phys. Chem.*, 1967, **71**, 4007.
[e] K. M. Kelly, G. D. Paterson and A. E. Tonelli, *Macromolecules*, 1977, **10**, 859.

In summary, statistical weight parameters elucidated from the NMR measurements on DME have been used successfully in the estimation of various conformation-dependent properties of the POE chain in the unperturbed state. However a value of $C_\infty = 6.9$ derived recently from the SANS studies[34] on the molten POE samples cannot be reproduced within a reasonable range of RIS parameters. At present, we have no plausible explanation for such a large enhancement in C_∞ in the melt.

2.5 THE RIS TREATMENT OF CONFORMATION-DEPENDENT PROPERTIES

2.5.1 Poly(Alkyl Vinyl Ether)s

2.5.1.1 Computational procedure

As illustrated in Figure 1, the energy calculation for polymethylene (PM) yields four $g^\pm g^\mp$ minima at the approximate angles (80°, 115°), (80°, −115°), (−80°, 115°) and (−80°, −115°). These minima can be incorporated in a five-rotational-state scheme comprising rotational states defined at 0°, 80°, 115°, −115° and −80°. This model represents more rigorously the characteristic feature of the

---

* In the present treatment, the dipole moment ratio $\langle \mu^2 \rangle / x$ is calculated for random-coil configurations of the chain. Since relevant experimental values are usually derived in thermodynamically good solvents, the respective zero subscript on $\langle \mu^2 \rangle_0$ is deleted. The excluded volume effect on the mean-square dipole moment is known to be less pronounced in the polymer system which involves an appropriate symmetry element for the disposition of the bond dipoles.[31, 32] Mattice *et al.*[33] have studied the effect for model chains generated by computer simulation.

conformational energy surface of the chain. Calculations carried out according to the five-state scheme yielded values of the characteristic ratio (7.4) only a little smaller than were obtained (7.6) using three-rotational states with comparable values of the parameters.[12] The three-state scheme includes only two $g^{\pm}g^{\mp}$ pairs in place of four pairs in the more realistic five-state model. Existing experimental results on the conformation-dependent properties can be treated satisfactorily by a three-state model by suitable choice of parameters. For a simple flexible chain such as PM, use of the three-state scheme as a computational device is fully justified. In this scheme E_{ω} may be regarded as an empirical parameter.

The presence of substituents on alternate skeletal carbon atoms proliferates steric interactions in the vinyl chains as compared with PM. Suter *et al.*[18] estimated the conformational energy surface of 2,4-dimethylpentane, a model compound for polypropylene (PP), by means of semiempirical energy expressions. Large displacements of high energy conformations from the perfectly staggered positions were found, which were taken into account in their five-state matrix scheme, which thereby offered a more exact representation of the energy contour map. The characteristic ratio C_{∞} of PP and its temperature coefficient were satisfactorily reproduced within the accuracy that these quantities are known. According to this model, C_{∞} decreases monotonically as the proportion of *meso* diads increases. Thus C_{∞} for the isotactic chain is somewhat smaller than for the atactic chain. As inspection of the three-state scheme should reveal, the same trend can be reproduced in this scheme by adjusting the statistical weight parameters for the second-order interaction somewhat arbitrarily.[35, 36] Accordingly the three-state scheme has been widely adopted in treating various vinyl polymer systems.[2]

A variety of vinyl polymers with polar side chains are known. In most cases, polar groups are incorporated in a 'flexible' side chain and, in the treatment of conformation-dependent properties, statistical mechanical averages must be taken over all configurations of side chains as well as skeletal bonds. The poly(alkyl vinyl ether)s (PAVE) provide an illustrative example (Figure 4).[37] The original Flory scheme has been extended to include statistical treatment of the flexible side chains. The method has been applied to the estimation of the unperturbed end-to-end distance, dipole moment and bond conformations of the PAVE chains. For PAVEs carrying asymmetric side chains such as $R = C^{*}HMeEt(S)$ (*cf.* Figure 4), a large enhancement in the optical rotatory power has been observed as compared with those of the low molecular weight model compounds. The effects arising from such asymmetric side chains have been discussed.[37]

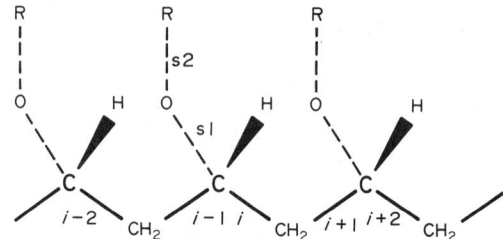

Figure 4   Schematic representation of an isotactic PAVE chain in its planar conformation. Rotational states t, g^{+}, g^{-} about bond s1 in the side chain are defined by the relative position of bond s2 in reference to bond i

Statistical weight matrices generally applicable to the *meso* and *racemo* diads of a vinyl polymer have been developed by Flory and his collaborators.[38] Adopting their notation, for PAVE, we have for the *meso* diad,

$$U''_m = \begin{bmatrix} \eta^2\omega'' & \eta & \eta\tau\omega' \\ \eta & \omega & \tau\omega' \\ \eta\tau\omega' & \tau\omega' & 0 \end{bmatrix} \tag{40}$$

and for the *racemo* diad,

$$U''_r = \begin{bmatrix} \eta^2 & \eta\omega' & \eta\tau\omega'' \\ \eta\omega' & 1 & \tau\omega \\ \eta\tau\omega'' & \tau\omega & 0 \end{bmatrix} \tag{41}$$

where rows and columns are indexed to rotational states in the order t, g and $\bar{g}$ for consecutive bonds i and $i+1$, respectively. The g state, in which CH is *syn* with respect to CH_2, is assigned a statistical weight of unity. The relative statistical weight for the t state in which CH is *syn* to O is expressed by η, and that for $\bar{g}$, in which CH is *syn* to both CH_2 and O, is represented by τ. Various second-order interactions taking place between groups separated by four bonds are distinguished by ω, ω' and ω''. Since these matrices are presented in a symmetrical form, distinctions between (*dd*) and (*ll*) (*meso*) and between (*dl*) and (*ld*) (*racemo*) are not required. The statistical weight matrix U' generally applicable to any vinyl chains in conjunction with the U'' matrices given above takes the form

$$U' = \begin{bmatrix} 1 & 1 & 1 \\ 1 & \omega & 1 \\ 1 & 1 & 0 \end{bmatrix} \tag{42}$$

The conformational energies required in the Boltzmann expression for the statistical weight factors have been estimated by the semiempirical energy calculations,[37] and the results are summarized in Table 7. Reliability of these parameters has been confirmed by testing against existing NMR data on 2,4-dimethoxypentane.[39]

Table 7 Separate Interaction Energy Parameters for the Skeletal Conformation of PAVE

Statistical weight factor	Interacting groups	Interaction energy parameter (kcal mol^{-1})[a]
η	$C^\alpha H\text{--}CH_2\text{--}C^\alpha H\text{--}O$	-0.44
τ	CH_2	0.5
ω''	$O\text{--}C^\alpha H\text{--}CH_2\text{--}C^\alpha H\text{--}O$	1.1
ω'	$CH_2\text{--}C^\alpha H\text{--}CH_2\text{--}C^\alpha H\text{--}O$	0.9–1.6
ω	$CH_2\text{--}C^\alpha H\text{--}CH_2\text{--}C^\alpha H\text{--}CH_2$	1.9–2.2

[a] 1 kcal = 4.2 kJ.

Statistical weights attributable to side chain conformations may be evaluated according to the scheme[37,40]

$$\beta(\delta, \varepsilon) = S_1^{(p)} \tag{43}$$

where δ and ε denote each of the rotational states about skeletal bonds, $i-1$ and i, on both sides of the methine carbon (*cf.* Figure 4). Each S_h in the serial product $S_1^{(p)}$ represents the statistical weight matrix for bond h ($1 \geq h \geq p$) in the side chain. Each element $u'_{\delta\varepsilon}$ of the U' matrix given in equation (42) is multiplied by the corresponding statistical weight factor $\beta(\delta, \varepsilon)$ to complete the expressions required for the vinyl polymer system with flexible side chains. Thus

$$U' = [u'_{\delta\varepsilon} \beta(\delta, \varepsilon)] \tag{44}$$

As is easily shown by inspection of a model, statistical weight matrices U'' should not be affected by the side chain conformations provided the assumption holds that interactions of higher orders are negligible (*cf.* Section 2.5.1.3). As a consequence of the conformational energy considerations described above, the g^- state is suppressed for the first side chain bond s1 (*cf.* Figure 4). With a symmetrical side chain OR, various conformations realizable through rotations about the second or a higher bond dictate equal statistical weights for the rotational states (t and g^+) permitted for bond s1. Side chain contributions, as assembled in a matrix scheme, are given by

$$\beta = \begin{bmatrix} 0 & 1 & 1 \\ 1 & 2 & 2 \\ 1 & 2 & 2 \end{bmatrix} \tag{45}$$

This simple expression should be generally applicable to PAVE chains having a symmetrical side chain such as $R = Me$, CH_2Me, $CHMe_2$ or CH_2CHMe_2. When the R group involves an asymmetric center, however, statistical weights associated with the t and g^+ states for bond s1 are no longer equivalent. Thus, any simple cancellation of statistical weights between these two states cannot be expected in principle. In general, the $\boldsymbol{\beta}$ matrix should assume either the form

$$\boldsymbol{\beta}_d = \begin{bmatrix} 0 & b & b \\ a & a+b & a+b \\ a & a+b & a+b \end{bmatrix} \tag{46}$$

or by symmetry

$$\boldsymbol{\beta}_l = \boldsymbol{\beta}_d^{\mathrm{T}} \tag{47}$$

depending upon the stereochemical configuration (d or l) at the main chain methine carbon. In equation (47), superscript T indicates a transpose. Parameters a and b represent the statistical weights ascribable to the asymmetric side chain when the s1 bond is in the t and g^+ state, respectively. Multiplication according to equation (44) yields a pair of U' matrices, i.e. U'_d and U'_l, for an asymmetric side chain. For a symmetric side chain, $a = b$, and naturally equations (46) and (47) reduce to equation (45).

The conformational partition function takes the form

$$Z = U_0 \left(\prod_{k=1}^{x-1} U'_k U''_k \right) U_x \tag{48}$$

where the U''_k matrices must be properly chosen depending upon the stereochemical character of the kth diad. As usual, special care must be taken for the terminal bonds of the chain. Corresponding to equation (48), the mean-square value of a conformation-dependent quantity (the second moment M^2) for a vinyl chain is given by

$$\langle M^2 \rangle = Z^{-1} g_0 \left(\prod_{k=1}^{x-1} g'_k g''_k \right) g_x \tag{49}$$

where g'_k and g''_k are the generator matrices defined for the first and second bond of the kth diad, respectively. When the moment m is defined for a single bond of the skeletal chain, pseudoelements of matrix g'_k (similarly g''_k) may be expressed as

$$g'_k = [u'_{\delta\varepsilon} G'_\varepsilon]_k \tag{50}$$

where

$$G'_\varepsilon = \begin{bmatrix} 1 & 2m^{\mathrm{T}}T & m^2 \\ 0 & T & m \\ 0 & 0 & 1 \end{bmatrix} \tag{51}$$

m^{T} being the transpose of m. Here T is the transformation relating the Cartesian coordinate system affixed to bond $i+1$ to that affixed to bond i, thus being defined as a function of the bond angle $(\pi - \theta)$ between these two bonds and the rotational states ϕ_ε of bond i. g_0 and g_x take care of the terminal bonds of the chain. Identification of m with the bond vector gives the mean square of the chain vector $\langle r^2 \rangle_0$ according to equation (49).

For a PAVE chain, bond dipoles may be assigned to the first and second bond of the side chain, corresponding to the group dipole moment of the ether linkage. The orientation of bond dipole m_1, being invariably defined in the coordinate system affixed to bond i, depends only on the skeletal conformation of the chain. The spatial arrangement of m_2 requires specification of rotational states (t, g^+, g^-) about bond s1 in addition to those of the backbone chain. For the purpose at hand, the bond moment m appearing in equation (51) should be replaced by the averaged quantity $\hat{m}$ over all side chain conformations in the coordinate system fixed to bond i. The mean dipole $\hat{m}$ thus defined must be derived for each combintion of rotational states (δ, ε) about bonds $i-1$ and i. The expression (50) now requires specification of two consecutive rotational states, and the generator matrix g'_k must read as

$$g'_k = [u'_{\delta\varepsilon} G'_{\delta\varepsilon}]_k \tag{52}$$

Thus the mean square of dipole moment $\langle \mu^2 \rangle$ can be calculated by equation (49) provided that proper account is taken of the terminal bonds.

2.5.1.2 *Results of calculation*

(i) Symmetric side chains

Characteristic ratios C_∞ calculated for Monte Carlo chains are plotted in Figure 5(a) as a function of the diad replication probability Pr, *i.e.* the probability of a *meso* placement.[2] As suggested by the conformational energy calculation, rotational states are taken to occur at $\phi_g = 110°$ and $\phi_t = 0$. Values of C_∞ vary from 16.1 for the isotactic ($Pr = 1.0$) to 12.7 for the syndiotactic chain ($Pr = 0.0$) through minimum values of *ca.* 5 in the range $Pr = 0.3$ to 0.5. The characteristic ratios have been experimentally estimated to be in the range 6 to 8 for moderately isotactic poly(methyl vinyl ether).[41] The results obtained for polymers with $R = Et$, Pr^i, or Bu^n on unfractionated samples, although less reliable, suggest that the ratios C_∞ are of a similar magnitude. According to the NMR studies, tacticities of these 'moderately' isotactic samples are probably in a range of $Pr = 0.7$ to 0.8.[37] The results of calculations are therefore in reasonable agreement with experimental observations.

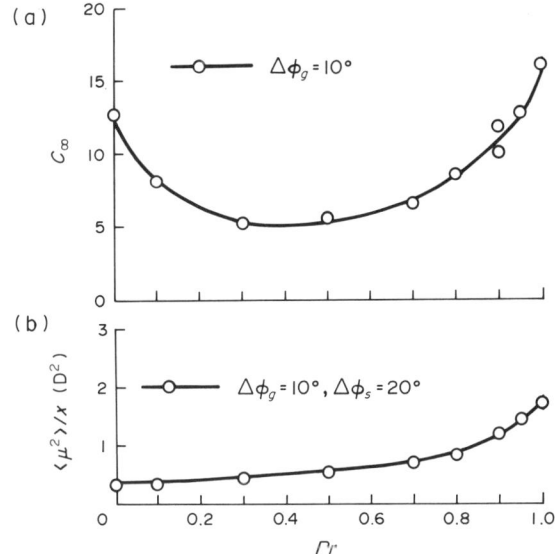

Figure 5 (a) Characteristic ratios C_∞ calculated for PAVEs having symmetric side chains. Computations were carried out using Monte Carlo chains of 200 units, ϕ_t being kept at 0. (b) Mean-square dipole moments per repeat unit $\langle \mu^2 \rangle / x$ calculated for Monte Carlo chains of 100 units. Displacements of rotational states are indicated in the figure for the main chain ($\Delta\phi_g$) and side chain (about s1) conformations ($\Delta\phi_s$), respectively, ϕ_t being kept at 0

The mean-square dipole moments $\langle \mu^2 \rangle / x$ calculated for Monte Carlo chains ($x = 100$) are shown in Figure 5(b), the moment μ being given in debye (1 debye $= 3.338 \times 10^{-30}$ C m). In treating the dipole moment of PAVEs, the spatial arrangement of the s2 bond (O–C) is important. A displacement of $\Delta\phi_s = 20°$ is assumed for the t and g^+ state about bond s1 as suggested from the conformational analysis[42a] as well as the crystallographic data.[42b] The g^- conformation involves very high steric interactions and therefore the state is assigned a nil weight. In contrast to the results shown for C_∞ in Figure 5(a), values of $\langle \mu^2 \rangle / x$ decrease monotonically from $Pr = 1.0$ to 0.0. The mean-square dipole moment $\langle \mu^2 \rangle / x$ observed for moderately isotactic poly(isopropyl vinyl ether)[43] (0.67 D^2) corresponds to that calculated for $Pr \simeq 0.7$. The higher value (1.22 D^2) observed for the atactic sample[43] is, however, at variance with the results presented in Figure 5(b). Observed dipole moments for poly(isobutyl vinyl ether) vary little with the tacticity in the moderately isotactic to atactic region. Experimental values[43,44] of $\langle \mu^2 \rangle / x = 0.95$ to 1.35 D^2 are reasonably approximated by the curve in the moderately isotactic range.

(ii) Asymmetric side chains

Calculations have been carried out for poly[(S)-1-methylpropyl vinyl ether] and poly[(S)-2-methylbutyl vinyl ether] using properly estimated conformational energy parameters. The characteristic ratios C_∞ obtained are shown in Figure 6(a). For the former polymer, large enhancement in C_∞ is predicted both in the isotactic ($Pr > 0.8$) and in the syndiotactic ($Pr < 0.2$) region.

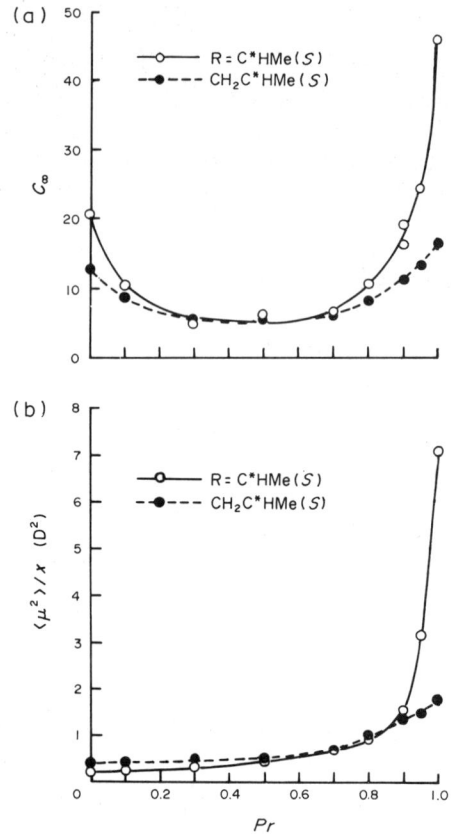

Figure 6 (a) Characteristic ratios and (b) mean-square dipole moments per repeat unit, calcultated for PAVEs carrying asymmetric side chains. Values of the parameters are the same as those given in the legend to Figure 5

Results for the latter polymer were found to be almost indistinguishable from those calculated for the symmetric side chain (*cf.* Figure 5a). Such an asymmetric effect may be best described by a parameter $q=(a-b)/b$, a and b being the parameters introduced in equation (46). With symmetric side chains for which $a=b$, $q=0$. Thus, q provides a measure of the 'asymmetric effect'. Values of q are estimated to be 1.63 for $R=C^*HMeEt(S)$ and 0.033 for $R=CH_2C^*HMeEt(S)$.

Variation of the mean-square dipole moment $\langle\mu^2\rangle/x$ with Pr is shown for the same polymers in Figure 6(b). These results should be compared with those for the symmetric side chain given in Figure 5(b). In the highly isotactic region ($Pr>0.9$), values of $\langle\mu^2\rangle/x$ increase very rapidly with Pr for $R=C^*HMeEt(S)$. In contrast to what is observed for C_∞ (*cf.* Figure 6a), no enhancement in $\langle\mu^2\rangle/x$ is predicted for the syndiotactic chain; in fact, $\langle\mu^2\rangle/x$ decreases slightly with increase in the q value. Experimental values of $\langle\mu^2\rangle/x$ are available for these polymers in the moderately isotactic to atactic region;[43] 0.90 (isotactic), 0.98 (atactic) for $R=C^*HMeEt(S)$ and 1.02 (isotactic), 1.06 (atactic) for $R=CH_2C^*HMeEt(S)$, respectively, the unit of μ being given in debye (1 debye $=3.338\times10^{-30}$ C m). Values obtained for the isotactic polymers are in agreement with those calculated for $Pr=0.8$. The slightly higher values measured for the atactic fractions are, however, somewhat at variance with the present analysis (*cf.* Figure 6b). As pointed out by Luisi *et al.*,[43] the dipole moment of PAVE is insensitive to the chemical structure of side chains, whether symmetric or asymmetric, in the moderately isotactic region. Distinct effects should be expected only for 'highly' isotactic polymers having high q values.

Fractions of left-handed (designated by fr^-) and right-handed conformation (fr^+) have been calculated for Monte Carlo chains with $x=200$ units in the conventional manner (Figure 7). For a symmetric side chain ($q=0$), $fr^+=fr^-$. Results for polymers having $R=C^*HMeEt(S)$ ($q=1.63$) and $CH_2C^*HMeEt(S)$ ($q=0.033$) are shown by circles ($\bigcirc$, $\bullet$) and triangles ($\triangle$, $\blacktriangle$), respectively. Disparity between fr^+ and fr^- is markedly high for the former polymer (0.013 to 0.868 at $Pr=1.0$), indicating that conformational rigidity imposed by branching at the β position in the side chain is very important. The effect is much smaller for the latter polymer. In each case values of fr are amazingly invariant over the entire range of Pr. As demonstrated in the analysis of optically active poly-1-alkenes[45–47] using the Whiffen–Brewster empirical rule,[48] the observed enhancement in the optical rotatory power of these polymer systems arises from the conformational asymmetry as described in terms of the disparity between fr^+ and fr^-. The optical rotation of PAVEs should obey the same principle. From the results illustrated in Figure 7, one would immediately expect that the optical activity enhancement would be very high for poly[(S)-1-methylpropyl vinyl ether].[45] It is interesting to note that the enhancement should not much depend on the stereoregularity of the chain. The effect for poly[(S)-2-methylbutyl vinyl ether] should be essentially the same but smaller in magnitude. Experimental observations reported by Pino and co-workers[49] are $[M]_D^{25} = +312$ for the former and $+5.6$ for the latter polymer. These values are to be compared with those observed for the corresponding model compounds;[49] $[M]_D^{25} = +36.13$ for (S)-1-methylpropyl ethyl ether and $+1.18$ for (S)-2-methylbutyl ethyl ether. In accordance with the preceding analysis on C_∞ and $\langle \mu^2 \rangle / x$, the tacticities of these moderately isotactic polymers may be in the range $Pr = 0.7$ to 0.8. An optical activity enhancement due to similar conformational asymmetry has been reported for poly-1-alkenes,[49,50] polyaldehydes[51] and polymethacrylates.[52]

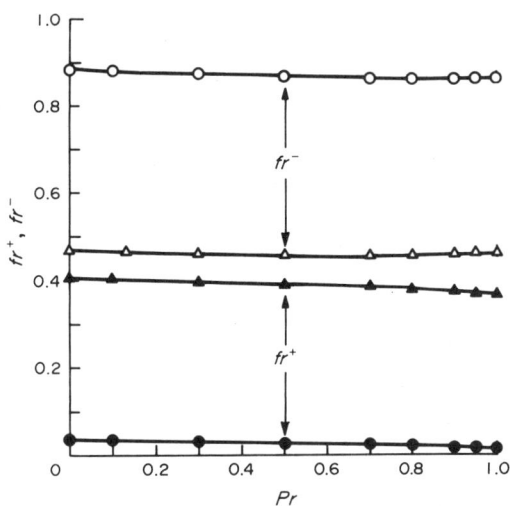

Figure 7 Fractions of left-handed (fr^-) and right-handed helical conformations (fr^+). The open ($\bigcirc$) and filled circles ($\bullet$) indicate the results obtained for R = C*HMeC$_2$H$_5$(S), and the open ($\triangle$) and filled triangles ($\blacktriangle$) for R = CH$_2$C*H MeC$_2$H$_5$(S). All calculations were carried out for Monte Carlo chains of 200 units

2.5.1.3 *Amendment for the possible conformational-space exclusion between adjacent side groups*

As inspection of a model should reveal, steric overlaps between the neighboring side chains may be substantial for PAVEs carrying bulky pendant groups such as R = CHMe$_2$ (*cf.* Figure 4). This problem has been studied by Suter *et al.*[53] for poly(4-methyl-1-pentene), a hydrocarbon analog of poly(isopropyl vinyl ether). In his treatment, all possible conformations are enumerated for the central unit of an isotactic triad sequence, the terminal groups being kept at the minimum energy estimated separately in the diad computation. The average energy was evaluated by numerical integration according to the trapezoidal rule on the basis of the potential energy surface thus derived. The 'net energy' was then estimated as the difference from that obtained for the conformation unperturbed by the inter-side-chain interactions. These space exclusions may occur cooperatively along the chain, and the exact treatment of such a long-range effect is very difficult in the present RIS scheme. The results of calculations for C_∞, in spite of the approximate nature of the amended treatment, showed substantial improvement in agreement with experiments.[53]

In this connection, it is appropriate to mention the work of Green *et al.*[54] on polyisocyanates carrying asymmetric side groups: poly[(S)-] and poly[(R),(S)-2,2-dimethyl-1,3-dioxolane-4-methyleneisocyanate]. In these polymer systems, steric interactions between the neighboring side chains are very severe. The results of light scattering measurements indicate that the backbone chain is highly extended when the polymer comprises only one optical antipode. Conformational-space exclusion taking place along the chain should play an important role in enhancement of the persistence length for the optically active polymer. Another example of the mutual space exclusion of the adjacent side chains has been reported from the deuterium NMR analysis for the α-helical poly(γ-benzyl-L-glutamate)s.[60]

2.6   CONFIGURATION OF POLYMERS IN BULK

2.6.1   Interpretation of the Neutron Scattering Data for Polyisobutylene

The analysis of the spatial configurations of macromolecular chains presented above is addressed primarily to an isolated molecule as it exists, for example, in a dilute solution. On theoretical grounds, the results obtained should be equally applicable to the molecules in the amorphous state. According to Flory's coil model,[55] the molecular conformations in bulk and in a Θ solvent should be identical. Recent applications of neutron scattering to studies of polymers in bulk have provided much valuable information inaccessible by other methods.[56] In particular, small angle scattering, applied to various polymers, has demonstrated that chain dimensions in the amorphous bulk state and in the melt are the same as those in Θ solvents, thus providing compelling evidence for the theoretical prediction that the chain configurations in amorphous and liquid polymers are substantially unperturbed. Nevertheless, one may still argue the possibility that the effects of local perturbations could somehow be compensated by alteration of the configuration of the chain as a whole in such a manner as to yield unperturbed overall dimensions. However, intermediate angle scattering carries information regarding correlation distances between scattering elements separated by comparatively short bond sequences within a polymer chain.

Flory and his co-workers[57,58] have developed a theoretical procedure for the RIS interpretation of small and intermediate angle X-ray and neutron scattering data. The scattering function P(s) for a system of independent molecules oriented at random is generally given by

$$P(s) = N^{-2} \left\langle \sum_i \sum_j (sr_{ij})^{-1} \sin(sr_{ij}) \right\rangle \tag{53}$$

where s represents the magnitude of the scattering vector for a given scattering angle, r_{ij} is the distance between the ith and jth scattering elements, N is the number of scattering elements, and the angle brackets denote the statistical mechanical average over all configurations of the molecule. To reduce the computational burden, the double sum in equation (53) may be replaced by a single sum such as

$$P(s) = N^{-2} \sum_k \left\langle h(r_k)(sr_k)^{-1} \sin(sr_k) \right\rangle \tag{54}$$

where $h(r_k)$ is the distance distribution function representing the number of distances r_{ij} falling between $r_k - \Delta/2$ and $r_k + \Delta/2$, Δ being an increment of r_k. Theoretical scattering functions can be calculated by the Monte Carlo method, the positions of individual atoms being treated explicitly. For a Monte Carlo configuration generated in a conventional manner, the positions of all scattering points are calculated by serial multiplication of the appropriate generator matrices for the chain vectors. Then the distance between every pair of scattering points is calculated and stored in the distance distribution function $h(r_k)$. The function P(s) is calculated according to equation (54) from these data. This process is repeated iteratively until a reasonable convergence is reached.

Flory and his collaborators[58] measured elastic neutron scattering from labelled polyisobutylene (PIB) at intermediate angles as well as at small angles, and compared the results of scattering measurements for the polymer in bulk and in dilute solution in a Θ solvent. Experiments in bulk were performed on (a) protonated PIB (PIB-H) dispersed in matrices containing fully deuterated PIB (PIB-d_8), and (b) the same PIB-H dispersed in hexadeuterio-PIB (PIB-d_6) in which only the methyl protons were deuterated. Neutron scattering experiments in a Θ solvent were carried out on (c) PIB-H dissolved in benzene-d_6. Experimental scattering curves were compared with theoretical scattering functions computed by a Monte Carlo method on the basis of torsional angles for an appropriate set of RIS parameters determined from the conformational analysis of PIB.[59] The scattering functions were calculated by assuming that the scattering points were situated on (i) all eight protons of the PIB-H monomer for samples (a) and (c), and (ii) the six methyl protons for sample (b). The agreement between the experimental and theoretical curves was excellent, both for bulk and solution, throughout the s range of the measurements. The maximum value $s = 0.6 \, \text{Å}^{-1}$ at which the measurements were performed corresponds to distances of *ca.* 10 Å in real space, which is the equivalent of four monomer units in the all-*trans* conformation. It follows from these observations that the RIS scheme, when properly prescribed (*cf. seq.*) precisely predicts the configuration of PIB chains in the bulk down to segments comprising four monomers. The analysis also demonstrates that the local configurations of individual chains in the bulk are random down to distances of *ca.* 10 Å; they are statistically unperturbed by intermolecular interactions which might have been thought to cause some local ordering.

Comparison of the measurements on samples (a) and (c) indicates that the experimental scattering curves are indistinguishable throughout the s range covered by the experiments. Since these scattering curves were measured using the same PIB-H molecules, it is scarcely conceivable that their correspondence is merely a coincidence. The agreement between the theoretical and experimental curves for sample (c) is also excellent, indicating that the theory predicts the chain configuration in the Θ solvent with the same accuracy (10 Å resolution) as in the bulk. These results, together with the observations of radii of gyration, provide decisive proof that chain configurations in amorphous polymers are essentially unperturbed, as predicted from theoretical considerations.[55] Analyses of the neutron diffraction data obtained from PIB-H in the partially deuterated matrix, PIB-d_6 (sample b), and X-ray scattering observed on PIB-H dissolved in *n*-heptane (sample d) also provide additional support to the conclusion stated above.

The RIS parameters used in the theoretical estimation of scattering functions of PIB are adapted from the results of the molecular force field calculation reported by Suter *et al.*[59] The bond lengths, bond angles and rotational angles elucidated by their analysis are summarized in Table 8. For both the diad bond pair and the pair flanking the substituted carbon, four different energy minima, characterized as t_+, t_-, g_+^+ and g_-^-, were identified at the bond rotation angles $+15°$, $-15°$, $+130°$ and $-130°$, respectively. Accordingly, the statistical weight matrices are given by a 4×4 scheme

$$U' = \begin{bmatrix} 0 & 0 & 1 & \xi \\ 0 & 0 & \xi & 1 \\ 1 & \xi & 0 & 0 \\ \xi & 1 & 0 & 0 \end{bmatrix} \tag{55}$$

and

$$U'' = \begin{bmatrix} 1 & 0 & 1 & 0 \\ 0 & 1 & 0 & 1 \\ 1 & 0 & 1 & 0 \\ 0 & 1 & 0 & 1 \end{bmatrix} \tag{56}$$

the parameter ξ being assigned a value of 0.0065, appropriate for 300 K. Matrices U' and U'' are applicable to the bond pair within the diad bounded by successive substituted carbon atoms.

Table 8 Geometrical Parameters Estimated for the PIB Chain, $\{CH_2C^sMe_2\}_x^a$

Parameter	Value
Bond lengths (Å)	
C–C	1.53
C–H	1.10
Bond angles (°)[b]	
$\angle C^sCC^s$	123
$\angle CC^sC$	109
Rotational angles for the skeletal bond (°)	
t_+	15
t_-	-15
g_+^+	130
g_-^-	-130

[a] U. W. Suter, E. Saiz and P. J. Flory, *Macromolecules*, 1983, **16**, 1317. [b] C^s denotes the substituted carbon.

In Table 9, radii of gyration ($\langle s^2 \rangle_w^{\frac{1}{2}}$) determined for PIB-H by neutron scattering in bulk as well as in solution are compared with those calculated by using the RIS parameter given above.[58] In these studies, parameters required in the conformational energy expressions were carefully adjusted so as to reproduce the torsional barrier heights of low molecular weight hydrocarbons such as Me–Me, Me–Et, Me–CHMe$_2$, Me–CMe$_3$ and Et–Et. Close agreement between experiment and calculation confirms the validity of conformational analysis on the basis of the molecular force field treatment.

Table 9   Radii of Gyration and Molecular Weights of PIB-H Determined by
Neutron Scattering[a]

Matrix or solvent	$\bar{M}_w$	Expl $\langle s^2 \rangle_w^{1/2}$ (Å)	Calcd $\langle s^2 \rangle_w^{1/2}$ (Å)
PIB-d_8	48000 ± 3000	68 ± 5[b]	$67–68$[c]
Benzene-d_6	52000 ± 4000	69 ± 5[b]	$67–68$[c]

[a] H. Hayashi, P. J. Flory and G. D. Wignall, *Macromolecules*, 1983, **16**, 1328; H. Hayashi and P. J. Flory, *Physica B + C (Amsterdam)*, 1983, **120**, 408. [b] Calculated from the observed values of $\langle s^2 \rangle_z^{1/2}$. [c] Calculated by assuming $C_\infty = 6.6–6.8$.

As is demonstrated in the examples presented in this chapter, various average properties of a polymeric chain can be interpreted in terms of this spatial configuration by using mathematical methods developed within the framework of the RIS scheme. The importance of these methods lies not so much in the facility with which they can be applied to the analysis of chemical and physical properties, but rather in the insights they afford to the understanding of the spatial configurations of polymeric chains. Through their use, a realistic comprehension of the spatial configurational characteristics has been achieved for polymeric chains of various kinds.

2.7 REFERENCES

1. P. J. Flory, 'Statistical Configuration of Macromolecular Chains', Nobel Prize Address, The Nobel Foundation, 1975; see also P. J. Flory, *Makromol. Chem., Macromol. Symp.*, 1986, **1**, 5.
2. P. J. Flory, 'Statistical Mechanics of Chain Molecules', Interscience, New York, 1969.
3. S. Mizushima, 'Structure of Molecules and Internal Rotation', Academic Press, New York, 1954.
4. M. Volkenstein, 'Configurational Statistics of Polymer Chains' (translated from the Russian, ed. S. N. Timasheff and M. J. Timasheff), Interscience, New York, 1963.
5. Yu. Ya. Gotlib, *Zh. Tekh. Fiz.*, 1959, **29**, 523.
6. T. M. Birshtein and O. B. Ptitsyn, 'Conformations of Macromolecules' (translated from the Russian, ed. S. N. Timasheff and M. J. Timasheff), Interscience, New York, 1966.
7. S. Lifson, *J. Chem. Phys.*, 1959, **30**, 964.
8. K. Nagai, *J. Chem. Phys.*, 1959, **31**, 1169.
9. C. A. J. Hoeve, *J. Chem. Phys.*, 1960, **32**, 888.
10. K. S. Pitzer and W. D. Gwinn, *J. Chem. Phys.*, 1942, **10**, 428; K. S. Pitzer, *J. Chem. Phys.*, 1946, **14**, 239; J. E. Kilpatrick and K. S. Pitzer, *J. Chem. Phys.*, 1949, **17**, 1064.
11. A. Morita and H. Watanabe, *Chem. Phys. Lett.*, 1979, **64**, 158; H. Watanabe and A. Morita, *J. Phys. Chem.*, 1985, **89**, 1787 and also see references cited.
12. A. Abe, R. L. Jernigan and P. J. Flory, *J. Am. Chem. Soc.*, 1966, **88**, 631.
13. U. Burkert and N. L. Allinger, 'Molecular Mechanics', ACS Monograph 177, Washington, DC, 1982.
14. M. Vásquez, G. Nemethy and H. A. Scheraga, *Macromolecules*, 1983, **16**, 1043 and references cited therein.
15. R. A. Scott and H. A. Scheraga, *J. Chem. Phys.*, 1966, **44**, 3054.
16. T. Oyama and K. Shiokawa, *Polym. J.*, 1981, **13**, 1145.
17. A. Abe, *Polym. J.*, 1982, **14**, 427.
18. U. W. Suter and P. J. Flory, *Macromolecules*, 1975, **8**, 765.
19. N. Saito, 'Kobunshi no Butsuri', Physics Society of Japan, Asakura, Tokyo, 1963; N. Saito, K. Okano, S. Iwayanagi and T. Hideshima, *Solid State Phys.*, 1963, **14**, 344.
20. A. Abe and K. Tasaki, *J. Mol. Struct.*, 1986, **145**, 309.
21. J. E. Mark and P. J. Flory, *J. Am. Chem. Soc.*, 1965, **87**, 1415; J. E. Mark and P. J. Flory, *J. Am. Chem. Soc.*, 1966, **88**, 3702.
22. A. Abe and J. E. Mark, *J. Am. Chem. Soc.*, 1976, **98**, 6468.
23. K. Tasaki and A. Abe, *Polym. J.*, 1985, **17**, 641.
24. (a) A. Abe, K. Tasaki and J. E. Mark, *Polym. J.*, 1985, **17**, 883; (b) H.-M. Li, B. Post and H. Morawetz, *Makromol. Chem.*, 1972, **154**, 89; A. Abe and K. Tasaki, *Macromolecules*, 1986, **19**, 2647.
25. W. J. Orville–Thomas, (ed.), 'Internal Rotation in Molecules', Wiley, New York, 1974.
26. Y. Sasanuma, Dissertation for Master Degree, Tokyo Institute of Technology, 1982.
27. T. Miyajima and T. Hirano, *J. Mol. Struct.*, 1984, **125**, 97; T. Hirano and T. Miyajima, *J. Mol. Struct.*, 1985, **126**, 141.
28. H. S. Gutowsky, G. G. Belford and P. E. McMahon, *J. Chem. Phys.*, 1962, **36**, 3353.
29. H. Wieser, W. G. Laidlaw, P. J. Krueger and H. Fuhrer, *Spectrochim. Acta, Part A*, 1968, **24**, 1055; T. Kitagawa and T. Miyazawa, *Bull. Chem. Soc. Jpn.*, 1968, **41**, 1976; J. P. Perchard, *J. Mol. Struct.*, 1970, **6**, 457.
30. K. Oyanagi and K. Kuchitsu, *Bull. Chem. Soc. Jpn.*, 1978, **51**, 2237.
31. J. Marchal and H. Benoit, *J. Chim. Phys. Phys.-Chim. Biol.*, 1955, **52**, 818; J. Marchal and H. Benoit, *J. Polym. Sci.*, 1957, **23**, 223; W. H. Stockmayer, *Pure Appl. Chem.*, 1967, **15**, 539; K. Nagai and T. Ishikawa, *Polym. J.*, 1971, **2**, 416; M. Doi, *Polym. J.*, 1972, **3**, 252.
32. S. C. Liao and J. E. Mark, *J. Chem. Phys.*, 1973, **59**, 3825.
33. W. L. Mattice and D. K. Carpenter, *Macromolecules*, 1984, **17**, 625; W. L. Mattice, D. K. Carpenter, M. D. Barkley and N. R. Kestner, *Macromolecules*, 1985, **18**, 2236; W. L. Mattice and A. C. Lloyd, *Macromolecules*, 1986, **19**, 2250.

34. J. Kugler, E. W. Fischer, M. Peuscher and C. D. Eisenbach, *Makromol. Chem.*, 1983, **184**, 2325.
35. P. J. Flory, J. E. Mark and A. Abe, *J. Am. Chem. Soc.*, 1966, **88**. 639.
36. A. Abe, *Polym. J.*, 1970, **1**, 232.
37. A. Abe, *J. Polym. Sci., Polym. Symp.*, 1976, **54**, 135; A. Abe, *Macromolecules*, 1977, **10**, 34.
38. P. J. Flory, P. R. Sundararajan and L. C. DeBolt, *J. Am. Chem. Soc.*, 1974, **96**, 5015.
39. K. Matsuzaki, K. Sakota and M. Okada, *J. Polym. Sci., Part A-2*, 1969, **7**, 1444.
40. For the treatment of side chain conformations, see also W. L. Mattice, Macromolecules, 1975, **8**, 644; W. L. Mattice, *Macromolecules*, 1977, **10**, 1171.
41. J. A. Manson and G. J. Arquette, *Makromol. Chem.*, 1960, **37**, 187.
42. (a) K. Tasaki, Y. Sasanuma, I. Ando and A. Abe, *Bull. Chem. Soc. Jpn.*, 1984, **57**, 2391; (b) I. W. Bassi, *Atti Accad. Naz. Lincei, Cl. Sci. Fis., Mat. Nat., Rend.*, 1960, **(8)29**, 193.
43. P. L. Luisi, E. Chiellini, P. F. Franchini and M. Orienti, *Makromol. Chem.*, 1968, **112**, 197.
44. M. Takeda, Y. Imamura, S. Okamura and T. Higashimura, *J. Chem. Phys.*, 1960, **33**, 631; H. A. Pohl and H. H. Zabusky, *J. Phys. Chem.*, 1962, **66**, 1390.
45. A. Abe, *J. Am. Chem. Soc.*, 1968, **90**, 2205.
46. A. Abe, *J. Am. Chem. Soc.*, 1970, **92**, 1136; A. Abe, *Rep. Prog. Polym. Phys. Jpn.*, 1970, **13**, 465.
47. U. W. Suter, S. Pucci and P. Pino, *J. Am. Chem. Soc.*, 1975, **97**, 1018.
48. D. H. Whiffen, *Chem. Ind. (London)*, 1956, 964; J. H. Brewster, *J. Am. Chem. Soc.*, 1959, **81**, 5475; J. H. Brewster, 'Topics in Stereochemistry', ed. N. L. Allinger and E. L. Eliel, Interscience, New York, 1967, vol. 2.
49. P. Pino and P. L. Luisi, *J. Chim. Phys. Phys.-Chim. Biol.*, 1968, **65**, 130 and references cited therein.
50. W. J. Bailey and E. T. Yates, *J. Org. Chem.*, 1960, **25**, 1800; P. Pino, F. Ciardelli, G. P. Lorenzi and G. Montagnoli, *Makromol. Chem.*, 1963, **61**, 207; P. Pino, *Adv. Polym. Sci.*, 1965, **4**, 393; S. Nozakura, S. Takeuchi, H. Yuki and S. Murahashi, *Bull. Chem. Soc. Jpn.*, 1961, **34**, 1673; M. Goodman, K. J. Clark, M. A. Stake and A. Abe, *Makromol. Chem.*, 1964, **72**, 131.
51. A. Abe and M. Goodman, *J. Polym. Sci., Part A-1*, 1963, 2193; A. Abe, K. Tasaki, K. Inomata and O. Vogl, *Macromolecules*, 1986, **19**, 2707.
52. Y. Okamoto, K. Suzuki, K. Ohta, K. Hatada and H. Yuki, *J. Am. Chem. Soc.*, 1979, **101**, 4763; Y. Okamoto, K. Suzuki and H. Yuki, *J. Polym. Sci., Polym. Chem. Ed.*, 1980, **18**, 3043.
53. H. Wittwer and U. W. Suter, *Macromolecules*, 1985, **18**, 403.
54. M. M. Green, R. A. Gross, C. Crosby, III and F. C. Schilling, *Macromolecules*, 1987, **20**, 992.
55. P. J. Flory, 'Principles of Polymer Chemistry', Cornell University Press, Ithaca, NY, 1953; P. J. Flory, *J. Chem. Phys.*, 1949, **17**, 303.
56. R. G. Kirste, W. A. Kruse and K. Ibel, *Polymer*, 1975, **16**, 120; G. D. Wignall, D. G. H. Ballard and J. Schelten, *Eur. Polym. J.*, 1974, **10**, 861.
57. Y. Fujiwara and P. J. Flory, *Macromolecules*, 1970, **3**, 288; D. Y. Yoon and P. J. Flory, *Macromolecules*, 1976, **9**, 294, 299.
58. H. Hayashi, P. J. Flory and G. D. Wignall, *Macromolecules*, 1983, **16**, 1328; H. Hayashi and P. J. Flory, *Physica B+C (Amsterdam)*, 1983, **120**, 408.
59. U. W. Suter, E. Saiz and P. J. Flory, *Macromolecules*, 1983, **16**, 1317; see also L. C. DeBolt and U. W. Suter, *Macromolecules*, 1987, **20**, 1425.
60. T. Yamazaki, Ph. D. Thesis, Tokyo Institute of Technology, 1988; T. Yamazaki and A. Abe, *Polymer J.*, 1987, **19**, 777.

3
Polymer Solutions

EDWARD F. CASASSA and GUY C. BERRY
Carnegie Mellon University, Pittsburgh, PA, USA

3.1 INTRODUCTION

Although analogies between polymerization reactions and those leading to well-defined, low molecular weight compounds would suggest the correctness of the idea that monomeric units can be joined by primary valence bonds to form chainlike macromolecular entities, the earliest direct evidence from typical organic polymeric products themselves was found in their properties in solution. These materials were found to disperse, albeit perhaps slowly, in solvents that were effective for low molecular weight species similar in chemical constitution to the monomeric units comprising the polymer. However, the solutions exhibited properties that were unprecedented from experience with ordinary solutions. The viscosity was unexpectedly high, and measurements of colligative properties revealed extreme deviations from ideality, *i.e.* enormous negative deviations from what

Raoult's law would require were the solute dissociated into monomeric units. Such phenomena provided evidence of the existence of macromolecular entities, and their stability indicated that they could not be loosely bound colloidal aggregates. Thus, the experimental information obtained from polymer solutions was crucial to the genesis of polymer science. In the ensuing half century, solution properties continued to be a focus of polymer physical chemistry, and they remain so today, despite the emergence of many aspects of the properties of the solid and amorphous states of polymeric materials as expanding fields of investigation. The questions stimulating interest in solution properties concern a range from some of the most fundamental theoretical problems to the utilitarian goal of molecular weight determination.

In the first stages of the quantitative physical chemical investigation of polymer solutions, the most serious efforts to interpret experimental behavior were based on the random flight model for chain conformations and the Flory–Huggins lattice model for thermodynamic properties. Soon Flory, among others, recognized the necessity of confronting the excluded volume problem, *i.e.* how to take proper account of the conformational restrictions on a flexible chain made up of volume-filling segments, no two of which can simultaneously occupy the same point in space. This fundamental property of real chains is of course missing from the random flight representation of the chain contour as the path of a randomly diffusing particle, in which self-intersections are allowed.

Progress along these lines to the early 1950s is summarized in Flory's classic, and still useful, book 'Principles of Polymer Chemistry'[1] and in Tompa's monograph 'Polymer Solutions'.[2] Yamakawa's 'Modern Theory of Polymer Solutions'[3] of 1971 affords another milestone as the definitive summary of the next stage, when more detailed probing of the very difficult excluded volume problem led to the elaboration of the so-called 'two-parameter' family of theories of dilute solution properties. In the 1970s physicists began to apply theoretical renormalization group techniques to polymer solutions. At first, this development seemed altogether discontinuous with the past and was typically expressed in language quite exotic to most polymer scientists; but in 'Scaling Concepts in Polymer Physics',[4] an important work published in 1979, de Gennes aimed at a rapprochement between disciplines by explaining the full range of the new ideas in terms that would be largely familiar to readers of Flory's work. A 1985 review by Oono[5] continues in the interdisciplinary spirit of de Gennes, bringing matters up to date and presenting some detailed representative calculations. That work and a more recent book by Freed,[6] 'Renormalization Group Theory of Polymer Solutions', provide excellent summaries of their authors' research contributions. A thick volume by des Cloizeaux and Jannink,[7] 'Les Polymères en Solution: leur Modélisation et leur Structure' is a treatise on the static properties of polymer solutions. Together with much else, it summarizes the accomplishments of the French school, of which the authors are prominent members.

Essentially, the older ideas are not discredited by the new; but the latter provide fresh insights into the earlier concepts, deepening and extending them, and putting many ideas into more precise and logically consistent form. A distinctly new understanding is achieved of moderately concentrated or 'semidilute' solutions. Much of the recent progress has been marked by application of elegant mathematical techniques.

This chapter gives a concise overview of the equilibrium properties of polymer solutions and of frictional behavior, insofar as it elucidates molecular conformational properties. Significant results that bear upon the coherent picture we try to present are stated, but many details of derivations have had to be omitted. In view of the subject matter of Chapter 1 of this volume, scaling and renormalization ideas are treated much less fully than would otherwise be appropriate. The discussion of experimental work is in no way comprehensive; the studies selected for consideration are intended as representative of correlations among data and of data with theory.

3.2 CONFORMATIONAL STATISTICS OF A SINGLE POLYMER CHAIN

3.2.1 The Unperturbed Chain[8-10]

Polymer chain structures range from those with a paucity of internal conformational states, such as the extreme of rodlike chains formed by *para*-catenated aromatic chains or helix structures stabilized by secondary intramolecular bonding, to chains with an abundance of conformational states due to rotational isomerism about valence bonds along the chain backbone. This review is mainly concerned with the latter category, the flexible chain polymers. Under certain conditions, elaborated below, interactions (contacts) among chain elements distantly connected along the chain contour can be neglected in describing the conformational states of flexible chains. The conceptual chain without such interactions is called an 'unperturbed' chain. An immediate consequence of this property is that the mean-square end-to-end distance of the chain scales with the number of repeat units N in the chain. In practice, in many cases, the averaged chain conformation obtaining in a dilute solution at a certain temperature called the theta temperature (or Flory temperature) Θ closely approximates that of the unperturbed chain. By *definition*, Θ is taken as the temperature at which the second virial coefficient in the osmotic equation of state vanishes for infinitely long chains (see below, Section 3.3). We denote the root-mean-square end-to-end distance R_L by $R_{L,0}$ for the unperturbed chain and by $R_{L,\Theta}$ for the chain under theta conditions, with the expectation that usually $R_{L,0} \approx R_{L,\Theta}$.

Methods of estimating $R_{L,0}$ from local microscopic chain states have focussed on the rotational isomeric state model.[1,9,11] The end-to-end vector L of a particular chain conformer is the sum of N bond vectors l_i representing each repeat unit

$$L = \sum_{i=1}^{N} l_i \tag{1}$$

and the average square over all allowed conformers is then

$$R_L^2 \equiv \langle L^2 \rangle = \langle L \cdot L \rangle = \sum_i^N \sum_j^N \langle l_i \cdot l_j \rangle \tag{2}$$

Here, and in all that follows, pointed brackets (angle brackets) indicate averages over chain conformations. The simplest, but unrealistic, specific case is the freely jointed chain with units of length l connected by universal swivels. Since there is no angular correlation between segment vectors, terms $\langle l_i \cdot l_j \rangle$ vanish unless $i = j$, with the result that

$$R_{L,0}^2 = N l^2 \tag{3}$$

For a free-rotation chain with fixed bond angle θ and bond length l, the cross terms in equation (2) do not vanish. After some algebra,[8] there is obtained, for large N

$$R_{L,0}^2 = [(1 - \cos\theta)/(1 + \cos\theta)]N l^2 \tag{4}$$

For the particular case of tetrahedral bond angles, θ is 109.5°, $\cos\theta$ is 1/3, and therefore

$$R_{L,0}^2 = 2N l^2 \tag{5}$$

At the next level of complication, a discrete number of rotational states, characterized by a set of rotation angles marking the potential energy minima and weighting factors (for energy differences), are assigned to each of the N bonds with fixed bond angle θ. This model gives the same result as the free-rotation chain if the rotational potential function is symmetrical (as is often the case) and the potential minima lie at the same energy.[8] When there are energy differences among allowed rotational states (as is usual), some configurations are favored over others, with the result that the numerical factor in equation (5) is altered.

In higher order models, the weighting factors g_i for the rotational states of bond i are allowed to depend on the states of neighboring bonds. This kind of model has been used to estimate $R_{L,0}^2$ and its temperature dependence for a multitude of chain structures. Usually, to keep calculations feasible, the interactions are counted only as far as nearest neighbor or next nearest neighbor groups along the chain.[9,12]

The quantity

$$C_N \equiv R_{L,0}^2/N l^2 \tag{6}$$

or its limit C_∞ for large N is often reported in rotational isomeric state calculations as a dimensionless parameter characterizing the chain conformation. For example, C_∞ is two for the free-rotation chain model. In practice, C_∞ is seldom this small, but more usually in the range four to ten,[13,14] indicating net effects of local steric interferences that extend the chain well beyond the free-rotation dimensions. The essential result is that when 'long-range' interactions can be neglected, C_N must be independent of N for large N. An alternative measure of chain dimensions is the ratio σ of the root-mean-square unperturbed end-to-end distance $R_{L,0}$ to the corresponding quantity for the free-rotation model. Yet another measure of chain size for a particular polymer is the characteristic ratio $R_{L,0}/M^{1/2}$, M denoting the molecular weight.

As Flory long ago postulated,[8] unperturbed dimensions are almost entirely a property of the polymer chain structure, specific solvent effects having proved very difficult to demonstrate.

Typically, the temperature dependence of $R_{L,0}^2$ is not large, but it may be either positive or negative. For example, $10^3 \, d \ln R_{L,0}^2/dT$ is -1.1 for polyethylene, 0.4 for atactic polystyrene, -0.1 to -0.3 for polyisobutylene, -0.2 for poly(ethyl acrylate), 2.2 for poly(hexyl methacrylate) and 0.5 for poly(vinyl alcohol).[13,14] A large compendium of chain dimension parameters is available in the 'Polymer Handbook'.[14] A 1969 monograph by Flory[9] gives a comprehensive summary of the application of the rotational isomeric state concept with near neighbor interactions to a great variety of real-chain model structures.

3.2.2 The Random Flight Chain

Though it has been by far the most fruitful idealization in polymer science, the representation of the conformation of a flexible unperturbed polymer chain by the path of a Brownian particle might seem impossibly extreme at the microscopic level of valence bond structure; but at a coarser-grained level of description, it does constitute a 'renormalized' model with physical significance. In this model, introduced long ago by Kuhn,[15] the real chain with its N bond vectors in any given conformation is replaced by a conceptual chain of $n = N/m$ segments of average length $b > l$. The new chain may be constructed by connecting the termini of bond vectors $1, m + 1, 2m + 1, 3m + 1$, *etc.* to form a sequence of n vectors that shows less correlation between successive steps than do the original bond vectors, and yet in a less detailed way still models the real chain. Then, as m is made larger, the resulting sequence of vectors is presumed to approach asymptotically a random flight conformation, in which successive steps are uncorrelated in both length and direction, with the result $R_{L,0}^2 \propto n \propto N$.

Fortunately, the mathematical analysis of Brownian motion in a system of dimensionality d^* is thoroughly understood.[16] We can take as fundamental the relation for the probability $P(\{r_n\}) \, d\{r_n\}$ that the configuration of a sequence of n d^*-dimensional Gaussian vectors $r_1, r_2, \ldots$, symbolized by $\{r_n\}$, lies in the differential element of configuration space $d\{r_n\}$

$$P(\{r_n\}) d\{r_n\} = \prod_{i=1}^{n} \tau_i dr_i \tag{7}$$

where

$$\tau_i = \left(\frac{d^*}{2\pi b^2}\right)^{d^*/2} \exp\left(-\frac{d^* r_i^2}{2b^2}\right) \tag{8}$$

b^2 denoting the mean-square length of the statistical chain step or segment. Integration of equation (7) over coordinates under the condition $\Sigma_i r_i = L$ gives

$$P_n(L) = \left(\frac{d^*}{2\pi n b^2}\right)^{d/2} \exp\left(-\frac{d^* L^2}{2n b^2}\right) \tag{9}$$

i.e. the distribution function for the end-to-end distance of the n-step random flight in unbounded d^*-dimensional space. We introduce d^* for generality, but, except as specifically noted, 3-space is assumed in all that follows.

Various averages over the chain conformation can be computed in standard fashion from the distribution function. The simplest one is the unperturbed mean-square end-to-end distance

$$R_{L,0}^2 \equiv \langle L^2 \rangle_0 = n b^2 \tag{10}$$

(or in general, $d^* n b^2/3$). It is apparent that these relations must apply equally to the separation L_{ij} of any pair of chain segments (or steps) i and j along the chain, so that

$$\langle L_{ij}^2 \rangle_0 = |i - j| b^2 \tag{11}$$

It will be noted that the constancy of $R_{L,0}^2/N l^2$ for the chain of bond vectors in a long chain is preserved in the random flight model.

The mean-square radius of gyration about the center of mass of the molecule is given by

$$R_{G,0}^2 \equiv \frac{1}{n} \sum_i \langle s_i^2 \rangle = \frac{1}{2n} \sum_i \sum_j \langle L_{ij}^2 \rangle = \frac{1}{6} n b^2 \tag{12}$$

where s_i is the vector from the center of mass to segment i.

Other averaged statistical dimensions of the random flight chain can be calculated, *e.g.* the mean projection of the chain onto a line, which is $(2nb^2/3\pi)^{1/2}$. Such measures of chain dimensions are discussed in a book by Volkenstein.[11] However, the root-mean-square molecular radius of gyration is of special significance because it is an entity measured directly by a scattering experiment. It is important to note that if $R_{G,0}^2$ is given by equation (12), the combination nb^2 is a physical entity derivable from experiment, but that the statistical model parameters n and b individually have no definite physical meaning. One way to fix the values of n and b is to let the stretched-out contour lengths $Nl \equiv L_c$ and nb of the valence bond chain and the random flight chain coincide, and then express n and b (or m and b/l) in terms of the experimental $R_{G,0}$ and L_c calculated from known bond lengths and angles and the degree of polymerization. However, this particular specification of n and b is not an essential feature of the model.

For branched chains the end-to-end distance is not a meaningful quantity, but the definition of the mean-square radius in equation (12) still holds. It is customary to express the effects of branching by comparing the branched species with the linear analog having the same molecular weight

$$R_{G,0}^2 = gnb^2/6 \tag{13}$$

The g factors have been computed by Zimm and Stockmayer[17] for various branched models made up of linear random flight subchains. For a star-branched polymer with f identical arms connected at a single node, the result is

$$g = \frac{3f - 2}{f^2} \tag{14}$$

An approximate result has been given[18] for a 'comb' structure with f identical branches each of n_b segments spaced uniformly along a backbone chain of n_s segments

$$g \approx \lambda + (1 - \lambda)^{7/3}\left(\frac{3f - 2}{f^2}\right) \tag{15}$$

where $\lambda = n_s/(n_s + fn_b)$ is the fraction of segments in the chain backbone. This approximation is in error by less than 1% when $f > 4$. When $\lambda \to 0$, g for a star molecule is recovered; and when $\lambda \to 1$, the correct limit, $g = 1$, for a linear chain is obtained. Equation (15) shows, as one would expect, that adding a few branches to a long linear chain will have little effect on R_G. Usually g is a slowly decreasing function of the number of branches. The greatest variation is perhaps seen in the star, for which g varies asymptotically as f^{-1} for large f.

3.2.3 The Wormlike Chain[10]

For some real chain structures $R_{L,0}^2/N$ may not attain its asymptotic constant limit, except at large degrees of polymerization, perhaps close to or beyond the upper limit that can be realized or utilized practically in the laboratory. Typical examples are certain polysaccharides, nucleic acids and helicoidal polypeptides. Chain structures of this kind are described as semiflexible or stiff (the latter term is sometimes misunderstood as connoting kinetic effects whereas only thermodynamic equilibrium is relevant in this context). The persistent or 'wormlike' chain model of Kratky and Porod[19] takes account of this behavior. With the free-rotation chain as the starting point, a characteristic persistence length ρ is defined in terms of the average projection of the end-to-end vector L upon the first bond vector l_1

$$\langle L \cdot l_1 \rangle = l^2 \frac{1 - (-\cos\theta)^N}{1 + \cos\theta} = l\rho\left[1 - \left(1 - \frac{L_c/\rho}{L_c/l}\right)^{L_c/l}\right] \tag{16}$$

where ρ is defined as $l/(1 + \cos\theta)$. It follows that

$$\lim_{N \to \infty} \langle L \cdot l_1 \rangle = l\rho[1 - \exp(-L_c/\rho)] \tag{17}$$

so that $\langle L \cdot l_1 \rangle$ tends to $l\rho$ for ρ/L_c small, *i.e.* for a chain with $N \gg m$, ρ tends to zero.

The wormlike chain model affords a continuous description of chain character ranging from 'flexible' (ρ/L_c small) to rodlike (ρ/L_c large). The unpertubed mean-square dimensions $R_{L,0}^2$ and

$R_{G,0}^2$ may be expressed in the forms

$$R_{L,0}^2 \;=\; 2L_c\rho \; W(\rho/L_c) \tag{18}$$

$$R_{G,0}^2 \;=\; (L_c\rho/3)\,[1 \;-\; 3(\rho/L_c) \;+\; 6(\rho/L_c)^2 \, W(\rho/L_c)] \tag{19}$$

where $W(y) = 1 - y[1 - \exp(-1/y)]$. Accordingly, in the flexible-chain limit, equations (18) and (19) reduce to

$$R_{L,0}^2 \;=\; 2L_c\rho \qquad R_{G,0}^2 \;=\; L_c\rho/3 \tag{20}$$

and in the rod limit to

$$R_{L,0}^2 \;=\; L_c^2 \qquad R_{G,0}^2 \;=\; L_c^2/12 \tag{21}$$

The length 2ρ is sometimes called the 'Kuhn length' of the chain. For the wormlike model, the ratio C_N, defined by equation (6), is

$$C_N \;=\; 2(\rho/l)\,W(\rho/L_c) \tag{22}$$

For large L_c/ρ, $W(\rho/L_c)$ tends to unity and $C_\infty = 2\rho/l$. Values of C_∞ for typical vinyl polymers give ρ/l in the range two to five.[13]

3.2.4 The Chain with Excluded Volume

3.2.4.1 *Expansion of molecular dimensions*

Unlike the local intramolecular interactions due to bond correlations, the excluded volume effect is long range in that it does not vanish for interactions between chain segments remotely connected through the chain contour. Indeed, interactions between segments far apart in this sense are dominant, even though two segments must naturally be in a conformation in close contact when they interact. The disallowance of intersecting conformations that would be permitted to a random flight means that the equilibrium dimensions of the real chain are larger than those for the corresponding unperturbed random flight. Fundamentally, the chain with excluded volume no longer obeys random flight statistics, and the various averaged chain dimensions are distorted from random flight values to different degrees. It is customary to express the change of a statistical dimension by a linear expansion factor α, *e.g.* for the root-mean-square end-to-end distance R_L and radius of gyration R_G

$$\alpha_L \;=\; R_L/R_{L,0} \qquad \alpha_G \;=\; R_G/R_{G,0} \tag{23}$$

It is important to remember that the effective segment–segment interaction potential in a polymer solution is by no means the same as for the hypothetical case of the chain in a vacuum. In a real system the net effect includes contributions from polymer–solvent and solvent–solvent interactions. In a given polymer + solvent system there may be a theta temperature (see above) at or near which there is a delicate balance among these effects such that the net excluded volume effect vanishes and random flight statistics prevail, *i.e.* R_L and R_G both become proportional asymptotically to the square root of the polymer molecular weight.

3.2.4.2 *Perturbation theory near the theta temperature*

For the regime of weak excluded volume interactions, it is natural to undertake a perturbation treatment of the deviation from random flight behavior. There are a number of derivations,[20] but the standard treatment is that due to Fixman,[21] which is inspired by the Ursell–Mayer cluster theory for the imperfect gas. The probability distribution of the linear chain is represented as in equations (7) and (8) but each configuration is weighted by a product of terms $\exp\{-U(R_{ij})/kT\}$, where kT has the usual meaning and $U(R_{ij})$ is the energy of interaction (more precisely the effective potential of average force) between segments i and j, a function of the scalar intersegmental distance R_{ij}

$$P(\{r_n\}) \;=\; C^{-1} \sum_{i<j}\sum \exp\!\left(-\frac{U(R_{ij})}{kT}\right)\!\prod_i \tau_i(r_i)\, dr_i \tag{24}$$

with the normalization

$$C = \int \ldots \int \sum_{i<j} \exp\left(-\frac{U(R_{ij})}{kT}\right) \prod_i \tau_i(r_i)\, dr_i \tag{25}$$

Equation (24) introduces the crucial approximation of the total interaction energy by the sum of all pairwise segment interaction energies. A new function

$$\chi_{ij} = 1 - \exp\{U(R_{ij})/kT\} \tag{26}$$

is substituted in equation (24) and the products are expanded. The short range character of segment–segment interactions is introduced by representing the χ functions as[20,22]

$$\chi_{ij} = -\beta\delta(\boldsymbol{R}_{ij}) \tag{27}$$

$$\chi_{ij}\chi_{jk}\chi_{ik} = -\beta_3\delta(\boldsymbol{R}_{ij})\delta(\boldsymbol{R}_{jk}) \tag{28}$$

where δ is the three-dimensional Dirac delta function, and β and β_3 are binary and ternary segment cluster integrals

$$\beta = -\int \chi_{ij}\, d\boldsymbol{R}_{ij} \tag{29}$$

$$\beta_3 = -\int\int \chi_{ij}\chi_{jk}\chi_{ik}\, d\boldsymbol{R}_{ij}\boldsymbol{R}_{jk} \tag{30}$$

These correspond, respectively, to the two-body and three-body integrals that appear in the second and third virial coefficients in the theory of the imperfect gas.

In a homopolymer, the chain segments are taken to be identical, and the χ functions and excluded volume integrals β are the same for all i,j. Since $U(\boldsymbol{R}_{ij})$ must in general comprise both attractive and repulsive terms, β is expected to depend on temperature, and may be zero at a particular temperature or at more than one temperature (see Section 3.3.4). By contrast, β_3 is expected to depend only weakly on temperature and to be positive, albeit possibly very small.[5,23,24]

As already mentioned, the second virial coefficient vanishes at a temperature Θ (or, more precisely at Θ_N, which may possibly depend weakly on N but tends to the Flory theta temperature as $N \to \infty$, see Section 3.3.2.2). The limit Θ represents a tricritical temperature[25,26,27] in the sense introduced by Griffiths and co-workers.[28] In Fixman's original derivation, like others of its genre, terms involving β_3 are suppressed as presumably negligible;[20] and with the single excluded volume integral β, the temperature at which β vanishes becomes synonymous with Θ_N (see Section 3.3.2.2). However, three-body effects manifested by a nonvanishing β_3 are intrinsically of interest, especially for T near Θ_N,[29] even if their effect on chain dimensions should ordinarily prove to be unimportant. Except as the context indicates otherwise, the ensuing discussion is consistent with a β that increases, and hence the solvent becomes 'better', as the temperature is raised. In this circumstance, Θ is the familiar 'upper' theta temperature associated with a solubility gap that appears as the temperature is lowered.

To compute an averaged dimension, say the mean-square separation of segments i and j of a chain, the distribution $P(\{r\}_n)$ according to equation (24) is multiplied by L_{ij}^2. Then integrations, made possible by the short range character of the χ functions, are performed over all coordinates except L_{ij}. The normalization factor C^{-1} is expressed as a geometric series. The result is obtained as a series in powers of β (together with terms in β_3 and, in principle, for higher order segment clusters) with coefficients containing increasingly complicated combinations of multivariate Gaussian probability distributions. For the end-to-end distance, the leading terms of the series are

$$R_L^2 = R_{L,0}^2 - \beta\sum_i\sum_j \int L^2[P_0(\boldsymbol{L}, O_{ij}) - P_0(\boldsymbol{L})P_0(O_{ij})]d\boldsymbol{L} + O(\beta_3) + O(\beta^2) \tag{31}$$

The functions P_0 are random flight probability densities: *i.e.* $P_0(\boldsymbol{L})$ for the end-to-end vector $\boldsymbol{L}$, $P_0(O_{ij})$ for contact between segments i and j, and $P_0(\boldsymbol{L}, O_{ij})$ for simultaneous occurrence of the two events. The probability densities are obtained using Fixman's generalization[21] of a theorem of Wang and Uhlenbeck. Finally the sums over segment indices are converted to integrals, in effect passing to a continuum model for long chains,[30] and the remaining integrations are done.

The 'bare' perturbation theory carried through in this way[22] gives a series for the mean-square end-to-end distance.

$$R_L^2 = R_{L,0}^2 [1 + (4/3)(z + 4n^{1/2}z_3) + \dots] \tag{32}$$

where

$$z \equiv \left(\frac{3}{2\pi b^2}\right)^{3/2} n^{1/2} \beta = \left(\frac{3}{2\pi n b^2}\right)^{3/2} n^2 \beta \tag{33}$$

$$z_3 \equiv (3/2\pi b^2)^3 \beta_3 \tag{34}$$

According to this result, R_L is greater than $R_{L,0}$ when $z = 0$ if $z_3 > 0$.

Recent refinements of perturbation theory incorporating effects significant at the theta temperature give more complicated expressions, *e.g.* a relation for linear chains with β small and n large[23]

$$\frac{R_L^2}{R_{L,0}^2} = 1 + \frac{4}{3}\left\{1 - \frac{3}{2}\left(\frac{a}{n}\right)^{1/2}\right\}\left\{z + 8\left[\left(\frac{n}{4a}\right)^{1/2} - 1\right]z_3\right\} - \left(4\pi - \frac{32}{3}\right)z_3 + \dots \tag{35}$$

with an additional 'cut-off' parameter a of order unity. The product of terms within braces becomes asymptotically proportional to $M^{1/2}$ for large M, whereas z_3 is independent of M. As is shown below, this product of terms vanishes when the second virial coefficient is zero; and then $R_L = R_{L,\Theta}$ is predicted to be (slightly) less than $R_{L,0}$ if $z_3 > 0$. The calculation of terms proportional to z_3 is delicate; certain alternative relations[27,31] are discussed in ref. 23. Analogous expressions for R_G^2 are also available, in which the coefficient 4/3 in equation (35) is replaced by 134/105 and the coefficient of the last term in z_3 is replaced by one that is very small (and negative).[26,31] These theoretical studies indicate that while the effect of z_3 may be uncertain and small, there is no reason to expect exact compensation of the interactions among chain segments as reflected in different properties at the same temperature.

Renormalization has been applied to express R_L^2 as a function of a renormalized three-body interaction parameter $\bar{z}_3$ for the case with β_3 not small. Freed's result[23] at the theta temperature is

$$\frac{R_{L,\Theta}^2}{R_{L,0}^2} = 1 - \left(4\pi - \frac{32}{3}\right)\frac{\bar{z}_3}{1 + 44\pi\bar{z}_3 \ln(2\pi n/a')} \tag{36}$$

where a' is of order unity. If, as expected, $\bar{z}_3$ is small, this result reduces in practice to the preceding bare expansion (equation 35).

In the limit of vanishing z_3, equations (32), (35) and (36) all reduce to the results of the two-parameter theory as discussed, for example, in Yamakawa's book[3] (see also below, Section 3.2.4.3). An assessment of z_3 can be made from measurements[32,33] of the third virial coefficient (Section 3.3.2.2) under conditions with $A_2 = 0$. It is indeed found to be small, of the order of 10^{-3} to 10^{-2} for the few systems studied.

With branched chains, on the other hand, the probability of ternary segment clustering is increased, with the possibility that Θ and $R_L/R_{L,0}$ or $R_G/R_{G,0}$ at Θ may depend sensibly on N and the branched-chain architecture.[34,35] Application of renormalization–group theoretical methods to assess this possibility remains to be done.

A recent derivation by Martin[36] of behavior near the theta point refines Fixman's perturbation treatment by replacement of the delta function representation of segment pair interactions with an attractive–repulsive potential of plausible form. The theory predicts that the chain at Θ is expanded relative to the unperturbed chain, but the ratio of $R_{L,\Theta}$ to $R_{G,\Theta}$ still has the unperturbed value. Other, earlier schemes that would allow the chain conformation to be perturbed at the theta point typically involve smoothed-density chain models[22,35,37,38] (Section 3.2.4.3) or may be largely phenomenological.[39] Explicitly or implicitly these approaches introduce some form of three-body segment interaction and/or relax the short range character of intersegmental potentials of average force. Although such treatments may offer additional free parameters, and consequent scope for data correlation, their physical content seems obscure.

3.2.4.3 Perturbation theory far from the theta temperature

Under good-solvent conditions, excluded volume contributions due to the binary cluster integral β far outweigh those involving β_3, allowing the latter to be neglected; and the celebrated two-parameter formalism making R_L and R_G functions only of unperturbed dimensions and z, or a similar parameter (see below), is valid. The perturbation development for small z is accomplished as already indicated, but the omission of β_3 contributions facilitates computation of higher order terms in the series expressions[20]

$$R_L^2/R_{L,0}^2 \equiv \alpha_L^2 = 1 + \sum_{j=1} a_{L,j} z^j \tag{37}$$

$$R_G^2/R_{G,0}^2 \equiv \alpha_G^2 = 1 + \sum_{j=1} a_{G,j} z^j \tag{38}$$

Fixman obtained two coefficients, $a_{L,1} = 4/3$ and $a_{L,2} = 2.076$; and for the mean-square radius of gyration, $a_{G,1} = 134/105$. It has also long been known[20] that $a_{G,2} = 2.082$. Although Fixman's derivation specifies a systematic technique for extracting higher terms of the expansions, the difficulty increases formidably with each successive term; and for 30 years, attempts to obtain anything more than one additional term were regarded as not feasible. Apparently, even $a_{L,3}$ was not calculated accurately.[20,40] Recently, however, Muthukumar and Nickel[41] have succeeded in determining exactly all the $a_{L,j}$ up to $j = 6$. In a variation of the procedure outlined above, they make use of a Fourier transform of the segment vector distribution $P(\{\mathbf{r}_n\})$ with respect to segment coordinates and a Laplace transform of the end-to-end displacement distribution with respect to the chain contour length in a way that remarkably facilitates the computation.

A more important result of the study by Muthukumar and Nickel is a proof that the series representation of R_L^2 is free of divergences in $\ln n$ to all orders. The same conclusion has been reached independently by des Cloizeaux.[42] Such divergent terms appear in intermediate results in the conventional perturbation calculation, and it had not been clear whether they would finally cancel out exactly for all powers of β.[43] Thus it had been questioned[44] whether equations (37) and (38) converge, even asymptotically. It is now clear that the expansions for R_L^2 and R_G^2 are divergent but asymptotic.

The perturbation series in equations (37) and (38) with only a few terms known obviously cannot be used directly except for very small values of z, thus only in a narrow temperature interval about $T = \Theta$, where z vanishes as β vanishes. For very long chains in a good solvent ($T > \Theta$), z becomes indefinitely large. For a solvent poorer than a theta solvent, z is negative; and at some temperature below Θ, the system becomes unstable. There follows a phase separation or, at extreme dilution and high polymer molecular weight, a coil-collapse transition.[79] There has been considerable theoretical speculation[45,46,47] about the reality and character of the latter phenomenon, but some careful experiments (static and dynamic light scattering) do indicate a precipitous fall in $R_{G,0}^2$ below Θ.[48,49]

As expected, α_L in equation (37) is not the same as α_G in equation (38), but the most significant feature of these relations is that the expansion factors are functions of the single excluded volume variable z. From the second form on the right hand side of equation (33), we see that z can be expressed in terms of two groups of statistical quantities nb^2 and $n^2\beta$. As mentioned above, the first group is simply related to an observable property; and the second is obtainable from thermodynamic measurements (see below, Section 3.3.4). Thus z is a function of two physically meaningful quantities, and the theory of the expansion factor outlined here is a 'two-parameter' theory[3] in the sense proposed by Stockmayer.[29]

The coefficient $a_{G,1}$ of z in the series expansion of α_G^2 is known for some nonlinear chain geometries. For a ring molecule it is $\pi/2$.[50] For regular star chains of functionality f it increases with f from the linear-chain value of 1.2762 ($f = 1, 2$) and asymptotically approaches $0.430 f^{1/2}$.[51] The calculation of α_G^2 to this order with identical branches attached at equal intervals along the backbone chain proves to be very cumbersome,[51] and the results contain irreducible finite multiple (single, double, triple, quadruple) sums of algebraically complicated arguments over branch numbers $1, 2, \ldots f$ as summation indices. Consequently machine calculations are necessary to obtain numerical values. Results are qualitatively very similar for α_G^2 averaged over a large population of combs which are homogeneous in molecular weight but with the f branches in each molecule placed randomly along the backbone. This 'random comb' model is more realistic than the regular comb in respect to comb species produced by ordinary synthetic chemical procedures, and the coefficient of z in the theoretical expression is notably simpler in that a closed analytical form is obtained.[52] The randomization of branch placements allows the troublesome multiple sums to be replaced by multiple integrals. The integrals are easy, but there are so many of them (several hundred) that a computer program for algebraic manipulation is a practical necessity to avoid errors.

3.2.4.4 *Approximate theories for good-solvent systems*

Since the series expansions for α_L^2 and α_G^2 are not practically useful, there is evident need for a different approach applicable to the regime of strong segment–segment interactions: *i.e.* β non-vanishing and z large. In fact, a treatment developed by Flory[53,54] for this purpose antedates the perturbation theory. A somewhat generalized version by Fixman[21] affords a concise presentation. Any configuration of a polymer chain is represented by a spherically symmetrical distribution of chain segments characterized by a radial density distribution $\rho(s)$. With any $\rho(s)$ there is associated an energy density at distance s from the center of mass proportional to the square of the segment density. Hence the total energy attributed to segment–segment excluded volume interactions for the configuration in question can be written

$$\frac{E}{kT} = \frac{\beta}{2}\int_0^\infty 4\pi s^2 \rho^2(s)\,\mathrm{d}s = \frac{2\pi n^2 \beta}{S^3}\int\left(\frac{s}{S}\right)^2 \mathrm{f}\left(\frac{s}{S}\right)\mathrm{d}\left(\frac{s}{S}\right) = \frac{n^2 \beta C'}{S^3} \tag{39}$$

where S is the radius of gyration of the configuration. The unknown dimensionless function $\mathrm{f}(s/S)$, and thus the constant C', depend on $\rho(s)$. The number of states for a given S is proportional to

$$P(S) = \mathrm{W}(S)4\pi S^2 \exp(-E(S)/kT) \tag{40}$$

The statistical weighting factor $\mathrm{W}(S)$ is assumed to be given by a Gaussian function

$$\mathrm{W}(S) = (3/2\pi R_{\mathrm{G},0}^2)^{3/2}\exp(-3S^2/2R_{\mathrm{G},0}^2) \tag{41}$$

Then to simplify matters, the expansion factor α at equilibrium is taken as the ratio of the most probable value of S [obtained by maximizing $P(S)$] to the most probable value for an unperturbed random flight, $(2R_{\mathrm{G},0}^2/3)^{1/2}$. The value of the constant C' can be fixed by assuming a Gaussian density distribution for $\rho(s)$ and normalizing it to yield the correct mean-square radius. This gives

$$\alpha^5 - \alpha^3 = (3/2)^{5/2}n^2\beta C'/R_{L,0}^3 = (3^4/2^{7/2})z = 7.16z \tag{42}$$

The numerical coefficient 7.16 is much larger than Flory's value of 2.60 from his slightly more complicated treatment, as it depends on different assumptions made in the derivations. Other variations on the method [*e.g.* use of a more accurate sum of n Gaussian functions for $\rho(s)$] lead to still other values for the coefficient.[55] Given the arbitrariness of the assumptions, no great significance can be attached to the numerical value, and it seems reasonable to follow a recommendation of Stockmayer[56] and let

$$\alpha_L^5 - \alpha_L^3 = (4/3)z \tag{43}$$

$$\alpha_G^5 - \alpha_G^3 = (134/105)z \tag{44}$$

thereby forcing agreement of the approximate closed forms with the 'exact' two-parameter first-order perturbation developments for the mean-square end-to-end separation and the mean-square radius of gyration as α_L and α_G approach unity.

Flory's derivation exemplifies the mean-field method. In the model, details of chain connectivity are obliterated, and any segment 'sees' only the averaged effects of a distribution of disconnected segments. Although equation (42) refers ostensibly to an expansion factor for the most probable value of the radius of gyration, the model is obviously so crude that equation (42) might be taken to represent the expansion of any statistical chain dimension. Thus, the conformation of the chain modified by the excluded volume effect can be approximated as a random flight with all dimensions expanded uniformly by a common factor α, *i.e.* the step length b can be replaced by αb.[53,54] Since the expansion of R_G^2 is directly related to experimental measurement, the uniform expansion factor is usually identified with α_G.

A notable aspect of Flory's treatment of the chain expansion in a good solvent is the asymptotic behavior for long chains

$$\alpha_G^2 \propto z^{2(2\nu-1)} \qquad R_G \propto n^\nu \tag{45}$$

where $\nu = 3/5$. Of course, with this model, the same proportionalities obtain for α_L. It is improbable that the asymptotic law could ever be realized experimentally, but it has been tested by direct enumeration of walks on various lattices[57–59] and by Monte Carlo simulations.[60–62] Although this is still not a closed subject,[63,64] these direct methods are consistent with the 3/5 power law for the end-to-end separation, within the inevitable hazards of sampling statistics and/or extrapolation to infinite chain length.[65] As discussed below, calculations based on scaling and renormalization–group methods also yield asymptotic power laws with v very close to 0.6, though usually slightly smaller.

What is essential in Flory's theory is the repulsive segment–segment (free) energy of interaction that scales as $n^2\beta T/S^3$ and the opposing entropic free energy of distortion of the equilibrium conformation that scales as $S^2 T/nb^2$. Minimizing the total free energy then gives the asymptotic form in equation (45). This argument is easily generalized to dimensionality d^* to obtain for the limiting exponent

$$v = \begin{cases} \dfrac{3}{d^* + 2} & d^* \leqslant 4 \\ 1/2 & d^* > 4 \end{cases} \tag{46}$$

This is a remarkable result; there is good reason to believe that it is almost correct despite the extreme simplifications made in the derivation. It shows that the strength of the excluded volume interactions decreases as the dimensionality increases until, in 4-space, the chain behaves as a random flight, *i.e.* self-intersections become so infrequent as to have negligible effect, giving $R_L^2 \propto N$.

Flory's theory leading to the form of equation (42) is only the first and most famous of many attempts, by a variety of techniques, to obtain expansion factors α_L or α_G for moderate and large values of z. A partial summary is given in ref. 40 (see also Chapter 1 in this volume). Other treatments are discussed in detail in ref. 20. Variants of Flory's theory give results that are conveniently expressed in the form

$$\alpha_G^5 - \alpha_G^3 = a_{G,1} z h(z/\alpha_G^3) \tag{47}$$

A derivation by Flory and Fisk,[66] which avoids or refines some of the approximations in the Flory treatment, gives a result that can be expressed by

$$h(y) = 0.508[1 + 0.969(1 + 10y)^{-2/3}] \tag{48}$$

which leads to the asymptotic form

$$\alpha_G^2 = (0.508 a_{G,1} z)^{2/5} = 0.841 z^{2/5} \tag{49}$$

in agreement with equation (45) with $v = 3/5$. Equation (48) has been used successfully to correlate experimental data covering a wide span of z.[67]

Another smoothed density model,[68] based on consideration of the ellipsoidal shape of a segment distribution averaged about any given end-to-end distance, leads to a different power law dependence

$$\alpha_L^3 - \alpha_L = (4/3)^{5/2} z\left(1 + \frac{1}{3\alpha_L^2}\right)^{-3/2} \tag{50}$$

which gives the (presumably) incorrect asymptotic dependence $v = 2/3$. A derivation by Yamakawa and Tanaka[20,69] gives an expression for α_L^2

$$\alpha_L^2 = 0.5364 + 0.4636(1 + 5.989z)^{0.4802} \tag{51}$$

The numerical parameters are obtained by matching the series expansion of the form of equation (51) with the known third-order perturbation expansion for α_G^2. The exponent 0.4802 corresponds to the asymptotic behavior $v = 0.6201$. Equation (51) is based on a revised value of $a_{L,3}$[41] and thus differs slightly from the version given by Yamakawa.[20] The Yamakawa–Tanaka expression for α_G^2 is obtained similarly by fitting the form of equation (51) to a third-order perturbation development for $R_{G,0}^2$, with the arbitrary assumption that $a_{G,3} = a_{L,3}$.

A quite different approach is to effect closure of the series in equations (37) and (38) by use of Padé approximants.[70] For example, the (2,2) approximant expression

$$\frac{\alpha_L^5 - 1}{z} = \frac{10}{3}\left(\frac{1 + 2.735z}{1 + 3.291z}\right) \tag{52}$$

proposed by Stockmayer[71] agrees with the third-order perturbation of equation (37), and asymptotically also gives $v = 3/5$ and the limit $\alpha_L^2/z^{2/5} = 1.50$, which is not far from a value of 1.64 deduced from enumeration of chain configurations on lattices.[59] Now, of course, one could refine this interpolation scheme by using more known coefficients of equation (37). The coefficients in equation (52) are slightly different from those actually given by Stockmayer because we have used the revised $a_{L,3}$.

Application of renormalization group methods and expansion of the expression for $P(\{r_n\})$ for dimension d^* in terms of the parameter $\varepsilon = 4 - d^*$ affords alternative expressions for α_G and α_L. The so-called epsilon expansion[72] is motivated by the finding that $\alpha_G = \alpha_L = 1$ exactly for $d^* = 4$. The procedure involves a double expansion of α_G or α_L in terms of ε (regarded as a continuous variable) and a z-like parameter $\tilde{z}$. Renormalization theory prescribes a method for analytical continuation of the results, valid near $d^* = 4$, to 3-space. Unfortunately, an additional β-like interaction parameter σ also appears explicitly, together with $\tilde{z}$. To order ε, one such treatment gives[73]

$$\alpha_L^2 = \left[\frac{\lambda_1}{(1 - \lambda_1)(1 + \sigma)} \right]^{\delta_1} [1 + d_1 \lambda_1] \tag{53}$$

$$\alpha_G^2 = \alpha_L^2 [1 + d_1 \lambda_1 / 12] \tag{54}$$

for $1 \leqslant \alpha_L^2 < 1.2$. Here, d_1, λ_1 and δ_1 are the special cases for $\varepsilon = 1$ (for 3-space) of the quantities

$$d_\varepsilon = -w_\varepsilon^* = -\frac{\varepsilon}{8} - \frac{21}{4}\left(\frac{\varepsilon}{8}\right)^2 + \cdots \tag{55}$$

$$\lambda_\varepsilon = \frac{4\tilde{z}/3w_\varepsilon^*}{1 + 4\tilde{z}/3w_\varepsilon^*} \tag{56}$$

$$\delta_\varepsilon = (1 + \varepsilon)/8 + \cdots \tag{57}$$

i.e. $\delta_1 = 1/4$ and $w_1^* = 1/8$, to order ε. Expansion of equation (53) for small $\tilde{z}$ gives

$$\alpha_L^2 = 1 + \left(\frac{1}{4} - d_1\right)\left(\frac{4\tilde{z}}{3w_1^*}\right) - \frac{3}{4}\left(d_1 + \frac{1}{8}\right)\left(\frac{4\tilde{z}}{3w_1^*}\right)^2 + \cdots \tag{58}$$

Here $4\tilde{z}/3w_1^*$ is set equivalent to either z or z/α_G^3 by comparison with the first-order perturbation theory (equation 37 to order z). The latter identification provides improved (but not perfect) agreement with the second-order perturbation theory.

Equation (53) does not give the expected power-law behavior of equation (46) for large $\tilde{z}$. Resort to asymptotic behavior in $\tilde{z}$ to order ε^2 is necessary to obtain the result:

$$\alpha_L^2 = (1 + d_1)(4\tilde{z}/3w_1^*)^{2(2\nu - 1)} \tag{59}$$

$$2\nu - 1 = w_1^* - \frac{3}{8}(w_1^*)^2 + \cdots = \frac{\varepsilon}{8} + 3\left(\frac{\varepsilon}{8}\right)^2 + \cdots \tag{60}$$

where w_1^* and ν are to be evaluated to order ε^2 (*i.e.* $w_1^* = 0.2070$ and $\nu = 0.586$).

A different result for α_L has been obtained with the use of the Fourier–Laplace transform of $P(\{R\})$ mentioned in connection with the perturbation analysis of α_L. In that treatment, α_L^2 is expressed as the ratio of certain inverse transforms, and perturbation series are developed for the arguments of these transforms, leading to equation (37). In an alternative calculation, analytical forms are found to represent the arguments and then used in a numerical evaluation of the inverse transforms to obtain α_L as a function of z. The analysis is completed by use of scaling arguments to guide construction of the needed analytical arguments (details are given above in Chapter 1 of this volume). The numerical results can be represented by the simple analytical relation[40]

$$\alpha_L^2 = \left(1 + \frac{a_{L,1}}{2\nu - 1} z + bz^2\right)^{2\nu - 1} \tag{61}$$

where $\nu = 0.5886$ (the 'best' value[40] for ν) and $a_{L,1} = 4/3$. The parameter b is taken as 11.062 to fit the asymptotic behavior found for large z

$$\alpha_L^2 = 1.531 z^{2(2\nu - 1)} = 1.531 z^{0.3544} \tag{62}$$

As may be seen, equation (61) is exact for both limits of large and small z; it is also close to the numerical results for all z. This function, with the above numerical parameters, gives values of α_L^2 between those from the original Flory theory with the coefficient 2.60 and the modified form with the coefficient $4/3$ (equation 43). An alternative analysis based on direct use of the sixth-order perturbation result of Muthukumar and Nickel[41] gives results in good quantitative agreement with equation (61).[74] The Padé approximant function in equation (52) gives α_L^2 differing from values from equation (61) by up to 12% for $\alpha_L \leqslant 2$. The deviation increases to 17% for $\alpha_L = 3$, and of course increases without limit as $z \to \infty$ because the limiting power laws are different. A [2,2] approximant for $[\alpha_L^{1/(2v-1)} - 1]/z$ with $v = 0.5886$ that fits the third-order perturbation gives α_L^2 nowhere differing from equation (61) by more than 1.2%.

Comparisons of a number of approximate theoretical functions for α_L^2 are shown in Figures 5 and 6 of Chapter 1 of this volume.

3.3 THERMODYNAMICS: DILUTE SOLUTIONS[75]

3.3.1 The Virial Series Expansion

In a general theory of solutions, McMillan and Mayer[76] demonstrated the formal equivalence between the pressure of a gas and the osmotic pressure of a solution. Hence the ratio of the osmotic pressure Π of a dilute solution to the concentration (number density) ρ of the solute can be expanded in a power series in ρ; and the coefficients of the series can be expressed, as in the theory of a real gas, in terms of cluster integrals determined by intermolecular potential energy functions. The only difference is, as already mentioned, that in the solution these potentials are effective potentials of average force, which include implicitly the effects of the solvent, modelled as a continuum.

In polymer science, weight-based concentrations are almost always more useful than molar units, and accordingly the osmotic equation of state is customarily written in the form

$$\Pi/RTc = M^{-1} + \sum_{2,3,\ldots} A_i c^{i-1} \tag{63}$$

with concentration c in units of mass per unit volume. The coefficient A_i is the ith virial coefficient, with $A_i M$ in units of (volume mass$^{-1})^{i-1}$, and R is the molar gas constant.

The generalization of the osmotic pressure relation equation (63) to a multicomponent solute is[77]

$$\Pi/RTc = 1/\bar{M}_n + A_2^{(OS)}c + O(c^2) \tag{64}$$

$$A_2^{(OS)} = \sum_J \sum_K A_{JK} w_J w_K \tag{65}$$

where w_J is the weight fraction of the Jth solute, $\bar{M}_n$ is the number-average molecular weight, and, as before, c is the total solute concentration. The coefficient A_{JK} with $K = J$ is the usual second virial coefficient measured on solutions of a single solute. The cross-coefficient $A_{JK} (J \neq K)$ represents interaction between unlike solute species J and K, and can be obtained experimentally by measurements on three concentration series including solutions of at least one binary solute composition (e.g. $w_J = 1$, $w_K = 1$, $w_J = w_K = 0.5$).

Light scattering from solutions depends on concentration fluctuations and thus can be related to thermodynamics by standard statistical mechanical methods. The ratio $c/R(\theta, c)$, where $R(\theta, c)$ is the reduced scattered intensity (or Rayleigh's ratio) due to the solute, is extrapolated to zero scattering angle θ. For a binary system, $R(0, c)$ is related very simply to the osmotic pressure by

$$\mathscr{K}c/R(0, c) = (RT)^{-1}(\partial\Pi/\partial c) = M^{-1} + 2A_2 c + 3A_3 c^2 + \cdots \tag{66}$$

where $\mathscr{K}$ is a collection of constants. In terms of the osmotic modulus, $K_{OS} \equiv c(\partial\Pi/\partial c)$, this relation becomes

$$\mathscr{K}c/R(0, c) = K_{OS}/RTc \tag{67}$$

The generalization to a multicomponent solute is nontrivial,[78, 79] but for a homologous polymer with a dispersion in molecular weight, the resulting virial expansion for light scattering is just

$$\mathscr{K}c/R(0, c) = 1/\bar{M}_w + 2A_2^{(LS)}c + O(c^2) \tag{68}$$

where $\bar{M}_w$ is the weight-average molecular weight and the measured second virial coefficient is the average

$$A_2^{(LS)} = \bar{M}_w^{-2} \sum_J \sum_K A_{JK} M_J M_K w_J w_K \tag{69}$$

The statistical mechanical problem of the virial coefficients is of course much more difficult for a polymeric solute than for a low molecular weight species, simply because of the enormous number of internal degrees of freedom possessed by the polymer chain. The multibody character of the problem is not obviated by passing to the dilute-solution limit; the connectivity of the polymer chain insures that multiple segment interactions are not diluted out relative to single segment–segment contacts between two molecules in a bimolecular cluster.

In the McMillan–Mayer formalism, the second virial coefficient is given by

$$A_2 = -\frac{N_A}{2VM^2} \int g_2\{1,2\} d\{1\} d\{2\} \tag{70}$$

$$g_2\{1,2\} = F_2\{1,2\} - F_1\{1\} F_1\{2\} \tag{71}$$

where V is the volume of the system and N_A is Avogadro's number. The molecular distribution functions $F\{1\}$, $F\{2\}$ denote normalized probability distributions for single molecules 1 and 2 in volume V with coordinates (external and internal) symbolized by $\{1\}$, $\{2\}$; and $F\{1,2\}$ is the joint distribution for the pair of molecules 1 and 2 with coordinates $\{1,2\}$. The single-molecule function $F_1\{x\}$ is identical with the probability density $P(\{r\})$ of equation (24) for all coordinates of molecule x. When there is no correlation between the pair of molecules (in configurations such that there are no intermolecular interactions), $F\{1,2\}$ equals the product $F\{1\} F\{2\}$ and the function $g_2\{1,2\}$ vanishes. Equation (70) can be taken as the fundamental point of departure for a theory of the second virial coefficient. Higher coefficients of order i can be expressed in terms of more complex integrals involving ith order distribution functions $g_i(1,2, \ldots i)$ *and* all the g functions of lower order.

3.3.2 Perturbation Theory

3.3.2.1 General formulation

A theory developed by Zimm,[80] like Fixman's treatment of the expansion factor discussed above (which it antedates), is based on a perturbation expansion for weak interactions. For interacting polymer chains, Zimm assumes that intermolecular intersegmental potentials of average force $U(r_{i_1 i_2})$ between segments i_1 in molecule 1 and i_2 in molecule 2 are additive

$$F_2\{1,2\} = F_1\{1\} F_1\{2\} \exp\left[-\sum_{i_1} \sum_{i_2} U(r_{i_1 i_2})/kT\right] \tag{72}$$

with the sums running over all n segments of molecules 1 and 2.

The χ functions, defined as in equation (26), but now for *intermolecular* segment pairs, are introduced and the resulting products are expanded. The χ functions for intramolecular pairs will appear in expansions of the single-molecule distribution functions $F_1\{1\}$, $F_1\{2\}$. As in the treatment of the chain expansion factor, the $U(r_{i_1 i_2})$ are of short range, so that the delta function representation can be assumed for integrals of the χ functions over coordinates. In fact, for a chemically homogeneous solute, the χ functions are exactly the same for intramolecular and intermolecular segment pairs since the potential of average force between two segments is the same whether the segments are in the same or different chains. The integrations over space coordinates lead to terms that are conveniently segregated into three classes in the perturbation series expression for the second virial coefficient of a solution of a monodisperse polymer

$$A_2 = \frac{N_A}{2M^2} \sum_{i_1} \sum_{i_2} [I + II + III] \tag{73}$$

$$I = \beta - \beta^2 n^{-2} \sum_{j_1} \sum_{j_2} P_0(O_{j_1 j_2}|O_{i_1 i_2}) + \beta^3 n^{-2} \sum_{j_1} \sum_{j_2} \sum_{k_1} \sum_{k_2} P_0(O_{j_1 j_2}, O_{k_1 k_2}|O_{i_1 i_2}) + \cdots \tag{74}$$

$$II = 2\beta^3 n^{-2} \sum_{j_1} \sum_{j_2} \sum_{s_1} \sum_{t_1} [P_0(O_{j_1 j_2}, O_{s_1 t_1}|O_{i_1 i_2}) - P_0(O_{j_1 j_2}|O_{i_1 i_2}) P_0(O_{s_1 t_1})] + \cdots \tag{75}$$

$$\text{III} \quad = \quad 2\beta_3 \sum_{j_1} P_0(O_{i_1 j_1}) \quad + \quad \ldots \tag{76}$$

The P_0s are probability densities, *e.g.* $P_0(O_{s_1 t_1})$ for a contact between segments s_1 and t_1, both in molecule 1, and $P_0(O_{j_1 j_2}, O_{k_1 k_2} | O_{i_1 i_2})$ for simultaneous intermolecular contacts j_1, j_2 and k_1, k_2 between chains 1 and 2, conditional upon an intermolecular contact already existing between segments i_1 and i_2 of the two chains. The subscript zero signifies that these are taken as random flight distributions. The summations are to be understood as limited to avoid multiple counting of contacts.

The series I contains the terms appearing in Zimm's original theoretical development. They concern successive orders of configurations with 1,2,3, . . . *etc.* intermolecular segment pairs simult-aneously in contact. The terms II, which were investigated later,[81,82] give contributions from configurations with both intermolecular and intramolecular segment pairs in contact.[75] Finally, III exhibits contributions from three-body segment interactions, *i.e.* an intermolecular contact pair also in contact with another segment from one of the two chains.[22] In principle, higher order, multibody interactions would be included in III. The excluded volume integrals β and β_3 have the same meaning as in Section 3.2.4.

The probability functions are evaluated, and the sums over segment indices are converted to integrals (as in the analogous procedure in the perturbation theory of the expansion factors α_L and α_G).

3.3.2.2 First-order perturbation near the theta temperature

The perturbation development to the first order in β and β_3 according to equation (73) gives the limiting behavior in the vicinity of $T = \Theta$

$$A_2 \quad = \quad 4\pi^{3/2}(N_A R_{G,0}^3/M^2)\,[z \quad + \quad 4z_3 n^{1/2} \quad + \quad \ldots] \tag{77}$$

where z and z_3 are the quantities defined in equations (33) and (34) and $n \propto M$ is the number of statistical chain segments. According to this result, $(z + n^{1/2} 4 z_3)$ is zero when A_2 vanishes; and then comparison with equation (32) gives $R_L = R_{L,\Theta} = R_{L,0}$. It should be noted that to the first-order perturbation approximation no terms II from equation (73) appear: the combination of two intermolecular and one intramolecular pair contacts only contributes to order β^3.

As in the treatment of R_L near the theta temperature, refinement of the perturbation treatment with ternary segment interactions for A_2 near zero leads to a slight revision[23]

$$A_2 \quad = \quad 4\pi^{3/2} \frac{N_A R_{G,0}^3}{M^2} \left\{ z \quad + \quad 8 z_3' \left[\left(\frac{n}{4a}\right)^{1/2} - 1 \right] \quad + \quad \ldots \right\} \tag{78}$$

where a is the cut-off parameter of order unity in equation (35). Now comparison with equation (35) shows that $R_{L,\Theta} < R_{L,0}$ at the theta temperature if $z_3 > 0$; by contrast, $R_{G,\Theta}^2$ would be close to $R_{G,0}^2$. However, since z_3 is independent of M, a finite z_3 does not invalidate the proportionality between $R_{L,\Theta}^2$ and M. According to equation (78), the temperature Θ_N for which A_2 vanishes at finite N may differ slightly from Θ, at least in principle, but the effect is expected to be small.

Under conditions such that $A_2 = 0$, the third virial coefficient is given by[83]

$$A_3 = (4\pi)^3 (N_A^2 R_{G,0}^6 / 3M^3) z_3 \quad + \quad \ldots \tag{79}$$

and, as already indicated, accurate experimental data with equation (79) can provide a first-order estimate of z_3. For example, light-scattering data on polystyrene solutions in cyclopentane give $z_3 \approx 0.012$, independent of T and M.[33] For this system, the behavior at large M gives $z \approx 0.005 M^{1/2}(1 - \Theta/T)$ for T near Θ.[67] Thus, $\Theta_N - \Theta \approx 8\Theta/M^{1/2}$ is small for polymers of high molecular weight.

3.3.2.3 Perturbation theory far from the theta temperature

As in the case of the chain expansion factors, terms of order z_3 become insignificant in the perturbation expansion of the second virial coefficient when $\beta \gg \beta_3$. Hence the two-parameter approximation becomes acceptable and only the I and II terms are needed in equation (73).

Consequently, evaluation of the Gaussian probability densities and summation (integration) over segment indices gives A_2 in the form

$$A_2 = A_2^{(L)} F(z) \tag{80}$$

here written as the product of a factor independent of molecular weight

$$A_2^{(L)} = \beta n^2 N_A / 2M^2 = 4\pi^{3/2} z N_A R_{G,0}^3 / M^2 \tag{81}$$

and a function of the excluded volume variable z

$$F(z) = 1 - b_1 z + b_2 z^2 + \ldots \tag{82}$$

In the context of the two-parameter theory, the form of equation (82) is valid for any chain architecture. For linear chains, the coefficient b_1 is $(32/15)(7 - 2^{5/2}) = 2.8654$, and for f-functional star chains[84]

$$b_1 f^{1/2} = 2.8654 + 0.9668(f - 1) + 0.2201(f - 1)^2 \tag{83}$$

from which it is evident that b_1 asymptotically approaches $0.2201 f^{3/2}$ at large f. Expressions for b_1 have also been obtained for regular and random combs.[85,86] As in the calculation of the linear approximation for $\alpha_L^2(z)$, randomization of comb branch placements along the backbone leads to mathematically simpler but quantitatively very similar results.[86] The effects of chain architecture on b_1 are correlated, at least roughly, by the mean-square-radius ratio g (equation 13); and as might be expected, b_1 increases with decreasing g, and thus with increasing intramolecular segment density.

First-order perturbation calculations of the mixed second virial coefficients A_{JK} have been carried out for chains differing in molecular weight but having the same architecture and chemical structure (so that the segment–segment excluded volume parameter β is the same for all polymer species), and for some combinations of topologically different chains, e.g. linear chains with rings, stars or combs, stars with stars of different functionality, stars with combs, etc.[87,88] The simplest result is that for two homologous linear chain species with a molecular weight ratio $\omega = M_K/M_J$[89]

$$A_{JK} = (N_A \beta n_J^2 / 2M_J^2)[1 - b_{1,JK} z_J + O(z_J^2)] \tag{84}$$

$$b_{1,JK} = (16/3)(1 + \omega^{1/2}) - (32/15\omega)[(1 + \omega)^{5/2} - \omega^{5/2} - 1] \tag{85}$$

where z_J is the excluded volume parameter z for polymer J.

To the double contact approximation, the three coefficients A_{JJ}, A_{JK} and A_{KK} for homologous linear chain solutes J and K are always of the same sign, and A_{JK} is intermediate in value between A_{JJ} and A_{KK}, according to equations (82) and (85). It follows that, to this approximation, the virial coefficient ($A_2^{(OS)}$ or $A_2^{(R)}$) for a ternary system containing solutes J and K is a monotone function of solute composition w_J.[90] On the other hand, if the two chain species are different topologically, e.g. linear and branched, and if the molecular weight ratio M_K/M_J is sufficiently far from unity, the cross-coefficient A_{JK} can be the largest of the three, and then the virial coefficient for the mixture (presuming the double contact approximation) must pass through a maximum as a function of the solute composition.

Just as the calculation of the triple contact (three contact pairs) term in the perturbation expansion for the expansion of the dimensions of a linear chain is a more formidable undertaking than calculation of the double contact term, a correct and complete determination of the coefficient of z^2 in equation (82) long proved elusive. The problem involves a complicated combination of types of simultaneous intermolecular and intramolecular contact configurations. Purely intermolecular triple contact pair configurations contribute $b_{2,\text{inter}} = 8.851$ to b_2 and these contributions (not calculated accurately) were at first assumed to afford a good approximate value for b_2. The various earlier efforts at partial and/or approximate evaluation of b_2 were reviewed in ref. 91 and a complete derivation, in part requiring numerical integration, was given. For 13 years the result, $b_2 = 14.278$, was thought to be definitive until Tanaka and Šolc[92] found a simple calculation error in ref. 91 and revised the result to the correct one, $b_2 = 13.928$. More recently, Muthukumar and Nickel[40] have shown that this result, still involving cumbersome numerical analysis in ref. 92, can be obtained quite simply in analytical form. In ref. 91 the derivation of b_2 was generalized to interactions between linear chains of unequal length, but these results are falsified quantitatively by the calculation error just mentioned. Ref. 92 corrects this also.

It is not difficult to formulate the two-parameter perturbation expansion for the third virial coefficient; but in view of the difficulties of determining the coefficients of the first few terms in the series for the second coefficient, it is not surprising that calculations for the third coefficient have not been carried very far. However, the leading term is known[29, 75, 93]

$$A_3/A_2^2 M \quad = \quad (4/3)\lambda z \quad + \quad O(z^2) \tag{86}$$

where $\lambda = 1.664$ for linear chains. This result demonstrates one consequence of the neglect of three-body interactions in the two-parameter formalism in that A_3 and A_2 vanish simultaneously. Of course equation (86) has at best only limited value, since β_3 cannot be neglected when A_2 is small (if three-body interactions have an effect) and the relation is not applicable unless z is small.

3.3.3 Approximate Theories for Good-solvent Systems

As in the case of the expansion factors α_G and α_L, the few available terms in the direct perturbation expansion of the molecular weight dependent function F(z) in equation (80) do not give a useful representation for z appreciably different from zero. Consequently a number of approximate closed form expressions for F(z), developed in much the same spirit as those for the expansion factor, have been proposed to analyze thermodynamic behavior in good solvents. A thorough discussion may be found in ref. 75. Usually, the closed form expressions are expressed in terms of a dimensionless 'penetration' function $\Psi(\bar{z})$ where $\bar{z} = z/\alpha_G^3$ and

$$\Psi(\bar{z}) \quad = \quad A_2 M^2/4\pi^{3/2} N_A R_G^3 \tag{87}$$

Experimentally, $\Psi(\bar{z})$ appears to tend toward a constant value Ψ_∞ for large z (*i.e.* large A_2). This limiting behavior provides one convenient criterion for theories of $\Psi(\bar{z})$, since it is independent of estimates of $\bar{z}$. Comparison with equations (80) and (81) gives

$$\Psi(\bar{z}) \quad = \quad (z/\alpha_G^3) F(z) \tag{88}$$

Consequently, for small z the perturbation theories (equations 38 and 82) give

$$\Psi(\bar{z}) \quad = \quad \bar{z}[1 \quad - \quad b_1\bar{z} \quad + \quad \ldots] \tag{89}$$

$$= \quad z[1 \quad - \quad (b_1 \quad + \quad 3a_{G,1}/2)z \quad + \quad \ldots] \tag{90}$$

Empirically it is found, perhaps surprisingly, that $A_2 M^2/R_{G,0}^3$ (or $A_2 M^{1/2}$) is very nearly proportional to $(\alpha_G^2 - 1)$ over a wide range of α_G.[94] Figure 1a, from light-scattering data of Berry,[67] illustrates this property for four narrow distribution polystyrenes in decalin over a 100 °C span and in toluene at room temperature. Figure 1b for two polystyrenes in cyclopentane[95] covers a much smaller range of α_G; but this system exhibits two theta temperatures, between which there lies a better-solvent region ($\beta > 0$). Thus the common linear relation holds, even as $(\alpha_G^2 - 1)$ and A_2 each increase from zero, pass through a maximum, and decline again to zero.

This behavior indicates, that the function G(z) in the expression

$$A_2 M^2 R_{G,0}^3 \quad = \quad 4\pi^{3/2} N_A(\alpha_G^2 - 1) G(z) \tag{91}$$

is nearly unity. It follows obviously that

$$\Psi(\bar{z}) \quad = \quad \frac{(\alpha_G^2 - 1)}{\alpha_G^3} G(z) \tag{92}$$

providing an expression for $\Psi(\bar{z})$ in terms of α_G if G(z) is constant. With the perturbation theories, we have

$$G(z) \quad = \quad (a_{G,1})^{-1}[1 \quad - \quad (b_1 \quad + \quad a_{G,2}/a_{G,1})z \quad + \quad \ldots] \quad = \quad (105/134) \quad - \quad 0.967z \quad + \quad O(z^2) \tag{93}$$

so that G(z) tends to a constant value for small z. In fact, experimental data show that G(z) is approximately equal to $1/a_{G,1}$ over a range of z for data on both linear and branched polymers.[96] If $G(z) \approx 1/a_{G,1}$ for all z, then the relation

$$\Psi(\bar{z}) \quad \approx \quad (\alpha_G^2 \quad - \quad 1)/a_{G,1}\alpha_G^3 \tag{94}$$

provides an implicit expression for $\Psi(\bar{z})$. None of the relations given above for α_G give constant $\Psi(\bar{z})$

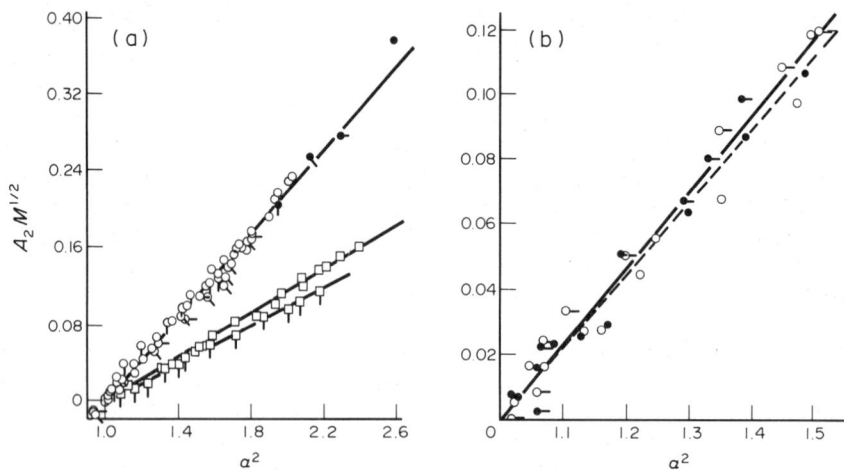

Figure 1 Plots illustrating proportionality between $A_2 M^{1/2}$ and $(\alpha^2 - 1)$. (a) Data for narrow distribution linear poly-styrenes ($0.6 < \bar{M}_w \times 10^{-6} < 4.4$) in decalin (open circles) at various temperatures from $10\,°C$ to $110\,°C$ and in toluene (filled circles) at $12\,°C$.[67] Data for two comb branched polystyrenes: 24 branches, $\bar{M}_w \times 10^4 = 24 \times 4.62 + 133$ (squares); 22 branches, $\bar{M}_w \times 10^4 = 22 \times 12.0 + 133$ (squares with pips).[96] (b) Data for two narrow distribution linear polystyrenes $\bar{M}_w \times 10^6 = 0.86$ (pips) and 1.97 (no pips) in cyclopentane.[95] Open circles are for temperatures from the limiting UCST ($\Theta_U = 19.6\,°C$) to *ca.* $70\,°C$. Filled circles are for temperatures from *ca* $70\,°C$ to the limiting LCST ($\Theta_L = 154.5\,°C$). The solid line is an empirical fit of all the data. The dashed line is the first-order perturbation result $A_2 M^{1/2}/(\alpha^2 - 1) = 0.219$ (reproduced by permission of Wiley, from *J. Polym. Sci., Part B: Polym. Phys.*, 1987, **25**, 685)

for large z since all have the asymptotic form $\alpha \propto z^{2\nu - 1}$. However, with equation (61), and the assumption that $\alpha_L \approx \alpha_G$, $(\alpha_G^2 - 1)/a_{G,1}\alpha_G^3$ *vs.* z exhibits a broad maximum for $z = 6.3$, giving a value of 0.30 ± 0.01 for $0.78 < \bar{z} < 2.0$ (or $1.5 < \alpha_G < 2.07$). Then $(\alpha_G^2 - 1)/a_{G,1}\alpha_G^3$ decreases slowly with increasing $\bar{z}$ (*e.g.* being 0.222 for $\alpha_G^2 = 10$, an unusually large value experimentally). Such a slow decrease in $\Psi(\bar{z})$ for large $\bar{z}$ is indicated in some experimental data.[97] Of course, G(z) could increase slowly with z in such fashion that $\Psi(\bar{z})$ increases monotonically to a limit Ψ_∞.

In the earliest 'smoothed density' model, Flory[98] represented a polymer molecule as a sphere containing a uniform density of chain segments, but having the correct molecular mass and radius of gyration R_G. Flory and Krigbaum[99] made the model more realistic by substituting a Gaussian segment distribution about the molecular center of mass. The interaction of two chains is then represented by a binary cluster integral, as in the theory of a real gas, with the local energy of interaction assumed to be proportional to the product of molecular segment densities at each point in the region of chain overlap. As is typical of smoothed density models, the effects of intramolecular interactions are accounted for by the expedient of expanding all chain dimensions uniformly by the factor α; and, as in other theories, the effect of this combination of intramolecular and intermolecular theories is to give the functional form derived for chains without intramolecular interactions but with the variable z replaced by $\bar{z} \equiv z/\alpha_G^3$. The Flory–Krigbaum expression for the second virial coefficient is a definite integral that can be evaluated numerically or graphically, but a quantitatively adequate analytical approximation

$$\Psi(\bar{z})_{\text{FKOm}} = [\ln(1 + 2b_1\bar{z})]/2b_1 \tag{95}$$

is a modification of one given by Orofino and Flory.[37] In their original form of equation (95), the numerical coefficient was 2.30 rather than $2b_1 = 5.730$; but it seems preferable to arbitrarily adopt the latter value to make the Taylor series expansion of equation (95) agree with equation (82) through the linear term.[29] In disagreement with indications from experiment, this expression fails to give a finite limit of $\Psi(\bar{z})$ for large $\bar{z}$.

Other approximate treatments of the second virial coefficient differ in the way the connectivity of the chain conformation is simplified to attain a tractable mathematical function. For example, the theory of Casassa and Markovitz[100] takes equation (73) as its starting point and replaces the conditional probabilities for multiple pairwise intermolecular segment contacts by products of single conditional pair contacts; thus $P(O_{j_1 j_2}, O_{k_1 k_2} | O_{i_1 i_2})$ is replaced by $P(O_{j_1 j_2} | O_{i_1 i_2})$ $P(O_{k_1 k_2} | O_{i_1 i_2})$, *etc.* With this simplification the sums in equation (73) can be done, assuming random flight chain statistics, save for the last pair over indices i_1 and i_2. This remaining double sum (or integral) can be evaluated numerically, but the final result is given to a good approximation by the simple form

$$\Psi(\bar{z})_{CM} = \frac{1 - \exp(-2b_1\bar{z})}{2b_1} \tag{96}$$

which was also obtained in a different way by Fixman.[101] Here the coefficient $2b_1$, which leads to $\Psi_\infty = 5.730^{-1} = 0.1745$, arises naturally from the model, with no arbitrary adjustment; and expansion of the exponential leads to a series expression that is correct through the linear term in $\bar{z}$. The factorization of multiple contact probabilities has the effect of averaging the segment distributions of interacting chains about an 'initial' contact of segments i_1 and i_2 before the final averaging over all choices of i_1 and i_2.

The functions in equations (95) and (96) are monotone, increasing with increasing $\bar{z}$, but $\Psi(\bar{z})_{FKO}$ increases much more rapidly than $\Psi(\bar{z})_{CM}$. These dependences appear to represent extremes in behavior; a number of other proposed approximate functions give plots *vs.* $\bar{z}$ that lie between these two cases.[75] It will be noted that these two functions also correspond to extremes in symmetry of models for the bimolecular cluster, *i.e.* the interpenetration of two soft spheres in the FKO model describes inherently asymmetric configurations of segments, while the CM model makes every configuration a spherically symmetrical superposition of the segments of the two participant chains.

The most sophisticated treatment of this kind is perhaps the 'differential equation' approach of Kurata *et al.*[75,82] Approximating probabilities for multiple intermolecular contacts, much as in the derivation of equation (96), they obtain a differential equation representation for $\partial A_2/\partial \beta$. This leads to a final result of the form

$$\Psi(\bar{z})_{KY} = \frac{1 - (1 + K\bar{z})^{-(2b_1 - K)/K}}{(2b_1 - K)} \tag{97}$$

where $K = (3b_{2,inter}/b_1) - 2b_1 = 3.536$ and $(2b_1 - K)/K = 0.6206$. These numbers are different from those given by Yamakawa[75] since they reflect the more recent revision of b_2 by Tanaka and Šolc.[92] With this K, equation (97) agrees with the second-order perturbation, and gives the limit $\Psi_\infty = (2b_1 - K)^{-1} \approx 0.4557$. However, by arbitrary adjustment of K, equations (95) and (96) are obtained as special cases, as is a simple form proposed by Stockmayer[29] with $K = b_1$

$$\Psi(\bar{z})_S = \bar{z}/(1 + b_1\bar{z}) \tag{98}$$

which affords a good representation of some experimental data.[102] This relation gives a finite limit $\Psi_\infty = 0.3490$. Alternatively, K could be chosen to force agreement of Ψ_∞ with experiment, *e.g.* $K \approx 2.40$ if $\Psi_\infty \approx 0.30$ to fit the data on polystyrene | decalin[67] shown in Figure 2.

This version of the theoretical function $\Psi(\bar{z})_{KY}$ is displayed in Figure 2 together with plots for the modified Flory–Krigbaum–Orofino function, equation (95), the 'original' version with 2.60 in place of b_1, equation (96) and equation (98). Over the range of $\bar{z}$ covered in the plot, equation (98) is in reasonable agreement with equation (94), if α_G is given according to equations (47) and (48). The calculation of $\bar{z}$ for the experimental data points is discussed below (Section 3.3.4).

The 'ε expansion' method mentioned above has also been applied to computation of the second virial coefficient to give a result to order ε^2 in 3-space[5,73,103]

$$\Psi(\bar{z})_{OF} = \begin{cases} w_1^*[\lambda_1 + k\lambda_1^2] & \bar{z} < 0.75 \tag{99} \\ w_1^*(1 + k) \approx 0.269 & \bar{z} > 0.75 \tag{100} \end{cases}$$

where $k = 2(7 + 24 \ln 2)/159 \approx 0.297$ and λ_1 is given by equation (56), with $w_1^* = 0.2070$ to order ε^2. Since λ_1 depends only on the z-like parameter $\bar{z}$, this result lies within the scope of two-parameter treatments.[104] Expansion of $\Psi(\bar{z})_{OF}$ for small $\bar{z}$ gives

$$\Psi(\bar{z})_{OF} = \frac{4\bar{z}}{3}\left[1 - \frac{4\bar{z}}{3}\left(\frac{1 - k}{w_1^*}\right)\left(\frac{4\bar{z}}{3}\right) + \cdots\right] \tag{101}$$

where $(1 - k)/w_1^* = 3.395$ to order ε^2. Equation (101) may be compared with equation (89) or equation (90) from the perturbation theory to evaluate $4\bar{z}/3$ as z or $\bar{z}$ respectively; either assignment gives the first-order perturbation, but neither provides perfect agreement to second order. From equation (100), it is evident that this theory belongs to the class for which A_2 is asymptotically proportional to $z^{-2/5}$.

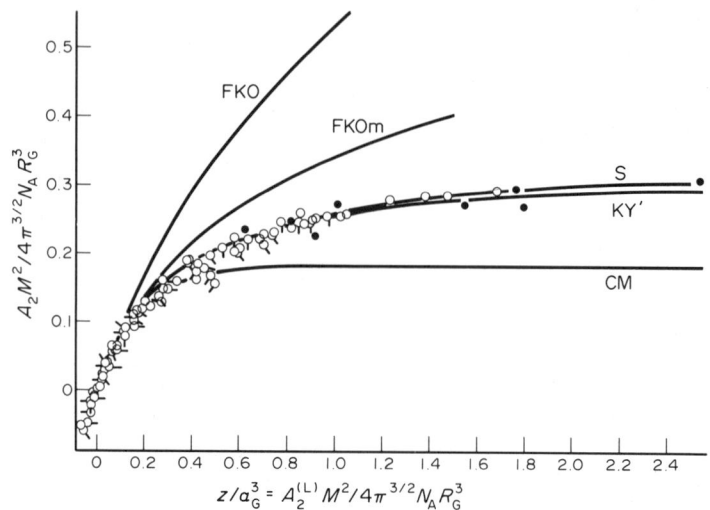

Figure 2 Penetration functions $\Psi(\bar{z})$ for several approximate theories as indicated by labels: (FKO) equation (95) with 2.60 in place of b_1; (FKOm) equation (95); (KY') equation (97) with $K = 2.40$; (S) equation (98); (CM) equation (96); Data are for polystyrene + decalin, as in Figure 1(a). Values of z/α_G^3 for the experimental points are calculated using equation (115) for z with $\Theta = \Theta_U$ and $B = B_U$

A renormalization group calculation by Burch and Moore[105] is based on analogies between lattice model field theories for magnetic systems and polymer solutions. By adjustment of a constant, their results agree with the perturbation expansion of $F(z)$ in equation (82) to order z. With the same value of the constant, $\Psi(\bar{z})$ in the good solvent range, $z > 0.75$, is given as an implicit function of $\bar{z}$ by the relations

$$\Psi(\bar{z})_{BM} = 0.209[1 - 0.0014z^{-6/5} + O(z^{-12/5})]/D \tag{102}$$

$$\bar{z}^{-1} \approx \alpha_G^3/z = 1.828\,z^{-2/5}D \tag{103}$$

$$D = [1 + 0.099z^{-6/5} + O(z^{-12/5})]^{3/2} \tag{104}$$

These relations give $\Psi(\bar{z})$ within 2% of that from equation (97) if K is chosen to make $\Psi_\infty = 0.209$.

The expectation that $\Psi(\bar{z})$ approaches a finite limit at large $\bar{z}$ is equivalent to the assertion that the second virial coefficient becomes asymptotically proportional to a volume R_G^3/M^2. Accepting also that $\alpha_G^{2\nu-1}/z$ asympotically approaches a constant value, it is a simple matter to show that A_2 must become proportional to $z^{-\mu}$ (or $M^{-\mu/2}$) where $\mu = 4 - 6\nu$. Thus if $\nu = 3/5$, A_2 is proportional to $z^{-2/5}$ (or $M^{-1/5}$). Most of the approximate closed forms that have been proposed for A_2, or $\Psi(\bar{z})$, behave in this fashion: equations (96) and (98) do so, as does the function from Flory's old uniform density sphere model; but the Flory–Krigbaum function equation (95) shows a logarithmic divergence. Although it is usually problematic whether experimental data actually represent asymptotic behavior, there is support for a limiting exponent μ of 0.40 from measurements on high molecular weight polymers in very good solvents.[97] Somewhat different values have been suggested from extrapolation of numerical data on relatively short lattice chains to infinite chain length: 0.5 from exact enumeration,[106] and 0.56 from Monte Carlo simulations.[107] From a Padé approximant formulation carried out in the same spirit as that mentioned above (Section 3.2.4.4) for the expansion factor, Tanaka[108] obtained an exponent of 0.4035 (later revised[92] to 0.4179 by virtue of the corrected coefficient b_2 in equation 82).

Mention was made above (Section 3.3.2.3) of calculations of mixed second virial coefficients $A_{JK}(J \neq K)$ to the double contact approximation. Greater interest attaches to relations among A_{JK}, A_{JJ} and A_{KK} for mixtures of homologous polymers in good-solvent systems, where an analysis calls for generalization of treatments like those that give equations (95) through to (98) to the case of species differing in molecular weight. The Flory–Krigbaum model[99] is one that has been so extended. It predicts that the cross-coefficient A_{JK} will always be the greatest of the three coefficients, provided the polymers differ in molecular weight by a factor of at least 4.24.[90] As a consequence, $A_2^{(OS)}(w)$ or $A_2^{(LS)}(w)$ will exhibit a maximum as a function of solute composition w (weight fraction) for mixtures of two sharp polymer fractions. A simpler scheme is to postulate that molecules of kinds J and K interact like effective hard spheres with A_{JJ} and A_{KK} the actual virial coefficients in the two binary systems with only one solute component.[109] With the further assumption (for convenience) that A_{JJ} is proportional to M^{-a}, the exponent $-a$ being regarded simply as an empirical constant, it is easy to show that for a mixture obeying this rule the cross-coefficient A_{JK} is again the greatest of the three coefficients with any physically reasonable values of a for good solvents (say 0.20 to 0.25) and a molecular weight ratio appreciably different from unity. The maxima in $A_2^{(OS)}$ and $A_2^{(LS)}$ according to this scheme are much more prominent than those predicted by the Flory–Krigbaum theory. In still another treatment, Tanaka and Šolc[92] have devised a Padé approximant expression that correctly reproduces three terms of the perturbation series in equation (84) for two linear chains of different length. Their results are qualitatively like those for the hard sphere interaction model in that the cross-coefficient is always larger than either of the other two.

None of these theoretical treatments would allow the cross-coefficient to be the least of the three; nor, therefore, could A_2 for the mixture ever exhibit a minimum at some composition. It may be noted that calculating the cross-coefficient as either the arithmetic or geometric mean of the other two (as has sometimes been arbitrarily done) would always make $A_2^{(OS)}(w)$ and $A_2^{(LS)}(w)$ monotone functions. Of the few reports of experimental maxima in $A_2(w)$ for solutions of binary mixtures of sharp polymer fractions, the most credible appear to be those of osmotic pressure measurements by Krigbaum and Flory[110] and by Noda *et al.*[111] The reality of the maxima in other data appears uncertain; but whether or not the maxima are real, it seems that the function $A_2^{(OS)}(w)$ is always concave downward, *i.e.* there is no experimental evidence of minima. Comparisons with the experiments suggest that the hard sphere interaction overestimates and the Flory–Krigbaum theory underestimates the prominence of any maxima. The averaging of the A_{JK} in the light-scattering virial coefficient $A_2^{(LS)}(w)$ causes any maximum for a mixture of two homologous fractions of widely different molecular weight to appear at very low content of the higher species, and its discernment seems to be subject to even more ambiguity than in the case of $A_2^{(OS)}(w)$[112] (an analysis of available data is given in ref. 109). An experimental maximum in $A_2^{(LS)}$ has been reported for mixtures of a branched and a linear polystyrene.[113]

Recently, Schäfer[114] challenged both the theories and the analysis of the experiments. He uses renormalization arguments to conclude that a maximum in $A_2^{(OS)}(w)$ or $A_2^{(LS)}(w)$ is not admissible. He also questions the practical possibility, without theoretical guidance, of extrapolating experimental data to zero concentration to obtain true second virial coefficients in systems containing long chains, *i.e.* his theoretical analysis indicates that needed data extend into the concentration regime where the simple virial series expansion is presumably no longer adequate.

The theoretical treatments of A_{JK} discussed here were motivated by the question of the effects of molecular weight dispersion on measured second virial coefficients. Once $A_{JK}(M_J, M_K)$ is available it is obvious in principle how to obtain $A_2^{(OS)}$ or $A_2^{(LS)}$ for any desired form of distribution. Detailed calculations using the hard-spherelike interaction model with the familiar Schulz–Zimm distribution[115–117] indicate that the ratio of virial coefficients $A_2^{(OS)}/A_2^{(LS)}$ increases without limit with the breadth of the distribution, becoming asymptotically proportional to $\bar{M}_w/\bar{M}_n$. For the 'most probable' distribution (a weight-average to number-average molecular weight ratio of two) and $a = 0.25$, $A_2^{(OS)}/A_2^{(LS)}$ is 1.255. Polydispersity increases both $A_2^{(OS)}$ and $A_2^{(LS)}$ relative to A_2 for the monodisperse polymer corresponding, respectively, to the number-average and weight-average molecular weights of the polydisperse polymer. For the parameters cited, the two ratios are 1.127 and 1.068. The first ratio increases without limit as the distribution is broadened; the second approaches a limit of 1.13 for large dispersions.

Tanaka and Šolc used both Schulz–Zimm and logarithmic–normal[117,118] distributions to calculate the effects of polydispersity on second virial coefficients for the Padé approximant expression. For the Schulz–Zimm function, $A_2^{(OS)}/A_2^{(LS)}$ increases from unity with increasing dispersion, as in the case of hard sphere interactions, but it approaches a finite limit; and the ratio of $A_2^{(OS)}$ to A_2 for the monodisperse species matching $\bar{M}_n$ passes through a maximum with increasing dispersion and may then become less than unity. The trends exhibited in these results suggest that careful conventional fractionation procedures applied to typical noncrystalline polymers should diminish the effects of polydispersity on the second virial coefficient to the extent that a solution of a polymer fraction can safely be treated as a binary system.

Since the third virial coefficient can seldom be determined with much accuracy from experimental data, there have been relatively few attempts at theoretical developments like those outlined above for A_2 in good-solvent systems. For hard spheres, the ratio $\gamma \equiv A_3/A_2^2 M$ is 5/8; and the simplest expedient is to assume that it is the same for interactions of polymer chains.[99] To improve on this approximation Stockmayer and Casassa[119] evaluated γ using a 'soft' potential chosen largely for its mathematical tractability. Variation of a parameter in the result makes γ increases from zero as A_2 increases and suggests that γ never exceeds 5/8. Their result can be put in the form

$$\gamma = (4/3)\,\kappa\,\Psi(\bar{z}) \tag{105}$$

where the parameter κ is expected to be close to unity (see equation 136). With this relation, γ tends to about 0.4 for large $\bar{z}$, in reasonable accord with some experimental estimates.[33,67]

Other theoretical estimates also make γ an increasing function of z but do not agree on the upper limit.[75] One example is the relation[75]

$$A_3/A_2^2 M = (4/3)\,[1 - \exp(-\lambda z/\alpha_G^3)] \tag{106}$$

This gives $A_3/A_2^2 M = 4/3$ in the limit of large $\bar{z}$.

A practical use of theoretical estimates of A_3 is to facilitate reliable extrapolation of osmotic and light-scattering data to zero concentration in determining molecular weights and second virial coefficients. Substituting A_3 in terms of γ and A_2 into equation (63) and taking the square root gives

$$\left(\frac{\Pi}{RTc}\right)^{1/2} = \frac{1}{M^{1/2}}\left[1 + \frac{1}{2}A_2Mc + \frac{1}{8}(4\gamma - 1)(A_2Mc)^2 + \dots\right] \tag{107}$$

It will be noted that if γ were $1/4$, the third term in equation (107) would vanish; and if the series truncated at this point were sufficient, a plot of data as $(\Pi/c)^{1/2}$ *vs.* c would be sensibly linear to higher concentration than the usual plot of Π/c *vs.* c. For this reason, the assumption that $\gamma = 1/4$ has been advocated as an adequate compromise for analysis of osmotic data.[110,120] The perhaps more realistic value of γ of *ca.* 0.4 would not indicate a significantly worse linear fit on the 'square root' plot.[110] Empirically it is found that plots of this kind for polymers in good solvents do show significantly less curvature than the conventional representation. The analogous square root plot for light scattering also gives a more nearly linear representation of data[67,121] as would be anticipated from the relation

$$\left(\frac{\mathscr{K}c}{R(0,c)}\right)^{1/2} = \frac{1}{M^{1/2}}\left[1 + A_2Mc + \frac{1}{2}(3\gamma - 1)(A_2Mc)^2 + \dots\right] \tag{108}$$

from equation (66) with A_3 expressed in terms of γ and A_2.

3.3.4 Correlation of Configurational and Thermodynamic Data in the Two-parameter Context[95,122]

Equations (33), (37), (80) and (81) constitute the basic relations applicable to the class of two-parameter theories for the radius of gyration and the second virial coefficient. Quite apart from specific knowledge of the functions $\alpha_G(z)$ and $F(z)$, or $\Psi(\bar{z})$, these relations assert that the effects of excluded volume on R_G and A_2 depend on the single excluded volume parameter z. Accordingly, correlations with experiment involve the problem of determining z itself, as well as $\alpha_G(z)$ and $F(z)$, from observable quantities. It is useful to rearrange equation (81) into the dimensionless grouping

$$A_2^{(L)}M^2 R_{G,0}^{-3} = 4\pi^{3/2}N_A z \tag{109}$$

This equation, together with equation (80), gives

$$A_2 M^2 R_{G,0}^{-3} = 4\pi^{3/2}N_A z F(z) \tag{110}$$

which expresses $zF(z)$ in terms of experimental quantities. Another useful correlation of similar form with experimental quantities is given by equation (87)

$$A_2 M^2 R_G^{-3} = 4\pi^{3/2}N_A \Psi(\bar{z}) \tag{111}$$

The two-parameter formalism makes no stipulations concerning the temperature dependence of the radius of gyration and of the second virial coefficient, except that both quantities must reflect any temperature dependence of the excluded volume integral β through the variable z. Without an explicit theory of the behavior of β, we can only assert general thermodynamic requirements. We recall that the osmotic pressure is given by $-(RT/V_1)\ln a_1$, with a_1 the activity of the solvent and V_1 its molar volume. Then using equation (63), we can write the chemical potential μ_1 of the solvent as a virial expansion

$$\mu_1 - \mu_1^0 = -RTV_1\,(c/M + A_2c^2 + \dots) \tag{112}$$

from which it is apparent that $RTV_1A_2c^2$ is a free energy quantity, which, considering equation (110) (or equation 77 if z_3 can be neglected), approaches $RTV_1A_2^{(L)}c^2$ as T approaches Θ. Like any free energy, $RTV_1A_2c^2$ must obey the Gibbs–Helmholtz relation; so that near Θ there is obtained the well known form[120]

$$A_2^{(L)} = 4\pi^{3/2}N_A B(1 - \Theta/T) \tag{113}$$

where

$$4\pi^{3/2}N_A B = -\Theta^{-1}(\partial A_2/\partial T^{-1})_{T=\Theta} = \Theta(\partial A_2/\partial T)_{T=\Theta} \tag{114}$$

If the theta temperature is accessible, a parameter B, *defined* by equation (114), is obtainable unambiguously from the slope at $T = \Theta$ of a plot of A_2 vs. T or $1/T$—without assumptions about a particular theory. Further, if the two-parameter theory implied by equations (80) and (81) (or equation 109) is accepted, z is also obtainable from experiment

$$z = \frac{1}{4\pi^{3/2}N_A}\left(\frac{R_{G,0}^2}{M}\right)^{-3/2} M^{1/2}A_2^{(L)} = \left(\frac{R_{G,0}^2}{M}\right)^{-3/2} M^{1/2}B\left(1 - \frac{\Theta}{T}\right) \tag{115}$$

From the definition of z, equation (33), it is evident that equation (115) expresses the statistical parameter βn^2 in terms of physical quantities.

For many years typical polymer + solvent systems were studied near and/or above a theta temperature in the region where the (positive) second virial coefficient increases with temperature, *i.e.* the solvent becomes better at higher temperature. It is easily shown that this behavior implies that both enthalpy and entropy of mixing are positive. The temperature dependence of $A_2^{(L)}$ was usually, rather uncritically, assumed to be given by equation (113) with the constant B evaluated at Θ, even at temperatures far above Θ. This assumption was justified by A_2 and R_G data, spanning wide ranges of molecular weight and temperature, that could be very successfully correlated by this scheme. One example is the system polystyrene + *cis*-decalin, where data analyzed in this way for molecular weights from 2×10^4 to 4.4×10^6 and temperatures from $10\,°C$ to $110\,°C$ ($\Theta = 12.2\,°C$) define a single curve, within experimental uncertainty, when plotted as $A_2M^{1/2}$ vs. z or as $\Psi(\bar{z})$ vs. z (or $\bar{z}$).[67] This correlation is shown by the data plotted in Figure 2.

Despite the success of correlations based on equation (113), there is no warrant for assuming that the temperature dependence of $A_2^{(L)}$ will be given at temperatures far from Θ by this expression with the B evaluated at Θ. It has now been demonstrated for many polymer + solvent systems that the virial coefficient first increases with temperature above Θ and then decreases, eventually to vanish again at another theta point. Indeed, it is now believed that this is a universal feature of solutions of long-chain molecules in low molecular weight solvents. Of course, near the high temperature theta point, a relation just like equation (113) must hold, but in terms of parameters valid at that temperature. As will be shown in subsequent discussion, the conventional and high temperature theta points are associated, respectively, with an upper critical phase separation (the solubility gap below the critical temperature) and a lower critical separation (the solubility gap above the critical temperature). Accordingly, we designate the two theta temperatures as Θ_U and Θ_L, and the corresponding B parameters as B_U and B_L.

Figure 3 shows the variation of the second virial coefficient with reciprocal temperature for polystyrene + cyclopentane near both theta points for this system. From the slopes of the two plots, the coefficients B_U and B_L are obtained. Then with R_G known, z and $\bar{z}$ are established within the range of validity of either of the two expressions for $A_2^{(L)}$.

For some range of temperatures between Θ_U and Θ_L, neither expression of the form of equation (113) can be valid; and z defined by equation (33) obviously cannot be given everywhere by equation (115) with a constant B. If data covering the temperature range between Θ_U and Θ_L are to be adequately analyzed, one needs not only the functions α_G^2 and $F(z)$ but also a theory relating $A_2^{(L)}$ with T over the full range, or at least a phenomenological connection between the two regimes about Θ_U and Θ_L. The latter is provided in a derivation by Eichinger,[123] and the former is addressed by Flory's 'liquid state' theory (and similar theories), which we shall consider briefly below (see Section 3.5.3.1).

Eichinger's treatment conforms to the limiting behavior of $A_2^{(L)}$ required by equation (113) at $T = \Theta_U$ and its analog at Θ_L. To represent behavior at intermediate temperatures by a continuous function, the temperature dependence of the free energy $-RTV_1A_2^{(L)}c^2$ (see equation 111) is explicitly characterized by a partial molar heat capacity of dilution $RTV_1c^2\partial^2(TA_2^{(L)})/\partial T^2$. This is expanded about Θ_U in a series in non-negative powers of $(T - \Theta_U)/\Theta_U$. Two integrations with respect to T between limits Θ_U and T yield a complicated series expression for $A_2^{(L)}$. This procedure

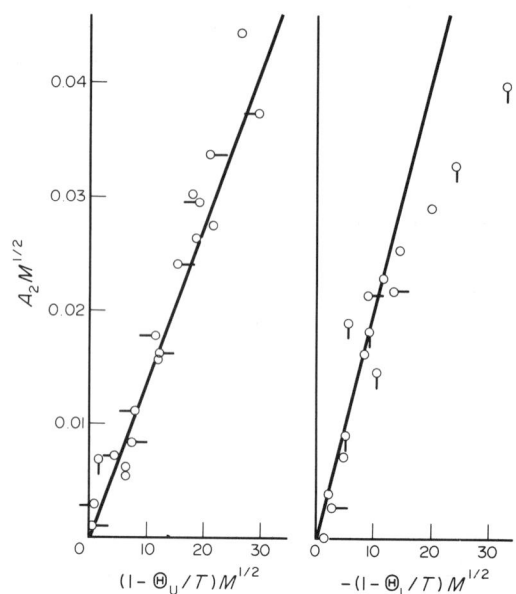

Figure 3 Temperature dependence of A_2 in the vicinity of the theta temperatures Θ_U and Θ_L for four polystyrenes ($\bar{M}_w \times 10^6 = 0.16, 0.41, 0.86$ and 1.97) in cyclopentane.[95] The initial slopes of the plots give B_U (left) and B_L (right) (reproduced by permission of Wiley, from *J. Polym. Sci., Part B: Polym. Phys.*, 1987, **25**, 681)

imposes agreement with equation (113) as $T \to \Theta_U$ and thus allows fitting with empirical values of Θ_U and the slope B_U. Details are given in ref. 95, which expands slightly on Eichinger's derivation. The first three terms of the series give

$$\frac{A_2^{(L)}}{4\pi^{3/2}N_A B_U} = 1 - \frac{\Theta_U}{T} - k_1\left[1 - \frac{\Theta_U}{T} + \ln\left(\frac{\Theta_U}{T}\right)\right] + \frac{k_2}{2}\left[\frac{T}{\Theta_U} - \frac{\Theta_U}{T} + 2\ln\left(\frac{\Theta_U}{T}\right)\right]$$

$$- \frac{k_3}{3}\left[\left(1 - \frac{T}{\Theta_U}\right)\left(\frac{T}{2\Theta_U} - \frac{\Theta_U}{T} - \frac{5}{2}\right) + 3\ln\left(\frac{\Theta_U}{T}\right)\right] \qquad (116)$$

Expressed in terms of the appropriate derivatives, the k_i would be the Taylor's series expansion coefficients for the heat capacity, but here they are treated as adjustable parameters to fit the available experimental data.

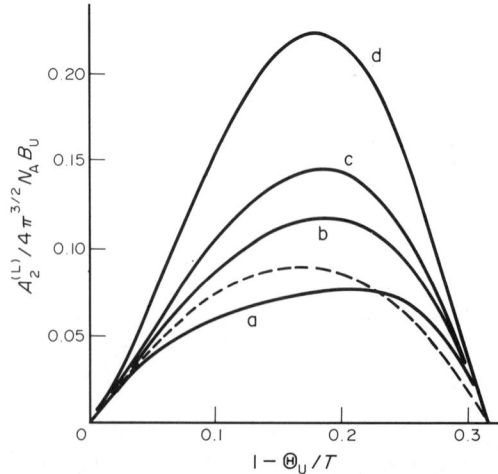

Figure 4 Temperature dependence of $A_2^{(L)}$ as given by equation (116) with various combinations of k_1, k_2 and k_3. All curves fit experimental values of Θ_U, B_U and Θ_L for polystyrene + cyclopentane. The dashed curve is Eichinger's approximation ($k_2 = k_3 = 0$). The four solid curves fit the experimental B_L as well as the three other experimental parameters; values of k_3 are: (a) -270; (b) 0; (c) 180; and (d) 680 (reproduced by permission of Wiley, from *J. Polym. Sci., Part B: Polym. Phys.*, 1987, **25**, 687)

Figure 4 illustrates application of equation (116) with the experimental parameters Θ_U, Θ_L, B_U and B_L of the system polystyrene + cyclopentane.[95] In the case worked out by Eichinger, the first order of refinement over equation (113), both k_2 and k_3 are set equal to zero, and k_1 is adjusted to make $A_2^{(L)}$ vanish at the experimental value of Θ_L. This gives the dashed curve in Figure 4, which does not match the experimental slope of A_2 at Θ_L. If, as in this case, data permit an estimate of B_L from dA_2/dT at Θ_L, k_1 and k_2 can be adjusted and only k_3 arbitrarily set to zero. Now the theoretical curve b is obtained, which matches the slopes at both theta temperatures. To proceed further, it can readily be shown that the experimental maximum A_2 must occur at the same temperature T_{max} as the maximum in $A_2^{(L)}$, and this additional item of information could, in principle, be used to fix a unique set of three coefficients for equation (116). However, it turns out in practice that the temperature at the calculated maximum is so insensitive to k_3 that it is not possible to use this datum to fix three k_i from the virial coefficient data. All the solid curves in Figure 4 conform to the same four experimental parameters for the temperature dependence of A_2 but correspond to vastly different choices of k_3 (together with different k_1, k_2 pairs to force conformity with data). An acceptable function for the polystyrene + cyclopentane system would have a maximum at *ca.* 70 °C. The limitations of experimental data would obviously render completely meaningless the retention of any further terms in equation (116).

Eichinger[123] reanalyzed Berry's[67] data on polystyrene + decalin using equation (116) with k_2 and k_3 both zero and k_1 calculated to give Θ_L at about $\Theta_U + 145$ °C: *i.e.* the gas–liquid critical temperature of *trans*-decalin. This value was chosen, for lack of other information, as representing a tentative upper limit for Θ_L. Then equation (115) yielded revised values of $z(T)$ and thence of $\bar{z}$ to permit correlation with experiment *via* equation (110) or equation (111). Eichinger's values of z are considerably different from Berry's based on equation (113) (smaller by as much as 35% at large z, as $A_2^{(L)}$ presumably levels off in approaching the temperature of the maximum); however, the *different* correlations with the experimental A_2 and R_G data appear equally good.

This analysis of the typical experimental data discussed here demonstrates agreement with the general requirements of two-parameter theory. Data on the radius of gyration and the second virial coefficient are in conformity with the functional forms in equations (109), (110) and (111): within the limitations of experiment, unique dependences on the excluded volume variable z are demonstrated over extended ranges of temperature and molecular weight. It is suggested that these dependences constitute common dilute solution behavior for a large class of typical polymer + solvent systems. Deviations might be expected near a theta point, but their detection makes stringent demands upon experiment. It will be recalled that small effects of three-body interactions on osmotic pressure, or light scattering, have been deduced from information on the third virial coefficient at the theta point. However, the *form* of equation (113) must hold near a theta point, irrespective of three-body contributions, although their involvement would affect the proper interpretation of the empirically determined B.

Two experimental findings on the polystyrene + cyclopentane system are worth noting.[95] The unperturbed radius of gyration is found to be the same at Θ_U and Θ_L, and thus the variation of $R_{G,0}$ with temperature is negligible. It also turns out that the parameters B_U and B_L are related by $B_U \Theta_L = -B_L \Theta_U$. It is not known how general this latter behavior might be since few data are available, accurate experimentation at temperatures in the range of Θ_L being a decidedly nontrivial undertaking.

3.3.5 Mixed-solvent Systems

Up to this point, the discussion has concerned systems with one or more polymer species dissolved in a pure solvent. However, thermodynamic behavior in the more general case of systems with multiple solvents involves additional effects that are of considerable interest. Here, for simplicity, only a three-component system of a polymer (component 2) in two low molecular weight solvents (components 1 and 3) is considered;[124] but an exhaustive analysis of the general multicomponent case is available.[125] The classification of solvents and solute is of course arbitrary, but it has utility with respect to specific experimental measurements. For instance, the operative principle in an osmotic measurement is that solvents pass through the membrane that confines a polymeric solute. The obvious effect of the additional thermodynamic degree of freedom in the three-component system is the possibility of selective interaction (preferential solvation or 'binding') of the solute with a solvent component. In an osmotic experiment this will be manifested by equilibrium solvent compositions that are different in the solute and solvent compartments of the osmometer. It is possible, though not always very useful, to formally redefine a solute component so as to include in it any excess (or deficiency) of a solvent component required to make the 'free' solvent mixture appear like a single liquid,[126] and so to picture the solute molecule as somehow bearing a local environment different from the overall composition. Without regard to the physical reality of such a visualization, the selective interaction is often expressed in the form[127,128]

$$\lambda = (\partial\phi_3/\partial c_2)_{T,\mu_1,\mu_3} \tag{117}$$

i.e. the dependence of the volume fraction of solvent 3 in the solvent mixture on the concentration c_2 of polymer, with the chemical potentials μ_1, μ_3 of the diffusible components in an osmotic experiment held fixed. Rigorously, ϕ_3 should be based on partial molar volumes in solution; but in practice, volumes of unmixed components are more often used. The volume-based composition variables in equation (117) would obviously not be particularly convenient for thermodynamic derivations since changing temperature or adding or removing a component changes the volume. The consequences of transformations to more rational concentration units, *e.g.* all based on a reference mass of one of the solvents, have been worked out in detail.[125]

The preferential interaction coefficient λ can be determined directly from the equilibrium distribution of a solvent across an osmotic membrane, but light scattering has been the more usual means of investigation from the time when Ewart *et al.* first showed that polystyrene 'binds' benzene from a mixed solvent of benzene and methanol.[129] The constant $\mathscr{K}$ in the light-scattering relation for a two-component system (equation 66) contains as a factor the square of the specific refractive index increment: *i.e.*

$$\mathscr{K} = \mathscr{K}'(\partial\tilde{n}/\partial c_2)_{p,T}^2 \tag{118}$$

where $\tilde{n}$ is the refractive index of the solution and c_2 is the polymer concentration c in the previous equation. The fact that the refractive increment can be measured independently (*e.g.* by differential refractometry) makes light scattering an absolute method for molecular weight determination. For a system with two solvents, the first equality of equation (66) is no longer valid. The correct expression can be written[128]

$$\frac{\mathscr{K}'c_2}{R(0,c)}\left[\left(\frac{\partial\tilde{n}}{\partial c_2}\right)_{p,T,\phi_3}\right]^2 = \frac{1}{\Omega^2 M} + O(c_2) \tag{119}$$

where M is the molecular weight of polymer and

$$\Omega = 1 + \frac{\lambda(\partial\tilde{n}/\partial\phi_3)_{p,T,c_2}}{(\partial\tilde{n}/\partial c_2)_{p,T,\phi_3}} \tag{120}$$

It follows from equations (119) and (120) that an analysis of light-scattering data on the three-component system that ignores the presence of two solvents and uses the conventional refractive increment for the polymer, measured at fixed ϕ_3, leads to an apparent molecular weight $M^\dagger = \Omega^2 M$, which depends on the preferential interaction and the refractive increments of solute and one (either) of the solvents. It turns out, however, that the light-scattering relation can be formulated, within inconsequential error, in terms of a refractive increment of the polymer defined by osmotic equilibrium[124,125]

$$\frac{\mathscr{K}'c_2}{R(0,c)}\left[\left(\frac{\partial\tilde{n}}{\partial c_2}\right)_{T,\mu_1,\mu_3}\right]^2 = \frac{1}{M} + O(c_2) \tag{121}$$

Since the refractive increment in equation (121) can be measured by comparing the refractive indices of the polymer solution and dialysate in equilibrium with it, the true molecular weight M is obtainable without ambiguity. Furthermore, by comparison of M and $M^\dagger$ the preferential interaction λ can be determined if the refractive increment $(\partial\tilde{n}/\partial\phi_3)_{p,T,c_2}$ is also measured.

Although preferential interactions have been studied for a long time, there does not seem to be a general framework for correlating the dependence of λ on composition and molecular weight. Some data suggest competitive equilibrium sorption of the two solvents by the polymer,[128] but such a scheme would not be applicable to well known instances where λ changes sign at some composition of the solvent mixture.[130] A recurrent suggestion is that λ should be correlated with the averaged density of polymer segments in the domain of the dissolved chain.[127,131,132]

3.4 THERMODYNAMICS: MODERATELY CONCENTRATED SOLUTIONS

The validity of the virial series expansion of the osmotic equation of state becomes questionable for moderately concentrated (or 'semidilute') solutions, in which the overall number density $N_A c/m_0$

of monomer units of molar weight m_0 is comparable to the number density $M/(R_G^0)^3 m_0$ of monomers in the domain of a single coil (the superscript zero denoting the infinite-dilution value of R_G). Thus the use of the virial expansion becomes problematic when the reduced concentration

$$\hat{c} \equiv N_A (R_G^0)^3 c/M \tag{122}$$

is of the order of unity. However, data on polystyrene and polyisobutylene solutions[32,33] with $A_2 = 0$ and $\hat{c} < 2$ can be fitted by simple quadratic formulas, as shown in Figure 5

$$\frac{\Pi M}{RTc} = 1 + A_3 M c^2 = 1 + \psi_3 \hat{c}^2 \tag{123}$$

or, in terms of the osmotic modulus, which is directly related to light scattering (Section 3.3.1)

$$\frac{K_{os}M}{RTc} = \frac{\mathscr{K} c M}{R(0,c)} = 1 + 3A_3 M c^2 = 1 + 3\psi_3 \hat{c}^2 \tag{124}$$

where $\psi_3 = A_3 M [M/N_A (R_G^0)^3]^2$. Hence, in this instance, contributions from higher-order virial terms are negligible. For polystyrene in cyclopentane, analysis of data using equation (124) to determine A_3 together with equation (79) gives $z_3 \approx 0.012$.[33]

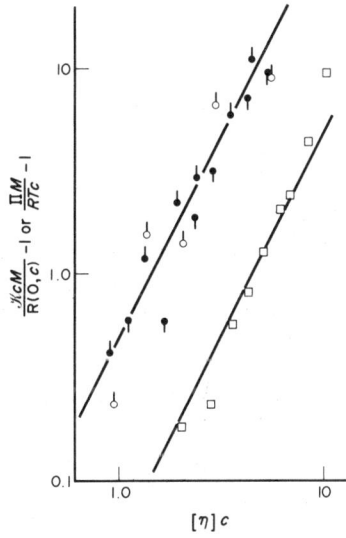

Figure 5 Bilogarithmic plot of $(\mathscr{K} c M/R(0,c) - 1)$ *vs.* $[\eta]c$ for two polystyrenes ($\bar{M}_w = 0.86 \times 10^6$, filled circles; $\bar{M}_w = 2.10 \times 10^6$, open circles) in cyclopentane at Θ_U (pips up) and Θ_L (pips down).[33] Squares represent osmotic data for polystyrene ($\bar{M}_n = 84 \times 10^3$) in benzene at Θ_U (24.5 °C).[32] The lines in the plot have a slope of two in conformity with the quadratic form of equation (124)

When A_2 is zero, R_G should be proportional to $M^{1/2}$ at all concentrations, and nearly equal to the unperturbed value. On the other hand, in a solution with A_2 large, R_G is greater than $R_{G,0}$ at small c and decreases with increasing c until it becomes equal to $R_{G,0}$ at some concentration c_s. Beyond c_s, R_G remains at $R_{G,0}$. Long ago, Flory[53] gave a convincing intuitive argument for the approach of α to unity at high polymer concentration; subsequent theory[133] and experiment[134,135] have provided ample confirmation. At infinite dilution, R_G is asymptotically proportional to M^ν, as discussed above. In the concentration range about $\hat{c} \approx 1$, however, $R_G/R_{G,0}$ is not expected to depend on chain length, even though R_G decreases with increasing c, so long as $R_G > R_{G,0}$. With these stipulations, the ratio α_c of R_G at concentration c to $R_{G,\Theta} \approx R_{G,0}$ may be expressed in the form[136]

$$\alpha_c^2 = \begin{cases} \alpha^2 [1 + (k_\alpha \hat{c})^2]^{-s/2} & c \leqslant c_s \\ 1 & c > c_s \end{cases} \tag{125}$$

where $s = (2\nu - 1)/(3\nu - 1)$; α is the limit of α_c at infinite dilution; $c_s \leqslant \rho$; and k_α is a constant. Estimates of α_c from viscosities[137] and from R_G determined by neutron scattering[135] are consistent with equations (125) with $k_\alpha \approx 6.8$. With this value, $k_\alpha \hat{c}$ is approximately equal to $[\eta]c$, where the intrinsic viscosity $[\eta]$ (see below, Section 3.6.1) depends implicitly on α (or $\hat{z}$). For example, the power s may be evaluated as an implicit function of α using equation (61). Typically, the ratio c_s/ρ is greater than 0.2, where ρ is the polymer density.

Renormalization group methods have been used to estimate α_c, with the result[103]

$$\alpha_c^2 = \alpha^2 \exp[-uZr(Z)] \tag{126}$$

where u is an interaction parameter that increases from zero for $\beta = 0$ to unity for large β, and

$$Z = 2\exp[-uF_0]A_2Mc \tag{127}$$

with $F_0 \approx 0.7$ and $r(Z) \approx 0.05(1 + Z)^{-1/2}$. A similar dependence of α_c/α on A_2Mc is given by an alternative treatment.[138]

The dependence of the osmotic pressure on concentration for solutions with A_2 large and $A_2Mc > 1$ reflects the variation of chain dimensions with c as well as the increased density of intermolecular interactions with increasing c. A simple scaling argument devised by des Cloizeaux[139] to represent these effects rests on the assumption that for $A_2Mc > 1$ (but $c < c_s$)

$$\Pi M/RTc \approx g(A_2Mc) = g(\psi_2 \hat{c}) \tag{128}$$

where $\psi_2 = A_2 M^2/N_A(R_G^0)^3$. The dependence on A_2Mc for A_2 large may be rationalized by the assertion that the higher virial coefficients A_j may each be represented in the form $A_j/A_2^j M =$ constant $(j = 3, 4, \dots)$. Then since ψ_2 is expected to be a constant for A_2 large, the condition that $\partial \Pi/\partial M \approx 0$ for $A_2Mc > 0$ gives the result $\partial \ln g/\partial \ln M = 1$; or with use of equation (122), the relation $\partial \ln g/\partial \ln \hat{c} = 1/(3v - 1)$. Thus if a power law $g \propto \hat{c}^{r-1}$ is assumed, it follows that $\Pi \propto c^r$, independent of M, with $r = 3(1 - s) = 3v/(3v - 1)$, so that $\Pi \propto c^{9/4}$ if $v = 3/5$.

Experimental data reveal power-law behavior for Π of the form expected, but r is often smaller than 9/4. It is interesting that an empirical power-law behavior for Π with $r = 2.27$ was reported nearly 40 years ago,[140] but the result was not then taken seriously, given the lack of a theoretical basis for such a seemingly peculiar result.[141]

A 'crossover' relation to express Π in the intermediate range of A_2Mc for large A_2 would obviously be desirable. One result, developed to reproduce the power-law form and virial expansion for $A_2Mc \gg 1$, gives convenient relations for the osmotic pressure and osmotic modulus[33]

$$\frac{\Pi M}{RTc} = 1 + A_2Mc\left(1 + \frac{\gamma_\infty}{p}A_2Mc\right)^p \tag{129}$$

$$\frac{K_{os}M}{RTc} \approx 1 + 2A_2Mc\left(1 + \frac{2\gamma_\infty}{p}A_2Mc\right)^p \tag{130}$$

Here, γ_∞ denotes the limiting value of A_3/A_2^2M for large A_2; and $p = (2 - 3v - \Delta)/(3v - 1 + \Delta)$, with $\Delta = \partial \ln(A_2M^2/R_G^3)/\partial \ln M$. It is expected that Δ will be nearly zero in a solution with A_2 large [for $\Delta = 0$, $p = r - 2 = (2 - 3v)/(3v - 1)$]. Comparisons of equations (129) and (130) with some experimental data[33,142,143] are given in Figure 6.

Renormalization group methods have been used to compute Π in the crossover regime, with results[103,144] that can be expressed in the form

$$\frac{\Pi M}{RTc} = 1 + \frac{Z}{2}\exp\{u[F_0 + F_1(Z)]\} \tag{131}$$

$$\frac{K_{os}M}{RTc} = Z \exp\{u[F_0 + F_2(Z)]\} \tag{132}$$

where u, F_0, and Z are as defined above; and in the range $1 < F_1 < 360$, $F_1 \approx 0.87(Z/2)^{0.87}$ and $F_2 \approx 0.87Z^{0.87}$. These relations tend to underestimate Π and K_{os} given in Figure 6. An alternative form for Π[138] gives estimates of $\Pi M/RTc$ and $K_{os}M/RTc$ as functions of A_2Mc, parametric in α (see Chapter 1 in this volume). The latter dependence appears to be stronger than that found experimentally.

A useful representation of Π or K_{os} in the crossover regime may be obtained by a type of generalized series expansion in powers of $\hat{c}$. It will be recalled that in the treatment of A_2 and A_3 with

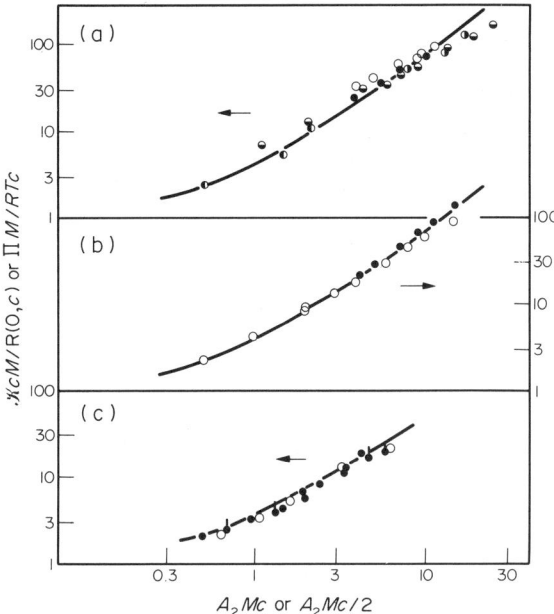

Figure 6 Bilogarithmic plots of $\mathscr{H}cM/R(0,c)$ (or $\Pi M/RTc$) vs. A_2Mc (or $A_2Mc/2$). (a) Light-scattering data for polystyrene (six polymers with $\bar{M}_w$ from 3.6×10^5 to 7.6×10^6) in benzene (a good solvent) at $15\,°C$.[33,142] The curve represents an empirical fit of the data to equation (130). (b) Light-scattering (open circles) and osmotic pressure (filled circles) data for poly(α-methylstyrene) in toluene (a good solvent).[143] The curve is an empirical fit of equations (129) and (130). (c) Light-scattering data for polystyrene; $\bar{M}_w = 8.62 \times 10^5$ (filled circles) and 1.50×10^6 (open circles) in cyclopentane at $40\,°C$ (no pips) and at $55\,°C$ (*i.e.* in an indifferent solvent not very far from Θ_U). The curve is an empirical fit to equation (130)

A_2 large, one of the approximations involved replacing $R_{G,0}$ by the expanded radius $R_G = R_{G,0}\alpha_G$ (see Section 3.3.3). If variation of R_G with concentration is considered, a further modification in the same spirit is to replace $R_{G,0}$ by $R_{G,0}\alpha_c$, where the concentration-dependent expansion factor α_c tends to unity as $\hat{c} \geqslant 1$. Thus, to order $\hat{c}^2$, one may express the osmotic modulus by the relation

$$K_{oS}M/RTc = (M/RT)\,(\partial\Pi/\partial c) = 1 + 2\psi_2(\hat{z}/\alpha_c^3)\,(\alpha_c/\alpha)^3\,\hat{c} + 3\psi_3(\hat{z}/\alpha_c^3, z_3)\,(\alpha_c/\alpha)^6\,\hat{c}^2 \tag{133}$$

Here, the functions ψ_2 and ψ_3 reduce to $\psi_2(\hat{z}/\alpha^3) = A_2M^2/N_A(R_G^0)^3$ and $\psi_3(\hat{z}/\alpha^3, z_3) = A_3M^3/N_A^2(R_G^0)^6$ respectively, as c tends to zero; and $\hat{z} = z + 4z_3n^{1/2}$ is an interaction parameter that vanishes when $A_2 = 0$ (see equation 77). In a good solvent (assuming z_3 to be small), we have

$$\psi_3(\hat{z}/\alpha^3, z_3) \approx (\kappa/3\pi^{3/2})\psi_2^3(\hat{z}/\alpha^3) + (4\pi)^3z_3 \tag{134}$$

where κ, defined in equation (105), is a function of $\hat{z}/\alpha^3$, with a limiting value κ_∞ for large $\hat{z}$. Moreover, $\psi_2(\hat{z}/\alpha^3)$ tends to the limiting value $\psi_2(\infty)$ for large $\hat{z}$. If the same form obtains in a moderately concentrated solution so that α can be replaced by α_c, we have in the limit of $\hat{z}$ and $\hat{c}$ large (but $c < c_s$)

$$K_{oS}M/RTc \approx \{(\kappa_\infty/3\pi^{3/2})\,\psi_2^3(\infty) + (4\pi)^3z_3\}\,(k_\alpha\hat{c})^{-3s}\hat{c}^2 \tag{135}$$

For $c > c_s$, $(k_\alpha\hat{c})^{-3s}$ is replaced by α^{-6}. Equation (135) reproduces the expected result that the osmotic pressure is independent of molecular weight in moderately concentrated solutions. This behavior has been cited above in scaling arguments that give a proportionality of $\Pi M/RTc$ with $\hat{c}^{(2-3s)}$. If $\nu = 3/5$, equation (129) gives $\Pi M/RTc \propto \hat{c}^{5/4}$, the result originally obtained by des Cloizeaux.[7]

The crossover behavior for small to intermediate $\hat{z}$ is represented by equation (133). Equation (61) may be used to represent α in terms of $\hat{z}$, and equation (97), with α_G replaced by α_c, is used for $\psi_2(\hat{z}/\alpha_c^3)$. With equation (105), we have

$$\kappa = [1 + k_3\psi_2(\hat{z}/\alpha_c^3)]^{-3/2} \tag{136}$$

where k_3 is nearly zero. Data on $K_{oS}M/RTc$ measured for polystyrene under conditions including a range of $A_2M^2/N_AR_G^3$ shown in Figure 6 may be fitted in this way.[136]

For the case with A_2 large, the above relations for Π or K_{OS} are expressed in terms of $A_2Mc = (A_2M^2/N_A R_G^3)\hat{c}$. Correlations of this type become suspect as $\hat{c}$ becomes large. The upper limit of c for the use of expressions such as equations (129) and (130) is uncertain; but the c^2 dependence expressed by equation (135) for large c must become inadequate as c approaches the density ρ of the neat polymer liquid. However, the high concentration range, very roughly defined by $c > c_s$, is the proper regime for the concepts underlying the venerable Flory–Huggins theory, which is discussed in the following section, to have some validity.

3.5 THERMODYNAMICS: CONCENTRATED SOLUTIONS

3.5.1 Lattice Treatments

3.5.1.1 The Flory–Huggins theory[120]

The extraordinary deviations from thermodynamic ideality exhibited by polymer solutions were addressed in the early 1940s by the celebrated liquid-lattice theory associated with the names of Flory[145] and Huggins[146] (Miller[147,148] must also be credited with an independent development). The theory is basically very simple, especially in the form due to Flory. The system is visualized as an array of N identical lattice cells with coordination number z. A cell may contain either a solvent molecule or a polymer chain segment; thus the segment is defined and the 'degree of polymerization' r is the ratio of the molecular volumes of polymer and solvent. In the combinatorial calculation the number of conformations (singly connected sequences of sites) is first calculated for a single chain on the empty lattice, ignoring self-intersections of indirectly connected segments. Then more polymer chains are added to the lattice one by one, and the number of conformations allowed to each chain is counted, taking account of loss of conformations due to occupancy of cells by molecules previously added. The computation is rendered tractable by representing the previously added molecules as a uniform distribution of chain segments. This approximation is a characteristic mean field maneuver which obliterates the detailed structure of the system. In this way, the total number of conformations Ω is counted for the complete system containing N_2 polymer chains. The remaining $N_1 = (N - rN_2)$ cells are filled with solvent. Then the combinatorial entropy of the system can be obtained as $k \ln \Omega$. However, what is really wanted is the entropy change of mixing of amorphous (disoriented) polymer with solvent. Thus the combinatory calculation is repeated for the N_2 polymer chains on a liquid lattice of rN_2 cells to find the number of arrangements Ω_0 of the neat polymer, and the entropy difference is calculated as $k \ln(\Omega/\Omega_0)$.

The most familiar presentation is given in Flory's 1953 book,[120] but in the 1970 Speirs lecture[149] Flory expressed the combinatorial result in a particularly illuminating form

$$\Omega = [(z - 1)^{(r-1)N_2}] \left[\frac{r}{e^{r-1}} \right]^{N_2} \left[\left(\frac{N}{N_1} \right)^{N_1} \left(\frac{N}{rN_2} \right)^{N_2} \right] \tag{137}$$

as a product of three independent quantities. The first factor in brackets enumerates the internal conformations of the polymer chains in the absence of intermolecular interferences. The second, intermolecular, factor effects a correction for the large proportion of conformations that are disallowed when undiluted chains are brought together on the lattice and compete for lattice sites. The third factor expresses the effect of dilution with solvent in increasing the total space available to each polymer chain, thus enhancing its conformational freedom.

It is important to observe that only the first term in equation (137) depends on chain architecture and the lattice coordination number, and that both it and the second term cancel out of the entropy of mixing, leaving only the third term equal to Ω/Ω_0. Consequently the theory would apply alike to solutions of flexible and stiff-chain polymers (except that above a certain concentration the latter would respond to packing constraints and minimize the free energy by separation of an ordered phase). The factorization of the combinatorial factors in equation (137) is the fundamental reason why the lattice calculation works at all, despite the extreme artificiality of picturing the chain as fitting a sequence of regular lattice sites with a definite coordination number. These aspects of the model simply disappear in the final result. The independence of the intermolecular factor also implies that the chain conformation should be independent of dilution: R_G should be the same in pure liquid polymer as in solution. Naturally this rationale would not hold for dilute solutions, for which the intermolecular factor in equation (137) is not valid.

Following equation (137), the combinatory entropy of mixing can be written in the very simple and familiar form

$$\Delta S_m(\text{combinatory}) = -k(N_1 \ln \phi_1 + N_2 \ln \phi_2) \tag{138}$$

where $\phi_1 = N_1/(N_1 + rN_2)$ and $\phi_2 = 1 - \phi_1$ are respectively the volume fractions of solvent and polymer. Obviously, the equivalence of polymer and solvent cells on the lattice implies that there is no volume change of mixing in the model.

To complete the Flory–Huggins treatment, noncombinatory effects of mixing are included in *ad hoc* fashion. It is supposed that mixing involves breaking polymer–polymer and solvent–solvent contacts or 'bonds' and forming the corresponding number of polymer–solvent contacts. Since $z - 2$ adjacent cells are available to a chain segment to form intermolecular contacts and the probability that any one of these cells is occupied by a solvent molecule is ϕ_1, the total number of heterocontacts per chain is $(z - 2)r\phi_1$. Then with $rN_2/N_1 = \phi_2/\phi_1$, the enthalpy of mixing arising from these contacts is written

$$\Delta H_m = (z - 2)wN_1\phi_2 = kT\chi N_1\phi_2 \tag{139}$$

where χ is an 'exchange' parameter, w being the energy change per polymer–solvent contact formed. (This χ is not to be confused with the function χ_{ij} defined by equation 26.)

Finally, the free energy of mixing is the sum of combinatory and noncombinatory contributions

$$\Delta G_m/kT = N_1 \ln \phi_1 + N_2 \ln \phi_2 + \chi N_1\phi_2 \tag{140}$$

or for a solution occupying one mole of lattice cells

$$\Delta G_m/RT = \phi_1 \ln \phi_1 + (\phi_2/m) \ln \phi_2 + \chi\phi_1\phi_2 \tag{141}$$

It is evident that for components of the same molar volume ($r = 1$) these relations revert to those for a regular solution.[150] An immediate generalization is to modify the added van Laar enthalpy term to include any local noncombinatory entropic contributions, which are counted in exactly the same way as the contact exchange energies. Thus it is useful to write χ as a sum of entropic and enthalpic free energy terms

$$\chi = \chi_H + \chi_S \tag{142}$$

By definition, a purely enthalpic χ would be inversely proportional to the absolute temperature; but, in general, the expressions

$$\chi_S = d(\chi T)/dT; \quad \chi_H = -Td\chi/dT \tag{143}$$

follow from the Gibbs–Helmholtz relation.

Partial differentiation of equation (140) with respect to N_1 at fixed p, T, N_2, gives the chemical potential μ_1 (per mol) of the solvent

$$(\mu_1 - \mu_1^0)/RT = \ln a_1 = \ln \phi_1 + (1 - r^{-1})\phi_2 + \chi\phi_2^2 \tag{144}$$

Differentiation with respect to N_2 gives an analogous equation for the chemical potential of the polymer.

From equation (144), the general osmotic pressure relation $\Pi V_1 = -RT \ln a_1$, and the fact that $\phi_2/rV_1 = c/M$, the virial expansion for a Flory–Huggins binary system is easily shown to be given by

$$\frac{\Pi}{RTc} = \frac{1}{M} + \frac{\bar{v}^2}{V_1}\left(\frac{1}{2} - \chi\right)c + \sum_j \frac{\bar{v}^j}{jV_1}c^j \tag{145}$$

where $\bar{v}$ is the (partial) specific volume of the polymer and $j = 2, 3, \ldots$.

Although equations (144) and (145) have sometimes been used uncritically to analyze the behavior of dilute polymer solutions, it was pointed out from the beginning that the Flory–Huggins model cannot be expected to apply to a dilute system.[98,120] One of the simplifications in the combinatorial analysis, the successive addition of polymer chains to a lattice containing a uniform distribution of chain segments, seems obviously more realistic for systems sufficiently concentrated for the chains to interpenetrate extensively than for very dilute solutions with the chains, on the average, occupying isolated domains separated by expanses of pure solvent. Although equation (145) is not an adequate representation of the behavior of a dilute polymer solution, it is consistent with the thermodynamic requirement that the solvent conform to Raoult's law as the solute concentration vanishes, and the limiting behavior near the theta temperature is correct. Essentially, the molecular weight invariant second virial coefficient in equation (145) refers to a system of disconnected chain segments. Thus it can be identified with the factor $A_2^{(L)}$ in equation (80). The theta condition, where the second virial coefficient and the excluded volume integral β vanish (neglecting β_3), corresponds to $\chi = 1/2$; in good-solvent systems χ is less than $1/2$. The higher virial coefficients in equation (145) are patently meaningless from the fact that they contain nothing relating to interactions in molecular clusters.

In a commonly used notation originated by Flory, the temperature dependence of the interaction parameter is given by

$$\frac{1}{2} - \chi = \psi_1\left(1 - \frac{\Theta}{T}\right) \tag{146}$$

so that, from equation (113)

$$\frac{1}{2} - \chi = 4\pi^{3/2}N_A\frac{V_1}{\bar{v}^2}B\left(1 - \frac{\Theta}{T}\right) = \frac{A_2^{(L)}V_1}{\bar{v}^2} \tag{147}$$

Now the excluded volume variable z can be written in terms of experimental quantities in the form

$$z = \frac{2}{N_A}\left(\frac{4\pi R_{G,0}^2}{M}\right)^{-3/2}\frac{\bar{v}^2}{V_1}\psi_1\left(1 - \frac{\Theta}{T}\right)M^{1/2} \tag{148}$$

which is equivalent to equation (115). The expressions in equations (146), (147) and (148) apply alike to the upper-critical and lower-critical regions. To associate the discussion specifically with one or the other, we add subsripts U or L (in keeping with Section 3.3.4) to the parameters Θ, B and ψ_1.

3.5.1.2   *Concentration dependence of the interaction parameter*

Experimental data on polymer solutions in good solvents have long been available to show that the simple ideas motivating the original Flory–Huggins theory are not adequate, even for the concentration range where the condition of uniform segment density is reasonably approximated. Empirically, the exchange parameter χ (*e.g.* as obtained at high polymer concentrations from measurements of vapor pressure lowering) is often far from independent of concentration; it may either decrease or increase, sometimes to values well above $1/2$, with increasing concentration. In systems where χ is found to be nearly independent of concentration, resolution of χ into its entropic and enthalpic components (by analysis of its temperature dependence) may show them individually to have large concentration dependences that tend to cancel. Moreover, the entropic term χ_S, which was originally supposed to be a small negative correction to a positive van Laar heat term (as it is for typical small-molecule regular solutions), usually turns out to be a dominant positive contribution.

There have been a number of attempts to improve on the original lattice theory by modification of the exchange term in the free energy to introduce a concentration dependence. The χ in equation (140) can be replaced by a concentration-dependent noncombinatory free energy parameter $g(\phi_2, T)$, whence the χ in equation (141) becomes

$$\chi(\phi_2, T) = g(\phi_2, T) - (1 - \phi_2)(\partial g/\partial \phi_2)_T \tag{149}$$

The function $g(\phi_2, T)$ has been represented in various ways as a series in powers of ϕ_2 or as a semiempirical closed form. While some justification may be cited for a chosen functional form, the concomitant introduction of adjustable parameters usually insures that agreement with experiment will be bettered.[151] Perhaps the most interesting effort of this kind is one in which Koningsveld *et al.*[152] devised a continuous free energy function to cover the entire concentration range while preserving the correct behavior for dilute solutions. They let $g(\phi_2, T)$ be given by a linear combination of functions g_{conc} and g_{dil} of temperature and concentration for the limiting cases, respectively, of concentrated and dilute solutions

$$g(\phi_2, T) = g_{conc} + P(g_{dil} - g_{conc}) \tag{150}$$

The parameter P is reasonably identified with the probability that the domain of a chain does not contain segments of the $N - 1$ other chains in the system. This probability is given approximately by $\exp(-\lambda\phi_2)$, where λ^{-1} is the value of ϕ_2 at which the N_2 coil domains, measured by R_G, would equal the total volume of the system if there were no overlap (a criterion essentially equivalent to making the reduced concentration $\hat{c}$ in Section 3.4 of order unity). Manipulations involving equations (80), (148) and (149) with equation (150) lead to an expression for $g(\phi_2, T)$ as a function of concentration, temperature, and molecular weight:

$$g(\phi_2, T) = g_{conc} + \psi_{1U}\left(1 - \frac{\Theta_U}{T}\right)[1 - F(z)]\frac{\exp(-\lambda\phi_2)}{1 + \lambda} \tag{151}$$

For g_{conc}, a semiempirical form to fit concentrated-solution data is used; realistic values of physical constants indicate that $\lambda \approx 0.5r^{1/2}$, and $F(z)$ is taken as one of the approximate closed-form expressions discussed earlier. Since the parameter λ can be determined objectively, though approximately, equation (151) has the virtue of conforming to the correct dilute-solution behavior at the limit — without introduction of new adjustable parameters.

An expression for the osmotic pressure that is useful at high polymer concentration (above c_s, as indicated in the previous section) is obtained directly from equation (144) if the logarithmic term is not expanded and if, for generality, χ is allowed to be concentration dependent:[2,149]

$$\frac{\Pi M}{RTc} = 1 - r\left[1 + \frac{\ln(1 - \phi_2)}{\phi_2} + \phi_2\chi(\phi_2)\right] \tag{152}$$

where

$$\chi(\phi_2) = \sum_{j=1} \chi_j\phi_2^{j-1} \tag{153}$$

The corresponding relation for the osmotic modulus is

$$\frac{K_{os}M}{RTc} = 1 + r\phi_2\left[\frac{1}{1 - \phi_2} - 2\chi(\phi_2) - \phi_2\frac{\partial\chi(\phi_2)}{\partial\phi_2}\right] \tag{154}$$

These forms have been found to give a good account of many osmotic pressure data over a wide range of concentration for $c > c_s$.[153,154,155] Often, only two terms are needed to represent Π, in keeping with the quadratic form in equation (135). Of course, in fitting data the χ_j coefficients are taken as dependent only on temperature. The success of this correlation implies that the Flory–Huggins combinatory entropy gives an adequate representation of reality at high concentrations, but it is unrevealing as to the form of the noncombinatory free energy.

3.5.2 Phase Separation[156,157]

As has been noted above, the familiar theta temperature Θ_U is associated with an upper-critical phase separation of a noncrystallizing polymer. The enthalpy and entropy of mixing are both positive (ψ_{1U} is positive); and below the critical temperature (the upper-critical solution temperature, UCST), the positive noncombinatory free energy $RT\chi\phi_2^2$ dominates the negative combinatory term. Thus there is a range of concentrations where the single-phase system is unstable relative to two liquid phases (two solutions) with different concentrations. Above the UCST the combinatory free energy of mixing wins out at all concentrations and there is only one phase. Application of the general thermodynamic requirements for a critical point in a binary mixture, *i.e.* that both $(\partial\mu_1/\partial\phi_1)_{p,T}$ and $(\partial^2\mu_1/\partial\phi_1^2)_{p,T}$ vanish, to the Flory–Huggins free energy function leads to two relations

$$1 - (1 - r^{-1} + 2\chi\phi_2)(1 - \phi_2) = 0 \tag{155}$$

$$1 - 2\chi(1 - \phi_2)^2 = 0 \tag{156}$$

which can be solved for the critical composition

$$\phi_{2U} = 1/(1 + r^{1/2}) \tag{157}$$

and, with equation (146), for the critical temperature T_U

$$\frac{1}{T_U} = \frac{1}{\Theta_U} + \frac{1}{\psi_{1U}\Theta_U}\left(\frac{1}{r^{1/2}} + \frac{1}{2r}\right) \tag{158}$$

Equation (158) indicates that a plot of T_U^{-1} for a series of polymers of different molecular weights against the expression in parentheses on the right hand side should be linear with positive intercept and slope, Θ_U^{-1} and $(\psi_{1U}\Theta_U)^{-1}$ respectively. Hence, the theta temperature Θ_U is identified with the limit of T_U at infinite molecular weight.[120,158] From studies of many polymer + solvent systems, it has been amply verified that experimental plots are linear and do extrapolate correctly to the temperature at which the second virial coefficient vanishes.[120] It has also been shown that the limiting behavior of equation (155) does not depend on the special form of the Flory–Huggins free energy though it does imply the validity of a virial expansion near the critical point. Despite the agreement of the limiting T_U with Θ_U, the entropy of mixing parameter ψ_{1U} derived from critical point data *via* equation (158) is usually larger than the value obtained unambiguously from the temperature dependence of the second virial coefficient at $T = \Theta_U$. This discrepancy is not surprising since interpretation of the limiting slope of T_U^{-1} *vs.* $r^{-1/2}$ as $(\psi_{1U}\Theta_U)^{-1}$ does depend on the incorrect third virial coefficient given by the Flory–Huggins model.[29]

The mapping of the critical points as a function of molecular weight given by equations (157) and (158) indicates the leading qualitative features of Flory–Huggins upper-critical phase diagrams. As the molecular weight increases in a homologous series, the critical points move to more dilute solutions and to higher temperatures, until at infinite molecular weight the critical point is at infinite dilution. Thus the solubility gap becomes more skewed toward low concentration at high molecular weight.

Once the parameters Θ_U and ψ_{1U} (or B_U) have been established, either from dilute solution measurements on a system or from the observed molecular weight dependence of critical points, the coexistence curve, or binodal, can be calculated from the thermodynamic requirement of uniformity of the chemical potential, derived from equation (141), of each component throughout the system. In Figure 7, calculated binodals are compared with the much cited phase separation data of Shultz and Flory[156,159] for the system polyisobutylene + diisobutyl ketone. The Flory–Huggins binodals are seen to give a qualitative picture of the phase separation, but they make the solubility gap much too narrow and the critical point tends to be at too low a concentration. Other systems exhibit comparable behavior. Figure 7 also shows the calculated spinodals (the metastability limits of a single phase); *i.e.* the loci of points for which $(\partial \mu_1/\partial \phi_2)_{p,T} = 0$, which is given directly by the quadratic relation in equation (156). More recent experimental studies on this system show, not unexpectedly, that the Flory–Huggins spinodals are also much too narrow.[160] Since the spinodal curve touches the binodal at the critical point, where the two are tangential, any displacement of the critical point is shared by the spinodal.

Proposals like those mentioned above to generalize the interaction parameter χ have usually been motivated by the goal of quantitatively improving the theoretical depiction of upper-critical phase

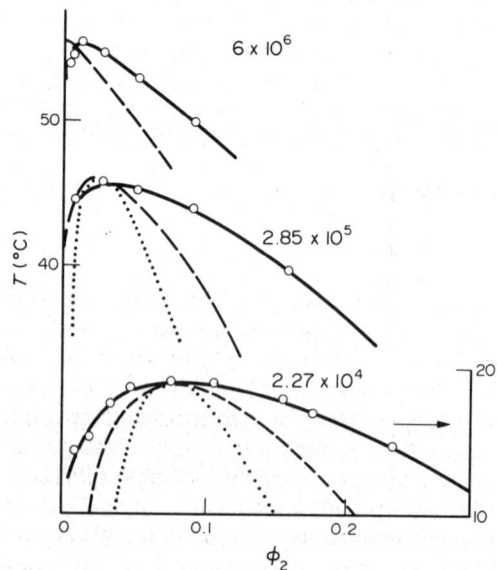

Figure 7 Precipitation temperatures (cloud points) *vs.* composition for polyisobutylene fractions with the indicated number average molecular weights in diisobutyl ketone.[159] Dashed curves are Flory–Huggins binodals with parameters Θ_U and ψ_{1U} determined from a plot of the measured critical points according to equation (158). The dotted curves are the corresponding Flory–Huggins spinodals (reproduced by permission of Wiley, from *J. Polym. Sci., Polym. Symp.*, 1976, **54**, 62)

diagrams. Koningsveld *et al.*[152] showed that equation (151) gives much better upper-critical spinodals than the original Flory–Huggins free energy function with a constant χ. Theoretical spinodals are far easier to calculate than binodals but, except for the facility of predicting the critical point from the maximum in the spinodal, there was less interest in them until the comparatively recent development of techniques for accurate experimental determinations (see Chapter 4 of this volume).[160]

Mean field calculations[161] suggest that coexistence curves of binary polymer solutions should obey a corresponding states principle near a critical point (ϕ_{2c}, T_c), such that $|\phi_2 - \phi_{2c}|/\phi_{2c}$ is a universal function of εr^b where $\varepsilon \equiv |T - T_c|/T_c$ and b is a universal exponent. However, the symmetry of this relation is evidently belied by experimental binodals, except perhaps for values of ε too small for experimental investigation. The discrepancy raises an interesting question as to whether the experimental data could satisfy a corresponding states principle expressed in terms of a new concentration variable, a 'symmetrical fraction' ψ that would make the binodals symmetrical about the critical 'concentration' ψ_c. Noting that volume fractions are related to other concentration variables defined on the interval [0, 1] by expressions of the form $\psi = \phi_2/[\phi_2 + R_c\phi_1]$, Sanchez[162] has defined the 'symmetrical fraction' ψ by taking R_c as an adjustable parameter to render any binodal T *vs.* ψ symmetrical about ψ_c. Upper-critical $(T_c = T_U)$ binodal data[163] for ten samples of polystyrene (with $1 < M \times 10^{-4} < 72$) in methylcyclohexane are found to conform very well to a scaling law

$$|\psi - \psi_c| = \psi_0(\varepsilon r^{0.313})^\beta \qquad \varepsilon = (T_U - T)/T_U \qquad (159)$$

where $\beta = 0.327$, and ψ_c and ψ_0 are constants independent of the degree of polymerization r. Thus $\varepsilon r^{0.313}$ *vs.* ψ defines a common coexistence curve. The experimental value of the critical exponent agrees well with renormalization group theoretical calculations for the Ising model.[164, 165] The exponent 0.102 of r may be universal, and it is much smaller than the value $1/4$ given by a mean field (Flory–Huggins) calculation.[162] The free parameter R_c used to symmetrize the data depends strongly on molecular weight, but any physical significance it might have is obscure.

Although phase separation studies on polymer solutions have mostly been concerned with the upper-critical separation and the UCST, the occurrence of the solubility gap widening above a lower-critical solution temperature (LCST) has been known for a long time. It was first reported for polar systems, in which miscibility could plausibly be attributed to formation of hydrogen-bonded complexes that dissociate as the temperature is increased. In such systems the LCST may be near room temperature and thus easily accessible. However, the LCST was eventually found in typical nonpolar systems at much higher temperatures, even in the neighborhood of the gas–liquid critical point of the solvent.[166] Despite the considerable difficulty of experimental work at temperatures well above the normal boiling point of the solvent, and such that incipient thermal degradation might be a severe limitation, enough systems have been studied to establish the prevalence of the LCST separation; and in a few favorable cases, such as polystyrene + cyclopentane,[167] polystyrene + acetone[168] at very low molecular weight, and polystyrene + cyclohexane,[169] it has been possible to study phase diagrams in detail in both the UCST and LCST regions. Representative data on the last system are shown in Figure 8.

The UCST and LCST diagrams display a roughly mirror-symmetric disposition: with increasing molecular weight, the LCST moves downward in temperature and toward lower concentration as the solubility gap widens, while the UCST moves upward. (In the extreme case of polystyrene + acetone above molecular weight 20 000, the UCST and LCST binodals merge, the phase diagram takes on an hour-glass shape, and there is no longer a region of complete miscibility.) Just as the UCST approaches Θ_U at the infinite molecular weight limit, so the LCST approaches Θ_L. While the enthalpy and entropy of mixing are both positive near the UCST, they are both negative near the LCST. The unexpectedly large and positive experimental entropy parameter χ_S for typical polymer systems in the UCST region has already been mentioned; furthermore, its temperature coefficient is large and positive. This behavior is an indication of a large *increase* in local order in the formation of unlike intermolecular contacts and of a trend making this effect more important as temperature is raised toward the LCST. Thus behavior near Θ_U already evidences trends that result in the appearance of the LCST separation at much higher temperature.

Equations (155) and (156) describe Flory–Huggins binodals for the LCST region just as they do for the upper-critical separation. The form of equation (158) with critical temperature T_L and a negative entropy parameter ψ_{1L} indicates the proper extrapolation to infinite chain length to obtain Θ_L. From the comparatively limited experimental information that is available, it appears that Flory–Huggins LCST binodals and spinodals are deficient in analogous fashion to the UCST functions.

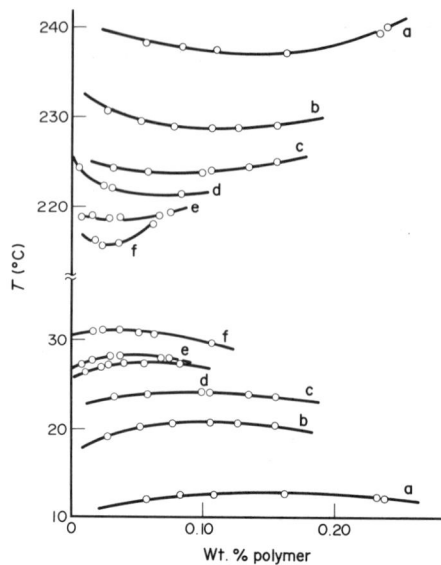

Figure 8 Cloud point curves from Saeki *et al.*[180] for a series of polystyrene fractions in cyclohexane showing both UCST and LCST regions. Molecular weights range from 37.0×10^3 (a) to 2.70×10^6 (f) (reproduced by permission of Wiley, from *J. Polym. Sci., Polym. Symp.*, 1976, **54**, 66)

By the 1960s, the growing body of experimental data provided compelling evidence that the lower-critical separation is a universal phenomenon of solutions of long-chain molecules in low molecular weight solvents. All systems exhibiting the familiar UCST will also have the LCST at some higher temperature, as will systems in very good solvents, which may have no UCST at any possible temperature for existence of a liquid phase. Theoretical support for this proposition is discussed below (Section 3.5.3.1).

The decreasing solubility of chemically homologous polymer species with increasing molecular weight, as illustrated by the phase diagrams in Figures 7 and 8, provides the basis for the standard techniques of fractionation of noncrystallizing polymers by incremental precipitation from a poor-solvent medium. For decades, before modern chromatographic methods had been fully developed, the great practical importance of precipitation fractionation focussed attention on the phase relationships of polymers with a molecular weight distribution dissolved in one or several solvents, and was undoubtedly a major impetus in the production of the voluminous literature on this topic. The effects of the form and breadth of the molecular weight distribution, the possibility of variation of polymer composition in a phase, and (in some cases) the occurrence of multiphase equilibria, make any complete description extremely complex. Despite its shortcomings, most studies have been based on the Flory–Huggins free energy function, usually (but not always) with concentration-independent interaction parameter(s) χ. Efforts to go beyond the basic ideas explored by Flory[156] and by Tompa, in a classic monograph,[2] were typically implemented by large-scale iterative computation.[170] However, in more recent years, careful study of the analytical properties of the Flory–Huggins model for this kind of multicomponent system have yielded important insights into the behavior of binodals and spinodals, and the associated critical points.[171–173] Analysis of experimental data is complicated by the circumstance that precipitation curves ('cloud point' curves obtained from turbidity measurements), as in Figures 7 and 8, are not equivalent to binodals if polydispersity is significant, *i.e.* horizontal lines connecting the two branches of a curve no longer represent coexisting phase compositions, and the maximum in the curve is not a critical point. For polymer systems that are chemically heterogeneous, the analysis of phase equilibrium is a still more formidable challenge, and only limited conclusions have been reported.[174]

3.5.3 Variants of the Lattice Theory

3.5.3.1 *Liquid state theory*

Like the two-parameter formalism, the Flory–Huggins theory contains nothing specific in its structure to account for the temperature dependence of the free energy of mixing that leads to the appearance of the lower-critical separation. Eichinger's treatment of $A_2^{(L)}$ will of course accommodate lower-critical phase diagrams in a reasonable phenomenological context, to the extent that $(1/2 - \chi)$ can be represented by the concentration-independent function $A_2^{(L)} V_1/\bar{v}^2$, but it too says nothing about the physics of the situation. In view of the accumulation of experimental facts, there was obvious need for a theoretical framework that could account for those facts in terms of the long-chain character of a polymer solute. It was also clear that a theoretical treatment with more specific physical content would be necessary if it were to have any predictive value concerning the transition from the upper-critical to the lower-critical regime and also account for the apparently system-specific dependences of χ on concentration.

The so-called liquid state theories, developed first by the 'Brussels School' of Prigogine and his associates[175] and later by Flory[149] meet these requirements. They may be regarded as combinations of the simple Flory–Huggins combinatorics with treatments of the exchange interactions in a fashion deriving from the old cell theory of liquids. Physical parameters of a pure liquid component, derived experimentally or theoretically, are incorporated naturally into a corresponding-states formalism. A solution is assumed to behave like a pure liquid with the corresponding-states reduction parameters determined by averages of the pure-component properties. The distinctly new idea introduced in the Prigogine theory is that of a 'free-volume' difference between solvent and polymer that arises directly from the catenation of the latter.[175-178] In the polymer, chain segments have lost some of the external degrees of freedom of homologous small molecules in a liquid. Consequently a polymer liquid is hindered in thermal expansion as compared with the small-molecule liquid; and as a polymer solution is heated, an increasing dissimilarity in the expansivities of the components results in an increasingly unfavorable noncombinatory entropy of mixing. Eventually, a solubility gap appears at the LCST. The observed decrease in the LCST with increasing molecular weight is consistent with this picture.

It is instructive to consider the approximate partition function for a single-component liquid system of N molecules that is the starting point of Flory's version of the theory

$$Z = Z_{\text{comb}} \, g^{rNc} [v^{1/3} - (v^*)^{1/3}]^{3rNc} \exp(rNs\eta/2vkT) \tag{160}$$

Here v is the volume of a segment and v^* is its 'hard-core' volume. The segment size and the degree of polymerization r are fixed as in the original Flory–Huggins model, except that a polymer segment is assigned the same hard-core volume as a solvent molecule and r is the ratio of hard-core molecular volumes. The parameter g is a geometrical factor, and c is Prigogine's parameter for the mean number of external degrees of freedom per segment. The 'free-volume' function in brackets is rigorous as a difference in lengths for a one-dimensional fluid, but here it is cubed as an approximation for a three-dimensional system. A segment has s contact sites for interaction with its neighbors, and η is an energy parameter for contact of two sites. Flory avoids specifying a potential energy function by simply introducing an interaction energy per segment $-s\eta/2v$ of the van der Waals form in the Boltzmann factor. This procedure can be defended on theoretical and empirical grounds and is preferable to accepting the unrealistic energy prescribed by the cell theory.[149] The factor Z_{comb} in equation (160) is just the combinatory partition function that leads to the Flory–Huggins entropy (except for the refinement that volume fractions ϕ_i are now defined precisely as segment fractions based on hard-core volumes). Thus the liquid state theory preserves the significant approximations of the original lattice combinatory theory. Save for the all-important parameter c, the partition function in equation (160) is one proposed long ago by Eyring and Hirschfelder.[179]

The partition function leads to a simple reduced equation of state

$$\frac{\tilde{p}\tilde{v}}{\tilde{T}} = \frac{\tilde{v}^{1/3}}{\tilde{v}^{1/3} - 1} - \frac{1}{\tilde{v}\tilde{T}} \tag{161}$$

where

$$\tilde{v} = v/v^* \qquad \tilde{T} = T/T^* = 2v^* ckT/s\eta \qquad \tilde{p} = p/p^* = 2p(v^*)^2/s\eta \tag{162}$$

The reduction parameters p^*, T^* and v^* for pure liquids can be deduced from measurements of density, thermal expansivity $(\partial \ln v/\partial T)_p$, and the thermal pressure coefficient $(\partial p/\partial T)_v$. Introduction of the parameter c means that all three of the reduction parameters, rather than only two, can be specified independently inasmuch as they are related by $p^*v^* = ckT^*$. The theory has been found to give reasonable fits to the isotherms of a number of fluids, both polymeric and nonpolymeric. Its deficiencies show up, however, in some drift of the reduction parameters with temperature.

Application of the partition function to polymer solutions is uncomplicated: it is assumed that the core volumes of the components are additive and that the intermolecular interaction energy is proportional to the contact surface area. With these assumptions, the mixture is treated as a single fluid with its effective $1/r$, s and c linear functions of the composition in terms of segment fractions. Mixing is assumed to be random, so that the probability that a segment of kind i will be found adjacent to any given site is equal to the site fraction θ_i; for example

$$\theta_2 \;=\; 1 \;-\; \theta_1 \;=\; \phi_2 s_2/(\phi_1 s_1 \;+\; \phi_2 s_2) \;=\; \phi_2 s_2/s \tag{163}$$

Here and in what follows, the quantities without subscripts refer to the mixture; and, as before, solvent and polymer are designated by subscripts 1 and 2. These prescriptions lead to

$$p^* \;=\; \phi_1 p_1^* \;+\; \phi_2 p_2^* \;-\; \phi_1 \theta_2 X_{12} \tag{164}$$

$$T^* \;=\; p^* \left(\frac{\phi_1 p_1^*}{T_1^*} \;+\; \frac{\phi_2 p_2^*}{T_2^*} \right)^{-1} \tag{165}$$

where X_{12} is an exchange parameter

$$X_{12} \;=\; [s_1/2(v^*)^2]\,[\eta_{11} \;+\; \eta_{22} \;-\; 2\eta_{12}] \tag{166}$$

the η_{ij} characterizing contact energies for the three kinds of contacts. The parameter X_{12} is strictly an energy density; it contains no entropy contribution. The (internal) energy of mixing is then $Nrv^*(\phi_1\theta_2/\tilde{v})X_{12}$.

The other thermodynamic properties follow in standard fashion from the partition function. In particular, unlike the original lattice treatment, the theory accommodates a nonvanishing excess volume V^E, the volume change on mixing

$$V^E/V \;=\; 1 \;-\; (\phi_1 \tilde{v}_1 \;+\; \phi_2 \tilde{v}_2)/\tilde{v} \tag{167}$$

The residual chemical potential μ_1^R of the solvent is

$$\mu_1^R \;\equiv\; \mu_1 \;-\; \mu_1^0 \;-\; RT[\ln(1 \;-\; \phi_2) \;+\; (1 \;-\; 1/r)\phi_2] \;=\; p_1^* V_1^* \left[3\tilde{T}_1 \ln\!\left(\frac{\tilde{v}_1^{1/3} \;-\; 1}{\tilde{v}^{1/3} \;-\; 1} \right) \right.$$

$$\left. +\; \frac{1}{\tilde{v}_1} \;-\; \frac{1}{\tilde{v}} \right] \;+\; \frac{V_1^*}{\tilde{v}}X_{12}\theta_2^2 \tag{168}$$

The quantity $(V_1^*/\tilde{v})X_{12}$ in the last term is a contact energy parameter, as in the original definition of χ in equation (139), but the new equation of state terms have no counterparts, even in the Flory–Huggins treatments with χ modified as described above.

To compare theoretical predictions with experimental data on mixtures, the parameters p^*, V^* and T^* can be obtained independently using equation of state data for the pure components and the mixing rules; and the segment–surface ratio s_2/s_1, required to calculate the site fraction θ_2, can be estimated reasonably from molecular structural data. The contact energy parameters are included in X_{12}, which is chosen as a constant to optimize fitting of an experimental thermodynamic property, usually the enthalpy of mixing or of dilution. If the best-fit reduction parameters of the components vary with temperature, comparisons with the mixture data are made using the optimal parameter values at each temperature. Analysis of data for a variety of systems shows the predictions of the theory to be semiquantitatively correct. Calculated excess volumes are correct in sign and magnitude. Calculated values of χ_S are usually positive and large, in accord with experiment; and, also in accord with experiment, χ and both its components, χ_H and χ_S, increase with increasing solute concentration. Of a variety of systems analyzed by Flory,[149] the only egregious deviations between theory and experiment were found with poly(dimethylsiloxane) solutions.

Flory was able to demonstrate that experimental values of X_{12} mostly fall into distinct groups characteristic of chemical composition, e.g. no more than 10 J cm^{-3} for aliphatic–aliphatic mixtures and larger (ca. 40 J cm^{-3}) for aliphatic–aromatic mixtures. The pattern of consistency and the fact that X_{12} values for nonpolar mixtures are invariably positive, supports the assignment of the empirical X_{12} to purely exchange interactions. The admission of X_{12} as a constant to be adjusted to fit data allows no more freedom than the choice of a purely enthalpic empirical χ (inversely proportional to the absolute temperature) in the simplest form of the Flory–Huggins theory. In some cases the fit between experiment and theory could be improved by introduction of an adjustable exchange entropy parameter Q_{12} in an additional volume-independent term $-V_1^*\theta_2^2 TQ_{12}$ on the right hand side of equation (168). The equation of state is not affected by this arbitrary addition, nor are the enthalpy and volume of mixing.

Although agreement with experiment is hardly complete, Flory suggests that practically it may be about optimal: 'instead of seeking a theory which is accurate in detail as well as comprehensive in scope, it may be more fruitful to adopt a simpler treatment of reasonable generality at sacrifice of accuracy of representation of individual cases. Departures from such a scheme may then be interpreted . . . with reference to molecular characteristics of the particular components.'[149]

By expanding the expression for the residual chemical potential, equation (168) in powers of ϕ_2 passing to the limit $\phi_2 = 0$ and setting $\mu_1^R/RT\phi_2^2$ equal to $1/2$ when $T = \Theta$, one obtains the expression

$$\frac{1}{\Theta} = \frac{1}{X_{12}}\left[\left(\frac{R}{2V_1^*\rho^2} + Q_{12}\right)\tilde{v}_1 - \frac{A^2\alpha_1 p_1^*}{2\rho^2}\right] \tag{169}$$

where

$$A = (1 - T_1^*/T_2^*)(p_2^*/p_1^2) - \rho X_{12}/p_1^* \tag{170}$$

α_1 is the thermal expansivity of the liquid solvent at low pressure, and $\rho = s_2/s_1$. With parameters to accommodate solution data in the appropriate temperature range, equation (169) gives the limiting UCST for a polymer in a poor solvent. If equation of state data for liquid polymer and solvent are available to high enough temperature, the equation can be used to predict the LCST. From such calculations on polystyrene + cyclohexane, Höcker *et al.*[155] determined Θ_L at 200 °C, which compares well with earlier experimental estimates and a more recent value of 213 °C.[169] For lack of p–V–T data on other polymers over the full temperature range from Θ_U to Θ_L, LCSTs cannot be calculated in this way for comparison with values from phase diagrams.[169,180]

The lack of data makes it impossible to use Flory's liquid state theory to obtain any meaningful estimate of $A_2^{(L)}$ of polystyrene + cyclopentane as a function of temperature for comparison with the Eichinger functions shown in Figure 4. In the only system for which any calculation can be made, polystyrene + cyclohexane, it appears that a plot of $A_2^{(L)}$ against $1 - \Theta/T$ would be sharply skewed to the right.[122] It may be noted that those Eichinger curves for polystyrene + cyclopentane in Figure 4 that are constrained to the experimental value of B_U are visibly skewed to the right.

3.5.3.2 Lattice models with vacancies

Another variant of the lattice model for fluids and fluid mixtures has been developed by Lacombe and Sanchez.[181-184] In their treatment the fluid lattice contains empty sites (holes), whose equilibrium concentration is temperature dependent. Exchange interaction energies are taken proportional to contact surfaces, as in Flory's theory, but hole–hole and hole–mer interactions are zero. 'Close-packed' molecular volumes, independent of temperature and pressure, appear in place of hard-core volumes. The holes provide the free volume, and there is no need to introduce a separation of internal and external molecular degrees of freedom or a parameter c. These considerations lead to a reduced equation of state for an r–mer fluid in the form

$$\frac{\tilde{p}\tilde{v}}{\tilde{T}} = \tilde{v}\ln\left(\frac{\tilde{v}}{\tilde{v}-1}\right) - \frac{1}{\tilde{v}\tilde{T}} - \left(1 - \frac{1}{r}\right) \tag{171}$$

with the reduced variables obtained, as usual, by multiplication by scaling parameters, *i.e.*

$$\tilde{v} = v/v^* \qquad \tilde{T} = T/T^* = kT/\varepsilon^* \qquad \tilde{p} = p/p^* = pv^*/\varepsilon^* \tag{172}$$

so that $p^*v^*/kT = 1$. Here, rv^* is the close-packed molecular volume, and ε^* is the fixed energy required to create a hole ($r\varepsilon^*$ is also the characteristic interaction energy per molecule in the absence of holes). For a long-chain polymer, the term $1/r$ on the right-hand side of equation (171) vanishes. Hence, in this limit polymer fluids satisfy a corresponding states principle. Equation (171) provides excellent fits of density data for a variety of polymeric liquids to a common set of reduced isobars.[184]

Formally, the equation of state for a mixture is identical with equation (171), with reduction parameters obtained by mixing rules consistent with the character of the model.[183] The close-packed volume V^* of the mixture is related to the averaged close-packed volume v^* per 'monomer' unit and the close-packed molecular volumes of pure components

$$\sum_i r_i^0 N_i v_i^* \equiv v^*\sum_i r_i N_i \equiv V^* \qquad v^* = \sum_i \phi_i^0 v_i^* \tag{173}$$

where N_i is the number of r_i-mers, superscript zero designates the pure state of each component, and the ϕ_i^0 are volume fractions calculated on the basis of the r_i^0. The residual chemical potential of component 1 in a binary mixture has the form

$$
\frac{\mu_1^R}{kT} = \frac{r_1^0}{\tilde{v}}\left[\chi + \lambda_{12}\left(1 - \frac{v_1^*}{v_2^*}\right)\right]\phi_2^2 + \frac{r_1^0 \varepsilon_{11}^*}{kT}\left\{-\frac{1}{\tilde{v}} + \tilde{p}_1\tilde{v}\right.
$$
$$
\left. + \tilde{T}_1\left[(\tilde{v} - 1)\ln\left(1 - \frac{1}{\tilde{v}}\right) - \frac{\ln(\tilde{v}\omega_1)}{r_1^0}\right]\right\} \tag{174}
$$

$$
\lambda_{12} = \frac{1}{\tilde{T}_1} - \frac{1}{\tilde{T}_2} + (\phi_1 - \phi_2)\chi \tag{175}
$$

where ω_1 is a constant, the number of configurations available to the r_1-mer in the pure close-packed liquid. Here the ϕ_1 and ϕ_2 are volume fractions based on the r_i in the mixture, and χ is the energy exchange parameter $(\varepsilon_{11}^* + \varepsilon_{22}^* - 2\varepsilon_{12}^*)/kT$. Despite its conceptual simplicity, this lattice-hole model gives a realistic account of the thermodynamic behavior of nondilute polymer solutions: it accommodates a volume change on mixing, and naturally gives both the UCST and LCST separations, indicating the universality of the latter if there is a large disparity between r_1 and r_2. At high pressure and/or low temperature, the reduced specific volumes approach the maximum value of unity and the Flory–Huggins chemical potential is recovered. There appears but one parameter χ, inversely proportional to temperature (thus purely enthalpic), to characterize a binary mixture; all the other parameters are obtainable independently from properties of the pure components.

The invariant hole size and energy of hole formation in the Lacombe–Sanchez model implies a dependance of the internal energy on the density that is strictly of the van der Waals form, as in Flory's theory. Another, somewhat more complicated, model developed by Simha and his collaborators[185–187] is similar to that of Sanchez and Lacombe in its use of a liquid lattice with vacant sites, but it also retains features of Prigogine's earlier cell model in the c parameter for external degrees of freedom and a lattice energy with a density dependence based on an effective (6–12) pair potential. Like the other theories, this one has been successful in correlating equation of state data both for neat polymer liquids and for nondilute solutions.

3.5.3.3   *Spin lattice models*

De Gennes[188,189] and des Cloizeaux[190] have noted a formal analogy between self-avoiding polymer chains disposed on a lattice and a system of n-component magnetic spins in the limit as $n \to 0$. Analysis of this model has provided important information on critical exponents at the limit of infinite molecular weight in good solvents; but apart from this universal behavior, it has not been possible to deduce properties dependent on molecular weight and solvent character. A serious deficiency of the magnetic representation is its imposition of a large and uncontrollable molecular weight distribution. This inherent property appears to be responsible for the inability of models of this type to generate the Flory–Huggins mean field result.

Recently Freed and co-workers[191,192] have modified the magnetic lattice representation by introducing a $2n$-component, $n \to 0$ model, where the spins have an internal symmetry label that controls the connectivity of the polymer segments in the chains and allows the molecular weight distribution to be specified. For low volume fractions, this formalism reduces to the Edwards–Freed[193] type of field theory for dilute and semidilute solutions; and in the mean field limit, to the Flory–Huggins theory. The importance of the latter result is that a cluster expansion can be formulated to generate rigorous corrections to the Flory–Huggins theory and their concentration dependence. Thus it overcomes a major weakness of the simple theory, its inability to generate such corrections. The new theory also indicates the source of the entropic contribution to the Flory–Huggins interaction parameter χ that experiment requires.

3.6   HYDRODYNAMIC PROPERTIES

3.6.1   Dilute Solutions

Hydrodynamic properties of dilute solutions of flexible-chain polymers are of interest in the context of this chapter because they are closely associated with chain dimensions. The concentration

(c) dependence of the solution viscosity η in dilute solutions may be expressed in the form[194]

$$\eta = \eta_s\{1 + [\eta]c + k'[\eta]^2c^2 + \ldots\} \tag{176}$$

$$k' = k'_0 - k'_1 A_2 M/[\eta]. \tag{177}$$

where η_s is the solvent viscosity and $[\eta]$ is the intrinsic viscosity (or limiting viscosity number): *i.e.* the dilute solution limit of the specific viscosity $(\eta - \eta_s)/\eta_s$ divided by c. The parameter k' is usually called the Huggins coefficient. The mutual diffusion coefficient $D_M(c)$ (a property determined in dynamic light-scattering experiments[121,195]) may be expressed in the analogous form[194]

$$D_M = D_T\{1 + k_D[\eta]c + \ldots\} \tag{178}$$

$$k_D = -c_0 + c_1 A_2 M/[\eta] \tag{179}$$

where D_T is the translational diffusion coefficient. Theoretical assessments of the numerical parameters $k'_0, k'_1, c_0,$ and c_1 of these equations are available.[194] For uncharged, flexible-chain polymers, k'_1 is positive, and k' decreases from its value $k'_0 \approx 1/2$ for $A_2 = 0$ to about 1/3 for A_2 large (by large A_2 we mean that the penetration function $\Psi(\bar{z})$ approximates its asymptotic value). In contrast, since both c_0 and c_1 are positive, k_D is negative for $A_2 = 0$, zero for $A_2 M/[\eta] = c_0/c_1$, and positive for larger $A_2 M/[\eta]$.

The intrinsic viscosity is related to R_G and the hydrodynamic radius R_H, defined as

$$R_H = kT/6\pi\eta_s D_T \tag{180}$$

It is convenient to express the intrinsic viscosity in the form

$$[\eta] = \pi N_A K R_G^2 R_H/M \tag{181}$$

because K turns out, empirically, to be a constant, usually between one and five, dependent on the shape of the molecule (or particle) contributing to $\eta(c)$.[196-198] For flexible-chain polymers in the unperturbed state, theoretical treatments give[194]

$$R_{H,0} = R_{G,0}\frac{3\pi^{1/2}X_0}{3 + 8X_0} \tag{182}$$

where X_0 is the unperturbed value of the draining parameter X. The latter depends on molecular weight through

$$\pi^{1/2}X = (L_c/R_G)(\zeta/6\pi\eta_s b) \tag{183}$$

Here, L_c is the chain contour length and ζ is the friction factor for a chain unit of length b. Data on R_H/R_G under conditions with $A_2 = 0$ (denoted by subscript Θ), offer an approximation to $R_{H,0}/R_{G,0}$, so that for large M

$$(R_G/R_H)_\Theta = (R_{G,0}/R_{H,0})(1 - a_H z_3 + \ldots) \tag{184}$$

According to one calculation,[83] a_H is 20.13, suggesting that $(R_G/R_H)_\Theta$ could be smaller than $R_{G,0}/R_{H,0}$, since z_3 is positive. Experimental data give $(R_G/R_H) \approx 1.27$ for a variety of systems for large M,[83] whereas $R_{G,0}/R_{H,0} = 1.505$, according to equation (182).

Calculations of $K_0(X_0)$, the value of K for the unperturbed chain, show that it depends only weakly on X_0,[194] being equal to $K_0(\infty) = 3.15$ for large M. Hence, at that limit, $[\eta]_0$ may be expressed conveniently as

$$[\eta]_0 = \Phi'(R_{G,0}^2/M)^{3/2}M^{1/2} \tag{185}$$

with $\Phi' = 3\pi^{3/2}N_A K_0(\infty)/8$. Values of $[\eta]/M^{1/2}$ from experiments under conditions with $A_2 = 0$ confirm this behavior over a wide range of M.[199] In fact, $[\eta]/M^{1/2}$ at $A_2 = 0$ appears to be independent of M over *too wide* a range, given the computed behavior of $K_0 R_{H,0}/R_{G,0}$ and the expected variation of $R_{G,0}^2/M$ at small M.[199]

In good solvents, where A_2 is large, both R_G and R_H increase relative to their unperturbed values, with a consequent increase in $[\eta]$ (unless K decreases). First-order perturbation treatments[194] of R_H and $[\eta]$ in the two-parameter context give

$$R_H(z, X_0) = R_{H,0}(X_0)(1 + h_1(X_0)z + \ldots) \tag{186}$$

$$[\eta]\,(z, X_0) \;=\; [\eta]_0(X_0)\,(1 \;+\; p_1(X_0)z \;+\; \ldots) \tag{187}$$

where, for large X_0, $h_1 = 0.069$ and $p_1(\infty) = 1.55$. With equations (38) and (181), these relations give

$$K(z, X_0) \;=\; K_0(X_0)\,\{1 \;+\; [p_1(X_0) \;-\; a_1 \;-\; h_1(X_0)]z \;+\; \ldots\} \tag{188}$$

Thus, for large X, K is expected to decrease with increasing z (or A_2), according to the first-order theory. By contrast, theoretical treatments using renormalization methods suggest that K may be nearly independent of A_2[197] so that excluded volume effects on $[\eta]$ are just those on $R_G^2 R_H$. The effects of excluded volume interactions on R_G have been considered in detail above (Section 3.2.4). The effects of excluded volume on R_H, and therefore on $[\eta]$, are more complex. Whereas the so-called 'nondraining' limit (large X) appears to apply under theta conditions over the range of M usually studied, excluded volume effects in good solvents can decrease $X \propto L_c/R_G$ to such an extent as to produce observable effects on R_H and $[\eta]$.[197,200,201] One treatment, based on renormalization methods, of the behavior for large A_2 gives[197]

$$R_H/R_{H,\Theta} \;=\; (R_G/R_{G,\Theta})^{X^*/(1 + X^*)} \tag{189}$$

where X^* is a draining parameter related to X. With this result, $[\eta]$ can be put in the form

$$[\eta]/[\eta]_\Theta \;=\; (R_G/R_{G,\Theta})^{3 - \delta} \tag{190}$$

where $\delta = 1/(1 + X^*) \approx 1/X^*$. In the limit $X^* \gg 1$, the familiar Flory–Fox expression[54,202]

$$[\eta]/[\eta]_\Theta \;=\; \alpha_G^3 \tag{191}$$

is recovered. Equation (190) may be written in the form

$$([\eta]/[\eta]_\Theta) \;-\; 1 \;=\; \Lambda(X, \alpha_G)\,(\alpha_G^3 \;-\; 1) \tag{192}$$

where $X = X_0/\alpha_G$, and the function $\Lambda(X, \alpha_G)$ depends only weakly on α_G. Of course, the Flory–Fox relation is recovered if $\Lambda = 1$. Behavior similar to that of equation (192) with $\Lambda(X, \alpha_G)$ independent of α_G has been reported[96,200] for linear and branched polymers in solution when A_2 is large, and is shown in Figure 9.

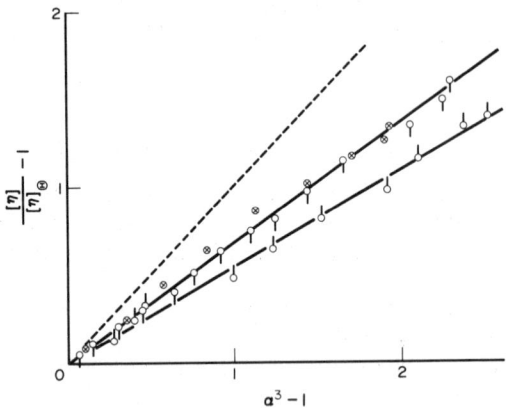

Figure 9 Plots showing linear relations between $[\eta]/[\eta]_\Theta$ and α_G^3 for branched polystyrene in decalin.[96] Data points are for two comb polymers [24 branches, $\bar{M}_w \times 10^4 = 24 \times 4.62 + 133$ (pips down); 22 branches, $\bar{M}_w \times 10^4 = 22 \times 12.0 + 133$ (pips up)] and for a star polymer [$\bar{M}_w \times 10^4 = 4 \times 40$ (cross within circles)]. The dashed line is the experimental function for linear chains (reproduced by permission of Wiley, from *J. Polym. Sci., Part A-2*, 1971, **9**, 710)

The empirical Mark–Houwink–Sakurada equation[203]

$$[\eta] \;=\; K_{MHS}M^\upsilon \tag{193}$$

is employed ubiquitously to correlate data on intrinsic viscosity and molecular weight. Values of K_{MHS} and υ have been tabulated for many polymer + solvent systems — notably in ref. 14. As is made evident by the preceding discussion, the parameters K_{MHS} and υ cannot be expected to remain constant over a wide range of M. At the theta temperature for large M, υ is $1/2$, so that $K_{MHS} = \Phi'(R_{G,0}^2/M)^{3/2}$ (*cf.* equation 185). However, for small M, $R_{G,0}^2/M$ may depend[13] on M, and X becomes small enough that hydrodynamic draining effects should impose an additional dependence on M. If equations (45) and (191) apply for high M in good solvents with A_2 large, it follows that $\upsilon \approx 3v - 1$. In practice, υ usually lies between $1/2$ and $4/5$ for flexible-chain polymers; but hydrodynamic draining effects can cause υ to be larger (also see below for rodlike chains).

In the nondraining limit with X large, the dependence of $[\eta]$ on R_G has been utilized to develop a number of approximate relations intended to deduce estimates of $R_{G,0}$ and $z/M^{1/2}$ from data on $[\eta]$ and M. These usually take the form of plots involving some combination of $[\eta]/M^{1/2}$ and $M^{1/2}$. For example, equation (44) together with the Flory–Fox expression gives a variant of the first relation proposed for this purpose[202]

$$\left(\frac{[\eta]}{M^{1/2}}\right)^{2/3} = \left(\frac{[\eta]_\Theta}{M^{1/2}}\right)^{2/3} + \frac{134}{105}\left(\frac{z}{M^{1/2}}\right)\left(\frac{[\eta]_\Theta}{M^{1/2}}\right)^{5/3}\left(\frac{M^{1/2}}{[\eta]}\right)M^{1/2} \tag{194}$$

In use of the ellipsoidal model of segment density distribution mentioned above in connection with equation (50), it has been suggested that the first-order perturbation expression equation (186) may in fact be valid for all z with the result[68]

$$\frac{[\eta]}{M^{1/2}} = \frac{[\eta]_\Theta}{M^{1/2}} + 1.55\left(\frac{z}{M^{1/2}}\right)M^{1/2} \tag{195}$$

so that a simple plot of $[\eta]/M^{1/2}$ *vs.* $M^{1/2}$ yields the desired parameters. Unfortunately in many instances, the plot is not linear. An alternative based on the form of equation (195) has been used to account for such nonlinearity.[204] A more fundamental criticism may be raised against the practice of estimating $[\eta]_\Theta/M^{1/2}$ and $z/M^{1/2}$ by procedures involving extrapolation to small M.[71,205] The interpretation of such extrapolations is compromised by the non-Gaussian chain behavior expected at low M and the effects of partial hydrodynamic draining that may intrude at small X.[206] For example, data on a ladder polymer give $\partial \ln[\eta]/\partial \ln M$ close to 0.9 under conditions for which $A_2 = 0$.[207] Moreover, a plot of $[\eta]/M^{1/2}$ *vs.* $M^{1/2}$ happens to be nearly linear with a nonzero slope. However, as Stockmayer pointed out,[71] the latter cannot be interpreted with equation (195). Rather, the observed behavior may correspond to either of the two effects mentioned above. Several models incorporating such behavior may be expressed in the form

$$M^{1/2}/[\eta] = \mathscr{B}_1 C_\infty [1 + (\mathscr{B}_2/M)^{1/2}] \tag{196}$$

where $\mathscr{B}_1$ and $\mathscr{B}_2$ are parameters to be evaluated according to the model selected.[208] In particular, the data on the ladder polymer mentioned above, taken over a wide range of M, may be fitted by relations for a wormlike chain model,[209] having nothing to do with effects of excluded volume on chain dimensions. Similar conclusions were reached on the behavior of cellulose esters.[208] Thus, caution must be used in the interpretation of experimental data with equation (195) or similar relations.

The wormlike chain alluded to in the preceding paragraph can be represented by equation (181) with $R_{G,0}$ given by equation (19). Both $R_{H,0}$ and K are functions of L_c/ρ and L_c/d_H, with d_H the effective hydrodynamic diameter of the chain segments. Numerical results[210] for $R_{H,0}$ may be approximated by the simple relation[211]

$$R_{H,0} \approx \frac{L_c}{2}\left[\left(\frac{27L_c}{16\rho}\right) + \left(\ln\frac{3L_c}{2a}\right)^2\right]^{-1/2} \tag{197}$$

which reproduces $R_{H,0}$ for the random flight and rodlike polymer chains, in the limits of small and large L_c/ρ, respectively. The function K in equation (181) may be approximated by[198]

$$K(L_c/\rho, d/\rho) \approx 1 + (1/2)[K(\infty, d/\rho) - 1]\{\tanh[0.25 \ln (L_c/340\rho)]\} \tag{198}$$

where $K(\infty, d/\rho)$ is near unity for $d \ll \rho$. Thus for a rodlike chain (L_c/ρ large and d_H/ρ small), equation (181) applies with $K = 1$. Over the span of M of usual interest experimentally, $[\eta]$ may be approximated by equation (193) with $\upsilon \approx 9/5$ and $K_{MHS} = 0.031\,\pi N_A d_H^{0.2} M_L^{-0.8}$ (cgs units), where $M_L = M/L$ is the molar mass per unit chain length.[212]

3.6.2 Moderately Concentrated Solutions

The dependence of $D_M(c)$ and $\eta(c)$ on c for concentrations large enough that $[\eta]c \approx 1$ is more complex than is given by equations (176) to (179). For example, with neglect of a factor $(1 - \phi_2)$, $D_M(c)$ may be expressed as[121,213]

$$D_M(c) = \frac{M}{N_A c} \frac{K_{OS}(c)}{\Xi(c)} \tag{199}$$

where $\Xi(c)$ is a molecular friction factor and, as before, K_{OS} is the osmotic modulus. Representation of Π and $\Xi(c)$, for $[\eta]c \ll 1$, by virial expansions gives equation (178). As discussed above (Section 3.4), a virial expansion is not appropriate for K_{OS} if $[\eta]c \gg 1$, but alternative relations are available (see equations 130, 132 or 135). Similarly, alternative expressions have been suggested for $\Xi(c)$. One of these is[136]

$$\frac{\Xi(c)}{\Xi_0} = \frac{MK_{OS}}{cRT}\left[\left(\frac{MK_{OS}}{cRT}\right)^{\hat{c}_1/2} \exp(-\hat{c}_0[\eta]c) + \frac{R_H}{\xi(c)}\right]^{-1} \frac{\zeta(c)}{\zeta(0)} \tag{200}$$

Here, $\zeta(c)$ is a monomeric friction factor, which may differ significantly from $\zeta(0)$ if c/ρ exceeds ca. 0.1;[137] $\hat{c}_0$ and $\hat{c}_1$ are constants expected to be close to c_0 and c_1 respectively in equation (178), and $\xi(c)$ is a correlation length given by scaling arguments[121,133]

$$\xi(c)/R_H = K_\xi([\eta]c)^{-r/3} \tag{201}$$

for large $[\eta]c$, where K_ξ is a constant (reported to be 5.26[136]) and $r = 3v/(3v - 1)$. The length $\xi(c)$ represents the distance required for the monomer pair correlation function to reduce to the uniform value at large monomer separation.[133] With equations (200) and (201), equation (199) gives

$$\frac{D_M(c)}{D_T} = \left[\left(\frac{MK_{OS}}{RTc}\right)^{\hat{c}_1/2} \exp(-\hat{c}_0[\eta]c) + \frac{([\eta]c)^{r/3}}{K_\xi}\right]\frac{\zeta(0)}{\zeta(c)} \tag{202}$$

Expressions for $\zeta(c)$ are discussed below. For $[\eta]c \gg 1$, $D_M(c)\zeta(c)/D_T\zeta(0)$ is proportional to c if $A_2 = 0$, or to $c^{3/4}$ if A_2 is large (and $v = 3/5$). Behavior of this kind is illustrated in Figure 10 with data for polystyrene in cyclohexane[136,214] and in tetrahydrofuran.[215,216]

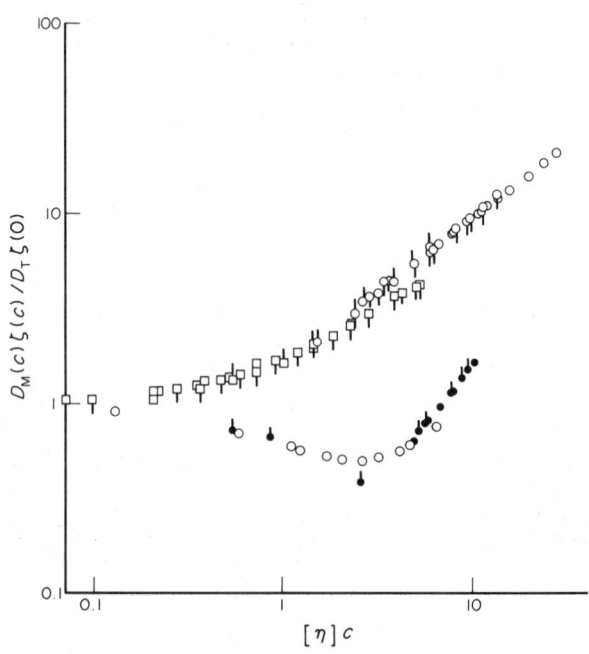

Figure 10 Bilogarithmic plot of $D_M(c)\zeta(c)/D_T\zeta(0)$ vs. $[\eta]c$ for polystyrene in a good solvent, tetrahydrofuran [upper curve: data from refs. 215 (circles) and 216 (squares)] and in cyclohexane at Θ [lower curve: data from refs. 214 (filled circles) and 136 (open circles)]

Over a range of concentrations with $[\eta]c < 1$, but not at infinite dilution, equation (176) for the viscosity is sometimes expressed in the form[137,217]

$$\eta = \eta_s\{1 + [\eta]c \exp(K_M k'[\eta]c)\} \tag{203}$$

in the hope that the exponentiation may (partially) account for the missing higher order terms (K_M would have to be unity to reproduce equation (176) for small c, but it is usually allowed to be adjustable). Indeed, plots of $\log(\eta - \eta_s)/c$ vs. $[\eta]c$ do tend to be linear over a range of concentration extending up to $[\eta]c \approx 10$ in solutions with A_2 large. The range of linearity becomes smaller as A_2 goes to zero, and deviations from linearity obtain in any case with increasing c. Moreover, the empirical behavior is best fitted with $K_M \neq 1$, and K_M varies with both M and A_2.[137] These deviations from equation (203) may result from the failure to include the effects of the decrease in coil dimensions with increasing concentration for solutions with A_2 large and/or the changing nature of hydrodynamic interactions. Thus, the hydrodynamic interactions are expected to change from the nondraining limit in dilute solutions (with large X) to screened interactions with increasing c (in a dilute solution, such screening would correspond to $[\eta]$ in the limit of small X).

According to a treatment by Muthukumar and Edwards,[218] (see Chapter 1 in this volume), the viscosity increment $(\eta - \eta_s)$ due to solute varies from $\eta_s[\eta]c$ for very dilute solutions to

$$\eta - \eta_s \approx k_\eta(c\alpha_c/\rho)^n \eta^{(R)} \tag{204}$$

for moderately concentrated solutions (if c is not too large — see below), where

$$\eta^{(R)} = N_A \hat{X} \zeta(c)/6 \tag{205}$$

Here k_η is a constant of order unity ($1/2\pi$ in ref. 218); $\hat{X} = c\alpha_c^2 R_{G,0}^2/m_a$; and m_a and ζ are the molar mass and friction factor per chain atom, respectively. The function $\eta^{(R)}$, with $\alpha_c = 1$, corresponds to a prediction of the Rouse theory.[219] The theoretical estimate $n = 1$ is larger than the value $n = 1/2$ observed for solutions at the theta temperature.[137] Equation (204) specifies a complex dependence on c and M, i.e.

$$\eta - \eta_s \propto Mc^{1+n}\alpha_c^{2+n} \approx Mc^p \tag{206}$$

where $p = 1 - s + n(1 - s/2)$. Thus if $s = 1/4$ (corresponding to A_2 large and M large), it follows that $p = (3/4) + (7/8)n$; or $p = 1$ if $n = 1/2$ and $p = 13/8$ if $n = 1$. Data on high molecular weight polymers in good solvents frequently give $p \approx 1$ in this regime.[220]

With increasing c, chain entanglements enforce reptational motion[213] among the chains, requiring modification of equation (204). This effect occurs when $cM\alpha_c^2 > \rho M_c$, where M_c is a number characteristic of the flexible-chain polymer. In this range of concentration, η is far greater than η_s; and the viscosity can be represented as[137,221-223]

$$\eta = \eta^{(R)}(c\alpha_c/\rho)^n E(cM\alpha_c^2/\rho M_c) \tag{207}$$

where $\eta^{(R)}$ is defined by equation (205) and

$$E(y) = [1 + y^{2.4\varepsilon}]^{1/\varepsilon} \tag{208}$$

Equation (207), which represents a modification of equation (204) (with $k_\eta = 1$) to account for chain entanglements, provides an empirical representation for concentrated solutions ($\alpha_c = 1$) and for pure polymer liquid ($\alpha_c = 1$, $c = \rho$), as well as for moderately concentrated solutions ($\alpha_c \geq 1$). Since $R_{G,0}^2 \propto M$, $\hat{X}$ is proportional to $cM\alpha_c^2$ and $\partial \ln \eta/\partial \ln M$ changes from 1.0 to 3.4 for $cM\alpha_c^2 \approx \rho M_c$. For many polymers, $\hat{X}_c$, the value of $\hat{X}$ for neat liquid ($M = M_c$), is a constant[220] (ca. 400×10^{17} in cgs units).

Most of the temperature dependence of η is given by $\zeta(c)$; the temperature dependence of the unperturbed chain dimensions may also contribute, but any effect is usually small. Semiempirical expressions for $\zeta(c)$ are available.[220] With equation (207), the dependence of η on c is complex, e.g.

$$\frac{\partial \ln \eta}{\partial \ln c} = a\left(1 + \frac{\partial \ln \alpha_c^2}{\partial \ln c}\right) + \frac{\partial \ln \zeta(c)}{\partial \ln c} \tag{209}$$

where a is 1.5 if $\hat{X} < \hat{X}_c$ and 3.9 for larger $\hat{X}$ (for $n = 0.5$). Inasmuch as $\partial \ln \alpha_c^2/\partial \ln c \leq 1$, the observed $\partial \ln \eta/\partial \ln c$ is diminished if α_c is larger than unity. Since $\hat{X}$ depends on $cM\alpha_c^2$ and α_c depends on $R_{G,0}^3 c/M \propto [\eta]c$, the scaling of $\eta/\zeta(c)$ with c and M is complicated.

For concentrated solutions, the dependence of ζ on c may be marked. In such cases, $\zeta(c)$ is often given by a Vogel relation[220]

$$\zeta(c) = \zeta_{\text{VOG}} = \zeta \exp\left(\frac{C}{\Delta + T - T_{\text{g}}}\right) \qquad (210)$$

where T_{g} is the glass transition temperature and ζ is a constant. Although T_{g} may vary substantially with c, both C and Δ tend to be nearly independent of c in concentrated solution.[220] In that case, $\zeta(c)$ will be determined by the variation of T_{g} with c. Equation (210) may be understood in the framework of free volume concepts. For moderately concentrated solutions, free volume effects become less pronounced, and the effects of an activated flow process must be included

$$\zeta(c) = \zeta_{\text{VOG}} + \zeta_{\text{ACT}} \qquad (211)$$

$$\zeta_{\text{ACT}} = \eta_{\text{s}}(T)p \exp(W/T) \qquad (212)$$

where p and W are parameters that may depend on c. However, for the lower end of the moderately concentrated regime, p is expected to be a constant, W is about zero, and $\zeta_{\text{ACT}} \gg \zeta_{\text{VOG}}$.[224] In that case, η/η_{s} is independent of T (unless α_{c} or $R_{\text{G},0}$ depends on T).

Examples of η/η_{s} for moderately concentrated solutions of polystyrene in cyclopentane at $T = \Theta_{\text{L}}$ and in the good solvent benzene are shown in Figure 11. The smaller value of $\partial \ln(\eta/\eta_{\text{s}})/\partial \ln(\phi_2\bar{M}_{\text{w}})$ in the latter solvent is attributed principally to the effect of $\partial \ln \alpha_{\text{c}}^2/\partial \ln c$ in equation (209) since $\partial \ln \zeta(c)/\partial \ln c$ is small for the data shown. The data obtained at $T = \Theta_{\text{L}}$ are fitted well by equation (207) with $n + 1/2$. Examples of η for concentrated solutions[222] $(c > c_{\text{s}})$ are given in Figure 12. In this case $\partial \ln \alpha_{\text{c}}/\partial \ln c$ is zero, but ζ depends markedly on c. Thus, plots of $\log \eta$ vs. $\log (\phi_2\bar{M}_{\text{w}}/m_{\text{a}})$ at fixed ϕ_2 are parallel, with slopes of 1.0 and 3.4 for $\hat{X} < \hat{X}_{\text{c}}$ and $\hat{X} > \hat{X}_{\text{c}}$ respectively, but displaced from one another by the effects of decreasing $\zeta(c)$ with decreasing c. As expected from equation (207), the value of $\hat{X}_{\text{c}}$ is independent of concentration.

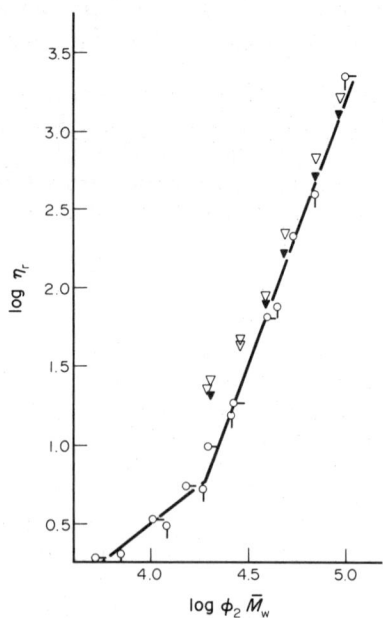

Figure 11 Relative viscosity $\eta_{\text{r}} = \eta/\eta_{\text{s}}$ as a function of $\phi_2\bar{M}_{\text{w}}$ for polystyrene solutions.[137] Open circles are for solutions in cyclopentane at Θ_{L}: $\bar{M}_{\text{w}} = 8.62 \times 10^5$ (pips right), 2.0×10^6 (pips down). Triangles indicate solutions ($\bar{M}_{\text{w}} = 2.0 \times 10^6$) in benzene at 15 °C (unfilled symbols) and 40 °C (filled symbols)

The form of equation (208) for $E(c\alpha_{\text{c}}^2 M/\rho M_{\text{c}})$ is consistent with certain calculations based on the reptation model of chain dynamics with intermolecular entanglement constraints. In its simplest version, application of the reptation model gives

$$E \approx 15 \, (\hat{X}/\hat{X}_{\text{c}})^2 \qquad (213)$$

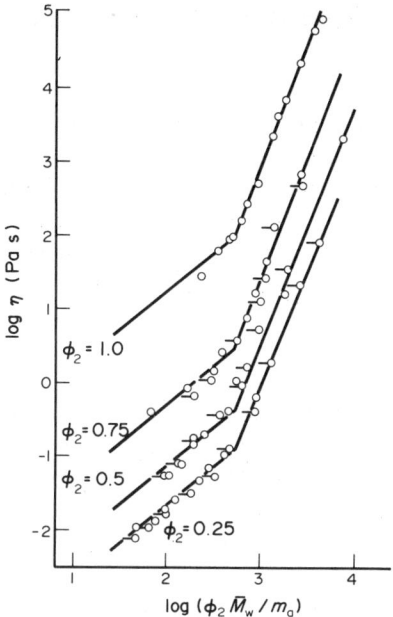

Figure 12 Log η at constant temperature *vs.* log $\phi_2 \bar{M}_w/m_a$ for solutions of poly(vinyl acetate) in diethyl phthalate (plain circles) and in cetyl alcohol (pips) at Θ_L for the latter solvent: $m_a = 43$.[222] The straight lines have slopes 1.0 and 3.4. The invariance with concentration of the value of $\phi_2 \bar{M}_w/m_a$ at the crossover point should be noted

for $\hat{X} > \hat{X}_c$.[225] In this model, a polymer chain deformed from its equilibrium conformation relaxes by reptation along a curvilinear tube formed by neighboring chains. Modification of this treatment to account for fluctuations in the contour length of the tube leads to the result[226]

$$ E \approx 9.6[1 - 1.04(\hat{X}_c/\hat{X})^{1/2}]^3 \, (\hat{X}/\hat{X}_c)^2 \qquad (214) $$

for $\hat{X} > \hat{X}_c$. This relation is essentially equivalent to equation (209) for $\hat{X} > 100\hat{X}_c$, a range that includes all the experimental data with one exception. Experimental data nearly always conform to the 3.4 power law for η. In a heroic effort, data were collected on a system with X/X_c up to 2600, the largest ratio yet studied.[227] The results revealed a possible deviation from equation (209) beginning at $X/X_c \approx 150$, with the deviation tending in the direction required by equation (213). Thus, if the pure reptation result ever applies, it does so only for polymer of extremely high molecular weight.

3.7 REFERENCES

1. P. J. Flory, 'Principles of Polymer Chemistry', Cornell University Press, Ithaca, NY, 1953.
2. H. Tompa, 'Polymer Solutions', Butterworths, London, 1956.
3. H. Yamakawa, 'Modern Theory of Polymer Solutions', Harper & Row, New York, 1971.
4. P.-G. de Gennes, 'Scaling Concepts in Polymer Physics', Cornell University Press, Ithaca, NY, 1979.
5. Y. Oono, *Adv. Chem. Phys.*, 1985, **61**, 301.
6. K. F. Freed, 'Renormalization Group Theory of Macromolecules', Wiley, New York, 1987.
7. J. des Cloizeaux and G. Jannink, 'Les Polymères en Solution: leur Modélisation et leur Structure', Editions de Physique, Paris, 1987.
8. P. J. Flory, ref. 1, chap. 10.
9. P. J. Flory, 'Statistical Mechanics of Chain Molecules', Wiley, New York, 1969.
10. H. Yamakawa, ref. 3, chap. 2.
11. M. V. Volkenstein, 'Configurational Statistics of Polymeric Chains' (S. N. Timasheff and M. J. Timasheff, translators), Interscience, New York, 1963, chap. 4.
12. T. M. Birshtein and O. B. Ptitsyn, 'Conformations of Macromolecules' (S. N. Timasheff and M. J. Timasheff, translators), Interscience, New York, 1966, chap. 5.
13. P. J. Flory, ref. 9, chaps. 1 and 2.
14. M. Kurata, Y. Tsunashima, M. Iwama and K. Kamada, in 'Polymer Handbook', 2nd edn., ed. J. Brandrup and E. H. Immergut, Wiley, New York, 1975, Table IV-1.
15. W. Kuhn, *Kolloid-Z.*, 1936, **76**, 258; 1939, **87**, 3.
16. S. Chandrasekhar, *Rev. Mod. Phys.*, 1943, **15**, 1; reprinted in 'Selected Papers on Noise and Stochastic Processes', ed. N. Wax, Dover, New York, 1954.

17. B. H. Zimm and W. H. Stockmayer, *J. Chem. Phys.*, 1949, **17**, 1301.
18. G. C. Berry, L. M. Hobbs and V. C. Long, *Polymer*, 1964, **5**, 31.
19. O. Kratky and G. Porod, *Recl. Trav. Chim. Pays-Bas*, 1949, **68**, 1106.
20. H. Yamakawa, ref. 3, chap. 3.
21. M. Fixman, *J. Chem. Phys.*, 1955, **23**, 1656.
22. H. Yamakawa, *J. Chem. Phys.*, 1966, **45**, 2606.
23. K. F. Freed, ref. 6, chap. 11, D.
24. B. Duplantier, *J. Phys., Lett. (Les Ulis, Fr.)*, 1980, **41**, L409; *J. Phys. (Les Ulis, Fr.)*, 1982, **43**, 991.
25. P. G. de Gennes, *J. Phys., Lett. (Les Ulis, Fr.)*, 1975, **36**, L55.
26. A. L. Kholodenko and K. F. Freed, *J. Chem. Phys.*, 1984, **80**, 900.
27. B. Duplantier, G. Jannink and J. des Cloizeaux, *Phys. Rev. Lett.*, 1986, **56**, 2080.
28. M. Blume, V. J. Emery and R. B. Griffiths, *Phys. Rev. A*, 1971, **4**, 1071.
29. W. H. Stockmayer, *Makromol. Chem.*, 1960, **35**, 54.
30. S. F. Edwards, *Proc. Phys. Soc., London*, 1965, **85**, 613.
31. B. J. Cherayil, J. F. Douglas and K. F. Freed, *J. Chem. Phys.*, 1987, **87**, 3089.
32. P. J. Flory and H. Daoust, *J. Polym. Sci.*, 1957, **25**, 429.
33. B. L. Hager, G. C. Berry and H.-H. Tsai, *J. Polym. Sci., Part B: Polym. Phys.*, 1987, **25**, 387.
34. G. C. Berry and E. F. Casassa, *J. Polym. Sci., Part D*, 1970, **4**, 1.
35. E. F. Casassa, *J. Polym. Sci., Part A-2*, 1970, **8**, 1651.
36. J. E. Martin, *Macromolecules*, 1984, **17**, 1263, 1275.
37. T. A. Orofino and P. J. Flory, *J. Chem. Phys.*, 1957, **26**, 1067.
38. A. Vrij, *J. Polym. Sci., Part A-2*, 1969, **7**, 1627.
39. F. Candau, P. Rempp and H. Benoit, *Macromolecules*, 1972, **5**, 627.
40. M. Muthukumar and B. G. Nickel, *J. Chem. Phys.*, 1987, **86**, 460.
41. M. Muthukumar and B. G. Nickel, *J. Chem. Phys.*, 1984, **80**, 5839.
42. J. des Cloizeaux, *J. Phys. (Les Ulis, Fr.)*, 1982, **43**, 1743.
43. K. F. Freed, *Adv. Chem. Phys.*, 1972, **22**, 1.
44. S. Aronowitz and B. E. Eichinger, *Macromolecules*, 1976, **9**, 377.
45. I. C. Sanchez, *Macromolecules*, 1979, **12**, 981.
46. I. M. Lifschitz, A. Yu. Grosberg and A. R. Khoklov, *Rev. Mod. Phys.*, 1978, **50**, 683.
47. A. L. Kholodenko and K. F. Freed, *J. Phys. A: Math. Gen.*, 1984, **17**, 2703.
48. E. Slagowski, B. Tsai and D. McIntyre, *Macromolecules*, 1976, **9**, 687.
49. I. Nishio, G. Swislow, S. T. Sun and T. Tanaka, *Nature (London)*, 1982, **300**, 243.
50. E. F. Casassa, *J. Polym. Sci., Part A*, 1965, **3**, 605.
51. G. C. Berry and T. A. Orofino, *J. Chem. Phys.*, 1964, **40**, 1614.
52. P. J. Solensky and E. F. Casassa, *Macromolecules*, 1980, **13**, 500.
53. P. J. Flory, *J. Chem. Phys.*, 1949, **17**, 303.
54. P. J. Flory, ref. 1, chap. 14.
55. E. F. Casassa and T. A. Orofino, *J. Polym. Sci.*, 1959, **35**, 553.
56. W. H. Stockmayer, *J. Polym. Sci.*, 1955, **15**, 595.
57. C. Domb, *J. Chem. Phys.*, 1963, **38**, 2957.
58. C. Domb and G. S. Joyce, *J. Phys. C*, 1972, **5**, 956.
59. C. Domb and A. J. Barrett, *Polymer*, 1976, **17**, 179.
60. S. Windwer, in 'Markov Chains and Monte Carlo Calculations in Polymer Science', ed. C. G. Lowry, Dekker, New York, 1970.
61. Z. Alexandrowicz and Y. Accad, *Macromolecules*, 1973, **6**, 251.
62. A. Baumgärtner, in 'Applications of the Monte Carlo Method in Statistical Physics', ed. K. Binder, Springer–Verlag, Berlin, 1984.
63. I. Majid, Z. V. Djordjevic and H. E. Stanley, *Phys. Rev. Lett.*, 1983, **51**, 1282.
64. D. C. Rapaport, *J. Phys. A: Math. Gen.*, 1985, **18**, 113.
65. G. Tanaka, *Macromolecules*, 1980, **13**, 1513.
66. P. J. Flory and S. Fisk, *J. Chem. Phys.*, 1966, **44**, 2243.
67. G. C. Berry, *J. Chem. Phys.*, 1966, **44**, 4550.
68. M. Kurata, W. H. Stockmayer and A. Roig, *J. Chem. Phys.*, 1960, **33**, 151.
69. H. Yamakawa and G. Tanaka, *J. Chem. Phys.*, 1967, **47**, 3991.
70. C. M. Bender and S. A. Orszag, 'Advanced Mathematical Methods for Scientists and Engineers', McGraw-Hill, New York, 1984.
71. W. H. Stockmayer, *Br. Polym. J.*, 1977, **9**, 89.
72. K. F. Freed, ref. 6, chap. 6.
73. K. F. Freed, ref. 6. chap. 8.
74. J. des Cloizeaux, *J. Phys. (Les Ulis, Fr.)*, 1981, **42**, 635.
75. H. Yamakawa, ref. 3, chap. 4.
76. W. G. McMillan and J. E. Mayer, *J. Chem. Phys.*, 1945, **13**, 276.
77. G. Scatchard, *J. Am. Chem. Soc.*, 1946, **68**, 2315.
78. J. G. Kirkwood and R. J. Goldberg, *J. Chem. Phys.*, 1950, **18**, 54.
79. W. H. Stockmayer, *J. Chem. Phys.*, 1950, **18**, 58.
80. B. H. Zimm, *J. Chem. Phys.*, 1946, **14**, 164.
81. M. Kurata and H. Yamakawa, *J. Chem. Phys.*, 1958, **29**, 311.
82. M. Kurata, M. Fukatsu, H. Sotobayashi and H. Yamakawa, *J. Chem. Phys.*, 1964, **41**, 139.
83. B. J. Cherayil, J. F. Douglas, and K. F. Freed, *J. Chem. Phys.*, 1985, **83**, 5293.
84. E. F. Casassa, *J. Chem. Phys.*, 1962, **37**, 2176.
85. E. F. Casassa, *J. Chem. Phys.*, 1964, **41**, 3213.
86. E. F. Casassa and Y. Tagami, *J. Polym. Sci., Part A-2*, 1968, **6**, 63.

87. E. F. Casassa, *J. Polym. Sci., Part C*, 1966, **15**, 299.
88. E. F. Casassa, *J. Chem. Phys.*, 1966, **45**, 2811.
89. H. Yamakawa and M. Kurata, *J. Chem. Phys.*, 1960, **32**, 1852.
90. E. F. Casassa, *Polymer*, 1960, **1**, 169.
91. Y. Tagami and E. F. Casassa, *J. Chem. Phys.*, 1969, **50**, 2206.
92. G. Tanaka and K. Šolc, *Macromolecules*, 1982, **15**, 791.
93. H. Yamakawa, *J. Chem. Phys.*, 1965, **42**, 1764.
94. B. H. Zimm, W. H. Stockmayer and M. Fixman, *J. Chem. Phys.*, 1953, **21**, 1716.
95. G. C. Berry, E. F. Casassa and P.-Y. Liu, *J. Polym. Sci., Part B: Polym. Phys.*, 1987, **25**, 673.
96. G. C. Berry, *J. Polym. Sci., Part A-2*, 1971, **9**, 687.
97. Y. Miyaki, Y. Einaga, T. Hirosye and H. Fujita, *Macromolecules*, 1977, **10**, 1356.
98. P. J. Flory, *J. Chem. Phys.*, 1949, **17**, 1347.
99. P. J. Flory and W. R. Krigbaum, *J. Chem. Phys.*, 1950, **18**, 1086.
100. E. F. Casassa and H. Markovitz, *J. Chem. Phys.*, 1958, **29**, 493.
101. M. Fixman, Ph.D. Thesis, Massachusetts Institute of Technology, 1953.
102. E. F. Casassa, *Pure Appl. Chem.*, 1972, **31**, 151.
103. K. F. Freed, *Acc. Chem. Res.*, 1985, **18**, 38.
104. J. F. Douglas and K. F. Freed, *Macromolecules*, 1984, **17**, 2344; 1985, **18**, 201.
105. D. J. Burch and M. A. Moore, *J. Phys. A: Math. Gen.*, 1976, **9**, 435.
106. D. S. McKenzie and C. Domb, *Proc. Phys. Soc., London*, 1967, **92**, 632.
107. A. Bellemans and M. Janssens, *Macromolecules*, 1974, **7**, 809.
108. G. Tanaka, *J. Polym. Sci., Polym. Phys. Ed.*, 1979, **17**, 305.
109. E. F. Casassa, *Polymer*, 1962, **3**, 625.
110. W. R. Krigbaum and P. J. Flory, *J. Am. Chem. Soc.*, 1953, **75**, 1775.
111. I. Noda, T. Kitano and M. Nagasawa, *J. Polym. Sci., Polym. Phys. Ed.*, 1977, **15**, 1129.
112. E. F. Casassa and W. H. Stockmayer, *Polymer*, 1962, **3**, 53.
113. T. P. Wallace and E. F. Casassa, *Polym. Prepr., Am. Chem. Soc., Div. Polym. Chem.*, 1970, **11**(1), 136.
114. L. Schäfer, *Macromolecules*, 1982, **15**, 652.
115. G. V. Schulz, *Z. Phys. Chem., Abt. B*, 1939, **43**, 25.
116. B. H. Zimm, *J. Chem. Phys.*, 1948, **16**, 1099.
117. G. C. Berry, in 'Encyclopedia of Materials Science and Engineering', ed. M. B. Bever, Pergamon Press, Oxford, 1986, p. 3759.
118. H. Wesslau, *Makromol. Chem.*, 1956, **20**, 111.
119. W. H. Stockmayer and E. F. Casassa, *J. Chem. Phys.*, 1952, **20**, 1560.
120. P. J. Flory, ref. 1, chap. 12.
121. G. C. Berry, in 'Encyclopedia of Polymer Science and Engineering' ed. H. Mark, C. G. Overberger, N. M. Bikales and G. Menges, Wiley, New York, 1987, vol. 8, p. 721.
122. E. F. Casassa, *J. Polym. Sci., Polym. Symp.*, 1976, **54**, 53.
123. B. E. Eichinger, *J. Chem. Phys.*, 1970, **53**, 561.
124. H. Eisenberg, 'Biological Macromolecules and Polyelectrolytes in Solution', Clarendon Press, Oxford, 1976, chaps. 2–4.
125. E. F. Casassa and H. Eisenberg, *Adv. Protein Chem.*, 1964, **19**, 287.
126. E. F. Casassa and H. Eisenberg, *J. Phys. Chem.*, 1960, **64**, 753; 1961, **65**, 427.
127. A. Dondos and H. Benoit, *Makromol. Chem.*, 1970, **133**, 119.
128. E. F. Casassa, *Polym. J.*, 1972, **3**, 517.
129. R. H. Ewart, C. P. Roe, P. Debye and J. R. McCartney, *J. Chem. Phys.*, 1946, **14**, 687.
130. J. M. G. Cowie, *Pure Appl. Chem.*, 1970, **23**, 355.
131. A. Dondos, *Eur. Polym. J.*, 1976, **12**, 435.
132. A. Dondos, *Macromolecules*, 1980, **13**, 1023.
133. P.-G. de Gennes, ref. 4, chap. 3.
134. R. G. Kirste and B. R. Lehnen, *Makromol. Chem.*, 1976, **177**, 1137.
135. M. Daoud, J. P. Cotton, B. Farnoux, G. Jannink, G. Sarma, H. Benoit, R. Duplessix, C. Picot and P. G. de Gennes, *Macromolecules*, 1975, **8**, 804.
136. S.-J. Chen and G. C. Berry, to be published.
137. B. W. Hager and G. C. Berry, *J. Polym. Sci., Polym. Phys. Ed.*, 1982, **20**, 911.
138. M. Muthukumar and S. F. Edwards, *J. Chem. Phys.*, 1982, **76**, 2720.
139. J. des Cloizeaux and G. Jannink, ref. 7, chap. 13.
140. C. R. Masson and H. W. Melville, *J. Polym. Sci.*, 1949, **4**, 337.
141. H. Tompa, ref. 2, p. 157.
142. H. Benoit and C. Picot, *Pure Appl. Chem.*, 1966, **12**, 545.
143. I. Noda, N. Kato, T. Kitano and M. Nagasawa, *Macromolecules*, 1981, **14**, 668.
144. T. Ohta and Y. Oono, *Phys. Lett. A*, 1982, **89**, 460.
145. P. J. Flory, *J. Chem. Phys.*, 1941, **9**, 660; 1942, **10**, 51.
146. M. L. Huggins, *J. Chem. Phys.*, 1941, **9**, 440; *Ann. N. Y. Acad. Sci.*, 1942, **43**, 1; *J. Phys. Chem.*, 1942, **46**, 151; *J. Am. Chem. Soc.*, 1942, **64**, 1712.
147. A. R. Miller, *Proc. Cambridge Philos. Soc.*, 1942, **38**, 109; 1943, **39**, 54.
148. A. R. Miller, 'The Theory of Solutions of High Polymers', Clarendon Press, Oxford, 1948.
149. P. J. Flory, *Discuss. Faraday Soc.*, 1970, **49**, 7.
150. E. A. Guggenheim, 'Mixtures', Clarendon Press, Oxford, 1952.
151. R. Koningsveld and L. A. Kleintjens, *Macromolecules*, 1971, **4**, 637.
152. R. Koningsveld, W. H. Stockmayer, J. W. Kennedy and L. A. Kleintjens, *Macromolecules*, 1974, **7**, 73.
153. P. J. Flory and H. Höcker, *Trans. Faraday Soc.*, 1971, **67**, 2258.
154. H. Höcker and P. J. Flory, *Trans. Faraday Soc.*, 1971, **67**, 2270.
155. H. Höcker, H. Shih and P. J. Flory, *Trans. Faraday Soc.*, 1971, **67**, 2275.

156. P. J. Flory, ref. 1, chap. 13.
157. E. F. Casassa, in 'Fractionation of Synthetic Polymers: Principles and Practices', ed. L. H. Tung, Dekker, New York, 1977, chap. 1.
158. E. F. Casassa, *Macromolecules*, 1975, **8**, 242.
159. A. R. Shultz and P. J. Flory, *J. Am. Chem. Soc.*, 1952, **74**, 4760.
160. K. W. Derham, J. Goldsborough and M. Gordon, *Pure Appl. Chem.*, 1974, **38**, 97.
161. I. C. Sanchez, *Macromolecules*, 1984, **17**, 967.
162. I. C. Sanchez, *J. Appl. Phys.*, 1985, **58**, 2871.
163. T. Dobashi, M. Nakata and M. Kaneko, *J. Chem. Phys.*, 1980, **72**, 6685, 6692.
164. J. C. Le Guillou and J. Zinn-Justin, *Phys. Rev. B: Condens. Matter*, 1980, **21**, 3976.
165. D. Z. Alpert, *Phys. Rev. B: Condens. Matter*, 1982, **25**, 4810.
166. P. I. Freeman and J. S. Rowlinson, *Polymer*, 1960, **1**, 20.
167. G. Allen and C. H. Baker, *Polymer*, 1965, **6**, 181.
168. K. S. Siow, G. Delmas and D. Patterson, *Macromolecules*, 1972, **5**, 29.
169. S. Saeki, N. Kuwahara, S. Konno and M. Kaneko, *Macromolecules*, 1973, **6**, 246.
170. R. Koningsveld and A. J. Staverman, *J. Polym. Sci., Part A-2*, 1968, **6**, 305, 325, 349.
171. K. Šolc, *Macromolecules*, 1970, **3**, 665; 1983, **16**, 236.
172. K. Šolc, L. A. Kleintjens and R. Koningsveld *Macromolecules*, 1984, **17**, 573.
173. K. Šolc and K. Battjes, *Macromolecules*, 1985, **18**, 220.
174. K. Šolc, *Macromolecules*, 1986, **19**, 1166.
175. I. Prigogine (with V. Mathot and A. Bellemans), 'The Molecular Theory of Solutions', North Holland, Amsterdam, 1957.
176. D. Patterson, *Macromolecules*, 1969, **2**, 672.
177. D. Patterson, *Rubber Chem. Technol.*, 1967, **40**, 1; *J. Polym. Sci, Part C*, 1968, **16**, 3379; *Pure Appl. Chem.*, 1972, **31**, 133.
178. D. Patterson and G. Delmas, *Discuss. Faraday Soc.*, 1970, **49**, 98.
179. H. Eyring and J. O. Hirschfelder, *J. Phys. Chem.*, 1937, **41**, 249.
180. S. Saeki, N. Kuwahara, S. Konno and M. Kaneko, *Macromolecules*, 1973, **6**, 589.
181. I. C. Sanchez and R. H. Lacombe, *Nature (London)*, 1974, **252**, 381.
182. I. C. Sanchez and R. H. Lacombe, *J. Phys. Chem.*, 1976, **80**, 2352.
183. R. H. Lacombe and I. C. Sanchez, *J. Phys. Chem.*, 1976, **80**, 2568.
184. R. H. Lacombe and I. C. Sanchez, *J. Polym. Sci., Polym. Lett. Ed.*, 1977, **15**, 71.
185. R. Simha and T. Somcynski, *Macromolecules*, 1969, **2**, 343.
186. R. Simha, *Macromolecules*, 1977, **10**, 1025.
187. R. K. Jain and R. Simha, *Macromolecules*, 1980, **13**, 1501.
188. P. G. de Gennes, *Phys. Lett. A*, 1972, **38**, 339.
189. P. G. de Gennes, ref. 4, chap. 10.
190. J. des Cloizeaux, *J. Phys. (Les Ulis, Fr.)*, 1975, **36**, 281.
191. K. F. Freed, *J. Phys. A: Math. Gen.*, 1985, **18**, 871.
192. M. G. Bawendi, K. F. Freed and U. Mohanty, *J. Chem. Phys.*, 1986, **84**, 7036.
193. S. F. Edwards and K. F. Freed, *J. Phys. C*, 1970, **3**, 739; 1970, **3**, 750.
194. H. Yamakawa, ref. 3, chap. 6.
195. B. J. Berne and R. Pecora, 'Dynamic Light Scattering', Wiley, New York, 1976, chap. 5.
196. G. Weill and J. des Cloizeaux, *J. Phys. (Orsay, Fr.)*, 1979, **40**, 99.
197. K. F. Freed, ref. 6, chap. 10.
198. G. C. Berry, *J. Polym. Sci., Part B: Polym. Phys.*, 1988, **26**, 1137.
199. M. Kurata and W. H. Stockmayer, *Fortschr. Hochpolym. Forsch.*, 1963, **3**, 196.
200. G. C. Berry, *J. Chem. Phys.*, 1967, **46**, 1338.
201. Y. Miyaki and H. Fujita, *Macromolecules*, 1981, **14**, 742.
202. P. J. Flory and T. G. Fox, *J. Am. Chem. Soc.*, 1951, **73**, 1904.
203. P. J. Flory, ref. 1, chap. 7.
204. G. C. Berry, *J. Polym. Sci., Part B*, 1966, **4**, 161.
205. P. J. Flory, *Makromol. Chem.*, 1966, **98**, 128.
206. P. J. Flory, ref. 9, chap. 4.
207. T. E. Helminiak, C. L. Benner and W. E. Gibbs, *Polym. Prepr., Am. Chem. Soc., Div. Polym. Chem.*, 1967, **8**(1), 284.
208. D. W. Tanner and G. C. Berry, *J. Polym. Sci., Polym. Phys. Ed.*, 1974, **12**, 941.
209. T. E. Helminiak and G. C. Berry, *J. Polym. Sci., Polym. Symp.*, 1978, **65**, 107.
210. H. Yamakawa and M. Fujii, *Macromolecules*, 1973, **6**, 407; 1974, **7**, 128.
211. R. Furukawa and G. C. Berry, *Pure Appl. Chem.*, 1985, **57**, 913.
212. G. C. Berry, P. M. Cotts and S.-G. Chu, *Br. Polym. J.*, 1981, **13**, 47.
213. P.-G. de Gennes, ref. 4, chap. 8.
214. J. Roots and B. Nyström, *Macromolecules*, 1980, **13**, 1595.
215. W. Brown and R. M. Johnsen, *Macromolecules*, 1985, **18**, 379.
216. W. Mandema and H. Zeldenrust, *Polymer*, 1977, **18**, 835.
217. R. Simha and L. Utracki, *J. Polym. Sci., Part A-2*, 1967, **5**, 853.
218. M. Muthukumar and S. F. Edwards, *Polymer*, 1982, **23**, 345.
219. J. D. Ferry, 'Viscoelastic Properties of Polymers', 3rd edn., Wiley, New York, 1980, chap. 9.
220. G. C. Berry and T. G Fox, *Adv. Polym. Sci.*, 1968, **5**, 261.
221. B. L. Hager, G. C. Berry and H. H. Tsai, *Polym. Prepr., Am. Chem. Soc., Div. Polym. Chem.*, 1978, **19**(2), 719.
222. G. C. Berry, H. Nakayasu and T. G Fox, *J. Polym. Sci., Polym. Phys. Ed.*, 1979, **17**, 1825.
223. D. S. Pearson, A. Mera and W. E. Rochefort, *Polym. Prepr., Am. Chem. Soc., Div. Polym. Chem.*, 1981, **22**(1), 102.
224. J.-O. Park and G. C. Berry, to be published.
225. M. Doi and S. F. Edwards, 'The Theory of Polymer Dynamics', Clarendon Press, Oxford, 1986, chap. 7.
226. M. Doi, *J. Polym. Sci., Polym. Phys. Ed.*, 1983, **21**, 667.
227. R. H. Colby, L. J. Fetters and W. W. Graessley, *Macromolecules*, 1987, **20**, 2226.

4

Phase Separation and Pulse-induced Critical Scattering

RUDY VAN DER HAEGEN
DSM Research BV, Geleen, The Netherlands
and
BERNARD W. READY
University of Essex, Colchester, UK

4.1 INTRODUCTION

The pulse-induced critical scattering (PICS) technique is designed to investigate the miscibility behaviour or compatibility in polymer solutions and in polymer blends over wide temperature (up to about 250 °C) and concentration ranges. It was developed during the 1970s at the Institute of Polymer Science at the University of Essex by Gordon and Ready in close cooperation with Koningsveld and Kleintjens of the DSM Research Laboratory.[1-4] The light scattering from small amounts of a polymer system in a laser beam, induced by a fast thermal pulse, is isothermally followed at two detection angles.

To illustrate the method, we recall the typical and schematic phase diagram for a polydisperse polymer solution (Figure 1), which is explained in detail in Section 4.2. It is the purpose of PICS experiments to determine the lines and special points in such diagrams. Although various polymer systems show a wide variety and often complexity of the phase diagram, this does not alter the principle of PICS as is explained in the following discussion. Fluctuations in density and especially in concentration which occur in the neighbourhood of the phase boundaries, as shown schematically in Figure 1, lead to dramatic changes in low angle light scattering, a phenomenon which is discussed in Section 4.3. This fluctuation scattering serves to locate the spinodal curves in polymer phase diagrams using the Debye–Scholte[5-7] extrapolation procedure (see also Sections 4.3 and 4.4). One may also observe the formation of droplets in certain regions of the phase diagram and the light scattering from such stable spherical particles can lead to the determination of the cloud point curve which is explained in Section 4.5. For the determination of spinodal loci, the great advantage of PICS over the classical light scattering lies in the speed, automation and accuracy of the required extrapolation. Four samples of a polymer system can be treated in one run and each experiment

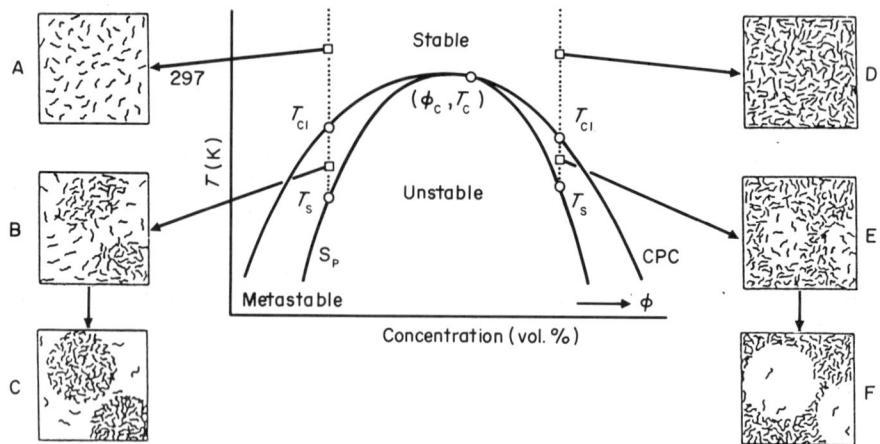

Figure 1 Typical phase diagram for a polydisperse polymer in solution showing upper critical solution temperature behaviour (UCST) with schematic molecular interpretations (see text of Section 4.2 for detailed explanation). CPC denotes the cloud point curve and S_P the spinodal curve in the temperature–concentration (volume percent of polymer in the solution) plane at ambient pressure. T_{cl} and T_s are points on the CPC and S_P curve respectively. The critical state is denoted by (ϕ_c, T_c). The PICS technique uses thermal pulses along the vertical broken lines bringing the dilute or concentrated system from a stable homogeneous region into the metastable part of the phase diagram, where the equilibrium fluctuation light scattering is measured. The solution is then put into the stable state again and the same operation is repeated several times bringing the temperature in the metastable part closer to T_s by about 0.03 °C each time

requires only a few milligrams of samples which may be expensive to obtain on a larger scale, *e.g.* polymer fractions prepared by chromatography.

PICS can be used for the characterization of polymers using phenomenological theories and basic aspects of the thermodynamic theory of light scattering, a subject much developed by Gordon and co-workers for the system polystyrene–cyclohexane,[1,8,9] exploiting the high sensitivity of measured spinodals by PICS to higher moments of the molecular weight distribution (MWD) of the polymer. Using semi-empirical fits of PICS data on liquid–liquid phase separation in solutions, useful structural and thermodynamic information on polymers can be obtained.[10-12] The mechanisms and kinetics of crystallization of polymer–solvent systems can be successfully studied.[13] Many structural variables of polymers such as copolymer composition, tacticity, crosslinking, branching or quantitative characterization of interacting molecular surface areas have a sensitive effect on their phase behaviour which has been revealed and studied by PICS experiments.[14-16]

Although the principle of obtaining the spinodal data or cloud points (as described in Sections 4.4 and 4.5 respectively) has not been changed since the simple Mark I PICS instrument was introduced,[1,17] a new version of the Mark III apparatus containing the most recent advances in software and hardware applications is now developed and commercially available.[38] The description of this modified Mark III instrument is given in Section 4.6.

In the last section, we present some examples of recent spinodal and cloud point data in various polymer solutions and polymer mixtures from PICS measurements by different authors, followed by some concluding remarks.

4.2 DISCUSSION OF THE PHASE DIAGRAM

Figure 1 presents a schematic picture of the phase boundaries in a polydisperse polymer solution having upper critical solution temperature (UCST) behaviour, together with the molecular interpretations of the observed phenomena, shown schematically in squares A–F as a function of temperature and concentration (the pressure is assumed to have a constant value). In the high temperature region of the phase diagram (above 297 K) the system is stable, which means that the polymer and solvent are compatible in every ratio, leading to one homogeneous concentrated or dilute phase (squares D and A in Figure 1 respectively). For a sample having a certain fixed polymer concentration, a decrease in temperature will encounter the miscibility gap and induce the formation of two-phase emulsions after some delay which is due to the need for droplets of a new phase to be nucleated. The onset of phase separation is the so-called cloud point of the solution and the cloud point temperature T_{cl} varies with concentration resulting in a cloud point curve (CPC). At sufficiently low viscosities gross separation into two phases will show up at the cloud point when the droplets formed have coalesced and settled. By sufficiently fast cooling of the system, the cloud point can be brought into the metastable region of the phase diagram. Here the onset of phase separation is foreshadowed by substantial concentration fluctuations as is shown in the middle squares B and E of Figure 1. These fluctuations also occur above the cloud point but are usually very weak unless they occur close to the critical concentration (see below). The closer one approaches the spinodal, the larger these fluctuations become, since at the spinodal temperature, which is the limit for metastable equilibrium, the energy barrier against diffusion into two phases vanishes. At and below the spinodal temperature, the system will spontaneously phase separate by a kinetic mechanism known as spinodal decomposition.[18] This spinodal decomposition process confers immense importance on the spinodal in technology, *e.g.* of metals,[19] ceramics and blends, as well as in polymer science. Approaching T_s from the metastable region in the phase diagram would allow the irregular fluctuation regions to become of infinite size if it were not stopped by the formation of spherical droplets (squares C and F, Figure 1) which form an emulsion. The stability of this emulsion is limited by a tendency of slow coalescence of droplets. Fluctuations in density and concentration give rise to inhomogeneities in the refractive index of the mixture which are detectable by low angle light scattering. By extrapolation to infinite scattering intensity, the spinodal temperature of the system is located.

Another special point in the phase diagram is the critical point, denoted by (ϕ_c, T_c), where the CPC and spinodal curve touch each other and share a common tangent as required by thermodynamics. Only for mixtures of exactly two chemical species, *e.g.* pure solvent plus completely homodisperse polyethylene, is the critical point situated at a common extremum of both the spinodal and binodal curve. In the case of UCST behaviour the extremum is a maximum; if it is a minimum, it is called a lower critical solution temperature (LCST). In polydisperse polymer solutions, even for polymers characterized by a narrow MWD, the critical point moves appreciably away from the extremum of the CPC towards the polymer rich side of the phase diagram as is shown in Figure 1.

The critical scattering phenomenon is due to the metastable inhomogeneities in the refractive index induced by the irregular concentration fluctuations schematically shown in squares B and E of Figure 1, which foreshadow the approach to phase separation. Debye *et al.*[20,21] have developed the appropriate thermodynamic theory for fluctuation scattering. From their theory it follows that in a first approximation and for a binary mixture the light-scattering intensity in the limit of zero scattering angle I_0 is inversely proportional to an appropriate measure of the curvature of the plot of molar Gibbs free energy G against the concentration ϕ at constant pressure. For a binary (or two-component) system, the second derivative $(\delta^2 G/\delta\phi^2)_{T,P}$ is the correct measure. This requires generalization to the Gibbs determinant for multicomponent systems (see below). The spinodal points are those where $(\delta^2 G/\delta\phi^2)$, or generally the Gibbs determinant, vanishes. In Figure 2, three

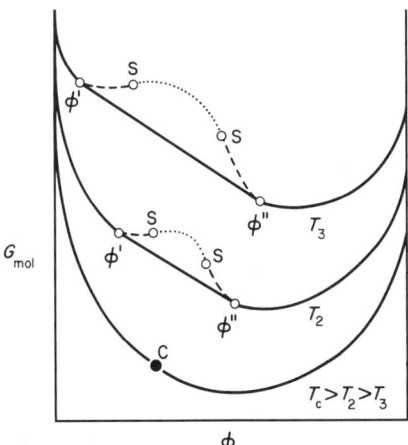

Figure 2 Schematic diagram of the molar Gibbs free energy G of a strictly binary polymer solution showing UCST demixing as a function of the volume fraction ϕ of the polymer for different successive temperatures (the pressure is kept constant). The highest temperature T_c is the critical temperature where, at the critical volume fraction ϕ_c, the radius of curvature approaches infinity. At lower temperatures, where phase separation is allowed, a loop in the free energy curve springs from the point denoted C. The dotted portions represent unstable regions of the two sample curves, while the broken-line portions represent metastable regions. S are the spinodal points (inflection points) and ϕ' and ϕ'' are the volume fractions of the polymer in the diluted and concentrated phase respectively, which are in equilibrium at the temperature under consideration

typical plots of the molar Gibbs free energy as a function of concentration are presented for different temperatures. Above the temperature T_c the sense of curvature of the free energy will be the same (convexity property) for all compositions ϕ, which means that the system is stable. At the temperature T_c the curvature is zero, *i.e.* the curvature vanishes. For temperatures below T_c each of the two upper plots shows two inflection points, marked S. These are spinodal points, since clearly the curvature vanishes there. Consider the broken portion of the curve. Small concentration fluctuations may lead to a separation into two regions, one more concentrated, the other more diluted than the original homogeneous phase. The free energy of the density fluctuating system then moves vertically away from the homogeneous system curve towards the cord spanning the two concentrations; the free energy is higher than that of the homogeneous system. Therefore the system is metastable and phase separation will not occur. In the dotted part of the curve, a local disturbance leading to regions of different concentration will lower the free energy because the cord spanning of these two concentrations lies below the original convex upward curve. This means that the system is unstable and the driving force for restoring homogeneity has vanished here. The inflection points S thus form the limit of the metastable region in the phase diagram, which is the spinodal. The two points ϕ' and ϕ'' share a common tangent to the free energy curve and are the volume fractions of two polymer solutions in equilibrium at that particular temperature. The locus of these points as a function of temperature is the binodal curve. C is the point where spinodal and binodal meet and share a common tangent at T_c and ϕ_c, which therefore are the critical temperature and concentration.

The general shape for the three plots in Figure 2 is dependent on the molecular weight of the two components in such a way that the greater the discrepancy in their size and shape, the greater the dissymmetries in the diagram will be, moving the miscibility gap towards the axis $\phi = 0$. Further, it has to be pointed out that for a critical point related to LCST demixing, the undulating curve would lie below the bottom curve. However, all the molecular features involved and explained here remain the same. Finally, it should be mentioned here that the simple measure $(\delta^2 G/\delta \phi^2)$ for the curvature is no longer a measure for the scattering intensity in the case of mixtures of more than two components. Consider a solution of a polymer in a single solvent. For an n-component polymer (usually n homodisperse polymer fractions) we have n independent volume fractions ϕ_i $(i = 1, \ldots, n)$ as coordinates leaving the solvent volume fraction determined by the normalization condition $\phi_0 + \sum_{i=1}^{n} \phi_i = 1$. The molar free energy G_{mol} may be thought of as a hypersurface in the n-dimensional space and the important quantity of scattering intensity and curvature is now, following Gibbs[22] stability analysis, the spinodal determinant $|\delta^2 G/\delta \phi_i \delta_j|_p$. This quantity vanishes whenever there is some direction in space where the free energy locally becomes flat (saddle point). The spinodal will now be located in composition–temperature space on the contours where the condition

$$J_s \equiv \left| \frac{\delta^2 G}{\delta \phi_i \delta \phi_j} \right|_{p,T} = 0 \tag{1}$$

holds, where ϕ_i is the volume fraction of the polymer species in the molecular weight distribution.

4.3   CRITICAL LIGHT SCATTERING

The thermodynamic theory of light scattering relates the light intensity I_θ of light scattered by a homogeneous solution at an angle θ to the incident beam to the free energy of mixing and the concentrations by the proportionality relation

$$I_\theta \propto TP(\theta)/J_s, \quad i,j = 1, 2, \ldots, n \tag{2}$$

where $P(\theta)$ is the properly averaged particle-scattering factor. From this it follows that light-scattering measurements contain information on: (i) the size and shape of particles through $P(\theta)$ (for an example, see ref. 23); (ii) the spinodal determinant J_s in homogeneous systems if the proportionality constant in equation (2) is determined or known;[24,25] and (iii) the spinodal determinant in heterogeneous systems, whose zones are the prime objective of the PICS technique, measuring intensities at constant sufficiently small θ. This leaves $P(\theta)$ a constant [merely used as $P(30°)$] to be absorbed in the proportionality factor, and $\theta = 30°$ is an adequate approximation to the limit as $\theta \to 0$.

The Gibbs spinodal determinant J_s in equation (2) vanishes at the spinodal temperature T_s. In the vicinity of T_s, the small value of the determinant is expanded by the Debye–Scholte theory in a

Taylor series, truncated after the linear term, in powers of the distance $\delta T = T - T_s$ from the spinodal temperature. The constant term is then proportional to $\sin^2(\theta/2)$

$$I_\theta \propto \frac{T}{[c_0 \sin^2(\theta/2) + c_1 \delta T]} \tag{3}$$

where c_0 and c_1 are constants. In the range $0.03\,°C < \delta T < 5\,°C$ and at an angle of about $30°$, the equation is approximated by

$$I_{30} \propto \frac{1}{\delta T} \tag{4}$$

This theoretical prediction is verified by the linearity of Debye–Scholte plots of ΔI_{30}^{-1} against temperature (Figures 3–5). From these figures it can be seen that the classical linear theory is found to be adequate for the present purposes. ΔI_{30} is here the difference in scattering intensities at $30°$ detection angle between solution and solvent (deducting the small solvent scattering effect is equivalent to eliminating the scatter due to density fluctuations in the solvent superimposed on the concentration fluctuations of interest).

Scholte used standard light-scattering equipment. His results on the polystyrene/cyclohexane system have been confirmed and extended to a wider range of polystyrene samples by PICS measurements.[1] Chu and co-workers[24, 26] use highly sensitive and specially constructed apparatus for locating spinodals based on a generalized non-linear extrapolation method.

4.4 LOCATION OF SPINODALS BY PICS

Measuring and collection of pairs of data points (ΔI_{30}, T) by automated and computer-controlled conditions is the aim of the PICS apparatus. Debye–Scholte extrapolation of such series of pairs of data points (see Figures 3, 4 and 5) lead to spinodal data pairs (ϕ, T_s) from which the phase diagram

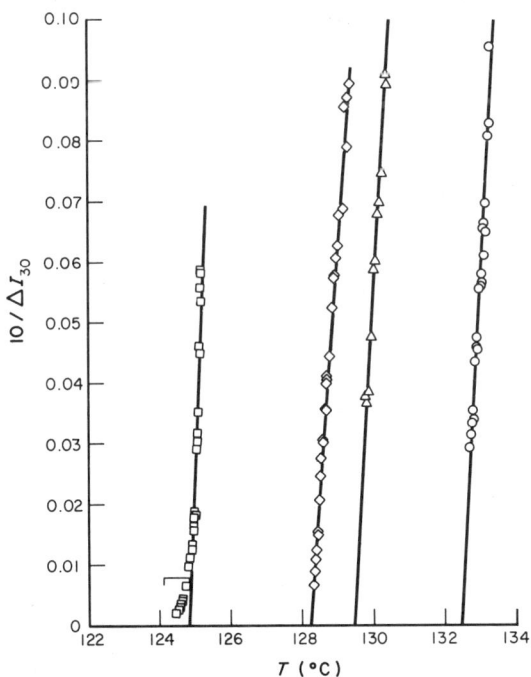

Figure 3 Typical Debye–Scholte plots for four solutions of a low density polyethylene in diphenyl ether recently measured (see also Figure 11). The ordinates are proportional to the reciprocal intensity difference of scattering between solution and solvent at $30°$ detection angle. The wt. % of polymer and the extrapolated spinodal temperature are (going from left to right): □ 14.58%, $T_s = 124.8\,°C$; ◇ 10.13%, $T_s = 128.2\,°C$; △ 8.66%, $T_s = 129.4\,°C$; and ○ 5.62%, $T_s = 132.4\,°C$. The regression lines to obtain the values for T_s were computer drawn. For one sample (□), a number of points have been included in the plot at the bottom part where the onset of phase separation is noticed. Such points are obviously left out of the extrapolation procedure.

Note the sensitivity to polymer concentrations and the precision at elevated temperatures of these measurements

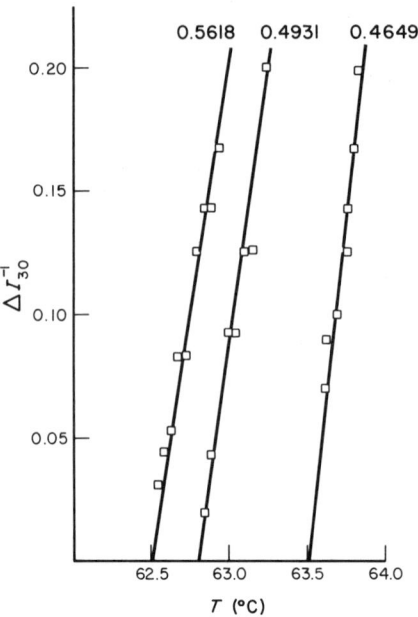

Figure 4 Debye–Scholte plots for three mixtures of low molecular weight polystyrene ($M_w = 2.10$ kg mol^{-1}) and polyisoprene ($M_w = 0.82$ kg mol^{-1}). The parameter values are weight fraction concentrations (data from ref. 8)

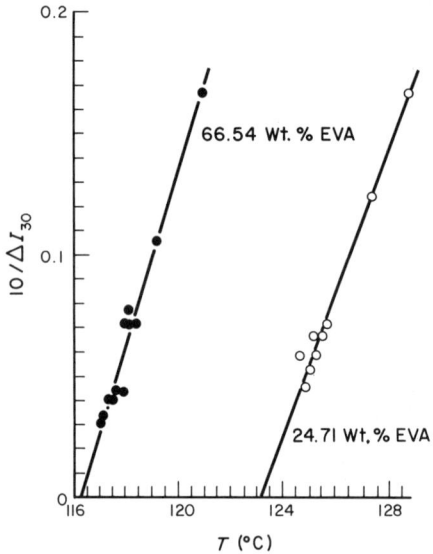

Figure 5 Debye–Scholte plots for two blends of poly(ethylene/vinyl acetate) [poly(EVA)] and a low molecular weight polyethylene with narrow MWD. The copolymer has a content of 7.5% VA and the concentrations of copolymer in the blend are indicated (measurements by H. Grooten)

can be constructed (for an example see Section 4.7). Although the principle of collecting spinodal data has not been changed since the most simple version of the apparatus,[17] the present commercially available and computerized equipment (described in Section 4.6) includes the most recent software and hardware applications and design.

About 15 mg of the sample of interest, usually a polymer solution or polymer mixture, is forced into a thin-walled glass capillary cell of 1 mm inner diameter. This cell has one lens-shaped end and after being filled with the sample of accurately known polymer concentration or ratio, the other end of the cell is sealed under vacuum. The cell thus obtained has a length of about 25 mm which allows it to be mounted in the sample carousel of the PICS apparatus, where four samples can be analyzed in one run. Although a very important step in the procedure, the preparation of the samples requires only a limited amount of skill and time.

Essentially, the technique consists of two operations, explained in the following for a polymer solution with UCST behaviour (see Figure 6a for a plot of the typical phase diagram). The system is first exposed to a fast thermal pulse, going from a temperature T_1 (typically 10 to 20 °C above the cloud point T_{cl}) to T_2 lying in, or just above, the metastable region of the phase diagram (between T_{cl} and the desired spinodal temperature T_s) (see Figures 6a and 6b). During the thermal pulse, the capillary cell is axially illuminated by a low power (about 1 mW) laser beam. Thermal equilibrium is reached within a few seconds. Due to the concentration fluctuations light is scattered from the sample and observed at selected detection angles of 30° and 90° by using light guides leading to photosensitive detectors (see Figure 6b) while the temperature is kept constant. The holding time at T_2 is at least about 10 s for polymer solutions but can be as high as 30 min for very viscous bulk mixtures. Intensities of scattered light are measured after immersion of the sample and again when metastable fluctuation equilibrium is established, which determines the necessary holding time (Figure 6c). The measured values for I_{30} and I_{90}, sampled at the proper times electronically, the temperature values and the heating conditions are processed by computer. Before nucleation and macroscopic growth of emulsion particles lead to phase separation, the sample is returned into the high temperature compartment of the PICS machine for immediate rehomogenization. The cells are thermally pulsed one by one in rotation. At successive pulses the temperature of the oil bath is changed by typically 0.01 to 0.03 °C towards the spinodal temperature.

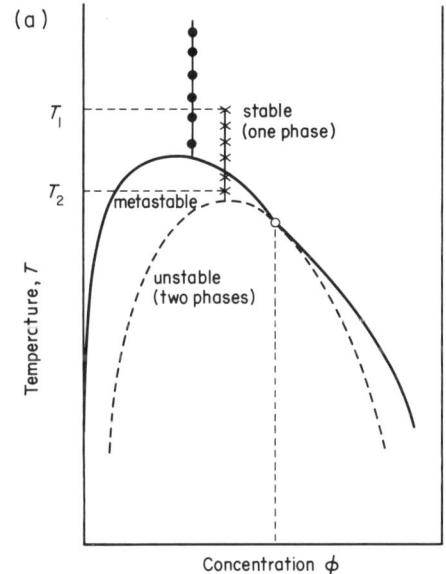

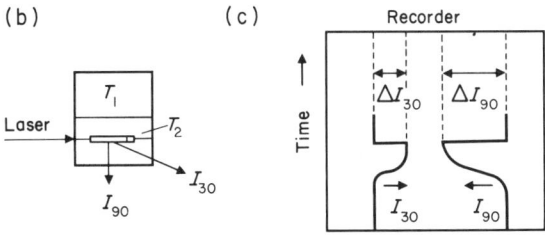

Figure 6 (a) Schematic phase diagram for a quasi-binary polymer solution showing upper critical solution temperature behaviour. The extrapolation ranges for determination of the spinodal temperature are compared for the methods developed by Scholte[6,7] and Chu[24] (●) and in PICS (×); – – – – spinodal curve, ——— cloud point curve, ○ critical point. (b) Principle of pulse-induced critical scattering. The cell containing the sample, mounted on a carousel, is quickly transferred from the stirred air box (at temperature T_1 above the cloud point of the solution) into the stirred oil bath (temperature T_2) in about 50 ms with the aid of a stepping motor. (c) Typical recorder measurement of equilibrium fluctuation light scattering at a temperature in the metastable part of the phase diagram. Scattering angles are 30° and 90° measured by photosensitive detectors through light guides placed in the oil bath

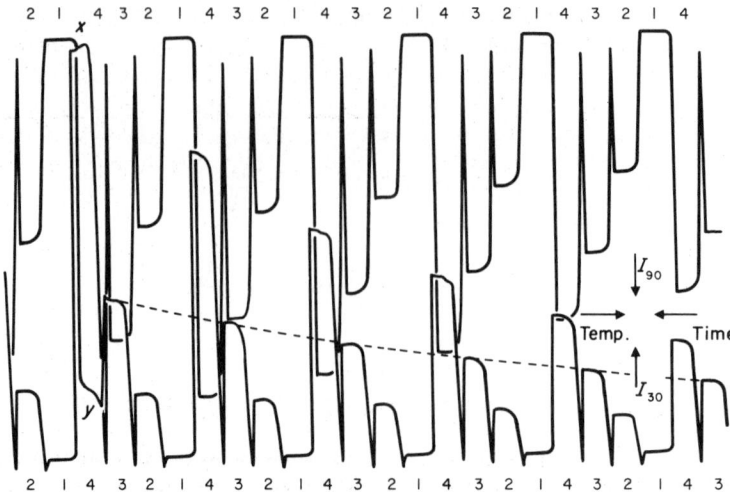

Figure 7 Recorder plot — to be read from right to left — of a PICS run with four tubes (numbered 1–4). Six successive pulses are shown for each of the four cells in rotation. Scattered intensities at 30° are plotted from the bottom upwards, 90° are plotted from the top downwards. For details see text of Section 4.4.

Figure 7 shows the light intensities recorded for four sample cells over a temperature interval of about 0.4 °C. The temperature linearly decreases from right to left over a total time of approximately six minutes. Cell 1 shows hardly any scattering while cell 2 is sufficiently close to its T_s to show some scattering. The effect is more pronounced in cells 3 and 4, which show a rapid increase in metastable fluctuation scattering. A small amount of superposed noise is visible in the case of cell 4. The slow-rise portions of the I_{30} pulses of cells 3 and 4 can be joined with a single temperature–intensity curve which is a proof for true equilibrium–metastable equilibrium fluctuation scattering. Cell 4 is so close to its T_s that the intensity pulses show a maximum (x in I_{30}, y in I_{90}) which is characteristic for the onset of the formation of droplets. The I_{90} scattering is observed to detect this phenomenon with great sensitivity and to exclude in this way I_{30} values masked by phase separation from the Debye–Scholte extrapolation procedure. Further, the I_{90} scattering intensity may also be helpful in the determination of cloud points (see Section 4.5). Apart from phase separation effects upon intensity measurements (x in Figure 7), there are other effects which will prevent the observation of indefinitely large I_{30} values as T_s is approached, *e.g.* the cut-off effect coming from the term $c_0 \sin^2(\theta/2)$ in equation (3). This term will thus raise $1/I_{30}$ near T_s. Such measurements, together with a small number of outlier points (occasioned by one of several possible causes — see end of Section 4.6) can be left out of the extrapolation procedure by the operator, or by the computer using statistical criteria. By the methods outlined here, spinodal temperatures in polymer systems showing partial miscibility can be determined with an accuracy of about 0.2 °C or better[1] (for example see Figures 3–5 and Section 4.7).

4.5   DETERMINATION OF CLOUD POINTS BY PICS

The cloud point temperature T_{cl} of a polymer solution or blend of known composition can be measured approximately by simply holding a sample in the laser beam while slowly cooling the oil bath starting from a temperature T_1 (Figures 6a and b) in the homogeneous region of the phase diagram. The slightest increase in the intensity of the scattered light observed then gives an indication that T_{cl} has been reached and that the formation of droplets has started. This standard practice of determining thermodynamic binodal or cloud points rapidly from the homogeneous side turns out to be somewhat troublesome merely due to hysteresis and baseline effects. However, this fast technique may be used to do experimental investigation on the miscibility behaviour of polymer systems for industrial purposes or to expose major structural differences between samples (for instance copolymers differing in composition of one of the two monomers, see Section 4.7). The accuracy of cloud point measurements thus obtained may typically be estimated to be about 1 °C. Relating thermodynamic measurements to molecular theories requires at least an uncertainty which is five times better on the location of the cloud point. Therefore Gordon *et al.* investigated the seeding technique for the determination of cloud points and tested it on polystyrene/cyclohexane solutions. In the seeding method, phase separation is approached from the heterogeneous side of the phase diagram using an emulsion carefully seeded to contain only minute droplets by a thermal step to a temperature just above the spinodal. The cloud point is then identified as the temperature where the small droplets are dissolved by heating when the scattering intensity returns to the low level characteristic for equilibrium fluctuation scattering. The method has been described in detail elsewhere[1] and the results show that supercooling can be serious in determining cloud points of dilute polymer solutions. One of the subjects of present investigation in PICS is how the 'seed' method applies to polymer mixtures.

Once the cloud point curve and spinodal curve are measured by PICS, the whole phase diagram can be constructed for the partially miscible polymer system under investigation, including a rough indication of the critical temperature and concentration where the two curves share a common tangent. It has been observed[27] that the spinodal curve does not always touch the CPC in the critical region but is slightly shifted downwards in temperature in the case of UCST behaviour over a temperature range of at most 0.5 °C. This is due to the high absorption in the critical region and it has been shown that this systematic discrepancy between spinodal and critical temperatures can be overcome by correcting the intensities measured by PICS for this absorption.[28]

4.6 DESCRIPTION OF THE PICS INSTRUMENT

The practical requirements of the pulse-induced critical scattering system are for instrumental control, data collection and data analysis. In its most recent form, it has therefore been convenient to integrate an IBM PC compatible computer into the design of the instrument (see Figure 8). A cruciform holder called a carousel is used to transport cells between two temperature-controlled baths. The rotary axis of the carousel is positioned within a stirred air bath at temperature T_1. One of the four arms of the carousel projects through an aperture to a stirred liquid bath at temperature T_2. A cell is mounted at the end of each arm and the carousel is positioned so that three cells lie within the air bath and one cell is completely immersed in the liquid bath at the lowest point of its carousel orbit. The other cells are immersed in turn, through a sequence of 90° rotational steps, to this same precise position in the liquid bath. Each remains there for a fixed period of time, and is then returned to the air bath by the following step.

The carousel is rotated by means of a stepping motor which has a single step angle of 0.9°. The motor is programmed with an optimum speed profile to rotate through 90° and to accurately position the immersed cell in line with three light guides. These carry the incident light beam and gather the scattered light at 30° and 90° from the incident beam. The light guides are mounted close to the cell position in order to reduce the amount of stray light interference as much as possible and to minimize dispersion due to dust or bubbles in the bath fluid. The cells are identified by rotary position detectors in order to aid automation and to ensure the valid location of the cell positions.

The liquid bath temperature is usually held constant or programmed to follow a slowly changing temperature profile. A wide range of power control is required and this can be achieved by utilizing up to six different power heaters under background control, or as digitally selected proportional controls. The air bath temperature is maintained roughly constant at a level necessary to keep the

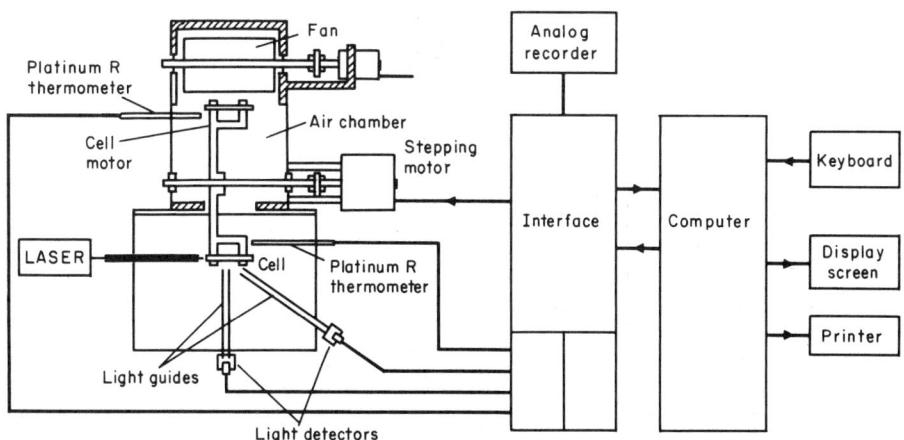

Figure 8 Simplified schematic diagram of the PICS machine

contents of all cells in a stable homogeneous state; this is done by a single digitally controlled heater. The programmed control selects half percentage power levels switched at zero power crossings over a two-second integration period. Temperature measurements over a range of 300 °C to a resolution of 0.01 °C are obtained by balancing a platinum resistance bridge using a 12-bit digital to analogue converter and then using a 12-bit analogue to digital converter with a ratiometric algorithm to interpolate the high resolution balance. Zero temperature coefficient resistors trimmed to give accurate calibration values for the platinum resistors at 0 and 100 °C are switched under program control to provide regular automatic self-calibration of the temperature-measuring systems. The intensity of scattered light is measured immediately after the cell arrives at its position in the liquid bath, while the true cell temperature is still close to that of the air bath. This 'background' scattering intensity is subtracted from the intensity measured after the cell has equilibrated at the liquid bath temperature and sufficient time has elapsed for the metastable (or possibly stable but weak) equilibrium fluctuations to become established.

When the measurement phase is complete the data analysis prompts the operator to select the useful temperature range for the measurements, including especially the valid metastable scattering region of the data peaks for each cell. This is quite simply done by examining the tabulation of scattered values and entering the corresponding temperature limits into the computer for each cell. The program then calculates the spinodal temperatures by constructing the Debye–Scholte plots for the data. During the initial fitting of the data, the program may list a small number of stray points, which are rejected by the computer as outliers on statistical grounds. Such occasional outliers might be due to various *random* causes, *e.g.* scattering from a dust particle in the stirred oil bath, when it passes through the laser beam in the small space between the sample cell and light guide. (The bath should therefore be kept as dust free as possible). *Systematic* outliers, recording too low an intensity, are generally seen at the low end of the Debye–Scholte plot. This is illustrated by the squares in Figure 3 and attributed to the cut-off effect described in Section 4.4.

4.7 EXAMPLES OF PICS MEASUREMENTS

We now illustrate the importance and sensitivity of PICS measurements by showing some recent examples of cloud point curves and spinodals detected in various polymer systems. In doing so, we do not want to go into detail about the molecular origin or thermodynamic theories behind these experimental findings because they are not relevant for the present purposes and are explained elsewhere. The sensitivity of measured cloud point curves by PICS towards structural differences in polymers such as copolymer composition is underlined by Figure 9 where measured phase

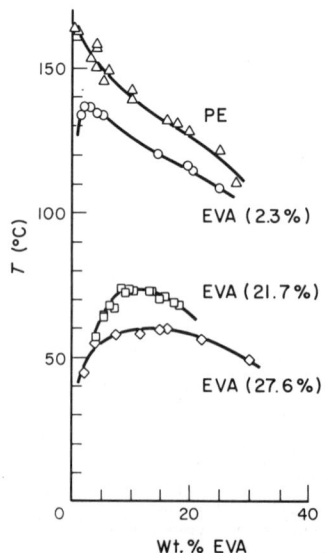

Figure 9 Cloud point curves measured by PICS for various quasi-binary mixtures of EVA copolymers in solution of 2-heptanone (content of VA in the copolymers indicated). PE denotes a linear low density polyethylene solution ($M_n = 7.9$; $M_w = 89$; $M_z = 730$ kg mol^{-1}) (curves drawn by hand; measurements taken from ref. 29)

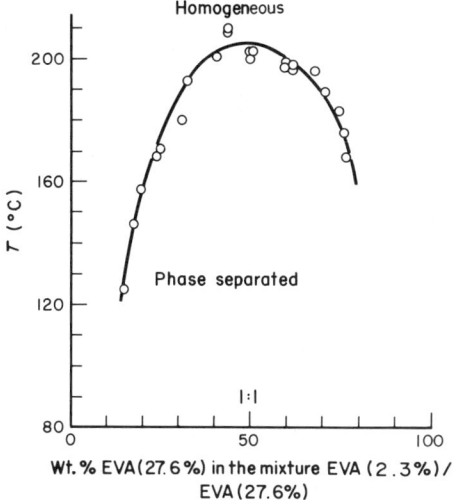

Figure 10 Cloud point curve, drawn by hand, for random ethylene/vinyl acetate copolymer mixtures having different contents of vinyl acetate (VA) (indicated) measured by PICS for various ratios of the copolymers. All samples have 40 wt % overall copolymer concentration, the solvent being 2-heptanone (60 wt %) (measurements taken from ref. 29)

boundaries for ethylene/vinyl acetate copolymers (EVA) differing in VA concentration are shown and are seen to fall remarkably rapidly with temperature as the weight percent of VA in the statistical copolymer increases[29] (2-heptanone is chosen to be the solvent here). Figure 10 shows a CPC for EVA mixtures, differing by 25.3% in VA concentration, of 40 wt% overall copolymer concentration in 2-heptanone.

The effect of chain branching on liquid–liquid phase relationships in solution has been experimentally and theoretically studied in detail for the system low density polyethylene (PE) in diphenyl ether.[11, 16, 30] In the same way as polydispersity, chain branching will also be extremely sensitively reflected in spinodal curves. One of the samples studied before (a PE fraction, labelled number 5, ref. 10, stored under dry and dark conditions) has recently been reinvestigated with PICS. This was a first attempt in a much more detailed experimental study[31] of spinodal curves in EVA copolymers in solution of diphenyl ether, PE being one of the homopolymers. In Figure 11, we compare the spinodal recently measured with PICS to the results obtained before. It turns out that the spinodal has shifted over 1 to 2 °C to lower temperatures, which suggests[30] that a higher degree of branching has appeared in the sample during ageing. Determining the mass molecular weight by static light

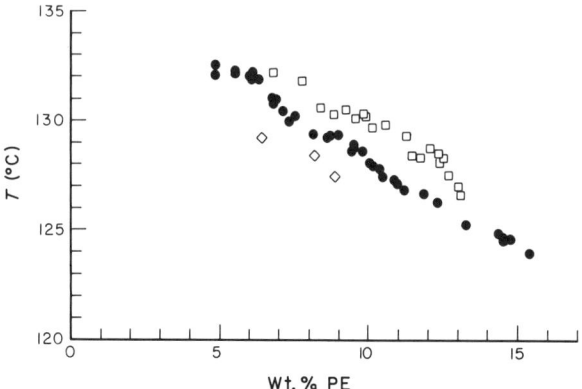

Figure 11 Spinodal measurements by PICS for a low density polyethylene fraction in diphenyl ether. The curve (□) was measured by Kleintjens[11, 30] (sample no. 5, $M_w = 70$ kg mol^{-1}). The second curve (●) was recently measured for the same sample ($M_w = 100$ kg mol^{-1}, see text, Section 4.7, higher degree of branching during ageing of the sample) by Van der Haegen and Van Opstal.[31] This again illustrates the sensitivity of the PICS technique towards structural changes in the polymer. All the samples measured here (□ and ●) contain a polymer stabilizer, dibutyl-*p*-cresol (DBPC) in a concentration of 2 g l^{-1} solvent. The influence of the concentration of DBPC on the spinodal decomposition of PE/diphenyl ether can be seen from the spinodal points (◇) where the concentration of DBPC is ten times higher (measurements by H. Roolant)

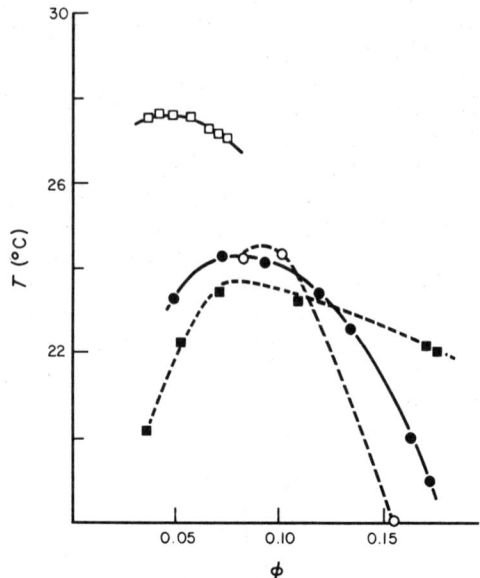

Figure 12 Spinodal points by PICS for branched polystyrene samples: ○ $M_n = 265$, $M_w = 618$; ● $M_n = 285$, $M_w = 770$; ■ $M_n = 306$, $M_w = 1.880$ kg mol^{-1}; measurements taken from ref. 32, dashed curves drawn by hand, □ PICS data for linear polystyrene at $M_w = 500$ kg mol^{-1} (data from ref. 33)

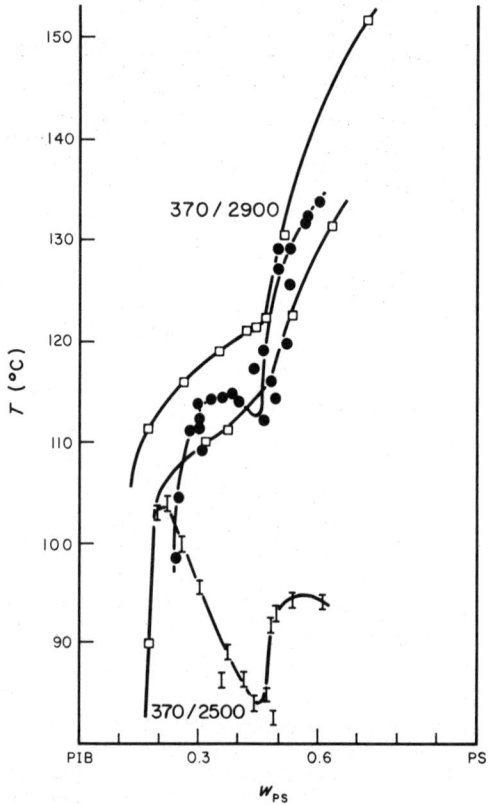

Figure 13 Spinodals (I, ●) and cloud point curves (□) for polyisobutene/anionic polystyrene (molar masses indicated in g mol^{-1}). Curves drawn by hand; w_{PS} = mass fraction polystyrene in the blend[15,34]

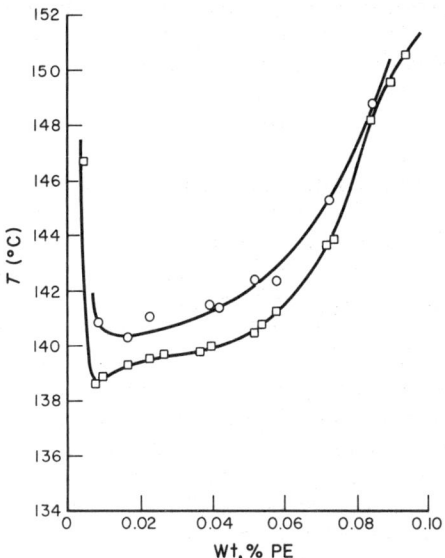

Figure 14 PICS measurements of spinodals ($\bigcirc$) and cloud points ($\square$) for a system with LCST behaviour. Linear polyethylene (PE) ($M_n = 8$, $M_w = 177$, $M_z = 990$ kg mol^{-1}) in *n*-hexane was measured under ambient pressure of the solutions (*ca.* 5 atm). Data by Hembury and Kleintjens[8]

scattering reveals a higher value for M_w (100 kg mol^{-1}) than before (70 kg mol^{-1}, ref. 11) but also gives an indication of a much higher degree of branching. A somewhat similar effect has been observed by PICS on the spinodal decomposition of branched polystyrenes in cyclohexane.[32] In this work, linear polystyrene having a narrow MWD has been γ-irradiated for several periods of time. From Figure 12, it can be seen that the location of the spinodal is hardly affected by crosslinkage, increasing the molar mass by almost a factor of ten.

An example of PICS spinodals and cloud point curves in polymer mixtures is given in Figure 13 for the polystyrene/polyisobutene system. Here, two-peaked spinodals are observed to go with cloud point curves having a shoulder.[15, 34] It can be noted from Figure 14 that, although we have described the phase diagrams (Figures 2 and 6a) of UCST behaviour, PICS can also be applied to polymer systems showing LCST behaviour, such as the PE/*n*-hexane system.[8]

4.8 CONCLUSIONS

It can be concluded that pulse-induced critical scattering is a very useful, elegant and fast technique for the determination of phase boundaries and spinodal curves in macromolecular systems, thus serving the understanding of molecular backgrounds of liquid–liquid phase separation and the refinement of statistical–thermodynamic modelling of these phenomena. The advantages of PICS over conventional light scattering, apart from it being less time-consuming, lies in the fact that the desired spinodal temperature is more closely approached so that the extrapolation of data to locate T_s is more accurate. Finally, note that most of the presented measurements by PICS were made possible through the use of the 'centrifugal homogenizer' (CH), developed by the same researchers mentioned in Section 4.1 and designed to fulfil the need for preparing homogeneous bulk mixtures up to very high viscosities, thus eliminating the very slow process of spontaneous diffusion. We refer to the literature for more details on the principle and on thermodynamic measurements with the CH.[35 – 37]

4.9 REFERENCES

1. K. W. Derham, J. Goldsbrough and M. Gordon, *Pure Appl. Chem.*, 1974, **38**, 97.
2. M. Gordon, P. Irvine and J. W. Kennedy, *J. Polym. Sci., Polym. Symp.*, 1977, **61**, 199.
3. M. Gordon, J. Goldsbrough, B. W. Ready and K. W. Derham, in 'Industrial Polymers', ed. J. H. S. Green and R. Dietz, Transcripta, London, 1973.
4. R. Koningsveld and L. A. Kleintjens, *J. Polym. Sci., Polym. Symp.*, 1977, **61**, 221.

5. P. Debye, *J. Chem. Phys.*, 1959, **31**, 680.
6. Th. G. Scholte, *J. Polym. Sci., Part A-2*, 1970, **8**, 841.
7. Th. G. Scholte, *J. Polym. Sci., Part C*, 1972, **39**, 281.
8. H. Galina, M. Gordon, B. W. Ready and L. A. Kleintjens, in 'Polymers in Solution', ed. W. C. Forsman, Plenum Press, New York, 1986.
9. H. Galina, M. Gordon, P. Irvine and L. A. Kleintjens, *Pure Appl. Chem.*, 1982, **54**, 365.
10. R. Koningsveld and L. A. Kleintjens, 'Macromolecular Chemistry — 8', ed. K. Sareela, Butterworths, London, 1973.
11. L. A. Kleintjens, H. M. Schoffeleers and R. Koningsveld, *Ber. Bunsenges. Phys. Chem.*, 1977, **81**, 980.
12. E. Nies, R. Koningsveld and L. A. Kleintjens, *Prog. Colloid Polym. Sci.*, 1985, **71**, 2.
13. D. M. Koenhen, C. A. Smolders and M. Gordon, *J. Polym. Sci., Polym. Symp.*, 1977, **61**, 93.
14. E. L. Atkin, L. A. Kleintjens, R. Koningsveld and J. L. Fetters, *Makromol. Chem.*, 1984, **185**, 377.
15. R. Koningsveld, L. A. Kleintjens and M. H. Onclin, *J. Macromol. Sci., Phys.*, 1980, **18**, 363.
16. L. A. Kleintjens, R. Koningsveld and M. Gordon, *Macromolecules*, 1980, **13**, 303.
17. J. M. G. Cowie, J. Goldsbrough, M. Gordon and B. W. Ready University of Essex, *Br. Pat.* 1 377 478 (1971) (*Chem. Abstr.*, 1972, **77**, 103 698); *Ger. Pat.* 161 555 (1971); *US Pat.* 3 807 865 (1974).
18. J. Goldsbrough, *Sci. Prog. (Oxford)*, 1972, **60**, 281.
19. H. E. Cook and J. E. Hilliard, *Trans. Metall. Soc. AIME*, 1965, **233**, 142.
20. P. Debye, H. Coll and D. Woermann, *J. Chem. Phys.*, 1960, **33**, 1746.
21. P. Debye, B. Chu and D. Woermann, *J. Chem. Phys.*, 1962, **36**, 1803.
22. 'The Scientific Papers of J. W. Gibbs', ed. J. W. Gibbs, Dover, New York, 1961, vol. 1.
23. W. Burchard and K. Kajiwara, *Proc. R. Soc. London, Ser. A*, 1970, **316**, 185.
24. B. Chu, F. J. Schoenes and M. E. Fisher, *Phys. Rev.*, 1969, **185**, 219.
25. Th. G. Scholte, *Eur. Polym. J.*, 1970, **6**, 1063.
26. B. Chu, *J. Chem. Phys.*, 1967, **47**, 3816.
27. H. Galina, M. Gordon, P. Irvine and L. A. Kleintjens, *IUPAC Int. Symp. Adv. Polym. Char.*, 1981.
28. E. Nies, Ph. D. Thesis, University of Antwerp, 1983.
29. R. Van der Haegen, in 'Integration of Fundamental Polymer Science and Technology', ed. L. A. Kleintjens and P. J. Lemstra, Elsevier, London, 1986.
30. L. A. Kleintjens, Ph. D. Thesis, University of Essex, 1979.
31. R. Van der Haegen, L. A. Kleintjens and L. Van Opstal, work in progress.
32. K. Kajiwara, W. Burchard, L. A. Kleintjens and R. Koningsveld, *Polym. Bull. (Berlin)*, 1982, **7**, 191.
33. P. Irvine and M. Gordon, *Macromolecules*, 1980, **13**, 761.
34. R. Koningsveld and L. A. Kleintjens, in 'Polymer Blends and Mixtures', ed. D. J. Walsh, J. S. Higgins and A. Macchonnachie, Martinus Nijhoff, Dordrecht, 1985.
35. M. Gordon, L. A. Kleintjens, B. W. Ready and J. A. Torkington, *Br. Polym. J.*, 1978, **10**, 170.
36. M. Gordon and B. W. Ready University of Essex, *US Pat.* 4 131 369 (1978); *Br. Pat.* 1 584 776 (1978).
37. R. Koningsveld, M. H. Onclin and L. A. Kleintjens, in 'Polymer Compatibility and Incompatibility, Principles and Practices', ed. K. Šolc, MMI Press, Harwood Academic, New York, 1982.
38. Further information and commercial conditions can be obtained from Stamicarbon BV, PO Box 53, 6160 AB Geleen, The Netherlands.

5
Polymer Blends

DAVID J. WALSH

E. I. Du Pont de Nemours & Co., Wilmington, DE, USA

5.1 INTRODUCTION

In the search for new polymer materials people have polymerized new monomers, or made new random, block or graft copolymers from existing monomers. A third alternative has been to blend existing polymers to produce materials with new properties. An obvious advantage of this approach is that it usually requires little or no capital expenditure relative to the production of new polymers. It is also possible to produce a range of materials with properties completely different from those of the blend constituents. These materials may be one phase or two phase. It is convenient to define miscibility as the ability to be mixed at a molecular level to produce one homogeneous phase. The term compatibility is often used interchangeably but is also used practically to mean 'able to be mixed to produce useful materials' and as such is often used to describe immiscible materials.

When two polymers are blended, by whatever method, the most likely result is a two phase material. The reason why two polymers are not usually miscible becomes apparent from simple thermodynamic considerations. A necessary (but not sufficient) criterion for miscibility is that the free energy of mixing ΔG_m be negative. This is given by

$$\Delta G_m = \Delta H_m - T\Delta S_m \tag{1}$$

where ΔH_m is the enthalpy of mixing, ΔS_m is the entropy of mixing and T the absolute temperature. In terms of the lattice theory developed by Flory and Huggins[1,2] the entropy of mixing is given by

$$\Delta S_m = -R(N_1 \ln \phi_1 + N_2 \ln \phi_2) \tag{2}$$

135

where N_i is the number of moles and ϕ_i the volume fraction of component i and R is the gas constant. The enthalpy of mixing is given by

$$\Delta H_m = RT\chi_{12}N_1\phi_2 = BV_1N_1\phi_2 = (v_1 + v_2)B\phi_1\phi_2 \tag{3}$$

where v_1 and v_2 are the actual volumes of the components, V_1 is the molar volume of component 1, B is an interaction energy density and χ_{12} is the interaction parameter (relative to 1 mole of component 1) which can be expressed as

$$\chi_{12} = BV_1/RT = z\Delta w_{12}N_A/RT \tag{4}$$

where N_A is Avogadro's number, z is the coordination number of the lattice and Δw_{12} is the energy for the formation of an unlike contact pair which can be expressed as

$$\Delta w_{12} = w_{12} - \tfrac{1}{2}(w_{11} + w_{22}) \tag{5}$$

where w_{12}, w_{11} and w_{22} are the energies of the respective pair interactions. In the case of interactions between nonpolar molecules w_{12} is usually less than the mean of w_{11} and w_{22} and, by the geometric mean assumption, is given by $(w_{11}w_{22})^{1/2}$. This means that the enthalpy of mixing is expected to be positive, *i.e.* unfavorable for mixing.

We now consider the entropy of mixing as expressed in equation (2). As the molecular weight of the polymers in the blend becomes high then the number of moles in the blend, N_1 and N_2, will become very small. As the molecular weights tend to infinity the number of moles tends to zero and therefore so does the entropy of mixing. Since the entropy of mixing is very small and the enthalpy of mixing is expected to be unfavorable, polymers are not expected to mix.

In more advanced theories extra contributions to the free energy of mixing are considered which take account of possible volume changes accompanying mixing, which are assumed to be zero in the above treatment. As formulated these contributions are, however, also unfavorable for mixing.

Mixing can, however, be predicted to occur under three circumstances. (a) If the polymers are not of very high molecular weight then the entropy will not be negligible and may outweigh an unfavorable enthalpy of mixing. Thus one might expect some oligomer mixtures to be homogeneous. (b) If the enthalpy of mixing is positive but very small, then a small entropy of mixing may be sufficient. This might occur if the two polymers are very similar physically and chemically. Thus, for example, copolymers of very similar compositions might be expected to be miscible. (c) If the enthalpy of mixing is negative then two polymers would be expected to be miscible. This might occur if, for example, there is a favorable interaction such as a hydrogen bond between the polymers.

The miscibility of polymers was once considered to be a rare occurrence, but over the last twenty years the number of miscible pairs known has increased by an order of magnitude or more and many of them have found practical applications. A survey of miscible polymers can be found in the literature.[3]

Most blends which are used as structural materials consist of two or more phases. This arises since the inclusion of rubbery particles within a glassy matrix can produce an improvement in impact and other properties without too large a decrease in modulus. Miscible polymers are used as polymeric plasticizers when a decrease in modulus is desired, and these are used, for example, in PVC blends.[4] One can also make homogeneous blends of two glassy polymers with different glass transition temperatures such that a range of glass transitions can be obtained giving a range of ease of processability and a corresponding range of high temperature use properties. An example of this is the blend of polystyrene with poly(2,6-dimethyl-1,4-phenylene oxide)[5] with the trade name Noryl, though this commercial blend may also contain a rubbery second phase.

5.2 PREPARATION OF BLENDS

The most common commercial method for producing polymer blends is mechanical mixing, due to its simplicity and low cost. In the laboratory, examinations of miscibility are often conducted using methods requiring a third component, such as casting from a common solvent, since these methods require little material or complicated equipment. A third method, used both commercially and in the laboratory, is mixing *via* reaction such as the *in situ* polymerization of a monomer containing a polymer in solution. Blend preparation has been reviewed in the literature.[6,7] Each method has its own advantages and problems.

5.2.1 Mechanical Mixing

The preferred industrial method of mechanical mixing is by use of a screw extruder. This has the advantage that it can be run continuously and produces product in a convenient form for further processing. Polymer, usually solid pellets, is fed into the rear of the screw and it is driven forward by the rotating screw through the barrel where it melts and mixes. Twin screws and other devices are often used to increase shear and improve mixing.

Another common mixer is the Banbury mixer. This consists of a mixing chamber containing two counter-rotating multilobed cams. The most immediately noticeable feature of these mixers is the large size of the electric motors relative to the sizes of the mixing chambers. This is a result of the power required to stir semimolten polymer. A small mixer with a capacity of just a couple of kilograms may be driven by a motor which is 1 m across. Similar mixers to these are available with volumes down to 30 cm³ and are therefore more convenient for laboratory use than extruders, the smaller ones of which generally require 1–2 kg of material.

Many other types of mixers exist, the commonest of which is probably the two-roll mill in which mixing takes place by repeatedly forcing the polymers through the gap between two rotating rollers. These machines are also commonly used for converting bulk polymer into a sheet which is easier to handle.

The production of homogeneous blends by mechanical mixing poses special problems. Low molecular weight liquid polymers mix to form one phase very easily but the low diffusion rate of high polymers makes the mixing process very slow. Prolonged mixing in the melt also causes problems due to the poor thermal stability of some polymers. There is also the fact that many miscible polymers phase separate at higher temperatures, as will be discussed later. In some cases, for example in the production of Noryl,[5] these problems can be overcome. In many cases complete miscibility may not be essential for the end use, for example in the case of many polymeric modifiers for PVC.[4]

5.2.2 Mixing from Ternary Systems

If two polymers are dissolved in a common solvent and the solvent allowed to evaporate to produce a film, then one might expect intimate mixing without mechanical work. However, if two polymers are miscible this method of preparation does not ensure a homogeneous blend since they may phase separate in the presence of the solvent. This occurs if there is a difference in the interactions between the solvent and each of the two polymers and has been called the $\Delta \chi$ effect.[8,9] Simply speaking, if one polymer interacts very strongly with the solvent then it will force out the other polymer from solution in order to maximize the number of favorable contacts. The result is a two phase region in the polymer/polymer/solvent three component phase diagram as shown in Figure 1. If a solution at A is evaporated to produce a blend at composition D the mixture will pass through the two phase region B–C, the mixture may not remix on leaving it, and a heterogeneous mixture may be formed. There are also problems in efficient removal of solvent, especially in thicker

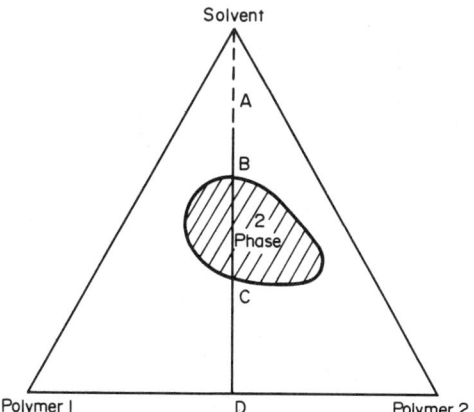

Figure 1 A hypothetical phase diagram for two polymers with a common solvent showing a two phase region. When solvent is evaporated from a mixed solution at composition A it enters the two phase region at B. When it leaves at C the phase-separated regions may have grown to such a size that remixing does not easily occur and the resultant blend at D is inhomogeneous

films. Commercially this method would not be attractive unless the blend was wanted in film form, for example in membrane formation.

A variation on this theme is freeze drying where one might be expected to maintain the well-dispersed state of the polymers if cooling is fast enough. It is, however, limited by the small number of solvents from which one can easily freeze dry, the requirement being a reasonably high vapor pressure above the crystalline solvent, which is usually fulfilled when the boiling point and freezing point are not too far separated. A second variation is mixing from emulsions, though this does not ensure such intimate mixing.

5.2.3 Mixing *via* Reaction

If a polymer is dissolved in a monomer and the monomer polymerized then a fine dispersion of the two polymers may be produced. There are many examples of the commercial application of this process for the production of rubber-toughened plastics, some of which have been described in detail.[10]

One material, the preparation of which has been studied in some detail, is high impact polystyrene (HIPS). Polybutadiene is dissolved in styrene monomer which is then polymerized free radically. When only a small amount of styrene has polymerized, phase separation takes place. The polystyrene/polybutadiene/styrene phase diagram has been studied and has the form shown in Figure 2.[11,12] Initially droplets of a polystyrene rich phase are formed within a polybutadiene rich phase (at A). As the polymerization proceeds the ratio of the two phases will change until the volume fraction of the polystyrene rich phase exceeds the volume fraction of the polybutadiene rich phase (between B and C). The mixture will then phase invert so that the polystyrene phase becomes continuous. The actual point of phase inversion depends on the viscosities of the phases and hence the molecular weights of the polymers, and on the stirring within the reactor. The resultant morphology is shown in Figure 3; polystyrene phases are trapped within the butadiene phases during phase inversion and the complicated morphology formed is responsible for many of the properties of these blends. The morphology and the material properties are also much influenced by grafting between the polymers which can occur during the polymerization (see Section 5.4).

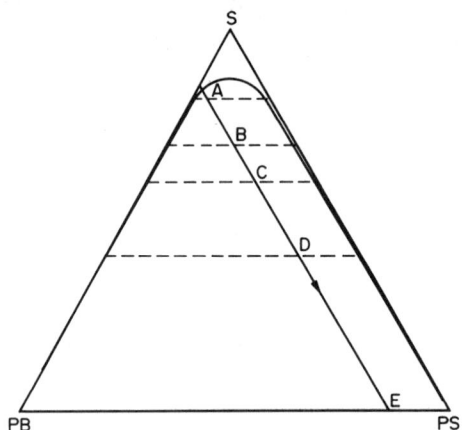

Figure 2   The ternary phase diagram for mixtures of styrene/polystyrene/polybutadiene (S/PS/PB). If styrene is polymerized in the presence of polybutadiene, phase separation takes place. Initially, at A, droplets of polystyrene rich phase form within a polybutadiene rich phase. Between B and C the mixture will phase invert and by D the continuous phase will be polystyrene rich

Many polymer blends are prepared by emulsion polymerization with sequential addition of different monomers. One example of this is the preparation of acrylonitrile/butadiene/styrene (ABS) plastics. Typically the butadiene is polymerized first in the presence of a crosslinking agent and the other two monomers added in a later stage.

Another example is the preparation of rubber-modified epoxy resins and other thermosets.[13] These and other blends where one or both of the components are crosslinked are often classified as semiinterpenetrating networks and interpenetrating networks. Their preparation, properties and applications have been reviewed.[14,15]

It is also possible to prepare homogeneous polymer mixtures by reaction. One example is the *in*

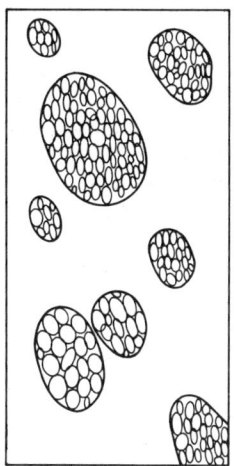

Figure 3 Representation of the morphology produced when styrene is polymerized in a mixture with polybutadiene. When phase inversion takes place droplets of polystyrene are trapped within the polybutadiene phase which is itself within a continuous polystyrene phase

situ polymerization of vinyl chloride in the presence of polymers which are miscible with poly(vinyl chloride) (PVC). This can be carried out during the suspension polymerization of PVC. However, just because the two polymers are miscible does not ensure homogeneous blends. The unpolymerized monomer acts as a solvent during the reaction and this can induce phase separation in the same way as for solvent casting.

In the case of the polymerization of vinyl chloride in the presence of poly(butyl acrylate) the phase diagram has the form shown in Figure 4.[16] Polymerization of a 50:50 mixture from A to B would produce a homogeneous blend, whereas if the poly(butyl acrylate) represented only 10 or 20% of the total, as would be more common were it present as a rubber modifier or processing aid, then the polymerization pathway would pass through the two phase region and a heterogeneous blend would be formed. Grafting reactions may also occur during the polymerization and these might be very pronounced if the initial polymer were one prone to grafting such as a copoly(ethylene/vinyl acetate).

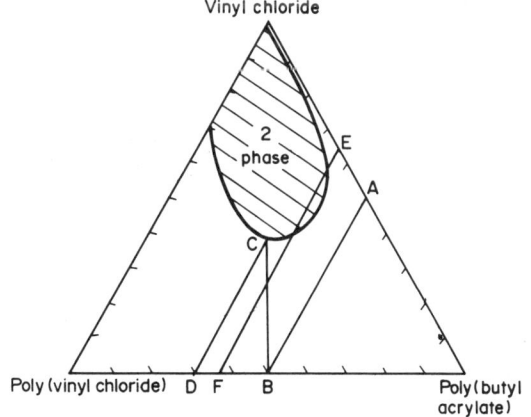

Figure 4 The three component phase diagram for mixtures of vinyl chloride, PVC and poly(butyl acrylate). A polymerization from E to F passes through the two phase region and an inhomogeneous blend results. A polymerization from A to B, followed by reswelling with vinyl chloride to C and repolymerization to D, avoids the two phase region and produces a homogeneous blend

5.3 MISCIBLE POLYMER BLENDS

5.3.1 Partial Miscibility and Phase Diagrams

Many polymers show a variation in miscibility with temperature. Low molecular weight polymers, having large favorable entropies of mixing and unfavorable enthalpies of mixing, are typically more miscible at higher temperatures and may phase separate on cooling, showing upper critical solution temperature behavior (UCST). High molecular weight polymers which form homogeneous mixtures are typically less miscible at higher temperatures and may phase separate on heating, showing lower critical solution temperature behavior (LCST).

In the temperature range where partial miscibility is observed, plots of ΔG_m against composition may have the form shown in Figure 5, which also shows the resultant phase diagram in the case of a UCST. At a temperature where miscibility is observed over the entire composition range (T_3 in Figure 5) the second differential of the free energy with respect to composition is always positive. This is the true thermodynamic criterion for miscibility.

At a temperature within the region where phase separation takes place (T_1 in Figure 5), compositions between ϕ_B and ϕ_B' can phase separate to reduce the overall free energy and give two phases at compositions ϕ_B and ϕ_B' which occur at the points where a common tangent can be drawn to the curve (not necessarily the minima) where

$$\frac{\delta}{\delta\phi_2}(\Delta G_m)_B = \frac{\delta}{\delta\phi_2}(\Delta G_m)_{B'} \qquad (6)$$

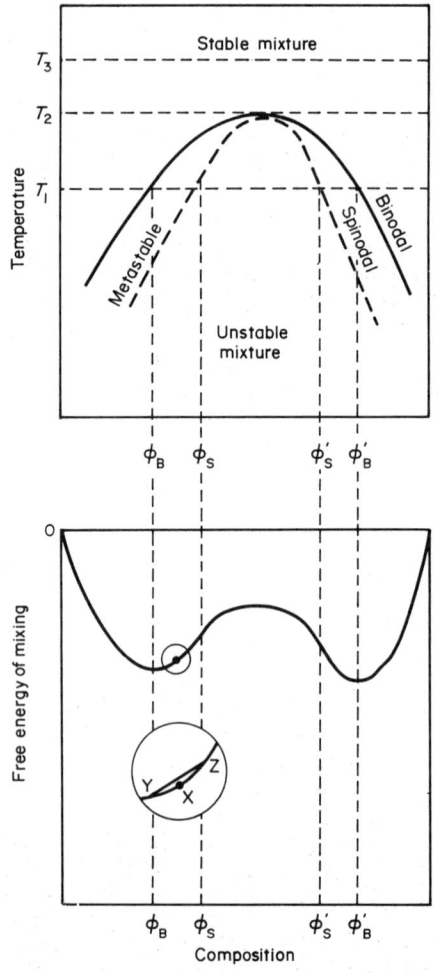

Figure 5 The phase diagram for a system showing upper critical behavior. At temperature T_3 mixtures are stable at all compositions whereas at T_1 phase separation can occur. The origins of this phase separation are shown in the lower diagram which shows the plot of free energy of mixing against composition at T_1. A metastable region exists between the binodal and the spinodal points (ϕ_B ϕ_B' and ϕ_S ϕ_S' respectively). The inset in the lower diagram shows that if composition X separates into Y and Z the average free energy is higher, even though complete phase separation to ϕ_B and ϕ_B' produces a lower free energy

At a temperature T_2 (the maximum for monodisperse polymers) the two points meet at the critical point and

$$\frac{\delta^2}{\delta\phi_2^2}(\Delta G_m) = \frac{\delta^3}{\delta\phi_2^3}(\Delta G_m) = 0 \tag{7}$$

Between ϕ_B and ϕ_S and between ϕ'_B and ϕ'_S a region of metastability occurs. This arises because any small fluctuation which occurs in composition produces an increase in free energy which acts as a barrier to phase separation. In the inset of Figure 5, if composition X phase separates to a mixture of Y and Z then the average free energy will be higher. Phase separation can only occur by a process of nucleation and growth. Between ϕ_S and ϕ'_S concentration fluctuations are immediately lower in free energy and phase separation can take place spontaneously by spinodal decomposition. The mechanism of spinodal decomposition involves the growth of a preferred concentration fluctuation wavelength. Thus there is a continuous change in the composition of the phases, while at the early stages the spacing remains constant, unlike the nucleation and growth process which involves the growth of phases having constant composition. These processes are described in Section 5.3.6. The spinodal occurs at the points of inflection of the free energy plot of Figure 5 where

$$\frac{\delta^2}{\delta\phi_2^2}(\Delta G_m) = 0 \tag{8}$$

On the resultant phase diagram of Figure 5 the line connecting points at various temperatures at composition ϕ_B satisfying equation (6) is the binodal. The line connecting points at composition ϕ_S satisfying equation (8) is the spinodal. In the case of high molecular weight polymers the same sort of phase diagram is found but it is commonly the other way up, with an LCST.

If we combine equations (1), (2) and (3) we obtain

$$\Delta G_m = RT(N_1 \ln \phi_1 + N_2 \ln \phi_2) + BV_1 N_1 \phi_2 \tag{9}$$

This equation, with B independent of temperature, can only be used to predict a UCST since the entropic contribution becomes more favorable at higher temperatures whereas the enthalpic contribution is constant. However if there is an entropic contribution to the interaction parameter such that

$$B = B_H - TB_S \tag{10}$$

then an LCST can be accommodated.

The origins of phase separation on heating can be attributed to three possible causes, more than one of which may be operative. (a) Contributions from the volume change on mixing to the free energy can become more unfavorable at higher temperatures. Free volume theories which address this problem will be discussed in Section 5.3.3. (b) The heat of mixing may be temperature dependent. This may especially be true if miscibility arises due to a specific interaction which may dissociate on heating. (c) There may be unfavorable entropy contributions which could arise from nonrandom mixing, for example associated with specific interactions. This may not necessarily result in a temperature dependence consistent with equation (10). The role of specific interactions in polymer miscibility will be discussed in Section 5.3.4.

As well as a temperature dependence of the interaction parameter there may also be a composition dependence. This is in part also addressed by the free volume theories. These theories, however, cannot explain all of the observed temperature and concentration dependences, and some investigators have chosen to describe the behavior in terms of semiempirical expressions by expanding the interaction term in these variables.[17]

One can compare various systems in terms of the sizes of the various contributions to the free energy of mixing: combinatorial entropy, free volume, enthalpy and entropy of interactions. Oligomer mixtures may often be very similar to polymer solutions. High polymer mixtures are very different in that the combinatorial entropy becomes very small. When partial miscibility is observed this may then be because all the other contributions are also very small and the free energy is very finely balanced. Examples of this which have been extensively studied[18,19] occur with mixtures of polystyrene and poly(α-methylstyrene) and with mixtures of polychlorostyrenes. In other cases partial miscibility may occur because two of the other terms are large and similar in size. The free energy of mixing may be changing quickly with temperature as a result of these terms and the combinatorial entropy may only play a minor role in the phase diagram at the extremes of composition. Examples of this are mixtures of polymers where there are strong specific interactions, as described in Section 5.3.4.

5.3.2 The Effect of Polydispersity

We have so far considered each polymer to be a single component, though it actually consists of a large number of different molecular weight species. This turns out to be important in many cases since when phase separation takes place the species with different molecular weights may not distribute themselves equally between the two phases. The phenomenon is therefore similar to fractionation in polymer solution phase separation, but involves both components. If we assume that the interaction parameter B is not molecular weight dependent then equation (9) becomes

$$\Delta G_m = RT(\Sigma N_{1i} \ln \phi_{1i} + \Sigma N_{2i} \ln \phi_{2i}) + BV_1 N_1 \phi_2 \qquad (11)$$

The effect of polydispersity has been considered by many authors[20,21] and it is found that the spinodal is just a function of $\bar{M}_w$, the critical point depends on $\bar{M}_w$ and on $\bar{M}_z$ and the binodal is dependent on the full distribution. The binodal can often have a complicated shape with shoulders as a result of polydispersity and/or a complex dependence of the interaction parameter on composition.

Many of the features caused by polydispersity can be understood by considering the simple case of a monodisperse polymer (A) mixed with two monodisperse fractions of another polymer (B1 and B2 with molecular weights in the order B2 > B1). The ternary phase diagram at one temperature is shown in Figure 6. A mixture at X does not phase separate into Y and Z but into V and W. We now consider a pseudobinary phase diagram at all temperatures for a mixture of A with a B1/B2 pseudocomponent with a ratio defined by point V. If we cool from the homogeneous melt to V, we meet the binodal, but the coexisting phase is not at composition U. It is in fact not in the plane of the paper at all since it has a different B1:B2 ratio. A line connecting all such coexisting phases is therefore often called the shadow curve.

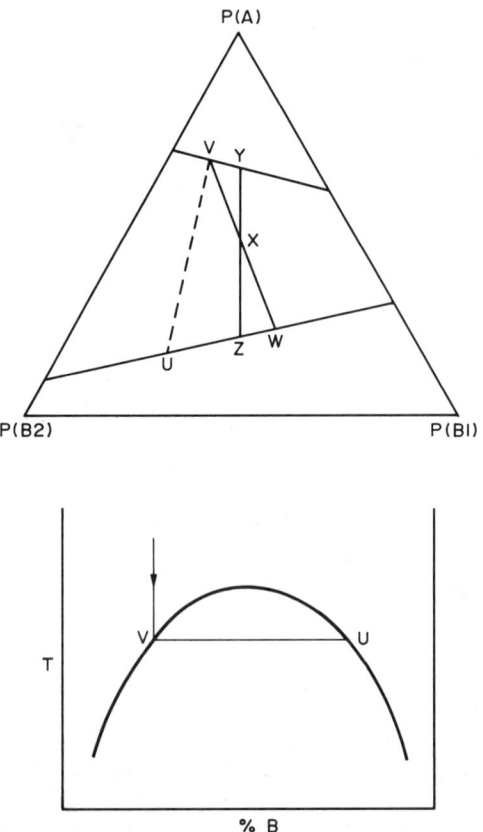

Figure 6 Hypothetical ternary phase diagram for a mixture of polymer A with another polymer B at two molecular weights, B1 and B2. A mixture at X will not phase separate into compositions Y and Z but into V and W with fractionation of polymer B. In the psuedobinary phase diagram below, on cooling a mixture to V one might expect the coexistent phase U to separate but in fact the coexisting phase is at W with a different distribution of B1 and B2

The effect of polydispersity is dealt with in much more detail in other chapters of this work (*e.g.* Volume 1, Chapter 15). In the case of polymer mixtures its effect will be expected to be greatest in oligomer mixtures and in high polymer mixtures when all the contributions to the free energy of mixing are small. In cases where the combinatorial entropy plays a minor role then the effect of polydispersity will also be less pronounced.

5.3.3 Free Volume Theories

Theories that account for free volume changes on mixing have already been dealt with in the chapter on polymer solutions (Volume 2, Chapter 3). In the case of polymer mixtures, because of the small combinatorial entropy change, these effects are of great importance.

The theory which has been applied most frequently to polymer mixtures is the equation of state theory of Flory and co-workers.[22,23] The theory prescribes the following equation for the pure components and their mixtures

$$\tilde{P}\tilde{v}/\tilde{T} = \tilde{v}^{1/3}/(\tilde{v}^{1/3} - 1) - 1/\tilde{T}\tilde{v} \tag{12}$$

where $\tilde{P} = P/P^*$, $\tilde{v} = v/v^*$ and $\tilde{T} = T/T^*$, and P^*, v^* and T^* are the hard core reduction parameters, which for the pure components can be found from thermal expansion and compressibility data.

For the mixture, the hard core volumes are considered to be additive and an interaction term X_{12} is introduced which reflects the difference in the chemical nature of the components. The hard core pressure and temperature are given by

$$P^* = \phi_1 P_1^* + \phi_2 P_2^* - \phi_1 \theta_2 X_{12} \tag{13}$$

and

$$P^*/T^* = \phi_1 P_1^*/T_1^* + \phi_2 P_2^*/T_2^*, \tag{14}$$

where ϕ_i are hard core volume fractions and θ_i are segment surface fractions given by

$$\phi_2 = \frac{M_2 V_2^*}{M_1 V_1^* + M_2 V_2^*}; \quad \phi_1 = 1 - \phi_2 \tag{15}$$

and

$$\theta_2 = \frac{(s_2/s_1)\phi_2}{(s_2/s_1)\phi_2 + \phi_1}; \quad \theta_1 = 1 - \theta_2 \tag{16}$$

where M_i are weight fractions and s_2/s_1 is the surface area per unit volume ratio, which can be best found by group contribution methods.

Using the combinatorial entropy of mixing taken from the Flory Huggins theory the chemical potential of mixing is calculated as

$$\Delta \mu_1/RT = \ln \phi_1 + (1 - r_1/r_2)\phi_2 + P_1^* V_1^*/RT[3\tilde{T}_1 \ln (\tilde{v}_1^{1/3} - 1)/(\tilde{v}^{1/3} - 1) + 1/\tilde{v}_1 - 1/\tilde{v} + \tilde{P}_1(\tilde{v} - \tilde{v}_1)]$$
$$+ V_1^* X_{12}\theta_2^2/TR\tilde{v} - V_1^* Q_{12}\theta_2^2/R \tag{17}$$

where V_1^* is the hard core molar volume of component 1. Q_{12} is a semiempirical interaction entropy parameter which reflects extra entropy contributions which might arise from, for example, non-random mixing due to specific interactions.

The reduction parameters must be obtained from PVT data of the pure components and this can be done directly by fitting to the equation of state if such data are available. If these data are not available a less satisfactory procedure is to use the known thermal expansion coefficient (α) to obtain $\tilde{v}$ via

$$\tilde{v} = \left[\frac{3 + 4\alpha T}{3 + 3\alpha T}\right]^3 \tag{18}$$

and to use the equation of state at zero pressure

$$\tilde{T} = \frac{\tilde{v}^{1/3} - 1}{\tilde{v}^{4/3}} \tag{19}$$

to obtain $\tilde{T}$. From the thermal pressure coefficient $\gamma = (\partial P/\partial T)_v$ one can then obtain P^* from

$$P^* = \gamma T \tilde{v}^2 \tag{20}$$

The thermal pressure coefficient is not often known but can be estimated from the solubility parameter δ by $\gamma \simeq \delta^2/T$.

McMaster[24] used a generalized version of this theory and simulated the spinodal and binodal curves of a hypothetical polymer mixture. He showed that it was possible to predict the occurrence of UCST and LCST phase diagrams and diagrams showing both simultaneously.

McMaster[24] and others[6] have examined the influence of the various parameters on miscibility. The main conclusions are as follows. (a) Polymers are predicted to be most miscible (larger range of temperature for miscibility) when they have similar expansion coefficients, which is equivalent to having similar values of T^*. (b) Changes in γ have less effect than changes in α. The size of the 'γ effect' depends on the 'α effect'. Higher average γ values reduce the range of miscibility. A difference between the γ values of the components affects the shape and moves the minimum position of the curve. (c) Increasing the molecular weights reduces the miscibility and alters the shape in the same way as would be expected in the simple Flory–Huggins treatment. (d) Introducing a negative or positive X_{12} makes the polymers more or less miscible respectively. The curves also become much flatter with large X_{12} values, as this composition independent factor then dominates the miscibility. (e) The value of s_2/s_1 has a large effect only in the presence of large X_{12} (or Q_{12}) terms, when it acts to tilt the curve to one side, the effect increasing as the value deviates from unity. (f) A negative Q_{12} makes the polymers less miscible.

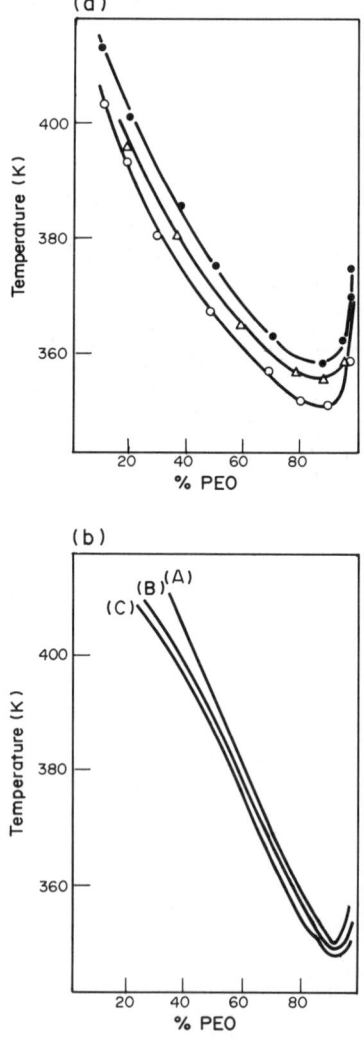

Figure 7 (a) Plots of the cloud point curves for poly(ethylene oxide)/polyether sulphone(PEO/PES) mixtures. The curves are for PEO of molar mass $4000\,\mathrm{g\,mol^{-1}}$ ($\bullet$), $20\,000\,\mathrm{g\,mol^{-1}}$ ($\blacktriangle$), and $200\,000\,\mathrm{g\,mol^{-1}}$ ($\bigcirc$). (b) Plots of the simulated spinodal curves for PEO/PES mixtures with PEO of molar mass 4000 (A), 20 000 (B), and $200\,000\,\mathrm{g\,mol^{-1}}$. (c) The spinodal curves were calculated using a value of $X_{12} = -40\,\mathrm{J\,cm^{-3}}$ and a value of $Q_{12} = -0.048\,\mathrm{J\,cm^{-3}\,k^{-1}}$ adjusted to fit the cloud point curve at its minimum

Many authors have used this theory to simulate the phase diagrams of real polymer mixtures.[25-28] In order to do this an estimate of the interaction parameter must be made. This has been done using inverse gas chromatography and by measuring heats of mixing of low molecular weight analogues as described in Section 5.3.7. In cases where the heats of mixing are favorable, a favorable (negative) value of X_{12} must be balanced by an unfavorable (negative) value of Q_{12} in order to simulate the phase diagram. This can be explained by an unfavorable entropy associated with the formation of specific interactions. An example of the fit which can be achieved is shown in Figure 7 for a blend of poly(ethylene oxide) with a polyether sulfone.[26] For this blend there are large free volume contributions. The interesting feature of this blend is that the phase diagram is very asymmetrical and one might be inclined to put this down to a molecular weight effect. However the phase diagram is not very molecular weight dependent. The asymmetry arises due to the difference in the surface areas per unit volume of the polymers. In other cases the interaction parameters suggested by heat of mixing measurements on low molecular weight analogues have proved to be too large (high and negative) and produced simulated phase diagrams which were much flatter bottomed than found experimentally.[27,28] This could be due to the fact that heats of mixing are highly temperature dependent due to dissociation of specific interactions as is described in Section 5.3.4. Specific interactions cannot be accounted for by a simple interaction parameter X_{12}.

The effect of pressure on phase diagrams has also been investigated and simulated.[29] The results of one experiment are shown in Figure 8 for oligomeric polystyrene/polybutadiene blends. The effect of pressure is selectively dependent on the volume change on mixing and the fact that the theory explains the data relatively well supports the conclusion that the free volume effects are adequately accounted for.

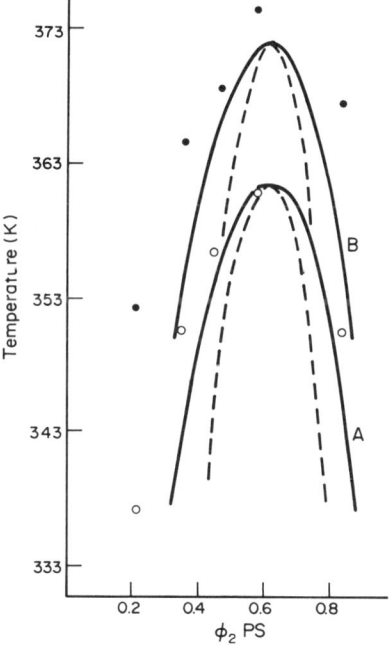

Figure 8 Simulated (——) binodals and (– – –) spinodals of polybutadiene ($\bar{M}_w = 2350$)/polystyrene ($\bar{M}_w = 1520$) mixtures at (A) 1 and (B) 1000 atm. The cloud point data at ($\bigcirc$) 1 and ($\bullet$) 1000 atm are also shown

Other free volume theories such as those of Sanchez and Lacombe[30] and Simha and Somcynski[31] are based on a lattice model and all or part of the free volume arises from vacancies on the lattice, unlike the Flory theory where free volume arises from an overall increase in molecular separations. Such theories are discussed in the chapter on polymer solutions (Volume 2, Chapter 3) and have not been much used in relation to polymer mixtures. Their use may well prove to be valuable since, especially using the theory of Simha and Somcynski, they much better describe the properties of the pure components.

5.3.4 Specific Interactions

As discussed earlier miscibility often arises due to a favorable (negative) heat of mixing. This infers the presence of some specific effect or interaction between the two components. The role of these interactions in mixtures has been reviewed in the literature.[6] Such an interaction can dominate the properties of a system and none of the theories so far discussed were designed to treat such systems. The interaction energy is just included within the interaction parameter and one must question whether the composition and temperature dependencies of the free energy as described by these parameters can be expected to reflect those of the interaction. Semiempirical expansions of the interaction parameter as used by some authors[17] may better describe this behavior.

Evidence for a specific interaction may be seen in, for example: (a) a favorable heat of mixing as discussed above; (b) if one of the components is a copolymer, an increase in the range of miscibility as the concentration of the interacting group is increased, though this may be counterbalanced by unfavorable changes in dispersive interactions (see Section 5.3.5); (c) a higher than expected mean glass transition temperature of the blend (see Section 5.3.7); and (d) direct spectroscopic evidence of the interaction.

The specific interaction which has received the most attention is the hydrogen bond. Studies of blends of PVC with polycaprolactone showed shifts of 4–6 cm^{-1} in the carbonyl band of the polycaprolactone relative to the pure polymer.[32,33] Frequency shifts have also been seen in blends of poly(methyl methacrylate) with poly(vinylidene fluoride).[34] In the case of PVC blends there has been some controversy about the exact nature of the interaction. It had originally been assumed that the interaction involved the methine hydrogen of PVC, (*i.e.* C=O · · · H–C–Cl) but mixing studies of THF with various chlorinated hydrocarbons show the heat of mixing to depend on the number of chlorines present, independent of whether they were present as CCl_2 or CCl_3 groups, *etc.*[35] It was concluded that in this case the chlorine atoms are involved in the interaction. IR studies using selectively deuterated PVC, however, suggest that the hydrogen of PVC is involved.[36]

Evidence from spectroscopic studies of other interactions is not plentiful. Polystyrene/poly(2,6-dimethyl-1,4-phenylene oxide) blends have been studied by IR and UV spectroscopy and interactions involving the aromatic rings proposed.[37,38] Studies of low molecular weight ethers with aromatic compounds have shown evidence of specific interactions and this has been extended to blends of polystyrene with poly(methyl vinyl ether).[39]

Weak interactions might be expected to dissociate on heating and this effect has been studied by IR spectroscopy.[40] Blends of copoly(ethylene/vinyl acetate) [copoly(E/VA)] with PVC and with chlorinated polyethylene were heated from room temperature through the region where phase separation takes place on heating. The carbonyl absorption of the copoly(E/VA) was followed. The peak is a sum of two absorptions, one shifted, the other unshifted. The maximum shows a gradual shift over the region studied as shown in Figure 9. This is attributed to a gradual dissociation of the specific interaction. The effect is very large compared to other contributions to the free energy and in

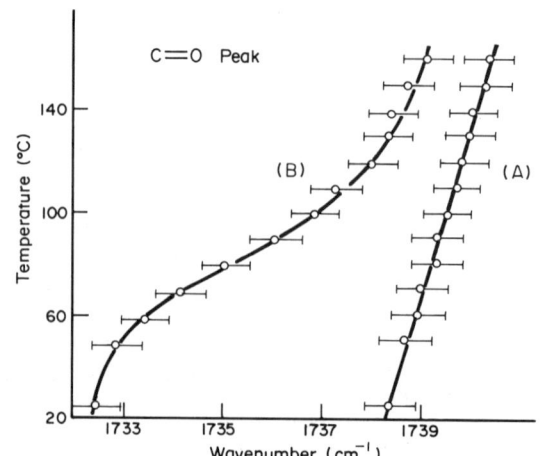

Figure 9 A plot of the frequency of maximum IR carbonyl absorption against temperature for (A) copoly(ethylene/vinyl acetate) and (B) a blend containing 80 wt.% chlorinated polyethylene and 20 wt.% copolymer. The shift is reduced at higher temperatures, which is attributed to a dissociation of the specific interaction between the polymers

effect the dissociation of the specific interaction is the main reason for the phase separation of this blend.

5.3.5 Miscibility of Copolymers

It has been noted that copolymers are more likely to be miscible with other polymers than homopolymers. One often observes what are called 'miscibility windows', a range of copolymer composition over which it is miscible with another polymer. For example, some ethylene/vinyl acetate copolymers are miscible with PVC whereas neither polyethylene nor poly(vinyl acetate) are miscible. Some have explained this as due to the contribution of 'cross terms', *i.e.* unfavorable interactions between different segments in the same polymer.[41-43] Such terms have often been considered in the literature and previously have been used in, for example, describing the solution properties of copolymers.[44] It is suggested that a copolymer is more likely to be miscible with other polymers since mixing can reduce the number of unfavorable contacts between the two types of segments.

The process can be visualized by reference to Figure 10. The (1) and (2) units of the copolymer mix with the units (3) of the other polymer. If the interaction energies between the various segments are B_{12}, B_{13} and B_{23} then the interaction energy for the blend will be given by

$$B_{\text{blend}} = B_{13}\phi_1' + B_{23}\phi_2' - B_{12}\phi_1'\phi_2' \tag{21}$$

Thus if B_{12} is positive and large enough to counteract B_{13} and B_{23}, the overall interaction term can be negative.

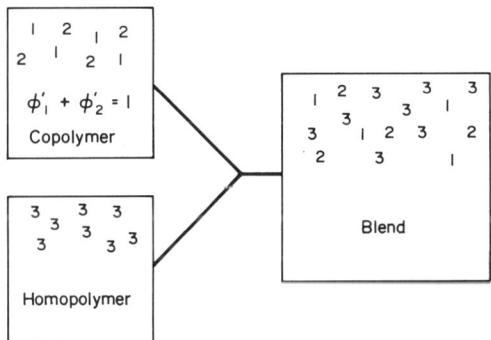

Figure 10 Schematic diagram for the mixing of a copolymer (1,2) with a homopolymer (3). If the units 1 and 2 of the copolymer have a sufficiently unfavorable interaction, they will preferentially mix with 3 to reduce the number of 1–2 contacts

One might therefore ask whether it is necessary to consider specific interaction in systems such as copoly(E/VA) with PVC. Firstly, it has been pointed out that as long as each segment/component can have attributed to it a single solubility parameter, then the use of the cross term cannot generate a negative heat of mixing.[43] Systems would have to be very nonideal to generate large negative heats of mixing. Secondly, there is direct spectroscopic evidence for specific interactions[40] and the estimated energies of interaction are considerable.

An alternative explanation for the window of miscibility which might apply in some cases has been suggested.[45] The specific interaction contributes to the heat of mixing but in some cases it may be outweighed by unfavorable dispersive forces. Crudely speaking, as one varies the ratio of the segments in a copolymer there is a point where the solubiliy parameter is equal to that of the other polymer. Around that value the specific interactions are able to produce an overall favorable interaction and the two polymers will be miscible.

The cross terms in the interaction parameter certainly exist but their relative importance in different systems will vary. In mixtures of nonpolar polymers where there is no specific interaction their effect is likely to be most important.

5.3.6   Kinetics of Phase Separation

Section 5.3.1 described the origins of the phase diagram for a polymer mixture as summarized in Figure 6. On cooling from the homogeneous region to a temperature below the binodal for a UCST (or heating for an LCST), phase separation can take place within the metastable region by a process of nucleation and growth. If the temperature is quickly dropped below the spinodal then fluctuations in concentration are stable and phase separation can take place by spinodal decomposition. The main features of the two phase separation processes are shown in Figure 11. In the nucleation and growth mechanism a nucleus having a composition close to that of one of the coexisting phases with concentration ϕ_B grows in size by the diffusion of one component down a concentration gradient. In the spinodal decomposition mechanism, in the initial stages periodic fluctuations in concentration increase in intensity while maintaining their periodicity, and the process occurs by diffusion up a concentration gradient.

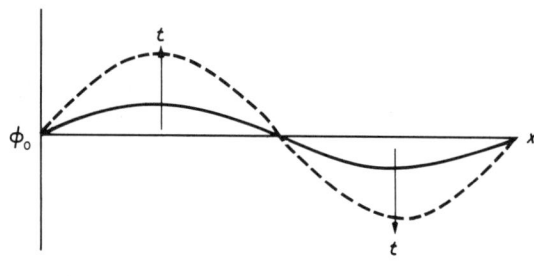

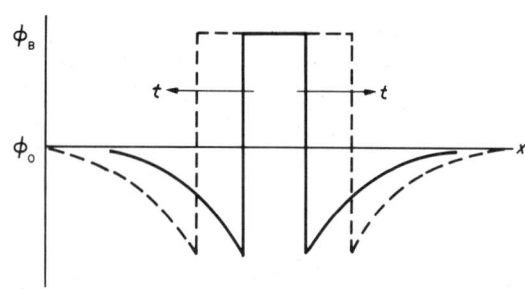

Figure 11   A schematic representation of the processes of spinodal decomposition (upper diagram) and nucleation and growth (lower diagram). In spinodal decomposition, fluctuations in composition away from the original (ϕ_0) increase in amplitude with time (t) at a constant wavelength. In nucleation and growth, nuclei are formed at the composition (ϕ_B) and the nuclei grow in size with time

It has been suggested that nucleation is less likely to take place in high molecular weight polymers though heterogeneous nucleation should not be ruled out. The model used to describe the spinodal decomposition process is that attributed to Cahn and Hilliard[46,47] and to Cook.[48] It has been questioned whether this model can adequately describe phase separation in the case of small molecules but mean field treatments are more appropriate in the case of polymers and so the model may be at least qualitatively correct.[49]

For a composition modulation of wavenumber β the difference in composition from the average at time t and position x is found to be

$$\phi - \phi_0 = \Sigma A(\beta) e^{R(\beta)t} \cos \beta x \tag{22}$$

where A is the Fourier coefficient of the component and $R(\beta)$ defines the rate of growth and is given by

$$R(\beta) = \tilde{D}(1 + 2K\beta^2/f'')\beta^2 \tag{23}$$

where f'' is the second derivative of the free energy with respect to composition, K is the gradient energy coefficient, a constant which reflects the influence of the concentration gradient on the local free energy, and $\bar{D}$ is an effective coefficient of diffusion which is proportional to f'' and is therefore negative in the unstable region.

If R(β) is negative then that particular modulation will not grow and from equation (23) the critical wavenumber is given by

$$\beta_c = -(f''/2K)^{1/2} \tag{24}$$

Equation (23) also predicts a sharp maximum at $\beta_c/2^{1/2}$ and thus the phase separation is dominated by this wavelength which gives rise to the periodicity of the structure.

Experimental studies have shown that the $\tilde{D}$ is negative as predicted.[50,51] They have also shown that $\tilde{D}$ is proportional to the temperature difference from the spinodal. This was predicted assuming the free energy could be described by the Flory–Huggins theory with an interaction term varying with temperature according to equation (10), though this is a very simplistic model and the fit may be fortuitous.

5.3.7 Properties of Homogeneous Mixtures and Criteria for Miscibility

Since many of the physical properties of homogeneous blends are also used as evidence for miscibility it is convenient to consider these two aspects at the same time. Such properties are also used in order to establish the phase diagrams of mixtures and this aspect will also be discussed.

5.3.7.1 Light scattering

Optical clarity is usually the first indication that two polymers are miscible though in itself it is not sufficient proof. It can be deceptive if the refractive indices of the two polymers are close, if the mixture has separated into two separate layers or if the separate phase domains are much smaller than the wavelength of light. The observation of an increase in turbidity or scattering has also often been used as a way of identifying the phase separation temperature of a mixture. This, of course, only indicates the point where the phases have grown to an extent where they can be detected under the particular circumstances, and one has no way of knowing whether phase separation has taken place close to the spinodal or the binodal.

An interesting related method is pulse-induced critical scattering (PICS) as this determines the spinodal curve directly.[52,53] In this technique the sample is brought for a short time into the temperature region where the mixture is metastable but is removed before phase separation can take place. It is possible to approach quite close to the spinodal. The fluctuations in concentration which occur in the metastable region can be detected by light scattering. This critical scattering intensity approaches infinity at the spinodal so that a plot of the reciprocal intensity against temperature extrapolates to the spinodal at zero. Due to the viscosity of high polymers this technique has so far been limited to oligomer mixtures showing UCST phenomena.

The use of scattering methods in studying the interaction between polymers will be discussed in the next section.

5.3.7.2 Measurement of interaction parameters

The strength of the interaction between two polymers can be considered as a criterion for miscibility since for high polymers the interaction parameter needs to be negative. Several reviews of the techniques can be found in the literature.[6,54] Most experimental methods for measuring the interaction parameter give the total interaction including both enthalpic and entropic parts or, in terms of the equation of state theory, $X_{12} - T\tilde{v}Q_{12}$. One noted exception to this is heat of mixing measurement, which by definition includes only enthalpic contributions.

For oligomer mixtures the heat of mixing can be measured directly. For high polymers one can measure the heat of mixing in the presence of a solvent and, knowing the heat of dissolution of the base polymers, extract a value by use of Hess's law. This process has problems associated with the accumulation of errors in the series of experiments. Alternatively one can measure the heats of mixing of low molecular weight analogues. This has problems due to differences in the chemical nature and density between the polymers and the analogues used.

The interaction parameter can be determined from a comparison of the interaction of a solvent with a blend and of the same solvent with the base polymers of the blend. Inverse gas chromatography (IGC) measures the retention of a known volatile solvent on a stationary phase consisting of, in this case, a polymer blend. The experiments are easy to carry out but there are problems associated with surface adsorption and diffusion limitation of the solvent in the polymers. Solvent vapor sorption measurement is a static method which is equivalent to IGC. The results of vapor sorption have been compared to IGC for PVC, polystyrene, and poly(methyl methacrylate) and the interaction parameters have been found to agree within experimental error.[55]

The depression of the melting point of a crystalline polymer in blends can be used to determine the interaction parameter between the polymers. In practice it is difficult to measure the true thermo-dynamic equilibrium melting point due to problems of inhomogeneity and diffusion limitation in heating measurements and of supercooling during cooling measurements, and results depend on the rate of heating and cooling. One also has to consider the effect of morphological changes on the melting point depression. A generalized form of the equation first derived by Flory has been given as[56]

$$\frac{\Delta H_u(T_m^0 - T_m)}{\phi_1 R T_m^0} - \frac{T_m}{M_1} - \frac{\phi_1 T_m}{M_2} = \frac{C}{R} - \chi'_{12}\phi_1 T_m \qquad (25)$$

where T_m^0 is the melting point of the pure polymer and T_m is that in the blend. This assumes that the morphological contribution is proportional to ϕ_1 with a proportionality constant C.

Thermodynamic data can be obtained from light, X-ray or neutron scattering techniques *via* the second osmotic virial coefficient A_2. At low values of the scattering vector Q (where $Q = (4\pi/\lambda) \times \sin(\theta/2)$, λ is the wavelength and θ the scattering angle) and for low concentrations of polymer 2 in component 1, the scattering intensity of the mixture minus solvent (polymer 1), $I(Q)$, is given by

$$\frac{KC_2}{I(Q)} = \frac{1}{\bar{M}_w}\left[1 + \frac{R_z^2 Q^2}{3}\right] + 2A_2 C_2 \qquad (26)$$

where C_2, $\bar{M}_w$ and R_z are the concentration, weight average molecular weight and z average radius of gyration of polymer 2. K is a contrast factor depending on the refractive indices for light, electron densities for X-rays or scattering lengths for neutrons. The Flory–Huggins interaction parameter is then obtained from

$$\chi_{12} = \frac{1}{P_1}\left[\frac{1}{2} - \frac{V_1 M_2^2 A_2}{V_2^2}\right] \qquad (27)$$

where P_1 is the degree of polymerization of component 1. This method has been used to obtain interaction parameters from neutron scattering.[57] Other treatments have allowed interaction parameters to be obtained over the whole range of concentration.[58]

5.3.7.3 *Glass transition temperature*

A miscible blend of two polymers shows a single glass transition temperature (T_g) which is generally between the T_g values of the individual polymers. The use and interpretation of T_g measurements have been reviewed.[59] As a criterion for miscibility it is superior to light scattering since it is sensitive to phase sizes greater than the order of 10 nm. Problems in sensitivity can however occur if the values of T_g for the components are less than about 20 K apart or if one constituent represents less than 10% of the total. Techniques also generally operate by scanning a property through the temperature range of the T_g which might cause problems if the miscibility is temperature dependent.

The most common techniques are differential-scanning calorimetry, the measurement of heat capacity as a function of temperature, and dynamic mechanical analysis, the measurement of storage and loss moduli as functions of temperature and possibly frequency. Other techniques are dielectric relaxation spectroscopy and dilatometry.

Several miscible blends studied by Hichman and Ikeda[60] exhibit a composition dependence of T_g which can be described by the Fox relationship

$$1/T_g = m_1/T_{g_1} + m_2/T_{g_2} \qquad (28)$$

where m_1 is the weight fraction of polymer 1 with T_{g_1}. Other relationships have also been developed.[61] Plots of T_g against composition for blends which have a strong specific interaction between the polymers show marked deviations from linearity.

5.3.7.4 *Other mechanical properties*

Homogeneous polymer mixtures find use in polymeric plasticizers and in producing blends with processing temperatures and heat distortion temperatures intermediate between those of the components. This is due to the averaging of glass transition described above.

The mechanical properties of homogeneous glassy polymer blends have been reviewed by Kambour.[62] To a first approximation the properties are additive. In many cases negative deviations in ductility are found which are believed to be due to the reductions in mobility associated with the volume contraction on mixing and interactions in the blends. Since these are necessary for mixing to take place, positive deviations are seen as unlikely. Toughened plastics are generally two phase materials.

5.4 TWO PHASE POLYMER BLENDS

The aspects of two phase polymer blends which most interest the physical scientist are those which concern the morphology and the interface. The interface is characterized by the interfacial tension and interfacial thickness and these affect the mixing process, morphology and properties. The morphology can be characterized by various techniques such as microscopy and scattering methods. A full description of the engineering and mechanical aspects of polymer blends is to be found in Volume 7, Chapter 4.

5.4.1 Characterization of Morphology

Optical and electron microscopy are the most informative techniques for studying phase morphology. The techniques are many and often specific for one class of system. A short overview has been presented by Shaw.[63] Detailed descriptions are given in Part 6 of Volume 1.

Scattering methods are described in Part 5 of Volume 1. For a detailed description of scattering methods applied to blends see a review by Stein.[64]

5.4.2 Interfacial Tension and Surface Profile

Extensive surveys of this subject are available in the literature including a book by Wu[65] and a review by Koberstein.[66] There are many theories of the polymer interface but none which is generally accepted.

The interfacial tension γ can be given in terms of the surface tensions of the components γ_i^0 and the work of adhesion W_a by

$$\gamma_{12} = \gamma_1^0 + \gamma_2^0 - W_a \tag{29}$$

This can be expressed as in the Good–Girifalco equation[67] as

$$\gamma_{12} = \gamma_1^0 + \gamma_2^0 - 2\phi(\gamma_1^0\gamma_2^0)^{1/2} \tag{30}$$

where the interaction parameter ϕ can be related in principle to molecular parameters though these are often not known for polymers. It varies for pairs of polymers between 0.8 and 0.97, tending to be higher for pairs of similar polarity.

Another approach is to divide the work of adhesion and surface tensions into two terms describing the polar (p) and dispersive (d, nonpolar) interactions respectively.[65,68] This gives two equations, the harmonic mean equation

$$\gamma_{12} = \gamma_1^0 + \gamma_2^0 - \frac{4\gamma_1^d\gamma_2^d}{\gamma_1^d + \gamma_2^d} - \frac{4\gamma_1^p\gamma_2^p}{\gamma_1^p + \gamma_2^p} \tag{31}$$

which is valid for two low energy materials, and the geometric mean equation

$$\gamma_{12} = \gamma_1^0 + \gamma_2^0 - 2(\gamma_1^d\gamma_2^d)^{1/2} - 2(\gamma_1^p\gamma_2^p)^{1/2} \tag{32}$$

which performs better for a high energy and a low energy material.

Lattice calculations have been made by Roe[69] and Helfand[70] which give the interfacial tension and interfacial thickness, a_1 (defined as the difference in composition of the phases divided by the gradient at the midpoint). These theories have the disadvantage that the result depends on the size and coordination number of the lattice which are unknown.

An alternative theory by Helfand,[71] having the advantage of not requiring assumptions of the lattice parameters, assumes that the segments interdiffuse to lower the free energy and gives

$$\gamma_{12} = \tfrac{2}{3} k T \alpha^{1/2} \left(\frac{\beta_1^3 - \beta_2^3}{\beta_1^2 - \beta_2^2} \right), \quad a_1 = 2 \left(\frac{\beta_1^2 + \beta_2^2}{2\alpha} \right) \tag{33}$$

where $\alpha = P_0 \chi$ is an interaction parameter, $\beta_i = \tfrac{1}{6} P_{0i} b_i^2$, b is the statistical segment length and P_0 the chain unit number density. This gives the following scaling relations

$$\gamma_{12} \propto \chi^{1/2}, \quad a_1 \propto \chi^{-1/2}, \quad \gamma_{12} \propto a_1^{-1} \tag{34}$$

which agree with the predictions of the lattice theory of Helfand.[70] Calculated values of interfacial tensions agree with experiment reasonably well.[65,72]

The square gradient approach of Cahn and Hilliard[46] has also been used but can only be tested close to the critical point of the mixture where interfacial tension data are not available. It has, however, been applied to demixed polymer solutions.[73]

Most interfacial tension measurements have been made by the pendant drop technique. A drop of a polymer melt is formed in another polymer melt. The shape is recorded photographically and the interfacial tension found from drop dimensions, *i.e.* maximum width and width at a height equal to the maximum width, using tabulated ratios found from numerical solutions of the equations of Bashforth and Adams.[65] The interfacial tension can also be found by a full analysis of the shape of the drop. The method is preferred over alternatives since equilibration is not a serious problem but accurate density measurements are required. Tabulated values of interfacial tensions are available in the literature[65,66] and range from about $1 \, \text{mN m}^{-1}$ for polymers similar in polarity such as polyethylene/polypropylene to $11 \, \text{mN m}^{-1}$ for dissimilar polymers such as polyethylene/poly-(methyl methacrylate).

Estimates of interfacial thicknesses have been made by analysis of electron micrographs or by scanning analytical electron microscopy.[66] X-Ray and neutron scattering methods have also been used but most extensively in the study of the interface in block copolymers. It is generally necessary to fit the scattering profile to a complete model of the morphology which contains a diffuse interface.

Additives can greatly reduce the interfacial tension between polymers and hence modify the mixing process and the properties of the blend. Block and graft copolymers are the most effective interfacial agents and can be added to the blend or formed during the mixing process by reactions between the base polymers.[74,75] There are theoretical treatments of the behavior of block copolymers at the polymer/polymer interface[76,77] but comparable experimental data is scarce.

5.4.3 The Blending Process

The blending of two polymers has been described by several authors[72,78] and is discussed in Part 2 of Volume 7. Here we will just describe how it interacts with the previous discussion of the interface. When two polymers are mixed in commercial equipment such as extruders and internal mixers they are subject to complex, nonuniform shear and elongational fields. Furthermore, polymers are viscoelastic. Qualitatively, however, the drop break up process can be understood in terms of a simple model.

Taylor's[79] classic work deals with the break up of a Newtonian liquid drop in a Newtonian matrix when subjected to a simple shear field. The drop deforms into an ellipse due to the viscous forces upon it and this deformation is resisted by the interfacial tension. The interfacial force depends on γ/a, the interfacial tension divided by the drop radius, and the viscous force depends on $\eta_m G$, the matrix viscosity multiplied by the shear rate. The ratio of these is the Taylor number

$$N_{\text{TA}} = \frac{\gamma}{\eta_m a G} \tag{35}$$

The critical condition for drop break up is approximately given by

$$\frac{G \eta_m a}{\gamma} > \frac{16p + 16}{19p + 16} \tag{36}$$

where p is the viscosity ratio η_d/η_m, η_d being the drop viscosity. Drop break up is therefore favored by low interfacial tension and by similar values for the drop and matrix viscosity and this agrees with experimental observations in blends. Others have stressed the importance of drop break up in extensional flow to generate very small particle dispersions or have considered the stability of a filament of one component in a similar way to Rayleigh's capillarity instability.

Interfacial agents such as block and graft copolymers are known to reduce the interfacial tension and hence are expected to increase the degree of dispersion in blends. This has been found to be the case both for added graft copolymers[80] and for cases where the graft copolymer is generated during the mixing process.[81]

5.5 REFERENCES

1. M. L. Huggins, *J. Am. Chem. Soc.*, 1942, **64**, 1712.
2. P. J. Flory, *J. Chem. Phys.*, 1942, **10**, 51.
3. O. Olabisi, L. M. Robeson and M. T. Shaw, 'Polymer–Polymer Miscibiliby', Academic Press, New York, 1979.
4. G. H. Hofmann, 'Polymer Blends and Mixtures', ed. D. J. Walsh, J. S. Higgins and A. Maconnachie, NATO ASI Series, Martinus Nijhoff, Dordrecht, 1985, chap. 7.
5. E. P. Cuzek, *US Pat.* 3 383 435 (1968).
6. D. J. Walsh and S. Rostami, *Adv. Polym. Sci.*, 1985, **70**, 119.
7. M. T. Shaw, 'Polymer Blends and Mixtures', ed D. J. Walsh, J. S. Higgins and A. Maconnachie, NATO ASI Series, Martinus Nijhoff, Dordrecht, 1985, chap. 4.
8. M. Bank, J. Leffingwell and C. Thies, *Macromolecules*, 1971, **4**, 43.
9. C. C. Hsu and J. M. Prausnitz, *Macromolecules*, 1974, **7**, 320.
10. C. B. Bucknall, 'Toughened Plastics'; Applied Science, London, 1977.
11. J. L. White and R. D. Patel, *J. Appl. Polym. Sci.*, 1975, **19**, 1775.
12. P. Gaillard, M. Ossenbach-Sauter and G. Riess, in 'MMI Press Symposium', ed. K. Solc, 1982, vol. 3, p. 289.
13. A. J. Kinloch, 'Polymer Blends and Mixtures', ed. D. J. Walsh, J. S. Higgins and A. Maconnachie, NATO ASI Series, Martinus Nijhoff, Dordrecht, 1985, chap 21.
14. L. H. Sperling, 'Interpenetrating Polymer Networks and Related Materials', Plenum Press, New York, 1981.
15. L. H. Sperling, 'Polymer Blends and Mixtures', ed. D. J. Walsh, J. S. Higgins and A. Maconnachie, NATO ASI Series, Martinus Nijhoff, Dordrecht 1985, chap 13.
16. D. J. Walsh and G. L. Cheng, *Polymer*, 1982, **23**, 1965.
17. H. Tompa, 'Polymer Solutions', Butterworth, London, 1956.
18. S. Saeki, J. M. G. Cowie and I. J. McEwen, *Polymer*, 1983, **24**, 60.
19. S. L. Zacharius, G. tenBrinke, W. J. MacKnight and F. E. Karasz, *Macromolecules*, 1983, **16**, 381.
20. R. Koningsveld and A. J. Staverman, *J. Polym. Sci., Part A-2*, 1968, **6**, 305.
21. R. Koningsveld and L. A. Kleintjens, 'Polymer Blends and Mixtures', cd. D. J. Walsh, J. S. Higgins and A. Maconnachic, NATO ASI Series, Martinus Nijhoff, Dordrecht, 1985, chap. 6.
22. P. J. Flory, R. A. Orwoll and A. Vrij, *J. Am. Chem. Soc.*, 1964, **86**, 3507.
23. B. E. Eichinger and P. J. Flory, *Trans. Faraday Soc.*, 1968, **64**, 2035.
24. L. P. McMaster, *Macromolecules*, 1973, **6**, 760.
25. O. Olabisi, *Macromolecules*, 1975, **8**, 316.
26. Z. Chai, S. Ruona, D. J. Walsh and J. S. Higgins, *Polymer*, 1983, **24**, 263.
27. D. J. Walsh and S. Rostami, *Polymer*, 1985, **26**, 418.
28. S. Rostami and D. J. Walsh, *Macromolecules*, 1984, **17**, 315.
29. S. Rostami and D. J. Walsh, *Macromolecules*, 1985, **18**, 1228.
30. I. C. Sanchez and R. H. Lacombe, *Macromolecules*, 1978, **11**, 1145.
31. R. Simha and T. Somcynsky, *Macromolecules*, 1969, **2**, 342.
32. O. Olabisi, *Macromolecules*, 1975, **8**, 316.
33. M. M. Coleman and J. Zarian, *J. Polym. Sci., Polym. Phys. Ed.*, 1979, **17**, 837.
34. M. M. Coleman, J. Zarian, D. F. Varnell and P. C. Painter, *J. Polym. Sci., Polym. Lett. Ed.*, 1977, **15**, 745.
35. J. Pouchly and J. Biros, *J. Polym. Sci., Polym. Lett. Ed.*, 1969, **7**, 463.
36. M. M. Coleman and P. C. Painter, *Appl. Spectrosc. Rev.*, 1984, **20**, 255.
37. S. T. Wellinghoff, J. L. Koenig and E. Baer, *J. Polym. Sci., Polym. Phys. Ed.*, 1977, **15**, 1913.
38. D. Lefebvre, B. Jasse and L. Monnerie, *Polymer*, 1981, **22**, 1616.
39. H. Yang, G. Haziioannou and R. S. Stein, *J. Polym. Sci., Polym. Phys. Ed.*, 1983, **21**, 159.
40. M. M. Coleman, E. J. Moskala, P. C. Painter, D. J. Walsh and S. Rostami, *Polymer*, 1983, **24**, 1410.
41. R. P. Kambour, J. T. Bendler and R. C. Bopp, *Polym. Blends, PRI Conf., 1981*, University of Warwick.
42. G. tenBrinke, F. E. Karasz and W. J. MacKnight, *Macromolecules*, 1983, **16**, 1827.
43. D. R. Paul and J. W. Barlow, *Polymer*, 1984, **25**, 487.
44. W. H. Stockmayer, L. D. Moore, Jr., M. Fixman and B. N. Epstein, *J. Polym. Sci.*, 1955, **16**, 517.
45. D. J. Walsh and G. L. Cheng, *Polymer*, 1984, **25**, 499.
46. J. W. Cahn and J. E. Hilliard, *J. Chem. Phys.*, 1959, **31**, 688.
47. J. E. Hilliard, in 'Phase Transformations', ed. H. L. Anderson, American Society for Metals, Metals Park, Ohio, 1970.
48. H. E. Cook, *Acta Metall.*, 1970, **18**, 297.
49. K. Binder, *J. Chem. Phys.*, 1983, **79**, 6387.
50. T. Nishi, T. T. Wang and T. K. Kwei, *Macromolecules*, 1975, **8**, 227.
51. H. L. Snyder, P. Meakin and S. Reich, *Macromolecules*, 1983, **16**, 757.
52. M. H. Onclin, L. A. Kleintjens and R. Koningsveld, *Makromol. Chem., Suppl.*, 1979, **3**, 197.

53. M. Gordon, 'Polymer Blends and Mixtures', ed. D. J. Walsh, J. S. Higgins and A. Maconnachie, NATO ASI Series, Martinus Nijhoff, Dordrecht, 1985, p. 429.

54. D. R. Paul, 'Polymer Blends and Mixtures', ed. D. J. Walsh, J. S. Higgins and A. Maconnachi, NATO ASI Series, Martinus Nijhoff, Dordrecht, 1985, chap. 1.

55. P. J. T. Tait and A. M. Abushihada, *Polymer*, 1977, **18**, 810.

56. T. K. Kwei and H. L. Frisch, *Macromolecules*, 1978, **11**, 1267.

57. A. Maconnachie, R. P. Kambour, D. M. White, S. Rostami and D. J. Walsh, *Macromolecules*, 1984, **17**, 2645.

58. M. Warner, J. S. Higgins and A. J. Carter, *Macromolecules*, 1983, **16**, 1931.

59. F. E. Karasz, 'Polymer Blends and Mixtures', ed. D. J. Walsh, J. S. Higgins and A. Maconnachie, NATO ASI Series, Martinus Nijhoff, Dordrecht, 1985, chap. 2.

60. J. J. Hickman and R. M. Ikeda, *J. Polym. Sci., Polym. Phys. Ed.*, 1973, **11**, 1713.

61. J. M. G. Cowie, *J. Macromol. Sci., Phys.*, 1980, **18**, 563.

62. R. P. Kambour, 'Polymer Blends and Mixtures'; ed. D. J. Walsh, J. S. Higgins and A. Maconnachie, NATO ASI Series, Martinus Nijhoff, Dordrecht, 1985, chap. 16.

63. M. T. Shaw, 'Polymer Blends and Mixtures', ed. D. J. Walsh, J. S. Higgins and A. Maconnachie, NATO ASI Series, Martinus Nijhoff, Dordrecht, 1985, chap. 3.

64. R. S. Stein, 'Polymer Blends', ed. D. R. Paul and S. Newman, Academic Press, New York, 1978, vol. 1, chap. 9.

65. S. Wu, 'Polymer Interface and Adhesion', Dekker, New York, 1982.

66. J. T. Koberstein, 'Polymer Interface Structure and Properties', in 'Encyclopedia of Polymer Science and Technology', 2nd edn., Wiley, New York, vol. 8, to be published.

67. R. J. Good and L. A. Girifalco, *J. Phys. Chem.*, 1960, **64**, 561.

68. S. Wu, *J. Polym. Sci., Part C*, 1971, **34**, 19.

69. R. J. Roe, *J. Chem. Phys.*, 1975, **62**, 490.

70. E. Helfand, *J. Chem. Phys.*, 1975, **63**, 2192.

71. E. Helfand, in 'Recent Advances in Polymer Blends, Grafts and Blocks', ed. L. H. Sperling, Plenum Press, New York, 1974.

72. S. Wu, *Polym. Eng. Sci.*, 1987, **27**, 335.

73. I. C. Sanchez, *Annu. Rev. Mater. Sci.*, 1983, **13**, 387.

74. H. T. Patterson, K. H. Hu and T. H. Grindstaff, *J. Polym. Sci., Part C*, 1971, **34**, 31.

75. N. G. Gaylord, *Adv. Chem. Ser.*, 1975, **142**, 76.

76. J. Noolandi and K. M. Hong, *Macromolecules*, 1982, **15**, 482; 1983, **16**, 1083.

77. L. Leibler, *Macromolecules*, 1982, **15**, 1283.

78. J. L. White, 'Polymer Blends and Mixtures', ed. D. J. Walsh, J. S. Higgins and A. Maconnachie, NATO ASI Series, Martinus Nijhoff, Dordrecht, 1985, chap. 22.

79. G. I. Taylor, *Proc. R. Soc. London, Ser. A*, 1932, **138**, 41; 1934, **146**, 501; 1954, **226**, 34.

80. W. M. Barentsen, D. Heikens and P. Piet, *Polymer*, 1974, **15**, 119; W. M. Barentsen and D. Heikens, *Polymer*, 1977, **18**, 69.

81. S. Cimmino, L. D'Orazio, R. Greco, G. Maglio, M. Malinconico, C. Mancarella, E. Martuscelli, R. Palumbo and G. Ragosta, *Polym. Eng. Sci.*, 1984, **24**, 48.

6

Chain Segregation in Block Copolymers

RICHARD A. BROWN, ANDREW J. MASTERS, COLIN PRICE and
XUE FENG YUAN
University of Manchester, UK

6.1   INTRODUCTION

The mixing of two different homopolymers leads in most cases to a two-phase blend. The main reason for this behaviour is that the combinatorial entropy of mixing is very small for chain molecules and specific interactions are required to achieve miscibility.[1] Such two-phase blends are generally characterized by opacity, distinct thermal transitions for each phase and poor mechanical properties. In block and graft copolymers the different types of chains (*i.e.* the component blocks or the grafts and backbone) are bonded together covalently, thereby preventing immiscibility from leading to phase separation on the macroscopic scale.

Microphase separation is the term used to describe the type of chain segregation which is able to occur in block and graft copolymers in the bulk state and in concentrated solution, and the structures generated by the packing of domains on the macrolattice are termed mesomorphic structures. In common with its use in colloid science, micellization is the term used to indicate molecular association of copolymers and block segregation in dilute solution. Their ability to undergo microphase separation and micellization endow block and graft copolymers with certain properties not possessed by homopolymers or simple statistical copolymers. They also enable block and graft copolymers to enhance significantly the mechanical properties of blends.

One of the earliest block copolymers to be prepared was reported in 1938 by Bolland and Melville[2] who showed that a poly(methyl methacrylate) film deposited on the walls of an evacuated tube could initiate the polymerization of 2-chloro-1,3-butadiene (chloroprene). Progress with block systems, however, was initially slow because of the difficulty in preparing reasonably homogeneous samples.

Bayer *et al.*[3,4] reported in the early 1950s on $(AB)_n$ multiblock networks formed by linking polyesters and other prepolymers through diisocyanate links.

In 1951, Wyandotte Chemical Company[5] introduced commercially poly(oxyethylene)-*block*-poly(oxypropylene)-*block*-poly(oxyethylene) copolymers (named Pluronics) which are non-ionic surfactants. In a paper on the synthesis and application of block copolymers in 1951, Mark[6] paid particular attention to the role of surfactants in aqueous media.

In 1954, Coleman[7] reported on how the introduction of poly(oxyethylene) blocks into poly(ethylene terephthalate) chains could modify the fibre properties of the latter giving improved hydrophilicity and dyeability; the work was inspired by the knowledge that silk fibroin had some block character. By increasing the polyether content, Charch and Shivers[8] extended the work to include elastomeric materials that proved to have good recovery properties. (These materials were the forerunners of the commercial poly(ether-*alt*-esters) developed some time later.)

The term 'grafting' seems to have been first introduced by Alfrey at an American Chemical Society meeting in Chicago in 1950, and brought into general use shortly afterwards through the publications of Mark.[9] In the late 1950s, Bateman[10] and Merritt[11] showed that graft copolymers consisting of a natural rubber backbone with poly(methyl methacrylate) grafts could exist in two physical forms (hard or soft) depending on the precipitation conditions. Merritt was able to account for the results in terms of domain formation. Thus, by this stage in the development of the subject, it was fully recognized that block and graft copolymers could micellize in solution and exist in a microphase-separated state in the bulk.

The Strasbourg school[12] were the first to investigate mesomorphic structures formed by block copolymers in a systematic manner using the technique of X-ray diffraction. The work commenced shortly after well-defined polystyrene-*block*-poly(oxyethylene) copolymers had been synthesized anionically using the 'living polymerization' method developed by Szwarc and co-workers.[13,14] Lyotropic mesophases formed by soaps were also being investigated at the time by X-ray diffraction and a connection between the two types of materials was quickly established, both forming similar structures.[15,16]

Ceresa[17] drew attention to the fact that out of 1400 copolymers only 5% of the products had been isolated with a reasonable degree of purity and only 20 species had been properly characterized. Exploitation of anionic polymerization, however, soon improved the situation markedly. The use of alkyllithium initiators, which were initially evaluated by Stavely and co-workers at Firestone for the synthesis of high-*cis* polyisoprene, in the preparation of block copolymers[18] from styrene and butadiene, and styrene and isoprene, is particularly noteworthy.

In 1965, Shell brought on the market a number of polystyrene-*block*-polybutadiene-*block*-polystyrene and polystyrene-*block*-polyisoprene-*block*-polystyrene copolymers. These ABA copolymers, in the chemically uncrosslinked state, show high tensile strength and rubber elasticity approaching that of vulcanized rubber. Whilst the copolymers contained a small amount of homopolymer contaminant, they were remarkably well-defined in copolymer terms. A fascinating article by Legge *et al.*[19] describes how the styrene end-blocks were first investigated as a means of modifying the cold-flow rheology of alkyllithium-initiated polybutadiene, but were soon recognized as a means of preparing 'thermoplastic elastomers'. A paper dealing with the structure/property relationships of these materials, which were physically crosslinked by virtue of glassy polystyrene domains formed through the aggregation of polystyrene end-blocks, was presented at the International Rubber Conference[20] held in the UK in 1967. The results presented at that meeting and at subsequent gatherings shortly afterwards, together with the samples of commercial and specially prepared block copolymers made freely available to academics by Shell, stimulated many groups to enter the block copolymer field.

Also in the mid-1960s, it was concluded by Cooper and Tobolsky[21] that segmented multiblock polyurethane elastomers, which were already commercial products (Estane), were physically cross-linked by virtue of the presence of segregated hard crystalline segments in a soft phase, rather than being crosslinked primarily by hydrogen bonding. The chain structure of this class of thermoplastic elastomer consists of segments of flexible aliphatic polyethers (or polyesters) alternating with stiff segments containing isocyanate linkages.

Segmented poly(ether-*alt*-ester) copolymers, which were marketed in 1972 by Du Pont (Hytrel), function on a similar basis to the segmented polyurethanes. In this case, soft polyether segments are linked to hard crystalline polyester groups.

As yet there are no significant commercial products in which graft copolymers constitute the major component, although a number of graft copolymers showing thermoplastic elastomer behaviour have been thoroughly investigated.[22] The most important use of graft copolymers and the area in which they have received most scientific attention is in rubber toughening of glassy polymers such as polystyrene and poly(styrene-*stat*-acrylonitrile). Graft copolymers which are formed *in situ* act as stabilizers in the early stages of the preparation (emulsion or suspension) and in the final product provide adhesion between the rubbery inclusions and glassy matrix by bridging the phase boundary.[23]

In this chapter we review the general topic of chain segregation in copolymers and the structures thereby generated in the bulk state and in solution. Physical properties and applications of block copolymers are covered in other chapters.

6.2 MICROPHASE SEPARATION

6.2.1 General Characteristics

A schematic outline of microphase separation in an ABA-type block copolymer in which both blocks are non-crystalline is given in Figure 1. In the case shown, the domains (which could be either spheres or cylinders viewed end-on) are composed of A blocks and the matrix of B blocks. The space in the domains and the matrix would be packed with A and B blocks respectively to a density commensurate with the bulk state. The speckled regions in the diagram draw attention to the presence of an interfacial region in such systems in which mixing of segments of A and B blocks can occur.

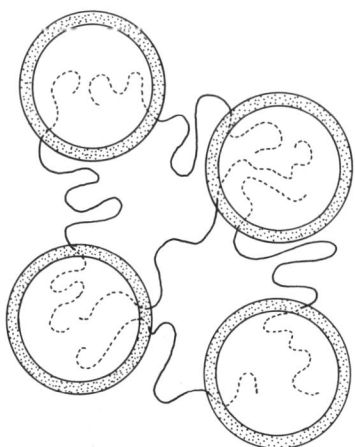

Figure 1 Schematic outline of microphase separation in an ABA block copolymer in which both components are non-crystalline (see text)

For relatively homogeneous AB, ABA and (AB)$_n$X copolymers (where the latter are 'star-shaped' systems with $n = 3$, 4, 5, *etc.*), in which both blocks are non-crystalline, it has been shown by transmission electron microscopy (TEM) and diffraction methods that three basic types of domain are possible.[24-30] These are spheres, cylinders and alternating lamellae. Because of the difficulty of reaching true equilibrium as opposed to metastable states, the structure and dimensions can be a function of the history (solution, mechanical and thermal) of the sample.[31-36] The problem arises because changes in the domain morphology can require chain transport *via* intermediate states of high energy (*e.g.* states in which the two types of block are molecularly mixed).[37] Whilst recognizing this problem, one can in general expect on theoretical grounds, and indeed find experimentally, a gradual change in structure of the type illustrated in Figure 2 as the composition of the copolymer is varied from predominantly B to predominantly A.

Increasing A content

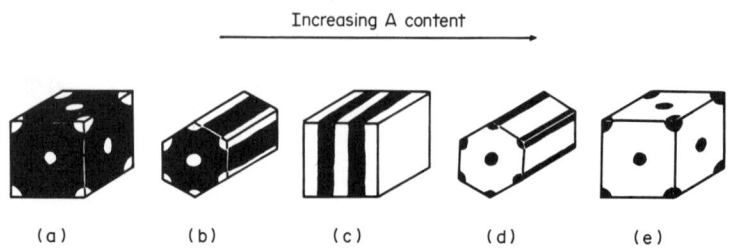

(a)   (b)   (c)   (d)   (e)

Figure 2   Effect of composition on block copolymer morphology: (a) spheres of A in matrix of B, (b) cylinders of A in matrix of B, (c) alternating A and B lamellae, (d) cylinders of B in matrix of A, (e) spheres of B in matrix of A (after Molau[38])

For equilibrium states, theoretical studies have related domain dimensions and lattice parameters, and the composition of transitional boundaries separating spherical, cylindrical and lamellar morphologies, to block types and molecular weight and to the nature and concentration of any diluent (see Section 6.3).

In practice, it is found that the degree of regularity of the domain structure depends very much on the physical treatment to which a sample has been subjected. When the structures are highly ordered it is found that lamellar domains are regularly alternating,[39] long cylindrical domains are hexagonally packed and spherical domains can be packed[40] in one of the three cubic forms: simple (sc),[41] face-centred (fcc)[42] or body-centred (bcc).[36] Other regular structures of a bicontinuous type appear also to be possible.[43]

Even when the domain packing is highly ordered the structures are nearly always 'polycrystalline' in nature, having a grain texture[34,36] rather than being 'single crystal'. However, a few examples of the 'single crystal' type have been reported.[40] A sample with a highly ordered domain morphology can easily be converted to one with only short-range order (and *vice versa*) by the appropriate physical treatment.

For the vast majority of AB, ABA and (AB)$_n$X systems which have been studied to date, the domain diameters for spheres and cyclinders and the thicknesses of lamellae fall within the range 5–100 nm. However, the length of cylindrical domains and the breadth and length of the lamellae can approach macroscopic dimensions in samples where the morphology is well developed.

With samples of relatively homogeneous AB- and ABA-type block copolymers, in which one or both types of block has crystallized, the domain morphology is frequently based on alternating lamellae[43-46] (schematically outlined in Figure 3). However, other structures are possible depending on the crystallization conditions and the structure of the crystallizing blocks.[47,48] Block copolymers containing crystalline blocks have in addition the possibility of developing the structural features (spherulites, fibrous textures, *etc.*)[49,50] associated with crystalline polymers in general.

In (AB)$_n$-type multiblock copolymers consisting of alternating short blocks, microphase separation can also occur, even if, as is often the case, the blocks are variable in length, provided at least one of the blocks is crystalline or contains highly polar groups. In such cases the sizes of the domains are smaller than those observed for the regular copolymer types discussed earlier and the domain packing in general shows only short-range order.[51]

Compared with block copolymers, relatively few studies of microphase separation have been carried out on graft copolymers. The main reason for this deficiency is the difficulty which exists in preparing well-defined graft copolymer samples. From the work that has been carried out, it appears that, depending on their chain structure and homogeneity, graft copolymers[29,52,59] can have domain morphologies varying from something approaching those exhibited by the regular AB-, ABA- and (AB)$_n$X-type systems to those of the segmented multiblock copolymers.

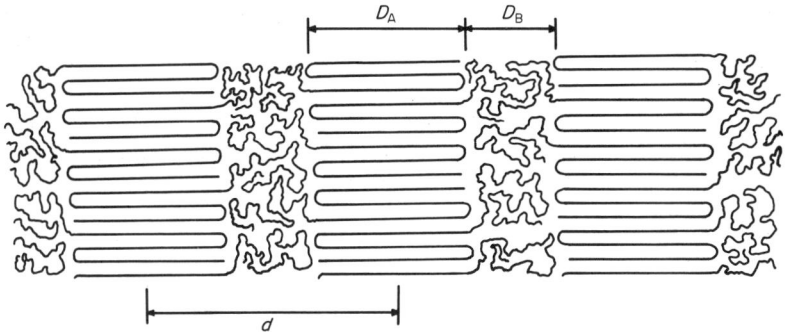

Figure 3 Schematic outline of microphase separation in an AB copolymer in which one component is crystalline; D_A and D_B are the thicknesses of the lamellae domains and d is the lattice repeat distance

6.2.2 Structural Studies

6.2.2.1 *Electron microscopy*

Electron microscopy in the transmission mode (TEM) has been extensively used for imaging domain structures since it has the resolution to probe the major features of interest. The disadvantages encountered in its use are those normally connected with the technique:

(a) The requirement with most samples to enhance contrast between microphases by staining or shadowing. The most widely used method has been staining with osmium tetroxide vapour (Kato's method)[54] which forms an adduct with organic double bonds, and enables contrast to be enhanced between polydiene chains (*e.g.* polybutadiene or polyisoprene) and unsaturated chains or between diene chains and simple aromatic polymers such as polystyrene. Whilst staining with osmium tetroxide does not appear to cause any significant morphological changes, results need to be viewed with great caution when assessing fine detail (*e.g.* thicknesses of interfacial regions). Shadowing with heavy metals has found only limited use, but provided useful supporting evidence of microphase separation in early studies.

More recently, further evidence for microphase separation in polydiene/polystyrene block copolymers using a technique which avoids staining was provided by Handlin and Thomas.[55] These workers used a defocusing technique to obtain phase contrast. The images obtained were in agreement with images obtained using osmium tetroxide staining, both for ordered and disordered domain structures.

(b) Only ultrathin specimens (usually 10–100 nm) can be studied by TEM. Specimens can be obtained by sectioning of bulk specimens using ultramicrotomy. Great care has to be taken to ensure structural changes do not occur during sectioning and specimen mounting. Rubbery specimens must be rendered sufficiently rigid by cooling or chemical hardening before sectioning. Alternatively, thin specimens can be cast from solution. In this case, however, the information obtained by TEM does not relate directly to the bulk state, although when solvents which are relatively good for both components are used such studies can provide a useful guide to structures which might be formed in the bulk state.

(c) Radiation damage can be a serious problem particularly when investigating fine detail.

(d) The method is essentially one for studying the solid state and the study of gels is excluded because of the vapour pressure of the solvent.

(e) Only small volumes of polymer are investigated by this method and assessments have to be made whether micrographs are 'representative'. However, when the structure has long-range periodicity, switching the microscope into a diffraction mode can be used to confirm lattice parameters.[27,29]

Electron micrographs of domain structures based on lamellae show black and white stripes or areas of uniform shade depending on the angle of section. Cylindrical domains show arrays of black dots on a white background or white dots on a black background depending on which component has been stained, or alternatively black and white stripes. Spherical domains also show black dots on a white background or white dots on a black background. It is possible to distinguish between the domain types by sectioning samples at different orientations. When spherical domains are present, particularly careful work is required to decide between the three possible types of lattice. Because of the many problems associated with interpreting electron micrographs, it is customary when dealing with periodic domain structures to seek supporting evidence from either small-angle X-ray or neutron diffraction.

In the case of solvent-cast ultrathin films, sectioning is not possible and structural assignments are generally aided by estimating the volume fractions of the components from the micrographs (assuming various structural models) and comparing the values obtained with the volume fraction determined from the chain structure.

Electron micrographs showing examples of microphase separation found in various copolymer systems stained with osmium tetroxide are given in Figures 4–9.

Figure 4 shows microphase separation in a polystyrene-*block*-polyisoprene-*block*-polystyrene film.[29] The periodic regularity of the structure has been enhanced by annealing the specimen after casting. The white domains are polystyrene cylinders viewed end-on in a polyisoprene matrix stained with osmium tetroxide. The electron micrograph reveals the specimen has a grain texture.

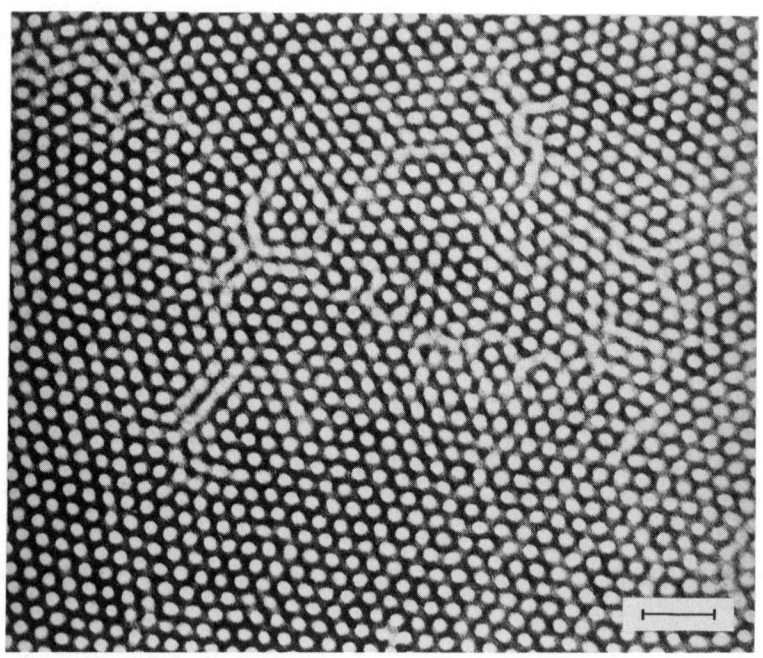

Figure 4 Electron micrograph of a film of a polystyrene-*block*-polyisoprene-*block*-polystyrene (12 500 : 35 500 : 12 500 M_n) copolymer cast from benzene solution (the molecular weights of the blocks are given inside the brackets). View shows polystyrene cylinders end-on in a polyisoprene matrix. Scale bar indicates 50 nm (reproduced from ref. 29 with permission of Society of Chemical Industry)

Figure 5 shows an electron micrograph of a polystyrene-*block*-polyisoprene film. This specimen was also annealed to enhance the structural regularity. In this case the domain morphology consists of polyisoprene cylinders viewed end-on in a polystyrene matrix. Again a grain texture is revealed.

Figure 6 shows microphase separation in a polystyrene-*graft*-polyisoprene copolymer film. Although this specimen was also subjected to a careful annealing procedure, the separation of the microphases is not as distinct as for the block systems and the domains appear slightly more irregular. The electron micrograph illustrates the point that in a graft copolymer the two components are not able to segregate as well, in general, as in a block copolymer.

Figure 7 shows an electron micrograph of an ultramicrotomed section of a polystyrene-*block*-polybutadiene-*block*-polystyrene copolymer in which the domain morphology consists of polystyrene cylinders in a polybutadiene matrix. The specimen was sectioned from a compression-moulded sheet.

The electron micrographs in Figures 8, 9(a) and 9(b) were obtained by Pedemonte *et al.*[36] for specimens ultramicrotomed from an extruded plug of a polystyrene-*block*-polyisoprene-*block*-polystyrene copolymer (K1107). Each of the specimens was cut perpendicular to the extrusion

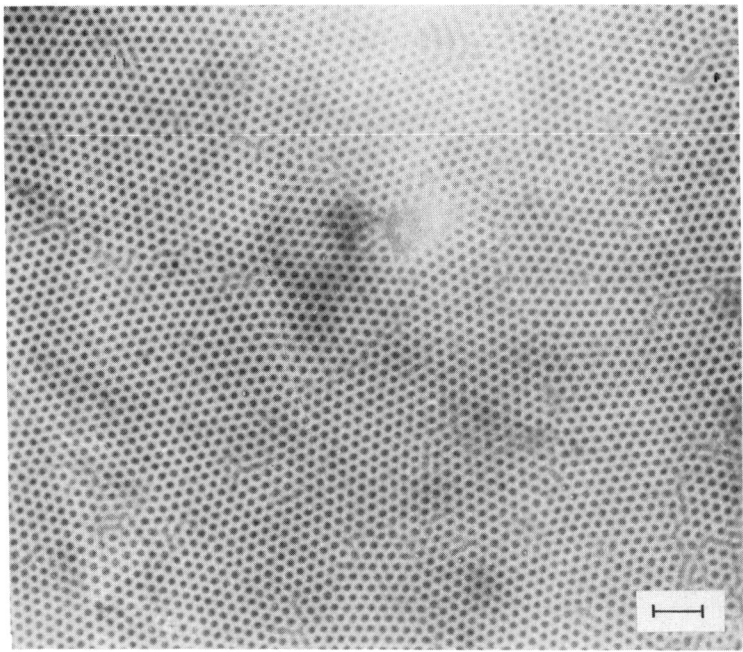

Figure 5 Electron micrograph of a film of a polystyrene-*block*-polyisoprene (34 000 : 10 000 M_n) copolymer cast from toluene solution. Scale bar indicates 100 nm. View shows polyisoprene cylinders end-on in a polystyrene matrix

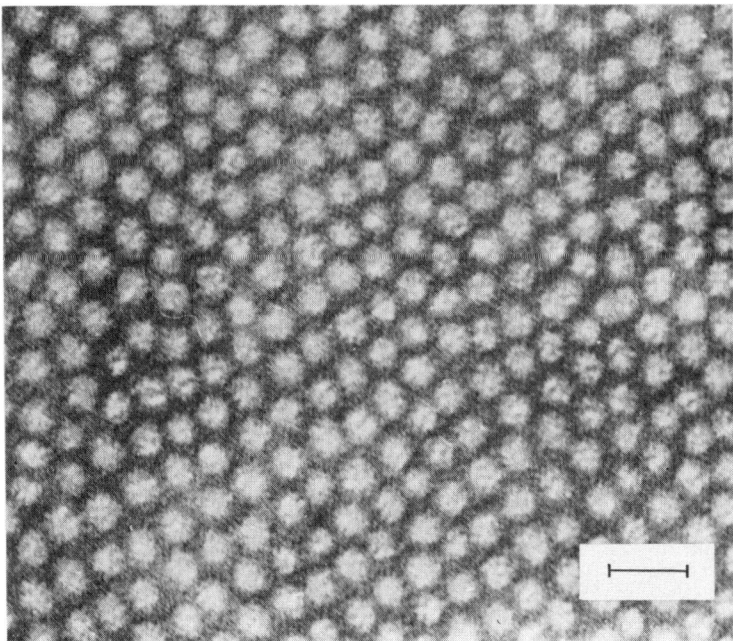

Figure 6 Electron micrograph of a film of a polystyrene-*graft*-polyisoprene (190 000 : 360 000 M_n) copolymer cast from hexane solution showing polystyrene domains in a polyisoprene matrix. Scale bar indicates 50 nm (reproduced from ref. 56 with permission of IPC Business Press Ltd.)

direction. The electron micrograph shown in Figure 8, revealing a very irregular domain structure, was obtained for a section which had been cut before the sample was annealed. The electron micrographs shown in Figures 9(a) and 9(b) were obtained for sections cut after annealing the plug at 150 °C for 6 days. They show spherical domains of polystyrene with bcc packing in a polyisoprene matrix.

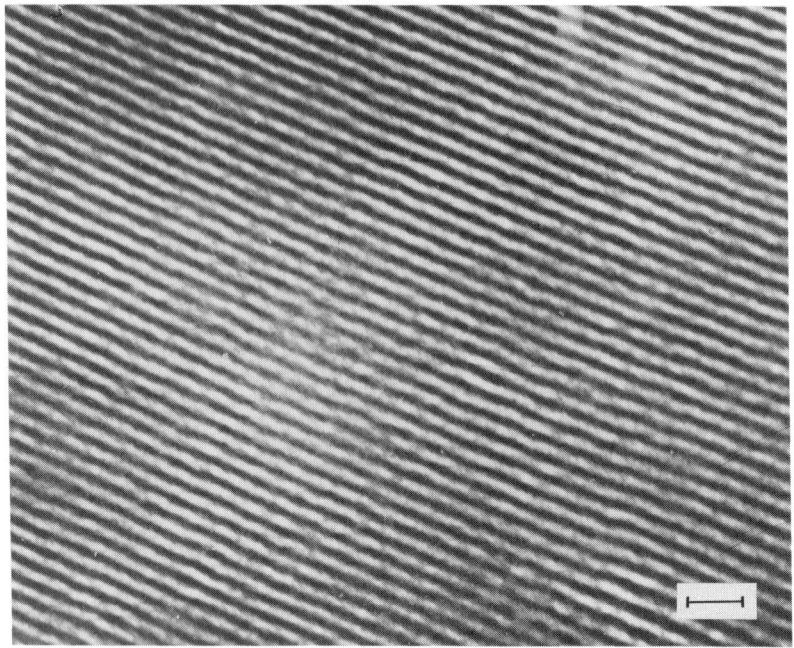

Figure 7 Electron micrograph ‾of an ultrathin section of a polystyrene-*block*-polybutadiene-*block*-polystyrene $(10\,000:71\,000:10\,000\ M_n)$ copolymer showing polystyrene cylinders in a polybutadiene matrix. Scale bar indicates 100 nm

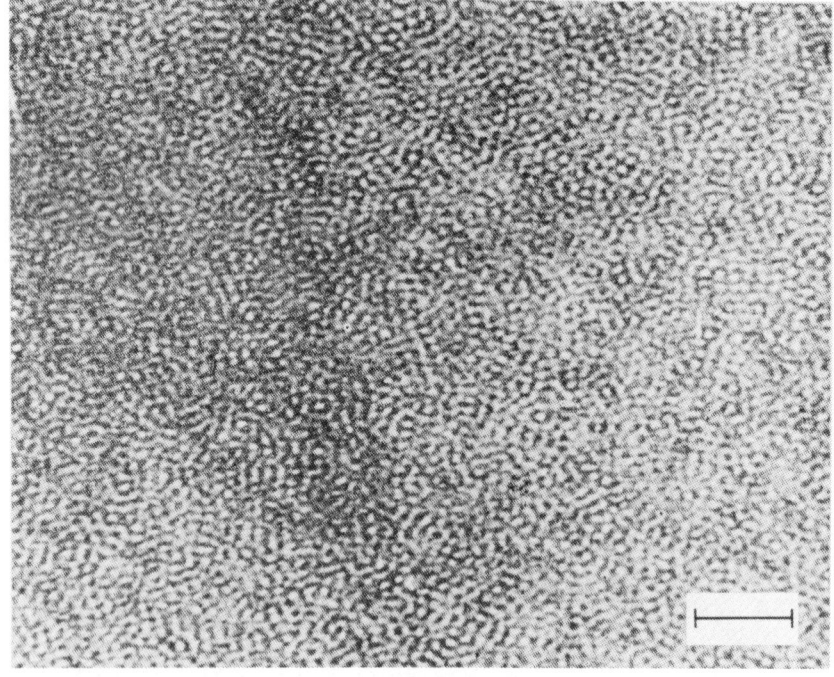

Figure 8 Electron micrograph of an unannealed ultrathin section of an extruded plug of a polystyrene-*block*-polyisoprene-*block*-polystyrene $(7500:135\,000:7500\ M_n)$ copolymer K1107 cut perpendicular to the extrusion direction. Scale bar indicates 0.2 μm (reproduced from ref. 36 with permission of IPC Business Press Ltd.)

Whilst multisectioning can be used to build up a fairly complete picture of microphase separation in systems consisting of uniform domains in periodic array, it is much less useful when applied to systems in which the domains have irregular features as in Figure 8. Recognizing this difficulty Spontak, Williams and Agard[57] have recently used a new procedure for extracting more detailed

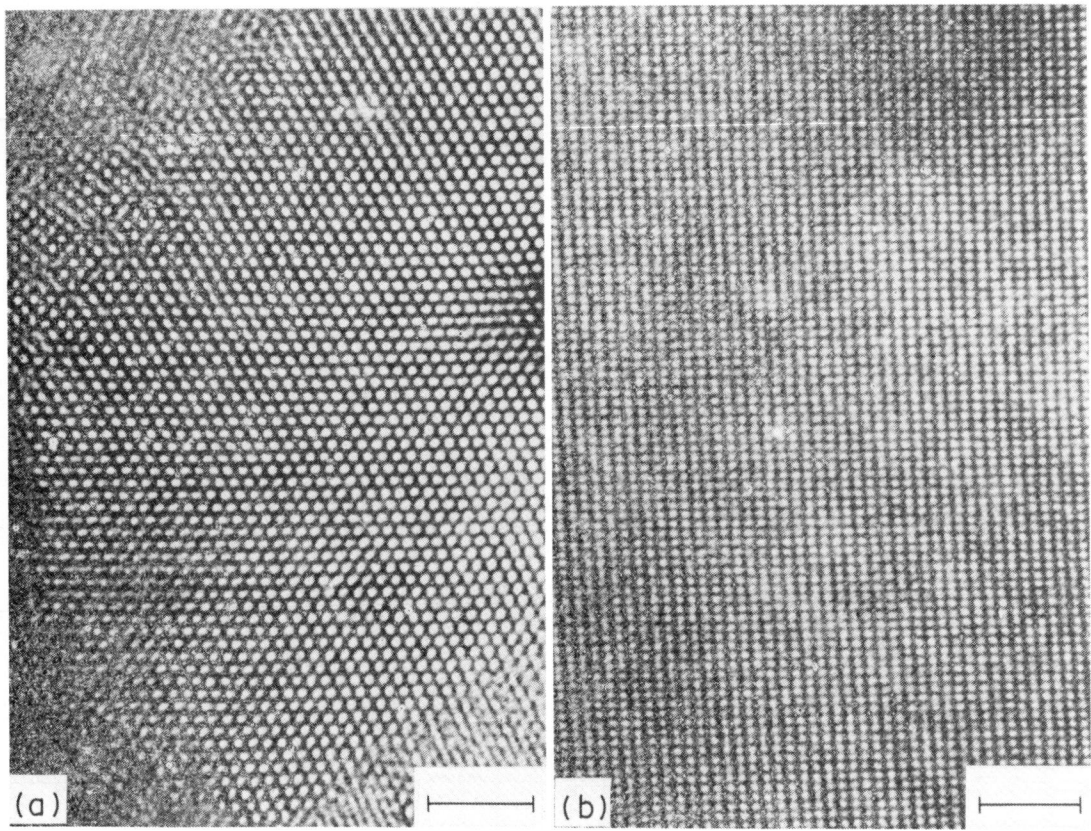

Figure 9 Electron micrographs of ultrathin sections of an annealed extruded plug of a polystyrene-*block*-polyisoprene-*block*-polystyrene (7500:135 000:7500 M_n) copolymer K1107 cut perpendicular to the extrusion direction: (a) view of 111 plane of spherical polystyrene domains in a polyisoprene matrix, (b) view of 110 plane of spherical polystyrene domains in a polystyrene matrix. Scale bars indicate 0.2 μm (reproduced from ref. 36 with permission of IPC Business Press Ltd.)

information from individual sections. Using TEM they obtained images (which are essentially two-dimensional projections) of a polystyrene-*block*-polybutadiene-*block*-polystyrene copolymer stained with osmium tetroxide. Images were obtained for a range of tilt angles ranging from $-61.5°$ to $48.5°$ in 5° increments. The two-dimensional images were then combined, after corrections had been made for variations in optical density, to give a single three-dimensional image when observed with a stereoscopic viewer. The analysis revealed two types of structure were present. The first type resembled cylinders in hexagonal packing with some deviations from geometric ideality. These structures were oriented normal to the surface of the film and extended through the thickness of the sample which was approximately 36 nm. The second type of structure consisted of irregular globular domains which were believed to be well removed from thermodynamic equilibrium. It was argued that cylindrical domains evolve from spherical domains in the development of the microphase structure.

The characterization of microphase separation in segmented multiblock copolymers by TEM has presented some very challenging problems. In studying polyester- and polyether-based polyurethanes and poly(ether-*alt*-esters), progress has been limited both by the lack of phase contrast and the small size of the domains. Mixed success has been obtained in attempts to enhance contrast by defocus techniques. Roche and Thomas[58] have demonstrated striking similarities between electron micrographs of films of segmented polyurethanes and films of carbon and polystyrene. The latter workers point out that several published micrographs which claim to show microphase separation in the polyurethanes had features attributable to spatially filtered noise structure, defocus and spherical aberration. Recently progress has been made by studying model segmented polyurethanes containing polybutadiene as the soft block[59-61] and by combining the results from TEM with information from other methods (*e.g.* small-angle X-ray scattering and transport measurements).

Many material science laboratories now have the possibility of carrying out scanning transmission electron microscopy (STEM) in addition to TEM. As yet, however, STEM has not played a significant role in studying block copolymer structures. Scanning electron microscopy (SEM) on the other hand has played a part in investigating crystalline textures.[62] Its lower resolution, however, has meant it has not been helpful in studying domains.

6.2.2.2 *Small-angle X-ray (SAXS) and neutron scattering (SANS)*

For coherent elastic scattering of monochromatic radiation at angle 2θ to the incident beam, the intensity of scatter is proportional to the squared product of a lattice factor, $L(s)$ and a structure factor $F(s)$, where s is a scattering vector of magnitude $s = 2\pi\lambda^{-1}\sin\theta$ and λ is the wavelength of the radiation. When the lattice has long-range periodicity, $L(s)$ consists of linear arrays of delta functions, each array being perpendicular to the particular set of lattice planes from which they are generated by Fourier transformation, and the spacing of the delta functions in each array is inversely proportional to the separation of the lattice planes.[63]

For mesomorphic structures $F(s)$ is the Fourier transform of the structure of an individual domain (or several domains if they constitute the repeat unit). In the case of X-rays, scattering structure is defined in terms of electron density, whilst for neutrons it is differences in scattering cross-sections, a nuclear characteristic, which is important. For the basic domain types (solid sphere, cylinder and lamella) the structure factors are rapidly oscillating functions[64] (*e.g.* for spheres $F(s)$ is a Bessel function of order 3/2 with the first two minima in reciprocal space occurring at $s = 4.43$ and 7.73).

The lattice factor can be considered to sample the structure factor at different points in reciprocal space and the observed diffraction pattern is essentially a two-dimensional projection of the square of this sampling. For samples which are 'polycrystalline' in nature (*i.e.* the grains are randomly oriented) the diffraction patterns are of the Debye–Scherrer type consisting of concentric rings. When the structural units are oriented along a particular axis the rings give way to broad spots, the distribution of which can reveal the orientation of the specimen.[65]

If the structure factor is sampled by the lattice factor at a point at which the former is zero, or very weak, then no diffraction peak will ensue. On the other hand, sampling in a region where the structure factor is strong can lead to a relatively strong peak. Thus, whilst an analysis of the positions of the diffraction peaks (or rings) can give the type of lattice and its parameters, the intensities of the spots (or rings) contain the information relating to the structure and dimensions of the domains.[64]

The type of lattice can be determined by comparing the observed sequence of d spacings (*i.e.* lattice-plane spacings) with characteristic sequences of model lattices (given in Table 1). The d spacings are determined through the Bragg equation, $n\lambda = 2d\sin\theta$, where the Bragg angle θ is half the scattering angle at which the diffraction peak is observed. Figure 10 shows a plot of X-ray scattering intensity against angle for the polystyrene-*block*-polybutadiene-*block*-polystyrene $(10\,000:71\,000:10\,000\ M_w)$ copolymer[34] sample referred to in Section 6.2.2.1 (the molecular weights of the blocks are given inside the brackets). The sequence of spacings $(d_2/d_1 = 0.560,\ d_3/d_1 = 0.368,\ d_4/d_1 = 0.338)$ suggested hexagonally packed cylinders; the absence of the second peak in the expected sequence for cylinders was assumed to be due to a weak structure factor in this region. The distance between the centres of adjacent cylinders was determined from the spacings to be 28.0 nm. The cylindrical assignment was in agreement with the results from electron microscopy.

If sufficient numbers of diffraction peaks are observed, simple inspection for the absence or weakening of particular peaks in an observed sequence of d spacings can provide a useful estimate of the radius (or thickness) of domains. To obtain accurate values, however, it is necessary to compare the whole of the observed intensity profile (after correcting for polarization, the Lorentz factor, *etc.*) with intensity profiles predicted by theoretical models.[66] When there are only a few peaks (which is almost always the case with spherical domains, and is often the case with cylindrical domains) then a fitting procedure must be used. With the scattering profile obtained for the polystyrene-*block*-polybutadiene-*block*-polystyrene $(10\,000:71\,000:10\,000\ M_w)$ copolymer this approach yielded a value of 12 nm for the cylindrical diameter.

So far in our consideration of small-angle diffraction behaviour we have not considered the effective of diffuse interfacial layers. Porod[67] showed that the tail (*i.e.* the asymptotic behaviour at high angles) of a scattering curve for an ideal two-phase system with sharp boundaries between the phases should have an intensity proportional to s^{-4} for a system studied with point collimation and proportional to s^{-3} when studied with infinite slit collimation.

Theory developed by Ruland[68] shows that diffuse boundaries give rise to negative deviations from Porod's law and these deviations can be used to determine the nature of such boundaries. However,

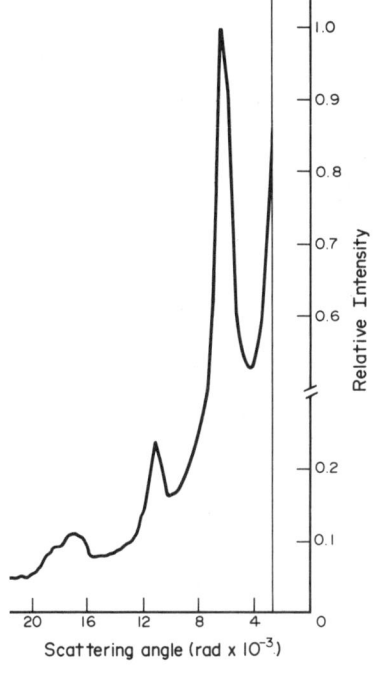

Figure 10 Variation of X-ray scattering intensity with angle for a poly-styrene-*block*-polybutadiene-*block*-polystyrene (10 000 : 71 000 : 10 000 M_n) copolymer having a domain morphology consisting of polystyrene cylinders in a polybutadiene matrix (reproduced from P. R. Lewis and C. Price, *Polymer*, 1971, **12**, 258 with permission of IPC Business Press Ltd.)

Table 1 Ratios of Consecutive Bragg Spacings for Different Model Morphologies

Morphologies	Ratios
Alternating lamellae	1 : 0.500 : 0.333 : 0.250 : 0.200
Hexagonally packed cylinders	1 : 0.577 : 0.500 : 0.378 : 0.333
Fcc packed spheres	1 : 0.866 : 0.612 : 0.521
Sc packed spheres or bcc packed spheres	1 : 0.706 : 0.577 : 0.500

positive deviations from Porod's law can be caused by density variations within phases. Hence an important aspect in studying boundary layers is the accurate subtraction of background scattering such as arises from natural density fluctuations within phases, impurities and from parasitic scattering. Hashimoto *et al.*[69, 70] have obtained the background by extrapolation to high angles. Vonk fitted the background to an empirical expression.[71]

For experiments carried out with point collimation the observed scattering intensity in the tail for systems in which the interface is a linear slope and for which positive deviations have been subtracted is given by[71]

$$I(s) = Ks^{-4}[1 - (\pi^2\delta^2 s^2)/3] \tag{1}$$

where K is a constant and δ is the thickness of the interfacial region in systems in which the structure has a grain texture. If, as is more realistic, the interfacial gradient is assumed to have a sigmoidal shape, then for point collimation and in the absence of positive deviations, the scattering intensity in the tail is given by

$$I(s) = Ks^{-4}\exp(-\sigma^2 s^2) \tag{2}$$

where σ is the standard deviation of the sigmoidal gradient. In relation to the characteristic thickness $\sigma = (2\pi)^{-1/2}\delta$ Tyagi *et al.*[72] have tabulated the various expressions which have been used to interpret scattering profiles for slit collimation assuming a sigmoidal gradient.

By studying negative deviations from Porod's law a number of workers have obtained interfacial thicknesses for domain structures. Hashimoto *et al.*[69, 70] using SAXS found the interfacial thickness for a polystyrene-*block*-polyisoprene copolymer (assuming perfectly oriented lamellae normal to the bulk surface so that $I(s) \propto s^{-2}$) fell within the range 1.9–2.2 nm. Using SANS in studies of polystyrene-*block*-polyisoprene and polystyrene-*block*-polyisoprene-*block*-polystyrene copolymers, Richards and Thomason[73] also obtained values for the interfacial thickness in the region of 2 nm.

Vonk, using SAXS, obtained a value of 1.5 nm for the thickness of the interfacial region in a poly(methyl methacrylate)-*block*-polyisoprene-*block*-poly(methyl methacrylate) copolymer.[71]

More recently Shibayama and Hashimoto[66] have carried out a very thorough investigation of interfacial thicknesses in two polystyrene-*block*-polyisoprene copolymers containing highly oriented lamellae. Arguments were presented for taking $I(s) \propto s^{-4}$ instead of $I(s) \propto s^{-2}$. In addition to using negative departure from Porod's law, they fitted theoretical scattering profiles to their experimental results to obtain the interfacial thicknesses. The two approaches gave results in good agreement as shown in Table 2.

Table 2 Structural Parameters for Lamellae Domains Determined by Small-angle X-ray Scattering[66]

Sample	$M_n/10^4$	wt % PS	d	δ^a	δ^b
L-2	3.1	40	24.8 ± 0.5	1.9 ± 0.3	1.9
L-7	10.5	46	47.6 ± 0.5	1.8 ± 0.3	1.8

[a] Determined by fitting theoretical scattering profiles to experimental results.
[b] Determined from the scattering tail by Porod's law; d is the repeat distance of the lattice perpendicular to the lamellae surfaces.

For systems having broad interfacial layers (*e.g.* systems swollen with a solvent or containing tapered blocks) the only reliable procedure for obtaining interfacial thicknesses is to fit the experimental results to theoretical scattering profiles. Annighoefer and Gronski[74] have studied the interfacial layers in a range of polystyrene-*block*-(polystyrene-*stat*-polyisoprene)-*block*-polyisoprene copolymers having lamellar morphologies by using SAXS and TEM. The statistical centre block ranged from 0–33% of the copolymer and the overall styrene content was 50%. By evaluation of electron micrographs of ultrathin sections stained with osmium tetroxide and by analysis of the total meridional SAXS intensity distribution using a one-dimensional paracrystalline model, the variation of the mean long period fluctuation of the lamellae thickness and the variation of the extent of the interfacial region were obtained in relation to the content of the copolymer centre block. The analysis showed that the volume fraction of the inerfacial region varied from 8.5 to 65%.

Elastic scattering has also been used to investigate block copolymers in the disordered state.[75,76] Results may be interpreted in terms of the random phase approximation, which allows one to predict spectra of thermal concentration fluctuations of copolymers in the disordered state up to the microphase transition.[76,77]

When studying domain morphologies there is little to choose between SAXS and SANS. The majority of such studies have naturally been carried out using SAXS because the equipment is more widely available. The great advantage of SANS over SAXS is the ability of SANS to prove chain structure factors (and thereby the average dimensions) of individual copolymer chains and blocks in the bulk state. Contrast in scattering experiments of this type is generated by dispersing small concentrations of suitably deuterated chains in a matrix of protonated polymer. The scattering theory underlying the method is similar to that developed for studying copolymers in dilute solutions by elastic light-scattering and small-angle X-ray scattering.

Unfortunately, there is a major problem associated with this application of SANS. The method relies on no perturbation being produced by the labelling with deuterium. However, a degree of perturbation is inevitable because deuterium is a heavier atom than hydrogen and has different vibrational characteristics. In some cases the effect can be sufficiently strong that it leads to segregation. Another problem is that to generate sufficient contrast it can be necessary to use higher concentrations than are desirable (*e.g.* 2–4 wt %). Because of such difficulties very few studies have been reported of the dimensions of copolymer blocks incorporated in mesophase structures. From the limited number of results available it is not possible to draw any general conclusions. However, it should be noted that, using SANS, Hadziioannou *et al.*[78] successfully studied the conformation of labelled polystyrene blocks in a polystyrene-*block*-polyisoprene copolymer with lamellae domains. Bates *et al.*[79] reported on labelled polybutadiene blocks in a polystyrene-*block*-polybutadiene copolymer with spherical domains. They noted a possible segregation effect arising from deuteration with one of their samples.

6.2.2.3 *Other methods*

Whilst electron microscopy, SAXS and SANS have provided most of the basic structural information on microphase separation, other methods have played a significant role.

Polarizing optical microscopy has been used to observe liquid-crystal-like textures of block copolymers which can be particularly distinctive when systems are swollen with solvent.[80,81] It has also been used to study birefringence and hence to provide a measurement of chain orientation.[82,83] Information on chain conformation has been obtained from IR and Raman spectroscopy.[84]

Wide-angle X-ray scattering (WAXS) has been used in the usual manner to characterize crystalline regions in microphase structure.[85] In addition to providing information on chain structure and packing the technique can be used to obtain the degree of crystallinity and the size of crystallites.

Low-angle light-scattering (LALS) has proved useful in studying some of the coarser features of interest in block copolymer morphology such as grain size, and sperulitic textures in systems containing crystallizable blocks.[86,87]

In addition to the major role it has played in the determination of the chemical structure of polymer chains, nuclear magnetic resonance (NMR) spectroscopy has provided information concerning microphase morphology and interfacial characteristics in both crystalline and non-crystalline systems.[88-92] The investigation of microphases by this approach is possible because the molecular motion of segments is very dependent on their location in the structure.

Mechanical,[93-97] electrical[98,99] and ultrasonic[100] moduli and loss factors for block copolymers are properties of great importance in their own right. In addition their determination has been used to assess the degree of block segregation in copolymers and also to provide information concerning interfacial regions. Of similar importance for probing block segregation has been differential scanning calorimetry (DSC).[101]

6.2.3 Stability and General Characteristics of Domain Structures

6.2.3.1 *Non-crystalline systems*

The most important factor in deciding the type of equilibrium domain structure adopted by a system is the weight fraction of the components. The molecular weight appears to play a secondary role in this respect, but strongly influences the location of the boundary defining the transition from homogeneous to microphase-separated states. This behaviour, which is predicted by theory[102] (see Section 6.4.2), is outlined schematically in Figure 11.

Reported studies of domain structures encompass a wide range of copolymer systems. By far the most comprehensive work, however, has been carried out on polystyrene/polydiene block copolymers.

Gallot[103] has considered the results obtained by his group and others. For polystyrene-*block*-polybutadiene copolymers he reports bcc polystyrene spheres up to approximately 20 wt % poly-

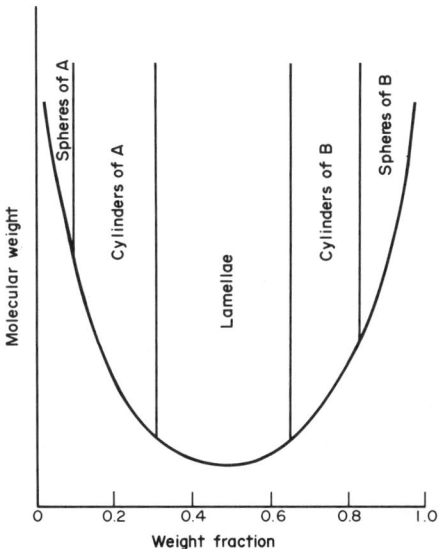

Figure 11 Phase diagram for the geometry, stability and microdomains of an AB-type block copolymer. The vertical lines indicate transitions from one type of domain geometry to another. The lower curves indicates the boundary between microdomain and homogeneous phase behaviour (after Helfand and Wasserman[102])

styrene, hexagonally packed polystyrene cylinders between approximately 20 and 40%, lamellae from 40 to 64%, polybutadiene cylinders from 64 to 82% and bcc polybutadiene spheres above 84%. Between 82 and 84% an orthorhombic structure was reported, but there is a good possibility in our view that this state is a metastable one.

Changes in molecular architecture can have an effect on domain structure and dimensions but is not as important as might be expected. Price et al.[29] showed that the structure and dimensions of linear and star-branched (polystyrene-block-polyisoprene)$_n$X copolymers are indpendent of the value of n for $n = 1$, 2, 3 and 4. This result has recently been confirmed by Alward et al.[43] However, changes in domain structure are observed when $n > 8$, in which case the formation of an ordered bicontinuous structure occurs.

After numerous experimental studies it now seems to have been reasonably well established that for AB- and ABA-type block copolymers containing polystyrene and polyisoprene (or polybutadiene) the microdomain size, D_i, and lattice repeat distance d, of periodically organized microdomains have the approximate molecular weight dependence[104-106]

$$D_i \sim (M_i)_n^{2/3} \tag{3}$$

$$d \sim (M)_n^{2/3} \tag{4}$$

where $(M_i)_n$ and $(M)_n$ are the number-average molecular weights of the blocks forming the domain and the overall number-average molecular weight of the copolymer respectively. However, in view of the difficulty in achieving thermodynamic equilibrium, further work is required in this area. It would also be of great value to widen the scope of the investigations to other block copolymers.

Interesting results have been obtained with respect to the uniformity of microdomain sizes.[106] For polystyrene-block-polyisoprene copolymers the distribution of microdomain size within a given sample was shown to be much narrower than the molecular size distribution; the heterogeneity index for sizes $(D_i)_w/(D_i)_n$ was typically 1.001, whilst that for the molecular weight distribution of the block copolymers was typically 1.1. Similar sharpening in size uniformity can be expected for other systems. The same situation exists for micelles in dilute solution.

The dependence of domain dimensions on temperature appears to be small from the few studies that have been made.[107] From studies of polystyrene-block-polydiene copolymers swollen with non-selective solvents, Shibayama et al.[37] show that in the equilibrium regime the dimensions decrease slightly with an increase in temperature (as expected from theory), whilst in the non-equilibrium regions the dimensions increase slightly with an increase in temperature, being governed by the coefficient of thermal expansion.

A number of workers have attempted to study the phase-boundary region for the microdomain–homogeneous-state transition which is expected to occur with polystyrene/polydiene copolymers on heating to relatively high temperatures ($> 150\,°C$). Distinct transitions have been observed by some workers,[108-112] but not others, and the picture is far from clear. There is undoubtedly a need for much more work in this area, particularly on systems which exhibit transition behaviour at somewhat lower temperatures, so that problems associated with thermally induced chemical changes do not arise.

When a non-selective solvent is added to a block copolymer, changes in domain size occur. Detailed studies of this behaviour have been carried out by Shiboyama et al.[113] on a polystyrene-block-polyisoprene copolymer exhibiting a lamellar microdomain structure at concentrations greater than 20 wt % copolymer. Two regimes in the concentration dependence of the lattice repeat distance d were identified: (a) in the low-concentration regime (< 70 wt % copolymer) d increases with increasing c, while (b) in the high-concentration regime (> 70 wt % copolymer) d tends to increase with increasing c which is consistent with earlier work reported by Skoulios et al.[114] and Douy et al.[115] It was proposed by Shibayama et al.[113] that in the low-concentration regime the system was able to attain thermodynamic equilibrium, in which case one expected increasing segregation power between the polymeric blocks as the amount of solvent increased. In the high-concentration regime the behaviour was governed by the inability of the system to reach equilibrium during the time scale of the experiment.

The presence of a solvent which is selectively poor for one of the blocks can be considered to alter the relative volume fractions or the components.[114] Adding such a solvent to a copolymer system might be expected therefore to modify the domain morphology. However, when one of the components is only poorly swollen, kinetic factors are likely to be very important and the expected changes may not occur.

6.2.3.2 *Crystalline systems*

Studies of microphase separation have been made on many regular block copolymers containing crystallizable blocks. The types of crystallizable blocks which have been examined include polyethers,[45,116-122] polyamides,[123,124] polypeptides[48,125] and polyesters.[126]

Copolymers containing poly(oxyethylene) and atactic polystyrene blocks have been studied in detail. Studies have been carried out on AB-, ABA- and (AB)$_n$-type systems.[45,116-122] At high temperature ($>100\,°C$) both components are rubbery and the domain morphology depends in the usual way on the volume fractions of the components. On cooling, the polystyrene microphase becomes glassy in the region of $100\,°C$ or below depending on the molecular weight of the polystyrene block. On further cooling the poly(oxyethylene) microphase crystallizes. When the volume fraction of poly(oxyethylene) is sufficiently large to ensure that in the melt state the morphology comprises either a poly(oxyethylene) matrix or lamellae domains then spherulitic textures with fibrous features[49] can be expected. Many of the systems studied show such textures. Their basic morphology is of the sandwich type with layers of polystyrene regularly spaced between lengths of folded poly(oxyethylene) chains.[12] Solution-grown single[127] crystals of polystyrene-*block*-poly(oxyethylene) also show chain folding, the lamellar crystals being encrusted with a covering of non-crystalline polystyrene. When the volume fractions of the components dictate that the polystyrene blocks form the matrix, then in the bulk state it appears the poly(oxyethylene) crystallizes within the disperse phase without any obvious rearrangement of the matrix.[47]

Less extensive studies have been carried out on poly(oxyethylene)/polydiene copolymers which exemplify systems in which the second component is rubbery.[21,122]

Studies have also been carried out on regular block copolymers in which both components are crystallizable. A good example[44] is provided by the system poly(oxyethylene)-*block*-poly(ε-caprolactone). As might be expected it appears that the first component to crystallize on cooling from the melt imposes restrictions on the second when its turn comes to crystallize.

To investigate the effect of restrictions imposed on crystallizable chains packing in adjacent domains Booth *et al.*[84,128] have studied a range of model completely crystalline oligo(ethylene glycol di-*n*-alkyl ether)s, $H(CH_2)_nO(CH_2CH_2O)_m(CH_2)_nH$ with $m=9$ or 15 and n in the range 1–30. On crystallization the block oligomers were shown to form layer structures in which the oxyethylene central-block and the *n*-alkyl end-blocks may be crystalline or non-crystalline (liquid crystalline) depending upon the block length and the crystallization conditions. Samples with $m=9$ or 15 and $12 < n < 21$ gave completely crystalline layer structures. Spectroscopic evidence indicates that the oxyethylene blocks, which in these samples crystallize first on cooling, were in a helical conformation as in poly(oxyethylene) homopolymer and the *n*-alkyl blocks crystallized in a planar zigzag conformation. X-Ray scattering from the long spacings of specimens which had been oriented by slow crystallization in capillaries gave many orders of reflections (*e.g.* up to order 30) The intensities of the reflections were analyzed by use of a model electron-density distribution for the layer structure. It was shown that the helical oxyethylene block was oriented normal to the layer end-plane, while the alkyl end-blocks were tilted relative to the layer end-plane at an angle in the range 70–54° depending on the sample.

In samples in which the *n*-alkyl blocks have the higher crystallization temperature, layer structures are again formed. In this case however the *n*-alkyl blocks in planar zigzag conformation are normal to the layer end-plane, whilst at ambient temperature the oxyethylene blocks are restricted to a liquid-crystalline state[84] in which the chains are in a predominantly planar zigzag rather than helical conformation. It is clear from the results that the relative cross-sectional areas of the chains play a major role in governing the state of the micophases.

Because of the complexities involved, the understanding of segmented multiblock polymers is far less advanced than that of the regular block polymers. Some of the clearest evidence of segmented multiblock polymers has been obtained for systems containing polybutadiene soft segments. A range of model polybutadiene/polyurethanes have recently been studied by Serrano *et al.*[60] using TEM, SAXS and sorption methods. The rubbery component is hydroxy-terminated polybutadiene end-capped with 2,4-toluene diisocyanate, whilst the glassy component is the reaction product of 2,4-toluene diisocyanate with 1,4-butanediol. The system offered the advantages of no hydrogen bonding or semicrystalline supramolecular structures, and the possibility of enhancing phase contrast by osmium tetroxide staining of the polybutadiene component. The results indicated the presence of phase separation on the scale of 10 nm in samples with 44 to 75 wt % hard segment, but there was no long-range order in the system. The scale of the morphology was found to be a major limiting factor, since to obtain planar images of individual domains it was necessary to use films approximately 10 nm thick and these proved extremely difficult to prepare. Solely from the TEM results it was not possible to make any morphological assignments. However, combining these results with those from the other techniques employed it was possible to suggest that cylinders of soft segment in a glassy matrix of hard segment was the morphology in 75 wt % hard segment materials whilst lamellae structures were established in the 52 wt % material.

The microphase dimensions and thickness of the diffuse boundary zone in a multiblock copoly(ether-*alt*-ester) based on poly(tetramethylene terephthalate) and on poly(oxytetramethylene) chains has been investigated by Droescher *et al.*[129] using SAXS. The system crystallizes into a pure crystalline ester microphase and a mixed non-crystalline phase. It was shown that SAXS results were consistent with a lamellar model comprising a crystalline core of thickness 3.5–4.5 nm, a diffuse boundary zone of width 0.9 nm, and a non-crystalline layer.

Using a combination of SAXS, TEM, DSC and dynamical mechanical measurements Tyagi *et al.*[72] have recently investigated a range of well-defined alternating segmented polysulfone/poly(dimethylsiloxane) copolymers. It was found that the thickness of the interfacial regions and the degree of phase separation was dependent on composition and block length. An advantage of the system was that there was sufficient contrast between the blocks to make staining unnecessary in the TEM studies.

6.3 MICELLIZATION

6.3.1 Early Studies

The reversible association of amphiphiles, which are molecules containing both hydrophobic and hydrophilic components, has attracted great attention since the early part of this century. One of the earliest workers to carry out quantitative studies on the subject was McBain,[130] and it was he who first used the term micelle to describe the colloidal products formed by the association of soap molecules. The basic structural features of simple soap micelles were worked out by Hartley[131,132] in the 1930s and 1940s: they consist of a compact globular hydrophobic core containing very little water surrounded by polar or ionic groups which protrude into the aqueous medium. Later it was established that other molecular shapes (*e.g.* worm-like micelles)[133] were also possible depending on the solution conditions and nature of the amphiphilic species.

As already mentioned in the introduction, when polymer chemists started to prepare amphiphilic block copolymers in the 1950s it was found that they also showed surfactant behaviour and exhibited micelle formation. It was also found that block copolymers (and graft copolymers) could form colloidal aggregates in organic solvents if these were selectively poor for one of the polymeric blocks.[134–138] These colloidal aggregates are also termed micelles.

6.3.2 Thermodynamic Background

The multistage process involved in micelle formation may be represented by a set of equilibria

$$A_1 + A_1 \underset{}{\overset{K_2}{\rightleftharpoons}} A_2$$

$$A_2 + A_1 \underset{}{\overset{K_3}{\rightleftharpoons}} A_3$$

$$A_3 + A_1 \underset{}{\overset{K_4}{\rightleftharpoons}} A_4$$

(6)

$$A_{N-1} + A_1 \underset{}{\overset{K_N}{\rightleftharpoons}} A_N$$

First, unassociated chains A_1 combine to form a dimer A_2 in a stage with equilibrium constant K_2 which is followed by further stages involving successive additions of block copolymer chains to the growing micelle. Formation of an N-mer involves multiple additions.

If all the equilibrium constants were equal, the molecular weight distribution of the micelles would have a most probable distribution. This extreme case is termed continuous association, or open association by Elias.[139] Soap micelles in hydrocarbon media apparently approach this type of behaviour. It has been shown, however, that when block copolymers having a narrow molecular weight distribution associate reversibly in either organic or aqueous media to form globular micelles the latter generally have a narrow size distribution. This means that only a narrow range of equilibrium stages have significantly high equilibrium constants.

When the micelles have a narrow size distribution it is useful from a thermodynamic standpoint to assume there is just a single stage equilibrium between unassociated chains and micelles with a fixed association number, m.

$$m\mathrm{A}_1 = \mathrm{A}_m \tag{7}$$

If it is also assumed the solution is ideal except for intramicellar interactions the equilibrium constant is given by

$$K_m = [\mathrm{A}_m]/[\mathrm{A}_1]^m = \exp(-m\Delta G^\ominus/RT) \tag{8}$$

where $\Delta G^\ominus$ is the free energy change per chain (expressed in molar units) of formation of micelles from unassociated chains.

For this limiting case, termed closed association,[139] the weight-average molecular weight of the total solute species in solution (micelles plus unassociated chains) at a total concentration of block copolymer c (in units of mol dm^{-3}), is related to the standard free energy of micellization by

$$\Delta G^\ominus = (m - 1)m^{-1}RT\ln c - RTm^{-1}\ln(M_\mathrm{w}M_1^{-1} - 1) + RT\ln(m - M_\mathrm{w}M_1^{-1})$$
$$+ m^{-1}RT\ln m - (m - 1)m^{-1}RT\ln(m - 1) \tag{9}$$

where M_1 is the molecular weight of the unassociated chains. If $m \gg 1$ and also the solution is at the critical micelle concentration (cmc; *i.e.* the concentration at which the presence of micelles is just detectable by an experimental method, such as light scattering), it follows from equation (9) that

$$\Delta G^\ominus \simeq RT\ln(\mathrm{cmc}) \tag{10}$$

If in addition m is independent of temperature, it follows from the above relations and the Gibbs–Helmholtz equation that[140-142]

$$\Delta H^\ominus \simeq R\,\mathrm{d}\ln(\mathrm{cmc})/\mathrm{d}T^{-1} \tag{11}$$

where $\Delta H^\ominus$ is the standard enthalpy of micellization.

5.3.3 Size Distribution of Copolymer Micelles

5.3.3.1 *Electron microscopy*

Price *et al.* have made wide use of electron microscopy in characterizing the size distribution of copolymer micelles. Two methods have been used to prepare specimens for examination by TEM.[143,144]

In the first method,[143] freeze etching, a drop of solution is rapidly frozen by shock-cooling with liquid nitrogen. Solvent is then allowed to evaporate off from a freshly microtomed surface of the droplet. Finally a replica is made of collapsed micelles left proud of the frozen surface. In one report results are reported for four copolymers: a polystyrene-*block*-polyisoprene copolymer, a (polystyrene-*block*-polyisoprene)$_4$Si star copolymer and two polystyrene-*graft*-polyisoprene copolymers; M_n, $M_\mathrm{w}/M_\mathrm{n}$ and the wt % polystyrene contents for the unassociated copolymers were respectively (51 000, 1.12, 25.2), (188 000, 1.10, 24.9), (550 000, 1.18, 29.2) and (420 000, 1.22, 38.1). Technical white oil, which is a selectively bad solvent for polystyrene, was used for the study. Within experimental error the micelles formed by the two block copolymers were monodisperse in size. The graft copolymer micelles on the other hand showed measurable size distributions.

The second method[144] involved allowing a drop of micellar solution to spread and evaporate on a carbon substrate. In one case it was applied to a polystyrene-*block*-polyisoprene (32 000 : 8600 M_n) copolymer having $M_\mathrm{m}/M_\mathrm{n} = 1.08$, which had also been successfully investigated by GPC. An electron micrograph of micelles which were stained in solution with osmium tetroxide prior to isolation is shown in Figure 12. The osmium tetroxide reacted selectively with alkenic bonds of the polyisoprene and served to enhance contrast between the two types of block. The electron micrograph showed that the micelles had a very narrow size distribution ($M_\mathrm{w}/M_\mathrm{n} \lesssim 1.02$). On final evaporation of the solvent during specimen preparation the micelles were thought to collapse to form particles having a density similar to that of bulk polymer. From shadowing it was concluded that the spherical micelles became approximately disc-shaped during isolation because they tended to flow and wet the carbon substrate. With some modification to allow for the involatility of the solvent medium the latter method was also used[145] to show that the micelles formed by a polystyrene-*block*-poly(ethylene/propylene) copolymer in the base lubricating oil had a narrow size distribution as first indicated by gel permeation chromatography (GPC).

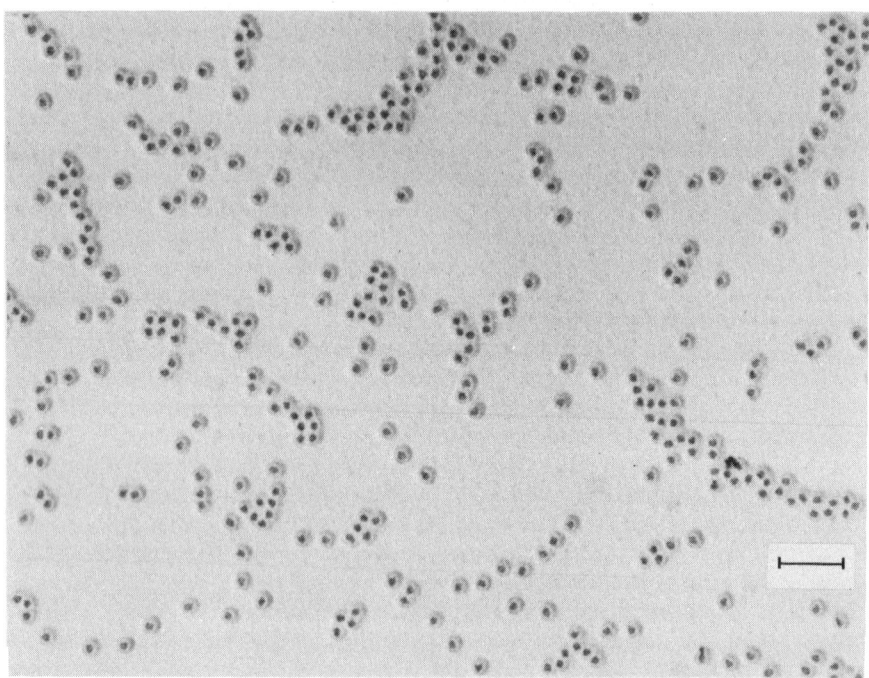

Figure 12   An electron micrograph of micelles isolated from a solution of a polystyrene-*block*-polyisoprene (32 000 : 8600 M_n) copolymer in *N,N*-dimethylacetamide. Scale bar indicates 100 nm (reproduced from ref. 144 with permission of Royal Society of Chemistry)

Using an evaporation technique Horii *et al.*[146] studied micelles formed by some poly(vinyl acetate)-*graft*-polystyene copolymers in ethyl acetoacetate, which is a poor solvent for polystyrene. It was possible, using dialysis, to replace the ethyl acetoacetate by either methanol or acetonitrile without affecting the colloidal stability of the micelles. The dried micelles were spherical when isolated from methanol or acetonitrile, but were somewhat flattened when prepared from the original solvent. Neither the nature of the solvent nor the initial concentration of the solution was found to influence the average diameter of the spheres. It seems likely that the polymeric content of the micelles remained fixed during dialysis. A broad distribution of micelle sizes was attributed by the authors to the polydispersity of the graft copolymers.

6.3.3.2   Ultracentrifugation

Tuzar *et al.*[147] have shown that sedimentation velocity analysis can be a very useful tool in the investigation of micelle formation by block copolymers. With this method one measures the velocity at which a solute species is displaced under the influence of a strong centrifugal force. The sedimentation velocity of a solute depends on its molecular weight, its buoyancy and friction factor and on the centrifugal force applied. Detailed studies were made of a polystyrene-*block*-polybutadiene-*block*-polystyrene copolymer for which $M_w = 1.4 \times 10^5$ g mol^{-1} in the unassociated state and the polystyrene content was 52 wt %. The copolymer was dissolved in the mixed selective solvent tetrahydrofuran/allyl alcohol. Tetrahydrofuran is a relatively good solvent for polystyrene whilst allyl alcohol is a rather strong precipitant for polybutadiene. The cores of the micelles would have consisted mainly of polybutadiene blocks swollen with a solvent mixture that was enriched with tetrahydrofuran by preferential absorption.

Interpretation of the results was much simplified since mixtures of tetrahydrofuran ($n_D^{25} = 1.407$, $\rho^{25} = 0.8820$) and allyl alcohol ($n_D^{25} = 1.412$, $\rho^{25} = 0.8469$) could be regarded as isorefractive and isopycnic. Three types of solution behaviour were observed: (i) with volume fractions of allyl alcohol up to 0.42 at 25 °C (or up to 0.44 at 35 °C) the solutions were clear and gave only a single maximum on the gradient curve corresponding to molecularly dissolved copolymer chains; (ii) on passing to higher volume fractions of allyl alcohol (0.44–0.50 at 25 °C and 0.46–0.52 at 35 °C) the solutions became turbid and a second, faster component appeared corresponding to compact copolymer micelles; it was deduced from the shape of the gradient curves that the micelles had a comparatively narrow distribution; and (iii) when the volume fraction of allyl alcohol in the mixture was 0.52 or more, very rapid sedimentation occurred due to macroscopic precipitation of the copolymer.

6.3.3.3 Gel permeation chromatography (GPC)

In this technique a porous rigid gel is used in the separation process. Fractionation occurs because the amount of solvent required to elute a particle through the gel is related to the hydrodynamic size of the particle; since large particles can only enter large pores they are eluted first. The chromatograms require correction for instrumental broadening; this correction is particularly important for particles with a narrow size distribution. When dealing with samples containing large particles care has to be taken that there are a sufficient number of permeable pores to provide adequate resolution.

In one case[144] studies were made of micelles formed by the polystyrene-*block*-polyisoprene (32 000 : 8600 M_n) copolymer having $M_w/M_n = 1.08$, referred to in the electron microscopy section on micelles (Section 6.3.3.1). The column packing material used in the studies was a poly(styrene–vinylbenzene) gel. Since the solvent used is a selectively bad solvent for polyisoprene, the spherical micelles formed by the block copolymer had a swollen polyisoprene core and an outer flexible fringe of polystyrene chains. At 26 °C the thermodynamic equilibrium was predominantly in favour of micelle formation. A single sharp peak at elution volume 126.5 cm³ (which was well within the range of resolution of the columns) recorded at this temperature indicated that micelles were able to enter the pores of the gel and elute through the columns as globular particles. The reverse flow method of correction of diffusional broadening was used to show that within experimental error the micelles were monodisperse in size ($M_w/M_n = 1.00$, experimental uncertainty = 0.02).

In the temperature range 50–65 °C the shape of the chromatograms indicated the presence of both micelles and free chains. As the temperature was raised the equilibrium shifted progressively in favour of the unassociated chain form.

More recently, additional studies involving the use of GPC in micelle characterization have been reported.[148,149] From GPC studies carried out on poly(oxyethylene) *n*-alkyl ethers in aqueous media,[150] it has been shown using a model proposed by Coll that the position and width (after correction for instrumental spreading) of the elution peak for micelles is not influenced by micelle dissociation provided the critical micelle concentration (cmc) is very low compared with the concentration of the solution injected into the columns.

That micelles can have narrower size distributions than the unassociated chains has been treated by Elias and Solc.[151,152] They have shown from statistical arguments that sharpening of the distribution should occur for associations of both end-to-end and segment-to-segment types. End-to-end association is when one group, or one collection of groups per molecule, is capable of association. Segment-to-segment association is when association may occur at more than one site, the number of sites being roughly proportional to the length of the chain. Micelle formation by copolymers having one block per molecule capable of association can be treated as end-to-end association.

6.3.4 Thermodynamic Studies

Standard free energies of micellization $\Delta G^{\ominus}$ and their enthalpy $\Delta H^{\ominus}$ and entropy $-T\Delta S^{\ominus}$ components have been determined by Price *et al.* for a number of block copolymers in organic solvents using light scattering, membrane osmometry and calorimetry.[153–158]

One of the systems studied[153] was a polystyrene-*block*-poly(ethylene/propylene) (37 300 : 59 700 M_n) copolymer in decane. Electron microscopy studies showed that the micelles formed by the block copolymer were spherical in shape and had a narrow size distribution. Since decane is a selectively bad solvent for polystyrene, the latter component formed the cores of the micelles. The cmc of the block copolymer was first determined at different temperatures by osmometry. Figure 13 shows a plot of π/cRT against c (where c is the concentration of the solution) for $T = 97.1$ °C. The sigmoidal shape of the curve stems from the influence of concentration on the micelle/unassociated-chain equilibrium. When the concentration of the solution is very low most of the chains are unassociated; extrapolation of the curve to infinite dilution gives M_n^{-1} of the unassociated chains. On increasing the concentration of the solution the osmotic pressure decreases rapidly over a narrow concentration range as expected for closed association. The arrow indicates the cmc. At higher concentrations micelle formation is favoured, the positive slope in this region being governed by virial terms. Similar shaped curves were obtained for other temperatures.

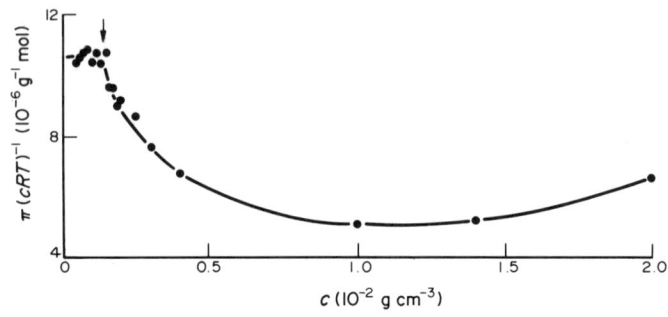

Figure 13 A plot of reduced osmotic pressure (π/cRT) against concentration c for a polystyrene-*block*-poly(ethylene/propylene) (37 000 : 59 700 M_n) copolymer at 97.1 °C. The arrow indicates the cmc (reproduced from ref. 153)

The slope of the linear plot of ln(cmc) against T^{-1} (see Figure 14) gave a value for $\Delta H^{\ominus}$ of -130 kJ mol^{-1}. The values of $\Delta G^{\ominus}$ and $T\Delta S^{\ominus}$ for $T = 86$ °C were -30 and 100 kJ mol^{-1} respectively; the standard states are ideally dilute with $c = 1$ mol dm^{-3}.

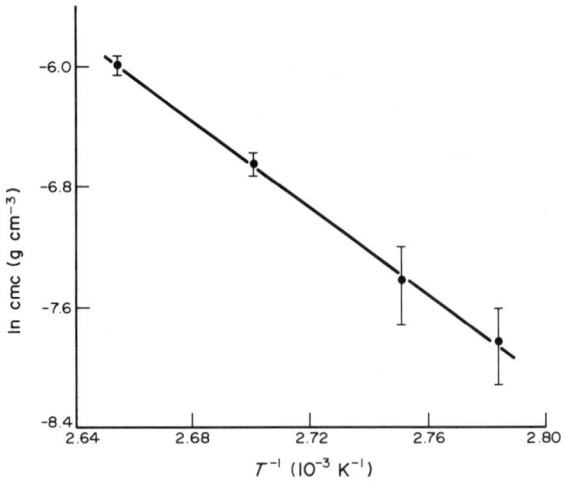

Figure 14 A plot of the logarithm of the cmc against the reciprocal of the absolute temperature. The data were obtained by osmometry for a polystyrene-*block*-poly(ethylene/propylene) (37 000 : 59 700 M_n) copolymer in decane (reproduced from ref. 153)

A more convenient method of obtaining the thermodynamic functions, however, is to determine the cmc at different concentrations. A plot of light-scattering intensity against concentration is shown in Figure 15 for a solution of concentration $c = 3.8 \times 10^{-5}$ g cm^{-3} and a scattering angle of 60°. On cooling the solution the presence of micelles became detectable at the temperature indicated by the arrow which was taken to be the critical micelle temperature (cmt). On further cooling the weight fraction of micelles increases rapidly leading to a rapid increase in scattering intensity at lower temperatures till the micellar state predominates. The slope of the linear plot of ln c against (cmt)$^{-1}$ shown in Figure 16, which is equivalent to the more traditional plot of ln (cmc) against T^{-1}, gave a value of $\Delta H^{\ominus} = -141$ kJ mol^{-1} which is in fair agreement with the result obtained by osmometry considering the difficulties in locating the cmc by the osmometric method. Direct calorimetric measurements gave a value of 138 kJ mol^{-1} for $\Delta H^{\ominus}$.

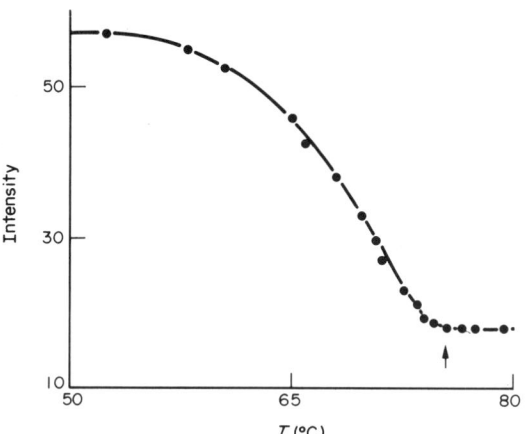

Figure 15 A plot of light-scattering intensity against temperature for a polystyrene-*block*-poly(ethylene/propylene) (37 000 : 59 700 M_n) copolymer in decane. The concentration of the solution was 3.85×10^{-5} g cm^{-3} and the angle of scatter 60°. The arrow indicates the cmt (reproduced from ref. 153)

Results obtained for a range of polymers are given in Table 3.[154,155,159] The first two sets of results were obtained using light scattering to determine the cmt. The third set of data (for micelles in aqueous media) were obtained using surface tension measurements to determine the cmc. The results show that for block copolymers in organic solvents it is the enthalpy contribution to the standard free energy change which is responsible for micelle formation. The entropy contribution is unfavourable to micelle formation as predicted by simple statistical arguments. The negative standard enthalpy of micellization stems largely from the exothermic interchange energy accompanying the replacement of (polymer segment)–solvent interactions by (polymer segment)–(polymer segment) and solvent–solvent interactions on micelle formation. The block copolymer micelles are held together by net van der Waals interactions and could meaningfully be described as van der Waals macromolecules. The combined effect per copolymer chain is an attractive interaction similar in magnitude to that posed by a covalent chemical bond.

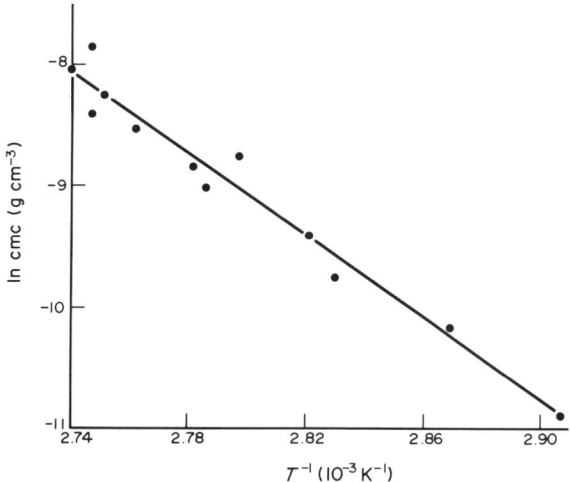

Figure 16 A plot of the logarithm of the cmc against temperature. The data were obtained by light scattering for a polystyrene-*block*-poly(ethylene/propylene) (37 000 : 59 700 M_n) copolymer in decane (reproduced from ref. 153)

Table 3 Thermodynamics of Micellization[a]

Polymer	M^b	M^c	$\Delta G^{\ominus}$ (kJ mol^{-1})	$\Delta H^{\ominus}$ (kJ mol^{-1})	$-T\Delta S^{\ominus}$ (kJ mol^{-1})
Polystyrene-*block*-polyisoprene copolymer[d]	24 600[e]	7000[e]	−21.4	−40.7	19.3
	46 000[e]	8650[e]	−23.6	−72.6	49.0
	26 800[e]	13 000[e]	−31.8	−86.4	54.6
	54 228[e]	12 700[e]	−30.5	−115.3	84.8
Polystyrene-*block*-poly(ethylene/propylene) copolymer[f]	146 000[g]	49 800[g]	−43.0	−181	138
	97 000[g]	37 300[g]	−41.5	−141	100
	110 000[g]	31 500[g]	−41.7	−103	61
Poly(oxyethylene) *n*-alkyl ethers[h]	1370[g]	150[g,i]	−24.6	10.9	−35.5
	1430[g]	210[g,i]	−26.8	16.9	−43.7
	1470[g]	250[g,i]	−28.8	24.5	−53.3
	1510[g]	290[g,i]	−30.2	32.0	−62.2
	1570[g]	350[g,i]	−31.4	39.4	−70.8

[a] Standard states are ideally dilute with $c = 1$ mol dm^{-3}.
[b] Overall average molecular weight of copolymer.
[c] Average molecular weight of polystyrene block.
[d] In hexadecane at $T = 40\,°C$.[154]
[e] $M = M_w$.
[f] In decane at $T = 80\,°C$.[155]
[g] $M = M_w$.
[h] In water at $T = 35\,°C$.[159]
[i] Average molecular weight of *n*-alkyl block.

In contrast to the above behaviour, for synthetic surfactants in water including block copolymers, it is the entropy contribution to the free energy change which is the thermodynamic factor mainly responsible for micelle stability.[159,160] Results for the thermodynamics of micellization of poly(oxyethylene) *n*-alkyl ethers (structural formula: $MeO(CH_2CH_2O)_{27}(CH_2)_nH$, where $n = 12, 14, 16, 18, 21$) in water are given in Table 3. Whilst a number of factors govern the overall magnitude of the entropy contribution, the fact that it is favourable to micelle formation arises largely from the structural changes[161] which occur in the water matrix when the hydrocarbon chains are withdrawn to form the micellar cores.

6.3.5 Characterization Studies in Dilute Solution

6.3.5.1 Methods

Provided the cmc is very low compared with the concentrations of the dilute solutions used in the study, the measurements may be interpreted as if the micelles were large macromolecules,[144] *i.e.* the effect of the micelle/unassociated-chain equilibrium on the results will not be significant. In such a case, light scattering and membrane osmometry can be applied in the normal way to determine the weight-average and number-average molecular weights respectively, capillary flow viscometry can be used to determine the intrinsic viscosity and photon correlation spectroscopy to determine the transitional diffusion coefficient.

6.3.5.2 Elastic light-scattering

As with copolymers in general, if the micelles are heterogeneous in composition then light scattering from a single solvent gives only an apparent molecular weight M_{app}.[162,163] The normal practice of using M_{app} values obtained in a range of solvents to establish the true value of M_w cannot be applied for micelles, however, because the nature of the solvent affects the aggregation number of the micelles. When both micelles and unassociated chains are present in solution a theoretical model of the association has to be invoked to interpret the results. Interpretation of viscosity measurements in this situation is particularly difficult.

Light-scattering results obtained for spherical micelles formed by polystyrene-*block*-poly(ethylene/propylene) (37 500 : 59 700 M_n) copolymer in decane at 30 °C are shown in Figure 17, where

$(Kc/R_\theta)_{\theta=0}$ is plotted against concentration.[153] Since the block copolymer was homogeneous in chemical composition

$$(Kc/R_\theta)_{\theta=0} = M_w^{-1} + 2A_2c + \ldots \ldots \ldots \tag{12}$$

Extrapolation of the results to zero concentration yielded a value for the weight-average molecular weight of the micelles of $6.6 \times 10^6 \text{ g m}^{-1}$ (the cmc of the system at $30\,^\circ\text{C}$ was $2.4 \times 10^{-8} \text{ g mol}^{-3}$).

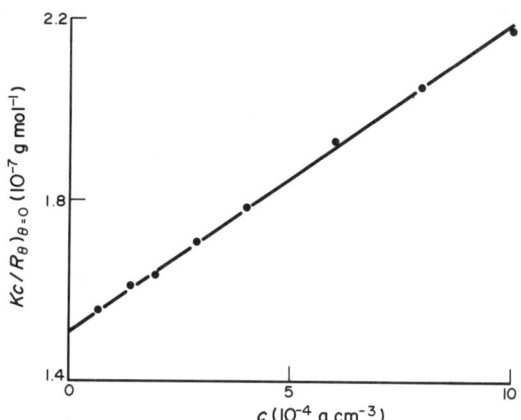

Figure 17 Values of $(Kc/R_\theta)_{\theta=0}$ obtained by extrapolating light-scattering data to zero angle plotted against concentration c for a polystyrene-*block*-poly(ethylene/propylene) ($37\,000:59\,700\ M_n$) copolymer in decane at $30\,^\circ\text{C}$ (reproduced from ref. 153)

The compact nature of the micelles formed in *n*-decane was reflected by the dependence of the dissymmetry ratio, $I(45^\circ)/I(135^\circ)$, on concentration.[153] As shown in Figure 18, the dissymmetry ratio fell below unity at $c = 0.8 \times 10^{-3} \text{ g cm}^{-3}$. This type of behaviour can be predicted quite well by assuming that on increasing the concentration the micelles pack together like hard spheres (*i.e.* the local packing can be described by radial distribution functions for hard spheres). The effective hard-sphere radius was set equal to 46.4 nm. The first part of the dissymmetry curve (indicated by the full line in Figure 18) was calculated using radial distribution functions obtained from an analytical expression predicted by the Percus–Yevick theory.[164] Radial distribution functions[165] obtained by Monte Carlo calculations were used to predict the behaviour at two higher concentrations (indicated by the squares in Figure 18); the two higher concentrations correspond to hard-sphere number densities of $0.298\ ^n\rho_0$ and $0.372\ ^n\rho_0$ where $^n\rho_0$ is the number density of a close-packed system. The repulsive potential between the micelles stems primarily from the large reduction in entropy when the relatively dense micellar fringes overlap.

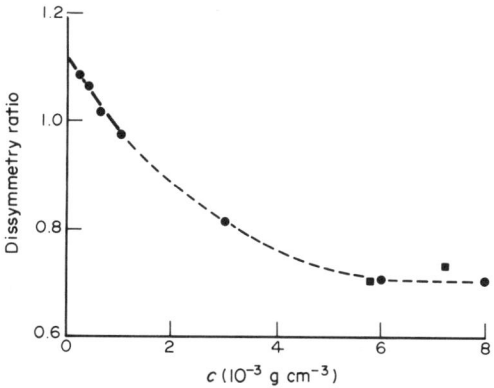

Figure 18 Plot of light-scattering dissymmetry ratio $[I(45^\circ)/I(135^\circ)$ against concentration c for a polystyrene-*block*-poly(ethylene/propylene) ($37\,000:59\,700\ M_n$) copolymer in decane at $30\,^\circ\text{C}$. The experimental points are indicated by circles (●). The full curve and squares (■) indicates the behaviour predicted theoretically using a hard-sphere model (reproduced from ref. 15)

The distribution of a particular polymer component in a micelle can only be obtained by light scattering by rendering the other component 'invisible' by using an isorefractive solvent. Tanaka *et al.*[166,167] have studied systems in which the core component was rendered invisible. Experimental results for polystyrene-*block*-poly(methyl methacrylate) micelles in a mixture of toluene/*p*-cymene were compared with calculated particle-scattering functions. Excellent agreement was obtained using a three-phase model consisting of a compact spherical core containing only poly(methyl methacrylate) chains, a mixed phase within which the poly(methyl methacrylate)/polystyrene junctions were located and an outer fringe consisting of polystyrene chains.

6.3.5.3 *Quasi-elastic light-scattering*

Brownian motion of particles in solution gives rise to a spectral distribution in the scattered light. From measurements on the line width of scattered laser light by photon correlation spectroscopy (PCS) the translational diffusion coefficient may be determined. Unlike the traditional method of investigating translational diffusion, PCS does not require a macroscopic concentration gradient and therefore can more readily be applied to investigate association processes. The method has been used to determine the translational diffusion coefficients, D_c, of micelles formed by a number of block copolymers in selectively bad solvents.[168-170]

If the micelles are treated as hydrodynamically equivalent spheres then the translational diffusion coefficient for the limiting case of infinitely dilute solution is given by the Stokes–Einstein relation

$$D_0 = kT/6\pi\eta R_d \qquad (13)$$

where R_d is the hydrodynamic radius and η the viscosity of the solvent. For micelles polydisperse in size distribution, PCS yields approximately a z-average diffusion coefficient and therefore a z-average reciprocal hydrodynamic radius $(R_D^{-1})_z$. Studies have been made on micelles having a very narrow size distribution in conditions where the association equilibrium overwhelmingly favoured micelle formation. For the polystyrene-*block*-polyisoprene copolymer in *N,N*-dimethylacetamide (DMA) for which the electron micrograph is shown in Figure 13, $R_D = 22.8$ nm.[168] Plots of D_c against temperature for the system were found to show fairly abrupt changes in regions where the micelles were beginning to break up to form free chains.

6.3.5.4 *Small-angle X-ray scattering*

SAXS may be usefully employed to investigate the molecular weight, overall size and phase structure of micelles.[171-176] Studies in general are limited to concentrations down to $c > 0.5 \times 10^{-2}$ g cm^{-3}.

One of the more detailed studies, which was reported by Pleštil and Baldrian,[171] deals with solutions of a polystyrene-*block*-polybutadiene (5700 : 59 000 M_n) copolymer in heptane, a selectively bad solvent for polystyrene. Measurements in heptane were made at four temperatures ranging from 18 to 50 °C. The analysis of results was complicated by the knowledge (gained from ultracentrifugation) that a significant proportion of the chains were unassociated.

Plots of log (1/c) against θ showed a shoulder appearing on the central peak at about 0.003 rad and a weak maximum within the range 0.0130 to 0.145 rad. Absolute intensity measurements taken below $\theta = 0.003$ rad were extrapolated to zero angle using Guinier's equation. The M_w of the micelles and the degree of swelling of the individual components were determined; the number of copolymer chains in the micelles varied from 190 (at 18 °C) to 100 (at 50 °C). It was argued that at angles greater 0.0055 rad the shape of the scattering curves was governed mainly by scattering from the micelle cores. For each of the four temperatures a two-phase concentric sphere model for the micelles was used to calculate the overall diameter of the micelles (55.8–53.2 nm) and the radius of gyration (19.1–18.0 nm) of the micelles, from absolute intensity measurements. Support for the model was provided by showing that an alternative treatment involving direct analysis of relative intensity measurements gave a similar set of values.

Roe *et al.* have used SAXS to investigate micelles formed by block copolymers solubilized in a homopolymer.[175,176]

6.3.5.5 *Small-angle neutron scattering*

SANS is potentially a more powerful technique for investigating micelle structure than X-ray scattering owing to the possibility of producing large contrasts by deuteration. Because of the difference in scattering cross-sections between hydrogen and deuterium, measurable scattering can be generated by dispersing a small concentration of fully deuterated polymer in a hydrogeneous matrix of *vice versa*. The technique offers a means of studying the multiphase structure of micelles and the average dimensions of copolymer chains and their component blocks in a micellar framework. It has been used to study micelles formed by a polystyrene-*graft*-poly(oxyethylene) copolymer in water.[177] It has also been used[178] in characterizing the dimensions of a polystyrene-*block*-poly(dimethylsiloxane) copolymer absorbed on dispersed polymer particles in the solvents heptane, 1,1,2-trichloro-1,2,2-trifluoroethane and silicone fluid.

6.3.5.6 *Dilute solution viscometry*

Capillary flow viscometry has been extensively used to investigate micellar solutions of block copolymers. Values of η_{sp}/c, where η_{sp} is the specific viscosity, have been reported for dilute micellar solutions of various block and graft copolymers. The effects of temperature and solvent[168,169,179,180] composition[137,181,182] have been investigated. In many cases the results obtained were extrapolated to infinite dilution to obtain the intrinsic viscosity, $[\eta]$. Such extrapolations, however, are only meaningful when the thermodynamic equilibrium overwhelmingly favours either micelles or unassociated chains.[183] For some systems η_{sp}/c has been found to increase sharply[184] (for solutions of fixed concentration) on passing from a predominantly micellar solution to one in which the chains are unassociated; in other cases decreases are observed. Results shown in Figure 19 are for the polystyrene-*block*-polyisoprene ($13\,000:38\,000\ M_n$) copolymer at $c = 6.49 \times 10^{-3}\ \mathrm{g\,cm^{-3}}$ and the polystyrene-*graft*-polyisoprene ($190\,000:360\,000\ M_n$) copolymer at $c = 4.97 \times 10^{-3}\ \mathrm{g\,cm^{-3}}$. Polystyrene-*block*-poly(ethylene/propylene) copolymers in base lubricating oils show a marked decrease in η_{sp}/c on raising the temperature through the transition region. The latter copolymers are used at concentrations of 1 to 2 wt % as viscosity index improvers for lubricating oils and their improver properties fall off rapidly on reaching the transition region at higher temperatures.

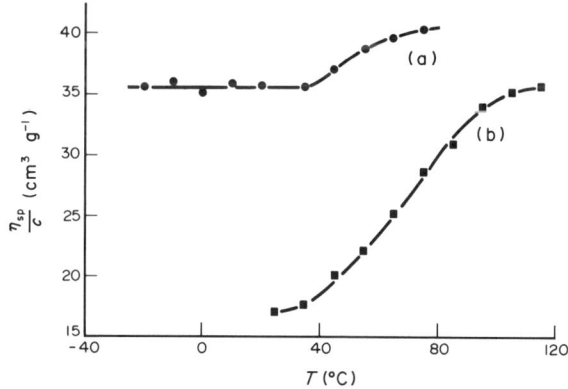

Figure 19 Plots of η_{sp}/c against temperature for copolymers in decane: (a) polystyrene-*block*-polyisoprene ($13\,000:38\,000\ M_n$) copolymer at $c = 6.49 \times 10^{-3}\ \mathrm{g\,cm^{-3}}$ and (b) polystyrene-*graft*-polyisoprene ($190\,000:360\,000\ M_n$) copolymer at $c = 4.97 \times 10^{-3}\ \mathrm{g\,cm^{-3}}$ (reproduced from ref. 184 with permission of Applied Science Publishers Ltd.)

6.3.6 Internal Viscosity

Micelles may have cores which are liquid-like, glassy or crystalline. For many block copolymer micelles in organic solvents, the cores are plasticized by the presence of solvent.

The fluidities of liquid-like cores may be investigated by comparisons of the mobilities of fluorescence[185] and electron spin resonance[186] probe molecules solubilized in micelles and in pure organic solvents. Use of NMR offers a method of studying the fluidity of micelle cores without the uncertainties introduced by the perturbing effects of probes.[187-189] The occurrence of colloidally stable aggregates with crystalline cores formed by poly(oxyethylene) blocks is suggested by results obtained for a polystyrene-*block*-poly(oxyethylene) copolymer in ethylbenzene.[190] Micelles formed by polystyrene-*block*-poly(oxyethylene) copolymers in water generally have glassy polystyrene cores at room temperature, but these become rubbery on heating above the T_g of the polystyrene blocks.[191]

6.3.7 Worm-like Micelles

Large changes in light-scattering dissymmetry indicating the formation of extended micelles were observed for synthetic surfactants in the early 1950s.[133] For such systems it appears that an increase in surfactant concentration always favours the formation of extended micelles rather than globular micelles.

Evidence for elongated micelles in dilute solutions of a block copolymer has been provided by Utiyama *et al.*[192] By elastic light-scattering they found that over a limited range of solvent composition a polystyrene-*block*-poly(methyl methacrylate) formed micelles with elongated polystyrene cores in the mixed solvent toluene/furfuryl alcohol. Evidence for the presence of large micelles over a very narrow solution range has been provided by Price *et al.*[193] and Tuzar *et al.*[194] for solutions of polystyrene-*block*-polybutadiene-*block*-polystyrene copolymers. Further evidence for the formation of worm-like or lamellar micelles over a narrow range has been presented by Mandema *et al.*[169] for polystyrene-*block*-poly(ethylene/propylene) copolymers in decane.

More recently[195] it has been shown by electron microscopy and light scattering that worm-like micelles can exist over wide concentration and temperature ranges in solutions of a polystyrene-*block*-polyisoprene (6500 : 22 800 M_n) copolymer in DMA (see Figure 20). However, it appears that they are metastable with respect to spherical micelles. It was possible to determine the intrinsic viscosity of the worm-like micelles which is 25.3 cm^3 g^{-1} at 30 °C in DMA. The spherical micelles were found to replace the worm-like micelles after taking the solutions through a thermal cycle which involved heating above the cmt. The intrinsic viscosity of spherical micelles was 5.8 cm^3 g^{-1} at 30 °C in DMA.

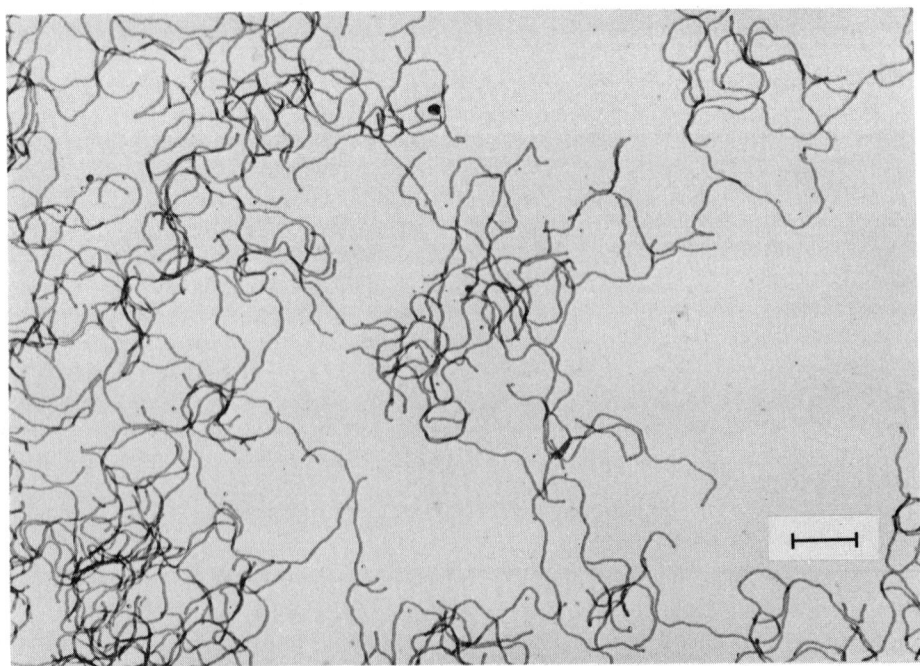

Figure 20 An electron micrograph of worm-like micelles formed by a polystyrene-*block*-polyisoprene (6500 : 22 800 M_n) copolymer in DMA. Scale bar indicates 1 μm (reproduced from ref. 195 with permission of IPC Business Press Ltd.)

6.3.8 Surface Activity

Block and graft copolymers may be absorbed at various surfaces: liquid/gas, liquid/liquid, solid/liquid, *etc.* The surface-active properties of copolymers (particularly block polymers) have made them useful as dispersants, emulsifiers, foam stabilizers, flocculants and wetting agents. These aspects of block copolymers do not fall within the scope of this review. An overview of these applications has been presented by Reiss,[196] who has also widely researched the area.[197]

Block copolymers containing both hydrophilic and hydrophobic components exhibit typical surfactant behaviour in aqueous media.[198,159] Adsorption of block copolymers at organic/air interfaces can also give measurable reductions in surface tension.

Surface tension (γ) measurements have shown that a polystyrene-*block*-polyisoprene (7000:24 600 M_w) copolymer[199] is positively adsorbed at a hexadecane/air interface. The changes which occur, however, are much smaller than those normally recorded for synthetic surfactants at the water/air interface. The linearity of plots of γ against $\ln c$ for the copolymer/hexadecane system just below the cmc (see Figure 21) indicates approximately constant surface concentration in this region. This means that, as with synthetic surfactants in water, the interface is already effectively saturated with a largely incompressible monolayer of solute at the stage when micelle formation is just detectable. The polystyrene blocks will be located in the upper part of the interface and the polyisoprene blocks in the lower. Both types of block can be expected to be in a coil form, but the polyisoprene blocks will be highly swollen with solvent and extend down into the hexadecane solution. Like the polystyrene blocks in the micellar cores, the polystyrene blocks at the interface will only be very sparingly swollen.

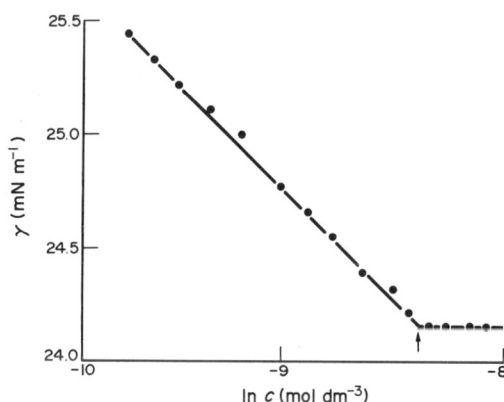

Figure 21 Plot of surface tension γ against the logarithm of the solution concentration c for a polystyrene-*block*-polyisoprene (37 000:59 700 M_n) copolymer in hexadecane. The arrow indicates the cmc (reproduced from ref. 199 with permission of the Society of Chemical Industry)

6.3.9 Solubilization of Homopolymers

From some of the earliest studies on block and graft copolymers it became clear that homopolymers could be held in colloidal suspension through the solubilizing action of copolymers. Attempts to fractionate homopolymers from copolymers by addition of non-solvents were frequently unsuccessful because of this effect. The behaviour is analogous to the stabilization of hydrophobic substances by surfactants in aqueous media. Several quantitative studies of homopolymer absorption in dilute solution have been reported.[200-202] The early results have been reviewed by Tuzar and Kratochvil.[203] It is found that the extent of homopolymer absorption depends strongly on the molecular weight of the homopolymer and the molecular weight of the blocks forming the micellar cores. A homopolymer of type A is only absorbed by an AB block copolymer having a core composed of A blocks if the molecular weight of the homopolymer is lower than that of the A blocks. The absorption of other species has also been studied.[204] The results demonstrate that the absorption process can be very selective with respect to composition.

6.4 THEORIES OF CHAIN SEGREGATION

In this section we set out to give a brief review of the theories describing various types of chain segregation. We begin by considering the single isolated chain. This is not only because it is a fascinating source of study in itself, with it possibly displaying a 'segregation' transition as the interaction between the two types of monomer becomes less favourable, but it is also a necessary ingredient in order to have a good understanding of micellization, when such quantities as the critical micelle concentration depend sensitively on the difference in free energy between chains in a micelle and free chains in solution. The next section is a review of the theory of copolymers in bulk and as to how and why they give rise to different structures in the phase diagram, and in the final part we turn to the problem of the formation of micelles and the theories proposed to understand the phenomenon.

6.4.1 The Isolated Chain

Because the interactions between different chains and between chains and solvent molecules are so complicated, it is normal to consider instead a model polymer in which all the microscopic details, such as bond angles, torsional potentials, solvent mediated forces between different monomers, *etc.*, are coarse grained into a few parameters which are then generally fixed empirically or alternatively from computer simulation.

Clearly, because of this coarse graining, it is only long wavelength properties of the polymer that can hope to be reproduced by the model and examples of these properties are thermodynamic quantities and the radius of gyration. Thus a single homopolymer chain is often represented as consisting of N beads linked together and the probability distribution function $P(r_1, r_2, \ldots, r_N)$ for finding the first bead at position r_1, the second at position r_2, and so on is given by[1,205,206]

$$P(r_1, r_2, \ldots, r_N) = P_0(r_1, r_2, \ldots, r_N) Z^{-1} \exp\left[-\beta \sum_{\substack{(i>j,\,j=1)}}^{N} \sum^{N-1} V(r_{ij}) - \beta \sum_{\substack{(i>j,\,j>k,\,k=1)}}^{N} \sum^{N-1} \sum^{N-2} W(r_i, r_j, r_k) \ldots \ldots \right] \quad (14)$$

where $\beta = (kT)^{-1}$, with k the Boltzmann constant and T the absolute temperature, $V(r_{ij})$ is the pair potential of mean force between beads i and j, $W(r_i, r_j, r_k)$ is the irreducible three-body potential and the dots represent irreducible potentials arising from higher order clusters. The configurational integral, denoted by Z, provides the normalization of $P(r_1, r_2, \ldots, r_N)$, and $P_0(r_1, r_2, \ldots, r_N)$ is the probability distribution function for the beads in the absence of any interbead interactions and is generally taken to be a normalized Gaussian distribution function, *viz.*

$$P_0(r_1, r_2, \ldots, r_N) = \prod_{i=1}^{N-1} (3/2\pi L^2)^{3/2} \exp(-3r_{i,i+1}^2/2L^2) \quad (15)$$

Here L is the Kuhn length which corresponds physically to that length of real polymer chain which separates statistically independent parts of that chain.

The next stage is very often to simplify equation (14) still further and to replace the interbead potentials with simple pseudo-potentials, generally involving Dirac delta functions. Thus a commonly used form is

$$P(r_1, r_2, \ldots, r_N) = P_0(r_1, r_2, \ldots, r_N) Z^{-1} \exp\left[-V_0 \sum_{\substack{(i>j,\,j=1)}}^{N} \sum^{N-1} \delta(r_i - r_j) \right.$$

$$\left. - W_0 \sum_{\substack{(i>j,\,j>k,\,k=1)}}^{N} \sum^{N-1} \sum^{N-2} \delta(r_i - r_j)\delta(r_i - r_k) \ldots \ldots \right] \quad (16)$$

where V_0 and W_0 are related to excluded volumes. One might attempt to relate these quantities to the fuller potentials used in equation (14) and so, for example, the quantity V_0 may be associated with the second virial coefficient between two free beads, *i.e.*

$$V_0 = \int [1 - \exp(-\beta V(r))] \, dr \quad (17)$$

but as stressed by Freed,[206] it is often better if one is working with equation (16) to fix V_0, W_0, *etc.* by comparing theoretical predictions directly with real systems, rather than go through equation (17) or analogous theory.

It is now pertinent to ask whether, after these drastic simplifications of the orginal problem, equation (16) gives an adequate representation of a real homopolymer even in the long wavelength limit. The answer would appear to be that if $V_0 > 0$, corresponding to a polymer in a good solvent, then equation (16) is adequate and in fact three-body and higher-order cluster terms may also be safely ignored (*i.e.* one can set $W_0 = 0$). Near the θ point, when $V_0 \simeq 0$, one should keep the three-body term and then one still gets good answers.[206] For poor solvents, though, when $V_0 < 0$, then it is unlikely that equation (16) will be physically reasonable at least for very negative values of V_0, for there is nothing in the Hamiltonian describing the hard-core repulsive force between the beads and thus the model can lead to collapsed polymer configurations of unphysically high density. Extensive studies of the single homopolymer chain have been made, both in good and bad solvents, but it would appear that the present understanding on the good solvent side is far in advance of that of the chain in a bad solvent when the chain takes up a much more compact structure.[207-209]

Turning now to a copolymer chain, one can follow the same simplifying procedures as sketched out above for the homopolymer chain but this time one must take into account the various types of interaction between different types of monomer. Thus for a diblock copolymer composed of N_A A beads with a Kuhn length L_A and N_B B beads with a Kuhn length L_B, the equivalent of equation (16) will involve quantities such as V_0^{AA}, V_0^{AB} and V_0^{BB}, referring to the excluded volumes between two A beads, an A and a B bead and two B beads. Most work to date has ignored the three-body terms but, in analogy with the homopolymer in a bad solvent problem, it is not clear how justifiable this is when one or more of the three V_0 values are negative.

Among the early work on the isolated copolymer problem was that of Froelich and Benoit[210] who investigated the situation of small excluded volumes so that departures from Gaussian configurations could be analyzed *via* a perturbation expression. However, they raised the possibility of more dramatic segregation effects for larger excluded volumes. This effect was explored by Edwards[211] using a self-consistent set of mean field equations. After taking the continuous chain limit of the discrete bead model outlined above, one obtains an A chain of length $N_A L_A$ attached to a B chain of length $N_B L_B$. Fixing the position of the join at r_0 and measuring lengths along the chain from that point, then Edwards' equation takes the form

$$\left(\frac{\partial}{\partial t} - \frac{L_A^2}{6} \nabla^2 - V_0^{AA} \rho_A - V_0^{AB} \rho_B \right) G_A(r, r_0; t) = \delta(t)\delta(r - r_0) \tag{18a}$$

and

$$\left(\frac{\partial}{\partial t} - \frac{L_B^2}{6} \nabla^2 - V_0^{AA} \rho_A - V_0^{AB} \rho_B \right) G_B(r, r_0; t) = \delta(t)\delta(r - r_0) \tag{18b}$$

where $G_A(r, r_0; t)$ is proportional to the probability density of finding a piece of A chain at a position r, a length $L_A t$ from the joint as measured along the chain, and $G_B(r, r_0; t)$ has a similar interpretation but for the B chain. $\rho_A(r)$ is the number density of A leads at point r, and is given in this particular mean field theory by

$$\rho_A(r) = \int_0^{N_A} dt\, P_A(r, r_0; t) \tag{19a}$$

where

$$P_A(r, r_0; t) = G_A(r, r_0; t) / \int dr_1\, G_A(r_{ij}, r_0; t) \tag{19b}$$

and a similar relation holds for $\rho_B(r)$.

In the special case that $L_A = L_B$, $N_A = N_B$ and $V_0^{AA} = V_0^{BB} = -V_0^{AB}$, Edwards showed one solution to the mean field equations was a Gaussian distribution for each block, yielding a spherically symmetric density profile centred on r_0. Further analysis, though, implied that when V_0^{AB} exceeded a critical positive value of order $L_A^3/N_A^{1/2}$, then another solution to the equations appeared with a dumb-bell-like density distribution in which the A chain concentrated in one lobe and the B chain in the other and which had a lower free energy. Thus a phase transition was predicted and was also assumed to take place under certain circumstances for the more general case, when for example the three excluded volumes were different in magnitude.

Although a single chain of finite length cannot really be expected to undergo a phase transition, presumably Edwards' analysis is really referring to an infinite chain limit. A real chain might be expected to undergo the transition continuously, but over a transition region that gets narrower as the chain length increases.

Computer simulations have also been carried out to give insight into this possible phenomenon. The method of Tanaka *et al.*[212-215] was to generate chain configurations as a walk on a cubic lattice, the model being that the A chain and B chain each did random walks but with the proviso that there could be no crossing of the A and B chains. Furthermore an interaction energy ε_{AB} was associated with A–B nearest neighbours, and to obtain final properties each configuration was weighted by the appropriate Boltzmann factor. Thus, in terms of excluded volume parameters, this model has $V_{AA} = V_{BB} = 0$, whilst V_{AB} can be made either positive or negative by altering the sign and magnitude of ε_{AB}. The chain lengths went up to 400 segments and up to 2000 configurations were generated for the averaging. The general conclusion for a diblock copolymer was that the 'segregated state' was very much like the distribution of the two halves of a homopolymer in a good solvent and making ε_{AB} more negative led to a condensed state — very similar to the coil–globule transition in a homopolymer.

The Monte Carlo simulation of Birshtein *et al.*[216] again used a cubic lattice, but this time no part of the chain could intersect any other part, and furthermore energies of ε_{AA}, ε_{BB} and ε_{AB} were associated with each (non-bonded) A–A, B–B and A–B nearest neighbour respectively. The case $N_A = N_B$ was studied and the maximum total chain length considered was 128. The results were presented in such a way that for given values of ε_{AA} and ε_{BB}, the variation with ε_{AB} could be seen. The general conclusion, *vis-a-vis* a segregation transition, was much the same as reached by Tanaka *et al.*, *viz.* for fixed values of ε_{AA} and ε_{BB}, in that varying ε_{AB} just led to behaviour that resembled that of a homopolymer.

Additional theoretical approaches include the biellipsoidal model calculations of Bendler *et al.*[217-219] in which the A and B chains are modelled as smooth ellipsoidal density clouds. The method is to generate an unrestricted (*i.e.* intersecting) random walk for each block and use this to obtain a Gaussian ellipsoidal density distribution for the two blocks. These two clouds are joined at their chain ends and then allowed to pivot around the bonding point. In order to obtain the orientation, a particular configuration is Boltzmann weighted by a mean field potential given by $\beta V = V_{AA} + V_{BB} + V_{AB}$ where

$$V_{AA} = \frac{1}{2} V_0^{AA} \int \rho_A^2(r)\, dr \tag{20a}$$

and similarly for V_{BB}, and

$$V_{AB} = V_0^{AB} \int \rho_A(r)\rho_B(r)\, dr \tag{20b}$$

The theory is in basic agreement with the simulation results of Tanaka *et al.* and the authors report that for $V_0^{AA} = V_0^{BB} = -V_0^{AB}$, the combination studied by Edwards, rapid conformational changes do occur.

Kurata and Kimura[220] present a smoothed-density treatment in which the probability distribution function for the end-to-end vector R of the total copolymer chain is given approximately by

$$P(R) = P_0(R)\exp[-\beta W(R)] / \int P_0(R)\exp(-\beta W(R)\, dR \tag{21a}$$

where

$$\beta W(R) = \sum_{i<j} V_0^{ij} \int P_{0i}(r|R)P_{0j}(r|R)\, dr \tag{21b}$$

Here the zero subscript indicates properties of a Gaussian chain, so $P_0(R)$ is the Gaussian end-to-end distance distribution function and $P_{0i}(r|R)$ the conditional Gaussian distribution function for finding segment i at a position r, given the end of the chain is at R. Furthermore, $V_0^{ij} = V_0^{AA}$, V_0^{AB} or V_0^{BB} depending on whether i and j are A or B segments. This theory for $N_A = N_B$ predicts the coil–globule collapse if V_0^{AB} is made more and more negative when $V_0^{AA} = V_0^{BB} = 0$, and in the Edwards case where $V_0^{AA} = V_0^{BB} = -V_0^{AB}$ predicts the collapsed dumb-bell form as did Edwards for sufficiently large and positive V_0^{AB}. When the theory is applied to the triblock ABA chain (with equal numbers of A and B segments) then a collapse transition is predicted in the $V_0^{AA} = V_0^{BB} = -mV_0^{AB}$ case, for $m < -2$, with the two A chains coming together.

More recent theoretical studies have been carried out using renormalization group techniques both for diblock and triblock copolymers, and good quantitative agreement is claimed between the theoretical results and Monte Carlo[221,222] simulations. These calculations have the great advantage of avoiding many of the approximations that are introduced in more intuitive theories, but unfortunately the theories are largely limited to polymers in good solvents and so do not, as yet, give information about the coil–globute or the segregation transition.

In conclusion, then, it seems that the problems of the single copolymer chain mirror those of the homopolymer chain — in good solvents, renormalization group calculations and scaling arguments have proved remarkably successful, but in poor solvents they are not fully understood. Clearly there is an acute shortage of computer simulation data and equally clearly there is a need for good theoretical insights into the collapsed states.

We finish off this section by showing some pictures of typical copolymer conformations obtained from Monte Carlo simulations using a modification of the program developed by Clark *et al.*[223–225] The chain does a self-avoiding walk on a diamond lattice, and interaction energies ε_{AA}, ε_{BB} and ε_{AB} are associated with non-bonded A–A, B–B and A–B neighbours respectively. Thus for a diblock AB copolymer in which each block contains 100 beads we illustrate the segregation effect which becomes more pronounced on passing from Figure 22 (a; $\varepsilon_{AA} = \varepsilon_{BB} = +1$, $\varepsilon_{AB} = -1$) to Figure 22 (b; $\varepsilon_{AA} = \varepsilon_{BB} = -1$, $\varepsilon_{AB} = +1$). A negative value of ε indicates an attraction and a positive value of ε a repulsion, and in the figures the A beads are black and the B beads are red. Figure 22 (c; $\varepsilon_{AA} = -1$, $\varepsilon_{BB} = \varepsilon_{AB} = +1$) illustrates the collapsed coil effect — a configuration of relevance for AB copolymers in a poor solvent for A and a good solvent for B, which is just the condition for micelles to form at a sufficiently high copolymer concentration. Finally, for a triblock ABA copolymer in which each A block contains 50 beads and the B block 100 beads, Figure 23 (a; $\varepsilon_{AA} = -1$, $\varepsilon_{BB} = +1$, $\varepsilon_{AB} = +1$) and 23 (b; $\varepsilon_{AA} = -1$, $\varepsilon_{BB} = +1$, $\varepsilon_{AB} = -0.5$) show some pictures indicating the possibility of a transition from two collapsed A coils at either end of the B chain to one collapsed A coil with a loop of B chain sticking out as one makes ε_{AB} more negative.

Experimentally it has proved extremly difficult to investigate the possibility of a collapsed coil effect in single block copolymer chains (*i.e.* the formation of so-called monomolecular micelles.). This is undoubtedly because the phenomenon is always likely to occur very near or under conditions where conventional micelle formation occurs. As pointed out by Cowie,[226] who has reviewed the experimental evidence available, many of the dilute solution studies are open to some flexibility of interpretation and a new approach to the experimental problem is required.

6.4.2 Microdomain Structure in Bulk Block Copolymers

In this section we review some of the theoretical work aimed at understanding why block copolymers form microdomain structures, and at trying to account for the geometries found and to construct the phase diagram. We shall exclude the even more complicated problem of the behaviour of copolymer–homopolymer blends, except to give some general references.[227] Roughly speaking there have been three complementary sorts of theoretical approach to these questions. The first approach is a direct attempt to split up the free energy of a phase into several clearly differentiated contributions, such as terms arising from surface tension effects, entropy loss due to the confinement of parts of the chains in domains, entropy loss due to the restricted space available for putting the junction between the two blocks and so on. The second approach again attempts to calculate the free energy of a phase but this time by means of solving self-consistent mean field equations for the conformations of a copolymer chain. The third approach considers the melt of block copolymers and then investigates those composition fluctuations that lead to the various microphase separation transitions. Early theories based on the direct free energy method have been reviewed by Keller,[228] but possibly among the most important of these was that due to Meier,[229–231] who was among the first to recognize the importance of ensuring that space was uniformly filled with polymer segments — the constant density constraint. Thus, in its more recent formulation the theory gives the free energy change for the transformation of a diblock copolymer system from a randomly mixed state to the domain state as

$$\Delta G = \Delta H_{mix} - T[\Delta S_J + \Delta S_{con}] \tag{22}$$

where ΔH_{mix} is the heat of mixing including surface tension terms, ΔS_J is the entropy change due to placing the A–B junction in the interfacial volume and ΔS_{con} is the entropy change on the chains being constrained to their domains.

Specializing to the case of lamellae, consider the geometry in Figure 24. The thickness of the interfacial region is denoted by δ and the phase is assumed to be strongly segregated so that

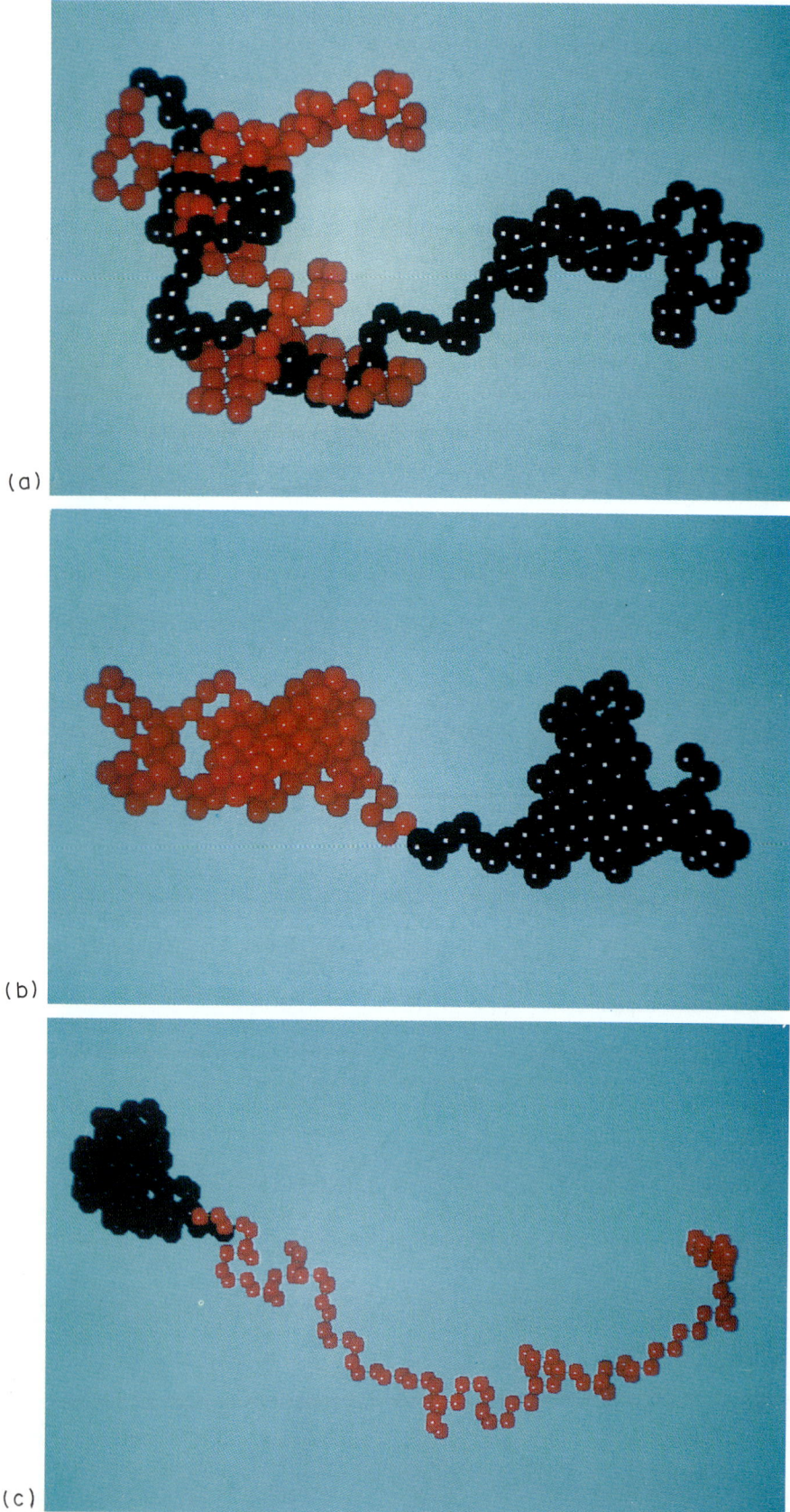

Figure 22   Typical conformations of a diblock **AB** copolymer taken from Monte Carlo simulations. The copolymers are (a) $N_A = N_B = 100$, $\varepsilon_{AA} = \varepsilon_{BB} = +1$, $\varepsilon_{AB} = -1$, (b) $\varepsilon_{AA} = \varepsilon_{BB} = -1$, $\varepsilon_{AB} = +1$ and (c) $\varepsilon_{AA} = -1$, $\varepsilon_{BB} = \varepsilon_{AB} = +1$. The A beads are black and the B beads are red

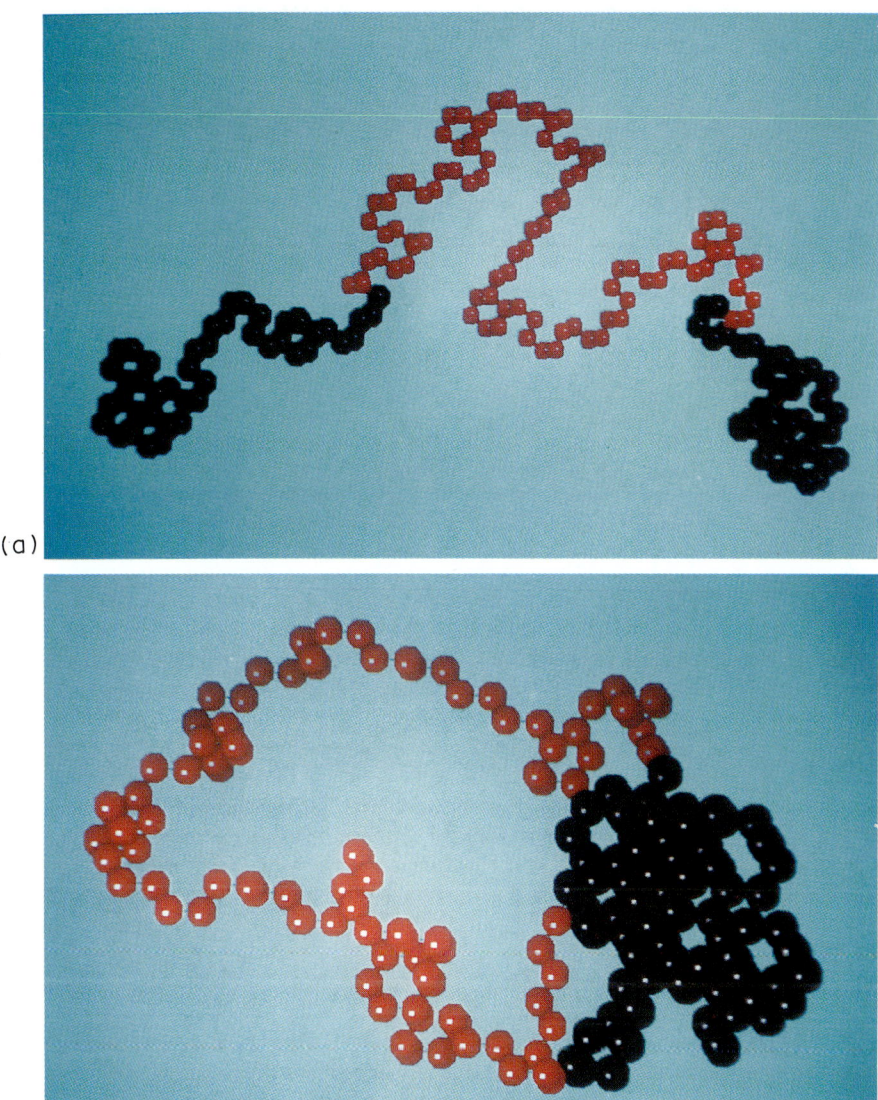

Figure 23 Typical conformations of a triblock ABA copolymer taken from Monte Carlo simulations. Each A block contains 50 beads and the B blocks 100 beads. The energy parameters are (a) $\varepsilon_{AA} = -1$, $\varepsilon_{BB} = +1$, $\varepsilon_{AB} = +1$ and (b) $\varepsilon_{AA} = -1$, $\varepsilon_{BB} = +1$, $\varepsilon_{AB} = -0.5$. The A beads are black and the B beads are red

$\delta \ll D_A, D_B$. Away from these interfacial regions the domains consist either of pure A or pure B at the density of the homopolymer melt $\rho_{0,A}$ or $\rho_{0,B}$ respectively. ΔH_{mix} may then be estimated as

$$\Delta H_{mix} = \gamma \xi - V \phi_A \phi_B \chi_{AB} kT \qquad (23)$$

where ξ is the total surface area of the domains, γ is the surface tension between the phases, V is the total volume of the system, ϕ_A and ϕ_B are the volume fractions of A and B respectively and χ_{AB} the Flory–Huggins interaction parameter. This second term, which really drives the transition so that unfavourable high energy A–B contacts are reduced, gives the enthalpy change on going from a randomly mixed state to the domains assuming no volume change, whilst the first term is the surface term. From space-filling considerations, we have

$$\xi D_A \rho_{0,A} = \xi D_B \rho_{0,B} = N_P \qquad (24)$$

where N_P is the total number of copolymer chains. Hence

$$\Delta H_{mix} = \gamma N_P (D_A \rho_{0,A})^{-1} - V \phi_A \phi_B \chi_{AB} kT \qquad (25)$$

and γ depends on the interactions between the blocks in the interfacial region and may be estimated, for example, in terms of χ_{AB} and δ.

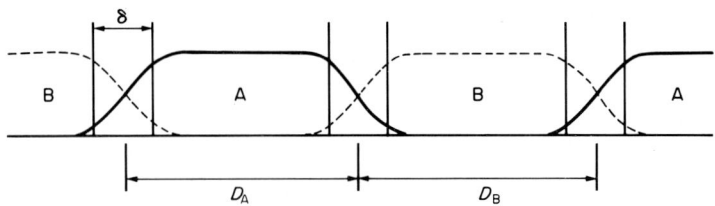

Figure 24   Density distribution of A and B segments in a block copolymer exhibiting microphase separation

ΔS_J arises from the entropy change in placing the A–B junctions within an interfacial volume (V_i) and is estimated as

$$\Delta S_J \;=\; N_p k \ln(V_i/V) \;\simeq\; N_p k \ln \left(\delta \rho_{0,A}/[D_A(\rho_{0,A} \;+\; \rho_{0,B})]\right) \tag{26}$$

Finally ΔS_{con} arises from the constraint that the blocks of the copolymer are restricted to domains. Because chains in the polymer melt behave as if they were ideal (*i.e.* as if there were no excluded volume effects), then it is frequently argued that Gaussian statistics should apply to these domains of pure A or B and provided the inequality[232] $L_k N_k^{1/2} < D_k < L_k N_k$ holds good (k = A or B) this leads to the result

$$\Delta S_{con} \;=\; k_A D_A^2 (N_A L_A^2)^{-1} \;+\; k_B D_B^2 (N_B L_B^2)^{-1} \tag{27}$$

where k_A and k_B are constants which may be estimated from the solution of the equation of motion for a random flight chain contained within a domain. If these results are combined together, if ΔS_J is neglected (generally it is a small term) and if the surface tension does not depends on N_A and N_B (again an excellent approximation), then the geometry of the phase is obtained by minimizing ΔG with respect to D_A (note from equation (24) that D_A and D_B are not independent), which gives the scaling law

$$D_A \;\sim\; (N_A \;+\; N_B)^{2/3} \tag{28}$$

at least for a constant ratio N_A/N_B. This and similar scaling results obtained from related arguments are found to be in good agreement with experimental findings.[232-233] (See Section 6.2.3.1.)

Meier, though, attempts more than this and estimates the values of γ and k_A and k_B, so as to obtain the full free energy of the lamellar phase, and uses similar arguments to obtain the free energy of the cylindrical and spherical domain phases, and thus compose a phase diagram. Essentially using Flory–Huggins theory and making some assumptions about the density profiles of A and B blocks across the interfacial region, Meier estimates γ as

$$\gamma \;=\; kT \chi_{AB} \delta/2 \tag{29}$$

Then ΔS_{con} is obtained from solving the random flight equations of motion for a polymer in a constrained domain, but adding in a requirement that the density profile within the domain as predicted by the model should be as uniform as possible. The free energy of the phase is then obtained by minimizing ΔG both with respect to D_A and δ, which incidentally yields an interfacial thickness that depends on N_A and N_B and hence a molecular-weight-dependent surface tension (*cf.* the assumption about γ in the scaling argument). Comparison of theoretical predictions with experiment shows good quantitative agreement as to the variation of domain size with molecular weight, but experimentally the interfacial dimensions varied much less with molecular weight than predicted.[231]

Finally this approach may be used to analyze cylindrical and lamellae domains and the general conclusion, in accord with experiment, is that as the fraction of A is increased one passes through the various states illustrated in Figure 2.

The attractions of this approach are firstly that it is very direct and physical and that for all its simplicity it seems to work well, even quantitatively, at predicting the nature of strongly segregated phases. It does, however, make several assumptions that should be tested and so now we consider the second type of approach based on self-consistent field theories. These methods have, in the main, been developed by Helfand *et al.*[102,234-237] and by Noolandi *et al.*[238,239] The idea is that a chain moves in a mean field due both to itself and the other chains, and this mean field not only contains within its contributions due to the unfavourable A–B interactions, but also most importantly, exercises the constraint that the system be at a constant overall density everywhere. This constraint was only treated approximately by Meier, but here it is put into effect exactly.

The mean field equations are written for the quantities $G_A(r, r_0; t)$ and $G_B(r, r_0; t)$, as discussed in the single-chain section where again r_0 is the position of the A–B join. Thus we have

$$\partial G_k / \partial t = (L_k^2 \nabla^2 G_k)/6 - \omega_k(r) G_k \tag{30}$$

where k = A or B and where one boundary condition is $G_k(r, r_0; t=0) = \delta(r - r_0)$. Other spatial boundary conditions also depend on the geometry of the phase under consideration but are of the form $n \cdot \nabla G_k(r, r_0; t) = 0$ on every reflection symmetry plane of the system, n being a normal to that plane. The mean field, $\omega_k(r)$, is given according to Helfand by

$$\omega_k(r) = \rho_{0,k}^{-1}\{\alpha[\tilde\rho_{k'}(r)/(\tilde\rho_A(r) + \tilde\rho_B(r))]^2 + (\kappa kT)^{-1}[\tilde\rho_A(r) + \tilde\rho(r) - 1]\} \tag{31}$$

where k' = B if k = A and *vice versa*, $\tilde\rho_k(r) = \rho_k(r)/\rho_{0,k}$, α is proportional to χ_{AB} and measures the A–B interaction relative to A–A and B–B interactions and κ is the compressibility. The limit $\kappa \to 0$ is generally taken, leading to the incompressibility condition

$$\tilde\rho_A(r) + \tilde\rho_B(r) = 1 \tag{32}$$

In the theory of Noolandi *et al.* a slightly different form for $\omega_k(r)$ emerges and the incompressibility condition is imposed using Lagrange multipliers.

To complete the set of equations a relation is needed between $\rho_k(r)$ and G_k which is given by

$$\rho_A(r) = N_P(VQ)^{-1} \int_0^{N_A} dt' \int_V dr_A dr_0 dr_B G_A(r_A, r; N_A - t') G_A(r, r_0; t') G_B(r_0, r_B; N_B) \tag{33a}$$

where

$$Q = V^{-1} \int_V dr_A dr_0 dr_B G_A(r_A, r_0; N_A) G_B(r_0, r_B; N_B) \tag{33b}$$

These equations may then be solved numerically in the geometry of interest, and the free energy of the phase may then be calculated in Helfand's[102] approach from

$$\Delta G / VkT = -NV^{-1} \ln Q - \alpha(N_A/\rho_{0,A})(N_B/\rho_{0,B})[(N_A/\rho_{0,A}) + (N_B/\rho_{0,B})]^{-2} \tag{34}$$

again the reference state being a hypothetical random bulk copolymer phase. Again a similar sort of expression for ΔG is given by Hong and Noolandi. Then, as before, the equilibrium geometry is obtained by minimizing ΔG (with respect to D_A, for example, in the lamellar phase).

A virtue of the approach is that a narrow interfacial thickness need not be assumed. Indeed, specializing to the case of a symmetric diblock polymer with $L_A = L_B$ and $\rho_{0,A} = \rho_{0,B}$, then the density profile for lamellae was plotted out for various values of $N_A\chi$ ($\chi = \alpha/\rho_{0,A}$ here), and for $N_A\chi = 10$ significant interpenetration of the block was observed. If, however, the interface is assumed to be narrow, then the theory may be simplified somewhat and the free energy cast into a form similar to that in equation (22). For the lamellar phase the procedure led to very similar results for the form of ΔH_{mix} and ΔS_J, but the big difference was in ΔS_{con}, where Helfand found approximately that

$$\Delta S_{con} \sim (D_A/N_A^{1/2})^{2.5} \tag{35}$$

again for a constant N_A/N_B. In other phases similar discrepancies emerged between this term as calculated from mean field theory and as calculated assuming random chain statistics. Presumably the difference comes about because if the chains all have their junction in a thin layer and if one demands the density in the domain really is uniform, then the chains cannot obey simple free-chain statistics. It does seem more work is needed, though, to fully resolve this difference.

These numerical studies again lead to predicted geometries and also a phase diagram showing the regions of stability as $(N_A + N_B)$ and N_A/N_B are varied. The geometries compare pretty well with a wide variety of experimental results,[240] whereas it appears the predicted phase boundaries tend to be at lower values of N_A/N_B than those observed experimentally. The homogeneous–microdomain transition was also investigated, but because in this limit the interface becomes extremly broad and fluctuations become very important, the validity of a mean field theory becomes doubtful. The theory, however, did still predict a dramatic change in structural features near experimentally observed temperatures.

The final part of this section deals with methods that consider composition fluctuations in the homogeneous copolymer melt in the spinodal region and the presentation closely follows that of Fredrickson and Helfand.[241] Consider a melt in which each copolymer chain has Nf A segments and $N(1-f)$ B segments, and let the Kuhn lengths in each block be the same (L) and, in the spirit of Flory–Huggins, let the density ρ be equal to L^{-3}. The central quantity in this theory, $\psi(r)$ is defined by

$$\psi(r) = \rho_A(r)/\rho - f \tag{36}$$

where $\rho_A(r)$ is the number density of A segments at r. Clearly $\psi(r)$ is proportional to a composition fluctuation in A segments (because of the incompressibility assumption this is the only independent composition fluctuation), and its ensemble average, $\bar{\psi}(r)$, is zero in the homogeneous melt but is a spatially periodic function in an ordered phase. Hence $\bar{\psi}(r)$ acts as an order parameter. For weak fluctuations, the Hamiltonian may be expanded in the form

$$\beta H(\psi) = (2!)^{-1} \int_q \gamma_2(q, -q)\psi(q)\psi(-q) + (3!)^{-1} \int_{q_1 q_2} \gamma_3(q_1, q_2, -q_1 - q_2)\psi(q_1)\psi(q_2)\psi(-q_1 - q_2)$$

$$+ (4!)^{-1} \int_{q_1 q_2 q_3} \gamma_4(q_1, q_2, q_3, -q_1 - q_2 - q_3)\psi(q_1)\psi(q_2)\psi(q_3)\psi(-q_1 - q_2 - q_3) + \dots$$

$$\tag{37}$$

where $\int_q = (2\pi)^{-3} \int dq$, all lengths are in units of L and $\psi(q)$ is given by

$$\psi(q) = \int dr\, e^{iq \cdot r} \psi(r) \tag{38}$$

An important property of $\gamma_2(q, -q)$ is that it has a minimum at $|q| = q^* \sim N^{1/2}$, and indeed equals zero on the spinodal. Thus near the spinodal the dominant fluctuations will be plane waves with wavevectors of a magnitude near q^*. Hence as the phase transition ought to occur at temperatures near the spinodal, γ_2 may be expanded about q^* to yield

$$\gamma_2(q, -q) \sim \tau + (q - q^*)^2 + \dots \tag{39}$$

where τ measures the distance from the spinodal. The higher order vertex functions, γ_3 and γ_4, do not show any such behaviour, and a simple approximation is to treat them as constants. Doing this reduces the Hamiltonian to that treated by Brazovskii.[242] Leibler's treatment[76] of the Hamiltonian was to introduce a potential $J(r)$ that coupled to ψ, and which thus created a non-zero value of $\bar{\psi}(r)$ and a free energy that depended on J, $F(J)$. However $\bar{\psi}(r)$ also depends on J, so this may be used to express the free energy of the system as a power series in $\bar{\psi}(r)$ [or $\bar{\psi}(q)$], and the vertex functions involved were calculated using a random phase approximation. This led to the result

$$\beta F(\bar{\psi}) \simeq \beta F_0 + (2!)^{-1} \int_q \gamma_2(q, -q)\bar{\psi}(q)\bar{\psi}(q) + (3!)^{-1} \int_{q_1 q_2} \gamma_3(q_1, q_2, -q_1 - q_2)\bar{\psi}(q_1)\bar{\psi}(q_2)\bar{\psi}(-q_1 - q_2)$$

$$+ (4!)^{-1} \int_{q_1 q_2 q_3} \gamma_4(q_1, q_2, q_3, -q_1 - q_2 - q_3)\bar{\psi}(q_1)\bar{\psi}(q_2)\bar{\psi}(q_3)\bar{\psi}(-q_1 - q_2 - q_3) + \dots \tag{40}$$

This corresponds to calculating the free energy from equation (37) using a saddle-point method, which is a mean field approximation, neglecting fluctuations about the extremum field.

Near the spinodal the important Fourier components of $\bar{\psi}(q)$ are those with $|q| = q^*$ and ordered phases may form by the superposition of plane waves of certain wavevectors of magnitude q^*. Leibler considered all the allowed structures and studied which could be stable at the transition point. He predicted that the structure of spherical domains that would appear just at the transition would be body-centred cubic (bcc) and he generated a phase diagram showing the regions of phase stability in terms of f and χN. He predicted the system would be homogeneous if $\chi N < 10.495$, and at the critical point ($\chi N = 10.495$, $f = 0.5$) a second-order phase transition to the lamellar phase would take place, though it was noted that the fluctuations ignored in the mean field treatment would change this into a first-order transition. At higher values of χN and for $f = 0.5$, there would be a first-order transition to the bcc phase, and then on increasing χN, transitions would occur first to a hexagonal (cyclindrical) phase and then to a lamellar phase. An independent study based on this sort of theoretical approach has been given by Yerukhimovich.[243]

The theory discussed is a weak segregation theory as it only takes into account values of q of magnitude close to q^*. Generalizations of the theory to cover the strong segregation limit were carried out by Ohta and Kawasaki[244] and Semenov.[245] In particular they took special care to include the small q limit of $\gamma_2(q, -q)$, which is proportional to q^{-2}. Ohta and Kawasaki thereby obtained the regions of phase stability in this limit, obtaining rather different boundaries from those obtained in the weak segregation limit. Another result they obtained was that in the lamellar phase $D_A \sim N^{2/3}$, which is the same result obtained from equation (22), with $\Delta S_{con} \sim D_A^2/N$ and again dropping the ΔS_J term. Clearly this scaling question requires further study!

Returning to the weak segregation limit, Fredrickson and Helfand[241] used a Hartree-like analysis of the Hamiltonian in equation (24) following the analysis of Brazovskii,[242] thereby including fluctuation corrections to Leibler's mean field result. Among their conclusions were that the fluctuations did give rise to a first-order phase transition from the disordered to lamellar phases at $f = 0.5$ and the location of the transition was at $\chi N = 10.4995 + 41.022 \, N^{-1/3}$. Furthermore, composition windows appeared through which a disordered phase could pass directly to each of the ordered phases by changing the temperature. And finally it was noted that Leibler's results were recovered in the limit $N \to \infty$.

This concludes the survey of theories about copolymer domains. Although there are matters of dispute it seems much is understood about these systems — in fact, as Helfand pointed out, the field is probably in a better state than that concerning single copolymeric chains.

6.4.3 Micelles

In this section we consider the formation of micelles when AB diblock copolymers are placed in a solvent that is good for A but bad for B. Many of the theories specialize to the case when the solvent is the homopolymer A. First of all we review some basic theory of micelle formation.[160] For a given number of copolymer chains in the solvent, there will exist a chemical equilibrium between isolated chains and all clusters of two or more chains. In dilute solution, where interactions between micelles and single copolymer chains may be neglected, one obtains the result

$$x_m = x_1^m \exp[\beta(mG_1^\ominus - G_m^\ominus)] \tag{41}$$

where x_m is the mole fraction of m-micelles, x_1 is the mole fraction of monomer, $G_m^\ominus$ is the standard free energy of an m-micelle, and $G_1^\ominus$ the standard free energy of an isolated chain monomer. Clearly the total overall mole fraction of copolymer, x, is given by

$$x = \sum m x_m \tag{42}$$

Hence, given $G_1^\ominus$ and $G_m^\ominus$, one can calculate the micelle distribution as a function of x and hence determine such quantities as the critical micelle concentration (cmc).

If the solution is not dilute this equation has to be modified to take into account non-ideality effects. This has been done *via* a virial expression for the case of amphiphilic micelles,[246,247] but the general approach of much of the theory done for copolymeric micelles has been to estimate the free energy of a distribution of monomers and micelles using a Flory–Huggins-like theory,[1] and then obtain the equilibrium distribution by minimization.

Most of the theories put forward are reminiscent of the 'direct free energy' approach as applied to microdomains in a bulk copolymer phase — the basic philosophies are similar though there are differences in detail.[232,238-239,248-250] Let us consider a spherical micelle containing N_P co-polymeric AB diblock chains immersed in a solvent S which is good for A and bad for B. Then a schematic model of the micelle is a spherical inner core of B and S of radius R_B surrounded by a corona containing A and S of radius R_A, and outside that pure S. The approximations often made are that the densities are uniform in the various regions; thus, denoting the core region by I, the corona by II and outside the corona by III, we have

$$\phi_B^I + \phi_s^I = 1, \qquad \phi_A^I = 0 \qquad \text{Region I}$$

$$\phi_A^{II} + \phi_s^{II} = 1, \qquad \phi_B^{II} = 0 \qquad \text{Region II}$$

$$\phi_s^{III} = 1, \qquad \phi_{A,B}^{III} = 0 \qquad \text{Region III}$$

where the ϕ values are volume fractions. We now estimate the free energy of this structure, $G_{N_P}^{\ominus}$, taking as reference states the pure solvent and a fictitious homogeneous melt of A–B copolymers in which the A–B interaction has been switched off (*i.e.* $\chi_{AB}=0$). If, as before, we let the Kuhn lengths of the A block be L_A, its number of segments be N_A and its number density in the pure homopolymer melt be $\rho_{0,A}$ (and similarly for B), then the geometry of the model requires

$$\frac{4}{3}\phi_B^I \pi R_B^3 = N_P N_B/\rho_{0,B} \tag{43a}$$

and

$$\frac{4}{3}\phi_A^{II}\pi[(R_A + R_B)^3 - R_B^3] = N_P N_A/\rho_{0,A} \tag{43b}$$

The total free energy then is estimated by

$$G_{N_P}^{\ominus} = G_{\text{mix}} + G_{\text{el}} + G_J + 4\pi R_B^2 \gamma \tag{44}$$

where G_{mix} is the free energy from mixing chains and solvent in the various regions. G_{el} is an elastic energy caused by the distortion of the chain lengths from their equilibrium values, G_J is the localization free energy arising from placing the A–B join at the interface between regions I and II and the final term is the surface energy term at the interface of regions I and II (because S is a good solvent for A, any such effects at the boundary of regions II and III are ignored).

G_{mix} may be estimated according to Flory–Huggin's theory to give

$$\beta G_{\text{mix}} = \frac{4}{3}\pi R_B^3[\alpha_{BS}\phi_B^I\phi_s^I + \phi_s^I \ln \phi_s^I] + \frac{4}{3}\pi((R_A + R_B)^3 - R_B^3)[\alpha_{AS}\phi_A^{II}\phi_s^{II} + \phi_s^{II}\ln\phi_s^{II}] \tag{45}$$

where $\alpha_{AS}=\chi_{AS}\rho_{0,s}$, χ_{AS} being the Flory–Huggins parameter and $\rho_{0,s}$ is the pure solvent number density and a similar relation holds for α_{BS}. Terms involving the entropy of mixing of the polymer have been neglected.

The elastic energy is normally estimated according to random flight statistics but various forms have been used. The A–B join is assumed to be in a narrow interface between the core and the mantle, and the average end-to-end distance of the A block is estimated as R_A and that of the B block as R_B. If we denote the ratios of the perturbed by unperturbed end-to-end distances by α_A and α_B for the A and B blocks respectively so that

$$\alpha_A = R_A/(N_A L_A^2)^{1/2} \tag{46}$$

and similarly for α_B, then several formulae have been used for G_{el}, *e.g.*

$$\beta G_{\text{el}} = \frac{3}{2}N_P[\alpha_P^2 - 1 - 2\ln\alpha_A + \alpha_B^2 - 1 - 2\ln\alpha_A] \tag{47a}^{229}$$

$$\beta G_{\text{el}} = \frac{3}{2}N_P[\alpha_A^2 + \alpha_B^2 + 2.3^{-3/2}(\alpha_A^{-1} + \alpha_B^{-1}) - 6] \tag{47b}^{248}$$

$$\beta G_{\text{el}} = \frac{3}{2}N_P[\alpha_A^2 + \alpha_A^{-2} + \alpha_B^2 + \alpha_B^{-2} - 4] \tag{47c}^{250}$$

Plausibility arguments exist for each of these formulae, but as the whole approach is so approximate anyway, perhaps it is impossible to argue convincingly which of these is the best — they all should give the same qualitative sort of behaviour. It is worth asking though whether it is correct to make use of results, no matter of what exact form, that make use of random flight statistics. The solution in the dense core is similar to that of spherical domains in a bulk copolymer phase, and there the solution of the mean field equations enforcing a constant density profile gave a very different form of this term than that obtained from Gaussian statistics, and as yet the correctness of the various approaches seems to be unclear. In the corona there is, of course, much solvent penetration so the uniform density constraint is not present (at least in the real micelle as opposed to the model!), so, for a θ-point solvent for A, random flight behaviour is expected. If, however, the solvent is good, one really should take into account excluded volume effects.

G_J is the entropic term arising from placing the A–B join at the interface, and is frequently neglected as a small term. Estimates for it are given by Meier[229] and Noolandi and Hong[248] (see also equation 26).

Finally one needs an expression for γ for the interfacial surface tension, and this has been calculated as a function of ϕ_B^I, ϕ_S^I, ϕ_A^{II} and ϕ_S^{II}, using mean field theory, by Hong and Noolandi.[238,239] In the special case when the solvent is the homopolymer A, and the Kuhn lengths of A and B chains are both L, the result of Helfand *et al.*[251,252] is $\beta\gamma = (\chi_{AB}/6)^{1/2} L^{-2}$.

One can now immediately find the most favourable geometry by minimizing $G_{N_P}^{\ominus}$ with respect to any free parameters (*e.g.* R_A and R_B). It then remains to estimate $G_1^{\ominus}$, the free energy of a single copolymer chain in the solvent, if we wish to use equations (41) and (42). As pointed out in section 6.4.1, this is no trivial problem! One would not wish to do a highly sophisticated treatment of the single chain, considering how crudely the theory treats the micelle, so among the simple options are (i) to assume a single chain forms a micelle on its own, and use the preceding equations with $N_P = 1$, or (ii) to assume a random configuration and to treat a dilute solution of copolymer chains in the solvent by Flory–Huggins theory. Most of the published work goes (implicitly) for option (ii), but it is an approximation of dubious quality. Unfortunately, though, quantities such as the cmc depend critically on the value used for $G_1^{\ominus}$.

Mostly, though, the method used to calculate the micelle parameters is firstly to assume a monodisperse micelle distribution (*i.e.* only one aggregation number, m for example) which is found to be the case experimentally to a very good approximation, and then to minimize the sum of G and a term corresponding to the entropy of mixing of the micelles themselves with respect to m as well as the other free parameters.[248-250]

For the case when the solvent is the homopolymer, one finds a cmc at low copolymer contents, even for small values of χ_{AB}. The core is almost entirely composed of B chains (*i.e.* $\phi_B^I \simeq 1.0$), and its radius depends primarily on N_B and not N_A, *viz.* $R_B \sim N_A^\mu N_B^\nu$ where $0.67 \lesssim \nu \lesssim 0.76$ and $-0.1 \lesssim \mu \lesssim 0$. The cmc has a dominant exponential dependence on $\chi_{AB} N_B$. Furthermore, good agreement (typically to within 10%) was obtained between theory and experiment for the values of R_A and R_B as N_A, N_B and the degree of polymerization of the homopolymer and the types of copolymer and homopolymer were varied. For the case of a distinct solvent S, Noolandi and Hong[247] again claim excellent agreement with experiment for micellar dimensions, though the estimate of the aggregation number as a function of temperature was not so good. This was possibly attributed to not taking into account the temperature dependence of χ_{AS}.

It thus seems that the basic physics of the process of micellization is well understood, but one can hardly expect the theories to be terribly quantitative. Some properties, such as the dimensions of the micelle, are not overly sensitive to the details of the approximation scheme, but other properties, such as cmc, the aggregation number and the thermodynamics of micelle formation are much more volatile in their behaviour. The theories presented all assumed a monodisperse micelle distribution, but in fact one can use the methods to calculate the full distribution from equations (42) and (43), and indeed the distribution does turn out to be narrowly peaked. One can also use the theories to estimate the relative stabilities of spherical micelles *vis-a-vis* non-spherical micelles, infinite cylinders and bilayers, and preliminary studies indeed indicate the possibility of infinite cylinders at copolymer concentrations less than the cmc.[253] The possible formation of these and other structures should be more thoroughly investigated.

The obvious deficiency in the theory is the assumption of constant density of A in region II — in fact this assumption is inconsistent with the assumption of A chains being anchored to the interface and behaving as random flight polymers. One obvious way of relaxing this is to introduce volume fraction as a continuous fraction of r, the distance from the micellar centre and write a free energy expression such as [253]

$$\beta G_{N_P}^{\ominus} = \int dr \left\{ \phi_s \ln \phi_s + \alpha_{AB}\phi_A\phi_B + \alpha_{AS}\phi_A\phi_s + \alpha_{BS}\phi_B\phi_s + \sum_{k,k'} \beta_{kk'}(d\phi_k/dr)(d\phi_{k'}/dr) \right\} + \beta G_{el}$$

$$(48)$$

where $\phi_s(r) + \phi_A(r) + \phi_B(r) = 1$, k, k' = A, B or S, and G_{el} is given by one out of equations (47a–c) with R_A and R_B estimated from moments of ϕ_A and ϕ_B. The term involving $\beta_{kk'}$ is a Cahn–Hilliard-like[254] term for inhomogeneous systems with typically $\beta_{kk'} \sim L_{k'}L_k\alpha_{kk'}$, and this produces a surface-tension-like term. Then $G_{N_P}^{\ominus}$ is found by minimizing this expression with respect to ϕ_A and ϕ_B. In practice a variational principle may be used employing trial functions but the minimization is numerically difficult, the theory is still very crude and it turns out the results and conclusions are not significantly different from those obtained using the simpler models. It would seem that calculations based on self-consistent mean field theories for the equation of motion of the copolymer chain in the presence of solvent[238,239] would remove a number of the approximations described above, and we look forward to seeing the results of such studies. Furthermore, such an approach would treat both the single chain and the micelle in what would hopefully be a unified way.

In conclusion it seems that there are still many outstanding theoretical problems in the field of copolymers. This includes problems in microdomain formation in the bulk phase, but more pressing are the areas of micelles and properties of the single chain which are much less well understood. Although more complete, more complex and increasingly sophisticated theories will no doubt be applied to these situations, it would appear that there is a real need for more computer simulations of the systems. Even for the single chain the quantity of simulation data is scarce, and although it is a difficult problem to tackle interfaces and many chain interactions, one feels that many useful results should be forthcoming from such studies, which may well give a good physical insight as to how real chains with real excluded volumes are behaving in their domains.

ACKNOWLEDGEMENTS

We would like to thank Dr. M. Lal for providing us with the single polymer chain Monte Carlo program and we are also grateful to Dr. J. Cox and Dr. P. Lawrence for help with the photography and computer graphics. The computer graphics package was Chem-X, developed and distributed by Chemical Design Ltd., Oxford, England.

6.5   REFERENCES

1. P. J. Flory, 'Principles of Polymer Chemistry', Cornell University Press, Ithaca, New York, 1953.
2. J. H. Bolland and H. W. Melville, in 'Proceedings of the Rubber Technology Conference, London', ed. T. R. Dawson and J. R. Scott, Heffer, London, 1938, p. 239.
3. E. Müller, O. Bayer, S. Peterson, H. F. Piepenbrink, H. F. Schmidt and E. Wienbrenner, *Angew. Chem.*, 1952, **64**, 523.
4. O. Bayer, E. Müller, S. Peterson, H. F. Piepenbrink and E. Windemuth, *Angew. Chem.*, 1950, **62**, 57.
5. L. G. Lunsted, *J. Am. Oil Chem. Soc.*, 1951, **28**, 294.
6. H. F. Mark, *Text. Res. J.*, 1953, **23**, 294.
7. D. Coleman, *J. Polym. Sci.*, 1954, **14**, 15.
8. W. H. Charch and J. C. Shivers, *Text. Res. J.*, 1959, **29**, 536.
9. H. F. Mark, *Rec. Chem. Prog.*, 1959, **12**, 139.
10. L. C. Bateman, *Ind. Eng. Chem.*, 1957, **49**, 704.
11. F. M. Merritt, *J. Polym. Sci.*, 1957, **24**, 467.
12. A. Skoulios, G. Finaz and J. Parrod, *C. R. Hebd. Seances Acad. Sci.*, 1960, **251**, 739.
13. D. H. Richards and M. Szwarc, *Trans. Faraday Soc.*, 1959, **55**, 1644.
14. M. Szwarc, 'Carbanions, Living Polymers and Electron Transfer Processes', Wiley, New York, 1968.
15. V. Luzzati, H. Mustacchi and A. Skoulios, *Discuss. Faraday Soc.*, 1958, **25**, 43.
16. A. Skoulios, *Adv. Colloid Interface Sci.*, 1967, **1**, 79.
17. R. J. Ceresa (ed.), 'Block and Graft Copolymers', Butterworths, Washington, DC, 1962.
18. F. W. Stavely, *Ind. Eng. Chem.*, 1956, **48**, 778.
19. N. R. Legge, S. Davison, H. E. de la Mare, G. Holden and M. K. Martin, *ACS Symp. Ser.*, 1985, **285**, 175.
20. G. Holden, E. T. Bishop and N. R. Legge, in 'Proceedings of the International Rubber Conference', MacLaren, London, 1968, p. 287.
21. S. L. Cooper and A. V. Tobolsky, *Text. Res. J.*, 1966, **36**, 800.
22. D. S. Campbell, in 'Developments in Block Copolymers – 2', ed. 1. Goodman, Elsevier Applied Science, London, 1985, chap. 6.
23. J. Mann and G. R. Williamson, in 'The Physics of Glassy Polymers', ed. R. N. Haward, Appled Science, London, 1973, chap. 8.
24. H. Hendus, K. H. Illers and E. Ropte, *Kolloid-Z. Z. Polym.*, 1967, **216**, 110.
25. M. Matsuo and S. Sagaya, in 'Colloid and Morphological Behavior of Block and Graft Copolymers', ed. G. E. Molau, Plenum Press, New York, 1971, p. 1.
26. T. Hashimoto, K. Nagatshi, A. Todo, H. Hasagawa and H. Kawai, *Macromolecules*, 1974, **7**, 364.
27. G. Kämpf, M. Hoffmann and H. Krömer, *Ber. Bunsenges. Phys. Chem.*, 1970, **70**, 851, 859.
28. B. Gallot, *Pure Appl. Chem.*, 1974, **38**, 1.
29. C. Price, T. P. Lally, A. G. Watson, D. Woods and M. T. Chow, *Br. Polym. J.*, 1972, **4**, 413.

30. A. Skoulios, in 'Advances in Liquid Crystals', ed. G. H. Brown, Academic Press, New York, 1975, vol. 1, p. 169.
31. M. Matsuo, *Jpn. Plast.*, 1968, **2**, 6.
32. A. Douy and B. Gallot, *C. R. Hebd. Seances Acad. Sci.*, 1972, **274**, 498.
33. P. R. Lewis and C. Price, *Nature (London)*, 1969, **223**, 494.
34. P. R. Lewis and C. Price, *Polymer*, 1971, **12**, 258.
35. P. R. Lewis and C. Price, *Polymer*, 1972, **13**, 20.
36. E. Pedemonte, A. Turturro, U. Bianchi and P. Devetta, *Polymer*, 1973, **14**, 145.
37. M. Shibayama, T. Hashimoto and H. Kawai, *Macromolecules*, 1983, **16**, 1434.
38. G. E. Molau, in 'Block Polymers', ed. S. L. Aggarwal, Plenum Press, New York, 1970, p. 102.
39. M. Shibayama and T. Hashimoto, *Macromolecules*, 1986, **19**, 740.
40. J. Dlugosz, A. Keller and E. Pedemonte, *Kolloid-Z. Z. Polym.*, 1970, **242**, 1125.
41. M. Shibayama, T. Hashimoto and H. Kawai, *Macromolecules*, 1983, **16**, 16.
42. G. Tsouladże and A. Skoulios, *J. Chim. Phys. Phys.-Chim. Biol*, 163, **60**, 626.
43. D. B. Alward, D. J. Kinning, E. L. Thomas and L. J. Lewis, *Macromolecules*, 1986, **19**, 215.
44. R. Perret and A. Skoulios, *Makromol. Chem.*, 1972, **162**, 147, 163.
45. M. Gervais and B. Gallot, *Makromol. Chem.*, 1973, **171**, 157.
46. M. Gervais and B. Gollot, *Makromol. Chem.*, 1977, **178**, 1577; 1977, **178**, 2071.
47. B. Lotz and A. A. Kovacs, *Polym. Prepr., Am. Chem. Soc., Div. Polym. Chem.*, 1969, **10**, 820.
48. T. Hayashi, in 'Developments in Block Copolymers — 2', ed. I. Goodman, Elsevier Applied Science, London, 1985, chap. 4.
49. R. G. Crystal, P. F. Erchardt and J. J. O'Malley, in 'Block Polymers', ed. S. L. Aggarwal, Plenum Press, New York, 1970, p. 179.
50. A. K. Fritzsche and F. P. Price, in 'Block Polymers', ed. S. L. Aggarwal, Plenum Press, New York, 1970, p. 249.
51. P. E. Gibson, M. A. Vallance and S. L. Cooper, in 'Developments in Block Copolymers — 1', ed. I. Goodman, Applied Science, London, 1982, chap. 6.
52. D. C. Evans, M. H. George and J. A. Barrie, *Polymer*, 1975, **16**, 690.
53. R. C. Thamm and W. H. Buck, *J. Polym. Sci., Polym. Chem. Ed.*, 1978, **16**, 539.
54. K. Kato, *J. Polym. Sci., Part B*, 1966, **4**, 35.
55. L. D. Handlin, Jr. and E. L. Thomas, *Macromolecules*, 1983, **16**, 1514.
56. C. Price, R. Singleton and D. Woods, *Polymer*, 1974, **15**, 117.
57. R. J. Spontak, M. C. Williams and D. A. Agard, *Polymer*, 1988, **29**, 387.
58. E. J. Roche and E. L. Thomas, *Polymer*, 1981, **22**, 333.
59. C. H. Y. Chem-Tsai, E. L. Thomas, W. J. MacKnight and N. S. Schneider, *Polymer*, 1986, **27**, 659.
60. M. Sorrano, W. J. MacKnight, E. L. Thomas and J. M. Ottino, *Polymer*, 1987, **28**, 1667, 1674.
61. R. R. Lagasse, *J. Appl. Polym. Sci.*, 1977, **21**, 2489.
62. S. D. Bruck, *J. Polym. Sci., Polym. Symp.*, 1979, **66**, 283.
63. H. Tadakoro, 'Structures of Crystalline Polymers', Wiley-Interscience, New York, 1979.
64. G. Oster and D. P. Riley, *Acta Crystallogr.*, 1952, **5**, 1, 272.
65. A. Keller, J. Olugosz, M. J. Folkes, E. Pedemonte, F. P. Scalisa and F. M. Willmouth, *J. Phys. (Orsay, Fr.)*, 1971, **32**, 295.
66. M. Shibayama and T. Hashimoto, *Macromolecules*, 1986, **19**, 740.
67. G. Porod, *Kolloid-Z.*, 1951, **124**, 83.
68. W. J. Ruland, *J. Appl. Crystallogr.*, 1971, **4**, 70.
69. T. Hashimoto, K. Nagatoshi, A. Todo, H. Hasegawa and H. Kawai, *Macromolecules*, 1974, **7**, 364.
70. T. Hashimoto, A. Todo, H. Itoi and H. Kawai, *Macromolecules*, 1977, **10**, 377.
71. C. G. Vonk, *J. Appl. Crystallogr.*, 1973, **6**, 81.
72. D. Tyagi, J. L. Hendrick, D. C. Webster, J. E. McGrath and G. L. Wilkes, *Polymer*, 1988, **29**, 833.
73. A. W. Richards and J. L. Thomason, in 'Proceedings of the IUPAC Macromolecule Symposium', IUPAC, Oxford, 1982, p. 708.
74. F. Annighoefer and W. Gronski, *Makromol. Chem.*, 1984, **185**, 2213.
75. H. Benoit, W. Wu, M. Benmouna, B. Mozer, B. Bauer and A. Lapp, *Macromolecules*, 1985, **18**, 986.
76. L. Leibler, *Macromolecules*, 1980, **13**, 1602.
77. I. Yasuhito and T. Hashimoto, *Kao Fen Tzu T'ung Hsun*, 1988, **29**, 135.
78. G. Hadzioannou, C. Picot, A. Skoulios, M. L. Ionescu, A. Mathis, R. Duplessix, Y. Gallot and J. P. Lingelser, *Macromolecules*, 1982, **15**, 263.
79. F. S. Bates, C. V. Berney, R. E. Cohen and G. D. Wignal, *Polymer*, 1983, **24**, 519.
80. A. Skoulios and G. Finaz, *J. Chim. Phys. Phys. -Chim. Biol.*, 1962, **59**, 473, 626.
81. J. C. Wittmann, B. Lotz, F. Candau and A. J. Kovacs, *J. Polym. Sci., Polym. Phys. Ed.*, 1982, **20**, 1341.
82. R. S. Stein and G. L. Wilkes, *J. Polym. Sci., Part A-2*, 1969, **7**, 1525.
83. M. J. Folkes and A. Keller, *Polymer*, 1971, **12**, 222.
84. S. G. Yeates and C. Booth, *Eur. Polym. J.*, 1985, **21**, 217.
85. G. Perego, M. Cesari and R. Vitali, *J. Appl. Polym. Sci.*, 1984, **29**, 1157.
86. J. T. Koberstein and R. S. Stein, *Polymer*, 1984, **25**, 171.
87. A. Misra and S. N. Garg, *J. Polym. Sci., Polym. Lett. Ed.*, 1986, **24**, 999.
88. L. W. Jelinski, J. J. Dumais and A. K. Engel, *Org. Coat. Appl. Polym. Sci. Proc.*, 1983, **248**, 102.
89. R. A. Assink and G. L. Wilkes, *Polym. Eng. Sci.*, 1977, **17**, 606.
90. L. W. Jelinski, J. J. Dumais and A. K. Engel, *Macromolecules*, 1983, **16**, 492.
91. B. Morèse-Séguéla, M. St. Jacques, J. M. Renaud and Prud'homme, *Macromolecules*, 1980, **13**, 100.
92. J. Schaefer, M. D. Sefcik, E. O. Stejskal and R. A. McKay, *Macromolecules*, 1980, **13**, 1121.
93. J. F. Beecher, L. Marker, R. D. Bradford and S. L. Aggarwal, *J. Polym. Sci., Part C*, 1969, **26**, 117.
94. T. L. Smith and R. A. Dickie, *J. Polym. Sci., Part C*, 1969, **26**, 163.
95. M. Shen and D. Kaelble, *J. Polym. Sci., Part B*, 1970, **8**, 149.
96. W. Maung and L. H. Williams, *Polym. Eng. Sci.*, 1985, **25**, 113.
97. T. Hashimoto, Y. Tsukahara and H. Kawal, *Polym. J.*, 1983, **15**, 699.
98. A. M. North, *J. Polym. Sci., Part C*, 1975, **50**, 345.

99. A. M. North and J. C. Reid, *Eur. Polym. J.*, 1972, **8**, 1129.
100. K. Adachi, A. M. North, R. A. Pethrick, G. Harrison and J. Lamb, *Polymer*, 1982, **23**, 1451.
101. J. L. Castles, M. A. Valance, S. L. Cooper and J. M. McKenna, *Polym. Mater. Sci. Eng.*, 1984, **51**, 387.
102. E. Helfand and Z. R. Wasserman, in 'Developments in Block Copolymers — 1', ed. I. Goodman, Applied Science, London, 1982, chap. 4.
103. B. R. M. Gallot, *Adv. Polym. Sci.*, 1978, **29**, 85.
104. T. Hashimoto, M. Shibayama and H. Kawai, *Macromolecules*, 1980, **13**, 1273.
105. T. Hashimoto, M. Fujimura and H. Kawai, *Macromolecules*, 1980, **13**, 1660.
106. T. Hashimoto, H. Tanaka and H. Hasegawa, *Macromolecules*, 1985, **18**, 1864.
107. A. E. Skoulios, in 'Developments in Block Copolymers — 1', ed. I. Goodman, Applied Science, London, 1982, chap. 3.
108. C. I. Chung and J. C. Gale, *J. Polym. Sci., Polym. Phys. Ed.*, 1976, **14**, 1149.
109. C. I. Chung and M. I. Lin, *J. Polym. Sci., Polym. Phys. Ed.*, 1978, **16**, 545.
110. E. V. Gouinlock and R. S. Porter, *Polym. Eng. Sci.*, 1977, **17**, 573.
111. C. I. Chung, H. L. Griesbach and L. Young, *J. Polym. Sci., Palym. Phys. Ed.*, 1980, **18**, 1237.
112. G. Hadziioannou and A. Skoulios, *Macromolecules*, 1982, **15**, 271.
113. M. Shibayama, T. Hashimoto, H. Hasegawa and H. Kawai, *Macromolecules*, 1983, **16**, 1427.
114. M. L. Ionescu and A. Skoulios, *Makromol. Chem.*, 1976, **177**, 257.
115. A. Douy, R. Mayer, J. Rossi and B. Gallot, *Mol. Cryst. Liq. Cryst.*, 1969, **7**, 103.
116. H. R. Thomas and J. J. O'Malley, *Macromolecules*, 1979, **12**, 323.
117. J. J. O'Malley, H. R. Thomas and G. M. Lee, *Macromolecules*, 1979, **12**, 996.
118. Y. Shimura and T. Hatakeyama, *J. Polym. Sci., Polym. Phys. Ed.*, 1975, **13**, 653.
119. J. J. O'Malley, *J. Polym. Sci., Polym. Symp.*, 1977, **60**, 151.
120. P. K. Seow, Y. Gallot and A. Skoulios, *Makromol. Chem.*, 1976, **177**, 199.
121. M. Gervais and B. Gallot, *Makromol. Chem.*, 1977, **178**, 1577.
122. E. Hirata, T. Ijitzu, T. Seon, T. Hashimoto and T. Kawai, *Polym. Prepr., Am. Chem. Soc., Div. Polym. Chem.*, 1974, **15**, 177.
123. D. Petit, R. Jerome and Ph. Teyssié, *J. Polym. Sci., Polym. Chem. Ed.*, 1979, **17**, 2903.
124. W. L. Hergenrother and R. J. Ambrose, *J. Polym. Sci., Polym. Chem. Ed.*, 1974, **12**, 2613.
125. A. Nakajima, K. Kugo and T. Hayashi, *Macromolecules*, 1979, **12**, 845.
126. R. W. M. van Berkel, S. A. G. de Graaf, F. J. Huntjens and C. M. F. Vrouenraets, in 'Developments in Block Copolymers — 1', ed. I. Goodman, Applied Science, London, 1982, chap. 7.
127. B. Lotz, A. Kovacs, G. A. Basset and A. Keller, *Kolloid-Z. Z. Polym.*, 1966, **209**, 115.
128. T. G. E. Swales, R. C. Domszy, R. L. Beddoes, C. Price and C. Booth, *J. Polym. Sci., Polym. Chem. Ed.*, 1985, **23**, 1585.
129. M. Droescher, U. Bandara and F. G. Schmidt, *Macromol. Chem. Phys. Suppl.*, 1984, **6**, 107.
130. J. W. McBain, *Trans. Faraday Soc.*, 1913, **9**, 99.
131. G. S. Hartley, *Kolloid-Z.*, 1939, **88**, 33.
132. G. S. Hartley, 'Aqueous Solutions of the Paraffin-Chain Salt', Hermann, Paris, 1936.
133. P. Debye and E. W. Anacker, *J. Phys. Colloid Chem.*, 1951, **55**, 644.
134. S. Schlick and M. Levy, *J. Phys. Chem.*, 1960, **64**, 883.
135. S. Krause, *J. Phys. Chem.*, 1961, **65**, 1618.
136. S. Ye. Bresler, L. M. Pyrkov, S. Ya. Frenkel, L. A. Laius and S. I. Klenin, *Vysokomol. Soedin.*, 1962, **4**, 250.
137. Y. Gallot, E. Franta, P. Rempp and H. Benoit, *J. Polym. Sci., Part C*, 1964, **4**, 473.
138. A. Dondos, P. Rempp and H. Benoit, *J. Chim. Phys., Phys.-Chim. Biol.*, 1963, **62**, 821.
139. H.-G. Elias, in 'Light Scattering from Polymer Solutions', ed. M. B. Huglin, Academic Press, London, 1972.
140. J. N. Phillips, *Trans. Faraday Soc.*, 1955, **51**, 561.
141. J. Th. G. Overbeek, *Chem. Weekbl.*, 1958, **54**, 687.
142. G. Stainsby and A. E. Alexander, *Trans. Faraday Soc.*, 1958, **46**, 587.
143. C. Price and D. Woods, *Eur. Polym. J.*, 1973, **9**, 827.
144. C. Booth, T. deV. Naylor, C. Price, J. Rajab and R. B. Stubbersfield, *J. Chem. Soc., Faraday Trans. 1*, 1978, **74**, 2352.
145. C. Price, A. L. Hudd and R. B. Stubbersfield, *Polymer*, 1980, **21**, 9.
146. F. Horrii, Y. Ikada and I. Sakurada, *J. Polym. Sci., Polym. Chem. Ed.*, 1974, **12**, 323.
147. Z. Tuzar, V. Petrus and P. Kratochvil, *Makromol. Chem.*, 1974, **175**, 3181.
148. C. Price, A. L. Hudd, C. Booth and B. Wright, *Polymer*, 1982, **23**, 650.
149. P. Spacek and M. Kubin, *J. Appl. Polym. Sci.*, 1985, **30**, 143.
150. S. G. Yeates, C. Price and C. Booth, *J. Colloid Interface Sci.*, 1986, **114**, 416.
151. H.-G. Elias, *J. Macromol. Sci., Chem.*, 1973, **A7** (3), 601.
152. K. Solc and H.-G. Elias, *J. Polym. Sci., Polym. Chem. Ed.*, 1973, **11**, 137.
153. C. Price, *Pure Appl. Chem.*, 1983, **55**, 1563.
154. C. Price, R. B Stubbersfield, S. El. Kafrawy and K. D. Kendall, *Br. Polym. J.*, to be published.
155. C. Price, E. K. M. Chan and R. B. Stubbersfield, *Eur. Polym. J.*, 1987, **23**, 649.
156. C. Price, C. Booth, P. A. Canham, T. deV. Naylor and R. B. Stubbersfield, *Br. Polym. J.*, 1984, **16**, 311.
157. C. Price, E. K. M. Chan, G. Pilcher and R. B. Stubbersfield, *Eur. Polym. J.*, 1985, **21**, 627.
158. C. Price, E. K. M. Chan, R. H. Mobbs and R. B. Stubbersfield, *Eur. Polym. J.*, 1985, **21**, 355.
159. H. H. Teo, S. G. Yeates, C. Price and C. Booth, *J. Chem. Soc., Faraday Trans. 1*, 1984, **80**, 1787.
160. J. M. Corkill, J. F. Goodman and J. R. Tate, *Trans. Faraday Soc.*, 1964, **60**, 996.
161. C. Tanford, 'The Hydrophobic Effect', Wiley-Interscience, New York, 1973.
162. W. Bushuk and H. Benoit, *Can. J. Chem.*, 1958, **36**, 1616.
163. W. H. Stockmayer, L. D. Moore, M. Fixman and B. N. Epstein, *J. Polym. Sci.*, 1955, **16**, 517.
164. A. C. Wright, *Discuss. Faraday Soc.*, 1970, **50**, 111.
165. F. H. Ree, Y.-T. Lee and T. Ree, *J. Chem. Phys.*, 1971, **55**, 234.
166. T. Tanaka, T. Kotaka and H. Inagaki, *Bull. Inst. Chem. Res., Kyoto Univ.*, 1977, **55**, 206.
167. T. Tanaka, T. Kotaka and H. Inagaki, *Polym. J.*, 1972, **3**, 327.
168. C. Price, J. D. G. McAdam, T. P. Lally and D. Woods, *Polymer*, 1974, **15**, 228.

169. W. Mandema, H. Zeldenrust and C. A. Emeis, *Makromol. Chem.*, 1979, **180**, 1521.
170. L. Oranli, P. Bahadur and G. Reiss, *Can. J. Chem.*, 1985, **63**, 2691.
171. J. Plěstil and J. Baldrian, *Makromol. Chem.*, 1975, **176**, 1009.
172. J. Plěstil and J. Baldrian, *Makromol. Chem.*, 1973, **174**, 183.
173. T. L. Bluhm and S. L. Malhotra, *Eur. Polym. J.*, 1986, **22**, 249.
174. S. L. Malhotra. T. L. Bluhm and Y. Deslandes, *Eur. Polym. J.*, 1986, **22**, 391.
175. R. J. Roe and D. Rigby, *Polym. Mater. Sci. Eng.*, 1984, **51**, 382.
176. D. Rigby and J. R. Roe, *Macromolecules*, 1984, **17**, 1778.
177. F. Candau, J.-M. Guenet, J. Boutillier and C. Picot, *Polymer*, 1979, **20**, 1227.
178. J. S. Higgins, J. V. Dawkins and G. Taylor, *Polymer*, 1980, **21**, 627.
179. S. Kranse and P. Reismiller, *J. Polym. Sci., Part A-2*, 1975, **13**, 663.
180. C. Price and D. Woods, *Polymer*, 1973, **14**, 82.
181. I. E. Climie and E. F. T. White, *J. Polym. Sci.*, 1960, **47**, 149.
182. A. A. Demin, G. D. Rudkovskaya, L. V. Dmitrenko, L. A. Ovsyannikova, T. A. Sokolova, G. V. Samsonov and I. N. Nikonova, *Vysokomol. Soedin., Ser. A*, 1974, **16**, 2706.
183. J. G. Watterson, H.-R. Lässer and H.-G. Elias, *Kolloid-Z. Z. Polym.*, 1972, **250**, 64.
184. C. Price, in 'Developments in Block Copolymers — 1', ed. I. Goodman, Applied Science, London, 1982, chap. 2.
185. M. Shinizky, A. C. Dianoux, C. Gitler and G. Weber, *Biochemistry*, 1971, **10**, 2106.
186. A. S. Waggoner, O. H. Griffith and C. R. Christenson, *Proc. Natl. Acad. Sci. USA*, 1967, **57**, 1198.
187. F. Heatley and A. Begum, *Makromol. Chem.*, 1977, **178**, 1205.
188. F. Candau, F. Heatley, C. Price and R. B. Stubbersfield, *Eur. Polym. J.*, 1984, **20**, 685.
189. J. Spevacek, *Makromol. Chem., Rapid Commun.*, 1983, **3**, 697.
190. E. Franta, *J. Chim. Phys.*, 1966, **63**, 595.
191. T. N. Khan, R. H. Mobbs, C. Price, J. R. Quintana and R. B. Stubbersfield, *Eur. Polym. J.*, 1987, **23**, 191.
192. H. Utiyama, K. Takenako, M. Mizumori, M. Fukuda, Y. Tsunashimo and M. Kurata, *Macromolecules*, 1974, **7**, 515.
193. P. A. Canham, T. P. Lally, C. Price and R. B. Stubbersfield, *J. Chem. Soc., Faraday Trans. 1*, 1980, **76**, 1857.
194. Z. Tuzar, A. Sikora, V. Petrus and P. Kratochvíl, *Makromol. Chem.*, 1977, **178**, 2743.
195. C. Price, E. K. M. Chan, A. L. Hudd and R. B. Stubbersfield, *Polym. Commun.*, 1986, **27**, 196.
196. D. J. Wilson, G. Hurtrez and G. Reiss, *NATO Adv. Study. Inst., Ser. E*, 1985, no. 89, 195.
197. G. Reiss, *Makromol. Chem., Suppl.*, 1985, **13**, 157.
198. K. Nakamura, R. Endo and M. Takada, *J. Polym. Sci., Polym. Phys. Ed.*, 1976, **14**, 1287.
199. C. Price, E. K. M. Chan, A. L. Hudd and R. B. Stubbersfield, *Br. Polym. J.*, 1986, **18**, 57.
200. Y. Ikada, F. Horrii and I. Sakurada, *J. Polym. Sci., Polym. Chem. Ed.*, 1973, **11**, 27.
201. Z. Tuzar and P. Kratochvíl, *Makromol. Chem.*, 1973, **170**, 177.
202. C. Price and R. B. Stubbersfield, *Eur. Polym. J.*, 1987, **23**, 177.
203. Z. Tuzar and P. Kratochvíl, *Adv. Colloid Interface Sci.*, 1976, **6**, 201.
204. R. Nagarajan, M. Barry and E. Ruckenstein, *Langmuir*, 1986, **2**, 210.
205. P.-G. de Gennes, 'Scaling Concepts in Polymer Physics', Cornell University Press, Ithaca, NY, 1979.
206. K. F. Freed, 'Renormalization Group Theory of Macromolecules', Wiley, New York, 1987.
207. M. A. Moore, *J. Phys. A: Math. Gen.*, 1977, **10**, 305.
208. I. Lifshitz, A. Grosberg and A. Kholkhov, *Rev. Mod. Phys.*, 1978, **50**, 683.
209. A. L. Kholodenko and K. F. Freed, *J. Phys. A: Math. Gen.*, 1984, **17**, 2703.
210. O. Froelich and H. Benoit, *Makromol. Chem.*, 1966, **92**, 224.
211. S. F. Edwards, *J. Phys. A: Math., Nucl., Gen.*, 1974, **7**, 332.
212. T. Tanaka, T. Kotaka and H. Inagaki, *Macromolecules*, 1976, **9**, 561.
213. T. Tanaka, T. Kotaka, K. Ban, M. Hattori and H. Inagaki, *Macromolecules*, 1977, **10**, 960.
214. T. Tanaka and T. Kotaka, *Polym. Prepr., Am. Chem. Soc., Div. Polym. Chem.*, 1979, **20**, 9.
215. T. Tanaka, M. Omoto and H. Inagaki, *Macromolecules*, 1979, **12**, 147.
216. T. M. Birshtein, A. M. Skvortsov and A. A. Sariban, *Macromolecules*, 1976, **9**, 888.
217. J. Bendler, K. Solc and W. Gobusch, *Macromolecules*, 1977, **10**, 635.
218. J. Bendler and K. Solc, *Polym. Eng. Sci.*, 1977, **17**, 8.
219. J. Bendler and K. Solc, *Polym. Prepr., Am. Chem. Soc., Div. Polym. Chem.*, 1977, **18**, 319.
220. T. Kimura and M. Kurata, *J. Polym. Sci., Polym. Phys. Ed.*, 1979, **17**, 2133.
221. J.-F. Joanny, L. Leibler and R. Ball, *J. Chem. Phys.*, 1984, **81**, 4640.
222. J. F. Douglas and K. F. Freed, *J. Chem. Phys.*, 1987, **86**, 4280.
223. M. Lal, K. A. Richardson, D. Spencer and M. A. Turpin, *ACS Symp. Ser.*, 1975, **8**.
224. A. T. Clark, M. Lal, M. A. Turpin and K. A. Richardson, *Faraday Discuss. Chem. Soc.*, 1975, **59**, 189.
225. A. T. Clark, Ph. D. Thesis, CNAA Awards, London.
226. J. M. G. Cowie, in 'Developments in Block Copolymers — 1', ed. I. Goodman, Applied Science, London, 1982, chap. 1.
227. J. Noolandi, *Ber. Bunsenges. Phys. Chem.*, 1985, **89**, 1147.
228. M. J. Folkes and A. Keller, in 'The Physics of Glassy Polymers', ed. R. N. Haward, Halsted Press, New York, 1973.
229. D. J. Meier, *J. Polym. Sci., Part C*, 1969, **26**, 81.
230. D. J. Meier, *Polym. Prepr., Am. Chem. Soc., Div. Polym. Chem.*, 1970, **11**, 400.
231. D. J. Meier, *NATO Adv. Study Inst., Ser. E*, 1985, no. 89, 173.
232. P. G. de Gennes, *Solid State Phys., Suppl.*, 1978, **14**, 1.
233. T. M. Birshtein and Ye. B. Zhulina, *Polym. Sci. USSR (Engl. Transl.)*, 1985, **27**, 1807.
234. E. Helfand, *Macromolecules*, 1975, **8**, 552.
235. E. Helfand and Z. R. Wasserman, *Macromolecules*, 1976, **6**, 879.
236. E. Helfand and Z. R. Wasserman, *Macromolecules*, 1973, **11**, 960.
237. E. Helfand and Z. R. Wasserman, *Macromolecules*, 1980, **13**, 994.
238. K. M. Hong and J. Noolandi, *Macromolecules*, 1981, **14**, 727.
239. J. Noolandi and K. M. Hong, *Macromolecules*, 1982, **15**, 482.
240. B. R. M. Gallot, *Adv. Polym. Sci.*, 1978, **29**, 85.

241. G. F. Fredrickson and E. Helfand, *J. Chem. Phys.*, 1987, **87**, 697.
242. S. A. Brazovskii, *Sov. Phys.—JETP (Engl. Transl.)*, 1975, **41**, 85.
243. I. Ya. Yerukhimovich, *Polym. Sci. USSR (Engl. Transl.)*, 1983, **24**, 2223.
244. T. Ohta and K. Kawasaki, *Macromolecules*, 1986, **19**, 2621.
245. A. I. Semenov, *Sov. Phys.—JETP (Engl. Transl.)*, 1985, **61**, 733.
246. A. Ben-Shaul and W. M. Gelbart, *J. Phys. Chem.*, 1982, **86**, 316.
247. W. M. Gelbart, A. Ben-Shaul, H. F. McMullen and A. Masters, *J. Phys. Chem.*, 1984, **88**, 861.
248. J. Noolandi and K. M. Hong, *Macromolecules*, 1983, **16**, 1443.
249. M. D. Whitmore and J. Noolandi, *Macromolecules*, 1985, **18**, 657.
250. L. Leibler. H. Orland and J. C. Wheeler, *J. Chem. Phys.*, 1983, **79**. 3550.
251. E. Helfand and Y. Tagami, *J. Polym. Sci., Part B*, 1971, **10**, 741.
252. E. Helfand and A. M. Sapse, *J. Chem. Phys.*, 1975, **62**, 1327.
253. A. J. Masters and C. Price, unpublished work.
254. J. W. Cahn and J. E. Hilliard, *J. Chem. Phys.*, 1958, **28**, 258.

7

Hydrodynamic Properties

JEAN-FRANCOIS JOANNY
Université Claude Bernard, Villeurbanne, France
and
S. JEAN CANDAU
Université Louis Pasteur, Strasbourg, France

7.1   INTRODUCTION

Many of the industrial applications of polymers are related to their unusual hydrodynamic and rheological behavior; a common way of increasing the viscosity of a liquid (*e.g.* an oil) is to introduce a polymeric additive. The relaxation times associated with the internal motions of polymers in solution (or in the melt) are much longer than the relaxation times for simple molecules, and can be as long as several seconds. This, in turn, leads to very high viscosities for polymer solutions and very slow motions of the macromolecules. The response of polymer solutions to strong hydrodynamic forces is even more unusual, and characteristic of non-Newtonian viscoelastic fluids. This non-Newtonian behavior, related to the internal elasticity of polymer chains, is the basis of such properties as drag reduction or shear thinning of polymer solutions.

In this chapter we describe the hydrodynamic properties of polymer solutions close to thermodynamic equilibrium, but ignore all the fascinating non-linear effects. We will also limit ourselves to neutral linear flexible macromolecules, although very similar theories have been applied to more complicated systems such as polyelectrolytes, branched polymers and rod-like polymers.

The classical approach to hydrodynamic properties of polymer solutions is described in detail in the books of Ferry,[1] Bird *et al.*,[2] Flory[3] and Yamakawa.[4] The non-draining Zimm description of dilute solutions is quite consistent with experimental results obtained by viscometry, concentration gradient diffusion (CGD) measurements and sedimentation. This classical approach, however, is not equally consistent for semi-dilute solutions.

In recent years, our understanding of hydrodynamic properties has been considerably improved by the development of new experimental techniques, the most important of these being quasi-elastic light scattering[5-6] (QELS), which is described in detail in Volume 1, Chapter 8 of this work. In a QELS experiment one measures the time correlation function of the intensity of light scattered by the solution for a given transfer wavevector $q = [(4\pi n)/\lambda] \sin \theta/2$, where n is the refractive index of the medium, λ is the wavelength of the scattered light and θ is the scattering angle. This intensity correlation function is directly proportional to the time correlation function of the concentration; at low wavevectors it allows investigation of the diffusion constant of the solution; at higher wavevectors it probes the internal motions of polymer chains. QELS has now become one of the easiest ways of characterizing a polymer solution (or any colloidal system) by its diffusion constant (or hydrodynamic radius).

Some more refined techniques have been developed to probe length or time scales which are not accessible by QELS. Amongst these are neutron spin echo (NSE) scattering,[7] which has allowed hydrodynamic measurements at the shorter length scales inaccessible by light-scattering, also pulsed field gradient nuclear magnetic resonance,[8] forced Rayleigh light-scattering[9] and photobleaching fluorescence recovery,[10] which have allowed a detailed measurement of self-diffusive motions of polymer molecules, chemically labelled with fluorescent dyes, in more concentrated solutions and in the melt.

On the theoretical side, the introduction of scaling laws,[11] by analogy with critical phenomena, has provided a more complete description of dilute and semi-dilute polymer solutions. Most physical properties of polymer solutions can, in certain asymptotic limits, be expressed as power laws of a few physical parameters, such as the degree of polymerization N, the concentration c or the shift from the Flory compensation temperature $(T-\theta)/\theta$. The exponents of these power laws are universal, in the sense that they do not depend explicitly on the chemical nature of the polymer and the solvent but on the fact that the polymer–solvent pair belongs to a certain universality class: linear polymers in a good solvent, randomly branched polymers in a θ solvent. Very few exact results are known for these exponents: a very useful, and usually very good, approximation is obtained by using a mean field Flory approximation. A more systematic approach uses the renormalization group theory, which provides expansions of these exponents. The critical exponents are not the only universal quantities in a polymer solution; some combinations of the amplitudes of the power laws are also universal numbers which can be determined both experimentally and theoretically. More generally, physical quantities can be expressed as scaling functions of some reduced variables, these scaling functions becoming power laws in certain asymptotic limits.

The scaling law approach, and renormalization group calculations, have been extensively developed for the static properties of polymer solutions. Many exponents and universal ratios are known very accurately, but the determination of full scaling functions is much harder and one often needs some reexponentiations (non-universal) to get results which are comparable with experiment.

The development of scaling laws for dynamic properties has been pioneered by de Gennes[11] and seems to provide a good description of hydrodynamic properties of dilute and semi-dilute solutions. Systematic dynamic renormalization calculations[13] have been performed for dilute solutions. Such calculations, however, do not seem to be as systematic for semi-dilute solutions, where further approximations are needed. Nevertheless, these approximations appear to yield quantitatively good results.

We shall use the dynamical scaling theory to describe the hydrodynamic properties of polymer solutions, focusing mainly on the expected universal behavior. We use a Flory approximation for the power law behavior, which turns out to be a much easier approach and allows a simple understanding of the important physical features often masked by a heavier formalism. For comparison with experiment we shall sometimes quote more detailed results obtained by renormalization group calculations. We will also discuss briefly the deviations from universal behavior related to crossover effects.

Scaling theory suggests using reduced variables (which are combinations of measurable physical quantities) for the presentation of experimental results. This allows elimination of the non-universal behavior due to the specific chemical nature of the polymer or solvent, which is not usually well known, and leads to universal scaling functions which contain more information than the determination of a single exponent. We shall use such a presentation as often as possible in this chapter.

In Section 7.2 we describe briefly the static scaling laws for polymers both in good and θ solvents. Section 7.3 is devoted to the discussion of the hydrodynamic properties of dilute solutions, which are often used to characterize polymers. The hydrodynamic properties of semi-dilute solutions are divided into two groups; collective properties and single chain properties, which are described in Sections 7.4 and 7.5 respectively.

7.2 A SHORT REMINDER OF SCALING LAWS IN POLYMER STATISTICS

Chain statistics and scaling concepts[11] are described in detail in Volume 2, Chapter 1. We shall summarize here the main results which are useful in understanding the hydrodynamics of polymer solutions. All these results are expressed in terms of scaling laws.

The basic idea is that, in the limit of infinite masses, a single polymer in solution is a self-similar fractal object. It is thus characterized by a fractal dimension d_f which relates the average radius of gyration R_G to the degree of polymerization N[14]

$$R_G \sim N^{1/d_f} \tag{1}$$

The fractal dimension is universal and depends only on the solvent quality:

(i) In a good solvent the polymer chains behave as self-avoiding walks and the fractal dimension is given accurately by the Flory formula $d_f = 5/3$ (for the real case of a three-dimensional polymer).[3] The best estimate for $v = 1/d_f$ is $v = 0.588$.[15] In the following, we will use the Flory approximation for the theoretical scaling results; however, for comparison with experiment, the more precise value will sometimes be needed.

(ii) In a θ solvent the polymer chains behave as random walks; the fractal dimension is $d_f = 2$.

(iii) In a poor solvent the polymer chains are collapsed; the fractal dimension is equal to the space dimension $d_f = 3$.

The other quantity of interest is the concentration correlation function. If the monomer concentration is c, we define the pair correlation function by $g(r) = \langle c(0)c(r)\rangle c^{-1}$. The structure factor $S(q)$ is the Fourier transform of the pair correlation function.

In the limit of low wavevectors ($qR_G \ll 1$), $S(q)$ is given by a Guinier approximation

$$S(q) = N\left(1 - \frac{q^2 R_G^2}{3}\right) \tag{2}$$

In the limit of large wavevectors, the structure of $S(q)$ reflects the fractal character of the polymer chain

$$S(q) \sim \frac{1}{q^{d_f}} \tag{3}$$

or $S(q) \sim q^{-5/3}$ in a good solvent and $S(q) \sim q^{-2}$ in a θ solvent.

The pair correlation function $g(r)$ is obtained by Fourier inversion and we approximate it by

$$g(r) \sim \frac{1}{r^{3-d_f}} e^{-r/R_G} \tag{4}$$

Upon increasing the concentration of a dilute solution, we cross the overlap concentration $c^* = N/[(4\pi R_G^3)/3]$. Above this concentration, in the so-called semi-dilute regime, different chains overlap and form a transient network.

The basic idea here is that, in this semi-dilute regime, chain ends are not important for the local properties, so that these properties can be expressed as scaling laws of the monomer concentration c. In particular, this imposes a scaling structure for the mesh size of the network formed by the chains[16]

$$\xi \sim c^{-3/4} \quad \text{good solvent} \tag{5}$$

$$\xi \sim c^{-1} \quad \theta \text{ solvent} \tag{6}$$

The characteristic length ξ separates two very distinct regimes: (a) at length scales smaller than ξ a given chain does not see its neighbors, and the statistics are those of a single chain, while (b) at length scales larger than ξ, the statistics of the solution are reminiscent of those of a melt, the excluded volume interactions are screened, and the chains are ideal ($d_f = 2$).

A good qualitative picture of a semi-dilute solution is provided by grouping the monomers in each chain into subunits called blobs. Each blob has a size ξ; the number of monomers g in a blob is given by equation (7).

$$g = \xi^{d_f} \quad \text{or} \quad \begin{matrix} g \sim c^{-5/4} & \text{good solvent} \\ g \sim c^{-2} & \theta \text{ solvent} \end{matrix} \tag{7}$$

Each chain is a Gaussian chain of blobs with a radius of gyration

$$R_G \sim N^{1/2} c^{-1/8} \quad \text{good solvent} \tag{8}$$

$$R_G \sim N^{1/2} \quad \theta \text{ solvent} \tag{9}$$

The semi-dilute solution can then be described as a melt of blobs.

At length scales smaller than ξ ($q\xi \gg 1$) the correlations are that of a single chain (equation 3). At larger length scales ($q\xi \ll 1$), ξ can be considered as the correlation length of the polymer solution, the structure factor being given by

$$S(q) \sim \frac{S(0)}{1 + q^2 \xi^2} \tag{10}$$

Finally, the pair correlation function can be approximated in a semi-dilute solution by

$$g(r) \sim \frac{1}{r^{3-d_f}} e^{-r/\xi} \tag{11}$$

7.3 HYDRODYNAMIC PROPERTIES OF DILUTE SOLUTIONS

We shall discuss the hydrodynamic properties of very dilute solutions first, where the motions of different chains can be studied independently. The essential microscopic quantity is the monomer friction ζ, describing the motion of a single monomer in the solvent. The monomer friction coefficient obviously depends critically on the solvent viscosity η_0. If the monomers can be assimilated to spheres with a hydrodynamic radius a, ζ is given by Stokes law ($\zeta = 6\pi\eta_0 a$). The problem is to relate the microscopic parameter to more macroscopic quantities, such as a chain mobility or diffusion constant.

For ideal Gaussian chains this has been done in great detail in two extreme limits; the free-draining limit (Rouse model), ignoring the hydrodynamic interactions, and the non-draining limit (Zimm model). These two models are quickly introduced and serve as a guide to the dynamical scaling theory in a good solvent which is reviewed in the following section. We shall discuss the scaling results from experimental methods such as internal relaxation, diffusion, sedimentation and viscometry.

7.3.1 Hydrodynamic Models for a Gaussian Chain: the Rouse and Zimm Models

7.3.1.1 *Free-draining dynamics of an ideal chain: the Rouse model*

The simplest hydrodynamic model for a polymer chain in solution assumes that the solvent drains freely into the polymer chain.[17,18] In this model, first studied by Rouse, the motions of different monomers are independent, *i.e.* the friction force on a given monomer depends solely on the velocity of that monomer and not on that of the other monomers.

Consider a chain of N monomers whose configuration is described by the position r_i of monomer i. The Rouse equation of motion is a classical Langevin equation

$$\zeta \frac{dr_i}{dt} = f_i + \phi_i(t) \tag{12}$$

The left hand side is the viscous friction force on monomer i. This force is balanced by the other forces f_i applied on monomer i. The force f_i has two contributions, $f_i = f_i^{ext} + f_i^{int}$; f_i^{ext} is the external force and f_i^{int} is the elastic force due to the neighboring monomers. For a Gaussian chain (for the sake of simplicity, the temperature is measured in units such that the Boltzmann constant is $k_B = 1$)

$$f_i^{int} = \frac{3T}{a^2} \frac{d^2 r_i}{di^2} \tag{13}$$

$\phi_i(t)$ is the Langevin random force. It has a Gaussian distribution such that

$$\langle \phi_i(t) \rangle = 0 \text{ and } \langle \phi_i(0) \phi_j(t) \rangle - 6T\zeta \delta(t)\delta_{ij} \tag{14}$$

The Rouse dynamics have been studied in great detail but we shall simply note the main results; details can be found in the books of Bird *et al.*[2] and Ferry.[1]

The equation of motion of the center of mass of the polymer chain, $R_G = \frac{1}{N} \sum_i r_i$, is obtained by summing equation (12) over all monomers

$$N\zeta \frac{dR_G}{dt} = F + \phi(t) \tag{15}$$

The sum of all internal forces vanishes; F is the total external force on the chain. This defines the mobility of the polymer chain $\zeta_{ch} = N\zeta$. The Einstein relation gives the chain diffusion constant

$$D_0 = \frac{T}{N\zeta} \tag{16}$$

The diffusion constant is, in the Rouse limit, inversely proportional to the molecular mass, this being true even when excluded volume effects are incorporated.

The internal motions of the chains can be analyzed in terms of Eigen modes. The characteristic time of the Rouse model is the longest relaxation time

$$T_R = \frac{a^2 \zeta}{3T\pi^2} N^2 = \frac{2\zeta N}{T\pi^2} R_G^2 \tag{17}$$

The last quantity of interest is the shear viscosity η, which can be obtained by taking into account the effect of a macroscopic solvent shear flow in equation (12) and then calculating the average shear stress

$$\eta = \eta_0 + \frac{\pi^2}{6} \frac{c}{N} T_R \tag{18}$$

The intrinsic viscosity is

$$[\eta] = \lim_{c \to 0} \frac{\eta - \eta_0}{c\eta_0} \sim R_G^2 \sim N \tag{19}$$

7.3.1.2 Non-draining dynamics of an ideal chain: the Zimm model

The Rouse free-draining limit is based on a local response of the monomers to external forces and ignores any long range hydrodynamic interactions. However, it is well known that the motion of each monomer creates a backflow velocity field in the solvent[19]

$$w(r) = T(r)v \tag{20}$$

$T(r)$ is the Oseen tensor defined by

$$T(r)v = \frac{3a}{4r}\left[v + \frac{(r \cdot v)r}{r^2} \right] \tag{21}$$

for a monomer idealized as a sphere of radius a.

The backflow field created by a monomer decays very slowly with distance and is the origin of the long range hydrodynamic interactions between monomers; each monomer moves in the flow field created by all the others.

The Langevin equation must be modified in order to take into account the hydrodynamic interactions[20,21]

$$\frac{dr_i}{dt} = \sum_j \zeta_{ij}^{-1} f_j + \frac{1}{\zeta} \phi_i(t) \tag{22}$$

The important difference between the Rouse and Zimm equations of motion comes from the non-locality of the mobility tensor

$$\zeta_{ij}^{-1} = \frac{1}{\zeta}\left[\delta_{ij} + (1 - \delta_{ij}) T(r_i - r_j) \right] \tag{23}$$

δ_{ij} being a Kronecker symbol.

The first term δ_{ij}/ζ is the local Rouse term; the Oseen tensor describes the hydrodynamic interactions. A frequently used approximation, introduced first by Kirkwood and Riseman, averages the Oseen tensor over the configurations of the polymer chain in its equilibrium Gaussian state.[22] This replaces the Oseen tensor by the preaveraged Oseen tensor

$$T_{ij} = \langle T(r_i - r_j) \rangle = \left\langle \frac{a}{|r_i - r_j|} \right\rangle \tag{24}$$

Within this preaveraging approximation, the Zimm equation of motion becomes linear and can be studied by the same methods as the Rouse equation. The main results are summarized here.

Summing the equation of motion over all monomers, we get the equation of motion of the center of mass under an external force F

$$\frac{dR_G}{dt} = \frac{F}{N\zeta}\left[1 + \frac{Na}{R_H} \right] \tag{25}$$

R_H is the hydrodynamic radius of the polymer chain defined as

$$\frac{1}{R_H} = \frac{1}{N^2} \sum_{ij} \left\langle \frac{1}{r_{ij}} \right\rangle \tag{26}$$

(for a Gaussian chain $R_H/R_G = (3\pi^{1/2})/8$).

This leads to a diffusion constant

$$D_0 = \frac{T}{6\pi\eta_0 R_H} \qquad (\zeta = a6\pi\eta_0) \tag{27}$$

The internal motions of the polymer chain can be analyzed in terms of Eigen modes. The Zimm time is the longest relaxation time

$$T_z = \frac{\zeta a^2}{2 T (3\pi)^{3/2}} N^{3/2} = \frac{2^{1/2}}{\pi^{3/2}} \frac{\zeta}{aT} R_G^3 \tag{28}$$

Introducing an external macroscopic shear flow, one can calculate the shear viscosity

$$\eta = \eta_0 + \frac{cT}{N} T_z \zeta\left(\frac{3}{2}\right) \tag{29}$$

$$[\eta] \sim \frac{R_G^3}{N} \sim N^{1/2} \tag{30}$$

where $\zeta(3/2)$ is the Riemann function, $\zeta(3/2) = \sum 1/n^{3/2}$.

The predictions of the Zimm model for a Gaussian chain have a very simple interpretation; the Zimm limit is a non-draining limit, the solvent flow does not penetrate into the polymer chain and, as far as hydrodynamic properties are concerned, a polymer chain can be considered as a hard sphere.[23] The diffusion constant is then given by Stokes law, and the intrinsic viscosity by the Einstein equation for dilute solutions of hard spheres,[19] the only difference being in the numerical factors.

7.3.2 Dynamical Scaling in Dilute Solutions

7.3.2.1 *Relaxation time, dynamic exponent*

In order to construct a scaling theory of polymer dynamics one needs to know, as an input, the variation of the characteristic correlation time T under rescaling of the length unit. This defines a dynamic exponent z such that

$$T \sim R_G^z \tag{31}$$

In the non-draining limit, polymer chains behave as hard spheres and the dynamic exponent is $z = 3$ leading to a variation of the characteristic time with the molecular mass

$$T_z \sim N^{3/d_f} \tag{32}$$

In a good solvent the fractal dimension is $d_f = 5/3$ thus $T_z \sim N^{9/5}$. In a θ solvent $d_f = 2$, T is the Zimm time $T_z \sim N^{3/2}$ (*cf.* equation 28). An obvious generalization of the Rouse model (*cf.* equation 17) leads to a dynamic exponent $z = 2 + d_f$. Clearly, the actual value of z lies in between these extreme limits. In fact, des Cloizeaux has given an explicit proof that $z \geqslant 3$.[24]

Several renormalization group calculations of the dynamic exponent now exist,[25-27] starting from a Langevin equation identical to equation (22), with either a preaveraged or a non-preaveraged Oseen tensor. These calculations provide an expansion of the exponents as a function of the small parameter $\varepsilon = 4 - d$, where d is the space dimension. So far all calculations are consistent with a value $z = d$ ($z = 3$ in three dimensions). We will thus consider here that, whatever the quality of the solvent (good or θ solvent), the dynamic exponent is $z = 3$ and that the characteristic time of a polymer coil is given by a Zimm formula (equations 31 or 32).

7.3.2.2 *Center of mass diffusion*

If the polymer has only one characteristic time T_z, then this time also represents the diffusion time over the characteristic length, *i.e.* the radius of gyration. The diffusion constant D_0 is, therefore, such that

$$R_G^2 = D_0 T_z \tag{33}$$

which gives the scaling structure of the diffusion constant in agreement with the hard sphere picture

$$D_0 \sim \frac{1}{R_G} \sim \frac{1}{N^{1/d_f}} \quad \text{or} \quad \begin{array}{ll} D_0 \sim N^{-3/5} & \text{good solvent} \\ D_0 \sim N^{-1/2} & \theta \text{ solvent} \end{array} \tag{34}$$

The renormalization group theory allows the calculation of not only exponents but also some universal ratios between different physical quantities. We define the hydrodynamic radius by

$$D_0 = \frac{1}{6\pi\eta_0 R_H} \tag{35}$$

The ratio $x_R = R_G/R_H$ is, in the limit of infinite mass, a universal quantity which should not depend on the system (polymer–solvent) but only on the solvent quality.

The value of the universal ratio x_R depends slightly on whether or not the Oseen tensor is preaveraged. For an ideal chain $x_R = 8/(3\pi^{1/2}) = 1.50$. In a good solvent a first order ε expansion gives $x_R = 1.562$,[28, 29] Akcasu and Han propose $x_R = 1.86$.[30]

It should be noted, however, that these asymptotic values are reached only in the limit of infinite masses and that, for real finite masses, crossover effects turn out to be of major importance. These crossover effects have been studied by Weill and des Cloizeaux and by Oono *et al.* and are discussed below (*cf.* Section 7.3.2.6).

The chain friction constant $\zeta_{ch} = 6\pi\eta_0 R_H$ is directly measured in sedimentation experiments;[31] the sedimentation coefficient s_0 is given by

$$s_0 = \frac{N}{\zeta_{ch}}(1 - \rho v_2)m = \frac{N}{6\pi\eta_0 R_H}(1 - \rho v_2)m \tag{36}$$

$$s_0 \sim N^{1 - d_f^{-1}} \quad \text{or} \quad \begin{array}{ll} N^{3/5} & \text{good solvent} \\ N^{1/2} & \theta \text{ solvent} \end{array} \tag{37}$$

where m is the mass of a monomer, v_2 is the polymer partial specific volume and ρ is the density of the solvent.

7.3.2.3 Intrinsic viscosity

In a non-draining limit, the scaling structure of the intrinsic viscosity is given by equation (30)

$$[\eta] \sim R_G^{3 - d_f} \quad \text{or} \quad \begin{array}{ll} [\eta] \sim N^{4/5} & \text{good solvent} \\ [\eta] \sim N^{1/2} & \theta \text{ solvent} \end{array} \tag{38}$$

Several universal numbers can be constructed using the intrinsic viscosity and we consider here two of these universal ratios

$$u_{\eta A} = \frac{A_2 N}{[\eta]} \tag{39}$$

(where A_2 is the second virial coefficient) and

$$u_{\eta f} = \frac{1}{6\pi R_H}(N[\eta])^{1/d} \tag{40}$$

For a θ Gaussian chain $u_{\eta A} = 0$ and $u_{\eta f} = 0.1236$. A first order ε expansion in a good solvent yields $u_{\eta A} = 1.196$ and $u_{\eta f} = 0.1297$.[28] These universal ratios being reached in the asymptotic limit of infinite masses, crossover effects might again, in practice, create deviations from the universal values.

7.3.2.4 Dynamical structure factor

The dynamical structure factor is essentially measured by QELS and NSE. These techniques are presented in detail in Volume 1, Chapters 6 and 8. We will discuss only the initial decay rate (first cumulant) $\Omega(q)$.

It is often convenient to study a wavevector dependent diffusion coefficient, rather than the first moment of the dynamical structure factor

$$D(q) = \frac{\Omega(q)}{q^2} \tag{41}$$

This diffusion constant $D(q)$ can be calculated within the framework of linear response theory by using a Kubo formula[32]

$$D(q) = \frac{1}{3} \frac{\sum\limits_{\alpha=1}^{3} \int_0^{+\infty} dt \int e^{i\mathbf{q}\cdot\mathbf{r}} \langle j_\alpha^{(0,0)} j_\alpha^{(r,t)} \rangle d\mathbf{r}}{\int \langle c(0) c(r) \rangle e^{i\mathbf{q}\cdot\mathbf{r}} d\mathbf{r}} \tag{42}$$

$j_\alpha(\mathbf{r}, t)$ is the α component of the local current at time t. It is related to the local velocity by $\mathbf{j} = c\mathbf{v}$. The denominator is easily calculated, the concentration correlation function $\langle c(0) c(r) \rangle = cg(r)$ being given by equation (4). In order to calculate the numerator, we use the decoupling scheme of the velocity and concentration correlations, as introduced for the dynamics of phase transitions by Kawasaki[33] and Ferrel.[34]

$$\langle c(0)c(r) v_\alpha(0)v_\alpha(r) \rangle = cg(r)\langle v_\alpha(0)v_\alpha(r) \rangle \tag{43}$$

The velocity correlation function is given by the Oseen tensor (equation 21).

$$\int_0^{+\infty} \langle v_\alpha(0,0) v_\alpha(r, t) \rangle dt = \mathbf{T}_{\alpha\alpha}(r) \frac{T}{\zeta} \tag{44}$$

We thus obtain

$$D(q) = \frac{\int e^{i\mathbf{q}\cdot\mathbf{r}} g(r) \dfrac{T}{6\pi\eta_0 r} d\mathbf{r}}{\int g(r) e^{i\mathbf{q}\cdot\mathbf{r}} d\mathbf{r}} \tag{45}$$

At zero wavevector $q = 0$, this formula is equivalent to the Kirkwood–Riseman formula (equation 27); $D(q=0)$ is the center of mass diffusion constant D_0.

At high wavevectors $(qR_G \gg 1)$, we can neglect the exponential decay in the pair correlation function $g(r) \sim 1/(r^{3-d_f})$ and obtain

$$D(q) = \frac{1}{2-d_f} \frac{\sin\dfrac{\pi}{2}(2-d_f)}{\sin\dfrac{\pi}{2}(d_f-1)} \frac{T}{6\pi\eta_0} q \tag{46}$$

The scaling law is the same in both a θ solvent and a good solvent, i.e. $\Omega \sim (q^3 T)/\eta_0$, typical of Zimm-like behavior.[21] As expected, the large q limit of the diffusion constant is independent of the polymer molecular mass as a local property. The prefactor in equation (46) is about 10% higher in a good solvent than in a θ solvent. A more precise study of the full dynamic structure factor along these lines has been made by Akcasu and co-workers using a Zwanzig–Mori generalized hydrodynamic approach.[30,35-38]

When q describes the whole wavevector range, the diffusion constant can be written in a scaling form

$$D(q) = D_0 \phi_D(qR_G) \tag{47}$$

The crossover function ϕ_D is universal and depends only on the solvent quality. It has the following limits: $\phi_D(0) = 1$, $\phi_D(x) \sim x$ when x goes to infinity, in order to be consistent with equation (46). Equivalently we can define a universal function for the initial decay rate $\Omega(q)$

$$\Omega(q) = \frac{T}{\eta_0} q^3 \phi_D'(qR_G) \tag{48}$$

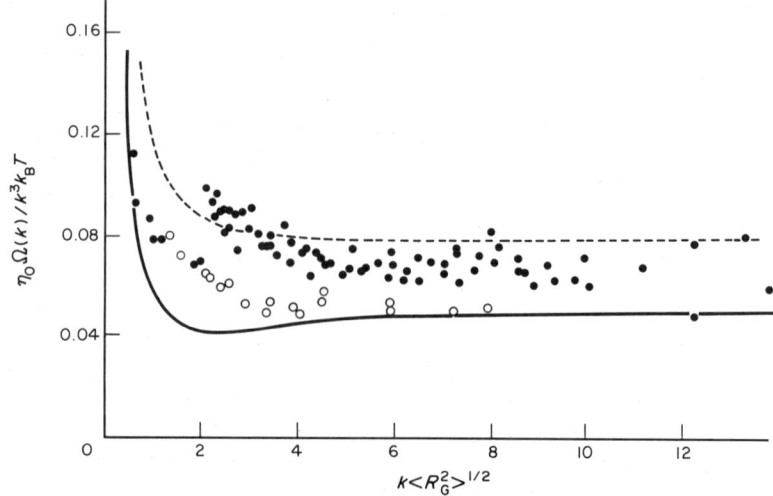

Figure 1   Universal plot of the reduced initial decay rate $[\eta_0\Omega(k)]/[k_B T k^3]$ as a function of the dimensionless wavevector kR_G. The solid curve is the renormalization group calculation of Lee *et al.*,[39] the broken curve is that of Akcasu and Benmouna.[36] Experimental points are: ○ Akcasu and Han (ref. 30) and ● Nemato *et al.* (ref. 39) for PS in toluene and THF (reproduced by permission of the American Institute of Physics from *Phys. Rev. A*, 1984, **30**, 968)

In a good solvent, the function ϕ'_D has been calculated by Lee *et al.*[39] using an ε expansion and by Akcasu *et al.*[37,38] using linear response theory. A comparison of their results, along with some experimental data, is shown in Figure 1.

7.3.2.5   Concentration effects, the virial regime

Up to now, we have only considered the hydrodyamic properties of a single isolated coil, ignoring the effect of other chains. In a dilute solution, the influence of the chain concentration is small and can be treated as a perturbation. The diffusion constant D and the sedimentation coefficient are

$$D = D_0(1 + k_D c + \ldots) \tag{49}$$

$$s^{-1} = s_0^{-1}(1 + k_s c + \ldots) \quad (c \text{ is the monomer concentration}) \tag{50}$$

In the non-draining limit, which is a good approximation for dilute solutions, polymer coils behave as hard spheres. For hard spheres of radius R there are discrepancies between the values of the hydrodynamic interaction coefficients calculated by different authors. A critical comparison of most of the published approaches can be found in refs. 40 and 41. The most commonly assumed values are $k_D = 1.56[(4\pi R^3)/3]$ and $k_s = 6.55[(4\pi R^3)/3]$.

For polymer chains we thus expect scaling laws

$$k_s \sim k_D \sim \frac{R_G^3}{N} \sim R_G^{3-d_f}, \quad k_s \sim \begin{cases} N^{4/5} & \text{good solvent} \\ N^{1/2} & \theta \text{ solvent} \end{cases} \tag{51}$$

It should also be noted that two types of interaction contribute to the value of k_D: hydrodynamic interactions and static excluded volume interactions. In a good solvent both contributions are of the same order of magnitude, so that, as for hard spheres, we expect positive values for k_D and k_s. Moreover, some universal ratios can be constructed from these numbers, such as $k_s/[\eta]$. In a θ solvent, the excluded volume interactions vanish (the second virial coefficient is $A_2 = 0$) and only the hydrodynamic interactions contribute. In this case, k_D is negative.[42] At higher concentrations, in the semi-dilute regime, the diffusion becomes cooperative, the diffusion constant goes through a minimum and then increases as a function of concentration.

As a consequence of the scaling law for the dynamical virial coefficients,[43] we expect universal behavior for s/s_0 and D/D_0 as a function of a variable proportional to c/c^* or $c[\eta]$.

7.3.2.6 Crossover effects between good and θ solvents

The crossover between good solvent and θ solvent is usually described in terms of thermal blobs which are subunits of radius ξ_T.[43] For length scales much less than ξ_T the chain is nearly Gaussian, while for length scales larger than ξ_T it exhibits the statistics of a good solvent. The thermal blob size ξ_T is related to the excluded volume parameter v (the virial coefficient between monomers, proportional to the reduced temperature $(T-\theta)/\theta$

$$\xi_T \sim \frac{1}{v} \tag{52}$$

A polymer chain can thus be viewed as a chain of thermal blobs in a good solvent with a radius

$$R_G \sim \left(\frac{N}{N_c}\right)^{3/5} \xi_T \tag{53}$$

where N_c is the number of monomers per thermal blob: $N_c \sim 1/v^2$.

The effective degree of polymerization, which should be large in the asymptotic limit, is thus N/N_c. This turns out to be particularly important for dynamical properties, such as R_H, that reach their asymptotic behavior at much higher degrees of polymerization than the static quantities. This was studied in detail by des Cloizeaux and Weill, who defined a temperature dependent exponent $v_D(T) = \dfrac{d\log R_H}{dN}$.[44] Close to the Flory temperature θ, $N/N_c < 1$, the chain statistics are Gaussian and $v_D = 0.5$. At higher temperatures excluded volume behavior is exhibited, $v_D = 0.6$. Figure 2 shows the variation of v_D between these two values based on the thermal blob model.

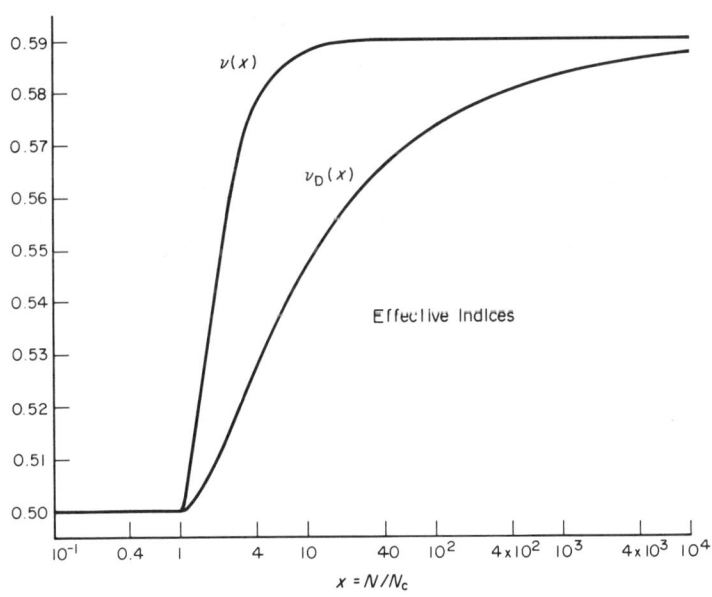

Figure 2 Variation of the effective indices $v(x)$ and $v_D(x)$ with respect to $x = N/N_c$. The asymptotic value is $v = 0.588$ in both cases (reproduced by permission of Commission des Publications Françaises de Physique from *J. Phys. (Orsay, Fr.)*, 1979, **40**, 99)

7.3.3 Experimental Results for Linear Chains in a Good Solvent

7.3.3.1 Scaling behavior

A number of experimental investigations of polymer–good solvent systems by means of hydrodynamic methods have shown weaker power law dependences than predicted by scaling arguments.[44] As mentioned before, higher molecular weights and higher excluded volume parameters are required for hydrodynamic parameters to become asymptotic than are necessary for static parameters.[44] Therefore, most of the data that will be subsequently discussed have been obtained in the crossover regime.

A comprehensive series of experiments has been carried out on toluene solutions of polystyrene (PS), with molecular weights ranging from 130 000 to 40×10^6 at $T = 20\,°C$.[45-53] The log–log plots of D_0, s_0 and $[\eta]$ as a function of the weight average molecular weight $\bar{M}_w$ are represented in Figures 3 and 4. The data can be fitted to straight lines obeying the following equations

$$D_0 (cm^2\,s^{-1}) = 3.64 \times 10^{-4}\, \bar{M}_w^{-0.577} \tag{54}$$

$$s_0 (s^{-1}) = 3.43 \times 10^{-15}\, \bar{M}_w^{0.413} \tag{55}$$

$$[\eta](cm^3\,g^{-1}) = 1.069 \times 10^{-2}\, \bar{M}_w^{0.724} \tag{56}$$

$$k_s (cm^3\,g^{-1}) = 8.08 \times 10^{-2}\, \bar{M}_w^{0.74} \tag{57}$$

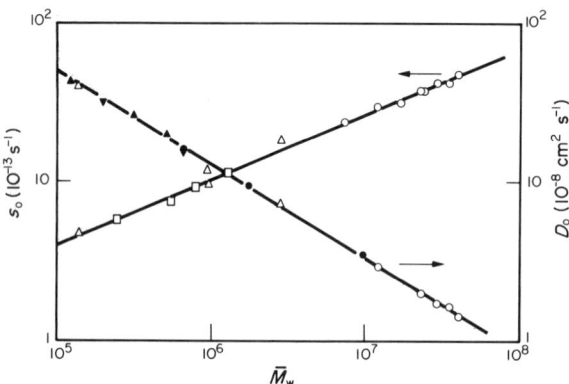

Figure 3 Sedimentation coefficient s_0 and diffusion constant D_0 as a function of molecular weight for polystyrene in toluene at $20\,°C$ (reproduced by permission of the American Chemical Society from *Macromolecules*, 1980, **13**, 657)

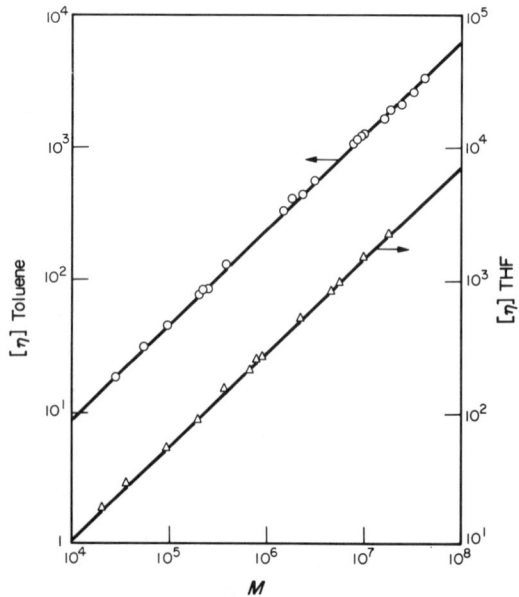

Figure 4 Intrinsic viscosity *vs.* molecular weight M for polystyrene in toluene (○) at $20\,°C$ and THF (△) at $25\,°C$ (reproduced by permission of the American Chemical Society from *Macromolecules*, 1979, **12**, 968)

For the same systems, the static parameters vary with $\bar{M}_w$ according to[46]

$$R_g(\text{nm}) = 1.107 \times 10^{-2} \bar{M}_w^{0.605} \tag{58}$$

$$A_2(\text{cm}^3 \, \text{mol} \, \text{g}^{-2}) = 6.36 \times 10^{-3} \bar{M}_w^{0.225} \tag{59}$$

where A_2 is the second virial coefficient.

Equations (54)–(59) are in good agreement with the scaling predictions which, for $v = 0.588$, are given by $D_0 \sim \bar{M}_w^{-0.588}$, $s_0 \sim \bar{M}_w^{0.412}$, $[\eta] \sim \bar{M}_w^{0.764}$, $k_s \sim \bar{M}_w^{0.764}$, $R_G \sim \bar{M}_w^{0.588}$, $A_2 \sim \bar{M}_w^{-0.236}$.

Furthermore, by combining equations (34), (37) and (38), one obtains the following relationships between the actual hydrodynamic exponents

$$b_{[\eta]} = 3v_D - 1 \tag{60}$$

$$b_s = 1 - v_D \tag{61}$$

where $b_{[\eta]}$ and b_s denote the exponents relative to viscosity and sedimentation experiments respectively. The above relationships have been well-verified experimentally, even for systems showing appreciable differences between theoretical and observed exponents.[3,4,31]

7.3.3.2 Universal ratios

Table 1 gives the values of the dimensionless parameters R_G/R_H, $u_{\eta A}$ and $u_{\eta f}$ obtained for solutions of polystyrenes of high molecular weight in toluene. The weight average molecular weights have been determined either from light-scattering experiments or by combining diffusion and sedimentation techniques and using Svedberg's relation[54]

$$\bar{M}_w = \frac{s_0}{D_0} \frac{RT}{1 - \rho v_2} \tag{62}$$

Table 1 Experimental Values of the Dimensionless Characteristic Parameters for Solutions of Polystyrene of High Molecular Weights in Toluene[d]

$10^6 \bar{M}_w$, LS^a	$10^6 \bar{M}_w$, SD^b	R_G/R_H	$u_{\eta A}$	$u_{\eta f}$
12.3	12.3	2	1.4[c]	0.13[c]
23.4	22.2	1.9	1.7	0.13
29.7	29.2	1.9	1.6	0.13
35.6	29.9	2.3	1.4[c]	0.14[c]
40.2	40.3	2	1.3	0.12

[a] Light scattering. [b] Sedimentation diffusion. [c] Calculated by using the values obtained from Equation (56) for $[\eta]$. [d] Data from refs 45 and 46.

Inspection of Table 1 shows that the experimental values of $u_{\eta f}$ approach closely the theoretical prediction of 0.1297 derived by Oono. It must be stressed that the determination of this parameter combines two hydrodynamic experiments. Similarly, the molecular weights obtained from Svedberg's relation between s_0 and D_0 are in very good agreement with light-scattering data.

On the other hand, the agreement between theory and experiment is not so satisfactory for parameters involving a static and a hydrodynamic experiment. This is the case for both $u_{\eta A}$ and R_G/R_H. With respect to R_G/R_H one must note that the experimental values are much closer to the prediction of Ackasu and Han (1.86) than to that of Oono (1.562). Values of R_G/R_H close to 2 have also been obtained, in the high molecular weight limit, for PS–*trans*-decalin at 40 °C[55] and poly (α-methylstyrene) in toluene.[56] The measured exponent of the power law, as a function of $\bar{M}_w$, is weaker for R_H than for R_G, therefore, as a general rule, R_G/R_H tends to increase with $\bar{M}_w$.

7.3.3.3 Dynamical structure factor

The dynamical structure factor $S(\mathbf{q},t)$ has been derived for various models, but no experimental attempts have been made to analyze the shape of $S(\mathbf{q},t)$ for polymer solutions in a good solvent. The interpretation of dynamic scattering experiments proceeds in terms of the initial slope of $S(q,t)$ or, equivalently, the q dependent diffusion coefficient $D(q)$, which should reduce to D_0 as $q \to 0$. The important feature is that, in the intermediate q range ($qR_G \gg 1$ and $qa \ll 1$), $D(q)$ is independent of the molecular weight of the polymer and approaches a q dependence for a Gaussian chain with hydrodynamic interactions ($D(q) \sim q^2$ if hydrodynamic interactions are not taken into account). The prefactor of the $D(q) \sim q$ power law depends slightly on the model.

Experimentally, $D(q)$ or, equivalently, $\Omega(q)$ can be obtained from a cumulant analysis of the autocorrelation function of scattered light. The q^3 dependence of $\Omega(q)$ in the intermediate q range has been observed for PS in various good solvents: benzene,[57] THF[38] and toluene.[37,38] The experimental results for the PS–toluene and PS–tetrahydrofuran systems are shown and compared with theory in Figure 1. For a detailed comparison between theory and experiment the reader should refer to a recent review by Schaefer and Han.[58]

7.3.4 Experimental Results for Linear Chains in a θ Solvent

7.3.4.1 Scaling behavior

The most complete investigation by means of hydrodynamic methods concerns the PS–cyclohexane system at $T = 35\,°C$. The experimental data fit the following molecular weight dependences:

$$D_0(\text{cm}^2\,\text{s}^{-1}) = 1.3 \times 10^{-4}\,\bar{M}_w^{-0.497} \tag{63}^{[59-61]}$$

$$s_0(\text{s}^{-1}) = 1.46 \times 10^{-15}\,\bar{M}_w^{0.501} \tag{64}^{[46]}$$

$$[\eta](\text{cm}^3\,\text{g}^{-1}) \simeq 8.4 \times 10^{-2}\,\bar{M}_w^{0.5} \tag{65}^{[62-64]}$$

$$k_s(\text{cm}^3\,\text{g}^{-1}) = 5.69 \times 10^{-2}\,\bar{M}_w^{0.5} \tag{66}^{[46]}$$

$$k_D(\text{cm}^3\,\text{g}^{-1}) = 4.1 \times 10^{-2}\,\bar{M}_w^{0.48} \tag{67}^{[59]}$$

$$R_G(\text{nm}) = 3.47 \times 10^{-2}\,\bar{M}_w^{0.5} \tag{68}^{[65]}$$

$$A_2 = 0 \tag{69}^{[66]}$$

These power laws are in excellent agreement with the theoretical predictions: $D_0^{-1} \sim s_0 \sim [\eta] \sim k_s \sim k_D \sim R_G \sim \bar{M}_w^{0.5}$.

In this respect, it must be noted that, in a θ solvent, the conditions to reach the asymptotic limit are much less restrictive than for a good solvent. A Gaussian conformation is ensured if the contour length is about ten times the persistence length.

The predicted scaling behavior of hydrodynamic parameters has been observed for many systems, mainly from sedimentation[67-70] and viscometry[3,71-78] experiments.

7.3.4.2 Characteristic ratios

As all hydrodynamic parameters follow the same power laws of $\bar{M}_w$, within experimental error, it is possible to calculate the characteristic ratios from the set of equations (63)–(69).

Thus for the system PS–cyclohexane at $T = 35\,°C$ one finds $u_{\eta A} = 0$, $u_{\eta f} = 0.121$ and $R_G/R_H = 1.51$. The agreement with theoretical predictions is excellent, so that in θ conditions the dilute polymer solutions exhibit a universal character expressed by the above ratios.

Furthermore, the variations of D/D_0 and s/s_0 with c/c^* can be represented by universal curves, independent of the molecular weight.

In Figure 5 D/D_0 is plotted *vs.* $c\sqrt{\bar{M}_w/m}$ (m is mass of the monomer) for PS–cyclohexane and PS–cyclopentane systems.[79] The minimum observed on this plot can be associated with the crossover between dilute and semi-dilute regimes.[42]

7.3.4.3 Dynamical structure factor

There have been several attempts[80-84] to analyze the shape of $S(q,t)$ from the normal mode development proposed by Pecora.[85-87] Pecora's model gives a normalized first-order correlation function $g^{(1)}(\tau)$ as a series of exponentials that can be approximated as a sum of two terms, for $qR_G \lesssim 2.5$.

$$|g^{(1)}(\tau)| \simeq \exp(-D_0 K^2 \tau)|S_0(q^2 R_G^2) + S_2(q^2 R_G^2)e^{-2\tau/\tau_1} + \ldots| \tag{70}$$

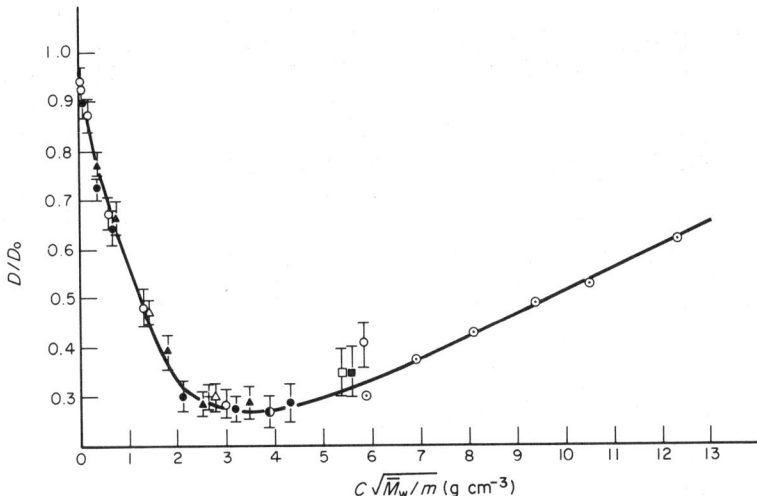

Figure 5 Normalized hydrodynamic diagram of polystyrene solutions at the θ temperature. Cyclohexane at 34.5 °C: $\triangle$, $\bar{M}_w = 20\,000$; $\blacktriangle$, $\bar{M}_w = 150\,000$; $\odot$, $\bar{M}_w = 10^6$ (data obtained by CGD experiments[124]); $+$, $\bar{M}_w = 2 \times 10^6$; $\square$, swollen network. Cyclopentane at 20.4 °C: $\bullet$, $\bar{M}_w = 130\,000$; $\bigcirc$, $\bar{M}_w = 570\,000$; $\mathbb{O}$, $\bar{M}_w = 10^6$; $\blacksquare$, swollen network (reproduced by permission of the American Chemical Society from *Macromolecules*, 1983, **16**, 71)

where τ_1 is the relaxation time of the first normal mode of the Gaussian coil and S_0 and S_2 are the dynamic form factors.[87]

A detailed QELS study was performed by Caroline and Jones[84] on PS–cyclohexane solutions, which led to the following relation for τ_1

$$\tau_1(s) = (7.7 \pm 0.3) \times 10^{-11} \, \bar{M}_w^{-(1.42 \pm 0.05)} \tag{71}$$

This result agrees satisfactorily with the theoretical prediction (equation 28), $\tau_1 \sim M^{1.6}$ corresponding to the non-free-draining case. Furthermore, the values of τ_1 approach closely those calculated from Zimm's theory.

In the high q domain the initial decay rate $\Omega(q)$ varies as q^3, as in the good solvent case, but with different magnitude. Figure 6 shows the variation of the normalized initial decay rate $\Omega(q)/(q^3 T/\eta_0)$ as a function of qR_G for PS solution in cyclohexane.[38]

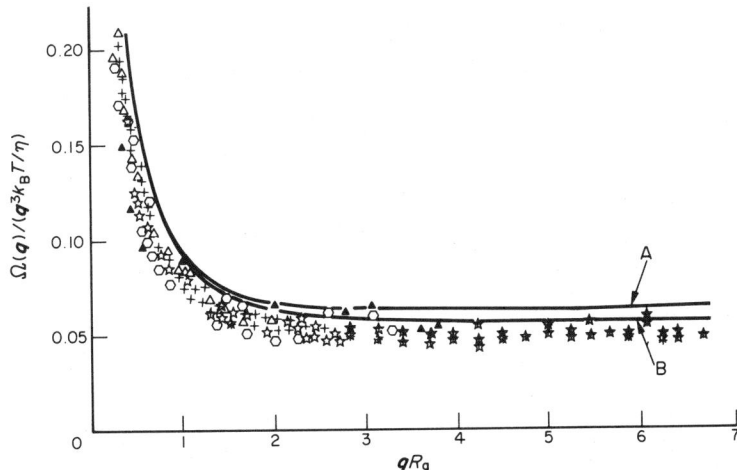

Figure 6 Variation of the reduced initial decay rate $[\Omega(q)\eta_0]/(k_D T q^3)$ as a function of qR_G for P S in cyclohexane at 35 °C. The curves A and B are the theoretical calculations of ref. 38 without and with preaveraging the Oseen tensor (reproduced by permission of the American Chemical Society from *Macromolecules*, 1981, **14**, 1080)

7.3.5   Crossover Between Good and θ Solvent

The θ conditions are characterized by $N/N_C \ll 1$ while good solvent conditions are reached in the limit $N/N_C \gg 1$. The transition from $N/N_C \ll 1$ to $N/N_C \gg 1$ can be achieved by changing the temperature, the quality of the diluent or the molecular weight of the polymer.

7.3.5.1   Scaling behavior

The analysis of the data in terms of power laws leads to actual exponents varying between θ and good solvent values, depending on N/N_C. For instance, the following exponents have been reported: $v_D = 0.55$ and $b_{[\eta]} = 0.69$ for the PS–Carbon tetrachloride system;[88] $v_D = 0.55$ and $b_{[\eta]} = 0.725$ for PS–benzene;[64,89] $v_D = 0.56$ and $b_s = 0.452$ for *n*-butanone;[46,90] $v_D = 0.559$ and $b_{[\eta]} = 0.714$ for PS–tetrahydrofuran.[45,91–93]

Again, it must be noted that the relations (61) and (62) between $b_{[\eta]}$ and v_D or b_s and v_D are generally well satisfied. Furthermore, for a given range of molecular weights, v_D was found to increase with temperature as illustrated by the following results obtained in the PS–cyclohexane system[60]

$T(°C) =$	30	35	40	50	60
v_D $=$	0.49	0.5	0.51	0.523	0.53

7.3.5.2   Characteristic ratios

In a crossover regime the characteristic ratios can no longer be considered as universal. This is illustrated by the variation of the ratio R_G/R_H with temperature for solutions of PS ($M_w = 14.7 \times 10^6$) in *trans*-decalin[55]

$T(°C)$	20	20.5(θ)	22.7	25	30	40
R_G/R_H	1.64	1.67	1.72	1.75	1.85	1.96

Also, the normalized diffusion coefficient D/D_0 cannot be represented by a universal curve as a function of c/c^* if c^* is determined from static measurements. On the other hand, Wiltzius *et al.*[94] have shown that, for PS–*n*-butanone and PS–toluene systems, D/D_0 is a single universal function of the reduced concentration $k_D c$ in both dilute and semi-dilute regimes.

7.3.5.3   Dynamical structure factor

Measurements of τ_1 using Pecora's analysis[85–87] have been performed on different systems using either polarized[80–83] or depolarized light-scattering.[88,95] The experimental values of τ_1 are in the region of the values predicted by the Rouse–Zimm model, but no molecular weight dependences are reported.

In the high q range, as mentioned before, $\Omega(q)$ varies as q^3 whatever N/N_C, the sole change being in the prefactor. Figure 7 shows the increase of the prefactor determined from NSE experiments[96] for PS–cyclohexane solutions when varying the solvent quality from θ to good solvent. This increase in the prefactor value has also been established by Han and Ackasu, using QELS experiments.[38]

7.4   HYDRODYNAMIC PROPERTIES OF SEMI-DILUTE SOLUTIONS: COLLECTIVE MODES

In very dilute solutions we have considered mainly single chain motions, and we have studied dynamic quantities in the limit of zero concentration. When the concentration is non-zero, one must take into account interactions between different chains, both static interactions and hydrodynamic interactions. If the concentration is smaller than the overlap concentration c^*, in the so-called virial regime, these effects are small and can be treated as perturbations. This is not the case, however, in a semi-dilute regime, where the motions become collective and thus renormalize the local monomer friction ζ which becomes concentration dependent.

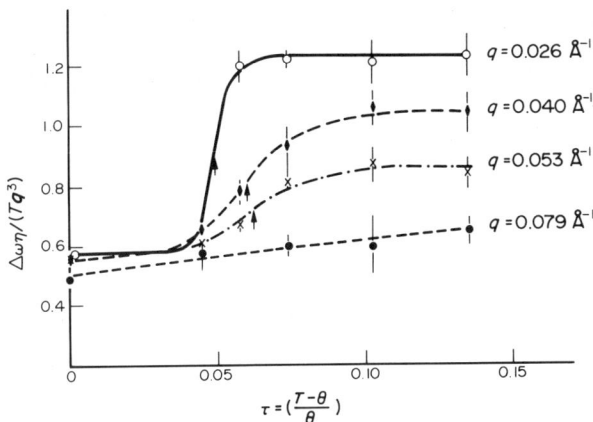

Figure 7 Measured reduced linewidth $\Delta\omega\eta/Tq^3$ *vs.* temperature for PS of molecular weight 6×10^4 in deuterated cyclohexane. The increase in temperature changes the solvent quality from a θ solvent ($\theta = 38\,°C$) to a good solvent (reproduced by permission of the American Institute of Physics from *Phys. Rev. Lett.*, 1980, **45**, 2121)

In the semi-dilute regime, two kinds of dynamical processes should be distinguished: collective processes involving cooperative motions of the solution and single chain processes involving the individual motion of a labeled chain. For instance, two independent diffusion constants can be defined, a cooperative diffusion constant, characterizing the relaxation of concentration fluctuations, and a self-diffusion constant for the motion of a labeled chain.

The dynamic collective modes of semi-dilute solutions both in good and θ solvents are considered here. Single chain motion and the related problem of macroscopic viscosity will be discussed in the next section.

We shall consider first the screening of hydrodynamic interactions, which turns out to be of major importance in a semi-dilute solution. We will then introduce the scaling laws of collective modes in semi-dilute solutions, emphasizing here the role of universal behavior as a function of the concentration variable. The experimental results are then discussed within the framework of this universal behavior.

7.4.1 Screening of Hydrodynamic Interactions

The main results for the dynamics of dilute solutions reflect the importance of hydrodynamic interactions: each moving monomer in the solvent creates a backflow field which decays very slowly with distance. In a semi-dilute solution, the interference between all these velocity fields induces a screening of the backflow field of a given monomer, which falls off exponentially after a characteristic distance λ. This idea was originally proposed by Edwards and Freed;[97,98] we shall briefly summarize their theory for ideal Gaussian chains.

Edwards and Freed start from two coupled equations for the solvent velocity $\boldsymbol{u}$ and the position $\boldsymbol{r}_{i\alpha}$ of monomer i of chain α

$$\rho_0 \frac{\partial \boldsymbol{u}}{\partial t} - \eta_0 \nabla^2 \boldsymbol{u} + \nabla P = \boldsymbol{F}_{\text{ext}}(\boldsymbol{r},t) + \sum_{\alpha i} \delta(\boldsymbol{r} - \boldsymbol{r}_{i\alpha}) \sigma_A^{i\alpha} \tag{72}$$

$$f_{i\alpha}^{\text{int}} = -\sigma_A^{i\alpha} + \boldsymbol{\phi}_{i\alpha}(t) \tag{73}$$

The solvent hydrodynamic equation is the Navier–Stokes equation; $\sigma_A^{i\alpha}$ is the friction force exerted by the monomer i on the solvent, P is the pressure and $\boldsymbol{F}_{\text{ext}}$ is the external force driving the solvent flow.

The second equation is very similar to the Rouse and Zimm equation of motion (Equations 12 and 22). The form of the friction force $\sigma_A^{i\alpha}$ between monomers and solvent is not specified here, it will be determined by the use of the no-slip Stokes condition

$$\boldsymbol{u}(\boldsymbol{r}_{i\alpha},t) = \frac{\mathrm{d}\boldsymbol{r}_{i\alpha}}{\mathrm{d}t} \tag{74}$$

$\phi_{i\alpha}$ is the Langevin random force and $f_{i\alpha}^{\text{int}}$ is the internal force due to neighboring monomers, given by equation (11).

Rather than the microscopic quantities such as u and $r_{i\alpha}$ we want to study more macroscopic quantities: the macroscopic velocity field v is obtained after averaging over all the chains' configurations. A tedious calculation, using a diagrammatic expansion in terms of the Green function of equation (72), leads to the macroscopic equation of motion

$$\rho_0 \frac{dv}{dt} - \eta_0 \nabla^2 v + \nabla P - \int dr' \int dt' \, \Sigma(r-r', t-t') v(r', t') = F_{\text{ext}}(r, t) \tag{75}$$

The tensor Σ accounts for the contribution of all chains to the stress tensor of the polymer solution. When $\Sigma = 0$ in the absence of polymer chains, the velocity v is related to the external force by

$$v(r) = \frac{1}{\zeta} \int T(r-r') F_{\text{ext}}(r') dr' \tag{76}$$

where T is the Oseen tensor defined in equation (21) and ζ the monomer friction $\zeta = 6\pi\eta_0 a$.

In the presence of polymer chains

$$v_\alpha(r) = \frac{1}{\zeta} \int G_{\alpha\beta}(r-r') F_{\text{ext}\beta}(r') dr' \tag{77}$$

where

$$G_{\alpha\beta}(r, r') = 6\pi a \int \frac{dk}{(2\pi)^3} \frac{1 - k_\alpha k_\beta / k^2}{k^2 - \Sigma(k)} e^{ikr} \tag{78}$$

Equation (78) describes a screening of the Oseen tensor. In the presence of the polymer chains the Oseen tensor $T(r)$ should be replaced by a screened tensor $G(r)$, which for small distances is approximately equal to

$$G(r) \simeq e^{-r/\lambda} T(r) \tag{79}$$

An explicit value of the screening length λ (or equivalently the tensor $\Sigma(r)$) was first given by Freed and Edwards, who derived a scaling law $\lambda \sim 1/c$ for Gaussian polymer chains (chains in a θ solvent).[97,98]

A description equivalent to equation (75) introduces a phenomenological non-local friction coefficient $\zeta(k)$ (in Fourier space) between monomer and solvent. This non-local friction coefficient is such that $-\Sigma(k) = c\zeta(k)$. In the limit of large wavevectors, $\Sigma(k)$, and thus $\zeta(k)$, reaches a finite value ζ, which represents the effective friction between a given monomer and the solvent.

If we consider motions over small characteristic distances (corresponding to large wavevectors) we can rewrite the equation of motion in a steady state with a local friction term

$$-\eta_0 \nabla^2 v + \nabla P + c\zeta v = F_{\text{ext}} \tag{80}$$

This equation relates the screening length of the hydrodynamic interactions λ to the effective monomer friction ζ

$$\lambda = \left(\frac{\eta_0}{c\zeta}\right)^{1/2} \tag{81}$$

Starting from this relation, De Gennes was the first to give a self-consistent argument showing that the hydrodynamic screening length λ depends only on the monomer concentration c and has the same scaling structure as the static correlation length;[99,100] $\lambda \sim c^{-3/4}$ (good solvent) and $\lambda \sim c^{-1}$ (θ solvent).[97,98]

A more direct calculation has been proposed by Muthukumar and Edwards, including excluded volume effects directly and using mode coupling arguments in equation (72), and leads to the same scaling laws for λ.[101,102]

Another important implication of the screening of the hydrodynamic interactions is that the effective friction of the monomers on the solvent is concentration dependent in a semi-dilute solution. The finite concentration of the surrounding monomers renormalizes the friction constant of the monomers, according to equation (81).

$$\zeta \sim \frac{6\pi\eta_0}{c\lambda^2} \sim c^{1/2} \quad \text{good solvent}$$

$$\sim c \quad \theta \text{ solvent} \tag{82}$$

7.4.2 Dynamical Scaling Theory in a Good Solvent

The proportionality between the hydrodynamic scaling length λ and the correlation length ξ means that, from a scaling point of view, there is only one characteristic length in a semi-dilute polymer solution, even if we consider hydrodynamic phenomena. This suggests that, as for dilute solutions, a dynamical scaling theory is possible. We therefore postulate that the existence of a single characteristic length implies a single characteristic time T_ξ. The scaling structure of this unit time is imposed by the value of the characteristic Zimm time in a dilute solution

$$T_\xi \sim \xi^z \sim \xi^3 \tag{83}$$

where ξ is the correlation length of the semi-dilute solution. We shall discuss the scaling predictions for the dynamical structure factor and sedimentation experiments in a semi-dilute solution in the following sections.

7.4.2.1 Dynamical structure factor[103]

This discussion will be limited to the behavior of the initial decay rate $\Omega(q)$ or the cooperative diffusion constant $D(q) = \Omega(q)q^{-2}$

The starting point is the Kubo equation with a Kawasaki–Ferrell decoupling approximation, as discussed earlier for dilute solutions[100–107]

$$D(q) = \frac{1}{3} \sum_{\alpha=1}^{3} \frac{\int_0^{+\infty} dt \int e^{iqr} cg(r) \langle v_\alpha(0,0)v_\alpha(r,t) \rangle dr}{\int cg(r) dr} \tag{84}$$

In a semi-dilute solution the pair correlation function is given by equation (11) and the velocity correlation function is given by the screened Oseen tensor

$$\int_0^{+\infty} dt \langle v_\alpha(0,0)v_\alpha(r,t) \rangle = \frac{T}{\zeta} \mathbf{G}_{\alpha\alpha}(r) \tag{85}$$

We shall discuss the long wavelength behavior ($q = 0$) of the diffusion constant first, then the short wavelength behavior ($q\xi \gg 1$).

7.4.2.2 Cooperative diffusion constant

The cooperative diffusion constant $D_{coop} = D(q=0)$ is obtained in the limit of zero wavevectors

$$D_{coop} \sim \frac{T}{\eta_0 \xi} \sim c^{3/4} \tag{86}$$

It scales as the inverse correlation length; we can then define a hydrodynamic correlation length ξ_H by

$$D_{coop} = \frac{T}{6\pi\eta_0\xi_H} \tag{87}$$

The scaling law, equation (86), suggests that the ratio $x_\xi = \xi/\xi_H$ is a universal number. From their mode coupling theory, Muthukumar and Edwards propose $x_\xi = 9/32$.[101]

Both the scaling law and the universal value of the ratio x_ξ are reached in the asymptotic limit. This notion of asymptotic limit is well defined in a dilute solution, where it corresponds to polymers of infinite masses (or more precisely to a number of thermal blobs in a chain $N/N_c \to +\infty$). It is, however, rather hard to achieve in practice for semi-dilute solutions, since it requires infinite masses (the ratio of the concentration c to the overlap concentration c^* should be infinite) and then a limit of zero concentration.

One type of correction to the scaling law, which exists even for very large polymer masses, is related to the fact that, in real experiments, the concentration of a semi-dilute solution is always finite (at least it is larger than the overlap concentration c^*) or, equivalently, that the number of monomers in a blob is not large enough to show excluded volume behavior. These corrections have received very little theoretical consideration. They should depend only on the monomer concentration and lead to slightly different observed scaling laws for ξ and ξ_H, similar to the different scaling laws observed experimentally for R_G and R_H in a dilute solution.

A second type of correction is related to the fact that, in real systems, the ratio c/c^* always has a finite value. These corrections thus describe the crossover from dilute to semi-dilute solutions and should depend on the variable c/c^*. We thus expect a scaling law

$$\frac{D_{coop}(c)}{D_0} = \frac{R_H}{\xi_H} = f_D\left(\frac{c}{c^*}\right) \tag{88}$$

D_0 is the diffusion constant in a dilute solution (given by equation 3) and c^* the overlap concentration. The scaling function f_D has the following limits imposed by asymptotic behavior in dilute and semi-dilute solutions

$$f_D(0) = 1 \quad \lim_{(c/c*) \to +\infty} f_D = (c/c^*)^{3/4} \tag{89}$$

In a semi-dilute solution there is, therefore, little hope of observing universal asymptotic behavior for the cooperative diffusion constant (as for any other dynamic quantity), the universal behavior being masked by crossover effects. As for dilute solutions, these crossover effects can be reduced if one uses, as a concentration unit, not the overlap concentration c^*, which is a static quantity, but a dynamic quantity with the same (theoretical) scaling behavior, such as the intrinsic viscosity $[\eta]^{-1}$ or the virial for diffusion k_D^{-1}.

Oono and Baldwin[104, 105] and Wiltzius *et al.*[94] suggest writing

$$\frac{D_{coop}(c)}{D_0} = \frac{R_H}{\xi_H} = f_D'(X) \tag{90}$$

where $X = ck_D\left[\exp + \dfrac{\varepsilon}{4}\right]$ up to first order in $\varepsilon = 4 - d$, d being the space dimension. Their first-order in ε expansion for the crossover function f_D' is, in the good solvent limit,

$$f_D'(X) = (1+X)^{-3/8\varepsilon}\left\{1 + \left[\left(1 + \frac{\varepsilon}{8}\right)X + \frac{\varepsilon}{4}\left(\ln\frac{1+X}{X} - 1\right)\right]\right\} \exp\frac{\varepsilon}{4}\left[\frac{1}{X} - \left(\frac{1}{X^2} - 1\right)\ln(1+X)\right] \tag{91}$$

7.4.2.3 Collective modes in the short wavelength limit ($q\xi \gg 1$)

In the limit of large wavevectors, the length scale probed is smaller than the blob size ξ and the portion of chain of interest exhibits individual chain behavior. In particular, the initial decay rate of the dynamical structure factor has the same value as in a dilute solution

$$\Omega(q) \sim \frac{T}{\eta_0}q^3 \tag{92}$$

as discussed in Section 7.3.2.4.

7.4.2.4 Sedimentation coefficient

The scaling result for the cooperative diffusion constant finds a simple interpretation in terms of blobs. The hydrodynamic screening length λ is of the order of the blob size ξ. As far as hydrodynamic properties are concerned, blobs behave as hard spheres (impenetrable by the flow field) and the cooperative diffusion constant is the Stokes–Einstein diffusion constant of a blob.

In an external flow field, the blobs behave independently, the friction constant of a blob being given by Stokes law

$$\zeta_b = 6\pi\eta_0\xi \tag{93}$$

The effective monomer friction constant is obtained by dividing the blob friction ζ_b by the number of monomers per blob $g = c\xi^3$

$$\zeta = \frac{6\pi\eta_0\xi}{g} = \frac{6\pi\eta_0}{c\xi^2} \tag{94}$$

a result equivalent to equation (82).

In a sedimentation experiment, one measures the sedimentation constant directly proportional to ζ^{-1}

$$s = (1 - \rho v_2)m\zeta^{-1} \tag{95}$$

As described earlier, m is the monomer mass, ρ the solvent density and v_2 the polymer specific volume.

A fluctuation dissipation theorem directly relates the sedimentation coefficient to the cooperative diffusion constant.[107]

$$\frac{s}{(1 - \rho v_2)m} = \zeta^{-1} = \frac{D_{coop}(c)}{\dfrac{\partial\Pi}{\partial c}} \tag{96}$$

Π being the osmotic pressure of the solution.

When increasing the concentration from a dilute solution to a semi-dilute solution, we expect scaling behavior for the sedimentation constant

$$\frac{s(c)}{s_0} = f_s(X) \tag{97}$$

$X \sim c/c^*$ being the same variable as in equation (90) and s_0 the sedimentation constant of a dilute solution.

The universal crossover function has the limiting behavior

$$f(0) = 1, \qquad \lim_{X \to +\infty} f_s(X) \sim X^{-1/2} \tag{98}$$

7.4.3 Dynamical Scaling Theory in a θ Solvent

Up to this point, we have treated in parallel the hydrodynamic properties of good and θ solvents, the only difference being the fractal dimension of the polymer in each type of solvent. However, while the scaling dynamic theory is well supported for polymer chains in a good solvent, this is not the case for a θ solvent, where, quite surprisingly, dynamical problems look more difficult to understand.

A first hint of this complication is found in dilute solutions. The theoretical models (Zimm or Rouse) that we introduced both consider phantom chains, they ignore the topological constraints due to the connections between polymer chains, and thus the effects of entanglements; this turns out to be a reasonable assumption if the number of entanglements or knots is small and is actually the case for an isolated polymer in a good solvent. However, in a Gaussian chain of N monomers (in a θ solvent), the number of knots scales as $N^{1/2}$: in a θ solvent, polymer chains are self-entangled.

Consideration of these knots is a formidable problem which has not been completely solved. It seems, however, that the scaling laws observed are not affected by entanglements which could play a role in the determination of some universal amplitudes.

In a semi-dilute solution in a good solvent, only entanglements between different chains exist; this is not the case in a θ solvent where both interchain entanglements and self-entanglements are present. This leads to the existence of two different relaxation modes and thus two different diffusion coefficients. In this section, we briefly discuss these two modes in a θ solvent and the crossover effects between good and θ solvents.

7.4.3.1   Isothermal diffusion coefficient, gel diffusion coefficient[108]

We have studied the hydrodynamic properties of polymer solutions by using microscopic equations as a starting point. An alternative way is to use a more macroscopic approach, very similar to the two fluid models for viscoelastic gels proposed by Tanaka, Hocker and Benedek.[109] In a semi-dilute solution, the polymer forms a transient network of elastic modulus E. The displacement of each point of the network is $u(r)$. The solvent is a continuum described by a velocity field $v(r)$.

The equation of motion of the network results from a balance between the friction force $c\zeta[(\mathrm{d}u/\mathrm{d}t) - v]$ and the elastic restoring force $E\nabla^2 u$

$$c\zeta\left(\frac{\mathrm{d}u}{\mathrm{d}t} - v\right) = E\nabla^2 u \tag{99}$$

Equation (99) yields the relaxation time of a perturbation u_q of wavevector q, $\tau_q = 1/Dq^2$, where the diffusion constant D is given by

$$D = \frac{E}{c\zeta} \tag{100}$$

Two different contributions should be distinguished in the elastic modulus E: the osmotic modulus and the shear modulus. In a good solvent these two moduli have the same scaling behavior (although the numerical factors can be very different). The modulus E scales as the osmotic pressure $E \sim T/\xi^3$ giving back a diffusion constant

$$D = \frac{T}{6\pi\eta_0\xi} \tag{101}$$

In a θ solvent, however, the two moduli have a different scaling structure which leads to two relaxation modes with very different relaxation times.[110] The osmotic modulus scales as the osmotic pressure $\sim T/\xi^3$ leading to a diffusion constant given by equation (101), *i.e* in a θ solvent $D \sim c$. The variation of this diffusion constant with concentration can be written as a universal function of the scaling variable $c/c^* \sim cN^{1/2}$, as for the case of a good solvent

$$\frac{D(c)}{D_0} = f_{D\theta}\left(\frac{c}{c^*}\right) \tag{102}$$

The sedimentation coefficient $s(c)$ is related to D by equation (96) and can thus also be written as a universal function of c/c^*. The gel shear modulus is related to the density of entanglements or to the number of two-body contacts in the solution ($\sim c^2$). The theory of rubber elasticity predicts, therefore, a shear modulus $E_{gel} \sim Tc^2$ leading to a gel diffusion coefficient independent of concentration[111]

$$D_{gel} \sim \frac{T}{6\pi\eta_0 a} \tag{103}$$

The entanglements relax with a characteristic time T_{rep}. At short wavevectors the relaxation is possible and the gel modulus does not contribute to the diffusion constant. At larger wavevectors a rubbery behavior[84] could be observed. A detailed comparison of the amplitude of these two modes, taking into account the relaxation phenomena, has been made by Brochard and De Gennes.[108,110]

7.4.3.2   Crossover between good and θ solvent

The crossover between good and θ solvents has been described for the static properties in terms of thermal blobs, as for dilute solutions.

For a semi-dilute solution in a good solvent, the correlation length ξ (the blob size) is much larger than the thermal blob size ξ_T. The statistics are Gaussian at length scales smaller than ξ_T and excluded-volume-like at length scales larger than ξ_T (and smaller than ξ, with ξ scaling as $c^{-3/4}$). In a θ solvent, the chain is Gaussian at all length scales and the correlation length ξ scales as c^{-1}.

We thus have two ways to induce a crossover from good solvent to θ solvent behavior: (i) increase the concentration, and thus decrease ξ until it becomes smaller than ξ_T; or (ii) decrease the temperature, and thus increase ξ_T until it becomes larger than ξ. The crossover region where the two length scales ξ_T and ξ are approximately equal is sometimes called a 'marginal solvent' regime.

If we omit the complication introduced by the gel modes, the collective diffusion constant might be calculated using a Kawasaki–Ferrell approximation (see equations 42 and 84), which always produces a hydrodynamic correlation length ξ_H proportional to the static correlation length ξ. Marginal solvents and the crossover between good and θ solvents have been reported in detail in ref. 112 and in the review of Schaefer and Han.[58]

7.4.4 Experimental Results for Semi-dilute Solutions in a Good Solvent

7.4.4.1 Cooperative diffusion

A straightforward method of studying the collective diffusion in semi-dilute solutions is to use the CGD technique. However, the cooperative diffusion coefficient D_{coop} is more usually determined by QELS from the characteristic frequency $\Omega(q) = D_{coop}\, q^2$.

Conflicting results and/or interpretations have been reported for QELS experiments; the major point of the controversy concerns the shape of the autocorrelation function of scattered light. While some authors fit the experimental autocorrelation function to a single exponential,[57,94,113–115] others claim that two modes, a 'fast' and a 'slow' mode, are present.[116–120] In the latter case, it is found that, as the concentration increases, the correlation function is increasingly dominated by the slow relaxation process when $qR_G \ll 1$. This slow mode was tentatively attributed to the self-diffusion of the chains, as the concentration dependence of the relaxation time was consistent with the reptation model.[117] However, the slow mode is between one and two orders of magnitude slower than self-diffusion as measured by pulsed field gradient NMR experiments.[118,119] Furthermore, ageing[121] and heating[122] of samples reduces the amplitude of the slow mode. Finally, one must keep in mind that stray scattering from dust would affect the measurements; dust-free samples are very difficult to prepare when working with semi-dilute solutions of high molecular weight polymers. Although the debate is not yet settled, there is a trend among the experimentalists to agree that, for solutions in very good solvents and at polymer volume fractions less than 0.1, the autocorrelation function is fitted by a single exponential.

The above considerations partly explain why conflicting values of D_{coop} were obtained from the same systems. Also, the conditions for reaching the asymptotic limit, discussed in Section 7.4.2.2., are more critical for semi-dilute solutions than for dilute solutions, so that the apparent exponents obtained by fitting power laws to experimental data depend on the range of concentrations investigated and on the quality of the diluent. The correct value of the exponent can be reached only at very low concentrations, which means that ultrahigh molecular weight polymers should be used

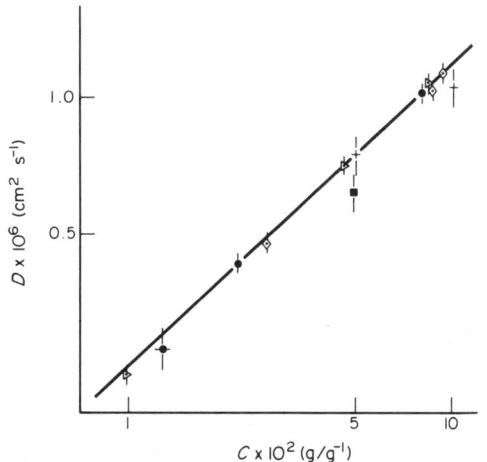

Figure 8 Diffusion coefficient D of semi-dilute PS–benzene solutions as a function of concentration. Results obtained from QELS experiments: $\diamondsuit$, $\bar{M}_w = 1.27 \times 10^6$; $\bullet$, $\bar{M}_w = 3.8 \times 10^6$; $\rhd$, $\bar{M}_w = 8.4 \times 10^6$; results obtained from CGD experiments: $+$, $\bar{M}_w = 3.2 \times 10^5$; $\blacklozenge$, $\bar{M}_w = 5.8 \times 10^6$; $\bullet$, $\bar{M}_w = 3.2 \times 10^5$. The straight line is the best fit obtained from QELS results (reproduced by permission of Springer from *Adv. Polym. Sci.*, 1982, **44**, 27)

The diffusion coefficient has been measured in semi-dilute solutions of PS in benzene by Adam *et al.*[123] who used both QELS and CGD experiments. The two methods give the same values for the diffusion coefficient, within experimental error, as shown in Figure 8. It must be noted that the QELS and CGD techniques, respectively, probe time scales much shorter and much longer than T_{rep}. The above results suggest that entanglements that are effective only at time scales shorter than T_{rep} do not play a major role in the cooperative diffusion process. Furthermore, the diffusion coefficient does not depend on the molecular weight of the polymer, and scales with the concentration as $c^{0.67 \pm 0.06}$. These results are similar to those obtained by Munch *et al.*[113] from QELS experiments performed on the same system, PS–benzene, and the results of Roots *et al.*,[124] who found for PS–toluene, by means of CGD experiments, that $D \sim {}^{0.70 \pm 0.01}$. A review of the experiments performed in various systems gives exponents for the power law fits to the data varying from 0.5 to 0.75;[58] that is, smaller than the static exponent.

Figure 9 is a direct comparison of the static and dynamic length scales for PS solutions in toluene and 2-butanone.[94] It shows that the ratio ξ_H/ξ is independent of M_w but increases with c, thus illustrating the difference between static and effective dynamic exponents.

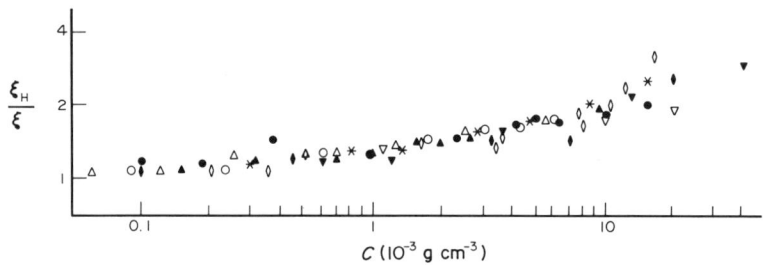

Figure 9 Ratio of the dynamic to the static length scale as a function of concentration for PS solutions of various molecular weights in toluene (open symbols) and 2-butanone (closed symbols and stars) (reproduced by permission of the American Institute of Physics from *Phys. Rev. Lett.*, 1984, **53**, 834)

7.4.4.2 Sedimentation

Sedimentation experiments also provide evidence of non-asymptotic behavior for semi-dilute solutions in a good solvent. Power law fits to the data provide exponents varying from 0.59 to 0.82,[125] while the theoretical scaling exponent is 0.54. These exponents are consistent with those obtained for D_{coop}. A detailed review on the cooperative diffusion and sedimentation results has been recently published by Nyström and Roots.[125]

7.4.4.3 Dynamical structure factor

The q dependence of the characteristic frequency $\Omega(q)$ in semi-dilute solutions has been measured both by QELS and by NSE methods.[115,126] Figure 10 shows the results obtained by NSE for a poly(dimethylsiloxane) (PDMS) of molecular weight $\bar{M}_w = 60\,000$ in deuterated benzene at 70 °C.[126]

In the dilute range, the $\Omega \sim q^3$ behavior is observed over the entire q range of the experiment. With increasing concentration, a crossover from $\Omega \sim q^3$ to $\Omega \sim q^2$ occurs at increasing q values. The data provide a direct experimental observation of the dynamic spatial crossover between Zimm-like relaxation at small dimensions (identical for dilute or semi-dilute regimes) and cooperative diffusion at larger dimensions in space. A similar crossover has been observed over a smaller q range by QELS for PS solutions in toluene and 2-butanone.[115]

7.4.5 Experimental Results for Semi-dilute Solutions in a θ Solvent

7.4.5.1 Light-scattering experiments

The results of QELS experiments in semi-dilute solutions in θ solvents are rather complicated. Three reasons can be invoked to explain the discordant conclusions reached by different experimentalists: (a) the difficulty of preparing clean homogeneous solutions; (b) the data analysis;

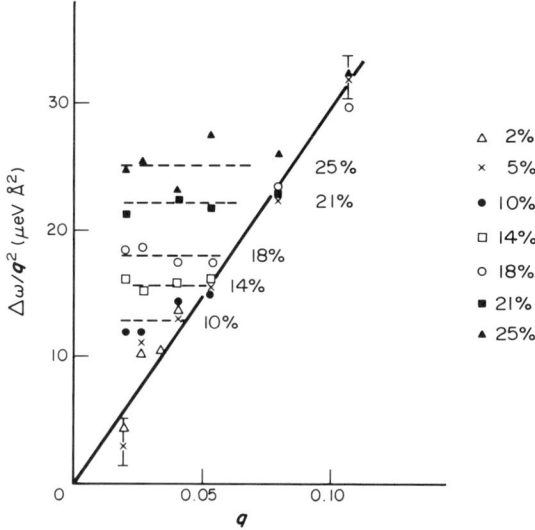

Figure 10 Concentration crossover in PDMS/d-benzene. Quasielastic line widths Ω/q^2 for different concentrations at 70 °C as a function of momentum transfer q. (reproduced by permission of Plenum Press from 'Scaling Phenomena in Disordered System', ed. R. Pynn and A. Skjeltorp, 1985)

and (c) the fact that the dynamic regime probed by QELS depends on the experimental conditions $(q, c, \bar{M}_w)$.

Tables 2 and 3 summarize the main conclusions obtained by several groups in the two limiting cases where the correlation time τ of collective fluctuations is, respectively, much longer and much shorter than T_{rep}.[127–131] We consider separately studies performed in the regime $\tau \simeq T_{rep}$.

The reptation time T_{rep} is a strongly increasing function of $\bar{M}_w$ and c (*cf.* Sections 7.5.1.2 and 7.5.2.2). Therefore the condition $\tau \gg T_{rep}$, corresponding to liquid-like behavior, is reached for relatively low molecular weight, low concentration and/or low q since τ^{-1} is expected to vary as q^2. Values of T_{rep} measured by Adam and Delsanti[132] for the PS–cyclohexane system (Section 7.5.2.2) can be used as references to determine the dynamical regime corresponding to experiment.

Inspection of Table 2 shows that in the limit $\tau \gg T_{rep}$ the autocorrelation function can be described by a single exponential with $\tau^{-1} \propto q^2$. There is a slight disagreement between authors concerning the concentration dependence of τ^{-1} but, as $(\tau q^2)^{-1}$ is reasonably close to D_{CGD}, one can infer from the results that it is the liquid-like behavior of the solutions that is actually probed in these experiments. Furthermore, Adam and Delsanti compared the concentration dependences of the diffusion coefficient $D = (\tau q^2)^{-1}$ obtained for the PS–cyclohexane system by QELS and CGD technique on the one hand and by combining intensity light-scattering and sedimentation experiments on the other

Table 2 Results of QELS Studies in Semi-dilute Solutions of PS in θ Solvents, $\tau \gg T_{rep}$

Solvent	Samples	Analysis	Shape of the auto-correlation function	Concentration and wavevector dependence	Ref.
cyclohexane	$\bar{M}_w = 4.22 \times 10^5$	Force fitting of a single exponential at different sampling time	Single exponential	$\tau^{-1} \propto q^2 c$	127
$= 34.5 °C$	$c < 7.37 \times 10^{-2} \, \text{g cm}^{-3}$	Heterodyne detection		$(\tau q^2)^{-1} \simeq D_{CGD}$ (within 10%)	
cyclopentane $= 20.4 °C$	$\bar{M}_w = 9.3 \times 10^4$ $c < 27 \times 10^{-2} \, \text{g g}^{-1}$	Bimodal analysis Homodyne detection	~ Single exponential	$\tau^{-1} \propto q^2 \, f(c)$ $f(c)$ increases weakly with c, $(\tau q^2)^{-1}$ $\sim 0.7 \, D_{CGD}$	128

Table 3 Results of QELS Studies in Semi-dilute Solutions of PS in θ Solvents, $\tau \ll T_{rep}$

Solvent	Samples	Analysis	Shape of the auto-correlation function	Concentration and molecular weight dependences	Ref.
Cyclohexane	$\bar{M}_w = 3\text{–}20.6 \times 10^6$ $c < 11 \times 10^{-2}$ g cm^{-3}	Force fitting of a single exponential at different sampling time Heterodyne detection	Multiexponential approximated by a bimodal shape	$\tau_{fast}^{-1} \propto q^2(a + b'c)\bar{M}_w^0$ $\tau_{slow}^{-1} \simeq T^{-1}$ $\propto q^0\, c^{-3.07}\, \bar{M}_w^{-3.8}$	127
Cyclopentane	$\bar{M}_w = 8 \times 10^6$ $c < 2.5 \times 10^{-2}$ g g^{-1} $\bar{M}_w = 1.5 \times 10^7$ $c < 4 \times 10^{-2}$ g g^{-1}	Bimodal analysis Homodyne detection	Bimodal	$\tau_{fast}^{-1} \propto q^2 c^0$ $\tau_{slow}^{-1} \propto q^2\, f(c)$ $f(c)$ decreases slightly as c increases	129
Cyclohexane	$\bar{M}_w = 20.6 \times 10^6$ $c < 9.54 \times 10^{-2}$ g cm^{-3}	Cumulants Heterodyne detection (flare + dusts)		$\langle \tau \rangle^{-1} \propto c^{0.94}$ (low q) $\langle \tau \rangle^{-1} \propto a' + b'c$ (high q)	130
Cyclopentane	$\bar{M}_w = 3.84 \times 10^6$ $c < 8.7 \times 10^{-2}$ g cm^3	Linear programming	Trimodal with two dominant modes	$\tau_{fast}^{-1} \propto c^{0.95}\, \bar{M}_w^0\, q^2$ $\tau_{slow}^{-1} \propto c^{-3.02}\, \bar{M}_w^{-3.1}\, q^0$	131

hand (*cf.* equation 96).[127] They obtained the following laws

(QELS+CGD) $\qquad\qquad\qquad D(\text{cm}^2\,\text{s}^{-1}) = (1.25 \pm 0.10) \times 10^{-6}\,c$ $\qquad\qquad$ (104)

(LS+Sedimentation) $\qquad\qquad D(\text{cm}^2\,\text{s}^{-1}) = 1.13 \times 10^{-6}\,c^{1.04}$ $\qquad\qquad\qquad$ (105)

Therefore in the liquid-like regime, experimental results are self-consistent and agree with the theoretical predictions.

In the gel-like regime ($\tau \ll T_{\text{rep}}$) the analysis of the data is more complicated because of the strong non-exponentiality of the autocorrelation function. All the authors agree that the autocorrelation function can be approximated by a bimodal shape. They also agree on the q^{-2} dependence of the fast relaxation time τ_{fast}. The major controversy concerns the q dependence of the slow mode. It must be noted that the results of refs. 127 and 131, relative to q, c and M_{w} dependences of the slow mode, are consistent with the theory of Brochard and de Gennes. Also, the experimental values of τ_{slow} are very close to those of T_{rep} (*cf.* Section 7.5.2.2). The linear dependence of τ_{fast}^{-1} with c is also predicted by the theory, and is due to the combined contributions of osmotic compressibility and shear modulus. Let us note that a log–log representation of τ_{fast}^{-1} *vs.* c leads to an apparent exponent of ~ 0.5 as actually observed by some authors.[129]

In the other two experimental studies reported, most of the data have been obtained in the regime $\tau \simeq \tau_{\text{rep}}$.[133,134] In both cases a bimodal distribution has been observed. The most interesting feature concerns the behavior of the slow relaxation time for the PS–cyclohexane system, given by

$$\tau_{\text{slow}}^{-1} \propto q^2 c^{-3} \bar{M}_{\text{w}}^{-2} \qquad\qquad (106)$$

The molecular weight and concentration dependences of $(\tau_{\text{slow}}\,q^2)^{-1}$ are exactly those expected for the self-diffusion coefficient. These results, however, are not yet well understood.

7.4.5.2 Sedimentation

Sedimentation experiments have been carried out by Nyström *et al.* in PS–cyclopentane,[135] hydroxypropyl cellulose–water [$\theta = 41\,°\text{C}$][136] and PS–*trans*-decalin[137] systems. The concentration dependences of the cooperative sedimentation can be described for $c < 7.5 \times 10^{-2}\,\text{g}\,\text{g}^{-1}$ by a power law with an exponent -1.0, in excellent agreement with the theoretical prediction. At higher concentration the sedimentation coefficient decreases more rapidly. This tendency is likely to be a manifestation of enhanced monomer–monomer friction, which increases with concentration. This result allows the estimation of the top limit of the concentration range corresponding to the semi-dilute regime. Also, c/c^* was found to be a reduced variable for both s_0/s^{70} and D/D_0^{79} as expected from the theory (*cf.* Section 7.4.3.1).

7.5 HYDRODYNAMIC PROPERTIES OF ENTANGLED POLYMER SOLUTIONS: REPTATION

Local collective motions in semi-dilute solutions are functions of a single variable, namely the monomer concentration c. This leads to a fairly universal behavior, at least in principle, with corrections to scaling playing a very important role in real experiments. If, however, motions on a larger length scale are considered, the effect of chain ends becomes important and the degree of polymerization of the polymers becomes a relevant variable for hydrodynamic phenomena. These hydrodynamic properties are dominated by entanglements, on a large length scale. They will be discussed here in the framework of the reptation model, which provides scaling laws in qualitatively good agreement with experiment. The very concept of scaling behavior is much more conjectural than for single chain dynamics or even static properties and has been limited, up to now, to the determination of exponents. We shall briefly describe the tube model introduced by Edwards for polymer melts, which is the basis of the so-called reptation model. Using scaling laws, this model can be generalized for semi-dilute solutions. We limit our study to three quantities: the reptation time, the self-diffusion coefficient of a labeled chain and the macroscopic viscosity of the solution. Finally, we briefly discuss the motion of labeled chains in a semi-dilute solution of a different length and the so-called tube renewal problem.

7.5.1 The Reptation Model

7.5.1.1 *The tube model*

Consider the motions of a tagged chain in a melt of identical chains. The test chain is entangled with the neighboring chains (Figure 11). It thus moves more easily parallel to itself than perpendicular. The perpendicular motion is hindered by the entanglements with neighboring chains which cannot be crossed.[138] In order to model the effect of the obstacles, Edwards suggested that the chain is trapped in a tube of small diameter (Figure 12) parallel to its contour.[139] The chain can only move by reptation inside the tube and has no transverse motion.[139-141] At each time, the chain leaves some parts of its tube and creates a new tube around its advancing end. In a first approximation, the environment of the chain is considered as quenched; the motion of the test chain is thus a one-dimensional diffusion inside the frozen tube.[139] This diffusive motion can be characterized by a mobility which gives the curvilinear velocity under an external force. The mobility μ_t is inversely proportional to the tube length, *i.e.* to the chain degree of polymerization N

$$\mu_t = \mu_0/N \tag{107}$$

μ_0 is a microscopic mobility depending on the local friction and on the tube diameter d.

Figure 11   Entanglements between neighboring chains

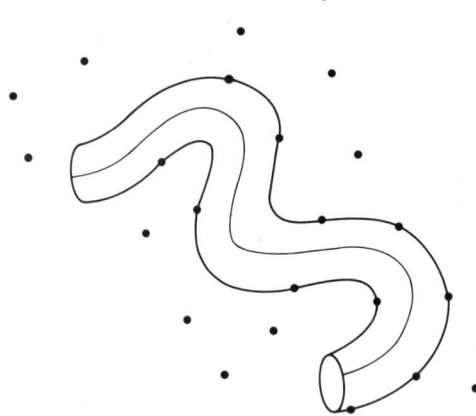

Figure 12   The tube model: the chain motion is hindered by obstacles made by the other chains, represented here as dots. It is confined to a tube of diameter d

The Einstein formula relates the tube mobility to a curvilinear diffusion constant

$$D_1 = T\mu_t = \frac{T\mu_0}{N} \tag{108}$$

The curvilinear displacement of a monomer is

$$l^2(t) \sim D_1 t \tag{109}$$

In a melt, polymer chains are Gaussian; if a monomer has a curvilinear displacement $l(t)$ it moves over a spatial distance $r(t) \sim l^{1/2}$. The Euclidian displacement of monomers by reptation is thus

(at short times)

$$r^4(t) \sim \frac{T\mu_0}{N} t \tag{110}$$

The reptation time T_{rep} is the time that a chain needs to disentangle from other chains; it is thus the time needed by a chain to entirely leave its tube. It is reached when $l(t) \sim N$ or $R_G \simeq r(t)$

$$T_{rep} \sim \frac{N^2}{D_f} \simeq N^3 \tau_0 \tag{111}$$

τ_0 being a microscopic time.

The reptation model outlined here has been studied in great detail by Doi and Edwards, who considered in particular the influence of the tube diameter (or the so-called entanglement mass) and the viscoelastic behavior of polymer melts;[142-144] this is the subject of Chapter 8.

7.5.1.2   *Reptation time in a semi-dilute solution*[11,100]

We have already mentioned that, for most purposes, a semi-dilute solution can be viewed as a melt of blobs of size ξ, each containing $g = c\xi^3$ monomers. It is thus reasonable to assume that the reptation model that we just described can be applied to a semi-dilute solution. The tube diameter is proportional to the blob size ξ in this model. The overall contour length of the tube is $(N\xi)/g$. The local mobility μ_0 is the blob mobility $\mu_0 = 1/(6\pi\eta_0\xi)$ and the tube diffusion constant is $D_1 = (Tg)/(6\pi\eta_0\xi N)$. This gives a reptation time for a polymer chain in a semi-dilute solution

$$T_{rep} = \frac{N^3 \xi^3 (6\pi\eta_0)}{g^3 T} \tag{112}$$

In a good solvent $T_{rep} \sim N^3 c^{3/2}$ and in a θ solvent $T_{rep} \sim N^3 c^3$.

As in a melt, the reptation time depends strongly on the molecular mass and, for high molecular mass polymers, can be as long as several seconds. As the concentration is decreased, the reptation time decreases and crosses over towards the Zimm time (the characteristic time of a dilute solution) at the overlap concentration $c = c^*$.

7.5.1.3   *Self-diffusion*[11,100]

At first a test chain has a one-dimensional motion inside its tube. After a reptation time T_{rep}, when the chain has disentangled from its neighbors and constructed a completely new tube independent of the original one, the motion becomes diffusive. The self-diffusion constant can be estimated by noting that the test chain has moved over a distance of the order of its radius of gyration during a time T_{rep}

$$R_g^2 = D_{self} T_{rep} \tag{113}$$

leading to a self-diffusion constant

$$D_{self} = \frac{T}{6\pi\eta_0\xi} \left(\frac{g}{N}\right)^2 \tag{114}$$

In a good solvent $D_{self} \sim N^{-2} c^{-7/4}$ and in a θ solvent $D_{self} \sim N^{-2} c^{-3}$.

Recently, the problem of self-diffusion of a chain in a semi-dilute good solvent solution has been studied by Hess using a direct approach, not based on the tube model.[145] The starting point is a microscopic Fokker–Planck equation governing the probability P that the solution has a given configuration at time t. This microscopic equation is studied within a Zwanzig–Mori projection operator scheme, all physical quantities being projected onto the test chain concentration. This leads to generalized Langevin equations for the concentration correlation function of the test chain and the self-diffusion constant. The generalized Langevin random force has two contributions in the equations, a hydrodynamic force and the short range excluded volume force with the neighboring chain. The excluded volume contribution is supposed to be dominant because of the screening of hydrodynamic interactions. The Langevin equation is then solved with a decoupling approximation.

If the test chain is infinite, the free energy of the polymer solution is invariant by translation of the chain along itself, the effect of excluded volume is, therefore, very different for motions parallel to the chain or transverse to it; this is the origin of a reptative motion. The reptation diffusion coefficient is obtained if the entanglement parameter ψ, proportional to the number of blobs per chain, is larger than a critical value; at low values of this parameter the self-diffusion constant is given by a Rouse approximation.

The crossover between these two extremes of behavior is found to be quite extensive. These results involve many approximations, which are not always well controlled, but they represent one of the first microscopic approaches to dynamics in semi-dilute solutions and they seem to confirm the more phenomenological tube model. It should also be noticed that the confinement of the chain in what would be the tube is not due to topological entanglements, as in Edwards approach, but rather to the excluded volume interactions with the other chains. The relative role of these two effects has not yet been studied.

7.5.1.4   Viscosity

Semi-dilute polymer solutions are viscoelastic liquids which have a very complex mechanical behavior, intermediate between viscous and elastic.[2] After a mechanical perturbation such as the application of stress, a semi-dilute polymer solution behaves as a solid over short time intervals, with a well-defined elastic modulus, and as a liquid over long time intervals, characterized by a macroscopic viscosity. The elastic modulus is related to the transient network made by the chains. This elastic behavior lasts until chains can disentangle and then the solution flows. The characteristic time separating viscous and elastic behavior is thus the reptation time.

A full description of the mechanical properties of the solution requires a study of its complex elastic modulus as a function of frequency. Such a description has been made in detail for polymer melts in the classic book by Ferry[1] (see also Volume 2, Chapter 8 of this work). We limit our discussion here to the macroscopic viscosity.

For a viscoelastic liquid with a single relaxation time, a Maxwell model can be made which relates the viscosity to the gel elastic modulus E and the stress relaxation time T_{rep}

$$\eta = E T_{rep} \tag{115}$$

For a dilute solution in a good solvent, the elastic modulus E scales as the osmotic pressure $E \sim T/\xi^3$, which leads to a viscosity[11,100]

$$\eta = (6\pi\eta_0)\left(\frac{N}{g}\right)^3 \sim N^3 c^{15/4} \tag{116}$$

The situation is less clear in a θ solvent where two elastic moduli have been defined: an osmotic modulus $E_0 = T/\xi^3$ and a gel shear modulus due to topological entanglements $E_{gel} \sim c^2$.[108]

If we use the gel shear modulus, as seems reasonable, in the Maxwell equation, we obtain

$$\eta \sim \eta_0 N^3 c^5 \tag{117}$$

This result, however, cannot be obtained directly by a scaling argument. In a semi-dilute θ solution, there are two characteristic lengths: the correlation length $\xi \sim 1/c$ (which represents the distance between three-body contacts and has been chosen here as the tube diameter) and the distance between entanglements ξ_2 (distance between two body contacts) which is the mesh size of the transient network. These two lengths play a role in the viscoelastic properties of semi-dilute θ solutions. Their relative importance is still a matter of controversy.

7.5.1.5   Tube renewal

So far, we have considered only the reptative motion of the test chains, assuming that their environment is frozen.[146] This pure reptation model should, therefore, be applied either to the motion of a linear chain in a gel or to the motions of a short chain in a matrix of infinite chains. If the test chain is embedded in a matrix of chains of the same length, then during the reptation motion the neighboring chains which build the tube also move and the tube is partially renewed. This tube renewal process, also called constraint release, should thus increase the self-diffusion constant, the effect being even higher for a chain in a matrix of shorter chains (although still entangled).

Several models have been proposed for tube renewal effects on the motion of a long chain of N monomers in a semi-dilute solution of P shorter chains.[147-150] The basic idea is that the entanglements constraining the chain inside its tube can switch if the neighboring chains move their extremities aside by reptation.

The tube itself is considered as a Rouse chain with an elementary jump frequency $W \sim 1/[T_{rep}(P)]$; this leads to a contribution to the diffusion constant due to tube renewal for a test chain of mass N[151]

$$D_{ren} = D_{rep}\left(\frac{g}{N}\right)^2\left(\frac{N}{P}\right)^3 \sim \frac{N^{-1}}{P^3}c^{-5/2} \quad \text{(good solvent)} \tag{118}$$

(g is the number of monomers per blob and D_{rep} the purely reptative diffusion constant). For sufficiently low values of P, therefore, tube renewal is indeed important and can dominate the diffusion process. The effect of tube renewal becomes less important as the mass of the matrix is increased.

7.5.2 Experimental Results

7.5.2.1 *Viscoelastic behavior in a good solvent*

In 1966, Onogi et al.[152] carried out viscometric experiments in semi-dilute toluene solutions of PS at 40 °C, with molecular weights ranging from 2.06×10^5 to 1.67×10^6 and concentrations $c \le 0.2 \text{ g g}^{-1}$. Their results show that the log η vs. log $\bar{M}_w$ and log η vs. log ρc curves (ρ = density of the solution) can be superimposed by shifts along the abscissa to give a smooth curve (Figure 13). The reduced abscissa is found to be $(c\rho)\bar{M}_w^{0.72}$. As the intrinsic viscosity $[\eta]$ of toluene solutions of PS is given by $[\eta] \sim \bar{M}_w^{0.72}$, it turns out that $(c\rho)[\eta]$ (or, equivalently, c/c^*) is a reduced variable of $[\eta]$, in good agreement with the theoretical prediction (equation 116). Furthermore, the straight portion of the master curve, in the semi-dilute regime, leads to the following power law

$$\eta \propto (c\rho[\eta])^{4.7} \propto (c\rho)^{4.7}\bar{M}_w^{3.4} \tag{119}$$

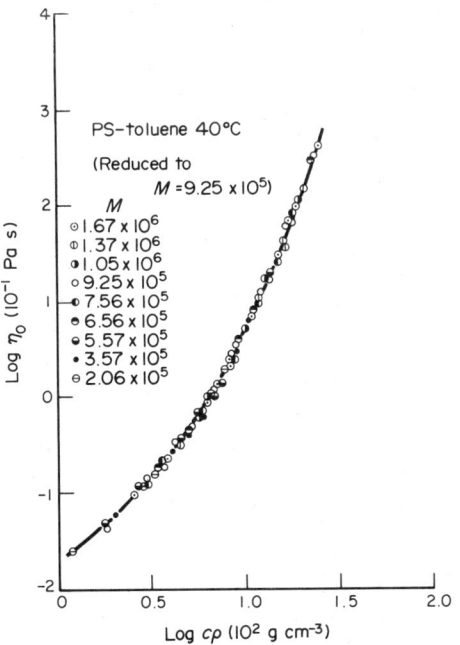

Figure 13 Viscosity of PS solutions in toluene at 40 °C as a function of concentration. The master curve is obtained by superposition of the curves relative to various molecular weights (*cf.* text) (reproduced by permission of Wiley from *J. Polym. Sci. Part C*, 1966, **15**, 381)

The value of the exponent relative to c is significantly higher than the theoretical value of 3.92 (assuming $v = 0.588$). It must be noted that an exponent of 3.4 for the molecular weight dependence of η has also been obtained for bulk polymers.[1,2]

Recent experiments by Adam and Delsanti in the PS–benzene system have confirmed that c/c^* is a reduced variable of η;[153] the experimental exponent of $\eta = f(c/c^*)$ is ~ 4.5 for $c/c^* \geq 10$. These authors also measured T_{rep} for the same systems. Figure 14 shows the variation of T_{rep}/T_z vs. c/c^*, where T_z was estimated from the intrinsic viscosity and c^* from osmotic pressure measurements. The ratio c/c^* is a reduced variable of T_{rep}/T_z as predicted from the combined equations (32) and (112) but the exponent of the power law fitting $T_{rep}/T_z = f(c/c^*)$ is 2.05 which is significantly higher than the theoretical value of 1.6. These deviations from the theory are not yet fully understood. The effect of local segmental frictional resistance has been invoked. This friction, which becomes important at polymer volume fractions larger than 0.1, does not affect c^* so that c/c^* should no longer be a reduced variable of the viscosity.[154] In fact, the effect of the local friction on the c/c^* dependence of the viscosity might remain in the limit of experimental accuracy. Furthermore the local friction would not explain the $\bar{M}_w^{3.4}$ dependence of η. It is noteworthy that the shear modulus obtained by combining viscosity and relaxation time measurement is independent of $\bar{M}_w$ and varies with c according to[153]

$$E(10^{-5}\text{N/cm}^2) = 8.32 \times 10^5 c^{2.36 \pm 0.2} \tag{120}$$

in very good agreement with the scaling prediction $E \propto c^{2.32}$ (see Section 7.4.3.1).

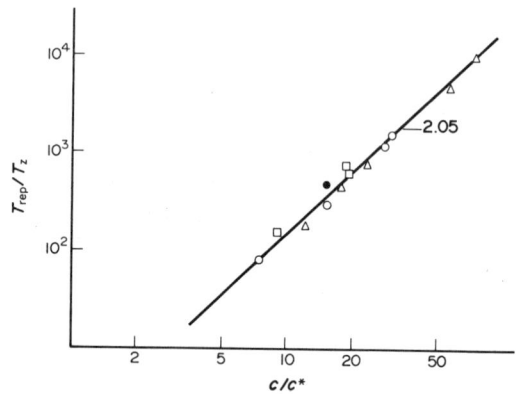

Figure 14 Variation of the reptation time with concentration in a semi-dilute solution of PS in toluene. The plot using reduced variables T_{rep}/T_z, c/c^* shows a universal behavior. Different symbols correspond to different molecular masses ranging from 1.7×10^5 to 2.07×10^7 (reproduced by permission of Commission des Publications Françaises de Physique from *J. Phys. (Orsay, Fr.)*, 1983, **44**, 1185)

The self-diffusion of labeled PS chains in semi-dilute solutions in benzene and tetrahydrofuran has been studied by forced Rayleigh light-scattering[151,155,156] and by pulse field gradient NMR techniques.[157] As a general rule, the logarithmic plots of $D_{self} \bar{M}_w^2$ against c did not give a single composite curve and showed no significant region with a slope of -1.85 as predicted by scaling theory. Effects of local friction were then taken into account by correcting the raw data by the multiplicative factor $D_{label}(\Phi = 0)/D_{label}(\Phi)$ where $D_{label}(\Phi)$ is the self-diffusion coefficient of the free label (unattached to the chain) in a solution of small unentangled polymer chains at volume fraction Φ. Figure 15 shows the variation of D_t/D_0 (where $D_t = D_{self}[D_{label}(\Phi = 0)]/[D_{label}(\Phi)]$) as a function of the reduced volume fraction Φ/Φ^* for PS chains labelled at each extremity with spiropyran molecules in solution in benzene. The data relative to different molecular weights and different concentrations lie on a universal curve. Tracer diffusion experiments, performed on labeled chains with index of polymerization N embedded in a semi-dilute solution of larger chains (index of polymerization P), showed that, for a given N, D_T decreases upon an increase of P. This result was attributed to a tube renewal effect. For $P/N > 5$, D_T becomes independent of P. In that case D_t/D_0 fits a power law with Φ/Φ^* in the semi-dilute range investigated ($\Phi/\Phi^* < 10$) with an exponent of -1.8, *i.e.* close to the theoretical value. This indicates that a pure reptative behavior of the labeled chain is observed in that case and that the matrix can be considered as completely frozen.

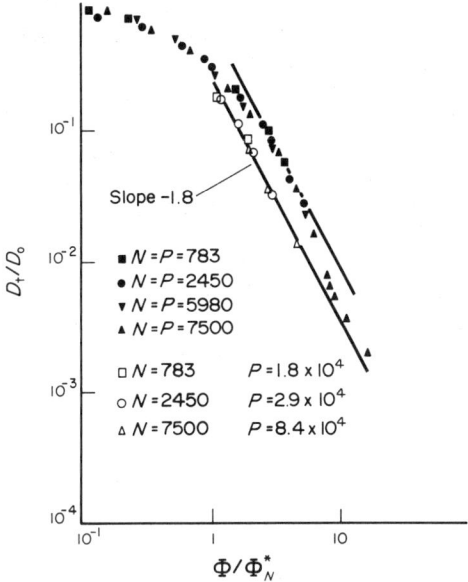

Figure 15 Normalized self diffusion constant D_t/D_0 as a function of the reduced volume fraction Φ/Φ_N^* for a PS chain of N monomers in a matrix of PS chains of P monomers in benzene. In the limit $N/P \ll 1$ a purely reptative behavior is observed (reproduced by permission of the American Institute of Physics from *Phys. Rev. Lett.*, 1985, **55**, 1078)

7.5.2.2 *Viscoelastic behavior in a θ solvent*

Measurements of viscosity and relaxation time were carried out by Adam and Delsanti in semi-dilute solutions of PS in cyclohexane.[132] The molecular weight range investigated was 2.89×10^6–2.06×10^7 and the concentration range 2.38×10^{-2}–10.85×10^{-2} g g^{-1}.

The following concentration and molecular weight dependences of η and T_{rep} were obtained

$$\eta \propto c^{5.14 \pm 0.16} \, \bar{M}_w^{3.75 \pm 0.5} \tag{121}$$

$$T_{rep} \propto c^{2.8 \pm 0.3} \, M_w^{3.75 \pm 0.5} \tag{122}$$

Both η and T_{rep} follow power laws of the concentration in satisfactory agreement with the scaling theory ($\eta \sim c^5$) and $T_{rep} \sim c^3$. On the other hand, the exponent of the power law of $\bar{M}_w$ is much higher than, not only the theoretical value of 3, but also, surprisingly enough, the value of 3.4 commonly found in the bulk. The ratio c/c^* is not a reduced variable of η and T_{rep} (Figure 16). Thus, it seems

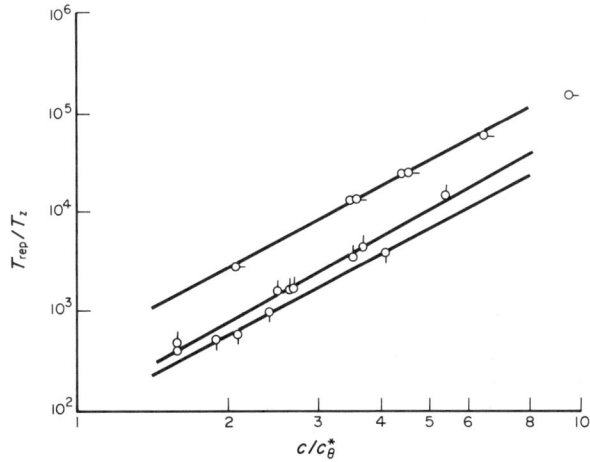

Figure 16 Variation of the reptation time with concentration in a semi-dilute solution of PS in cyclohexane at $T = 35\,°C$. No universal behavior is observed, different molecular masses yield different curves when plotted in reduced variables (reproduced by permission of Commission des Publications Françaises de Physique from *J. Phys. (Orsay, Fr.)*, 1984, **45**, 1513)

that the scaling approach does not satisfactorily describe the viscoelastic behavior of semi-dilute solutions at θ temperature. This is likely to be linked to the fact that two correlation lengths are necessary to describe the properties of these systems.

Measurements of D_{self} in semi-dilute solutions at θ temperature by forced Rayleigh light-scattering are scarce because they present considerable experimental difficulties. It seems that the photoexcited label molecules modify locally the quality of the solvent and, consequently, the polymer concentration. As a result, a phase grating, related to the formation of a modulation of the polymer concentration, appears in the solution. The long time relaxation of the polymer concentration modulation is driven by the diffusion of the labeled chains over the interfringe spacing, allowing the determination of D_{self}. Deschamps and Leger have recently measured the self-diffusion coefficient of PS chains labelled by spiropyran in solution in cyclopentane at θ temperature.[158,159] Three molecular weights (261 800, 657 000 and 861 000) were investigated in a concentration range $c \leq 0.2 \, \mathrm{g\,g^{-1}}$. The following variation of D_{self} with $\bar{M}_{w}$ and c was obtained

$$D_{self} \propto \bar{M}_{w}^{-2 \pm 0.1} \, c^{-3 \pm 0.1} \tag{123}$$

Contrary to the good solvent case a weak dependence of D_{self} on the molecular weight of unlabeled chains is observed. These results are in agreement with reptation and scaling predictions, assuming the tube width to be equal to the correlation length of the θ solution, $\xi \sim c^{-1}$. However, in this study no corrections were made to account for monomer–monomer friction. Such a correction would increase the exponent of the concentration dependence of D_{self}.

7.6 CONCLUSION

We have reviewed the main hydrodynamic properties of both dilute and semi-dilute polymer solutions in a good or θ solvent.

In dilute solutions, the classical Zimm non-draining description of polymers behaving as hard spheres, impenetrable by the flow field, is well confirmed and extended by the dynamical scaling theory. The agreement between theory and experiment is reasonably good for very long macro-molecular chains, not only for exponents but also for some universal amplitudes and even some scaling functions. The deviations observed can often be related to the finite molecular weight of the chains.

The situation is not as clear in a semi-dilute solution. In a good solvent a scaling theory has been developed in the asymptotic limit of infinite chains and small concentrations. Although re-normalization group calculations are not as systematic as for dilute solutions, they provide several quantitative results. Experimentally, however, the asymptotic conditions are reached only for very long chains and, in practice, corrections to the asymptotic results mask the expected scaling results. On the whole, for collective modes, the agreement between theory and experiment is fair. The motions of a tagged chain and the macroscopic viscosity are in good agreement with the reptation model, with the same restrictions as in the melt associated with the mass dependence of the viscosity, which shows an exponent of 3.4, higher than the predicted value of 3. Two main reasons have been invoked: tube renewal (constraint release) and fluctuations of the tube. The issue of the adequacy of the reptation theory, augmented by these corrections to describe the rheological properties of melts and concentrated solutions, remains an open question, which could be answered only by more detailed experiments. Another uncertain point in semi-dilute solutions is the role of the friction between monomers which has not been included in the theories, but which is often referred to as the origin of the discrepancies between experiments and the reptation model for the concentration dependences.

Our understanding of the hydrodynamic properties of semi-dilute solutions in a θ solvent or a marginal solvent is not as good as that for good solvents. No clear picture emerges from the experimental results. For the collective properties, the exact shape of the correlation function and the origin of the slow fluctuations are still a subject of discussion, although some differences between experiments can be explained by the different time scales probed. The behavior of the macroscopic viscosity is also not well understood; even the mass exponent seems higher than the value of 3.4 observed in melts and good solvents. The hydrodynamic behavior in θ solvents is thus very controversial. This might be related to the difficulty of making clear homogeneous θ solvent solutions with high molecular weight polymers, because of the vicinity of the phase separation in a poor solvent which could lead to aggregation between different chains. Also a dynamical scaling theory can be made in a θ solvent, but it is much less reliable, due to the existence of two characteristic lengths: the correlation length ξ and the distance between entanglements ξ_2, which is related to the existence of knots or self-entanglements of polymer chains in a θ solvent. A deeper understanding of the knots is thus probably needed to explain the experimental results.

We have limited this review to the discussion of a few experiments where the solution is close to thermodynamic equilibrium. There exist, however, several other ways to investigate hydrodynamic properties of polymer solutions.

A whole set of experiments, not discussed here, deals with some flow experiments where the polymer moves in a well-defined macroscopic solvent flow: in a shear[160] flow, one observes the flow birefringence which gives information very similar to the viscometry experiments (stress optic theorem). The behavior of polymer solutions in an elongational flow[161-163] is quite different; at low velocity gradients the viscosity of the solution increases with the velocity gradient instead of decreasing as in a shear flow. At higher elongational gradients, linear chains undergo an abrupt transition[161] between a coil conformation (with only slight perturbations to the equilibrium conformation) to a stretched conformation where the chains are extended ($R_G \sim N$). Experimentally, this leads to a very strong elongational flow birefringence which is often localized (on the main axis of a Taylor roll cell, for example). Elongational gradients also play a major role in the behavior of polymer solutions entering a porous medium[164-166] (a model membrane or a porous rock for practical applications). Individual chains are stretched in the convergent flow and enter the pores more easily as the solvent flow is increased.

Other powerful tools in the study of polymer properties are 'computer experiments'. These allow a good control of the physical parameters which can be varied almost at will, but one is limited by the polymer mass which cannot be as high as in real experiments. Monte Carlo simulations[167] are quite successful in studying the static properties of dilute solutions. The study of hydrodynamic properties of solutions requires the use of molecular dynamics and is only at an early stage even for dilute solutions.[170,171] The most important computer body of work[168,169] has been concerned with melts in order to test the reptation ideas. Although some results are not far from the reptation predictions, there seems to be no conclusive result in favor of the reptation theory.[179]

In conclusion we would like to mention several other chain properties, which we have not considered, but which could play an important role in hydrodynamic behavior.

We have studied flexible chains, the rigidity of the chemical backbone often being characterized by a persistence length l_p. The effect of the persistence length can be included in the scaling analysis[112] and enhances the marginal solvent behavior. As far as hydrodynamic properties are concerned, the local friction constant ζ obviously depends critically on the persistence length. At higher concentrations, the tube diameter of the reptation model and thus the viscosity depend strongly on the chain rigidity in a way which is not well understood. An extreme limit is that of rigid rods, where the persistence length is much larger than the chain contour length. Their behavior in a dilute solution is well understood. The reptation of rods in a semi-dilute solution is still a matter of controversy.[172,173]

Branching also plays a major role in the dynamics of semi-dilute solutions. The side branches prevent any reptative motion and the motion is considerably slower than that of a linear chain. This has been studied in detail only for star polymers.[150,174,175] Similar problems exist with the reptation of cyclic polymers.[150,176]

The existence of electric charges on the chemical backbone modifies considerably the properties of the polymers because of the long range character of the electrostatic interactions. Polyelectrolytes are of major importance in biology. While their static properties are now better understood,[177] little is known about their hydrodynamic behavior. At low concentrations the specific viscosity of a polyelectrolyte solution varies according to Fuoss law $[\eta] \sim 1/\sqrt{c}$;[178] there does not seem to be any satisfactory explanation of this law.

7.7 REFERENCES

1. J. Ferry, 'Viscoelastic Properties of Polymers', Wiley, New York, 1981.
2. B. Bird, O. Hassager, F. Curtiss and H. Armstrong, 'Dynamics of Polymeric Liquids', Wiley, New York, 1974.
3. P. J. Flory, 'Principles of Polymer Chemistry', Cornell University Press, Ithaca, NY, 1971.
4. H. Yamakawa, 'Modern Theory of Polymer Solutions', Harper and Row, New York, 1971.
5. B. Berne and R. Pecora, 'Dynamic Light Scattering', Wiley, New York, 1977.
6. G. Benedek, in 'Polarisation Matière et Rayonnement', jubilé de A. Kastler, P.U.F. Paris, 1969.
7. F. Mezei, 'Neutron Spin Echo', Springer, Berlin, 1980.
8. P. Stilbs, *J. Colloid Interface Sci.*, 1982, **87**, 385.
9. V. Degiorgio, M. Corti and M. Giglio, 'Light Scattering in Liquids and Macromolecular Solutions', Plenum Press, New York, 1980.

10. B. A. Smith and H. McConnell, *Proc. Nat. Acad. Sci. USA*, 1978, **75**, 2759.
11. P. G. de Gennes, 'Scaling Concepts in Polymer Physics', Cornell University Press, Ithaca, NY, 1985.
12. J. des Cloizeaux and G. Janninck, 'Polymères en Solution, Leur Modelisation et Leur Structure', Editions de Physique, Paris, 1987.
13. Y. Oono, in 'Advances in Chemical Physics', ed. I. Prigogine and S. A. Rice, Wiley, New York, 1985, vol. LXI.
14. B. Mandelbrojt, 'The Fractal Geometry of Nature', Freeman, San Francisco, 1982.
15. J. C. Le Guillou and J. Zinn Justin, *Phys. Rev. Lett.*, 1977, **39**, 95.
16. M. Daoud, J. P. Cotton, B. Farnoux, G. Janninck, G. Sarma, H. Benoit, R. Duplessix, C. Picot and P. G. de Gennes, *Macromolecules*, 1975, **8**, 804.
17. P. E. Rouse, *J. Chem. Phys.*, 1953, **21**, 1272.
18. P. G. de Gennes, *Physics (Long Island City, NY)*, 1967, **3**, 37.
19. L. Landau and E. M. Lifschitz, 'Mécanique des Fluides', Edition Mir, Moscow, 1971.
20. B. H. Zimm, *J. Chem. Phys.*, 1956, **24**, 269.
21. E. Dubois Violette and P. G. de Gennes, *Physics (Long Island City, NY)*, 1967, **3**, 181.
22. J. G. Kirkwood and J. Riseman, *J. Chem. Phys.*, 1948, **16**, 565.
23. P. Debye and A. M. Bueche, *J. Chem. Phys.*, 1948, **16**, 573.
24. J. des Cloizeaux, *J. Phys., Lett. (Orsay, Fr.)*, 1978, **39**, 151.
25. D. Jasnow and M. Moore, *J. Phys., Lett. (Orsay, Fr.)*, 1978, **38**, 467.
26. G. F. Al Noaimi, G. C. Martinez–Mekler and C. A. Wilson, *J. Phys., Lett. (Orsay, Fr.)*, 1979, **39**, 373.
27. Y. Oono and K. Freed, *J. Chem. Phys.*, 1981, **75**, 1009.
28. Y. Oono and M. Kohmoto, *J. Chem. Phys.*, 1983, **78**, 520.
29. Y. Oono, *J. Chem. Phys.*, 1983, **79**, 4629.
30. A. Z. Akcasu and C. Han, *Macromolecules*, 1979, **12**, 276.
31. C. Tanford, 'Physical Chemistry of Macromolecules', Wiley, New York, 1961.
32. R. Kubo, *Rep. Prog. Phys.*, 1966, **29**, 255.
33. K. Kawasaki, *Ann. Phys. (NY)*, 1970, **61**, 1.
34. R. Ferrel, *Phys. Rev. Lett.*, 1970, **24**, 1169.
35. A. Z. Akcasu and H. Gurol, *J. Polym. Sci., Polym. Phys. Ed.*, 1976, **14**, 1.
36. M. Benmouna and A. Z. Akcasu, *Macromolecules*, 1978, **11**, 1187.
37. A. Z. Akcasu, M. Benmouna and C. Han, *Polymer*, 1980, **21**, 866.
38. C. Han and A. Z. Akcasu, *Macromolecules*, 1981, **14**, 1080.
39. A. Lee, P. R. Baldwin and Y. Oono, *Phys. Rev. A.*, 1984, **30**, 968.
40. G. K. Batchelor, *J. Fluid Mech.*, 1976, **74**, 1.
41. B. U. Felderhof, *J. Phys. A: Math. Gen.*, 1978, **11**, 929.
42. W. Wan and S. L. Whittenburg, *Polym. Prep., Am. Chem. Soc., Div. Polym. Chem.*, 1986, **27**, 221.
43. B. Farnoux, F. Boue, J. P. Cotton, M. Daoud, G. Janninck, N. Nierlich and P. G. de Gennes, *J. Phys. (Orsay, Fr.)*, 1978, **39**, 77.
44. G. Weill and J. des Cloizeaux, *J. Phys. (Orsay, Fr.)*, 1979, **40**, 99.
45. G. Meyerhoff and B. Appelt, *Macromolecules*, 1979, **12**, 968.
46. B. Appelt and G. Meyerhoff, *Macromolecules*, 1980, **13**, 657.
47. G. Meyerhoff, *Z. Phys. Chem. (Frankfurt/Main)*, 1955, **4**, 335.
48. H. Baumann, Ph.D. Thesis, University of Mainz, 1964.
49. A. Burmeister, Ph.D. Thesis, University of Mainz, 1973.
50. K. Nachtigall and G. Meyerhoff, *Z. Phys. Chem. (Frankfurt/Main)*, 1961, **30**, 17.
51. G. Büldt, Ph.D. Thesis, University of Mainz, 1972.
52. S. Newman and F. Eirich, *J. Colloid Sci.*, 1950, **5**, 541.
53. J. Raczek and G. Meyerhoff, *Ber. Bunsenges Phys. Chem.*, 1979, **83**, 381.
54. T. Svedberg and K. O. Pederson, in 'Die Ultrazentrifuge', Steinkopferlag, Dresden, 1940.
55. T. Nose and B. Chu, *Macromolecules*, 1979, **12**, 1122.
56. J. C. Selser, *Macromolecules*, 1981, **14**, 346.
57. M. Adam and M. Delsanti, *Macromolecules*, 1977, **10**, 1229.
58. D. Schaeffer and C. Han, in 'Dynamic Light Scattering', ed. R. Pecora, Plenum Press, New York, 1985.
59. T. A. King, A. Knox, W. I. Lee and J. D. G. McAdam, *Polymer*, 1973, **14**, 151.
60. M. J. Pritchard and D. Caroline, *Macromolecules*, 1981, **14**, 424.
61. H. J. Cantow, *Makromol. Chem.*, 1959, **30**, 169.
62. G. C. Berry, *J. Chem. Phys.*, 1967, **46**, 1338.
63. W. R. Krigbaum and P. J. Flory, *J. Polym. Sci.*, 1953, **11**, 37.
64. Y. Einaga, Y. Miyaki and H. Fujita, *J. Polym. Sci., Polym. Phys. Ed.*, 1979, **17**, 2103.
65. D. Decker–Freyss, Ph.D. Thesis, University of Strasbourg, 1968.
66. C. Strazielle and H. Benoit, *Macromolecules*, 1975, **8**, 203.
67. M. Abe, K. Sakato, T. Kageyama, M. Fukatsu and M. Kurata, *Bull. Chem. Soc. Jpn.*, 1968, **41**, 2330.
68. K. Sakato and M. Kurata, *Polym. J.*, 1970, **1**, 260.
69. I. Noda, K. Mizutani and T. Kato, *Macromolecules*, 1977, **10**, 618.
70. B. Nyström, J. Roots and R. Bergman, *Polymer*, 1979, **20**, 157.
71. V. H. Lutje and G. Meyerhoff, *Makromol. Chem.*, 1963, **68**, 180.
72. P. J. Flory, L. Mandelkern, J. B. Kinsinger and W. B. Shultz, *J. Am. Chem. Soc.*, 1952, **74**, 3364.
73. M. Matsumoto and Y. Ohyanagi, *J. Polym. Sci.*, 1960, **46**, 441.
74. T. G. Fox, Jr. and P. J. Flory, *J. Am. Chem. Soc.*, 1951, **73**, 1909.
75. W. W. Everett and J. F. Foster, *J. Am. Chem. Soc.*, 1959, **81**, 3464.
76. H. L. Wagner and P. J. Flory, *J. Am. Chem. Soc.*, 1952, **74**, 195.
77. W. R. Krigbaum and L. H. Sperling, *J. Phys. Chem.*, 1960, **64**, 99.
78. A. Haug and G. Meyerhoff, *Makromol. Chem.*, 1962, **53**, 91.
79. J. P. Munch, G. Hild and S. Candau, *Macromolecules*, 1983, **16**, 71.

80. W. N. Huang and J. E. Frederick, *Macromolecules*, 1974, **7**, 34.
81. T. A. King, A. Knox and J. D. G. McAdam, *Chem. Phys. Lett.*, 1973, **19**, 351.
82. J. D. G. McAdam and T. A. King, *Chem. Phys.*, 1974, **6**, 109.
83. T. A. King and M. R. Treadaway, *J. Chem. Soc., Faraday Trans. 2*, 1976, **72**, 1473.
84. D. Caroline and G. Jones, in 'Light Scattering in Liquids and Macromolecular Solutions', ed. V. Degiorgio, M. Corti and M. Giglio, Plenum Press, New York, p. 97.
85. R. Pecora, *J. Chem. Phys.*, 1964, **40**, 1604.
86. R. Pecora, *J. Chem. Phys.*, 1965, **43**, 1562.
87. R. Pecora, *J. Chem. Phys.*, 1968, **49**, 1032.
88. D. R. Bauer, J. I. Brauman and R. Pecora, *Macromolecules*, 1975, **8**, 443.
89. M. Adam and M. Delsanti, *J. Phys. (Orsay, Fr.)*, 1976, **37**, 1045.
90. T. A. King, A. Knox and J. D. G. McAdam, *Polymer*, 1973, **14**, 293.
91. W. Mandema and Z. Zeldenrust, *Polymer*, 1977, **18**, 835.
92. T. L. Yu, H. Reihanian and A. M. Jamieson, *Macromolecules*, 1980, **13**, 1590.
93. M. E. McDonnell and A. M. Jamieson, *J. Macromol. Sci., Phys.*, 1977, **B13**, 67.
94. P. Wiltzius, H. R. Haller, D. S. Cannell and D. W. Schaeffer, *Phys. Rev. Lett.*, 1984, **53**, 834.
95. C. Han and H. Yu, *J. Chem. Phys.*, 1974, **61**, 2650.
96. D. Richter, B. Ewen and J. B. Hayter, *Phys. Rev. Lett.*, 1980, **45**, 2121.
97. S. F. Edwards and K. F. Freed, *J. Chem. Phys.*, 1974, **61**, 1189.
98. K. F. Freed and S. F. Edwards, *J. Chem. Phys.*, 1974, **61**, 3626.
99. P. G. de Gennes, *Macromolecules*, 1976, **9**, 587.
100. P. G. de Gennes, *Macromolecules*, 1976, **9**, 594.
101. M. Muthukumar and S. F. Edwards, *Polymer*, 1982, **23**, 345.
102. M. Muthukumar, *J. Phys. A: Math. Gen.*, 1981, **14**, 2129.
103. D. Richter, K. Binder, B. Ewen and B. Stuhn, *J. Phys. Chem.*, 1984, **88**, 6618.
104. Y. Oono, P. Baldwin and T. Ohta, *Phys. Rev. Lett.*, 1984, **53**, 2149.
105. Y. Oono and P. Baldwin, *Phys. Rev. A*, 1986, **33**, 3391.
106. M. Daoud and G. Jannink, *J. Phys. Lett. (Orsay, Fr.)*, 1980, **41**, 217.
107. M. M. Kops–Werkhoven, A. Vrij and M. Lekkerkerker, *J. Chem. Phys.*, 1983, **78**, 2760.
108. F. Brochard and P. G. de Gennes, *Macromolecules*, 1977, **10**, 1157.
109. T. Tanaka, L. O. Hocker and G. Benedek, *J. Chem. Phys.*, 1973, **59**, 5151.
110. F. Brochard, *J. Phys. (Orsay, Fr.)*, 1983, **44**, 39.
111. P. G. de Gennes, *J. Phys., Lett. (Orsay, Fr.)*, 1974, **35**, 133.
112. D. W. Schaeffer, J. F. Joanny and P. Pincus, *Macromolecules*, 1980, **13**, 1280.
113. J. P. Munch, S. Candau, J. Herz and G. Hild, *J. Phys. (Orsay, Fr.)*, 1977, **38**, 971.
114. J. P. Munch, P. Lemaréchal, S. Candau and J. Herz, *J. Phys. (Orsay, Fr.)*, 1977, 38, 1499.
115. P. Wiltzius and D. S. Cannell, *Phys. Rev. Lett.*, 1986, **56**, 61.
116. A. M. Jamieson, H. Reihanian, J. G. Southwick, T. L. Yu and J. Blackwell, *Ferroelectrics*, 1980, **30**, 267.
117. E. Amis and C. C. Han, *Polymer*, 1982, **23**, 1403.
118. W. Brown, R. M. Johnsen and P. Stilbs, *Polym. Bull. (Berlin)*, 1983, **9**, 305.
119. W. Brown, *Polymer*, 1984, **25**, 680.
120. D. Hwang and C. Cohen, *Macromolecules*, 1985, **18**, 1681.
121. P. Mathiez, C. Mouttet and G. Weisbuch, *J. Phys. (Orsay, Fr.)*, 1980, **41**, 519.
122. S. Candau, I. Butler and T. A. King, *Polymer*, 1983, **24**, 1601.
123. M. Adam, M. Delsanti and G. Pouyet, *J. Phys., Lett. (Orsay, Fr.)*, 1979, **40**, 435.
124. J. Roots and B. Nyström, *Macromolecules*, 1980, **13**, 1595.
125. B. Nyström and J. Roots, *J. Macromol. Sci., Rev. Macromol. Chem.*, 1980, **C19**, 35.
126. B. Ewen, D. Richter, J. B. Hayter and B. Lehner, *J. Polym. Sci., Polym. Lett.*, 1982, **20**, 233.
127. M. Adam and M. Delsanti, *Macromolecules*, 1985, **18**, 1760.
128. W. Brown and R. M. Johnsen, *Macromolecules*, 1985, **18**, 379.
129. W. Brown, *Macromolecules*, 1986, **19**, 387.
130. A. M. Hecht, H. B. Bohidar and E. Geissler, *J. Phys., Lett. (Orsay, Fr.)*, 1984, **45**, 121.
131. P. Stepànek, C. Konak and J. Jakes, *Polym. Bull. (Berlin)*, 1986, **16**, 73.
132. M. Adam and M. Delsanti, *J. Phys. (Orsay, Fr.)*, 1984, **45**, 1513.
133. E. Amis, C. Han and Y. Matsushita, *Polymer*, 1984, **25**, 650.
134. T. Nose and B Chu, *Macromolecules*, 1979, **12**, 590.
135. J. Roots and B. Nyström, *J. Polym. Sci., Polym. Phys. Ed.*, 1981, **19**, 479.
136. J. Roots and B. Nyström, *Polymer*, 1979, **20**, 148.
137. B. Nyström and J. Roots, *Eur. Polym. J.*, 1978, **14**, 551.
138. W. W. Graessley, *Adv. Polym. Sci.*, 1974, **16**, 1.
139. S. F. Edwards, *Proc. Phys. Soc., London*, 1967, **92**, 9.
140. P. G. de Gennes, *J. Chem. Phys.*, 1971, **55**, 572.
141. P. G. de Gennes and L. Léger, *Annu. Rev. Phys. Chem.*, 1982, **33**, 49.
142. M. Doi and S. F. Edwards, *J. Chem. Soc., Faraday Trans. 2*, 1978, **74**, 1789, 1802, 1818; 1978, **75**, 32.
143. M. Doi, *J. Polym. Sci., Polym. Phys. Ed.*, 1980, **18**, 1005.
144. M. Doi, *J. Polym. Sci., Polym. Phys. Ed.*, 1983, **21**, 667.
145. W. Hess, *Macromolecules*, 1986, **19**, 387.
146. K. E. Evans and S. F. Edwards, *J. Chem. Soc., Faraday Trans. 2*, 1981, **10**, 1891, 1929.
147. M. Daoud and P. G. de Gennes, *J. Polym. Sci., Polym. Phys. Ed.*, 1979, **17**, 1971.
148. J. Klein, *Macromolecules*, 1978, **11**, 852.
149. W. W. Graessley, *Adv. Polym. Sci.*, 1982, **47**, 67.
150. J. Klein, *Macromolecules*, 1986, **19**, 105.
151. M. F. Marmonier and L. Léger, *Phys. Rev. Lett.*, 1985, **55**, 1078.

152. S. Onogi, S. Kimura, T. Kato, T. Masuda and N. Miyanaga, *J. Polym., Sci., Part C*, 1966, **15**, 381.
153. M. Adam and M. Delsanti, *J. Phys. (Orsay, Fr.)*, 1981, **42**, 1135.
154. M. Adam and M. Delsanti, *J. Phys. (Orsay, Fr.)*, 1982, **43**, 549.
155. L. Léger, H. Hervet and F. Rondelez, *Macromolecules*, 1981, **14**, 1732.
156. J. A. Wesson, I. Noh, T. Kitano and H. Yu, *Macromolecules*, 1984, **17**, 782.
157. E. D. von Meerwall, E. J. Amis and J. D. Ferry, *Macromolecules*, 1985, **18**, 260.
158. H. Deschamps and L. Léger, *Macromolecules*, 1986, **19**, 2760.
159. H. Deschamps, Ph.D. Thesis, University of Paris, 1983.
160. H. Janeschitz-Kriegel, 'Polymer Melt Rheology and Flow Birefringence', Springer, Berlin, 1983.
161. P. G. de Gennes, *J. Chem. Phys.*, 1974, **60**, 5030.
162. M. R. Mackley and A. Keller, *Philos. Trans. R. Soc. London, Ser. A*, 1975, **278**, 29.
163. R. Cressely, R. Hocquart and O. Scrivener, *Opt. Acta*, 1979, **26**, 1173.
164. S. Daoudi and F. Brochard, *Macromolecules*, 1978, **11**, 751.
165. G. Chauveteau, M. Tirrell and A. Omari, *J. Colloid Interface Sci.*, 1984, **100**, 41.
166. T. D. Long and J. L. Anderson, *J. Polym. Sci., Polym. Phys. Ed.*, 1984, **22**, 1261.
167. A. Baumgärtner, *Annu. Rev. Phys. Chem.*, 1984, **35**, 419.
168. K. Kremer, *Macromolecules*, 1983, **16**, 1632.
169. J. M. Deutsch, *Phys. Rev. Lett.*, 1982, **49**, 926.
170. G. Grest and K. Kremer, *Phys. Rev. A*, 1986, **33**, 3628.
171. D. Ceperley, H. Kalos and J. Lebowitz, *Phys. Rev. Lett.*, 1978, **41**, 313.
172. M. Doi and S. F. Edwards, *J. Chem. Soc., Faraday Trans. 2*, 1978, **74**, 918.
173. J. A. Odell, A. Keller and E. D. T. Atkins, *Macromolecules*, 1985, **18**, 1443.
174. P. G. de Gennes, *J. Phys. (Orsay, Fr.)*, 1975, **36**, 603.
175. M. Doi and N. Y. Kozuu, *J. Polym. Sci., Polym. Lett. Ed.*, 1980, **18**, 775.
176. J. M. Deutsch and M. Cates, *J. Phys. (Orsay, Fr.)*, 1986, **47**, 2121.
177. T. Odijk, *Macromolecules*, 1979, **12**, 688.
178. F. Oosawa, 'Polyelectrolytes', Dekker, New York, 1971.
179. Note added in proof: in the recent book by M. Doi and S. F. Edwards, 'The Theory of Polymer Dynamics', Oxford University Press, 1986, hydrodynamic properties of polymer liquids and melts are discussed in great detail; in particular the reptation model is thoroughly reviewed.

8

Nonlinear Viscoelastic Behavior

R. BYRON BIRD and CHARLES F. CURTISS
University of Wisconsin-Madison, Madison, WI, USA
ROBERT C. ARMSTRONG
Massachusetts Institute of Technology, Cambridge, MA, USA
and
OLE HASSAGER
Danmarks tekniske Højskole, Lyngby, Denmark

8.1   EXAMPLES OF VISCOELASTIC BEHAVIOR IN POLYMERIC LIQUIDS

Because of their complex structure the mechanical behavior of polymeric materials is not well described by the classical constitutive equations: Hooke's law (for elastic solids) or Newton's law (for viscous liquids). Polymeric materials are said to be 'viscoelastic' inasmuch as they exhibit both viscous and elastic responses. This viscoelastic behavior has played a key role in the development of the understanding of polymer structure.[1] Viscoelasticity is also important in the understanding of various measuring devices needed for rheometric measurements.[2-4] In the fluid dynamics of polymeric liquids, viscoelasticity also plays a crucial role.[5-13] Also in the polymer-processing industry it is necessary to include the role of viscoelastic behavior in careful analysis and design.[14,15] Finally there are important connections between viscoelasticity and flow birefringence.[16]

We begin by looking at twelve experiments (see Figure 1) in which the behavior of Newtonian fluids is contrasted with that of polymeric fluids. These experiments, and many more, show that the differences between Newtonian flow and polymer-liquid flow are not just small quantitative deviations, but rather major qualitative differences.

The twelve experiments of Figure 1 help to develop intuition about how polymeric liquids behave:

(a) The surface of a Newtonian liquid near a rotating rod shows a depression near the rod, whereas the polymeric liquid climbs the rod. This rod-climbing is called the 'Weissenberg effect'.[17-20]

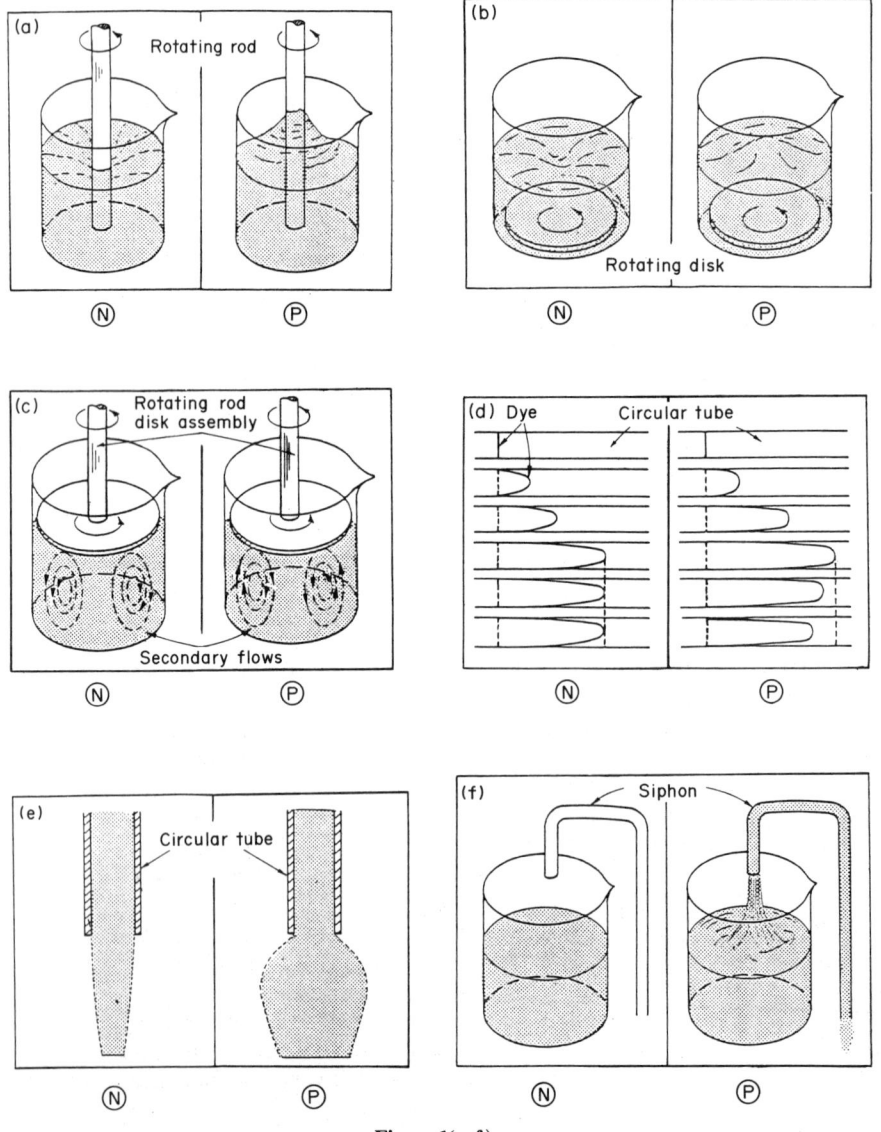

Figure 1(a–f)

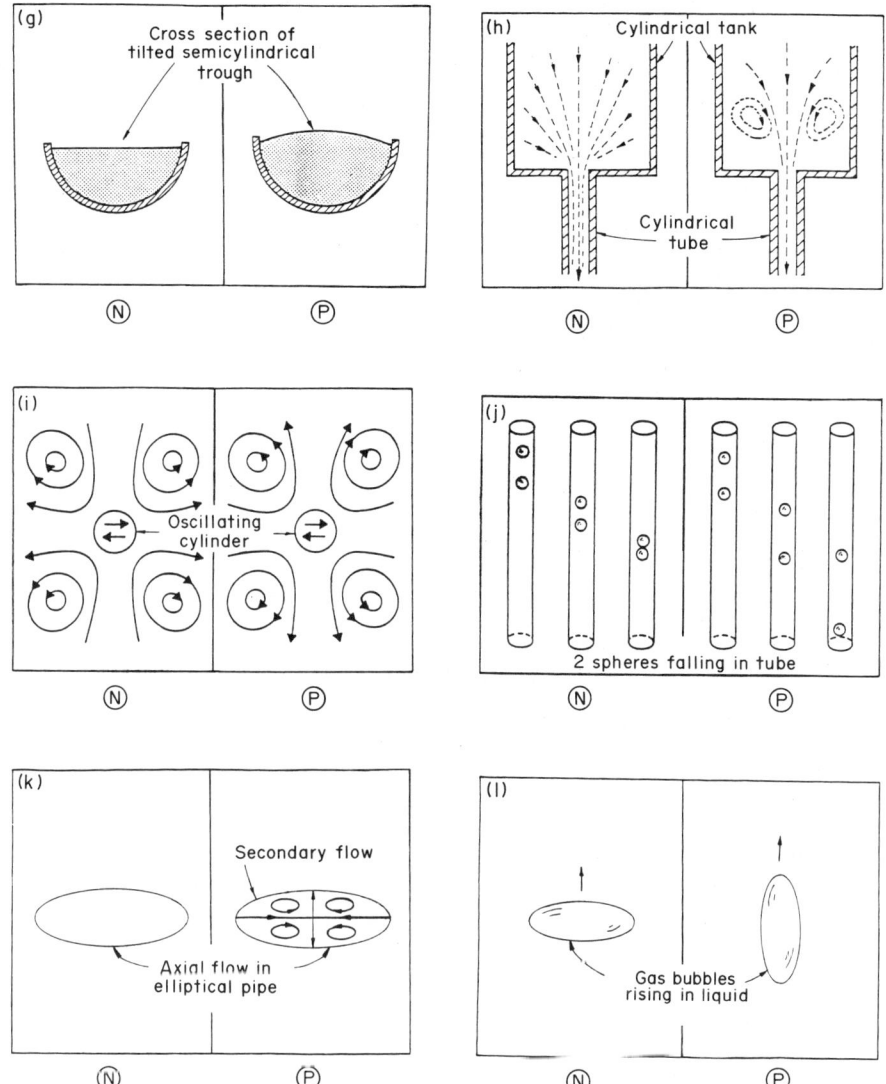

Figure 1 Comparison of the response of Newtonian N fluids and polymeric P fluids in twelve experiments. The behavior of polymeric fluids is qualitatively different from that of Newtonian fluids

(b) A rotating disk at the bottom of a beaker produces a dimple in the surface of a Newtonian liquid, whereas the polymeric liquid shows a lump.[20,21]

(c) A rotating disk placed at the surface of a Newtonian liquid in a beaker causes a primary flow in the tangential direction, but superposed on this primary motion is a weak secondary flow in which the fluid flows outward near the disk (because of centrifugal force) and then back upward near the axis of the beaker. For polymeric liquids the secondary flow goes in the opposite direction.[22–24]

(d) The motion of a liquid in a tube can be followed by watching a streak of dye that is inserted before the motion begins; six successive snapshots of the streak are shown. When the pump is turned off at the fourth snapshot, the Newtonian liquid comes to rest, but the polymeric liquid 'recoils' as shown in the fifth and sixth snapshots. This illustrates the 'memory' of the polymeric liquids. Because they do not return to their initial configuration, as a rubber band would after being stretched, we say that polymeric liquids have a 'fading memory'.[2,25]

(e) Polymeric liquids, on emerging from a tube or slit, exhibit a very substantial 'extrudate swell' (sometimes called 'die swell'); the cross-sectional area can increase by as much as a factor of nine (Newtonian fluids can exhibit a very small extrudate swell at very low flow rates; see ref. 6, chapter 8; ref. 2, p. 243; refs. 26, 27).

(f) For Newtonian fluids, siphons work only as long as the upstream end of the tube is beneath the surface of the liquid; when the tube is removed the siphoning action ceases. For polymeric fluids siphoning continues even if there is a gap of several centimeters between the liquid surface and the end of the tube.[28]

(g) When a liquid flows down a tilted trough of semicircular cross section in laminar flow, the Newtonian liquid surface is flat (except for the existence of meniscus effects), whereas the surface of polymeric liquids is slightly convex.[29,30]

(h) When fluids flow from a large-diameter tube into a small-diameter tube in slow flow, the polymeric liquid exhibits a vortex upstream from the small tube; fluid particles trapped in this vortex do not move on into the small tube.[31,32]

(i) When a cylinder oscillates in a plane containing the cylinder axis, a steady secondary motion is produced in a Newtonian fluid. The direction of the secondary motion in a polymeric liquid is just opposite to that in Newtonian liquids.[33]

(j) When two spheres are dropped one after the other into a Newtonian liquid, the second sphere always catches up with the first one and collides with it. In polymeric liquids the same thing happens if the second sphere is dropped very soon after the first one. However, if we wait longer than a critical time interval, then the spheres tend to move apart.[34]

(k) In axial flow through an elliptical pipe, Newtonian fluids in laminar flow show no secondary motions. Polymeric liquids, on the other hand, exhibit secondary flows with a division of the pipe cross-section into four cells.[47]

(l) When a tiny gas bubble rises in a Newtonian liquid it is oblate; in a polymeric liquid, however, the bubble assumes a prolate shape.[35-39]

Some of these experiments have been extensively studied, and many of them are now reasonably well understood (see ref. 5, chapter 2 for an extensive discussion of these and other experiments, along with additional literature citations and interpretations).

It should be evident from the twelve experiments cited above that polymeric liquids are not Newtonian liquids. It is a major challenge to find a constitutive equation that is capable of describing the flow of polymers; rheometric measurements, continuum mechanics and kinetic theories have all contributed to the development of constitutive equations. In the sections that follow a survey of these topics is given. It will be seen that in rheometric experiments only one or two components of the stress tensor can be measured in a few well-defined flows; this is not sufficient information to deduce the constitutive equation. Continuum mechanics provides some general guidelines as to constitutive relations within the limits of certain restrictions made at the outset; continuum mechanics also gives the key kinematic tensors and their interrelations. Kinetic theory explores the link between molecular structure and rheological responses; in any kinetic theory one has to begin by introducing a molecular model, and the kinetic theory results are necessarily restricted by the lack of realism of the model. Clearly no one of the above approaches — rheometry, continuum mechanics, kinetic theory — can possibly provide all the information needed to describe the mechanical behavior of polymeric materials. Generally speaking, kinetic theory can be used to suggest a form for the constitutive equation, which can be written in such a way that standard kinematic tensors developed in continuum mechanics appear, and rheometric measurements can be used to determine the model parameters that appear in the constitutive equation.

8.2 DEFINITIONS OF MATERIAL FUNCTIONS MEASURED IN RHEOMETRIC EXPERIMENTS

Although the experiments described in the foregoing section are very helpful for developing our intuition about the behavior of viscoelastic fluids such as polymers, they are not suitable for obtaining and cataloging information about specific polymeric materials. For the characterization of polymers it is necessary to make careful measurements of stresses in systems where the velocity or displacement field is known within rather strict limits. These 'rheometric experiments' provide information about one or more of the stress components as functions of shear rate, frequency, or of other controllable variables; these functions are generally referred to as 'material functions', since they are different for each material. Once these material functions have been measured, they can be used to test various empirical or molecular expressions for the stress tensor (that is, the constitutive equation), or they can be used to establish the values of the parameters that appear in these stress-tensor expressions.

There are two broad classes of rheometric experiments that have been developed: shear flows and shear-free flows. Within each category one can speak of steady flows and various unsteady flows; the latter can include step-function experiments, sinusoidal experiments and others. We now discuss these idealized flows and the material functions that are commonly defined. For a much more complete discussion, illustrated with experimental data on diverse polymeric systems, see ref. 5, chapter 3.

8.2.1 Shear Flows

The various idealized shear-flow experiments are shown in Figure 2. These experiments can be reasonably well approximated in a cone-and-plate viscometer.[2,4,5] This device can be programmed to give a wide variety of time-dependent shear flows, and the velocity gradient is known to be very nearly constant in the gap between the cone and plate. We now discuss some of the material functions that can be measured.

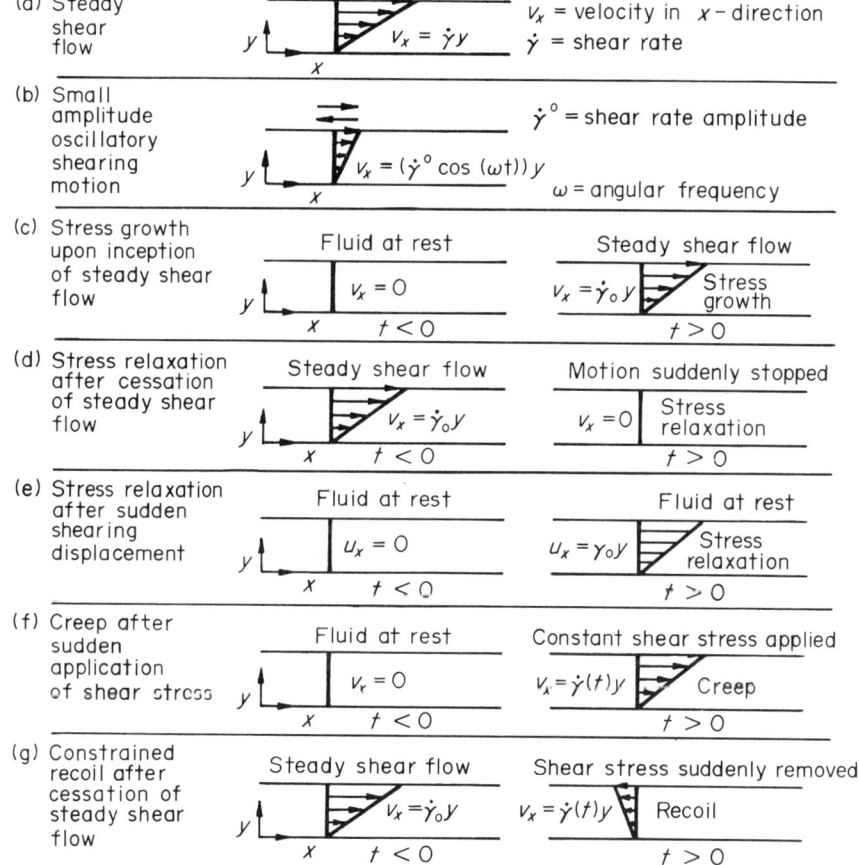

Figure 2 Various types of unidirectional shear flow experiments used in rheometry. Here v_x is the x-component of velocity and u_x is the x-component of the displacement (adapted from R. B. Bird, R. C. Armstrong and O. Hassager, 'Dynamics of Polymeric Liquids', 1st edn., Wiley, New York, 1977, vol. 1, p. 142)

8.2.1.1 *Steady-state shear flow*

This flow is shown in Figure 2(a) where the velocity distribution is given by $v_x = \dot{\gamma} y$, $v_y = 0$, $v_z = 0$ and $\dot{\gamma} = dv_x/dy$ is a constant. For this flow it is possible to measure a shear stress τ_{yx}, a 'first normal stress difference' $\tau_{xx} - \tau_{yy}$, and a 'second normal stress difference' $\tau_{yy} - \tau_{zz}$. These three quantities are in general strong functions of the shear rate $\dot{\gamma} = |dv_x/dy|$. It is conventional to define three 'viscometric functions', namely the (non-Newtonian) viscosity η (equation 1), the first normal stress coefficient Ψ_1 (equation 2) and the second normal stress coefficient Ψ_2 (equation 3), as follows

$$\tau_{yx} = -\eta\dot{\gamma} \tag{1}$$

$$\tau_{xx} - \tau_{yy} = -\Psi_1\dot{\gamma}^2 \tag{2}$$

$$\tau_{yy} - \tau_{zz} = -\Psi_2\dot{\gamma}^2 \tag{3}$$

Both η and Ψ_1 are known to be positive for amorphous polymers and Ψ_2 is negative and smaller in magnitude than Ψ_1. Most values of $-\Psi_2/\Psi_1$ reported in the literature are in the range between about 0.01 and 0.3. Both η and Ψ_1 decrease with the shear rate $\dot{\gamma}$, as may be seen in the sample experimental data of Figure 3.

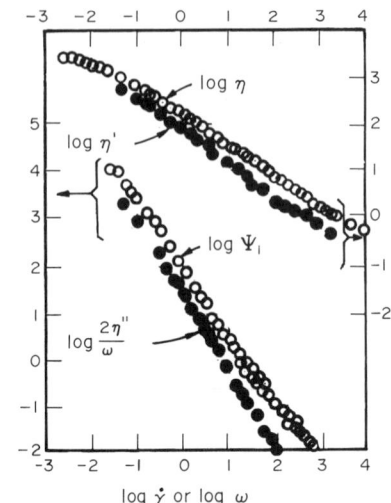

Figure 3   Material functions for a 1.5% polyacrylamide (Separan AP30) solution in a 50/50 mixture (by weight) of water and glycerine. The functions $\eta(\dot{\gamma})$, $\eta'(\omega)$, and $\eta''(\omega)$ are in Pa s, and $\Psi_1(\dot{\gamma})$ is in Pa s^2; both $\dot{\gamma}$ and ω are in s^{-1} (data are taken from J. D. Huppler, E. Ashare and L. A. Holmes, *Trans. Soc. Rheol.*, 1967, **11**, 159–179)

The dramatic decrease in the non-Newtonian viscosity with shear rate was one of the first properties to be measured extensively for polymeric liquids, because of its great importance in the description and prediction of polymer-melt flow in industrial problems. Clearly if the viscosity of a fluid is varying by several orders of magnitude in a flow field, this is an important effect that cannot be ignored.

The normal stress coefficients can be interpreted in the following way. The positive Ψ_1 causes the fluid to act as though it has an extra tension in the direction of the streamlines; the negative Ψ_2 can be thought of as an extra tension perpendicular to the streamlines and in the plane parallel to the two containing surfaces (see Figure 4). The tension corresponding to the negative Ψ_2 is generally smaller than the tension corresponding to the positive Ψ_1. Some of the experiments discussed in Section 8.1 can be interpreted qualitatively in terms of the normal stress coefficients: experiments (a), (b), (c), and (k) depend on both the first and second normal stress coefficients, with the former playing the predominant role; experiment (g) is a direct result of the fact that the second normal stress coefficient is negative. A number of other observed polymer flow phenomena can be interpreted qualitatively in terms of the normal stress coefficients.[5]

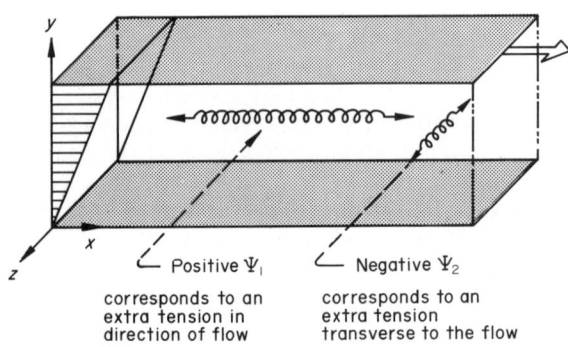

Positive Ψ_1
corresponds to an
extra tension in
direction of flow

Negative Ψ_2
corresponds to an
extra tension
transverse to the flow

Figure 4   Interpretation of Ψ_1 and Ψ_2 in the simple shear flow $v_x = \dot{\gamma}y$ as extra tensions, pictured here by springs

8.2.1.2 Unsteady-state shear flows

Many time-dependent flows have been studied by polymer chemists. A particularly important one is the small-amplitude sinusoidal shear flow (Figure 2b) where $v_x = \text{Re}\{\dot{\gamma}^0 e^{i\omega t}\}y$, $v_y = 0$, $v_z = 0$, in which $\dot{\gamma}^0$ is in general a complex quantity and ω is the frequency of oscillation; the notation $\text{Re}\{\ \}$ means 'the real part of'. Then because the amplitude of the oscillation is vanishingly small, the shear stress is also sinusoidal: $\tau_{yx} = \text{Re}\{\tau_{yx}^0 e^{i\omega t}\}$, where τ_{yx}^0 is complex. We now define a complex viscosity η^* by

$$\tau_{yx}^0 = -\eta^* \dot{\gamma}^0 \tag{4}$$

and $\eta^* = \eta' - i\eta''$ is a function of the frequency ω. In this sinusoidal shear flow experiment the normal stresses can also be measured and they oscillate with a frequency of 2ω about a nonzero mean. In Figure 3 we show data for the real and imaginary components of η^* for the same fluid for which the viscometric functions are plotted. For many fluids it has been found that there are simple relations between the viscometric properties and the components of the complex viscosity: the Cox–Merz rule[40] states that η is the same function of $\dot{\gamma}$ as $\eta' f^{0.5}$ is of ω; the Laun rule[41] states that Ψ_1 is the same function of $\dot{\gamma}$ as $(2\eta''/\omega)f^{0.7}$ is of ω. In these rules the quantity f is $1 + (\eta''/\eta')^2$; they can be used for making estimates of the viscometric functions if complex viscosity data are available.

The stress growth experiment of Figure 2(c) involves the study of the time evolution of the stresses when a fluid is brought instantaneously from a state of rest at $t = 0$ to a state of steady-state shear flow; this is an idealized experiment which presumes that in the experiment one can effectively minimize inertial effects and achieve the linear velocity profile within an acceptably short time interval. One can then define the 'growth functions' associated with the shear stress and the two normal-stress differences for $t > 0$ as follows

$$\tau_{yx} = -\eta^+(t, \dot{\gamma}_0)\dot{\gamma}_0 \tag{5}$$

$$\tau_{xx} - \tau_{yy} = -\Psi_1^+(t, \dot{\gamma}_0)\dot{\gamma}_0^2 \tag{6}$$

$$\tau_{yy} - \tau_{zz} = -\Psi_2^+(t, \dot{\gamma}_0)\dot{\gamma}_0^2 \tag{7}$$

These three functions of time and imposed shear rate have been successfully measured in cone-and-plate instruments.

The stress relaxation experiment in Figure 2(d) involves the measurement of the time evolution of the stresses after a steady-state shear flow has been suddenly stopped at time $t = 0$. Here again it is necessary to design the experiment in such a way that inertial effects are minimized. One can then define three 'relaxation functions' as follows for $t > 0$

$$\tau_{yx} = -\eta^-(t, \dot{\gamma}_0)\dot{\gamma}_0 \tag{8}$$

$$\tau_{xx} - \tau_{yy} = -\Psi_1^-(t, \dot{\gamma}_0)\dot{\gamma}_0^2 \tag{9}$$

$$\tau_{yy} - \tau_{zz} = -\Psi_2^-(t, \dot{\gamma}_0)\dot{\gamma}_0^2 \tag{10}$$

These functions have also been measured by cone-and-plate rheometry. Sample experimental data for the stress growth and stress relaxation functions can be found in chapter 3 of ref. 5, as well as for the other rheometric experiments shown in Figure 2. The analysis of the rheometric equipment used in these tests is discussed in chapter 10 of ref. 5 as well as in ref. 4.

8.2.2 Shear-free Flows

Several shear-free flows are shown in Figure 5. In each of these flows there are only normal stresses (and no shear stresses). These flows are much more difficult to maintain than shear flows and as a consequence they have not been so extensively studied. They are, however, extremely important, since the data obtained from them can provide crucial tests of continuum and molecular theories.

8.2.2.1 Steady-state shear-free flows

Shear-free flows are described by the velocity field $v_x = -\frac{1}{2}\dot{\varepsilon}(1+b)x$, $v_y = -\frac{1}{2}\dot{\varepsilon}(1-b)y$, $v_z = \dot{\varepsilon}z$ where $0 \le b \le 1$ and $\dot{\varepsilon}$ is the elongation rate, which is a constant for steady shear-free flows. There are three special shear-free flows that correspond to the three diagrams shown in Figure 5: elongational

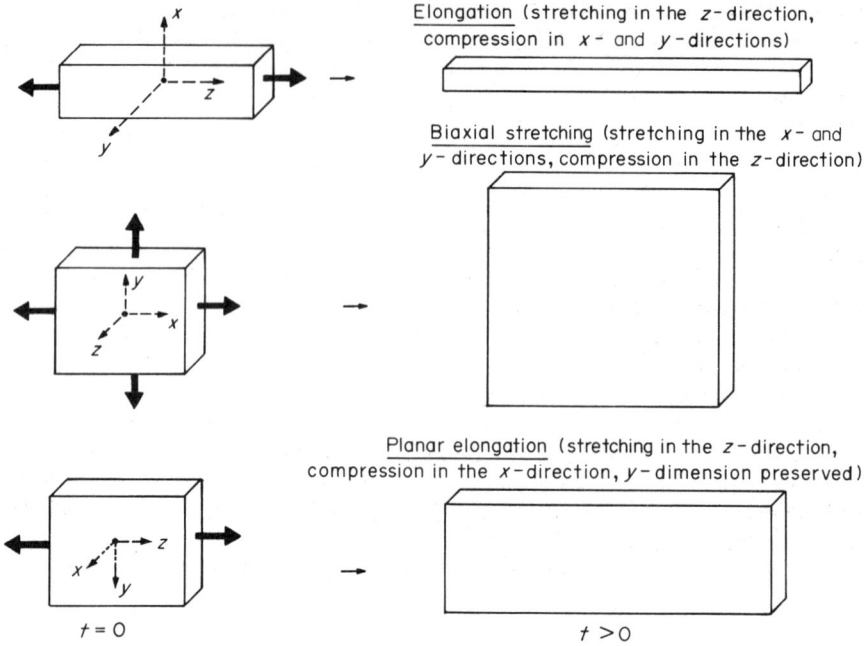

Figure 5 Shear-free deformations at constant volume

flow $b=0$, $\dot{\varepsilon}>0$; biaxial stretching flow $b=0$, $\dot{\varepsilon}<0$; and planar elongation flow $b=1$. For the shear-free flows one can define two material functions associated with two of the normal-stress differences

$$\tau_{zz}-\tau_{xx}=-\bar{\eta}_1\dot{\varepsilon} \tag{11}$$

$$\tau_{yy}-\tau_{xx}=-\bar{\eta}_2\dot{\varepsilon} \tag{12}$$

The two functions $\bar{\eta}_1$ and $\bar{\eta}_2$ depend on the parameter b as well as on the elongation rate $\dot{\varepsilon}$ in steady shear-free flows. When $b=0$, the function $\bar{\eta}_2$ is zero, and $\bar{\eta}_1$ is replaced by the symbol $\bar{\eta}$, which is called the 'elongational viscosity'. The elongational viscosity describes the resistance to elongational flow if $\dot{\varepsilon}$ is positive and the resistance to biaxial stretching if $\dot{\varepsilon}$ is negative (the terms 'extensional viscosity' and 'Trouton viscosity' have also been used for $\bar{\eta}$).

There is some question as to whether or not it is possible to attain and preserve a steady-state flow in these shear-free flow experiments, and hence only a limited amount of experimental data has been reported (for some sample data see ref. 5, chapter 3). The elongational viscosity for positive elongation rates has been extensively investigated by kinetic theories.[42] According to the dilute-solution kinetic theories the most realistic molecular models suggest that the elongational viscosity should increase monotonically from the zero-elongation-rate value of $3\eta_0$ and then level off at some rather high value (here η_0 is the zero-shear-rate viscosity). Some polymer melt kinetic theories, both network and reptation theories, predict that the elongational viscosity increases with elongation rate, goes through a maximum, and then decreases; these theories suggest that elongational viscosity is less dependent on elongation rate than the (shear) viscosity is on shear rate.

8.2.2.2 *Unsteady-state shear-free flows*

Because of the difficulty of attaining steady-state shear-free flows, it is thought that it may be preferable to study some of the unsteady-state flows experimentally. One can, of course, study the shear-free analogs of any of the unsteady-state experiments depicted in Figure 2 for shear flows. For example, one may define growth functions analogously to equations (5)–(7)

$$\tau_{zz}-\tau_{xx}=-\bar{\eta}_1^{+}\dot{\varepsilon}_0 \tag{13}$$

$$\tau_{yy}-\tau_{xx}=-\bar{\eta}_2^{+}\dot{\varepsilon}_0 \tag{14}$$

The material functions $\bar{\eta}_1^+$ and $\bar{\eta}_2^+$ depend on both t and $\dot{\varepsilon}_0$, and of course on the parameter b that specifies the type of shear-free flow. For elongational flow, with $b=0$ and $\dot{\varepsilon}_0$ positive, $\bar{\eta}_1^+$ becomes $\bar{\eta}^+$, the 'elongational stress growth function.' This quantity has been measured for a number of polymer melts. Further information on elongational properties can be found in several extensive reviews.[43,44]

8.3 KINEMATIC TENSORS USED IN CONTINUUM MECHANICS AND KINETIC THEORY

In this section definitions are given of the main kinematic tensors (see ref. 5, chapter 9) that are needed for the continuum mechanics description of viscoelastic materials as well as for the presentation of kinetic theory results. Some of these tensors are defined naturally in terms of the velocity field of the material, whereas others are defined easily in terms of the displacement functions that describe the motion of fluid particles. Inasmuch as the velocity field and the displacement functions are themselves interrelated, it is possible to interrelate the two groups of kinematic tensors. Here the emphasis is on working definitions of the kinematic tensors and not on their derivation from the motion of a convected coordinate system, which is a standard starting point for the discussion of continuum mechanics;[2,3,5] an important basic reference for the kinematics and dynamics of continuous media is a paper by Oldroyd.[45]

8.3.1 Kinematic Tensors Derived from the Velocity Field

We designate the velocity field as a function of position r and time t by $v = v(r, t)$. The components of the velocity gradient tensor ∇v and its transpose $(\nabla v)^\dagger$ are given by

$$(\nabla v)_{ij} = \frac{\partial}{\partial x_i} v_j; \qquad (\nabla v)_{ij}^\dagger = \frac{\partial}{\partial x_j} v_i \qquad (15)$$

Here x_i is the ith cartesian coordinate (*i.e.* $x_1 \equiv x$, $x_2 \equiv y$, $x_3 \equiv z$), and v_i is the ith cartesian component of the velocity vector (*i.e.* $v_1 \equiv v_x$, $v_2 \equiv v_y$, $v_3 \equiv v_z$). Then the rate of strain tensor (or rate of deformation tensor) $\dot{\gamma}$ is defined as

$$\dot{\gamma} = \nabla v + (\nabla v)^\dagger \qquad (16)$$

In some equations it is useful to use the symbols $\gamma^{(1)}$ and $\gamma_{(1)}$ instead of $\dot{\gamma}$ for the rate of strain tensor. Higher order rate of strain tensors are defined by the operations[45]

$$\gamma^{(n+1)} = \frac{D}{Dt} \gamma^{(n)} + \{(\nabla v) \cdot \gamma^{(n)} + \gamma^{(n)} \cdot (\nabla v)^\dagger\} \quad (n \geq 1) \qquad (17)$$

$$\gamma_{(n+1)} = \frac{D}{Dt} \gamma_{(n)} - \{(\nabla v)^\dagger \cdot \gamma_{(n)} + \gamma_{(n)} \cdot \nabla v\} \quad (n \geq 1) \qquad (18)$$

in which D/Dt is the 'substantial derivative' (or 'material derivative') given by $D/Dt = \partial/\partial t + (v \cdot \nabla)$; the $\gamma^{(n)}$ are often referred to as the Rivlin–Ericksen tensors,[46] and the operations on the right-hand side of equations (17) and (18) are referred to as 'convected derivatives'.[45] A table of these and other kinematic tensors for unsteady-state shear and shear-free flows is given in appendix C of ref. 5.

8.3.2 Kinematic Tensors Derived from the Displacement Functions

In a flow field a particle is located at position x_1, x_2, x_3 (abbreviated as x) at the current time t; the same particle is located at position x'_1, x'_2, x'_3 (abbreviated as x') at some previous time t'. Then the motion can be described by giving the relations

$$x'_i = x'_i(x, t, t'); \quad \text{or} \quad x_i = x_i(x', t', t) \qquad (19)$$

which are referred to as the 'displacement functions'.[45] It is possible in principle to obtain the velocity field from the displacement functions and *vice versa*, although analytical relations are not always easy to find.

The displacement functions can be used to define the components of two 'displacement gradient tensors'

$$\Delta_{ij}(x, t, t') = \partial x'_i(x, t, t')/\partial x_j \tag{20}$$

$$E_{ij}(x, t, t') = \partial x_i(x', t', t)/\partial x'_j \tag{21}$$

Note that the x, t in the arguments of Δ and E serve as a 'particle label' *i.e.* the fluid element located at position x at time t. Next, two 'finite strain tensors', widely used in continuum mechanics, can be defined

$$B^{-1} = \{\Delta^\dagger \cdot \Delta\} = \text{Cauchy strain tensor} \tag{22}$$

$$B = \{E \cdot E^\dagger\} = \text{Finger strain tensor} \tag{23}$$

For some purposes it is more convenient to use two closely related 'relative (finite) strain tensors'

$$\gamma^{[0]}(x, t, t') = \{\Delta^\dagger \cdot \Delta\} - \delta \tag{24}$$

$$\gamma_{[0]}(x, t, t') = \delta - \{E \cdot E^\dagger\} \tag{25}$$

in which δ is the unit tensor, whose components are given by the Kronecker delta ($\delta_{ij} = 1$ if $i = j$; $\delta_{ij} = 0$ if $i \neq j$). These tensors give the strain at time t' with respect to the state of the material at the current time t. The relative strain tensors just defined are convenient because they simplify to the 'infinitesimal strain tensor' $\gamma(x, t, t')$ in the limit of infinitesimally small displacement gradients. Successive differentiations of the relative strain tensors with respect to t' give sets of higher order quantities

$$\gamma^{[n+1]} = \frac{\partial}{\partial t'} \gamma^{[n]} \quad (n \geq 0) \tag{26}$$

$$\gamma_{[n+1]} = \frac{\partial}{\partial t'} \gamma_{[n]} \quad (n \geq 0) \tag{27}$$

The kinematic tensors designated with indices in brackets [ ] (derived from the displacement functions) are related to the kinematic tensors with indices in parentheses ( ) (derived from the velocity field) as indicated in Figure 6. The kinematic tensors with indices in brackets depend on two times — the current time t and the past time t' — and they appear in the integrands of time integrals in integral constitutive equations; the kinematic tensors with indices in parentheses depend on the current time t only, and they appear in differential constitutive equations.

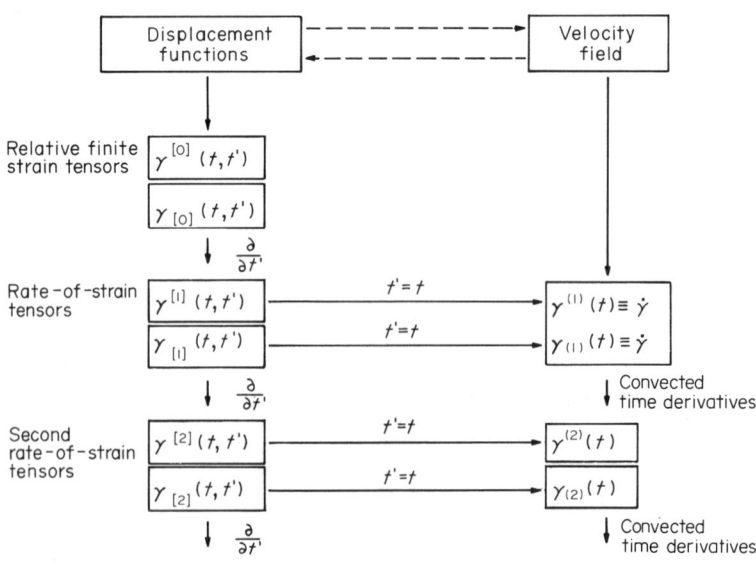

Figure 6 Relations among the kinematic tensors. Note that all the kinematic tensors can be obtained from the relative (finite) strain tensors

8.3.3 Scalar Invariants Derived from Kinematic Tensors

The scalar invariants of certain kinematic tensors play important roles in continuum mechanics, constitutive equations and kinetic theories. Of particular interest are the invariants of the rate of strain tensor and of the finite-strain tensors. There are many ways to define these invariants, and we give only those definitions that are used in later sections; in addition, much research has been done on the definition of joint invariants of several tensors.[46,47]

Three scalar invariants of the rate of deformation tensor may be defined as follows

$$I = \operatorname{tr} \dot{\gamma} = \Sigma_i \dot{\gamma}_{ii} \tag{28}$$

$$II = \operatorname{tr} \dot{\gamma}^2 = \Sigma_i \Sigma_j \dot{\gamma}_{ij} \dot{\gamma}_{ji} \tag{29}$$

$$III = \operatorname{tr} \dot{\gamma}^3 = \Sigma_i \Sigma_j \Sigma_k \dot{\gamma}_{ij} \dot{\gamma}_{jk} \dot{\gamma}_{ki} \tag{30}$$

The invariant I is always zero when the assumption of incompressibility is made. For the shear flows discussed in Section 8.2.1, both I and III are zero and $II = \dot{\gamma}_{yx}^2 = (dv_x/dy)^2$. For the shear-free flows of Section 8.2.2, the invariant I is zero, $II = (6 + 2b^2)\dot{\varepsilon}^2$ and $III = (6 - 6b^2)\dot{\varepsilon}^3$. Another quantity often used is the magnitude of the rate of strain tensor $\dot{\gamma} = \sqrt{(1/2)II}$ which is defined to be a positive quantity; for shear flows $\dot{\gamma} \equiv |\dot{\gamma}_{yx}|$, which is called the 'shear rate'.

Three scalar invariants of the Finger strain tensor may defined in the following way

$$I_1 = \operatorname{tr} \boldsymbol{B} \tag{31}$$

$$I_2 = \frac{1}{2}[(\operatorname{tr} \boldsymbol{B})^2 - \operatorname{tr} \boldsymbol{B}^2] = \operatorname{tr} \boldsymbol{B}^{-1} \tag{32}$$

$$I_3 = \det \boldsymbol{B} \tag{33}$$

When the material is assumed to be incompressible, the third invariant I_3 is found to be unity. Note that I_1 and I_2 are just the traces of the Finger and Cauchy strain tensors respectively. For shear flows $I_1 = I_2 = 3 + \gamma_{yx}^2$, where $\gamma_{yx}(t, t') = \int_t^{t'} \dot{\gamma}_{yx}(t'')dt''$; for shear-free flows $I_1 = \lambda_x^2 + \lambda_y^2 + \lambda_z^2$ and $I_2 = \lambda_x^{-2} + \lambda_y^{-2} + \lambda_z^{-2}$ where $\lambda_x = \exp[\frac{1}{2}(1+b)\varepsilon]$, $\lambda_y = \exp[\frac{1}{2}(1-b)\varepsilon]$ and $\lambda_z = \exp(-\varepsilon)$, with $\varepsilon(t, t') = \int_t^{t'} \dot{\varepsilon}(t'')dt''$.

One use of the invariants I_1 and I_2 is demonstrated in Figure 7, where the range of experimentally accessible combinations of these invariants is shown. Note that simple shearing motions correspond to the line of slope unity, and that the shaded regions correspond to various shear-free motions. It can be seen from this diagram why shear-free flows are so important in the testing of constitutive equations and kinetic theory results. The asymptotes of the boundaries of the shaded regions for large I_1 and I_2 are $I_1 = \frac{1}{4}I_2^2$ for uniaxial elongation, and $I_2 = \frac{1}{4}I_1^2$ for biaxial stretching.

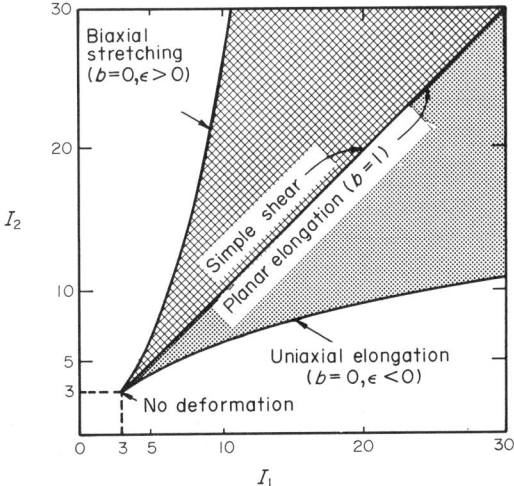

Figure 7 Chart in I_1, I_2-space showing the accessible regions for incompressible materials. The comprehensiveness of rheometric testing depends on how much of the shaded region is sampled by the experiments

8.4 CONSTITUTIVE EQUATIONS FOR POLYMERIC LIQUIDS

A major challenge in polymer rheology is to devise constitutive equations that can be used, along with the equations of change, to solve flow and heat-transfer problems. The equations of change for incompressible fluids are

$$\text{Continuity:} \quad (\nabla \cdot \boldsymbol{v}) = 0 \tag{34}$$

$$\text{Motion:} \quad \rho \frac{D\boldsymbol{v}}{Dt} = -[\nabla \cdot \boldsymbol{\pi}] + \rho \boldsymbol{g} \tag{35}$$

$$\text{Energy:} \quad \rho \hat{C}_p \frac{DT}{Dt} = -(\nabla \cdot \boldsymbol{q}) - (\boldsymbol{\pi} : \nabla \boldsymbol{v}) \tag{36}$$

Here ρ is the fluid density, $\boldsymbol{g}$ is the gravitational acceleration, $\hat{C}_p$ is the heat capacity at constant pressure (per unit mass), and $\boldsymbol{q}$ is the heat flux (taken to be given by Fourier's law $\boldsymbol{q} = -\lambda \nabla T$, where λ is the thermal conductivity). It is convenient to split the (total) stress tensor $\boldsymbol{\pi}$ into two parts $\boldsymbol{\pi} = p\boldsymbol{\delta} + \boldsymbol{\tau}$ where p is an isotropic pressure, $\boldsymbol{\delta}$ is the unit tensor, and $\boldsymbol{\tau}$ is the (extra) stress tensor, which is that part of the total stress tensor that vanishes when the fluid is at rest.

For incompressible Newtonian fluids the stress tensor $\boldsymbol{\tau}$ is given by

$$\boldsymbol{\tau} = -\mu \dot{\boldsymbol{\gamma}} \tag{37}$$

where μ is the viscosity, which is a constant for a given temperature, pressure and composition. For polymeric liquids the Newtonian fluid model is insufficient, as was demonstrated by the experiments in Section 8.1. For polymeric liquids the stress tensor cannot be linear in the velocity gradients and it must in general involve time derivatives and time integrals. Some of the more popular constitutive equations are considered here: (i) the retarded motion expansion: this can be thought of roughly as a Taylor series about the Newtonian fluid, and is useful in flow systems in which the velocity gradients and their time derivatives are small; (ii) the Criminale–Ericksen–Filbey equation: this equation is applicable only to steady-state shear flow, and can be obtained by collapsing the retarded motion expansion for this particular flow; (iii) the general linear viscoelastic model: this equation is applicable to arbitrary unsteady flows as long as the displacement gradients are infinitesimally small; and (iv) nonlinear viscoelastic models: these equations are empirical equations or equations obtained from molecular theories that one hopes are applicable to all types of flows; some of these contain time derivatives of the stress ('differential models' or 'rate models') and others contain time integrals ('integral models'). One can, of course, attempt to make some very minimal set of assumptions about the fluid response, and continuum mechanicists have devoted considerable time and effort to developing very general theories.[3,45-48] In the development of constitutive equations it is generally assumed that the stress tensor is symmetric and that it should be rheologically invariant,[45] in the sense that the stress tensor should not have any unwanted dependence on the orientation of a fluid element in space; we also assume that the fluids are incompressible. The constitutive equations below all satisfy these requirements, except for the linear viscoelastic models, which are not rheologically invariant; however, they can be obtained from rheologically invariant models by a suitable limiting process.

8.4.1 Retarded Motion Expansion

The retarded motion expansion (ref. 5, chapter 6) written through terms of third order, is[46]

$$\boldsymbol{\tau} = -[b_1 \boldsymbol{\gamma}_{(1)} + b_2 \boldsymbol{\gamma}_{(2)} + b_{11} \{\boldsymbol{\gamma}_{(1)} \cdot \boldsymbol{\gamma}_{(1)}\} + b_3 \boldsymbol{\gamma}_{(3)} + b_{12} \{\boldsymbol{\gamma}_{(1)} \cdot \boldsymbol{\gamma}_{(2)} + \boldsymbol{\gamma}_{(2)} \cdot \boldsymbol{\gamma}_{(1)}\} + b_{1:11} (\boldsymbol{\gamma}_{(1)} : \boldsymbol{\gamma}_{(1)}) \boldsymbol{\gamma}_{(1)} + \cdots] \tag{38}$$

The term containing the constant b_1 is the first-order term and corresponds to the Newtonian model of equation (37). The terms containing the constants b_2 and b_{11} are second-order terms, and if the expansion is truncated after these terms, one speaks of the 'second-order fluid'. Similarly, the terms containing the constants b_3, b_{12}, and $b_{1:11}$ are third-order terms, and if the expansion is truncated after these terms, the resulting constitutive equation is termed the 'third-order' fluid. This series systematically displays the deviations from Newtonian behavior as the velocity gradients and their time derivatives increase. There have been only very limited measurements of the constants other than b_1, which is identical to the zero-shear-rate viscosity η_0. However, for a wide range of molecular models, it is found[49] from kinetic theories of dilute solutions and melts that the second-order constants are negative, with $|b_2| > |b_{11}|$; for the third-order constants $b_3 > b_{12} > b_{1:11}$.

This model has been found to be useful in studying the slow motion of viscoelastic fluids around particles, droplets and bubbles, and in predicting the directions of secondary flows in rotating systems. Furthermore it is very helpful in providing a framework for the presentation of kinetic theory results obtained by perturbation theories. The retarded motion expansion is not, however, useful for most industrial flow problems, in which large velocity gradients or rapid time responses are generally encountered.

8.4.2 The Criminale–Ericksen–Filbey Equation[50]

If the retarded motion expansion is written for steady shear flows, there is a considerable simplification inasmuch as $\gamma_{(n)} = 0$ for $n \geq 3$. As a result the series 'collapses' to the form

$$\tau = -\eta\gamma_{(1)} + \frac{1}{2}\Psi_1\gamma_{(2)} - \Psi_2\{\gamma_{(1)} \cdot \gamma_{(1)}\} \tag{39}$$

in which η, Ψ_1, and Ψ_2 are the 'viscometric functions' discussed in Section 8.2.1.1, these being functions of the shear rate $\dot{\gamma}$. Hence once the viscometric functions have been determined by means of some rheometric experiment(s), the flow in all steady shear flows can be obtained (*e.g.* in tube flow, slit flow, and axial, tangential and helical flow between cylinders).

If the normal-stress terms in equation (39) are omitted, one obtains a much simpler constitutive equation, known as the 'generalized Newtonian fluid'.[51,52] This equation has been used for half a century by engineers in design work; it takes into account the change of viscosity with shear rate, which is the salient physical effect that needs to be included in steady-state flow problems involving computation of flow rates *vs.* pressure drops and torques *vs.* angular velocities. There are many empirical expressions available for the non-Newtonian viscosity function $\eta(\dot{\gamma})$ containing a small number of constants. The most popular of these has certainly been the 'power law' model:[53,54] $\eta(\dot{\gamma}) = m\dot{\gamma}^{n-1}$, in which m and n are constants characteristic of each fluid, with n having values between 0 and 1. A function describing the viscosity curves more realistically is the Carreau model:[55] $\eta/\eta_0 = [1 + (\lambda\dot{\gamma})^2]^{(n-1)/2}$, in which η_0 is the zero-shear-rate viscosity, λ is a time constant, and n is a dimensionless quantity between 0 and 1. Similar empirical functions for the normal stress coefficients can also be used in equation (39).

8.4.3 Linear Viscoelastic Models

The constitutive equation for the Newtonian fluid is linear in both stress and velocity gradients. If it is desired to retain a linear relation between the stress and rate of strain tensors, but allow for time derivatives, then one can write a relation of the form

$$\tau + \lambda_1 \frac{\partial}{\partial t}\tau = -\eta_0\left(\dot{\gamma} + \lambda_2 \frac{\partial}{\partial t}\dot{\gamma}\right) \tag{40}$$

in which λ_1 is called the 'relaxation time' and λ_2 is called the 'retardation time'; η_0 is the zero-shear-rate viscosity as before. Equation (40) is called the 'Jeffreys model';[56] if $\lambda_2 = 0$, it reduces to the 'Maxwell model'.[57] Of course one could add second, third and higher derivatives of the stress tensor and the rate of strain tensor and obtain a more complex model with additional time constants. These are all called 'linear viscoelastic models' (see ref. 5, chapter 5).

The Maxwell, Jeffreys and more complicated models can be integrated once with respect to time to obtain equations of the form

$$\tau = -\int_{-\infty}^{t} G(t-t')\dot{\gamma}(t')dt' \tag{41}$$

the form of the 'relaxation modulus' $G(t-t')$ depending on the form of the original equation. For example, for the Maxwell model $G(s) = (\eta_0/\lambda_1)\exp(-s/\lambda_1)$, in which the exponential provides a representation of the 'fading memory' of the fluid; the stress tensor is affected more strongly by kinematics of the recent past than by kinematic events of the distant past. Equation (41) can be further integrated by parts to give an equation of the form

$$\tau = + \int_{-\infty}^{t} M(t-t')\gamma(t,t')dt' \tag{42}$$

in which $\gamma(t,t')$ is the infinitesimal strain tensor (a symmetrized displacement gradient tensor) and the 'memory function' $M(t-t')$ depends on the form of the original equation; for example for the Maxwell model $M(s) = (\eta_0/\lambda_1^2)\exp(-s/\lambda_1)$. In general $M(t-t') = \partial G(t-t')/\partial t'$. Equations (41) and (42) with the functions $G(s)$ and $M(s)$ unspecified are referred to as the 'general linear viscoelastic model', and this model is the most general linear relation that can exist between the stress and kinematic tensors.

The general linear viscoelastic model has been widely used for the analysis of experiments in which the polymer sample is never allowed to stray very far from its initial condition.[1] For example it is not difficult to show that the complex viscosity is related to the relaxation modulus by $\eta^*(\omega) = \int_0^\infty G(s) e^{-i\omega s} ds$. The linear viscoelastic models discussed above have formed the starting point from which various empirical nonlinear viscoelastic models have been derived.

8.4.4 Nonlinear Viscoelastic Models (Differential Type)

Many nonlinear viscoelastic models of the differential type (ref. 5, chapter 7) have been proposed during the last three or four decades. Nonlinearities can be introduced by modifying differential linear viscoelastic models such as that in equation (40) in several ways: (a) replacement of time derivatives by convected derivatives (to satisfy the requirement of rheological invariance); (b) inclusion of additional terms that are nonlinear in stresses and/or velocity gradients; and (c) replacement of constants such as the zero-shear-rate viscosity and the time constants by functions of invariants of the rate of strain tensor.

An example of (a) is the 'convected Jeffreys model' or 'Oldroyd B model'[45]

$$\tau + \lambda_1\tau_{(1)} = -\eta_0(\gamma_{(1)} + \lambda_2\gamma_{(2)}) \tag{43}$$

in which $\tau_{(1)} = D\tau/Dt - \{(\nabla v)^\dagger \cdot \tau + \tau \cdot \nabla v\}$ is a convected derivative of the stress tensor. This constitutive equation predicts that all three viscometric functions are constants, with $\eta = \eta_0$, $\Psi_1 = 2\eta_0(\lambda_1 - \lambda_2)$, and $\Psi_2 = 0$. It can describe a number of the unsteady shear-flow properties qualitatively, but not quantitatively. It yields a steady-state elongational viscosity that goes to infinity at a finite value of the elongation rate. The model is thus not satisfactory for describing a wide range of rheological phenomena even qualitatively.

An example of (b) is the Oldroyd eight-constant model[58]

$$\tau + \lambda_1\tau_{(1)} + \frac{1}{2}\lambda_3\{\gamma_{(1)}\cdot\tau + \tau\cdot\gamma_{(1)}\} + \frac{1}{2}\lambda_5(\text{tr }\tau)\gamma_{(1)} + \frac{1}{2}\lambda_6(\tau:\gamma_{(1)})\delta = -\eta_0[\gamma_{(1)} + \lambda_2\gamma_{(2)} + \lambda_4\{\gamma_{(1)}\cdot\gamma_{(1)}\} + \frac{1}{2}\lambda_7(\gamma_{(1)}:\gamma_{(1)})\delta] \tag{44}$$

which contains a zero-shear-rate viscosity η_0 and seven time constants $\lambda_1, \lambda_2, \ldots, \lambda_7$. This model contains equation (43) as well as the second-order-fluid model. It does give shear-rate-dependent viscometric properties and is generally an improvement over equation (43). With some of the constants set equal to zero it has been widely used for trial flow calculations and rheological studies; however, inasmuch as the shapes of the material functions are not very well described, it cannot be regarded as a generally satisfactory model.

As an example of (c) we cite the White–Metzner model[59]

$$\tau + \frac{\eta(\dot{\gamma})}{G}\tau_{(1)} = -\eta(\dot{\gamma})\gamma_{(1)} \tag{45}$$

This modification of the convected Maxwell model contains one constant G (an elastic modulus) and the non-Newtonian viscosity function $\eta(\dot{\gamma})$. It describes the shear-rate dependence of the viscosity perfectly and the first normal stress coefficient rather well. In steady elongational flow it gives an infinite elongational viscosity, and does not simplify properly in the linear viscoelastic limit. Nonetheless it has been found to be useful in exploratory flow calculations aimed at assessing the interaction of shear thinning and memory.

One can object to the above models on the ground that they cannot describe the complete spectrum of relaxation times that one observes, for example, in the sinusoidal oscillatory shear experiment. This objection can be overcome by taking a superposition of any of the above models, thereby obtaining models of considerable flexibility. Rather than resorting to this kind of empiricism, however, it is probably preferable to seek guidance from molecular theories as to potentially successful forms for constitutive equations.

8.4.5 Nonlinear Viscoelastic Models (Integral Type)

In linear viscoelasticity the integral model (ref. 5, chapter 8) of equation (42) is the most general linear relation between the stress and infinitesimal strain tensors. Nonlinearities may be introduced into this model in several ways: (a) by replacing the infinitesimal strain tensor by one of the relative (finite) strain tensors given in equations (24) and (25); (b) by allowing the memory function to contain the invariants of the finite strain tensors; (c) by introducing products of the finite strain tensors, or, equivalently, use both of the strain tensors of equations (24) and (25); or (d) by including double, triple, and other multiple integrals containing various multiple products of strain tensors.

An example of (a) is the Lodge elastic liquid[2]

$$\tau = \int_{-\infty}^{t} M(t-t')\gamma_{[0]}(t')dt' \tag{46}$$

which simplifies to the general linear viscoelastic model of equation (42) in the limit of infinitesimal displacement gradients. This model gives constant viscometric functions: $\eta = M_1$, $\Psi_1 = M_2$ and $\Psi_2 = 0$, where $M_j = \int_0^\infty M(s)s^j ds$, and hence is not capable of describing a wide range of polymers. It can, of course, describe linear viscoelastic phenomena exactly and is regarded as useful for describing slight departures from linear viscoelasticity. The model has been very helpful as a starting point for developing useful empirical constitutive equations.

As an example of (b) the Wagner model[60] can be cited; this is obtained as an extension of the Lodge elastic liquid

$$\tau = \int_{-\infty}^{t} M(t-t')h(I_1,I_2)\gamma_{[0]}(t')dt' \tag{47}$$

Here $h(I_1,I_2)$, the 'damping function', is a function of the invariants of the Finger strain tensor given in equations (31) and (32); the damping function is determined by requiring the constitutive equation to describe shear and elongational flow data. Extensive comparisons with experimental data[60,61] show that this rather simple empiricism is extremely useful. Equation (47) gives a value of zero for the second normal stress coefficient.

A constitutive equation that includes the features in both (b) and (c) above is the factorized K-BKZ equation[62-65]

$$\tau = \int_{-\infty}^{t} M(t-t')\left[\frac{\partial W(I_1,I_2)}{\partial I_1}\gamma_{[0]} + \frac{\partial W(I_1,I_2)}{\partial I_2}\gamma^{[0]}\right]dt' \tag{48}$$

in which W, the 'potential function', is a function of the invariants of the Finger strain tensor in equations (31) and (32). This model has enormous flexibility and is capable of giving realistic descriptions of a wide range of rheological phenomena, depending on the choices of the memory function M and the potential function W. The K-BKZ equation has been compared extensively against experimental data and in addition it is found to contain as special cases the results of several kinetic-theory-based constitutive equations. It is regarded by many polymer rheologists as the most useful single-integral constitutive equation. It would be highly desirable to have good experimental determinations of the surface $W(I_1,I_2)$.

An example of (d) is the 'memory-integral expansion', which results from a Fréchet expansion of the stress tensor given as a general functional of the strain history; the general term in this expansion is an n-fold integral involving all various n-tuple products of the relative finite strain tensors

$$\tau = \int_{-\infty}^{t} M_1(t-t')\gamma'_{[0]}dt' + \frac{1}{2}\int_{-\infty}^{t}\int_{-\infty}^{t} M_2(t-t',t-t'')\{\gamma'_{[0]}\cdot\gamma''_{[0]} + \gamma''_{[0]}\cdot\gamma'_{[0]}\}dt''dt' + \dots \tag{49}$$

Here $\gamma''_{[0]}$ is shorthand for $\gamma_{[0]}(t,t'')$. This can be regarded as an expansion about the Lodge rubberlike liquid, which in turn includes the general linear viscoelastic model. By expanding the strain tensors in equation (49) about time t, the retarded-motion expansion of equation (38) is obtained, with the retarded-motion constants (the b's) given as integrals over the kernel functions (the M's). The memory integral expansion is useful for general continuum mechanics arguments and derivations, but it is not particularly useful for flow calculations; no experiments have been devised to determine the values of M beyond M_1, which is the memory function of linear viscoelasticity.

8.5 KINETIC THEORIES FOR DILUTE SOLUTIONS OF POLYMERS[66-72]

In order to calculate the viscoelastic properties of fluids it is necessary to have (a) a mechanical model to represent the polymer molecule, and (b) a kinetic theory that relates the stress tensor to the mechanical model.

Many mechanical models have been used for kinetic theory calculations. The simplest models are the dumbbell models:[73,74] two beads connected by a spring ('the elastic dumbbell') for flexible polymer molecules, and two beads joined by a rigid rod ('the rigid dumbbell') for rodlike polymers. Although these models are extremely crude, they have been very helpful for generating simple constitutive equations that can be used in polymer fluid dynamics. More realistic are the many-bead models: the freely jointed bead–spring chain (Rouse[75-77] and Zimm[78,79] models), the freely jointed bead–rod chain (the Kramers model[80,81]), the multibead rod,[82] the freely rotating bead–rod chain (the Kirkwood–Riseman model[83,84]), and the rotational isomeric state model used by Flory.[85] In addition to these bead–rod–spring models, a number of other models have been used — porous spheres, ellipsoids, strings, *etc.* — but these will not be discussed here.

Many of the kinetic theories published so far can be regarded as specialized forms of a general phase-space kinetic theory as shown in Figure 8. In the general phase-space theory one begins with the equations of motion of all the beads in the system and the Liouville equation in the complete phase space of the liquid. Then by a systematic development one performs a sequence of operations that leads to the main starting point for making kinetic theory calculations: (i) the equation of motion for the beads of a single polymer molecule and the equation of continuity for the configurational distribution function of a single molecule: these two equations can then be combined

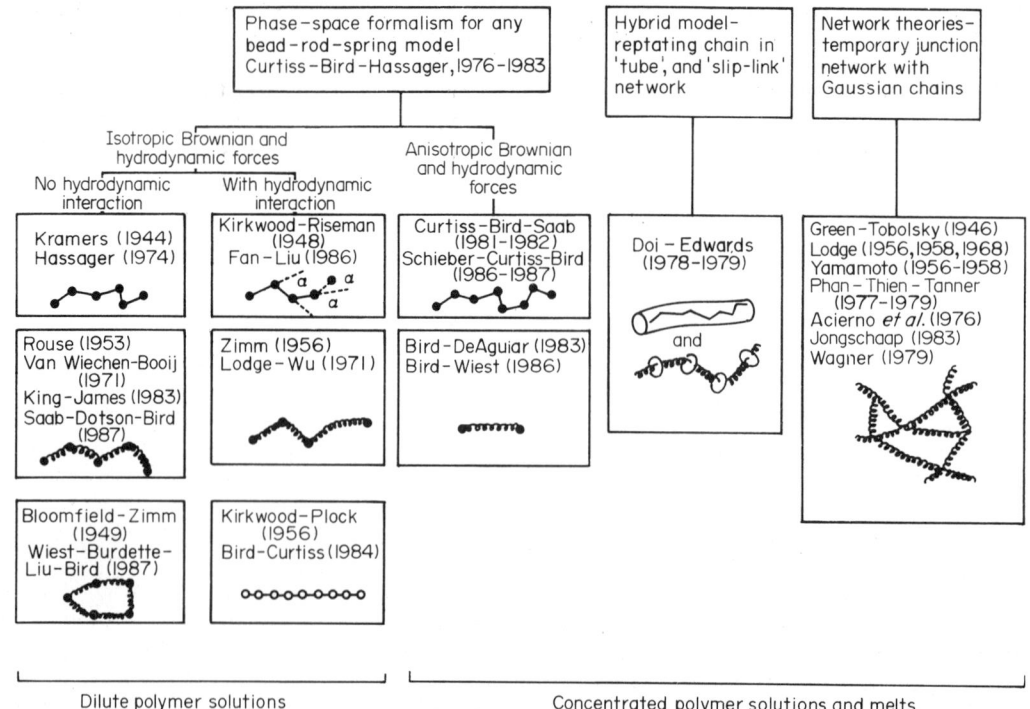

Figure 8 Models and theories. The chart summarizes molecular models and outlines the development of kinetic theories of polymer liquids. The sketches show some simple physical pictures that form the basis of theoretical models describing molecular motion in flowing systems

to give the 'diffusion equation' for the distribution function, and (ii) the expression for the stress tensor, given as integrals over the configurational distribution function.[86,87]

8.5.1 Elastic Dumbbell Models

The notation used for the elastic dumbbell model is shown in Figure 9. There are n dumbbells per unit volume, dissolved in a solvent with viscosity η_s. The imposed velocity distribution for the solution is given by $v = v_0 + [\kappa \cdot r]$, in which v_0 is independent of position, $\kappa = (\nabla v)^\dagger$, is a position-independent traceless tensor, and r is the position vector; such a velocity distribution is referred to as 'homogeneous' since the velocity gradients are constant throughout the fluid.

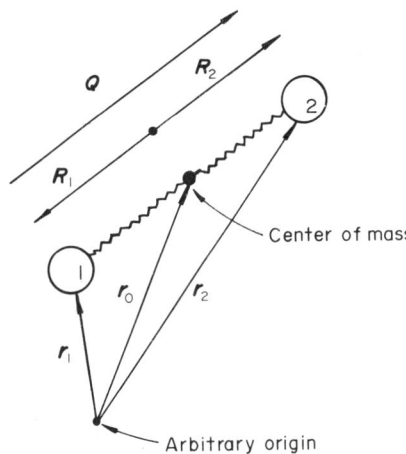

Figure 9 Elastic dumbbell formed by joining two identical beads with a spring

The equation of motion for bead v can now be written as follows

$$F_v^{(h)} + F_v^{(b)} + F_v^{(\phi)} + F_v^{(e)} = 0 \quad (v = 1,2) \tag{50}$$

The 'mass times acceleration' term is omitted, an assumption that is common to virtually all kinetic theories for polymers. The four forces in equation (50) are respectively the hydrodynamic, Brownian, spring and external forces; in this discussion we neglect the external forces entirely (see ref. 66, chapters 13 to 18, where these forces are included). The expressions for the three remaining forces are

$$F_v^{(h)} = -[\zeta \cdot ([\dot{r}_v] - v_v - v_v')] \tag{51}$$

$$F_v^{(b)} = -\frac{1}{\Psi}\left(\frac{\partial}{\partial r_v} \cdot [m(\dot{r}_v - v)(\dot{r}_v - v)]\Psi\right) \tag{52}$$

$$F_v^{(\phi)} = -\frac{\partial}{\partial r_v}\phi \tag{53}$$

where $[\![\ ]\!]$ is a momentum-space average. A brief discussion of each of these forces is now given.

(a) The hydrodynamic force on bead v is proportional to the difference between the bead velocity (appropriately averaged with respect to the velocity distribution) and the velocity of the fluid in the vicinity of the bead; v_v is the imposed fluid velocity, and v_v' is the perturbation of the fluid velocity because of the motion of the other bead ('hydrodynamic interaction'). To indicate that the force on the bead is not necessarily collinear with the velocity difference, a friction tensor ζ has been included; in most theories the friction tensor ζ is taken to be a multiple of the unit tensor $\zeta\delta$, where ζ is called the 'friction coefficient'.

(b) The Brownian motion force involves the configurational distribution function $\Psi(r_1, r_2, t)$, which gives the number of dumbbells per unit volume that have beads located at r_1 and r_2 at time t. The Brownian force depends on the divergence of a momentum flux, averaged over all the velocity space. If it is assumed, as is done in most kinetic theories, that the velocity distribution is Maxwellian, then the Brownian force contribution becomes $F_v^{(b)} = -kT\,\partial\ln\Psi/\partial r_v$.

(c) The spring force term depends on the potential energy ϕ of the spring. Inasmuch as the forces on the two beads must exactly balance, it is convenient to define a 'connector force' as $F^{(c)} = F_1^{(\phi)} = -F_2^{(\phi)}$. When hydrodynamic interaction is neglected, when the friction tensor is isotropic, and when the velocity distribution is Maxwellian, the equations of motion for the beads become

$$-\zeta([\dot{r}_\nu] - v_0 - [\kappa \cdot r_\nu]) - kT\frac{\partial}{\partial r_\nu}\ln\Psi + F_\nu^{(\phi)} = 0 \quad (\nu = 1, 2) \tag{54}$$

When these two equations are subtracted one from the other, the equation for the internal motion is obtained

$$[\dot{Q}] = [\kappa \cdot Q] - \frac{2kT}{\zeta}\frac{\partial}{\partial Q}\ln\psi - \frac{2}{\zeta}F^{(c)} \tag{55}$$

Here the vector Q describes the orientation and stretching of the dumbbell; in going from equation (54) to equation (55), the configurational distribution function has been written as $\Psi(r_1, r_2, t) = n\psi(Q, t)$ to indicate that the distribution of orientations is independent of position.

The equation of continuity for the configurational distribution function is given by

$$\frac{\partial \psi}{\partial t} = -\left(\frac{\partial}{\partial Q} \cdot [\dot{Q}]\psi\right) \tag{56}$$

When the last two equations are combined, the 'diffusion equation' is obtained

$$\frac{\partial \psi}{\partial t} = -\left(\frac{\partial}{\partial Q} \cdot \left([\kappa \cdot Q]\psi - \frac{2kT}{\zeta}\frac{\partial}{\partial Q}\psi - \frac{2}{\zeta}F^{(c)}\psi\right)\right) \tag{57}$$

When this equation is solved for given initial and boundary conditions, the distribution of configurations is then obtained for the prescribed velocity gradient field $\kappa(t)$.

The phase-space kinetic theory gives a general expression for the total stress tensor. For the elastic dumbbell this may be written as

$$\pi = \pi_s - n\langle QF^{(c)}\rangle + nm\sum_{\nu=1}^{2}\langle(\dot{r}_\nu - v)(\dot{r}_\nu - v)\rangle \tag{58}$$

Here the angular brackets $\langle\ \rangle$ indicate an average in the phase space of a single dumbbell. The first term is the solvent contribution to the stress tensor, the second term is the contribution resulting from the tensions in the springs and the third term is a result of the momentum flux associated with the bead motion. If the Maxwell velocity distribution is assumed, then the last term becomes an isotropic contribution. The (extra) stress tensor can now be written in several equivalent forms:

Kramers[80]
$$\tau = -\eta_s\dot{\gamma} - n\langle QF^{(c)}\rangle + nkT\delta \tag{59}$$

Modified Kramers
$$\tau = -\eta_s\dot{\gamma} + n\langle\sum_{\nu=1}^{2}R_\nu F_\nu^{(\phi)}\rangle + nkT\delta \tag{60}$$

Kramers–Kirkwood[68, 80]
$$\tau = -\eta_s\dot{\gamma} - n\langle\sum_{\nu=1}^{2}R_\nu F_\nu^{(h)}\rangle \tag{61}$$

Giesekus[88]
$$\tau = -\eta_s\dot{\gamma} + \frac{1}{4}n\zeta\langle QQ\rangle_{(1)} \tag{62}$$

Note that the first three of the above expressions are somewhat similar in appearance, but that they involve different forces. The fourth expression is obtained by combining equation (59) with the second moment of equation (57). Now that the main equations have been presented, a number of special cases and modifications can be discussed.

8.5.1.1 Hookean dumbbells

If the spring force law is taken to be Hookean, with $F^{(c)} = HQ$, where H is the spring constant, then it is possible to eliminate $\langle QQ\rangle$ between equations (59) and (62) to obtain the complete constitutive equation for the dilute solution of elastic dumbbells[114]

$$\tau_p + \lambda_H \tau_{p(1)} = -nkT\lambda_H \gamma_{(1)} \tag{63}$$

in which $\lambda_H = \zeta/4H$ is the time constant for the fluid, and $\tau_p = \tau - \tau_s$ is the polymer contribution to the stress tensor. This result can also be written in the form of the convected Jeffreys model of equation (43), in which the three constants are identified as: $\eta_0 = \eta_s + nkT\lambda_H$, $\lambda_1 = \lambda_H$, and $\lambda_2 = [\eta_s/(\eta_s + nkT\lambda_H)]\lambda_H$. Using standard continuum mechanics techniques, equation (63) may be rewritten as

$$\tau = -\eta_s \dot{\gamma} + \int_{-\infty}^{t} \left\{ \frac{nkT}{\lambda_H} \exp[-(t-t')/\lambda_H] \right\} \gamma_{[0]}(t, t') dt' \tag{64}$$

which is of the form of equation (46), namely the Lodge elastic liquid.

For this model it is thus possible to obtain the complete constitutive equation without actually finding the configurational distribution function. This function is, however, known[79] to be

$$\psi(Q, t) = \frac{(H/2\pi kT)^{3/2}}{\sqrt{\det \alpha}} \exp[-(H/2kT)(\alpha^{-1}:QQ)] \tag{65}$$

$$\alpha(t) = \delta - \frac{1}{\lambda_H} \int_{-\infty}^{t} \exp[-(t-t')/\lambda_H] \, \gamma_{[0]}(t, t') dt' \tag{66}$$

It is interesting to note that the diffusion equation contains the transpose of the velocity gradient tensor, but the solution is given in terms of one of the relative finite strain tensors. The tensor α plays an important role in the changes of the thermodynamic functions that occur when a polymer solution goes from a state of equilibrium to a state of flow. The changes in internal energy and entropy are[89,90]

$$\Delta U = \frac{1}{2} nkT \, \mathrm{tr} \, (\alpha - \delta) \tag{67}$$

$$\Delta S = nk \, \ln \sqrt{\det \alpha} \tag{68}$$

In a steady-state shear flow with shear rate $\dot{\gamma}$ these quantities become $\Delta U = nkT\lambda_H^2 \dot{\gamma}^2$ and $\Delta S = nk \ln\sqrt{1 + \lambda_H^2 \dot{\gamma}^2}$. Note also that the stretching of the dumbbells is given by

$$\langle Q^2 \rangle / \langle Q^2 \rangle_{eq} = 1 - \mathrm{tr}(\tau_p/3nkT) = \frac{1}{3} \mathrm{tr} \, \alpha \tag{69}$$

Hence once a flow problem has been solved for the constitutive equation for the elastic dumbbells, the molecular stretching and thermodynamic properties are known.

Because the constitutive equation for the dilute solution of Hookean dumbbells does not describe the shear-rate dependence of the viscometric functions and the elongational viscosity goes to infinity at a finite value of the elongation rate, there have been many attempts to modify the model in order to get better agreement with experiments.[74] A great deal has been learned about the relation between model structure and rheological properties. Although some of the modified dumbbell models do lead to useful equations for fluid dynamicists, one must keep in mind that dumbbell models do not contain enough internal degrees of freedom to describe completely the molecular motions, and in particular those involving very short time constants; in some flow situations involving several different time scales, it will be important to have models containing a wide spectrum of relaxation times, and for these purposes a full chain model will be required.

8.5.1.2 Finitely extensible nonlinear elastic (FENE) dumbbells

Hookean dumbbells are infinitely extensible, and real polymer molecules are not; this suggests that finitely extensible springs ought to be used, and consequently the FENE dumbbell model has been much studied.[91-97] The spring force is taken to be

$$F^{(c)} = \frac{HQ}{1 - (Q^2/Q_0^2)} \quad (Q \le Q_0) \tag{70}$$

in which Q_0 is the maximum allowable extension of the dumbbell. For this model the Kramers form of the stress tensor (equation 59) is

$$\tau = -\eta_s \dot{\gamma} - nH \left\langle \frac{QQ}{1-(Q^2/Q_0^2)} \right\rangle + nkT\delta \tag{71}$$

It is not possible to eliminate $\langle QQ \rangle$ between equations (71) and (62). However, it is possible if in equation (71) the average of the ratio is replaced by the ratio of averages plus an extra isotropic term to compensate partially for the error introduced (here $b = HQ_0^2/kT$)

$$\tau = -\eta_s \dot{\gamma} - nH \left(\frac{\langle QQ \rangle}{1-\langle Q^2/Q_0^2 \rangle} \right) + (1-\varepsilon b)nkT\delta \tag{72}$$

The constant ε is chosen to be $2/[b(b+2)]$ so that the trace of equation (72) is true at equilibrium. Now it is possible to eliminate the two averages $\langle QQ \rangle$ and $\langle Q^2/Q_0^2 \rangle$ from equations (71), (62), and the trace of (72) to get the following constitutive equation for the polymer contribution to the stress tensor[96,97]

$$Z\tau_p + \lambda_H \tau_{p(1)} - \lambda_H [\tau_p - (1-\varepsilon b)nkT\delta] D \ln Z/Dt = -(1-\varepsilon b)nkT\lambda_H \gamma_{(1)} \tag{73}$$

$$Z = 1 + (3/b)[(1-\varepsilon b) - \text{tr}(\tau_p/3nkT)] \tag{74}$$

Note that the resulting constitutive equation is nonlinear in the stresses. When b is infinite (and hence $Z = 1$), equation (73) simplifies to the convected Maxwell equation. This equation can describe the shear-rate dependence of the viscosity and first normal stress coefficient (see Figure 10) and gives an elongational viscosity that becomes large but finite. For the FENE dumbbell it has thus been possible to go from a mechanical model to an approximate constitutive equation, and to determine the parameters in the constitutive equation from rheometric data; it has been further possible to use the constitutive equation with these parameters to describe nontrivial flow problems, such as rod climbing, squeezing flow, torque on a rotating sphere and velocity overshoot at the inception of Couette flow.[98,99] It has been verified that the approximation introduced in equation (72) is not a serious one;[97] however it has also been established that when a similar approximation is used for a dumbbell with a friction coefficient that depends on Q, spurious 'hysteresis' effects are obtained.[100]

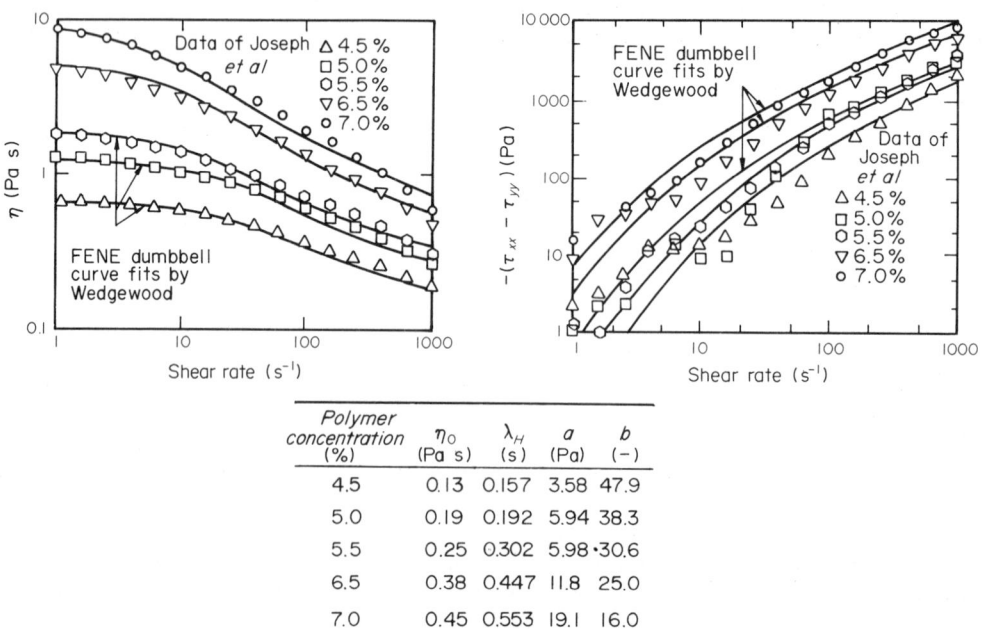

Polymer concentration (%)	η_0 (Pa s)	λ_H (s)	a (Pa)	b (−)
4.5	0.13	0.157	3.58	47.9
5.0	0.19	0.192	5.94	38.3
5.5	0.25	0.302	5.98	30.6
6.5	0.38	0.447	11.8	25.0
7.0	0.45	0.553	19.1	16.0

Figure 10 PMMA solution data reported by D. D. Joseph, G. S. Beavers, A. Cers, C. Dewald, A. Hoger and P. T. Than, *J. Rheol.*, 1984, **28**, 325, along with FENE dumbbell curve fits by L. E. Wedgewood (see L. E. Wedgewood and R. B. Bird, in 'Integration of Fundamental Polymer Science and Technology', ed. L. A. Kleintjens and P. J. Lemstra, Elsevier, Amsterdam, 1986, p. 337). The FENE-dumbbell parameters are also shown, where $a = nkT$ and $\eta_0 = \eta_s + nkT\lambda_H b/(b+5)$

8.5.1.3 Hydrodynamic interaction

In equation (51) the idea of hydrodynamic interaction was introduced, but this effect was neglected in subsequent equations. To include this effect it is necessary to insert an expression for, say, v'_1. A commonly used expression is $v'_1 = [\mathbf{\Omega} \cdot F_2^{(h)}]$, in which $\mathbf{\Omega}$ is the Oseen–Burgers tensor, introduced into the polymer kinetic theory by Kirkwood and his collaborators[68]

$$\mathbf{\Omega} = \frac{1}{8\pi\eta_s Q}\left(\boldsymbol{\delta} + \frac{QQ}{Q^2}\right) \tag{75}$$

This accounts approximately for the disturbance at bead '1' resulting from the motion of bead '2'; a more accurate description of hydrodynamic interaction is given by the Rotne–Prager–Yamakawa tensor.[101,102] When the above modification of the theory is made, equation (55) is changed to

$$[\![\dot{Q}]\!] = [\boldsymbol{\kappa} \cdot Q] + \left[(\boldsymbol{\delta} - \zeta\mathbf{\Omega}) \cdot \left(-\frac{2kT}{\zeta}\frac{\partial}{\partial Q}\ln\psi - \frac{2}{\zeta}F^{(c)}\right)\right] \tag{76}$$

The inclusion of $\mathbf{\Omega}$ in equation (76) complicates the theory considerably. In most of the polymer literature it has been commonplace to replace the Oseen tensor by its 'equilibrium averaged' value $\langle\mathbf{\Omega}\rangle_{eq} = \int \mathbf{\Omega}\psi_{eq}\,dQ = \Omega\boldsymbol{\delta}$, which is isotropic. Then all results for the Hookean dumbbell have to be modified only by replacing ζ^{-1} everywhere by $\zeta^{-1} - \Omega$ and $\lambda_H = \zeta/4H$ by $\tilde{\lambda}_H = 1/[4H(\zeta^{-1} - \Omega)]$. This minor modification does not allow for the description of the shear-rate dependence of η and Ψ_1, and Ψ_2 is still equal to zero. It may be remarked in passing that the translational diffusivity for Hookean dumbbells with an Oseen–Burgers equilibrium-averaged hydrodynamic interaction is $D_{tr} = (kT/2\zeta)(1 + \zeta\Omega)$.

Instead of carrying out equilibrium averaging, one can introduce a 'consistently averaged hydrodynamic interaction', in which the averaging is not performed with the equilibrium distribution function, but rather with a distribution function that is consistent with the local flow field.[103] This leads to shear-rate dependence for all the viscometric functions, and also to a nonzero (but positive) value of the second normal stress coefficient.

Another procedure is to assume that the Oseen–Burgers tensor is the equilibrium-averaged value, multiplied by a truncated Taylor series in the polymer contribution to the stress tensor; such an assumption leads, for Hookean dumbbells, directly to the Giesekus constitutive equation[115]

$$\tau_p + \tilde{\lambda}_H \tau_{p(1)} - \frac{a}{nkT}\{\tau_p \cdot \tau_p\} = -nkT\tilde{\lambda}_H\dot{\gamma} \tag{77}$$

in which a is an empirical constant not provided by any kinetic theory arguments.[104] This result leads to shear-rate-dependent viscometric functions and a negative second normal stress coefficient; however, the model allows for infinite stretching of the dumbbells in shear and elongational flows.[105]

Still another way to account for hydrodynamic interaction is to presume that the internal motions of the polymer molecule disturb the local flow field so that, on average, a molecule 'sees' an effective velocity gradient tensor $\boldsymbol{\kappa} - \xi\dot{\gamma}/2$ instead of $\boldsymbol{\kappa}$. The parameter ξ is an empirical quantity to be determined by data-fitting. Inclusion of this parameter does lead to shear-rate dependence of the viscometric functions for Hookean dumbbells, but the elongational viscosity still goes to infinity.[106-108]

8.5.1.4 Internal viscosity

It has been suggested[109-111] that, because of the uncoiling process in the stretching of a polymer molecule in a flow field, it would be more realistic to include a linear dashpot in parallel with a Hookean spring in a dumbbell model. That is, the connector force should be represented as

$$F^{(c)} = HQ + K(Q/Q)\dot{Q} \tag{78}$$

in which H is the spring constant and K is the dashpot constant. This kind of force law involves a nonconservative force, and therefore the usual methods of statistical mechanics cannot even be used to get the equilibrium configurational distribution function. Apparently all researchers who have used this model have tacitly replaced equation (78) by

$$\boldsymbol{F}^{(c)} = H\boldsymbol{Q} + K(\boldsymbol{QQ}/Q^2)\boldsymbol{\cdot}[\![\dot{\boldsymbol{Q}}]\!] \tag{79}$$

That is, a velocity-averaged dashpot is in fact used. For this model, which is then partly mechanical and partly statistically averaged, equation (55) becomes

$$[\![\dot{\boldsymbol{Q}}]\!] = \left[\left(\boldsymbol{\delta} - \frac{1}{(\zeta/2K)+1}\frac{\boldsymbol{QQ}}{Q^2}\right)\boldsymbol{\cdot}\left([\boldsymbol{\kappa}\boldsymbol{\cdot}\boldsymbol{Q}] - \frac{2kT}{\zeta}\frac{\partial}{\partial \boldsymbol{Q}}\ln\psi - \frac{2}{\zeta}H\boldsymbol{Q}\right)\right] \tag{80}$$

This model gives a shear-rate-dependent viscosity and a finite limiting value for $\eta'(\omega)$ as $\omega \to \infty$.[112,113]

8.5.1.5   *Anisotropic friction and Brownian motion*

If one allows for anisotropic frictional forces by retaining the friction tensor $\boldsymbol{\zeta}$ in equation (51), and allowing for anisotropic Brownian motion by allowing the Maxwellian velocity distribution to be skewed (so that $\boldsymbol{F}_v^{(b)} = -(kT/\Psi)[(\partial/\partial r_v)\boldsymbol{\cdot}\boldsymbol{\xi}^{-1}\Psi]$), then the diffusion equation and stress tensor expressions become

$$\frac{\partial\psi}{\partial t} = -\left(\frac{\partial}{\partial\boldsymbol{Q}}\boldsymbol{\cdot}\left[\boldsymbol{\kappa}\boldsymbol{\cdot}\boldsymbol{Q}\psi - 2kT\boldsymbol{\zeta}^{-1}\boldsymbol{\cdot}\left[\frac{\partial}{\partial\boldsymbol{Q}}\boldsymbol{\cdot}\boldsymbol{\xi}^{-1}\psi\right] - 2\boldsymbol{\zeta}^{-1}\boldsymbol{\cdot}\boldsymbol{F}^{(c)}\psi\right]\right) \tag{81}$$

$$\boldsymbol{\pi} = \boldsymbol{\pi}_{\mathrm{s}} - n\langle\boldsymbol{Q}\boldsymbol{F}^{(c)}\rangle + 2nkT\langle\boldsymbol{\xi}^{-1}\rangle \tag{82}$$

If, as shown in Table 1, simple postulates are made for the anisotropic tensors (case I is due to Giesekus[114,115]), then a variety of constitutive equations can be obtained, including some which are special cases of the Oldroyd model.

Still other postulates for the anisotropy tensors can be made. If it is assumed that the anisotropy tensors are of the form $A\boldsymbol{\delta} + B\boldsymbol{QQ}/Q^2$, then one is led to the 'encapsulated dumbbell model'.[116,117] If the FENE spring connector is used, then an approximate constitutive equation can be obtained, and some comparisons of this model have been made with experimental data on concentrated systems.[117]

Table 1   Constitutive Equations from Dumbbell Models[a]

Postulates for the anisotropic tensors		Time constant	Constitutive equation	Form of constitutive equation
$\boldsymbol{\zeta}\boldsymbol{\zeta}^{-1}$	$\boldsymbol{\xi}^{-1}$	λ_H	$\boldsymbol{\tau}_{\mathrm{p}} + \lambda_H\boldsymbol{\tau}_{\mathrm{p}(1)} =$	
I $\boldsymbol{\delta} - \dfrac{a}{nkT}\boldsymbol{\tau}_{\mathrm{p}}$	$\boldsymbol{\delta} - \dfrac{b}{nkT}\boldsymbol{\tau}_{\mathrm{p}}$	$\dfrac{\zeta(1+2b)}{4H(1+b)}$	$-\dfrac{nkT\lambda_H}{1+2b}\boldsymbol{\gamma}_{(1)} + \dfrac{a}{nkT}\{\boldsymbol{\tau}_{\mathrm{p}}\boldsymbol{\cdot}\boldsymbol{\tau}_{\mathrm{p}}\}$	Giesekus (equation 77)
II $\boldsymbol{\delta} - \dfrac{a\zeta}{4H}\dot{\gamma}$	$\boldsymbol{\delta} - \dfrac{b\zeta}{4H}\dot{\gamma}$	$\dfrac{\zeta}{4H}$	$nkT\lambda_H\{(1+b)\boldsymbol{\gamma}_{(1)} + ab\lambda_H\boldsymbol{\gamma}_{(1)}\boldsymbol{\cdot}\boldsymbol{\gamma}_{(1)} + 2b\lambda_H\boldsymbol{\gamma}_{(2)}\}$ $+\tfrac{1}{2}a\lambda_H\{\boldsymbol{\gamma}_{(1)}\boldsymbol{\cdot}\boldsymbol{\tau}_{\mathrm{p}} + \boldsymbol{\tau}_{\mathrm{p}}\boldsymbol{\cdot}\boldsymbol{\gamma}_{(1)}\}$	Oldroyd (equation 44)
III $\boldsymbol{\delta} - \dfrac{a\zeta}{4H}\dot{\gamma}$	$\boldsymbol{\delta} - \dfrac{b}{nkT}\boldsymbol{\tau}_{\mathrm{p}}$	$\dfrac{\zeta(1+2b)}{4H(1+b)}$	$-\dfrac{nkT\lambda_H}{1+2b}\boldsymbol{\gamma}_{(1)} + \dfrac{a\lambda_H(1+b)}{2(1+2b)}\{\boldsymbol{\gamma}_{(1)}\boldsymbol{\cdot}\boldsymbol{\tau}_{\mathrm{p}} + \boldsymbol{\tau}_{\mathrm{p}}\boldsymbol{\cdot}\boldsymbol{\gamma}_{(1)}\}$	Oldroyd (equation 44)
IV $\boldsymbol{\delta} - \dfrac{a}{nkT}\boldsymbol{\tau}_{\mathrm{p}}$	$\boldsymbol{\delta} - \dfrac{b\zeta}{4H}\boldsymbol{\gamma}_{(1)}$	$\dfrac{\zeta}{4H}$	$-nkT\lambda_H[(1+b)\boldsymbol{\gamma}_{(1)} + 2b\lambda_H\boldsymbol{\gamma}_{(2)}]$ $+\dfrac{a}{nkT}\{\boldsymbol{\tau}_{\mathrm{p}}\boldsymbol{\cdot}\boldsymbol{\tau}_{\mathrm{p}}\} + ab\lambda\{\boldsymbol{\gamma}_{(1)}\boldsymbol{\cdot}\boldsymbol{\tau}_{\mathrm{p}} + \boldsymbol{\tau}_{\mathrm{p}}\boldsymbol{\cdot}\boldsymbol{\gamma}_{(1)}\}$	Oldroyd (equation 44) plus $\{\boldsymbol{\tau}_{\mathrm{p}}\boldsymbol{\cdot}\boldsymbol{\tau}_{\mathrm{p}}\}$ term

[a] R. B. Bird and J. M. Wiest, *J. Rheol.*, 1985, **29**, 519.

8.5.2 Rigid Rodlike Molecules

The rigid dumbbell model and the multibead rod models are shown in Figure 11. For these models[73,82] we use a distribution function f(u, t), which gives the probability that a molecule is in the orientation u at time t, where u is the unit vector designating the orientation.

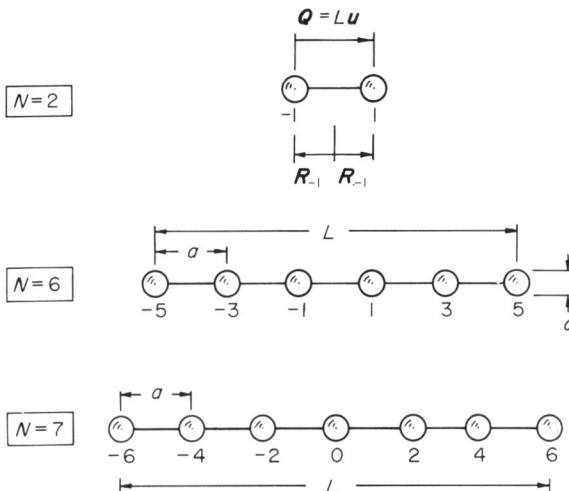

Figure 11 Rigid dumbbell and multibead rod models. The unit vector u gives the direction from $-v$ to $+v$, and $R_v = [Lv/2(N-1)]u = (va/2)u$ gives the position of bead v with respect to the center of mass. The ratio d/a is called ξ (with $0 \le \xi \le 1$); if $\xi = 1$ the model is the 'osculating multibead rod'

8.5.2.1 Hydrodynamic interaction neglected

The diffusion equation for the distribution function is (in the absence of hydrodynamic interaction)

$$\frac{\partial f}{\partial t} = \frac{1}{6\lambda}\left(\frac{\partial}{\partial u} \cdot \frac{\partial}{\partial u} f\right) - \left(\frac{\partial}{\partial u} \cdot [\kappa \cdot u - \kappa : uuu] f\right) \tag{83}$$

in which $\lambda = \zeta L^2/12kT$ is the time constant for the solution, and $\partial/\partial u$ is a gradient operator in the θ, ϕ space describing the orientation of the molecule. The stress tensor may be written in several different forms:

Kramers
$$\tau = -\eta_s \dot{\gamma} - 3nkT\langle uu \rangle - 6nkT\lambda\kappa:\langle uuuu \rangle + nkT\delta \tag{84}$$

Modified Kramers
$$\tau = -\eta_s \dot{\gamma} - 3nkT(2/L^2)\sum_{\substack{v=-1\\v\neq0}}^{+1}\langle R_v R_v \rangle - 6nkT\lambda\kappa:(8/L^4)\sum_{\substack{v=-1\\v\neq0}}^{+1}\langle R_v R_v R_v R_v \rangle + nkT\delta \tag{85}$$

Kramers–Kirkwood
$$\tau = -\eta_s \dot{\gamma} - n\sum_{\substack{v=-1\\v\neq0}}^{+1}\langle R_v F_v^{(h)} \rangle \tag{86}$$

Giesekus
$$\tau = -\eta_s \dot{\gamma} + 3nkT\lambda\langle uu \rangle_{(1)} \tag{87}$$

in which the $\langle \ \rangle$ brackets indicate an average with respect to the distribution function f. To extend the above results to multibead rods it is necessary only to replace the time constant $\lambda = \zeta L^2/12kT$ by the time constant $\lambda_N = \zeta L^2 N(N+1)/72(N-1)kT$, and to replace the sums from -1 to 1 by sums from $-(N-1)$ to $(N-1)$. Note that many of the results have a rather different form from the corresponding expressions for elastic dumbbells; this is the result of the fact that the rigid models contain 'constraints'.

For rigid dumbbells and multibead rods no general solution to the $f(u, t)$ equation has been found, and a complete constitutive equation has not been obtained. A complete solution is available for elongational flow, but for shear flow only a perturbation solution about the equilibrium state is available; the latter has been carried out up to 40th-order terms[118] and it has been determined that the series converges for $\lambda\dot{\gamma} < 0.81$. When hydrodynamic interaction is neglected, the second normal stress coefficient is exactly zero. A complete numerical solution is also available for shear flow.[119] It is found that the viscosity and first normal stress coefficient decrease with increasing shear rate, and that the elongational viscosity increases monotonously. In addition the retarded-motion constants have been determined up to fourth order, and the kernel functions of the memory-integral expansion through third order.[120]

8.5.2.2 *Hydrodynamic interaction included*

If the Rotne–Prager–Yamakawa hydrodynamic interaction is included in the rigid dumbbell model, it is found that the equation for the distribution function is identical to equation (83) except that the time constant λ must be replaced by

$$\lambda_2^{(1)}(h, \xi) = (\zeta L^2/12kT)\left[1 - h\left(1 + \frac{1}{6}\xi^2\right)\right]^{-1} \tag{88}$$

in which $h = \zeta/8\pi\eta_s L$ is the hydrodynamic interaction parameter, and $\xi = d/L$ (see Figure 11). Similarly, the Kramers form of the stress tensor has the same form as equation (84) but with the time constant λ replaced by

$$\lambda_2^{(2)}(h, \xi) = (\zeta L^2/12kT)\left[1 - 2h\left(1 - \frac{1}{6}\xi^2\right)\right]^{-1} \tag{89}$$

It is possible to extend the calculation to multibead rods by replacing $\lambda_2^{(1)}$ and $\lambda_2^{(2)}$ by $\lambda_N^{(1)}$ and $\lambda_N^{(2)}$, and by replacing the sums from -1 to $+1$ by $-N$ to $+N$. The definitions of these multibead time constants are not given here, since they are available elsewhere, along with a tabulation of the time constants up to $N = 70$ for the special case of osculating multibead rods (see ref. 66, section 14.6). It is found that the second normal stress coefficient is positive for $N = 2$, 3, 4, 5 and 6, but that it is negative for $N \geq 7$. We note in passing that for rigid dumbbells with hydrodynamic interaction (both Oseen–Burgers and Rotne–Prager–Yamakawa) the translational diffusivity is $D_{tr} = (kT/2\zeta)[1 + (4/3)h]$.

8.5.3   Bead–Spring Chain and Ring Models

In order to describe rheological properties that depend on a spectrum of relaxation times the dumbbell models are inadequate, and it is necessary to use mechanical models with many beads (ref. 66, chapter 15). The bead–spring chain model of Figure 12 has been the subject of much study;[75–79,121–123] this model poses no particular problems, since the chain geometry is well understood and the matrices needed to facilitate derivations have been presented systematically (ref. 66, tables 15.1-1 and 15.4-1).

8.5.3.1   *Hydrodynamic interaction neglected*

If hydrodynamic interaction is neglected, if a Maxwellian velocity distribution is assumed for the beads and if the friction tensor is taken to be isotropic, then the diffusion equation for the distribution function $\psi(\mathbf{Q}^{N-1}, t)$ is

$$\frac{\partial\psi}{\partial t} = -\Sigma_j\left(\frac{\partial}{\partial\mathbf{Q}_j}\cdot\left\{[\boldsymbol{\kappa}\cdot\mathbf{Q}_j]\psi - \frac{1}{\zeta}\Sigma_k A_{jk}\left[kT\frac{\partial}{\partial\mathbf{Q}_k}\psi + \mathbf{F}_k^{(c)}\psi\right]\right\}\right) \tag{90}$$

in which A_{jk} is the jk element of the Rouse matrix ($A_{jk} = 2$, if $j = k$; $A_{jk} = -1$ if $j = k \pm 1$; and $A_{jk} = 0$ otherwise); the Rouse matrix describes the coupling between the motions of adjacent beads due to the springs. This equation can be solved[77–79] by the method of separation of variables, after the 'normal coordinates' $\mathbf{Q}_k'$ have been introduced by $\mathbf{Q}_j = \Sigma_k\Omega_{jk}\mathbf{Q}_k'$, where $\Omega_{jk} = \sqrt{2/N}\sin(jk\pi/N)$. The equation corresponding to each normal mode turns out to be identical in form to equation (57) for the elastic dumbbell, but with a time constant characteristic for that normal mode.

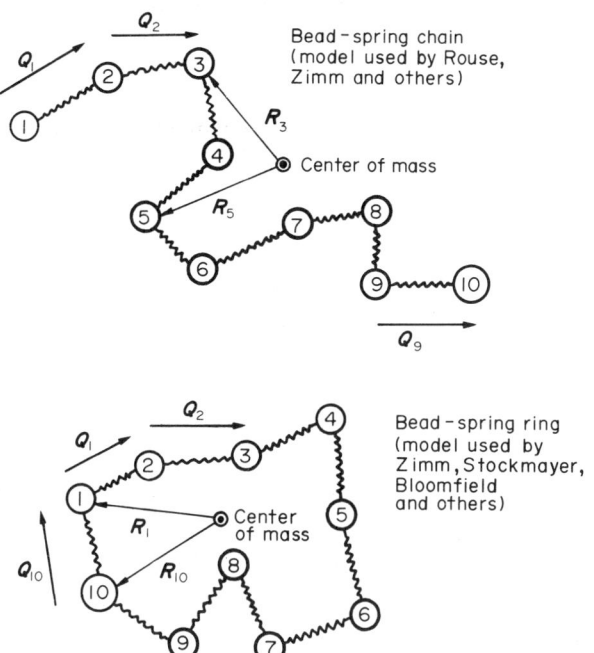

Figure 12 Bead–spring chain and ring models for $N=10$

The stress tensor expression can be written analogously to equations (59) to (62), thus:

Kramers
$$\tau = -\eta_s \dot{\gamma} - n \sum_{k=1}^{N-1} \langle Q_k F_k^{(c)} \rangle + (N-1)nkT\delta \tag{91}$$

Modified Kramers
$$\tau = -\eta_s \dot{\gamma} + n \sum_{v=1}^{N} \langle R_v F_v^{(\phi)} \rangle + (N-1)nkT\delta \tag{92}$$

Kramers–Kirkwood
$$\tau = -\eta_s \dot{\gamma} - n \sum_{v=1}^{N} \langle R_v F_v^{(h)} \rangle \tag{93}$$

Giesekus
$$\tau = -\eta_s \dot{\gamma} + \frac{1}{2} n\zeta \sum_{j=1}^{N-1} \sum_{k=1}^{N-1} \langle C_{jk} Q_j Q_k \rangle_{(1)} \tag{94}$$

In the last expression C_{jk} is the jk element of the Kramers matrix, which is the inverse of the Rouse matrix.

The constitutive equation for the Rouse chain (the bead–spring chain with Hookean springs) can then be written in either differential or integral form[77,79]

$$\tau = -\eta_s \dot{\gamma} + \Sigma_j \tau_j; \qquad \tau_j + \lambda_j \tau_{j(1)} = -nkT\lambda_j \dot{\gamma} \tag{95}$$

$$\tau = -\eta_s \dot{\gamma} + \int_{-\infty}^{t} \{nkT\Sigma_j \lambda_j^{-1} [\exp -(t-t')/\lambda_j]\} \gamma_{[0]}(t, t')dt' \tag{96}$$

in which the time constants are $\lambda_j = (\zeta/2Ha_j)$, where the $a_j = \sin^2(j\pi/2N)$ are the eigenvalues of the Rouse matrix. That is, in the differential form the constitutive equation contains a superposition of convected Maxwell models, and in the integral form it is of the form of the Lodge elastic liquid model. The viscometric functions are predicted to be constant

$$\eta - \eta_s = \frac{1}{3} nkT(\zeta/4H)(N^2 - 1) \tag{97}$$

$$\Psi_1 = \frac{2}{45} nkT(\zeta/4H)^2 (N^2 - 1)(2N^2 + 7) \tag{98}$$

$$\Psi_2 = 0 \tag{99}$$

The complex viscosity components are considerably more realistic for this model than for the elastic dumbbell. The elongational viscosity, however, goes to infinity at some finite elongation rate.

If the chain undergoes ring closure (see Figure 12) the theory is somewhat more complicated; early investigations[124-127] showed that for large N the viscosity should be one-half the value for the straight chain, and dilute-solution measurements corroborate this.[128,129] More recently the complete constitutive equation has been obtained[130] for all N, and it is found that for rings one replaces 1/3 by 1/6 in equation (97) and $(2N^2 + 7)/45$ by $(N^2 + 11)/180$ in equation (98) to account for ring closure. The constitutive equation is of the same form as equation (95) or (96), but with time constants given by $\lambda_j = (\zeta/2Hg_j)$ with $g_j = 4\sin^2(j\pi/N)$.

8.5.3.2 *Hydrodynamic interaction included*

When equilibrium-averaged hydrodynamic interaction is accounted for, the constitutive equation has the same form as equations (95) or (96), but with time constants that have to be calculated numerically by finding the eigenvalues of a modified Rouse matrix (that is, modified to take into account all the pairwise hydrodynamic interactions within the chain);[78,79] the inclusion of equilibrium-averaged hydrodynamic interaction allows for a very good description of linear viscoelastic data for some fluids over a wide range of frequencies.[1,131] The linear viscoelastic properties have also been obtained for complete hydrodynamic interaction (that is, without equilibrium averaging),[132-134] and in addition it has been shown that the ratio Ψ_2/Ψ_1 is about -0.01 if no preaveraging is done.[135]

Many other variations on the bead-spring chain have been worked on: inclusion of the 'excluded volume effect',[136] branched polymers,[137] addition of 'chain stiffness' to the model,[138] inclusion of 'internal viscosity',[139] and incorporation of nonlinear springs.[140]

8.5.4 Bead-Rod Chain Models[80-85]

The two models shown in Figure 13 have not been studied as much as the bead-spring chains, inasmuch as the constraints of constant bond lengths and constant bond angles require the use of generalized coordinates in the kinetic theory.

8.5.4.1 *The Kramers chain*

The freely jointed bead-rod chain (Kramers model) is of interest because it has a large number of internal degrees of freedom and is also finitely extensible. Many equilibrium properties have been

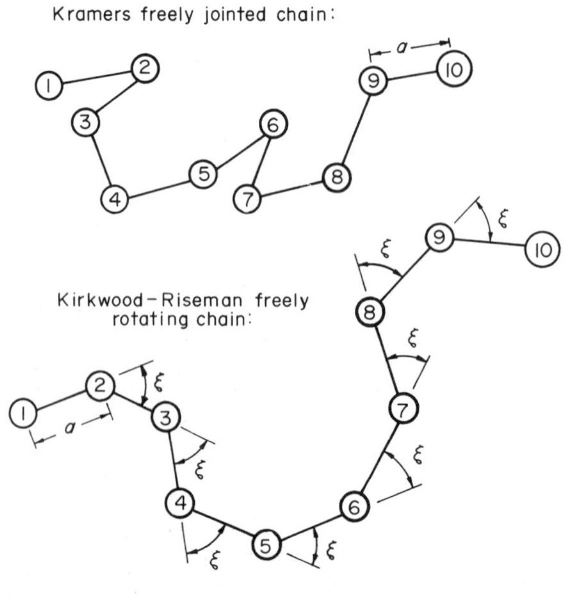

Figure 13 Bead-rod chain models for $N = 10$; the rod length is a

calculated for this chain assuming that the equilibrium distribution function is identical to the random walk,[81,85] even though it is known that the random walk distribution is inconsistent with the equilibrium statistical mechanics;[80,141-144] however, for very large N the differences are probably inconsequential.

For no hydrodynamic interaction the zero-shear-rate viscometric functions are known[80,81]

$$\eta_0 - \eta_s = \frac{1}{36} n\zeta a^2 (N^2 - 1) \tag{100}$$

$$\Psi_{1,0} = \frac{1}{16\,200} \frac{n\zeta^2 a^4}{NkT} (N^2 - 1)(10N^3 - 12N^2 + 35N - 12) \tag{101}$$

$$\Psi_{2,0} = 0 \tag{102}$$

Equation (101) is derived for the random walk and is believed to be good within about 2%; exact expressions for $\Psi_{1,0}$ have been obtained[143] for $N = 3$, 4 and 5. The elongational viscosity has also been worked out for large N

$$\frac{\bar{\eta} - 3\eta_s}{3(\eta_0 - \eta_s)} = \begin{cases} 1 + \dfrac{N^2 \zeta a^2 \dot{\varepsilon}}{90kT} + \cdots & \text{small } \dot{\varepsilon} \\[4mm] N - \dfrac{24kT}{N\zeta a^2 \dot{\varepsilon}} + \cdots & \text{large } \dot{\varepsilon} \end{cases} \tag{103}$$

in which $\dot{\varepsilon}$ is the elongation rate;[81] hence the molecular theory predicts that the elongational viscosity increases to some high value. There are no direct measurements to verify this result. The components of the complex viscosity have been obtained,[145] and it is found that $\eta'(\omega) - \eta_s$ approaches a nonzero value in the limit of infinite frequency, in contrast to the zero limiting value for the Rouse chain.

8.5.4.2 The Kirkwood–Riseman chain

Although this model was used in the development of a kinetic theory for polymer solutions,[83] very few calculations have actually been made for it. The equilibrium distribution function and some derived properties have been computed assuming a random walk (taking into account the fixed bond angles).[85] However, when the canonical distribution function of statistical mechanics is used, it is found that the deviations from the random-walk calculations increase with increasing N; the calculations include the mean-square end-to-end distance, the zero-shear-rate viscosity, and the zero-shear-rate first normal stress coefficient for $N = 3$–8, and the elongational viscosity curve for $N = 3$ and 4. From the latter calculation a tentative conclusion has been drawn that chain stiffness lowers the upper limiting value of the elongational viscosity.[84]

8.6 KINETIC THEORIES FOR UNDILUTED POLYMERS

For undiluted polymers and concentrated solutions, there are two types of theories: (a) the 'single-chain theories' or 'reptation theories', in which one focuses on the motions of one polymer molecule in the fluid as it moves in some kind of mean field provided by the surrounding polymer molecules; and (b) the 'network theories', in which one visualizes the fluid as a loosely joined network in which the network junctions have a distribution of lifetimes. The chain theories are similar in structure to the dilute solution theories, and one has to make some kinds of assumptions about how the surrounding molecules affect the hydrodynamic drag and the Brownian motion. The network theories are similar in structure to the kinetic theory of rubber elasticity, and one has to make some kinds of assumptions about the junction kinetics.

8.6.1 Single-chain models[146-158]

The first attempt to develop a kinetic theory for concentrated systems using a mean-field approach was that of Doi and Edwards.[69,146] They studied the motion of a polymer molecule (modeled as a Kramers chain) that is constrained in a 'tube', using the theory of fluctuations to apply to the backwards-and-forwards motion (*i.e.* 'reptation'[147]) of the polymer along its own backbone. In this way they obtained a diffusion equation for each portion of the chain. This result is used to evaluate the stress tensor, using the stress-tensor formula from rubber elasticity, and applying it to what they call a slip-link network; in this part of the theory the molecules are imagined to be Rouse chains, whose contour lengths are maintained to be constant by the use of Maxwell demons. The resulting constitutive equation predicts shear-rate dependence for the viscometric functions, an elongational viscosity that is monotone decreasing, and rather realistic curves for the components of the complex viscosity; the ratio $-\Psi_2/\Psi_1$ is found to be 2/7. However, the viscosity function decreases so rapidly with shear rate (approaching a high-shear rate asymptote of $\eta \sim \dot{\gamma}^{-3/2}$) that it is not possible to get good curve fits of experimental data. It has further been shown that over a wide range of rotational speeds the Weissenberg effect is not described.[19] Attempts have been made to modify the theory by allowing the 'tube' diameter to change during the deformation.[148]

The Curtiss–Bird theory[149] for concentrated systems is structured quite differently, taking as the starting point the general phase-space formalism as shown in Figure 8. The polymer molecule is modeled throughout as a Kramers chain, and the restricted motion of the molecule is described by means of an anisotropic Stokes law expression for the links and an anisotropic Brownian motion. Despite the great differences between the Doi–Edwards and Curtiss–Bird theories, some of the key results are very similar. For example, both theories lead to a diffusion equation for a segment of the polymer chain as follows[146,149]

$$\frac{\partial f}{\partial t} = \frac{1}{\lambda}\frac{\partial^2 f}{\partial \sigma^2} - \left(\frac{\partial}{\partial u} \cdot [\kappa \cdot u - \kappa : uuu] f\right) \tag{104}$$

Here $f(u, \sigma, t)\mathrm{d}u$ is the probability that the polymer chain has an orientation u at a fractional distance σ down the chain at time t. This equation can be solved to give[149]

$$f(u, \sigma, t) = \frac{1}{4\pi}\int_{-\infty}^{t}\frac{P(\sigma, t-t')}{(1 + \gamma^{[0]}:uu)^{3/2}}\,\mathrm{d}t' \tag{105}$$

in which

$$P(\sigma, s) = \frac{4\pi}{\lambda}\sum_{\alpha, \text{odd}}\alpha(\sin \pi\alpha\sigma)\exp(-\pi^2\alpha^2 s/\lambda) \tag{106}$$

where $\lambda = N^{3+\beta}\zeta a^2/2kT$ is the longest relaxation time for the fluid; the parameter β is the 'chain-constraint exponent' that accounts for the increase in the drag coefficient with molecular weight (the Doi–Edwards theory time constant differs by a factor of 2 and has $\beta = 0$). The stress tensor expression is[149]

$$\tau = NnkT\left[\frac{1}{3}\delta - \int_0^1 \langle uu\rangle\,\mathrm{d}\sigma - \varepsilon\lambda\kappa:\int_0^1 \sigma(1-\sigma)\langle uuuu\rangle\,\mathrm{d}\sigma\right] \tag{107}$$

in which the $\langle\ \rangle$ averages are taken with respect to f, and the parameter ε is the 'link-tension coefficient' which enters into the nonisotropic Stokes law expression ($\varepsilon = 1$ corresponds to an isotropic Stokes law, whereas $\varepsilon = 0$ corresponds to a cylindrically symmetrical drag law; the Doi–Edwards results are not recovered by setting $\varepsilon = 0$ because of the difference in the definition of the time constant, and because of a front factor of 3). From the last two equations one can obtain the constitutive equation[149]

$$\tau = NnkT\left[\frac{1}{3}\delta - \int_{-\infty}^{t}\mu(t-t')A(t,t')\,\mathrm{d}t' - \varepsilon\lambda\kappa:\int_{-\infty}^{t}v(t-t')B(t,t')\,\mathrm{d}t'\right] \tag{108}$$

in which there are two functions dealing with the properties of the fluid

$$\mu(s) = -\left(\frac{\lambda}{2}\right)\mathrm{d}v/\mathrm{d}s \tag{109}$$

$$v(s) = \frac{16}{\pi^2\lambda}\Sigma_{\alpha, \text{odd}}\frac{1}{\alpha^2}e^{-\pi^2\alpha^2 s/\lambda} \tag{110}$$

and two tensors that describe the flow field

$$A = \frac{1}{4\pi} \int [1 + (\gamma^{[0]}: uu)]^{-3/2} uu \, du \tag{111}$$

$$B = \frac{1}{4\pi} \int [1 + (\gamma^{[0]}: uu)]^{-3/2} uuuu \, du \tag{112}$$

Equation (108) gives a non-Newtonian viscosity that goes asymptotically as $\dot{\gamma}^{-1}$ for large shear rates, and gives $-\Psi_2/\Psi_1 = (2/7)(1-\varepsilon)$. Rod-climbing at low rotational rates is predicted[19] as long as $\varepsilon > 1/8$. The zero-shear-rate viscosity is proportional to $M^{3+\beta}$ and the zero-shear-rate normal stress coefficients are proportional to $M^{6+2\beta}$.

Equation (108) has been tested extensively by comparisons with experimental data[150–152,157] on systems that contain carefully fractionated polymers. It is possible to choose values of N, λ and ε to fit the experimental data on viscometric functions (see Figure 14); the description of time-dependent behavior is somewhat less satisfactory. From the data fitting one can obtain information about the dependence of N, λ and ε on concentration and molecular weight. Equation (108) has been used to solve somewhat more difficult flow problems by numerical methods.[153,158]

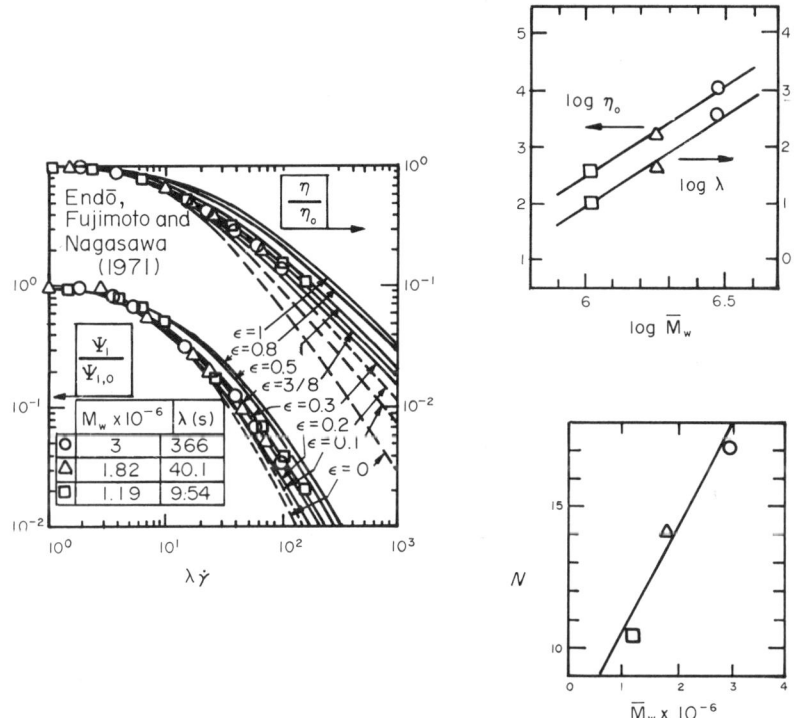

Figure 14 Data on $\eta(\dot{\gamma})$ and $\Psi_1(\dot{\gamma})$ for 7% solutions of poly(α-methylstyrene) in Kanechlor (by H. Endō, T. Fujimoto and M. Nagasawa, *J. Polym. Sci. Part A*, 1971, **2**, 345) and Curtiss–Bird theory curves computed by H. H. Saab (see ref. 152). The values of η_0 (in Pa s), λ, and N are shown and $\varepsilon = \frac{3}{8}$ provides a good fit for the data

The Curtiss–Bird theory has been extended to include polydispersity effects;[154,155] extensive data comparisons for the steady-state shear compliance and various nonlinear rheological properties further support the inclusion of the link-tension coefficient in the theory.

In addition to the Curtiss–Bird theory for Kramers bead–rod chains, a simpler theory for FENE dumbbells has also been worked out and compared with experimental data.[116,117] The simplicity of this model enables one to see the structure of the theory without the mathematical complications.

8.6.2 Network Models

The kinetic theory of permanently cross-linked rubber (see refs. 2, 60, 61, 159–174, also ref. 66, chapter 20) has been developed extensively and compared with experimental data; the treatise of

Treloar[159] provides a thorough and readable account of this subject. A review[161] of the network theory of amorphous polymer melts traces the development from the initial Green–Tobolsky theory[160] through the various refinements and embellishments on the basic network model.

For both solid rubber and polymeric liquids one envisions a coherent network of interconnected chains, in which the network junctions move affinely; thermal motions of the junctions are ignored. The segments between junctions are modeled as Gaussian chains, that is Kramers bead–rod chains obeying a random-walk distribution. For liquids one makes some additional assumptions in the model: segments may be created or lost during the flow, newly created chains immediately assume a Gaussian distribution and segments may be characterized by a single index i that indicates the 'complexity' of the segment.

No phase-space theory has been worked out for the network theory. Instead the derivations are carried out at once in terms of the configuration space of a single segment. The equation of motion for a typical segment is tantamount to a statement of affine motion of the junctions

$$\dot{Q} = [\kappa \cdot Q] \tag{113}$$

in which Q gives the length and orientation of a segment between two junctions. The continuity equation may be written

$$\frac{\partial \Psi_{iN}}{\partial t} = -\left(\frac{\partial}{\partial Q} \cdot \dot{Q}\Psi_{iN}\right) + L_{iN}(Q,t) - \frac{\Psi_{iN}}{\lambda_{iN}(Q,t)} \tag{114}$$

Here $\Psi_{iN}\,dQ$ is the number of segments per unit volume at time t that have complexity i and $N-1$ equivalent random links, and that have end-to-end vectors in the range dQ about Q. The last two terms describe the creation and loss of segments resulting from the creation and loss of junctions. The number per unit volume of iN segments of length Q that are created per unit time at time t is $L_{iN}(Q,t)$; the probability per unit time that an iN segment leaves the network is $1/\lambda_{iN}(Q,t)$. Combination of the last above two equations then gives the first-order 'convection equation' for the segments

$$\frac{\partial \Psi_{iN}}{\partial t} = -\left(\frac{\partial}{\partial Q} \cdot [\kappa \cdot Q]\,\Psi_{iN}\right) + L_{iN}(Q,t) - \frac{\Psi_{iN}}{\lambda_{iN}(Q,t)} \tag{115}$$

Various network theories result from making different assumptions for the segment creation and loss.

The stress tensor expression is obtained by summing the stresses resulting from the tensions in the springs in the network. This gives

$$\tau = -\Sigma_i \Sigma_N H_N(\langle QQ \rangle_{iN} - \langle QQ \rangle_{iN,\,eq}) \tag{116}$$

where the averages are calculated with respect to the distribution function obtained from the solution to equation (115), and H_N is the Hookean spring constant corresponding to a Gaussian chain of $N-1$ links.

From the above starting points constitutive equations may be obtained by making various assumptions about the creation and loss terms. For a cross-linked rubber all the λ_{iN} are infinite, and the L_{iN} are proportional to the Dirac delta function $\delta(t-t_0)$. This leads to

$$\tau = G_0 \gamma_{[0]}(t, t_0) \tag{117}$$

which has been used extensively in the theory of finite deformations of solids.[162,163] For polymer melts, the simplest assumptions that one can make are that the creation rates are independent of time, and that the loss rates are constant. This leads ultimately to a constitutive equation of the form

$$\tau = \int_{-\infty}^{t} \{\Sigma_j(\eta_j/\lambda_j^2)\exp[-(t-t')/\lambda_j]\}\,\gamma_{[0]}(t,t')\,dt' \tag{118}$$

This 'Lodge network model' result is a special case of the Lodge elastic liquid, in that the memory function is a sum of exponentials; it is also of the same form as the constitutive equation for the Rouse and Zimm models, except that here the constants λ_j and η_j are free parameters to be determined from the experimental data. If these quantities are both taken to be proportional to $M^{3+\beta}$, then the zero-shear-rate viscosity is proportional to $M^{3+\beta}$ and the first normal-stress coefficient is proportional to $M^{6+2\beta}$, just as in the Curtiss–Bird theory. Note that equation (118) can also be written

$$\tau = \Sigma_j \tau_j; \qquad \tau_j + \lambda_j \tau_{j(1)} = -\eta_j \dot{\gamma} \tag{119}$$

that is, as a sum of convected Maxwell models.

Because of the fact that equation (118) does not give shear-rate dependence for the viscometric functions, many other postulates have been made for the segment creation and loss rates, and some other modifications have been made in the theory. If equation (113) is replaced by $\dot{Q} = [(\kappa - \frac{1}{2}\xi\dot{\gamma}) \cdot Q]$ (*cf.* last paragraph of Section 8.5.1.3), then the constitutive equation becomes

$$\tau = \Sigma_j \tau_j; \qquad \tau_j + \lambda_j \tau_{j(1)} + \frac{1}{2}\lambda_j \xi \{\dot{\gamma} \cdot \tau_j + \tau_j \cdot \dot{\gamma}\} = -\eta_j (1 - \xi)\dot{\gamma} \tag{120}$$

which is a superposition of Oldroyd models. This leads to shear-rate dependence of the viscometric functions, but still allows the elongational viscosity to become infinite.[107,169-171]

By letting the creation and loss rates be functions of time, still better descriptions of the rheological properties can be obtained. The three most successful theories to date are those of Acierno *et al.*[172] (as rewritten by Jongschaap,[173] Phan-Thien and Tanner,[169-171] and Wagner and Stephenson).[60,61,174] The Phan-Thien and Tanner constitutive equation has the form

$$\tau = \Sigma_j \tau_j; \qquad Y_j \tau_j + \lambda_j \tau_{j(1)} + \frac{1}{2}\lambda_j \xi \{\dot{\gamma} \cdot \tau_j + \tau_j \cdot \dot{\gamma}\} = -\eta_j (1 - \xi)\dot{\gamma} \tag{121}$$

in which two proposals have been made for Y_j

$$Y_j = \begin{cases} 1 - (\varepsilon_1 \lambda_j / \eta_j) \operatorname{tr} \tau_j \\ \exp(-(\varepsilon_2 \lambda_j / \eta_j) \operatorname{tr} \tau_j) \end{cases} \tag{122}$$

where ε_1 and ε_2 are constants; the upper form in equation (122) gives a bounded monotone-increasing function for the elongational viscosity, whereas the lower form goes through a maximum (the latter behavior has been observed for some polymer melts). Note that there is some similarity with equations (73) and (74).

8.7 REFERENCES

1. J. D. Ferry, 'Viscoelastic Properties of Polymers', 3rd edn., Wiley, New York, 1980.
2. A. S. Lodge, 'Elastic Liquids', Academic Press, New York, 1964.
3. A. S. Lodge, 'Body Tensor Fields in Continuum Mechanics with Applications to Polymer Rheology', Academic Press, New York, 1974.
4. K. Walters, 'Rheometry', Wiley, New York, 1975.
5. R. B. Bird, R. C. Armstrong and O. Hassager, 'Dynamics of Polymeric Liquids', Wiley, New York, 2nd edn., vol. 1, 1987.
6. R. I. Tanner, 'Engineering Rheology', Oxford University Press, 1985.
7. W. R. Schowalter, 'Mechanics of Non-Newtonian Fluids', Pergamon, Oxford, 1978.
8. R. R. Huilgol, 'Continuum Mechanics of Viscoelastic Liquids', Wiley, New York, 1975.
9. Y. Tomita, 'Reorojii: Hisenkei Ryūtai no Rikigaku', Corona, Tōkyō, 1975.
10. G. Astarita and G. Marrucci, 'Principles of Non-Newtonian Fluid Mechanics', McGraw-Hill, New York, 1974.
11. M. J. Crochet, A. R. Davies and K. Walters, 'Numerical Simulation of Non-Newtonian Flow', Elsevier, Amsterdam, 1984.
12. J. F. Hutton and J. R. A. Pearson, 'Theoretical Rheology', Wiley, New York, 1974.
13. M. Yamamoto, 'Buttai no Henkeigaku', Seibundō-Shinkōsha, Tōkyō, 1972.
14. Z. Tadmor and C. G. Gogos, 'Principles of Polymer Processing', Wiley, New York, 1979.
15. J. R. A. Pearson, 'Mechanics of Polymer Processing', Elsevier, London, 1985.
16. J. Janeschitz-Kriegl, 'Polymer Melt Rheology and Flow Birefringence', Springer, New York, 1983.
17. K. Weissenberg, *Nature (London)*, 1947, **159**, 310.
18. F. H. Garner and A. H. Nissan, *Nature (London)*, 1946, **158**, 634.
19. A. S. Lodge, J. D. Schieber and R. B. Bird, *J. Chem. Phys.*, 1988, **88**, 4001.
20. G. R. Böhme, R. Voss and W. Warnecke, *Rheol. Acta*, 1985, **24**, 22.
21. K. Kirschke, Polymères et Lubrification, Colloques Internationaux de CNRS, No. 233, Editions de CNRS, Paris, 1974, p. 137.
22. C. T. Hill, *Trans. Soc. Rheol.*, 1972, **16**, 213.
23. J. P. Nirschl and W. E. Stewart, *J. Non-Newtonian Fluid Mech.*, 1984, **16**, 233.
24. J. M. Kramer and M. W. Johnson, Jr., *Trans. Soc. Rheol.*, 1972, **16**, 197.
25. N. N. Kapoor, M. S. Thesis, University of Minnesota, Minneapolis, 1964.
26. R. I. Tanner, *J. Polym. Sci., Part A*, 1970, **8**, 2067.
27. J. Vlachopoulos, M. Horie and S. Lidorikis, *Trans. Soc. Rheol.*, 1972, **16**, 669.
28. D. F. James, *Nature (London)*, 1966, **212**, 754.
29. R. I. Tanner, *Trans. Soc. Rheol.*, 1970, **14**, 483.
30. A. S. Wineman and A. C. Pipkin, *Acta Mechanica*, 1966, **2**, 104.
31. H. Giesekus, *Rheol. Acta*, 1968, **7**, 127.

32. D. V. Boger, *AIChE J.*, 1978, **869**, 992; 1979, **25**, 152.
33. C. F. Chang and W. R. Schowalter, *J. Non-Newtonian Fluid Mech.*, 1979, **6**, 47.
34. M. J. Riddle, C. Narvaez and R. B. Bird, *J. Non-Newtonian Fluid Mech.*, 1977, **2**, 23.
35. H. Giesekus, *Z. Angew. Math. Phys.*, 1978, **58**, T26-T36.
36. P. Brunn, *J. Non-Newtonian Fluid Mech.*, 1980, **7**, 271.
37. O. Hassager, *Nature (London)*, 1979, **279**, 402.
38. C. Bisgaard and O. Hassager, *Rheol. Acta*, 1982, **21**, 537.
39. C. Bisgaard, *J. Non-Newtonian Fluid Mech.*, 1983, **12**, 283.
40. W. P. Cox and E. H. Merz, *J. Polym. Sci.*, 1958, **28**, 619.
41. H. M. Laun, *J. Rheol.*, 1986, **30**, 459.
42. R. B. Bird, *Chem. Eng. Commun.*, 1982, **16**, 175.
43. C. J. S. Petrie, 'Elongational Flows', Pitman, London, 1979.
44. C. J. S. Petrie and J. M. Dealy, in 'Rheology', ed. G. Astarita, G. Marrucci and L. Nicolais, Plenum Press, New York, 1980, vol. 1, p. 171.
45. J. G. Oldroyd, *J. Non-Newtonian Fluid Mech.*, 1984, **14**, 9.
46. R. S. Rivlin and J. L. Ericksen, *J. Ration. Mech. Anal.*, 1955, **4**, 323.
47. R. S. Rivlin, in 'Research Frontiers in Fluid Dynamics', ed. R. J. Seeger and G. Temple, Wiley, New York, 1965.
48. R. D. Coleman, H. Markovitz and W. Noll, 'Viscometric Flows of Non-Newtonian Fluids', Springer, New York, 1966.
49. R. B. Bird, in 'Viscoelasticity and Rheology', ed. A. S. Lodge, M. Renardy and J. A. Nohel, Academic Press, New York, 1985.
50. W. O. Criminale, Jr., J. L. Ericksen and G. L. Filbey, Jr., *Arch. Ration. Mech., Anal*, 1958, **1**, 410.
51. M. Reiner, 'Deformation, Strain and Flow', Interscience, New York, 1960.
52. K. Hohenemser and W. Prager, *Z. Angew. Math. Mech.*, 1932, **12**, 216.
53. W. Ostwald, *Kolloid-Z.*, 1924, **36**, 99.
54. A. de Waele, *Oil Colour Chem. Assoc. J.*, 1923, **6**, 33.
55. P. J. Carreau, Ph.D. Thesis, University of Wisconsin, 1968.
56. H. Jeffreys, 'The Earth', Cambridge University Press, 1929, p. 265.
57. J. C. Maxwell, *Philos. Trans. R. Soc. London, Ser. A*, 1867, **157**, 49.
58. J. G. Oldroyd, *Proc. R. Soc. London, Ser. A*, 1958, **245**, 278.
59. J. L. White and A. B. Metzner, *J. Appl. Polym. Sci.*, 1963, **7**, 1867.
60. M. H. Wagner, *Rheol. Acta*, 1979, **18**, 33.
61. M. H. Wagner, T. Raible and J. Meissner, *Rheol. Acta*, 1979, **18**, 427.
62. A. Kaye, College of Aeronautics, Cranfield, Note No. 134, 1962.
63. B. Bernstein, E. A. Kearsley, and L. J. Zapas, *Trans. Soc. Rheol.*, 1963, **7**, 391.
64. B. Bernstein, *Int. J. Nonlinear Mech.*, 1969, **4**, 183.
65. L. J. Zapas and J. C. Phillips, *J. Rheol.*, 1981, **25**, 405.
66. R. B. Bird, C. F. Curtiss, R. C. Armstrong, and O. Hassager, 'Dynamics of Polymeric Liquids', 2nd edn., Wiley, New York, vol. 2, 1987.
67. H. Yamakawa, 'Modern Theory of Polymer Solutions', Harper and Row, New York, 1971.
68. J. G. Kirkwood, 'Macromolecules', ed. P. L. Auer, Gordon and Breach, New York, 1967.
69. M. Doi and S. F. Edwards, 'Theory of Polymer Dynamics', Oxford University Press, 1986.
70. N. Saitō, 'Kōbunshi Butsurigaku', Shōkabō, Tokyo, 1968.
71. M. Fixman and W. H. Stockmayer, *Annu. Rev. Phys. Chem.*, 1970, **21**, 407.
72. M. C. Williams, *AIChE J.*, 1975, **21**, 1.
73. R. B. Bird, H. R. Warner, Jr. and D. C. Evans, *Fortschr. Hochpolym.-Forsch.*, 1971, **8**, 1.
74. G. G. Fuller and L. G. Leal, *J. Non-Newtonian Fluid Mech.*, 1981, **8**, 271.
75. P. E. Rouse Jr., *J. Chem. Phys.*, 1953, **21**, 1272.
76. D. H. King and D. F. James, *J. Chem. Phys.*, 1983, **78**, 4743.
77. P. H. van Wiechen and H. C. Booij, *J. Eng. Math.*, 1971, **5**, 89.
78. B. H. Zimm, *J. Chem. Phys.*, 1956, **24**, 269.
79. A. S. Lodge and Y. Wu, *Rheol. Acta*, 1971, **10**, 539.
80. H. A. Kramers, *Physica (Amsterdam)*, 1944, **11**, 1.
81. O. Hassager, *J. Chem. Phys.*, 1974, **60**, 4001.
82. R. B. Bird and C. F. Curtiss, *J. Non-Newtonian Fluid Mech.*, 1984, **14**, 85.
83. J. G. Kirkwood and J. Riseman, *J. Chem. Phys.*, 1948, **16**, 565.
84. X.-J. Fan and T. W. Liu, *J. Non-Newtonian Fluid Mech.*, 1986, **19**, 303.
85. P. J. Flory, 'Statistical Mechanics of Chain Molecules', Wiley, New York, 1969.
86. C. F. Curtiss, R. B. Bird and O. Hassager, *Adv. Chem. Phys.*, 1976, **35**, 31.
87. C. F. Curtiss and R. B. Bird, *Physica A (Amsterdam)*, 1983, **118**, 191.
88. H. Giesekus, *Rheol. Acta*, 1962, **2**, 50.
89. G. C. Sarti and G. Marrucci, *Chem. Eng. Sci.*, 1973, **28**, 1053.
90. H. C. Booij, *J. Chem. Phys.*, 1984, **80**, 4571.
91. A. Peterlin, *Makromol. Chem.*, 1961, **44-46**, 338.
92. H. R. Warner, Jr., *Ind. Eng. Chem. Fundam.*, 1972, **11**, 379.
93. R. C. Armstrong, *J. Chem. Phys.*, 1974, **60**, 724, 729.
94. R. L. Christiansen and R. B. Bird, *J. Non-Newtonian Fluid Mech.*, 1977/78, **3**, 161.
95. R. C. Armstrong, S. K. Gupta and O. Basaran, *Polymer Eng. Sci.*, 1980, **20**, 466.
96. R. I. Tanner, *Trans. Soc. Rheol.*, 1975, **19**, 37.
97. X.-J. Fan, *J. Non-Newtonian Fluid Mech.*, 1985, **17**, 125.
98. Y. Mochimaru, *J. Non-Newtonian Fluid Mech.*, 1983, **12**, 135.
99. L. E. Wedgewood and R. B. Bird, in 'Integration of Fundamental Polymer Science and Technology', ed. L. A. Kleintjens and P. J. Lemstra, Elsevier, Amsterdam, 1986.
100. X.-J. Fan, R. B. Bird and M. Renardy, *J. Non-Newtonian Fluid Mechanics*, 1985, **18**, 255.
101. J. Rotne and S. Prager, *J. Chem. Phys.*, 1969, **50**, 4831.

102. H. Yamakawa, *J. Chem. Phys.*, 1970, **53**, 436.
103. H. C. Öttinger, *J. Chem. Phys.*, 1986, **84**, 4068.
104. R. B. Bird and J. M. Wiest, *J. Rheol.*, 1985. **29**, 519.
105. J. M. Wiest and R. B. Bird, *J. Non-Newtonian Fluid Mech.*, 1986, **22**, 115.
106. R. J. Gordon and W. R. Schowalter, *Trans. Soc. Rheol.*, 1972, **16**, 79.
107. M. W. Johnson and D. Segalman, *J. Non-Newtonian Fluid Mech.*, 1977, **2**, 255.
108. R. B. Bird, *J. Non-Newtonian Fluid Mech.*, 1979, **5**, 1.
109. W. Kuhn and H. Kuhn, *Helv. Chim. Acta*, 1945, **28**, 1533.
110. R. Cerf, *Fortschr. Hochpolym. Forsch.*, 1959, **1**, 382.
111. A. Peterlin, *J. Polym. Sci., Part A-2*, 1967, **5**, 179.
112. H. C. Booij and P. H. van Wiechen, *J. Chem. Phys.*, 1970, **52**, 5056.
113. C. W. Manke and M. C. Williams, *J. Rheol.*, 1986, **30**, 19.
114. H. Giesekus, *Rheol. Acta*, 1966, **5**, 29.
115. H. Giesekus, *J. Non-Newtonian Fluid Mech.*, 1982, **11**, 69; 1983, **12**, 367.
116. R. B. Bird and J. R. DeAguiar, *J. Non-Newtonian Fluid Mech.*, 1983, **13**, 149.
117. J. R. DeAguiar, *J. Non-Newtonian Fluid Mech.*, 1983, **13**, 161.
118. S. Kim and X.-J. Fan, *J. Rheol.*, 1984, **28**, 117.
119. W. E. Stewart and J. P. Sørensen, *Trans. Soc. Rheol.*, 1972, **16**, 1.
120. R. C. Armstrong and R. B. Bird, *J. Chem. Phys.*, 1973, **58**, 2715.
121. H. H. Saab and R. B. Bird, *J. Chem. Phys.*, 1987, **86**, 3032.
122. H. H. Saab and P. J. Dotson, *J. Chem. Phys.*, 1987, **86**, 3039.
123. H. C. Öttinger, *J. Chem. Phys.*, 1987, **86**, 3731; erratum, *ibid.*, 1987, **87**, 1460.
124. B. H. Zimm and W. H. Stockmayer, *J. Chem. Phys.*, 1949, **17**, 1301.
125. V. Bloomfield and B. H. Zimm, *J. Chem. Phys.*, 1966, **44**, 315.
126. M. Fukatsu and M. Kurata, *J. Chem. Phys.*, 1966, **44**, 4539.
127. S. Imai, *J. Chem. Phys.*, 1969, **51**, 1732.
128. J. Roovers, *J. Polym. Sci., Polym. Phys. Ed.*, 1985, **23**, 1117.
129. J. Roovers, *Macromolecules*. 1985. **18**, 1359.
130. J. M. Wiest, S. R. Burdette, T. W. Liu and R. B. Bird, *J. Non-Newtonian Fluid Mech.*, 1987, **24**, 279; T. W. Liu and H. C. Öttinger, *J. Chem. Phys.*, 1987, **87**, 3131.
131. K. Osaki, J. L. Schrag and J. D. Ferry, *Macromolecules*, 1972, **5**, 144.
132. M. Fixman, *J. Chem. Phys.*, 1966, **45**, 785, 793.
133. C. W. Pyun and M. Fixman, *J. Chem. Phys.*, 1965, **42**, 3838.
134. H. D. Stidham and M. Fixman, *J. Chem. Phys.*, 1968, **48**, 3092.
135. R. H. Shafer, *Macromolecules*, 1976, **9**, 895.
136. N. W. Tschoegl, *J. Chem. Phys.*, 1964. **40**, 473.
137. R. L. Sammler, Ph.D. Thesis, University of Wisconsin-Madison, 1985.
138. I. Noda and J. E. Hearst, *J. Chem. Phys.*, 1971, **54**, 2342.
139. E. R. Bazua and M. C. Williams, *J. Polym. Sci., Polym. Phys. Ed.*, 1974, **12**, 825.
140. H. C. Öttinger, *J. Non-Newtonian Fluid Mech.*, 1987, **26**, 207; L. E. Wedgewood and H. C. Öttinger, *ibid.*, 1988, **27**, 245.
141. M. Gottlieb and R. B. Bird, *J. Chem. Phys.*, 1976, **65**, 2467.
142. M. Gottlieb, *Computers and Chemistry*, 1977, **1**, 155.
143. C. F. Curtiss and R. B. Bird, *J. Non-Newtonian Fluid Mech.*, 1977, **2**, 392.
144. N. van Kampen, *Appl. Sci. Res.*, 1981, **37**, 67.
145. N. Fixman and J. Kovac, *J. Chem. Phys.*, 1974, **61**, 4950.
146. M. Doi and S. F. Edwards, *J. Chem. Soc., Faraday Trans. 2*, 1978, **74**, 1789, 1802, 1818; 1979, **75**, 38.
147. P.-G. de Gennes, *J. Chem. Phys.*, 1971, **55**, 572.
148. G. Marrucci and J. J. Hermans, *Macromolecules*, 1980, **13**, 380.
149. C. F. Curtiss and R. B. Bird, *J. Chem. Phys.*, 1981, **74**, 2016, 2026.
150. R. B. Bird, H. H. Saab and C. F. Curtiss, *J. Phys. Chem.*, 1982, **86**, 1102.
151. R. B. Bird, H. H. Saab and C. F. Curtiss, *J. Chem. Phys.*, 1982, **77**, 4747.
152. H. H. Saab, R. B. Bird and C. F. Curtiss, *J. Chem. Phys.*, 1982, **77**, 4758.
153. X.-J. Fan and R. B. Bird, *J. Non-Newtonian Fluid Mech.*, 1984, **15**, 341.
154. J. D. Schieber, C. F. Curtiss and R. B. Bird, *Ind. Eng. Chem. Fundam.*, 1986, **25**, 471.
155. J. D. Schieber, *J. Chem. Phys.*, 1987, **87**, 4917, 4928.
156. R. B. Bird and C. F. Curtiss, *Ned. Tijdschr. Natuurk., A*, 1981, **47**, 133.
157. D. A. Bernard and J. Noolandi, *Macromolecules*, 1983, **16**, 1358.
158. D. S. Malkus and B. Bernstein, *J. Non-Newtonian Fluid Mech.*, 1984, **16**, 77.
159. L. R. G. Treloar, 'The Physics of Rubber Elasticity', 3rd edn., Oxford University Press, London, 1975.
160. M. S. Green and A. V. Tobolsky, *J. Chem. Phys.*, 1946, **14**, 80.
161. A. S. Lodge, R. C. Armstrong, M. H. Wagner and H. H. Winter, *Pure Appl. Chem.*, 1982, **54**, 1349.
162. R. S. Rivlin, *Philos. Trans. R. Soc. London, Ser. A*, 1948, **240**, 459, 491, 509.
163. R. S. Rivlin, in 'Rheology', ed. F. R. Eirich, Academic Press, New York, 1956, vol. 1, p. 351.
164. A. S. Lodge, *Trans. Faraday Soc.*, 1956, **52**, 120.
165. A. S. Lodge, *Rheol. Acta*, 1968, **7**, 379.
166. M. Yamamoto, *J. Phys. Soc. Jpn.*, 1956, **11**, 413; 1957, **12**, 1148; 1958, **13**, 1200.
167. F. Wiegel, *Physica (Amsterdam)*, 1969, **42**, 156.
168. F. Wiegel and F. Th. de Bats, *Physica (Amsterdam)*, 1969, **43**, 33.
169. N. Phan-Thien and R. I. Tanner, *J. Non-Newtonian Fluid Mech.*, 1977, **2**, 353.
170. N. Phan-Thien, *J. Rheol.*, 1978, **22**, 259.
171. R. I. Tanner, *J. Non-Newtonian Fluid Mech.*, 1979, **5**, 103.
172. D. Acierno, F. P. La Mantia, G. Marrucci, and G. Titomanlio, *J. Non-Newtonian Fluid Mech.*, 1976, **1**, 125.
173. R. J. J. Jongschaap, *J. Non-Newtonian Fluid Mech.*, 1981, **8**, 183.
174. M. H. Wagner and S. E. Stephenson, *J. Rheol.*, 1979, **23**, 489.

9

Rubber Elasticity and Characterization of Networks

JEAN-PIERRE QUESLEL
Manufacture Michelin, Clermont-Ferrand, France
and
JAMES E. MARK
University of Cincinnati, OH, USA

9.1 INTRODUCTION TO RUBBER ELASTICITY

Substances which exhibit rubber elasticity can, by definition, sustain large deformations and then return to their original state when the deformation is removed.[1-8] This high extensibility combined with a capacity for full recovery is exhibited by nearly all polymers when in the disordered state above the glass transition temperature. However, under prolonged stress, linear or even branched polymers such as raw rubber will flow like a viscous liquid. Therefore, a cross-linking process is necessary to confer a permanence of shape to the polymer and remove the tendency to flow. Hence, in order for a material to exhibit high elasticity, three molecular conditions must be met: (i) the material must consist of polymeric chains; (ii) the chains must be joined into a network structure; and (iii) the chains must have a high degree of flexibility.

The first requirement arises from the fact that the molecules in a rubber or elastomeric material must be able to alter their arrangements and extensions in space in response to an imposed stress, and only a long-chain molecule has the very large number of spatial arrangements of very different extensions required. This versatility is illustrated in Figure 1, which depicts a random spatial arrangement of a relatively short polymer chain. In this random arrangement, the chain extension (as measured by the end-to-end separation) is quite small. For even such a short chain, the extension could be increased approximately fourfold by simple rotations about skeletal bonds, without any need for distortions of bond angles or bond lengths.

The second characteristic cited above, a network structure, is required to obtain the elastomeric recoverability. It is obtained by joining pairs of segments from different chains, approximately one out of a hundred, thereby preventing the extended polymer chains from irreversibly sliding by one another. When primary molecules are joined intermolecularly by means of a chemical reaction, the first stage can be viewed as the formation of a giant tree devoid of cyclic (intramolecular) connections. Additional cross-links would engage units intramolecularly into a three-dimensional network[9-16] which is no longer soluble in any solvent, although good solvents would swell the network structure.[17-21]

The functionality ϕ of the cross-links or junctions, *i.e.* the number of chains emanating from each of them, depends on the chemical process used to link the chain molecules. The network junctions are provided ordinarily by covalent bonds. In some instances, however, physical association of chain segments, for example in crystallites, may serve the same function as permanent chemical linkages.[22] Difunctional junctions are of no interest since they lead only to chain extension. The two most important types of networks are (i) the tetrafunctional (ϕ equals four), almost invariably obtained

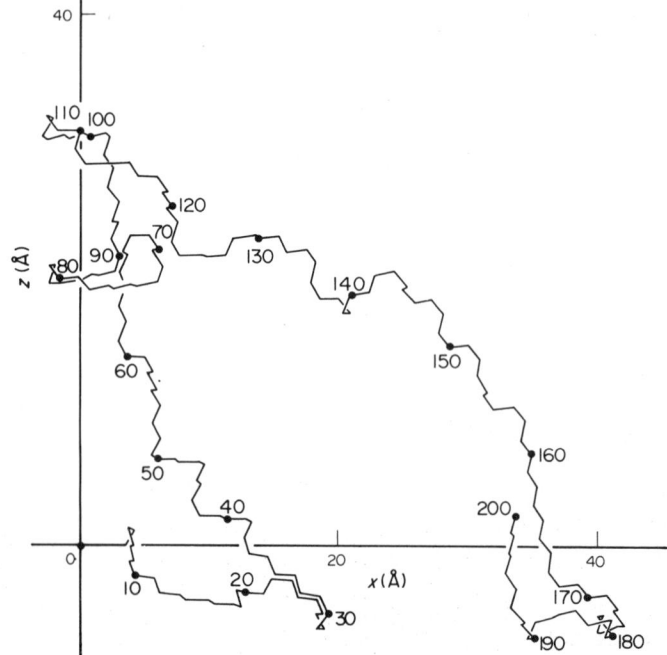

Figure 1 A computer-generated conformation of a polymethylene chain of 200 C–C bonds. The projection shown is in the plane of the first two bonds (reproduced by permission of Wiley from the article 'Elasticity', J. P. Queslel and J. E. Mark, 'Encyclopedia of Polymer Science and Engineering', 2nd edn., 1986, vol. 5, p. 365)

Figure 2 Sketch of an elastomeric tetrafunctional network showing the dense interpenetration of chains and junctions (dots)

upon joining two segments from different chains, as in the case of sulfur vulcanization,[23-25] peroxide[26] or irradiation cross-linking); and (ii) the trifunctional (ϕ equals three) obtained for example when forming a polyurethane network by end-linking hydroxyl-terminated chains with a triisocyanate.[27] A typical tetrafunctional network structure is illustrated in Figure 2, in which the cross-links are represented by dots.

The last characteristic specifies that the different spatial arrangements should not be hindered by constraints such as those which might result from inherent rigidity of the chains, extensive chain crystallization or the very high viscosity characteristic of the glassy state.

In 1805 Gough reported his observations on the essential thermoelastic properties of natural rubber.[27] When a rubber band is stretched rapidly, *i.e.* adiabatically, its temperature rises and, under tension, it contracts on heating and expands on cooling. Thus, stretched rubber possesses a negative thermal expansion coefficient in the direction of deformation, whereas a positive thermal expansion coefficient is displayed by low molecular weight systems as well as by undeformed polymers. These experiments were followed by the classic works of Joule[28,29] and Lord Kelvin[30] in the 1850s. Their correlation of the thermal effects accompanying elongation of rubber with its thermoelastic behavior represented one of the first applications of the second law of thermodynamics which, combined with Gough's observations, permits one to deduce that the entropy of rubber must decrease with extension. The change in the internal energy U accompanying the stretching of an elastic body[1,2,31,32] is given by equation (1), where the infinitesimal work of uniaxial extension $f \, dL$ is simply added to the work of compression. T is the absolute temperature, S the entropy, p the pressure, V the volume, f the applied force and L the length. The differential of the Gibbs free energy F is defined in equation (2). Equation (3) can then be obtained as one of the Euler relationships. On the assumption that the rapid (adiabatic) stretching is essentially reversible, the Gough experiment cited gives a positive value for the coefficient $(\partial T/\partial L)_{S,p}$. Since $(\partial T/\partial L)_{S,p}$ is equal to $-(\partial S/\partial L)_{T,p}/(\partial S/\partial T)_{L,p}$, and since the heat capacity $T(\partial S/\partial T)_{L,p}$ is clearly positive, it follows that $(\partial S/\partial L)_{T,p}$ is negative. This decrease of entropy is the source of the retractive force.

$$dU = T \, dS - p \, dV + f \, dL \tag{1}$$

$$dF = -S \, dT + V \, dp + f \, dL \tag{2}$$

$$(\partial f/\partial T)_{p,L} = -(\partial S/\partial L)_{p,T} \tag{3}$$

In 1932, Meyer, Von Susich and Valkó[33] advanced the essential insight into the molecular mechanism of high elasticity, roughly a decade after the nature of macromolecules had been established by Staudinger. They attributed high elasticity to the tendency of the deformed coiling chain molecule to come back to its original random shape, such as the one illustrated in Figure 1. This undeformed state is the most probable state having the maximum number of configurations and thus the maximum entropy. This molecular mechanism for the retractive force was thus in perfect agreement with the observations of Gough and Joule and with the thermodynamic analysis of Kelvin.

The thermodynamic behavior of rubber shows, therefore, a close analogy with that of an ideal gas, the entropy of which decreases during compression. In fact, the elastic equation of state for an ideal rubber is similar to the molecular form of the equation of state for an ideal gas. The stress replaces the pressure, and the number of network chains the number of gas molecules.

On the other hand, rubbers and metals differ greatly with regard to their deformation mechanisms.[31,34] Metals and minerals are formed of atoms arranged in a three-dimensional crystalline lattice, acted on by powerful valence forces operating at relatively short ranges. Deformation of such materials involves changes in interatomic distance, and this requires forces and energy changes that are very large. Hence the elastic modulus of these materials is very high. After a few percent deformation, slippage between adjacent crystalline regions occurs (at the yield point) and the deformation increases much more rapidly than the stress, becoming irreversible or plastic. The primary effect of stretching a metal short of this yield point is an increase in energy caused by changing the distance of separation between the metal atoms. The sample recovers its original length when the force is removed, since this process corresponds to a decrease in energy. Heating increases oscillations about the minimum in the asymmetric potential energy curve and thus causes the usual volume expansion. Differences in thermoelastic behavior and the origins of elasticity for a metal, a rubber and a gas are summarized in Figures 3 and 4, respectively.[31]

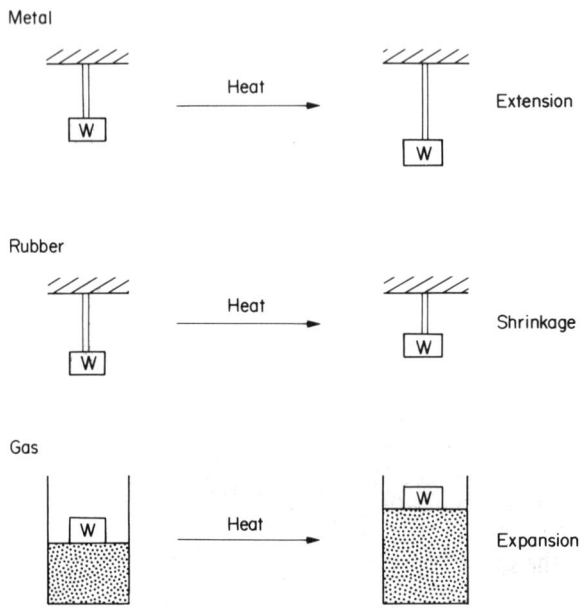

Figure 3 Results of thermoelastic experiments carried out on a typical metal, rubber and gas (reproduced by permission of the American Chemical Society from J. E. Mark, Audio Course Manual 'Physical Chemistry of Polymers', 1986)

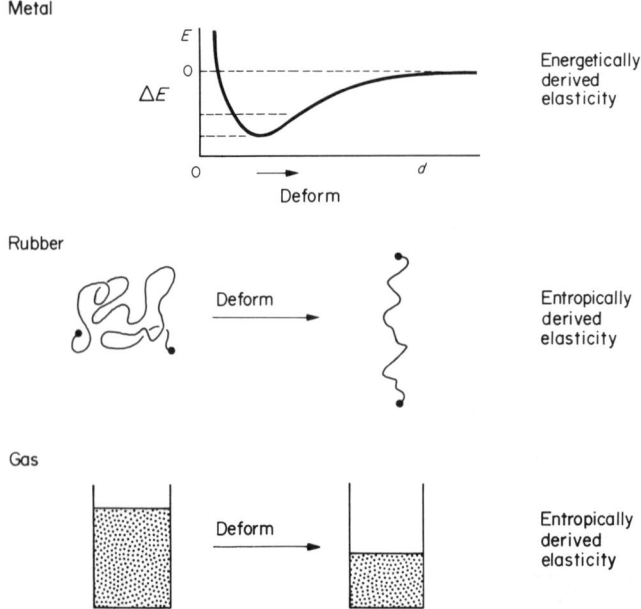

Figure 4 Sketches explaining the observations described in Figure 3 in terms of the molecular origin of the elastic force or pressure (reproduced by permission of the American Chemical Society from J. E. Mark, Audio Course Manual, 'Physical Chemistry of Polymers', 1986)

In marked contrast to the enthalpic deformation of metals, the stretching of elastomers does not involve any significant change in the interatomic distances and therefore the forces required are considerably lower. A representative stress–strain curve for an elastomer is given in Figure 5. Its most important features are a very low initial modulus ($\sim 10^{-5}$ times that of a typical metal), extremely high extensibility and reasonably high strength. The area under the curve is frequently of considerable interest since it is proportional to the work of deformation. Its value up to the rupture point is thus a measure of the toughness of the material.

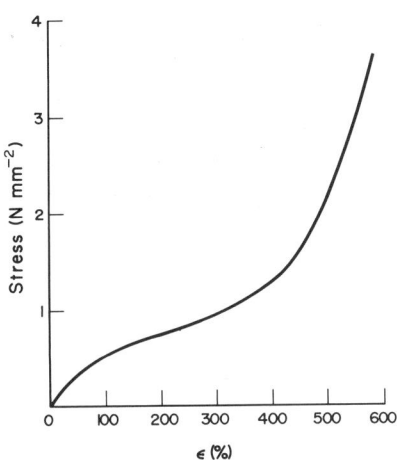

Figure 5   Typical stress–extension curve of a cross-linked rubber. Stress is defined as the force per unit cross-sectional area of the undeformed sample. Extension ε is the ratio of the increase in length to the original length

9.2   STATISTICS AND ELASTICITY OF A POLYMER CHAIN

As illustrated in Figure 2, elastomeric networks consist of chains joined by multifunctional junctions. As early as 1934, it was suggested by Guth and Mark[35] and by Kuhn[36] that the elastic retractive force exhibited by rubber upon deformation arises from the entropy decrease associated with the diminished number of conformations available to deformed polymer chains. It is, therefore, of primary interest to study the statistics of a polymer chain[8,37,38] and to establish the elastic equation of state for a single chain.

Characterization of the chemical constitution of a polymeric network requires descriptions at two levels. It must be characterized first at the macro level, where the overall topology, as expressed by the chain length between junctions, molecular weight distribution, cross-link density and chain connectivity, is taken into account. Secondly, it is necessary at a micro level to consider the constitution of the units comprising the polymer chains and the bonds connecting them to one another. Elucidation of the relationships between structure and properties necessitates inquiry into the spatial arrangement of the atoms comprising the chain skeleton and its pendant groups. In chain molecules, torsional rotations about skeletal bonds, as well as bond lengths and bond angles, must be taken into account. Torsional angles may assume wide ranges of values, and the spatial configuration of a chain molecule depends decisively on these angles, which specify its conformation. This set of variables is very large for a long-chain polymer, being equal (approximately) to the number of skeletal bonds. Consequently, the polymer molecule has access to an enormous array of conformations. Owing to its linearity, it may be treated as a one-dimensional mechanical system and such a system is amenable to exact methods. The large number of variables (*i.e.* torsional angles) required to specify a conformation notwithstanding, averages over these variables may be readily evaluated by dealing with them sequentially. Consequently, statistical mechanical averages of the properties of a polymer molecule of specified structure may be treated quite rigorously.

As an example of chemical structure and torsional angles, a sketch of a 1,4-polybutadiene chain in the planar all-*trans* conformation is shown in Figure 6.[39] Rotational angles ϕ about bonds of the chain skeleton are measured relative to the planar conformation shown, for which $\phi = 0°$. X-Ray diffraction studies[40,41] on *cis*- and *trans*-polybutadiene indicate that CH_2–CH_2 bonds such as bond $(i+2)$ connecting C_{i+1} and C_{i+2} in Figure 6 assume the *trans* conformation in the crystalline state. The rotational states accessible to these bonds in the amorphous state or in solution are concluded

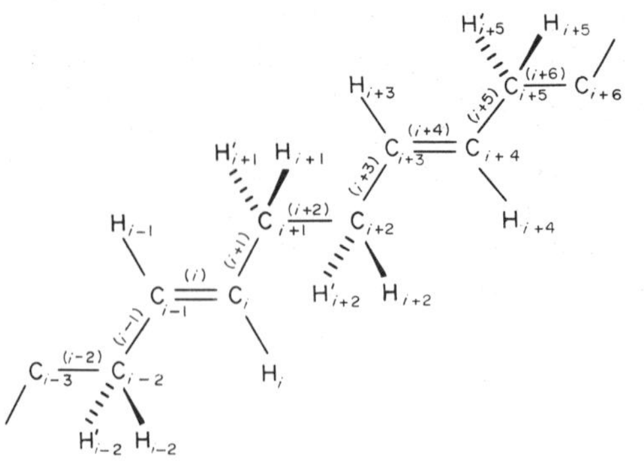

Figure 6 The *trans*-1,4-polybutadiene chain in the planar, all-*trans* conformation. This (highly disfavored) form is shown for convenience in enumerating atoms and bonds of the chain (reproduced by permission of the American Chemical Society from Y. Abe and P. J. Flory, *Macromolecules*, 1971, **4**, 219)

to be *trans* ($\phi = 0°$) or *gauche* ($\phi \simeq \pm 120°$) by analogy with CH_2–CH_2 bonds of normal alkanes and their derivatives. These analogues generally assume both *trans* and *gauche* conformations in the gaseous or liquid state but adopt exclusively the *trans* conformation in the crystalline state. According to studies on the rotational isomerism of propylene and its homologs, the rotational energy minima for the C–C bond adjacent to the double bond in the sequence C–C–C=C occurs at the *skew* ($\phi \simeq \pm 60°$) and *cis* ($\phi = 180°$) conformations.[42,43]

The rotational isomeric state method (matrix generation technique), associated with conformational energy calculations, is a useful technique for determining configurational averages for real chains, including the polybutadiene chain described above.[8,37,44–49] The procedure is applicable to chains of any length, to copolymers of any specified composition and sequence and to asymmetric (*e.g.* vinyl) chains of any stereochemical configuration and sequence, tactic or atactic.[50–55] Spatial configurations of macromolecules are represented in terms of a set of discrete rotational states. Properties of chain molecules that depend on configuration include the dimensions of the spatial configuration measured by the chain displacement vector *r* connecting its ends, or by any of various products that may be formed from *r*, the radius of gyration, dipole moments and optical anisotropies.

To establish a useful equation of state for the mechanical behavior of a rubber network, it is necessary to predict the most probable overall dimensions of the molecules under the influence of various externally applied forces. An interesting approach to rubber elasticity consists of simulating network chain configurations (and thus the distribution of end-to-end distances) by the rotational isomeric state technique cited above.[56,57] Based on the actual chemical structure of the chains, it enables one to circumvent the limitations of the Gaussian distribution function in the high deformation range. Nonetheless, the Gaussian distribution function of the end-to-end distance is very useful. It is obtained from a simple hypothetical model, the so-called freely jointed chain, which can be treated either exactly or at various levels of approximation.

9.2.1 Statistics of the Freely Jointed Chain

This hypothetical chain is considered to consist of *n* bonds of length *l* joined in sequence without any restrictions whatever on the angles between successive bonds, and with zero volume in the sense that different bonds do not interfere with one another in space. The corresponding chain configuration problem is similar to that of random flight. Specifically, the configuration of the freely jointed chain resembles the path described by a diffusing particle such as a gas molecule.[58]

The most appropriate quantity to specify the overall chain configuration is the vector distance *r* between the chain ends, called the displacement length (which must be distinguished from the fully extended or contour length *nl*). The fully extended conformation is, of course, only one of a great many conformations. It is more meaningful to consider an average size of the macromolecule, such as the mean-square end-to-end distance, $\langle r^2 \rangle$. It is equivalent in statistics to consider a molecule at different times or an assembly of *N* molecules at the same time. The mean-square end-to-end distance which describes this assembly is defined by equation (4), where r_i is the end-to-end distance of the *i*th chain. The vector r_i is the sum of the bond or link vectors I_j of the chain as shown in equation (5) and r_i^2 is the scalar product of $\mathbf{r}_i$ with itself, as shown in equation (6). If the chain is assumed to be freely jointed and volumeless, any two links can assume any orientation with respect to one other, as already stated. Therefore the second term of equation (6) is zero, resulting in equation (7). The mean-square end-to-end distance of a freely orienting chain is deduced from equations (4) and (7),[59-62] as shown in equation (8).

$$\langle r^2 \rangle = (1/N) \sum_{i=1}^{N} r_i^2 \tag{4}$$

$$r_i = \sum_{j=1}^{n} \mathbf{l}_j \tag{5}$$

$$r_i^2 = \sum_{j=1}^{n} l_j^2 + 2 \sum_{k<j} \mathbf{l}_k \cdot \mathbf{l}_j \tag{6}$$

$$r_i^2 = nl^2 \tag{7}$$

$$\langle r^2 \rangle = nl^2 \tag{8}$$

Another interesting quantity is the probability for a chain to have a given end-to-end distance. The problem is to calculate the probability that, if one extremity of vector $\mathbf{r}$ is fixed at the origin of the coordinates, the other lies in the infinitesimal volume $dxdydz$ centered around the point x, y, z, as illustrated in Figure 7.[2,35,36,63] This problem will be solved for the one-dimensional case first. The projection l_x on the x axis made by a single link of the chain has the value $l \cos \psi$ if this link lies at an angle ψ to the x direction. The probability $P(l_x)dl_x$ that this projection has a value lying between l_x and $l_x + dl_x$ is given by the fraction of the solid angle accessible to the bond which gives a value of the projection in the correct range. For a freely jointed chain, all angles are accessible and equation (9) is obtained. The extreme values of l_x are $+l$ for bonds lying in the positive direction of the x axis, i.e. $\psi = 0°$, and $-l$ for $\psi = \pi$. The mean value of l_x is zero and the mean square $\langle l_x^2 \rangle$ is obtained by integrating the product of l_x^2 and $P(l_x)dl_x$ from $-l$ to $+l$, as shown in equation (10).

$$P(l_x)dl_x = 2\pi \sin\psi \, d\psi / 4\pi = (1/2)\sin\psi \, d\psi \tag{9}$$

$$\langle l_x^2 \rangle = (1/2) \int_0^{\pi} l^2 \cos^2\psi \, \sin\psi \, d\psi = l^2/3 \tag{10}$$

For a chain consisting of a large number of links, every link can be assumed to contribute a length $l/3^{1/2}$ to its component along the x axis. The total chain length on the x axis is the number n_+ of link projections in the positive direction minus that in the negative direction multiplied by the average length of each component, as shown in equation (11). The probability of finding a particular value of x is the probability of a random flight of n_+ positive steps and n_- negative steps with $n = n_+ + n_-$. This probability is given in equation (12).

$$x = (n_+ - n_-)l/3^{1/2} \tag{11}$$

$$W(n_+, n_-) = (1/2)^n n! / n_+! n_-! \tag{12}$$

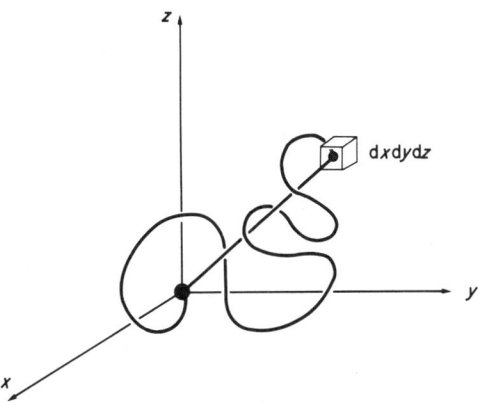

Figure 7 Schematic representation of a random chain in a Cartesian coordinate system

For an unrestricted chain, a bond does not make exactly the root-mean-square contribution to the projection and the real distribution of x is continuous. It can be shown[2,7,37] that the probability $W(x)dx$ that x assumes a value between x and $x+dx$ is the Gaussian distribution, provided that x is much smaller than the value nl corresponding to full extension of the chain. It is thus given by equation (13), with β as defined in equation (14). The generalization in three dimensions leads to equation (15), where r^2 is given in equation (16). Equation (15) gives the probability that if one end of a random chain is located at the origin of a Cartesian coordinate system, then the other end can be found in the volume element $dxdydz$ a distance r away. The probability of finding the free chain end anywhere at a distance r from the origin is expressed by the radial distribution function given in equation (17) (where r is now a scalar quantity and not a vector). The quantity $4\pi r^2 dr$ in equation (17) is the volume of a spherical shell of thickness dr located a distance r from the origin.

$$W(x)dx = (\beta/\pi^{1/2})\exp(-\beta^2 x^2)\,dx \tag{13}$$

$$\beta \equiv [3/(2nl^2)]^{1/2} \tag{14}$$

$$W(x,y,z)dxdydz = W(x)W(y)W(z)dxdydz$$

$$= (\beta/\pi^{1/2})^3\exp(-\beta^2 r^2)dxdydz \tag{15}$$

$$r^2 = x^2 + y^2 + z^2 \tag{16}$$

$$W(r)dr = (\beta/\pi^{1/2})^3\exp(-\beta^2 r^2)4\pi r^2 dr \tag{17}$$

The most probable value of r, occurring at the maximum in $W(r)$, is not zero and is found to be equal to $1/\beta$. Figure 8 shows a plot of the radial distribution function $W(r)$ for a six-link random chain with $l=1$ in arbitrary units.[1] The mean-square end-to-end distance is the second moment of the radial distribution function, and can thus be calculated from equation (18), which yields the results given in equations (8) and (19).

$$\langle r^2 \rangle = \int_0^\infty r^2 W(r)dr / \int_0^\infty W(r)dr \tag{18}$$

$$\langle r^2 \rangle = 3/(2\beta^2) = nl^2 \tag{19}$$

The Gaussian distribution function is really valid only when the chains are far below full extension. Its greatest shortcoming is the fact that it predicts zero probability only for $r=\infty$ instead of for all r in excess of that for full extension. In the region of high deformation, the appropriate non-Gaussian statistics treatment provides the general distribution function given in equation (20),[64-66] where $\mathscr{L}^{-1}$ is the inverse of the Langevin function defined in equation (21). Equation (21) can be expanded in the series shown in equation (22). The Gaussian distribution (equation 17) is recovered for $r \ll nl$. A comparison of the Gaussian and inverse Langevin distributions is shown in Figure 8.[1]

$$W(r)dr = \text{const. } \exp[-\int_0^r \mathscr{L}^{-1}(r/nl)dr/l]4\pi r^2 dr \tag{20}$$

$$\mathscr{L}(u) = \coth u - 1/u \tag{21}$$

$$W(r)dr = \text{const. } \exp(-n\{(3/2)(r/nl)^2 + (9/20)(r/nl)^4 + (99/350)(r/nl)^6 + \ldots\})4\pi r^2 dr \tag{22}$$

9.2.2 The Freely Rotating Chain

If the angles between successive bonds in a chain are held fixed, but the rotation of the bond is free, the resulting chain is termed freely rotating. Unlike the case of the freely jointed chains, the average of the scalar product of vectors $\boldsymbol{l}_i$ and $\boldsymbol{l}_j$ appearing in equation (6) is no longer zero but is given by the relationship shown in equation (23), where θ is the angle between successive bonds. The transverse projection of $\boldsymbol{l}_j$ or $\boldsymbol{l}_i$ averages to zero because of the free rotation. Substitution in equation (6) yields equation (24) for large n.[5,59,60] In the particular case of a tetrahedrally bonded chain, $\theta = 109.5°$ and the result, given in equation (25), is twice the value for a freely jointed chain.

$$\langle \boldsymbol{l}_i \cdot \boldsymbol{l}_j \rangle = l^2 \cos^{|j-i|}\theta \tag{23}$$

$$\langle r^2 \rangle = nl^2(1-\cos\theta)/(1+\cos\theta) \tag{24}$$

$$\langle r^2 \rangle = 2nl^2 \tag{25}$$

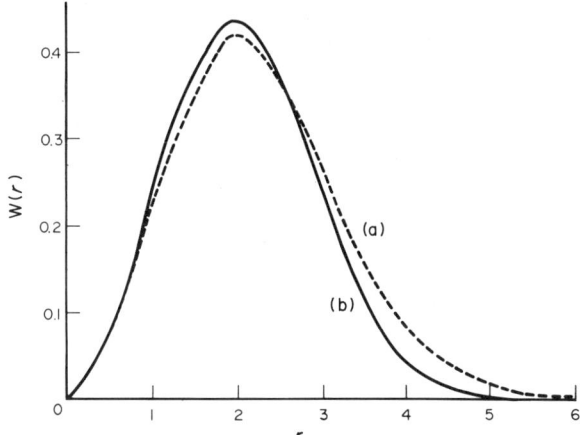

Figure 8 Radial distribution function $W(r)$ for a six-link random chain, with $l=1$; (a) Gaussian limit, (b) inverse Langevin distribution (reproduced by permission of Clarendon Press from L. R. G. Treloar, 'The Physics of Rubber Elasticity', 3rd edn., 1975)

9.2.3 Real Polymer Chains

In addition to valence angle restrictions, real chains have intrachain hindrances that restrict each bond to a small number of discrete rotational states. For large n, equation (26) results, where $\langle \cos \phi \rangle$ is the average value of $\cos \phi$, ϕ being the angle of rotation of a single bond around the direction of the preceding one.[67,68]

$$\langle r^2 \rangle = nl^2 \left(\frac{1-\cos\theta}{1+\cos\theta} \right) \left(\frac{1+\langle\cos\phi\rangle}{1-\langle\cos\phi\rangle} \right) \tag{26}$$

9.2.3.1 *Persistence vector, characteristic ratio and Kuhn step length*

The moments of the end-to-end vector r afford the most universal quantities with which to characterize the average configuration of a chain. The configurational average of r itself is called the persistence vector and is defined by $a = \langle r \rangle$. It is informative with regard to the characteristics of the spatial configurations of chain molecules of a given kind. Vectors r and a are defined in a reference frame composed of the first two bonds of the chain in the manner shown in Figure 9(a).[38] The x axis is taken along the first bond, and the y axis in the plane of the first two bonds, its direction making an acute angle with the second bond. The z axis, perpendicular to this plane, completes a right-handed Cartesian coordinate system. The persistence vector is then given by equation (27), where $\langle l_i \rangle$ is the configurational average of bond vector l_i in the reference frame of the initial bond defined in Figure 9(a), *i.e.* the components of $\langle l_i \rangle$ are the projections, averaged over all configurations of the chain, of bond vector l_i on the respective axes of this coordinate system.

$$a = \sum_{i=1}^{n} \langle l_i \rangle \tag{27}$$

The persistence vectors vividly portray the correlations between bonds of the chain. These correlations are due to the combined effects of narrowly restricted bond angles, preferences for one or more rotational states over others and the interdependence of rotations about neighboring bonds. The persistence of a freely jointed chain is just the first bond. Transverse components are necessarily null.

Increase in the mean-square end-to-end distance $\langle r^2 \rangle$, expressed in equations (19), (24) and (26), corresponds to new restrictions imposed on the chain configuration. The mean-square end-to-end distance for real chains consisting of a large number of primary valence bonds exceeds that of their random flight analogs, as represented in equation (28), where C_∞ is the characteristic ratio. This coefficient C_∞ represents the degree to which a real molecule departs from the freely jointed model. It may be deduced from equations (25) and (28) that C_∞ equals two for a tetrahedrally

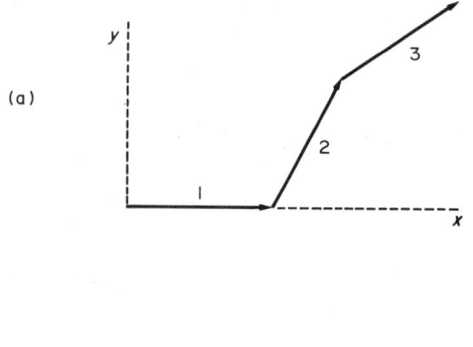

(a)

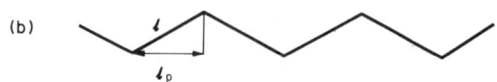

(b)

Figure 9   (a) Coordinate system for the persistence vector defined by the first two bonds of a chain (reproduced by permission of Pergamon Press from P. J. Flory, *Pure Appl. Chem.*, 1980, **52**, 241), (b) Fully extended chain conformation without distortion of bond angles or deformation of bond lengths; l is a bond vector and l_p its projection on the chain axis

bonded, freely rotating chain. The characteristic ratio depends on the chemical structure of the molecule, and generally also on temperature (because different rotational states generally correspond to different energies).[8,69-73] Values may be calculated by the rotational isomeric state method, or from measurements of solution viscosity or radiation scattering, using either light or thermal neutrons.[74] It is usually determined from intrinsic viscosity $[\eta]_\theta$ measurements in a θ solvent[2] by making use of equation (29). The molecular weight M can be obtained from light-scattering measurements, gel permeation chromatography or viscosity measurements. It is preferable that the molecular weight distribution be narrow. The quantity Φ is a universal constant approximately equal to 2.5×10^{21} when the other quantities are in their customary units.

$$\langle r^2 \rangle = C_\infty \, nl^2 \tag{28}$$

$$C_\infty = ([\eta]_\theta \, M / \Phi)^{2/3} / (nl^2) \tag{29}$$

The statistical distribution of end-to-end distances for any chain, whatever its geometrical structure, reduces to the Gaussian form if the number of rotatable links is sufficiently large. This is true even when rotations about single bonds are restricted by energy barriers. The effect of the chain structure on statistical properties is therefore represented simply by the numerical value of the parameter β in equation (17). Since this parameter also determines the value of the mean-square length through equation (19), it follows that if the mean-square end-to-end distance can be calculated, the Gaussian distribution is then completely determined. Therefore it is always possible to find, for a real chain, a corresponding freely jointed chain which will have the same statistical properties. The equivalent freely jointed chain has the same contour length and the same mean-square displacement length $\langle r^2 \rangle$ as the real chain.[8,75,76] These two conditions yield equations (30) and (31), respectively, for the number n_e of equivalent random links and for their length l_e. As shown in equation (32), combination of equations (28) and (31) gives the Kuhn step length l_e in terms of the characteristic ratio C_∞ and the average skeletal bond length l.

$$n_e = n^2 l^2 / \langle r^2 \rangle \tag{30}$$

$$l_e = \langle r^2 \rangle / (nl) \tag{31}$$

$$l_e = C_\infty l \tag{32}$$

A fully extended chain, without distortion of bond angles or deformation of bonds, is represented in Figure 9(b). In this conformation, the value of the end-to-end distance is nl_p, where l_p is the length of the bond vector projected on the chain axis. If the equivalent random chain is now required to have the same end-to-end distance in full extension and also the same mean-square displacement length as the real chain,[1] then equations (33) and (34) result. The maximum extension ratio λ_{max} of unperturbed macromolecules may be defined as the ratio of the fully extended length nl_p to the root-mean-square end-to-end distance in the isotropic state. Equation (35) then results,[77] and λ_{max} becomes proportional to the square root of the number of chain links.

$$n_e = n l_p^2 / (C_\infty l^2) \tag{33}$$

$$l_e = C_\infty l^2 / l_p \tag{34}$$

$$\lambda_{max} = (n/C_\infty)^{1/2} l_p / l \tag{35}$$

9.2.3.2 Chain dimensions in the amorphous bulk state

The spatial configuration of macromolecules is of central importance to understanding rubber elasticity. The high elastic extensibility that characterizes elastomers is a direct consequence of the randomness and diversity of the configurations accessible to the chains. In the amorphous bulk state, chains are densely packed without any possibility of segment overlapping. At one time there was considerable controversy concerning the perturbation of configurations by intermolecular inter-actions and the degree of order existing in amorphous polymers.[2, 11, 78, 79] Flory postulated[2] that chains exhibit the unperturbed dimensions characteristic of their behavior in θ solvents, and this was confirmed by small angle neutron-scattering experiments.[80-88] Changes in conformation such as those accompanying deformation do not, therefore, cause changes in intermolecular energy. Thus, the elastic response of a polymeric network originates within the chains and not, to a significant extent, from interactions between them, a statement strongly supported by thermoelasticity measurements.[88-94]

9.2.4 Elasticity of a Single Polymer Chain

According to the Boltzmann principle, the entropy S is proportional to the logarithm of the number of configurations Ω available to the system, as shown in equation (36), where k is the Boltzmann constant. In a deformed network, the chain end-to-end vector r is fixed by the positions of the cross-linking points at its ends. The number of conformations Ω is then proportional to the probablity $W(x, y, z)$ in equation (15). Equation (37) for the chain entropy is then deduced,[1, 2] where c is an arbitrary constant. The element dw of work required to move one end of the chain from a distance r to a distance $r + dr$ with respect to the other is given by equation (38). The quantity $-dw$ is also equal to the change dA in the Helmholtz free energy, as shown in equation (39). Equation (40) is then obtained by substitution of equations (37) and (38) into equation (39), and equation (41) from equation (19). Thus, a molecule with its ends fixed at specified points is acted on by a tensile force in the direction of the line joining its ends and proportional to the length of that line. It behaves as a spring of modulus $3kT/\langle r^2 \rangle$. When the displacement length r becomes sufficiently large, the Gaussian distribution function is no longer an adequate representation of $W(x, y, z)$ (equation 15) and it is necessary to use the Langevin function (equation 20) or its expansion. In this case, the force is given by equation (42), where r/nl is the fractional extension of the chain, *i.e.* the ratio of its displacement length to its contour length.

$$S = k \ln\Omega \tag{36}$$

$$S = c - k\beta^2 r^2 \tag{37}$$

$$dw = -f dr \tag{38}$$

$$-dw = dA = -T dS \tag{39}$$

$$f = 2kT\beta^2 r \tag{40}$$

$$f = 3kTr/\langle r^2 \rangle \tag{41}$$

$$f = (kT/l) \mathcal{L}^{-1}(r/nl) \tag{42}$$

9.3 TOPOLOGICAL STRUCTURE OF POLYMERIC NETWORKS

Network topology affects all of the elastomeric properties, including (i) equilibrium properties such as the modulus, ultimate strength, maximum extensibility and degree of swelling; and (ii) dynamic mechanical properties such as viscoelastic losses. The pore or mesh size is a fundamental quantity which characterizes the structure of the insoluble polymer network, and can be taken to be the chain molecular weight M_c between two consecutive cross-links. Another important way of characterizing a network is in terms of its degree of cross-linking, *i.e.* the number density of junctions joining the chains into a permanent structure, and of junction functionality ϕ.

Polymer networks are conveniently characterized in the elastomeric state, which is exhibited at temperatures above the glass-to-rubber transition temperature T_g. In this state, the large ensemble of configurations accessible to flexible chain molecules by Brownian motion is very amenable to statistical mechanical analysis.[9] Polymers with relatively high values of T_g such as polystyrene[95] or elastin[96,97] are generally studied in the swollen state to lower their values of T_g to below the temperature of investigation. It is also advantageous to study network behavior in the swollen state since this facilitates the approach to elastic equilibrium, which is required for application of rubber elasticity theories based on statistical thermodynamics.[89,98,99]

9.3.1 Network Formation and the Cross-linking Process

For several centuries, raw rubber was used for sport balls, waterproofing cloth and footwear, but it was never considered to be entirely satisfactory for any of these purposes. It melted in the summer heat and became brittle in the winter cold. It also became sticky if exposed to many organic solvents. Useful rubber products became practical only through the discovery of vulcanization by Goodyear and Hancock in 1843.[100] When rubber is vulcanized or cured with sulfur, it becomes insoluble in hydrocarbon solvents and loses its thermoplasticity. This cross-linking results in a three-dimensional stable network of polymer chains linked together by mono-, di- and poly-sulfide bridges. Organic peroxides can also be used to cross-link both saturated and unsaturated polymers.[101-108] After peroxide or irradiation cross-linking, chains are joined by carbon-to-carbon bonds formed between segments of two of the polymer chains.

In general, information on how the polymer network was formed is of little help in establishing structure–property relationships for elastomeric materials, since the cross-links required in its construction are usually introduced between chain segments in an indiscriminate, essentially uncontrolled manner.[109-113] As a result, there is no independent knowledge of the crucially important average molecular weight M_c between cross-links and the distribution about this average. Furthermore, it is impossible to vary the cross-link functionality ϕ, since the joining of a pair of segments from two chains will almost invariably give a junction of functionality four. Finally, there is little reliable information on imperfections in the network structure, for example the numbers and average lengths of ineffective and possibly detrimental dangling chains (which are attached to the network at only one end). An obvious solution to the problem is the preparation of model elastomers, *i.e.* polymer networks having controllable and independently known structural characteristics.[114-122] Such model networks may be prepared by very specific chemical reactions in which the cross-links are introduced in a carefully controlled manner. For example, poly-(dimethylsiloxane) (PDMS) networks of this type have been synthesized by end-linking hydroxyl-terminated PDMS chains by means of silicates, and vinyl-terminated or vinyl-substituted PDMS chains with silanes containing active hydrogen atoms. In this approach, chemical analysis and viscometric and gel permeation chromatographic (GPC) measurements are carried out on the chains prior to their cross-linking, in order to determine the number-average molecular weight $\bar{M}_n$ between the potential cross-linking sites. In the case of chains with reactive groups only at the ends, the GPC measurements provide the distribution of $\bar{M}_n$ as well. Networks then prepared by exclusively and exhaustively reacting these groups with a multifunctional cross-linking agent have a known structure in that (i) the molecular weight M_c between cross-links is simply $\bar{M}_n$; (ii) the distribution of M_c is also that of $\bar{M}_n$; (iii) the functionality of the cross-links is the same as that of the cross-linking agent; and (iv) the incidence of dangling-end network imperfections is very small. Also, known numbers of dangling chains of molecular weight $\bar{M}_n$ can be introduced by using less than the stoichiometrically required amount of end-linking agent.[123]

9.3.2 Description of Network Structure

The permanent structure required for elastic recovery is obtained by joining N linear chains by chemical cross-linkages. A polymer network may then be characterized by the number μ of its junctions, their functionality (or average functionality) and by the number ν_{ends} of chain ends.[9] The number ν of chains in the network, including those with only one end attached, is given by equation (43). The effective number of chains ν_e is less than ν owing to imperfections due to free chain ends.

For a perfect network with no chain ends, it is given by equation (44).

$$v = (1/2)(\phi\mu + v_{ends}) \tag{43}$$

$$v_e = v = \phi\mu/2 \tag{44}$$

A quantity that characterizes the network with greater generality, regardless of the nature of its imperfections, is the cycle rank ξ, or number of independent circuits it contains.[9,10] It may be defined alternatively as the minimum number of scissions required to reduce the network to a spanning tree, *i.e.* a unified structure comprising all the chains and containing no closed circuits or loops. According to graph theory, chain ends and junctions are called vertices and chains which join them are edges. The total number of vertices and edges are denoted, respectively, by ψ and v. In the spanning tree they are related by equation (45). The network is formed by coalescence of ξ pairs of vertices, reducing their number as shown in equation (46). In the case of a perfect network, the number of junctions μ equals the number of vertices. Equations (47) and (48) can then be established by making use of equations (44) and (46).

$$\psi = v + 1 \tag{45}$$

$$\psi = v - \xi + 1 \simeq v - \xi \tag{46}$$

$$\xi/V_0 = (\phi/2 - 1)\mu/V_0 = (1 - 2/\phi)\rho/M_c \tag{47}$$

$$\xi = (1 - 2/\phi)v \tag{48}$$

In a randomly cross-linked network, the number density of chain ends is related to the number-average molecular weight $\bar{M}_n$ of the N primary linear chains and to the polymer density ρ by equation (49). The vertices in the network consist of the junctions and the $2N$ chain ends. This case is described by equation (50), and combination of equations (43), (46), (49) and (50) yields equation (51). The number of junctons μ^* in the spanning tree is obtained by setting ξ equal to zero in equation (51), resulting in equation (52). The number of junctions μ_e which are elastically effective is the difference between μ and μ^* and is related to ξ by equation (53).

$$v_{ends}/V_0 = 2N/V_0 = 2\rho/\bar{M}_n \tag{49}$$

$$\psi = \mu + 2N \tag{50}$$

$$\xi = (\phi/2 - 1)\mu - N \tag{51}$$

$$\mu^* = N/(\phi/2 - 1) \tag{52}$$

$$\xi = (\phi/2 - 1)\mu_e \tag{53}$$

In a tetrafunctional cross-linking process, formation of an elastically effective junction will create two effective chains. Generalization of this to a ϕ-functional process leads to equation (54), a relationship between v_e and μ_e which is similar to that for perfect networks (equation 44). Combination of equations (53) and (54) then yields equation (55).

$$v_e = \phi\mu_e/2 \tag{54}$$

$$\xi = v_e - \mu_e \tag{55}$$

Scanlan[110] and Case[125] have used another criterion to identify the chains and junctions which should control the elastic properties of networks. In their scheme, a junction is elastically active if three or more of its arms are independently attached to the network. A strand is elastically active if it is attached at both ends to an active junction. However, it can be shown[112,125] that equation (55) still holds if v_e and μ_e are replaced by the numbers of active chains v_a and junctions μ_a, respectively. This is stated in equation (56). In the phantom network model, the elastic modulus depends only on the cycle rank and therefore is not influenced by the choice of criterion. Furthermore, there is no difference between the two criteria when applied to perfect networks. All the equations derived above by graph-theoretical arguments hold for odd as well as even values of ϕ and may be adapted to networks of variable functionality merely by replacing ϕ by its average $\bar{\phi}$.[124,126,127]

$$\xi = v_a - \mu_a \tag{56}$$

It is also possible to derive expressions for ξ and v_e for model networks with a known number of dangling ends of known length.[17] Experimentally, these networks can be formed by mixing, in a stoichiometric ratio, ϕ-functional agents, a weight fraction W_1 of di-end-reactive chains of weight $\bar{M}_{n1}$ and a weight fraction W_2 of mono-end-reactive chains of molecular weight $\bar{M}_{n2}$. The two weight fractions are, of course, related by equation (57). The resulting cycle rank density ξ/V_0 (where V_0 is a reference volume) and effective chain density[17] ν_e/V_0 are given by equations (58) and (59).

$$W_1 + W_2 = 1 \tag{57}$$

$$\xi/V_0 = (1 - 2/\phi)\rho W_1/\bar{M}_{n1} - \rho W_2/(\phi M_{n2}) \tag{58}$$

$$\nu_e/V_0 = \rho W_1/\bar{M}_{n1} - \rho W_2/(M_{n2}(\phi - 2)) \tag{59}$$

The elastic properties of rubbers depend on the network molecular structure and on the microscopic response of this structure to a macroscopic constraint. Studies of network formation by computer simulation,[128] sol–gel statistics[12–14,112] or determination of network structure by extraction and random degradation[129] may help to analyze results of mechanical testing. Generally, the probability parameters involved in the statistical theories of network formation are determined by comparison of theoretical predictions with experimental values of the sol fraction.[12–14]

Although model networks are very useful for testing elasticity theories, common applications involve statistically cross-linked polymers. The cycle rank of these networks can be determined from mechanical measurements through models based on statistical mechanics. Nevertheless, a relationship between the cycle rank, the molecular weight M_c and the number of junctions is needed to fully characterize the network. For regular networks formed by random tetrafunctional cross-linking of linear chains of molecular weight $\bar{M}_n$ and having no defects other than chain ends, equations (60) and (61) have been proposed.[15] Similarly, the topologies of networks formed by random tetra-functional cross-linking of star polymers having A arms of number-average molecular weight $\bar{M}_n$ have been investigated, with equations (62)–(64) as a result.[130]

$$\xi/V_0 = (\rho/2M_c)(1 - 3M_c/\bar{M}_n) \tag{60}$$

$$\mu/V_0 = (\rho/2M_c)(1 - M_c/\bar{M}_n) \tag{61}$$

$$\xi/V_0 = (\rho/2M_c)[1 - (A+2)M_c/A\bar{M}_n] \tag{62}$$

$$\mu/V_0 = (\rho/2M_c)[1 - (A-2)M_c/A\bar{M}_n] \tag{63}$$

$$\bar{\phi} = 4 + 2(A-4)(2 - A + A\bar{M}_n/M_c)^{-1} \tag{64}$$

9.4 MECHANICAL TESTING OF RUBBERS

Analysis of the deformation of an elastomer resulting from application of a force, or of the force required to achieve a specified deformation, is one of the important ways used to characterize polymeric networks (in addition to chemical analysis and swelling measurements). The results are generally expressed in terms of stress and strain so as to be independent of specimen geometry. Stress is the force f per unit original cross-sectional area and its units are Newtons per square meter ($N\,m^{-2}$). Other units are often used and are easily converted by the relationships between megapascals (MPa), Newtons per square millimeter ($N\,mm^{-2}$), pounds per square inch (psi), and kilograms of force per square centimeter ($Kgf\,cm^{-2}$) given in equations (65) and (66).

$$MPa \equiv N\,mm^{-2} \equiv psi \times 0.00689 \tag{65}$$

$$MPa \equiv Kgf\,cm^{-2} \times 0.09806 \tag{66}$$

Strain is the deformation per unit original dimension and is therefore dimensionless. The ratio of stress to strain is called the modulus, and a material is said to be 'Hookean' when its modulus is independent of strain (which is typical of a metal spring or wire). Elastomers are Hookean only in the range of very small deformations. The physical properties of rubbers are highly sensitive to specimen preparation and are also dependent on testing method and conditions (*e.g.* strain rate and temperature).[5,34,131–134] Standardization, however, has been achieved through the American Society for Testing and Materials (ASTM),[131,136] the International Standards Organization (ISO) and the British Standards Institution (BSI).[133,134,137,138] Description of network structure from mechanical testing is possible through the kinetic theory of rubber elasticity. It is based on statistical mechanics and therefore requires that measurements be made at thermodynamic equilibrium. This restrictive condition is fraught with difficulty. Dense chain entanglements induce a large spectrum of relaxation times, and very low frequency components may preclude complete attainment of equilibrium under normal experimental conditions. One possible way to circumvent this difficulty consists of measurements on swollen samples.

9.4.1 Uniaxial Extension

The experimental mechanical techniques most commonly used for network characterization are uniaxial extension and compression,[1,31,99,139-144] and also biaxial strain.[145-150] A sketch of a rubber sample under extension is shown in Figure 10(a). The nominal stress σ is defined as the ratio of the force f to the cross-sectional area A_0 of the undeformed specimen, and the strain ε as the ratio of the length change ΔL to the original length L_0. These definitions are given in equations (67) and (68). The deformation is also often expressed in terms of the extension ratio λ defined in equation (69). The cross-sectional area of the specimen varies with deformation. A true stress, defined as the ratio of the force to the real deformed area, is also frequently used.

$$\sigma = f/A_0 \tag{67}$$

$$\varepsilon = \Delta L/L_0 \tag{68}$$

$$\lambda = 1 + \varepsilon \tag{69}$$

Young's modulus E is the ratio of nominal stress to strain, as shown in equation (70). However, vulcanized rubbers do not obey Hooke's law (as is shown in Figure 5), so E is not a constant. The stress–strain relationship is generally assumed to be linear over small tensile or compressive strains, and Young's modulus is usually defined as the slope of the stress–strain curve in this range of deformation.[151-153] Hardness measurement is another way of determining values of this modulus.[154-158] It is noteworthy that the slope of the stress–strain curve in the tensile and compressive regions is slightly different.[151]

$$E = \sigma/\varepsilon \tag{70}$$

A typical apparatus for uniaxial extension measurements at thermodynamic equilibrium is shown schematically in Figure 11.[88] Specifically, an elastomeric strip is mounted between two clamps, the lower one fixed and the upper one attached to a movable force gauge. A recorder is used to monitor the output of the gauge as a function of time, in order to obtain equilibrium values of the force f suitable for theoretical analysis. The sample is generally protected with an inert atmosphere, such as nitrogen, to prevent network degradation, particularly in the case of measurements carried out at elevated temperatures. Both the sample cell and surrounding constant temperature bath are glass, thus permitting use of a cathetometer or travelling microscope to obtain values of the strain by measurements of the distance between two lines marked on the central portion of the test sample. The cross-sectional area is generally obtained by means of a micrometer. Stress–strain measurements are generally made using a sequence of increasing values of the elongation, with some inclusions of values out of sequence to test for reversibility.[159-161] For networks studied in the swollen state, a final variable is the volume fraction v_2 of polymer in the swollen network.[98,99,162]

Compression measurements are much more difficult than elongation measurements since friction between the sample and the compression plates generally causes serious sample distortion

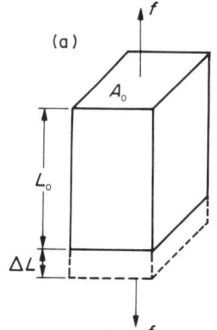
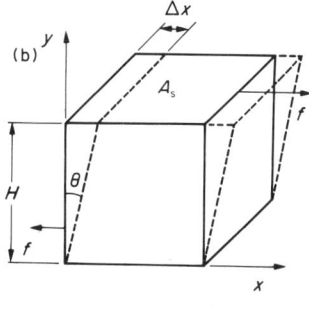

Figure 10 Sketch of a sample subjected to (a) uniaxial extension and (b) simple shear

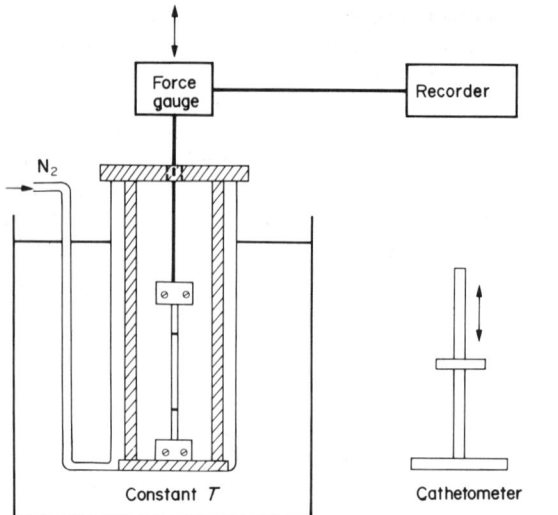

Figure 11 Schematic diagram of a typical apparatus used to measure the uniaxial stress as a function of strain for an elastomer (reproduced by permission of Wiley from J. E. Mark, *J. Polym. Sci., Macromol. Rev.*, 1976, **11**, 135)

('barrelling').[163-165] For this reason, equivalent experiments using equibiaxial extension,[146] including inflation of sheets,[166] are frequently carried out. Indeed, simultaneous prediction of elongation and compression mechanical behaviors is a stringent test for molecular models of rubberlike elasticity.[89,144,166]

9.4.2 Simple Shear

Simple shear is illustrated in Figure 10(b). The application of the force f results in the deformation Δx. The height H and the surface area A_s are kept constant. The shear stress σ_s and strain γ are defined in equations (71) and (72), and the shear modulus G in equation (73). The only practical difficulty in testing in shear is the necessity to bond the rubber test piece to rigid members in order to apply the shearing force. The stress–strain curve in simple shear is approximately linear up to relatively large strains.

$$\sigma_s = f/A_s \tag{71}$$

$$\gamma = \Delta x/H = \tan\theta \tag{72}$$

$$G = \sigma_s/\gamma \tag{73}$$

The shear and Young's moduli are related through equation (74),[167,168] where μ_p is Poisson's ratio of the change of volume dV accompanying the infinitesimal deformation $d\varepsilon$. It is defined in equation (75). Unfilled elastomers have a Poisson's ratio of *ca.* 0.49 and are thus nearly incompressible. For incompressible bodies, μ_p equals 0.5, and equation (74) then simplifies to equation (76).

$$E = 2G(1 + \mu_p) \tag{74}$$

$$\mu_p = (1/2)[1 - (1/V)(dV/d\varepsilon)] \tag{75}$$

$$E = 3G \tag{76}$$

9.5 THERMOELASTICITY

As already mentioned in Section 9.1, the elastic force is largely entropic in origin. More specifically, it arises primarily from the fact that stretching an elastomer decreases the entropy of its constituent chains, since the probability of any specific spatial arrangement or configuration of a chain decreases significantly with increase in its end-to-end separation r.[2] Such changes in r, however, can cause a change in the energy as well as the entropy of a network chain since the dimensional changes require changes in conformation. Also, it is known that different conformational states (generally *trans*, *gauche* positive and *gauche* negative) usually correspond to different energies, because of different steric and Coulombic interactions between proximate parts of the chain.[8] Thus the elastic force may also have a significant energetic component, although it is predominantly entropic in origin.

Thermodynamic analysis of rubber elasticity enables one to resolve the elastic force into entropic and energetic components, thereby elucidating the origin of rubberlike elasticity in general. It also indicates that intermolecular interactions do not affect the force and must be independent of the extent of the deformation and thus of the spatial configurations of the chains. Since the spatial configurations are independent of intermolecular interactions, the amorphous chains must be in random unordered configurations, the dimensions of which should be the unperturbed values.[2, 88]

Another important aspect of rubberlike elasticity, in which thermodynamic analysis is of the foremost importance, is the efficiency of storage of elastic energy. It has obvious bearing on the important problem of heat build-up in cyclic deformations. Since the energy change ΔU for elastic deformations is small, the work of deformation w must appear largely as a heat effect of magnitude Q. If the deformation is carried out relatively rapidly, as it is in most applications, then the process is approximately adiabatic and the heat generated can raise the temperature of the elastomer significantly. In cyclic deformations (including the flexing of automobile tires), retraction phases following elongation (or compression) phases do not give temperature decreases sufficiently large to cancel the increases. The processes are highly irreversible in the thermodynamic sense, which means that w (elongation), which is positive, has to be larger than the reversible amount, and w (retraction), which is negative, will be smaller than its corresponding reversible value. Thus, the elastic energy is stored with less than perfect efficiency. The net (positive) w per cycle can produce a sizable (negative) Q which, if not passed from the system to surroundings, can cause significant heat build-up. Such increases in temperature will affect the stability of an elastomeric material in any application involving cyclic deformations.

9.5.1 Thermoelastic Equations

The thermodynamic quantity of primary interest is the energetic contribution f_e to the elastic force f exhibited by a deformed polymer network. From the change in the Helmholtz free energy A given by equation (77) and from the change in the internal energy U (equation 1), the elastic force due to change in length L at constant temperature and volume may be expressed as the sum of a contribution f_e from the internal energy and a contribution from the entropy.[90, 169-172] This is stated in equation (78), with expressions for the energetic and entropic components given in equations (79) and (80).

$$dA = -p\,dV + f\,dL - S\,dT \tag{77}$$

$$f = f_e + f_s \tag{78}$$

$$f_e = (\partial U/\partial L)_{T,V} \tag{79}$$

$$f_s = -T(\partial S/\partial L)_{T,V} \tag{80}$$

Application of Maxwell's relation to $(\partial A/\partial L)_{T,V}$ and $(\partial A/\partial T)_{V,L}$ yields equation (81). The entropy component may then be related to the temperature coefficient of the retractive force at constant volume and length by combination of equations (80) and (81). The result is given in equation (82). Furthermore, it is now possible to express the fraction f_e/f of the total force which is of energetic origin from equations (78) and (82), as shown in equation (83). The ratio f_e/f may thus be obtained from force–temperature measurements at constant length and volume. Such thermoelastic studies have been carried out,[89, 171, 173, 174] but, because of severe experimental difficulties encountered in meeting the constant volume requirement, most experiments are conducted at constant pressure.

$$(\partial f/\partial T)_{V,L} = -(\partial S/\partial L)_{T,V} \tag{81}$$

$$f_s = T(\partial f/\partial T)_{V,L} \tag{82}$$

$$f_e/f = 1 - (T/f)(\partial f/\partial T)_{V,L}$$

$$= -T(\partial \ln(f/T)/\partial T)_{V,L}$$

$$= (1/T)(\partial \ln(f/T)/\partial(1/T))_{V,L} \tag{83}$$

Transformation of equation (83) to a form suitable for analysis of such isobaric data requires recourse to an appropriate elastic equation of state. Such an equation for simple elongation, valid for both swollen and unswollen samples, is obtained through the statistical mechanical treatment of phantom Gaussian networks.[9] It is given in equation (84). The extension α is measured relative to the length L_v^i of the unstretched sample when its volume is fixed at the same volume V as occurs in the stretched state, as specified in equation (85), where L is the length of the deformed sample at volume V. The quantity k is the Boltzmann constant, $\langle r^2 \rangle_v^i$ the mean-square end-to-end distance for a typical network chain in the isotropic state of volume V, and $\langle r^2 \rangle_0$ the corresponding value for the free chain in the absence of constraints imposed by network junctions. The dimensions $\langle r^2 \rangle_v^i$ depend on the volume of the network while $\langle r^2 \rangle_0$ (which is a characteristic of the polymer) depends upon bond lengths, bond angles, conformational energies and, accordingly, on temperature.[175] By making use of equations (83) and (84), an expression for f_e/f is obtained for measurements at constant length and pressure. It is given in equation (86), where b is the bulk coefficient of thermal expansion defined in equation (87). Similarly, equation (88) gives the relationship for measurements at constant pressure and deformation.

$$f = (\xi kT/L_v^i)(\langle r^2 \rangle_v^i / \langle r^2 \rangle_0)(\alpha - \alpha^{-2}) \tag{84}$$

$$\alpha = L/L_v^i \tag{85}$$

$$f_e/f = -T(\partial \ln(f/T)/\partial T)_{L,p} - bT/(\alpha^3 - 1) \tag{86}$$

$$b = (\partial V/\partial T)_p/V \tag{87}$$

$$f_e/f = -T(\partial \ln(f/T)/\partial T)_{p,\alpha} + bT/3 \tag{88}$$

It was assumed in the derivation of these equations that the stored elastic free energy resides exclusively within the chains of the network. Also, equation (84) has been derived specifically for a network of Gaussian chains. Compliance with the Gaussian distribution is not vitiated by dependence of the energy on the bond conformations,[176] and the statistical distribution of end-to-end distances for all flexible and long polymer chains must be approximately Gaussian. Consequently, the form of the isotherm relating stress to strain is unaffected by generalization to include non-ideal polymer chains for which the energy depends on the conformation.

Equations (83) and (84) provide a molecular interpretation of the thermoelastic data through equation (89). This equation establishes the relationship between the purely thermodynamic quantity f_e/f and its molecular counterpart of $d\ln\langle r^2 \rangle_0/dT$, which can be interpreted in terms of the rotational isomeric state theory of chain configurations.[44] It permits comparison of the change of the unperturbed dimensions $\langle r^2 \rangle_0$ obtained by thermoelastic measurements on polymer chains in the bulk (in network structures) with that obtained by viscosity measurements on chains of the same polymer, essentially isolated in dilute solution.[177]

$$f_e/f = Td\ln\langle r^2 \rangle_0/dT \tag{89}$$

9.5.2 Thermoelastic Experiments

A typical apparatus used for thermoelastic studies of networks in elongation and at thermodynamic equilibrium is shown in Figure 11.[88] Equations (83), (86) and (88) were derived to interpret thermoelastic data obtained at constant volume and length, constant length and pressure, and constant pressure and deformation, respectively. Equation (83) was obtained without recourse to a model of rubberlike elasticity. However, it is necessary to impose a sizeable hydrostatic pressure to keep the specimen volume constant with change in temperature.[89,171,173,174] This constancy of volume may be achieved also by the study of the network at swelling equilibrium with a diluent such that the change in polymer–solvent interactions with temperature exactly offsets the effect of thermal expansion.[178]

Development of equations (86) and (88) required adoption of an elastic equation of state, which is justified by the fact that measurements at constant pressure are much easier to make than those at constant volume. The correction term $bT/3$ in equation (88), required to interpret data obtained at constant deformation, is itself independent of deformation, an advantage which is somewhat counterbalanced by the necessity to adjust the deformed length L of the specimen at each new temperature to keep α constant, since the original length L_v^i in equation (85) is itself varying with temperature. Correction of measurements at constant length requires the term $bT/(\alpha^3 - 1)$, which can be made negligibly small by choice of sufficiently large values of α in elongation, thereby also suppressing the phenomenon of the thermoelastic inversion[1] encountered at small elongations. Values of f_e/f obtained from constant pressure experiments, and treated according to equations (86) or (88), are generally in good agreement with those obtained at constant volume.[88]

Some typical stress–strain curves,[88,170] obtained at constant pressure and length on amorphous polyethylene, are shown in Figure 12. The stress is generally found to vary linearly with temperature.

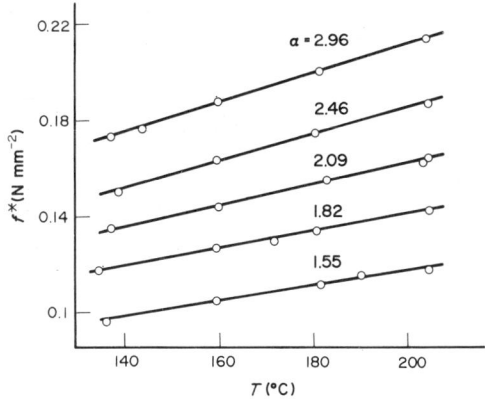

Figure 12 Stress–temperature curves at constant pressure and length for an amorphous polyethylene network in the unswollen state. Values of the elongation α and of the nominal stress f^* are calculated using the rest length and the undeformed cross-section, respectively, at the highest temperature investigated (reproduced by permission of Wiley from J. E. Mark, *J. Polym. Sci., Macromol. Rev.*, 1976, **11**, 135)

Equivalent thermoelastic data may be obtained by measurements in compression, torsion or shear,[1,179,180] from a series of stress–strain isotherms,[165,181–183] or from measurements of sample length as a function of temperature at constant force.[184] Additional information on the thermodynamic behavior of rubbers may be derived from stretching calorimetry[185–187] and dilatometry.[188,189]

The treatment of the data shown in Figure 12 by use of equation (86) indicates that the energetic contribution to the elastic force is large and negative.[170] The thermodynamic ratio f_e/f may be interpreted in molecular terms through its equality to $T\mathrm{d}\ln\langle r^2\rangle_0/\mathrm{d}T$ (equation 89). Specifically, deformation of a network requires the polymer chains to switch to less compact conformational states. In the case of amorphous polyethylene, the less compact states are the *trans* (planar) conformations, which are of lowest energy. Thus f_e/f is negative. In the case of poly-(dimethylsiloxane), the less compact states are the *gauche* (non-planar) conformations, of higher energy, and therefore f_e/f is positive.[8,88,90]

The ratio f_e/f does not depend on sample deformation during cross-linking, degree of crystallinity or presence of diluent during cross-linking.[88] There is no apparent effect from the degree of network swelling (dilation), and values of the temperature coefficient $\mathrm{d}\ln\langle r^2\rangle_0/\mathrm{d}T$ obtained from thermoelastic studies are in good agreement with values obtained from viscosity–temperature measurements on chains of the same polymer dispersed in a solvent at infinite dilution. This supports the assumption[2] that intermolecular interactions do not vary significantly with deformation and thus do not contribute to the elastic force. The retractive force and its energetic component are essentially intramolecular.

9.6 STATISTICAL THEORY OF RUBBER ELASTICITY

9.6.1 Affine and Phantom Gaussian Models

Relationships between the structure of elastomeric networks and their mechanical properties have been theoretically investigated, from the very first, by statistical mechanics.[1–3,5–9,31,32,64,65,88,109,111,117,124,190–197] These statistical mechanical theories of rubber elasticity are based on two fundamental postulates, supported by thermoelastic and neutron scattering experiments: (i) molecular chain configurations are random in undeformed amorphous polymers; and (ii) the elastic response of the network originates within the chains and not, to a significant extent, from interactions between them.

The high deformability of an elastomer, and the elastic force generated by deformation, stem from the essentially unlimited number of configurations accessible to long molecular chains. This versatility has been characterized in numerous theoretical investigations on typical polymer chains. They have shown that the Gaussian approximation for the distribution of the end-to-end distance should be quite satisfactory for chains consisting of 100 or more bonds.[9,198,199] Non-Gaussian effects generally exist only in the very high region of deformation, where the limited extensibility of the network chains becomes important.

The internal state of the network system can be specified in terms of the positions of the cross-linkages. Deformation of the rubber transforms the arrangements of these points, and hence alters the internal state of the network. Any such state can be represented by the system of vectors connecting neighboring junctions, *i.e.* the system of vectors each of which connects the two ends of a chain.

The first step in the molecular theory of rubberlike elasticity is the derivation of the statistical properties of a single chain. It is necessary to know quantitatively the free energy of the molecule as a function of its end-to-end distance r and a quantitative relation between this distance and the force necessary to maintain it. The second step consists in calculating the free energy of the network as a function of the macroscopic parameters which characterize the deformation. Since the network is formed by connecting single polymeric chains through the cross-links, the derivation ought to start from the expression for the free energy of a single chain. Moreover, in order to obtain the retractive force for such a system, it is necessary to assume some correlation between the macroscopic dimensions of the sample and the set of r values. The simplest assumption is that the displacement of network junctions should be affine (linear) in the macroscopic strain and, hence, that the transformation of the distribution of chain vectors r should likewise be affine.[9,200-205] With such an assumption, the elastic free energy of the system of v_e effective chains relative to the isotropic state is given by equation (90), where I_1 is the first invariant of the tensor of deformation, defined in equation (91), and the quantities λ_1, λ_2 and λ_3 are the principal extension ratios that specify the strain relative to an isotropic state of reference volume V_0. Their product is given in equation (92). In the state of reference, the mean-square end-to-end distance of the chains is assumed to be the corresponding mean-square end-to-end distance for the undeformed chains without cross-links, *i.e.* $\langle r^2 \rangle_0$. The quantity V is the volume of the deformed sample, and V_0 is the volume at which the cross-linking was carried out.[160,206]

$$\Delta A_{el}(\text{aff}) = (v_e kT/2)(I_1 - 3) - (v_e - \xi)kT \ln(V/V_0) \qquad (90)$$

$$I_1 = \lambda_1^2 + \lambda_2^2 + \lambda_3^2 \qquad (91)$$

$$\lambda_1 \lambda_2 \lambda_3 = V/V_0 \qquad (92)$$

The elastic equation of state for simple elongation along the axis denoted 1 takes the familiar neo-Hookean form shown in equation (93), where L_0 is the length of the specimen in the isotropic state of volume V_0. Because of the symmetry of the elongation, λ_2 and λ_3 are equal and may be calculated through equation (92). The uniaxial extension ratio λ is thus given by equation (94).

$$f = (v_e kT)[\lambda - V/(V_0 \lambda^2)] \qquad (93)$$

$$\lambda = \lambda_1 = L/L_0 \qquad (94)$$

Alternatively, the force is given by equation (95), where L_v^i is related to L_0, V, and V_0 through equation (96). Combination of equations (85), (95) and (96) yields the relationship between λ and α given in equation (97). The force per unit cross-sectional area A_0 of the undeformed sample of reference volume V_0 is deduced from equations (95) and (96) to be that shown in equation (98). In the deformation of swollen networks, A_0 is the cross-sectional area of the dry sample and V_0/V the volume fraction v_2 of polymer in the swollen system. The extension ratio α is then measured with respect to the isotropic swollen state.

$$f = (v_e kT/L_v^i)(V/V_0)^{2/3}(\alpha - \alpha^{-2}) \qquad (95)$$

$$L_v^i = L_0(V/V_0)^{1/3} \qquad (96)$$

$$\alpha = \lambda(V_0/V)^{1/3} \qquad (97)$$

$$f/A_0 = (v_e kT/V_0)(V_0/V)^{-1/3}(\alpha - \alpha^{-2}) \qquad (98)$$

James and Guth[207,208] considered a network of chains whose only action is to assert forces on the junctions to which they are attached. The chains are devoid of other material characteristics. In this idealized 'phantom' network, chains may move freely through one another, and are assumed to fluctuate around their mean positions due to Brownian motion. The instantaneous distribution of chain vectors is not affine in the strain because it is the convolution of the distribution of the mean vectors (which is affine) with the distribution of fluctuations (which is Gaussian and independent of the strain).[9,209] Gaussian statistics for undeformed phantom networks (denoted by the subscript 0) lead to the relationships given in equations (99) and (100),[9,127,210] between the mean-square values $\langle r^2 \rangle_0$ of the chain end-to-end length, fluctuations $\langle (\Delta r)^2 \rangle_0$ in the chain dimensions and fluctuations

$\langle(\Delta R)^2\rangle_0$ in the positions of junctions from their mean locations (where the brackets denote ensemble averages).

$$\langle(\Delta r)^2\rangle_0 = 2\langle r^2\rangle_0/\phi \tag{99}$$

$$\langle(\Delta R)^2\rangle_0 = [(\phi-1)/\phi(\phi-2)]\langle r^2\rangle_0 \tag{100}$$

The elastic free energy of a phantom network of Gaussian chains was obtained rigorously by Flory[9] and is valid for networks of any functionality, irrespective of their structural imperfections. It is given in equation (101). The elastic equation of state for phantom networks may then be expressed by equation (102). Equation (84) is then recovered, as expected, because of the relationship shown in equation (103).

$$\Delta A_{el}(\text{ph}) = (\xi kT/2)(I_1-3) \tag{101}$$

$$f/A_0 = (\xi kT/V_0)(V_0/V)^{-1/3}(\alpha-\alpha^{-2}) \tag{102}$$

$$\langle r^2\rangle_v^i/\langle r^2\rangle_0 = (V/V_0)^{2/3} \tag{103}$$

9.6.2 Theoretical Predictions and Comparison with Experiment

In view of the expressions for the elastic equation of state for affine networks (equation 98) and phantom networks (equation 102), it is customary to plot the reduced nominal stress $[f^*]$ measured in uniaxial extension (or compression) against reciprocal extension α^{-1}.[211] The reduced nominal stress is defined in equation (104), where the volume fraction v_2 of polymer in the swollen system is given in equation (105).

$$[f^*] \equiv (f/A_0)v_2^{1/3}/(\alpha-\alpha^{-2}) \tag{104}$$

$$v_2 = V_0/V \tag{105}$$

According to the affine model (equation 98), $[f^*]$ should be a constant equal to the right-hand side of equation (106). In the case of the phantom model, this constant modulus should be given by equation (107). Combination of equations (53), (54), (106) and (107) then yields equation (108).

$$[f^*]_{\text{aff}} = v_e kT/V_0 \tag{106}$$

$$[f^*]_{\text{ph}} = \xi kT/V_0 \tag{107}$$

$$[f^*]_{\text{ph}} - (1-2/\psi)[f^*]_{\text{aff}} \tag{108}$$

These theoretical predictions are shown schematically in Figure 13 in the region of moderate deformation. In the case of the two limits, the affine deformation and the non-affine deformation in the phantom limit, the reduced stress should be independent of α. According to equation (108), the value for the phantom limit should be reduced, however, by the factor $(1-2/\phi)$ in the case of a ϕ-functional network, as illustrated by the case $\phi=4$. Deviations from theory have been experimentally observed; specifically $[f^*]$ generally decreases significantly upon increase in α. This decrease is interpreted as resulting from a gradual change from more nearly affine to phantom behavior, as portrayed by the illustrative curve shown for $\phi=4$.[212,213] The usual range of elongations covered experimentally is shown by the heavier line. Furthermore, exhaustive investigations[89,98] have shown that the reduced stress decreases with dilation by swelling and approaches a limiting value with either increase in elongation or swelling. A typical example of these observations is represented in Figure 14.[214] Experimental data may be approximately fitted by straight lines, according to the Mooney–Rivlin phenomenological equation.[139] Their slopes decrease with increase in swelling. These observations have suggested that the limiting value of the reduced stress at high elongation or high dilation (swelling) is a fundamental characteristic of a given network. It became apparent later[213] that this quantity was the phantom network modulus $[f^*]_{\text{ph}}$.

Additional deviations from theory occur at high deformation, specifically increases of the reduced stress with elongation. In the case of natural rubber they are due largely, if not entirely, to strain-induced crystallization.[215] In a few studies of bimodal networks which do not crystallize under stress, the observed increases in modulus have been shown to be due to limited chain extensibility.[216] At very high elongations the number of spatial configurations decreases drastically, the entropy plummets and the force increases correspondingly. Such non-Gaussian behavior arising from limited chain extensibility occurs at approximately 60–70% of maximum chain extension. This upturn in the reduced stress at high strain for an amorphous network cannot be explained by the simple network theory of rubber elasticity based on the Gaussian statistical treatment of the long-

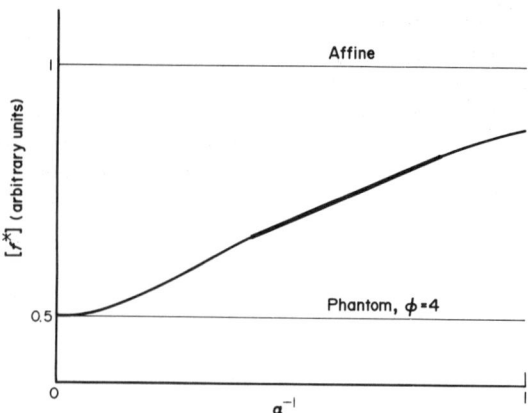

Figure 13 A schematic diagram qualitatively showing theoretical predictions for the reduced stress as a function of reciprocal elongation α^{-1} (reproduced by permission of Elsevier from 'Rubber Elasticity' by J. E. Mark in 'Frontiers in Materials Technology', Amsterdam, 1985)

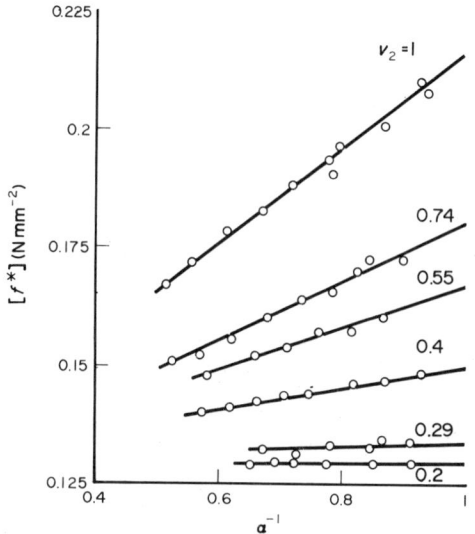

Figure 14 Plot of the reduced stress against the reciprocal of the extension ratio α^{-1} for a natural rubber vulcanizate. Values of the volume fractions v_2 of polymer in the networks swollen in benzene are indicated (reproduced by permission of The Royal Society of Chemistry from S. M. Gumbrell, L. Mullins and R. S. Rivlin, *Trans. Faraday Soc.*, 1953, **49**, 1495)

chain molecule and an approximation which is valid only at extensions of the molecule well below its maximum. A number of network theories based on more exact non-Gaussian chain statistics have been developed.[64-66,111,145,191,192,217] The theory of James and Guth[65] leads to the stress–strain relationship shown in equation (109) for a non-Gaussian affine network in uniaxial extension, where n is the number of randomly joined links in a chain. The inverse Langevin function $\mathscr{L}^{-1}$ may be approximated by the series expansion given in equation (110).

$$f/A_0 = (1/3)v_e kTn^{1/2}[\mathscr{L}^{-1}(\lambda n^{-1/2}) - \lambda^{-3/2}\mathscr{L}^{-1}(\lambda^{-1/2}n^{-1/2})] \tag{109}$$

$$\mathscr{L}^{-1}(t) = 3t + (9/5)t^3 + (297/175)t^5 + \dots \tag{110}$$

9.6.3 Statistical Mechanics of Real Networks

Recent investigations in the statistical mechanical analysis of elastomeric networks have involved the relationship between molecular chain deformation and macroscopic strain, in particular the failure of the assumption of affine transformation of the chain vectors under strain as the main source of disparity between theory and experiment.

In polymer networks, the region of space pervaded by a given chain is shared with many other chains and junctions. A tetrafunctional, statistically cross-linked *cis*-1,4-polyisoprene network, with a typical average molecular weight between junctions of 5000 g mol^{-1}, contains of the order of 5×10^{19} junctions cm^{-3}. The characteristic ratio for the *cis*-1,4-polyisoprene chain is 5.0 and, according to equation (28), the radius $\langle r^2 \rangle_0^{1/2}$ of the volume occupied by such a chain will be 56 Å. From equations (99) and (100), the mean radius of the fluctuation domain $\langle (\Delta r)^2 \rangle_0^{1/2}$ in a phantom network will be 40 Å and the average fluctuation distance $\langle (\Delta R)^2 \rangle_0^{1/2}$ of a junction from its mean position will be 34 Å. These large fluctuations may be drastically impeded in a real network by the random interspersion of the ϕ chains (emanating from a junction) with the neighboring chains. In the preceding numerical application, there will be approximately 37 junctions in the domain occupied by each chain. This degree of interpenetration, *i.e.* the number of junctions in the domain of a given network chain, is associated with the deviation of the properties of a real network from the phantom limit. Strong overlapping of chain configurations and intermolecular steric hindrances of chain motion (commonly termed 'entanglements') have been recognized as the origin of departures from both affine and phantom predictions. Different formalisms were proposed to include these intermolecular effects in the analysis of rubberlike elasticity at thermodynamic equilibrium.[16,32,212,218-224] In the one due to Flory and Erman,[11,124,213,225-227] entanglements are embodied as domains of constraints acting as restrictions on junctions fluctuations. Their theory was successful in accounting for the relationships of stress to strain for all varieties of strains, for typical elastomers throughout ranges accessible to experiment,[19,99,116,144,149,150,228-231] and is very amenable for use in network characterization. The constraints due to entanglements are introduced in the Flory and Erman model by means of spring-like forces acting on the fluctuating junctions, and a center of constraint is defined for each junction. The constraint domains can be initially represented as spheres, which are transformed to ellipsoids by the deformation. In uniaxial extension, the main axes of these ellipsoids are along the direction of stretching. Thus, fluctuations increase in the direction along which the stress is measured. The behavior of real networks then tends to that of phantom networks, with large fluctuations in the limit of infinite deformation. The decrease in reduced force with increasing elongation, observed experimentally in uniaxial tensile tests of amorphous rubbers (Figure 14), results from the transformation of fluctuations with strain in the manner stated above, and is indeed predicted by this molecular model of the constraints. The model based on Gaussian statistics does not account for the upturn exhibited by the reduced force at high strain, however, and has to be used for interpretation of data obtained only in the range of moderate deformations and at thermodynamic equilibrium.

The principal parameter κ of the theory is defined as the ratio of the mean-square radius of the fluctuations of the junctions in the phantom network $\langle (\Delta R)^2 \rangle_0$, to the mean-square radius $\langle (\Delta s)^2 \rangle_0$ of the Gaussian domains of constraints in the undistorted network. This is shown in equation (111). The parameter κ has been postulated to depend on the degree on interpenetration in the network, and this is well supported by experiments.[15,19,228,230,232,233] The parameter κ and the number of junctions μ in the volume V_0 of the state of reference should be related through equation (112),[11,230] where I is an interpenetration parameter which is frequently *ca.* 0.5[15,19] For sufficiently long chains, the mean-square end-to-end length $\langle r^2 \rangle_0$ of the (unperturbed) network chains is given by equation (113) derived from equation (28), where N is the average number of bonds per repeat unit, l^2 the mean-square skeletal bond length and m_0 the monomer molecular weight. A second parameter ζ is believed to reflect possible inhomogeneities in the network topology, which can perturb the affine transformation of constraint domains with macroscopic deformation.

$$\kappa = \langle (\Delta R)^2 \rangle_0 / \langle (\Delta s)^2 \rangle_0 \tag{111}$$

$$\kappa = I \langle r^2 \rangle_0^{3/2} (\mu / V_0) \tag{112}$$

$$\langle r^2 \rangle_0 = C_\infty M_c N l^2 m_0^{-1} \tag{113}$$

The molecular theory of elasticity outlined above predicts an expression for the elastic free energy change ΔA_{el}, which is the sum of the elastic free energy $\Delta A_{el}(\text{ph})$ for a phantom network, and a term ΔA_c which accounts for entanglement constraints. This is expressed by equation (114). The contribution $\Delta A_{el}(\text{ph})$ is given in equation (101) and the second term can be written as shown in equation (115), with B_t and g_t defined in equations (116) and (117), respectively. In the case of uniaxial extension of a swollen network, the elastic free energy expression, equation (114), yields equation (118) as a relationship between the reduced nominal stress and the uniaxial deformation α. The function K is defined in equation (119), with $\dot{B}_t$ and $\dot{g}_t$ defined in equations (120) and (121), respectively.

$$\Delta A_{el} = \Delta A_{el}(ph) + \Delta A_c \tag{114}$$

$$\Delta A_c = (\mu_e kT/2) \sum_{t=1,2,3} \{(1+g_t)B_t - \ln[(B_t+1)(g_tB_t+1)]\} \tag{115}$$

$$B_t = (\lambda_t - 1)(1 + \lambda_t - \zeta\lambda_t^2)(1+g_t)^{-2} \tag{116}$$

$$g_t = \lambda_t^2[\kappa^{-1} + \zeta(\lambda_t - 1)] \tag{117}$$

$$[f^*] \equiv (\xi kT/V_0)\{1 + (\mu_e/\xi)[\alpha K(\alpha^2 v_2^{-2/3}) - \alpha^{-2}K(\alpha^{-1}v_2^{-2/3})](\alpha - \alpha^{-2})^{-1}\} \tag{118}$$

$$K(\lambda_t^2) = B_t[\dot{B}_t(B_t+1)^{-1} + g_t(\dot{B}_tg_t + \dot{g}_tB_t)(g_tB_t+1)^{-1}] \tag{119}$$

$$\dot{B}_t = B_t\{[2\lambda_t(\lambda_t-1)]^{-1} + (1-2\zeta\lambda_t)[2\lambda_t(1+\lambda_t-\zeta\lambda_t^2)]^{-1} - 2\dot{g}_t(1+g_t)^{-1}\} \tag{120}$$

$$\dot{g}_t = \kappa^{-1} - \zeta(1 - 3\lambda_t/2) \tag{121}$$

The deformation ratios λ_t, $t = 1$, 2, 3, characterizing the deformation of the swollen network relative to the unswollen, undeformed state are related to α by equations (92), (94) and (97). The ratio μ_e/ξ depends on the functionality ϕ through equation (53). In the limit of vanishing constraints, equations (122) apply. The reduced stress then becomes equal to the phantom modulus $[f^*]_{ph}$.

$$\kappa = 0, \quad \kappa\zeta = 0, \quad K(\lambda_t^2) = 0 \tag{122}$$

The opposite extreme corresponds to the affine transformation of chain vectors which is reached in the limits given in equation (123). The nature of real network deformation is between affine and phantom behavior and is characterized by the set of parameters κ and ζ, κ being itself linked to the network topology through the interpenetration concept (equations 112 and 113).

$$\kappa \to \infty, \quad \zeta = 0, \quad K(\lambda_t^2) = 1 - \lambda_t^{-2} \tag{123}$$

In the case of perfect networks, combination of equations (47), (107), (112) and (113) yields equation (124). Thus, κ depends on the inverse square-root of the phantom modulus and is independent of swelling.[228,230,232] The factor N_A is Avogadro's number, which appears in equation (124) since μ in equation (112) is the number of junctions and not the number of moles of junctions. In the case of randomly cross-linked networks, use of equation (61) yields equation (125). In the case of networks formed by random cross-linking of star polymers, equation (63) is used instead of equation (61) to derive the expression for κ.[130] The other parameter ζ is the result of the relationship between κ and network inhomogeneities and its magnitude is estimated by experiment.

$$\kappa = 2I(N_A\rho C_\infty Nl^2 m_0^{-1}\phi^{-1})^{3/2}[kT(\phi-2)[f^*]_{ph}^{-1}]^{1/2} \tag{124}$$

$$\kappa = (IN_A\rho/2)(C_\infty Nl^2 m_0^{-1})^{3/2}M_c^{1/2}(1 - M_c/\bar{M}_n) \tag{125}$$

9.6.4 Characterization of Elastomeric Networks by Uniaxial Extension or Compression

The theoretical equations presented above can be used to interpret stress–strain measurements in uniaxial extension and thus to fully characterize elastomeric networks. In this regard, equations (124) and (125) are of particular interest since they relate the parameter κ, quantifying the entanglement constraints in the Flory and Erman model, to the polymer microstructure and conformational properties and to the network topology.[130,233] An illustrative analysis of stress–strain data due to Queslel, Thirion and Monnerie[234] is reported below.

Specifically, stress–strain relationships in uniaxial extension for networks of *cis*-1,4-polyisoprene (Shell IR 307) were obtained by extrapolation of relaxation measurements to infinite time. Employed for this purpose was a BKZ constitutive equation, with a memory function derived from the Helfand and Pearson theory. This theory involves the random-walk statistics for a chain in an obstacle net[235] in the case of the randomly distributed networks already considered by Curro and Pincus.[236] Three series of networks were investigated, each series being characterized by its precursor molecular weight. The influence of cross-link density was studied by varying the amount of dicumyl peroxide (dicup) used as cross-linking agent. Designation of vulcanizates, number-average molecular weight $\bar{M}_n$ of the precursor polymer and the quantity of peroxide incorporated in the rubber are reported in Table 1. Figures 15 and 16 show the equilibrium reduced forces obtained by the relaxation and extrapolation method described above as a function of the inverse of the deformation ratio, for the nine vulcanizates listed. The usual decrease in reduced force with increasing deformation was observed for all samples. The deformation was kept small enough to avoid any limited extensibility upturn. Experimental reduced forces were fitted by least-squares analysis with theoretical curves calculated from equation (118). The interpenetration parameter I was set equal to 0.5, and the two independent parameters ζ and M_c were chosen for the least-squares fitting. The other quantities ζ, κ, $[f^*]_{ph}$ and μ/V_0 were then calculated from M_c through equations (60), (61), (107) and (125) with knowledge of the independently determined values of $\bar{M}_n$. For the IR 307 polymer (92% *cis*, 5% *trans*), typical values of 5.0 for C_∞, 2.18 Å^2 for l^2 and 3.94 for N[72] lead to equation (126) for the relationship between $\langle r^2 \rangle_0$ and M_c, which are expressed in cm^2 and g mol^{-1}, respectively.

$$\langle r^2 \rangle_0 = 3.8 \times 10^7 \, M_c/N_A \qquad (126)$$

The predictions of the constrained junction theory are consistent with all the experimental data, and the range of κ is in agreement with values previously reported.[99,116,228] According to the interpenetration concept, κ should be a decreasing function of the cross-link density. Effectively, κ regularly decreases with increasing amount of dicumyl peroxide, as expected.

Beside determination of the mesh size M_c, the characterization of network structure by analysis of mechanical properties with a molecular model of elasticity gives some insight into the chemical cross-linking process. The number density of junctions μ/V_0 determined above may be compared with the number density of junctions μ_t/V_0 calculated from knowledge of the initial amount x (weight percent) of peroxide, assuming full decomposition of peroxide and a 1:1 cross-linking efficiency. The quantity μ_t/V_0 is then related to x through equation (127). The quantity μ_t/V_0 and the ratio μ/μ_t are reported in Table 1. The real density of junctions is about three times higher than the

Table 1 Characterization of the *cis*-Polyisoprene Vulcanizates[a]

Sample	$10^{-5} \bar{M}_n$ (g mol^{-1})	Dicup[b] x	$[f^*]_{ph}^c$ (N mm^{-2})	κ	ζ	$10^{-3} M_c$ (g mol^{-1})	$10^5 \mu/V_0$ (mol cm^{-3})	$10^5 \mu_t/V_0$ (mol cm^{-3})	μ/μ_t
1 IR 360	3.60	1.0	0.2350	4.6	0	4.69	9.58	3.34	2.9
2 IR 360	3.60	0.8	0.2113	4.9	0.011	5.19	8.63	2.67	3.2
1 IR 245	2.45	0.8	0.1967	4.9	0.017	5.44	8.17	2.67	3.1
2 IR 245	2.45	0.6	0.1592	5.4	0.011	6.62	6.68	2.02	3.3
3 IR 245	2.45	0.35	0.0930	6.8	0	10.72	4.06	1.18	3.4
1 IR 125	1.25	1.2	0.2452	4.3	0	4.21	10.46	4.00	2.6
2 IR 125	1.25	0.6	0.1445	5.3	0.035	6.67	6.46	2.01	3.2
3 IR 125	1.25	0.45	0.0989	6.0	0.010	9.08	4.65	1.51	3.1
4 IR 125	1.25	0.3	0.0747	6.6	0.016	11.22	3.69	1.01	3.7

[a] Reproduced by permission of Butterworth from J. P. Queslel, P. Thirion and L. Monnerie, *Polymer*, 1986, **27**, 1869. [b] Weight percent in bulk.
[c] At $T = 303$ K.

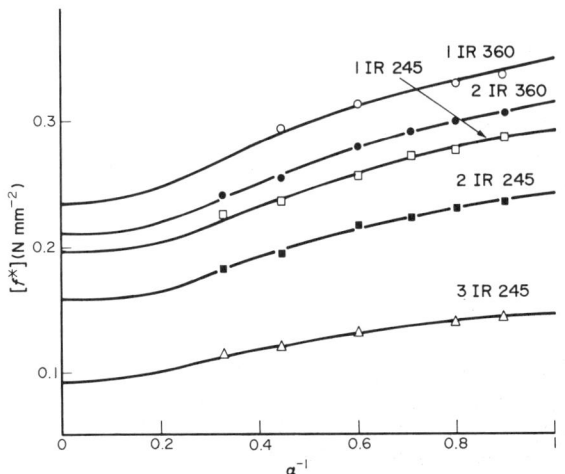

Figure 15 Plot of equilibrium reduced forces *vs.* reciprocal of the extension ratio. Experimental points were obtained by extrapolation of relaxation measurements to infinite time. Continuous curves are predictions of the constrained junction theory of rubber elasticity calculated with parameters listed in Table 1 (reproduced by permission of Butterworth from J. P. Queslel, P. Thirion and L. Monnerie, *Polymer*, 1986, **27**, 1869)

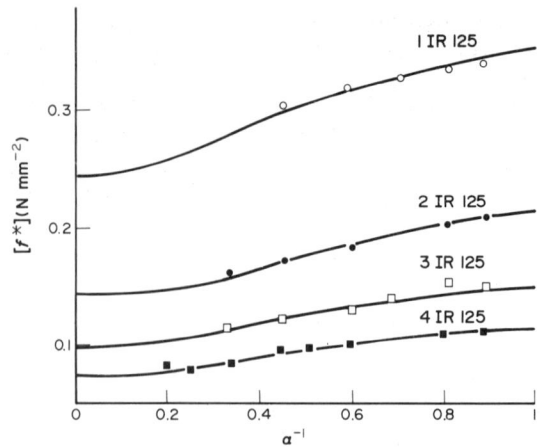

Figure 16   Plot of equilibrium reduced force *vs.* reciprocal of the extension ratio. Same symbols as in Figure 15 (reproduced by permission of Butterworth from J. P. Queslel, P. Thirion and L. Monnerie, *Polymer*, 1986, **27**, 1869)

calculated density. For this IR 307 polymer, one mole of dicumyl peroxide has thus created about three moles of C–C cross-links.

$$\mu_t/V_0 = \rho x/[270(100 + x)] \tag{127}$$

9.6.5   Entanglement Models in Rubber Elasticity

In the Flory and Erman model, entanglements are represented as domains of constraints which act on junction fluctuations. It is a mean-field theory since entanglements are seen as averages. In addition to stress–strain isotherms, this concept has been successfully applied to network chain orientation[237-239] and birefringence.[230,240] Other mean-field theories are based on the tube concept (in which each network chain is confined within a tube), which was introduced by de Gennes[241] and Doi and Edwards[242,243] in the viscoelasticity area.[243-245] In slip-link models, each network chain threads its way through a number of small rings representing entanglements.[221,246-248] These discrete theories have the disadvantage of introducing another topological parameter, which is the number density of entanglements. Combination of non-localization of trapped entanglements and restrictions on the fluctuations of permanent junctions has also been proposed for the departures from neo-Hookean behavior in real networks.[220,249] All these theories predict that the entanglement contribution in uniaxial extension tends to zero in the range of infinite deformation, since the freedom of junctions to fluctuate, or of slip-links to slide along the chains, increases with deformation in the direction of the macroscopic stretching.[247,249-251]

A phenomenological concept of trapped entanglements was introduced by Langley[16,218] and applied in the range of small uniaxial extensions.[116,120-122,250-256] According to this model, a fraction of the entanglements present in the melt are said to be trapped during network formation. The contribution of these entanglements to the equilibrium elastic modulus is assumed to be a fraction of the non-equilibrium plateau modulus as determined in viscoelasticity measurements.[257-259] This fraction is the trapping probability calculated through probability theories which simulate network formation, such as the Macosko–Miller branching theory.[14,118,260-265]

9.7   PHENOMENOLOGICAL THEORY OF RUBBERLIKE ELASTICITY

The phenomenological theory of rubberlike elasticity is based on continuum mechanics. It provides a mathematical structure from which, in principle, the deformation produced within a vulcanized elastomer by applied surface and bulk forces can be calculated. In the theory, the material is idealized by the assumption that it is perfectly elastic, isotropic in the undeformed state and incompressible.[211,266-269] The most general form of the strain energy function (which vanishes at zero strain) is the power series shown in equation (128) where the C_{ijk} coefficients have units of $N\,m^{-2}$. The quantities I_1, I_2, and I_3 are the three strain invariants, I_1 being defined by equation (91), and I_2 and I_3 by equations (129) and (130), respectively. It was found that a simple form of equation

(128) can reproduce the observed dependences.[270-273] The specific relationship is given in equation (131), where m is a parameter which can be adjusted to obtain the best fit of the data, and the ratio of C_3 to C_2 is 0.5. For uniaxial extension, the reduced nominal stress is expressed by equation (132). The Mooney–Rivlin equation is obtained for $m = 0$ and has been extensively used to fit experimental data in uniaxial extension.[116, 139, 143, 214, 274] Values of m were found to be between 0 and 0.5.[270, 273] Generally only a portion of typical experimental curves can be approximated by a straight line (equation 132), as shown on Figure 14. Also, equation (132) cannot account at all for data obtained in compression.[144]

$$w/V_0 = \sum_{i,j,k=0}^{\infty} C_{ijk}(I_1-3)^i (I_2-3)^j (I_3-1)^k \tag{128}$$

$$I_2 = \lambda_1^2 \lambda_2^2 + \lambda_2^2 \lambda_3^2 + \lambda_3^2 \lambda_1^2 \tag{129}$$

$$I_3 = \lambda_1^2 \lambda_2^2 \lambda_3^2 \tag{130}$$

$$w/V_0 = C_1(I_1-3) + C_2(I_2 I_3^{-1+(m/2)} - 3) - C_3 \ln I_3 \tag{131}$$

$$[f^*] \equiv 2C_1 + 2C_2 v_2^{(4/3)-m} \alpha^{-1} \tag{132}$$

9.8 SWELLING OF ELASTOMERIC NETWORKS

Swelling phenomena have always been closely associated with the use and the study of polymeric materials. For instance, less expensive industrial compounds may be obtained if rubber is mixed with a diluent (oil). Insoluble network polymers are conveniently characterized in the swollen state since this facilitates the approach to equilibrium.

9.8.1 Thermodynamic Equations of Swelling

Swelling is the result of two thermodynamic phenomena: an increase of entropy of the network–solvent system by introduction of small molecules as diluent, and a decrease of entropy of the polymer chains by the isotropic dilation. The Flory–Huggins lattice theory accounts for the first effect, *i.e.* the mixing of polymer and solvent.[2, 275, 276] Use of their relationship, and a statistical mechanical expression for the Gibbs free energy change with dilation, lead Flory and Rehner to a relationship between swelling and degree of cross-linking for an affine network[109, 204, 277] which has been widely used to characterize a variety of elastomers.[103] Refined swelling–structure relationships for real networks can be obtained by making use of the Flory and Erman model.[17-21, 232, 278]

The change ΔA of the Gibbs free energy due to mixing a solvent 1 with a polymer 2 is given by equation (133), where A_M, A_1 and A_2 are the free energies of the systems polymer–solvent, pure solvent and pure polymer, respectively. Chemical potentials of solvent in the uncross-linked polymer, μ_1^u, and in the pure state, μ_1°, are obtained by differentiation of ΔA with respect to the number of moles of solvent n_1 at constant n_2 of polymer. This is specified by equation (134). The difference $\mu_1^u - \mu_1^\circ$ is related to the vapor pressures of solvent p_1^u over the solution and p_1° of pure solvent. This is specified by equation (135). The liquid lattice theory of Flory and Huggins[275, 276] yields equation (136) for $\mu_1^u - \mu_1^\circ$, where v_2 is the volume fraction of polymer in the mixture, χ the polymer–solvent interaction parameter,[273, 279-281] and x the ratio of molar volumes of polymer, V_2, and solvent, V_1. The explicit expressions for x are given in equation (137), where $\bar{M}_1$ and $\bar{M}_2$ are the number-average molecular weights of solvent and polymer, respectively, and ρ_1 and ρ_2 their densities. The term $1/x$ in equation (136) is small and equal to zero for a network (which is a 'molecule' of infinite molecular weight).

$$\Delta A = A_M - A_1 - A_2 \tag{133}$$

$$(\partial \Delta A/\partial n_1)_{n_2 T, p} = \mu_1^u - \mu_1^\circ \tag{134}$$

$$\mu_1^u - \mu_1^\circ = RT \ln(p_1^u/p_1^\circ) \tag{135}$$

$$\mu_1^u - \mu_1^\circ = RT[\ln(1-v_2) + \chi v_2^2 + v_2(1-1/x)] \tag{136}$$

$$x = V_2/V_1 = (\bar{M}_2/\bar{M}_1)(\rho_1/\rho_2) \tag{137}$$

The difference between the chemical potential of solvent in the cross-linked polymer, μ_1^c, and in the pure state, μ_1°, is related to the vapor pressure p_1^c of solvent in the swollen network, and in the pure state, p_1° by equation (138). Combination of equations (135) and (138) at constant volume fraction of solvent then yields equation (139). The quantity $\mu_1^c - \mu_1^\circ$ results from an elastic contribution $(\mu_1^c - \mu_1^\circ)_{el}$ and a mixing contribution $(\mu_1^c - \mu_1^\circ)_{mix}$ approximately equal to $\mu_1^u - \mu_1^\circ$ for large x. It is generally assumed that these two contributions are separable. Introducing μ_1^u on the left side of equation (139) leads directly to equation (140). Swelling is a three-dimensional deformation (dilation) of the network and thus $(\mu_1^c - \mu_1^\circ)_{el}$ is related to the elastic free energy change ΔA_{el} by equation (141).

$$\mu_1^c - \mu_1^\circ = RT\ln(p_1^c/p_1^\circ) \tag{138}$$

$$\mu_1^c - \mu_1^u = RT\ln(p_1^c/p_1^u) \tag{139}$$

$$(\mu_1^c - \mu_1^\circ)_{el} = RT\ln(p_1^c/p_1^u) \tag{140}$$

$$(\mu_1^c - \mu_1^\circ)_{el} = (\partial\Delta A_{el}/\partial n_1)_{T,p} \tag{141}$$

The activities of solvent over the uncross-linked polymer, a_1^u, and over the network, a_1^c, are related to the vapor pressures by equations (142) and (143), respectively. Combination of equations (140)–(143) at constant volume fraction v_2 of polymer then gives equation (144). If solvent is present in excess, absorption equilibrium is reached when the chemical potential of solvent in the swollen network is equal to that of pure solvent outside the network, *i.e.* $\mu_1^c = \mu_1^\circ$. Hence equation (145) must apply, where $(\mu_1^c - \mu_1^\circ)_{mix}$ is given by equation (136) with $1/x = 0$.

$$a_1^u = p_1^u/p_1^\circ \tag{142}$$

$$a_1^c = p_1^c/p_1^\circ \tag{143}$$

$$RT\ln(a_1^c/a_1^u) = (\partial\Delta A_{el}/\partial n_1)_{T,p} \tag{144}$$

$$\mu_1^c - \mu_1^\circ = (\mu_1^c - \mu_1^\circ)_{mix} + (\mu_1^c - \mu_1^\circ)_{el} = 0 \tag{145}$$

The volume fraction of polymer at equilibrium (maximum) swelling is designated v_{2m}. Then, equations (144) and (146) may be used to test the phenomenological or molecular theories of elasticity, and to characterize elastomeric networks by differential solvent vapor sorption and swelling equilibrium measurements, respectively.

$$RT[\ln(1-v_{2m}) + \chi v_{2m}^2 + v_{2m}] + (\partial\Delta A_{el}/\partial n_1)_{T,p} = 0 \tag{146}$$

9.8.2 Techniques for Swelling Measurements

9.8.2.1 Swelling equilibrium

This simple experiment consists of completely immersing a sample in solvent and waiting until swelling equilibrium occurs. Then v_{2m} is determined by reweighting it. An accurate value is obtained only with samples which have been extracted to remove soluble material.

9.8.2.2 Determination of solvent activities and interaction parameters

The parameter χ can be obtained by several techniques including osmometry, vapor sorption, gas–liquid chromatography, freezing point depression of solvent, swelling equilibrium, intrinsic viscosity and critical solution temperatures.[279,280] The differential solvent vapour sorption technique enables one to measure interaction parameters and activities of solvent in uncross-linked and cross-linked polymers and is thus very useful for testing the different rubber elasticity theories through equation (144).[282,283] A typical apparatus is illustrated in Figure 17.[281] Specifically, a Cahn RG electrobalance is housed in stainless steel case B. Polymer samples on quartz sample pans are suspended on both sides of the balance with 32-gauge nichrome wires. A strip-chart recorder is used to display portions of the electrobalance read-out. Uncross-linked and cross-linked samples of approximately equal weight are placed on the two sides of the electrobalance. A glass weighing chamber K communicates with the balance case, and a quartz spring is attached to a removable chamber cap D. An uncross-linked polymer sample is suspended on the quartz spring, and all dry sample weights are accurately determined. Reservoir C, which serves to cushion sudden pressure changes and acts as a mercury trap, is connected to the chamber K and to one arm of the mercury manometer connected to a vacuum line. A solvent reservoir G can be opened to the system *via* a mercury float valve F. A similar valve E connects all parts of the apparatus to the vacuum line. A water bath is used as a thermostat for the apparatus. Measurements of the displacement of the quartz spring and manometer are made with a cathetometer, through a glass window. Valve F is opened to allow solvent vapor into the system. The amount of solvent introduced is controlled by controlling the temperature of the solvent reservoir. Valve F is closed after each addition of vapor and polymer samples are then allowed to equilibrate. The integral and differential sorptions are observed on the respective balances and the pressure of solvent vapor is recorded. It is then possible to calculate the volume fraction of polymer in the cross-linked and uncross-linked swollen samples, v_2^c and v_2^u respectively, at a solvent pressure $p_1^c = p_1^u$ and thus to know p_1^c and p_1^u at equal volume fractions $v_2^c = v_2^u$ and to calculate the ratio of activities $a_1^c/a_1^u = p_1^c/p_1^u$. From knowledge of p_1°, p_1^u, v_2^u and x, $\chi(v_2^u)$ is deduced through equations (135)–(137).

9.8.3 Phenomenological Approach to Swelling

Swelling is equivalent to an isotropic deformation of extension ratio λ given in equation (147). From equations (91) and (129)–(131), the stored elastic energy takes the form given in equation (148). The derivative of the elastic energy occurring on the right side of equation (144) may then be developed as shown in equation (149). The derivative of λ^2 with respect to n_1 is obtained from equation (147), the result being given in equation (150). A phenomenological isotropic swelling relationship, equation (151), is then obtained by combination of equations (144) and (148)–(150). Use of the Mooney–Rivlin free energy expression, *i.e.* equation (148) with $m=0$ and without the logarithmic term, leads to the alternative expression given in equation (152). Experimentally, it has been observed that there is a maximum in the dependence of $\lambda \ln(a_1^c/a_1^u)$ on λ^2.[281,282] Equations (151) and (152) cannot account for this maximum.[116,281] However, the phenomenological approach may provide a simple way to determine the interaction parameter if an expression for the stored elastic energy obtained from uniaxial and biaxial experiments is used in equation (146) with the value of v_{2m} from a swelling equilibrium measurement. The parameter χ will be obtained as the solution of this equation.

$$\lambda = \lambda_1 = \lambda_2 = \lambda_3 = [(n_1 V_1 + V_0)/V_0]^{1/3} = v_2^{-1/3} \tag{147}$$

$$w = 2C_1 V_0 [\lambda^2 - 1 - (1/2)\ln\lambda^2] + 3C_2 V_0 [(\lambda^2)^{(3m/2)-1} - 1] \tag{148}$$

$$(\partial\Delta A_{el}/\partial n_1)_{T,p} = (\partial\Delta A_{el}/\partial\lambda^2)_{T,p}(\partial\lambda^2/\partial n_1)_{T,p} \tag{149}$$

$$(\partial\lambda^2/\partial n_1)_{T,p} = 2V_1/3\lambda V_0 \tag{150}$$

$$\lambda\ln(a_1^c/a_1^u) = 2(C_1 V_1/RT)[1 - 2/\lambda^2 - (C_2/2C_1)(2-3m)\lambda^{3m-4}] \tag{151}$$

$$\lambda\ln(a_1^c/a_1^u) = 2(C_1 V_1/RT)[1 - C_2/(C_1\lambda^4)] \tag{152}$$

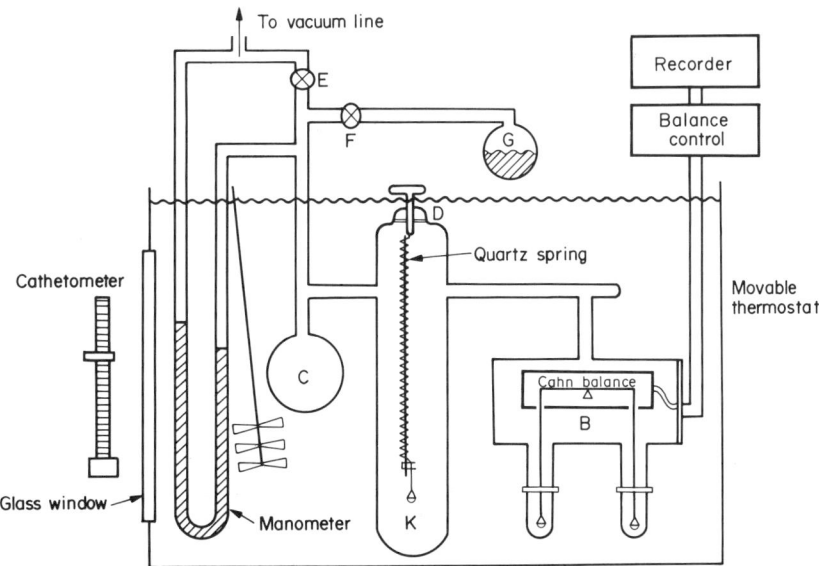

Figure 17 Schematic diagram of apparatus for the measurement of differential solvent vapor sorption (reproduced by permission of Wiley from L. Y. Yen and B. E. Eichinger, *J. Polym. Sci., Polym. Phys. Ed.*, 1978, **16**, 121)

9.8.4   Statistical Mechanical Energy of Dilation and Network Characterization

The derivative of the elastic free energy with respect to λ^2 in the Flory and Erman constrained junction analysis is the sum of the corresponding derivatives of $\Delta A_{el}(\text{ph})$ and ΔA_c.[18,116] These two contributions are given in equations (153) and (154), respectively. The statistical mechanical form of equation (144) is then the relationship shown in equation (155). Equation (146) then gives equation (156) at swelling equilibrium, where ξ is expressed in moles and the ratio μ_e/ξ is calculated through equation (53).

Swelling equations for phantom networks are recovered through $\kappa = 0$ and $K(\lambda^2) = 0$. Equations (157) and (158) give the specific relationships. For affine networks, $\kappa = \infty$ and $K(\lambda^2) = 1 - \lambda^{-2}$. This leads to equations (159) and (160). A plot of $\lambda \ln(a_1^c/a_1^u)$ vs. $\lambda^2 (\lambda = v_2^{-1/3})$ for the system poly(dimethylsiloxane)–benzene is given in Figure 18.[281] A similar curve was obtained by Gee and co-workers.[282] Data are compared with theoretical predictions given by equations (151), (152), (157) and (159).

$$(\partial \Delta A_{el}(\text{ph})/\partial \lambda^2)_{T,p} = (3/2)\xi kT \tag{153}$$

$$(\partial \Delta A_c/\partial \lambda^2)_{T,p} = (3/2)\mu_e kT K(\lambda^2) \tag{154}$$

$$\lambda \ln(a_1^c/a_1^u) = (\xi/V_0)V_1[1 + (\mu_e/\xi)K(\lambda^2)] \tag{155}$$

$$\ln(1 - v_{2m}) + \chi v_{2m}^2 + v_{2m} = -(\xi/V_0)V_1 v_{2m}^{1/3}[1 + (\mu_e/\xi)K(v_{2m}^{-2/3})] \tag{156}$$

$$\lambda \ln(a_1^c/a_1^u) = (\xi/V_0)V_1 \tag{157}$$

$$\ln(1 - v_{2m}) + \chi v_{2m}^2 + v_{2m} = -(\xi/V_0)V_1 v_{2m}^{1/3} \tag{158}$$

$$\lambda \ln(a_1^c/a_1^u) = (\xi/V_0)V_1[1 + (\mu_e/\xi)(1 - \lambda^{-2})] \tag{159}$$

$$\ln(1 - v_{2m}) + \chi v_{2m}^2 + v_{2m} = -(\xi/V_0)V_1 v_{2m}^{1/3}[1 + (\mu_e/\xi)(1 - v_{2m}^{2/3})] \tag{160}$$

The simple statistical mechanical models, *i.e.* affine and phantom, or the phenomenological approach cannot reproduce the maximum in the plot of $\lambda \ln(a_1^c/a_1^u)$ vs. λ^2.[116,281] This maximum is predicted, however, by the constrained junction theory[18,21,278,284] but, as shown in Figure 19, agreement with experiment is only qualitative.[17,20] This has to lead to questioning of the separability of mixing and elastic contributions in equation (145). This postulate may be valid only in the limit of swelling equilibrium.[19]

Characterization of elastomeric networks by swelling equilibrium measurements may take advantage of the applicability of the constrained junction theory already demonstrated for mechanical testing. Use of the interpenetration concept (equation 124 or 125), and of topological expressions for ξ (equations 47 or 60) cause equation (156) to depend only on M_c and χ for a perfect network, or on M_c, $\bar{M}_n$, and χ for a randomly cross-linked network. An example of such an application was given by Erman and Baysal.[19] Two cross-linked polystyrene networks were

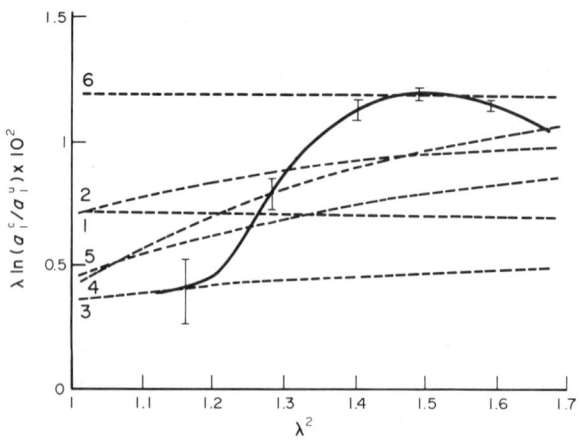

Figure 18   Plot of $\lambda \ln(a_1^c/a_1^u)$ vs. λ^2 as a smooth function for poly(dimethylsiloxane)–benzene at 25 °C (solid curve). Dashed lines are calculated from the reduced elastic component of chemical potential $(\mu_1^c - \mu_1^u)_{el}/RT$ given by various elasticity equations. Curve 1: equation (157), $(\xi/V_0)V_1 = 0.007$; curves 2 and 3: equation (159), $\mu_e/\xi = 1$, $(\xi/V_0)V_1 = 0.007, 0.0035$; curve 4: equation (152), $C_1 V_1/RT = 0.007$, $C_2 V_1/RT = 0.005$; curves 5 and 6: equation (151), $m = 1/2, 1$, $C_1 V_1/RT = 0.007, 0.0044$, $C_2 V_1/RT = 0.005, 0.0072$, respectively (reproduced by permission of Wiley from L. Y. Yen and B. E. Eichinger, *J. Polym. Sci., Polym. Phys. Ed.*, 1978, **16**, 121)

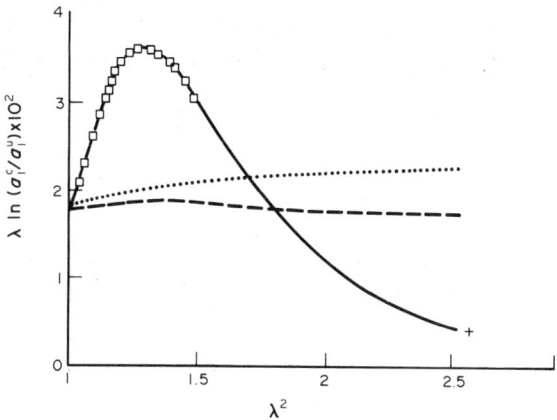

Figure 19 Plot of $\lambda \ln(a_1^c/a_1^u)$ *vs.* λ^2 as a smooth function for poly(dimethylsiloxane)–benzene at 30 °C (solid curve). The + point represents the swelling equilibrium point. Dashed and dotted curves represent equation (155) ($\mu_e/\xi = 1$, $(\xi/V_0)V_1 = 0.017$) with parameters $\kappa = 10$, $\zeta = 0.4$, and $\kappa = 10$, $\zeta = 0$, respectively (reproduced by permission of the American Chemical Society from R. W. Brotzman and B. E. Eichinger, *Macromolecules*, 1982, **15**, 531)

prepared by polymerization of mixtures of styrene, benzoyl peroxide, and *p*-divinylbenzene. Samples with dimensions of *ca.* $8 \times 3 \times 1$ mm^3 were machined from the dry glassy cross-linked networks. Two pieces of thin wires were placed approximately 6 mm apart on each sample, forming fiducial marks from which linear deformation upon swelling was measured. Samples were stored horizontally in a constant temperature bath (± 0.02 °C) and measurements of length were made in the immersed state with a travelling microscope. The degree of cross-linking was obtained by measuring the degree of equilibrium swelling of the network in toluene. The χ parameter for polystyrene–toluene at 22–30 °C was obtained from the data compiled by Orwoll,[280] which is given in equation (161). For polystyrene (at 25 °C), the parameter κ is given by equation (162), where the numerical coefficient is in units of N$^{1/2}$ mm^{-1}. Equation (162) was obtained from equations (112) and (113) assuming a perfect tetrafunctional network (equation 47 with $\phi = 4$), with $C_\infty = 8.5$, $m_0 = 100$ g mol^{-1}, $1 = 1.53 \times 10^{-8}$ cm, and $N = 2$. The results of swelling measurements of v_{2m} in toluene and calculations of the phantom modulus $\xi kT/V_0$ by combination of equations (119), (156) and (162) and of the molecular weight M_c from equation (47) are reported in Table 2. The parameter ζ was set equal to zero and the values of κ were calculated from the phantom moduli through equation (162).

$$\chi = 0.455 - 0.155 v_2 \qquad (161)$$

$$\kappa = 1.334(\xi kT/V_0)^{-1/2} \qquad (162)$$

Subsequent measurements of swelling ratios were carried out in a toluene–methanol mixture (75/25 wt%). By making use of the cycle rank values determined above, values of the interaction parameter for the polystyrene–toluene–methanol system were calculated at different temperatures by means of equation (156). Least-squares analysis of the data gave equation (163) for the dependences of χ on v_2 and T.

$$\chi = 0.4103 + 16.03/T + (0.5054 + 5.60/T)v_2 \qquad (163)$$

Table 2 Characterization of Polystyrene Networks by Swelling Equilibrium Measurements in Toluene, and Application of the Constrained Junction Theory[a]

Sample	v_{2m}[b]	κ	$\xi kT/V_0$ (N mm^{-2})[c]	$10^{-4} M_c$ (g mol^{-1})
1	0.067	10	0.0163	7.6
2	0.056	12	0.0110	11.3

[a] Reproduced by permission of the American Chemical Society from B. Erman and B. M. Baysal, *Macromolecules*, 1985, **18**, 1696. [b] In toluene. [c] At $T = 298$ K.

9.8.5 Swelling of Deformed Networks

For stress–strain measurements on swollen networks, it is advantageous to immerse the samples completely in excess solvent. This avoids errors due to solvent vaporization during the stretching of a swollen network exposed to the air. However, the swelling capacity of an elastomer is not constant with deformation.[1,17,204,230,285] A swelling equilibrium equation for this case can also be obtained by making use of the constrained junction model,[17,230] as follows.

An unswollen sample is deformed uniaxially. The principal extension ratio along the stretching direction is denoted $\lambda_{||}$. The sample is then immersed in solvent, and swelling occurs in directions normal to the stretching axis. This is accompanied by a change in lateral dimensions. The quantities $\lambda_\perp$ are the two principal extension ratios in the lateral directions with respect to the unswollen sample of volume V_0. The product of the three principal extension ratios is the inverse of the volume fraction of polymer v'_2 in the swollen network, as shown in equation (164). The elastic free energy ΔA_{el} can now be differentiated with respect to n_1, with the result shown in equation (165), where $\lambda_\perp$ is introduced since it is the extension ratio which varies with swelling at constant $\lambda_{||}$. By making use of equation (164), equation (166) may be obtained. The phantom free energy change may be expressed by equation (167), which leads to equations (168) and (169). Equation (146) then becomes equation (171), while for an affine network it is described by equation (172).

$$\lambda_{||}\lambda_\perp^2=(n_1 V_1+V_0)/V_0=v'^{-1}_2 \tag{164}$$

$$(\partial\Delta A_{el}/\partial n_1)_{T,p,\lambda_{||}}=(\partial\Delta A_{el}/\partial\lambda_\perp^2)_{T,p,\lambda_{||}}(\partial\lambda_\perp^2/\partial n_1)_{T,p,\lambda_{||}} \tag{165}$$

$$(\partial\lambda_\perp^2/\partial n_1)_{T,p,\lambda_{||}}=V_1/V_0\lambda_{||}=\lambda_\perp^2 v'_2 V_1/V_0 \tag{166}$$

$$\Delta A_{el}(ph)=(1/2)\xi kT(\lambda_{||}^2+2\lambda_\perp^2-3) \tag{167}$$

$$(\partial\Delta A_{el}(ph)/\partial\lambda_\perp^2)_{T,p,\lambda_{||}}=\xi kT \tag{168}$$

$$(\partial\Delta A_c/\partial\lambda_\perp^2)_{T,p,\lambda_{||}}=2(\partial\Delta A_c/\partial\lambda_t^2)_{T,p,\lambda_{||}}=\mu_e kT K(\lambda_\perp^2) \tag{169}$$

$$\ln(1-v'_2)+\chi v'^2_2+v'_2=-(\xi/V_0)V_1\lambda_\perp^2 v'_2[1+(\mu_e/\xi)K(\lambda_\perp^2)] \tag{170}$$

$$\ln(1-v'_2)+\chi v'^2_2+v'_2=-(\xi/V_0)V_1\lambda_{||}^{-1} \tag{171}$$

$$\ln(1-v'_2)+\chi v'^2_2+v'_2=-(\xi/V_0)V_1\lambda_{||}^{-1}[1+(\mu_e/\xi)(1-\lambda_{||}v'_2)] \tag{172}$$

It is possible to predict whether an already stretched network swells more or less than an undeformed network at equilibrium. The threshold is obtained by comparing equations (158) and (171), (160) and (172), with $v'_2=v_{2m}$. The solution is $\lambda_{||}=v_{2m}^{-1/3}$. The physical explanation was given by Treloar.[1] If the dry sample is uniaxially deformed, with an extension ratio $\lambda_{||}$ lower than the isotropic deformation $v_{2m}^{-1/3}$, then when swelling occurs, the clamps act as restrictions to swelling deformation and the net result is a compression. Thus a deformed network swells more than an isotropic network if $\lambda_{||}>v_{2m}^{-1/3}$.[17] If a sample is stretched when immersed completely in excess solvent, a new swelling equilibrium occurs at each extent of deformation. Generally, the force f is monitored as a function of α, the ratio of the length L_V of the deformed sample of volume V to the initial length L_V^i at the same volume. In the expression of the reduced force (equation 104), v_2 is now replaced by v'_2. Thus, to calculate $[f^*]$ and α, it is necessary to know v'_2 and L_V^i at each extent of deformation. The extension ratios $\lambda_{||}$ and $\lambda_\perp$ are defined with respect to the unswollen state. An example is given in equation (173), where $L_{V_0}^i$ is the length of the dry isotropic sample. By putting on the sample two ink marks parallel to the stretching axis and two along the lateral direction, it is possible to measure $\lambda_{||}$ and $\lambda_\perp$ with a cathetometer, and then to calculate v'_2 from equations (174) and (175).

$$\lambda_{||}=L_V/L_{V_0}^i \tag{173}$$

$$v'_2=(\lambda_{||}\lambda_\perp^2)^{-1} \tag{174}$$

$$L_V^i=v'^{-1/3}_2 L_{V_0}^i \tag{175}$$

9.9 RUPTURE OF ELASTOMERIC NETWORKS

The nominal tensile strength f_r/A_0 is the maximum tensile stress which a material is capable of sustaining. The maximum extensibility is the corresponding rupture strain λ_r. These ultimate tensile properties of elastomeric networks are very sensitive to temperature and test conditions, but can be characterized by a failure envelope which is generally represented by a plot of $\log(f_r T_0/A_0 T)$ vs. $\log(\lambda_r-1)$ where T and T_0 are respectively the test temperature and a reference temperature.[286,287]

This envelope may be used to describe relaxation, creep or constant strain rate measurements. A change in strain rate or temperature only shifts a point along the failure envelope, which is thus dependent only on the structural characteristics of the elastomer. The ultimate properties of rubbers are mainly governed by their viscoelastic properties, and reduced master curves can be obtained for tensile strength and strain as a function of time to break. The failure process is a non-equilibrium one, developing with time and involving the consecutive rupture of the molecular chains. The ultimate properties can then be predicted from creep measurements.[288-291]

9.10 STRAIN-INDUCED CRYSTALLIZATION IN ELASTOMERS

Elastomeric chains of sufficient regularity may crystallize under some conditions.[292] The deformation of a macromolecule is accompanied by a decrease in its conformational entropy. Its tendency to crystallize is therefore increased. The oriented system will crystallize even above the melting point of the isotropic system (and is thus 'strain-induced crystallization'). The crystallinity reaches a value which depends on the degree of deformation and the crystallization temperature. The melting point $T_m = \Delta H_m / \Delta S_m$, given by the ratio of the enthalpy of fusion to the entropy of fusion, is thus increased by the network deformation.

Crystallization, when reversibly induced in the deformation process, is of great importance with regard to the elastomeric and ultimate properties of the network.[215,293-298] It was found to be the origin of the upturn in the reduced nominal stress observed for some polymer networks in the region of very high elongation.[1] Strain-induced crystallites act as both physical cross-links of very high functionality and filler particles and, by thus reinforcing the network, greatly increase its ultimate properties. A typical example is natural rubber. During an experiment carried out at constant temperature, the modulus decreases abruptly but subsequently increases as the elongation increases. A decrease in modulus could result from the fact that the crystallites are oriented along the direction of stretching, and the chain sequences within the crystallites are in regular highly extended conformations. The straightening and aligning of portions of the network chains thus decrease the deformation in the remaining amorphous regions, with an accompanying decrease in the modulus.[215]

9.11 FILLERS IN ELASTOMERS

Elastomers are often compounded with finely divided solids to reinforce the rubber and to reduce costs. The most important fillers are carbon blacks, silica and silicates, clays and whiting (calcium carbonate).[299-304] The particles are the source of reinforcement through their interactions with the rubber, among themselves and with the chemistry of the cross-linking process. Abrasion resistance, tear strength and tensile strength are simultaneously improved. However, hysteresis, heat build-up and compression set (permanent deformation) are also known to increase as the reinforcing ability of the filler becomes more pronounced.

One of the most important features of the addition of fine particulate fillers is the increased stiffness and strength of the vulcanized rubbers. Theoretical studies have derived mainly from the formally analogous problem of the increase in viscosity caused by a suspension of solid particles. In this approach, the vulcanized rubber surrounding the filler particles is considered to be an isotropic elastic continuum having the same elastic properties as the rubber without filler. The rubber is also assumed to adhere to the filler surface, and by considering stresses in the rubber in the neighborhood of particles, expressions can be derived relating the elastic modulus E of the filled rubber to the modulus E_0 of the matrix. Taking into account the interactions between neighboring filler particles, Guth and Gold have obtained equation (176) for Young's modulus of the filled rubber as a function of the volume concentration c of filler.[305,306] Young's modulus of vulcanized natural rubber filled with different amounts of a coarse MT black are plotted *vs.* concentration of filler in Figure 20.[301] The full line in the figure is the theoretical curve given by equation (176). Good agreement is observed. For finer particle size carbon blacks, the actual increase of rubber modulus is much more than this equation predicts. Guth[307] introduced a shape factor f_N, the ratio of the largest dimension of the particle to the shortest, and thus derived equation (177). In a more refined treatment,[308] an effective filler concentration, augmented by rubber occluded in the concavities of the carbon black primary structure aggregates,[309] was substituted into equation (176).

$$E = E_0(1 + 2.5c + 14.1c^2) \tag{176}$$

$$E = E_0(1 + 0.67f_N c + 1.62f_N^2 c^2) \tag{177}$$

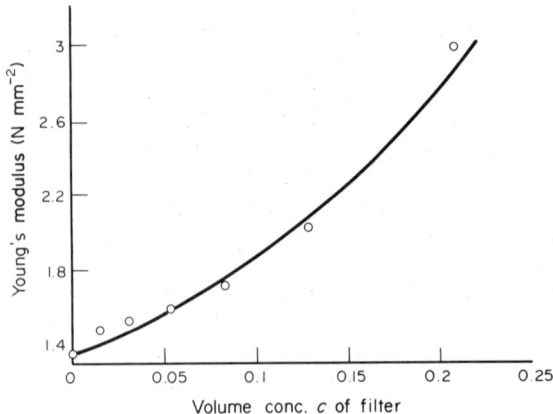

Figure 20 Young's modulus of rubbers containing MT carbon black filler as a function of filler concentration. The solid curve is predicted by equation (176) (reproduced by permission of Wiley from L. Mullins and N. R. Tobin, *J. Appl. Polym. Sci.*, 1965, **9**, 2993)

A filled vulcanizate may also be characterized by swelling.[302] Most fillers do not swell and, if the matrix is restricted by the filler through attachments on the filler surface, the swelling of the rubbery matrix will decrease as the filler loading increases. Typical stress–strain isotherms for different composites, containing some 20% by volume of various fillers tested in uniaxial extension, are shown in Figure 21.[300] Modulus and tensile strength are enhanced by carbon black having strong interactions with the polymer (curve A). Graphitization eliminates these links (D), and unfilled networks or those filled with inert filler have lower values of the modulus and tensile strength (B and C). There is no reversibility in these curves, and a significant hysteresis loop is observed. The stress softening phenomenon occurring during the second stretching is due to breakage of short chains extending between two filler aggregates (the 'Mullins effect')[310] and to slippage of chain segments along the filler surface.[311] The tensile strength of a carbon black filled vulcanizate is high, since strain energy may be dissipated by additional relaxation mechanisms introduced by such fillers.[300, 304, 312] As in the case of unfilled vulcanizates, it is useful to plot the reduced nominal stress *vs.* the reciprocal elongation (equation 104 and Figures 14–16). The curves still show a linear relation over a range of low and moderate extensions, but only for low concentrations of filler.[301] The increase in modulus due to the incorporation of filler results from the consequent distortion of

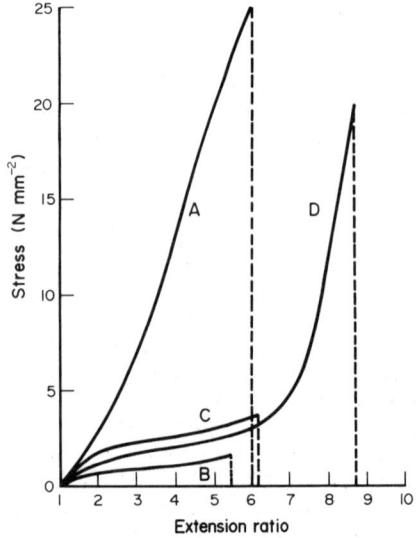

Figure 21 Schematic stress–strain curves in uniaxial extension; A, rubber filled with highly reinforcing carbon black; B, unfilled rubber; C, rubber filled with coarse filler of low activity; D, rubber filled with graphitized carbon black (reproduced by permission of Huthig und Wepf from G. Kraus, *Angew. Makromol. Chem.*, 1977, **60/61**, 215)

the stress pattern in the rubber matrix. The matrix is prevented from deforming uniformly by adhesion of the rubber to the surfaces of the particles, and thus the overall apparent strain is less than the strains occurring locally in the elastomeric matrix. The strain is increased by a factor X which takes into account both this disturbance of the strain distribution and the absence of deformation in the fraction of the material composed of filler. It appears reasonable to consider that the local strains are, on average, X times greater than the overall strain, X being thus the strain amplification factor. For several deformations and for spherical particles, this factor is the ratio of modulus E to E_0 occurring in equation (176). When the reduced curves for low concentrations of filler are now calculated with the amplified strain, they become closer to the curve corresponding to the unfilled vulcanizate.

9.12 REFERENCES

1. L. R. G. Treloar, 'The Physics of Rubber Elasticity', 3rd edn., Clarendon Press, Oxford, 1975.
2. P. J. Flory, 'Principles of Polymer Chemistry', Cornell University Press, Ithaca, NY, 1953.
3. A. V. Tobolsky, 'Properties and Structure of Polymers', Wiley, New York, 1960.
4. R. Houwink, 'Elasticity, Plasticity, and Structure of Matter', Cambridge University Press, London, 1937.
5. F. Bueche, 'Physical Properties of Polymers', Interscience, New York, 1962.
6. J. W. S. Hearle, 'Polymers and Their Properties', Wiley, New York, 1982, vol. 1.
7. J. J. Aklonis and W. J. MacKnight, 'Introduction to Polymer Viscoelasticity', 2nd edn., Wiley, New York, 1983.
8. P. J. Flory, 'Statistical Mechanics of Chain Molecules', Wiley–Interscience, New York, 1969.
9. P. J. Flory, *Proc. R. Soc. London, Ser. A*, 1976, **351**, 351.
10. P. J. Flory, *Macromolecules*, 1982, **15**, 99.
11. P. J. Flory, *Br. Polym. J.*, 1985, **17**, 96.
12. A. Charlesby, *Proc. R. Soc. London, Ser. A*, 1954, **222**, 542.
13. A. Charlesby and S. H. Pinner, *Proc. R. Soc. London, Ser. A*, 1959, **249**, 367.
14. C. W. Macosko and D. R. Miller, *Macromolecules*, 1976, **9**, 199.
15. J. P. Queslel and J. E. Mark, *J. Chem. Phys.*, 1985, **82**, 3449.
16. N. R. Langley, *Macromolecules*, 1968, **1**, 348.
17. J. P. Queslel and J. E. Mark, *Adv. Polym. Sci.*, 1985, **71**, 229.
18. P. J. Flory, *Macromolecules*, 1979, **12**, 119.
19. B. Erman and B. M. Baysal, *Macromolecules*, 1985, **18**, 1696.
20. M. Gottlieb and R. J. Gaylord, *Macromolecules*, 1984, **17**, 2024.
21. R. W. Brotzman and B. E. Eichinger, *Macromolecules*, 1983, **16**, 1131.
22. A. Y. Coran, in 'Science and Technology of Rubber', ed. F. R. Eirich, Academic Press, New York, 1978.
23. B. Ellis and G. N. Welding, *Rubber Chem. Technol.*, 1964, **37**, 563.
24. V. A. Shershnev, *Rubber Chem. Technol.*, 1982, **55**, 537.
25. M. L. Studebaker, *Rubber Chem. Technol.*, 1966, **39**, 1359.
26. L. D. Loan, *Rubber Chem. Technol.*, 1967, **40**, 149.
27. J. Gough, *Mem. Proc.—Manchester Lit. Philos. Soc.*, 2nd Ser., 1805, **1**, 288.
28. J. P. Joule, *Philos. Mag.*, 1857, **14**, 227.
29. J. P. Joule, *Philos. Trans. R. Soc. London, Ser. A*, 1859, **149**, 91.
30. Lord Kelvin, *Q. J. Math.*, 1857, **1**, 57.
31. J. E. Mark, *J. Chem. Educ.*, 1981, **58**, 898.
32. B. E. Eichinger, *Ann. Rev. Phys. Chem.*, 1983, **34**, 359.
33. K. H. Meyer, G. Von Susich and E. Valkó, *Kolloid Z.*, 1932, **59**, 208.
34. A. R. Payne and J. R. Scott, 'Engineering Design with Rubber', MacLaren, London, 1960.
35. E. Guth and H. Mark, *Monatsh. Chem.*, 1934, **65**, 93.
36. W. Kuhn, *Kolloid Z.*, 1934, **68**, 2.
37. M. V. Volkenstein, 'Configurational Statistics of Polymer Chains', translation from the Russian edition, 1959, by S. N. and M. J. Timasheff, Interscience, New York, 1963.
38. P. J. Flory, *Pure Appl. Chem.*, 1980, **52**, 241.
39. Y. Abe and P. J. Flory, *Macromolecules*, 1971, **4**, 219.
40. G. Natta and P. Corradini, *Angew. Makromol. Chem.*, 1956, **68**, 615.
41. P. Corradini, *J. Polym. Sci., Part B*, 1969, **7**, 211.
42. J. E. Mark, *J. Am. Chem. Soc.*, 1966, **88**, 4354.
43. J. E. Mark, *J. Am. Chem. Soc.*, 1967, **89**, 6829.
44. P. J. Flory, *Macromolecules*, 1974, **7**, 381.
45. T. M. Birshtein and O. B. Ptitsyn, 'Conformations of Macromolecules', translation from the Russian edition, 1964, by S. N. and M. J. Timasheff, Interscience, New York, 1966.
46. K. Nagai, *J. Chem. Phys.*, 1963, **38**, 924.
47. K. Nagai, *J. Chem. Phys.*, 1964, **40**, 2818.
48. P. J. Flory, *Proc. Natl. Acad. Sci. USA*, 1964, **51**, 1060.
49. P. J. Flory, *Proc. Natl. Acad. Sci. USA*, 1973, **70**, 1819.
50. P. J. Flory, J. E. Mark and A. Abe, *J. Am. Chem. Soc.*, 1966, **88**, 639.
51. Y. Fujiwara and P. J. Flory, *Macromolecules*, 1969, **2**, 315.
52. Y. Fujiwara and P. J. Flory, *Macromolecules*, 1970, **3**, 280.
53. A. Abe and P. J. Flory, *Macromolecules*, 1971, **4**, 230.
54. P. J. Flory, *J. Chem. Phys.*, 1972, **56**, 862.

55. P. J. Flory and Y. Abe, *J. Chem. Phys.*, 1971, **54**, 1351.
56. J. E. Mark and J. G. Curro, *J. Chem. Phys.*, 1983, **79**, 5705.
57. J. G. Curro and J. E. Mark, *J. Chem. Phys.*, 1984, **80**, 4521.
58. Lord Rayleigh, *Philos. Mag.*, 1919, **37**, 321.
59. H. Eyring, *Phys. Rev.*, 1932, **39**, 746.
60. F. T. Wall, *J. Chem. Phys.*, 1943, **11**, 67.
61. P. Debye, *J. Chem. Phys.*, 1946, **14**, 636.
62. B. H. Zimm and W. H. Stockmayer, *J. Chem. Phys.*, 1949, **17**, 1301.
63. S. Chandrasekhar, *Rev. Mod. Phys.*, 1943, **15**, 1.
64. W. Kuhn and F. Grun, *Kolloid Z.*, 1942, **101**, 248.
65. Yu. Ya. Gotlib, *Vysokomol. Soedin.*, 1964, **6**, 389.
66. M. C. Wang and E. Guth, *J. Chem. Phys.*, 1952, **20**, 1144.
67. R. A. Jacobson, *J. Am. Chem. Soc.*, 1932, **54**, 1513.
68. W. H. Hunter and G. H. Woollett, *J. Am. Chem. Soc.*, 1921, **43**, 135.
69. N. Hadjichristidis, X. Zhongde and L. J. Fetters, *J. Polym. Sci., Polym. Phys. Ed.*, 1982, **20**, 743.
70. F. J. Ansorena, L. M. Revuelta, G. M. Guzman and J. J. Iruin, *Eur. Polym. J.*, 1982, **18**, 19.
71. K. Kajiwara and W. Burchard, *Macromolecules*, 1984, **17**, 2669.
72. J. Mays, N. Hadjichristidis and L. J. Fetters, *Macromolecules*, 1984, **17**, 2723.
73. J. T. Gotro and W. W. Graessley, *Macromolecules*, 1984, **17**, 2767.
74. F. A. Bovey, 'Chain Structure and Conformation of Macromolecules', Academic Press, New York, 1982.
75. W. Kuhn, *Kolloid Z.*, 1936, **76**, 258.
76. W. Kuhn, *Kolloid Z.*, 1939, **87**, 3.
77. P. Smith, R. R. Matheson and P. A. Irvine, *Polym. Commun.*, 1984, **25**, 294.
78. A. E. Tonelli, 'Polymer Conformation and Configuration', in 'Encyclopedia of Polymer Science and Technology', ed. N. Bikales, 2nd edn., Wiley, New York, 1977.
79. P. J. Flory, *Pure Appl. Chem.*, 1984, **56**, 305.
80. J. S. Higgins and R. S. Stein, *J. Appl. Crystallogr.*, 1978, **11**, 346.
81. A. Maconnachie and R. W. Richards, *Polymer*, 1978, **19**, 739.
82. J. A. Hinkley, C. C. Han, B. Mozer and H. Yu, *Macromolecules*, 1978, **11**, 836.
83. S. Candau, J. Bastide and M. Delsanti, *Adv. Polym. Sci.*, 1982, **44**, 27.
84. M. Beltzung, J. Herz and C. Picot, *Macromolecules*, 1983, **16**, 580.
85. R. Ullman, *J. Polym. Sci., Polym. Lett. Ed.*, 1983, **21**, 521.
86. M. Beltzung, C. Picot, P. Rempp and J. Herz, *Macromolecules*, 1982, **15**, 1594.
87. L. H. Sperling, *Polym. Eng. Sci.*, 1984, **24**, 1.
88. J. E. Mark, *J. Polym. Sci., Macromol. Rev.*, 1976, **11**, 135.
89. G. Allen, M. J. Kirkham, J. Padget and C. Price, *Trans. Faraday Soc.*, 1971, **67**, 1278.
90. J. E. Mark, *Rubber Chem. Technol.*, 1973, **46**, 593.
91. P. J. Flory, *Pure Appl. Chem.*, 1972, **8**, 1.
92. P. J. Flory, A. Ciferri and R. Chiang, *J. Am. Chem. Soc.*, 1961, **83**, 1023.
93. J. E. Mark and P. J. Flory, *J. Am. Chem. Soc.*, 1964, **86**, 138.
94. R. Chiang, *J. Phys. Chem.*, 1966, **70**, 2348.
95. J. Bastide, C. Picot and S. Candau, *J. Polym. Sci., Polym. Phys. Ed.*, 1979, **17**, 1441.
96. C. A. J. Hoeve and P. J. Flory, *Biopolymers*, 1974, **13**, 677.
97. J. E. Mark, *Biopolymers*, 1976, **15**, 1853.
98. G. Gee, *Trans. Faraday Soc.*, 1946, **42**, 585.
99. B. Erman, *Br. Polym. J.*, 1985, **17**, 140.
100. I. J. Sjothun and G. Alliger, 'History and Technology of Elastomer Vulcanization', in 'Vulcanization of Elastomers', ed. G. Alliger and I. J. Sjothun, Reinhold, London, 1964.
101. I. Ostromeslenski, *Zh. Russ. Fiz.-Khim. O-va., Chast Khim.*, 1915, **47**, 1467.
102. L. O. Amberg and W. D. Willis, *Rubber Age (N.Y.)*, 1955, **77**, 83.
103. C. G. Moore and W. F. Watson, *J. Polym. Sci.*, 1956, **19**, 237.
104. B. M. E. Van der Hoff, *Rubber Chem. Technol.*, 1965, **38**, 560.
105. K. Hummel and G. Kaiser, *Rubber Chem. Technol.*, 1965, **38**, 581.
106. W. Scheele and E. Rohde, *Rubber Chem. Technol.*, 1966, **39**, 768.
107. L. D. Loan, *J. Appl. Polym. Sci.*, 1963, **7**, 2259.
108. E. M. Barral, R. Hawkins, A. A. Fukushima and J. F. Johnson, *J. Polym. Sci., Polym. Symp.*, 1984, **71**, 189.
109. P. J. Flory, *Chem. Rev.*, 1944, **35**, 51.
110. J. Scanlan, *J. Polym. Sci.*, 1960, **43**, 501.
111. K. Dusek and W. Prins, *Adv. Polym. Sci.*, 1969, **6**, 1.
112. D. S. Pearson and W.W. Graessley, *Macromolecules*, 1978, **11**, 528.
113. G. Friedmann and J. Brossas, *Eur. Polym. J.*, 1984, **20**, 1151.
114. J. E. Mark, *Rubber Chem. Technol.*, 1981, **54**, 809.
115. J. E. Mark, *Adv. Polym. Sci.*, 1982, **44**, 1.
116. J. P. Queslel and J. E. Mark, *Adv. Polym. Sci.*, 1984, **65**, 135.
117. K. Dusek, *Adv. Polym. Sci.*, 1982, **44**, 164.
118. K. Dusek, *Rubber Chem. Technol.*, 1982, **55**, 1.
119. P. Rempp and J. E. Herz, *Angew. Makromol. Chem.*, 1979, **76/77**, 373.
120. J. P. Queslel and J. E. Mark, *J. Polym. Sci., Polym. Phys. Ed.*, 1984, **22**, 1201.
121. M. Gottlieb, C. W. Macosko, G. S. Benjamin, K. O. Meyers and E. W. Merrill, *Macromolecules*, 1981, **14**, 1039.
122. D. S. Pearson and W. W. Graessley, *Macromolecules*, 1980, **13**, 1001.
123. A. L. Andrady, M. A. Llorente, M. A. Sharaf, R. R. Rahalkar, J. E. Mark, J. L. Sullivan, C. U. Yu and J. R. Falender, *J. Appl. Polym. Sci.*, 1981, **26**, 1829.
124. P. J. Flory, *Polym. J.*, 1985, **17**, 1.

125. L. C. Case, *J. Polym. Sci.*, 1960, **45**, 397.
126. J. P. Queslel and J. E. Mark, *Polym. Bull. (Berlin)*, 1984, **12**, 311.
127. W. W. Graessley, *Macromolecules*, 1975, **8**, 186.
128. Y.-K. Leung and B. E. Eichinger, *J. Chem. Phys.*, 1984, **80**, 3877, 3885.
129. M. Hoffman, *Makromol. Chem.*, 1982, **183**, 2191, 2213, 2237.
130. J. P. Queslel and J. E. Mark, *Polym. J.*, 1986, **18**, 263.
131. L. E. Nielsen, 'Mechanical Properties of Polymers', Reinhold, New York, 1962.
132. R. J. Moseley, R. A. Amos and J. R. Scott, 'Physico-Mechanical Testing of Unvulcanized and Vulcanized Rubbers', MacLaren, London, 1962.
133. J. R. Scott, 'Physical Testing of Rubbers', MacLaren, London, 1965.
134. R. P. Brown, 'Physical Testing of Rubbers', Applied Science, London, 1979.
135. ASTM Standards, 'Rubber, Electrical Insulation', Part 11, American Society for Testing and Materials, Philadelphia, 1961.
136. ASTM Standards, 'Plastics — Methods of Testing', Part 27, American Society for Testing and Materials, Philadelphia, 1961.
137. R. P. Brown (ed.), 'RAPRA Guide to Rubbers and Plastics Test Equipment', Rubber and Plastics Research Association, London, 1973.
138. R. P. Brown, *Elastomerics*, 1983, **115**, 46.
139. J. E. Mark, *Rubber Chem. Technol.*, 1975, **48**, 495.
140. J. E. Mark, *Makromol. Chem., Suppl.*, 1979, **2**, 87.
141. J. E. Mark, *Rubber Chem. Technol.*, 1982, **55**, 762.
142. L. R. G. Treloar, *Trans. Faraday Soc.*, 1944, **40**, 59.
143. T. L. Smith, *J. Polym. Sci., Part C*, 1967, **16**, 841.
144. M. Gottlieb and R. J. Gaylord, *Polymer*, 1983, **24**, 1644.
145. L. R. G. Treloar and G. Riding, *Proc. R. Soc. London, Ser. A*, 1979, **369**, 261.
146. S. Kawabata, M. Matsuda, K. Tei and H. Kawai, *Macromolecules*, 1981, **14**, 154.
147. M. Matsuda, S. Kawabata and H. Kawai, *Macromolecules*, 1981, **14**, 1688.
148. M. Matsuda, S. Kawabata and H. Kawai, *Macromolecules*, 1982, **15**, 160.
149. B. Erman, *J. Polym. Sci., Polym. Phys. Ed.*, 1981, **19**, 829.
150. L. R. G. Treloar, *Br. Polym. J.*, 1982, **14**, 121.
151. O. H. Yeoh, *Plastics and Rubber Processing and Applications*, 1984, **4**, 141.
152. E. Galli, *Plast. Compd.*, 1982, **5**, 14.
153. G. B. McKenna and L. J. Zapas, *Polymer*, 1983, **24**, 1502.
154. J. R. Scott, *Trans. Inst. Rubber Ind.*, 1935, **11**, 224.
155. J. R. Scott, *J. Rubber Res. Inst. Malays.*, 1948, **17**, 145.
156. A. N. Gent, *Trans. Inst. Rubber Ind.*, 1958, **34**, 46.
157. P. B. Lindley, 'Engineering Design with Natural Rubber', Technical Bulletin, Malaysian Rubber Producers' Association, UK, 1979.
158. A. N. Gent and P. B. Lindley, *Proc.-Inst. Mech. Eng.*, 1959, **173**, 111.
159. J. E. Mark and J. L. Sullivan, *J. Chem. Phys.*, 1977, **66**, 1006.
160. M. A. Llorente and J. E. Mark, *J. Chem. Phys.*, 1979, **71**, 682.
161. M. A. Sharaf, J. E. Mark, B. Gunesin, M. Julemont and P. Teyssie, *Polym. Eng. Sci.*, 1986, **26**, 162.
162. B. Erman, W. Wagner and P. J. Flory, *Macromolecules*, 1980, **13**, 1554.
163. M. J. Forster, *J. Appl. Phys.*, 1955, **26**, 1104.
164. F. P. Wolf, *Polymer*, 1972, **13**, 347.
165. R. Y. S. Chen, C. U. Yu and J. E. Mark, *Macromolecules*, 1973, **6**, 746.
166. H. Pak and P. J. Flory, *J. Polym. Sci., Polym. Phys. Ed.*, 1979, **17**, 1845.
167. P. Meares, 'Polymers: Structure and Bulk Properties', Van Nostrand, New York, 1965.
168. I. S. Sokolnikoff, 'Mathematical Theory of Elasticity', McGraw-Hill, New York, 1956.
169. P. J. Flory, A. Ciferri and C. A. J. Hoeve, *J. Polym. Sci.*, 1960, **45**, 235.
170. A. Ciferri, C. A. J. Hoeve and P. J. Flory, *J. Am. Chem. Soc.*, 1961, **83**, 1015.
171. G. Allen, U. Bianchi and C. Price, *Trans. Faraday Soc.*, 1963, **59**, 2493.
172. M. C. Shen, D. A. McQuarrie and J. L. Jackson, *J. Appl. Phys.*, 1967, **38**, 791.
173. G. Allen, G. Gee, M. C. Kirkham, C. Price and J. C. Padget, *J. Polym. Sci., Part C*, 1968, **23**, 201.
174. C. Price, J. C. Padget, M. C. Kirkham and G. Allen, *Polymer*, 1969, **10**, 573.
175. A. Ciferri, *J. Polym. Sci.*, 1961, **54**, 149.
176. P. J. Flory, C. A. J. Hoeve and A. Ciferri, *J. Polym. Sci.*, 1959, **34**, 337.
177. P. J. Flory, *Pure Appl. Chem., Macromolecular Chem.-8*, 1973, **33**, 1.
178. C. A. J. Hoeve and P. J. Flory, *J. Polym. Sci.*, 1962, **60**, 155.
179. L. R. G. Treloar, *Polymer*, 1969, **10**, 291.
180. T. Y. Chen, P. Ricica and M. Shen, *J. Macromol. Sci., Chem.*, 1973, **7**, 889.
181. C. U. Yu and J. E. Mark, *Polymer*, 1975, **16**, 326.
182. R. H. Becker, C. U. Yu and J. E. Mark, *Polym. J.*, 1975, **7**, 234.
183. R. L. Anthony, R. H. Caston and E. Guth, *J. Phys. Chem.*, 1942, **46**, 826.
184. J. A. Barrie and J. Standen, *Polymer*, 1967, **8**, 97.
185. Y. K. Godovsky, *Polymer*, 1981, **22**, 75.
186. Y. K. Godovsky, *Polym. Sci. USSR (Engl. Transl.)*, 1977, **19**, 2709.
187. H.-G. Kilian, G. W. H. Höhne, P. Trögele and H. Ambacher, *J. Polym. Sci., Polym. Symp.*, 1984, **71**, 221.
188. R. G. Christensen and C. A. J. Hoeve, *J. Polym. Sci., Part A-1*, 1970, **8**, 1503.
189. C. Price and G. Allen, *Polymer*, 1973, **14**, 576.
190. D. J. Williams, 'Polymer Science and Engineering', Prentice-Hall, Englewood Cliffs, NJ, 1971.
191. A. N. Gent and M. Shen, in 'Science and Technology of Rubber', ed. F. R. Eirich, Academic Press, New York, 1978, pp. 1 and 155 respectively.

192. W. R. Krigbaum and R.-J. Roe, *Rubber Chem. Technol.*, 1965, **38**, 1039.
193. K. J. Smith, Jr. and R. J. Gaylord, *Rubber Age (N.Y.)*, 1975, **107**(6), 31.
194. W. W. Graessley, *Adv. Polym. Sci.*, 1974, **16**, 1.
195. R. G. C. Arridge and P. J. Barham, *Adv. Polym. Sci.*, 1982, **46**, 67.
196. L. R. G. Treloar, *Rep. Prog. Phys.*, 1973, **36**, 755.
197. J. E. Mark, *J. Educ. Mod. Mater. Sci. Eng.*, 1982, **4**, 733.
198. D. Y. Yoon and P. J. Flory, *J. Chem. Phys.*, 1974, **61**, 5366.
199. P. J. Flory and V. W. C. Chang, *Macromolecules*, 1976, **9**, 33.
200. E. Guth and H. M. James, *Ind. Eng. Chem., Ind. Ed.*, 1941, **33**, 624.
201. H. M. James and E. Guth, *J. Chem. Phys.*, 1943, **11**, 455.
202. F. T. Wall, *J. Chem. Phys.*, 1943, **11**, 527.
203. L. R. G. Treloar, *Trans. Faraday Soc.*, 1943, **39**, 241.
204. P. J. Flory and J. Rehner, Jr., *J. Chem. Phys.*, 1943, **11**, 512.
205. W. Kuhn, *J. Polym. Sci.*, 1946, **1**, 380.
206. R. M. Johnson and J. E. Mark, *Macromolecules*, 1972, **5**, 41.
207. H. M. James, *J. Chem. Phys.*, 1947, **15**, 651.
208. H. M. James and E. Guth, *J. Chem. Phys.*, 1947, **15**, 669.
209. P. J. Flory, *Rubber Chem. Technol.*, 1975, **48**, 513.
210. W. W. Graessley, *Macromolecules*, 1975, **8**, 865.
211. M. Mooney, *J. Appl. Phys.*, 1940, **11**, 582.
212. G. Ronca and G. Allegra, *J. Chem. Phys.*, 1975, **63**, 4990.
213. P. J. Flory, *Polymer*, 1979, **20**, 1317.
214. S. M. Gumbrell, L. Mullins and R. S. Rivlin, *Trans. Faraday Soc.*, 1953, **49**, 1495.
215. J. E. Mark, *Polym. Eng. Sci.*, 1979, **19**, 254, 409.
216. A. L. Andrady, M. A. Llorente and J. E. Mark, *J. Chem. Phys.*, 1980, **73**, 1439.
217. L. R. G. Treloar, *Trans. Faraday Soc.*, 1954, **50**, 881.
218. N. R. Langley and K. E. Polmanteer, *J. Polym. Sci., Polym. Phys. Ed.*, 1974, **12**, 1023.
219. P. J. Flory, *J. Chem. Phys.*, 1977, **66**, 5720.
220. W. W. Graessley and D. S. Pearson, *J. Chem. Phys.*, 1977, **66**, 3363.
221. W. W. Graessley, *Adv. Polym. Sci.*, 1982, **47**, 67.
222. R. J. Gaylord, B. Joss, J. T. Bendler and E. A. DiMarzio, *Br. Polym. J.*, 1985, **17**, 126.
223. S. F. Edwards, *Br. Polym. J.*, 1985, **17**, 122.
224. R. Ullman, *J. Polym. Sci., Polym. Symp.*, 1985, **72**, 39.
225. B. Erman and P. J. Flory, *J. Chem. Phys.*, 1978, **68**, 5363.
226. P. J. Flory and B. Erman, *Macromolecules*, 1982, **15**, 800.
227. P. J. Flory, 'Contemporary Topics in Polymer Science', ed. E. M. Pearce and J. R. Schaefgen, Plenum Press, New York, 1977, vol. 2, p. 1.
228. B. Erman and P. J. Flory, *Macromolecules*, 1982, **15**, 806.
229. R. W. Brotzman and J. E. Mark, *Macromolecules*, 1986, **19**, 667.
230. B. Erman and P. J. Flory, *Macromolecules*, 1983, **16**, 1607.
231. B. Erman, *J. Polym. Sci., Polym. Phys. Ed.*, 1983, **21**, 893.
232. J. P. Queslel and J. E. Mark, *Polym. Bull. (Berlin)*, 1983, **10**, 119.
233. J. P. Queslel and J. E. Mark, *Eur. Polym. J.*, 1986, **22**, 273.
234. J. P. Queslel, P. Thirion and L. Monnerie, *Polymer*, 1986, **27**, 1869.
235. E. Helfand and D. S. Pearson, *J. Chem. Phys.*, 1983, **79**, 2054.
236. J. G. Curro and P. Pincus, *Macromolecules*, 1983, **16**, 559.
237. B. Erman and L. Monnerie, *Polym. Commun.*, 1985, **26**, 167.
238. B. Erman and L. Monnerie, *Macromolecules*, 1985, **18**, 1985.
239. J. P. Queslel, B. Erman and L. Monnerie, *Macromolecules*, 1985, **18**, 1991.
240. B. Erman and P. J. Flory, *Macromolecules*, 1983, **16**, 1601.
241. P. G. De Gennes, *J. Phys., Lett. (Orsay, Fr.)*, 1974, **35**, L-133.
242. S. F. Edwards, *Proc. Phys. Soc., London*, 1967, **92**, 9.
243. S. F. Edwards, *Br. Polym. J.*, 1977, **9**, 140.
244. G. Marrucci, *Macromolecules*, 1981, **14**, 434.
245. R. J. Gaylord, *Polym. Bull. (Berlin)*, 1982, **8**, 325.
246. G. Marrucci, *Rheol. Acta.*, 1979, **18**, 193.
247. R. C. Ball, M. Doi, S. F. Edwards and M. Warner, *Polymer*, 1981, **22**, 1010.
248. P. Thirion and T. Weil, *Polymer*, 1984, **25**, 609.
249. J. P. Queslel, *Rubber Chem. Technol.*, 1984, **57**, 145.
250. K. O. Meyers, M. L. Bye and E. W. Merrill, *Macromolecules*, 1980, **13**, 1045.
251. K. A. Kirk, S. A. Bidstrup, E. W. Merrill and K. O. Meyers, *Macromolecules*, 1982, **15**, 1123.
252. L. M. Dossin and W. W. Graessley, *Macromolecules*, 1979, **12**, 123.
253. E. M. Vallés and C. W. Macosko, *Macromolecules*, 1979, **12**, 673.
254. C. W. Macosko and G. S. Benjamin, *Pure Appl. Chem.*, 1981, **53**, 1505.
255. M. Gottlieb, C. W. Macosko and T. C. Lepsch, *J. Polym. Sci., Polym. Phys. Ed.*, 1981, **19**, 1603.
256. E. M. Vallés, E. J. Rost and C. W. Macosko, *Rubber Chem. Technol.*, 1984, **57**, 55.
257. J. D. Ferry, 'Viscoelastic Properties of Polymers', 2nd edn., Wiley, New York, 1970.
258. N. R. Langley and J. D. Ferry, *Macromolecules*, 1968, **1**, 353.
259. S. Granick, S. Pedersen, G. W. Nelb, J. D. Ferry and C. W. Macosko, *J. Polym. Sci., Polym. Phys. Ed.*, 1981, **19**, 1745.
260. D. R. Miller and C. W. Macosko, *Macromolecules*, 1976, **9**, 206.
261. D. R. Miller, E. M. Vallés and C. W. Macosko, *Polym. Eng. Sci.*, 1979, **19**, 272.
262. M. A. Bibbo and E. M. Vallés, *Macromolecules*, 1982, **15**, 1293.
263. M. Ilavsky and K. Dusek, *Polymer*, 1983, **24**, 981.

264. K. Dusek, *Macromolecules*, 1984, **17**, 716.
265. E. M. Vallés and C. W. Macosko, *Rubber Chem. Technol.*, 1976, **49**, 1232.
266. R. S. Rivlin, *Philos. Trans. R. Soc. London, Ser. A*, 1948, **241**, 379.
267. P. J. Blatz and N. W. Tschoegl, *Trans. Soc. Rheol.*, 1974, **18**, 145.
268. A. Ziabicki and J. Walasek, *Br. Polym. J.*, 1985, **17**, 116.
269. N. W. Tschoegl and C. Gurer, *Macromolecules*, 1985, **18**, 680, 687.
270. L. Mullins, *J. Appl. Polym. Sci.*, 1959, **2**, 257.
271. C. Booth, G. Gee, G. Holden and G. R. Williamson, *Polymer*, 1964, **5**, 343.
272. B. M. E. Van Der Hoff, *Polymer*, 1965, **6**, 397.
273. P. J. Flory and Y.-I. Tatara, *J. Polym. Sci., Polym. Phys. Ed.*, 1975, **13**, 683.
274. J. E. Mark, *J. Polym. Sci., Part C*, 1970, **31**, 97.
275. P. J. Flory, *J. Chem. Phys.*, 1942, **10**, 51.
276. M. L. Huggins, *J. Chem. Phys.*, 1941, **9**, 440.
277. P. J. Flory and J. Rehner, Jr., *J. Chem. Phys.*, 1943, **11**, 521.
278. R. W. Brotzman and B. E. Eichinger, *Macromolecules*, 1982, **15**, 531.
279. B. E. Eichinger and P. J. Flory, *Trans. Faraday Soc.*, 1968, **64**, 2035, 2053, 2061, 2066.
280. R. A. Orwoll, *Rubber Chem. Technol.*, 1977, **50**, 451.
281. L. Y. Yen and B. E. Eichinger, *J. Polym. Sci., Polym. Phys Ed.*, 1978, **16**, 117, 121.
282. G. Gee, J. B. M. Herbert and R. C. Roberts, *Polymer*, 1965, **6**, 541.
283. G. Allen, P. Egerton and D. J. Walsh, *Eur. Polym. J.*, 1979, **15**, 983.
284. R. W. Brotzman and B. E. Eichinger, *Macromolecules*, 1981, **14**, 1445.
285. G. Gee, *Trans. Faraday Soc.*, 1946, **42**, 33.
286. T. L. Smith, *Polym. Eng. Sci.*, 1977, **17**, 129.
287. T. L. Smith, *J. Appl. Phys.*, 1964, **35**, 27.
288. F. Bueche and J. C. Halpin, *J. Appl. Phys.*, 1964, **35**, 36.
289. J. C. Halpin, *J. Appl. Phys.*, 1964, **35**, 3133.
290. J. C. Halpin and F. Bueche, *J. Appl. Phys.*, 1964, **35**, 3142.
291. J. C. Halpin, *J. Polym. Sci., Part C*, 1967, **16**, 1037.
292. L. Mandelkern, 'Crystallization of Polymers', McGraw-Hill, New York, 1964.
293. A. N. Gent, *J. Polym. Sci., Part A-2*, 1966, **4**, 447.
294. R. S. Stein, *Polym. Eng. Sci.*, 1976, **16**, 152.
295. G. Allegra, *Makromol. Chem.*, 1980, **181**, 1127.
296. F. De Candia, C. Romano, R. Russo and V. Vittoria, *J. Polym. Sci., Polym. Phys. Ed.*, 1982, **20**, 1525.
297. K. J. Smith, Jr., *J. Polym. Sci., Polym. Phys. Ed.*, 1983, **21**, 45, 55.
298. G. Allegra and M. Bruzzone, *Macromolecules*, 1983, **16**, 1167.
299. G. Kraus, *Rubber Chem. Technol.*, 1965, **38**, 1070.
300. G. Kraus, *Angew. Makromol. Chem.*, 1977, **60/61**, 215.
301. L. Mullins and N. R. Tobin, *J. Appl. Polym. Sci.*, 1965, **9**, 2993.
302. B. B. Boonstra, *Polymer*, 1979, **20**, 291.
303. Z. Rigbi, *Adv. Polym. Sci.*, 1980, **36**, 21.
304. G. Kraus, *J. Appl. Polym. Sci.: Appl. Polym. Symp.*, 1984, **39**, 75.
305. E. Guth and R. Simha, *Kolloid Z.*, 1936, **74**, 266.
306. E. Guth and O. Gold, *Phys. Rev.*, 1938, **53**, 322.
307. E. Guth, *J. Appl. Phys.*, 1945, **16**, 20.
308. A. I. Medalia, *Rubber Chem. Technol.*, 1973, **46**, 877.
309. A. I. Medalia, *J. Colloid Interface Sci.*, 1970, **32**, 115.
310. L. Mullins, *Rubber. Chem. Technol.*, 1969, **42**, 339.
311. A. N. Gent, *J. Appl. Polym. Sci.*, 1974, **18**, 1397.
312. T. S. Chow, *J. Polym. Sci., Polym. Phys. Ed.*, 1982, **20**, 2103.

10

Glass Formation and Glassy Behavior

GREGORY B. McKENNA

National Bureau of Standards, Gaithersburg, MD, USA

10.1 INTRODUCTION

The scientific study of glass-forming substances has been important for well over half a century. The continuing interest in the behavior of these materials was confirmed by a simple computer search of the literature from 1968 to 1987 on 'glasses', which resulted in over 125 000 citations, and 'polymer glasses', which resulted in over 10 000 citations.[1] A further glimpse of the importance of glasses as an area of scientific inquiry can be obtained by browsing through the books, special journal issues and symposium proceedings which have appeared over the past decade and half.[2-10]

In spite of the large body of literature which has accumulated on glasses, there is still much room for increased understanding of the physics of glass formation and the nature of the glassy state. There are underlying thermodynamic questions which are still unanswered, such as: 'Is the glass transition a true thermodynamic transition (second-order transition) or is it a *kinetic* phenomenon which saves the thermodynamic "catastrophe"?'. Such questions will be addressed in the appropriate sections of this chapter. First, however, we present some of the phenomenology of the glass transition itself.

The basic event of the transition from liquid-like to glass-like behavior in glass-forming substances, including polymers, is observed to be a kinetic phenomenon under ordinary experimental conditions.[2-10] For example, as depicted in Figure 1, when isochronal volume measurements are

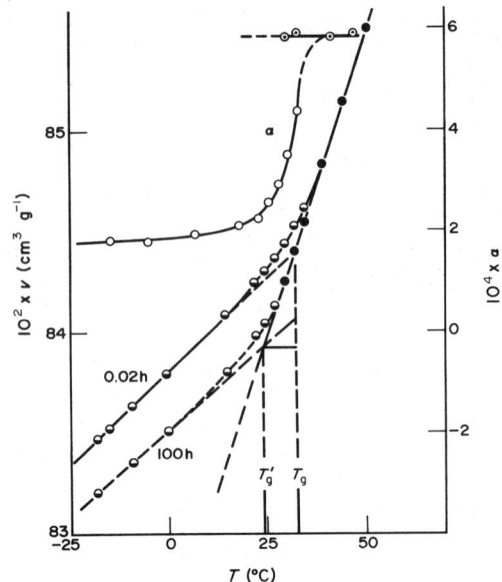

Figure 1 Specific volume v *vs.* temperature T for a poly(vinyl acetate) polymer. Plot shows effect of duration of experiment on glass temperature, *i.e.* $T'_g = T_g$ (100 h) $< T_g$ (0.02 h). Also depiected is α *vs.* T showing the discontinuity in thermal expansion coefficient at T_g (after ref. 11, with permission)

taken at different temperatures (at constant pressure), the volume–temperature line for the material begins to depart from the equilibrium line at a temperature at which the characteristic time τ for the molecular motions leading to volume recovery is longer than the time scale of the experiment. Therefore, as the time scale of measurement increases, the position of the departure from equilibrium moves towards lower temperatures. As shown in Figure 1, the intersection of the extrapolated glassy and liquid lines defines the glass transition temperature T_g or T'_g at the relevant annealing time.

Although the above description is a kinetic one phenomenologically, there is a change in the thermodynamics associated with the change from glassy behavior to liquid behavior. In Figure 2 it is shown how the heat capacity of a polymer glass changes upon going through the glass transition

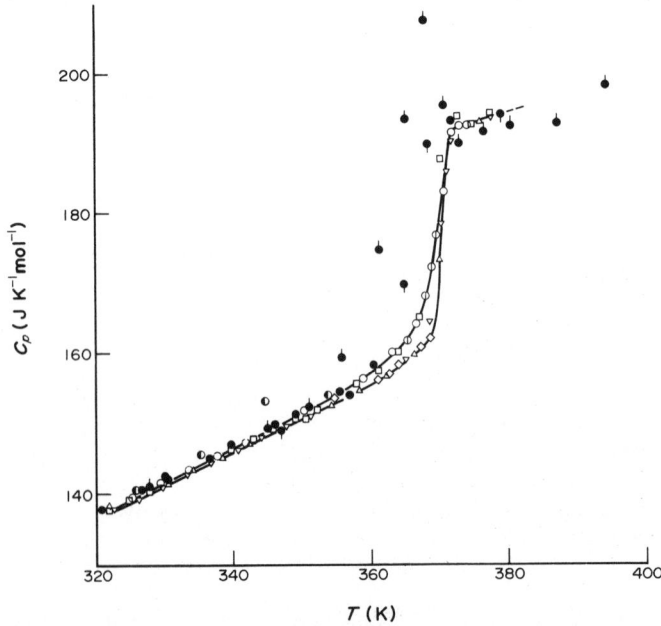

Figure 2 C_p *vs.* T for polystyrene showing the specific heat discontinuity at T_g. The points represent data for different samples and thermal histories (after ref. 12, with permission)

temperature. Interestingly, although T_g is a function of the experimental time-scale, the value of the change of the heat capacity ΔC_p is not.[13] Similar behavior is found for $\Delta \alpha$, the change in the coefficient of thermal expansion in going from the liquid to glassy states (see Figure 1). The quasi-discontinuous behavior of C_p and α at T_g suggests that, although the glass transition itself is a kinetic event under ordinary experimental conditions, there is an underlying thermodynamic transition, as discussed subsequently.

Returning to the kinetics, when a liquid is cooled to below the glass transition temperature and the cooling is arrested at some temperature T, the volume departure from equilibrium $\delta \; [= (v - v_\infty)/v_\infty]$ of the non-equilibrium glass evolves spontaneously towards equilibrium, as depicted in Figure 3 (where v is the specific volume of the glass and v_∞ the value in equilibrium). Associated with these changes in glass 'structure', one observes changes in the mechanical properties of the glass.[14] As shown in Figure 4, the creep compliance of a glassy polymer shifts along the time axis with aging time t_e after a quench from above to below T_g. This process, whereby the mechanical properties change as the volume of the glass recovers, is known as physical aging[14] and it affects both the linear and non-linear viscoelastic behavior as well as the ultimate properties of the glass.[14,15]

The above paragraphs cover what the following sections of this chapter will describe in some detail. We will examine first the thermodynamics of the glass transition, because an understanding of the equilibrium state is essential to any understanding of the non-equilibrium state.[16] This discussion will include both the configurational-entropy model, which is a true thermodynamic model, as well as the free-volume models which allow prediction of quasi-static thermodynamic

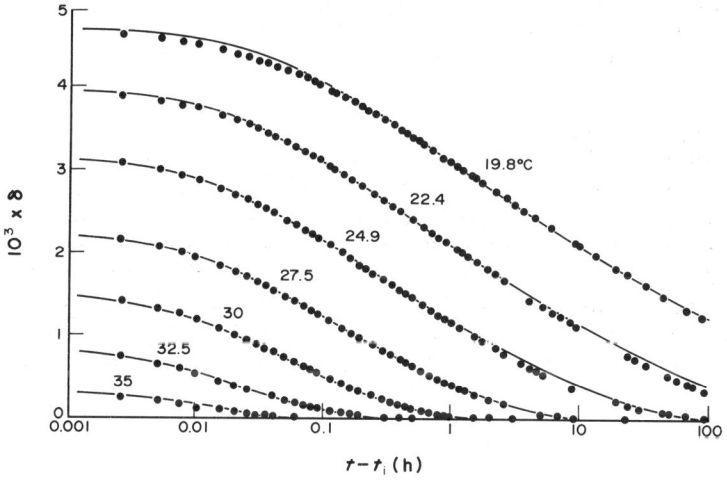

Figure 3 Isothermal contraction of glucose after quenching from $T_0 = 40\,°C$ to different temperatures T, as indicated (after ref. 13, with permission)

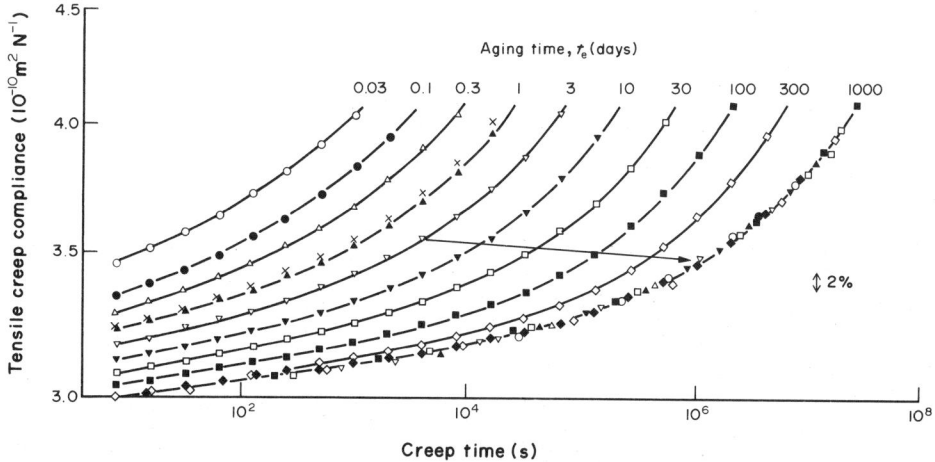

Figure 4 Creep compliance of poly(vinyl chloride) quenched from $90\,°C$ ($T_g + 10\,°C$) to $20\,°C$ at various aging times after the quench (after ref. 14, with permission)

properties. The discussion of free volume will include a description of the models used to describe the equilibrium behavior of a liquid near to the glass transition. We will then examine the kinetics of glasses. The discussion of the kinetics will emphasize the phenomenological equations[17-21] developed recently, which have been highly successful in describing the kinetics of the glass in the non-equilibrium state. Finally, we will discuss briefly the phenomenon of physical aging,[14] which has proved to be of much interest recently and which has important technological implications.

10.2 THERMODYNAMICS OF THE GLASS TRANSITION

10.2.1 The Kauzmann Paradox

Much of what has been written about the thermodynamics of glasses revolves around the question of how to resolve the Kauzmann[22] paradox. Kauzmann[22] examined the thermodynamic behavior of supercooled glass-forming liquids by extrapolating the equilibrium properties, *e.g.* volume (*cf.* Figure 1), enthalpy, entropy, *etc.*, to low temperatures. He found that 'a very startling result' was obtained. 'Not very far below the glass-formation temperature, but still far above 0 K, the extrapolated entropy (or other property) of the liquid becomes less than that of the crystalline solid.' Figure 5 shows the entropy differences between the supercooled liquid and crystalline phases for several low molecular weight glass-formers after Kauzmann's original work.[22] (It is interesting to note that for boron trioxide (B_2O_3) the entropy difference does not extrapolate to negative values.)

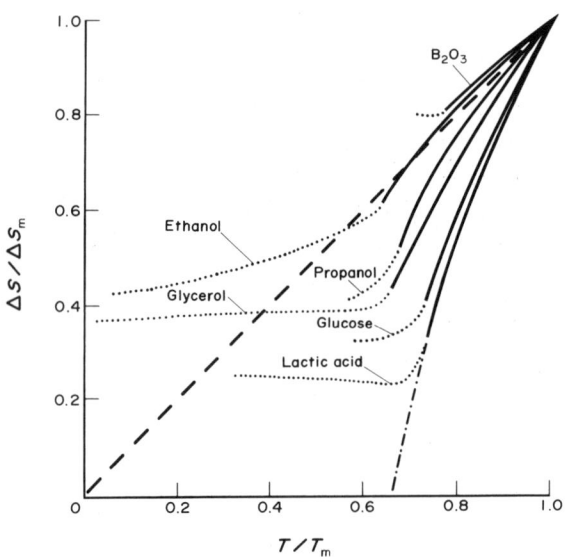

Figure 5 Differences in entropy between supercooled liquid and crystalline phases for different glass-forming materials, as indicated. $\Delta S/\Delta S_m$ = difference in entropy expressed as fraction of entropy of fusion; T/T_m = temperature expressed as fraction of melting temperature (after ref. 22)

Kauzmann argued that the appropriate comparison to be made is between the properties of a thermodynamically stable crystalline phase and a metastable glassy phase. His arguments imply that there is no 'thermodynamic' glass transition. He extended a liquid model by Mott and Gurney[23] to explain glass formation in terms of the entropy due to the orientational configurations of 'tiny' crystallites which make up the liquid state, which contributes to the appearance of a 'pseudo-critical' temperature T_K, at which the rate of crystallization of the supercooled liquid becomes slow enough that glass formation occurs over ordinary experimental time-scales. The model was crude and quantitatively incorrect; however, it did correctly portray the trend in the thermodynamic properties of the supercooled liquid discussed previously, *i.e.* the liquid properties become smaller than the crystal-state properties above 0 K.

10.2.2 Thermodynamic Transitions and Glasses

10.2.2.1 Definitions of first- and second-order transitions

Kauzmann[22] pointed out the difficulties involved in making a thermodynamic interpretation of the glass transition phenomenon and opted for a kinetic solution to the apparent paradox discussed above. However, the question of whether or not the glass transition is a true second-order transition is still one which is widely discussed in the literature. Before describing the two most prominent models of the glass transition, *i.e.* the Gibbs–DiMarzio[24] configurational-entropy model and the free-volume models,[25-29] we will attempt to clarify a point of confusion which arises from attempts to experimentally prove or disprove the glass transition event as a true *thermodynamic transition*. We remind the reader that, regardless of the underlying thermodynamics, the experimentally observed glass transition is a kinetic phenomenon.

According to Ehrenfest[31], a *first-order transition* is one for which the free energy as a function of any given state variable (V, P, T) is continuous, but the first partial derivatives of the free energy with respect to the relevant state variables are discontinuous. Thus, if the Gibbs free energy G at the transition temperature is continuous, but $(\partial G/\partial T)_P$ and $(\partial G/\partial P)_T$ are discontinuous, we have a first-order transition. At a typical first-order transition point, such as fusion or vaporization, there is a discontinuity in entropy S, volume V and enthalpy H, since[32]

$$\left[\frac{\partial G}{\partial T}\right]_P = -S \tag{1}$$

$$\left[\frac{\partial G}{\partial P}\right]_T = V \tag{2}$$

$$\left[\frac{\partial(G/T)}{\partial(1/T)}\right]_P = H \tag{3}$$

Similarly, a *second-order transition* is characterized by a discontinuity in the second partial derivatives of the free energy function with respect to the relevant state variables, but by continuity in both the free energy and its first partial derivatives. Thus, there is no discontinuity in S, V or H at the transition temperature but there is in the heat capacity C_p, the compressibility κ and the thermal expansion coefficient α, since[32, 33]

$$-\left[\frac{\partial^2 G}{\partial T^2}\right]_P = \left[\frac{\partial S}{\partial T}\right]_P = \frac{C_p}{T} \tag{4}$$

$$\left[\frac{\partial^2 G}{\partial P^2}\right]_T = \left[\frac{\partial V}{\partial P}\right]_T = -\kappa V \tag{5}$$

$$\frac{\partial}{\partial T}\left[\left[\frac{\partial(G/T)}{\partial(1/T)}\right]_P\right]_P = \left[\frac{\partial H}{\partial T}\right]_P = C_p \tag{6}$$

$$\left[\frac{\partial}{\partial T}\left[\frac{\partial G}{\partial P}\right]_T\right]_P = \left[\frac{\partial V}{\partial T}\right]_P = \alpha V \tag{7}$$

Some typical phenomena which have been described as second-order transitions are order–disorder transitions in metal alloys, onset of ferromagnetism, onset of ferroelectricity and onset of superconductivity.[32, 34, 35]

In Figure 6 are depicted schematically the behaviors of the Gibbs free energy function and its first and second derivatives for first-order, second-order and glass transitions.[32, 34, 35] The important thing to note from this figure is that, formally, the glass transition event appears to be a second-order transition. However, the nature of the glass transition is somewhat different from what is normally observed and classified as a second-order transition. Interestingly, the kinetics involved in the transitions differ, *e.g.* for glasses, an increase in the time scale of the experiment reduces the observed transition temperature, while for other second-order transitions this is not the case.[35]

10.2.2.2 Pressure dependence of the transition temperature

For a first-order transition from a phase α to a phase β, the continuity of the free energy function and the equilibrium condition result in the well-known Clausius–Clapeyron[36] relation (equation 8)

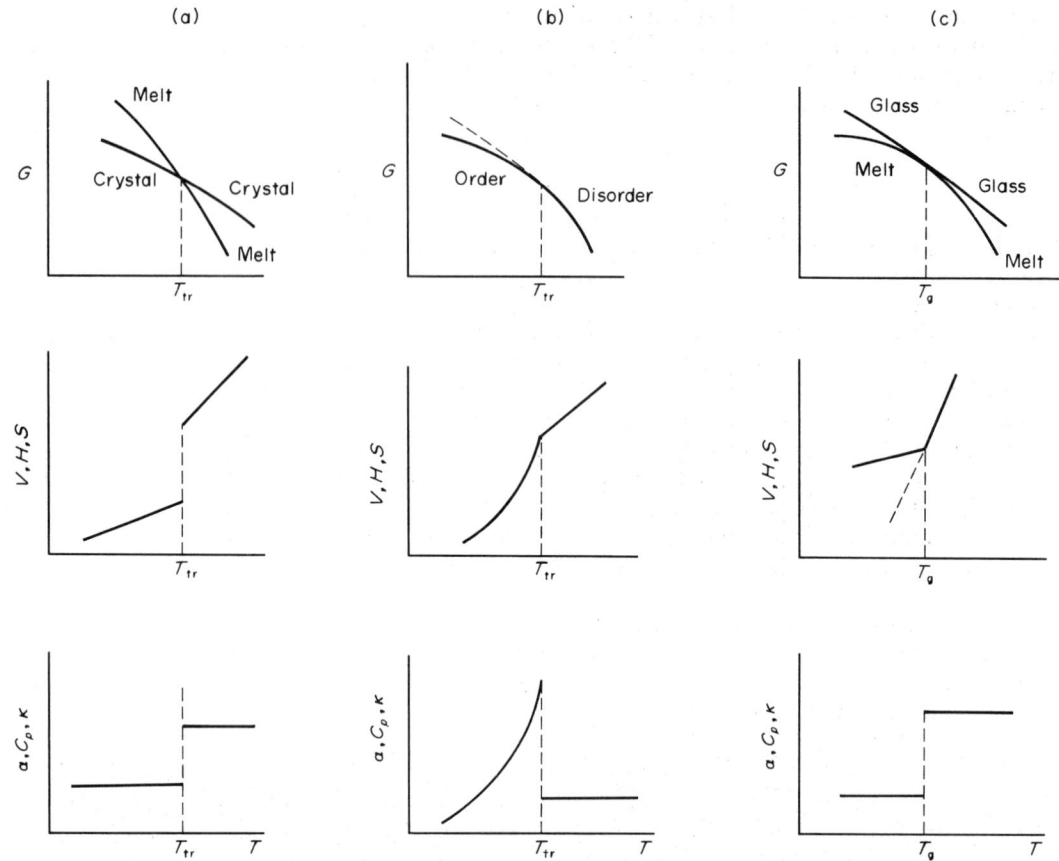

Figure 6   Schematic representation of changes with temperature of the free energy and its first and second derivatives for: (a) first order, (b) classical second-order and (c) glass transitions (after ref. 35, with permission)

$$\left[\frac{dP}{dT}\right]_{tr} = \frac{S_\beta - S_\alpha}{V_\beta - V_\alpha} = \frac{\Delta S}{\Delta V} = \frac{\Delta H}{T \Delta V} \qquad (8)$$

where the t_r refers to the transition line.

For a second-order transition, we will show the development of two similar relationships, one which results from the volume continuity at the transition and the other from the entropy continuity. In the first instance, we can write that the total differential for the volume in the liquid l and the glass g are the same.

$$dV_1 = \alpha_1 V_1 dT - \kappa_1 V_1 dP = dV_g = \alpha_g V_g dT - \kappa_g V_g dP \qquad (9)$$

Regrouping, and because $V_1 = V_g$ at the transition, we can write

$$\left[\frac{dT}{dP}\right]_{tr} = \frac{\Delta \kappa}{\Delta \alpha} \qquad (10)$$

Similarly, for the entropy, we write

$$dS_1 = \frac{C_{pl}}{T} dT - \alpha_1 V_1 dP = dS_g = \frac{C_{pg}}{T} dT - \alpha_g V_g dP \qquad (11)$$

Regrouping terms, and noting that $S_g = S_1$ and $V_g = V_1$ at the transition

$$\left[\frac{dT}{dP}\right]_{tr} = \frac{\Delta \alpha V T}{\Delta C_p} \qquad (12)$$

Expressions (10) and (12) will always be true for any material which undergoes an apparent second-order transition. Furthermore, if the volume and entropy surfaces for the transition are the same, as they must be thermodynamically, (10) and (12) can be equated and we arrive at the Prigogine–Defay[30] ratio, R

$$R = \frac{\Delta\kappa\Delta C_p}{TV(\Delta\alpha)^2} = 1 \qquad (13)$$

Together, equations (10), (12) and (13) have led to much discussion over the thermodynamics of the glass transition event. This is because, depending upon how the experiments are carried out, one can attain agreement, *i.e.* $R = 1$, or not with the prediction for a second-order thermodynamic transition. In this author's view, much of the discussion is unnecessary because the experiments measure a kinetic phenomenon (perhaps a manifestation of an underlying thermodynamic event) and, therefore, cannot test whether or not there is a true thermodynamic glass transition.

In the discussion which follows, we examine the thermodynamic data which have been used to 'test' the validity of relations (10), (12) and (13) and describe how the lack of agreement arises only when 'mixed' data are used. By mixed, we refer to data obtained on glasses with different formation histories. This effect has been recognized by other workers,[37-39] but we emphasize it here because we feel that experiments other than PVT or SPT types should be developed in order to define the thermodynamics of glasses. Such experiments will be described when we discuss theoretical models of glasses.

Typical PVT data for a glass-forming polymer are shown in Figure 7. These data were obtained by isobarically cooling the polymer liquid (melt) from above to below the glass transition temperature at a constant rate. Numerical fitting of the data allowed computation of the appropriate thermodynamic parameters, *i.e.* $\Delta\kappa_1$, $\Delta\alpha_1$ and dT_g^1/dP for these[37] 'variable-formation glasses'. In Figure 8 is depicted a similar plot for glasses formed at atmospheric pressure (1 atm, 101 325 Pa), 'constant-formation glasses'. In this instance, all pressure changes occurred in the glassy state after cooling to the appropriate temperatures. Again $\Delta\kappa_2$, $\Delta\alpha_2$ and dT_g^2/dP were calculated from these data. There are two points which can be made. First, the PVT data of Figures 7 and 8 are not the same, indicating a path dependence of the PVT surface. Second, within each set of data, *i.e.* Figure 7 or Figure 8, $dT_g/dP = \Delta\kappa/\Delta\alpha$, because of the tautological development of this relationship (equation 10). However, $dT_g^1/dP \neq dT_g^2/dP$. This is shown in Figure 9, which also includes data for a glass formed at 800 bar (8×10^7 Pa) (constant-formation glass). Interestingly, dT_g/dP for the latter is

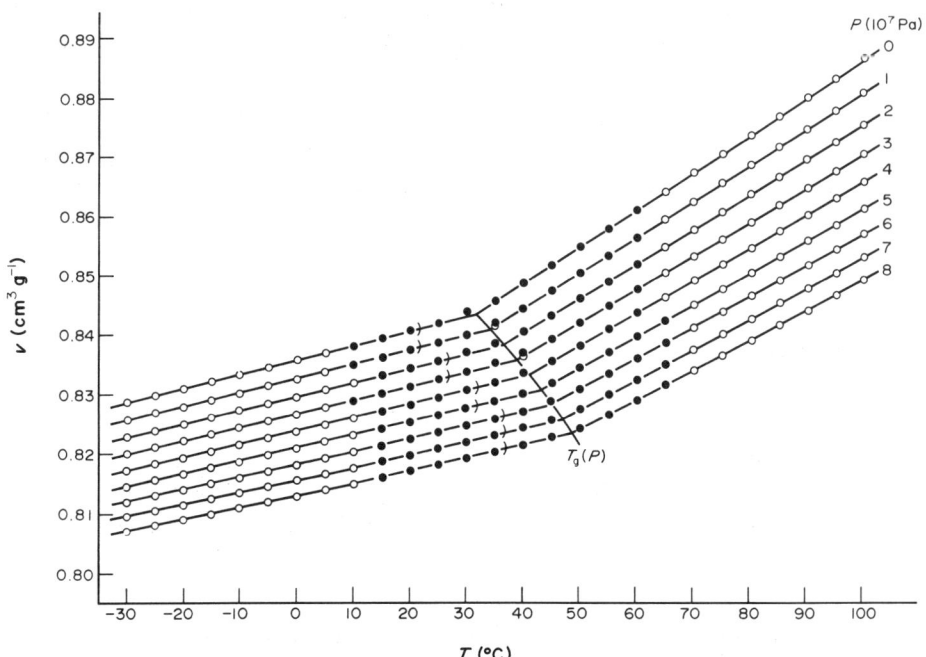

Figure 7 Specific volume v of poly(vinyl acetate) *vs.* temperature T at different pressures, as indicated, for variable-formation glass; note definition of $T_g(P)$ for this type of glass (after ref. 37, with permission)

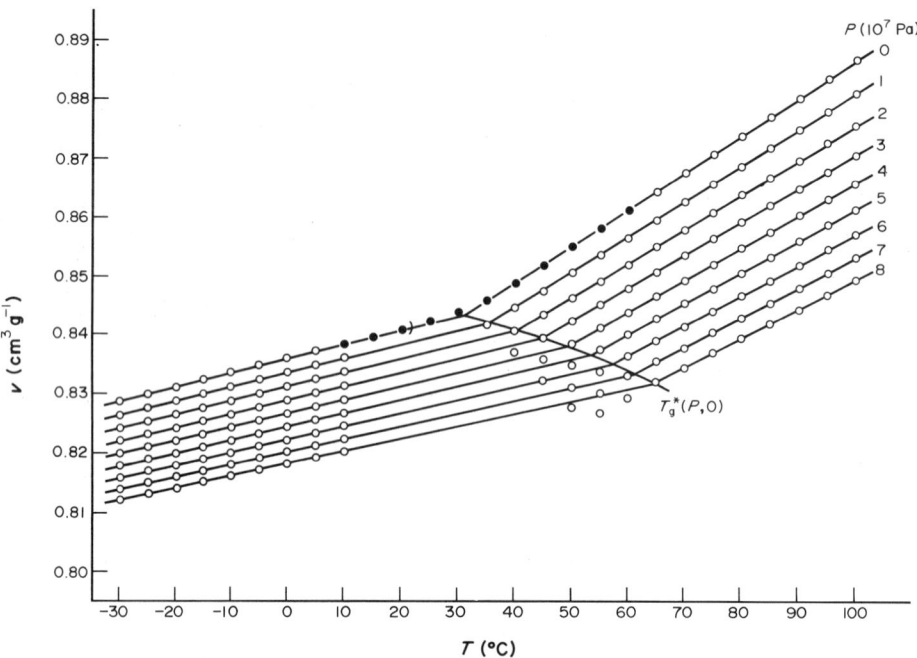

Figure 8 Specific volume v of poly(vinyl acetate) *vs.* temperature T at different pressures, as indicated, for isobaric glass formed at 1 atm; note definition of $T_g^*(P, 0)$ for this type of glass (after ref. 37, with permission)

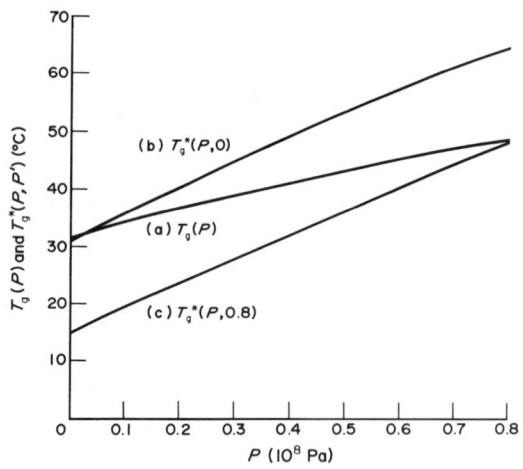

Figure 9 Transition map showing T_g and T_g^* *vs.* pressure P for different formation histories: (a) variable formation ($P' = P$), as in Figure 7; (b) isobaric glass formed at 1 atm ($P' = 0$), as in Figure 8; and (c) isobaric glass formed at 800 bar ($P' = 0.8$ kbar). Note that $dT_g(P)/dP < dT_g^*(P, 0)/dP \approx dT_g^*(P, 0.8)/dP$ (after ref. 37, with permission)

approximately equal to that of the glass formed at atmospheric pressure. McKinney and Goldstein's[37] data for the three glasses are shown in Table 1.

The fact that dT_g/dP depends upon the path of glass formation shows that if one mixes measurements on glasses formed differently then one can expect an apparent violation of the tautology represented by equation (10). Thus it is no surprise that there are conflicting reports in the literature concerning its validity. For example, Zoller[40] finds that $dT_g/dP = \Delta\kappa/\Delta\alpha$ for four different polymers. His data were obtained for variable-formation glasses. On the other hand, Oels and Rehage[41,42] show that $dT_g/dP \neq \Delta\kappa/\Delta\alpha$. They would have found a contradiction within their own data had they carried out an analysis of their PVT data, obtained for variable-formation glasses. However, they obtained their values for $\Delta\kappa$ by direct measurement of κ on glasses cooled isobarically to appropriate temperatures followed by incremental changes in pressure. Although this is not the

Table 1   Tabulation of Thermodynamic Parameters for Poly(vinyl acetate) Glasses Formed under Different Conditions

Variable-formation glass

P (bar)	T_g (°C)	v (cm³ g⁻¹)	$\alpha_1 \times 10^4$ (°C⁻¹)	$\alpha_g \times 10^4$ (°C⁻¹)	$\Delta\alpha \times 10^4$ (°C⁻¹)	$\kappa_1 \times 10^5$ (bar⁻¹)	$\kappa_g^* \times 10^5$ (bar⁻¹)	$\Delta\kappa^* \times 10^5$ (bar⁻¹)	dT_g/dP (°C bar⁻¹)	$T_g^* \Delta\alpha$
0	31.50	0.84358	7.118	3.031	4.088	4.990	3.902	1.088	0.0266	0.1245
100	34.08	0.84098	6.997	3.000	3.996	4.838	3.834	1.003	0.0251	0.1228
200	36.53	0.83840	6.850	2.969	3.881	4.690	3.766	0.924	0.0238	0.1202
300	38.85	0.83585	6.685	2.937	3.748	4.543	3.696	0.848	0.0226	0.1169
400	41.05	0.83334	6.509	2.904	3.606	4.397	3.624	0.773	0.0214	0.1133
500	43.13	0.83085	6.328	2.869	3.459	4.249	3.552	0.697	0.0202	0.1094
600	45.08	0.82839	6.148	2.834	3.314	4.098	3.478	0.620	0.0187	0.1055
700	46.86	0.82596	5.973	2.797	3.176	3.942	3.403	0.539	0.0170	0.1016
800	48.46	0.82355	5.807	2.759	3.048	3.780	3.326	0.454	0.0149	0.0980

1 atm glass

P (bar)	T_g^* (°C)	v (cm³ g⁻¹)	$\alpha_1 \times 10^4$ (°C⁻¹)	$\alpha_g \times 10^4$ (°C⁻¹)	$\Delta\alpha \times 10^4$ (°C⁻¹)	$\kappa_1 \times 10^5$ (bar⁻¹)	$\kappa_g \times 10^5$ (bar⁻¹)	$\Delta\kappa \times 10^5$ (bar⁻¹)	dT_g^*/dP^a (°C bar⁻¹)	$T_g^* \Delta\alpha$
0	30.68	0.84309	7.117	2.795	4.322	4.981	2.896	2.085	0.0482	0.1313
100	35.42	0.84177	6.996	2.771	4.225	4.856	2.880	1.976	0.0468	0.1304
200	40.05	0.84042	6.840	2.740	4.100	4.743	2.869	1.874	0.0457	0.1284
300	44.57	0.83905	6.664	2.704	3.960	4.637	2.864	1.773	0.0448	0.1258
400	49.00	0.83765	6.480	2.662	3.818	4.532	2.864	1.668	0.0437	0.1230
500	53.29	0.83620	6.300	2.614	3.686	4.420	2.868	1.552	0.0421	0.1203
600	57.39	0.83468	6.136	2.559	3.576	4.298	2.877	1.421	0.0397	0.1182
700	61.20	0.83308	5.998	2.499	3.499	4.158	2.889	1.270	0.0363	0.1170
800	64.61	0.83137	5.894	2.433	3.461	3.999	2.903	1.096	0.0317	0.1169

800 bar glass

P (bar)	T_g^* (°C)	v (cm³ g⁻¹)	$\alpha_1 \times 10^4$ (°C⁻¹)	$\alpha_g \times 10^4$ (°C⁻¹)	$\Delta\alpha \times 10^4$ (°C⁻¹)	$\kappa_1 \times 10^5$ (bar⁻¹)	$\kappa_g \times 10^5$ (bar⁻¹)	$\Delta\kappa \times 10^5$ (bar⁻¹)	dT_g^*/dP^a (°C bar⁻¹)	$T_g^* \Delta\alpha$
0	14.75	0.83361	7.085	2.944	4.141	4.849	2.833	2.016	0.0487	0.1192
100	19.44	0.83240	7.010	2.918	4.092	4.668	2.809	1.860	0.0454	0.1197
200	23.87	0.83115	6.885	2.890	3.996	4.511	2.783	1.728	0.0433	0.1187
300	28.12	0.82986	6.726	2.861	3.865	4.371	2.756	1.615	0.0418	0.1164
400	32.25	0.82856	6.542	2.832	3.711	4.244	2.728	1.516	0.0409	0.1133
500	36.30	0.82726	6.348	2.802	3.546	4.126	2.700	1.425	0.0402	0.1097
600	40.29	0.82596	6.153	2.771	3.382	4.010	2.672	1.338	0.0396	0.1060
700	44.21	0.82466	5.968	2.741	3.228	3.894	2.643	1.251	0.0388	0.1024
800	48.02	0.82335	5.804	2.709	3.095	3.773	2.614	1.159	0.0374	0.0994

[a] $dT_g^*/dP = \Delta\kappa/\Delta\alpha$.

exact procedure used by McKinney and Goldstein[37] for the glasses formed at atmospheric or 800 bar pressure, it is similar enough that one could predict Oels and Rehage's result. In particular, they found[37] that $\Delta\kappa/\Delta\alpha$ was considerably greater than dT_g/dP. This, of course, does point up an interesting problem in defining the properties of a glass. Although the glasses formed by Oels and Rehage[41] were individually isobarically formed glasses, the *PVT* surface corresponds to the variable-formation glasses of McKinney and Golstein[37] or Zoller.[40] However, upon incrementally changing pressure isothermally once the glass is formed, Oels and Rehage reproduced the compressibility surface (κ) corresponding to the constant-formation (isobaric) glasses.

In addition to the path-dependence problem of glass formation, it is also possible that both the McKinney and Goldstein[37] *PVT* surfaces for constant-pressure glasses and the Oels and Rehage[41] incremental compressibility measurements reflect the fact that the volume obtained upon changing the pressure shows kinetic effects similar to those that occur after a temperature change (recall Figure 3).

One final point concerning the path dependence of the *PVT* behavior of polymer glasses is the difference between isobaric glass formation (the variable-formation glass) and isothermal glass formation. In the latter, the melt is compressed until a glass is obtained. If there were no path dependence of behavior, a plot of the isobaric data as *V* vs. *P* at different temperatures would give the same result as isothermal measurements. This is not observed, as shown in Figure 10, where isobaric *PVT* data of Oels and Rehage[41] are replotted and compared with their isothermal *PVT* data.

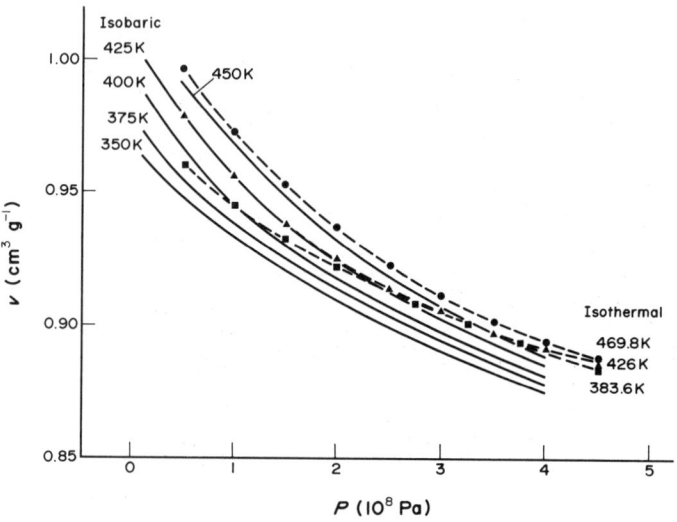

Figure 10   Specific volume *v* vs. pressure *P* for polystyrene through the glass transition range. Note that the isothermal data and the isobaric data do not show the same behavior (data from ref. 41)

The above discussion puts into perspective the question of the validity of the 'volume' equation (10), *i.e.* $dT_g/dP = \Delta\kappa/\Delta\alpha$ for glass-forming systems. However, the use of the Ehrenfest classification for second-order thermodynamic transitions also leads to the 'entropy' equation (12), *i.e.* $dT_g/dP = T_g V_g \Delta\alpha/\Delta C_p$. We note that it has frequently been reported that this equation is usually valid,[43-55] although discrepancies have been noted.[41,43,44] In a way it is surprising that equation (12) is so often found to be valid for the glass transition event which, as we have noted previously, is a kinetic phenomenon over normal experimental time-scales. Equation (12) is derived using the Maxwell relation (see equations 7 and 11).

$$\left[\frac{\partial S}{\partial P}\right]_T = \left[\frac{\partial V}{\partial T}\right]_P \tag{14}$$

We have already seen that an isothermally formed glass is different from one which is isobarically formed, and that the transition temperature from isobarically obtained data differs from that for isothermally obtained data. One may then question the meaning of equation (14) under such conditions.

However, for nearly all the reported results, equation (12) is a reasonable representation of the data. This can be seen in Figure 11, taken from a compilation by Angell and Sichina,[43] in which T_g vs. P was calculated using equation (12) and compared with experimental data. With only the exceptions of poly(vinyl chloride) and $ZnCl_2$ the agreement is quite good. The reasons for this agreement may stem from the fact that $\Delta\alpha$ does not vary much with formation history[37] (see Table 1, for example) and that the dT_g/dP values were more than likely obtained either from differential scanning calorimetry (DSC) scans under the pressure used to determine $\Delta C_p(P)$[43,44,54] or from data on variable-formation glasses.[40] In fact, McKinney and Goldstein[37] report that equation (13) is valid, *i.e.* $R = 1$, for their variable-formation glasses. Obviously, the same heat capacity measurements resulted in disagreement for the constant-pressure glasses.[37] We also note that $ZnCl_2$ glasses[43,44] show a larger than usual kinetic effect on T_g, *i.e.* $dT_g/d \log \tau \simeq -13.5$ K per decade, compared with the normally found value of about -3 K per decade,[56-57] where τ is the time scale of the experiment.

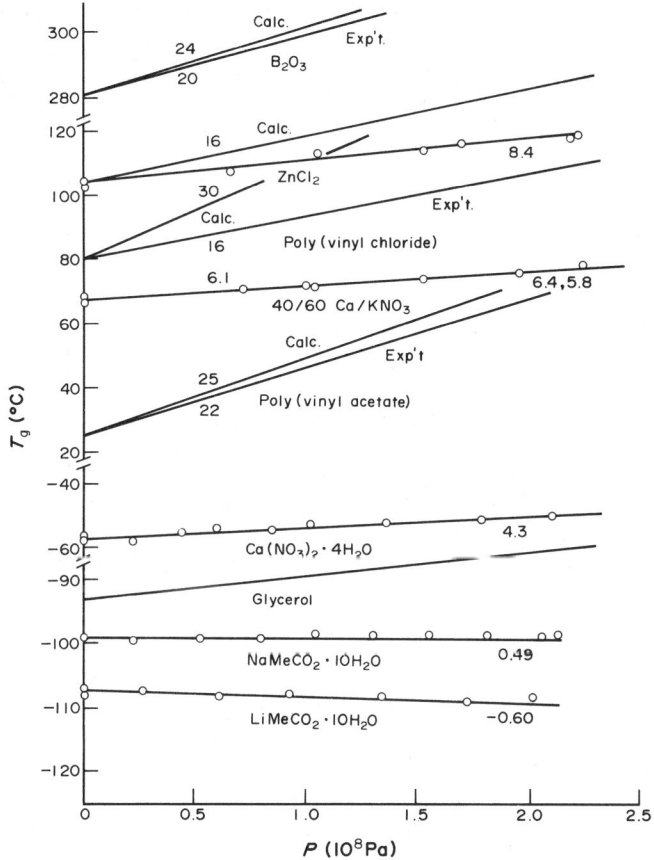

Figure 11 T_g *vs.* pressure P from different glass-forming systems, as indicated. Calculations are based on equation (12) (after ref. 43, with permission)

In summary, dT_g/dP depends upon the thermal and pressure history of the glass. With few exceptions, it is found that, when experiments are run in a self-consistent manner, equation (10) is valid. Equation (12) is also generally found to be valid, although this may arise from self-consistency in the experiments. Therefore, although the glass transition event which is observed experimentally is a kinetic phenomenon, the Prigogine–Defay ratio (equation 13) is equal to unity. In fact, such a finding, given the nature of the thermodynamic constructions used to derive equations (10), (12) and (13), is not surprising. If the glass transition event is a kinetic manifestation of an underlying second-order transition, then theoretical models of the underlying transition must be tested against predictions other than relations (10), (12) and (13). We will describe these models and the molecular parameters which most strongly influence the glass transition. As we will see, some of these are unique to polymers, although the glassy state is essentially ubiquitous. First, however, we make some comments on ordering-parameter descriptions of glassy thermodynamics.

10.2.2.3   *Ordering-parameter description of glassy thermodynamics*

As indicated in the previous section, there has been considerable discussion of the value of the Prigogine–Defay[30] ratio (equation 13), and the validity of the Ehrenfest[31] relations (equations 10 and 12). We also indicated that much of the problem arises from the fact that the results for dT_g/dP have been compared with ratios of $\Delta\kappa/\Delta\alpha$ or $T_g V\Delta\alpha/\Delta C_p$ determined from different paths. The use of such discrepancies in discussing the nature of the glass transition (for example, entropy- or volume-controlled) appears to this author to be inappropriate. However, such discrepancies have been treated and interpreted within the context of an ordering-parameter description of the glass transition as a 'freezing-in' process. Here we will describe this approach and how it relates to the non-equilibrium glassy state

The most important contribution to the extension of ordering parameters to describe the glass transition is that of Davies and Jones.[53] The basic idea is that the non-equilibrium thermodynamic state of the glass is described, for example, by a free energy function $F(V, T, Z_i)$, where the Z_i are internal variables which determine the non-equilibrium state of the glass in addition to the normal thermodynamic variables.

For a material defined by a single ordering parameter Z, Davies and Jones[53] showed that the glassy state was defined at constant Z, while the equilibrium liquid state was defined at $(\partial F/\partial Z)_{V,T} = 0$, e.g. $\alpha_g = (1/V)(\partial V/\partial T)_{P,Z}$ and $\alpha_l = (1/V)(\partial V/\partial T)_{P,\partial F/\partial Z = 0}$. From this point of view, the glass is defined by a line of constant Z, and in the glassy state Z is 'frozen-in'.

Importantly, Davies and Jones[53] found that, for a single ordering-parameter glass, the Ehrenfest relations (equations 11 and 12 are valid, and the Prigogine–Defay[30] ratio is unity. However, upon defining a material with multiple ordering-parameters, they[53] found the following inequality for the Prigogine–Defay[30] ratio

$$R = \frac{\Delta\kappa\Delta C_p}{T(\Delta\alpha)^2} \geq 1 \tag{15}$$

As discussed previously, it has often been reported that for real glasses $R \geq 1$ but that $dT_g/dP = T_g V\Delta\alpha/\Delta C_p$ (equation 12). While we have recognized this as being due to the path dependence of the glass transition temperature, the analysis of Davies and Jones[53] does not address this problem specifically, although they did subsequently examine the kinetics of glasses within this context. Furthermore, the entire approach of Davies and Jones[53] has recently been debated and discussed in a series of works by DiMarzio and co-workers,[58-62] Goldstein,[63,64] and Oels and Rehage.[41] Of particular interest is the point of view of DiMarzio[60-62] that the Davies and Jones[53] analysis leads to a contradiction, and, additionally, the Prigogine–Defay[30] ratio is unity even for systems having more than one ordering parameter. Goldstein[63,64] does not entirely agree, but notes[64] that, if DiMarzio's conclusions are correct, the ordering-parameter approaches taken previously to describe the thermodynamic aspect of the glass transition 'will be invalidated'.

Without going into this argument further, we note that the true importance of ordering parameters in describing glassy behavior rests in their use in the development of models of kinetics in the glassy state, which will be discussed in Section 10.3. Here we add that Davies and Jones,[53] DiMarzio[60-62] and Oels and Rehage[41] have commented upon the importance of dealing with the time-dependent behavior of the ordering parameters in order to correctly account for the glass transition behavior.

10.2.3   Models of the Glass Transition

As noted previously, the experimentally observed glass transition is dominated by kinetics. However, it is legitimate to ask whether, independent of the time scale, there exists a purely thermodynamic glass transition. It is in this context that the Gibbs–DiMarzio[24-26] theory was developed and eventually provided a theoretical basis for the resolution of the Kauzmann[22] paradox. Using a lattice model, Gibbs and DiMarzio[24-26] found that below a temperature T_2 the configurational entropy was zero. Thus, it was argued that a true second-order thermodynamic transition at T_2 underlies the kinetically observed glass transition at T_g. In the following sections we will describe the Gibbs–DiMarzio theory and, in the process of describing tests of this theory, survey the influence of various molecular parameters on the glass transition temperature. Subsequent sections will deal with free-volume models of the glass transition in the same manner.

10.2.3.1 Gibbs–DiMarzio theory

The Gibbs–DiMarzio[24–26] theory of the glass transition results from an application of the Flory–Huggins[65,66] lattice model of a polymer system. DiMarzio[16] has argued that the use of the lattice model to study polymeric, as opposed to simple, liquids is more promising, because in polymers it is possible to form glasses from systems which have no underlying crystalline phase, which makes arguments about the existence of metastable glasses (which are strongly dependent on kinetic questions) irrelevant to the equilibrium thermodynamic state described by the lattice model. Also, the study of polymer glasses must include such parameters as the effects of chain length, chain stiffness and chain connectivity, with the result that the theory can be compared to a richer class of experimental phenomena than is possible for simple liquids. These latter become important here because, as we noted in the previous discussion of equations (10), (12) and (13), experiments other than the PVT type are necessary to test models of the thermodynamics of glasses.

The parameters relevant to the lattice model are readily described by reference to Figure 12. The polymer chains of degree of polymerization X (DP = X) have many configurations which fit onto the lattice of coordination number Z. Each chain has a lowest energy shape, and the more the shape deviates from it, the greater the internal energy of the molecule. Gibbs and DiMarzio[24–26] assumed a simple form for this energy, in terms of the fraction f of bonds 'flexed' out of the lowest energy state and the energy associated with the flexed and unflexed potential wells, ε_2 and ε_1. They also allowed for n_0 vacant sites on the lattice. This latter results in a hole energy, proportional to the number of intermolecular or van der Waals bonds (bond energy α) broken by introduction of the vacancies into the lattice. Then there are two energies, $\Delta\varepsilon = \varepsilon_2 - \varepsilon_1$ and α, and two parameters, f and n_0, in the model. These latter two are controlled by the energies.[16]

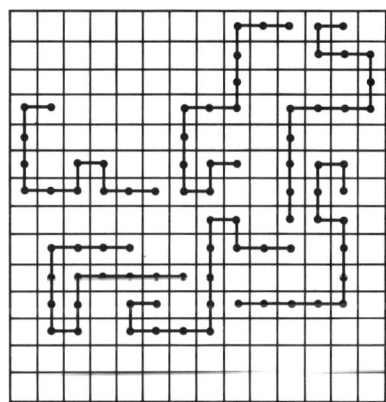

Figure 12 Representation of the Gibbs–DiMarzio model. A lattice of coordination number Z is used to calculate the number of complexions for a system of n_x molecules each of length X. Energy α is associated with holes and energy $\Delta\varepsilon$ is associated with flexing of a bond from its low energy shape (after ref. 16, with permission)

The calculation of the partition function can be done by the standard Flory–Huggins lattice method.[67] The lattice model predicts the existence of a true second-order transition at a temperature T_2. This is shown schematically in Figure 13 for the entropy–pressure–temperature equation of state. As can be seen, the transition occurs at a critical value of the entropy (zero configurational entropy) and the Kauzmann[22] paradox is resolved for thermodynamic reasons rather than kinetic ones, *i.e.* one is simply not permitted to extrapolate high temperature behavior through the glass transition. Rather, as the material is cooled, a break in the S–T (or V–T) curves occurs because of a second-order transition.

The physical reason for the existence of the transition at T_2 is then easily understood. As one cools the system, at a given pressure, the number of allowed arrangements for the molecules decreases for two reasons. In the first place, the number of holes decreases (a volume decrease) and the configurational entropy due to permuting holes and chains decreases. Secondly, the configurational entropy of the molecules decreases, because the chains favor low energy states (shapes) at lower temperatures. The $T(P)$ transition line defines the point where the total configurational entropy first becomes zero. It is a direct result of the lattice model that this occurs at a finite non-zero temperature, *i.e.* T_2. Gibbs and DiMarzio[24–26] hypothesized that the $T(P)$ line represents the thermodynamic glass transition in experiments of long time-scale.

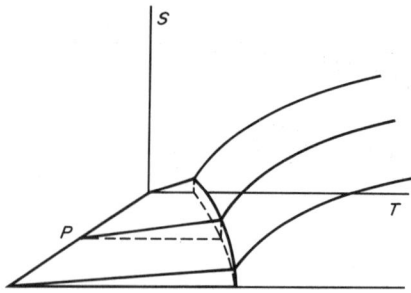

Figure 13   Schematic representation of the *SPT* surface calculated from the Gibbs–DiMarzio lattice model. The mean-field calculation on the lattice model results in a second-order transition in the Ehrenfest sense (after ref. 16, with permission)

The Gibbs–DiMarzio[24–26] theory allows the glassy state as a metastable state above the lowest-energy crystalline state for crystallizable materials, and resolves the Kauzmann paradox even for crystallizable materials. However, more importantly, it defines the glassy state as a 'fourth state of matter' in the case of materials, such as atactic polymers, which cannot crystallize. Glass formation is associated with the condition that the configurational entropy goes to zero. Once one allows that the glass transition temperature T_g, measured kinetically, is a reflection of the true thermodynamic transition at T_2, the model makes specific predictions, many of which have been tested experimentally. We now turn to the question of the ability of the Gibbs–DiMarzio theory to describe (or predict) experimental data.

(i) The influence of molecular weight on T_g

It is well-known that, for most linear polymers, the glass transition varies with molecular weight,[28,68–78] although exceptions to this rule have been reported for polymers with ionic[72] or hydroxyl[79,80] end-groups as well as for polyisoprene.[81] In fact, it is often reported that, over a limited range of molecular weight, T_g varies inversely as $\bar{M}_n$, the number-average molecular weight, and there are several treatments which give such a reciprocal relationship.[69,82,83] However, at low molecular weights, this relationship breaks down.[69,78,84] It is here that one can differentiate between the Gibbs–DiMarzio theory and other treatments.

The molecular weight dependence of T_g obtained by Gibbs and DiMarzio can be written[69,80,84] as

$$
\left(\frac{x}{x-3}\right)\left(\frac{\ln v_0}{1-v_0}\right) + \left(\frac{1+v_0}{1-v_0}\right)\ln\left[\frac{(x+1)(1-v_0)}{2x\,v_0} + 1\right] + \left(\frac{\ln 3(x+1)}{x}\right)
$$

$$
= \frac{-2\dfrac{\Delta\varepsilon}{kT_g}\exp\dfrac{-\Delta\varepsilon}{kT_g}}{\left(1+2\exp\dfrac{-\Delta\varepsilon}{kT_g}\right)} - \ln\left(1+2\exp\dfrac{-\Delta\varepsilon}{2kT_g}\right) \tag{16}
$$

where x is twice the degree of polymerization, $\Delta\varepsilon$ is the flex energy, v_0 is the volume fraction of holes, T_g is the glass temperature (associating T_2 with T_g) and k is the Boltzmann constant. Figures 14 and 15 show fits of equation (16) to experimental data[70] for poly(vinyl chloride) over a range of molecular weights from 540 to 45 000, using values of $v_0 = 0.025$ and $\Delta\varepsilon = 6.34\,\text{kJ mol}^{-1}$. The expression (15) fits the data quite well, even down to a molecular weight of 540 ($x = 9$). Because $\Delta\varepsilon$ is fixed by the value of T_g at high (infinite) molecular weight this is really a one-parameter fit, *i.e.* v_0 must be determined. v_0 has been related[69,70] to the 'free volume' for an infinite molecular weight polymer at T_g (free-volume concepts will be discussed in a later section). In reality, v_0 is related to the hole energy α in the lattice model.

As mentioned previously, the lattice model used by Gibbs and DiMarzio is readily generalized to other systems. An interesting prediction is made by the theory for chains which have been formed into uncatenated rings. Guttman and DiMarzio[85] have calculated that, unlike for linear chains, the glass temperature for rings should increase with decreasing molecular weight. There is little data for the glass temperatures of small rings; however, the data of Clarson *et al.*[246] for poly(dimethyl siloxane) (PDMS) rings do show the predicted increase (see Figure 16). The agreement is only

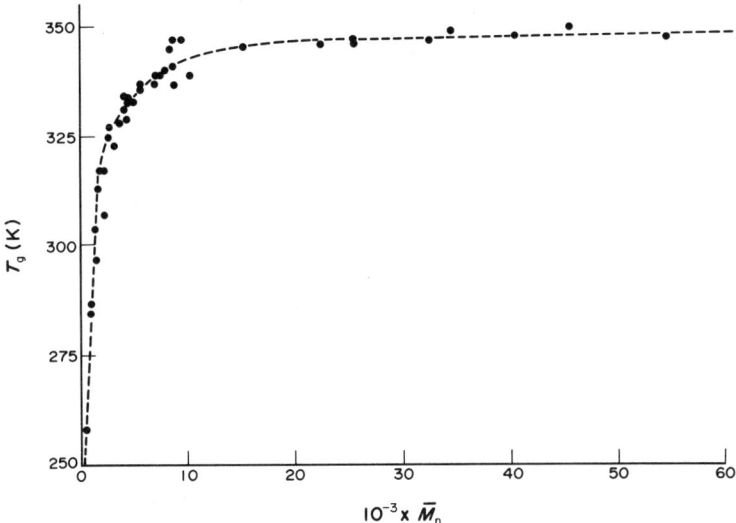

Figure 14 Change in T_g with number-average molecular weight for poly(vinyl chloride), the dashed line is the Gibbs–DiMarzio prediction (after ref. 70, with permission)

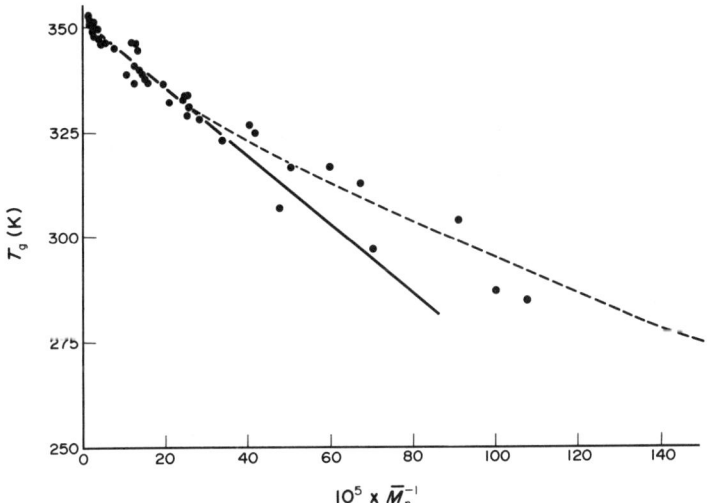

Figure 15 Change in T_g with reciprocal number-average molecular weight. Solid line, linear dependence on $\bar{M}_n^{-1}$; dashed line, Gibbs–DiMarzio prediction (after ref. 70, with permission)

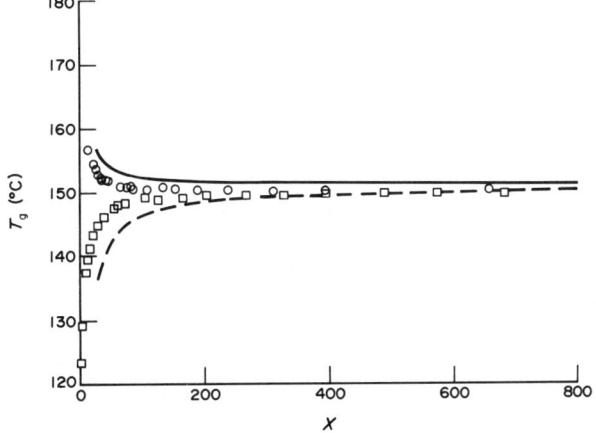

Figure 16 Glass transition temperature *vs.* degree of polymerization X for (○) cyclic and (□) linear poly(dimethyl siloxane). The lines represent the predictions of Gibbs–DiMarzio theory (after ref. 85, with permission)

qualitative for PDMS, but this is perhaps not surprising as the model does not represent the data for linear PDMS very well. The reason for this is unknown.

(ii) Prediction of the heat capacity change at T_g

As a thermodynamic model, the lattice model of Gibbs and DiMarzio allows the calculation of the change in specific heat capacities at constant volume, ΔC_v, or constant pressure, ΔC_p, once the parameters $\Delta\varepsilon$ and α (or $\Delta\varepsilon$ and v_0) are known. Havlicek et al.[86] have compiled comparisons between the predicted and experimental heat capacities for various polymers and find excellent agreement. Interestingly, they found that values of $\Delta\varepsilon$ and α increase linearly with T_g of the polymer and are not greatly different one from the other. Their data also show that the fractional free volumes at T_2 and T_g, i.e. $v_0(T_2)$ and $v_0(T_g)$, increase as T_g of the polymer increases. Their interpretation was that the lattice model parameters are physically meaningful. Berry,[87] on the other hand, argues that the parameters are not truly meaningful, but finds 'surprisingly' that $v_0(T)$ appears to have a physical sense.

DiMarzio and Dowell[88] have added a term for the contribution from lattice vibrations to the original model. A comparison is made of their predictions with experimental results in Table 2. The agreement for these polymers is excellent.

Table 2 Comparison of Gibbs-DiMarzio Predictions and Experimentally Observed Values of the Specific Heat Discontinuity ΔC_p at T_g for Different Polymers (After Ref. 88)

Polymers	ΔC_p theoretical[a] ($J\,g^{-1}\,K^{-1}$)	ΔC_p experimental ($J\,g^{-1}\,K^{-1}$)
Polyethylene	0.59	0.60
Polypropylene	0.51	0.48
Poly(isobutylene)	0.43	0.40
Poly(vinyl chloride)	0.36	0.30
Poly(vinyl acetate)	0.43–0.47	0.41
Poly(methyl methacrylate)	0.40–0.45	0.30
Polystyrene	0.31	0.34
Poly(α-methylstyrene)	0.31	0.32
Polycarbonate	0.28–0.33	0.24
Poly(ethylene terephthalate)	0.29–0.35	0.33

[a] When two numbers are given, it is because the number of flexes per monomer unit is uncertain for these polymers.

(iii) Change in T_g due to crosslinking

An important characteristic of high polymers is their ability to exhibit rubbery behavior above the glass transition temperature. When crosslinks are introduced into the polymer to form a permanent network rubber, the glass transition temperature can change significantly.[74,89–96] DiMarzio[59] showed that the effect of crosslinking could be calculated quantitatively for real systems. Setting the configurational entropy to zero at T_2 he found that[90]

$$\ln(1 + 2e^{-\Delta\varepsilon/kT}) + f\Delta\varepsilon/kT - 1 + \frac{T\Delta\alpha}{1 + 2T\Delta\alpha} - \frac{3X}{2} + \frac{3}{4}X\ln f + \frac{3}{4}X\ln(A'X) = 0 \quad (17)$$

where $\Delta\varepsilon$, k, T and f are as before, $\Delta\alpha$ is the change in coefficient of expansion from glass to liquid $(\alpha_l - \alpha_g)$, X is the crosslink density expressed as the number of chains per mole of segments and A' is a number independent of the material.

DiMarzio was able to greatly simplify equation (17) and obtained

$$\frac{T(\chi) - T(0)}{T(0)} = \frac{KM\chi/\gamma}{1 - KM\chi/\gamma} \quad (18)$$

where $T(\chi)$ is the glass temperature of the polymer with χ crosslinks per gram, M is the molecular weight of a residue and γ is the number of flexible bonds per residue. K is a constant, independent of material, approximately equal to 1.3×10^{-23}.

In Figure 17 we have replotted the data of Wood[92] for dicumyl-peroxide-crosslinked natural rubber for T_g vs. parts-per-hundred peroxide, and compare the predictions from equation (18) using

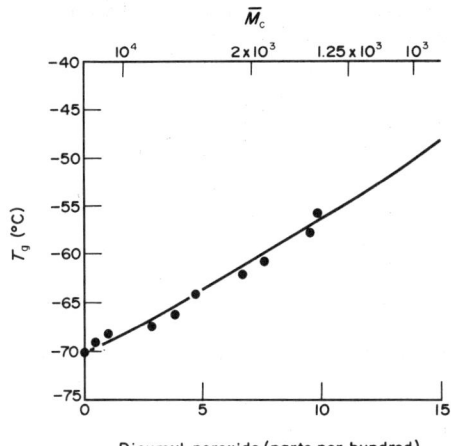

Figure 17 Change in T_g with crosslink density for peroxide-crosslinked natural rubber. The line is calculated from equation (18) (data from ref. 92)

$K = 1.3 \times 10^{-23}$ and $M/\gamma = 22.7$, as suggested by DiMarzio. Values for χ were obtained using Wood's[97] calculations for the number of crosslinks per cm^3 for different amounts of dicumyl peroxide added to the rubber. The agreement is quite good over the range of the data. We must point out here that these data are quite different from those used by DiMarzio in his original discussion. In that case he used data for sulfur-crosslinked rubber,[59] for which T_g increases much more rapidly than for the peroxide-crosslinked systems. In fact, while the data he used fit an equation of the form of equation (18), the parameter K differed by an order of magnitude from that which DiMarzio reported to be a universal number. This may simply reflect the complex nature of sulfur-crosslinked systems.

(iv) Effect of deformation on T_g

Another interesting aspect of the Gibbs–DiMarzio[24–26] theory is that it predicts an effect of deformation on the glass transition temperature due to changes in the entropy of the system upon straining the network. For uniaxial extension, the result in terms of the small-strain modulus G, measured at a temperature T_0, is[59]

$$\frac{T(\lambda)}{T(1)} = \exp\left[\frac{G}{2\Delta C_p T_0}[I_1 - 3]\right] \tag{19}$$

where $T(\lambda)$ is the transition temperature at a stretch λ, $T(1)$ is the transition temperature at zero deformation ($\lambda = 1$), and $I_1 = \lambda_1^2 + \lambda_2^2 + \lambda_3^2$ ($= \lambda^2 + 2/\lambda$ in extension) is the first invariant of the deformation tensor. This equation results from the classical network theory for entropy of a deformed chain,[98] and could presumably be improved upon, but the prediction is interesting, *i.e.* that the glass transition temperature increases exponentially with the deformation invariant. There are few results available on the effects of deformation on T_g, but that from the most thorough investigation by Gee *et al.*[99] is in reasonable agreement with the model. As shown in Figure 18, the transition temperature increases with increasing elongation.[99] Of significance in this respect is the prediction that, because the configurational entropy always decreases upon deformation, independently of the mode of deformation (*i.e.* tension, simple compression or shear), the glass temperature should always increase when a rubber sample is deformed.

We note here that other data have been reported which disagree[100–102] with those of Gee *et al.*,[99] as well as data which agree.[100, 102, 103] This is a subject which should be investigated further.

(v) Compositional dependence of T_g

The glass temperatures of polymer–diluent systems, polymer–polymer blends and copolymers vary with composition.[57, 105–120] The configurational-entropy model has proven relatively successful in describing compositional effects on T_g. However, it is limited in the sense that, while one

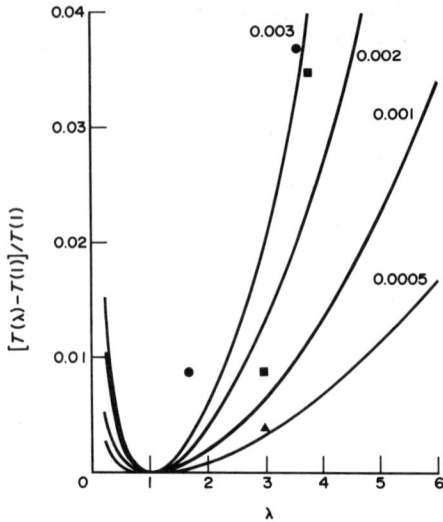

Figure 18   Relative change in glass-transition temperature *vs.* stretch λ for various values of $G/2\Delta C_p T(1) = Y$, as indicated. Experimental data are from Gee *et al.*[99] Circles are for natural rubber, for which $Y = 0.0032$; squares are for GRS rubber, for which $Y = 0.0012$; the triangle is for Hycar, for which $Y = 0.005$, (after ref. 59, with permission)

can calculate $T_g(T_2)$ for a given composition by assuming that the configurational entropy for the system is zero, analytical expressions for dependence of T_g on composition are not available from the Gibbs–DiMarzio model.[108,110]

For random mixing of polymer and solvent, the Gibbs–DiMarzio theory predictions for the glass temperature are quite good, as shown in Figure 19. Gordon *et al.*[108] have used the configurational-entropy model, but with an alternative method of calculation of the point of vanishing configurational-entropy, to derive expressions for the compositional dependence of T_g of regular solutions. They find that, in terms of the mole fraction x, T_g of a binary mixture can be expressed as

$$T_g(x) = \frac{x\,T_{g1} + (1 - x)\,T_{g2}(\Delta C_{p2}/\Delta C_{p1})}{x + (1 - x)(\Delta C_{p2}/\Delta C_{p1})} \tag{20}$$

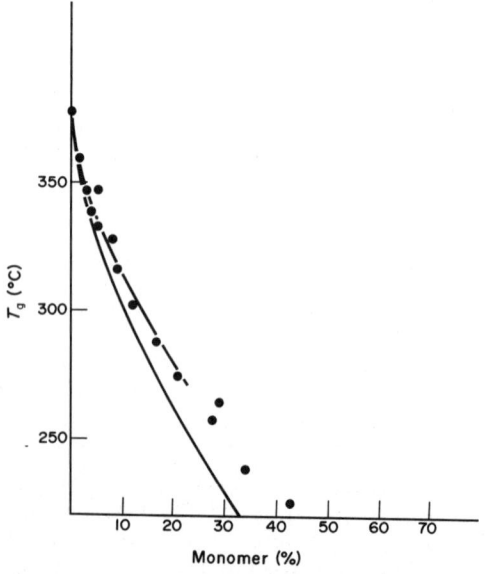

Figure 19   Variation of glass transition temperature with composition for solutions of polystyrene in styrene monomer. The lines represent calculations from the Gibbs–DiMarzio lattice model: upper line calculation allows for 'holes' while the lower one does not (after ref. 25, with permission)

which fits the data quite well.[108] More recently, Chow[106] has used the lattice model to successfully describe the change in T_g for non-random mixing of polymer solutions.

DiMarzio and Gibbs[112] have derived equations which successfully describe the variation of T_g with copolymer composition

$$T_g = B_A T_{gA} + B_B T_{gB} \tag{21}$$

where B_A and B_B are the fractions of rotatable bonds of polymer A and B, respectively, in the copolymer. Equation (21) results from a simplified derivation and is only valid when the stiffness energies used in the lattice model are not too different. De la Campa *et al.*[111] have shown that the Gibbs–DiMarzio model can be used to describe the behavior of poly(ethylene terephthalate-*co*-diethylene terephthalate).

Other molecular parameters affect the glass temperature of polymers, such as tacticity[81, 121] and blend composition.[113, 114] These have been successfully treated, at least partially, by the entropy model. In addition, it is observed that the T_g of crosslinked systems is more dramatically reduced upon addition of diluent than in uncrosslinked systems.[95, 106] This is also accounted for by considerations of configurational entropy.

Finally, we note that, over and above the Gibbs–DiMarzio approach described here, the composition dependence of T_g has been thoroughly treated by Couchman and Karasz[115–117] from a classical thermodynamic point of view. They claim that simple considerations of the thermodynamics of mixing and the glass temperature results in rules similar to those described above for the composition dependence of T_g, and that agreement with experiment is not truly support for the Gibbs–DiMarzio model. This point of view has been strongly disputed by Goldstein.[118]

10.2.3.2 Free-volume concepts

(i) Free volume and molecular mobility

Historically, the free-volume concept was developed to explain the non-Arrhenius dependence of the fluidity or viscosity of liquids on temperature.[27–29, 122–138] The most well-known formulation of this is the so-called Doolittle[29, 124–126] equation

$$\ln \eta = \ln A + B(v - v_f)/v_f \tag{22}$$

where η is the viscosity, v is the specific volume and v_f is the free volume. Doolittle[29, 124–126] recognized early that the definition of the 'free space' v_f was difficult, as was the 'limiting specific volume' v_0, where $v_f = v - v_0$. Without describing his definitions (we will subsequently use modern notation) we find that equation (22) describes quite well the viscosity–temperature behavior of many simple liquids.

The free-volume concept was subsequently introduced to describe the melt behavior of polymers in the work of Fox and Flory[127–129] and that of Williams, Landel and Ferry.[137] These latter found that, when the free volume was assumed to increase linearly with temperature (equation 23), the equation which is now known as the WLF equation could be derived from the Doolittle equation (22).

$$v_f = v_g[0.025 + \Delta\alpha(T - T_g)] \tag{23}$$

The WLF equation was developed empirically to describe the change in viscoelastic properties of polymers above the glass transition.[137] Starting from the principle of thermorheological simplicity, *i.e.* that the spectrum of relaxation (or retardation times) changes with temperature by a simple shift along the time axis, it was found that the shift factor for the viscosity could be written as[57]

$$a_T = (\eta_0 T_0 \rho_0)/(\eta_{00} T \rho) \tag{24}$$

or

$$\log a_T = \log(\eta_0/\eta_{00}) + \log(T_0 \rho_0/T\rho) \tag{25}$$

where η_0 and ρ are the viscosity and density at the temperature T of interest and η_{00} and ρ_0 are the viscosity and density at the reference temperature T_0, normally taken in the mid-range of the data. The correction term $T_0 \rho_0/T\rho$ is small and is often ignored in the reduction of the viscosity data.[57] Williams, Landel and Ferry[137] found that $\log a_T$ could be well represented by the empirical equation

$$\log a_T = - C_1^0(T - T_0)/(C_2^0 + T - T_0) \tag{26}$$

where C_1^0 and C_2^0 are constants which, while originally thought to be universal, are now recognized to depend on the polymer.

As noted previously, the WLF equation (26) and the Doolittle equation are found to be same if one treats the free volume as varying linearly with temperature. If we adopt Ferry's[57] notation that the fractional free-volume is $f = v_f/v$ and the temperature dependence of f is

$$f = f_0 + \alpha_f(T - T_0) \tag{27}$$

then, from equations (22), (24) and (27), and ignoring the $T_0\rho_0/T\rho$ correction, we find

$$\log a_T = \frac{-(B/2.303 f_0)(T - T_0)}{f_0/\alpha_f + T - T_0} \tag{28}$$

and we can then identify the constants in the WLF equation (26) with those in equation (28) and obtain

$$C_1^0 = \frac{B}{2.303 f_0} \tag{29}$$

$$C_2^0 = f_0/\alpha_f \tag{30}$$

$$f_0 = \frac{B}{2.303 C_1^0} \tag{31}$$

$$\alpha_f = \frac{B}{2.303 C_1^0 C_2^0} \tag{32}$$

If we identify the reference temperature as the glass transition temperature, then equation (31) becomes

$$f_g = \frac{B}{2.303 C_1^g} \tag{33}$$

Because B is an arbitrary parameter, the free-volume fraction cannot be determined directly from these equations. As noted previously, Doolittle[29,124-126] was concerned with the definition of free volume and in Figure 20 is shown a possible definition for the free-volume fraction based on the concept of an occupied volume. When one considers B to be unity, the value of the free-volume fraction at T_g falls[57] between 0.013 and 0.034 for many polymers, and the 'universal' value originally proposed by Williams, Landel and Ferry[137] was taken to be 0.025 (see equation 23; also the discussion about the free-volume fraction derived from the Gibbs–DiMarzio[24-26] lattice model).

We note that the Doolittle/WLF equations predict a rapid increase of viscosity with decreasing free volume (or as T_g is approached). Another empirical equation used to describe this is the

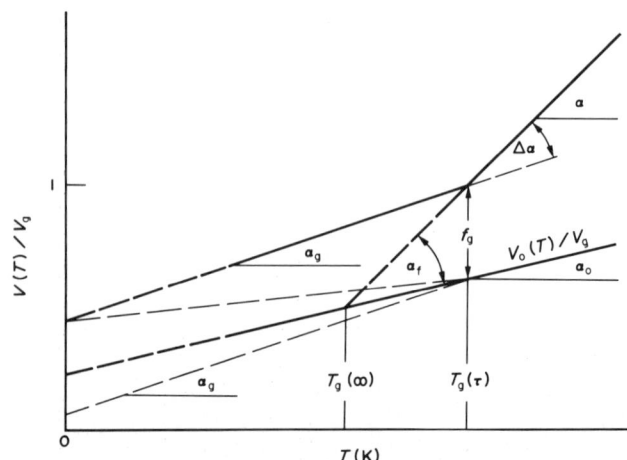

Figure 20 Schematic variation of total specific volume $V(T)/V_g$, occupied volume $V_0(T)$ and free volume at $T_g(f_g)$ with temperature T for a supercooled liquid. α_g is the coefficient of expansion (total) of the glass, α is the coefficient of expansion of the liquid, α_f is the coefficient of expansion of the free volume and α_0 is the coefficient of expansion of the occupied volume. $\Delta\alpha = \alpha - \alpha_g$ is the difference in slope between liquid and glassy states. $T_g(\tau)$ is the glass transition temperature at some time τ (or equivalent cooling rate) and $T_g(\infty)$ is the glass transition temperature at infinite time (after ref. 13, with permission)

Vogel,[139] Fulcher,[140] Tamann and Hess[141] (VFTH) equation

$$\log \eta = A + B/(T - T_\infty) \tag{34}$$

where T_∞ is often found to be approximately 50 °C below the conventionally measured glass temperature and has been associated with the T_2 of the Gibbs–DiMarzio[24-26] theory. The three equations (22), (28) and (34) are formally equivalent and describe the data quite well over a large range of viscosities, as shown in Figure 21. We remind the reader that these equations are valid[57] from approximately $T_g + 10 \,°C \le T_g < T_g + 100 \,°C$.

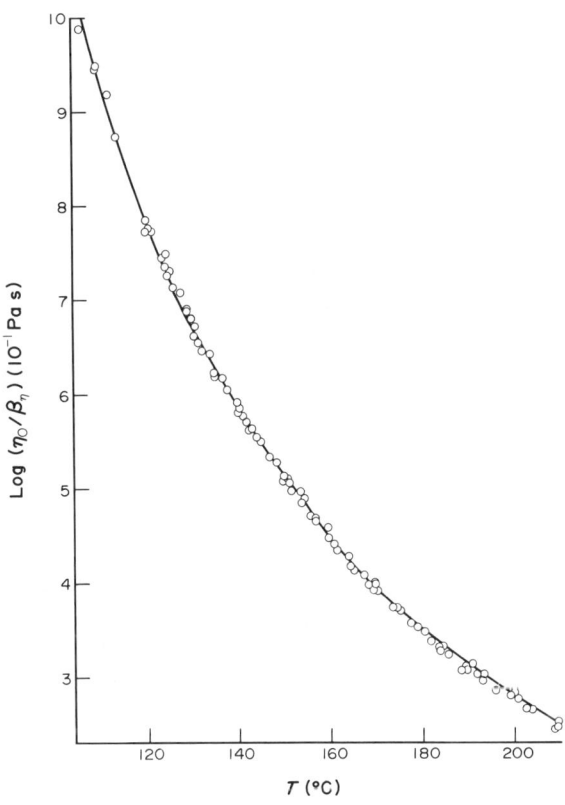

Figure 21 Logarithm of reduced viscosity η_0/β_η *vs.* temperature T for melts of cyclic polystyrene molecules showing rapid increase as T_g ($\approx 100 \,°C$) is approached. The line represents the VFTH equation with $T_0 = 43.5 \,°C$ (after ref. 245)

The empirical use of the free volume (in particular the WLF model) to describe the mobility of molecular fluids has been extended to include pressure effects. Then one can write that the shift factor is related to temperature and pressure through the free volume with the result[142-145]

$$\log a_{T,P} = \frac{(B/2.303 f_0)[T - T_0 - \theta(P)]}{f_0/\alpha_f(P) + T - T_0 - \theta(P)} \tag{35}$$

where

$$\theta(P) = C_3^0(P) \ln \left[\frac{1 + C_4^0 P}{1 + C_4^0 P_0} \right] - C_5^0(P) \ln \left[\frac{1 + C_6^0 P}{1 + C_6^0 P_0} \right] \tag{36}$$

and

$$C_3^0(P) = 1/k_r \alpha_f(P) \tag{37}$$

$$C_4^0 = k_r K_r^* \tag{38}$$

$$C_5^0(P) = 1/k_\phi \alpha_f(P) \tag{39}$$

$$C_6^0 = k_\phi/K_\phi^* \tag{40}$$

where the superscript on the C's refers to the reference temperature T_0; P_0 is the reference pressure and C_3^0 and C_5^0 depend on the experimental pressure P. f_0 is the free-volume fraction at the reference temperature and $\alpha_f(P)$ is the pressure-dependent coefficient of thermal expansion of the free volume. K_r^* is the bulk modulus of the rubber at T_0 and zero pressure, k_r is the Bridgeman pressure coefficient of the bulk modulus and K_ϕ^* and k_ϕ are the corresponding analogs for the occupied volume. The equations (35)–(40) were shown to successfully describe the temperature- and pressure-dependences of the viscoelastic shifts for a series of lightly crosslinked rubbers.[142–145]

In spite of this success, there is still some argument about the role of free volume or fractional free-volume in determining the mobility of liquids. Some fluids can apparently be described by the WLF-type equations over reasonably large ranges of temperature and pressure, but in other cases the description results in unreasonable values for the free-volume parameters.[146] Also, under conditions where the free-volume fraction is held constant by increasing pressure when the temperature is increased, the viscosity of the fluid is not constant.[146–150] Thus it is argued that an energy of activation is required in the free-volume model.

Much of the argument surrounding free-volume models results from the difficulty, recognized by Doolittle,[29,124–126] of defining both the free volume and the occupied volume. In a sense, they are fitting parameters for the data in the absence of a specific model which gives *a priori* the free volume. A point of importance here is that equations for molecular mobility, similar in form to those which can be derived from the free-volume concepts (equations 26, 28, 34 and 35), can be obtained from the configurational-entropy model of Gibbs and DiMarzio. Adam and Gibbs[202] carried out this extension of the Gibbs–DiMarzio theory by assuming that the relaxation time in the system is inversely proportional to the configurational entropy.

(ii) Free volume and the glass transition

The argument described above, that the free volume controls the molecular mobility, leads to a kinetic resolution of the Kauzmann[22] paradox. As the free volume collapses with decreasing temperature, molecular mobility decreases until the time required for molecular rearrangement is longer than the time scale of the experiment and equilibrium is not reached. In fact, when one considers the viscosity (or the shift factor $a_{T,P}$) as a measure of the time required for molecular rearrangement, the free-volume models give a critical temperature T_∞ (and pressure) at which the viscosity goes to infinity (see equations 22, 24–26, 28, 34 and 35, Figures 20 and 21) which corresponds to zero fractional free-volume. Thus the material never reaches equilibrium and the Kauzmann[22] problem is resolved for kinetic reasons. The value of T_∞ is found[57] to be approximately 50 °C below the conventionally measured T_g.

The concept of free volume is quite descriptive and, as discussed above, able to describe quite well the mobility of many simple and polymeric fluids. The quantitative interpretation of the free and occupied volumes, however, is not obvious and has been worried about since Doolittle.[29,124–126] Major breakthroughs in arriving at a molecular basis for free volume were based on ideas of cooperativity[131–133,151,152] of motion of neighboring molecules. According to Cohen and Turnbull,[153,154] molecular transport in simple liquids occurs when a molecule moves into voids having a greater size than some critical value v^*. The voids are created by the redistribution of free volume arising from the cooperative motion of neighboring atoms. According to the Cohen and Turnbull model,[153,154] there is no thermodynamic change in phase in going from the liquid state to the glassy state; rather, the glass transition exists because the free volume of the amorphous phase falls below some characteristic value which is small enough to increase the relaxation times (decrease molecular mobility) enough that one does not attain the equilibrium state. This underlying assumption of the nature of the glass temperature was also advanced by Fox and Flory[28,68,127–129] and entered into the thinking of Williams, Landel and Ferry[137] in their interpretation of the viscosity of polymers near T_g and their development of equation (26). The Cohen–Turnbull free-volume model arrives at the empirical Doolittle equation (22) from two simple assumptions: (1) molecular transport occurs only when voids having a volume greater than some critical value v^* form by the redistribution of the free volume; and (2) no energy is required for free-volume redistribution.

Because of the success of the empirical free-volume relations in describing the behavior of glass-forming liquids, there have been many attempts since the Cohen and Turnbull free-volume model to quantify the concept and make the free-volume physics more than a convenient way to correlate data. The reader is referred to the literature for a general look[1–11] at the various models and also for some specific developments.[155–163] However, due to space limitations, we limit our discussion to the cell model of Simha and Somcynsky[164,165] and the extensive developments of this model which have been carried out over the years by Simha and co-workers.[166–177]

iii) The hole model of Simha and Somcynsky

Simha and Somcynsky[164, 165] developed a cell model for polymer liquids which has been highly successsful in describing the PVT behavior of melts and glasses. The partition function is computed from a model which assumes that the polymer repeat units are confined to cells, the centers of which are in a regular lattice. The classical cell-model, which generally underestimates the entropy, is extended to include empty cells (holes), which adds a mixing term to the entropy. A square-well approximation to the Lennard-Jones and Devonshire cell potential is used and the canonical partition function can be evaluated to obtain the Helmholtz free energy A.

The model is generalized to include chain molecules (polymers) in addition to simple liquids by factorizing the free volume of a molecule into $3c$ elements, where $3c$ is the number of external degrees of freedom. Then, applying the equilibrium condition that

$$(\partial A/\partial y)_{V,T} = 0 \tag{41}$$

for a system of N molecules of degree of polymerization s occupying a fraction y of sites, one obtains

Equation of State
$$\tilde{P}\tilde{V}/\tilde{T} = [1 - 2^{-1/6}y(y\tilde{V})^{-1/3}]^{-1} + (2y/\tilde{T})(y\tilde{V})^{-2}[1.011(y\tilde{V})^{-2} - 1.2045] \tag{42}$$

Internal energy
$$\tilde{U} = (y/2)(y\tilde{V})^{-2}[1.011(y\tilde{V})^{-2} - 2.409] \tag{43}$$

Entropy
$$\tilde{S} = -s/c[(1 - y)/y]\ln(1 - y) + 3\ln[(y\tilde{V})^{1/3} - 2^{-1/6}y] + \ln V^* + \text{constant} \tag{44}$$

and, from the relation (41), the equation of constraint

$$(s/3c)[(s - 1)/s + y^{-1}\ln(1 - y)] = (y/6\tilde{T})(y\tilde{V})^{-2}[2.409 - 3.033(y\tilde{V})^{-2}]$$
$$+ [2^{-1/6}y(y\tilde{V})^{-1/3} - 1/3][1 - 2^{-1/6}y(y\tilde{V})^{-1/3}]^{-1} \tag{45}$$

where $3c/s$ is a characteristic flexibility parameter. The reduced variables are given by

$$\tilde{V} = V/Nsv^* = v/v^* \tag{46}$$

$$\tilde{T} = ckT/(qz\varepsilon^*) \tag{47}$$

$$\tilde{P} = Psv^*/(qs\varepsilon^*) \tag{48}$$

where ε^* and v^* are the characteristic energy and volume per segment, qz is the number of nearest neighbor sites per chain equal to $s(z-2)+2$, and z is the coordination number. Additionally we define three scaling parameters

$$P^* = P/\tilde{P} \tag{49}$$

$$V^* = V/\tilde{V} \tag{50}$$

$$T^* = T/\tilde{T} \tag{51}$$

The Simha–Somcynsky[164, 165] hole theory develops the free-volume model quantitatively by identifying the Doolittle free-volume fraction f, described previously, with the hole fraction $h = 1 - y$. Then the liquid state PVT behavior can be described using the hole theory, as shown in Figures 22 and 23. Figure 22 shows isobaric data as $\log \tilde{V}$ vs. $\log \tilde{T}$ for several polymer melts and Figure 23 shows reduced isotherms for polystyrene melts as $\tilde{P}\tilde{V}/\tilde{T}$ vs. $1/\tilde{V}$. Importantly we note that the hole theory is descriptive of the data rather than predictive. However, fitting to data then allows extraction of physically meaningful parameters, such as the hole fraction (free volume).

Application of the Simha–Somcynsky[164, 165] model to the glass transition allows some interesting observations about the hole fraction to be made. Because this model treats the glass transition as wholly kinetic, the transition event itself cannot be predicted *a priori* from the model but must arise from kinetic arguments. However, the glassy state can be described by the model. It is relevant to note that, in the hole model, unlike other free-volume treatments, there is no assumption that the free-volume fraction remains constant below T_g, although it is constant along the T_g^* vs. P line, where T_g^* refers to the glass temperature for constant-formation glasses, described previously.

In Figure 24, agreement between experiment and the hole model is excellent in the melt. The open circles correspond to the data for the glass (constant formation) and liquid. Below T_g, the solid curve for each pressure gives the theoretical values of $\tilde{V}$ corresponding to a constant value of $h = 1 - y = h(T_g(P))$. This constraint condition replaces the equilibrium condition (equation 44). The dashed lines represent the equilibrium liquid lines predicted by the theory. The large differences

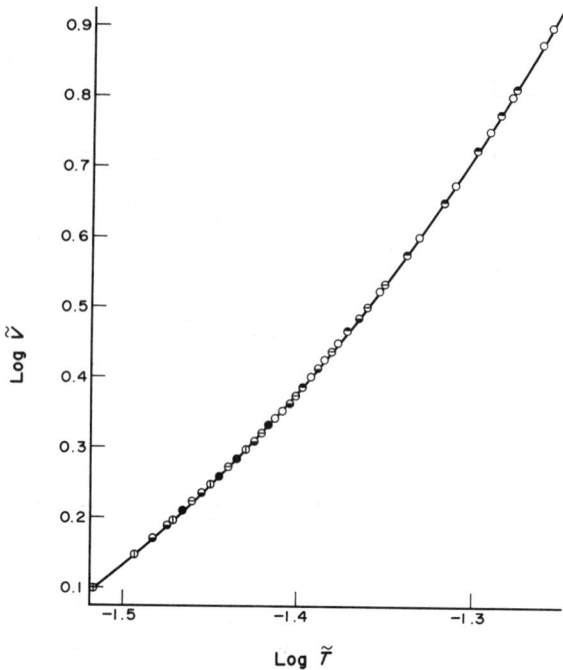

Figure 22   Logarithm of reduced volume *vs.* logarithm of reduced temperature for representative polymer melts and *n* alkane liquid. The line is calculated from equations (42) and (45) (after ref. 169, with permission)

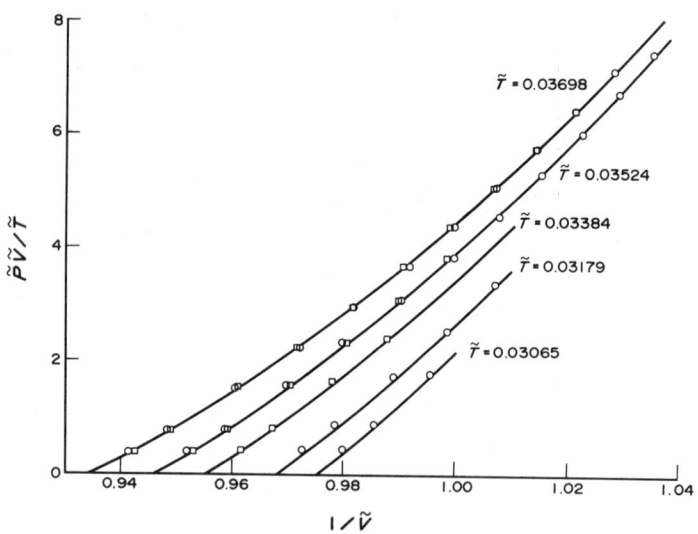

Figure 23   Reduced isotherms for polymer melts. The lines are calculated from equations (42) and (45) (after ref. 171, with permission)

between the slopes in the glassy region of the experimental volume–temperature curves and those for constant h (theoretical) led Simha and co-workers[170-173, 178] to the conclusion that $h = h(T_g)$ in the glass is an oversimplification.

An alternative procedure adopted by Quach and Simha[171] results in Figure 25. Here the glassy *PVT* data are used to calculate the fraction of holes as a function of temperature in the glass. The open circles show values of h calculated in this manner. Below T_g the solid line represents the assumption that the hole fraction in the glass remains constant at $h = h(T_g)$. Thus Figure 25 suggests that the free-volume fraction varies with temperature below T_g and the structure of the glass is not completely frozen. Finally, we note that the free-volume fraction at T_g given by the Simha-

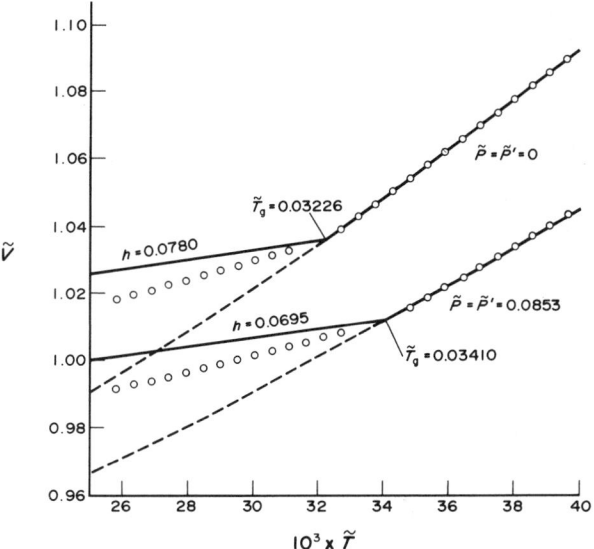

gure 24 Reduced volume *vs.* reduced temperature for poly(vinyl acetate) at two different reduced pressures. Solid lines are ▎culated from Simha–Somcynsky[164, 165] theory assuming that the hole fraction is frozen at T_g. Dashed lines are for the equilibrium liquids below T_g (after ref. 173, with permission)

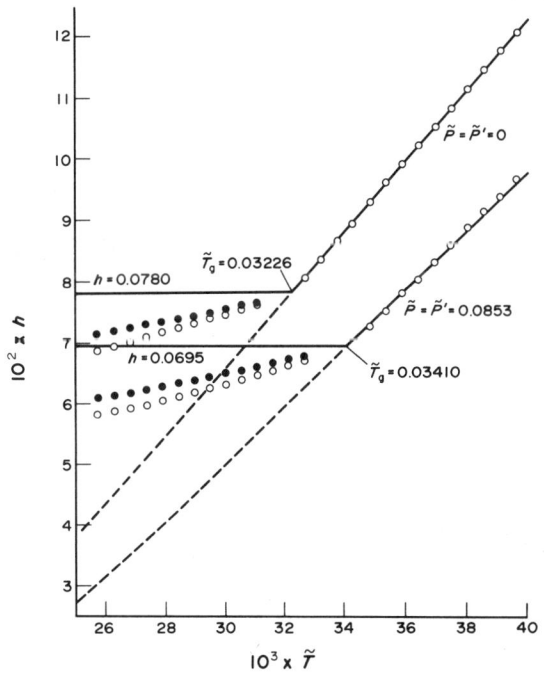

gure 25 Temperature dependence of hole fraction above and below the glass transition in poly(vinyl acetate). The points ▎present two different calculation schemes to determine h. These results show that h is not constant (*i.e.* frozen) below T_g (after ref. 170, with permission)

▎mcynsky[164, 165] model differs markedly from that used by Ferry and co-workers[57] (WLF-type), ▎ that which can be derived from the Gibbs–DiMarzio[24-26] lattice model. The Simha–Somcynsky ▎odel results in values of $h = f \simeq 0.07$, while the WLF empiricism and the Gibbs–DiMarzio lattice ▎odel give values of the free-volume fraction of about 0.025, as mentioned previously. The ▎screpancy with the WLF approach may come from the arbitrary choice of a value of unity for ▎arameter B in the Doolittle equation.

The Simha–Somcynsky model is successful in describing much of the thermodynamics (PVT) c glass-forming substances. Its one major drawback is that it does not develop the kinetics, an therefore the transition phenomenon, *a priori*. However, the model is useful as a quantitative free volume description of glassy behavior. As we will see, the cell-model limitations on prediction of th entropy (or energy) equation of state still hold for the Simha–Somcynsky description of the glass

In the following sections we will examine the ability of the free-volume models, with emphasis o the Simha–Somcynsky[164,165] model where possible, to describe the dependence of T_g on differer experimental parameters.

(iv) The influence of molecular weight on T_g

As discussed previously, there is a significant dependence of the glass transition temperatur on polymer molecular weight (see Figures 13 and 14) for linear chains. Fox and Flory[127-129] firs suggested that an empirical relation in which T_g varied inversely with number-average molecula weight, $\bar{M}_n$, could describe the dependence for $\bar{M}_n > 3000$

$$T_g = T_g^\infty - B/\bar{M}_n \tag{5.}$$

where T_g^∞ is the glass temperature for infinite molecular weight polymer and B is a constant. Thi relationship can be understood if the glassy state is taken to be an 'iso-free-volume' state. If we take linear chain, as $\bar{M}_n$ decreases the concentration of free ends increases and so does the total fre volume. Assuming that the free volume (or fractional free-volume) is constant at T_g for samples c different molecular weight, and that each terminal group contributes an excess free-volume θ, relation of the form of equation (52) is readily obtained.[70,82]

Let the excess free-volume per linear molecule be 2θ and the free volume per unit volume be equa to $2\theta\rho/N_A\bar{M}_n$, where ρ is density and N_A Avagadro's constant. Assuming constant free-volume, at T and a coefficient of expansion for the free volume α_f (see Figure 20), one can write

$$f = f_g + \alpha_f(T - T_g) \tag{5.}$$

As the excess free-volume introduced by the chain ends must be compensated for by therma contraction from T_g^∞ to $T_g^{\bar{M}_n}$; *i.e.*

$$2\theta\rho N_A/\bar{M}_n = \alpha_f(T_g^\infty - T_g) \tag{54}$$

then

$$T_g = T_g^\infty - \frac{2\theta\rho N_A}{\alpha_f\bar{M}_n} \tag{5.}$$

which is of the same form as equation (52) with $B = 2\theta\rho N_A/\alpha_f$.

Equation (52) or (55) describes the experimental data quite well except at low molecular weight (large values of $\bar{M}_n^{-1}$), and has been used extensively to describe the T_g–$\bar{M}_n$ relationship fo polymers. We note, however, that Gibbs and DiMarzio[25] have pointed out that the glass is not a tru iso-free-volume state, and the free-volume description of Simha and Somcynsky[164,165] also pre cludes[173] such a description of the glass transition.

The free-volume model does not explain the lack of molecular weight dependence of the T_g c certain polymers[72,79-81] mentioned previously, and obviously the increase in T_g at low molecula weights for ring-like molecules are not readily explicable in the context of a simple free-volum model.

(v) Heat capacity change at T_g

There has been little use of the free-volume models to describe or predict thermodynami functions other than the PVT equation of state for glasses, because these models are primaril concerned with molecular mobility. However, the cell model of Simha and Somcynsky[164,165] ha been used[170,178] to predict the heat capacities of poly(vinyl acetate) in the liquid and glassy state Table 3 shows a comparison between theory and experiment for C_p and C_v in the liquid and glass states, as well as for ΔC_p, ΔC_v and ΔS at the glass transition. Although the values of C_p and C obtained from the model are less than 20% of the observed ones, values for C_p–C_v are reasonabl close to those actually observed. The values for ΔC_p, ΔC_v and ΔS obtained from the model are les

Table 3 Comparison of Heat Capacity Data at $T_g = 304$ K for Poly(vinyl acetate) with Predictions from Simha–Somcynsky[164, 165] Cell Model (After Ref. 178)[a]

	Liquid		*Glass*
C_p $(T = T_g)$ (experimental)	1.77		1.27
C_p (theoretical)/C_p (experimental)	0.19		0.08
ΔC_p (experimental)		0.500	
ΔC_p (theoretical)		0.234	
ΔC_v (experimental)		0.310	
ΔC_v (theoretical)		0.039	
$S_g - S$ $(T = 273$ K) (experimental)		0.055	
$S_g - S$ $(T = 273$ K) (theoretical)		0.023	

[a] All units in $J\,g^{-1}\,K^{-1}$.

than half of the experimental values. This poor agreement with the experimental values undoubtedly lies in the previously mentioned problem that cell models predict low entropies. Thus, in spite of obtaining excellent agreement with the *PVT* behavior of the glass, the *SPT* behavior is not well described by the hole model.

(vi) Change in T_g due to crosslinking

The increase in glass transition temperature with increasing crosslink density is readily understood within the framework of the free-volume concept if one considers the fact that the introduction of crosslinks into the polymer involves the exchange of van der Waals' bonds for shorter covalent bonds. This results in a decrease in the specific volume (and, presumably, the free volume) of the polymer. Fox and Loshaek[74] and Loshaek[91] have treated the crosslinking effect on T_g from this point of view. They arrived at an expression for the difference between the glass transition of the crosslinked polymer $T_g(\rho)$ and the uncrosslinked polymer $T_g(0)$

$$T_g(\rho) - T_g(0) = \left(\frac{m_c \Delta \alpha_c T_g(0)}{\alpha_c - B_c - m_c \Delta \alpha \rho} \right) m_c \rho \tag{56}$$

where ρ is the number of crosslinks per gram, $\Delta \alpha = \alpha - \alpha_c$ is the difference between the coefficient of thermal expansion of the monomer, α, and that of the uncrosslinked high molecular weight polymer above T_g, α_c. B_c is the coefficient of expansion of the glassy high molecular weight polymer and m_c is the monomer molecular weight.

Equation (56) results in a linear dependence of T_g vs. ρ, and has been found to describe the change in T_g with crosslinking for many polymers. Figure 26 shows the results for poly(styrene-co-divinylbenzene) obtained by Fox and Loshaek.[74] We note here that some workers claim that T_g

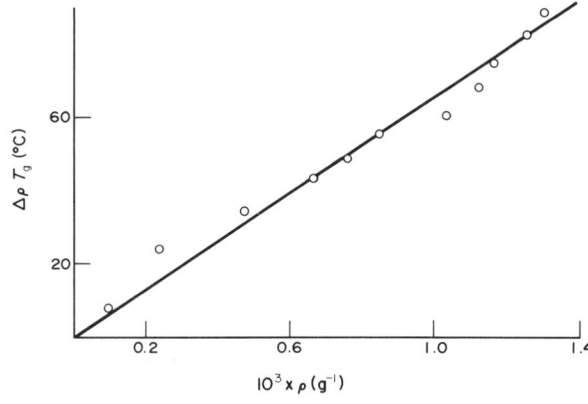

Figure 26 Change in glass temperature *vs.* crosslink density for poly(styrene-co-vinylbenzene) $\Delta\rho T_g = T_g(\rho) - T_g(0)$ and $\rho = $ crosslink density in crosslinks g^{-1}. The straight line is calculated from equation (56) (after ref. 74, with permission)

exhibits an exponential dependence on crosslink density, as predicted by the Gibbs–DiMarzio theory,[94] but data are difficult to obtain over a large range of crosslink densities, and distinguishing between a straight line and an exponential or other non-linear curve?[93] is difficult to do.

The free-volume models have also been used successfully to describe the effects of crosslinking on molecular mobility.[179,180] The WLF equation is generally invoked in these cases, in which case the free volume can be used to describe the mobility above T_g[180] and the effect of changing T_g on cure kinetics.[179,180]

(vii) Effect of deformation on T_g

The fact that the volume of a material changes upon deformation suggests that the free volume also changes. Thus, if one associates the sign of the global volume change with the change in free volume, the glass transition should shift to lower temperatures when the volume increases and to higher temperatures when the volume decreases. Thus, in uniaxial compression T_g should increase because the volume decreases. On the other hand, the volume in uniaxial extension increases and therefore T_g should decrease. In shearing deformations there should be no change in T_g.

These free-volume arguments give quite different predictions for the effect of deformation upon the glass transition temperature than does the Gibbs–DiMarzio theory discussed previously. Unfortunately, there have been few experiments designed to test the effects of deformation on the glass transition temperature. Furthermore, as noted by Johnston and Shen,[100] the change in T_g expected for tensile deformations is only of the order of 1–2 K for 100% elongation, and determinations of T_g are often imprecise and the elongations relatively small. Perhaps the best measurements are those of Gee et al.[99] cited previously (Figure 18), which indicate that T_g increases upon extension. However, Mason,[101] Witte and Anthony[102] and Stevens and Ivey[103] have observed decreases in T_g upon stretching. On the other hand, in addition to Gee et al.,[99] Oshimo and Kusumoto[104] have observed increases in T_g with stretching. This author knows of no measurements of T_g changes due to uniaxial compression or shearing deformations. Such experiments may elucidate the free volume vs. entropy and the thermodynamic vs. kinetic bases of the glass transition.

(viii) Compositional dependence of T_g

As mentioned previously, the glass temperature of polymer–diluent systems, polymer–polymer blends and copolymers depends upon the composition. The free-volume theories have been used to describe this behavior with varying degrees of success. First we examine the question of polymer–diluent systems.

Following Ferry,[57] we first remind the reader that addition of low molecular weight diluents depress the glass temperature, as depicted in Figure 27 (see also Figure 18). Jenckel and Heusch[181] first arrived at an equation in which the T_g depression is linear in the weight fraction, w_1, of the diluent

$$T_g(w_1) \; = \; T_g(0) \; - \; kw_1 \tag{57}$$

where $T_g(0)$ is the glass temperature of the undiluted polymer and k is a constant depending on the solvent. However, as seen in Figure 27, the actual dependence on weight fraction of diluent is nonlinear. This non-linear T_g depression has been attributed to the introduction of additional free-volume upon addition of the diluent,[57] as would be expected if the fractional free-volume of the diluent (f_1) exceeds that of the polymer (f_2). Normally, the T_g of the polymer is greater than that of the diluent, i.e. $T_{g1} < T_{g2}$, which corresponds to $f_1 > f_2$. Equation (57) results when the free volumes of the diluent and polymer are additive. In the linear range, f depends upon w_1 in a linear fashion. Then

$$f(T, w_1) \; = \; f_2(T) \; + \; \beta w_1 \tag{58}$$

or in terms of the volume fraction v_1

$$f(T, v_1) \; = \; f_2(T) \; + \; \beta' v_1 \tag{59}$$

where β and β' are constants, $\beta' = \beta \rho_1 / \rho_2$. If T_g represents an iso-free-volume-fraction state, then, from equation (57), $k = \beta / \alpha_f$. Alternatively,[182]

$$\beta' \; = \; \alpha_{f1}(T \; - \; T_{g1}) \; - \; \alpha_{f2}(T \; - \; T_{g2}) \tag{60}$$

where α_{f1} and α_{f2} are the expansion coefficients of the fractional free-volume defined previously (see Figure 20).

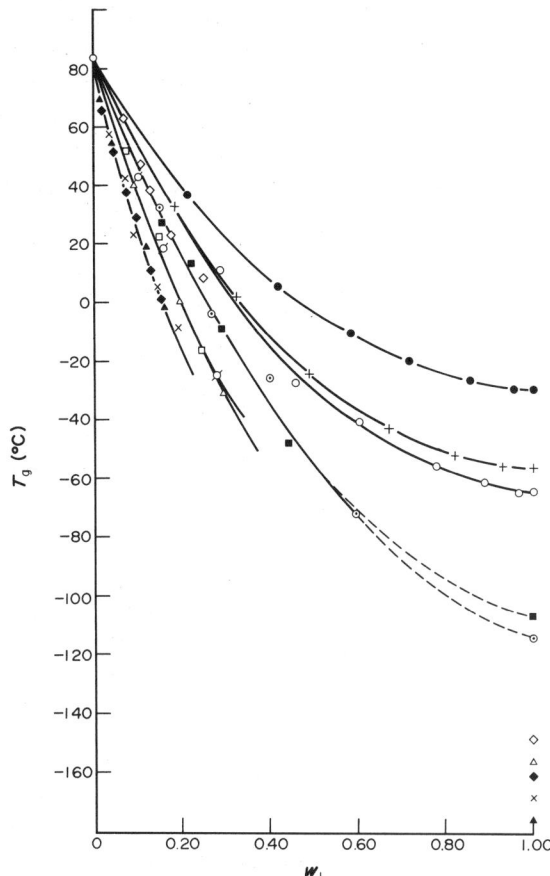

Figure 27 Glass transition temperatures of polystyrene solutions with various low molecular weight diluents: ●, β-naphthyl salicylate; +, phenyl salicylate; ○, tricresyl phosphate; ⊙ methyl salicylate; ■, nitrobenzene; ◇, chloroform; ×, methyl acetate; ◆, ethyl acetate; ▲, carbon disulfide; □, benzene; ⌀, toluene; △, amyl butyrate (after ref. 181, as cited in ref. 57, with permission)

If one assumes that the fractional free-volumes are not additive,[183] the curvature in Figure 27 can be accounted for and one writes the total fractional free volume as

$$f = v_1 f_1 + v_2 f_2 + k'_v v_1 v_2 \tag{61}$$

where k'_v is a negative-valued interaction parameter of the order of 10^{-2}. This fits well with the observation that the total specific volume depends in a similar way on weight fraction[181]

$$1/\rho = w_1/\rho_1 + w_2/\rho_2 + k_v w_1 w_2 \tag{62}$$

which implies that the occupied volumes, not the free volumes, are additive. Equation (61) leads to the following compositional dependence of T_g

$$T_g(v_1) = \frac{v_1 \alpha_{f1} T_{g1} + v_2 \alpha_{f2} T_{g2} + k'_v v_1 v_2}{v_1 \alpha_{f1} + v_2 \alpha_{f2}} \tag{63}$$

Kelley and Bueche[184] derived similar equations for the composition dependence of polymer–diluent systems from free-volume considerations and also showed acceptable fits to the data. The parameters used to fit equation (63) to the data are generally found to be in the range expected from free-volume arguments. Of particular interest is the fact that at small diluent-contents, one finds that equation (61) corresponds to[138]

$$\beta' = f_1 - f_2 + k'_v \tag{64}$$

and, since f_2 and k'_v are small, $\beta' \approx f_1$. Then, f_1 is in the range 0.1 to 0.3, which is the correct magnitude of the fractional free-volume of small-molecule liquids far above T_g. The above development shows the power of free-volume models to describe the observed experiments.

The dependence of the glass transition temperature on copolymer composition has been treated by Gordon and Taylor[185] using a volume approach. In their development for 'ideal' copolymers, Gordon and Taylor[185] assume additivity of volumes of the two components in both rubbery and glassy states. Then two equations result

$$T > T_g \quad v_R = w_1 v_{R1} + w_2 v_{R2} \tag{65}$$

$$T > T_g \quad v_G = w_1 v_{G1} + w_2 v_{G2} \tag{66}$$

where v_R and v_G are the specific volumes of the copolymer in the rubbery and glassy states, respectively, w_i is the weight fraction of the ith component and v_{Ri} and v_{Gi} are the specific volumes of the rubbery and glassy phases of the ith component. Assuming linear expansion coefficients in the liquid and glassy states for each component α_{li} and α_{gi}, one arrives at an equation for the concentration dependence of T_g for the copolymer[91,185]

$$T_g(w_2) = \frac{w_1 T_{g1} k + w_2 T_{g2}}{w_1 + k w_2} \tag{67}$$

where $k = \Delta\alpha_2/\Delta\alpha_1$ and $\Delta\alpha_i = \alpha_{li} - \alpha_{gi}$.

In Figure 28 is shown the excellent agreement obtained between equation (67) and the experimental data for poly(butadiene-*co*-styrene). Loshaek[91] reports equally good results for poly(methyl methacrylate-*co*-glycol dimethacrylate)s. Non-additivity of volumes would, of course, alter these results, as equations (65) and (66) would require additional terms. We note that equation (67) is identical to equation (63) with $k'_v = 0$.

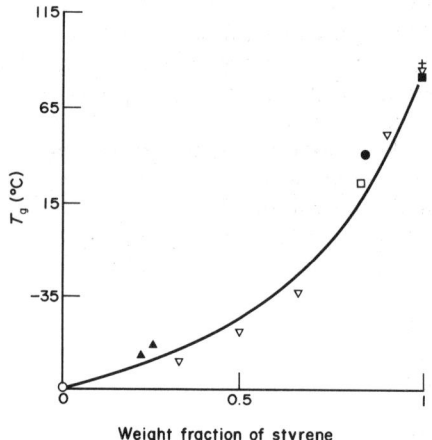

Figure 28   Glass transition temperature *vs.* weight fraction styrene for poly(butadiene-*co*-styrene) The line is calculated from equation (67) (after ref. 185, with permission)

Similar equations based on free-volume arguments have been proposed for block copolymers.[186] Again the agreement is quite satisfactory. For polymer blends, equations of the same sort have been successfully used to describe the change in T_g with blend composition.[113,187] For these latter systems, however, the experimental measurement of T_g may dramatically affect the agreement obtained. Figure 29 shows T_g *vs.* composition for mixtures of poly(2,6-dimethyl-1,4-phenylene oxide) (PPO) with polystyrene (PS) obtained by different experimental methods.[188] As can be seen, the shape of the curve depends dramatically on the method of measurement.

10.3   KINETICS OF GLASS-FORMING SYSTEMS

10.3.1   Introduction

In the previous section the thermodynamics of the glass transition and the glassy state were described. From the discussion there, it is obvious that a complete understanding of the nature of

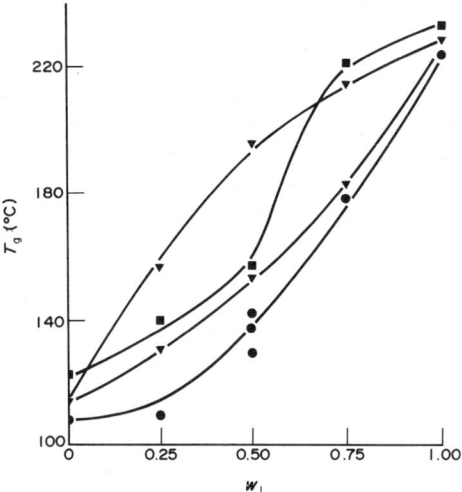

Figure 29 Glass transition temperature, measured by different techniques (●, DSC; ▼, mechanical; ■, dielectric), of blends of poly(phenylene oxide) and polystyrene *vs.* weight fraction poly(phenylene oxide) *w*. (after ref. 188, with permission)

glasses remains elusive for the major reason that, regardless of whether or not underlying the glass transition there is a true second-order thermodynamic transition, the overwhelming features of observed glassy behavior are kinetic.

The phenomenology of the kinetics of glasses has been studied and known for many years. However, it is only recently that the mathematical formalism has become available to describe the major features of the known behavior, which shows a rich array of phenomena. In the sections which follow, we describe the phenomenology of the kinetics of glassy systems and the equations which have been developed to explain their kinetic behavior. We subsequently show how the statistical mechanical models of glasses have been adapted to the mathematical formalism which describes the kinetics. We will also examine some recent developments in the use of computer simulations to look at the kinetics of glasses and, finally, we will briefly describe the effects of volume recovery on the mechanical behavior of glasses, *i.e.* physical aging.

10.3.2 Phenomenology

10.3.2.1 Kinetics of glass formations

The glass transition event manifests itself as a change in the temperature dependence of some property (*e.g.* volume V, enthalpy H) of a liquid upon cooling. As shown in Figure 1, below some critical temperature range, the volume (enthalpy, entropy) of the undercooled liquid departs from equilibrium. The glass transition temperature T_g is determined by the time scale of the experiment t_e, and corresponds to the temperature at which the characteristic time τ for the molecular rearrangements leading to structural (volume, enthalpy, entropy) recovery is longer than the time scale of the experiment, *i.e.* $\tau > t_e$. Therefore, as t_e increases, the position of the departure from equilibrium moves towards lower temperatures. The intersection of the extrapolated glassy and liquid lines defines $T_g(t_e)$. This definition was used by Tool[189,190] to define the fictive temperature, T_F, which has come to be used to quantify the structure of the glass at an arbitrary temperature in terms of the structure of the equilibrium liquid at T_F (or T_g). This concept, while not entirely valid, is useful and will be referred to subsequently.

The variation of T_g with the experimental time scale t_e, or rate of cooling q, for a large number of glass-forming systems is[57]

$$\frac{-dT_g}{d\log t_e} \simeq \frac{dT_g}{d\log q} \approx 3\,K \tag{68}$$

although, as indicated in Section 10.2, there are systems which show a rate sensitivity which is much higher.

10.3.2.2 Kinetics in the glassy state

(i) Intrinsic isotherms

Upon cooling the equilibrium liquid below the glass transition range, the thermodynamic properties of the glass (V, H, *etc.*) continue to evolve toward equilibrium as shown in Figure 3. Kovacs[11] has referred to the isothermal contraction curves resulting upon interruption of cooling at different temperatures T_1 as intrinsic isotherms. There are two interesting features of the curves in Figure 3. First, the departure from equilibrium, $\delta = (v - v_\infty)/v_\infty$ does not vary linearly with T_1. Second, the times required to reach equilibrium increases dramatically as δ_1, the initial departure from equilibrium, increases (or T_1 decreases).

(ii) Asymmetry of isothermal recovery

If the glass is allowed to equilibrate (reach true thermodynamic equilibrium) at a temperature $T_0 = T - \Delta T$ or $T_0 = T + \Delta T$, then the volume recovery upon changing the temperature to T will differ depending upon whether or not one is expanding or contracting towards equilibrium. This asymmetry of approach to equilibrium (depicted in Figure 30) is well understood qualitatively. When the approach towards equilibrium is from above, the material has a structure which corresponds to the liquid having higher mobility than the equilibrium mobility. Thus, the contraction is autoretarded. Conversely, the expansion towards equilibrium from below is autocatalytic, because mobility increases as the departure from equilibrium decreases towards an equilibrium state of higher mobility.

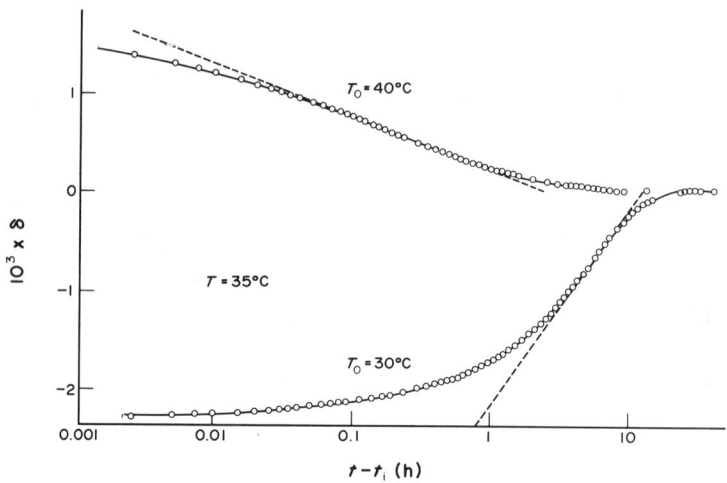

Figure 30 Expansion and contraction isotherms for poly(vinyl acetate) glass at $T = 35\,°C$ after heating or cooling from $T_0 = T \pm 5\,°C$. This plot shows the asymmetry of the expansion and contraction isotherms (after ref. 13, with permission)

(iii) Memory effects

One of the (apparently) most complicated features of the kinetic behavior of glasses arises when the material is allowed to recover isothermally for a period of time t_1 insufficient to reach equilibrium, and then heated to a higher temperature and allowed to recover. As shown in Figure 31, the volume departure from equilibrium can cross over the actual equilibrium and exhibits a maximum which depends upon the actual thermal history applied to the sample. These have been referred to as crossover or memory effects. As will be seen, they arise from the fact that the response function (*e.g.* for volume recovery) exhibits behavior equivalent to a multiplicity of retardation mechanisms.

(iv) Uniform heating through the transition range

A common experiment is to study the glass transition behavior by DSC. In such experiments, a sample is brought into the glassy state either by an unknown route, *i.e.* 'as received', or by a controlled thermal history. During the experiment, the heat flow into the sample required to

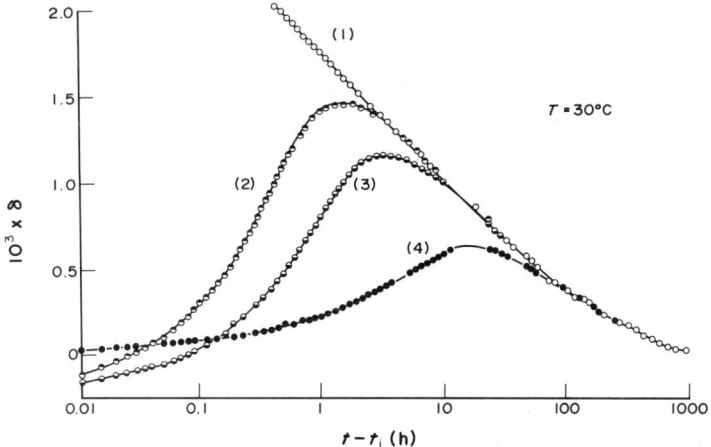

Figure 31 Isothermal evolution at $T = 30\,°C$ for poly(vinyl acetate) showing 'memory' effect: (1) quench from $40\,°C$ to $30\,°C$; (2) quench from $40\,°C$ to $10\,°C$ for 160 h followed by rapid heating to $30\,°C$; (3) quench from $40\,°C$ to $15\,°C$ for 140 h followed by rapid heating to $30\,°C$; and (4) quench from $40\,°C$ to $25\,°C$ for 90 h followed by rapid heating to $30\,°C$ (after ref. 13, with permission)

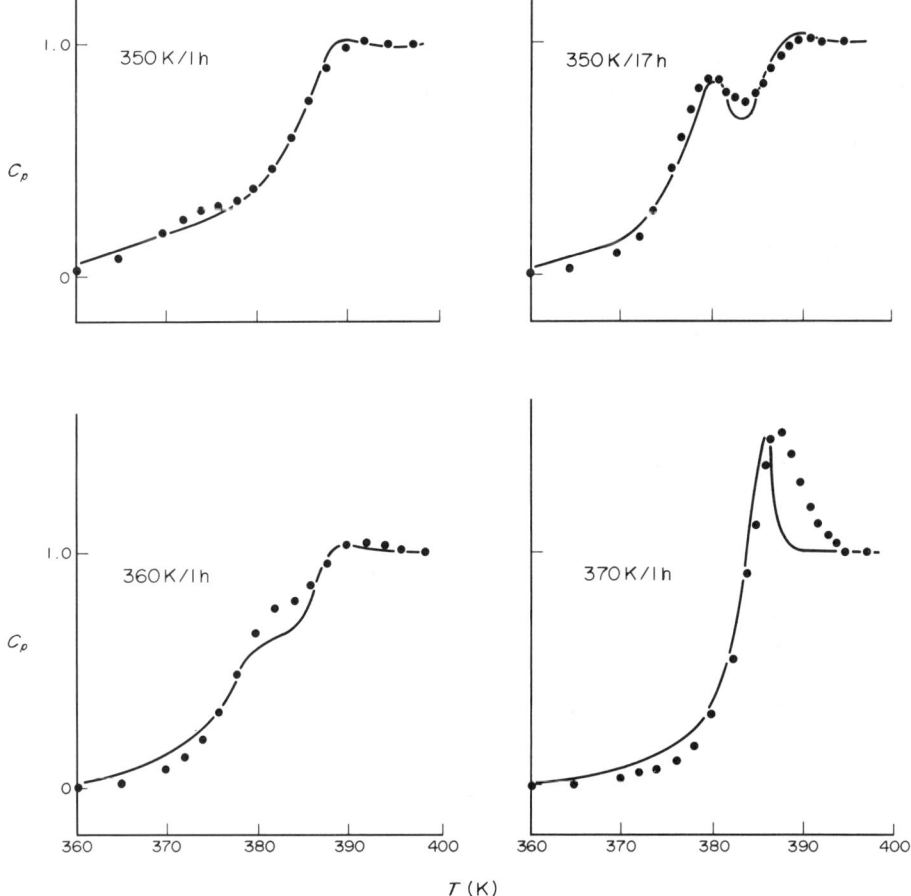

Figure 32 Heat capacity C_p from DSC measurements *vs.* temperature for poly(methyl methacrylate) (PMMA) annealed at T_e for a time t_e as indicated (*i.e.* T_e/t_e). The glasses were cooled at 40 K min^{-1} and reheated at 10 K min^{-1} after annealing. Note the 'features' in these DSC curves (after ref. 204, with permission)

maintain the desired temperature is measured. In an ideal case, the change in heat capacity can be measured on samples which have been cooled slowly and reheated immediately at approximately the same rate. However, upon choosing 'poor' experimental conditions, the DSC trace can exhibit complicated features, as shown in Figure 32. The peaks in Figure 32 have been variously interpreted as having specific molecular significance. However, as we shall see, they can arise simply from the kinetics of response to a complicated thermal history. Analogous features would be observed in volume measurements if α were followed as a function of temperature upon heating.[11,19]

10.3.3 Phenomenological Equations

10.3.3.1 *Requirements of the models*

From the above discussion one can see that the kinetics of glasses exhibit a rich variety of behaviors which offer a severe test of any model. Until the development by Narayanaswamy[17] in 1971 of a mathematical formalism analogous to that of reduced variables used in viscoelasticity theories, there was no model which incorporated all of the 'ingredients' necessary to explain the major features of the phenomenology described in the previous section.

The pieces necessary to describe the behavior, in addition to the viscoelasticity formalism, are (a) a non-exponential response function which can be represented in terms of multiple retardation mechanisms characterized by a distribution of retardation times τ_i; and (b) a departure from equilibrium, e.g. δ, related to T and t by a linear superposition principal but with each retardation time τ_i dependent upon the temperature and on the instantaneous state (structure) of the glass.

Several developments based upon these ideas were made by Narayanaswamy,[17] Kovacs and co-workers[18,19,191-194] and Moynihan and co-workers.[20,195,196] The result has been a tremendous improvement in our ability to understand and describe the kinetic behavior described above. In fact, once the model is developed, a single non-linear equation describes all of the major features of the phenomenology. In the next section, we will develop this equation following the approach of Kovacs and co-workers,[18] the so-called KAHR model, because of this author's familiarity with the model. We note that the resulting equations are formally identical with those of Narayanaswamy[17] and Moynihan and co-workers,[20,195,196] the differences being not in the underlying assumptions of the models, i.e. (a) and (b) above, but rather in the specific routes and approximations used to obtain the equations.[19] An excellent treatment of the Narayanaswamy model and applications is given by Scherer.[5]

10.3.3.2 *The Kovacs–Aklonis–Hutchinson–Ramos (KAHR) model*

(i) Single-parameter approach

In this section and the next will arrive at a set of equations which describe the volume recovery under isobaric conditions of model glasses. The KAHR model of Kovacs, Aklonis, Hutchinson and Ramos[18] formally uses a multiple ordering-parameter model to define the 'viscoelastic' response function for the glass. Here we take a somewhat different approach, summarized by Kovacs,[19] which avoids the specific formalism of ordering parameters and results in the same set of equations. It is important to remark that the ordering-parameter approach to describing the kinetics of glasses is interesting and has been increasingly formalized, not only in the KAHR model but also by Moynihan and co-workers.[195,196] Here we wish to focus on understanding the important parameters which yield the rich kinetic phenomenology described in the previous sections.

In early work, Tool[189,190] and Davies and Jones[53] examined the structural recovery of glasses assuming that the departure from equilibrium depends upon a single ordering-parameter. Accordingly, the recovery was described by a single retardation time τ_i. Here we define as the ordering parameter the volume departure from equilibrium expressed as a dimensionless variable δ

$$\delta = (\upsilon - \upsilon_\infty)/\upsilon_\infty \tag{69}$$

where υ is the instantaneous specific volume and υ_∞ the equilibrium value at the same T and (P). Then at any temperature the rate of recovery at constant P is[11,53,191,197]

$$-\frac{d\delta}{dt} = q\Delta\alpha + \delta/\tau \tag{70}$$

where $q = dT/dt$, t is the time and $\Delta\alpha = (\alpha_1 - \alpha_g)$ is the excess expansion coefficient of the liquid with respect to the glass. The enthalpy recovery can also be described by equation (70) if we refer to the enthalpy departure as $\delta_H = H - H_\infty$ and substitute the excess heat capacity $\Delta C_p = (C_{p_1} - C_{p_g})$ for $\Delta\alpha$. In addition, equation (2) can be cast in terms of the fictive temperature T_F to give[19,189,190]

$$dT_F/dt = (T - T_F)/T \tag{71}$$

where $T_F = T + \delta/\Delta\alpha$ and $T_F = T + \delta_H/\Delta C_p$ for volume and enthalpy recoveries, respectively. (Tool used the fictive temperature to define the structure of the glass. In simple cooling experiments T_F is the intersection of the extrapolated glassy and the extrapolated liquid lines, and is equivalent to T_g (see Figure 1, for example). For more complicated histories T_F can be redefined and the reader is referred to ref. 5 for an excellent discussion of T_F.)

The solution of equation (70) depends upon the functionality of the retardation time τ. This consequently determines how well actual glassy behavior is described by the one-parameter model. If one assumes that τ depends upon temperature alone (in isobaric experiments), then glass formation and the uniform heating through the glass transition range can be qualitatively described.[18,197] However, this model predicts that isothermal recovery proceeds *via* a simple exponential decay of δ in contradiction with observation. Also a dependence of τ on T alone cannot account for the asymmetry of approach depicted in Figure 31.

Tool,[189,190] who recognized this shortcoming, was able to account for the non-linearity and asymmetry of the isothermal approach to equilibrium by assuming that τ depends, in addition to T and (P), on the instantaneous state of the glass, *i.e.* on T_F or δ. Introduction of a δ-dependent τ into equation (70) provides qualitative agreement with the isothermal recovery of glasses. In particular, the intrinsic isotherms[11] are reasonably well described in such a model, as is the asymmetry of isothermal approach towards equilibrium.[5]

However, one-parameter models cannot even qualitatively describe the memory effects depicted in Figure 31. These effects arise precisely because of the contributions of at least two independent retardation mechanisms involving two or more retardation times. In this context, two or more ordering parameters are required to describe the kinetics. (We note that this is true within the context of a discrete spectrum of retardation times. Moynihan *et al.*[20] have pointed out the usefulness of a non-exponential decay function rather than the multiplicity of mechanisms inherent in the ordering-parameter approach.)

(ii) Multiparameter approach

The multiparameter treatment of the KAHR[18] model can be viewed as a simple extension of the single-parameter model. The development is formalized by defining the thermodynamic state of the glass system at any T and P by a set of N ordering parameters, Z_i. Both δ (the departure) and $\Delta\alpha$ are partitioned among the Z_i, with each of the latter being associated with a unique retardation time τ_i. It is furthermore assumed that the contribution of each Z_i to δ is independent of the other ordering parameters. Then, analogous to equation (70), we have

$$-\frac{d\delta_i}{dt} = q\Delta\alpha_i + \delta_i/\tau_i \quad (1 \leq i \leq n) \tag{72}$$

where $\Delta\alpha_i = g_i\Delta\alpha$ is the weighted contribution of the ith retardation mechanism to $\Delta\alpha$ with

$$\sum_{i=1}^{n} g_i = 1 \tag{73}$$

The KAHR model assumes that each τ_i depends on the total departure from equilibrium (*i.e.* on δ not on δ_i) which couples the set of equations (72). Furthermore, the total departure is the sum of the individual departures, *i.e.*

$$\delta = \sum_{i=1}^{n} \delta_i \tag{74}$$

The solution to equation (72) now depends upon the specific temperature (pressure) and structure dependence of the τ_i. These dependencies are put into the KAHR model in a manner equivalent to the time–temperature superposition principles of viscoelasticity theory.[57,198] Then KAHR assume that, by a change in temperature or δ, each retardation time is shifted by the same amount and that the amount of shift due to a change in temperature is independent from that due to the departure (δ) from equilibrium (*i.e.* structure). As a result of this, the spectrum of retardation times simply shifts along the time axis but does not change shape, *i.e.*

$$\tau_i(T, \delta) \;=\; \tau_{i,r} a_T a_\delta \tag{75}$$

where a_T and a_δ are the appropriate shift factors for the spectrum at any T and δ relative to $\tau_{i,r}$ in the reference state at equilibrium. Then simply

$$a_T \;=\; \tau_i(T, 0)/\tau_i(T_r, 0) \tag{76}$$

$$a_\delta \;=\; \tau_i(T, \delta)/\tau_i(T, 0) \tag{77}$$

where T_r is a reference temperature and $\delta = 0$ denotes equilibrium. The set of paired values $(g_i, \tau_{i,r})$ defines the retardation spectrum, $G(\tau_{i,r})$ of the system in a reference equilibrium state at T_r. Invariance of the shape of the spectrum is assured by assuming the g_i are independent of T and δ. Then one can write the thermal history dependence of δ in a compact form by using a reduced-time variable z

$$z \;=\; \int_0^t \frac{\mathrm{d}t'}{a_T a_\delta} \tag{78}$$

which reduces the instantaneous rates of change of structure (δ_i) at any T and δ to those obtained in equilibrium at T. The relevant expression for $\delta(t)$ is then given by the convolution integral

$$\delta(z) \;=\; -\Delta\alpha \int_0^z \mathrm{R}(z \,-\, z')\frac{\mathrm{d}T}{\mathrm{d}z}\,\mathrm{d}z' \tag{79}$$

where $\mathrm{R}(z)$ is the normalized retardation (recovery) function for the sytem

$$\mathrm{R}(z) \;=\; \sum_{i=1}^{N} g_i \exp(-z/\tau_{i,r}) \tag{80}$$

the value of which changes from unity to zero as z (or t) increases from zero to infinity.

Importantly, we now have a formalism which allows one to describe the behavior of a hypothetical glass once one has determined a_T, a_δ, $\Delta\alpha$ and $\mathrm{R}(z)$. All of these material functions must be determined from experiment. This implies further assumptions about the model which we will deal with in the next section, where we show how well the model agrees with the phenomenology presented previously. Before leaving this section, we note that two assumptions which have been made in the model are unnecessary. First is the invariance of the shape of the retardation function with changes in T and δ. The other is that the parameters g_i are independent of the magnitude of T or δ (or δ_H), which results in $\Delta\alpha$ (or ΔC_p) being a constant. For isobaric experiments this is perhaps unimportant. The extension of this formalism to include pressure histories as well may require a pressure dependence in the g_i (which weights $\Delta\alpha_i$) as well as another set of weighting functions for the pressure equivalent of $\Delta\alpha$, *i.e.* $\Delta\kappa$.[18] In the formalism of equation (79) $\Delta\kappa$ would no longer be a constant. Then equation (79) would exhibit non-linearities due to a magnitude dependence of $\Delta\alpha$ (or $\Delta\kappa$) in addition to those due to the structure-dependent retardation spectrum. This will not be dealt with further.

(iii) A comparison of the phenomenological equations with experiment

Equation (79) is a general equation in which the recovery function $\mathrm{R}(z)$ and the shift factors a_T and a_δ are unknown. $\Delta\alpha$ is unknown but easily measured. In the KAHR[18] model, $\mathrm{R}(z)$ is a function which can be expressed by a spectrum of retardation times (equation 80). A more easily visualized function is the so-called Kolrausch[199]–Williams-Watts[200] (KWW) function introduced to describe glassy kinetics by Moynihan *et al.*[20, 195, 196]

$$\mathrm{R}(z) \;=\; \exp(-z/\tau_r)^\beta \tag{81}$$

where τ_r denotes a characteristic retardation time at T_r and β is a parameter describing the width of the 'spectrum', $0 \le \beta \le 1$. In the discussion which follows we will use the KWW function for $\mathrm{R}(z)$, *i.e.* equation (81), rather than the sum of exponentials, equation (80).

The forms of a_T and a_δ used in the KAHR model are

$$a_T = \exp[-\Theta(T - T_r)] \tag{82}$$

$$a_\delta = \exp[-(1 - x)\Theta\delta/\Delta\alpha] \tag{83}$$

where Θ and x are material parameters, and x is a partitioning parameter ($0 \leq x \leq 1$) which determines the relative contributions of temperature and structure to τ_i. When $x = 1$, $a_\delta = 1$, *i.e.* there is no structural dependence of the retardation times and they depend only on temperature. This parameter has also been referred to as a non-linearity parameter,[201] since the non-linearity is introduced into the model due to the structure dependence of τ_i. The forms for a_T and a_δ introduced into the KAHR model are those originally put forward by Tool,[189,190] but expressed in terms of δ rather than T_F. Similar expressions can be obtained from free volume,[11,191] configurational entropy[202] or activation enthalpies.[17,20,195] These expressions reduce to equations (82) and (83) when the temperature interval chosen is not too wide.[18,19]

The problem of determining the adequacy of equation (79) to describe the data is one of determining $\Delta\alpha$, (or ΔC_p), x and Θ, and $R(z)$. The difficulty is that the non-linear nature of equation (79) makes unique determination of these parameters and functions difficult. However, upon assuming reasonable values for $\Delta\alpha$, Θ and x and a two-box representation of the distribution of retardation times, KAHR[18] were able to solve equation (79) numerically for various thermal histories. The results obtained reproduce[18,19] all of the features shown in Figures 1, 3 and 30–32, *i.e.* the specific non-linearity and asymmetry of the isothermal approach curves, memory effects and the peaks obtained on constant heating through the transition range, although in the latter α was used rather than C_p.

It is interesting to note here that the combination of a broad distribution of retardation times, combined with a structure-dependent shift factor for these times explains some extremely complicated features of glassy behavior. A set of experiments which are of particular interest are depicted in Figure 32, *i.e.* uniform heating through the transition range. The peaks exhibited in such DSC traces of heat capacity *vs.* temperature have been related to specific relaxation mechanisms in polymers. However, great care must be used in interpretation of such peaks, because in non-isothermal experiments the glassy kinetics themselves can give rise to such peaks. As an example, Hodge and Berens[203] have carried out calculations using the Narayanaswamy[17]–Moynihan[20] form of equation (79) for different conditions. As seen in Figures 33(a) and (b) the annealing time and temperature of the glass can affect considerably the position and magnitudes of the peaks. A full analysis and discussion of this is presented by Hodge and Berens,[201,203-205] by Kovacs and co-workers[18,19,191-194,197] and Tribone *et al.*[206] One question remains: 'How well can the models describe the experimental data?'. The answer is very well — but! The 'but' implies that there are still some things which have not been completely explained. Kovacs and co workers[18,19,191-194,197] treat the problem as one of developing a rigorous test, not only of the phenomenology predicted by the model, but also of the underlying physical assumption of the model. In this case, it is necessary to determine by independent measurements the values of $\Delta\alpha$, x, Θ and $R(z)$, and to then compare actual data with the model predictions. The non-linear nature of equation (79) makes such independent determinations difficult. However, KAHR[18] were able to give methods for determining all of the parameters in the model except x, the structure-partitioning or non-linearity term. Although Lagasse *et al.*[207] claim to have found a method to obtain x, this has been disputed by Kovacs.[208] Without x, one is left to examine the model by curve-fitting procedures. Several groups have adopted this approach with exceptional success.[201,203-206,209-212]

Finally, we note that the extensive testing of the phenomenological equations, while indicating satisfactory[5] agreement with the experimental results for most 'practical' applications, has shown less than satisfactory agreement from a scientific point of view, for two reasons. The first is that the values of the parameters required for optimal fit to the data change from one set of experiments to the next. In particular the shape of the distribution of retardation times, as measured by β in equation (81) and the structure dependence of the retardation times as measured by the non-linearity parameter x, change with thermal history.[5,201,206,209-212] In Figure 34, for example, a DSC curve calculated using parameters determined from one thermal history is compared with the data for a different history. Although the calculated behavior is similar to that observed, the agreement is not perfect. Better fits are obtained by allowing both x and β to vary with history, which then violates some underlying assumptions about the model, *i.e.* that the shape of the spectrum is invariant and, more importantly, the current retardation times depend only upon the instantaneous departure from equilibrium. The second difficult problem in the description of the data arises in the fits to the asymmetric-approach curves (see Figure 30). Although one can fit these curves quite well, a close

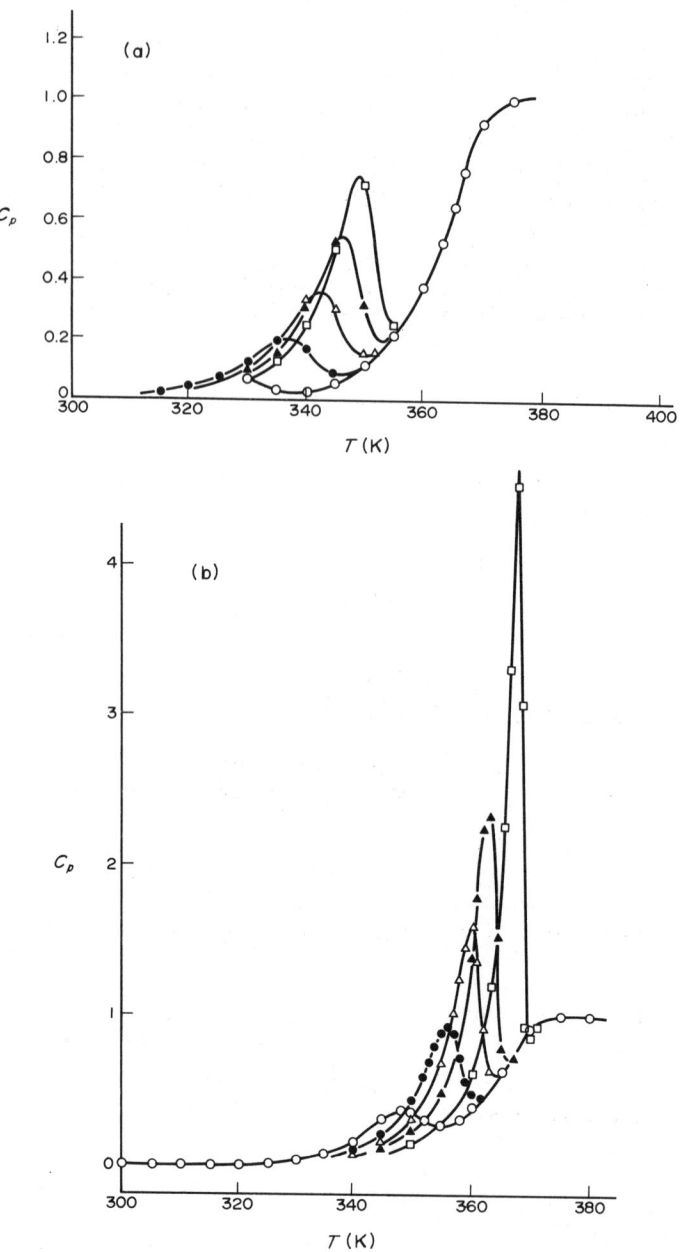

Figure 33 DSC curves calculated using a Narayanaswamy[17]–Moynihan[20]-type phenomenological model for glasses subjected to different annealing times t_e. ($\bigcirc$, 1 h; $\bullet$, 1 day; $\triangle$, 1 week; $\blacktriangle$, 1 month; $\square$, 1 year) after quenching to different temperatures: (a) $T_e = 40\,°C$; (b) $T_e = 60\,°C$. This demonstrates how complex features in DSC curves can be reproduced by phenomenological equations (after ref. 201, with permission)

scrutiny of the fit in the light of the so-called 'τ-effective paradox' (see next section) implies that there is an 'ingredient' missing in the underlying physics of the phenomenological models. We turn now to this paradox.

(iv) The τ-effective paradox

Upon bringing a glass from equilibrium at a temperature $T_0 = T_1 \pm \Delta T$ to the new temperature T_1, the volume (enthalpy) recovery towards equilibrium exhibits either autocatalytic or autoretarded behavior depending upon whether the approach towards equilibrium is from below or above. We previously referred to this behavior as the asymmetry of approach in T-jump experiments (Figure 30). As described above, the phenomenological equations of the Narayanaswamy-

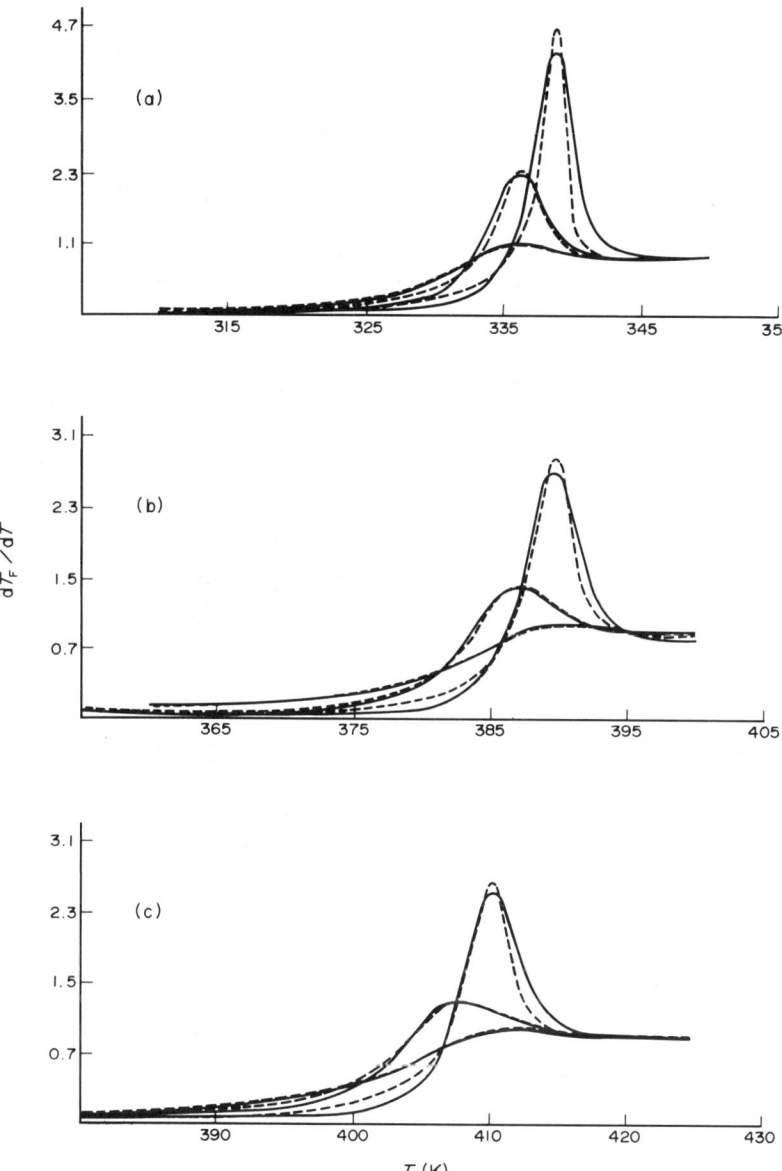

Figure 34 Comparison of experimental (solid) and calculated (dashed) enthalpy recovery curves: (a) isotactic PMMA, $T_e = 311$ K; (b) atactic PMMA, $T_e = 362$ K; (c) syndiotactic PMMA, $T_e = 380$ K. Annealing temperatures are approximately 15 K below T_g. The curves represent $t_e = 0, 6000$ and $60\,000$ s, with peak height increasing with annealing time (after ref. 206, with permission)

KAHR–Moynihan-type model this behavior quite well. However, in earlier work, Kovacs[13] examined this behavior through an effective retardation time τ_{eff} defined as

$$\tau_{eff}^{-1} = -\frac{1}{\delta}\left(\frac{d\delta}{dt}\right) \tag{84}$$

where t is the time and δ is the departure from equilibrium. τ_{eff} is a derivative parameter which makes fitting its behavior an extremely demanding test of any model.[18,19,21,213] In Figure 35 are depicted τ_{eff} plots for the expansion and contraction isotherms reported by Kovacs[13] for poly(vinyl acetate). The data on the right-hand side of the figure ($\delta > 0$) come from contraction experiments in which the sample was brought from equilibrium at $T_0 = T_1 + \Delta T$ to $T = T_1$. The left-hand side of the figure ($\delta < 0$) shows the response in the expansion experiments in which the sample was brought from equilibrium at $T_0 = T_1 - \Delta T$ to $T = T_1$. The curves are labeled in families by the given T_1 value, and the values for $T_1 \pm \Delta T$ are given near each individual curve. These data, of course, reflect the asymmetric approach to equilibrium. In discussing the families of curves, we want to keep in mind that each set of contraction and expansion isotherms corresponds to the same value of T_1, *e.g.* 35 °C. Of considerable interest is that the effective retardation times in the approaches from $\delta > 0$ (contraction) merge near $\delta = 0$, while from $\delta < 0$ (expansion) there is no corresponding merging of the τ_{eff} values. For example, at $T_1 = 35$ °C the responses to the temperature jumps are markedly different: as equilibrium is approached, at $\delta \approx 10^{-4}$, $[\tau_{eff}]_{30°C}/[\tau_{eff}]_{32.5°C} \approx 5$, *i.e.* there is a 'gap' between the curves. This behavior has been commented on previously by Kovacs *et al.*,[18,191] Ng and Aklonis[213] and others,[19,214] and is surprising because, near equilibrium, one would not expect such a difference in an effective retardation time, since the effects of the previous history should have faded away. In the original KAHR[18] paper, plots of $\log \tau_{eff}$ *vs.* δ were calculated for reasonable values of the material parameters and, as seen in Figure 36, for all of the curves, contraction and expansion, the values of τ_{eff} converge at $\delta = 0$. Furthermore, Ng and Aklonis[218] have analyzed many sets of parameters, and they were unable to reproduce the effect (even qualitatively) except with extremely unrealistic values of the material parameters. To this author's knowledge, there is only one model which accounts for the 'τ-effective paradox' — the coupling model of Ngai.

(v) The coupling model of Ngai and the τ-effective paradox

An interesting model of relaxation phenomena has appeared recently.[21,215-223] The appeal of the model from the point of view of glassy relaxations is that it can describe the τ-effective paradox

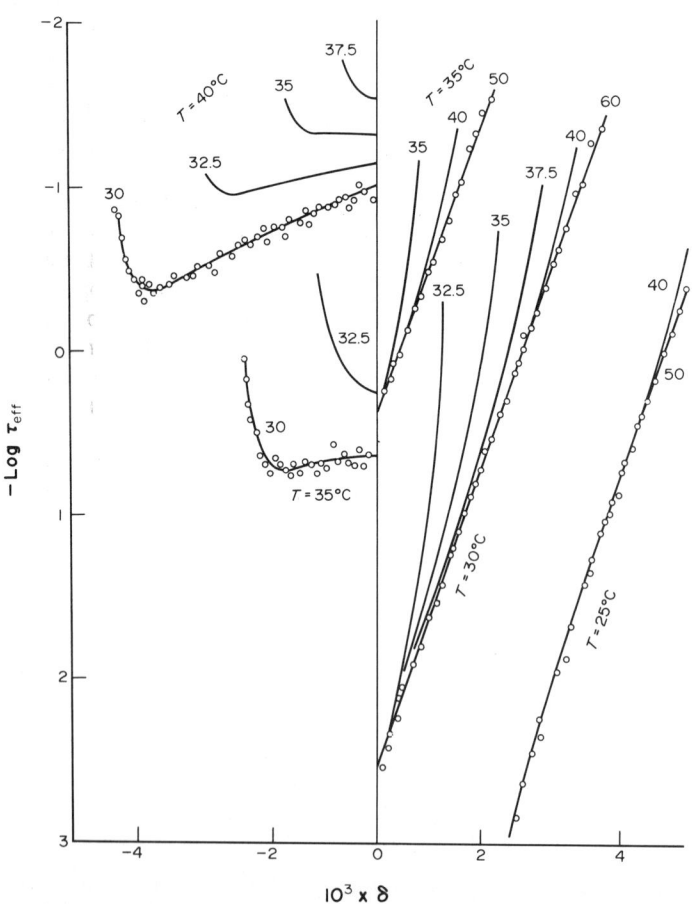

Figure 35 $\log \tau_{eff}$ *vs.* departure from equilibrium δ for expansion ($\delta \leq 0$) and contraction behaviors ($\delta \geq 0$) of a poly(vinyl acetate) glass. Experiments were conducted by equilibrating the samples at the temperatures indicated on each curve and jumping the temperature to that indicated with each family of curves. Thus, volume contraction experiments were carried out at $T = 30$ °C after equilibration at temperatures of 32.5, 35, 37.5, 40 and 60 °C. Importantly, note that while contraction curves merge at $\delta = 0$, the expansion curves do not. The latter is referred to as the 'expansion gap' (after ref. 13, with permission. Note that the figure is corrected courtesy of A. J. Kovacs)

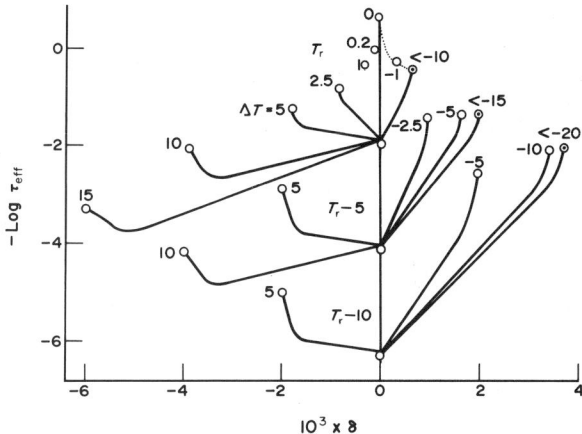

Figure 36   Calculations of log τ_{eff} vs. departure from equilibrium δ. Calculations are based on the KAHR model. This figure is a simulation using the KAHR model of the experimentally observed behavior depicted in Figure 35. T_r is the temperature of the volume contraction experiment and ΔT indicates the equilibration temperature. Note that the expansion gap is not reproduced (after ref. 18, with permission)

without recourse to additional assumptions beyond those used in the Narayanaswamy–KAHR–Moynihan models, other than that relaxation in 'complex' systems proceeds as described by the coupling model of Ngai.[21,215-223] Thus, the basic physics of glassy recovery remains the same, *i.e.* it is necessary to have a retardation function (spectrum) which exhibits non-exponential decay and retardation times which depend upon the instantaneous state (structure) of the glass. As mentioned previously, the shape of the retardation spectrum may change during recovery (*i.e.* as a function of the structure of the glass), but this is not the missing 'ingredient' required to explain the τ-effective paradox. The piece that is missing is the correct description of the relaxation process in 'complex' systems as given by the coupling model.

There is insufficient space here to derive the coupling model in detail, therefore we present the salient features of the model and show how it describes the isothermal recovery behavior of glasses in response to a T-jump experiment. In particular, we will pay attention to the expansion experiment, *i.e.* $\delta < 0$ where a 'gap' exists between the curves for log τ_{eff} vs. δ as δ approaches zero.

According to the coupling model, relaxation parameters (*e.g.* ordering parameters or mechanisms in the KAHR model) do not exhibit simple exponential decay. Rather, the physics of complex systems requires cooperativity among the relaxing species which acts to slow the simple constant relaxation rate into a time-dependent rate. Then one finds that the relevant variable initially decays at a constant rate, $W_0 = \tau_0^{-1}$, but at times larger than a characteristic time $t_c \equiv W_c^{-1}$ the rate becomes explicitly time-dependent, *i.e.* of the form $W = \tau_0^{-1} (\omega_c t)^{-n}$ where $0 < n < 1$. This is the basic result of the coupling model which can be summarized as

$$W(t) = \tau_0^{-1}, \quad \omega_c t \ll 1 \tag{85}$$

$$W(t) = \tau_0^{-1}(\omega_c t)^{-n}, \quad \omega_c t \gg 1 \tag{86}$$

The fact that the relaxing species exhibit a time-dependent rate in the complex environment has some important consequences. The relaxation rate of the macroscopic variable ϕ becomes

$$\frac{d\phi}{dt} = -W(t)\phi \tag{87}$$

and integration of equations (85) and (86) gives

$$\phi(t) = \phi(0)\exp(-t/\tau_0), \quad \omega_c t \ll 1 \tag{88}$$

$$\phi(t) = \phi(0)\exp\left[-\int_0^t dt' \, \tau_0^{-1}(\omega_c t')^{-n}\right], \quad \omega_c t \gg 1 \tag{89}$$

The slower-than-exponential decay of ϕ in equation (89) reflects the coupling of the primitive relaxing-species to its complex environment. Ngai and co-workers[217,218] have been able to determine ω_c for several polymers and find that it is typically in the range 10^9–$10^{10}\,s^{-1}$.

The form of equation (89) for equilibrium conditions, when τ_0, ω_c and n are constant, becomes

$$\phi(t) \;=\; \phi(0)\exp[-(t/\tau^*)^{1-n}], \quad w_c t \;\gg\; 1 \tag{90}$$

where

$$\tau^* \;=\; [(1 \;-\; n)\omega_c^n\tau_0]^{\frac{1}{1-n}} \tag{91}$$

is the effective relaxation time of the complex system (*i.e.* that observed experimentally), and is related to τ_0, the relaxation time of the primitive species (uncoupled). We remark that equations (90) and (91) contain the same parameter n, thus the measured relaxation time τ^* is related to the shape of the decay function (spectrum), which leads to testable predictions because n, which describes material behavior, must be the same in both equations (90) and (91). The model has proven successful in many instances.[217-223]

Importantly for the structural recovery of glasses, the model predicts an equilibrium decay function which is of KWW form ($\beta = 1 - n$), see equation (81), for even a single primitive species. Thus the requirement of a non-exponential decay function is fulfilled by the model. Although the other models use a broad relaxation function to describe behavior, they neither make the prediction of equation (91), nor can the general equation (89) result from them. In general, n and τ^* can be functions of T and δ; however, we treat only the case where τ^* is a function of δ, *i.e.* $\tau^* = \tau^*(T, \delta(t'))$. Then, rewriting equation (86) in terms of τ^* and identifying the macroscopic variable ϕ with the departure from equilibrium δ, we find that, for isothermal volume recovery, equation (89) becomes

$$\delta(t) \;=\; \delta_0\exp\left[-\int_0^t dt'(1 \;-\; n)(\tau^*)^{-1}(t'/\tau^*)^{-n}\right], \quad \tau^* \;=\; \tau^*(T, \delta) \tag{92}$$

There are then three important facets to equation (92). First, at equilibrium, the decay function is non-exponential (equation 90). Second, the observed retardation time depends upon δ and T. Third, both n and τ^* appear inside the integral, which implies that the history-dependent change of both these parameters is inherent in the model and so they affect the volume-recovery response differently than in the Narayanaswamy–KAHR–Moynihan-type models. For the same isothermal recovery after a T-jump, we can compare equation (92) with the response[21] for the KAHR[18] model

$$\delta(t) \;=\; \delta_0\sum_{i=1}^N g_i\exp\left(-\int_0^t \tau_i^{-1} dt'\right), \quad \tau_i \;=\; \tau_i(T, \delta) \tag{93}$$

and the Narayanaswamy[17]–Moynihan[20] form

$$\delta(t) \;=\; \delta_0\exp\left(-\int_0^t \tau^{-1} dt'\right)^\beta, \quad \tau \;=\; \tau(T, \delta) \tag{94}$$

As can be seen, equation (92) is different from (93) and (94), which is a direct result of the coupling model.[21] Rendell *et al.*[21] have solved equation (92) for the T-jump experiment. As can be seen from Figure 37, the expansion curve is reproduced, as is the 'gap' as δ approaches zero. The lower portion of Figure 37 shows the shift factor $a(T, \delta)$ which was used to fit the data compared with independent measurements obtained by Kovacs, Stratton and Ferry[224] (KSF).

An important result of the coupling model is that the expansion gap is directly related to n, the coupling parameter in the model. In fact, this is a severe test of the coupling model, as the value of n which fits the shape of the equilibrium retardation spectrum, must be that which determines the magnitude of the 'gap'. This is, in fact, what Rendell *et al.*[21] found.

We remark that the coupling-model predictions for contraction, *i.e.* $\delta > 0$, while reasonable at small ΔT, fit less well at large ΔT. This was attributed[21] to finite cooling-rate effects, not accounted for in equation (92). Furthermore, the coupling-model predictions have not been compared with other experiments, *e.g.* memory effect, uniform heating through the transition, *etc.*, and ultimate confirmation of the model awaits such quantitative comparison. The results presented above as an explanation of the τ-effective paradox are encouraging.

10.3.4 Free Volume and the Kinetics of Glasses

10.3.4.1 Introduction

The phenomenological models described in the previous section contain the essential ingredients needed to describe most of the kinetic behavior of glasses, *i.e.* a retardation function exhibiting non-

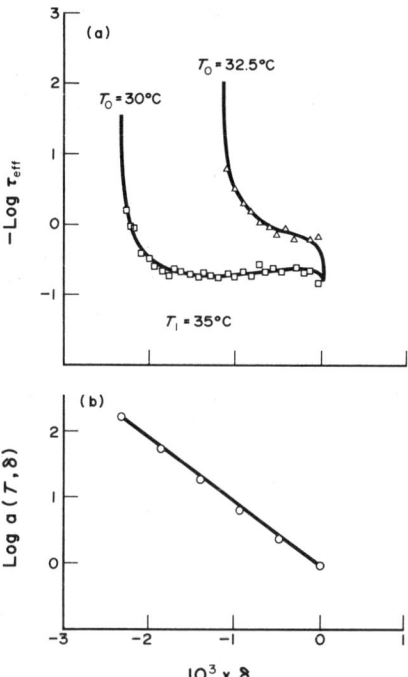

Figure 37 (a) Calculation of expansion side of the τ-effective plot using the Ngai coupling model. The expansion gap is reproduced. (b) shows the shift factor $a(T, \delta)$ used in the model *vs.* the experimental data of Kovacs, Stratton and Ferry[230] (after ref. 21, with permission)

exponential decay and a dependence of retardation times on the instantaneous state (structure) of the glass. The Narayanaswamy–KAHR–Moynihan models are virtually identical, while the coupling model of Ngai results in a somewhat different phenomenological equation, not by changing the 'essential ingredients' above, but by treating the relaxation process itself in a new way. In the next section we examine the use of a detailed model of free volume to describe glassy kinetics.

10.3.4.2 The Robertson–Simha–Curro (RSC) model

Robertson, Simha and Curro (RSC)[225-227] have developed a molecular theory of the kinetics of structural recovery of glasses. It is based upon a stochastic model originally proposed by Robertson,[228-231] and makes use of the previously discussed Simha–Somcynsky[164,165] cell model to calculate the free-volume function and its fluctuations. As we will show, the theory develops expressions for a spectrum of retardation times and the time-dependent distribution of the free volume.

RSC assume that the changes in measured volume associated with structural recovery are changes in free volume. The free volume is defined using the Simha–Somcynsky[164,165] equation of state (recall Section 10.2.3.2.iii)

$$\tilde{P}\tilde{V}/\tilde{T} = [1 - 2^{-1/6}y(y\tilde{V})^{-1/3}]^{-1} + (2y/\tilde{T})(y\tilde{V})^{-2}[1.011(y\tilde{V})^{-2} - 1.2045] \tag{42}$$

where the $\tilde{P}$, $\tilde{V}$ and $\tilde{T}$ are reduced variables and y is the fraction of occupied cells. Then fractional free-volume f is related to y by

$$f = 1 - y \tag{95}$$

At equilibrium, the free volume can be determined by minimization of the free energy

$$(s/3c)[(s - 1)/s + y^{-1}\ln(1 - y)] = (y/6\tilde{T})(y\tilde{V})^{-2}[2.409 - 3.003(y\tilde{V})^{-2}]$$
$$+ [2^{-1/6}y(y\tilde{V})^{-1/3} - 1/3][1 - 2^{-1/6}y(y\tilde{V})^{-1/3}]^{-1} \tag{45}$$

As RSC note, equation (45) is not valid when the polymer is not in equilibrium, and a different condition is needed to reduce the number of variables in equation (42).

To compute changes in free volume during recovery, the theory requires that one examine changes throughout the entire sample. This is because local changes occur at rates that depend upon the local free-volume, which can vary from point to point because of thermal fluctuations. (We note here that this assumption, due originally to Robertson,[228-231] is at variance with the assumption in the phenomenological models where the 'local' rate depends on the global departure $\delta = \Sigma \delta_i$). At equilibrium, one can write an appropriate expression for the mean-squared fluctuation in free volume[225,227] $\langle \delta f^2 \rangle = \langle \delta y^2 \rangle$:

$$N_s \langle \delta f^2 \rangle = [1 + (2/y)\ln(1 - y) + 1/(1 - y)]/y^2 + \frac{(2^{1/2} y \tilde{V})^{2/3} - (8/3)y(2^{1/2} y \tilde{V})^{1/3} + 3y^2}{3y^2[2^{1/2} y \tilde{V}^{1/3} - y]^2}$$
$$+ (1/3 \, y\tilde{T})[6.066(y\tilde{V})^{-4} - 2.409(y\tilde{V})^{-2}] \tag{96}$$

where N_s represents the size of the regions of interest and is the nominal number of chain units in those regions. Equation (96) represents the fact that the mean-square fluctuation in free volume is a constant independent of volume size.[225] At equilibrium, equation (96) is valid. In the 'frozen' glass, away from equilibrium, one must use the measured volume with equations (42) and (96) to determine the free volume. Also, although the value of N_s is the number of chain units required for local segmental rearrangements, and should be approximately 26 for vinyl-type polymers,[227] RSC treat it as an adjustable parameter in their computations.

In the RSC model there is a discrete free-volume distribution function, which allows one to write a set of stochastic equations in terms of the probability $w_i(t_0)$ of finding a local region out of equilibrium in the ith state at time t_0, i.e. it contains a fractional free-volume f_i. RSC go on to define a probability matrix $P_{ij}(t - t_0)$ which gives the probability that a region known to be in the ith state at time t_0 will be in the jth state after some infinitesimal time interval Δt. Assuming that a region undergoing a change from state i to state j will pass through all intervening states, the transition probabilities become

$$P_{i,i-1}(\Delta t) = \Delta t \lambda_i^- + 0(\Delta t) \tag{97}$$

$$P_{i,i+1}(\Delta t) = \Delta t \lambda_i^+ + 0(\Delta t) \tag{98}$$

$$P_{ii}(\Delta t) = 1 - \Delta t(\lambda_i^- + \lambda_i^+) + 0(\Delta t) \tag{99}$$

where $0(\Delta t)$ denotes a quantity which approaches zero faster than does Δt. λ_i^+ is the upward transition rate from state i and λ_i^- is the downward rate. At equilibrium, microscopic reversibility requires that

$$\xi_i \lambda_i^+ = \xi_{i+1} \lambda_{i+1}^- \tag{100}$$

where the ξ_i are the state occupancies given by the binomial distribution.[227]

In the model, the upward and downward transition rates are expected to depend on both the local free-volume and on the free volume of the neighbors. This is because any increase or decrease in local free-volume must be obtained from or passed on to a neighboring region. RSC[227] note that in a strict cell or lattice model the free volume would be required to diffuse to or from the surface for any net change to occur in the global free-volume of the system. This is one aspect of the free-volume models which is disturbing. Measurements by Braun and Kovacs,[232] using bulk and powdered samples, showed no effect of specimen geometry on the observed kinetics. RSC feel that, for real systems, net changes in free volume are 'probably liberated and absorbed by changes in various local strains of the system'. This problem is also discussed by Curro et al.,[233] who used a different model to describe the kinetics of recovery in glasses. In any event, there is no correlation between free-volume content of neighboring regions in equilibrium, and RSC further assume that the correlation remains weak away from equilibrium. Then the combined free-volume of the neighboring regions can be approximated by the average global free-volume $\bar{f}$. The free volume which affects the transition rate in the ith region of interest is

$$\hat{f}_i = [f_i + (Z - 1)f]/Z \tag{101}$$

where Z is the number of regions needed to liberate or absorb the net free-volume changes in the region of interest.

The rates of transition, λ_i^+ and λ_i^-, are assumed to have the same dependence on local free-volume as their global dependence. Therefore, RSC are able to use the Doolittle or WLF forms, described previously, to represent the structure dependence of the transition rates

$$\lambda_i^- = R\tau_g^{-1}(\xi_{i+1}/\xi_i)^{1/2}\beta^{-2}\exp\left(2.303\,C_1\frac{f_i - f_g}{\alpha_1 C_2/T^* + f_i - f_g}\right) \tag{102}$$

$$\lambda_i^+ = \lambda_{i+1}^-(\xi_{i+1}/\xi_i) \tag{103}$$

where R is an adjustable parameter, $(\xi_{i+1}/\xi_i)^{1/2}$ enforces the miscroscopic reversibility condition (equation 100) and β is the step between adjacent free-volume states.

With the above assumptions, RSC were able to compute volume-recovery curves which mimic the major features of the glassy phenomenology. In Table 4 are shown the parameters used in the calculations. Figure 38 shows the asymmetry of the calculated approach curves compared with the experimental data of Kovacs[11] for poly(vinyl acetate). In Figure 39 is shown a plot of $\log\tau_{\text{eff}}$ vs. δ calculated from the model compared with the data for poly(vinyl acetate). As was the case for the phenomenological models, the τ-effective paradox is not resolved.

An interesting aspect of the RSC model, however, is the prediction that the retardation spectrum changes in shape as recovery proceeds. As shown in Figure 40, it appears to narrow as equilibrium is

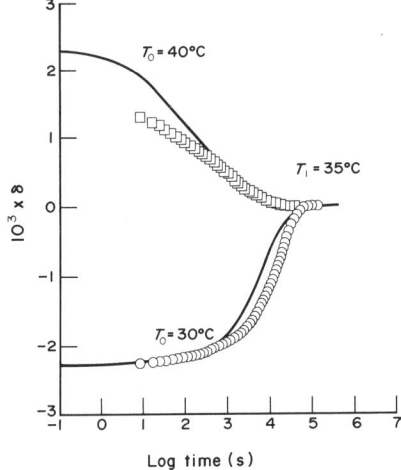

Figure 38 The departure from equilibrium δ vs. log time following steps in temperature to 35 °C from 40 °C (□) and 30 °C (○) for poly(vinyl acetate). The curves are computed from the RSC model (after ref. 225, with permission)

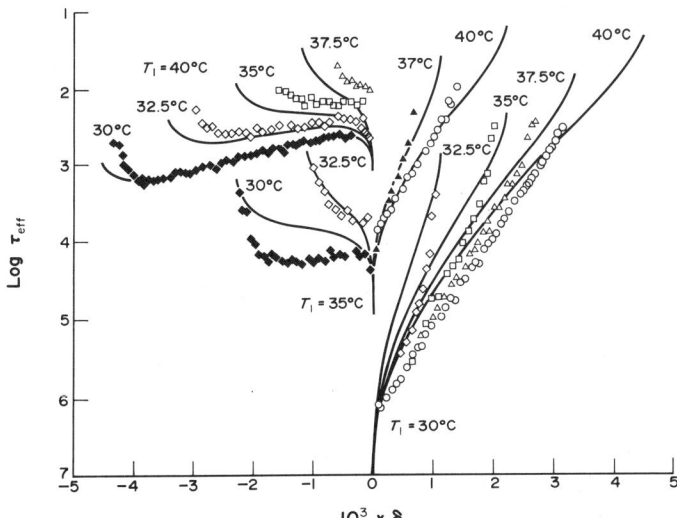

Figure 39 Computed curves (from RSC model) and experimental data for poly(vinyl acetate) for τ_{eff} vs. departure from equilibrium δ. Values of T_r and T_0 are indicated (after ref. 225, with permission)

Table 4 Parameters Used to Calculate the Kinetics of Volume Recovery in the Robertson–Simha–Curro Model (After Ref. 225)

$N_s = 26$	Number of segments in free-volume region
$C_1 = 15.6$ $\Big\}$	Time–temperature shift parameters (ref. 57)
$C_2 = 46.8$ K	
$T_g = 308$ K	Glass transition temperature at 100 kPa
$f_g = 0.08134$	Free volume at T_g and 100 kPa
$\tau_g = 3600$ s	Retardation time at T_g
$Z = 13$	Size-ratio controlling free volume

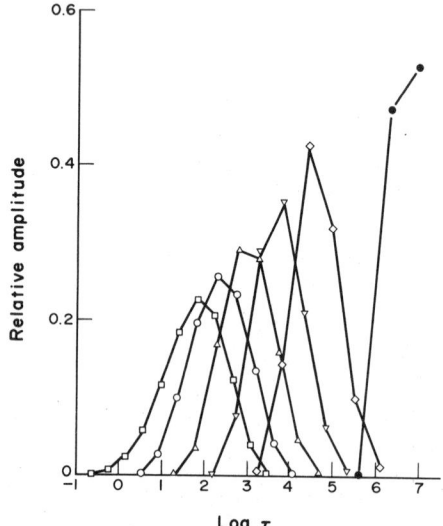

Figure 40 Retardation time spectra at stages during the volume recovery of poly(vinyl acetate) following a temperature step from 40 °C to 30 °C. Progression from the initial spectrum (□) immediately after the step to the final spectrum (●) shows narrowing as volume recovery proceeds (after ref. 225, with permission)

approached in the contraction experiment. This is attributed to a change in the distribution of the free volume as equilibrium is approached. Such behavior may mean that a more complicated free-volume treatment is required than the one presented here, as the observed experimental behavior appears to support a broadening rather than a narrowing of the spectrum.[209–212] (Recall that both β and x changed in the phenomenological Narayanaswamy[17]–Moynihan[20] model.)

The free-volume model of Robertson, Simha and Curro allows one to examine the role of free-volume fluctuations on the kinetics of glassy recovery and provides a detailed model of the physics of such processes. An interesting extension of this type of calculation might be a non-equilibrium Flory–Huggins lattice calculation, *i.e.* an extension of the Gibbs–DiMarzio theory to the non-equilibrium properties of glasses. This is particularly interesting since the Simha–Somcynsky cell model, as mentioned previously, does not correctly describe the energetics of glasses but only the *PVT* surface.

10.3.5 Computer Simulations

There has been considerable recent interest[234–244] in carrying out computer simulations of the behavior of glassy systems with the intention of understanding the kinetics by varying certain parameters. It is interesting to note that, as implied in the phenomenological and free-volume models, the computer simulations also imply that an important component in glassy kinetics is cooperativity of motion. A detailed treatment of the models is beyond the range of this article; however, an interesting finding was made recently by Fredrickson[236] when he compared the kinetics from the computer simulation with those given by the Narayanaswamy[17]–Moynihan[20] phenomenological model. He found that, for certain cases in which the phenomenological equations did not describe the simulation, the deviations were similar to those observed in real experiments. However, Frederickson[236] does note that the simulations are versatile enough that one may be able to simulate almost anything. Apparently more work is required before simulations reach the stage of sophistication where new physical insights will be obtained.

10.3.6 Physical Aging

In the above sections we have described the kinetics of the structural recovery of glasses from both phenomenological and microscopic viewpoints. As the structure (free volume, configurational entropy, *etc.*) of the non-equilibrium glass evolves in response to thermal history, the mechanical, dielectric, *etc.* properties also change. The changes in properties accompanying the structural changes of glasses have been known for some time,[224] but were intensively studied by Struik[14] and labeled 'physical aging' in his extensive treatise on the subject. A lengthy review of physical aging is outside the scope of this article and the reader is referred to Struik's[14,247] work and the more recent work of Plazek and Berry[248] for reviews of the subject. However, a few comments about aging are worth making.

The picture of aging formulated by Struik[14] is based on the postulate that, as the volume of the glass decreases with increasing aging-time after a quench (*cf.* Figure 3), the change in the glassy structure changes the viscoelastic spectrum governing the material response (*e.g.* mechanical, dielectric) by shifting it along the time axis by an amount a_{t_e} where t_e denotes the time after the quench. Such behavior can be seen in the creep-compliance data presented in Figure 4 for poly(vinyl chloride) quenched from above T_g to 20 °C. Here we see that, shortly after the quench, the compliance is greater than at longer aging times t_e. The creep curves can be superimposed by a shift along the time axis, *i.e.* the retardation spectrum shifts but does not change shape. The amount of shift per decade of time, μ, was found to be approximately equal to unity, *i.e.*

$$\mu = \frac{\mathrm{d}\log a_{t_e}}{\mathrm{d}\log t_e} \approx 1 \tag{104}$$

for many systems in the linear viscoelastic range.

Importantly, for this type of analysis to work, the duration of the mechanical measurement must be short relative to the total aging time, so that the tests are carried out at fixed glassy structure,[19] *i.e.* $t_1/t_e \leq 0.10$. Under such conditions, one can consider the measurement of the mechanical or dielectric response to be a probe of the glassy structure. In fact, Kovacs[19] has proposed associating a_{t_e} from the mechanical or dielectric measurements with the a_δ required for the phenomenological theories as a means of independently determining the non-linearity parameter x discussed previously (see equation 83).

Struik[14] also investigated the change with aging of the large stress (or deformation) properties of glasses. Along with other workers he reports two additional phenomena: (i) the viscoelastic spectrum

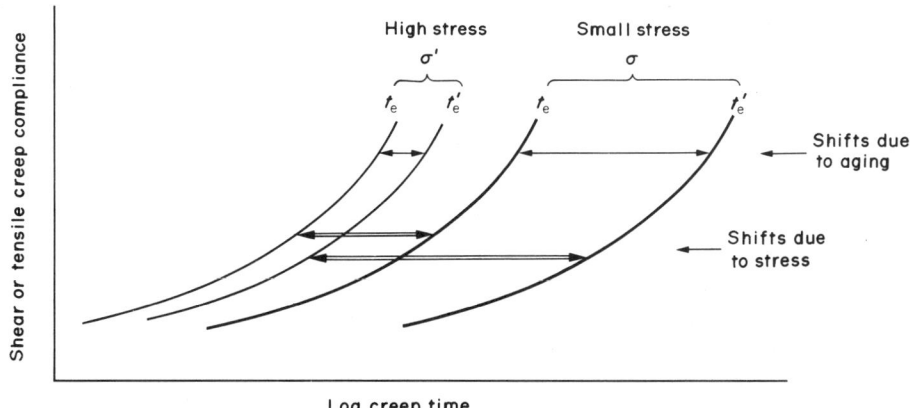

Figure 41 Schematic representation of the relative effects of aging on the small-stress and high-stress creep compliance of a glass (after ref. 14, with permission)

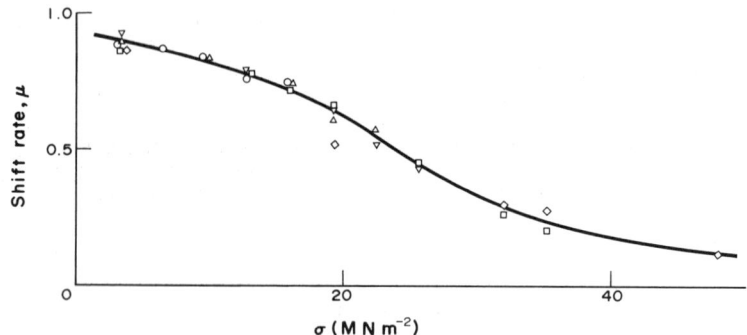

Figure 42 Variation of the rate of aging, μ (*i.e.* double logarithmic rate of change of the shift factor) with maximum shear stress for PVC quenched from 90 °C to various temperatures: $\bigcirc$, 60; $\triangle$, 50; $\bigtriangledown$, 40; $\square$, 20; $\lozenge$, −20 °C (after ref. 14, with permission)

is shifted[14, 15, 248−250] to shorter times by large stresses or deformations (see Figure 41); and (ii) the rate of aging μ decreases dramatically as the level of stress (or deformation) increases[14, 251, 252] (see Figure 42).

Struik[14] attributed the above effects to 'erasure' of the aging due to the mechanical deformation, possibly due to change in the structure of the glass created by the mechanical energy dissipated by the system. This contention has been supported by work of Smith and co-workers[253] and Yee and co-workers.[254, 255] However, Myers *et al.*[256] claim the opposite result, *i.e.* an enhancement of the aging response due to the applied deformation, and McKenna and Zapas[257, 258] have argued against the Struik interpretation on the grounds that non-linear memory effects explain, from simple viscoelasticity considerations, many of the observed 'rejuvenation' or 'erasure' phenomena.

Although the basic phenomenon of physical aging, *i.e.* the impact of structural changes in a glass on the viscoelastic or dielectric response, are well understood, particularly in the linear range,[14, 247−248, 259−261] there are details of behavior, such as aging effects near subglass relaxations[262−264] and changes in spectral shape,[264−267] which need to be clarified. Finally, the coupling between aging effects, large mechanical deformations, the structure of the glass and non-linear viscoelastic properties has become an area of great interest recently,[251−263, 268−275] and merits considerable research effort in the future.

ACKNOWLEDGEMENTS

The author appreciates many fruitful discussions during the writing of this article with E. A. DiMarzio. Thanks are also due to the Polymers Division and the Institute for Materials Science and Engineering of the National Bureau of Standards, which made the time available to produce this work.

10.4 REFERENCES

1. A computer search of 'Chemical Abstracts' (American Chemical Society, Washington, DC) using DIALOG (DIALOG Information Services) for 1967 to May 1987 resulted in the numbers quoted.
2. R. N. Haward, 'The Physics of Glassy Polymers', Wiley, New York, 1973.
3. 'Physical Aging Processes in Molecular and Atomic Glasses', NATAS Conference, 1983, Special Issue of *Polym. Eng. Sci.*, 1984, **24** (14).
4. M. R. Tant and G. L. Wilkes, *Polym. Eng. Sci.*, 1981, **21**, 874.
5. G. W. Scherer, 'Relaxation in Glass and Composites', Wiley, New York, 1986.
6. M. Goldstein and R. Simha (eds.), *Ann. N. Y. Acad. Sci.*, 1976, **279**.
7. J. M. O'Reilly and M. Goldstein (eds.), *Ann. N. Y. Acad. Sci.*, 1981, **371**.
8. C. A. Angell and M. Goldstein (eds.), *Ann. N. Y. Acad. Sci.*, 1986, **484**.
9. K. L. Ngai and G. B. Wright (eds.), 'Relaxations in Complex Systems', U. S. Government Printing Office, Washington, DC, 1984 (available from National Technical Information Service, 5285 Port Royal Road, Springfield, VA 22161, USA).
10. P. H. Gaskell (ed.), 'The Structure of Non-Crystalline Materials', Taylor and Francis, London, 1977.
11. A. J. Kovacs, *J. Polym. Sci.*, 1958, **30**, 131.
12. S. S. Chang, *J. Polym. Sci., Polym. Symp.*, 1984, **71**, 59.
13. A. J. Kovacs, *Fortschr. Hochpolym.-Forsch.*, 1964, **3**, 394.
14. L. C. E. Struik, 'Physical Aging in Amorphous Polymers and Other Materials', Elsevier, Amsterdam, 1978.

15. J. M. Crissman and G. B. McKenna, *J. Polym. Sci., Polym. Phys. Ed.*, 1987, **25**, 1667.
16. E. A. DiMarzio, *Ann. N. Y. Acad. Sci.*, 1981, **371**, 1.
17. O. S. Narayanaswamy, *J. Am. Ceram. Soc.*, 1971, **54**, 491.
18. A. J. Kovacs, J. J. Aklonis, J. M. Hutchinson and A. R. Ramos, *J. Polym. Sci., Polym. Phys. Ed.*, 1979, **17**, 1097.
19. A. J. Kovacs, *Ann. N. Y. Acad. Sci.*, 1981, **371**, 38.
20. C. T. Moynihan, P. B. Macedo, C. J. Montrose, P. K. Gupta, M. A. DeBolt, J. F. Dill, B. E. Dom, P. W. Drake, A. J. Easteal, P. B. Elterman, R. P. Moeller, H. Sasabe and J. A. Wilder, *Ann. N. Y. Acad. Sci.*, 1976, **279**, 15.
21. R. W. Rendell, J. J. Aklonis, K. L. Nagi and G. R. Fong, *Macromolecules*, 1987, **20**, 1070.
22. W. Kauzmann, *Chem. Rev.*, 1948, **43**, 219.
23. N. F. Mott and R. W. Gurney, *Trans. Faraday Soc.*, 1939, **35**, 364.
24. J. H. Gibbs, *J. Chem. Phys.*, 1956, **25**, 185.
25. J. H. Gibbs and E. A. DiMarzio, *J. Chem. Phys.*, 1958, **28**, 373.
26. E. A. DiMarzio and J. H. Gibbs, *J. Chem. Phys.*, 1958, **28**, 807.
27. A. J. Batschinski, *Z. Phys. Chem. (Leipzig), 1913*, **84**, 643.
28. T. G. Fox, Jr. and P. J. Flory, *J. Appl. Phys.*, 1950, **21**, 581.
29. A. K. Doolittle, *J. Appl. Phys.*, 1951, **22**, 1031.
30. I. Prigogine and R. Defay, 'Thermodynamique Chimique', Edition Desoer, Liège, Belgium, 1950.
31. P. Ehrenfest, *Proc. K. Ned. Akad. Wet.*, 1933, **36**, 153.
32. R. A. Swalin, 'Thermodynamics of Solids', 2nd edn., Wiley, New York, 1972.
33. J. J. Aklonis and W. J. MacKnight, 'Introduction to Polymer Viscoelasticity', Wiley, New York, 1983.
34. P. G. Shewmon, 'Transformations in Metals', McGraw-Hill, New York, 1969.
35. G. Rehage and W. Borchard, in 'The Physics of Glassy Polymers', ed. R. N. Haward, Wiley, New York, 1973, p. 54.
36. F. T. Gucker and R. L. Seifert, 'Physical Chemistry', Norton, New York, 1966, p. 386.
37. J. E. McKinney and M. Goldstein, *J. Res. Natl. Bur. Stand., Sect. A*, 1974, **78A**, 331.
38. G. Gee, *Polymer*, 1966, **7**, 177.
39. E. Passaglia and G. M. Martin, *J. Res. Natl. Bur. Stand., Sect. A*, 1964, **68A**, 273.
40. P. Zoller, *J. Polym. Sci., Polym. Phys. Ed.*, 1982, **20**, 1453.
41. H.-J. Oels and G. Rehage, *Macromolecules*, 1977, **10**, 1036.
42. G. Rehage and H.-J. Oels, *High Temp.— High Pressures*, 1977, **9**, 545.
43. C. A. Angell and W. Sichina, *Ann. N. Y. Acad. Sci.*, 1976, **279**, 53.
44. C. A. Angell, E. Williams, K. J. Rao and J. C. Tucker, *J. Phys. Chem.*, 1977, **81**, 238.
45. M. Goldstein, *J. Phys. Chem.*, 1973, **77**, 667.
46. J. M. O'Reilly, *J. Polym. Sci.*, 1962, **57**, 429.
47. J. E. McKinney, *Ann. N. Y. Acad. Sci.*, 1976, **279**, 88.
48. I. Havlíček, *Macromolecules*, 1981, **14**, 1595.
49. A. J. Staverman, *Rheol. Acta*, 1966, **5**, 283.
50. R.-J. Roe and A. E. Tonelli, *Macromolecules*, 1978, **11**, 114.
51. A. Quach and R. Simha, *J. Phys. Chem.*, 1972, **76**, 416.
52. M. Goldstein, *J. Chem. Phys.*, 1963, **39**, 3369.
53. R. O. Davies and G. O. Jones, *Adv. Phys.*, 1953, **2**, 370.
54. E. Williams and C. A. Angell, *J. Phys. Chem.*, 1977, **81**, 232.
55. K. D. Pae, C.-L. Tang and E.-S. Shin, *J. Appl. Phys.*, 1984, **56**, 2426.
56. C. T. Moynihan, A. J. Easteal, M. A. DeBolt and J. Tucker, *J. Am. Ceram. Soc.*, 1976, **59**, 12.
57. J. D. Ferry, 'Viscoelastic Properties of Polymers', 3rd edn., Wiley, New York, 1980.
58. E. A. DiMarzio, J. H. Gibbs, P. D. Fleming III and I. C. Sanchez, *Macromolecules*, 1976, **9**, 763.
59. E. A. DiMarzio, *J. Res. Natl. Bur. Stand., Sect. A*, 1964, **68A**, 611.
60. E. A. DiMarzio, *J. Appl. Phys.*, 1974, **45**, 4143.
61. E. A. DiMarzio, *Macromolecules*, 1977, **10**, 1407.
62. E. A. DiMarzio, in 'Relaxations in Complex Systems', ed. K. L. Ngai and G. B. Wright, U. S. Government Printing Office, Washington, DC, 1985, p. 43 (available from National Technical Information Service, 5285 Port Royal Road, Springfield, VA 22161, USA).
63. M. Goldstein, *J. Appl. Phys.*, 1975, **46**, 4153.
64. M. Goldstein, *Macromolecules*, 1977, **10**, 1407.
65. P. J. Flory, 'Principles of Polymer Chemistry', Cornell University Press, Ithaca, NY, 1953.
66. M. L. Huggins, *Ann. N. Y. Acad. Sci.*, 1942, **43**, 1.
67. A. R. Miller, 'The Theory of Solutions of High Polymers', Oxford University Press, London, 1948.
68. T. G. Fox and P. J. Flory, *J. Polym. Sci.*, 1954, **14**, 315.
69. R. B. Beevers and E. F. T. White, *Trans. Faraday Soc.*, 1960, **56**, 744.
70. G. Pezzin, F. Zilio-Grandi and P. Sammartin, *Eur. Polym. J.*, 1970, **6**, 1053.
71. R. B. Beevers, *J. Polym. Sci., Part A*, 1964, **2**, 5257.
72. A. Eisenberg and T. Sasada, in 'Proceedings of the Conference on Physics of Non-Crystalline Solids', ed. J. A. Prins, North-Holland, Amsterdam, 1965, p. 99.
73. R. Larrain, L. H. Tagle and F. R. Draz, *Polym. Bull. (Berlin)*, 1981, **4**, 487.
74. T. G. Fox and S. Loshaek, *J. Polym. Sci.*, 1955, **15**, 371.
75. E. V. Thompson, *J. Polym. Sci., Part A-2*, 1966, **4**, 199.
76. T. Hatakeyama and M. Serizawa, *Polym. J.*, 1982, **14**, 51.
77. F. Bastard and B. Jasse, *Polym. Bull. (Berlin)*, (1982), **7**, 331.
78. K. Ueberreiter and G. Kanig, *J. Colloid Sci.*, 1952, **7**, 569.
79. B. E. Read, *Polymer*, 1962, **3**, 529.
80. J. A. Faucher, *J. Polym. Sci., Part B*, 1965, **3**, 143.
81. J. M. Widmaier and G. C. Meyer, *Macromolecules*, 1981, **14**, 450.
82. F. J. Bueche, 'Physical Properties of Polymers', Interscience, New York, 1962.
83. T. Somcynsky and D. Patterson, *J. Polym. Sci.*, 1962, **62**, 151.

84. J. M. G. Cowie and P. M. Toporowski, *Eur. Polym. J.*, 1968, **4**, 621.
85. C. M. Guttmann and E. A. DiMarzio, *Macromolecules*, 1988, in press.
86. I. Havĺíček, V. Vojta, M. Ilavský and J. Hrouz, *Macromolecules*, 1980, **13**, 357.
87. G. C. Berry, *Macromolecules*, 1980, **13**, 550.
88. E. A. DiMarzio and F. Dowell, *J. Appl. Phys.*, 1979, **50**, 6061.
89. G. M. Martin and L. Mandelkern, *J. Res. Natl. Bur. Stand. (U. S.)*, 1959, **62**, 141.
90. K. Uebrreiter and G. Kanig, *J. Chem. Phys.*, 1950, **18**, 399.
91. S. Loshaek, *J. Polym. Sci.*, 1955, **15**, 391.
92. L. A. Wood, *J. Res. Natl. Bur. Stand., Sect. A*, 1972, **76A**, 51.
93. J. M. Charlesworth, *J. Macromol. Sci., Phys.*, 1987, **B26**, 105.
94. A. Schiraldi, P. Baldini and E. Pezzati, *J. Therm. Anal.*, 1985, **30**, 1343.
95. T. S. Ellis, F. E. Karasz and G. ten Brinke, *J. Appl. Polym. Sci.*, 1983, **28**, 23.
96. X. Jin, T. S. Ellis and F. E. Karasz, *J. Polym. Sci., Polym. Phys. Ed.*, 1984, **22**, 1701.
97. L. A. Wood, *J. Res. Natl. Bur. Stand. (U. S.)*, 1979, **84**, 353.
98. L. R. G. Treloar, 'The Physics of Rubber Elasticity', 3rd edn., Oxford University Press, Oxford, 1975.
99. G. Gee, P. N. Hartley, J. B. M. Herbert and H. A. Lanceley, *Polymer*, 1960, **1**, 365.
100. W. V. Johnston and M. Shen, *J. Polym. Sci., Part A-2*, 1969, **7**, 1983.
101. P. Mason, *Trans. Faraday Soc.*, 1959, **55**, 1461.
102. R. S. Witte and R. L. Anthony, *J. Appl. Phys.*, 1951, **22**, 689.
103. J. R. Stevens and D. G. Ivey, *J. Appl. Phys.*, 1958, **29**, 1390.
104. K. Oshima and H. Kusumoto, *J. Chem. Phys.*, 1956, **24**, 913.
105. R. K. Chan, K. Pathmanathan and G. P. Johari, *J. Phys. Chem.*, 1986, **90**, 6538.
106. T. S. Chow, *Ferroelectrics*, 1980, **30**, 139.
107. G. ten Brinke, F. E. Karasz and T. S. Ellis, *Macromolecules*, 1983, **16**, 244.
108. J. M. Gordon, G. B. Rouse, J. H. Gibbs and W. M. Risen, Jr., *J. Chem. Phys.*, 1977, **66**, 4971.
109. C. A. Angell, J. M. Sare and E. J. Sare, *J. Phys. Chem.*, 1978, **82**, 2622.
110. E. A. DiMarzio and J. H. Gibbs, *J. Polym. Sci., Part A*, 1963, **1**, 1417.
111. J. G. de la Campa, J. Guzmán, J. de Abajo and E. Riande, *Makromol. Chem.*, 1981, **182**, 3163.
112. E. A. DiMarzio and J. H. Gibbs, *J. Polym. Sci.*, 1959, **40**, 121.
113. H. A. Schneider and M.-J. Brekner, *Polym. Bull. (Berlin)*, 1985, **14**, 173.
114. P. Zoller and H. H. Hoehn, *J. Polym. Sci., Polym. Phys. Ed.* 1982, **20**, 1385.
115. P. R. Couchman and F. E. Karasz, *Macromolecules*, 1978, **11**, 117.
116. P. R. Couchman, *Polym. Eng. Sci.*, 1984, **24**, 135.
117. P. R. Couchman, *Macromolecules*, 1978, **11**, 1156.
118. M. Goldstein, *Macromolecules*, 1985, **18**, 177.
119. A. E. Tonelli, *Macromolecules*, 1977, **10**, 633.
120. A. E. Tonelli, *Macromolecules*, 1977, **10**, 716.
121. J. C. Wittmann and A. J. Kovacs, *J. Polym. Sci., Part C*, 1969, **16**, 4443.
122. J. H. Hildebrand, 'Viscosity and Diffusivity: A Predictive Treatment', Wiley, New York, 1977.
123. J. H. Hildebrand, *Science (Washington, D. C.)*, 1971, **174**, 490.
124. A. K. Doolittle, *J. Appl. Phys.*, 1951, **22**, 1471.
125. A. K. Doolittle, *J. Appl. Phys.*, 1952, **23**, 236.
126. A. K. Doolittle and D. B. Doolittle, *J. Appl. Phys.*, 1957, **28**, 901.
127. T. G. Fox, Jr. and P. J. Flory, *J. Am. Chem. Soc.*, 1948, **70**, 2384.
128. T. G. Fox, Jr. and P. J. Flory, *J. Polym. Sci.*, 1954, **14**, 315.
129. T. G. Fox, Jr. and P. J. Flory, *J. Phys. Chem.*, 1951, **55**, 221.
130. A. Bondi, *J. Phys. Chem.*, 1954, **16**, 929.
131. F. Bueche, *J. Chem. Phys.*, 1953, **21**, 1850.
132. F. Bueche, *J. Chem. Phys.*, 1956, **24**, 418.
133. F. Bueche, *J. Chem. Phys.*, 1959, **30**, 738.
134. F. Bueche, *J. Chem. Phys.*, 1962, **36**, 2940.
135. J. I. Frenkel, 'Kinetic Theory of Liquids', Clarendon Press, Oxford, 1946, p. 93.
136. H. Eyring, *J. Chem. Phys.*, 1936, **4**, 283.
137. M. L. Williams, R. F. Landel and J. D. Ferry, *J. Am. Chem. Soc.*, 1955, **77**, 3701.
138. J. D. Ferry and R. A. Stratton, *Kolloid-Z.*, 1960, **171**, 107.
139. H. Vogel, *Phys. Z.*, 1921, **22**, 645.
140. G. S. Fulcher, *J. Am. Ceram. Soc.*, 1925, **8**, 339.
141. G. Tammann and W. Hesse, *Z. Anorg. Allg. Chem.*, 1926, **156**, 245.
142. R. W. Fillers and N. W. Tschoegl, *Trans. Soc. Rheol.*, 1977, **21**, 51.
143. W. K. Moonan and N. W. Tschoegl, *Macromolecules*, 1983, **16**, 55.
144. W. K. Moonan and N. W. Tschoegl, *Int. J. Polym. Mater.*, 1984, **10**, 199.
145. W. K. Moonan and N. W. Tschoegl, *J. Polym. Sci., Polym. Phys. Ed.*, 1985, **23**, 623.
146. J. Koppelmann, in 'Proceedings of the International Congress on Rheology, 4th', ed. E. H. Lee and A. L. Copley, Wiley, New York, 1965, vol. 3, p. 361.
147. P. B. Macedo and T. A. Litovitz, *J. Chem. Phys.*, 1965, **42**, 245.
148. N. H. Nachtrieb and J. Petit, *J. Chem. Phys.*, 1956, **24**, 746.
149. R. E. Hoffman, *J. Chem. Phys.*, 1952, **20**, 1567.
150. J. D. McKenzie, *J. Chem. Phys.*, 1958, **28**, 1037.
151. R. M. Barrer, *Trans. Faraday Soc.*, 1942, **38**, 322.
152. R. M. Barrer, *Trans. Faraday Soc.*, 1943, **39**, 48.
153. M. H. Cohen and D. Turnbull, *J. Chem. Phys.*, 1959, **31**, 1164.
154. D. Turnbull and M. H. Cohen, *J. Chem. Phys.*, 1961, **34**, 120.
155. N. Hirai and H. Eyring, *J. Appl. Phys.* 1958, **29**, 810.
156. N. Hirai and H. Eyring, *J. Polym. Sci.*, 1959, **37**, 51.

157. R. P. Smith, *J. Polym. Sci., Part A-2*, 1970, **8**, 1337.
158. M. H. Cohen and G. S. Grest, *Phys. Rev. B: Condens. Matter*, 1979, **20**, 1077.
159. G. S. Grest and M. H. Cohen, *Phys. Rev. B: Condens. Matter*, 1980, **21**, 4113.
160. G. S. Grest and M. H. Cohen, *Adv. Chem. Phys.*, 1981, **48**, 455.
161. M. H. Cohen and G. S. Grest, *Ann. N. Y. Acad. Sci.*, 1981, **371**, 199.
162. Y. Lipatov, *Adv. Polym. Sci.*, 1978, **26**, 63.
163. R. N. Haward, *J. Macromol. Sci., Rev. Macromol. Chem.*, 1970, **C4**, 191.
164. R. Simha and T. Somcynsky, *Macromolecules*, 1969, **2**, 342.
165. T. Somcynsky and R. Simha, *J. Appl. Phys.*, 1971, **42**, 4545.
166. V. S. Nanda, R. Simha and T. Somcynsky, *J. Polym. Sci., Part C*, 1966, **12**, 277.
167. R. Simha, *Polym. Eng. Sci.*, 1980, **20**, 82.
168. P. S. Wilson and R. Simha, *Macromolecules*, 1973, **6**, 903.
169. R. Simha and P. S. Wilson, *Macromolecules*, 1973, **6**, 908.
170. J. E. McKinney and R. Simha, *Macromolecules*, 1976, **9**, 430.
171. A. Quach and R. Simha, *J. Appl. Phys.*, 1971, **42**, 4592.
172. J. E. McKinney and R. Simha, *J. Res. Natl. Bur. Stand., Sect. A*, 1977, **81A**, 283.
173. J. E. McKinney and R. Simha, *Macromolecules*, 1974, **7**, 894.
174. R. Simha, *Macromolecules*, 1977, **10**, 1025.
175. R. Simha and C. E. Weil, *J. Macromol. Sci., Phys.*, 1970, **B4**, 215.
176. S. T. J. Peng, R. F. Landel, J. Moacanin, R. Simha and E. Papazoglou, *J. Rheol.*, 1987, **31**, 125.
177. E. Papazoglou and R. Simha, *J. Rheol. (N. Y.)*, 1987, **31**, 135.
178. R. Simha, *Ann. N. Y. Acad. Sci.*, 1976, **279**, 2.
179. C. Feger and W. J. MacKnight, *Macromolecules*, 1985, **18**, 280.
180. M. Gordon and W. Simpson, *Polymer*, 1961, **2**, 383.
181. E. Jenckel and R. Heusch, *Kolloid-Z.*, 1953, **130**, 89.
182. L. J. Garfield and S. E. Petrie, *J. Phys. Chem.*, 1964, **68**, 1750.
183. G. Braun and A. J. Kovacs, in 'Proceedings of the Conference on Physics of Non-Crystalline Solids', ed. J. A. Prins, North-Holland, Amsterdam, 1965, p. 303.
184. F. N. Kelley and F. Bueche, *J. Polym. Sci.*, 1961, **50**, 549.
185. M. Gordon and J. S. Taylor, *J. Appl. Chem.*, 1952, **2**, 493.
186. I. J. McEwen and A. F. Johnson, in 'Alternating Copolymers', ed. J. M. G. Cowie, Plenum Press, New York, 1985, p. 239.
187. J. A. Manson and L. H. Sperling, 'Polymer Blends and Composites', Plenum Press, New York, 1976.
188. W. J. MacKnight, J. Stoelting and F. E. Karasz, *Adv. Chem. Ser.*, 1971, **99**, 29.
189. A. Q. Tool, *J. Am. Ceram. Soc.*, 1946, **29**, 240.
190. A. Q. Tool, *J. Res. Natl. Bur. Stand. (U. S.)*, 1946, **37**, 73.
191. A. J. Kovacs, J. M. Hutchinson and J. J. Aklonis, in 'The Structure of Non-Crystalline Materials', ed. P. H. Gaskell, Taylor and Francis, London, 1977, p. 153.
192. J. M. Hutchinson and A. J. Kovacs, in 'The Structure of Non-Crystalline Materials', ed. P. H. Gaskell, Taylor and Francis, London, 1977, p. 167.
193. A. J. Kovacs and J. M. Hutchinson, *J. Polym. Sci., Polym. Phys. Ed.*, 1979, **17**, 2031.
194. J. M. Hutchinson, J. J. Aklonis and A. J. Kovacs, *Polym. Prepr., Am. Chem. Soc., Div. Polym. Chem.*, 1975, **16** (2), 94.
195. M. A. DeBolt, A. J. Esteal, P. B. Macedo and C. T. Moynihan, *J. Am. Ceram. Soc.*, 1976, **59**, 16.
196. C. T. Moynihan and A. V. Lesikar, *Ann. N. Y. Acad. Sci.*, 1981, **371**, 151.
197. J. M. Hutchinson and A. J. Kovacs, *J. Polym. Sci., Polym. Phys., Ed.*, 1976, **14**, 1575.
198. H. Leaderman, 'Elastic and Creep Properties of Filamentous Materials and other High Polymers', Textile Foundation, Washington, DC, 1943.
199. F. Kolrausch, *Pogg. Ann. Phys.*, 1847, **12**, 393.
200. G. Williams and D. C. Watts, *Trans. Faraday Soc.*, 1970, **66**, 80.
201. I. M. Hodge and A. R. Berens, *Macromolecules*, 1982, **15**, 762.
202. G. Adam and J. H. Gibbs, *J. Chem. Phys.*, 1965, **43**, 139.
203. I. M. Hodge and A. R. Berens, *Macromolecules*, 1981, **14**, 1598.
204. I. M. Hodge, *Macromolecules*, 1983, **16**, 898.
205. I. M. Hodge and A. R. Berens, *Macromolecules*, 1985, **18**, 1980.
206. J. J. Tribone, J. M. O'Reilly and J. Greener, *Macromolecules*, 1986, **19**, 1732.
207. R. R. Lagasse, R. E. Cohen and A. Letton, *J. Polym. Sci., Polym. Phys. Ed.*, 1982, **20**, 375.
208. J. M. Hutchinson and A. J. Kovacs, *J. Polym. Sci., Polym. Phys. Ed.*, 1983, **21**, 2419.
209. I. M. Hodge and G. S. Huvard, *Macromolecules*, 1983, **16**, 371.
210. W. M. Prest, Jr., F. J. Roberts, Jr. and I. M. Hodge, in 'Proceedings of the North American Thermal Analysis Society Conference, 12th, Williamsburg, VA', ed. J. C. Buck, 1983, p. 119 (available from NATAS, A-1 Business Service, 219 Park Avenue, Scotch Plains, NJ, USA).
211. W. M. Prest, Jr., F. J. Roberts, Jr., D. M. Teegarten and K. M. Sheriden, in 'Proceedings of the North American Thermal Analysis Society Conference, 12th, Williamsburg, VA', ed. J. C. Buck, 1983, p. 251 (available from NATAS, A-1 Business Service, 219 Park Avenue, Scotch Plains, NJ, USA).
212. R. W. Rendell, T. K. Lee and K. L. Ngai, *Polym. Eng. Sci.*, 1984, **24**, 1104.
213. D. Ng and J. J. Aklonis, in 'Relaxations in Complex Systems', ed. K. L. Ngai and G. B. Wright, U. S. Government Printing Office, Washington, DC, 1985, p. 53 (available from National Technical Information Service, 5285 Port Royal Road, Springfield, VA 22161, USA).
214. M. Goldstein and M. Nakonecznyj, *Phys. Chem. Glasses*, 1965, **6**, 126.
215. K. L. Ngai, *Comments Solid State Phys.*, 1979, **9**, 127.
216. K. L. Ngai, *Comments Solid State Phys.*, 1980, **9**, 141.
217. R. W. Rendell and K. L. Ngai, in 'Relaxations in Complex Systems', ed. K. L. Ngai and G. B. Wright, U. S. Government Printing Office, Washington, DC, 1985, p. 309 (available from National Technical Information Service, 5285 Port Royal Road, Springfield, VA 22161, USA).
218. K. L. Ngai and G. Fytas, *J. Polym. Sci., Polym. Phys. Ed.*, 1986, **24**, 1683.

Glass Formation and Glassy Behavior

219. K. L. Ngai and D. J. Plazek, *J. Polym. Sci., Polym. Phys. Ed.*, 1985, **23**, 2159.
220. G. B. McKenna, K. L. Ngai and D. J. Plazek, *Polymer*, 1985, **26**, 1651.
221. R. W. Rendell, K. L. Ngai and G. B. McKenna, *Macromolecules*, 1987, **20**, 2250.
222. K. L. Ngai and D. J. Plazek, *J. Polym. Sci., Polym. Phys. Ed.*, 1986, **24**, 619.
223. K. L. Ngai, R. W. Rendell, A. K. Rajagopal and S. Teitler, *Ann. N. Y. Acad. Sci.*, 1986, **484**, 150.
224. A. J. Kovacs, R. A. Stratton and J. D. Ferry, *J. Phys. Chem.*, 1963, **67**, 152.
225. R. E. Robertson, R. Simha and J. G. Curro, *Macromolecules*, 1984, **17**, 911.
226. R. Simha, J. G. Curro and R. E. Robertson, *Polym. Eng. Sci.*, 1984, **24**, 1071.
227. R. E. Robertson, R. Simha and J. G. Curro, *Macromolecules*, 1985, **18**, 2239.
228. R. E. Roberston, *J. Polym. Sci., Polym. Symp.*, 1978, **63**, 173.
229. R. E. Robertson, *J. Polym. Sci., Polym. Phys. Ed.*, 1979, **17**, 597.
230. R. E. Robertson, *J. Appl. Phys.*, 1978, **49**, 5048.
231. R. E. Robertson, *Ann. N. Y. Acad. Sci.*, 1981, **371**, 21.
232. G. Braun and A. J. Kovacs, *Phys. Chem. Glasses*, 1963, **4**, 152.
233. J. G. Curro, R. R. Lagasse and R. Simha, *Macromolecules*, 1982, **15**, 1621.
234. G. H. Fredrickson and H. C. Andersen, *J. Chem. Phys.*, 1985, **83**, 5822.
235. G. H. Fredrickson and S. A. Brawer, *J. Chem. Phys.*, 1986, **84**, 3351.
236. G. H. Fredrickson, *Ann. N. Y. Acad. Sci.*, 1986, **484**, 185.
237. J. J. Fernández, C. Z. Andérico and T. S. J. Streit, *J. Appl. Phys.*, 1982, **53**, 7991.
238. C. Z. Andérico, J. F. Fernández and T. S. J. Streit, *Phys. Rev. B: Condens. Matter*, 1982, **26**, 3824.
239. A. Baumgärtner, *J. Chem. Phys.*, 1980, **73**, 2489.
240. R. D. de la Batie, J.-L. Viovy and L. Monnerie, *J. Chem. Phys.* 1984, **81**, 567.
241. S. M. Rekhson, D. M. Heyes, C. J. Montrose and T. A. Litovitz, *J. Non-Cryst. Solids*, 1980, **38/39**, 403.
242. C. A. Angell, *Ann. N. Y. Acad. Sci.*, 1981, **371**, 136.
243. L. V. Woodcock, *Ann. N. Y. Acad. Sci.*, 1981, **371**, 274.
244. H. J. Raveché, *Ann. N. Y. Acad. Sci.*, 1976, **279**, 36.
245. G. B. McKenna, G. Hadziioannou, P. Lutz, G. Hild, C. Strazielle, C. Straupe, P. Rempp and A. J. Kovacs, *Macromolecules*, 1987, **20**, 498.
246. S. J. Clarson, K. Dodgson and J. A. Semlyen, *Polymer*, 1985, **26**, 930.
247. L. C. E. Struik, *Polymer*, 1980, **21**, 962.
248. D. J. Plazek and G. C. Berry, in 'Glass: Science and Technology, Volume 3, Viscosity and Relaxation', ed. D. R. Uhlmann and N. J. Kriedl, Academic Press, Orlando, FL, 1986, p. 363.
249. S. Matsuoka, S. J. Aloisio and H. E. Bair, *J. Appl. Phys.*, 1973, **44**, 4265.
250. G. C. Berry, *J. Polym. Sci., Polym. Phys. Ed.* 1976, **14**, 451.
251. G. B. McKenna and A. J. Kovacs, *Polym. Eng. Sci.*, 1984, **24**, 1138.
252. G. B. McKenna, in 'Relaxations in Complex Systems', ed K. L. Ngai and G. B. Wright, U. S. Government Printing Office, Washington, DC, 1985, p. 129 (available from National Technical Information Service, 5285 Port Royal Road, Springfield, VA 22161, USA).
253. T. L. Smith, T. Ricco, G. Levita and W. K. Moonan, *Plast. Rubber Process. Appl.*, 1986, **6**, 81.
254. A. F. Yee, R. J. Bankert, K. L. Ngai and R. W. Rendell, in 'Advances in Rheology 3. Polymers', ed. B. Mena, A. Garcia-Rejon and C. Rangel-Nafaile, Universidad Nacional Autonoma de Mexico, 1984, p. 231.
255. A. F. Yee, R. J. Bankert, K. L. Ngai and R. W. Rendell, *J. Polym. Sci., Polym. Phys. Ed.*, 1988, in press.
256. F. A. Myers, F. C. Cama and S. S. Sternstein, *Ann. N. Y. Acad. Sci.*, 1976, **279**, 94.
257. G. B. McKenna and L. J. Zapas, *J. Polym. Sci., Polym. Phys. Ed.*, 1985, **23**, 1647.
258. G. B. McKenna and L. J. Zapas, *Polym. Eng. Sci.*, 1986, **26**, 725.
259. D. L. Huntson, W. T. Carter and J. L. Rushford, in 'Developments in Adhesives — 2', ed. A. J. Kinloch, Applied Science, Barking, 1981, p. 125.
260. L. Guerdoux, R. A. Duckett and D. Froelich, *Polymer*, 1984, **25**, 1392.
261. S. Matsuoka, *J. Rheol. (N. Y.)*, 1986, **30**, 869.
262. J. Heijboer, *Ann. N. Y. Acad. Sci.*, 1976, **279**, 104.
263. G. P. Johari, in 'Relaxations in Complex Systems', ed. K. L. Ngai and G. B. Wright, U. S. Government Printing Office, Washington, DC, 1985, p. 17 (available from National Technical Information Service, 5285 Port Royal Road, Springfield, VA 22161, USA).
264. B. E. Read and G. D. Dean, *Polymer*, 1984, **25**, 1679.
265. N. G. McCrum, *Polymer*, 1984, **25**, 309.
266. C. K. Chai and N. G. McCrum, *Polymer*, 1984, **25**, 291.
267. N. G. McCrum, *Polymer*, 1984, **25**, 299.
268. G. B. McKenna and L. J. Zapas, *Polymer*, 1985, **26**, 543.
269. T. S. Chow, *J. Rheol. (N. Y.)*, 1986, **30**, 729.
270. R. M. Shay, Jr. and J. M. Caruthers, *J. Rheol. (N. Y.)*, 1986, **30**, 781.
271. B. D. Snow, P. Potnis and D. C. Bogue, *J. Rheol. (N. Y.)*, 1984, **28**, 517.
272. B. D. Snow and D. C. Bogue, *J. Rheol. (N. Y.)*, 1984, **28**, 533.
273. R. F. Landel and S. T. J. Peng, *J. Rheol. (N. Y.)*, 1986, **30**, 741.
274. J. T. Bendler, *AIP Conf. Proc.*, 1985, **137**, 227.
275. J. T. Bendler, D. G. LeGrand and W. V. Olszewski, in 'Transport and Relaxation in Random Materials', ed. J. Klafter, R. J. Rubin and M. F. Shlesinger, World Scientific, Singapore, 1986, p. 240.

11

Crystallization and Melting

LEO MANDELKERN

Florida State University, Tallahassee, FL, USA

11.1 PROLOGUE

The crystallization of flexible chain molecules, of sufficient structural regularity, is widely observed in both naturally occurring macromolecular systems as well as those of synthetic origin. Wide-angle X-ray diffraction patterns have established that long range three-dimensional order is a characteristic of polymer crystallinity and is thus similar to crystals formed by low molecular weight substances. A major difference exists, however, since for low molecular weight species individual molecules occupy equivalent points in each unit cell while in polymers the chain repeating units play a corresponding role. The special feature of polymer structure, which manifests itself in all aspects of polymer crystallization, is the covalent connectivity of chain atoms and repeating units. This connectivity of hundreds to thousands of chain atoms sets the crystallization of polymers apart from other types of molecular systems. It is the underlying reason for the differences that are observed. For example, for the flexible type chain molecules the crystallization process is rarely if ever complete. For homopolymers, depending upon molecular weight and crystallization conditions, the extent of crystallization can range from 30 to 90%.[1,2] Because of the basic structural differences that exist it should be apparent that one cannot tacitly assume that polymers and low molecular weight substances display the same crystallization behavior and properties. Each aspect of the crystallization phenomenon needs to be examined separately to see what correlations actually exist. We shall find that for many aspects of the problem there are strong elements of similarity between the different molecular systems; there are, however, also major differences.

The study of crystalline polymers closely parallels the development of polymer science itself.[3] Very early, following the discovery by von Laue and by Bragg that simple crystalline substances diffract X-rays, a variety of polymers were subject to similar analysis. Irrespective of the specific polymers and the structures derived some important principles were established, which in retrospect may seem quite obvious. It was found that (i) normal bond distance, bond angles and other elements of structure are the general rule; (ii) the chemical repeating unit plays a role that is analogous to that of molecules in crystals of low molecular weight organic compounds; (iii) it is not uncommon to find more that one chain passing through the unit cell; and (iv) the unit cells are usually composed of from one to eight chain units. The realization that a unit cell need not contain a complete molecule was a very important development. It was crucial in the determination of crystal structure, in understanding some rudimentary features of polymer crystallization and in the development of the general macromolecular hypothesis.[3,4]

From the initial study of crystallographic structures the investigations of crystalline polymers branched into other directions. Particular emphasis was placed on the melting and crystallization processes. The very early pioneering work of Wood and Bekkedahl,[5,6] who studied the crystallization behavior of natural rubber, pointed out some of the differences as well as similarities between crystallization and melting of polymers and low molecular weight substances. The thermodynamic aspects of the melting of polymers will be given a major emphasis in this chapter. The early studies[3,5,6] made it apparent that the morphological and structural features, characteristic of the crystalline state, play an important role in governing the fusion process. Hence this subject area will also be discussed here. To properly understand the origin of the structural features considerations must also be given to the mechanisms and kinetics of crystallization.[7] This aspect of the problem will only be discussed briefly.

It is well-established experimentally that polymer crystallization from a pure melt only takes place, over a meaningful time scale, at temperatures well below the melting temperature. For a long chain hydrocarbon, such as $C_{94}H_{190}$, crystallization takes place very rapidly at temperatures just infinitesimally below its melting temperature. In contrast, for linear polyethylene of even modest molecular weight crystallization only occurs at reasonable rates at about 20 °C below the melting temperature.[8] For most other crystallizable polymers the undercooling required is even greater.[8] There are several reasons for this behavior which are related to the high interfacial free energy associated with the basal plane of the crystallites and the difficulty in extracting ordered sequences of sufficient length from the highly entangled randomly disordered melt to insure thermodynamic stability. Coupled with the fact that the crystallization process is rarely, if ever, complete a polycrystalline, morphologically complex system develops. The detailed structure that evolves is a consequence of the crystallization mechanism. Consequently the detailed structure of the crystalline state must be given serious consideration in analyzing fusion and the thermodynamic properties of the crystalline state.

The basic fundamental structural feature of bulk crystallized homopolymers, as well as co-polymers with surprisingly high co-unit content, is a lamellar-like crystallite.[9-18] This type of crystallite was originally observed in crystals formed by homopolymers from dilute solution by several different groups of investigators.[19-22] They are also widely observed in bulk crystallized systems and are the typical mode of crystallization. Typical thin section transmission electron micrographs, which depict the very general characteristics of the lamellae, are given in Figure 1. In these micrographs the crystallites, which reject the stain, are the light regions. The non-crystalline, interfacial and interlamellar regions absorb the stain and are represented by the darker region in the micrographs. Although the lamellae are usually limited to thicknesses of from 100–200 Å when formed from dilute solution, they can be as large as several thousand Ångstroms in homopolymers crystallized in bulk under atmospheric pressure.[9-17] The chain axes are preferentially (but not exactly) oriented normal to the wide faces, or basal planes, of the lamellar crystallite. Except for very low molecular weights the crystallite thickness is always much less than the extended chain length. The trajectory of the chains, after they emerge from the crystallite of origin, is not only a matter of interest but is also important in determining properties. In dilute solution, where the crystallites are isolated from one another, the chains have to return to the crystallite of origin. Hence, some type of 'folding' must take place.* For more concentrated systems, and for pure polymers, there is also the distinct possibility of chains escaping from the crystallite of origin and joining other crystallites. These different situations have to be sorted out.

With the discovery of the lamellar form of polymer crystallites, it was widely heralded, with minor exceptions,[23-28] that the chains crystallized in a regularly folded array with complete adjacent reentry.[29-32] Thus the basal plane was presumed to have a very smooth interfacial structure. For bulk crystallized systems a sequence of chain units would be allowed to escape on occasion 'as a defect' and join a neighboring crystallite. The concept of regular chain folding, where the chains

---

* Although the expression 'folding' is a legitimate one, its use without clear definition has lead to a great deal of confusion and ambiguity. The specific type of structures that are involved need to be clearly specified.

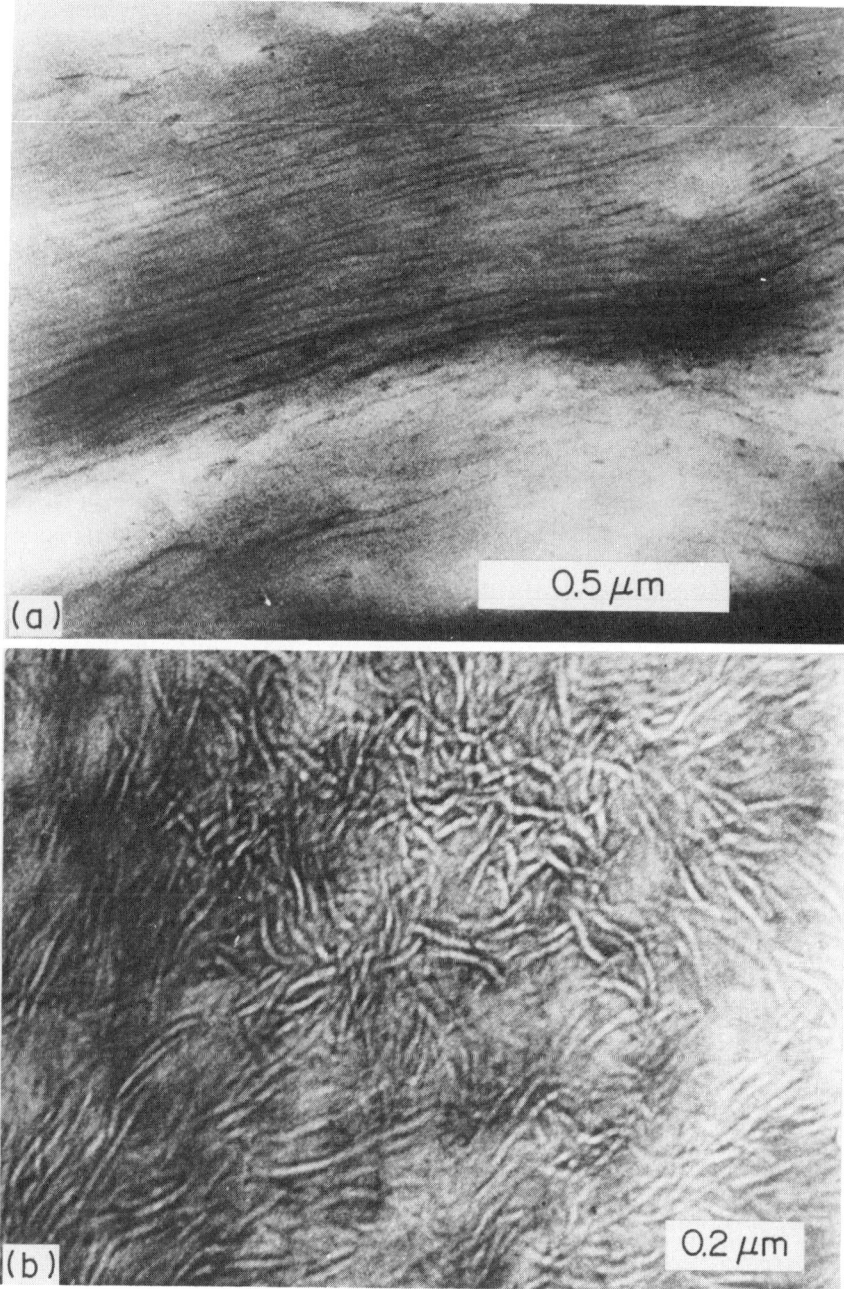

Figure 1 Thin section, stained electron micrographs of crystalline linear polyethylene fractions. (a) $M = 2.78 \times 10^4$ isothermally crystallized 128.8 °C; (b) $M = 1.89 \times 10^5$ rapidly crystallized. Lamellar crystallites light regions; interlamellar regions dark[17]

within the crystallites resembled firehoses, was so widely advertized that it became a matter of faith. In retrospect there seem to have been two major reasons that this view was once held. One reason was the *ad hoc* identification of the well-established lamellar crystallites with regularly folded chains. The apparent geometric surface regularity, as perceived by the electron microscope, for either bulk or dilute solution formed crystallites cannot be taken as evidence for sharp folds with adjacent reentry. The present resolving power of the electron microscope, $\simeq 50$–$100\,\text{Å}$, under the usual circumstances is not sufficient to preclude the existence of a disordered overlayer. Although the gross crystallite morphology is clearly lamellar the molecular morphology, or the chain structure, cannot be defined by the microscopic techniques currently available. Therefore, despite its aesthetic appeal there is no scientific basis for the identification of the lamellar crystallites with a set of regularly folded chains.

Another reason that the concept of regularly folded chains was promulgated for such a long time was the strong efforts that were made to legitimize the concept by application of nucleation theory.[33,34] Extensive studies of the temperature coefficient in the vicinity of the melting temperature of the overall crystallization rate in bulk and from solution,[35-39] of the spherulite growth rate in bulk*[40-42] and lamellar growth rate in solution[43-46] have firmly established that polymer crystallization is a nucleation controlled process.[8] This conclusion has turned out to be of extraordinary importance in the study of crystalline polymers. It is fundamental in understanding both the crystallization process and resultant properties since crystallite thicknesses are related to the critical size of the nucleus. As will be discussed in a more detailed analysis of nucleation theory later in this chapter, this conclusion is reached on the most general theoretical ground.[8,46,47] The strong dependence of the crystallite thickness of the undercooling at which the crystallization is conducted arises from the same source.[8] Because of the generality of these conclusions they do not depend on the form, structure or chain disposition within the nucleus. Virtually all conceivable nuclei structures will obey the required kinetics and size relations. Hence, based solely on kinetic studies, it is not legitimate to assume a specific molecular form for the nucleus and to allow it to develop into a mature crystallite. The procedure described was used to develop what is called the 'kinetic theory of chain folding' where it was assumed that the nuclei are comprised of regularly folded chains that subsequently grow into mature crystallites which have the same chain structure.[33] A specific nucleation path was arbitrarily chosen so that the two-dimensional Gibbs-type coherent nucleus would be stable at temperatures infinitesimally above the crystallization temperature. It has been pointed out on many occasions that the argument for regularly folded chains, based on kinetic studies and nucleation theory, is a circular one.[7,48-51] The nucleus structure must be established independently of kinetic arguments before any unique and thus meaningful molecular information can be obtained. There is in fact no substantive basis, either experimental or theoretical, to support the view that the lamellar crystallites of long chain molecules crystallized in bulk, or even from solution, are comprised of regularly folded chains where the chains make sharp hairpin curves to yield the optimum level of crystallinity. Fortunately, these views are being rapidly abandoned.[28] When the basic ideas which lead to the concept of regularly folded chains are recognized one can then proceed to discuss the total subject in an objective manner. However, to keep matters in perspective, it is very important to recognize that nucleation processes play a very important role in polymer crystallization.

The lamellar-like crystallites need not be identified with regularly folded or pleated chains. Their gross morphological forms are compatible with other type chain structures as is clearly indicated by the differential staining exhibited by the electron micrograph of Figure 1. A variety of techniques demonstrate that a disordered overlayer is even associated with the lamellar crystallites formed in dilute solution.[7,24,52,53] For bulk crystallization some of the chains could traverse a crystallite only once and then join a nearby crystallite; while others would return to the crystallite of origin. All of these general structural features are quite compatible with the gross morphology as revealed by electron and other types of microscopy. There is then the fundamental question as to why a lamellar crystallite habit is characteristic of polymer crystallization.

The basic reason for lamellar formation resides in the different spatial requirements of polymer chains in the interfacial zone and in the region of complete isotropy between lamellae.[23,54,55] The continuity of a long chain molecule, because of covalent bonding, imposes severe constraints on the transition between the perfect order of the crystalline region and the disorder of the isotropic, or liquid-like, region. Because of these restraints the transition between the two states cannot occur abruptly. This fact is a major distinguishing feature between polymer crystallization and the crystallization of small molecules. A significant proportion of the flux of chains emanating from the 001 crystal face is dissipated by chains returning to the crystallite of origin, but these do not necessarily have to be in juxtaposition.† Therefore, in order for a crystallite to grow laterally a significant amount of chain bending or folding must occur which will necessarily involve an expenditure of free energy. (cf. seq.) However, this loss in free energy can be easily overcome by lamellar growth. There is then a straightforward process by which well-developed lamellae will form. It is not necessary to invoke any inherent propensity of chain molecules to fold in a regular array nor to manipulate nucleation theory characteristic of monomeric substances.

---

* We cite here just a few of the large number of examples that exist in the literature which show essentially the same result.
† These considerations apply to a lamellar crystallite of very large, or in effect unlimited, lateral dimensions. However, these concerns do not apply to a nucleus of critical size.[8] Estimates of the interfacial free energies involved indicate that for polyethylene, for example, the nucleus will consist of from 20 to 100 sequences in the undercooling range in which crystallization is usually conducted.

There is also an abundance of experimental evidence which makes clear that the lamellae formed during bulk crystallization are not comprised of arrays of regularly folded chains.[7,27,56–59] There are two main types of studies that have been directed to this problem. One involves a study and analysis of the dependence of many different properties on molecular weight and crystallization conditions.[25,27,60–64] The analysis of the wide range in values that can be attained in a very systematic way for a given property precludes the existence of a regularly folded chain structure. Small-angle neutron scattering experiments with deuterated polymers cocrystallized with the protonated counterpart have also provided very strong and compelling evidence of a non-adjacent type reentry.

The small-angle neutron scattering studies have shown quite generally that for samples crystallized from the pure melt the radius of gyration of the chains does not change upon crystallization.[65–70] These results indicate that the molecules crystallize with a distribution of mass elements that is similar to that characteristic of the melt. They must, therefore, be distributed over several lamellae. Analysis of the experimental scattering curves obtained at intermediate angles shows that the results are incompatible with any significant amount of adjacent reentry.[68,71–77] Analysis of wide-angle neutron scattering yield the same conclusion.[78,79]

There is a substantial body of experimental evidence which indicates the existence of a significant interfacial region, or interphase, wherein some vestiges of order remain.[64,80–82] For linear polyethylene the relative interfacial content increases from about 5 to 15% as the molecular weights approach several million.[64] For random copolymers the interfacial region can represent as much as 20–25% of the system.[64,83] In a major theoretical advance, Flory, Yoon and Dill[54,55] calculated the structure of the equilibrium interphase associated with polyethylene crystallites.* In this calculation attention is focused on the region between the perfect crystal and the isotropic liquid-like phase. A lattice technique was used. Each segment of the chain is assigned to one of the sites of a cubic lattice. The dimension of a lattice site for a polyethylene segment is $\simeq 4.5$ Å. In the initial calculations the chain axes are taken to be oriented normal to the 001 basal plane of the crystallite. One of the axes of the lattice is set parallel to the chain axis in the crystal. There is then a set of lattice planes (or set of sites) parallel to the 001 plane which is separated by a distance of about 4.5 Å. With this model and taking into account the natural constraint of chain continuity, the conservation of segments entering and leaving a given plane, and the free energy of making a bend or fold, it is possible to calculate the chain trajectory as it emerges from the crystal face. The problem of flux dissipation is then analyzed in a straightforward manner. A schematic, two-dimensional view of the interphase is given in Figure 2. The layers, from the crystal surface outward, are labeled 0, 1, 2, ... The flux of chains emanating from the crystalline phase is very large. It can only be dissipated (neglecting chain ends) by reversals in the chain direction, so that the chains can return to the crystallites from which they initially emerged.

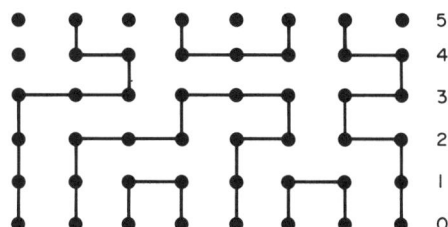

Figure 2 Schematic cross-sectional view of the interphase in a lamellar semicrystalline polymer. The layers are numbered from the crystal surface[54]

According to these calculations, in order for the flux of chains to be dissipated, about 70% of the chains return to the crystal face from which they emanate when the chain axes are oriented normal to the basal plane, without passing beyond the third layer. However, only a small portion of these, less than 20%, are adjacent folds in the first layer. Most of the reentry from the first three lattice layers adjoining the crystal surface occur at sites which are nearby, but not adjacent to, the emerging site. Regular folding, or adjacent reentry, will be relatively dilute according to these calculations. The vestiges of parallel order in the interfacial region will be largely dissipated in the third and fourth

---

* The term equilibrium is used advisedly since it represents the structure associated with the lamellar crystallites. For true equilibrium the crystallite thickness is comparable to the extended chain length[84] so that the problem of lamellar structure is not a matter of concern.

layers. Therefore, an interphase of 10–15 Å is indicated, a value which stands in very good agreement with experimental results.[64,81,85] Thus, the crystalline order is dissipated and a state of isotropy is reached about 10–15 Å from the crystallite surface.

From the foregoing analyses, it is also possible to calculate the interfacial free energies, σ_e, associated with the basal plane of the mature crystallites. Allowing for a reasonable variation in the parameters involved, σ_e is calculated to be in the range 50–60 erg cm^{-2} (50–60 × 10^{-7} J cm^{-2}). This theoretical estimate is in good accord with that determined for low molecular weight polyethylene, $M = 5600$, which displays a typical 'folded' type lamellar morphology. Such low molecular weight systems are best suited for comparison because the influence of chain entanglements, and other topological features, on the interfacial free energy will be minimized. These results show that the main reason for the relatively large interfacial free energy associated with the 001 face of polymer crystals is the persistence of order for a significant distance beyond the crystal boundary. For higher molecular weights the extensive entanglement and knotting of chains in the initial melt and subsequent non-crystalline region will prevent the attainment of the equilibrium interphase. The interphase will then extend further beyond the crystal surface so that there will be a concomitant increase in the interfacial free energy. Experimental estimates of the interfacial free energy associated with the basal plane of the mature crystallites of higher molecular weight chains is in the range 90–300 erg cm^{-2} (90–300 × 10^{-7} J cm^{-2}).[9,86,87a]

The calculations discussed above are based on the assumption that the chain axes within the crystallite are normal to the basal plane of the lamellae. However, for polyethylene, which is the only polymer that has been studied in detail, there is a systematic deviation from this orthogonality which has very important implications for the interfacial structure. For crystallization from the melt at low undercoolings the angle of inclination between the chain axis and the basal plane is about 19°. It increases to about 40° for crystallization at large undercoolings.[17,87a] The surface density of chains in the first layer of the interphase will obviously be reduced by the tilting. As a consequence the need for adjacent reentry to reduce the flux of chains will be diminished. For polyethylene the need for adjacent folds becomes negligible when the tilt angle is greater than 25°,[54] which is a commonly observed situation. Concomitantly the interfacial free energy will be reduced significantly.

In polyethylene the inclination of the chain axes removes from serious concern any extensive adjacent reentry. Since variations in the chain tilt are usually accomplished by changing the crystallization temperature[17,87a] the unique and important situation can develop where the interfacial free energy characteristic of the mature crystallites will depend on the crystallization temperature. In addition to the tilt angle the need for, and propensity of, adjacent reentry will depend on the energetics involved in making a sharp bend.[54] These structures usually involve *gauche* bond conformations which are energetically unfavorable for polyethylene. However, for other polymers, such as poly(ethylene oxide) for example, the *gauche* states of certain bonds are more stable.[87b] For such polymers, with all other factors being equal, adjacent reentry will not be disfavored as much in dissipating the chain flux. Calculations which do not take into account chain tilt and the free energy required to make a 'fold'[87c] do not conform to physical reality and, therefore, will only have very limited validity, if any at all.

The brief review of the basic crystallite structure has centered on the nature of the lamellar habit. There are several points that should be emphasized. The molecular basis for the formation of lamellae can be developed in a natural way, based on the conformational properties of long chains. It is not necessary to invoke any inherent propensity of chains to fold or to arbitrarily draw upon monomeric nucleation theory. The structure of the equilibrium interphase has been quantitatively described. Although the number of chains that return to the crystallite from which they emanate is relatively large, the number that return in adjacent positions is small. These equilibrium considerations will be tempered in general by kinetic factors and by the melt structure at high molecular weights. The amount of adjacent reentry will be further reduced while the residual melt structure will enhance the value of the interfacial free energy.

Although the lamellar crystallite is a universal characteristic of homopolymer crystallization differences in detailed structure are found with respect to changes in molecular weight and crystallization conditions. Thin section electron microscopic studies have revealed that when linear polyethylene is isothermally crystallized at low undercoolings large, geometrically well-developed crystallites form with low molecular weight samples.[14–17] When the molecular weight is increased the lamellae become curved and segmented internally, while their lateral extent is reduced. For the highest molecular weights studied, of the order of several million, the lateral dimensions of the lamellae become small. When the crystallization temperature is lowered it is found for all molecular weights that the lamellae become more curved and their lateral extent is severely reduced. Well-defined lamellae are also observed for compositional fractions of random ethylene copolymers containing as much as 3.2 mol % branch points.[18] The lamellae of the copolymers of lower co-unit content are quite flat. However, their lateral extent is restricted when compared with homopolymer crystallites of the same molecular weight. The lamellae of the higher co-unit content copolymers are very curved, their lateral extent is restricted and they are severely segmented. However, their lamellar characteristics are quite discernible. For copolymers which contain a larger concentration of structural irregularities, although crystalline regions can be detected by electron microscopy, lamellae are no longer observed.[18]

From small-angle neutron scattering, from studies of properties and from theory it is possible to develop a schematic representation of a rudimentary crystallite and associated regions. A highly schematic representation of crystallities is given in Figure 3. This schematic representation is designed to illustrate the major structural features. It is not concerned with fine detail. There are three major regions, each of which have different chain conformations. These are the crystalline region, the interfacial region or interphase, and what has been commonly called the amorphous or liquid-like region. The crystalline region represents the three-dimensional ordered structure with the typical lamellae-like habit. The chain structure, represented by the vertical lines, can be either planar zigzag or a variety of known helical structures.[88] In homopolymers the lamellae are typically several hundred to several thousand Ångstroms thick and several microns in the lateral direction. The imperfection levels within the interior of the crystallites formed by polymers are no different than those found in corresponding low molecular weight compounds. There is no experimental evidence for an excessive concentration of such defects. The crystallite or core thickness is controlled by nucleation requirements. This does not mean that the crystallite thickness is identical with that of a critical size nucleus. Under many circumstances the crystallites thicken with time.[89-92] The rate of crystallite thickening depends on molecular weight and temperature.[89,90] The base thickness is, however, set by the nucleation requirements.

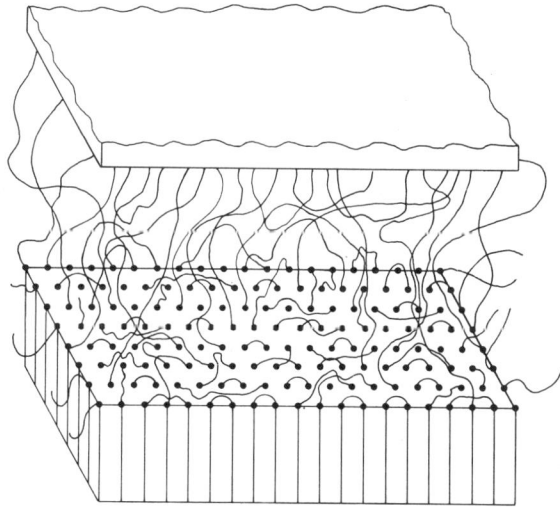

Figure 3 Schematic rendering of isotropic, interlamellar region of semicrystalline polymers

The structure of the interphase has already been discussed in some detail. It is a diffuse region that extends some 10–15 Å above the basal plane. Illustrated in the schematic representation are sequences in adjacent reentry, neighboring returns, long loops as well as sequences of units which escape the interfacial region. The chain dispositon within the interphase leaves some vestiges of order.[54]

The other major region, which is the non-crystalline one, reflects the structure of the sequences of repeating units which connect lamellae. Excluding the interphase, it represents the structure of the interlamellar region. Spectral and thermodynamic studies show that the units in this region are in essentially random conformation and thus liquid-like in character.[57,60-62,64,93-96] The quantitative nature of the degree of crystallinity,[57,58] glass formation of semicrystalline polymers[57] and low frequency Raman spectroscopy[97,98] all substantiate the liquid-like, isotropic character of this region. The condition of isotropy, which though in itself is very important, obscures some important structural details. The pure melt is of course completely isotropic. Statistical arguments, as well as deductions from the neutron scattering experiments make clear that the topological arrangements of the chain in the non-crystalline region must resemble those that existed in the initial melt. In the melt the molecules are profusely entangled with one another; the extent of this entanglement is molecular weight dependent. With minor exceptions, the entangled units are excluded from the crystallites and hence are trapped and concentrated in the interlamellar regions. This topological effect manifests itself quite vividly in the influence of molecular weight on crystallization kinetics and the level of crystallinity that can be attained at a given temperature. From these considerations, one concludes that neighboring lamellae are lavishly interconnected by sequences of chain units in random conformation. Many different types of connections can be envisaged, some of which are illustrated in Figure 3. A chain can traverse one lamellae to another, either directly or after entangling and intertwining with other chains. Connections can also be provided by entanglements involving chains that originate from neighboring lamellae and that also return to the crystallite of origin after traversing a portion of the adjacent non-crystalline region. Such a connection has been postulated earlier[99] and recently established by mathematical analysis.[100] All the connecting structures illustrated, as well as others that can be envisaged, are isotropic. Hence they have not as yet been discriminated by the usual type of measurement. A very important challenge is to quantitatively describe the structure of the isotropic interlamellar region. Although the condition of isotropy establishes the general structure of the interlamellar region, and can account for many properties through their dependence on the degree of crystallinity,[58] it is not adequate in itself to explain other features of the crystalline state such as mechanical and ultimate properties[101] and the influence of molecular weight on the level of crystallinity that can be attained.[63]

With this background and discussion of the major features of structure and morphology of semicrystalline polymers, we can now direct our attention to the major subjects that will be discussed in this chapter.

11.2 INTRODUCTION

We shall limit ourselves in this chapter to a discussion of quiescent crystallization of flexible chain molecules. The very important and interesting problem of the crystallization of liquid-crystal forming polymers will not be discussed. Major attention will be directed to bulk crystallization, although the thermodynamic aspects of polymer–diluent mixtures will also be considered. The important area involving the crystallization and melting of deformed systems, *i.e.* strain induced crystallization, will not be treated here because of space limitations. The major focus here will be on the melting of polymers, and related thermodynamic features. However, this task cannot be accomplished without some understanding of the kinetics and mechanisms of crystallization as well as the general aspects of the structure and morphology of crystalline polymers. There are many facets to the subject of crystalline polymers. It has been pointed out that it is often treacherous to consider one aspect of polymer crystallization in isolation from the total subject.[7] This review is not intended to be a compendium of all experimental and theoretical results. The emphasis will be on a discussion of basic principles and the consequences thereof. Experimental results have been selected that either illustrate these principles or point out areas in need of further theoretical analysis. Since this selection must necessarily be arbitrary many experimental results have not be given their due consideration.

Before examining the fusion of long chain molecules in detail it is necessary to clearly distinguish between several different molecular situations. The nature of a homopolymer is clear in that, except for end groups, the chemical and structural repeating units are identical along the chain. There are situations where the nature of the end group can be important.

From the point of view of crystallization behavior the definition of copolymer can be rather subtle. There can be several different types of structural irregularities incorporated into a chain. There can of course be two chemically dissimilar monomers or co-units. However, depending on the nature of the repeating units, there can also be geometric or stereoisomers, and head–head as opposed to head–tail structures. Branch points and crosslinks are other sources of structural irregularities. Thus it is possible to have copolymer behavior from a crystallization point of view with chemically identical repeating units. Also of concern, for all types of polymers, are the molecular weight and molecular weight distribution which play crucial roles in crystallization mechanisms and related properties. It is important to note that no matter how well-fractionated a system of long chain molecules may be, no-one has yet been able to produce a system where all the molecules have exactly the same chain length.

We shall start by considering the general theoretical aspects of the fusion process in terms of the different types of molecular systems just described. The comparison between theory and experiment will bring in naturally the influence of structure and morphology.

11.3 FUSION OF POLYMERS

11.3.1 Character of the Fusion Process

The melting–crystallization process of small molecule systems is formally described as a first-order phase transition, with the appropriate laws being applicable. For a one-component system, at constant pressure, the transformation temperature is independent of the relative abundance of either of the two phases being maintained in equilibrium. Melting is infinitely sharp. The characteristic temperature of equilibrium is defined as the melting temperature. For the above conditions to be experimentally satisfied an almost perfect internal arrangement of the crystalline phase is required. Moreover, crystals of large size are required to minimize any excess contribution to the free energy change caused by the surfaces or junctions between the two phases. Deviations from these idealized conditions will inevitably lead to a broadening of the melting range.

The criteria set forth for an idealized first-order phase transition apply equally well to the melting–crystallization of all substances. Before considering whether it is valid to apply these classical ideas to polymers it is instructive to examine the fusion process of *n*-hydrocarbons that are low molecular weight chain molecules. The results of a dilatometric study of the fusion of pure $C_{44}H_{90}$ and $C_{94}H_{190}$ are presented in Figure 4.[102] For both these pure compounds the chain lengths are uniform. The complete molecule participates in the crystal structure and molecular crystals are formed. For both compounds the fusion process is relatively sharp. The melting temperature, representing the termination of fusion, is very clearly defined and can be determined with only a very small uncertainty. Upon close scrutiny, however, differences can be observed in the fusion of the two compounds. The $C_{44}H_{90}$ melts almost exactly according to theoretical expectations. The fusion process is relatively sharp and takes place within less than 0.25 °C. On the other hand, although the chemically pure $C_{94}H_{190}$ gives a well-defined melting temperature, the fusion process is much broader. In this case melting takes place over a 1.5–2 °C temperature interval. It is well accepted that this compound also is undergoing a first-order phase transition on melting. Here the broadening of the fusion range cannot be attributed to chemical impurities (or an added second component). It is reasonable to assume that morphological or structural 'impurities' are the cause for the broadening.

An interesting comparison can be made between the fusion process of a normal alkane and a low molecular weight linear polyethylene of comparable molecular weight. Such a comparison is given in Figure 5, for the fusion of $C_{44}H_{90}$ (M = 618) and a linear polyethylene fraction $\bar{M}_n = 725$, $\bar{M}_w/\bar{M}_n$

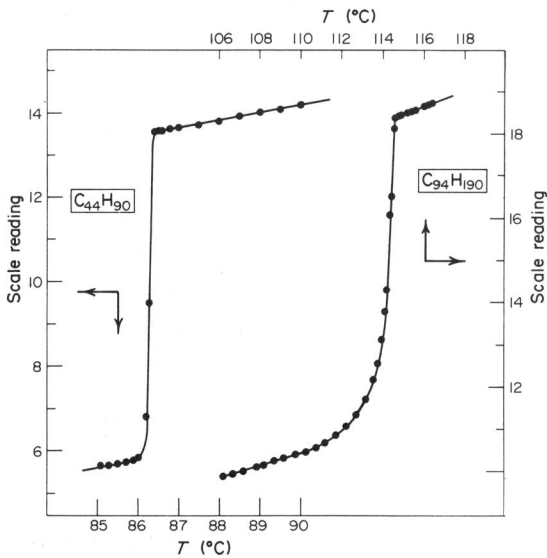

Figure 4 Fusion of *n*-hydrocarbons. Plot of dilatometric scale reading against temperature for $C_{44}H_{90}$ and $C_{94}H_{190}$[102]

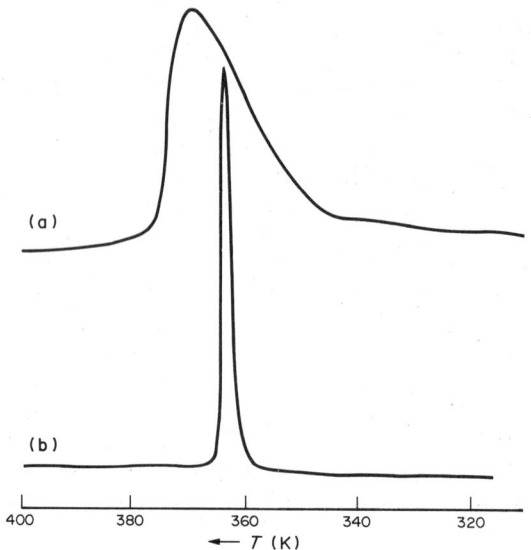

Figure 5 Differential Scanning Calorimetric Endotherms: (a) low molecular weight polyethylene $M = 725$; (b) n-hydrocarbon $C_{44}H_{90}$[103]

$= 1.1$.[103] The differences between the two endotherms is quite striking. The normal hydrocarbon displays classical behavior. While melting is clearly evident for the polymer, the fusion process is quite broad, reflecting the lack of uniformity in chain length even for such a narrow molecular weight distribution.

Long chain molecules that are packed in perfect array in crystallites of sufficiently large dimensions represent a state that can be termed crystalline. The fact that a chain molecule may penetrate many unit cells is of no consequence in these considerations. Accordingly, melting should be a first-order transition and the important consequences of this proposition would follow. If this condition is not satisfied by experiment and detailed molecular theory the foregoing premise will have to be discarded. The extent to which this condition is fulfilled must ultimately be judged in terms of the narrowness of the fusion process and the sharpness and reproducibility of the melting temperature. There are inherent difficulties in classifying a transition by the shape of the fusion curve. A detailed investigation of the nature of the fusion and the characteristics of the transformation temperature is required.[104]

We examine this question by analyzing the melting of the long chain linear polyethylenes. The fusions of these polymers are known to be typical of other crystalline polymers[1] and offer a continuity to the melting of the low molecular weight homologs illustrated in Figure 4. In order to study the fusion process and to determine the nature of the transformation, procedures must be adopted that ensure conditions satisfying equilibrium are attained. Experience has taught us that these requirements are best fulfilled when the transformation from the liquid to crystalline state is carried out slowly at temperatures close to the melting temperature or by protracted annealing at elevated temperatures of the already formed crystalline phase. Attention must also be given to the molecular constitution of the chains. For homopolymers this involves the molecular weight for fractions and the molecular weight distribution for polydisperse systems.

Utilizing the linear polyethylenes as model systems we next examine the influence of molecular constitution and crystallization conditions on the fusion process. Figure 6 illustrates the influence of crystallization conditions on the fusion of a very polydisperse linear polyethylene ($\bar{M}_n = 1.2 \times 10^3$; $\bar{M}_w = 1.5 \times 10^5$).[105] In one case, represented by the open circles, the sample was slowly cooled from the melt to room temperature. The solid circles give the results when the sample was crystallized at 130 °C for 40 days then cooled to room temperature prior to fusion. In both examples melting is relatively sharp and the last traces of crystallinity disappear at a well-defined temperature, which is taken in the conventional manner to represent the melting temperature. Although the melting temperature is the same for both samples close scrutiny of the plots in Figure 6 indicates that the isothermally crystallized samples melt over a significantly narrower temperature interval. The clear implication of these results is that if more stringent crystallization conditions could be employed fusion would become increasingly sharper. Thus, for two molecularly identical systems the crystalliz-

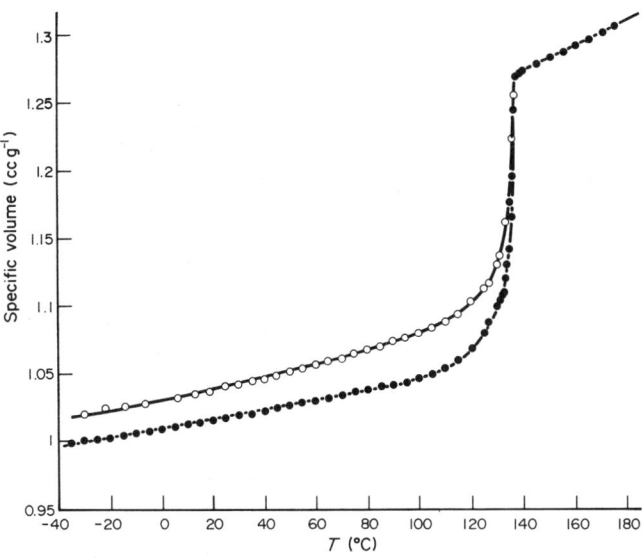

Figure 6 Specific volume–temperature relations for an unfractionated linear polyethylene sample. Slowly cooled from melt ○; isothermally crystallized at 130 °C for 40 days then cooled to room temperature ●[105]

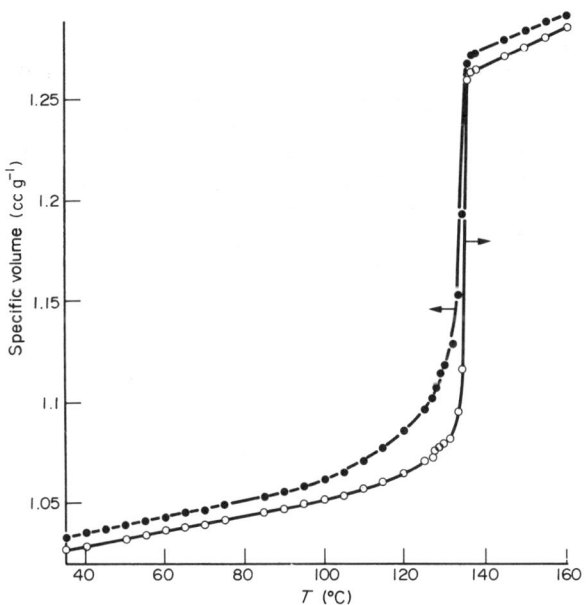

Figure 7 Specific volume–temperature relation for linear polyethylene samples. Samples initially crystallized at 131.3 °C for 40 days. Unfractionated polymer ●; fraction $M_n = 32\,000$ ○[106]

ation conditions, and presumably the resulting morphological forms, influence the course of fusion although the melting temperatures themselves are clearly defined.

Varying the molecular constitution of the chain also influences the fusion. A comparison between the unfractionated polymer of Figure 6 and a molecular weight fraction, $M = 3.2 \times 10^4$ crystallized under extreme isothermal conditions, is given in Figure 7.[106] For both samples the melting temperatures are the same and clearly defined. It is evident in Figure 7 that the melting of this fraction is appreciably sharper than the whole polymer. For the fraction, 80% of the transformation occurs over a 2 °C range; over the same temperature interval the change is only 35–40% for the polydisperse system. Molecular weight fractions apparently form a more perfectly developed crystalline state with a concomitant sharper fusion process. However, more detailed studies have shown that even with fractions there is a profound influence of molecular weight.[60] Figure 8

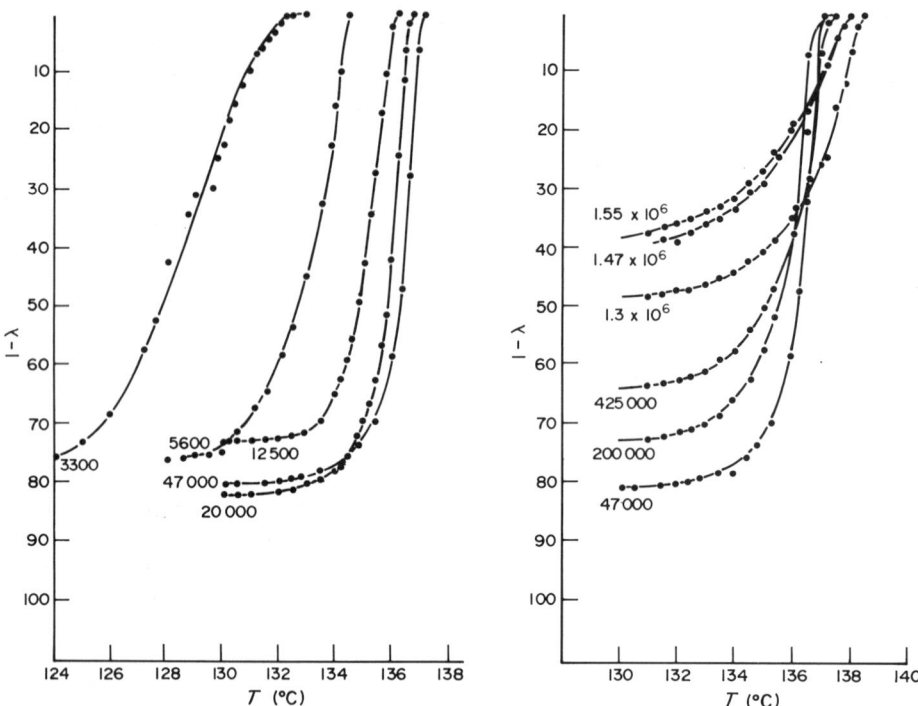

Figure 8 Plot of degree of crystallinity, $1 - \lambda$, as a function of temperature after crystallization at high temperature for polyethylene fractions of the molecular weights indicated[60]

illustrates some results for molecular weight fractions ranging from 3.3×10^3 to 1.55×10^6 which were isothermally crystallized and never cooled subsequent to the initiation of fusion. The two lowest molecular weight fractions melt very broadly. This result is easily explained by the high concentration of end groups and their exclusion from the crystal lattice (*cf. seq.*) We have in effect a significant impurity concentration and the expected broadening of the melting range. As the end group concentration becomes insignificant an appreciable sharpening of the melting takes place. For molecular weights in the range 1.25×10^4 to 4.7×10^4 about 80–90% of the transformations occur over only a 2 °C interval. The curves begin to closely resemble the one in Figure 4 for $C_{94}H_{190}$ and thus the behavior expected for a first-order phase transition of a pure substance. However, for higher molecular weights, including examples not shown here, the curves broaden significantly with increasing chain length. It has also been found for the high molecular weights that if the level of crystallinity is restricted the melting range narrows considerably and becomes comparable to that observed for the lower molecular weight species. The factors involved in the broadening appear to be associated with the increasing level of crystallinity although for high molecular weights the absolute level of crystallinity that can be attained is relatively low.[63,107] The reasons for the broadening appear to be structural and morphological in charactrer. It could be caused by a distribution of crystallite sizes and the influence of the structure of the interfacial regions.

The results described above for linear polyethylene give clear evidence that it is possible to develop fusion curves in homopolymers that are comparable to those obtained for low molecular substances. In all cases the temperature at which the last traces of crystallinity disappear is clearly and accurately defined. Molecular weight polydispersity as well as structural and morphological irregularities tend to broaden the fusion range. However, even for monomeric substances the fusion process can be broadened by rapid cooling from the melt and the freezing in of non-equilibrium states. It is not unexpected, therefore, that these processes will be exaggerated in polymers. For long chain molecules, in order to develop a sharper fusion process the chain length must be highly uniform. In addition, the stringent crystallization conditions that are necessary cannot be easily employed because of kinetic restraints. The differences in the fusion process between polymers and low molecular weight substances is, therefore, one of degree rather than of kind. We can, therefore, conclude that the melting of crystalline polymers is a first-order phase transition and all of the dictates of this transition should be followed. The consequences of this conclusion are profound and have far-reaching implications. Before pursuing these consequences it is of interest to examine the fusion of other than pure one-component systems.

Impurities, or second-components, can be incorporated into polymeric systems in several ways. They can be built directly into the chain as a co-unit in a copolymer or added as a monomeric diluent. Although the fusion of copolymers will be considered in quantitative detail in a subsequent section, it is appropriate at this point to consider the general aspects of the fusion of random type copolymers and compare the results with those of homopolymers just described. The results for copolymers are generally the same, irrespective of the specific chemical or physical nature of the structural irregularities, as long as they are in random sequence distribution. A typical example is given in Figure 9 for three different polybutadienes.[108] Only the 1,4-*trans* units crystallize and the mole fractions of these samples range from 0.64 to 0.81. The fusion process has obviously broadened considerably when compared to that of pure homopolymers. The transformation occurs over a very wide temperature interval and the melting curves are sigmoidal in character. Such fusion curves are typical of random copolymers and are theoretically expected (see below). As the concentration of crystallizable units decreases the melting range becomes progressively broader and the level of crystallinity decreases. However, even the situation illustrating a melting temperature can be definitely established.

The addition of a low molecular weight diluent, which does not enter the crystal lattice, will lower the melting temperature and broaden the melting range of a crystalline polymer. A quantitative analysis of the melting point depression for such systems will be given subsequently. At present we are only interested in analyzing the fusion range, which can be accomplished by just assuming phase equilibrium.[106,109] If the diluent is uniformly distributed throughout the non-crystalline phase then the volume fraction of polymer in this phase, v_2', will vary with the degree of crystallinity, $1 - \lambda$, according to

$$v_2' = \frac{v_2 \lambda}{1 - v_2 + v_2 \lambda} \tag{1}$$

where v_2 is the nominal composition of the system. At phase equilibrium the chemical potential of the polymer unit in the pure crystalline phase will be equal to that in the liquid phase at the composition given by equation (1). By utilizing the Flory–Huggins theory for the chemical potential of the liquid phase it is found that[106]

$$\frac{1}{T_\lambda} - \frac{1}{T_m^\circ} = \frac{R}{\Delta H_u} \frac{V_u}{V_1}(v_1' - \chi_1 v_1'^2) \tag{2}$$

Here T_λ is the equilibrium temperature for a degree of crystallinity, $1 - \lambda$, corresponding to a volume

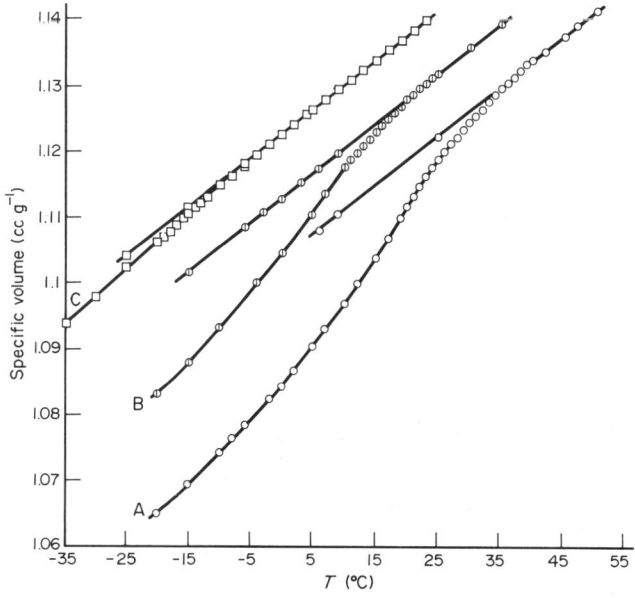

Figure 9 Plot of specific volume against temperature for polybutadienes of varying concentration of crystallizing *trans* 1,4 units. Curve A, $X_A = 0.81$; curve B, $X_A = 0.73$; curve C, $X_A = 0.64$. Curves B and C are arbitrarily displaced along the ordinate[108]

fraction of diluent v_1'. T_m° is the equilibrium melting temperature of the pure undiluted polymer; ΔH_u is the enthalpy of fusion per repeating unit, V_u and V_1 are the molar volume of the repeating unit and diluent respectively and χ_1 is the thermodynamic polymer–solvent interaction parameter.[110] Equation (2) specifies the equilibrium degree of crystallinity at temperature T_λ for the nominal composition v_2.

Experimental studies by Chiang and Flory[106] of the specific volume–temperature relation of polyethylene–diluent mixtures are well suited to test the thesis that phase equilibrium can be maintained at finite levels of crystallinity.* The experimental results and their analysis are given in Figure 10. The solid points in this figure represent the experimental observations, while the dashed lines are calculated from equation (2) using values for ΔH_u and χ_1 that were determined independently. The agreement between the theoretical expectations and the experimental observations is excellent in the composition range studied, for up to about 50% crystallinity. For systems containing smaller amounts of diluent, slight deviations between theory and experiment are observed.[106] These differences can be attributed to the enhanced difficulties of establishing equilibrium. This formal thermodynamic analysis demonstrates rather conclusively that equilibrium can be established between the two distinct phases even when appreciable levels of crystallinity are developed. One of these phases has the thermodynamic properties of the liquid mixture (at the appropriate composition) while the other has that of the pure crystalline polymer.

11.3.2 Homopolymers of Uniform Chain Length

11.3.2.1 *Equilibrium*

Fundamental to understanding the crystallization behavior of polymers, whether the interest be melting, kinetics or morphology, is knowledge of the equilibrium melting temperature. The

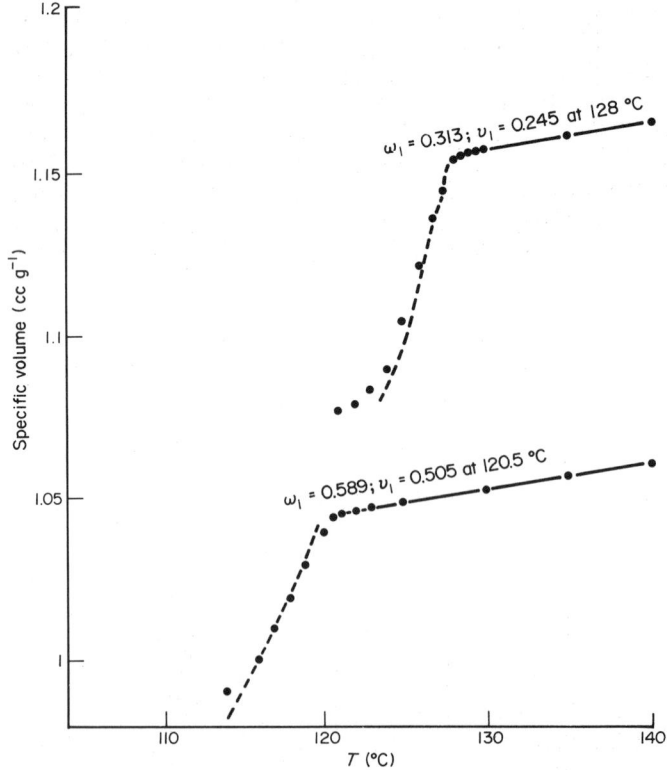

Figure 10 Relationship of specific volume to temperature for mixtures of polyethylene ($M = 50\,000$) with α-chloronapthalene of the indicated weight (w_1) and volume (v_1) fractions. Dashed line represents calculated equilibrium between the crystalline and liquid phases[106]

---

* These experiments were conducted under carefully controlled crystallization conditions. For this system the specific volume is easily converted to the degree of crystallinity.

equilibrium melting temperature, T_m°, of a crystalline polymer is the melting temperature of a perfect crystal formed by infinite molecular weight chains.[1,84] The melting temperature under equilibrium conditions of a chain of finite molecular weight, T_m, is also an important quantity. There are several reasons that the equilibrium melting temperature is important. One of these is that it reflects the conformational character of the chain. Another is that when the idealized equilibrium melting temperatures are compared with those obtained for real systems important morphological and structural information can be deduced. It is also the key parameter in influencing crystallization rates because it establishes the undercooling and thus controls the very important nucleation processes. By definition T_m° cannot be determined by direct experiment. For chains of finite molecular weight morphological complexities and kinetic restrictions placed on the crystallite dimensions make the direct determination of the equilibrium melting temperature virtually impossible. To determine these quantities experimentally recourse must be made to extrapolative procedures. The basis for these extrapolations, and their reliability, will be discussed in detail subsequently.

It is possible to theoretically calculate the equilibrium melting temperature of chains of finite molecular weight and extrapolate these results to T_m°.[84,111,112] To accomplish these calculations models have to be assumed with respect to the crystalline state. In the simplest case all the chains are assumed to be of precisely the exact chain length. In analogy to the n-hydrocarbons molecular crystals are formed as is illustrated in Figure 11. In this figure the vertical straight line represents the ordered sequence conformation. In the case of polyethylene it is planar zigzag. In the crystalline state depicted the molecules are placed end to end so that the terminal groups are juxtaposed in successive layers of the lattice. The end groups are paired, one to another, so that the sequence of chain units from one molecule to the next is perpetuated through successive layers of the lattice. The requirement of *exactly* the same chain length is crucially important if this model is to be valid. This condition cannot be satisfied by polymers, no matter how well fractionated. Hence there is a limitation, which must be clearly recognized, in applying the results of this analysis to polymers of finite chain length. However, it is valid for the analysis of T_m° utilizing thermodynamic data for low molecular weight compounds which satisfy the above structural features. Such data are available for the n-hydrocarbons[111,113-117] and have more recently become available for the n-fluorocarbons.[118]

It had been assumed for a long time that the melting temperatures of the n-hydrocarbons could be explained by representing both the enthalpy and entropy of fusion by a linear expression comprised of a term proportional to the number of carbon atoms, n, and an additive content representing the end group contribution.[113] Extrapolation could then be made to chains of infinite molecular weight. Flory and Vrij[111] pointed out that although the enthalpy of fusion could be reasonably taken as a linear function of n, the situation was quite different with respect to the entropy of fusion. Upon melting, the end pairing represented in Figure 11 is destroyed. This disruption leads to a contribu-

Figure 11 Schematic representation of a molecular crystal wherein the end groups are paired

tion to the entropy of fusion in addition to the usual disordering characteristics of the liquid state. The terminal segment of a chain can now be paired with any of the cn segments of another molecule, where c is a constant. An additional contribution to the entropy of fusion results which is of the form $R \ln cn$.[111]

The molar free energy of a chain of n repeating units at any arbitrary temperature T can be expressed as[111]

$$n\Delta G_n = n\Delta G_u(T) + \Delta G_e(T) - RT \ln n \tag{3}$$

In this equation $\Delta G_u(T)$ represents the free energy of fusion of a repeating unit in the limit of infinite chain length at temperature T. $\Delta G_e(T)$ is the end group contribution which is assumed to be independent of n. The constant $R \ln c$ is incorporated into ΔG_e. This latter term plays a role analogous to that of an interfacial free energy.

The temperature dependence of ΔG_u and ΔG_e can be accounted for by performing a Taylor series expansion around the equilibrium melting temperature. By expanding ΔG_u to second order one obtains[111]

$$\Delta G_u(T) = \Delta G_u - \Delta S_u[T - T_m^\circ] - \Delta C_p/(2T_m^\circ)[T - T_m^\circ]^2 \tag{4}$$

Here the quantities ΔG_u and ΔS_u represent the values of these functions at T_m°. By defining $\Delta T \equiv T_m^\circ - T$ and noting that at T_m°, $\Delta G_u = 0$ and $\Delta S_u = \Delta H_u/T_m^\circ$ equation (4) reduces to*

$$\Delta G_u(T) = \Delta H_u \Delta T/T_m^\circ - \Delta C_p(\Delta T)^2/2T_m^\circ \tag{5}$$

Expanding ΔG_e to first order yields

$$\Delta G_e(T) = \Delta G_e - \Delta S_e[T - T_m]$$
$$\Delta G_e(T) = \Delta H_e - T\Delta S_e \tag{6}$$

Both ΔG_u and ΔG_e could obviously be expanded to as high an order as desired. In this way any extremes in their temperature dependence would be accounted for. However, the second and first order expansions that are illustrated here are adequate for most purposes encountered. By inserting equations (5) and (6) into equation (1) the following free energy of fusion results

$$n\Delta G_n = n[\Delta H_u \Delta T/T_m^\circ - \Delta C_p(\Delta T)^2/2T_m^\circ] + \Delta H_e - T\Delta S_e - RT \ln n \tag{7}$$

At the melting temperatures of an n-mer, $\Delta G_n = 0$ and $T = T_m$. After rearrangement, the melting temperature of a series of homologues of length n is given by[111]

$$n\Delta H_u \Delta T/R - n\Delta C_p(\Delta T)^2/2R - T_m T_m^\circ(\ln n) = [T_m^\circ/R](T_m \Delta S_e - \Delta H_e) \tag{8}$$

This is the equation that Flory and Vrij used to analyze the experimental melting points of the n-alkanes that were available to them.[111]

The melting temperature data of the n-alkanes were organized into a compilation made by Broadhurst.[113]† The other parameters necessary to perform the analysis were also available. The value of ΔH_u had been obtained by the melting point depression method.[122,123] A satisfactory estimate of ΔC_p could be made from specific heat measurements of the n-alkanes.[111] The major objective in this analysis is to determine the value of T_m° from the melting point data of the n-alkanes. The utilization in this analysis of the melting temperature data for polymers of finite molecular weight[111,112] would not be appropriate. No matter how well these samples were fractionated the requirement of uniformity of chain length would not be met.

Utilizing the parameters cited, the left-hand side of equation (8) can be calculated for each n-alkane for an assumed value of T_m°. Flory and Vrij[111] found that with the available data the best value of T_m° for linear polyethylene was 418.5 K. Another way to assess the validity of the analysis is to directly calculate from equation (8) the melting point of the n-alkanes. The solid line in Figure 12 represents the calculated values of T_m plotted against n for the n-alkanes up to C_{100}.[112] Here, T_m° was taken to be 145.5 °C and the best values of the other parameters were used. The experimental

---

* It is necessary that the complete function ΔG_u be expanded and not the individual quantities ΔH_u and ΔS_u independently. Incorrect equations and serious errors result when this latter procedure is followed.[112,119,120]

† Of interest here, in relation to the polyethylenes, is the melting of the orthorhombic crystalline form. Some of the lower members of the n-alkane series undergo a transition from the orthorhombic to hexagonal form prior to melting.[111,113] For this case the melting temperature for the metastable orthorhombic form can be calculated in a straightforward manner.[11] Contrary to other suggestions[121] the resulting melting temperatures can be quite properly used in the analysis.

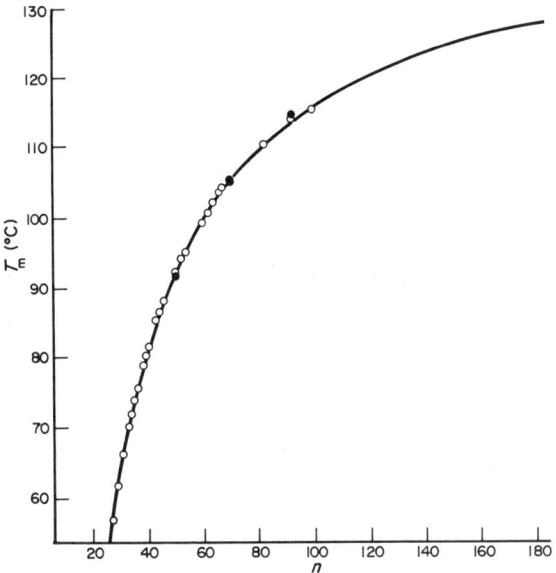

Figure 12 Melting temperatures of *n*-alkanes (up to C_{100}) as a function of chain length. Experimental points taken from Broadhurst compilation.[113] Solid curve calculated from Flory–Vrij analysis[112]

points are represented by the open circles. It is quite evident that the Flory–Vrij analysis gives an excellent representation of the *n*-alkane melting temperatures for this particular database.

The analysis can also be assessed by comparing the observed and calculated enthalpies of fusion.[111,112] By expanding ΔH_u as a function of temperature one obtains an expression for the enthalpies of fusion of the *n*-alkanes. Thus

$$n\Delta H_n = n\Delta H_u - n\Delta C_p \Delta T + \Delta H_e \tag{9}$$

where ΔH_n is the enthalpy of fusion of an *n*-mer. A comparison of the observed and calculated enthalpies of fusion is given in Table 1.[111,112] The values taken for ΔC_p and ΔH_e are the same as were used in the melting point analysis. Extraordinarily good agreement is obtained between theory and experiment so that the theoretical analysis is also confirmed in this manner. It should be noted that there is a range of 100 °C in the melting temperatures of the sample represented in Table 1. The agreement obtained indicates that the temperature expansion used to represent the enthalpy of fusion is more than adequate for present purposes. We conclude that the Flory–Vrij analysis quantitatively explains all the enthalpy of fusion data that is currently available for the *n*-alkanes.

Table 1 Enthalpies of Fusion of *n*-Alkanes

| | ΔH_u(cal mol^{-1} *of CH$_2$ units*) | | |
n	*Observed*	*Equation (9)*[a]	*Ref.*
15	700	699	136
19	750	751	136
25	800	800	136
29	805	822	136
30	795	826	137
43	800	870	137
100	924	928	138

[a] Calculated assuming $\Delta H_u = 980$ cal mol^{-1} (4.1 kJ mol^{-1}).[112]

Recently, synthetic methods have been developed which have enabled the preparation of *n*-alkanes containing up to 390 carbon atoms ($M = 5408$).[114,115,116] Although the main thrust of these studies were in other aspects of crystallization behavior, melting temperatures of these compounds were reported. The melting temperatures were determined by differential scanning calorimetry and were identified with the maximum in the endothermic peak. It has been shown that with proper caution thermodynamically meaningful melting temperatures, comparable to those obtained by classical calorimetry, can be obtained for the lower molecular weight n-alkanes by differential calorimetry.[112,124] Sufficient detail has not as yet been given for the high molecular weight n-alkanes (n > 100) to assess the thermodynamic significance of the reported melting temperatures. Nevertheless, with this important cautionary note, it is of interest to analyze these results in terms of the Flory–Vrij theory and the experimental data reported previously for the lower molecular weight compounds.

A compilation of the melting temperatures of the pertinent n-alkanes is given in Figure 13.[114-117] The most extensive melting point data is that given by Wegner and Lee, which covers the range n = 44 to 216. In this data set there is a considerable overlap in carbon number with the Broadhurst compilation. Starting with n = 48 the melting temperatures reported are about two degrees lower than the Flory–Vrij calculation which, as we have already noted, agrees very well with the older, conventional and accepted values for the melting temperature of the n-alkanes. Melting temperatures for the new compounds n = 72 and 96 show the same difference in melting temperatures. This disagreement is maintained up to n = 216, the highest molecular weight n-hydrocarbon that they found to form extended chain crystallites. The results of Ungar et al.[114] cover the range n = 102 to n = 390. The melting temperatures that they report for n = 102 and 150 are in good agreement with the Flory–Vrij prediction. However, as the carbon number increases the observed melting temperatures become progressively lower than the expected values so that for n = 390 the difference between theory and observation is about four degrees. We should note that the melting temperature for n = 294, which was reported to contain a significant impurity concentration,[114] falls on a smooth curve for this data set. The results of Takamijawa et al.[117] are in very good agreement with theory for n = 60 to n = 120. However, a significant deviation at n = 160 is again noted.

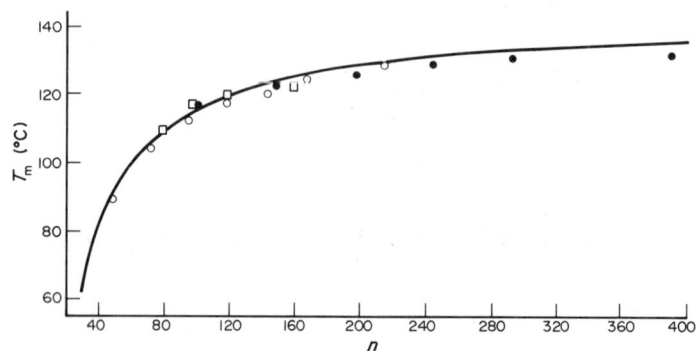

Figure 13   Melting temperatures of n-alkanes as a function of chain length. Solid curve calculated from Flory–Vrij analysis. Experimental results: ●;[114] ○;[115,116] □[117]

There is cause for concern as to why there is a deviation in melting temperatures, between theory and experiment, for n > 120. There are several possibilities that need to be explored in more detail before any definitive conclusions can be reached with regard to this discrepancy. One major concern is the determination of thermodynamically significant melting temperatures by differential scanning calorimetry. This general problem has been discussed in detail previously.[112,124] Although appropriate melting temperatures can be obtained by this technique, special care and procedures need to be adopted. There is also the question of the purity of the compounds with respect to chain length and chemical effects within the chain. The aforementioned points need to be clarified in detail before theoretical revisions are undertaken. It would be extraordinarily helpful in resolving these difficulties if the enthalpies of fusion were reported for the higher molecular weight n-alkanes so comparison could be made with the predictions of equation (9).

Starkweather has analyzed the melting of a set of normal perfluoroalkanes.[118] Although the samples were limited to the interval n = 6 to 24, when compared to the much larger range that is available for the n-alkanes, important information concerning the melting of the corresponding homopolymer, poly(tetrafluoroethylene), was obtained by applying the Flory–Vrij analysis. The measured enthalpies of fusion of the low molecular weight homologues and the enthalpy of fusion per repeating unit of the infinite chain were related by equation (9). The value extrapolated from the n-perfluoroalkane data for T_m° is in close accord with the estimate from experimental studies. The importance of including the $R \ln n$ term in the total entropy of fusion was also demonstrated. Failure to include this term lowers the expected T_m° by about 40 °C which is not acceptable from an experimental point of view.

The above discussion has been an idealized one, even for chains of uniform length. Only two states were considered; the extreme of perfect crystalline order on one hand and a completely disordered molten liquid on the other. It can be hypothesized that under favorable circumstances the molecular crystals may undergo partial melting, with disruption of the planar array of end groups, prior to final melting. This possibility has also been considered.[111] It was concluded that for such partial melting under equilibrium the melting temperature T_m would have to be very close to T_m°. This condition would require very large values of n.[111] This analysis can explain the narrow melting range observed for the n-alkanes. It also predicts that when n is very large, although molecular crystals would be stable at low temperatures, partial melting to crystallites having a thickness $\zeta < n$ should occur below T_m. A connection can then be made between the melting of an idealized system of large n, which has been described above, to that for a real polymer system. A schematic model of this crystallite is represented in Figure 14. It is generated from the molecular crystal model illustrated in Figure 11 by unpairing the terminal sequences and melting a sequence of m units. Thus a core crystal results which is $\zeta = n - m$ units in length.

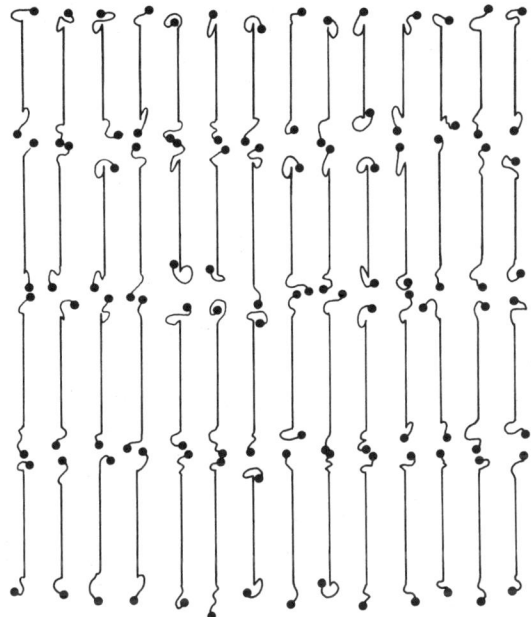

Figure 14 Schematic representation of a non-molecular crystal where the end sequences are in disordered conformation

11.3.3 Real Polymer Systems

11.3.3.1 Equilibrium

The situation depicted in Figure 14 where a long chain polymer participates in a crystal of length ζ, which is less than n, has also been analyzed.[84] This model is applicable to polymer fractions and to specific polydisperse systems. The theoretical relations developed predict stable forms of crystallites of length $\zeta (< n)$ over a finite temperature range below the melting temperature. (The treatment is, however, limited to low levels of crystallinity.) In analyzing this model cognizance must be taken of the fact that the chain ends, and the contiguous non-crystalline sequences are excluded from the crystal lattice. Thus, for polymers of finite molecular weight there is in essence a 'built in' set of impurities. Flory,[84] taking cognizance of these facts, developed a quantitative description of the semicrystalline state of unoriented polymers. It has turned out to be an important analysis not only for the description of the equilibrium condition, but it can also be used when the crystallites are of non-equilibrium size. The results can also be adopted to develop nucleation theory appropriate to long chain molecules.[35,36,125,126] It is, therefore, not necessary that nucleation theory of monomeric systems be adopted in an *ad hoc* manner to polymeric systems.

Relying on a lattice model, Flory[84] considered the general case of N homopolymer molecules, each comprised of exactly x repeating units, which were mixed with n_1 molecules of a low molecular weight species. The composition of the mixture is characterized by the volume fraction of polymer

v_2. Since in general the diluent will be structurally different from the polymer repeating unit it is assumed to be excluded from the crystal lattice.* The objective of the calculation is to calculate the free energy of fusion.

The configurational properties of the semicrystalline polymer are conveniently described by using a lattice with coordination number Z. The size of the lattice cell is chosen to accommodate one segment, the volume of the segment being so chosen as to equal the volume of the diluent. The number of segments per structural unit is z. Definite regions in the lattice must be reserved for occupancy by the crystallite. There are v crystallites, each having an average length of ζ repeating units and a cross section of σ chains. The total number of crystalline sequences, m, is equal to $v\sigma$; $m\zeta$ is the total number of chain units which participate in the crystallization. The major problem to be solved is the calculation of the configurational entropy of the semicrystalline polymer. When the entropy of the completely disordered mixture is subtracted from this quantity the entropy of fusion, ΔS_f, results. This quantity is expressed by[84]

$$\Delta S_f/xN = (1-\lambda)\Delta S_u - R[(V_u/V_1)(1-v_2)/v_2 + 1/x]\ln[1-v_2(1-\lambda)] - R[(1-\lambda)/\zeta]\{\ln v_2 D + \ln[(x-\zeta+1)/x]\} \quad (10)$$

In equation (10) λ is the fraction of non-crystalline (amorphous) polymer and is equal to $(xN-\zeta m)/xN$. The entropy of fusion per repeating unit, ΔS_u, is formally defined as kz $\ln[(Z-1)/e]$. The parameter D is defined as $(Z-1)/ze$. V_u and V_1 are the molar volumes of the repeating unit and diluent respectively. In calculating ΔS_f it was assumed that sharp boundaries exist between the crystalline and amorphous regions. At the same time it was recognized that this is a physically untenable situation and that some degree of order must persist for some distance beyond the crystallite boundary. The effect of the diffuseness of the boundary on the configurational entropy can be formally accounted for by redefining the parameter D. Equation (10) represents a melting and dilution process. However, even in the absence of diluent ΔS_f depends on the degree of crystallinity, $1-\lambda$, and the crystallite thickness ζ. It is, therefore, not an inherent property of the polymer chain itself. On the other hand, the entropy of fusion per structural unit, ΔS_u, is characteristic of a given polymer irrespective of the actual characteristics of the crystallite.

The enthalpy of fusion consists of a contribution arising from the melting of the crystallites and the mixing of these previously crystalline segments with the amorphous mixture. The former contribution can be expressed as $\zeta m \Delta H_u$, where ΔH_u is the enthalpy of fusion per repeating unit. The effect of the lower energy that would be expected at the crystallite boundary can also be incorporated into the parameter D. The heat of mixing can be expressed in the van Laar form as is customary in polymer solution theory. The free energy change on fusion, ΔG_f, can then be expressed as [84]

$$\Delta G_f/xN = (1-\lambda)(\Delta H_u - T\Delta S_u) + RT\{[(V_u/V_1)(1-v_2)/v_2 + 1/x]\ln[1-v_2(1-\lambda)] + [(1-\lambda)/\zeta]\ln v_2 D$$
$$+ \ln(x-\zeta+1)/x + \chi_1^*(1-v_2)^2(1-\lambda)/(1-v_2+v_2\lambda)\} \quad (11)$$

Here χ_1^* is related to the conventional thermodynamic interaction parameter χ_1 by the relation

$$\chi_1^* = \chi_1(V_u/V_1) \quad (12)$$

The discussion of the quantities which comprise the parameter D makes clear that it plays a role analogous to that of an interfacial free energy. It can be redefined as

$$-\ln D = \frac{2\sigma_e}{RT} \quad (13)$$

so as to correspond to convention notation. Here σ_e is the interfacial free energy characteristic of the equilibrium crystallite.

The equilibrium, or most stable semicrystalline state, characterized by the equilibrium values $\lambda = \lambda_e$; $\zeta = \zeta_e$, is obtained when ΔG_f assumes its maximum value. Accordingly it is found that ζ_e is given by

$$-\ln v_2 D = \zeta_e/(x-\zeta_e+1) + \ln[(x-\zeta_e+1)/x] \quad (14)$$

and λ_e by

$$1/T - 1/T_m^\circ = R/\Delta H_u\{[(V_u/V_1)(1-v_2)+v_2/x]/[1-v_2(1-\lambda_e)] + 1/(x-\zeta_e+1) - \chi_1^*\{(1-v_2)/[1-v_2(1-\lambda_e)]\}^2\} \quad (15)$$

---

* An assumption of this type is always necessary in multicomponent–multiphase systems. For polymer–diluent mixtures this assumption has been found to be valid for virtually all systems studied.

In the absence of diluent equations (14) and (15) reduce to

$$-\ln D = \frac{2\sigma_e}{RT} = \zeta_e/(x - \zeta_e + 1) + \ln[(x - \zeta_e + 1)/x] \tag{16}$$

and

$$1/T - 1/T_m^\circ = R/\Delta H_u[1/x\lambda_e + 1/(x - \zeta_e + 1)] \tag{17}$$

Here T_m°, the equilibrium melting temperature of the pure polymer of infinite chain length, is equated with the ratio $\Delta H_u/\Delta S_u$. The fact that ζ_e is independent of λ_e is a consequence of approximations made in calculating ΔS_f. Physically significant values of ζ_e occur only when $v_2 D$ is less than unity.

Equation (17) gives the expectation that even for chains of exactly the same length fusion will occur over a finite temperature range in contrast to the melting of pure monomeric species. Experimental observations support this conclusion, particularly in the low molecular weight range. These results are seen in Figure 8 for the two lowest molecular weight species. The theory also predicts that the fusion interval will decrease as the molecular weight increases. For the linear polyethylenes a very perceptible sharpening of the fusion occurs up to molecular weights of about 5×10^4. However, the experimental evidence, in Figure 8, also indicates that at the higher molecular weights 10^5–10^6 the melting range broadens appreciably with increasing chain length. This disparity can be attributed to the extreme difficulty in achieving the equilibrium conditions, $\zeta \rightarrow \zeta_e$ and $\lambda \rightarrow \lambda_e$, for very high molecular weights.

Despite the fact that the melting of a semicrystalline polymer will occur over a finite temperature range, even under equilibrium conditions, as λ_e approaches unity $d\lambda_e/dT \neq 0$. Therefore, the last traces of crystallinity will disappear at a well-defined temperature. Above this temperature $d\lambda_e/dT = 0$. Thus the thermodynamic melting temperature $T_{m,e}$ is defined. At $\lambda_e = 1$, when $T \equiv T_{m,e}$ we find that

$$\frac{1}{T_{m,e}} - \frac{1}{T_m^\circ} = \frac{R}{\Delta H_u}\left(\frac{1}{x} + \frac{1}{x - \zeta_e + 1}\right) \tag{18}$$

$$2\sigma_e = RT_{m,e}\left\{\frac{\zeta_e}{x - \zeta_e + 1} + \left(\frac{x - \zeta_e + 1}{x}\right)\right\} \tag{19}$$

Equations (18) and (19) represent the dependence of the equilibrium melting temperature on chain length for a pure polymer system which adheres to the crystallite model of Figure 14. Here σ_e represents the interfacial free energy associated with the basal plane of the equilibrium crystallite. $T_{m,e}$ is the equilibrium melting temperature of a crystallite of size ζ_e which is comprised of chains of x units. It should not be identified *a priori* with the melting temperature of a molecular crystal formed by chains of the same size. Equation (18) can be rewritten in more compact form as

$$1/T_m - 1/T_m^\circ = (R/\Delta H_u)(1 + b)/x \tag{20}$$

where $b = [1 - (\zeta_e - 1)/x]^{-1}$. If σ_e is independent of chain length, b will be approximately constant, unless x is very small. Under these circumstances

$$\frac{1}{T_m} - \frac{1}{T_m^\circ} \propto \frac{1}{x} \tag{21}$$

This inverse relationship, for a constant interfacial free energy, between melting temperature and chain length contrasts markedly to the results obtained for molecular crystals. Equation (21) is a consequence of the fact that the ends are unpaired in this model.

It is tempting to apply equation (21) to experimental data to extrapolate to the equilibrium melting temperature of the infinite chain. However, the conditions that lead to equation (21) need to be satisfied. The experimental quantities that are usually determined are T_m and x for a given molecular weight. Equations (18) and (19) make clear that only given this information T_m° cannot be determined without some arbitrary assumptions being made with respect to either ζ_e or σ_e. The relationship expressed by equation (21) is derived from equations (18) and (19) under the assumption that σ_e is independent of x. One has to establish whether this assumption is valid over the molecular weight range where data are available that satisfy the requirements of theory. A basic requirement that needs to be satisfied is that the crystalites attain equilibrium size. This condition requires that the crystallite thickness be comparable to the extended chain length. Moreover, the crystallite must also be of uniform thickness. The molecular weight range over which extended chain crystals can be formed is limited. Suitable data that are available for analysis are restricted to very low molecular weight fractions of polyethylene[89] and poly(ethylene oxide).[112,119,120,127]*

The analyses of the appropriate available data for linear polyethylene and poly(ethylene oxide) are summarized in Tables 2 and 3.[89,112] In order to perform the calculation it is necessary to assume a value for T_m°. For polyethylene we have taken $145 \pm 1\,°C$;[111] for poly(ethylene oxide) $76\,°C$ and $80\,°C$.[112] Although the precise values of the parameters deduced will depend on the value of T_m° the trends with molecular weight are essentially unaffected. Very similar results are obtained with both polymers. Over the molecular weight range for which appropriate data are available σ_e varies three- to four-fold. Put another way, the parameter b in equation (20) varies by about a factor of two over the accessible molecular weight range appropriate for analysis. Booth et al.[127] performed a similar with the low molecular weight poly(ethylene oxides) and found σ_e values of the same magnitude as reported here, which also increased with molecular weight. It is evident that equation (21) cannot be used with the experimental data presently available to extrapolate to the equilibrium melting temperature.

Table 2 Parameters[a] Governing Fusion of Linear Polyethylene[87]

M_n	x	$T_m(°C)$	ζ_e	ζ_e/x	$\sigma_e(\text{cal mol}^{-1})$	$x-\zeta_e$
1586	113	124.5	95 ± 1	0.84 ± 0.01	1298 ± 200	18
2221	159	126.0	140 ± 2	0.88 ± 0.01	2024 ± 200	19
3769	269	132.0	242 ± 3	0.90 ± 0.01	2551 ± 300	27
5600	400	134.2	368 ± 4	0.92 ± 0.01	3485 ± 500	32

[a] Uncertainties calculated assuming $T_m = \pm 1\,°C$.

Table 3 Parameters[a] Governing Fusion of Poly(ethylene oxide)[112]

M_n	x	T_m (°C)	$T_m = 80\,°C$				$T_m = 76\,°C$			
			ζ_e	ζ_e/x	σ_e (cal mol^{-1})	$x-\zeta_e$	ζ_e	ζ_e/x	σ_e (cal mol^{-1})	$x-\zeta_e$
1110	25	43.3	23	0.90	1413	2	22	0.88	1186	3
1350	31	46.0	28	0.91	1734	3	28	0.90	1447	3
1890	43	52.7	39	0.91	1995	4	38	0.89	1588	5
2780	63	57.6	58	0.93	2567	5	57	0.90	1954	6
3900	89	60.4	84	0.94	3410	5	82	0.92	2523	7
5970	136	63.3	129	0.95	4776	7	127	0.93	3389	9
7760	176	64.3	169	0.96	6080	7	166	0.94	4261	10

[a] Melting temperature data from ref. 119.

Some interesting parameters have resulted from this analysis. The increase in σ_e with chain length is caused by the first term on the right of equation (19). It arises from a consideration of the number of ways the equilibrium sequence of crystallization units ζ_e can be chosen from among the x units if the chain ends are excluded. On this basis it can be expected that σ_e will reach a limiting value at some higher molecular weight. The ratio ζ_e/x increase with x and appears to be reaching a limiting value which is the order of 0.95. The difference between x and ζ_e, given in the last columns of Tables 2 and 3, increases with molecular weight. This trend would be expected to continue at still higher molecular weights.

We are now in a position to compare the theoretical melting temperature–molecular weight relation for molecular crystals with the unpaired, disordered interface model that we have just discussed. In order to make this comparison values have to be taken for the parameters ΔG_e and σ_e. The comparison between the two models is given by the curves in Figure 15 where ΔG_e has been taken from the Flory–Vrij analysis and for the unpaired model σ_e values of 1200 and 4600 cal mol^{-1} (5.0 and 19.2 kJ mol^{-1}) were considered. Curve A in Figure 15 is a repeat and continuation of the solid curves in Figures 12 and 13. The figure makes clear that the stability of the model will depend on the values of the corresponding interfacial free energy parameter. We have found that in the low

* Crystallization under conditions of high pressure and high temperature usually produces thick crystallites for very high molecular weights. The thicknesses of these crystallites are not, however, comparable to the extended chain length.[128,129]

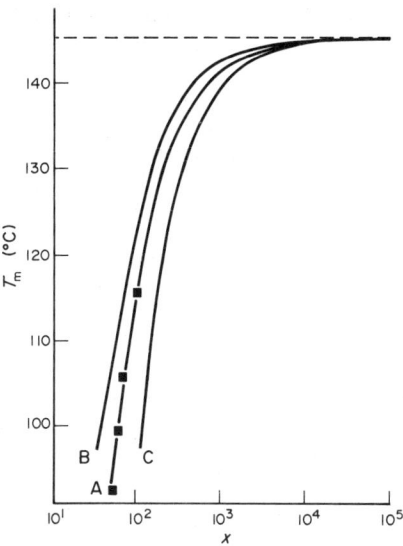

Figure 15 Plot of melting temperature as a function of the number of carbon atoms in the chain. Curve A, Flory–Vrij anaysis. Curves B and C, theoretical calculation for equations (18) and (19) with $\sigma_e = 1200$ and 4600 respectively. ■, values for *n*-alkanes

molecular weight range, σ_e has relatively low values which increase with molecular weight. It is, therefore, theoretically possible that for low x the unpaired model, as represented by low molecular weight fractions, would have a melting temperature somewhat greater than the corresponding *n*-alkanes. As σ_e increases to the range of 2000–3000 cal mol^{-1} (8–12 kJ mol^{-1}) the melting temperatures of the unpaired crystallites and the molecular crystals of the *n*-alkanes would be comparable to one another. At higher chain length, with a further increase in σ_e, the molecular crystal model would become more stable.

To examine the validity of these conclusions requires the study of extremely narrow molecular weight distributed samples which are crystallized in such a manner so that fractionation is avoided.[130,131] These exacting conditions are difficult to fulfill. However, there are data available for molecular weight fractions, in the appropriate range, with which a comparison can be made. A summary of the results, for $x = 40$ to 400, is given in Figure 16. The solid curve again represents the Flory–Vrij relation for *n*-hydrocarbons. The short, vertical lines represent the results for the low molecular weight polyethylene fractions, with the number average molecular weight being used.[60,87,132] Within the limits of the uncertainties expressed above, the general expectations are fulfilled. For $x \leq 160$ the melting temperatures of the polymers are comparable to or slightly greater than the corresponding *n*-hydrocarbons. For the two best fractions, $x = 75$ and 113 respectively, the polymers have slightly higher melting temperatures. As x increases to 400 the melting temperatures of the polymers are less than those expected for *n*-hydrocarbons. It is interesting to note that the melting temperatures of the polymers and the newly synthesized *n*-hydrocarbons are very close to one another.

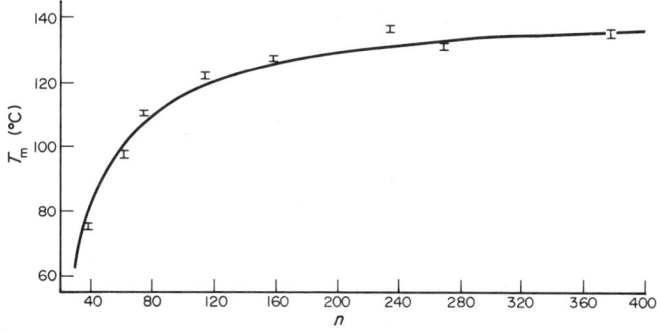

Figure 16 Plot of melting temperature against chain length for low molecular weight polyethylene fractions[60,112,130]

Polydisperse systems where the molecules vary in length according to the 'most probable distribution' defined by

$$w_x = (1-p)^2 \, p^{x-1} \qquad (22)$$

can also be treated by the methods outlined.[84] Here w_x is the weight fraction of the species comprised of x repeating units. In this case the only change that occurs in the expression for the entropy of fusion, at low levels of crystallinity, is the replacement of the chain length x by the number average degree of polymerization, $\bar{x}_n$. Consequently at the melting temperaure for this specific heterogeneous system

$$\frac{1}{T_m} - \frac{1}{T_m^\circ} = \frac{2R}{\Delta H_u \bar{x}_n} \qquad (23)$$

The quantity $2/\bar{x}_n$ represents the mole fraction of the non-crystallizing terminal units. Since they are distributed at random equation (23) can also be derived from the condition for phase equilibrium with the impurities being restricted to the liquid phase.[133] The melting temperature expressed by equation (23) is characteristic of very long crystalline sequences. These will be formed from the larger molecular weight species even for a very low number average molecular weight. Such sequences will be rare. Therefore, their melting will be difficult to determine experimentally.

Evans, Mighton and Flory[134] have studied the melting temperatures of poly(decamethylene adipate) samples that were prepared in such manner as to possess the 'most probable' molecular weight distribution. Samples were prepared which were terminated in the conventional manner with hydroxyl and carboxyl end groups as well as with bulky end groups such as benzoate, α-naphthoate and cyclohexyl. The melting temperature–molecular weight results for the hydroxyl–carboxyl terminated samples are plotted in Figure 17 according to equation (23). It is clear from the figure that the functional form of this equation is obeyed over the complete molecular weight range studied. Moreover, the value for ΔH_u obtained from the slope of the straight line in Figure 17 is in very good agreement with the value obtained by other methods.[134] The melting temperatures for the polyesters with other terminal groups are virtually identical to those in Figure 17. It would hardly be expected that these bulky end groups would participate in the crystal lattice. Thus, these results substantiate the underlying basis for the development of equation (23), namely the exclusion of terminal groups from the lattice.

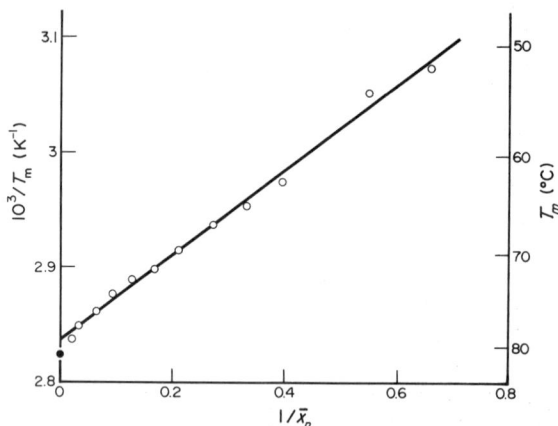

Figure 17 Plot of $1/T_m$ against $1/\bar{x}_n$ for poly(decamethylene adipate)s terminated with hydroxyl and carboxyl end groups[134]

Booth and co-workers have adopted the Flory treatment to a polydisperse system which has an exponential molecular weight distribution.[135] For this distribution

$$w(x) = \frac{b^{(a+1)}}{a!} x^a e^{-bx} \qquad (24)$$

here $b = a/\bar{x}_n$, $a = \bar{x}_n/(\bar{x}_w - \bar{x}_n)$ where $\bar{x}_n$ and $\bar{x}_w$ are the number average and weight average chain length respectively. In this case the equilibrium melting temperature is given by

$$T_m = T_m^\circ \left(1 - \frac{2\sigma_e}{\Delta H_u \zeta_e} \right) \Big/ \left[1 + \frac{RT_m^\circ}{\Delta H_u} \left(\frac{1}{x_n} - \frac{\ln I}{\zeta_e} \right) \right] \qquad (25)$$

where

$$I = \frac{b^{(a+1)}}{a!} \int_{\zeta_e}^{\infty} x^{(a-1)} e^{-bx}(x - \zeta_e + 1) dx \tag{26}$$

This analysis is based on the inherent assumption that cocrystallization of all species occurs, *i.e.* there is no fractionation or segregation.

Calculations based on equation (25) indicate that for a fixed value of σ_e the equilibrium melting temperature of the poly(ethylene oxides), and presumably other polymers as well, will be very sensitive to the width of this type of distribution. Significant changes in the melting temperature can occur. For example, for σ_e equal to 1500 cal mol^{-1} (6.3 kJ mol^{-1}) and $\bar{M}_n = 1000$, T_m increases by about 10 °C as $\bar{M}_w/\bar{M}_n$ increases from one to two. When $\bar{M}_n$ is increased to 6000, T_m increases by about three degrees for the same change in the width of the distribution.[133] Thus, for equilibrium crystallization (cocrystallization and extended chain crystallites) the results, and their interpretation in the low molecular weight range will be dependent on the molecular weight distribution. A further complication is then added to the analysis of the data in Figure 16.

11.3.4 Fusion of Copolymers

The introduction into a crystallizable polymer chain of units which are chemically or structurally different will be expected to alter its crystallization behavior. Some of the different types of chemical structural units that are involved have already been described. It is important to recall that they are very often quite subtle in character.* Problems involving the crystallization and melting of copolymers cannot in general be formulated uniquely since two phases and at least two species are involved. The disposition of the species among the phases needs to be specified; it cannot be established *a priori* by theory. This is a general problem that is not unique to polymeric systems. When the co-units are excluded from the crystal lattice the thermodynamic and structural situation is clear. However, if the co-unit is located within the lattice several different situations need to be distinguished. The co-unit can either enter the crystal lattice as an equilibrium requirement, or be located within the lattice as a non-equilibrium defect. These are obviously distinctly different situations. If not distinguished from one another the problem can become confused. In some cases these situations are difficult to distinguish. The more general problem of deciding whether a unit does or does not enter the lattice in real systems can also be difficult.

The equilibrium theory for the fusion of copolymers has been developed in most detail for the case where the crystalline phase remains pure. We consider a chain consisting of A and B units where the B units are not crystallizable, *i.e.* they are excluded from the crystal lattice. The problem is more complex than just having a set of isolated impurities. In the initial molten state, prior to crystallization, the units occur in a specified sequence length distribution determined by the copolymerization kinetics, *i.e.* the relative reactivity rates. The crystalline state is comprised of crystallites of varying lengths. The length of a crystalline sequence is expressed by ζ, the number of A units that traverse the crystallite from one end to the other. The development of a crystallite in the chain direction is restricted by the occurrence of a non-crystallizing B unit. The lateral development of crystallites of length ζ is governed by the concentration of sequences of sufficient length in the residual melt and the free energy decrease that occurs upon the crystallization of a sequence of ζ A units. These concepts have been formulated more quantitatively.[84, 140]

Following Flory[140] we consider the probability p_ζ that a given A unit in the non-crystalline amorphous region is located within a sequence of at least ζ such units. We also let w_j represent the probability that a unit chosen at random from the amorphous region is an A unit and also a member of a sequence of j A units terminated at either end by B units. We then consider the probability that the specific A unit selected is followed in a given direction by at least $\zeta - 1$ similar units. We can then write[140]

$$P_{\zeta,j} = \frac{(j - \zeta + 1)w_j}{j} \quad j \geq \zeta$$

$$P_{\zeta,j} = 0 \quad\quad\quad j < \zeta \tag{27}$$

$$P_\zeta = \sum_{j=\zeta}^{\infty} P_{\zeta,j} = \sum_{j=\zeta}^{\infty} \frac{(j - \zeta + 1)w_j}{j}$$

---

* The influence of intermolecular crosslinks needs to be considered separately. Although clearly structural irregularities, their influence on the crystallization behavior requires special treatment since the resultant properties depend on the state in which the crosslinks were introduced.[139]

Solving for successive values of ζ yields the difference equation

$$w_\zeta = \zeta(P_\zeta - 2P_{\zeta+1} + P_{\zeta+2})$$

(28)

Equations (27) and (28) represent quite general properties of the non-crystalline regions. They do not depend on the presence or absence of crystallites.

In the completely molten polymer, prior to the development of any crystallinity

$$w_\zeta^\circ = \frac{X_A \zeta v_\zeta^\circ}{N_A}$$

(29)

here N_A is the total number of A units in the copolymer, X_A the corresponding mole fraction, and v_ζ° the number of sequences of A units initially present in the melt. If we assume that the probability, p, of an A unit being succeeded by another A unit is independent of the number of preceding A units then equation (29) can be expressed as*

$$w_\zeta^\circ = \frac{\zeta X_A (1-p)^2 p^\zeta}{p}$$

(30)

By combination of equations (29) and (30) we find that

$$P_\zeta^\circ = X_A p^{\zeta-1}$$

(31)

for the completely molten polymer.

For the crystalline polymer under thermodynamic equilibrium the probability P_ζ^e that in the non-crystalline region a given A unit is located in a sequence of at least ζ such units is given by

$$P_\zeta^e = \exp\left(-\frac{\Delta G_\zeta}{RT}\right)$$

(32)

Here ΔG_ζ is the standard free energy of fusion of a sequence of ζA units from a crystallite ζ units long. ΔG_ζ can be expressed as

$$\Delta G_\zeta = \zeta \Delta G_u - 2\sigma_e$$

(33)

Crystallites are assumed to be sufficiently large in the direction transverse to the chain axis so that the contribution of the excess lateral surface free energy to equation (33) can be neglected. Equation (32) can be expressed as

$$P_\zeta^e = \frac{1}{D} \exp(-\zeta\theta)$$

(34)

where

$$\theta = \frac{\Delta H_u}{R}\left(\frac{1}{T} - \frac{1}{T_m^\circ}\right)$$

(35)

and

$$D = \exp\left(-\frac{2\sigma_e}{RT}\right)$$

(36)

If crystallites of length ζ, $\zeta+1$ and $\zeta+2$ are present, and are in equilibrium with the melt, then equations (28) and (34) can be combined to give

$$w_\zeta^e = \zeta D^{-1}[1 - \exp(-\theta)]^2 \exp(-\zeta\theta)$$

(37)

Equation (37) expresses the residual concentration in the melt of sequences of A units which are ζ units long.

The necessary and sufficient conditions for crystallization can be stated as

$$P_\zeta^\circ > P_\zeta^e$$

(38)

---

* The sequence propagation probability p can be related to the comonomer reactivity ratio in a straightforward manner. The problem can be further generalized to the case where p is influenced by the penultimate group.[141,142]

for one or more values of ζ. The condition $w_\zeta^\circ > w_\zeta^e$, for one or more values of ζ, is a necessary but not sufficient condition for crystallization. Equation (30) for the completely molten polymer and equaton (37) for the equilibrium crystalline polymer are functions of the sequence length ζ. There is therefore a critical value $\zeta = \zeta_{cr}$ at which these two distributions are equal. This condition is specified by

$$\zeta_{cr} = \frac{-\{\ln(DX_A/p) + 2\ln[(1-p)/(1-e^{-\theta})]\}}{\theta + \ln p} \tag{39}$$

For values of $\zeta < \zeta_{cr}$, w_ζ^e is greater than w_ζ°; for $\zeta > \zeta_{cr}$ the converse holds. Thus, ζ_{cr} represents the limiting size above which crystallites can exist at equilibrium. Smaller values of ζ cannot be maintained at equilibrium.

The theory, as has been outlined up to this point, allows for an estimate to be made of the fraction of A units that are crystalline at temperatures below the melting temperature. This estimate is obtained by summing all the sequences of A units that participate in crystallites. This procedure leads to a slight overestimate of the degree of crystallinity since sequences greater than ζ units in length can participate in crystallites which are only ζ units long. If w_ζ^c is the concentration of sequences of ζ units involved in a crystallite then

$$w_\zeta^c = w_\zeta^\circ - w_\zeta^e \tag{40}$$

The fraction of A units in the crystalline state, w_c, is given by

$$w_c = \sum_{\zeta=\zeta_{cr}}^{\infty} w^c = \sum_{\zeta=\zeta_{cr}}^{\infty} (w_\zeta^\circ - w_\zeta^e) \tag{41}$$

Using the expressions for w_ζ° and w_ζ^e from equations (29) and (37) one obtains

$$w_c = \frac{X_A}{p}(1-p)^2 p^{\zeta_{cr}}\{p(1-p)^{-2} - e^{-\theta}(1-e^{-\theta})^{-2} + \zeta_{cr}[(1-p)^{-1} - (1-e^{-\theta})^{-1}]\} \tag{42}$$

Equation (42) is an expression for the degree of crystallinity as a function of the reduced temperature (θ) as it depends on the copolymer structure embodied in the parameter p and the interfacial free energy σ_e.

Theoretical curves of the degree of crystallinity as a function of temperature can be constructed from equation (42) by assignment of the appropriate parameters. Typical plots are given in Figure 18 for random copolymers, $p \equiv X_A$, of different co-unit contents using the indicated parameters.[143] At comparable temperatures the equilibrium degree of crystallinity that can be attained is severely reduced as the concentration of the non-crystallizing B units increases. Fusion of copolymers is expected to occur over a broad temperature range in marked contrast to what is expected and actually observed for homopolymers. For random copolymers small, but significant, amounts of

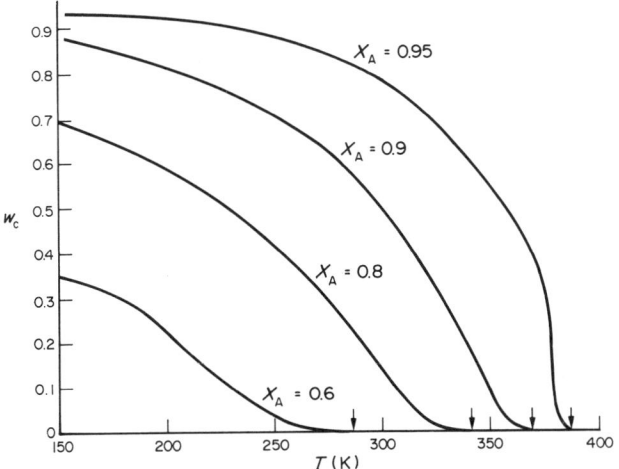

Figure 18 Theoretical plot, according to equation (42), of the crystalline fraction, w_c, as a function of temperature for random type copolymers of indicated composition. Short vertical arrows indicate melting temperatures of each copolymer. For case illustrated, $T_m^\circ = 400\,\text{K}$, $\Delta H_u = 10^3\,\text{cal mol}^{-1}$ ($4 \times 10^3\,\text{J mol}^{-1}$) and $\ln D = -1$[143]

crystallinity persist for an appreciable temperature interval before the transformation is complete. The breadth of this interval increases substantially with the concentration of non-crystallizable co-units. Although the absolute amount of crystallinity that persists is small it is a significant amount from the point of view of theory as well as its influence on properties.

On cursory examination the diffuse nature of the curves depicted in Figure 18 do not appear to be typical of a first-order phase transition. However, these types of melting curves are natural consequences of the constitution of random copolymers and the requirement that the B units are restricted to the non-crystalline phase. At the melting temperature, w_c and all its derivatives vanish. A true discontinuity then exists. There is a true thermodynamic melting temperature representative of the disappearance of crystallites composed of very long sequences.

The unique features of the fusion of random copolymers are basically consequences of the broad distribution of sequence length. There are several specific factors involved. These are simultaneously the influence of crystallites of finite size which vary with temperature, an impurity effect, and the changing sequence distribution in the residual melt. At a given temperature only those sequences whose length exceeds ζ_{cr} can participate in the crystallization process; ζ_{cr} being a function of temperature. As the temperature changes, the composition and sequence distribution of the amorphous portion of the system will also change. These factors govern the fraction of A units that can participate in the crystallization and thus provide a natural explanation for the diffuse melting curves. These factors will be severely exacerbated for the real non-equilibrium situations which are usually encountered. As the melting temperature is approached ζ_{cr} assumes very large values. At the same time the concentration of sequences that exceed this critical length becomes quite small. This requirement accounts for the expected persistence of smal amounts of crystallinity at temperatures just below the melting temperature.

The major conclusions of the theoretical development, as embodied in Figure 18, are substantiated by experiment. Random copolymers are well known to melt over a broad temperature range, irrespective of the specific nature of the structural irregularity involved or the homopolymer from which they are derived. The parent homopolymers which display these features include polyesters, polyamides, polyalkenes, fluorocarbon polymers and crystallizable vinyl polymers to cite but a few examples.[114] In many instances the structural irregularities can be subtle in nature so that the copolymeric character of the chain being studied is not recognized. A typical example of the melting of random copolymers has been illustrated in Figure 9 with a set of polybutadienes. These melting curves are typically sigmoidal and the transformation occurs over a very wide temperature interval. The melting range becomes broader with decreasing concentration of crystallizing units. Small amounts of crystallinity persist at temperatures just below the melting temperature. In accord with the theoretical prediction this final portion of the melting curve encompasses a larger temperature interval as the crystallizing unit concentration decreases. The level of crystallinity also decreases substantially with decreasing concentration of crystallizable units. Consequently the change that occurs on melting can be very small and can very often be undetected. There is an inherent difficulty in detecting crystallinity and in determining the melting temperature of a random copolymer which has a relatively high concentration of non-crystallizing structural units. Yet the persistence of even very small amounts of crystallinity can influence properties in a significant manner.

Graessley and co-workers[145] were able to crystallize a polybutadiene sample which only contained 56 mol % of the 1,4-*trans* units. Depending on molecular weight, the level of crystallinities that developed ranged from 2 to 5%. Thus, copolymers which contain a very high non-crystallizing content can be crystallized not only in principle but also in practice. At low temperatures, where these small amounts of crystallinity develop, anomalies were observed in the viscoelastic behavior of these polybutadienes.[145] Presumably, other physical properties will also be affected.

Besides describing the nature of the fusion process, relations involving the melting temperature can also be derived.[84, 140] Returning to the inequality of equation (33) and substituting the expressions for P_ζ° and P_ζ^c from equations (31) and (33) respectively we obtain[140]

$$\frac{X_A}{p} p^\zeta > \frac{1}{D} \exp(-\theta\zeta) \tag{43}$$

Except for the special case of copolymers which exhibit a very strong tendency for alternation, $1/D$ is greater than X_A/p. Thus the inequality of equation (43) becomes

$$\frac{1}{T} - \frac{1}{T_m^\circ} > -\frac{R}{\Delta H_u} \ln p \tag{44}$$

A limiting temperature must therefore exist above which the inequality of equation (44) cannot be fulfilled. At higher temperatures crystallization cannot occur. This limiting temperature is the equilibrium melting temperature T_m of the copolymer. Therefore

$$\frac{1}{T_m} - \frac{1}{T_m^\circ} = -\frac{R}{\Delta H_u} \ln p \tag{45}$$

Equation (45) can be derived in an alternative manner by treating the copolymer as an ideal binary mixture.[146] In analogy to the classical statistical derivation of Raoult's law in three dimensions we seek the number of distinguishable ways, Ω, in which the sequences of A and B units can be arranged in the linear chain. Following a straightforward combinatorial analysis[146,147] equation (45) is regenerated.

The derivation of equation (45) is based on the concept that total equilibrium prevails throughout the system. This assumption implies that longitudinal crystallite growth is impaired only by the occurrence of non-crystallizable units along the chain; the lateral development of a crystallite is restricted only by the specified thermodynamic conditions. The parameter D, related to the end surface free energy, does not explicitly appear in equation (45). The equilibrium melting temperature equation does not, therefore, account for the role of the interphase between the crystalline and non-crystalline regions. In effect, therefore, a sharp demarcation is assumed between the two regions which is not a realistic situation. Appropriate cognizance must be taken of the departure from equilibrium in analyzing actual experimental data and the theoretical equation must be modified accordingly.

The most significant consequence, or prediction, of equation (45) is that the melting temperature of a copolymer, in which only one type unit is crystallizable, depends only on the sequence propagation probability p and not directly on composition. This is a rather unusual result and is unique to long chain molecules. Considering the major categories of copolymer structure we find that: for a random copolymer $p = X_A$; for a block copolymer $p \gg X_A$; and for an alternating copolymer $p \ll X_A$. Many real systems will not fit these conditions exactly but will fall between the specifications cited above. Equation (45) implies that, depending on the sequence arrangement, very large differences can be obtained between the melting temperatures of copolymers of exactly the same composition. This most significant prediction has been amply supported by experiment. Some dramatic examples are found with copolyesters.[148,149] A typical example is given in Figure 19 where the melting temperature–composition relations are given for block and random copolymers of ethylene terephthalate with different comonomers. As predicted by theory the melting temperature does not depend on the specific chemical nature of the co-units. While the melting temperatures of the random copolyesters decrease monotonically with co-unit content, in contrast the melting temperatures of the block copolymers are constant in these examples up until co-unit content of

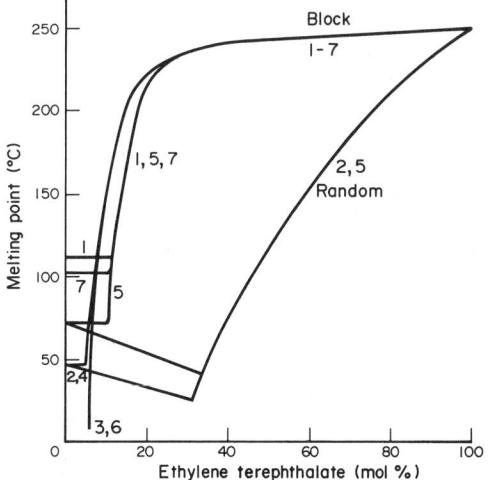

Figure 19 Melting temperature–composition relations for block copolymers of ethylene terephthalate with; (1) ethylene succinate; (2) ethylene adipate; (3) diethylene adipate; (4) ethylene azelate; (5) ethylene sebacate; (6) ethylene phthalate; (7) ethylene isophthalate. For comparative purposes data for random copolymers with ethylene adipate and with ethylene sebacate are also given[149]

about 80%. These results are the theoretical expectations. Consequently one can observe melting point differences close to 200 °C for the same copolymer composition. These differences are reflected in major changes in many physical and mechanical properties. It has been reported[150] that the melting temperatures of block copolymers of ethylene terephthalate/butylene terephthalate are lowered by about 20 °C for only 20 weight % of co-unit. The reason for this apparent deviation from theory needs further investigation.

The influence of sequence distribution on melting temperatures need not be as drastic as is illustrated in Figure 19. There can also be other important, but more subtle effects. Figure 20 illustrates such an example. Here the melting temperature–composition relations for several different polyethylene-based statistical copolymers are given.[83] For present purposes it is sufficient to limit consideration to copolymers that contain less than three mol % of branched groups. A significant feature of these results is that hydrogenated polybutadiene copolymers with ethyl, propyl and acetate side groups all obey the same melting temperature–composition relation. The sequence distributions of these copolymers are close to random.[83,151,152] Fractions of copoly-(ethylene/1-octene) also follow the same melting temperature–composition relation. On the other hand, the melting temperatures of copoly(ethylene/1-butene) fractions, the open rectangles in Figure 20, are significantly higher than those of the other copolymers when compared at the same composition. The difference is about 5 °C for 0.5 mol % side groups and increases to 10 °C at about 3 mol %. These melting temperature differences cannot be attributed to the chemical nature of the side group since both copoly(ethylene/1-butene) and hydrogenated polybutadiene contain ethyl branches. On the basis of equation (45) we can conclude that the melting point difference is a result of different sequence distributions. In the composition range of present interest, *i.e.* the order of a few mol % co-unit, only very small differences in the sequence propagation parameter p suffice to cause the melting point differences of the magnitude observed here. For example, for a random copolymer $p = 0.980$ for 2 mol % side groups. For 2 mol % branches in copoly(ethylene/1-butene), based on the observed melting temperature, the calculated value of p would be equal to 0.9875. This small but significant difference in the parameter p is adequate to explain the melting point differences. Therefore, as theory teaches us and experiment strongly confirms, neither the chemical nature of the co-unit nor its nominal composition directly determines the melting temperature of a copolymer.

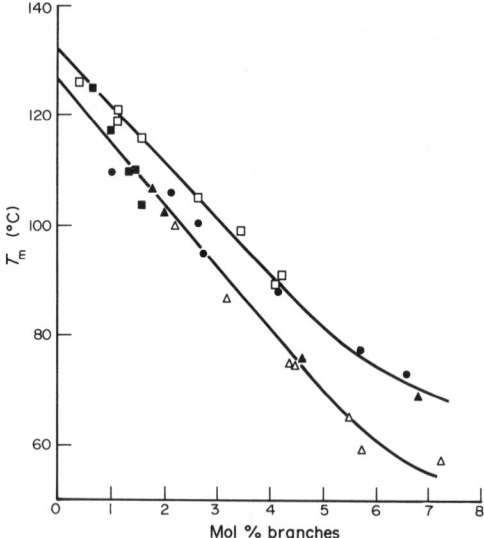

Figure 20 Melting temperature T_m of rapidly crystallized fractions of polyethylene-based copolymers as determined by differential scanning calorimetry: hydrogenated polybutadiene with ethyl groups $\triangle$; copoly(ethylene vinyl acetate) $\bullet$; diazoalkane copolymer with propyl side groups $\blacktriangle$; copoly(ethylene/1-butene) $\square$; copoly(ethylene/1-octene) $\blacksquare$ [83]

For a variety of random copolymer types a linear relation is invariably observed between $1/T_m$ and $-\ln X_A$. Although the proper functional form is obtained with respect to sequence distribution, equation (45) is not quantitatively obeyed. The value of ΔH_u that is deduced from equation (45) is always much less than that determined by other methods. The reason for this disparity appears to lie in the inability of a real random copolymer system to achieve the conditions stipulated by the equilibrium theory. Theory requires that the melting temperature recorded represent the disappearance of very long sequences of A units. As has been indicated in the theoretical fusion curves of Figure 18, even at equilibrium, the concentration of such sequences is relatively low. They will be difficult to develop at any reasonable crystallization rates. Their detection will also require very sensitive expermental methods. Therefore, even under very careful crystallization conditions the recorded melting temperatures of random copolymers will be less than the theoretical expectation. This difference will become larger as the co-unit content increases and will result in an apparent lower enthalpy of fusion.

The small thicknesses of the crystallites that are actually formed bring into focus the contribution of the interfacial structure. Since the crystallites formed by random copolymers have a definite lamellar habit, even for surprisingly high co-unit contents,[18] a sharp boundary cannot exist between crystalline and non-crystalline sequences. The structure of the interfacial region, and thus the interfacial free energy, has to be taken into account in analyzing the melting temperature of random copolymers. Because of the limitations on the thickness and lateral extent of the lamellae the distinct possibility exists that a high concentration of co-units will accumulate in the interfacial region. Consequently, the composition of the isotropic, or liquid-like region, will not reflect the nominal composition.

When there is a strong tendency for the comonomeric units to alternate, that is when $p \ll X_A$, a very large depression of the melting temperature is predicted by equation (45). This expectation is based on the requirement that only the A units crystallize and that the same crystal structure, corresponding to the homopolymer, is present over the whole composition. This condition will obviously be very difficult to fulfill. If this situation is met at all it would only be at very high concentrations of A units. What is usually observed is that with a strong, regular alternation of A and B units long regularly ordered sequences develop which crystallize in a completely new structure which is different from that of the homopolymers. The AB repeat can then be treated as the crystallizing unit in a random copolymer.[153]

Typically for crystallizable alternating copolymers, the maximum melting temperature occurs at equal molar ratios of the two structural repeating units.[153–157] The melting temperature–composition relation of the new structure follows the functional form of a random copolymer as described by equation (45).[153,157] An example is given in Figure 21 for *alt*-copoly(ethylene/carbon monoxide). Here $1/T_m$ is plotted against $\ln X$ where X represents the fraction of crystallizable units (*i.e.* CH_2CH_2CO). The expected straight line results. Similar results were also found[157] for *alt*-copoly (ethylene/chlorotrifluoroethylene). The equal molar (ethylene/carbon monoxide) copolymer has a crystal structure that is similar to polyethylene. The unit cell is orthorhombic but has a repeat of six planar zigzag chain atoms and larger a and shorter b axes. At lower concentrations of carbonyl units, zero to 13.2%, it has been reported that there was essentially no change in the melting temperature with composition. At high ethylene unit concentrations the c axis becomes that of

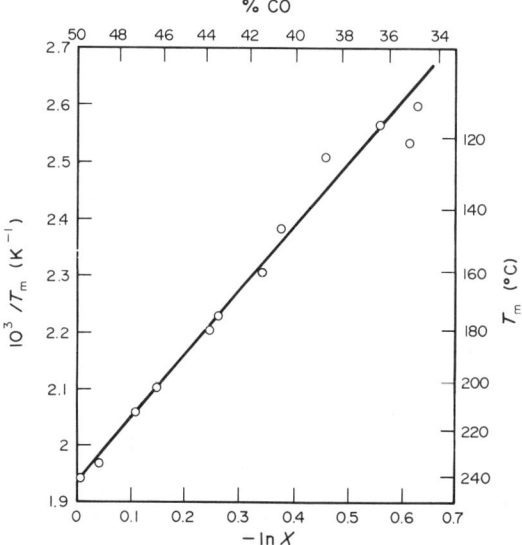

Figure 21 Melting temperature–composition relation for alternating copolymer of ethylene and carbon monoxide[153]

polyethylene with a continuous change in the *a* and *b* dimensions. The melting temperature behavior is not that of a random copolymer and the data suggest inclusion of co-units on a non-equilibrium basis.

In the discussion up to now of copolymer crystallization the specific assumption has been made that the co-units are excluded from the lattice. As was pointed out earlier an *a priori* assumption of this type has to be made in multicomponent–multiphase problems. The case where the co-unit enters the lattice is of considerable theoretical importance. Properties are also significantly altered. There are two distinctly different situations that can exist when the co-unit enters the lattice. The co-unit can enter as an equilibrium requirement or in a quite different manner as a non-equilibrium impurity or defect. These are quite different situations, both theoretically and practically.

Before considering the role played by the co-unit a decision has to be made as to whether it actually enters the lattice. This question cannot be answered in any simple manner. The classical procedure of establishing a true phase diagram, *i.e.* the composition of the solidus in equilibrium with the liquidus, is not easily accomplished with polymers. There are thus no general principles that can be invoked. The analysis is hampered by the fact that equation (45) is not quantitatively obeyed even if the sequence distribution has been established independently. If equation (45) were quantitatively obeyed then, given the enthalpy of fusion per repeating unit and the sequence distribution, a straightforward decision could be made as to whether the co-unit enters the lattice. In the absence of this basic guiding principle each situation has to be considered individually.

Analysis of the lattice parameters has been a very popular method of determining the purity of the crystal lattice. Extensive work in this area has been carried out with polyethylene-based copolymers.[82,158–162] The usual assumption that is made is that the expansion of the lattice reflects the inclusion of the co-unit. However, Bunn[163] has pointed out that this is not necessarily a unique interpretation. The crystallite thicknesses of these copolymers are relatively small, of the order of 100 Å or less depending on co-unit content.[83] With such thin crystallites and a thick interfacial region the strain that develops could easily cause the lattice expansion. Hence, the analysis of the lattice parameters does not necessarily yield definitive information with respect to the problem at hand.

Another method involves the utilization of chemical reactions where the non-crystalline region is presumed severed from the crystalline core. The selective oxidation of polyethylene-based copolymers, which has been extensively studied, is an example of this method.[165–168] The crystalline core is presumed to remain behind as a residue which can then be analyzed for purity by many different methods. The problem here, as well as in other chemical methods, is establishing the reactivity of the interfacial zone and the need to decide whether or not it is penetrated by the reacting agent. If not easily removed it would be construed to be part of the crystalline core. As a consequence conflicting results have been obtained by this and other chemical methods.[169,170] In a very specific example crystals formed from a dilute solution of copolymers of L-lactide and the racemic DL-lactide were selectively hydrolyzed and analyzed by optical rotation.[171] It was concluded that there was a significant partitioning of the D units between the crystalline and non-crystalline phases. However, this method was not selective for the melt crystallized copolymer.

Melting temperature relations, based on the liquidus composition, can be helpful in settling the partitioning issues in specific cases. An example is given in Figure 22 for a set of polyethylene-based copolymers with different types of branches. The dashed line in the figure represents the theoretical expectation for a random copolymer. As we have found for all other systems the actual observed melting temperatures are significantly below the theoretical expectations. Of particular importance in this figure, however, is the fact that although the ethyl, propyl and acetate branches fall on the same line the copolymers containing directly bonded methyl groups have much higher melting temperatures. There is in fact a maximum in the melting temperature–composition relation.[146] Richardson, Flory and Jackson pointed out[172] that a proportion of the methyl groups enter the lattice on an equilibrium basis. In contrast, co-units entering the lattice as non-equilibrium defects will invariably cause a lowering of the melting temperature. It should not be surprising, therefore, that the structural features[83] and mechanical properties[173] of the methyl branched copolymer are very similar to those of the parent homopolymer as contrasted with the copolymers that contain larger side groups. The melting point–composition data of Figures 20 and 22 indicate that the ethyl and larger branches behave in a similar manner. It is highly unlikely that the very large side groups would enter the lattice,[83,146] so that on this basis the ethyl groups would also be excluded.

The melting temperature–composition relations for poly(oxymethylene)-based copolymers depend on the method of synthesis.[174–176] Inoue[174] prepared copolymers, containing up to 10 mol % co-units, from trioxane with dioxalane, 1,3-dioxane, dioxepane and epichlorohydrin using boron trifluoride etherate as catalyst. The melting temperatures decreased monotonically with the

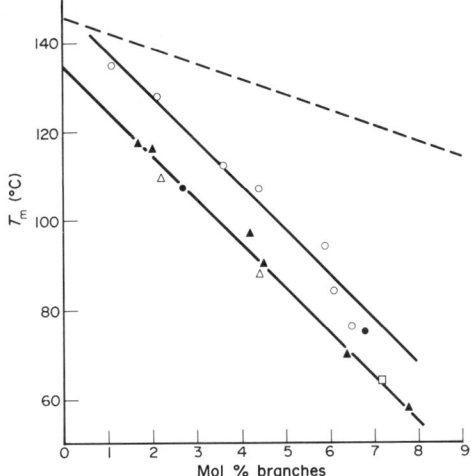

Figure 22 Dilatometric determined melting temperature, T_m, of polyethylene-based copolymers plotted against mol % branches. Dashed line represents equilibrium theory for random copolymers. Diazoalkane copolymers with methyl branches ○, ethyl branches □, propyl branches ▲; hydrogenated polybutadiene with ethyl branches △; copoly(ethylene vinyl acetate) ● [83]

co-unit content and were independent of the chemical nature of the comonomer. The density–composition relations were also independent of the chemical nature of the co-units and were of the form expected for a random copolymer where the co-units are excluded from the lattice. Conventionally prepared copoly(oxymethylene/oxyethylene)s also showed similar behavior.[175]

Droscher et al.[176] prepared chemically similar poly(oxymethylene)-based copolymers (with 1,3-dioxalane, 1,3-dioxane, 1,3-dioxepane and 1,3,6-trioxocane as respective comonomers) which were of extended crystal form and presumed to be completely crystalline. For this series of copolymers, the melting temperature–composition relation at low co-unit content depended on the chemical nature of the comonomer used. The characteristic curve observed by Inoue[174] was not found. It was presumed that an appreciable concentration of co-units enter the lattice. An interesting feature developed at higher co-unit concentration. Depending on the co-unit type, the melting temperatures became invariant with composition, while the normal poly(oxymethylene) unit cell was maintained.

Poly(ethylene oxide) copolymers which contain up to 4 mol % of either butylene oxide or styrene oxide behave in a conventional manner.[177] In these cases the results are typical of random copolymers where the co-units are excluded from the crystal lattice.

In some specific cases scattering methods have been used to determine the distribution of the B units between the phases. Fischer and co-workers[178] combined the analysis of the integrated intensity of small-angle neutron and X-ray scattering to determine the extent of partitioning of the co-units of a random copolymer of 1,3,5-trioxane and 1,3-dioxolane and of chlorinated polyethylene. A two phase model was assumed to describe the structure of the semicrystalline copolymer. Each phase was taken to be homogeneous and separated by sharp boundaries. This assumption, which is an inherent part of the analysis, is contrary to theoretical expectations and to the relatively large interfacial region that is usually associated with crystalline random copolymers.[64, 83] This method of analysis would, therefore, not be expected to give a quantitative measure of the partitioning. However, it should be adequate to allow for a decision as to whether all the B units are confined to the isotropic, liquid-like region. If the small-angle X-ray scattering patterns can be fitted to a model of only two distinct phases the implication is that the electron density (and hence density) of the interfacial region is close, or identical, to one of the other phases.* All of the copolyethers studied showed some degree of partitioning, *i.e.* the B units were not restricted to the non-crystalline region.

Roe and Gieniewski[179, 180] studied the absolute intensity of the small-angle X-ray scattering of a series of randomly chlorinated polyethylenes. When only the X-ray method is used, in addition to the assumption with respect to the boundary separation of the phases, the densities of the crystalline and noncrystalline regions need to be supplied. It was found in this study that the chlorine atoms enter the lattice with the extent of partitioning becoming more accentuated the more rapid the rate of crystallization.

---

* Extensive analysis of the degree of crystallinity of linear polyethylene and random polyethylene-based copolymers has shown that the density of the interfacial region is very close to that of the crystallite core.[58]

As would be expected, for many copolymers, there are specific co-units, or side groups, which enter the crystal lattice. Still to be established in most of these cases is whether the unit is behaving as a true crystal defect or enters the lattice as an equilibrium requirement. This distinction is important in developing a proper analysis. For the polyethylene-based copolymers smaller side groups, such as Me, Cl, O, enter the lattice to some extent. It has, however, only been established that the Me group does so on an equilibrium basis.

Cocrystallization, or isomorphous replacement, is observed in copolymers and is very similar to the phenomenon found in low molecular weight systems. For polymers the two different co-units are covalently linked and are thus not separable as in the monomeric case. When there is complete isomorphic replacement, the X-ray diffraction patterns indicate the presence of only one crystal structure over the complete composition range. The melting temperature–composition relations for such systems are different from those for ordered, alternating or random copolymers where the crystalline phase remains pure.

Examples of isomorphic replacement are found in a variety of copolymer types such as copolyamides,[181-186] copolyesters,[187] vinyl copolymers[188-190] and diene copolymers.[191] It is difficult to formulate a definitive set of rules that establishes the comonomer structural factors which favor cocrystallization. There are, however, some interesting examples that illustrate various principles. Edgar and Hill[181] noted that the distances between the carboxyl groups in terephthalic and adipic acid are almost identical. On this basis, partial substitution in the crystal lattice of these two diacids was anticipated. This was found to be the case for copoly(hexamethylene adipamide/terephthalamide). The melting temperatures of these random copolymers increase monotonically with increasing concentration of aromatic units. No minimum in the melting point–composition relation is observed, as would be expected if the comonomer crystallized separately. It can, therefore, be concluded that the cocrystallization of the two co-units has occurred. Having co-units of identical length is not, however, a sufficient requirement for cocrystallization. Copolyesters correspondig to the copolyamides just discussed do not cocrystallize. Tranter[185] has described several copolyamide systems which appear to satisfy the above requirement, some of which cocrystallize and some of which do not.

An example of cocrystallization among vinyl copolymers can be found in copoly(styrene/p-fluorostyrene).[188] The stable crystal modifications of the homopolymers polystyrene and poly(p-fluorostyrene) are quite different. In the former case the ordered chain conformation is a threefold helix while in the latter it is a four-fold one. Nevertheless, the copolymers are crystalline over the complete composition range, and the melting temperature is essentially a linear function of composition. These data by themselves provide adequate evidence of cocrystallization.

Hachiboshi *et al.*[187] were able to crystallize random copolymers of ethylene terephthalate/ethylene isophthalate over the complete composition range. The wide-angle X-ray patterns of these copolymers change systematically with co-unit content and it is concluded that the two units can cocrystallize and can form a new unit cell. A complete phase diagram, involving both the solidus and liquidus composition, was determined and is illustrated in Figure 23. The solidus was determined by

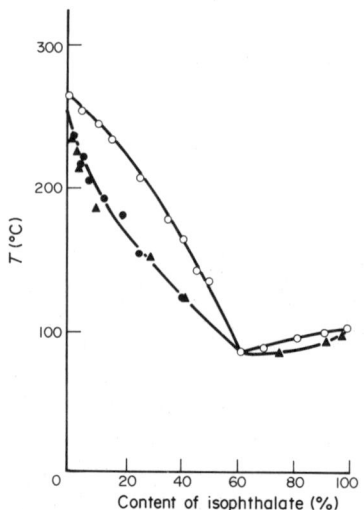

Figure 23   Phase diagram for ethylene terephthalate/isophthalate copolymers[187]

assuming the additivity of the lattice spacing. The phase diagram of Figure 23 is essentially a classical one, in terms of low molecular weight systems. It even contains an azeotropic point. Polymer crystallization, therefore, is not atypical as has been found in virtually all the examples. For low molecular weight systems at the azeotropic point the liquid and solid must have the same composition. For random copolymers, the comparable requirement would be that the sequence propagation probability must be the same in both phases.

When the co-units are partitioned between the two phases on an equilibrium basis, *i.e.* when the crystalline phase does not remain pure, the melting temperature–composition relations should be classical. For long chain molecules, therefore, the composition variable will be replaced by the sequence propagation probability. A variety of melting temperature–composition relations can be generated. Co-units entering the lattice on a non-equilibrium basis will behave as true defects. The melting temperature will then be lower than that predicted. Attempts have been made to calculate the effects of such internal defects on the melting temperature.[192–197] All these calculations have assumed a regularly folded-chain lamellar structure. The co-unit, or defect, was treated as if it were an isolated impurity, as in a momeric system. These assumptions do not correspond to the experimental situation and thus the conclusions are of limited validity. In this problem the crucial matter is the sequence distribution within the lattice as compared to the distribution in the non-crystalline region. Specific assumptions have to be made for each non-equilibrium situation. The limiting relation that has been proposed can be derived on the most general grounds.[133]

It is evident that the study of crystalline copolymers presents some very difficult as well as intriguing problems. The theoretical development of equilibrium relations and phase diagrams can be accomplished in principle when it is recognized that one must be concerned with the sequence propagation probability and not the composition. One major difficulty is that although the liquidus of the phase diagram can be determined quite easily, the solidus is extremely difficult to establish in most cases. In addition, it is virtually impossible to establish equilibrium on an experimental basis. As we have enumerated above there is a large number of important contributions to the deviations from equilibrium. The major ones are the finite crystallite size (relative to the equilibrium requirement), possible defected structures within the lattice and most importantly the structure of the interphase.

11.3.5 Polymer–Diluent Mixtures

To analyze the fusion of mixtures of homopolymers and low molecular weight diluents we return to equations (11), (14) and (15). Equation (11) was derived by utilizing lattice techniques. In effect, the Flory–Huggins[198,199] expression for the free energy of mixing was used. We are, therefore, limited at this point to situations where the polymer segments and diluent are uniformly distributed in space. Dilute solutions are therefore excluded from the present discussion but will be considered separately subsequently. It has also been assumed that the diluent is excluded from the crystal lattice. This condition has been satisfied by almost all the polymer–diluent mixtures that have been studied.

Equation (14) indicates that the melting range depends on the value of $v_2 D$. Hence, the fusion range will be expected to broaden with increasing diluent concentration. This expectation has been observed in numerous systems.[106,200] The equilibrium melting temperature of the polymer–diluent mixture, T_m, is obtained when $\lambda_e = 1$. Hence from equation (15) we obtain

$$1/T_m - 1/T_m^\circ = R/\Delta H_u \{(V_u/V_1)(1-v_2) + (1/x)[v_2 + x/(x-\zeta_e+1)] - \chi_1^*(1-v_2)^2\} \tag{46}$$

For large x, equation (46) reduces to

$$1/T_m - T_m^\circ = (R/\Delta H_u)(V_u/V_1)[(1-v_2) - \chi_1(1-v_2)^2] \tag{47}$$

Equation (47) is very similar to the classical expression for the depression of the melting temperature of low molecular weight binary systems, *i.e.* the freezing point depression equation. The only difference results from the expression for the activity of the crystallizing polymeric component in the molten phase. Consequently equation (47) can also be derived by the application of the criterion for phase equilibrium.[2,201]

According to polymer solution theory the chemical potential of a repeating unit in a high molecular weight polymer is[201]

$$\mu_u - \mu_u^\circ = RT(V_u/V_1)(v_1 - \chi_1 v_1^2) \tag{48}$$

where μ_u° is the chemical potential of the pure liquid polymer. The difference between the chemical potential of the crystalline unit in the polymer μ_u^c and that of the pure liquid polymer can be written as

$$\mu_u^c - \mu_u^\circ = -\Delta G_u = -(\Delta H_u - T\Delta S_u) \tag{49}$$

Since $T_m^\circ = \Delta H_u / \Delta S_u$, equation (52) becomes

$$\mu_u^c - \mu_u^\circ = \Delta H_u(1 - T/T_m^\circ) \tag{50}$$

When the stipulation for phase equilibrium

$$\mu_u^c - \mu_u^\circ = \mu_u - \mu_u^\circ \tag{51}$$

is invoked at the melting temperature of the polymer–diluent mixture equation (47) is once again obtained.

According to equation (47) the depression of the melting temperature, for a given polymer, depends on the volume fraction of diluent in the mixture and on its thermodynamic interaction with the polymer. Other quantities being equal a larger depression of the melting temperature should be observed with good solvents (smaller values of χ_1) than with a poor one. The depression will be larger with diluents of smaller molar volume; the larger the value of ΔH_u the smaller will be the expected melting point depression. ΔH_u, which represents the enthalpy of fusion per mole of repeating units, is an inherent and characteristic property of the repeating unit of the crystalline polymer. It does not depend on the level of crystallinity or other morphological features. It should not be identified with the enthalpy of fusion obtained by direct calorimetric measurements. The quantity ΔH_u can be gotten from experiment by means of equation (47).[200] One can thus obtain a very important characteristic parameter of the crystalline polymer.

A large number of experiments, involving a wide variety of different polymers, has conclusively demonstrated that equation (47) is quantitatively obeyed.[200] Moreover, for a given polymer the same value of ΔH_u is obtained irrespective of the chemical and structural nature of the diluent used. The experimental verification of equation (47) is further evidence that the fusion of crystalline polymers can be treated as a first-order phase transition.

An interesting situation develops when sufficiently poor solvents are used as diluents. With the initial addition of diluent a relatively small depression of the melting temperature is observed. However, above a critical diluent concentration the melting temperature remains invariant with further dilution, until the very dilute range is reached.[202–206] When this invariance in melting temperature is observed the melt always consists of two immiscible liquid phases. Hence, at the melting temperature three phases must be coexisting in equilibrium. As a consequence of the phase rule, since the pressure is constant, the melting temperature must remain invariant with composition.

When analyzing melting temperature–composition relations according to equation (47) the underlying assumption is made that the crystallite structure and size is the same over the complete composition range being studied. For non-equilibrium crystallites it is also assumed that the interfacial free energy remains constant. There are no major problems when the polymer is crystallized external to the diluent and the melting temperature is determined at a given concentration. However, when the crystallization takes place within the polymer–diluent mixtures, prior to determining the melting temperature, complications can develop. When limited to concentrated solutions, i.e. $v_2 \geq 0.50$, there is no serious difficulty. In more dilute solutions, however, the crystallite structure and size change and eventually reach those typical of dilute solution. There will be an inconsistency in the analyses unless care is taken to ensure that the melting temperatures of the same type crystallites are measured.[24]

If the added component is a good solvent, and a wide concentration range is studied, a large change in melting temperature will result. Under these circumstances the tacit assumption that ΔS_u and ΔH_u are independent of temperature will no longer be correct and appropriate corrections need to be made. The entropic contribution to χ_1 will also have to be taken into account.[207,208] Equation (47) will still quantitatively describe the equilibrium but appropriate corrections, which can be applied in a purely formal manner, need to be made.

The analysis of the binary systems have so far has been restricted to the very common case where the crystalline phase remains pure. However, other situations which are similar to those found in low molecular weight systems also exist. For example, for the poly(2,6-dimethyl-1,4-phenylene oxide)–methylene chloride mixture the diluent enters the lattice.[209,210] The analysis of the collagen–water system has lead the conclusion that water was present within the lattice.[211] Several binary mixtures have also been studied where the pure diluent can also crystallize. Hence, a classical eutectic mixture results.[212-215] Schematic representations of the different types of phase diagrams that can be developed by polymer–diluent mixtures have been given by Papkov.[216]

In the analysis that has been given the tacit assumption has been made that the polymer segments are uniformly distributed throughout the solution. Under these circumstances the use of the Flory–Huggins expression for the free energy of mixing, and thus the chemical potential of the species, is valid. However, a dilute solution of flexible chain molecules is characterized by a non-uniform polymer segment distribution throughout the medium. The use of the Flory–Huggins free energy function is no longer valid. An exception occurs with very poor solvents since the molecules can interpenetrate one another freely.

In order to develop a theory for crystallization from dilute solution it is necessary to have an expression for the chemical potential of the polymer species. In the general thermodynamic theory of dilute solutions the chemical potential of the solvent species can be expressed in virial form. The equilibrium requirement, equation (57) leads[217,218] to an expression for the melting point or dissolution temperature, in dilute solution, the limiting form of which is[219]

$$\frac{1}{T_m} - \frac{1}{T_m^\circ} = \frac{R}{\Delta H_u} \frac{V_u}{V_1} \left[-\frac{\ln v_2}{x} + (1 - 1/x)v_1 - \chi_1 v_1^2 \right] \tag{52}$$

In this expression x represents the number of segments per molecule.

A very similar analysis of the problem has been given by Beech and Booth[220] who did not, however, have occasion to actually calculate the melting point relation. Pennings,[221] following a similar procedure, did not arrive at equation (52) because of differences in expressing the higher virial coefficients. Sanchez and DiMarzio[222] offered an empirical relation to resolve the problem. There is no need to adopt arbitrary methods since the problem can be resolved in a straightforward manner.

Equation (52) is very similar in form to equation (47) which was derived for more concentrated systems of high molecular weight. In fact, as $x \to \infty$, equation (52) reduces to equation (47). Deviation from the limiting form of equation (47) would only be expected at extremely high dilutions and low molecular weights. Numerical analysis indicates that, even in the low molecular weight range, deviation from the limiting form requires that v_2 be much less than 0.05. Moreover, data at this and slightly higher concentrations cannot be extrapolated in any simple manner to give meaningful results in the dilute range.[223]

11.3.6 Polymer–Polymer Mixtures

Crystallization can also take place from polymer mixtures. One common example involves two chemically different polymeric species that are completely miscible in the melt from which only one component crystallizes. The melting temperature–composition relation for this case has been formulated by Nishi and Wang,[224] who utilized the free energy of mixing expression given by Scott[225] following the Flory–Huggins lattice treatment. From this free energy expression the chemical potential per mole of repeating unit in the melt can be expressed as

$$\mu_{2u} - \mu_{2u}^\circ = \frac{RTV_{2u}}{V_{1u}} \left[\frac{\ln v_2}{x_2} \right] + \left(\frac{1}{x_2} - \frac{1}{x_1} \right)(1 - v_2) + \chi_{12}(1 - v_2)^2 \tag{53}$$

The V_u's are the molar volumes of the respective repeating units, x_2 and x_1 the number of segments per molecule for each of the chains, and χ_{12} is the polymer–polymer interaction parameter. Here v_2 is the volume fraction in the mixture of the crystallizing component.

The melting temperature reaction was found to be

$$\frac{1}{T_m} - \frac{1}{T_m^\circ} = -\frac{RV_{2u}}{\Delta H_u V_{1u}} \chi_{12}(1 - v_2)^2 \tag{54}$$

There are several interesting consequences of equation (54). Since the melting point depression is a colligative property the influence of a high molecular weight diluent will be expected to be relatively small as predicted. This expectation is widely observed experimentally.[226-230]

If χ_{12} is assumed to only involve enthalpic interactions (or only a very small temperature interval is involved) then to a good approximation[207]

$$\chi_{12} = \frac{BV_{1u}}{RT} \tag{55}$$

so that equation (54) becomes

$$1 - T_m/T_m^\circ = -\frac{BV_{2u}}{\Delta H_u}(1 - v_2)^2 \tag{56}$$

Thus, in principle, the parameter B can be obtained from measurement of the melting point depression; it is an important parameter in determining miscibility of the liquid phase. Even if equilibrium conditions prevail, the small melting point depression observed precludes an accurate determination of the magnitude of B. However, its sign, which is important, should be accessible from this type of experiment.

Shultz and McCullough[231] have extended the analysis to the ternary system consisting of a low molecular weight diluent and two chemically different polymers only one of which crystallizes. The analysis was successfully applied to a range of experimental data for the system poly(2,6-dimethyl-1,4-phenylene oxide), polystyrene and toluene.

Binary systems, involving two chemically identical polymers that differ only in molecular weight, are also of interest. The major influence on melting temperatures will be in the low molecular weight range. It is clear that equation (53) will no longer be appropriate, although it has been used on several occasions to analyze crystallization from chemically identical polymer mixtures.[232-234] The fact that the thermodynamic interaction parameter enters equation (53) makes clear *a priori* that it will not be applicable to mixtures of the same polymer, which constitute an athermal system. The problem has been formulated in terms of the Flory–Huggins expression for the free energy of mixing a set of chemically identical species with a low molecular weight diluent.[235] The analysis indicates that significant changes in melting temperatures will only take place with low molecular weight species in a mixture of low number average molecular weight.[236]

11.3.7 Other Thermodynamic Aspects

The analysis of the melting of polymer–diluent mixtures enables the quantity ΔH_u to be determined. This is an important characteristic of the chain. Heats of fusion, as well as latent volume changes, observed for long chain molecules are comparable to the values characteristic of molecular crystals formed by low molecular weight substances. We have already noted that there is a natural continuity in the values of the enthalpy of fusion, from the *n*-hydrocarbons to linear polyethylene. From the enthalpy of fusion and the melting temperature the entropy of fusion per repeating unit can be obtained. It is then possible to examine these parameters in terms of chain structure. Such an analysis has been given in detail elsewhere and will not be repeated here.[237,238] As a general rule it has been found that the main influence of chain structure on the melting temperature is through its conformational properties and thus the entropy of fusion. The entropy of fusion thus plays a key role in determining the location of the melting temperature. Rotational isomeric state theory has been successfully applied to the analysis of the entropy of fusion at constant volume.[239-242d]

Phase equilibrium between the crystalline and disordered state has been demonstrated by the application of the Clapeyron equation to the relationship between melting temperature and applied hydrostatic pressure.[237,242] In a one-dimensional analogue to the Clapeyron equation, applicable to oriented fibrous systems, the relationship between tensile force and length can be established. It is found that[244]

$$\left[\frac{\partial(f/T)}{\partial(1/T)}\right]_{p,\,eq} = \frac{\Delta H}{\Delta L} \tag{57}$$

Here f is the applied tensile force and ΔH and ΔL are the changes in enthalpy and length that occur upon the fusion of the entire fiber at constant T, p and f. In this case the coexistence of two phases is expected and has been actually demonstrated for a variety of fibrous systems.[244-246] The equilibrium force is analogous to the negative of the pressure and the length to the volume in the conventional pressure–volume equilibrium. As a consequence of the coexistence and different equations of state for the two phases, and their coexistence in equilibrium, contractility and tension development can be demonstrated for many fibrous macromolecular systems.[246-249]

11.3.8 Non-equilibrium Aspects of Fusion

The discussion given has had as its primary focus the equilibrium aspects of crystallization and melting. The thermodynamics of fusion, utilizing the principle of a first-order phase transition, is the basic foundation upon which the understanding of all aspects of the crystallization behavior of polymers is developed. The major deviations from equilibrium that are observed for copolymer melting have already been discussed. Similar problems should be expected for homopolymer and polymer–diluent mixtures. For homopolymers, for example, one of the equilibrium conditions is that the crystallites be extended, i.e. ζ_e must be comparable to the chain length.[84] This requirement will obviously be very difficult to achieve for polymers of any appreciable molecular weight when consideration is given to entanglements and to the profuse interdispersion of chains in the melt.[99,250] Crystallization kinetic studies bear out this concern. For example, the n-hydrocarbon $C_{94}H_{190}$ can only be supercooled infinitesimally below its melting temperature. Crystallization is rapid. Extended chain molecular crystals are formed and the transformation from the liquid to crystalline state is complete. On the other hand for the corresponding polymer, linear polyethylene, no matter how well fractionated and even for low molecular weights undercoolings of the order of 20 °C are necessary in order for crystallization to proceed at a reasonable rate.* Molecular crystals are never formed, The transformation is rarely if ever complete[63] and only the very low molecular weight species form extended chain crystals.[89] Accompanying the restricted crystallite thickness are contributions from the interfacial region and other aspects of morphological structure.

The major deviations from equilibrium to be considered here will be concerned with the limited crystallite size and the contribution from the interfacial region. Extended chain crystals have been demonstrated to form with low molecular weight fractions of linear polyethylene[45b,89,232] and poly(ethylene oxide).[119,127,129] Raman LAM and small-angle X-ray scattering studies have shown that for linear polyethylene of molecular weight less than about 3000 extended chain crystals having narrow size distributions are formed. In this molecular weight range the melting temperature is independent of the crystallization temperature. These type samples can be properly used to analyze the equilibrium conditions. For molecular weights of about 8000 or greater, there is a continuous increase in the melting temperature with crystallization temperature and extended chain crystals are not formed. For molecular weights greater than 10^4 the crystallite thickness is always very much less than the extended chain length.[9,251] Molecular weight fractions of linear polyethylene in the range from about 3000–6000 show intermediate behavior which depends on the crystallization temperature.

The free energy of fusion of a non-equilibrium crystallite is obtained from equation (11) prior to maximizing the function. Although generally applicable to pure polymers and to polymer–diluent mixtures, for simplicity, we focus attention here on the pure system. For a crystallite ζ repeating units thick that has ρ sequences in cross section, the free energy of fusion for the pure system, $v_2 = 1$, can be written as

$$\frac{\Delta G_f}{xN} = \frac{\zeta\rho}{xN}\Delta G_u + RT\left[\ln\left(1 - \frac{\zeta\rho}{xN}\right) + \frac{\zeta\rho}{xN}\frac{1}{\zeta}\left[\ln D + \ln\frac{(x-\zeta+1)}{x}\right]\right] \tag{58}$$

It is assumed that the crystallite is sufficiently large in the directions normal to the chain axis so that the influences of the lateral surface free energies can be neglected. By expanding $\ln\left(1 - \frac{\zeta\rho}{xN}\right)$ equation (58) becomes

$$\frac{\Delta G_f}{\zeta\rho} = \Delta G_u - \frac{2\sigma_{ec}}{\zeta} - \frac{RT}{x} + \frac{RT}{\zeta}\left[\ln\left(\frac{x-\zeta+1}{x}\right)\right] \tag{59}$$

where $\ln D \equiv -\frac{2\sigma_{ec}}{RT}$. Here σ_{ec} represents the interfacial free energy of the basal plane associated with the non-equilibrium crystallite. It cannot be *a priori* identified with the interfacial free energy of the equilibrium crystallite since the corresponding surface structures are not necessarily the same. In equation (59) the first term represents the bulk free energy of fusion for the $\zeta\rho$ units involved. The second term represents the excess free energy due to the interfacial contribution of the chains emerging from the 001 crystal face (the basal plane). The last two terms result from the finite length of the chain. They only make significant contributions at the low molecular weights. The first of

---

* The recent synthesis of n-alkanes of high carbon numbers[114,115] raises the very interesting questions as to whether these compounds can be supercooled appreciably and if the transformation will be complete under all crystallization conditions.

these represents the entropy gain which results from the increased volumes available to the ends of the molecule after melting. The last term results from the fact that only a portion of the units of a given chain are included in the crystallite. It represents the entropy gain that arises from the number of different ways a sequence of ζ units can be located in a chain x units long with the stipulation that terminal units are excluded from the lattice.

At the melting temperature T_m^* of the non-equilibrium crystallite $\Delta G_f = 0$, so that equation (59) becomes

$$\frac{1}{T_m^*} - \frac{1}{T_m^\circ} = \frac{R}{\Delta H_u}\left[\frac{2\sigma_{ec}}{RT_m^*\zeta} - \frac{1}{\zeta}\left(\frac{x-\zeta+1}{x}\right) + \frac{1}{x}\right] \qquad (60)$$

Equation (60) represents the relation between the melting temperature and crystallite thickness ζ for a chain of length x. The crystallite thickness ζ is not constrained to its equilibrium value and σ_{ec} is characteristic of the particular interface that is developed under the specific set of crystallization conditions. The melting temperature depression in equation (60) is reckoned from the equilibrium melting temperature of the infinite chain.*

For high molecular weights equation (60) reduces to

$$\frac{1}{T_m^*} - \frac{1}{T_m^\circ} = \frac{2\sigma_{ec}}{\Delta H_u T_m^* \zeta} \qquad (61)$$

or

$$T_m^* = T_m^\circ[1 - 2\sigma_{ec}/\Delta H_u\zeta] \qquad (62)$$

which is the classical Gibbs–Thompson expression for the melting of crystals of finite size. It is somewhat ironical that non-equilibrium crystallites of high molecular weight chains obey the same melting point relation as do monomeric substances. However, corrections need to be made for lower molecular weight chain molecules.†

Equation (62) can be used, in principle, to determine T_m° since T_m^* can be measured as a function of ζ. A plot of T_m^* against $1/\zeta$ should be linear and extrapolate to T_m° as $\zeta \to \infty$.[233,252,253] However, the use of equation (62) requires that σ_{ec} be the same for each sample. This condition may be very difficult to fulfill. For example, the experimentally deduced σ_{ec} values for linear polyethylene are a function of molecular weight. These values range from 2–3000 cal mol^{-1} (8–12 kJ mol^{-1}) of sequences emerging from the basal plane (64–95 erg cm^{-2} or 64–95 mJ m^{-2}) at lower molecular weights to about 8000 cal mol^{-1} (32 kJ mol^{-1}) at molecular weights of 10^5 and greater.[9,86] Even for a given molecular weight, changes in σ_{ec} are to be expected with changes in the crystallization temperature. This is the main technique by which the crystallite thickness is varied for a homopolymer. For linear polyethylene the chain tilt, *i.e.* the angle between the chain axis and the basal plane, increases with decreasing crystallization temperature.[89] Concomitantly there should be a substantial decrease in the interfacial free energy. These factors, which lead to varying values of σ_{ec} with changes in crystallization temperature and molecular weight, need to be given serious consideration when designing experiments to use equation (61) for the determination of T_m°. Although there are reports of linear relations[233,252,253] when T_m^* is plotted against $1/\zeta$ it is clear that these concerns have not been taken into account in analyzing the data. For example, when applying this method of extrapolation to poly(ethylene oxide) an equilibrium value of T_m° of 69 °C was deduced.[233] This is a substantially lower value of T_m° than has been directly observed for this polymer[254] or that has been extrapolated by other methods.[255,256]

Although a discussion of the properties of crystals formed in dilute solution is not a major topic in this chapter, they are of great importance, both as entities in themselves and in relation to bulk crystallized systems. Of particular importance in the present context is the level of crystallinity that is attained and the structure of the non-crystalline region. The lamellar-like platelets that are formed in dilute solution and the chain orientation within the crystallites is well established.[21,22] Based almost solely on the observation of lamellae, and the concept of regularly folded chains, a completely

---

* The melting temperature T_m^* can be related to the equilibrium melting temperature for the chain of length x by introducing equation (20) into equation (60)

† Strictly speaking equation (60) only applies to the case where a chain supplies only one sequence to a crystallite. This equation will thus be valid for the lower molecular weights where the last two terms are important. For higher molecular weights equation (61) can be shown to be an excellent approximation to the situation where many sequences from the same chain participate in a given crystallite. This also includes the hypothetical model where the chains are regularly folded within the crystallite. In this case ζ is identified with the length of each sequence and an additional interfacial free energy is added to account for the folds.

crystalline system was promulgated. Early physical chemical measurements on such systems indicated that this situation could not be true.[24] The concept that crystals formed in dilute solution are not completely crystalline has received further quantitative support from a myriad of thermodynamic, spectroscopic, relaxation and scattering (small angle neutron as well as wide and small angle X-ray) experiments.[57,257,258,259] Depending on the crystallization conditions approximately 20–30% of the chain units are located in the disordered overlayer. Raman internal mode analysis indicates that the major portion of the disordered region is isotropic with only about 5% of the material being assigned to the interfacial zone.[260] This result is in good accord with theoretical expectations for solution formed crystals.[55]

There is interest in comparing the interfacial free energy between solution formed crystals and those formed in the bulk. The theoretical calculations of Flory, Yoon and Dill[54,55] indicate that they should not differ from one another by more than about 10%. To make the comparison, utilizing equation (61), it is necessary that the two sets of crystallites have the same, narrow crystallite thickness distribution and that reliable melting temperatures be obtained. This latter condition requires that any melting–recrystallization process, which could be occurring, needs to be analyzed in detail.[261] A quantitative comparison has been made for narrow distributed systems, with a crystallite thickness centered about 150 Å. Careful determination of the melting temperatures indicates that they do not differ by more than 1–2 °C.[261] Thus, the interfacial free energy associated with the respective basal planes are the same and found to be 2385 ± 40 cal mol^{-1} (10.0 ± 0.2 kJ mol^{-1}) of sequences, corresponding to 87 erg cm^{-2} (87 mJ m^{-2}). The theoretical expectations of similar interfacial free energies for the two crystallization modes is confirmed in this size range. However, the values are somewhat higher than predicted for reasons that have been previously discussed. It is important that these comparisons be extended to higher size ranges with the restraint, however, of narrow size distributions.

11.4 POLYMER NUCLEATION

As was pointed out in the prologue, a very well established feature of polymer crystallization is that it is a nucleation controlled process. This conclusion is based on the strong negative temperature coefficients that are observed for the crystallization process and the relation of the crystallite thickness to the undercooling at which the crystallization takes place. This principle is based on the most general grounds and is independent of either the shape of the nucleus or the disposition of chain units within a given structure.

An expression for the steady state nucleation rate for all substances, which is applicable to all nuclei types and shapes, has been derived by Fisher and Turnbull. It is given by[46]

$$\dot{N} = N_0 \exp\left(-\frac{E_D + \Delta G^*}{RT}\right) \tag{63}$$

Here E_D is the free energy of activation for transport across the liquid–nucleus interface and ΔG^* is the free energy necessary to form a nucleus of critical size. The steady state nucleation rate can then be incorporated into the expression for different type crystallization processes, as for example the overall crystallization rate or spherulite growth rate.[270] The problem then is to calculate ΔG^* for nuclei that contain long chain molecules where even at equilibrium molecular crystals are not formed, *i.e.* not all the chain units participate in the crystallization. The expression for the free energy of fusion given by equation (11) lends itself naturally to the calculation of ΔG^*. The problem of the dissipation of the chain flux, which plays an important role in the development of mature crystallites and controls the structure of the interphase,[23,54] does not influence the development of nuclei since their lateral dimensions are small.[262] This important point was not recognized in an earlier development where, for space-filling reasons, it was concluded that the nuclei were made up of regularly folded chains.[263] It has also been postulated that the nucleus must be comprised of regularly folded chains since this particular structure would have a lower free energy.[33] Calculations have made abundantly clear that this assumption is incorrect.[54]

In order to calculate the free energy of forming a nucleus it is necessary to define its shape. For simplicity, we will consider a cylindrical nucleus comprised of a bundle of crystalline sequences since only one new surface is introduced.* Since we are effectively dealing with a small crystallite the

---

* All other geometries, including regularly folded chains, can be treated equally well by the methods outlined here. The results differ only in the value of certain constants while the physical significance is not sensibly affected.

contribution of the lateral surface free energy needs to be taken into account. Hence under these circumstances

$$\Delta G = 2\pi^{1/2}\rho^{1/2}\zeta\sigma_u + 2\rho\sigma_{en} + \frac{RT}{x}\zeta\rho - \rho\zeta\Delta G_u - \rho RT \ln\left[\frac{x-\zeta+1}{x}\right] \tag{64}$$

Here σ_u is the interfacial free energy characteristic of the lateral surface of the nucleus and σ_{en} is the interfacial free energy for the nucleus. We must carefully distinguish between σ_{en} and the other interfacial free energies that have been introduced earlier. When the complete free energy function is treated the development that follows can also include heterogeneous polymers, polymer–diluent mixtures and copolymers. We restrict consideration here to homogeneous pure systems.

The surface described by equation (64) does not have a maximum or minimum but does contain a saddle point. The coordinates of the saddle point are defined by $\left(\frac{\partial\Delta G}{\partial\rho}\right)_\zeta = 0$ and $\left(\frac{\partial\Delta G}{\partial\rho}\right)_\rho = 0$. These coordinates define the dimensions ζ^* and ρ^* of the critical size nucleus. They are given by

$$\rho^{*1/2} = \frac{2\pi^{1/2}\sigma_u}{\Delta G_u - \dfrac{RT}{x} - \dfrac{RT}{(x-\zeta^*+1)}} \tag{65}$$

and

$$\frac{\zeta^*}{2}\left[\Delta G_u - \frac{RT}{x} + \frac{RT}{x-\zeta^*+1}\right] = 2\sigma_{en} - RT \ln\left(\frac{x-\zeta^*+1}{x}\right) \tag{66}$$

It follows that

$$\Delta G^* = \pi^{\frac{1}{2}}\zeta^*\rho^{*\frac{1}{2}}\sigma_u \tag{67}$$

Equation (67) is identical in form with the equation obtained for a nucleus composed of small molecules arranged in the same geometric array. The values of ζ^* and ρ^* are molecular weight dependent with a particularly strong dependence in the low molecular weight range. The implications of these results have been discussed in detail elsewhere.[89, 125, 126] However, if x is very large and $\zeta^* \ll x$ equations (65) and (66) reduce to that for an infinite molecular weight chain. These expressions are identical to those for a critical size nucleus composed of low molecular weight species. Under these conditions

$$\zeta^* = \frac{4\sigma_{en}}{\Delta G_u}; \quad \rho^* = \frac{4\pi\sigma_u^2}{\Delta G_u^2} \tag{68}$$

and

$$\Delta G^* = \frac{8\pi\sigma_u^2\sigma_e}{\Delta G_u^2} \tag{69}$$

The source of the strong negative temperature coefficient in the crystallization rate, in the vicinity of T_m, becomes apparent when we recall that ΔG_u is proportional to ΔT. When equation (69) is inserted into the expression for the steady state nucleation rate, equation (63), the strong dependence on temperature results. The critical dimensions ρ^* and ζ^* will also be very dependent on the undercooling at which crystallization takes place. Since lateral growth will be rapid, vestiges of the original ρ^* are usually lost as the lamellae develop in directions normal to the chain axis. Growth or thickening in the chain direction has now been clearly demonstrated.[89–92] Depending on temperature and molecular weight, it is usually a relatively slow process.[89,90] Although the crystallite thicknesses cannot be directly identified with ζ^* they are related to nucleation requirements.

At large undercoolings this strong temperature dependence of the rate is lost. At temperatures well removed from the melting temperature the activation energy term becomes important and a maximum in the crystallization rate results.[262,264] As the glass temperature is approached the crystallization rate tends toward zero.[262,263]

Equations (67) and (68) have been developed for a three dimensional nucleus which is formed homogeneously by statistical fluctuations. Although homogeneous nucleation will admittedly be a difficult event, the analysis given above serves as an ideal model or base for considering real systems. It should be clear that the key results, the inverse dependence of ΔG^* on ΔG_u^2, will not depend on the specific geometry of the nucleus, the chain arrangements within the nucleus and for high molecular weights the fact that polymers are involved at all. Nuclei formed heterogeneously, for both small molecules and polymers alike, have the same temperature dependence as in the homogeneous case.[265,266] However, depending in great detail on the structure of the heterogeneity involved, the critical free energy for nucleus formation can be reduced.[266]

Another basic nucleation process is the type that has been described by Gibbs.[267] In this model the nucleus is deposited coherently on an already existing crystal face. This type nucleus has been assumed to be involved in lamellar growth of crystallites of long chain molecules. Quite generally, utilizing equation (64), the free energy of forming such a nucleus can be expressed as

$$\Delta G = 2\zeta\sigma_u + 2\rho\sigma_{en} - \zeta\rho\Delta G_u + \frac{RT}{x}\zeta\rho - \rho RT\ln\left(\frac{x-\zeta+1}{x}\right) \tag{70}$$

The first three terms in equation (70) represent the expression for the infinite chain, that is the same for small molecule systems. The last two terms represent the contribution of the finite chain. The coordinates of the saddle point of the surface described by equation (70) are given by[126]

$$\rho^* = \frac{2\sigma_u}{\Delta G_u - \dfrac{RT}{x} - \dfrac{RT}{x-\zeta^*+1}} \tag{71}$$

$$\zeta^* = \frac{2\sigma_{en} - RT\ln\left(\dfrac{x-\zeta^*+1}{x}\right)}{\Delta G_u - \dfrac{RT}{x}} \tag{72}$$

$$\Delta G^* = \frac{2\sigma_u\left[2\sigma_{en} - RT\ln\left(\dfrac{x-\zeta^*+1}{x}\right)\right]}{\Delta G_u - \dfrac{RT}{x}} = 2\sigma_u\zeta^* \tag{73}$$

For high molecular weights these expressions reduce to

$$\Delta G^* = \frac{4\sigma_{en}\sigma_u}{\Delta G_u}; \quad \zeta^* = \frac{2\sigma_{en}}{\Delta G_u}; \quad \rho^* = \frac{2\sigma_u}{\Delta G_u} \tag{74}$$

which are also the classical results for a Gibbs type nucleus made up of small molecules. These results are independent of the shape of the nucleus and the chain disposition within it. As equation (74) attests it is even independent of the molecular system involved. ΔG^*, ρ^* and ζ^* again display very strong temperature dependence which reflects itself in crystallization rates and in crystallite dimensions.

The Gibbs type nucleus has some unique features which merit special concern. A three-dimensional nucleus can gain stability, $\Delta G < 0$, and develop into a mature crystallite with only lateral growth taking place. Growth in the chain direction is not necessary in this case in order to develop a thermodynamically stable crystallite.[262] In contrast a Gibbs type nucleus cannot become stable without growth in the chain direction, irrespective of the amount of lateral growth that takes place after nucleation. Put another way, this nucleus is unstable at temperatures that are infinitesimally above the crystallization temperature. This problem can be overcome quite naturally by allowing for growth along the chain direction. Stability would then be ensured. However, when regularly folded chains are assumed to comprise the nucleus the problem cannot be resolved this simply. This difficulty has been circumvented for this model by allowing for fluctuations in the thickness of the nucleus.[33] The increased average thickness that results from this analysis only marginally enhances the stability of the nucleus and cannot account for the relationship between the melting temperature and crystallization temperature.[268]

11.5 DETERMINATION OF EQUILIBRIUM MELTING TEMPERATURE

Except for very low molecular weight chains, where extended crystals are formed, ζ will invariably be very much less than ζ_e. Thus, a direct determination of the equilibrium melting temperature for chains of moderate to high molecular weight poses almost insurmountable difficulties.* Except for

---

* It was thought at one time that the crystallization of linear polyethylene at high pressures and high temperatures lead to extended chain crystals even for high molecular weights.[271] It was subsequently found that this was not the case. The crystallite thicknesses were of the same order as when crystallized under atmospheric pressure for long times at low undercoolings.[128,129] Hence direct measurements with such samples do not lead to the value for the equilibrium melting temperature.

some specialized techniques which are not generally applicable[269,270] some type of extrapolative method has to be used. The nature of the problem is exemplified by Tables 4 and 5 which give respectively the directly determined melting temperatures of linear polyethylenes and those determined by different extrapolation methods. For bulk crystallization at atmospheric pressure the highest directly observed value for $T_m = 139.0\,°C$ while other crystallization procedures have lead to reports of higher values of varying degree. In Table 5 the two lowest extrapolated values for T_m° can be discarded since they are significantly lower than directly observed values in Table 4. The remaining extrapolated values fall in the range of 141–146 °C. Although this difference may appear to be small it is significant from both a theoretical point of view[112] and in the analysis of crystallization kinetics. In principle the determination of the melting temperatures as a function of crystallite thickness should lead to an extrapolated value for the equilibrium melting temperature. The difficulties that are associated with this method have already been discussed.

Table 4   Directly Determined Melting Temperatures for Linear Polyethylene

Sample	Crystallization Method	T_m (°C)	Ref.
$M = 4.9 \times 10^5$	131.3 °C, atmospheric	138.5[a]	106
$M = 5.7 \times 10^5$	130.0 °C, atmospheric	139.0[a]	9
$M = 4.7 \times 10^5$	130.0 °C, atmospheric	138.5[a]	274
$M = 6.6 \times 10^5$	132.1 °C, atmospheric	138.6[a]	293
Unfractionated	Solution stirring	146.0[a]	270
Unfractionated	High pressure	145.0[b]	294
Polymethylene	High pressure	141.5	271
Unfractionated	High pressure	139.2[c]	252

[a] Dilatometry. [b] DTA extrapolated to zero heating rate. [c] Rapid heating on hot stage.

Table 5   Extrapolated T_m° for Linear Polyethylene

T_m° (°C)	Method	Ref.
145.1 ± 1	Extrapolation of *n*-alkanes	111
146.0	Extrapolation of *n*-alkanes and polyethylene	120
145.5	$T_m - T_c$ analysis	295
146.0	$T_m - T_c$ analysis	274
145.8	$T_m - 1/L$ (solution crystals)	296
141.6	Specific heat of *n*-alkanes	297
142.0	$T_m - 1/L$ (bulk crystallized)	298
143.5	$T_m - 1/L$ (bulk crystallized)	252
141.0	$T_m - 1/L$ (bulk and solution)	253
138.9	$T_m - 1/L$ (bulk crystallized)	299
138.1	$T_m - 1/L$ (solution crystals)	300

Another method, which has found widespread use, is determining the melting temperature T_m^* as a function of the crystallization temperature T_c. Analysis of these data can in principle lead to a value for T_m°. The essence of this method lies in relating the crystallite thickness to the critical dimensions of a nucleus and assuming that no structural changes occur during fusion. A plot of T_m^* against T_c is made. An extrapolation to the intersection with straight line $T_m^* = T_c$ gives the equilibrium melting temperature. In many ways this method is deceivingly simple and should be analyzed in some detail. The analysis given here is for homopolymer fractions but can be extended in a straightforward way to copolymers and polymer–diluent mixtures.[272] The melting temperature T_m^* for a crystallite of thickness ζ is given by equation (60). The critical nucleus size in the chain direction ζ^* is given by equation (66). When the usual approximation, $\Delta G_u(T_c) = \Delta H_u(T_m^\circ - T_c)/T_m^\circ$, is made then equations (60) and (66) can be combined to give an expression for T_m^* in terms of T_c, ζ, ζ^*, σ_{en} and σ_{ec}.* When x becomes large then[273,274]

$$T_m^* = \frac{a}{2n} T_c + T_m^\circ \left(1 - \frac{a}{2n}\right) \tag{75}$$

* In this development we have assumed that the lateral dimensions are sufficiently large that the corresponding interfacial free energies can be neglected. This assumption is made to simplify the analysis that is given here. It is possible to consider the contributions of all the surfaces involved.[273] Although in using equation (75) we have tacitly assumed a three-dimensional nucleus the problem can also be formulated for the Gibbs type nucleus. For this case ζ must increase beyond ζ^* for thermodynamic stability to be achieved.

Here $a = \sigma_{ec}/\sigma_{en}$ and $n = \zeta/\zeta^*$. Thus it is not necessary to assume the identity of either the interfacial energies between the mature crystallite and nuclei or their thicknesses. For the very special case for x large and where simultaneously $\zeta = \zeta^*$ and $\sigma_{en} = \sigma_{ec}$ equation (75) reduces to

$$T_m^* = \tfrac{1}{2}(T_m^{\circ} + T_c) \qquad (76)$$

Equation (76) is the expression originally given by Hoffman and Weeks[275] who made the explicit assumption that the lamellar crystallites were made up of regularly folded chains. As the general derivation indicates, these are restrictive conditions. In fact lamellar crystallites are not even required for equation (76) to be valid. Based on either equations (75) or (76) a plot of T_m^* against T_c should be linear and the intersection with the 45° line gives T_m°. However, only when equation (76) is applicable will the slope of the straight line be 0.5.

Before this analysis is compared with experimental results it is instructive to examine the influence of molecular weight in more detail. An example is given in Figure 24 of the expected dependence of T_m^* on T_c for different molecular weights. The curves illustrated were calculated according to equations (60) and (66) using linear polyethylene as a model. For illustrative purposes, the simplifying assumption was made that $\sigma_{en} = \sigma_{ec}$ and $\zeta = \zeta^*$. These assumed equalities do not sensibly alter the general conclusion drawn from this model calculation. The point at which the curves for each value of x intersect the straight line $T_m^* = T_c$ gives the equilibrium melting temperature for that particular chain length. The character of the curves in Figure 24 are quite different from one another depending on molecular weight. At high molecular weights the T_m^* *vs.* T_c plot is linear with a slope of 0.5. At the very low molecular weights, $x \leq 500$, the plots are only slightly curved and can be represented by a straight line with a slope close to zero. At the intermediate molecular weights the curvature is quite marked. In this range a linear extrapolation of data that is accessible for analysis could lead to major errors in determining the equilibrium melting temperature.

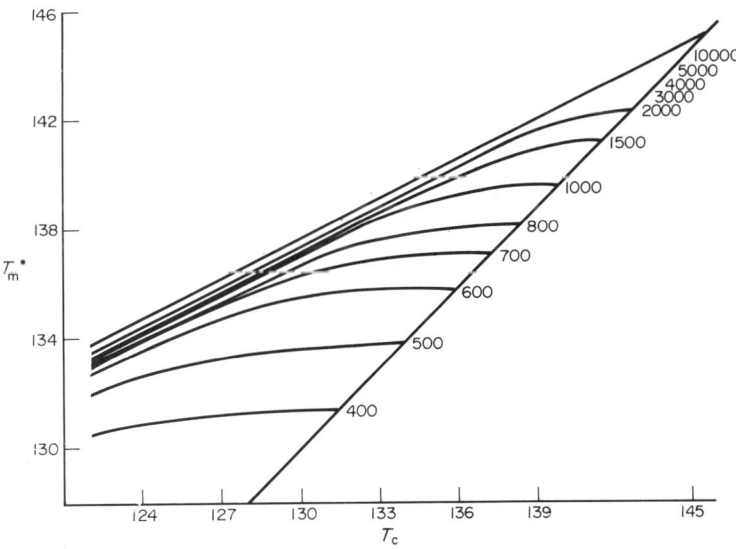

Figure 24 Theoretical plot of T_m^* against T_c for indicated chain lengths of linear polyethylene. For this example $\zeta/\zeta^* = 1$ and $p_e = \sigma_{ec} = \sigma_{en} = 4600$ cal mol^{-1} (19 kJ mol^{-1})[274]

When actual experimental data is analyzed further complications can develop. For low molecular weight polyethylene and poly(ethylene oxide) T_m^* is essentially independent of T_c[89,119] as would be expected from the curves of Figure 24. Extended chain crystals are formed and the extrapolation is straightforward. Unusual T_m^* *vs.* T_c relations are found for polyethylenes of 3000–5000 molecular weight and poly(ethylene oxide) in a similar range. Here there is a well-defined temperature at which there is a jump in the crystallite thickness accompanied by a discontinuity in T_m^*.[89] Above this temperature extended chain crystals are formed. Hence it is important that experiments be conducted in this latter temperature range to insure a consistent extrapolation.

Another, and more serious, complication develops for high molecular weights as is evidenced by the plot in Figure 25 for linear polyethylene fractions. The problem involves the level of crystallinity

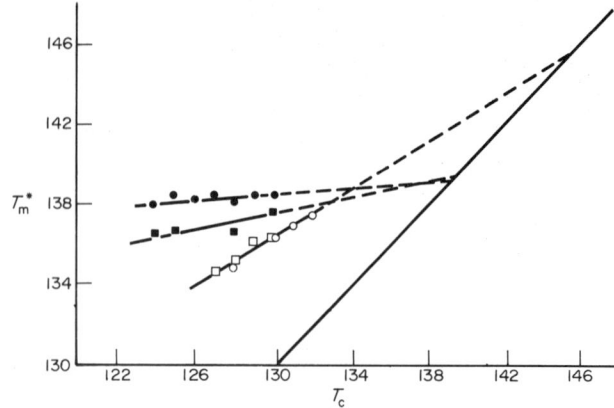

Figure 25 Plot of T_m^* against T_c for two high molecular weight fractions of linear polyethylene. $M = 1.0 \times 10^6$: low level of crystallinity □; high level of crystallinity ■. $M = 4.7 \times 10^5$: low level of crystallinity ○; high level of crystallinity ●[274]

in the samples for which T_m^* is determined. The low level of crystallinity data yield a linear relation, with slope 0.6 and an extrapolated T_m° of 146 °C in accord with theoretical expectations.[111,112] However, at higher levels of crystallinity the slope is reduced to the range 0.1–0.2, and the extrapolated melting temperature is reduced to 139 °C. A clearly significant difference in melting temperatures is observed, which is of theoretical and practical importance. These results are not limited to polyethylene. Beech and Booth,[256] for a high molecular weight fraction of poly(ethylene oxide), found that samples with low levels of crystallinity gave results which extrapolated to a significantly higher value for T_m°.

The reasons for the differences that the observed reside in the underlying restraints to the analyses. It is required that the ratio of crystallite to nucleus thickness be small, definite and maintained at each of the crystallization temperatures. There will then be a definite and systematic change in thickness with T_c, which is the underlying premise of the theory. These conditions are best met by the lowest level of crystallinity that can be determined experimentally. However, as crystallization proceeds the level of crystallinity increases and concomitantly the crystallites thicken.[89,90,92] In these circumstances there is no longer a major discrimination of crystallite thickness with crystallization temperature. Therefore, although for a given T_c, T_m^* increases with crystallinity there is no significant change with crystallization temperature. Hence the slope of the T_m^* vs. T_c plot is close to zero and the extrapolated melting temperature is less than that deduced from low crystallinity measurements. In order for this extrapolative procedure to be effective it must be limited to low levels of crystallinity. With proper concern and caution reliable values of T_m° can be obtained by this method. The underlying theory does not depend on the crystallite morphology or the nature of the interfacial structure.

11.6 EPILOGUE

We have described some of the basic thermodynamic and structural features which characterize the crystalline state of flexible chain molecules. This is a necessary foundation if one is to develop an understanding of the properties of such systems on a molecular level. A great deal of effort has been expended in the direction of understanding properties. Space does not allow for a detailed discussion of these works here. However, a brief summary of the progress in this area is appropriate. In this endeavor it is necessary to identify and quantitatively describe the principal independent structural variables which describe the crystalline state. These parameters, either individually or in combination, govern a given property. The important independent variables that have been identified are: the degree of crystallinity; the structure and topology of the residual non-crystalline region; the crystallite thickness distribution; the structure and relative amount of the interfacial region; the structure of the lamellar crystallites; and the supermolecular structure.[58] These factors depend in a very systematic manner on molecular weight, molecular weight distribution, the structural regularity of the chain, and on the crystallization conditions.[58] By control of the elements of molecular constitution these structural features can be varied over very wide limits and their influence on properties assessed. A very general, but important, conclusion that has been reached is that a very wide range in values for a given property can be achieved. Utilizing this strategy the factors which control a variety of thermodynamic properties[60-63], IR absorption,[93] Raman internal modes,[64] [13]C relaxation times,[276] dynamic mechanical properties,[277,278] force–length curves and ultimate properties,[279] to cite but a few examples, have been established.

The structural and morphological features that evolve are the result of the crystallization mechanisms which can, in principle, be elucidated by crystallization kinetic studies. Despite the complex structure of the melt, where the chains are intertwined and entangled, polymer crystallization can be formally treated as a concurrent nucleation and growth process. The classical analysis of Göler-Sachs[280] and Avrami[281] apply. The only modification necessary for polymers is to account for the fact that the transformation is rarely if ever complete. The extent of the transformation, or level of crystallinity that is attained, is very dependent on molecular weight[58,63,83] and chain structure.

The importance of nucleation processes in polymer crystallization has already been emphasized. Equation (63) for the steady state nucleation rate can be used for homogeneous and heterogeneous three-dimensional nucleation, Gibbs type nucleation as well as other nucleation processes that can be conceived. All that is required is the insertion of the appropriate value for ΔG^*. In an analysis of the nucleation of small molecules on an already formed substrate (nucleation followed by growth) Hillig[282] analyzed two extreme situations. These were named Regimes I and II when applied to polymers.[41] In Regime I, Hillig allowed the growth step to sweep completely across the face of the crystal and a pause occurs before the next layer is nucleated. In other words the rate of lateral growth of a nucleus across the face of the crystal is very much faster than the nucleation rate. In Regime II new growth steps are allowed to nucleate before the previous layer has filled the substrate. Detailed analysis of this problem has shown that the temperature coefficients of the crystal growth rates in Regimes I and II will differ by a factor of two.[41,283,284]

The demarcation between Regimes I and II, with the required change in temperature coefficients, has been experimentally observed for several different polymers by studies of either the overall crystallization rate or the spherulite growth rate.[41,63,285,286] Either of these techniques is adequate to establish the demarcation. It is not necessary that direct visualization be made of either spherulite or lamellar growth.[63] The temperature of demarcation is the same irrespective of the kinetic method used.[285] This is a particularly advantageous situation since spherulites are usually only formed in a restricted molecular weight range.[287-289] These results also point out that there is no special theoretical significance in studying spherulite growth rates.

It was originally thought that this changes in the nucleation/growth relations reflected changes in the supermolecular structure that took place above and below the demarcation temperature of Regimes I and II.[41,287] However, more detailed study has shown that there is no relation between the temperature at which there is a change from Regime I and II kinetics and changes in the supermolecular structure.[285] Although the change from Regime I to II is of kinetic and mechanistic importance its reflection on morphology and properties is still to be clarified.

Since this analysis is very formal it obviously does not require any specification of the structure of the nucleus. It clearly is not concerned with regularly folded chains or adjacent reentry.[41,298] No assumptions, direct or indirect, are required with respect to the interfacial structure. Neither does reptation theory have any direct bearing on this phenomenon and classification of Regimes.

When the undercooling is sufficiently large multiple nucleation events will take place on the same layer and growth will take place on both sides of the nuclei. The space, or niche, between nuclei will become small and comparable to the chain width. Lateral growth of nuclei will thus be retarded. This situation, assumed to occur at a discrete undercooling, has been termed Regime III.[290,291] The temperature coefficient of crystal growth turns out to be the same as for Regime I. The situation where the nucleation rate is much larger than the later growth rate had been analyzed previously.[292] This case corresponds to the deposition of stable nuclei on the substrate with either minimal or no further lateral growth. The temperature dependence of this process is the same as Regime I. Regime III is physically very similar to the case just described. Therefore, it is not surprising that they have similar temperature coefficients.

The formal of phenomenological development of crystallization kinetics in polymers has reached a level of understanding and sophistication that is very similar to that achieved for low molecular substances. A detailed connection between the kinetics and crystallization mechanisms with the independent structural variables that describe the semicrystallize state still remains to be made. This endeavor presents a major challenge to the further understanding of semicrystalline polymers.

ACKNOWLEDGEMENT

The support of this review by the National Science Foundation Polymer Program Grant DMR 83-14679 to Florida State University is gratefully acknowledged.

11.7 REFERENCES

1. L. Mandelkern, 'Crystallization of Polymers', McGraw-Hill, New York, 1964, p. 20ff.
2. L. Mandelkern, *Chem. Rev.*, 1956, **56**, 903.
3. L. Mandelkern, *J. Macromol. Sci., Chem.*, 1981, **15**, 1211.
4. P. J. Flory, 'Principles of Polymer Chemistry', Cornell University Press, 1953, p. 22ff.
5. L. A. Wood and N. Bekkedahl, *J. Appl. Phys.*, 1946, **17**, 362.
6. L. A. Wood and N. Bekkedahl, *J. Res. Nat. Bur. Stand.*, 1946, **36**, 489.
7. L. Mandelkern, *Faraday Discuss Chem. Soc.*, 1979, **68**, 310.
8. Ref. 1, p. 215ff.
9. L. Mandelkern, J. M. Price, M. Gopalan and J. G. Fatou, *J. Polym. Sci., Part A-2*, 1966, **4**, 385.
10. R. Eppe, E. W. Fischer and H. A. Stuart, *J. Polym. Sci.*, 1959, **34**, 721.
11. D. C. Bassett and A. H. Hodge, *Proc. R. Soc. London, Ser. A*, 1981, **377**, 25.
12. D. C. Bassett, A. H. Hodge and R. H. Olley, *Faraday Disc. Chem. Soc.*, 1979, **68**, 218.
13. G. Strobl, M. Schneider and I. G. Voigt-Martin, *J. Polym. Sci., Polym. Phys. Ed.*, 1980, **18**, 1368.
14. I. G. Voigt-Martin, E. W. Fischer and L. Mandelkern, *J. Polym. Sci., Polym. Phys. Ed.*, 1980, **18**, 2347.
15. G. M. Stack, L. Mandelkern and I. G. Voigt-Martin, *Polym. Bull. (Berlin)*, 1982, **8**, 421.
16. I. G. Voigt-Martin and L. Mandelkern, *J. Polym. Sci., Polym. Phys. Ed.*, 1981, **19**, 1769.
17. I. G. Voigt-Martin and L. Mandelkern, *J. Polym. Sci., Polym. Phys. Ed.*, 1984, **22**, 1901.
18. I. G. Voigt-Martin, R. Alamo and L. Mandelkern, *J. Polym. Sci., Polym. Phys. Ed.*, 1986, **24**, 1283.
19. R. Jaccodine, *Nature (London)*, 1955, **176**, 301.
20. P. H. Till, *J. Polym. Sci.*, 1957, **24**, 301.
21. A. Keller, *Philos. Mag.*, 1957, **2**, 1171.
22. E. W. Fischer, *Z. Naturforsch., Teil A*, 1957, **12**, 753.
23. P. J. Flory, *J. Am. Chem. Soc.*, 1962, **84**, 2857.
24. J. B. Jackson, P. J. Flory and R. Chiang, *Trans. Faraday Soc.*, 1963, **59**, 1906.
25. L. Mandelkern, *Polym. Eng. Sci.*, 1967, **7**, 232.
26. L. Mandelkern, *Polym. Eng. Sci.*, 1969, **9**, 255.
27. L. Mandelkern, *Acc. Chem. Res.*, 1976, **9**, 81.
28. P. J. Flory, *Faraday Discuss. Chem. Soc.*, 1979, **68**, 489.
29. A. Keller, *Makromol. Chem.*, 1959, **34**, 1; *Polymer*, 1962, **3**, 393.
30. P. H. Lindenmeyer, *Science*, 1956, **147**, 1256.
31. J. D. Geil, 'Polymer Single Crystals', Wiley–Interscience, New York, 1963.
32. J. D. Hoffman, *SPE Trans.*, 1964, **4**, 315.
33. J. D. Hoffman and J. K. Lauritzen, *J. Res. Nat. Bur. Stand., Sect. A.*, 1960, **64**, 73; 1961, **65**, 297.
34. F. P. Price, *J. Chem. Phys.*, 1964, **35**, 1884.
35. L. Mandelkern, F. A. Quinn, Jr. and P. J. Flory, *J. Appl. Phys.*, 1955, **26**, 443.
36. L. Mandelkern, *J. Appl. Phys.*, 1955, **26**, 443.
37. L. Mandelkern, *Polymer*, 1964, **5**, 637.
38. J. G. Fatou, E. Riande and R. Garcia Valdecasas, *J. Polym. Sci., Polym. Phys. Ed.*, 1975, **13**, 2103.
39. C. Devoy, L. Mandelkern and L. Bourland, *J. Polym. Sci., Part A-2*, 1970, **8**, 869.
40. P. J. Flory and A. D. McIntyre, *J. Polym. Sci.*, 1955, **18**, 592.
41. J. D. Hoffman, L. J. Frolen, G. S. Ross and J. J. Lauritzen, *J. Res. Nat. Bur. Stand.*, 1975, **A79**, 671.
42. J. H. Magill, in 'Treatise on Materials Science and Technology', ed. J. M. Schultz, Academic Press, New York, 1977, vol. 10A, p. 268ff.
43. D. J. Blundell and A. Keller, *J. Macromol. Sci., Phys.*, 1968, **B2**, 301.
44. A. Keller and E. Pedemonte, *J. Cryst. Growth*, 1973, **18**, 111.
45. (a) M. Cooper and R. St. J. Manley, *Macromolecules*, 1975, **8**, 219; (b) W. M. Leung, R. St. J. Manley and A. R. Panaras, *Macromolecules*, 1985, **18**, 760.
46. D. Turnbull and J. C. Fischer, *J. Chem. Phys.*, 1949, **17**, 71.
47. D. Turnbull, in 'Solid State Physics', Academic Press, New York, 1956, vol. 3.
48. F. P. Price, in 'Nucleation', ed. A. C. Zettlemoyer, Marcel Dekker, New York, 1969, p. 401.
49. P. D. Calvert and D. R. Uhlmann, *J. Appl. Phys.*, 1972, **43**, 944.
50. H. G. Zachmann, *Pure Appl. Chem.*, 1974, **38**, 79.
51. I. C. Sanchez, *J. Macromol. Sci., Rev. Macromol. Chem.*, 1974, **C10**, 1.
52. L. Mandelkern, in 'Progress in Polymer Science', ed. A. D. Jenkins, Pergamon Press, Oxford, 1970, vol. 2.
53. L. Mandelkern, in 'Annual Review of Materials Science', ed. R. A. Huggins, Anuual Reviews, Palo Alto, CF, 1976, vol. 6, p. 119.
54. P. J. Flory, D. Y. Yoon and K. A. Dill, *Macromolecules*, 1984, **17**, 862.
55. D. Y. Yoon and P. J. Flory, *Macromolecules*, 1984, **17**, 868.
56. L. Mandelkern, *J. Polym. Sci., Polym. Symp.*, 1975, **50**, 457.
57. L. Mandelkern, in 'Characterization of Materials in Research: Ceramics and Polymers', Syracuse University Press, 1975.
58. L. Mandelkern, *Polym. J.*, 1985, **17**, 337.
59. P. J. Flory, in 'Structural Order in Polymers', ed. F. Ciardelli and P. Gusti, Pergamon Press, Oxford, 1981, p. 3.
60. J. G. Fatou and L. Mandelkern, *J. Phys. Chem.*, 1965, **69**, 417.

61. L. Mandelkern, A. L. Allou, Jr. and M. Gopalan, *J. Phys. Chem.*, 1968, **72**, 309.
62. L. Mandelkern, *J. Phys. Chem.*, 1971, **75**, 3909.
63. E. Ergoz, J. G. Fatou and L. Mandelkern, *Macromolecules*, 1972, **5**, 147.
64. M. Glotin and L. Mandelkern, *Colloid Polym. Sci.*, 1982, **260**, 182.
65. B. Crist, W. Graessley and G. D. Wignall, *Polymer*, 1982, **23**, 1561.
66. E. W. Fischer, M. Stamm, M. Dettenmaier and P. Herschenroeder, *Polym. Prepr., Am. Chem. Soc., Div. Polym. Chem.*, 1979, 20 (**1**), 219.
67. G. Lieser, E. W. Fischer and K. Ibel, *J. Polym. Sci., Polym. Lett. Ed.*, 1975, **13**, 39.
68. J. Schelten, D. G. H. Ballard, G. D. Wignall, G. Longman and W. Schmatz, *Polymer*, 1976, **27**, 751.
69. D. G. H. Ballard, P. Cheshire, G. W. Longman and J. Scheltern, *Polymer*, 1978, **19**, 379.
70. J. M. Guenet, *Polymer*, 1981, **22**, 313.
71. D. Y. Yoon and P. J. Flory, *Polym. Bull. (Berlin)*, 1981, **4**, 692.
72. P. J. Flory, *Pure Appl. Chem.*, 1984, **56**, 305.
73. D. Y. Yoon and P. J. Flory, *Faraday Discuss Chem. Soc.*, 1979, **68**, 288.
74. J. D. Hoffman, C. M. Guttman and E. A. DiMarzio, *Faraday Discuss. Chem. Soc.*, 1979, **68**, 197.
75. D. M. Sadler and R. Harris, *J. Polym. Sci., Polym. Phys. Ed.*, 1982, **20**, 561.
76. E. W. Fischer, K. Hahn, J. Keigler and R. Bom; *J. Polym. Sci., Polym. Phys. Ed.*, 1984, **22**, 1491.
77. E. W. Fischer, *Polym. J.*, 1985, **17**, 307.
78. M. Stamm, *J. Polym. Sci., Polym. Phys. Ed.*, 1982, **20**, 235.
79. G. D. Wignall, L. Mandelkern, C. Edwards and M. Glotin, *J. Polym. Sci., Polym. Phys. Ed.*, 1982, **20**, 245.
80. R. Kitamaru, F. Horii and S. H. Hyon; *J. Polym. Sci., Polym. Phys. Ed.*, 1977, **15**, 821.
81. R. Kitamaru, F. Horii and K. Marayama, *Macromolecules*, 1986, **19**, 636.
82. C. G. Vonk and A. P. Pijpers, *J. Polym. Sci., Polym. Phys. Ed.*, 1985, **23**, 2517.
83. R. Alamo, R. Domszy and L. Mandelkern, *J. Phys. Chem.*, 1984, **88**, 6587.
84. P. J. Flory, *J. Chem. Phys.*, 1949, **17**, 223.
85. G. R. Strobl and W. J. Hagedorn, *J. Polym. Sci., Polym. Phys. Ed.*, 1978, **16**, 1181.
86. J. M. Schultz, W. H. Robinson and G. M. Pound, *J. Polym. Sci., Part A-2*, 1967, **5**, 511.
87. (a) G. M. Stack, L. Mandelkern and I. G. Voigt-Martın, *Macromolecules*, 1984, **17**, 321; (b) P. J. Flory, 'Statistical Mechanics of Chain Molecules' Interscience, New York, 1969, p. 165ff; (c) J. A. Marqusee and K. A. Dill, *Macromolecules*, 1986, **19**, 2420.
88. Ref. 1, p. 1ff.
89. G. M. Stack, L. Mandelkern and I. G. Voigt-Martin, *Polym. Bull. (Berlin)*, 1982, **8**, 421.
90. G. M. Stack, Ph.D. Dissertation, Florida State University, August 1983.
91. J. Dlugosz. C. Fraser, D. Grubb. A. Keller. J. Odell and P. Goggin, *Polymer*, 1976. **17**, 471.
92. P. J. Barham, R. A. Chivers, D. A. Jarvis, J. Martinez-Salajar and A. Keller, *J. Polym. Sci., Polym. Lett. Ed.*, 1981, **19**, 539.
93. T. Okada and L. Mandelkern, *J. Polym. Sci., Part A-2*, 1967, **5**, 239.
94. J. F. Jackson and L. Mandelkern, in 'Analytical Calorimetry', ed. Johnson & Porter, Plenum Press, New York, 1968, vol. 1, p. 1.
95. R. Komoroski, J. Maxfield, F. Sakaguchi and L. Mandelkern, *Macromolecules*, 1977, **10**, 550.
96. R. Komoroski, J. Maxfield and L. Mandelkern, *Macromolecules*, 1977, **10**, 545.
97. R. G. Snyder, N. E. Schlotter, R. Alamo and L. Mandelkern, *Macromolecules*, 1986, **19**, 621.
98. L. Mandelkern, R. Alamo, W. L. Mattice and R. G. Snyder, *Macromolecules*, 1986, **19**, 2404.
99. P. J. Flory and D. Y. Yoon, *Nature (London)*, 1978, **272**, 226.
100. R. C. Lacher, L. Bryant, L. N. Howard and D. W. Sumners, *Macromolecules*, 1986, **19**, 2639.
101. R. Popli and L. Mandelkern, *J. Polym. Sci., Polym. Phys. Ed.*, 1988, in press.
102. S. H. Kim and L. Mandelkern, unpublished results.
103. F. Smith, E. Chan and L. Mandelkern, unpublished results.
104. J. E. Mayer and S. F. Streeter, *J. Chem. Phys.*, 1939, **7**, 1019.
105. L. Mandelkern, *Rubber Chem. Technol.*, 1949, **32**, 1392.
106. R. Chiang and P. J. Flory, *J. Am. Chem. Soc.*, 1961, **83**, 2857.
107. P. Calvert, *J. Polym. Sci., Polym. Phys. Ed.*, 1979, **17**, 1341.
108. L. Mandelkern, M. Tryon and F. A. Quinn, Jr., *J. Polym. Sci.*, 1956, **19**, 77.
109. Ref. 1, p. 47.
110. P. J. Flory, 'Principles of Polymer Chemistry', Cornell University Press, 1953, p. 508ff.
111. P. J. Flory and A. Vrij, *J. Am. Chem. Soc.*, 1963, **85**, 3548.
112. L. Mandelkern and G. M. Stack, *Macromolecules*, 1984, **17**, 871.
113. M. G. Broadhurst, *J. Res. Nat. Bur. Stand., Sect. A*, 1962, **66**, 241.
114. G. Ungar, J. Stejny, A. Keller, I. Bidd and M. C. Whiting, *Science*, 1985, **229**, 386.
115. K. S. Lee and G. Wegner, *Makromol. Chem., Rapid Commun.*, 1985, **6**, 203.
116. K. S. Lee and G. Wegner, to be published.
117. K. Takamijawa, Y. Ogawa and T. Oyama, *Polym. J.*, 1982, **14**, 441.
118. H. W. Starkweather, Jr., *Macromolecules*, 1986, **19**, 1131.
119. C. P. Buckley and A. J. Kovacs, *Prog. Colloid Polym. Sci.*, 1975, **58**, 44.
120. N. J. Hay, *J. Polym. Sci., Polym. Chem. Ed.*, 1976, **14**, 2845.
121. O. V. Romankevich and S. Ya Frenkel, *Vysokomol. Soedin., Ser. A*, 1978, **20**, 2417.
122. F. A. Quinn, Jr. and L. Mandelkern, *J. Am. Chem. Soc.*, 1958, **80**, 3178.
123. L. Mandelkern, *Rubber Chem. Technol.*, 1959, **32**, 1392.
124. L. Mandelkern, G. M. Stack and P. J. Mathieu, *Anal. Calorim.*, 1984, **5**, 223.
125. J. G. Fatou, C. Howard and L. Mandelkern, *J. Phys. Chem.*, 1964, **68**, 3386.
126. L. Mandelkern, J. G. Fatou and C. Howard, *J. Phys. Chem.*, 1965, **69**, 956.
127. D. R. Beech, C. Booth, C. J. Pickles, R. R. Sharpe and J. R. S. Waring, *Polymer*, 1972, **13**, 246.
128. L. Mandelkern, M. R. Gopalan and J. F. Jackson, *J. Polym. Sci., Part B*, 1967, **5**, 1.

129. J. F. Rabolt and C. H. Wang, *Macromolecules*, 1983, **16**, 1698.
130. L. Mandelkern, *Faraday Discuss Chem. Soc.*, 1979, **68**, 414.
131. R. H. Glaser and L. Mandelkern, *J. Polym. Sci., Polym. Phys. Ed.*, 1988, **26**, 221.
132. L. Mandelkern, E. Chan and A. Prasad, unpublished observations.
133. P. J. Flory, L. Mandelkern and H. K. Hall, *J. Am. Chem. Soc.*, 1951, **73**, 2532.
134. R. D. Evans, H. R. Mighton and P. J. Flory, *J. Am. Chem. Soc.*, 1950, **72**, 2018.
135. D. R. Beech, C. Booth, D. V. Dodgson, R. R. Sharpe and J. R. S. Waring, *Polymer*, 1972, **13**, 73.
136. A. A. Schaerer, C. J. Busso, A. F. Smith and L. Skinner, *J. Am. Chem. Soc.*, 1955, **77**, 2017.
137. W. M. Mazee, *Recl. Trav. Chim. Pays-Bas*, 1948, **67**, 197.
138. N. J. Hay, *J. Polym. Sci., Polym. Lett. Ed.*, 1970, **8**, 395.
139. Ref. 1, p. 146ff.
140. P. J. Flory, *Trans. Faraday Soc.*, 1955, **51**, 848.
141. B. D. Coleman, *J. Polym. Sci.*, 1958, **31**, 155.
142. Ref. 1, p. 98ff.
143. Ref. 1, p. 80.
144. Ref. 1, p. 82ff.
145. R. H. Colby, G. E. Milliman and W. W. Graessley, *Macromolecules*, 1986, **19**, 1261.
146. C. H. Baker and L. Mandelkern, *Polymer*, 1966, **7**, 7.
147. R. J. Orr, *Polymer*, 1961, **2**, 74.
148. D. H. Coffey and T. J. Meyrick, *Proc. Rubber Technol. Conf., 3rd*, 1954, 170.
149. J. F. Kenney, *Polym. Eng. Sci.*, 1968, **8**, 216.
150. A. Misra and S. N. Garg, *J. Polym. Sci., Polym. Phys. Ed.*, 1986, **24**, 983.
151. J. C. Randall, *J. Polym. Sci., Polym. Phys. Ed.*, 1975, **13**, 1975.
152. T. M. Krigas, J. M. Carella, M. J. Struglueski, B. Crist, W. W. Graessley and F. C. Schilling, *J. Polym. Sci., Polym. Phys. Ed.*, 1985, **23**, 509.
153. H. D. Starkweather, Jr., *J. Polym. Sci., Polym. Phys. Ed.*, 1977, **15**, 247.
154. Y. Chatani, T. Takizawa, S. Murahashi, Y. Sakata and Y. Nishimma, *J. Polym. Sci.*, 1961, **55**, 811.
155. P. Colombo, L. E. Kuracka, J. Fontana, R. N. Chapman and M. Steinberg, *J. Polym. Sci., Part A-1*, 1966, **4**, 29.
156. H. D. Starkweather, Jr., *J. Polym. Sci., Polym. Phys. Ed.*, 1973, **11**, 587.
157. C. Garbuglio, M. Ragazzini, O. Pilato, D. Carcaro and G. B. Cendalli, *Eur. Polym. J.*, 1967, **3**, 137.
158. C. G. Vonk, *J. Polym. Sci., Part C*, 1972, **38**, 429.
158a. G. Kortleve, C. A. F. Tuijnman and C. G. Vonk, *J. Polym. Sci., Part A-2*, 1972, **10**, 123.
159. E. R. Walter and F. P. Reding, *J. Polym. Sci.*, 1956, **21**, 501.
160. R. M. Eichhorn, *J. Polym. Sci.*, 1958, **31**, 197.
161. P. R. Swan, *J. Polym. Sci.*, 1962, **56**, 409.
162. C. H. Baker and L. Mandelkern, *Polymer*, 1966, **7**, 71.
163. J. E. Preedy, *Br. Polym. J.*, 1973, **5**, 13.
164. C. W. Bunn, in 'Polyethylene', ed. A. Renfrew and P. Morgan, Illife, London, chap. 5.
165. D. J. Cutler, P. J. Hendra, M. E. A. Cudby and H. A. Willis, *Polymer*, 1977, **18**, 1005.
166. T. N. Bowmer and J. H. O'Donnell, *Polymer*, 1977, **18**, 1032.
167. P. J. Holdsworth and A. Keller, *Makromol. Chem.*, 1969, **125**, 82.
168. J. Vile, P. J. Hendra, H. A. Willis, M. E. A. Cudby and G. Gee, *Polymer*, 1984, **25**, 1173.
169. C. G. Vonk, 'Polyethylene Golden Jubilee Conference', Plastics and Rubber Institute, London, 1983, vol. D2.1.
170. R. Seguela and F. Rietsch, *J. Polym. Sci., Polym. Lett. Ed.*, 1986, **24**, 29.
171. E. W. Fischer, H. J. Sterzel and G. Wegner, *Kolloid Z. Z. Polym.*, 1973, **251**, 980.
172. M. J. Richardson, P. J. Flory and J. B. Jackson, *Polymer*, 1963, **4**, 221.
173. G. Capaccio and I. M. Ward, *J. Polym. Sci., Polym. Phys. Ed.*, 1984, **22**, 475.
174. M. Inoue, *J. Appl. Polym. Sci.*, 1964, **8**, 2225.
175. H. Wilski, *Makromol. Chem.*, 1971, **150**, 209.
176. M. Droscher, K. Hertwig, H. Reimann and G. Wegner, *Makromol Chem.*, 1976, **177**, 1695.
177. F. T. Simon and J. M. Rutherford, Jr., *J. Appl. Phys.*, 1964, **35**, 82.
178. U. Kalepky, E. W. Fischer, P. Herchenroder, G. Lieser and G. Wegner, *J. Polym. Sci., Polym. Phys. Ed.*, 1979, **17**, 2117.
179. R. J. Roe and C. Gieniewski, *J. Cryst. Growth*, 1980, **48**, 295.
180. R. J. Roe and C. Gieniewski, *Macromolecules*, 1973, **6**, 212.
181. O. B. Edgar and R. Hill, *J. Polym. Sci.*, 1952, **8**, 1.
182. A. J. Yu and R. D. Evans, *J. Am. Chem. Soc.*, 1959, **81**, 5361.
183. A. J. Yu and R. D. Evans, *J. Polym. Sci.*, 1960, **42**, 249.
184. M. Levine and S. C. Temin, *J. Polym. Sci.*, 1961, **49**, 241.
185. T. C. Tranter, *J. Polym. Sci., Part A*, 1964, **2**, 4289.
186. F. P. Prince, E. M. Pearce and R. J. Fredericks, *J. Polym. Sci., Part A-1*, 1970, **8**, 3533.
187. M. Hachiboshi, T. Fukuda and S. Kobayashi, *J. Macromol. Sci., Phys.*, 1960, **35**, 94.
188. G. Natta, *Makromol. Chem.*, 1960, **35**, 94.
189. G. Natta, P. Corradini, D. Sianesi and D. Morero, *J. Polym. Sci.*, 1961, **51**, 527.
190. G. Natta, G. Allegra, I. W. Bassi, D. Sianesi, G. Caporiccio and E. Torti, *J. Polym. Sci., Part A*, 1965, **3**, 4263.
191. G. Natta, L. Porri, A. Carbonaro and G. Lugli, *Makromol. Chem.*, 1962, **53**, 52.
192. R. K. Eby, *J. Appl. Phys.*, 1966, **34**, 2442.
193. J. P. Colson and R. K. Eby, *J. Appl. Phys.*, 1966, **37**, 3511.
194. I. C. Sanchez and R. K. Eby, *J. Res. Nat. Bur. Stand., Sect. A*, 1973, **77**, 353.
195. E. Helfand and J. I. Lauritzen, Jr., *Macromolecules*, 1973, **6**, 631.
196. I. C. Sanchez and R. K. Eby, *Macromolecules*, 1975, **8**, 638.
197. I. C. Sanchez, *J. Polym. Sci., Polym. Symp.*, 1977, **59**, 109.
198. M. L. Huggins, *J. Phys. Chem.*, 1942, **46**, 151; *Ann. N.Y. Acad. Sci.*, 1942, **41**, 1; *J. Am. Chem. Soc.*, 1942, **64**, 1712.
199. P. J. Flory, *J. Chem. Phys.*, 1942, **10**, 51.

200. Ref. 1, p. 38ff.
201. Ref. 110, p. 568ff.
202. L. Mandelkern, *Rubber Chem. Technol.*, 1959, **32**, 1392.
203. A. M. Bueche, *J. Am. Chem. Soc.*, 1952, **74**, 65.
204. P. J. Flory, L. Mandelkern and H. K. Hall, *J. Am. Chem. Soc.*, 1951, **73**, 2532.
205. R. B. Richards, *Trans. Faraday Soc.*, 1946, **42**, 10.
206. A. Nakajima, H. Fujiwara and F. Hamada, *J. Polym. Sci., Part A-2*, 1966, **4**, 507.
207. Ref. 110, p. 507ff.
208. A. Nakajima and F. Hamada, *Kolloid Z. Z. Polym.*, 1966, **205**, 55.
209. A. Factor, G. E. Heinsohn and L. H. Vogt, Jr., *J. Polym. Sci., Polym. Lett.*, 1969, **7**, 205.
210. A. R. Shultz and C. R. McCullough, *J. Polym. Sci., Part A-2*, 1969, **7**, 1577.
211. P. J. Flory and R. R. Garrett, *J. Am. Chem. Soc.*, 1958, **80**, 4836.
212. P. Smith and A. J. Pennings, *Polymer*, 1974, **15**, 413.
213. P. Smith and A. J. Pennings, *J. Mater. Sci.*, 1976, **11**, 1450.
214. P. Smith and A. J. Pennings, *J. Polym. Sci., Polym. Phys. Ed.*, 1977, **15**, 523.
215. C. C. Gryte, H. Berghmans and G. Smets, *J. Polym. Sci., Polym. Phys. Ed.*, 1979, **17**, 1925.
216. S. P. Papkov, *Vysokomol. Soedin., Ser. A*, 1978, **20**, 11, 2517.
217. Ref. 110, p. 533.
218. P. J. Flory and W. R. Krigbaum, *J. Chem. Phys.*, 1950, **18**, 1086.
219. Ref. 1, p. 50.
220. D. R. Beech and C. Booth, *Polymer*, 1972, **13**, 355.
221. A. J. Pennings, in 'Characterization of Macromolecular Structure', National Academy of Science, 1968, p. 214.
222. I. C. Sanchez and E. A. DiMarzio, *Macromolecules*, 1971, **4**, 677.
223. W. M. Leung, R. St. J. Manley and A. R. Panaras, *Macromolecules*, 1985, **18**, 746.
224. T. Nishi and T. T. Wang, *Macromolecules*, 1975, **8**, 909.
225. R. L. Scott, *J. Chem. Phys.*, 1949, **17**, 279.
226. D. R. Paul, J. W. Barlow, R. E. Bernstein and D. C. Warhmund, *Polym. Eng. Sci.*, 1978, **18**, 1225.
227. B. S. Morra and R. S. Stein, *J. Polym. Sci., Polym. Phys. Ed.*, 1982, **20**, 2243.
228. P. B. Rim and J. P. Runt, *Macromolecules*, 1984, **17**, 1520.
229. J. M. Jonza and R. S. Porter, *Macromolecules*, 1986, **19**, 1946.
230. G. C. Alfonso and T. P. Russell, *Macromolecules*, 1986, **19**, 1143.
231. A. R. Shultz and C. R. McCullough, *J. Polym. Sci., Part. A-2*, 1972, **10**, 307.
232. P. Smith and R. St. J. Manley, *Macromolecules*, 1979, **12**, 483.
233. S. Z. D. Cheng and B. Wunderlich, *J. Polym. Sci., Polym. Phys. Ed.*, 1986, **24**, 577.
234. P. Smith and K. Gardner, *Macromolecules*, 1985, **18**, 1222.
235. Ref. 110, p. 509, 513.
236. L. Mandelkern, E. K. Chan and F. A. Smith, in preparation.
237. Ref. 1, chap. 5.
238. L. Mandelkern, in 'Physical Properties of Polymers', ed. J. E. Mark, American Chemical Society, Washington, 1984, p. 155ff.
239. A. E. Tonelli, *J. Chem. Phys.*, 1970, **52**, 4749; 1971, **54**, 4637; 1972, **56**, 5533.
240. A. E. Tonelli, *Macromolecules*, 1972, **5**, 563.
241. (a) P. R. Sundarajan, *J. Appl. Sci.*, 1978, **22**, 1391; (b) Y. Takahashi and J. E. Mark, *J. Am. Chem. Soc.*, 1976, **98**, 756; (c) A. Abe, *Macromolecules*, 1980, **13**, 546; (d) P. Corradini, *J. Polym. Sci., Polym. Symp.*, 1975, **50**, 327.
242. B. Wunderlich, 'Macromolecular Physics', Academic Press, New York, 1980, vol. 3, p. 91ff.
243. P. J. Flory, *J. Am. Chem. Soc.*, 1956, **78**, 5222.
244. J. F. M. Oth and P. J. Flory, *J. Am. Chem. Soc.*, 1958, **80**, 1297.
245. P. J. Flory and O. K. Spurr, Jr., *J. Am. Chem. Soc.*, 1961, **83**, 1308.
246. L. Mandelkern, *Annu. Rev. Phys. Chem.*, 1964, **15**, 421.
247. L. Mandelkern, D. E. Roberts, A. F. DiOrio and A. S. Posner, *J. Am. Chem. Soc.*, 1959, **81**, 4148.
248. A. Katchalsky, S. Lifson, I. Michaelis and H. Zurich, in 'Size and Shape of Contractile Polymers', ed. A. Wasserman, Pergamon Press, New York, 1960, p. 1.
249. Ref. 1, chap. 7.
250. P. J. Flory and D. Y. Yoon, *Faraday Discuss. Chem. Soc.*, 1979, **69**, 389.
251. F. R. Anderson, *J. Appl. Phys.*, 1964, **35**, 64.
252. R. G. Brown and R. K. Eby, *J. Appl. Phys.*, 1964, **35**, 1156.
253. K. H. Illers and H. Hendus, *Makromol. Chem.*, 1968, **113**, 1.
254. R. C. Allen, M. S. Thesis, Florida State University, 1980.
255. N. J. Hay, *J. Chem. Soc., Faraday Trans.*, 1972, **68**, 656.
256. D. R. Beech and C. Booth, *J. Polym. Sci., Polym. Lett. Ed.*, 1970, **8**, 731.
257. L. Mandelkern, in 'Progress in Polymer Science', ed. A. D. Jenkins, Pergamon Press, Oxford, 1970, vol. 2, p. 165.
258. L. Mandelkern, in 'Annual Review of Materials Science', ed. R. A. Huggins, Annual Reviews, Palo Alto, CF, 1976, vol. 6, p. 119.
259. L. Mandelkern, *Faraday Discuss Chem. Soc.*, 1979, **68**, 454.
260. A. Peacock and L. Mandelkern, *Polym. Bull. (Berlin)*, 1986, **16**, 529.
261. A. Alamo and L. Mandelkern, *J. Polym. Sci., Polym. Phys. Ed.*, 1986, **24**, 2087.
262. Ref. 1, chap. 8.
263. C. F. Frank, in 'Growth and Perfection of Crystals', ed. R. H. Doremus, B. W. Roberts and D. Turnbull, Wiley, New York, 1958, p. 529.
264. J. H. Magill, in 'Treatise on Materials Science and Technology', ed. J. M. Schultz, Academic Press, New York, 1977, p. 3.
265. D. Turnbull, *J. Chem. Phys.*, 1950, **18**, 198.
266. C. Devoy and L. Mandelkern, *J. Chem. Phys.*, 1970, **7**, 3827.
267. J. W. Gibbs, 'Collected Works', Longmans Green and Co., New York, 1928, p. 325.

268. L. Mandelkern, *Faraday Discuss. Chem. Soc.*, 1979, **68**, 375.
269. R. Koningsveld and A. J. Pennings, *Recl. Trav. Chim. Pay-Bas*, 1964, **83**, 552; A. J. Pennings and A. M. Kiel, *Kolloid Z.*, 1965, **205**, 160.
270. A. M. Rijke and L. Mandelkern, *J. Polym. Sci., Part A-2*, 1970, **8**, 825.
271. B. Wunderlich and T. Arakawa, *J. Polym. Sci., Part A*, 1964, **2**, 3694.
272. M. H. Theil and L. Mandelkern, *J. Polym. Sci., Part A-2*, 1970, **8**, 957.
273. Ref. 1, p. 321ff.
274. M. Gopalan and L. Mandelkern, *J. Phys. Chem.*, 1967, **71**, 3833.
275. J. D. Hoffman and J. J. Weeks, *J. Res. Nat. Bur. Stand., Sect. A*, 1962, **66**, 13.
276. D. E. Axelson, L. Mandelkern, R. Popli and P. Mathieu, *J. Polym. Sci., Polym. Phys. Ed.*, 1983, **21**, 2319.
277. R. Popli and L. Mandelkern, *Polym. Bull. (Berlin)*, 1983, **9**, 260.
278. R. Popli, M. Glotin, L. Mandelkern and R. S. Benson, *J. Polym. Sci., Polym. Phys. Ed.*, 1984, **22**, 407.
279. R. Popli and L. Mandelkern, *J. Polym. Sci., Polym. Phys. Ed.*, 1987, **25**, 441.
280. F. von Göler and G. Sachs, *Z. Phys.*, 1932, **77**, 281.
281. M. Avrami, *J. Chem. Phys.*, 1939, **7**, 1103; 1940, **8**, 212; 1941, **9**, 177.
282. W. B. Hillig, *Acta Metall.*, 1966, **14**, 1868.
283. J. I. Lauritzen, Jr., *J. Appl. Phys.*, 1973, **44**, 4353.
284. I. C. Sanchez and E. A. DiMarzio, *J. Res. Nat. Bur. Stand., Sect. A*, 1972, **76**, 213.
285. R. C. Allen and L. Mandelkern, *Polym. Bull. (Berlin)*, 1987, **17**, 473.
286. R. Alamo, J. G. Fatou and J. Guzman, *Polymer*, 1982, **23**, 374; 1981, **23**, 379.
287. J. Maxfield and L. Mandelkern, *Macromolecules*, 1977, **10**, 1141.
288. R. C. Allen and L. Mandelkern, *J. Polym. Sci., Polym. Phys. Ed.*, 1982, **20**, 1465.
289. L. Mandelkern, M. Glotin and R. S. Benson, *Macromolecules*, 1981, **14**, 22.
290. J. D. Hoffman, C. M. Guttman and E. A. DiMarzio, *Faraday Discuss. Chem. Soc.*, 1979, **68**, 177.
291. J. D. Hoffman, *Polymer*, 1983, **24**, 3.
292. I. C. Sanchez and E. A. DiMarzio, *J. Chem. Phys.*, 1971, **55**, 893.
293. E. Ergoz, Ph.D. Dissertation, Florida State University, Dec. 1970.
294. J. D. Hoffman, *Faraday Discuss. Chem. Soc.*, 1979, **68**, 470.
295. J. J. Weeks, *J. Res. Nat. Bur. Stand., Sect. A*, 1963, **67**, 331.
296. H. E. Bair, T. W. Hreseby and R. Salovey, *Polym. Prepr., Am. Chem. Soc., Div. Polym. Chem.*, 1968, **9**, 795.
297. B. Wunderlich and G. Czornyj, *Macromolecules*, 1977, **10**, 960.
298. D. C. Bassett, A. M. Hodge and R. H. Olley, *Proc. R. Soc. London, Ser. A*, 1981, **377**, 39.
299. W. D. Varnell, I. R. Harrison and J. I. Wang, *J. Polym. Sci., Polym. Phys. Ed.*, 1981, **19**, 1577.
300. J. Runt, I. R. Harrison and S. Dobson, *J. Macromol. Sci., Phys.*, 1980, **B17**, 99.

12

Crystallization and Morphology

ALUN S. VAUGHAN
Central Electricity Research Laboratories, Leatherhead, UK
and
DAVID C. BASSETT
University of Reading, UK

12.1 INTRODUCTION

The crystalline high polymers represent a class of materials which, in the solid state, exhibit a complex hierarchical morphology in which structural ordering occurs over a wide range of dimensional levels. These range from spacings of the order of angstroms between neighbouring chains within a crystal lattice through the tens of nanometres to microns scale dimensions of a single crystal, up to complex polycrystalline aggregates which, in extreme cases, can extend for several millimeters. Thus, from an academic point of view, the topic of polymer morphology provides vast scope for both experimental and theoretical study. In addition, the structure of these materials is also of great importance commercially, since many important properties are fundamentally dependent upon microstructure. Thus a satisfactory understanding of material microstructure represents an essential prerequisite to a comprehensive understanding of macroscopic properties. However, as stated above, polymers are structurally complex materials and for this reason the necessary morphological understanding can frequently only be obtained through a diverse range of complementary experimental routes. For example, whilst conventional X-ray diffraction techniques can give detailed information concerning the molecular packing within the crystal lattice, little can generally be deduced about the precise spatial arrangement of crystallites within the bulk. Similarly, the transmission electron microscope frequently represents an ideal tool for the examination of crystal habits, but is difficult to apply to the measurement of parameters such as degree of crystallinity or lamellar thickness, which are best obtained by diffraction or thermal techniques which lead directly to values statistically averaged throughout the bulk. Thus to provide as complete a picture as possible of the arrangement of molecules and crystallites within a polymer, a range of complementary techniques are essential. Indeed, historically, many of the major advances in the understanding of polymeric materials have paralleled the application of novel methods of experimental examination.

12.1.1 Historical Survey

Much of the early work on the structure of crystalline polymers involved the study of naturally occurring systems by the techniques of X-ray diffraction.[1] Such experiments generally result in patterns in which both sharp spots and diffuse haloes are evident, indicating that polymers do not attain perfect crystalline ordering. Other macroscopic properties, such as the density and melting behaviour of these materials, are also indicative of only limited crystallinity. In an attempt to explain these observations, a model was proposed in which the polymer molecules, which were known to be long thread-like entities, were incorporated into a two-phase structure composed of ordered regions, tens of nanometres in size, embedded in a disordered matrix. The polymer chains were then imagined to meander from crystallite to crystallite *via* the amorphous phase as shown in Figure 1.[2] For many years this so-called fringed-micelle model represented the accepted view of the structure of crystalline polymers, since it was reasonably successful in explaining many of the observed phenomena.

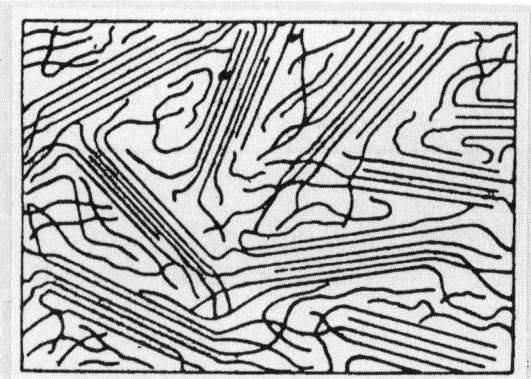

Figure 1   Schematic representation of the fringed-micelle model (after Bunn[3])

However, in 1945 Bunn and Alcock[4] reported an additional scale of ordering, much greater than that present within the fringed-micelle model, following a study of polyethylene in the polarizing optical microscope. These entities, termed spherulites (see Figure 2), were several microns in diameter and were shown to consist of an approximately spherical array of equivalent radiating crystalline units. The existence of such structures within crystalline polymers highlighted some of the deficiencies of the fringed-micelle model since it was by no means obvious how the small crystallites of the model could give rise to these structures which were several orders of magnitude larger, particularly since detailed optical analysis showed the molecular chain axis within spherulites to be tangential rather than radial as would have been expected from a fringed-micelle type structure.

This difficulty of accommodating spherulitic ordering within a structure of the type shown in Figure 1 remained until the 1950s when, following the development of catalysts whereby linear, stereoregular polymers could be synthesized, individual crystals of polyethylene were first grown from dilute solution.[5-7] Examination of these crystals in the then newly developed first generation of commercial transmission electron microscopes (TEM) showed them to be sheet-like in nature, ~ 100 Å in thickness and with other dimensions extending to several microns (see Figure 3). What was most startling, however, was that within these crystals the molecular chain axis was approximately perpendicular to the plane of the lamella. On the basis of these observations, Keller concluded that for molecules ~ 1 μm long to be incorporated within a crystal ~ 100 Å thick in the way

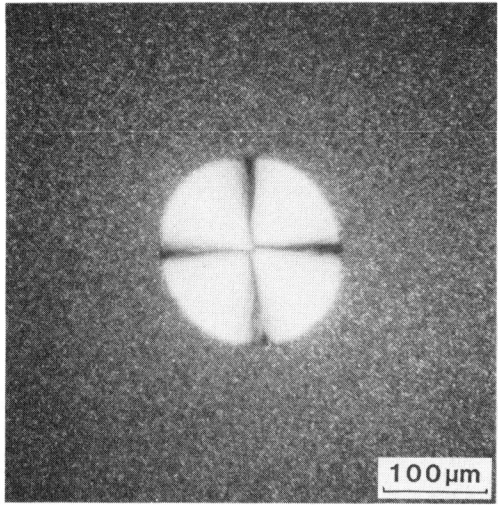

Figure 2 A spherulite of the γ-crystal form of poly(vinylidene fluoride), crystallized at 173 °C

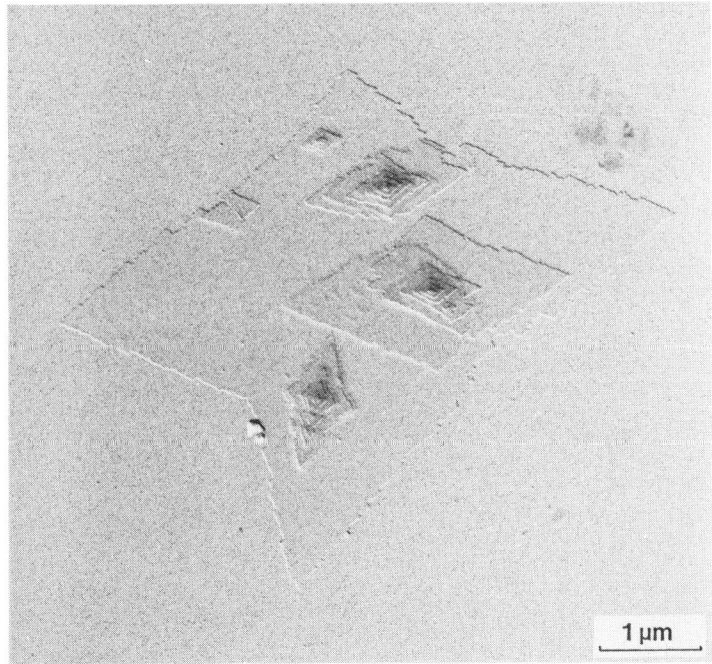

Figure 3 Polyethylene lamellae, crystallized from solution

described above, they must traverse the crystal many times, this being accomplished by the chain folding back on itself at the crystal surface. Thus the contemporary view of chain-folded lamellar crystallization in polymers was established, nearly 20 years after the concept had first been tentatively suggested by Storks.[8]

The economic importance of the new stereoregular polymers, together with the experimental possibilities afforded by electron microscopy, led to a considerable research effort in the early 1960s into the structure of linear crystalline polymers. However, much of this was directed towards examination of lamellae and lamellar aggregates grown from solution. *Via* this crystallization route, a wide range of crystal habits and morphological forms were identified, depending upon the particular polymer and crystallization conditions employed. Much of the interest in solution crystallization, however, stemmed from the experimental difficulties associated with the preparation of bulk melt-crystallized samples for examination in the TEM. Since crystallization from solution generally gives rise to individual lamellae or simple aggregates of crystals, such structures are amenable to the transmission of electrons either directly or following simple sample preparation procedures,[9] due to lamellar thicknesses and, as a result, specimen thicknesses generally being limited to a scale of the order of hundreds of angstroms. For melt-crystallized materials to be examined directly in the TEM, specimens of a similar thickness must be prepared, either by crystallization of thin films or else by sectioning, following crystallization in the bulk. However, in some circumstances the restricted geometries and modified diffusion fields, which are an inevitable consequence of crystallization within a 1000 Å thick film, give rise to morphological forms which may differ considerably from those present within the bulk.[10] Similarly, sectioning of bulk-crystallized specimens is not entirely free of pitfalls, since the preparation of samples which are not only ~1000 Å thick but also undeformed is by no means trivial.[11] Therefore both these sample preparation routes present potential limitations or difficulties. In addition, even when samples of a suitable thickness are available, direct TEM examination is not free of problems, due to the fundamental nature of the interaction between high energy electrons and polymeric materials.[12] The difficulties which are encountered may be classified under two headings. Firstly, radiation damage, which can give rise to changes in crystal structure, mass transport, enhancement of structural features and, ultimately, the complete destruction of crystallinity, and secondly, inherently low contrast, due to the elemental composition of most polymers. As a result of these problems, much of the early work on melt-crystallized systems was conducted using the optical microscope. However, the inherent resolution limitations of this instrument means that only large scale structures, such as spherulites, can be examined directly. Thus, lamellar architectures had to be deduced from indirect observations, taking into account knowledge gained from simpler systems. As a result of the experimental difficulties outlined above, two less direct methods of TEM examination were developed, whereby representative melt-crystallized morphologies could be studied.

The first class of techniques, which has its origins in similar biological preparative methods,[13] involves staining the sample under investigation prior to examination, such that contrast is enhanced by the selective addition of electron dense material.[14] In the case of polyethylene, for example, treatment with chlorosulfonic acid[15,16] converts the polymer into some chemically cross-linked material, with chlorine and sulfur atoms attached to the lamellar surfaces. These regions then give rise to the contrast observed in the electron microscope, whereby lamellae viewed edge-on are seen outlined in black, as shown in Figure 4. This technique also has the effect of improving the stability

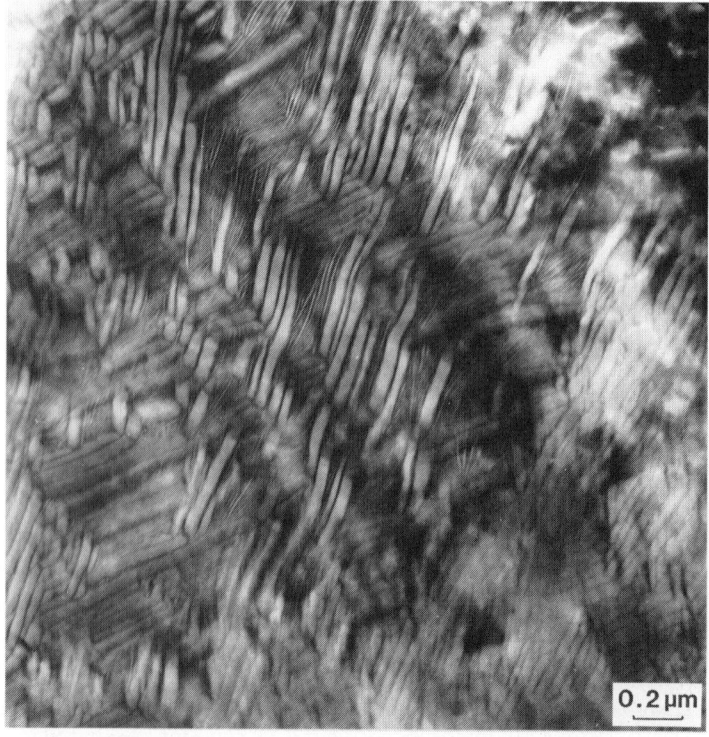

Figure 4   Sample of polyethylene crystallized at 130 °C and stained with chlorosulfonic acid, showing ridged lamellae

of the sample in the electron beam and reducing the problems associated with sectioning.[18] Other staining reagents that have been used in the investigation of polymers include phosphotungstic acid,[19] tin chloride[20] and osmium tetroxide.[21–23] Techniques such as these involve the examination of the distribution of heavy, electron dense atoms within the microstructure. As such they are subject to geometrical limitations whereby unambigous image interpretation is only possible when lamellae are viewed in an edge-on projection, where the difference in concentration of staining medium between amorphous and crystalline regions can be readily identified (see Figure 4).

The second important experimental approach has long been employed in many aspects of materials science to study the surface topography of samples which are opaque to electrons, and involves the production of a cast or replica in some electron transparent material which faithfully reproduces the structural features present in the surface of the sample.[24] Initially in polymers this approach was limited to either free growth surfaces or else to internal surfaces exposed by fracturing the specimen. However the structural features present in both free growth and fracture surfaces may not be truly representative of the morphology present within the bulk, as a result of molecular fractionation/segregation effects in the former cases and through the influence of the underlying morphology on the crack propagation route in the latter, such that the appearance is biased towards certain morphological features. However, early investigations of polymer microstructure using replication techniques were limited to these rather specific areas, since attempts to devise an etchant for polymers whereby more typical surfaces could be examined had met with only limited success. Nevertheless, in 1978 an etching reagent based on potassium permanganate dissolved in sulfuric acid was reported by Olley *et al.*,[25] which selectively removes disordered material present at the surface such that structural features are revealed. Following its initial application to polyethylene (see Figure 5), permanganic etching has subsequently been applied successfully in various guises to a wide range of different polymers.[25–29] Together with standard replication techniques, permanganic etching has, over the last 10 years, proved to be a powerful tool which permits the study of melt-crystallized morphologies in unprecedented detail without the projection limitations of staining techniques. Nevertheless staining and etching should be considered complementary since each has both strengths and limitations. Some other articles relevant to the selective etching of polymeric materials are listed in refs. 30 to 32.

From the brief preceding introduction it is clear that as far as melt-crystallized polymers are concerned, the structure can be considered in terms of three fundamental units, namely: (1) the

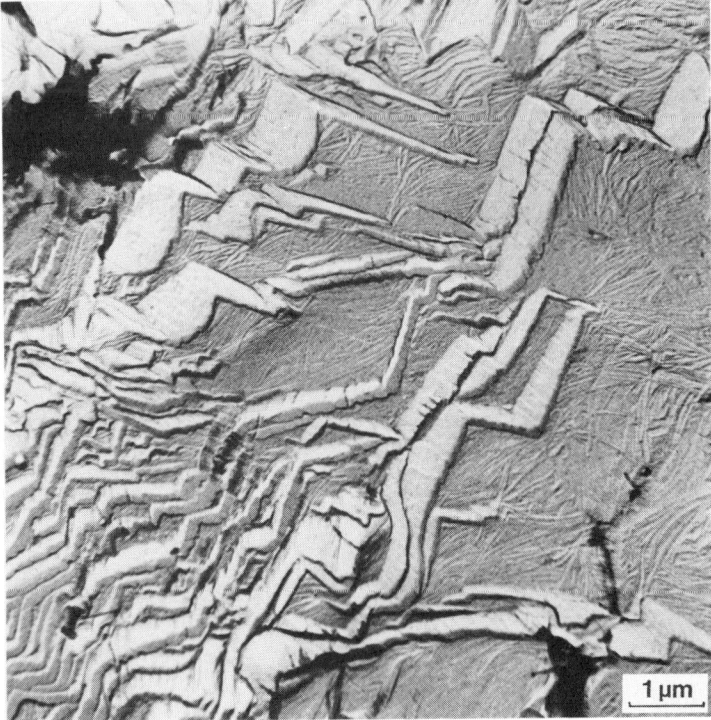

Figure 5 Sample of polyethylene crystallized at 130 °C and etched. The ridged morphology is similar to that shown in Figure 4

molecule; (2) the lamellar crystal; and (3) spherulites and lamellar aggregates. These three basic elements will now be considered in order.

12.2 MOLECULAR REQUIREMENTS FOR CRYSTALLINITY

For crystallinity to be observed in a given polymeric system, the constituent molecules must be capable of adopting an ordered configuration in which adjacent chains lie parallel to one another on well-defined lattice sites. Implicit in this statement are two elements, first, that the molecules in the melt are capable of relative motion and, secondly, that they themselves are ordered. Whilst the first of these factors merely serves to preclude highly cross-linked materials such as epoxy resins, the degree to which linear systems may or may not be capable of crystallization depends upon the precise degree of regularity present within the molecules in question. In this context the degree of regularity may be considered under three headings: (1) chemical regularity; (2) geometrical regularity; and (3) spatial regularity.

12.2.1 Chemical Regularity

Polymer molecules consist of long sequences of comparatively simple chemical units bonded together. Thus the simplest carbon-based polymer molecule is polyethylene which, ignoring end groups, simply consists of a long sequence of CH_2—CH_2 units.

In this idealized form molecules of polyethylene exhibit chemical regularity along the chain such that the first of the above requirements for crystallinity is met. Such a system in which all the repeat units between end groups are identical is termed a homopolymer. Alternatively, molecules may be synthesized in which all the monomer units are not identical. Such systems are termed copolymers.

In general, copolymers do not possess the necessary chemical regularity along the chain for them to be able to crystallize fully. In the case of random copolymers the degree to which crystallization is possible will generally depend on the relative number and similarity of the different monomer units present within the chain. For example the idealized polyethylene molecule described above is never attained in practical, commercial materials. So-called linear polyethylene, produced by Ziegler–Natta catalysis generally has two to three side branches per 1000 carbon atoms,[33] and is capable of giving materials with ~90% crystallinity.[34] Branched polyethylene, produced by a high pressure route, however, has ~30 side branches per 1000 carbon atoms[33] and, as a result, only exhibits crystallinity values up to ~50%,[34, 35] since most branches, being excluded from crystalline regions,[36] serve to limit the degree to which crystallization can occur. The above example serves to illustrate how comparatively small amounts of co-units, in this case alkylene units generated during the polyethylene synthesis, can reduce the degree to which crystallization is possible.

Block copolymers, being made up of well-ordered subchains, apparently possess the necessary prerequisites for crystallinity. However, except for a few systems[37-41] in which the synthesis is such that subchains are generated with narrow length distributions, crystallinity is precluded by the variation in length of the blocks between the copolymer molecules.

12.2.2 Geometrical Regularity

In the case of polymers made up of units of the type CH_2—CR_2 a possible irregularity may arise as a result of the absence of symmetry about the C—C bond. Generally the polymerization proceeds such that the CRR group from one monomer unit is bonded to the CH_2 unit from the next, referred to as head-to-tail addition. Occasionally however, during polymerization one CRR group becomes bonded to another to form a defect in the molecular structure which is termed a head-to-head defect. As in the copolymer case, the extent to which such defects may or may not affect the ability of a molecule to crystallize will depend upon the frequency with which head-to-head defects occur. In general, however, it is found experimentally that such defects, which tend to be present in polymers prepared by a free radical addition mechanism,[42, 43] occur only infrequently and therefore do not have a major bearing on whether or not a particular system will crystallize.[44]

12.2.3 Spatial Regularity

Spatial or stereoregularity represents an important criterion for crystallization in polymers which contain a pendant side group attached to the backbone *via* an asymmetric carbon atom. Different

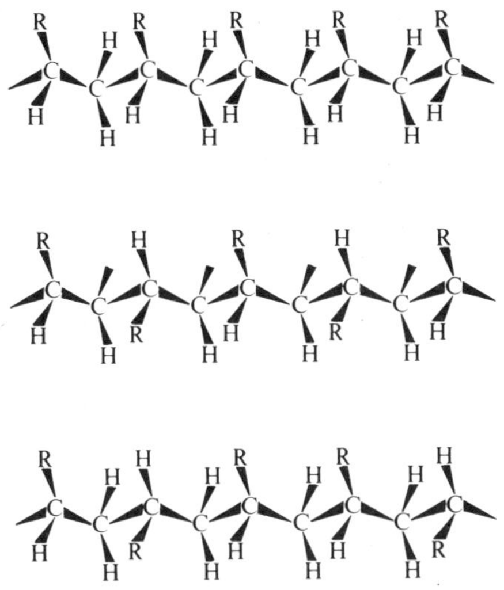

Figure 6 Schematic representation of isotactic, syndiotactic and atactic vinyl polymers

configurations of a polymer molecule composed of a sequence of CH_2—CHR units, differing only with respect to the relative spatial placement of the pendant side group, R, are shown in Figure 6. Isotactic and syndiotactic polymers are spatially ordered and are crystallizable. Atactic polymers, having a spatially disordered configuration, cannot be arranged into a regular lattice and, as a result, are generally unable to crystallize. Poly(vinyl alcohol), however, constitutes a system which is atactic but which, nevertheless, exhibits crystallinity.[45] Examples of common commercially important polymers in which the crystallinity is dependent upon stereoregularity are polypropylene and polystyrene, where the side groups R in Figure 6 are Me and Ph respectively.

In practice, perfect tacticity is not achieved and some stereoirregular placements occur within otherwise isotactic and syndiotactic molecules. As for chemical and geometrical regularity, the importance of such defects with respect to the degree to which a material will crystallize depends upon the frequency distribution of the stereoregular sequences present.

12.3 LAMELLAE

12.3.1 General Features

Individual polymer lamellae were first reported by Keller,[5] Fischer[6] and Till[7] in 1957 following the successful crystallization of linear polyethylene from a dilute solution in xylene. In their simplest form such structures are found to grow as flat sheets with dimensions typically of the order of 100 Å thick by several microns wide. Examination of these crystals by electron diffraction shows that the c-axis of the unit cell, and hence the chain axis of the molecule, is generally close to perpendicular to the plane of the lamella. Thus since, under suitable conditions, crystals can grow as isolated entities, it is necessary that each of the constituent molecules traverse the lamella many times, the upper and lower crystal surfaces being described as fold surfaces, within which the molecules fold back upon themselves.

A full discussion of the nature of lamellar folding goes beyond the scope of this chapter, since questions such as whether or not a molecule leaving a crystal returns to an adjacent site, or whether a fold contains a large or small number of carbon atoms have, over the years, been highly controversial (see for example, Volume 2, Chapter 11). Nevertheless polymer crystals can be modelled to a first approximation by simply considering them to consist of two components with different properties, *i.e.* a well-ordered crystalline interior plus interlamellar regions which incorporate the fold planes without specifying the nature of these surfaces further. In this way many experimental observations such as the presence of both sharp and diffuse X-ray reflections and variations in sample density and enthalpy of fusion may be explained. However, this is not to say that the precise nature of the non-crystalline interlamellar regions is unimportant. Indeed in the case of mechanical properties, for example, the frequency and conformation of tie molecules, which bridge the non-crystalline regions between adjacent lamellae, are of great consequence with respect to the ultimate properties of the material, since such molecules give cohesion between crystallites in addition to that due to van der Waals forces.

Following crystallization of a given polymeric system, the precise lamellar architecture observed depends upon a large number of different parameters. These involve the molecular considerations described in the preceding section, the molecular weight distribution present in the sample and the physical conditions pertaining during growth. Nevertheless, in most circumstances, crystallization results in polycrystalline lamellar aggregates such as spherulites, axialites and hedrites. Whilst such structures will be discussed in some detail in Section 12.4, it is first necessary to consider the individual lamellae of which these complex objects are composed.

12.3.2 Monolayer Crystals

In polymeric systems, individual lamellae only grow under very specific crystallization conditions. Monolayers are most easily generated by crystallizing a polymer from solution at low super-coolings,[17] whereby crystals are often formed with well-defined habits, bounded by simple, low index crystallographic growth faces. For example, in polyethylene, crystallization from a dilute solution in xylene gives rise to crystals which in their simplest form are lozenge shaped, as shown in

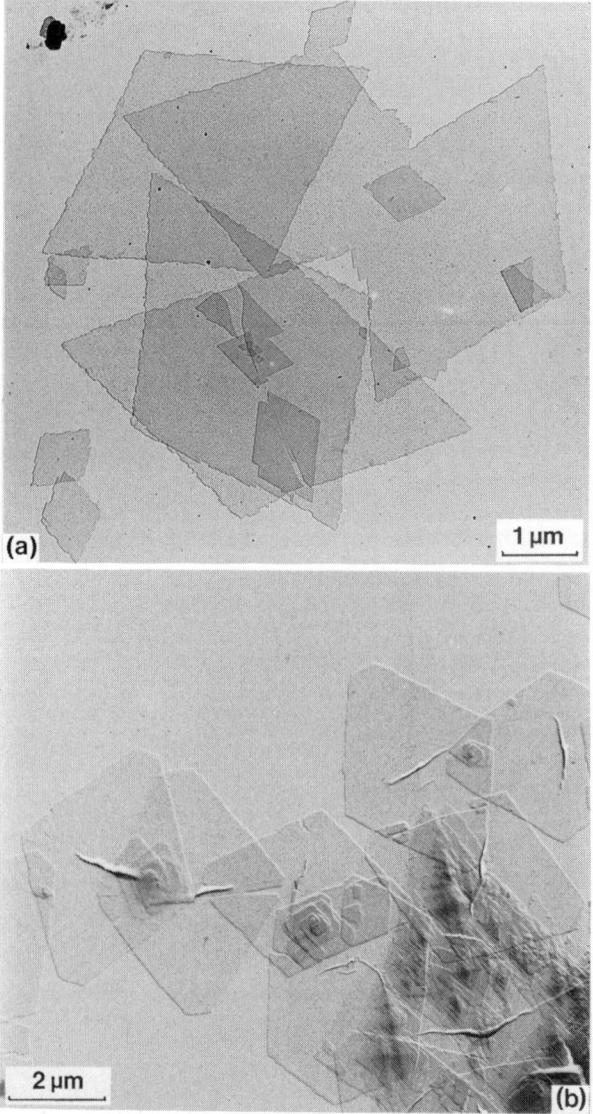

Figure 7   Polyethylene lamellae crystallized from solution: (a) lozenge shaped (b) truncated lozenge shaped

Figure 7(a), with {110} growth faces.[46] At higher crystallization temperatures this form becomes modified through the development of additional {100} faces (see Figure 7b) to give lamellae which are termed truncated lozenges.[46] Being bounded by low index crystallographic growth faces, the habits of such monolayers necessarily reflect the symmetry of the underlying unit cell, which in the case of polyethylene is normally orthorhombic. Thus poly(oxymethylene)[47] and isotactic polystyrene,[48] having hexagonal crystal structures, tend to produce hexagonal crystals and, similarly, poly-(4-methylpentene-1),[49] having tetragonal symmetry, tends to give rise to lamellae which are square. Examples of monolayer crystals from these systems are shown in Figure 8.

Monolayer polymer lamellae, being individual entities are often referred to as single crystals. However, strictly speaking such objects are more accurately described as multiple twins, since detailed studies show that the well-defined crystal habits described above are associated with each lamella being subdivided into a number of distinct crystallographic regions, each corresponding to a particular growth face. This occurs as a result of the presence of the molecular folds, which can introduce a structural complexity into polymer crystals that is not found in crystals of low molecular weight materials, since the orientation of the folds on the surface, *i.e.* the orientation of the fold planes, varies from one region of a crystal to another, even though the lattice of the bulk is

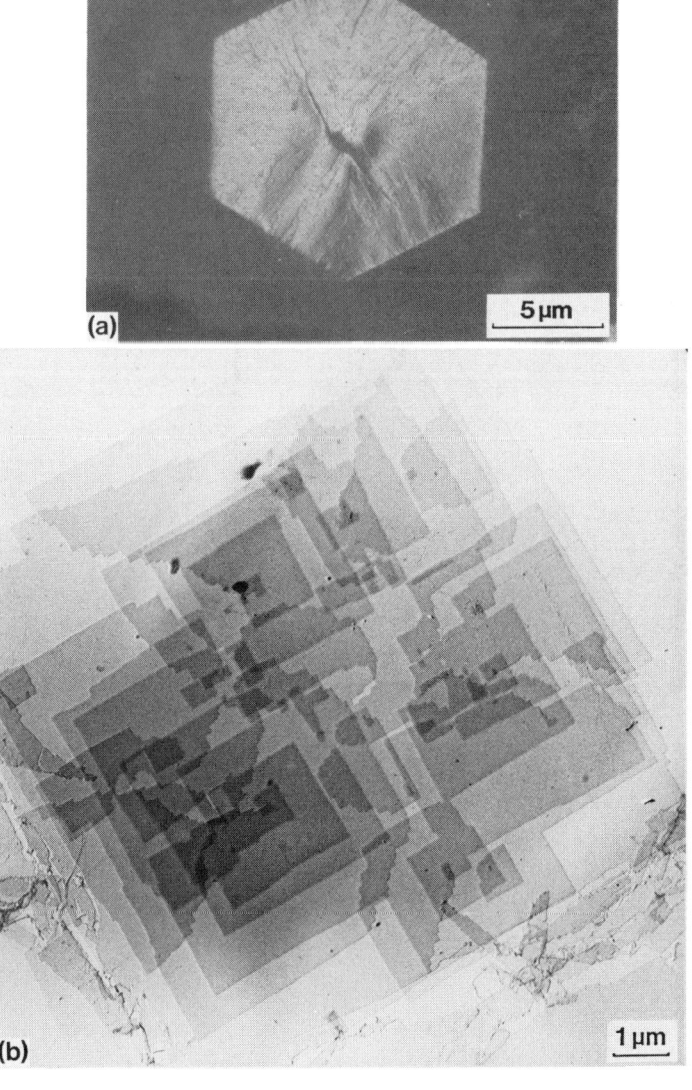

Figure 8 Solution-crystallized lamellae of (a) poly(oxymethylene), (b) poly(4-methylpent-1-ene)

continuous. These regions, consisting of portions of a crystal in which the fold planes have a common orientation, are termed sectors or fold domains.[50, 51] This is not to say, however, that even in well-sectored crystals grown from dilute solutions, all molecular folds lie parallel to the appropriate growth face, but rather that under such conditions folding does not constitute a random reversal of molecular direction, since then all subcell planes would be affected equally and, as a result, sectorization would not be observed.

Under certain circumstances this tendency for molecular folding to occur along specific planes can lead to a slight distortion being introduced into the crystal lattice whereby nominally equivalent crystallographic planes become differentiated into fold and non-fold planes. In the cases of polyethylene, poly(oxymethylene) and poly(4-methylpentene-1) the difference in lattice spacing between fold plane and 'equivalent' non-fold plane spacings is ~ 0.01 Å, as measured using moiré techniques.[47] Despite the fact that this value represents a very small lattice distortion indeed, it is still sufficient to give rise to lattice mismatches at sector boundaries such that otherwise planar lamellae become distorted into shallow conical habits. Alternatively non-planar habits can be generated as a result of the need to pack adjacent folds together. For example, in polyethylene, where the separation of adjacent stems in the fold plane is comparatively small, lamellae are formed in which the molecules are inclined to the basal plane of the crystal by 30° and more, thereby increasing the area available to accommodate each fold.

The presence of molecular folds and the need to pack chain conformations together have a bearing not only on the molecular structure within a lamella, but also on the ultimate lamellar form.

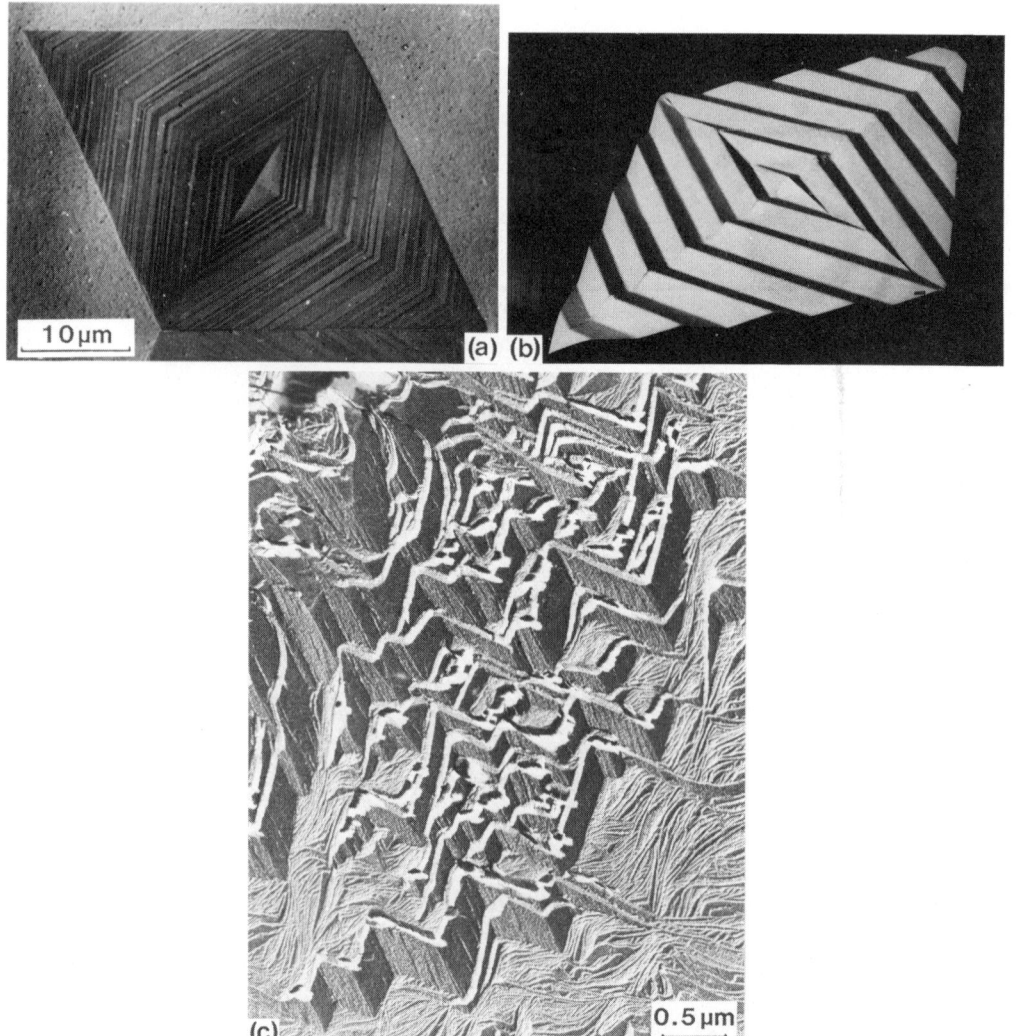

Figure 9 Ridged non-planar lamellar forms in polyethylene: (a) crystallized from solution; (b) a schematic representation of (a); (c) crystallized from the melt

Thus in polyethylene complex ridged habits can be generated through periodic reversals in the sense of staggering of the molecular folds.[52] Such structures are shown in Figure 9. In the case of both melt and solution growth, the molecular chain axis is such that it bisects the dihedral angle between adjacent ridge portions.

The above discussion is primarily concerned with monolayers crystallized from solution since, historically, much of the work on single lamellae was conducted *via* this crystallization route. However, it is possible under appropriate conditions to grow individual lamellae from the melt. Keith, Vadimsky and Padden[48] reported observing some monolayers on crystallizing isotactic polystyrene from a blend containing 90% atactic polystyrene. Whilst these workers described the above situation as crystallization from solution, it may be argued that under conditions where both solute and solvent are polymeric, melt crystallization is a more apt description. Single lamellae were also observed by Edwards and Phillips, in thin films of isotactic polystyrene[22] and *cis*-poly-isoprene[23] crystallized from the melt. In both these cases the observations were made during investigations into spherulitic nucleation processes. As such the samples were crystallized for very short times to yield small lamellar aggregates made up of several layers, together with some individual lamellae. In the case of *cis*-polyisoprene the individual lamellae extended to only ∼0.1 μm, whilst in the polystyrene case, lamellar dimensions were of the order of 0.5 μm.

As part of a similar investigation into nucleation processes in polyethylene it has recently been shown that under suitable conditions of temperature and molecular weight, individual lamellae can also be grown in that system.[53] The resulting crystals are interesting for a number of reasons. Firstly, as shown in Figure 10, these lamellae have an unusual elliptic habit which is bounded by curved growth faces. Hence these edges must be rough on a molecular level, an observation which has considerable implications for conventional secondary nucleation kinetic theory.[54-56] Also, being so structurally simple, the observation of the way in which these lamellae begin to develop into three-dimensional entities has provided an interesting insight into the mechanism by which banded spherulites form. These observations will be discussed later. Similar lamellar habits to that shown in Figure 10 have also been reported recently following a detailed investigation into the crystallization of polyethylene from a range of solvents at high temperatures.[57-59]

All the lamellae so far described share the common feature of exhibiting well-defined crystal habits, even those in which growth faces are curved. In systems such as linear polyethylene, which

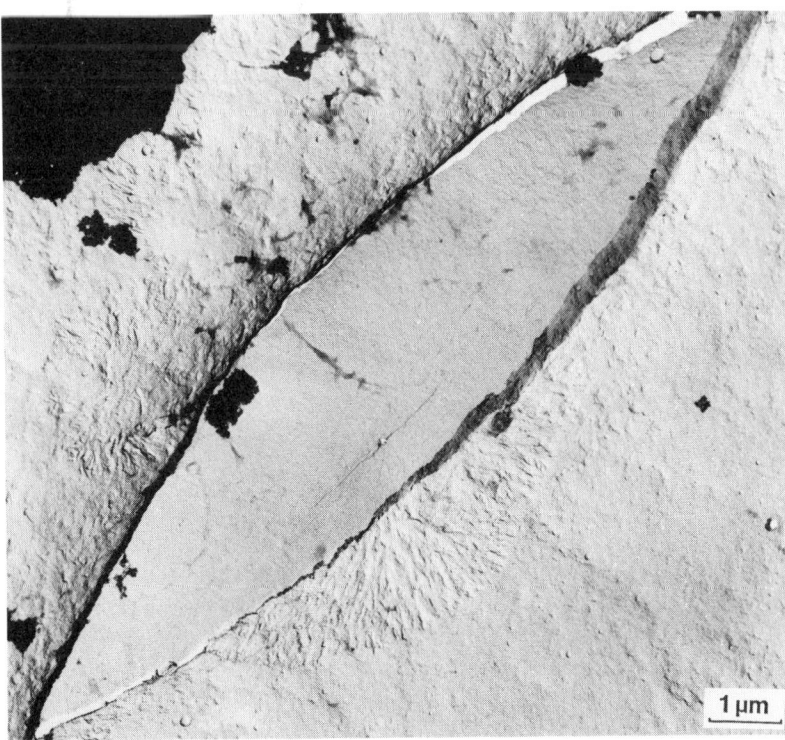

Figure 10 Isolated elliptic lamella of polyethylene. Rigidex 140–60 crystallized from the melt at 130 °C

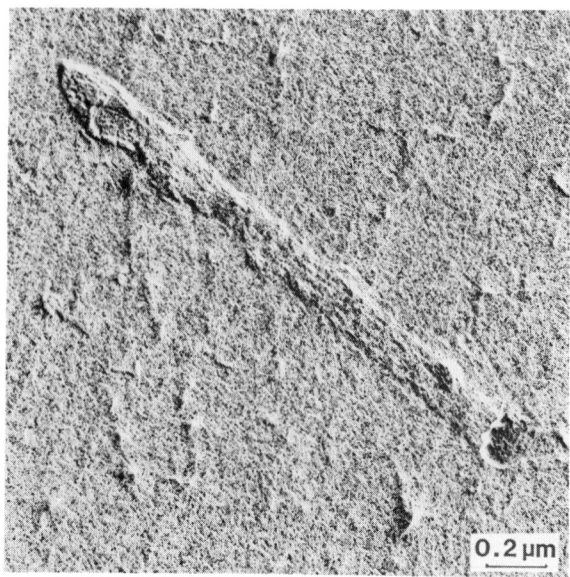

Figure 11   Melt-crystallized poly(aryl ether ether ketone) (PEEK) showing the irregular crystal habit

are capable of crystallizing to a high degree, this is generally the case. However in copolymers, branched materials or systems of low tacticity, ill-defined irregular crystal habits are often observed.[60-63] Figure 11 shows a micrograph of poly(aryl ether ether ketone) (PEEK; poly[bis(1,4-phenoxy)-1,4-benzoyl), a polymer in which crystallinity values are typically of the order of 30%, and in which crystals are formed which do not exhibit simple geometrical lateral shapes. Similar habits to this have been observed in many other systems where the degree to which crystallization can proceed is limited by molecular considerations.

The above discussion shows that even when considering individual lamellae, a wide range of structures can be identified depending upon both the particular polymer being studied and the crystallization conditions employed. Nevertheless in polymeric systems, generally speaking, isolated single lamellae represent a morphological form far removed from that which is normally encountered. Even in the case of crystallization from dilute solution, under most conditions objects develop which consist of more than one lamella. Such multilayered structures will now be considered.

12.3.3   Multilayer Crystals

In most circumstances, multilayered lamellar aggregates result from the action of screw dislocations. Such defects constitute an important phenomenon in polymer morphology since they give rise to the proliferation of layers which, together with splaying, can generate sheaf-like structures and ultimately spherulites. However, here discussion will be limited to the simple structures, composed of only a few layers, that develop readily on crystallization from solution, or from the melt under certain conditions.

In the simplest case such structures have the appearance of a spiralling ramp, as shown in Figure 12. Here consecutive terraces are in crystallographic register and the general crystal habit is preserved in successive layers. Examination of polyethylene samples crystallized from alkanic solvents has shown that the stress fields which accompany spiral growth are small. Thus it seems likely that the shears involved in the development of screw dislocations, where the Burgers vector is parallel to the stems of the folded molecule and equal in magnitude to the long period of the crystal, are not elastic strains but rather result from molecular fold conformations which are staggered.[44,64] Often, where crystallization occurs slowly, multilayered structures such as that in Figure 12b are generated by screw dislocations located near the centre of the crystal, so suggesting that spiral development was initiated as a result of the nucleation process. However, when crystal growth proceeds rapidly, *i.e.* at large supercoolings, screw dislocations may be generated profusely at crystal boundaries. Figure 13 shows the characteristic dendritic form of polyethylene lamellae crystallized from solution at temperatures below ~80 °C.[46] In this object large numbers of growth spirals can be

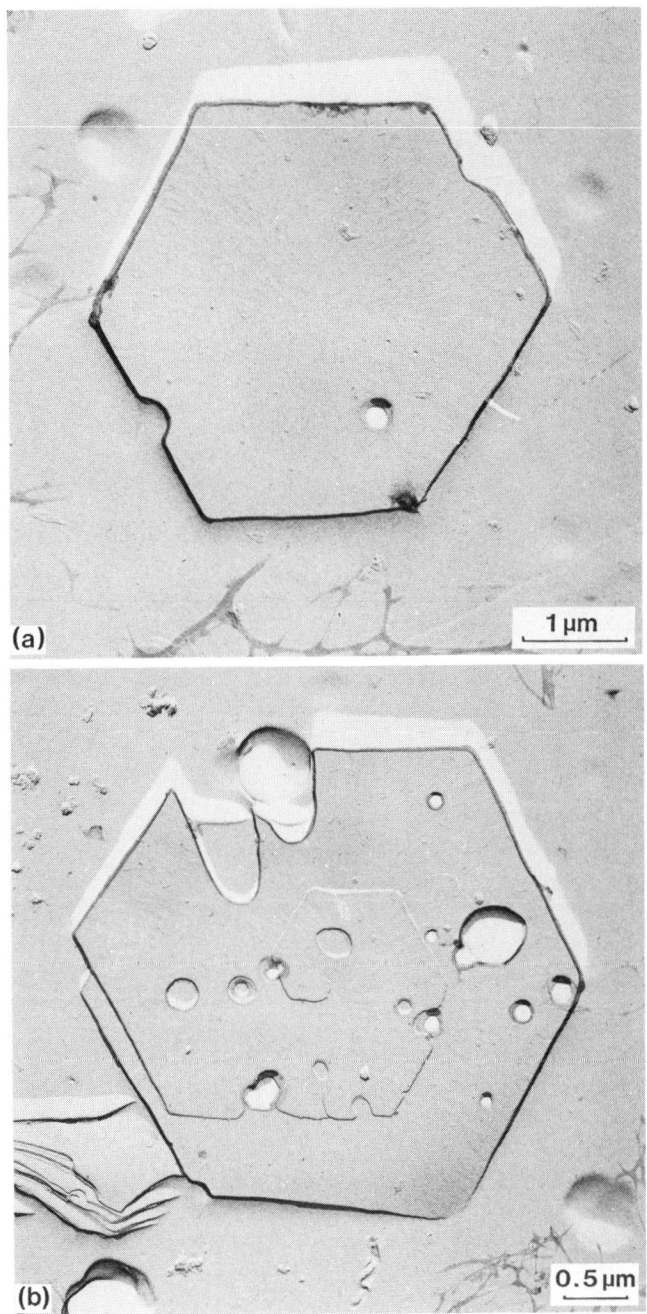

Figure 12 Hexagonal lamellae of isotactic polystyrene crystallized from the melt at 228 °C, illustrating preservation of crystal habit (a) and in the spiral form (b)

seen. As in the polystyrene sample illustrated in Figure 12, the general crystal habit is preserved in successive layers of each spiral, despite the tendency of growth faces to break up under these crystallization conditions. Figure 14 also illustrates this point. In this case however, the polyethylene sample shown was decorated with evaporated low molecular weight polyethylene after crystallization and then shadowed with a gold/palladium alloy. This technique, which results in the growth of small crystals of polyethylene preferentially oriented perpendicular to the fold planes in each sector,[65] shows how not only the general habit, but also that the sectorization is preserved during screw dislocation growth.

Other features found in multilayered crystals involve the interpenetration of a pair of screw dislocations of opposite hand and the relative rotation and splaying apart of successive terraces in

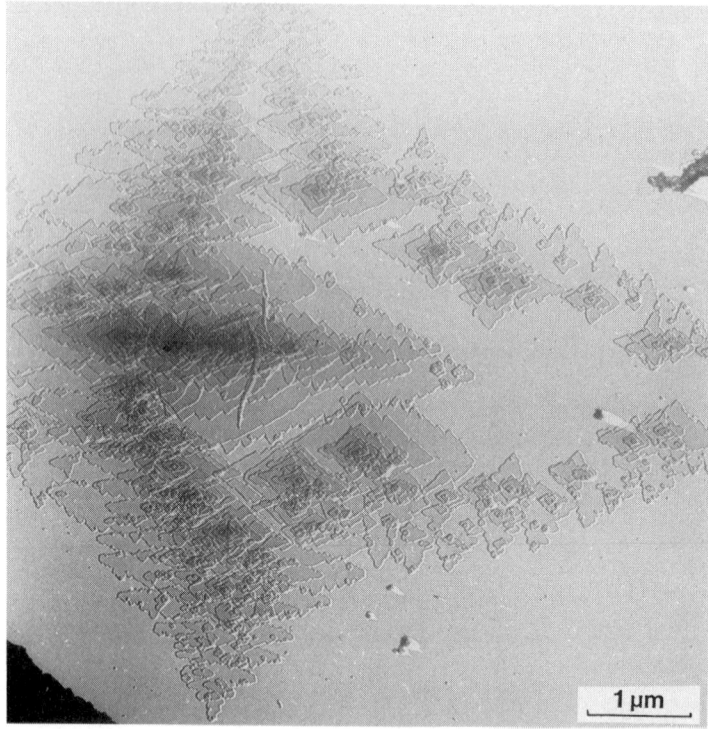

Figure 13   Polyethylene dendrite crystallized from solution at low temperature

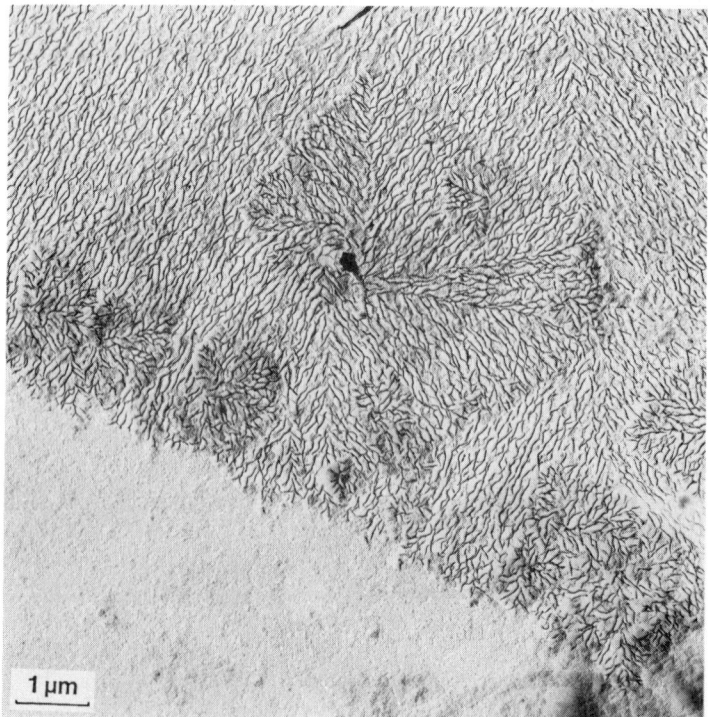

Figure 14   Decorated polyethylene lamella showing sectorization and a screw dislocation

the spiral. By superimposing such processes onto the basic elements described above, more complex structures can be generated (see Khoury and Passaglia[33]). A full account of such effects would, however, go beyond the scope of this chapter. The discussion above highlights the basic elements concerned with the proliferation of lamellae through the action of screw dislocations and, as will be demonstrated shortly, these basic principles are of considerable significance in relation to the growth of complex lamellar aggregates such as spherulites.

12.4 LAMELLAR AGGREGATES—SPHERULITES AND AXIALITES

The preceding section described how, during lamellar growth, additional layers may be formed through the action of screw dislocations. Under conditions where growth is slow and the number of such defects limited, well-ordered so-called multilayered crystals are formed. However, such structures are only generated under specific growth conditions such as at moderate solution concentrations, or from melts which are either highly impure or else only slightly supercooled. More generally, polymers tend to form complex lamellar aggregates which are either sheaf-like or spherical in outline. Such structures are of great interest from an industrial point of view, being typical of melt crystallization. As such, spherulites have been studied extensively with a view to understanding their origins and structure since, as far as many macroscopic properties are concerned, the precise arrangement of lamellae within spherulites and of spherulites within the bulk are of great importance. For example, it has long been appreciated that as crystallization proceeds, so the shorter, less crystallizable components of the melt tend to be rejected, such that they become concentrated between spherulites.[66] As such, interspherulitic regions constitute sites of potential weakness, *via* which crack propagation or dielectric breakdown may occur. For reasons such as this spherulitic size represents an important parameter for bulk property control.

In the following sections the development of spherulitic morphologies will be discussed both from a historical point of view and in the light of recent investigations.

12.4.1 Spherulites, Axialites and Hedrites

The most commonly studied structure in polymer morphology is the spherulite. This is because spherulites constitute the morphological form which is most readily grown from the melt, and also because of the ease by which such structures may be viewed using the polarizing optical microscope. However spherulites are not exclusive to polymeric materials.[67] Indeed such structures may be

Figure 15 A saccharine spherulite

grown in many organic and inorganic small molecular systems where crystallization proceeds from a melt which is both viscous and impure. Under these conditions approximately spherical poly-crystalline aggregates develop, with a variety of internal microstructures, depending upon the materials involved. An example of a non-polymeric spherulitic structure is shown in Figure 15.

In addition to spherical entities, during an investigation of the growth of apatite in a glass–ceramic system sheaf-like objects were observed.[68] Similar structures can be identified in polymeric systems, where they are believed to represent the precursors to full spherulitic development. Such entities have variously been described as axialites or hedrites, depending upon the crystallization conditions.

The term axialite was first coined by Bassett, Keller and Mitsuhashi[69] to describe multilayered objects grown from relatively concentrated solutions of polyethylene in xylene. The lamellae in these objects, which are bounded by low index faces, diverge, to a first approximation, from a single central axis and, as a result, the overall appearance is either polygonal or sheaf-like depending upon the direction in which they are viewed. Although similar objects crystallized from the melt were originally termed hedrites,[51] it is the term axialite which is now more commonly applied and it is this which will be used here to describe sheaf-like structures which have not developed a spherical envelope.

12.4.2 Spherulites and Crystal Growth

The formation of spherulites is a feature of crystal growth which is still not well understood. That is reflected in part in the different views taken by different authors as to what constitutes the essential nature of a spherulite.[27, 70–77] Three principal views are to be found. If one starts from the classic optical observation of a spherulitic section between crossed polars then invariance of the Maltese cross with rotation of the specimen implies that, to the resolution of the microscope, all radii are equivalent. In practice this is something of an approximation, partly because differences of orientation (extinction with respect to the radius) can often be detected in coarse textures and, less trivially, because the central sheaf-like stages following nucleation are ignored. Some authors consider that spherical symmetry, with all radii equivalent, is essential to a spherulite and draw a distinction between the final, large object and its initial, non-spherical stages. An alternative view, which we support, is to recognize that all stages require a branching and splaying microstructure to maintain radial orientation and fill space. Accordingly the essence of spherulitic growth would be the production of such a microstructure. All stages of spherulitic development fall within this definition but the earliest would not have attained full spherical symmetry. The differences between these two points are largely semantic; they have no significant consequence in the understanding of spherulitic growth but need to be appreciated to avoid confusion.

There is, however, a third point of view associated especially with the theory of Keith and Padden.[66, 78] According to these authors spherulites are 'fibrous' and they have developed a theory of how 'fibres' would be expected to form from typical polymer melts. On the other hand evidence of internal lamellar organization has been used to show that such 'fibres' are not generally present in polymer spherulites.[71] Although this disagreement turns in part on what is meant by a 'fibre', matters of fact are also involved.

Early optical studies established that there is radial lattice continuity in spherulites.[79] Continuity in the equivalent radial units implies that they are long and narrow, in which sense they can be termed 'fibres'. But Keith and Padden go further, identifying 'fibres' as a result of habit degeneration due to cellation in crystal growth. This would give 'fibres' a specific character and predicted cross-section, different from what has been found. Moreover evidence for the proposed habit degeneration is either weak or absent. Nevertheless, it has been held that spherulites contain lamellae in 'fibrous' units in contrast to axialites which have not reached a sufficient size for the lamellar habit to degenerate.[80] The new evidence fails to support this hypothesis which, in our view, is not a satisfactory basis by which to designate objects as spherulites.

The original proposal of Keith and Padden[66, 78] was an ingenious attempt to carry over to polymers ideas of crystal growth which were being successfully applied to segregation in alloys.[81] In particular they drew attention to the fact that polymers were polydisperse and did not crystallize homogeneously from the melt. They showed directly, in polyethylene, that uncrystallizable additives became concentrated to some extent along radii but especially between spherulites. With hindsight it is now apparent that segregation is especially prominent in polyethylene, and that other factors besides segregation are also involved in spherulitic growth, notably in low crystallinity materials. Recent work has reinforced earlier doubts that segregation is the cause of spherulitic growth; it appears rather to be an effect. Nevertheless with the wealth of new information now available at the lamellar level, as opposed to previous optical studies, it is worth re-examining spherulitic growth from first principles.

The salient facts of spherulitic growth are its geometry (by which we mean the eventual spherical envelope and equivalence of radial units produced from a space-filling, branching and splaying microstructure) and the linear increase in radius with time under isothermal conditions. It was pointed out by Keith and Padden that this constant growth rate was inconsistent with the solution to the diffusion equation for an expanding sphere. They inferred that this meant growth conditions were constant with time and, in the knowledge that added impurities coarsened the texture eventually to the extent of separating crystalline units at optical resolutions, proposed that the relevant diffusion fields were local and related to impurity segregation. This is the central premise of their theory but it is not a necessary one from the experimental evidence. The isothermal lateral growth rate of polymer lamellae, including those from the melt, is itself constant. One need look no further to account for the constant growth rate of spherulites. Moreover, the separated radial units seen in thin films of heavily doped spherulites are not necessarily rod-like units. They may approach this condition in isotactic polypropylene[26] where lamellar widths are normally large fractions of a micron but in isotactic polystyrene where lamellar widths can exceed 10 μm, one is simply seeing groups of lamellae edge on.[71,82] Furthermore the texture of spherulites grown from highly-doped melts is different from typical textures. Doping, for example, with 50% atactic polymer produces groups of, say, five or six lamellae separated by atactic polymer not unlike a two-dimensional projection of the Keith–Padden model. Typical spherulites have the different texture produced by splaying of *individual* dominant lamellae. This appears to be the most general texture found throughout spherulites of many polymers. The features of heavily-doped melts appear to be additional phenomena possibly diffusion related.

We have now touched on the remaining feature of spherulites, namely the scale, measured tangentially, of texture within spherulites. Optical measurements, not least the extensive series due to Keith and Padden,[66,78] show finer textures at lower crystallization temperatures, coarser and more open ones for growth nearer the melting point. Keith and Padden interpreted this as resulting from local diffusion and the segregation of impurities. By analogy with cellular microstructures produced in the linear solidification of metallic and other alloys the expectation is of 'fibres' (*i.e.* cells) of polymer lamellae separated by regions with a high concentration of 'impurities' (*i.e.* cell walls). The width of these cells was predicted, as in metallic alloys, to be of order δ, the characteristic length $= D/G$. This would be measured by the lamellar width whereas the lamellar thickness, l, would be set by chain-folding and $\delta/l \gtrsim 1$.

This prediction that lamellar widths should vary as $\sim \delta$ is the only quantitative one made by Keith and Padden. However, measurements of Bassett and Hodge[83–85] on polyethylene showed that this was not the case and that lamellar width varied much less, remaining within a factor of 3 μm over the achievable crystallization range; Keith and Padden have now confirmed that this is so.[77]

On the general question of whether the scale of internal texture is set by diffusion, two points are to be made. Firstly a spherulitic framework of individual dominant lamellae which branch and splay is qualitatively different from the cellulation model assumed by Keith and Padden. It has later crystallizing species located between individual dominants rather than concentrated at 'fibre' boundaries. Secondly extensive measurements of the scale of internal texture, namely the separation of adjacent dominants, denoted λ, have been made in isotactic polystyrene.[82] These do not scale with the characteristic length, δ. Nor would they be expected to unless branching and splaying were controlled by diffusion. In the theory of Keith and Padden branching is viewed as a statistical process, with branching angles decreasing in probability from zero, but only those growth centres separated by $\sim \delta$ are viable. In practice branching angles are much greater, $\sim 25°$, and once a dominant lamella has begun to grow it generally continues until impingement on another, *i.e.* its viability depends on geometry rather than on the scale of a diffusion field.

A final point concerns the reality of the instability of lamellar habit once the size exceeds the characteristic length. The only direct evidence of which we are aware is of a composite figure of four different lamellae of isotactic polystyrene.[80] These were interpreted as showing the sequence of development of an individual lamella. However, if one does observe the development of an individual lamella with time, similar shapes can be reproduced but they are found to grow with constant shapes. Similar experiments on poly(4-methylpentene-1) show that lamellae of this polymer also retain their equiaxial habit to dimensions considerably exceeding the characteristic length. We conclude that there is no clear evidence in favour of habit changes due to cellulation.

In practice spherulites are constructed of radial sequences of lamellae often linked by screw dislocations. In high symmetry polymers such as polystyrene and poly(4-methylpentene-1) this is how radial continuity is achieved, rather than by a change in habit to a 'fibre'. This factor was neglected by Keith and Padden[67] but its importance has recently been recognized.[77]

In summary, therefore, textural examination of spherulites with lamellar resolution does not, in our opinion, support the hypothesis that spherulites are a general consequence of cellulation during crystal growth. The actual position appears to be that concluded previously,[71] *i.e.* that the lateral textural dimension is generally set by the branching and splaying of individual dominant lamellae. Additional ordering occurs in highly doped systems which in some respects might be governed by the characteristic length. Although there has been controversy on this topic the recent recognition that ordering associated with segregation in whole polymers (*i.e.* with a substantial low molecular weight component) is different from fractions suggests that the facts are now removing the basis of disagreement.[71,77]

12.4.3 Axialites

Sheaf-like axialites have been identified in a number of systems, under particular crystallization conditions. Such structures are observed where the nucleation density is high and the volume available for growth is sufficiently limited that mature spherulites are unable to develop,[86] or in situations where growth occurs slowly, such as from impure melts[87] or at temperatures approaching the system's equilibrium melting point.[26,27,70] Since, as far as the first of these two situations is concerned, the resulting objects clearly differ from spherulites only in extent, growth having been curtailed through impingement with neighbouring structures, further discussion here will be limited to the larger sheaf-like structures which are commonly observed at high crystallization temperatures.

Figure 16 shows the detailed lamellar microstructure of sheaf-like axialites in a number of systems, all of which have been studied at the University of Reading. Since the Polymer Physics Group at this institution has taken a leading role in the investigation of melt-crystallized morphologies, much of that which follows, both within this and subsequent sections, is necessarily concerned with a review of its work. Micrographs such as those shown illustrate the great variety of morphological forms which are observed within polymeric systems, together with their structural complexity. Also these micrographs exemplify the ever-present problems associated with the differentiation of fundamental underlying principles from the superimposed features which relate to a particular polymer and the particular crystallization conditions. For example, despite the chemical simplicity of polyethylene, this polymer exhibits a great variety of lamellar forms, depending upon the molecular weight distribution of the crystallizing material and the conditions employed.[83-85] A combination of these

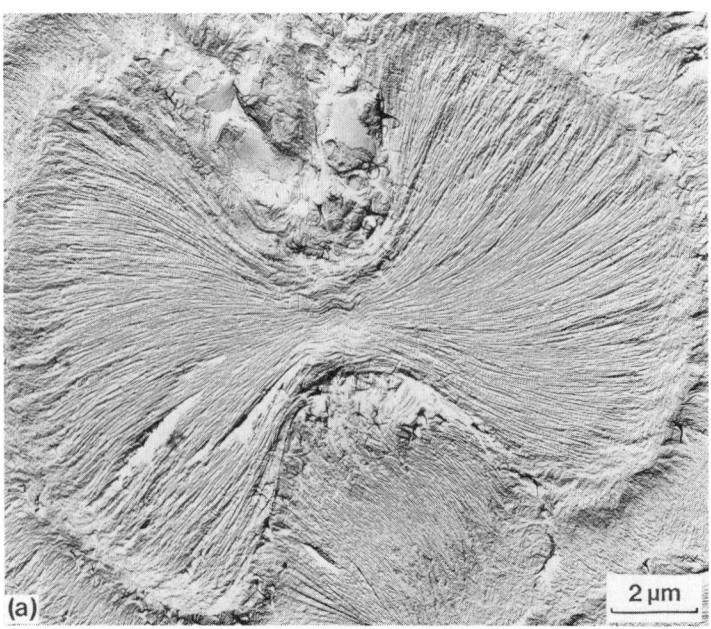

Figure 16(a)

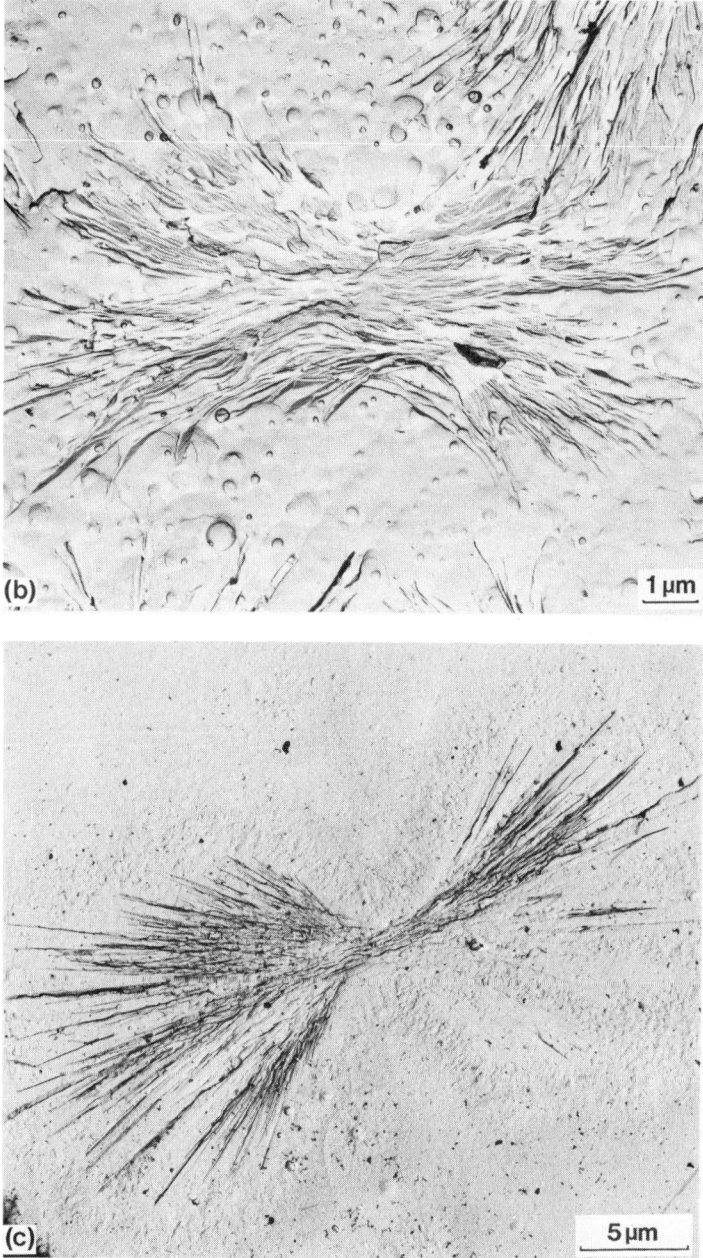

Figure 16 Sheaf-like lamellar aggregates crystallized from the melt in (a) a blend of linear and low density polyethylene at 125 °C, (b) isotactic polystyrene at 220 °C, (c) isotactic polypropylene at 160 °C

morphological complexities with the difficulty associated with quenching polyethylene to 'freeze in' the structure present at any given time within the melt means that micrographs of polyethylene are often difficult to interpret. Polypropylene presents similar problems, particularly at low crystallization temperatures as a result of copious twinning,[26, 33, 89] which gives rise to the characteristic so-called cross-hatched structure of this polymer. At high temperatures, however, cross-hatching is suppressed and structures of the type shown in Figure 16(c) are seen.[26] Objects such as this appear axialitic in structure, possessing orthogonal views which are sheaf-like and polyhedral and with splaying restricted to a single axis (see Figure 17). However, the very limited lateral development of the lath-like lamellae, which are characteristic of α-polypropylene, means that the objects shown are effectively two dimensional, as a result of which classic splaying behaviour is to be expected. Under certain circumstances the appearance of polypropylene, like polyethylene, can become complicated

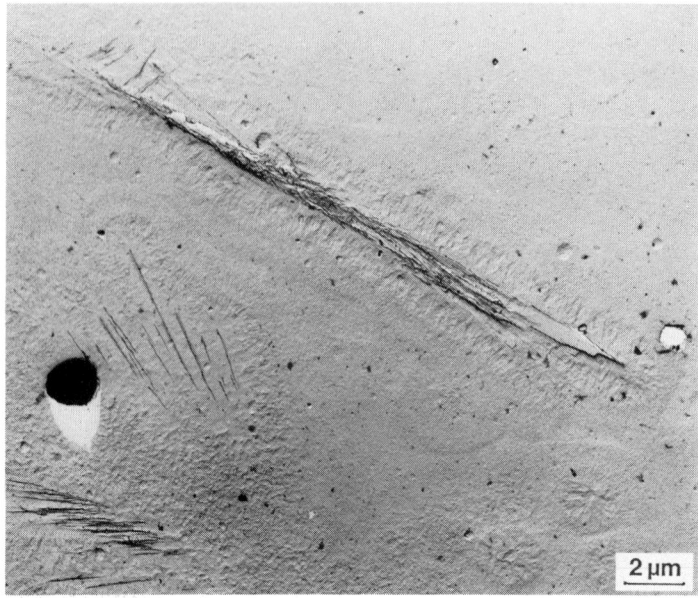

Figure 17 Lath-like projection of a lamellar aggregate in polypropylene crystallized from the melt at 160 °C

as a result of crystallization during quenching. The 'fringe' surrounding the polypropylene lamellae seen in Figures 16(c) and 17 result from such development.

Isotactic polystyrene, however, constitutes a system which has none of the problems and limitations described above. Not only do polystyrene lamellae tend to adopt a simple, planar hexagonal habit under a wide range of crystallization conditions, such that complex hierarchical morphologies are not observed, but also the melt is easily quenched into the glassy state. Thus polystyrene constitutes a system which is structurally simple, and, therefore, amenable to investigation.

Figure 18 shows three orthogonal projections of a sheaf-like structure in isotactic polystyrene crystallized at 228 °C. This temperature, which is only about 14 °C below the equilibrium melting point, represents a temperature at which crystallization proceeds very slowly, the objects in

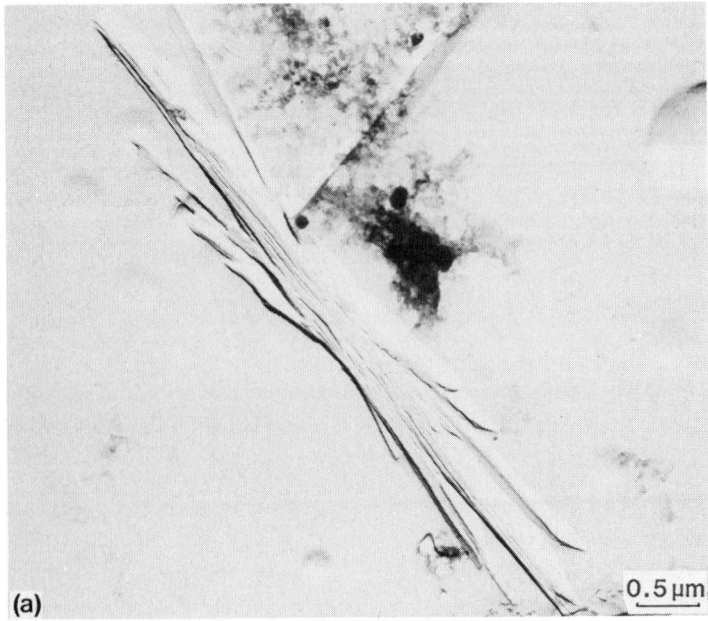

Figure 18(a)

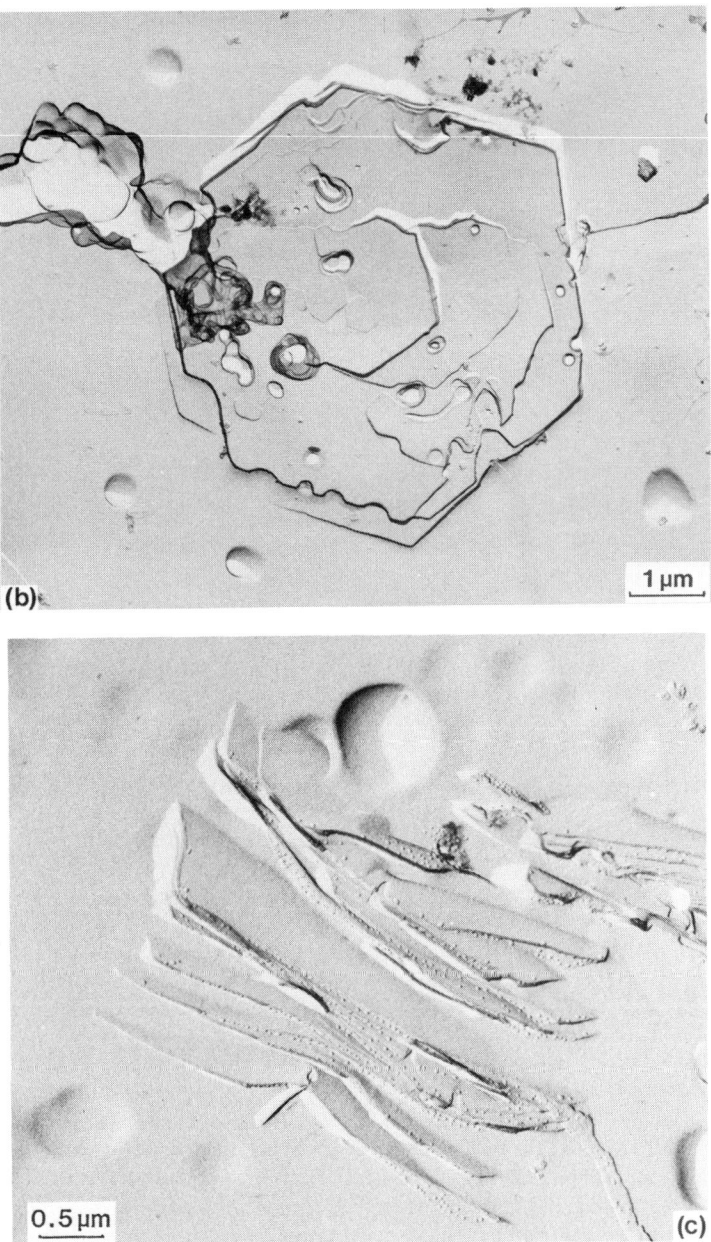

Figure 18 Isotactic polystyrene crystallized at 228 °C showing lamellar aggregates similar to idealized axialities: (a) sheaf-like view, (b) hexagonal view, (c) edge-on view

Figure 18 having taken eight days to grow. As a result of this low growth rate, the objects shown exhibit a simple structure, similar to that of an idealized axialite, as previously defined. However, as the crystallization temperature is reduced, so the observed structures undergo significant changes both with respect to their complexity and axial nature.

Figure 19 shows a sample of isotactic polystyrene crystallized at 210 °C, in which the objects seen are more representative of the type of structure usually referred to as axialite. However, these sheaf-like structures are not axialites in the strictest sense, since their constituent lamellae do not splay about a single axis. This fact is clearly demonstrated when the etched surface intersects with an object in a plane well removed from its centre. In Figure 20(a) an object is viewed in such a projection. Here the appearance is such that it is not possible to identify a unique axis about which lamellae diverge from one another. Rather, what is shown by the structure seen in Figure 20(a) is that lamellae splay apart about all possible axes. This conclusion is confirmed by the structure shown in Figure 20(b) which could also not be generated if splaying were restricted to a single axis.

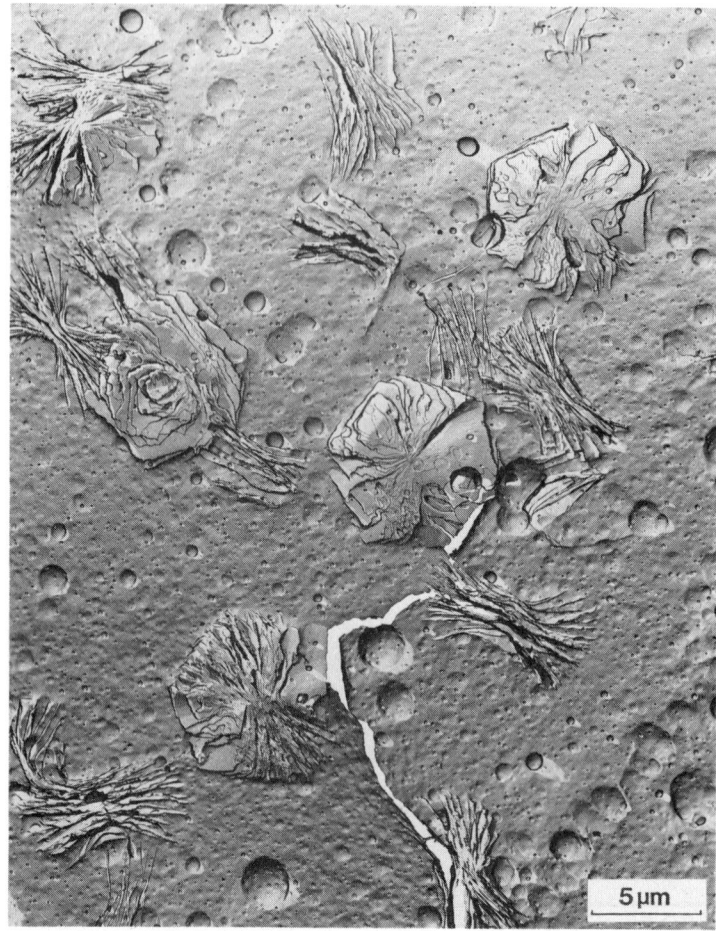

Figure 19 Isotactic polystyrene crystallized at 210 °C showing structures in a range of orientations

In systems such as polystyrene where lamellae grow as well-developed two-dimensional sheets this conclusion is not surprising. Lamellae branch through the action of screw dislocations, which give rise to spiral structures containing a number of layers. As shown in Figure 21, successive layers then tend to splay apart. Thus under circumstances where the lamellar habit is well developed in two dimensions and where, as previously demonstrated, successive spiral layers are also well developed in two dimensions, there is no reason on symmetry grounds why the forces which give rise to splaying should be restricted so that lamellae diverge solely about some unique axis. In addition the fact that screw dislocations may develop on any of the lamellar growth faces will also promote full three-dimensional development.

From this it is clear that as sheaf-like structures grow, lamellae will branch and splay apart such that eventually all solid angles will be filled and a near spherical envelope attained. Thus it may be argued that the differentiation of axialites from spherulites is an artificial one, since given sufficient time and volume, real, non-idealized axialites will eventually develop spherical envelopes as a direct consequence of lamellar splaying not being confined to one axis. Nevertheless particular morphological forms have been associated with particular growth conditions and transitions from sheaf-like to spherical morphologies have been observed at distinct temperatures. In polyethylene, for example, an axialite/spherulite transition has been observed at ~ 127 °C,[70] a temperature close to that at which the growth kinetics change from regime I to regime II, and the association of particular growth kinetics with each of these structural forms has been suggested. In isotactic polystyrene similar morphological observations have also been made.[82,87] At temperatures above ~ 200 °C sheaves are observed, whilst below this value spherulites are seen. However, analysis of the growth kinetics of this system has not revealed any distinct change from regime II growth kinetics on raising the crystallization temperature.[87,90] Nevertheless, to a large extent, whether sheaves or spheres are observed under particular growth conditions depends upon the frequency with which branching

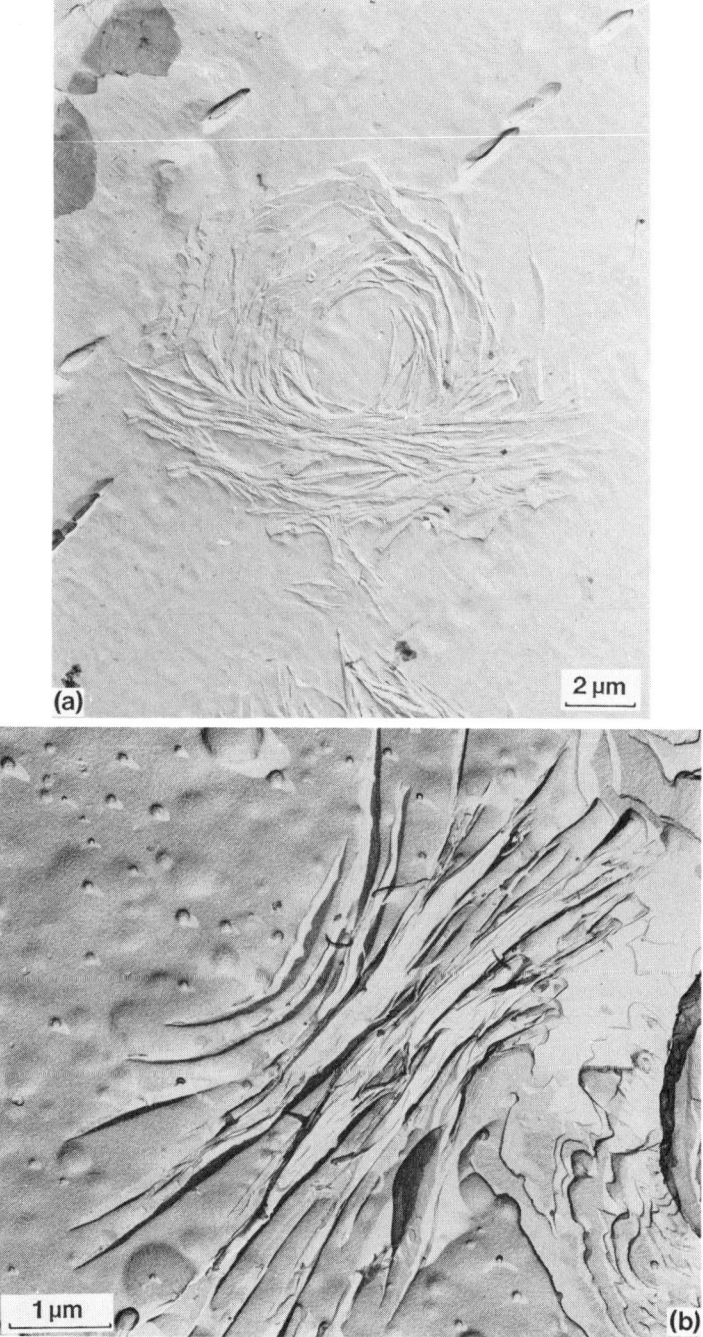

Figure 20 Micrographs of sheaf-like structures in polystyrene crystallized at 210 °C for an etched surface which intersects with the lamellar aggregates well away from the centre of the axialite: (a) etched plane parallel to the hexagonal view shown in Figure 18(b); (b) etched plane parallel to the sheaf view shown in Figure 18(a)

occurs. Since branching stems from the presence of screw dislocations, which themselves result from staggered molecular fold conformations, it would seem likely that the precise arrangement of molecular stems on the crystal growth surface should be important in determining the morphological form observed. Nevertheless from the preceding argument the authors believe that in general spherulites and axialites differ by degree rather than in any more fundamental respect.

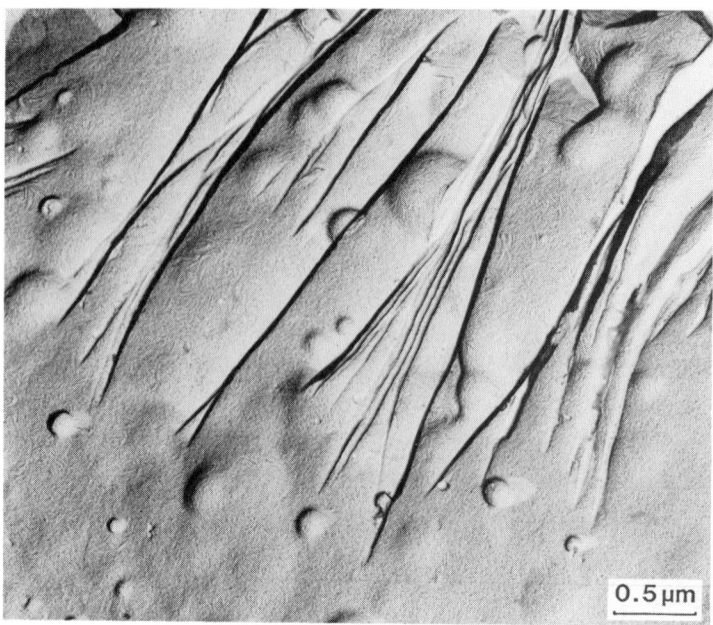

Figure 21   Lamellar branching in isotactic polystyrene crystallized at 220 °C

12.4.4   Spherulites — Structure and Growth

In comparison with those objects so far discussed, spherulites are highly complex lamellar aggregates which can exhibit a wide number of different features, depending upon the system and crystallization conditions considered. For this reason the spherulite architectures observed in a number of systems will be described, and the features which are common to several systems will be discussed.

12.4.4.1   *Polyethylene*

The material for which melt-crystallized morphologies were first studied in detail was polyethylene, largely as a result of this being the system on which the techniques of chlorosulfonation and permanganic etching were developed.

The first samples examined were crystallized at high temperatures, where the resulting structures are at their most simple. It was found that in a linear fraction ($\bar{M}_w = 2.6 \times 10^4$, $\bar{M}_w/\bar{M}_n = 1.3$) crystallized for 20 days at 130.6 °C, lamellae primarily grow in the form of ridged sheets, within which the molecules are inclined to the basal planes at $\sim 38°$. Thus the fold surfaces of these crystals approximately correspond to the {201} planes of the polyethylene subcell. When viewed down their growth direction (the crystallographic *b*-axis) the appearance of the lamellar aggregates which develop under the above conditions is as shown in Figure 22. This micrograph shows an array of ridged lamellae ~ 500 Å thick, separated from one another by regions of very much thinner lamellae. By examining the edge of a spherulite it can be shown that the thick lamellae develop first and in so doing enclose columns of melt which may subsequently crystallize either isothermally (near the centre of a spherulite) or on quenching (near the growth tips). In this way, the overall form of the structures observed are dictated by the first-forming, thicker lamellae which are, as a result, termed dominant.[83] The thinner later-forming lamellae which crystallize within the pre-established dominant framework only play a passive infilling role in terms of the overall development of the gross morphological forms which are observed. As a result these lamellae have been termed subsidiary.[83] This is not to say, however, that subsidiary lamellae are of secondary importance, but merely that the shape of the gross morphological features observed in practice is dictated by the development of the dominants. Indeed many important macroscopic parameters are fundamentally influenced by the way in which subsidiary crystallization occurs.

It has long been supposed that the component of the melt which crystallizes first corresponds in common circumstances to the longest and most crystallizable molecules in the distribution.

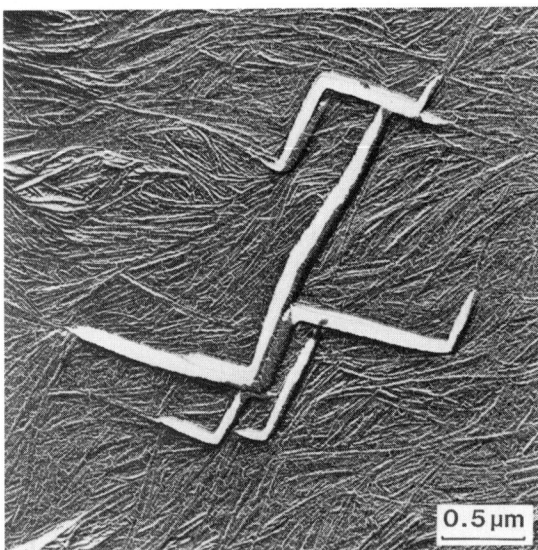

Figure 22 Spherulite tip viewed down the radial crystallographic *b* direction, showing ridged lamellae: polyethylene crystallized at 130·6 °C

In polyethylene this has been demonstrated clearly, using solvent extraction[84] and blending techniques.[91,92] Thus the longer molecules tend to be located within the population of dominant lamellae, shorter molecules in regions which crystallized isothermally within the dominant framework, and shorter molecules still in regions which were unable to crystallize prior to quenching. In this way molecular fractionation occurs, whereby molecules of a particular length and degree of regularity tend to become concentrated at specific locations within the overall structure. Such placement of particular molecular fractions at particular sites can exert a considerable influence on macroscopic properties. For example, polyethylene regions in which lower molecular weight material is concentrated are less able to resist crack growth and provide easy routes by which cracks may propagate through the bulk, so leading to premature failure. However in the case of polyethylenes in which an appreciable quantity of branched molecules are present, molecular segregation occurs primarily on the basis of branch content, since branched structures are not easily incorporated into a crystal lattice. As a result, the average separation between adjacent branches sets an approximate upper limit for the lamellar thickness. Thus long-branched molecules introduced into an otherwise linear material tend to become concentrated in what would otherwise be regions of exclusively low molecular weight polymer: locally the average length of segregated molecular species is increased with, as a result, an improvement in overall stress-cracking resistance.

The above describes how molecular fractionation occurs in a melt-crystallized polymer, and the effect that such a process can have under appropriate conditions. However, polyethylene fractionation does not only give rise to the concentration of particular molecular components at particular locations within the bulk, but also, under certain circumstances, can give rise to a variety of different lamellar habits.

Previously the morphology of a narrow fraction of polyethylene crystallized at a very high temperature was described, in which the basic lamellar unit is the ridged sheet. However, as the crystallization temperature is reduced a transition is observed from a ridged lamellar habit to one of planar sheets.[83–85] A similar change is observed if the molecular weight is varied at constant temperature. However this change, which effectively involves an increase in distance between consecutive reversals of direction of the ridge, represents only a transitory change in morphology, from ridges to the more general S-shaped lamellar habit which is characteristic of melt-crystallized polyethylene under most conditions. Such lamellae are shown in Figure 23 where again a spherulite is viewed down its radial growth direction so that, as in Figure 22, a number of individual, dominant lamellae are seen separated by regions of quenched melt. However, unlike the ridged structures previously described, S-shaped lamellae do not exhibit a habit with more than one reversal of direction, *i.e.* curvatures of more than one wavelength are not observed. Nevertheless C-shaped structures, corresponding to one half of an S, are seen. A possible explanation for these curved habits was proposed by Keith and Padden.[93] They suggested that when lateral growth faces and fold

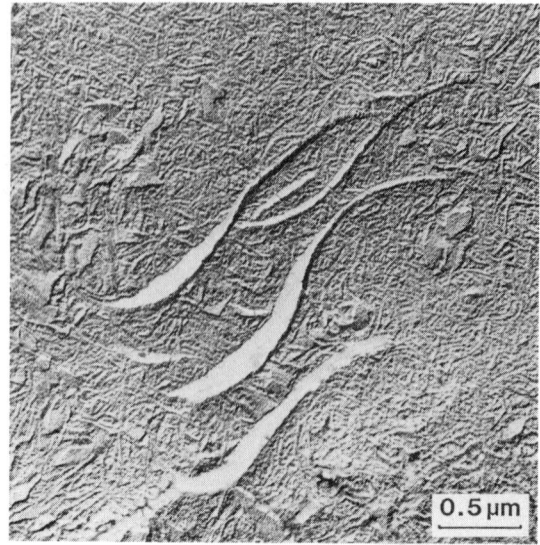

Figure 23 Spherulite tip viewed down the radial crystallographic *b* direction showing S-shaped lamellae: polyethylene crystallized at 128.1 °C

surfaces are not orthogonal, as is the case in polyethylene, then different degrees of disorder develop at the opposite fold planes of one lamella, such that the stresses on opposing surfaces are unequal. These disparate stresses, they proposed, then give rise to the bending moments which result in the observed non-planar crystal habits.

In the above discussion a number of features of spherulitic growth have been described, which, it is believed, are common to all the polymeric systems so far studied. These main points are worthy of further emphasis at this point. Growth proceeds by the initial formation of an array of dominant lamellae, composed of the longest and most regular molecules present within the melt. Following

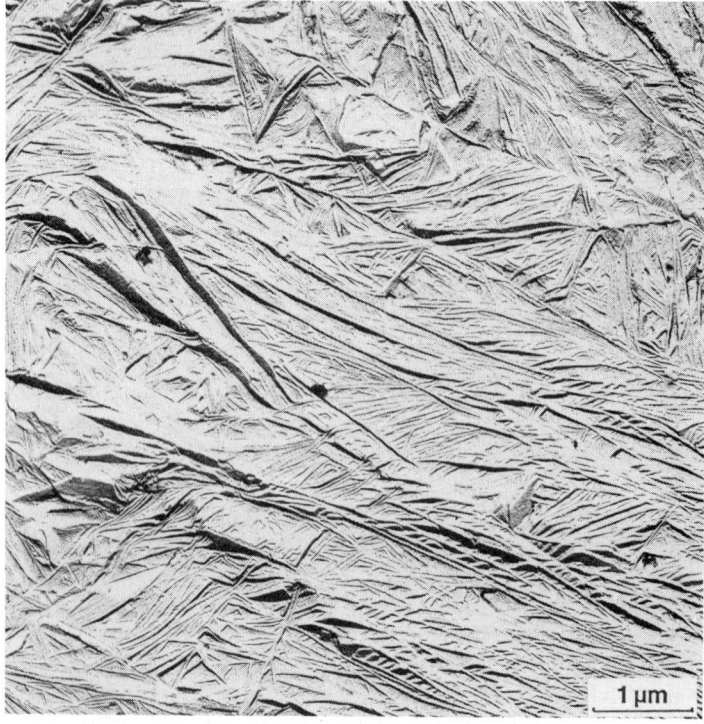

Figure 24 Blend of linear and low density polyethylene (1:1) crystallized at 125 °C showing a range of different lamellar habits

establishment of the dominant framework, subsequent subsidiary crystallization occurs at interstitial sites, to a degree dictated by the crystallization conditions and the precise molecular composition of the melt. For example crystallization of blends of linear and low density polyethylene (LDPE) at temperatures above that at which LDPE can crystallize results in only limited subsidiary isothermal development, no further crystallization being possible prior to quenching. A micrograph of such a sample is shown in Figure 24, where isothermally crystallized lamellae can be seen separated by quenched LDPE. Fractionation effects such as this lead to particular molecular components being sited at particular locations. Thus complex hierarchical microstructures can be generated, as shown, in which curved, planar and ridged lamellae can all be seen.

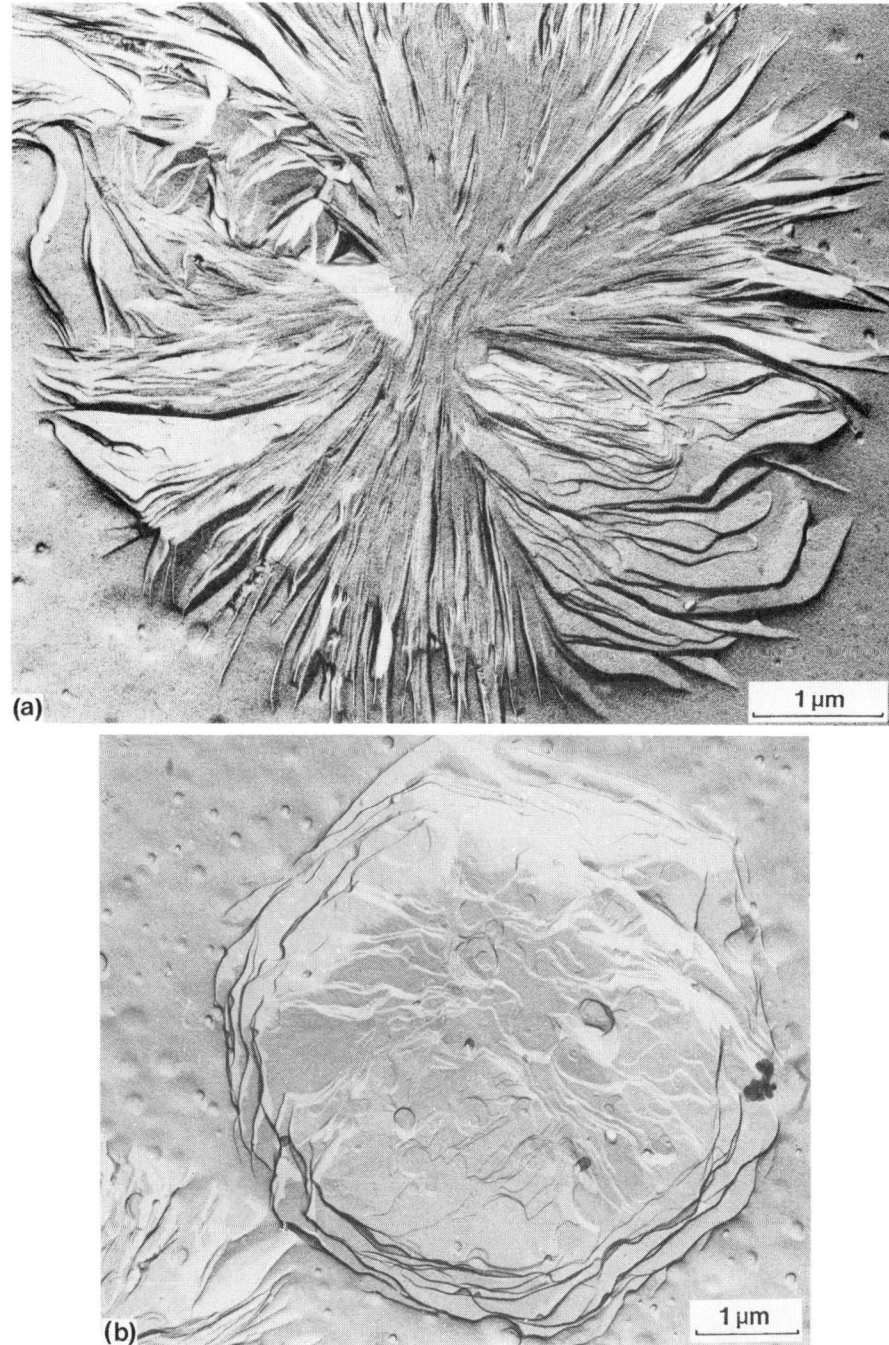

Figure 25 Micrographs of isotactic polystyrene crystallized at 200 °C showing spherulites in the (a) sheaf-like and (b) hexagonal projections

12.4.4.2   *Isotactic polystyrene*

In Section 12.4.3, the sheaf-like structures which develop in isotactic polystyrene at crystallization temperatures of 210 °C and above, were discussed. Figure 25 shows two micrographs of a sample of isotactic polystyrene crystallized at 200 °C for 40 mins. From Figure 25(a) it can be seen that in this sample development has proceeded far enough for spherical objects to be observed. These spherulites are composed of basically hexagonal lamellae, shown in Figure 25(b), which have grown out, branched and splayed apart in three dimensions until an approximately spherical envelope has been attained. This is not to say, however, that all the lamellae in the objects shown in Figure 25 are perfectly hexagonal, since clearly this cannot be so, on geometrical grounds. However, branching occurs through the formation of screw dislocations which give rise to the proliferation of layers in a spiral form. As discussed in Section 12.3, in the absence of external influences, successive terraces of the spiral reflect the symmetry of the parent lamellar crystal. Thus the objects shown in Figure 25,

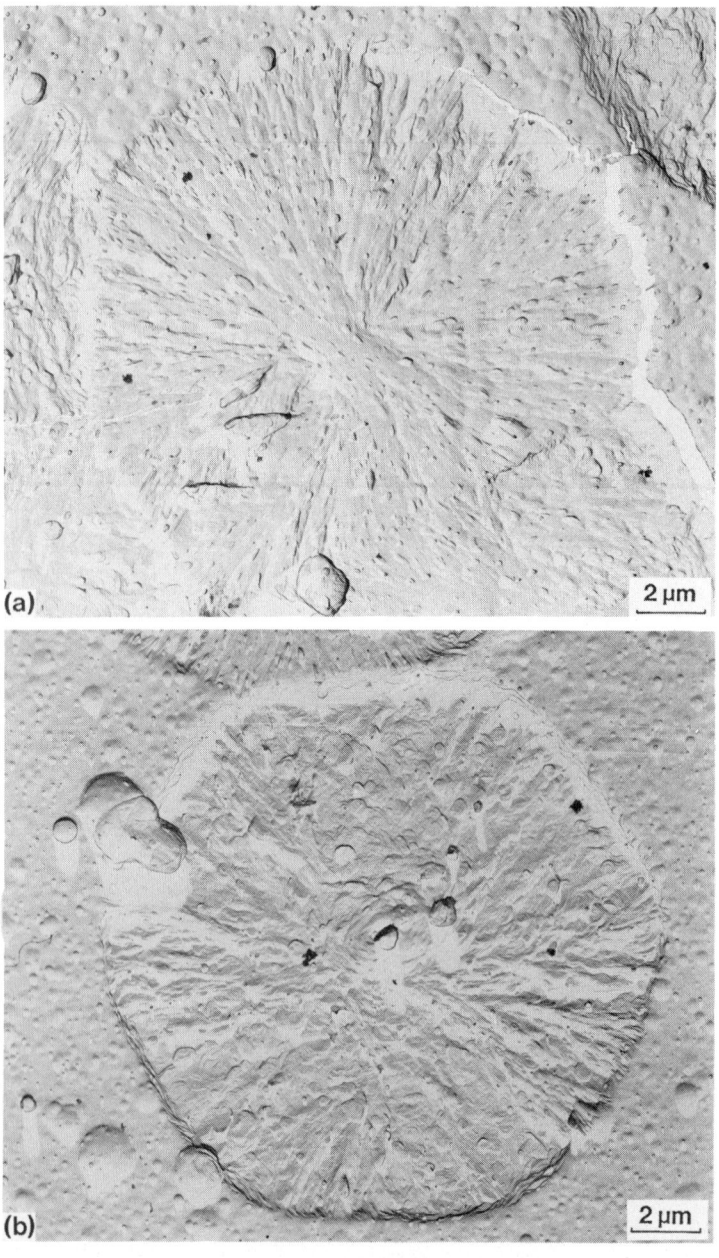

Figure 26   Micrographs of isotactic polystyrene crystallized at 190 °C showing spherulites in the (a) sheaf-like and (b) hexagonal projections

although not made up of a large number of complete, perfectly hexagonal crystals, are composed of lamellae the shapes of which reflect their mechanism of formation (generation by screw dislocations) together with the symmetry of the parent crystal habit. Thus aggregates are formed which, when viewed down the chain direction, can be seen to be hexagonal in shape, being composed of lamellae with wide, approximately planar growth facets.

Similar features are observed in more mature objects such as those shown in Figure 26, which were crystallized from the melt at 190 °C. Figure 26(a) shows the densely packed lamellar structure of such a spherulite together with, at the centre, its sheaf-like origins. Also, as in the objects previously

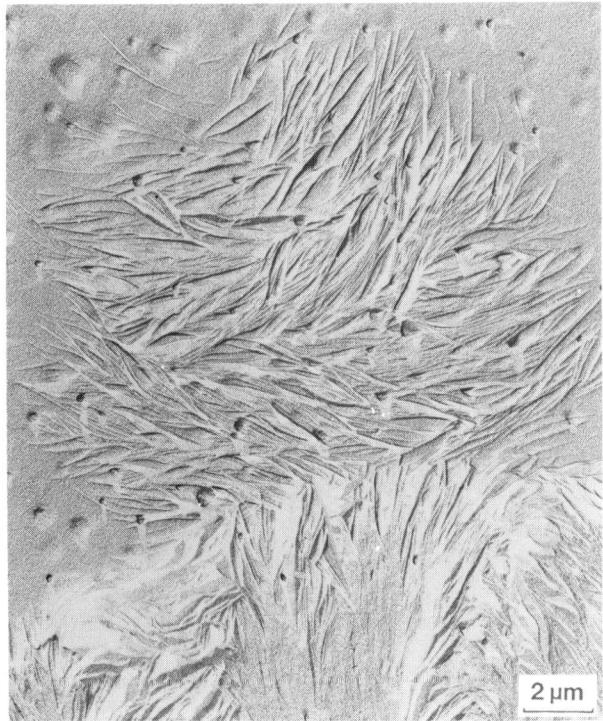

Figure 27 Spherulite of isotactic polystyrene crystallized at 190 °C, viewed along its radial growth direction

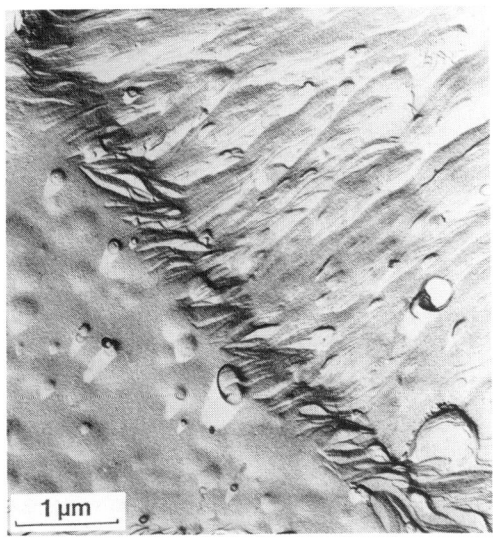

Figure 28 Growth tips of a spherulite of isotactic polystyrene crystallized at 190 °C, showing dominant/subsidiary structure

described at 200 °C, spherulites crystallized at 190 °C have a hexagonal appearance when viewed down the molecular chain axis (see Figure 26b). Examination of such structures along a radial direction shows that, as in polyethylene, growth proceeds by a dominant/subsidiary mechanism (see Figure 27). However, limited attempts to study fractionation in polystyrene[27] suggest that, whilst this does occur, such effects are not as marked as in polyethylene. Also, since in polystyrene only one lamellar habit is ever seen at a given crystallization temperature, no marked differences between dominant and subsidiary lamellae are observed. Thus dominant and subsidiary lamellae can only be clearly differentiated near the edge of a spherulite, where individual dominant lamellae can be seen at the tips of the structure separated from one another by regions of quenched melt. Such an arrangement is shown in Figure 28 together with the very dense array of lamellae which subsequently develops on subsidiary crystallization.

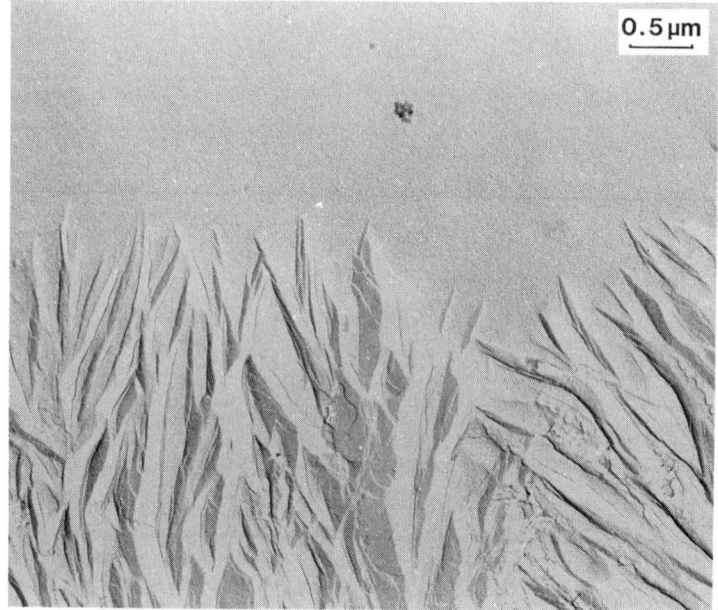

Figure 29 Spherulitic growth tips in a 1:1 blend of isotactic and atactic polystyrene crystallized at 190 °C

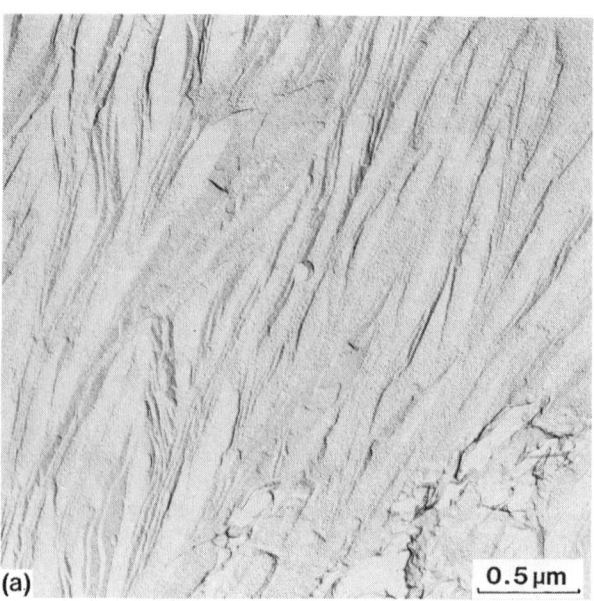

Figure 30(a) Radiating lamellar bundles separated by non-crystalline regions. A 1:1 blend of isotactic and atactic polystyrene crystallized at 190 °C

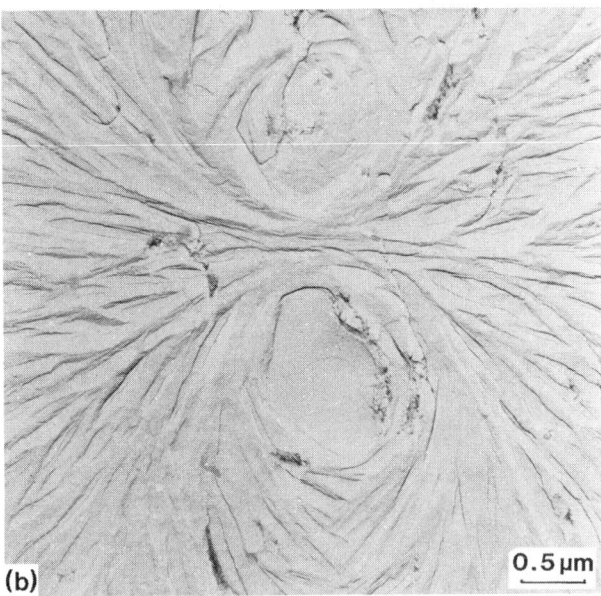

Figure 30(b)   Spherulitic structure in a 1:1 blend of isotactic and atactic polystyrene crystallized at 190 °C

The above mechanism of growth also occurs in blends of isotactic and atactic polystyrene, but with rather different consequences with respect to the overall appearance of the structure. Figure 29 shows the growth tips of a spherulite crystallized at 190 °C from a blend containing equal masses of the two polymers. In these circumstances the isotactic material present in the interstitial regions of melt between the established dominant lamellae undergoes subsidiary crystallization, leaving the atactic material incorporated within the spherulite in distinct, well-defined regions. Thus, as shown in Figure 30, the general appearance of such spherulites is one of an array of fibrous lamellar bundles, separated from one another by non-crystalline regions. Although the thickness of these fibres along the chain directions is limited to about five lamellae, their width is only limited by the criteria discussed above for the completely isotactic case. Indeed as mentioned in Section 12.2, crystallization from highly doped melts has been successfully employed to generate simple multi-layered crystals of isotactic polystyrene with a well-defined hexagonal habit.

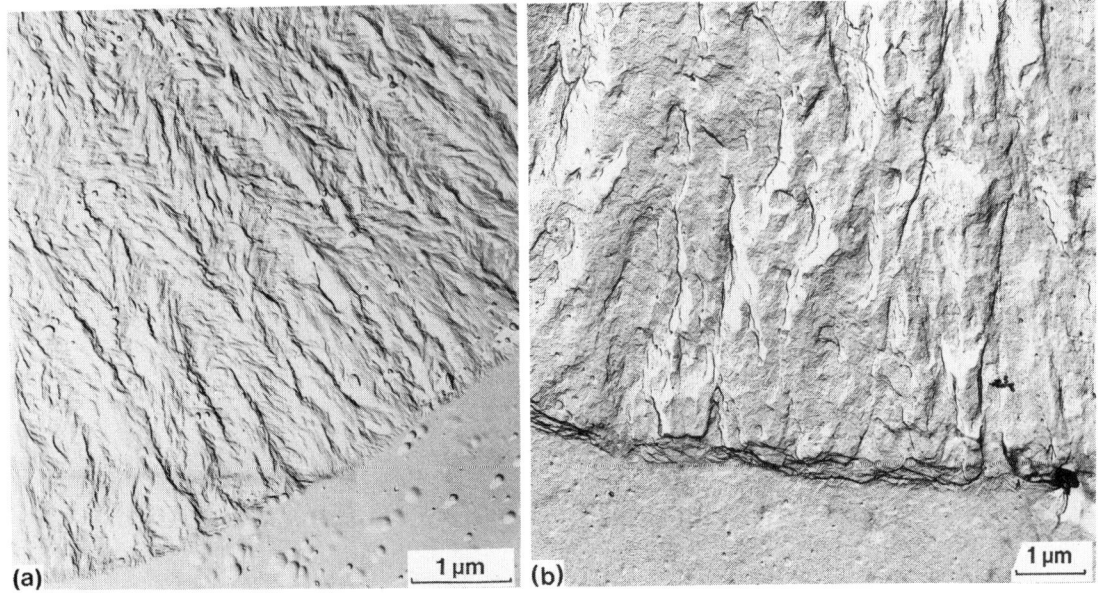

Figure 31(a,b)   Spherulitic edge structure in polystyrene: (a) 100% isotactic melt crystallized at 190 °C; (b) 100% isotactic melt crystallized at 162 °C

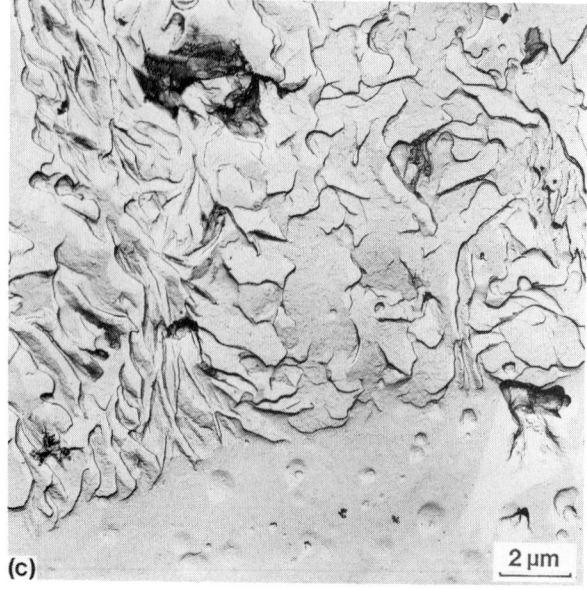

(c)   2 μm

Figure 31(c)   Spherulitic edge structure in polystyrene crystallized at 162 °C; 20% isotactic, 80% atactic melt

All the spherulitic morphologies so far described, both in polyethylene and polystyrene, have been based on individual dominant and infilling subsidiary lamellae and, as such, differ greatly from expectations based on lamellar fibrillation due to cellulation where a hexagonal shape would be unstable. However, in polystyrene, significant changes are observed at crystallization temperatures lower than those so far considered. The structural changes seen suggest that some additional process becomes operative which serves to limit widths of lamellar aggregates, so giving rise to an additional dimension of ordering intermediate between the lamellar and the spherulitic. A comparison of the edge structure in spherulites grown at 162 °C and 190 °C is given in Figure 31. In samples crystallized at 162 °C there appears to be a radially elongated texture. In terms of individual lamellae, fine detail suggests that this is linked to their proliferation, especially through screw dislocations, leading to an overall elongation along the radius. This effect appears merely to be superimposed onto the dominant/subsidiary growth mechanism previously described, as shown in Figure 32 where individual lamellae can again be seen at the spherulite growth tips, and bears no

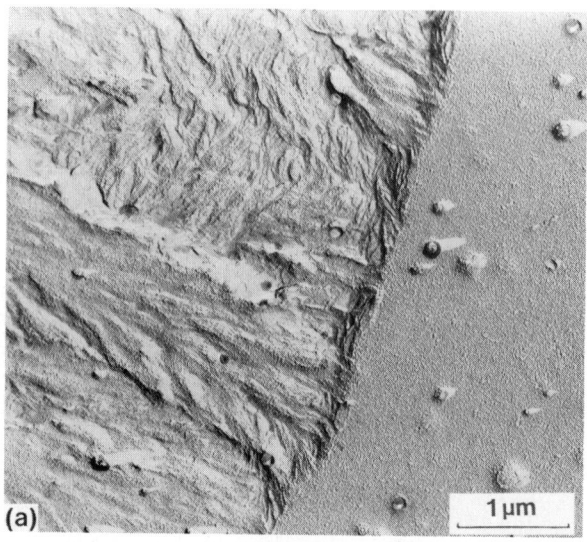

(a)   1 μm

Figure 32(a)

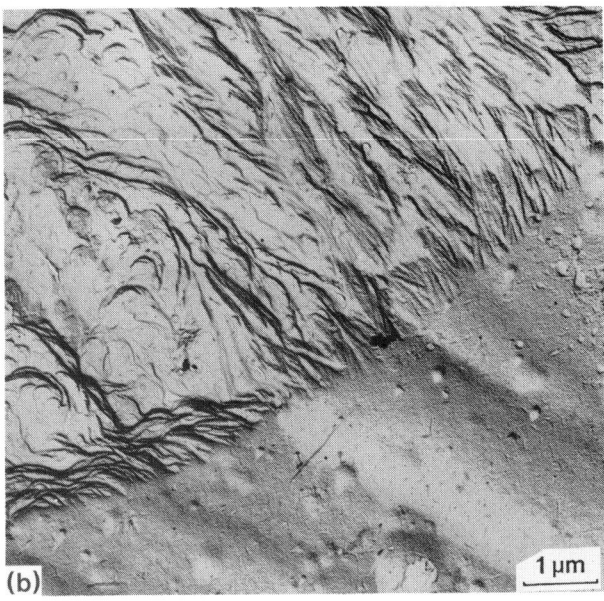

Figure 32 Dominant/subsidiary structure in polystyrene crystallized at 162 °C: (a) isotactic polystyrene; (b) 1:1 blend of isotactic and atactic polystyrene.

obvious relation to diffusion-generated instabilities. Thus it would seem that even where lamellae do become radially elongated, this process is of only secondary importance to spherulitic growth rather than having any fundamental significance.

12.4.4.3 Polypropylene

Under most conditions polypropylene crystallizes into the monoclinic α form which is based on narrow lath-like lamellae. Frequently this gives rise to very complex morphologies as a result of the

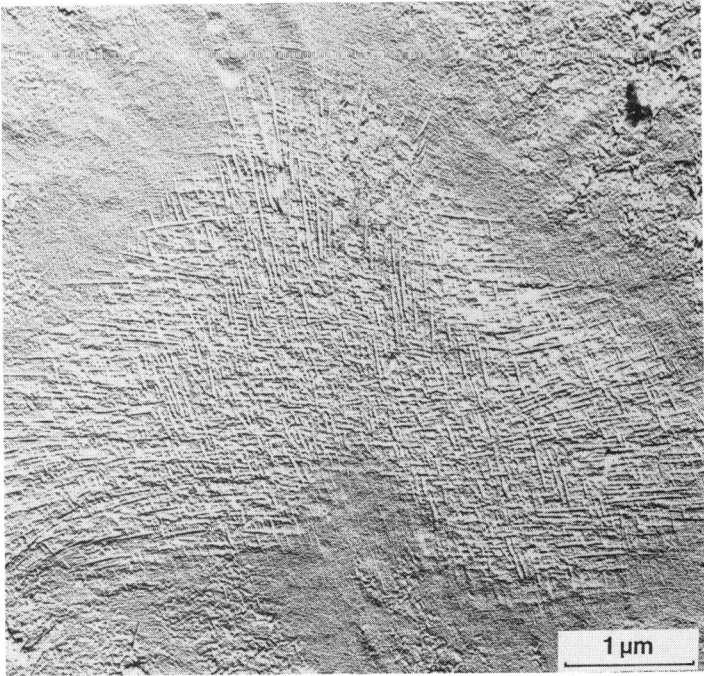

Figure 33 Isotactic polypropylene crystallized from the melt at 150 °C, showing cross-hatching

frequent twinning which occurs, this being known as cross-hatching.[88] A comparatively simple cross-hatched structure is shown in Figure 33. By examination of similar structures to this, but crystallized from solution, it has been shown that in one component of the twinned structure laths grow outward along a^* with the chain axis (c-axis) perpendicular. The second, identical component of the twinned structure is then inclined at 80.7° to the first, the crystallographic b-axis being common to both populations.[89] At high crystallization temperatures however, above about 155 °C, cross-hatching is suppressed, such that simple sheaf-like structures are formed initially (see Figure 16c), which eventually develop into spherulites. Examination of Figure 34 once again reveals an open framework of dominant lamellae near the extremities of the structure, together with a very much more compact arrangement of lamellae nearer the centre, which is generated by subsidiary crystallization. At lower temperatures dominant/subsidiary growth still occurs; however by then the resulting structures become very much more complicated due to copious cross-hatching. Figure 34(b) shows a spherulite crystallized at 150 °C within which this characteristic cross-hatched lamellar arrangement can be seen. However, this feature is not found distributed equally throughout the structure, rather it tends to occur preferentially at specific locations. For example in the object

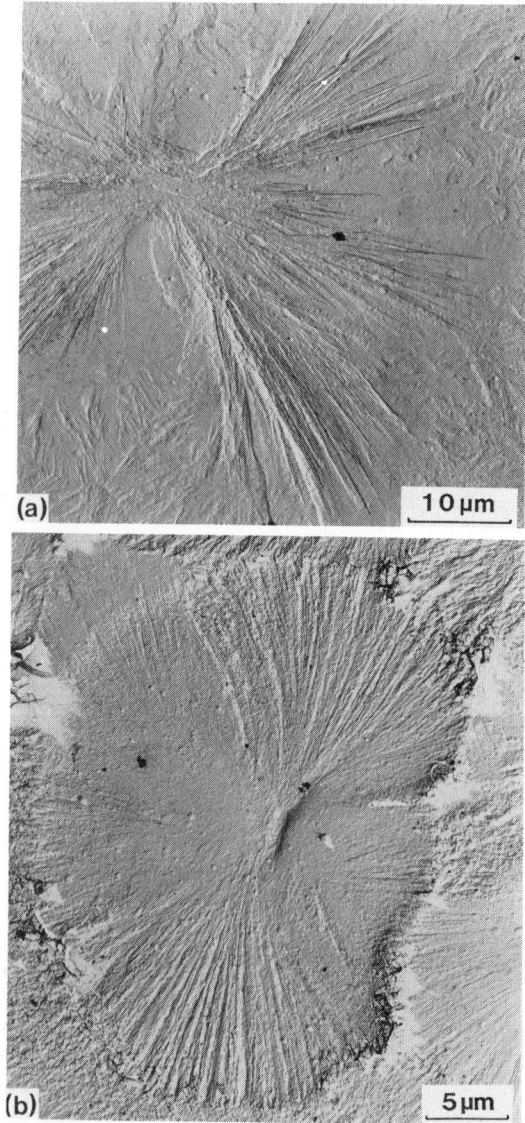

Figure 34 Spherulites of polypropylene: (a) when crystallized at 159.5 °C, cross-hatching is absent due to the high crystallization temperature; (b) when crystallized at 150 °C, cross-hatching develops, particularly on either side of the central sheaf

shown in Figure 34(b) the 'eyes' on either side of the distinct sheaf-like region are profusely cross-hatched, in contrast to the sheaf itself where the dense array of radiating laths tends to suppress the twinning process.

As in polystyrene, the above objects, which were crystallized at comparatively high temperatures, are all based on individual dominant, subsidiary and cross-hatching lamellae. However under conditions where growth is sufficiently fast another textural feature becomes superimposed onto the prevailing pattern. This is shown in Figure 35 where the interior morphology of a polypropylene spherulite crystallized at 130 °C can be seen. The appearance of this textural element, based on many lamellae, is remarkably similar to that of certain lamellar eutectic alloys which develop under conditions of constitutional supercooling.[94,95] In polypropylene these structures may well have their origin in similar diffusion-related processes, and their dimension would be expected to be of considerable relevance to the Keith and Padden theory of spherulite growth.[67] However as in polystyrene, the significance of such a textural pattern would appear to be limited due to the restricted range of conditions under which it is observed, since spherulitic structures develop under conditions where cellulation is not apparent. Rather the process which seems to be generally applicable is that of branching and splaying apart of dominant lamellae followed by in-filling, cross-hatching and subsidiary crystallization.

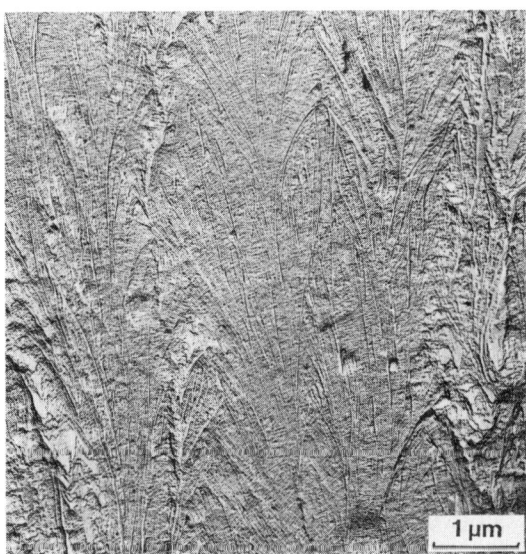

Figure 35 Interior morphology of a polypropylene spherulite crystallized at 130 °C showing the additional textural element intermediate in scale between the spherulite and the lamella

12.4.4.4 *Banded spherulites*

In all the spherulites so far described the structure is based on lamellar crystals radiating outwards from the centre. In this way the early optical observation of polymer spherulites being birefringent entities with radial symmetry may be explained. However in addition to the characteristic Maltese cross extinction pattern which is seen when such an object is viewed in plane-polarized light, in certain circumstances spherulites may be seen which exhibit an additional pattern of concentric rings. Examples of such spherulites, which are termed banded, are shown in Figure 36. More detailed optical examination reveals that this characteristic extinction pattern occurs as a result of the optic axes of the crystalline units spiralling systematically and cooperatively about the spherulite radius. That is, as the lamellae grow outwards from the spherulite's centre, so they twist about the radius and adopt a helical configuration. This has been confirmed by X-ray microbeam diffraction and TEM, whereby it has been shown that the dark concentric bands correspond to regions of the structure where lamellae are viewed flat on, whilst in the bright birefringent region they are seen in an edge-on projection. Thus whilst the general structural significance of banded extinction patterns is understood, the reason why lamellae should adopt such a conformation is still unclear.

The first polymer in which the lamellar structure of banded spherulites was studied in any detail was polyethylene, a system in which such structures are primarily composed of lamellae with a non-

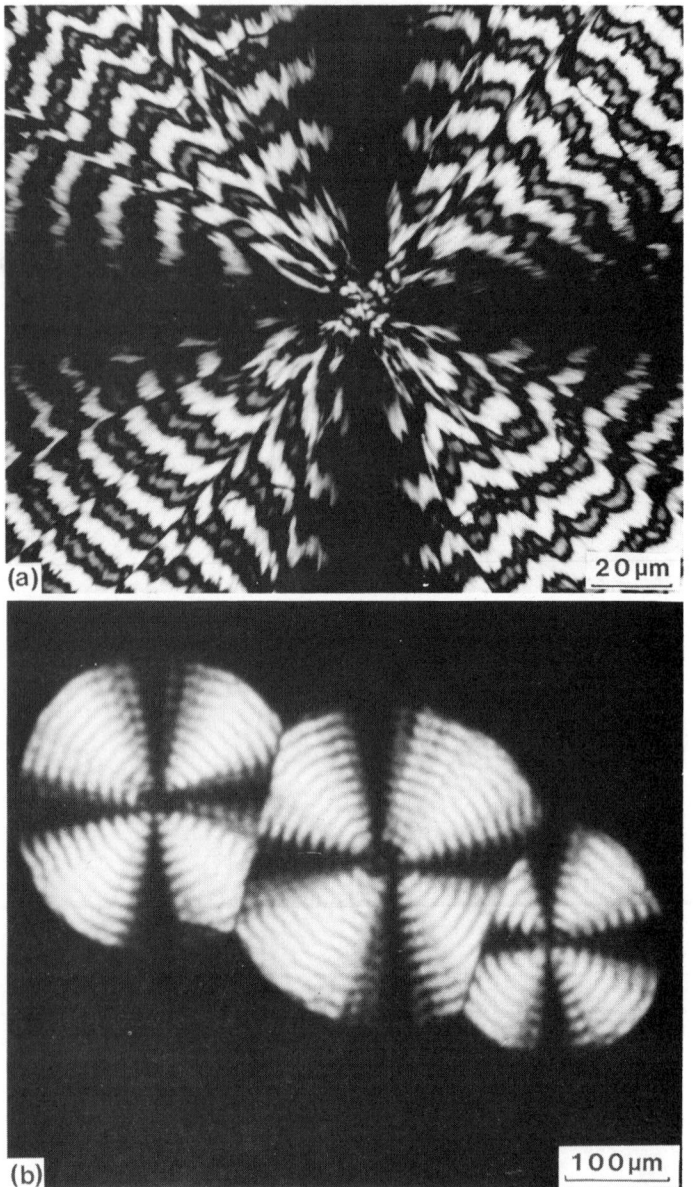

Figure 36 Optical micrographs of banded spherulites — plane-polarized, transmitted light: (a) poly(hydroxybutyrate); (b) poly(vinylidene fluoride)

planar S-shaped profile. As previously described, Keith and Padden[93] proposed that such a lamellar habit develops in response to disparate surface stresses, which originate in the inclined molecular stem conformation characteristic of polyethylene lamellae. These workers also argued that, when accompanied by asymmetrical development, as a result of diffusion processes or impingement limiting lateral growth, such forces would give rise to lamellae adopting a smoothly twisted helicoidal form. However, experimentally, no evidence was found in polyethylene for such uniformly twisted lamellae; rather what were observed were lamellae with only slight twisting for about one third of a band period followed by a region in which more abrupt orientational changes occurred (see Figure 37).[85] Nevertheless Keith and Padden reconciled this apparent discrepancy with the suggestion that the observed non-uniform twisting is not a feature of the initial dominant growth but rather results from in-filling crystallization.

Poly(vinylidene fluoride), PVF$_2$, is another polymer in which banded spherulites are observed.[96] A recent study of the lamellar structure of these objects[97] reveals that, as in polyethylene, lamellae do not adopt a smoothly twisted conformation but rather undergo major re-orientations at regular

Figure 37 A banded spherulite of polyethylene crystallized at 122 °C, showing lamellar re-orientation approximately every third of a band period

intervals. The lamellar structure of a banded spherulite of the α crystal form of PVF_2 is shown in Figure 38. However, unlike polyethylene, the lamellae of α-PVF_2 grow in the form of planar sheets, as can be seen in Figure 39 where a spherulite is viewed along its radius. Whilst no evidence is as yet available concerning the orientation of molecular stems within the lamellae of α-PVF_2 banded spherulites, examination of solution-grown crystals by electron diffraction shows the stems to be perpendicular to the lamellar fold surface.[98] This, together with the cross-sectional profiles of the lamellae shown in Figure 39 and the apparently flat lamellae seen in certain regions of Figure 38 is difficult to reconcile with the model for banded spherulites outlined above.

An alternative model for banded spherulites has been proposed by Schultz and Kinloch,[99] which is based on the inherent lamellar twist perpendicular to the axis of a screw dislocation. By considering the cumulative effect of a sequence of screw dislocations with the same hand or sign, a theoretical expression was derived which provided a good fit to the experimental data of Lindenmeyer and Holland[100] for the twisting of crystals in a fraction of polyethylene crystallized at several temperatures. However, this model also fundamentally depends on molecular chain stems being inclined relative to the lamellar normal, since it is proposed that when a screw dislocation is formed at the growth tip of a lamella composed of inclined molecular stems then two energetically non-degenerate structural possibilities arise. In one case the sign of the screw dislocation is such that the accompanying molecular staggering brings adjacent folds closer together whilst in the other, the folds are effectively moved further apart, *i.e.* the fold density decreases. Thus, screw dislocations of opposite sign involve different changes in surface free energy. Since the lower energy form will tend to predominate, the lamellar twist associated with the screw dislocations will tend to accumulate, such that a twisted conformation results. In polyethylene, where molecules are inclined to the lamellar normal, changes in molecular orientation have been observed to be associated with the

Figure 38 A banded spherulite of poly(vinylidene fluoride) crystallized at 163 °C

Figure 39 A banded spherulite of poly(vinylidene fluoride) crystallized at 163 °C, viewed along a direction close to the radial crystallographic *b* direction. Lamellae can be seen to be inherently planar rather than S-shaped as in polyethylene

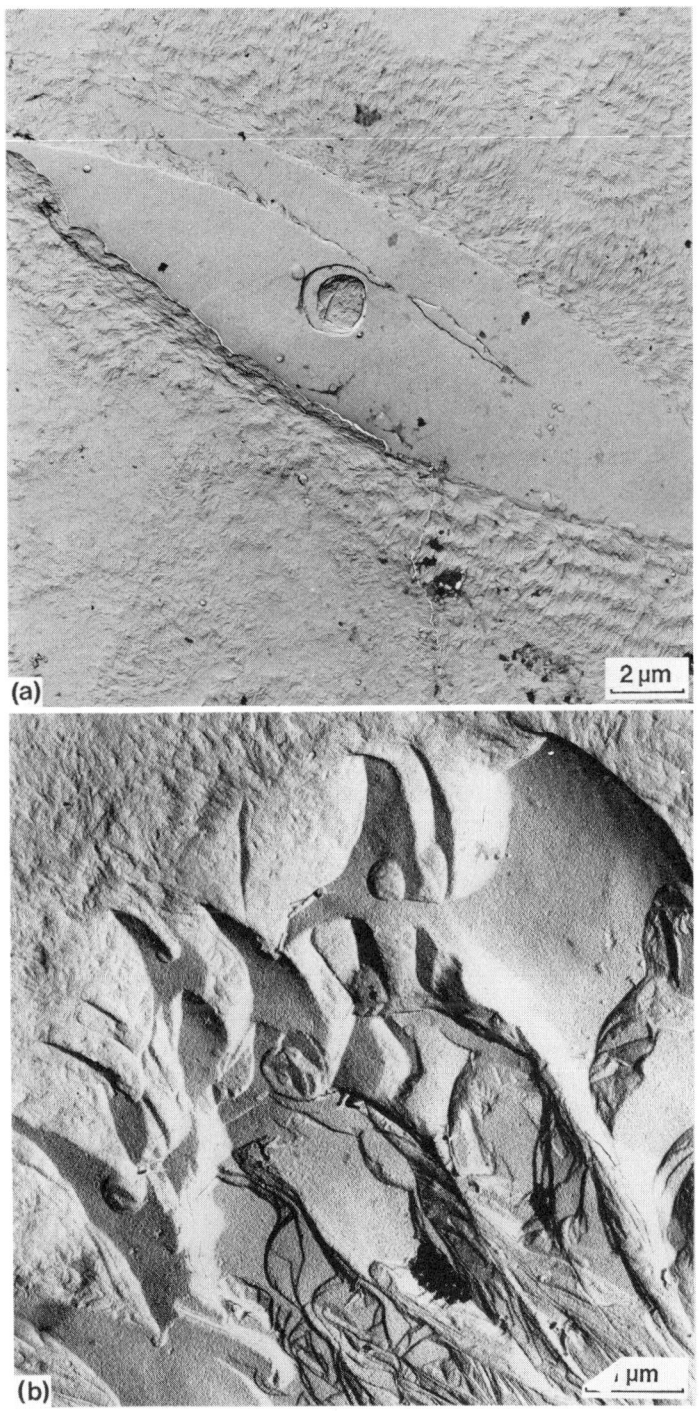

Figure 40 Lamellar re-orientation at screw dislocations in polyethylene: (a) a simple lamellar aggregate crystallized at 130 °C; (b) the edge of a spherulite grown at 122 °C

presence of screw dislocations, as shown in Figure 40. As such, in that system, the model of Schultz and Kinloch is appealing. However, whether the difference in energy between left- and right-handed screw dislocations is sufficient to account for the morphological effects observed is, at this point, unclear. Also, there is apparently no reason for screw dislocations of opposite sign to be energetically non-degenerate in the case of α-PVF$_2$, where no evidence exists for inclined molecular stems.

12.4.4.5 *Spherulite nucleation*

The nucleation step of spherulite growth is known to be associated with the presence of inhomogeneities in the melt. The importance of foreign bodies, such as catalyst particles remaining after synthesis, or nucleating agents, such as talc, deliberately introduced to influence spherulite size and ultimate properties, is readily appreciated. However recent work suggests that under certain circumstances an additional nucleation process may operate.[101]

Figure 41 shows a large number of small lamellar aggregates of isotactic polystyrene. The sample chosen was crystallized at 220 °C from a high molecular weight fraction, prepared by a dissolution/precipitation process, such that it was reasonably free of catalyst residues and other foreign bodies. Whilst the sheaf-like aggregates of hexagonal lamellae shown are not dissimilar to those previously described in samples of polystyrene crystallized at high temperatures, they do differ from more mature structures in a number of important ways. First, due to their limited size, in-filling crystallization from within the interstitial regions between the first forming lamellae has not begun, and differentiation into dominant and subsidiary lamellae is impossible. Secondly all the lamellae within a given object are of about the same size. This implies that in each structure several lamellae began growing at the same moment, rather than one forming first followed by the generation of additional layers through the action of screw dislocations. This is not to say that screw dislocations do not play a part in the early stages of lamellar growth, but rather that an additional process may operate which can give rise to multilamellar aggregates directly.

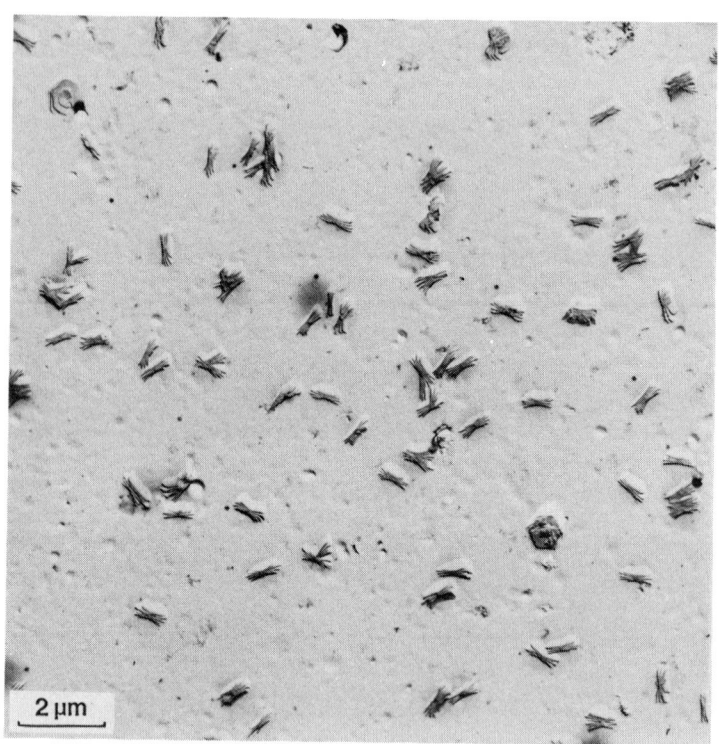

Figure 41   Small nuclei in isotactic polystyrene crystallized at 220 °C for 81 min

Examination of even less mature objects than those shown in Figure 41 reveals the source of this effect. The lamellar aggregates shown in Figure 42 can be seen to consist of an array of parallel, presumably chain-folded lamellae, connected together by a central thread or core. Comparison with similar shish-kebab morphologies suggests that the structures shown result from chain-folded crystallization about a core which contains some chains which have adopted an extended conformation.[102,103] Whereas in the case of shish-kebabs, such extended molecular conformations are introduced and sustained through the imposition of some external stress field,[104–107] in the case of the polystyrene samples shown, which were not deliberately strained, it has been suggested that the extended molecular conformations were established during production of the samples and were then

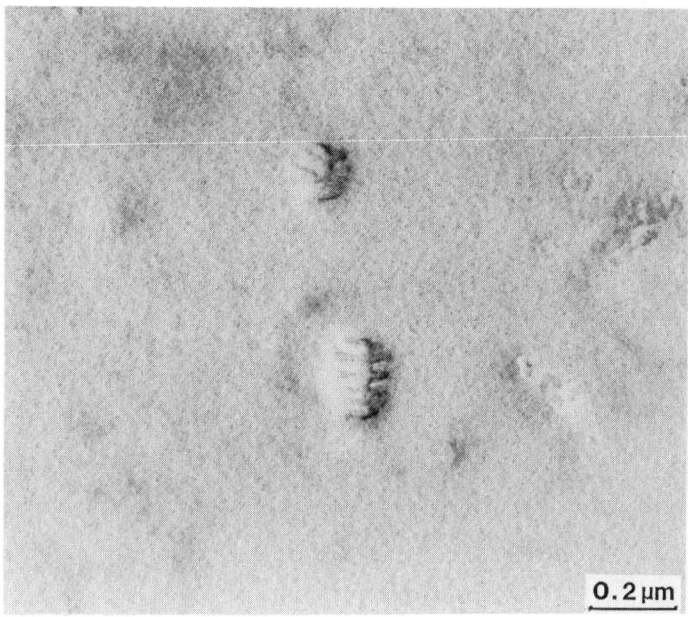

0.2 μm

Figure 42 Small nuclei exhibiting internal structure, *i.e.* a core plus lamellar overgrowths: isotactic polystyrene crystallized at 220 °C for 17 min

unable to relax prior to crystallization, despite heating the polymer to ~ 20 °C above its equilibrium melting point, as a result of the action of molecular entanglements.

In the case of linear polymers, this process has been demonstrated to be highly dependent on the molecular weight distribution, since it is the longer molecules which are most prone to entanglement.[108] However, in crystallizable cross-linked systems where a molecular network already exists, the above nucleation process should be of considerable consequence, since in such materials extended molecular conformation may occur without the prerequisite of entanglement. Indeed in the case of cross-linked polyethylene, recent results demonstrate that the nucleation density does indeed increase with increased cross-linking.[86]

12.5 CONCLUSION

In this chapter we have attempted to cover the basic features which are important in the morphology of crystalline polymers. This has been approached both from a historical point of view and in the light of recent work. Inevitably many areas have not been covered, such as the morphology of drawn polymers, gels and materials crystallized at high pressures. Nevertheless, despite these omissions, it is intended that the preceding account should cover the area of bulk, melt-crystallized polymers in a way which provides both the polymer specialist with an up to date review of current understanding in the field of polymer morphology, and the non-specialist with an introduction to the vast range of structural forms that are observed in crystalline polymers.

ACKNOWLEDGEMENTS

The authors are indebted to Dr A. M. Hodge, Mr R. H. Olley, Ms D. Patel, Dr I. A. M. al Raheil and Mr M. J. Blissett for provision of figures.

12.6 REFERENCES

1. W. T. Astbury, 'Fundamentals of Fibre Structure', Oxford University Press, Oxford, 1933.
2. O. Gerngross, K. Herrmann and W. Abitz, *Z. Phys. Chem., Abt. B*, 1930, **10**, 371.
3. C. W. Bunn, in 'Fibres from Synthetic Polymers', ed. R. Hill, Elsevier, Amsterdam, p. 240.
4. C. W. Bunn and T. C. Alcock, *Trans. Faraday Soc.*, 1945, **41**, 317.

5. A. Keller, *Philos. Mag.*, 1957, **2**, 1171.
6. E. W. Fischer, *Z. Naturforsch., Teil A*, 1957, **12**, 753.
7. P. H. Till, Jr., *J. Polym. Sci.*, 1957, **24**, 301.
8. K. H. Storks, *J. Am. Chem. Soc.*, 1938, **60**, 1753.
9. D. C. Bassett, *Philos. Mag.*, 1961, **6**, 1053.
10. A. J. Lovinger and H. D. Keith, *Macromolecules*, 1979, **12**, 919.
11. A. M. Glauert and R. Phillips, in 'Techniques for Electron Microscopy', ed. D. H. Kay, Blackwell, Oxford, 1965, p. 213.
12. D. T. Grubb, *J. Mater. Sci.*, 1974, **9**, 1715.
13. A. M. Glauert, in 'Techniques for Electron Microscopy', ed. D. H. Kay, Blackwell, Oxford, 1965, p. 254.
14. D. T. Grubb, in 'Developments in Crystalline Polymers — 1', ed. D. C. Bassett, Applied Science, London, 1982, p. 1.
15. G. Kanig, *Kolloid Z. Z. Polym.*, 1973, **251**, 782.
16. S. G. Burnay and G. W. Groves, *J. Mater. Sci.*, 1977, **12**, 1139.
17. D. C. Bassett, 'Principles of Polymer Morphology', Cambridge University Press, Cambridge, 1981.
18. E. L. Thomas, in 'Structure of Crystalline Polymers', ed. I. H. Hall, Elsevier, London, 1984, p. 79.
19. C. W. Hock, *J. Polym. Sci., Part A-2*, 1967, **5**, 471.
20. A. C. Reimschuessel and D. C. Prevorsek, *J. Polym. Sci., Polym. Phys. Ed.*, 1976, **14**, 485.
21. E. H. Andrews, *Proc. R. Soc. London, Ser. A*, 1964, **277**, 562.
22. B. C. Edwards and P. J. Phillips, *Polymer*, 1974, **15**, 491.
23. B. C. Edwards and P. J. Phillips, *J. Polym. Sci., Polym. Phys. Ed.*, 1975, **13**, 2117.
24. D. E. Bradley, in 'Techniques for Electron Microscopy', ed. D. H. Kay, Blackwell, Oxford, 1965, p. 96.
25. R. H. Olley, A. M. Hodge and D. C. Bassett, *J. Polym. Sci., Polym. Phys. Ed.*, 1979, **17**, 627.
26. D. C. Bassett and R. H. Olley, *Polymer*, 1984, **25**, 935.
27. D. C. Bassett and A. S. Vaughan, *Polymer*, 1985, **26**, 717.
28. R. H. Olley, D. C. Bassett and D. J. Blundell, *Polymer*, 1986, **27**, 344.
29. R. M. Gohil and P. J. Phillips, *Polymer*, 1986, **27**, 1687.
30. R. H. Olley and D. C. Bassett, *Polymer*, 1982, **23**, 1707.
31. R. H. Olley, *Sci. Prog. (Oxford)*, 1986, **70**, 17.
32. H. S. Bu, S. Z. D. Cheng and B. Wunderlich, *Polym. Bull.*, 1987, **17**, 567.
33. F. Khoury and E. Passaglia, in 'Treatise on Solid State Chemistry', ed. N. B. Hannay, Plenum, New York, 1976, vol. 3, p. 335.
34. E. Passaglia and H. K. Kevorkian, *J. Appl. Polym. Sci.*, 1963, **7**, 119.
35. Y. H. Kao and P. J. Phillips, *Polymer*, 1986, **27**, 1669.
36. D. C. Bassett, in 'Developments in Crystalline Polymers — 1', ed. D. C. Bassett, Applied Science, London, 1982, p. 115.
37. B. Lotz and A. J. Kovacs, *Kolloid Z. Z. Polym.*, 1966, **209**, 97.
38. B. Lotz, A. J. Kovacs, G. A. Bassett and A. Keller, *Kolloid Z. Z. Polym.*, 1966, **209**, 115.
39. A. J. Kovacs, J. A. Manson and D. Lévy, *Kolloid Z. Z. Polym.*, 1966, **214**, 1.
40. R. Perret and A. Skoulios, *Makromol. Chem.*, 1972, **162**, 147.
41. R. Perret and A. Skoulios, *Makromol. Chem.*, 1972, **162**, 163.
42. N. C. Billingham and A. D. Jenkins, in 'Polymer Science', ed. A. D. Jenkins, North Holland, Amsterdam, 1972, vol. 1, p. 1.
43. N. C. Billingham and A. D. Jenkins, in 'Polymer Science', ed. A. D. Jenkins, North Holland, Amsterdam, 1972, vol. 1, p. 121.
44. J. B. Lando and W. W. Doll, *J. Macromol. Sci., Phys.*, 1968, **B2**, 205.
45. J. A. Brydson, 'Plastic Materials', Butterworth, London, 1975.
46. D. C. Bassett and A. Keller, *Philos. Mag.*, 1962, **7**, 1553.
47. D. C. Bassett, *Philos. Mag.*, 1964, **10**, 595.
48. H. D. Keith, R. G. Vadimsky and F. J. Padden, *J. Polym. Sci., Part A-2*, 1970, **8**, 1687.
49. F. Khoury and J. D. Barnes, *J. Res. Nat. Bur. Stand., Sect. A*, 1972, **76**, 225.
50. D. C. Bassett, F. C. Frank and A. Keller, *Nature (London)*, 1959, **184**, 810.
51. P. H. Geil, 'Polymer Single Crystals', Wiley, London, 1963.
52. D. C. Bassett, F. C. Frank and A. Keller, *Philos. Mag.*, 1963, **8**, 1739.
53. I. A. M. al Raheil, R. H. Olley and D. C. Bassett, *Polymer*, 1988, **29**, 1539.
54. J. D. Hoffman and J. I. Lauritzen, *J. Res. Nat. Bur. Stand., Sect. A*, 1961, **65**, 297.
55. J. D. Hoffman, G. T. Davis and J. I. Lauritzen, 'Treatise on Solid State Chemistry', ed. N. B. Hannay, Plenum Press, New York, 1976, vol. 3, p. 497.
56. J. D. Hoffman, *Polymer*, 1983, **24**, 3.
57. S. J. Organ and A. Keller, *J. Mater. Sci.*, 1985, **20**, 1571.
58. S. J. Organ and A. Keller, *J. Mater. Sci.*, 1985, **20**, 1586.
59. S. J. Organ and A. Keller, *J. Mater. Sci.*, 1985, **20**, 1602.
60. V. F. Holland, S. B. Mitchell, W. L. Hunter and P. H. Lindenmeyer, *J. Polym. Sci.*, 1962, **62**, 145.
61. P. J. Holdsworth and A. Keller, *J. Polym. Sci., Part B*, 1967, **5**, 605.
62. A. Nakajima and S. Hayashi, *Kolloid Z. Z. Polym.*, 1969, **228**, 12.
63. A. J. Lovinger and D. D. Davis, *Polym. Commun.*, 1985, **26**, 322.
64. H. D. Keith, *J. Appl. Phys.*, 1964, **35**, 3115.
65. J. C. Wittman and B. Lotz, *J. Polym. Sci., Polym. Phys. Ed.*, 1985, **23**, 205.
66. H. D. Keith and F. J. Padden, *J. Appl. Phys.*, 1964, **35**, 1270.
67. H. D. Keith and F. J. Padden, *J. Appl. Phys.*, 1963, **34**, 2409.
68. P. R. Carpenter, M. Campbell, R. D. Rawlings and P. S. Rogers, *J. Mater. Sci. Lett.*, 1986, **5**, 1309.
69. D. C. Bassett, A. Keller and S. Mitsuhashi, *J. Polym. Sci., Part A*, 1963, **1**, 763.
70. J. D. Hoffman, L. J. Frolen, G. S. Ross and J. I. Lauritzen, *J. Res. Nat. Bur. Stand., Sect. A*, 1975, **79**, 671.
71. D. C. Bassett, *CRC Crit. Rev. Solid State Mater. Sci.*, 1984, **12**, 97.
72. H. D. Keith and F. J. Padden, Jr., *Polymer*, 1986, **27**, 1463.
73. D. C. Bassett and A. S. Vaughan, *Polymer*, 1986, **27**, 1472.

74. N. Goldenfeld, *J. Cryst. Growth*, 1987, **84**, 601.
75. H. D. Keith and F. J. Padden, Jr., *J. Polym. Sci., Polym. Phys. Ed.*, 1987, **25**, 229.
76. H. D. Keith and F. J. Padden, Jr., *J. Polym. Sci., Polym. Phys. Ed.*, 1987, **25**, 2265.
77. H. D. Keith and F. J. Padden, Jr., *J. Polym. Sci., Polym. Phys. Ed.*, 1987, **25**, 2371.
78. H. D. Keith and F. J. Padden, Jr., *J. Appl. Phys.*, 1964, **35**, 1286.
79. A. Keller, *J. Polym. Sci.*, 1959, **39**, 151.
80. H. D. Keith, *J. Polym. Sci., Part A*, 1964, **2**, 4339.
81. J. W. Rutter and B. Chalmers, *Can. J. Phys.*, 1953, **31**, 15.
82. A. S. Vaughan, D. C. Bassett and R. H. Olley, in 'Morphology of Polymers', ed. B. Sedlacek, de Gruyter, Berlin, 1986, p. 387.
83. D. C. Bassett and A. M. Hodge, *Proc. R. Soc. London, Ser. A*, 1981, **377**, 25.
84. D. C. Bassett, A. M. Hodge and R. H. Olley, *Proc. R. Soc. London, Ser. A*, 1981, **377**, 39.
85. D. C. Bassett and A. M. Hodge, *Proc. R. Soc. London, Ser. A*, 1981, **377**, 61.
86. P. J. Phillips and Y. H. Kao, *Polymer*, 1986, **27**, 1679.
87. A. S. Vaughan, Ph. D. Thesis, University of Reading, 1984.
88. F. Khoury, *J. Res. Nat. Bur. Stand., Sect. A*, 1966, **70**, 29.
89. A. J. Lovinger, *J. Polym. Sci., Polym. Phys. Ed.*, 1983, **21**, 97.
90. T. Suzuki and A. J. Kovacs, *Polym. J.*, 1970, **1**, 82.
91. J. M. Rego Lopez and U. W. Gedde, *Polymer*, 1988, **29**, 1037.
92. J. M. Rego Lopez, M. T. Conde Brana, B. Terselius and U. W. Gedde, *Polymer*, 1988, **29**, 1045.
93. H. D. Keith and F. J. Padden, Jr., *Polymer*, 1984, **25**, 28.
94. G. A. Chadwick, *Prog. Mater. Sci.*, 1963, **10**, 97.
95. G. Chalmers, 'Principles of Solidification', Krieger, Melbourne, 1964.
96. A. J. Lovinger, *J. Polym. Sci., Polym. Phys. Ed.*, 1980, **18**, 793.
97. A. S. Vaughan and D. C. Bassett, to be published.
98. A. J. Lovinger, in 'Developments in Crystalline Polymers — 1', ed. D. C. Bassett, Applied Science, London, 1982, p. 195.
99. J. M. Schultz and D. R. Kinloch, *Polymer*, 1969, **10**, 271.
100. P. H. Lindenmeyer and V. F. Holland, *J. Appl. Phys.*, 1964, **35**, 55.
101. A. S. Vaughan and D. C. Bassett, *Polymer*, 1988, **29**, 1397.
102. T. Nagasawa and Y. Shimomura, *J. Polym. Sci., Polym Phys. Ed.*, 1974, **12**, 2291.
103. J. Petermann, M. Miles and H. Gleiter, *J. Polym. Sci., Polym. Phys. Ed.*, 1979, **17**, 55.
104. A. J. Pennings, *J. Polym. Sci., Polym. Symp.*, 1977, **59**, 55.
105. J. Dlugosz, D. T. Grubb, A. Keller and M. B. Rhodes, *J. Mater. Sci.*, 1972, **7**, 142.
106. D. T. Grubb and A. Keller, *J. Polym. Sci., Polym. Lett. Ed.*, 1974, **12**, 419.
107. E. H. Andrews, *Proc. R. Soc. London, Ser. A*, 1964, **277**, 562.
108. A. Lee and R. P. Wool, *Macromolecules*, 1987, **20**, 1924.

13

Oriented Polymers

SHUNJI NOMURA

Kyoto Institute of Technology, Japan

13.1 INTRODUCTION

Polymeric chain molecules contain atomic groups, covalently bonded in the chain direction. In most polymers the covalent bonds have approximately the same strength as the C–C bond in diamond, which is the strongest material in nature. However, the forces perpendicular to the chain direction are relatively weak intermolecular forces such as van der Waals forces and hydrogen bonding. The large difference in strength between the intramolecular and intermolecular forces generates a large mechanical anisotropy. The elastic modulus of a polyethylene (PE) crystal is approximately 240 GPa in the chain direction due to stretching of bonds and distortion of valence angles, and about 4 GPa perpendicular to the chain direction due to van der Waals interactions.[1] Even in the case of poly(vinyl alcohol), which is intermolecularly hydrogen bonded, the moduli are 255 GPa and 9 GPa, parallel and perpendicular to the chain direction respectively.[1] The mechanical anisotropy of polymer chains is well utilized in synthetic polymeric fibers. The highest elastic modulus reported for a PE fiber is about 150 GPa obtained for a sample produced by drawing single-crystal mats of ultrahigh molecular weight PE.[2]

The anisotropy of polymer molecules is manifested not only in mechanical properties but also in various kinds of physical and chemical properties. The anisotropy of the polymer molecules does not necessarily lead to anisotropy in the bulk properties of the polymeric materials as a whole, because the microscopic anisotropy is cancelled out when the polymer molecules are randomly oriented. The relationship between the microscopic and macroscopic anisotropy can be expressed in terms of the intrinsic anisotropy of the structural units such as the crystal unit cell and the segment of the noncrystalline polymer chain, and the shape and degree of orientation of the structural units. The sharper the orientation or the more anisotropic the physical properties of the structural units, the greater is the anisotropy shown by the physical properties in the bulk polymeric materials.

Early studies of physical anisotropy were focused upon form birefringence and form dichroism in native fibers.[3-5] Wiener[6,7] developed theories for the form birefringence and form dichroism of a two-phase system. One of his theories predicts that a two-phase system, where rodlets are regularly embedded in a matrix, gives form birefringence[6] even if both rodlets and matrix are entirely optically isotropic and absorb polarized light to the same extent. On the other hand, selective absorption by one of the two components in this system results in form dichroism.[7] Möhring expressed the total birefringence of swollen cellulosic fibers in terms of the sum of the intrinsic birefringence of the anisotropic rodlets within the isotropic matrix and Wiener's form birefringence.[8]

Preston carried out the first quantitative investigation of dye dichroism related to micellar orientation in cellulosic fibers and films and made the first study of dichroic ratio in the anisotropic absorption of polarized visible rays.[9] A series of his works[9-13] has covered various kinds of optical anisotropies, such as absorption, reflection and fluorescence as well as birefringence, in order to characterize the average degree of molecular orientation in the system. The average degrees of molecular orientation derived from the different optical anisotropies were compared with each other and also with that obtained from X-ray diffraction, in order to evaluate crystalline and noncrystalline contributions to the average degrees of molecular orientation. Morey[14-18] also quantitatively investigated polarization of fluorescence in relation to molecular orientation, and defined the degree of molecular orientation in terms of the polarization of fluorescence. However, in his method as well as that of Preston,[12] the incident radiation was not polarized, in contrast to a recent fluorescence polarization method where polarized radiation is employed, as will be discussed later.

Extensive studies on the optical anisotropy of hydrated cellulosic gel in relation to deformation have been reported by Kratky[17-27] and by Hermanns.[23,28-43] Kratky studied the deformation mechanism of fibrous materials by means of X-ray diffraction and birefringence, and postulated two kinds of deformation models for the swollen gel,[19,21] in order to explain the orientation mechanism of the structural units in terms of the orientation distribution (but not in terms of the average degree of orientation of the structural units). Hermanns carried out studies on the same lines as Kratky, and introduced one of the orientation factors as a measure of the average degree of orientation.[35]

Müller investigated the relationship between the molecular orientation and the mechanical and optical anisotropies in fibrous as well as crosslinked rubberlike materials.[44] The mechanical and optical properties of a crosslinked rubber network under deformation were theoretically analyzed with the so-called freely jointed equivalent chain model by Kuhn and Grün[45] and Treloar.[46] Mechanical anisotropy in preoriented crystalline polymers was discussed by Raumann[47-49] and by Ward et al.[50-55] on the basis of the infinitesimal anisotropic elastic theory. Ward also expressed the mechanical and optical anisotropies for preoriented systems in terms of the orientation of the aggregation model,[50,55] and further extended his studies to the viscoelastic anisotropy in oriented crystalline polymers.[56-58] Similar kinds of studies on the mechanical and optical anisotropies as well as on swelling anisotropy were performed by Nomura et al.[59,60] and Maeda et al.[61] for two-phase systems of oriented crystalline polymers.

Optical anisotropy has been studied not only to characterize the optical anisotropy of oriented polymeric materials but also to evaluate their molecular orientation. The methods of measuring molecular orientation, together with much of the underlying theory, were developed during the 1960s and 1970s. Detailed discussion on the measurements of molecular orientation may be found in several reviews.[62-65]

In this chapter, firstly a mathematical representation of orientation distribution functions of structural units will be discussed in terms of an expansion of the distribution functions in a series of generalized spherical harmonics and generalized orientation factors. Secondly, the deformation mechanism of polymer spherulites will be shown to be one of the areas where the above theory can be applied. Thirdly, the relationship between the optical anisotropy in oriented polymeric materials and the orientation of the structural units will be described in general by several types of average degrees of molecular orientation. Finally, the mechanical anisotropy in oriented polymeric materials will be discussed in terms of orientation of the structural units.

13.2 QUANTITATIVE DESCRIPTION OF ORIENTATION IN ORIENTED POLYMER SYSTEMS

Physical anisotropies are related to various kinds of average degrees of orientation, hence it is not possible to discuss physical anisotropies in terms of a single average degree of orientation, *i.e.* not only the average degree of orientation but also the shape of orientation has to be taken into account. A mathematical representation of the orientation distribution function of the crystalline and noncrystalline structural units will be given as an expansion of the distribution function in terms of a series of spherical harmonics,[66-68] and each coefficient of the expanded series will be discussed in relation to the average degrees of orientation of the structural units, the so-called orientation factors independently defined by several authors.[35,69-74]

13.2.1 Series Expansion of the Orientation Distribution Function

The orientation of the structural units within the space of a film specimen can be specified by using three Euler angles, ϕ, θ and η, as shown in Figure 1.

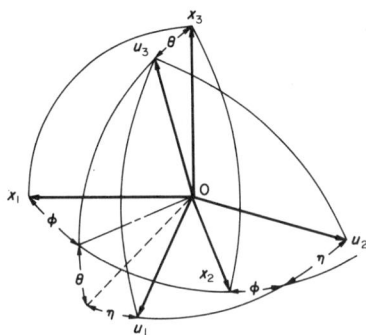

Figure 1 Euler angles ϕ, θ and η specifying the orientation of Cartesian coordinates $0-u_1 u_2 u_3$ fixed within the structural unit with respect to other Cartesian coordinates $0-x_1 x_2 x_3$ fixed within the space of the film specimen

$0-x_1 x_2 x_3$ are Cartesian coordinates which are fixed within the space of the specimen and coincide with axes of symmetry. $0-u_1 u_2 u_3$ are another set of Cartesian coordinates fixed within the structural units and these also coincide with axes of symmetry. In most cases, it is convenient to choose the x_3 axis along the machine or draw direction of the specimen and the u_3 axis along the polymer chain direction. The angles, θ and ϕ, are respectively the polar and azimuthal angles of the u_3 axis with respect to $0-x_1 x_2 x_3$, and η specifies the rotation of the unit around its own u_3 axis. The orientation distribution function of the structural units within the specimen space may be normalized as given by equation (1), where ξ is defined as equal to $\cos \theta$.

$$\int_0^{2\pi} \int_0^{2\pi} \int_0^{\pi} w(\cos\theta, \phi, \eta)\sin\theta \, d\theta \, d\phi \, d\eta = \int_0^{2\pi} \int_0^{2\pi} \int_{-1}^{1} w(\xi, \phi, \eta) \, d\xi \, d\phi \, d\eta = 1 \tag{1}$$

Unfortunately, we do not have any direct measurement of the orientation distribution function $w(\xi, \phi, \eta)$. We can only measure the orientation distribution function of an axis r_j fixed within the structural unit with respect to the specimen space. The orientation of a vector r_j may be specified by polar and azimuthal angles θ_j and ϕ_j with respect to $0-x_1 x_2 x_3$ as shown in Figure 2 and by polar and azimuthal angles Θ_j and Φ_j with respect to $0-u_1 u_2 u_3$ as shown in Figure 3. The orientation distribution function of the vector r_j within the specimen space may also be normalized as given by equation (2) where ζ_j is defined as equal to $\cos \theta_j$.

$$\int_0^{2\pi} \int_0^{\pi} q_j(\cos\theta_j, \phi_j) \sin\theta_j \, d\theta_j \, d\phi_j = \int_0^{2\pi} \int_{-1}^{1} q_j(\zeta_j, \phi_j) \, d\zeta_j \, d\phi_j = 1 \tag{2}$$

Two sets of angles, (θ_j, ϕ_j) and (Θ_j, Φ_j), referring to the identical vector r_j, can be related to each other by the coordinate transformation between $0-u_1 u_2 u_3$ and $0-x_1 x_2 x_3$ as equation (3), where (t_{ki}) is a linear transformation operator, and t_{ki} is the direction cosine between the u_k and x_i axes and the function of ϕ, θ and η.

$$\begin{pmatrix} \sin\theta_j \ \cos\phi_j \\ \sin\theta_j \ \sin\phi_j \\ \cos\theta_j \end{pmatrix} = (t_{ki}) \begin{pmatrix} \sin\Theta_j \ \cos\Phi_j \\ \sin\Theta_j \ \sin\Phi_j \\ \cos\Theta_j \end{pmatrix} \tag{3}$$

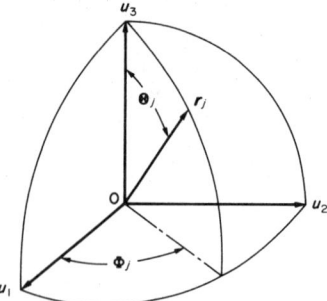

Figure 2   Polar and azimuthal angles Θ_j and Φ_j specifying the orientation of a given jth axis of the structural unit with respect to the Cartesian coordinates $0-u_1 u_2 u_3$

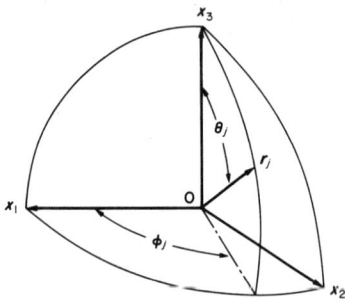

Figure 3   Polar and azimuthal angles θ_j and ϕ_j specifying the orientation of the jth axis of the structural unit with respect to the Cartesian coordinates $0-x_1 x_2 x_3$ fixed within the space of the film specimen

According to Roe's analyses, two kinds of orientation distribution functions given by equations (1) and (2) may be expanded into infinite series of normalized and generalized spherical harmonics as equations (4) and (5).[67,68] $Z_{lmn}(\xi)$ and $\Pi_l^m(\zeta_j)$ in equations (4) and (5) are the normalized associated Legendre's polynomials as given by equations (6) and (7). $P_{l-m}^{(m-n,\,m+n)}(x)$ and $P_l^m(x)$ in equations (6) and (7) are Jacobi's polynomial and an associated Legendre's polynomial (not normalized) respectively.

$$w(\xi, \phi, \eta) = \sum_{l=0}^{\infty} \sum_{m=-l}^{l} \sum_{n=-l}^{l} W_{lmn} Z_{lmn}(\xi) \exp\{-i(m\phi + n\eta)\} \tag{4}$$

$$q_j(\zeta_j, \phi_j) = \sum_{l=0}^{\infty} \sum_{m=-l}^{l} Q_{lm}^j \Pi_l^m(\zeta_j) \exp(-im\phi_j) \tag{5}$$

$$Z_{lmn}(x) = \left[\frac{2l+1}{2} \frac{(l-m)!}{(l-n)!} \frac{(l+m)!}{(l+n)!} \right]^{1/2} (1/2)^m (1-x)^{(m-n)/2} (1+x)^{(m+n)/2} P_{l-m}^{(m-n,\,m+n)}(x) \tag{6}$$

$$\Pi_l^m(x) = Z_{lm0}(x) = \left[\frac{2l+1}{2} \frac{(l-m)!}{(l+m)!} \right]^{1/2} P_l^m(x) \tag{7}$$

In turn, the coefficients of the expanded series W_{lmn} and Q_{lm}^j can be obtained from equations (8) and (9) respectively. Clearly, from equations (8) and (9), the coefficients W_{lmn} and Q_{lm}^j are the averages of the distribution functions with respect to the lmnth and lmth orders of spherical harmonics respectively. In other words, the coefficients are the lmnth and lmth averages of the orientation distribution functions, and are a sort of general definition of the orientation factors as will be discussed later.

$$W_{lmn} = \frac{1}{4\pi^2} \int_0^{2\pi} \int_0^{2\pi} \int_{-1}^{1} w(\xi, \phi, \eta) Z_{lmn}(\xi) \exp\{i(m\phi + n\eta)\} \, d\xi \, d\phi \, d\eta \tag{8}$$

$$Q_{lm}^j = \frac{1}{2\pi} \int_0^{2\pi} \int_{-1}^{1} q_j(\zeta_j, \phi_j) \Pi_l^m(\zeta_j) \exp(im\phi_j) \, d\zeta_j \, d\phi_j \tag{9}$$

Applying a generalization of the Legendre addition theorem to equation (3), equation (10) can be obtained. Multiplying both sides of equation (10) by $w(\xi, \phi, \eta) \times q_j(\zeta_j, \phi_j)$, integrating over the range of the five angles and taking the relations of equations (8) and (9) leads to equation (11) which gives the important relation between W_{lmn} and Q_{lm}^j. Equation (11) relates not only the coefficients W_{lmn} and Q_{lm}^j to each other but also provides a relationship between the distribution functions $w(\xi, \phi, \eta)$ and $q_j(\zeta_j, \phi_j)$. Thus the coefficients W_{lmn} as well as the distribution function $w(\xi, \phi, \eta)$ can be estimated from the coefficient Q_{lm}^j or the distribution function $q_j(\zeta_j, \phi_j)$, which are measurable quantities as discussed above.

$$\Pi_l^m(\zeta_j)\exp(im\phi_j) = \left[\frac{2}{2l+1}\right]^{1/2} \sum_{n=-l}^{l} Z_{lmn}(\xi)\exp\{i(m\phi + n\eta)\} \, \Pi_l^n(\cos\Theta_j)\exp(in\Phi_j) \tag{10}$$

$$Q_{lm}^j = 2\pi \left[\frac{2}{2l+1}\right]^{1/2} \sum_{n=-l}^{l} W_{lmn} \, \Pi_l^n(\cos\Theta_j)\exp(in\Phi_j) \tag{11}$$

Finally, let us consider the orthogonal biaxial symmetry of the distribution function $q_j(\zeta_j, \phi_j)$ and the orthorhombic symmetry of the structural unit. The coefficients W_{lmn} and Q_{lm}^j have the symmetries given by equations (12) and (13). Consequently, the distribution functions can be expressed by the expansion series only for the case when l, m and n are all even numbers, and equation (11) can be rewritten as equation (14).

$$Q_{lm}^j \begin{cases} = Q_{l\bar{m}}^j \neq 0 & \text{when both } l \text{ and } m \text{ are even} \\ = 0 & \text{when either } l \text{ or } m \text{ is odd} \end{cases} \tag{12}$$

$$W_{lmn} \begin{cases} = W_{l\bar{m}n} = W_{lm\bar{n}} = W_{l\bar{m}\bar{n}} \neq 0 & \text{when } l, m \text{ and } n \text{ are all even} \\ = 0 & \text{when } l, m \text{ or } n \text{ is odd} \end{cases} \tag{13}$$

$$Q_{lm}^j = 2\pi \left[\frac{2}{2l+1}\right]^{1/2} \left[W_{lm0} \, \Pi_l(\cos\Theta_j) + 2 \sum_{n=2}^{l} W_{lmn} \, \Pi_l^n(\cos\Theta_j)\cos n\Phi_j \right] \tag{14}$$

13.2.2 Generalization of the Orientation Factor

As has been discussed with reference to equations (8) and (9), the expansion coefficients W_{lmn} and Q_{lm}^j may be understood as averages of the orientation distribution functions $w(\xi, \phi, \eta)$ and $q_j(\zeta_j, \phi_j)$ with respect to the lmnth and lmth orders of the spherical harmonics, i.e. the lmnth and lmth order moments of the orientation distribution functions, which have been termed 'orientation factors'. Therefore the lmnth and lmth orders of the orientation factors can be defined in terms of equations (15) and (16) respectively.[68]

$$F_{lmn} = \left[\frac{2}{2l+1} \frac{(l+m)! \, (l+n)!}{(l-m)! \, (l-n)!}\right]^{1/2} 4\pi^2 W_{lmn} \tag{15}$$

$$F_{lm}^j = \left[\frac{2}{2l+1} \frac{(l+m)!}{(l-m)!}\right]^{1/2} 2\pi Q_{lm}^j \tag{16}$$

The $l00$th and $l0$th orders of the orientation factors, F_{l00} and F_{l0}^j, as well as W_{l00} and Q_{l0}^j, specify the average orientation of the u_3 axis and the vector r_j, both with respect to the reference x_3 axis, i.e. the orientation distribution with respect to either the polar angle θ or θ_j respectively. The $lm0$th and lmth orders of the orientation factors F_{lm0} and F_{lm}^j specify the average biaxial orientation of the u_3 axis and the vector r_j with respect to the $0-x_1x_2x_3$ system, respectively. Furthermore, F_{l0n} specifies the $l0n$th order of the average orientation of the structural unit $0-u_1u_2u_3$ with respect to the x_3 axis, i.e. the average biaxial orientation of the x_3 axis with respect to the $0-u_1u_2u_3$ coordinates.

Some of the generalized orientation factors defined by equations (15) and (16) have been independently reported by several authors. One of the second order orientation factors F_{200} has been proposed by Hermanns and Platzek[35] to relate the birefringence of fibrous materials to the uniaxial orientation of cylindrical–symmetric structural units, and F_{20}^j has been defined by Stein and Norris[69] to characterize the uniaxial orientation of polyethylene crystals. F_{220} and F_{22}^j are also the second order orientation factors defined by Nomura *et al.*[72] for relating the absorption dichroism of a material to the biaxial orientation of the structural units. The fourth order of the orientation factors F_{400} and F_{40}^j have been defined by Kimura *et al.*[73] and by Stein and Read[74] for relating the absorption/emission dichroism of a material to the uniaxial orientation of the structural units.

Some of the generalized orientation factors may be written in more concrete form for the orthogonal-biaxial symmetry of the orientation distribution and the orthorhombic symmetry of the structural unit as shown by equations (17)–(28), the triangular brackets $\langle\ \rangle$ indicating averages.

$$F_{200} = 4\pi^2(2/5)^{1/2}\,W_{200} = (3\langle\cos^2\theta\rangle - 1)/2 \tag{17}$$

$$F_{220} = 4\pi^2\,12/(15)^{1/2}\,W_{220} = 3\langle\sin^2\theta\cos 2\phi\rangle \tag{18}$$

$$F_{202} = 4\pi^2\,12/(15)^{1/2}\,W_{202} = 3\langle\sin^2\theta\cos 2\eta\rangle \tag{19}$$

$$F_{222} = 4\pi^2\,24(2/5)^{1/2}\,W_{222} = 6\langle(1+\cos^2\theta)\cos 2\phi\cos 2\eta\rangle \tag{20}$$

$$F_{400} = 4\pi^2(2)^{1/2}/3\,W_{400} = (35\langle\cos^4\theta\rangle - 30\langle\cos^2\theta\rangle + 3)/8 \tag{21}$$

$$F_{420} = 4\pi^2\,4(5)^{1/2}\,W_{420} = (15/2)\langle(7\cos^2\theta - 1)\sin^2\theta\cos 2\phi\rangle \tag{22}$$

$$F_{440} = 4\pi^2\,16(35)^{1/2}\,W_{440} = 105\langle\sin^4\theta\cos 4\phi\rangle \tag{23}$$

$$F_{20}^j = 2\pi(2/5)^{1/2}\,Q_{20}^j = (3\langle\cos^2\theta_j\rangle - 1)/2 \tag{24}$$

$$F_{22}^j = 2\pi\,12/(15)^{1/2}\,Q_{22}^j = 3\langle\sin^2\theta_j\cos 2\phi_j\rangle \tag{25}$$

$$F_{40}^j = 2\pi(2)^{1/2}/3\,Q_{40}^j = (35\langle\cos^4\theta_j\rangle - 30\langle\cos^2\theta_j\rangle + 3)/8 \tag{26}$$

$$F_{42}^j = 2\pi\,4(5)^{1/2}\,Q_{42}^j = (15/2)\langle(7\cos^2\theta_j - 1)\sin^2\theta_j\cos 2\phi_j\rangle \tag{27}$$

$$F_{44}^j = 2\pi\,16(35)^{1/2}\,Q_{44}^j = 105\langle\sin^4\theta_j\cos 4\phi_j\rangle \tag{28}$$

13.2.3 Graphic Representation of the State of Orientation of Structural Units

There have been several types of graphical illustrations to represent the state of orientation of structural units. Stein has proposed orthogonal equilateral triangle coordinates for the uniaxial symmetric system[70] and Desper has proposed equilateral triangle coordinates for the orthogonally biaxial symmetric system.[75]

Let us consider the second moments of the orientation distribution of the *j*th vector with respect to the x_i axis, *i.e.* $\langle\cos^2\theta_{ij}\rangle$, for which equation (29) may be obtained. Any point within the triangle coordinates, as shown in Figure 4 in terms of $(\langle\cos^2\theta_{1j}\rangle, \langle\cos^2\theta_{2j}\rangle, \langle\cos^2\theta_{3j}\rangle)$, satisfies equation

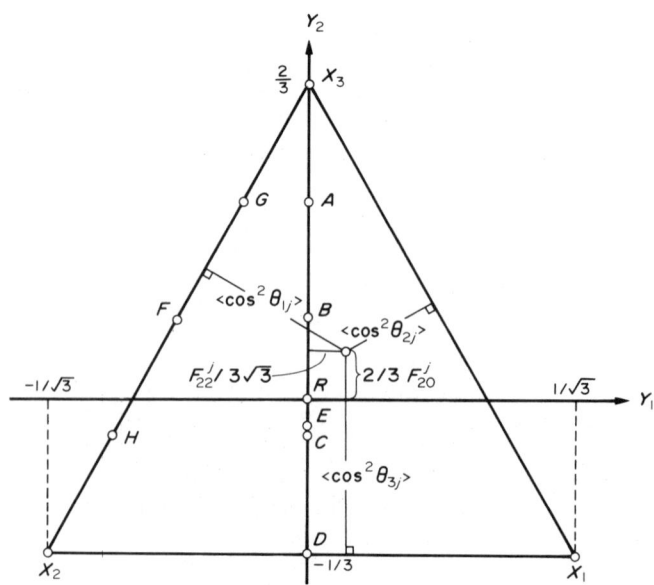

Figure 4 Graphical representation of the state of orthogonal biaxial orientation in terms of the second order of the orientation factors

(29) and represents the state of orientation of the orthogonally biaxial symmetric system.

$$\sum_{i=1}^{3} \langle \cos^2 \theta_{ij} \rangle = 1 \tag{29}$$

On the other hand, the second-order orientation factors F_{20}^j and F_{22}^j are given by equations (30) and (31). Thus the point $(\langle \cos^2 \theta_{1j} \rangle, \langle \cos^2 \theta_{2j} \rangle, \langle \cos^2 \theta_{3j} \rangle)$ is coincident with the point $(F_{22}^j/(3)^{3/2}, 2F_{20}^j/3)$ in the orthogonal coordinate system $0-Y_1 Y_2$, shown in Figure 4, when the origin of the coordinates is taken as the center of the triangle and the $0Y_1$ and $0Y_2$ axes are taken as F_{20}^j and F_{22}^j coordinates respectively.[72] The orthogonal coordinates are therefore also considered to be one representation of the state of orientation of the orthogonally biaxial symmetric system. The states of orientation represented by the points X_1, X_2, X_3, A through H, and R in Figure 4, which are rather extreme models of uniaxial or orthogonally biaxial symmetric systems, are illustrated in Figure 5 and tabulated in Table 1 in terms of the code of the model and the values of the second and fourth order orientation factors.

$$F_{20}^j = (3\langle \cos^2 \theta_{3j} \rangle - 1)/2 \tag{30}$$

$$F_{22}^j = 3(\langle \cos^2 \theta_{1j} \rangle - \langle \cos^2 \theta_{2j} \rangle) \tag{31}$$

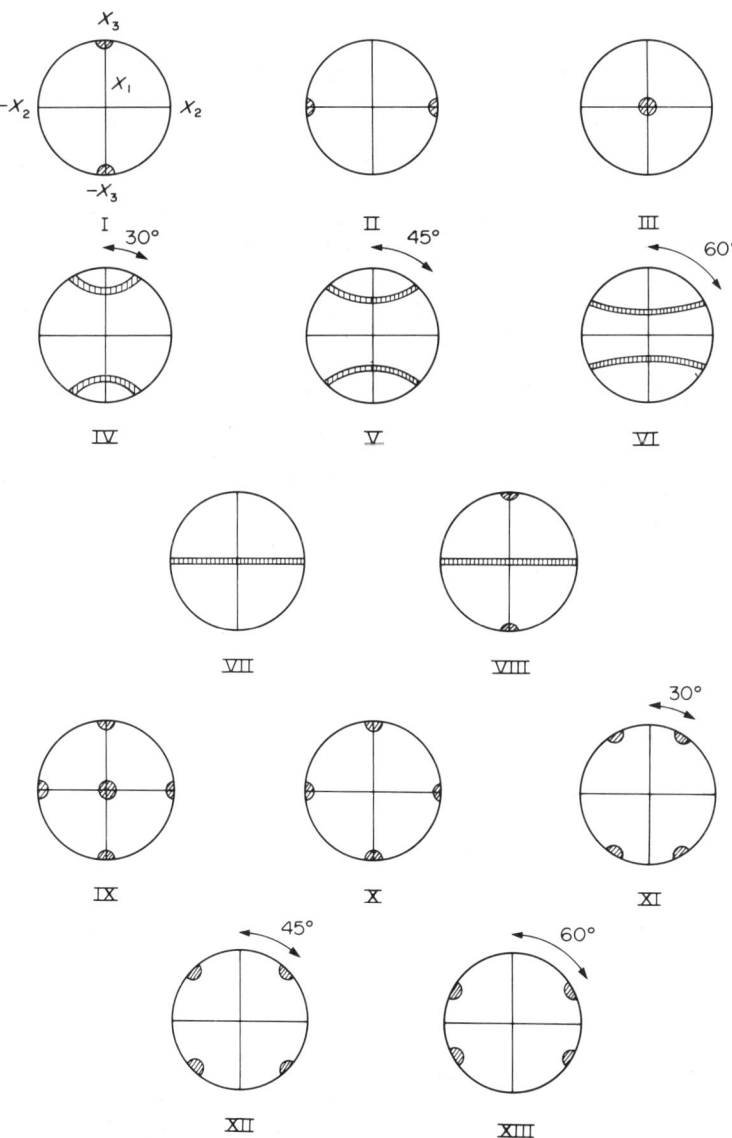

Figure 5 Pole figures with respect to the x_1 axis, illustrating the orientation distributions for various types of extreme models for uniaxially and orthogonally biaxial symmetric systems

Table 1　Model Systems and Orientation Factors

Orientation distribution	F^j_{20}	$\dfrac{F^j_{22}}{3}$	F^j_{40}	$\dfrac{2F^j_{42}}{15}$	$\dfrac{F^j_{44}}{105}$	Location in Figure 4	Location in Figure 6
Random orientation	0	0	0	0	0	R	R
Ideal parallel model							
I　　$\theta_j=0$	1	0	1	0	0	X_3	X_3
II　　$\theta_j=\pi/2,\ \phi_j=\pi/2$	$-1/2$	-1	$3/8$	1	1	X_2	X_2
III　　$\theta_j=\pi/2,\ \phi_j=0$	$-1/2$	1	$3/8$	-1	-1	X_1	X_1
Uniaxial model (Dirac delta function)							
IV　　$\theta_j=\pi/6,\ \phi_j$; random	$5/8$	0	$3/128$	0	0	A	A
V　　$\theta_j=\pi/4,\ \phi_j$; random	$1/4$	0	$-13/32$	0	0	B	B
VI　　$\theta_j=\pi/3,\ \phi_j$; random	$-1/8$	0	$-37/128$	0	0	C	C
VII　　$\theta_j=\pi/2,\ \phi_j$; random	$-1/2$	0	$3/8$	0	0	D	D
VIII　　$\dfrac{3}{11}(\text{I})+\dfrac{8}{11}(\text{VII})$; (bimodal)	$-1/11$	0	$6/11$	0	0	E	E
Biaxial model							
IX　　$\dfrac13(\text{I})+\dfrac13(\text{II})+\dfrac13(\text{III})$; (trimodal)	0	0	$7/12$	0	$2/3$	R	Q
X　　$\dfrac12(\text{I})+\dfrac12(\text{II})$; (bimodal)	$1/4$	$-1/2$	$11/16$	$1/2$	$1/2$	F	F
XI　　$\theta_j=\pi/6,\ \phi_j=\pi/2$	$5/8$	$-1/4$	$3/128$	$-17/16$	$1/16$	G	G
XII　　$\theta_j=\pi/4,\ \phi_j=\pi/2$	$1/4$	$-1/2$	$-13/32$	$-5/4$	$1/4$	F	K
XIII　　$\theta_j=\pi/3,\ \phi_j=\pi/2$	$-1/8$	$-3/4$	$-37/128$	$-9/16$	$9/16$	H	H

Let us further consider representations of the state of orientation provided by the fourth-order orientation factors. Figure 6 shows two projections of the Cartesian coordinates of F_{40}^j, F_{42}^j and F_{44}^j, in which the states of orientation corresponding to the extreme models from (I) to (XIII) in Table 1 and the random orientation model are again plotted. As seen in Figure 6, one can distinguish differences in states of orientation between the random orientation model and model (IX) and between models (X) and (XII), which cannot be distinguished in Figure 4 by using the second-order orientation factors. Apparently, using higher-order orientation factors enables the state of orientation to be described in more detail.

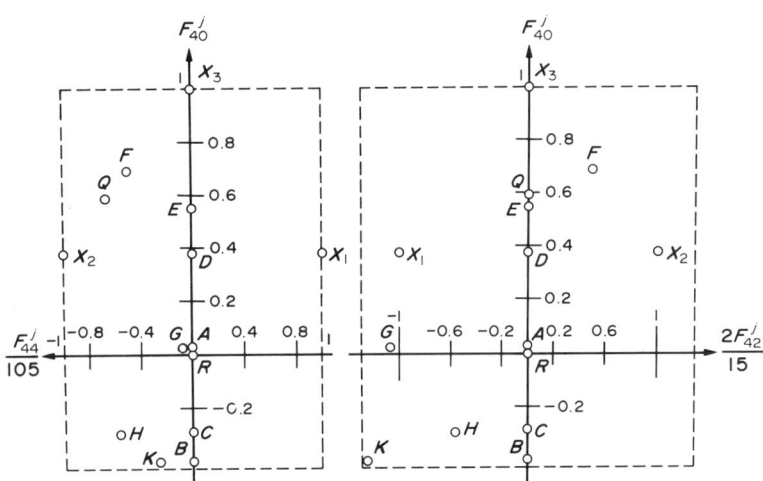

Figure 6 Two projections of Cartesian coordinates illustrating the state of orthogonally biaxial orientation in terms of the fourth order of the orientation factors, F_{40}^j, $2/15F_{42}^j$ and $F_{44}^j/105$

In the orthogonal coordinates in Figure 4, when the point $(F_{220}/(3)^{3/2}, 2F_{200}/3)$ is plotted instead of $(F_{22}^j/(3)^{3/2}, 2F_{20}^j/3)$, the state of orthogonal biaxial orientation of the u_3 axis may be represented. Furthermore, when one plots the point $(F_{202}/(3)^{3/2}, 2F_{200}/3)$, the point has the same significance as that given by Stein for the uniaxial orientation of polyethylene crystallites.[70]

13.2.4 Estimation of the Orientation Distribution Function from Finite Series

The orientation distribution functions for the structural units cannot be directly determined from any experimental sources, except for the crystalline units from X-ray diffraction measurements. Usually the orientations of structural units are obtained as several kinds of averages, *i.e.* the second and/or fourth orders of the orientation factors from optical measurements such as birefringence and absorption/emission dichroism.

Therefore, the estimation of the orientation distribution function $q_j(\zeta_j, \phi_j)$ from the finite series of the spherical harmonics and the error in the estimation as compared to the real distribution function must be discussed quantitatively. For simplicity, the discussion will be limited mainly to the uniaxially symmetric system. The average square error between the real distribution function and the function estimated from the finite series of the expansion is given by equation (32). Let us evaluate the error quantitatively by using a uniaxially symmetric orientation distribution function proposed by Kratky as a floating rod model in the affine matrix,[17] given by equation (33), where λ is the stretching ratio of the bulk specimen along the symmetric axis.

$$\sigma_k = \int_0^{2\pi} \int_{-1}^{1} \left\{ q_j(\zeta_j, 0) - \sum_{l=0}^{k} Q_{l0}^j \Pi_l(\zeta_j) \right\}^2 d\zeta_j d\phi_j = 2\pi \int_{-1}^{1} \{q_j(\zeta_j, 0)\}^2 d\zeta_j - 2\pi \sum_{l=0}^{k} \{Q_{l0}^j\}^2 \tag{32}$$

$$q_j(\cos\theta_j, 0) = \frac{1}{4\pi} \frac{\lambda^3}{\{\lambda^3 - (\lambda^3 - 1)\cos^2\theta_j\}^{3/2}} \tag{33}$$

Figure 7 shows the change of the first term on the right hand side of equation (32) and the error incurred with the increase of F_{20}^j accompanying the increase in λ. For random orientation,

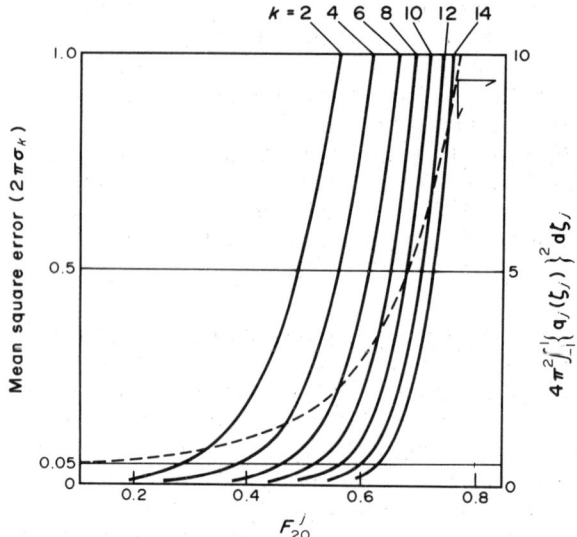

Figure 7 Changes of the first term of the right hand side of equation (32) and the average square error for k varying from 2 to 14, both with the increase of F_{20}^{j}

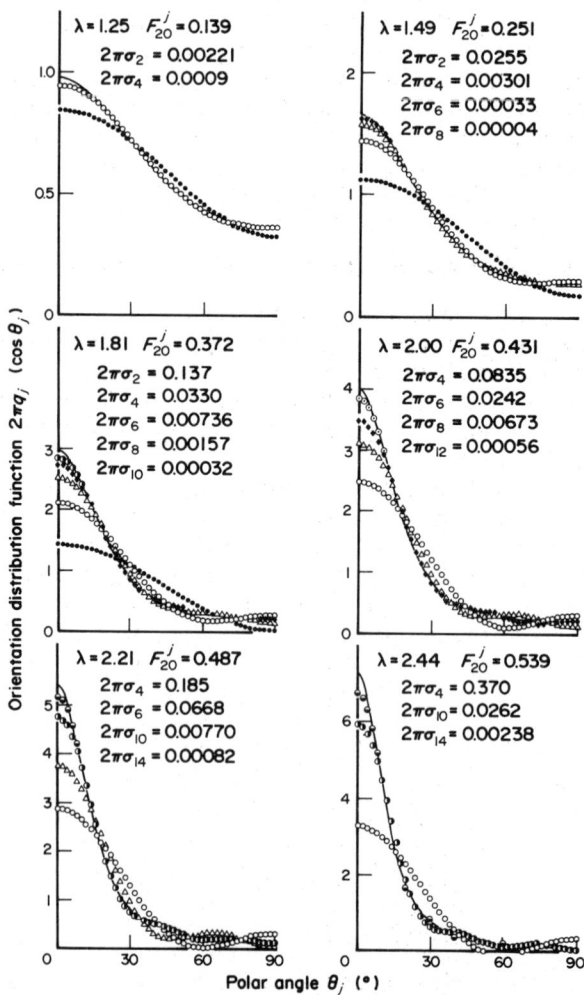

Figure 8 Comparison of the Kratky distribution function at various stretch ratios with the calculated distribution function from the expansion of the Kratky function in a finite series of spherical harmonics, where ● corresponds to $k=2$, ○ to $k=4$, △ to $k=6$, ◆ to $k=8$, ◑ to $k=10$, ⊙ to $k=12$, and ◓ to $k=14$

$q_j(\zeta_j, 0) = 1/(4\pi)$ and $(Q^j_{k0})^2 = 1/(8\pi^2)$ or 0 for $k = 0$ or $k \neq 0$. This means that the average square error is zero even when k is taken as low as zero. On the other hand, the sharper the distribution function, the larger the value of the integrand on the right-hand side of equation (32) and hence the larger the error itself; that is, the sharper the orientation distribution, the higher the order of k which should be taken into account to approach the real distribution.

Figure 8 shows a comparison of Kratky's distribution function at various stretching ratios, indicated by the solid curve, with the distribution function calculated from finite series of the expansion. Again it may be seen that the larger the stretching ratio, *i.e.* the larger the value of F^j_{20}, the higher the moments necessary to approach the real distribution function.

13.3 DEFORMATION MECHANISM OF POLYMER SPHERULITES

One of the most interesting applications of the theory of Krigbaum and Roe[66,67] described above, is a computer simulation of the deformation mechanism of spherulitic crystalline texture.[76-79] The deformation mechanism of crystalline polymers has been extensively studied by a number of authors using different optical techniques in order to investigate the mechanism in terms of the change in crystalline texture during deformation, especially for polyethylene specimens.[80-101]

Some of the investigations were performed for a bulk-crystallized specimen of spherulitic texture by comparing the orientation behavior of crystallites within the texture during the deformation of the specimen, as observed by X-ray diffraction, with those predicted from a deformation model for the spherulitic texture, so as to characterize the mechanism in terms of the model parameters. Earlier quantitative investigations were concerned with a comparison of the second-order orientation factors for crystallographic axes of the crystallites.[83,84,86-89] However, as described above, these factors are merely averages of the orientation distributions of the axes, and are not necessarily sufficient to deduce detailed information on the deformation mechanism of the spherulitic texture. This shortcoming has been greatly reduced by Nomura *et al.*,[76,77] who used the orientation distribution functions of particular reciprocal lattice vectors. A further improvement is to use the orientation distribution function of the crystallites themselves, as discussed below.

13.3.1 Experimental Results on Uniaxial Orientation Behavior of Polyethylene Crystals

Figure 9 shows the uniaxial orientation distribution functions $q_j(\zeta_j, 0)$, determined from X-ray diffraction measurement, for a high-density polyethylene specimen stretched to an extension ratio of 1.4.[78] Twelve different jth reciprocal lattice vectors were observed. With $q_j(\zeta_j, 0)$ for $j = 1$ to 12, the coefficients Q^j_{l0} were calculated from equation (9) for any higher orders of l, and then the coefficients W_{l0n} were calculated from the simultaneous equations of equation (11) with respect to j by the weighted least-square method up to $l = 18$.

Figure 10 compares the observed orientation distribution functions with those calculated for the respective reciprocal lattice vectors at the extension ratio of 1.4. That is, Q^j_{l0} were calculated in turn from W_{l0n}, which were initially determined by the weighted least-squares method, by using equation (11), and further $q_j(\zeta_j, 0)$ were calculated from the recalculated Q^j_{l0} by the use of equation (5).

As can be seen in Figure 10, fairly good agreement between the observed and calculated distribution functions was obtained even for the less accurately measured reciprocal lattice vectors, such as those of the (311), (120) and (002) crystal planes. This suggests a high reliability of the W_{l0n} thus determined. A mean-square error between the calculated and recalculated Q^j_{l0} may be defined as in equation (34). The values of the mean-square error are 0.020–0.036 for four specimens with different extension ratios, again indicating the high reliability of W_{l0n}. These values of R are much smaller than those obtained in the studies of uniaxially stretched polyethylenes by Krigbaum and Roe[102] (10%) and by Matsuo *et al.*[99] (30%).

$$R = \left[\sum_{j,l} \rho_j \{(Q^j_{l0})_{\text{cal}} - (Q^j_{l0})_{\text{recal}}\}^2 \right] \Big/ \left[\sum_{j,l} \rho_j \{Q^j_{l0})_{\text{cal}}\}^2 \right] \qquad (34)$$

Figure 11 shows the change in the uniaxial orientation distribution function of crystallites, $w(\xi, 0, \eta)$ with the extension ratio. This function was calculated using equation (4) with a finite series expansion, with the coefficients W_{l0n} determined from equation (11), l taking a maximum value of 18. As can be seen from Figure 11, two populous regions appear in the orientation distribution; *i.e.* one at θ a little larger than 30° and η of 90°, and the other at θ a little smaller than 30° and η of 0°. It seems

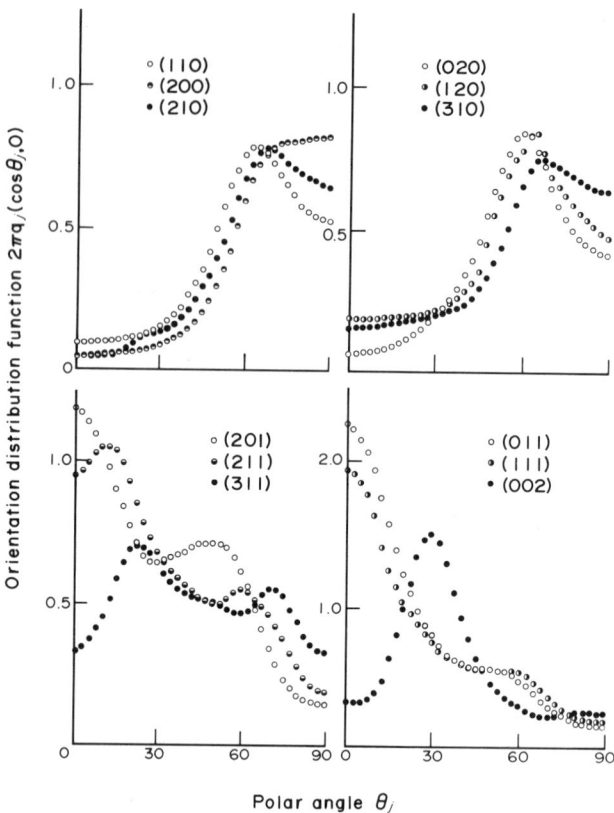

Figure 9   Uniaxial orientation distribution functions $q_j(\zeta_j, 0)$ for the jth reciprocal lattice vectors determined from X-ray diffraction intensity distributions at a given extension ratio of 1.4

that the former population appears at a later stage than, but at a much greater rate than, the latter population, suggesting that different mechanisms of spherulitic deformation yield the two populations.[78]

Actually, as was pointed out in previous studies on the deformation mechanism of polyethylene spherulites,[76, 77, 86] the appearance of two populations in $w(\xi, 0, \eta)$ must be explained in terms of two types of preferential reorientation of crystallites (crystal grains) within orienting lamellae being accompanied by the unaxial deformation of spherulites; that is, rotations of the crystal grains about their own crystal a and b axes respectively, in association with lamellar shearing and untwisting mostly at the polar and equatorial zones of the spherulites, both so as to align their crystal c axes (molecular axes) toward the stretching direction (x_3 axis) as a result of the straining of tie-chain molecules between adjacent lamellae. The above concept of rotations of the crystal grains about the crystal a and b axes does not necessarily invoke a specific nature for the plasticity of the polyethylene crystal, but a hypothetical concept of paracrystallinity of the crystal lamella; that is, the lamella must consist of 'aggregates of microparacrystallite domains' (crystal grains)[103] or of the so-called crystal mosaic blocks, capable of rotating at the grain boundaries.

13.3.2   Calculation of the Orientation Distribution Function from the Spherulite Deformation Model

The orientation distribution function of crystallites can be calculated using a spherulite deformation model. By comparing the calculated result with that observed, the deformation mechanism can be evaluated in terms of the model parameters. Figure 12 shows the orientation of a set of Cartesian coordinates $0-v_1 v_2 v_3$ fixed within a crystal lamella relative to the Cartesian coordinates $0-x_1 x_2 x_3$ fixed within the bulk specimen in terms of Euler angles ϕ', θ' and η'. The Cartesian coordinates $0-v_1 v_2 v_3$ may be fixed within a segment of crystal lamella in such a way that the v_1 and v_3 axes are respectively normal to the lamellar surface and parallel to the lamellar axis.

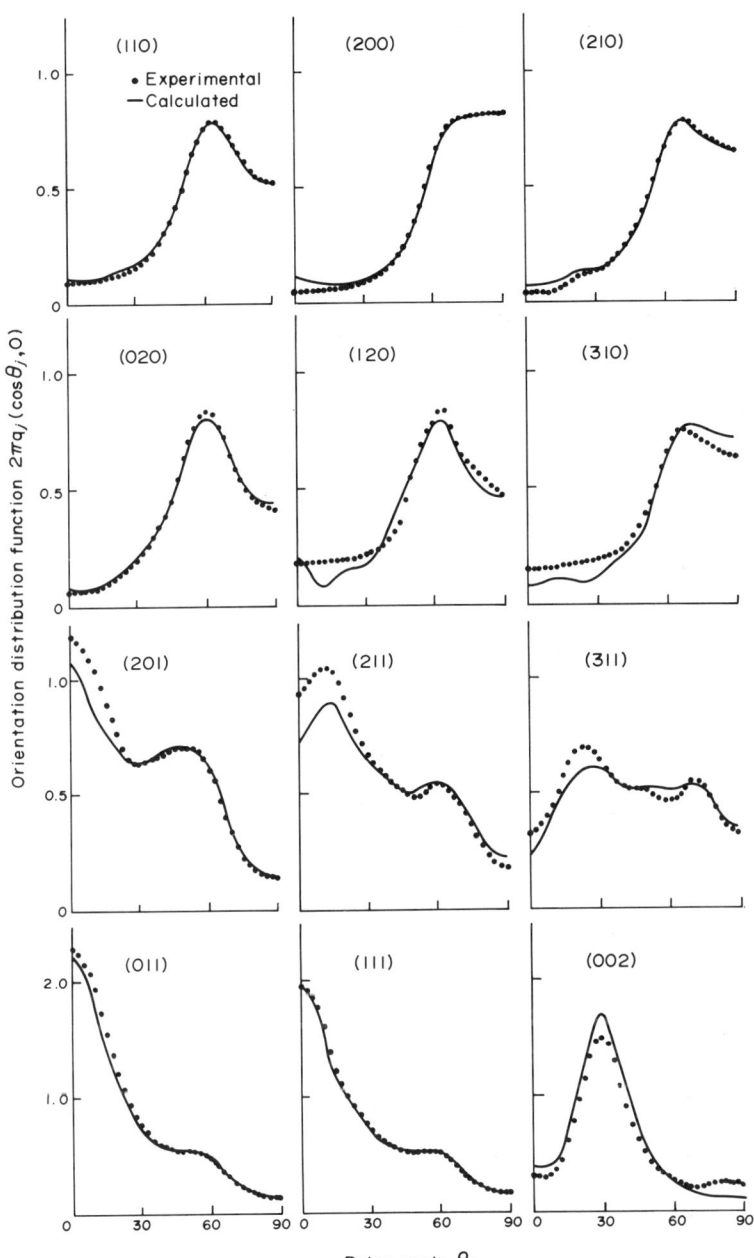

Figure 10   Comparison of observed orientation distribution functions $q_j(\zeta_j, 0)$ (circles) for the jth reciprocal lattice vectors with those calculated (solid curves) at a given extension ratio of 1.4

Assuming the unaxial deformation of the polyethylene spherulite to follow an affine deformation hypothesis keeping its volume constant, as has been postulated by several authors,[76,77,81-84,92,93,99,104] the number of crystallites orienting in a range from θ' to $(\theta' + d\theta')$ is given by equation (35), where N_0 is the total number of crystallites within a spherulite and λ is the extension ratio of the spherulite along the x_3 axis.

$$\frac{N(\theta')}{\dot{N}_0} = \frac{\lambda^3}{2\{\lambda^3 - (\lambda^3 - 1)\cos^2 \theta'\}^{3/2}} \qquad (35)$$

Let us formulate the following four types of reorientation mechanism of the crystal grains within the orienting lamellae.

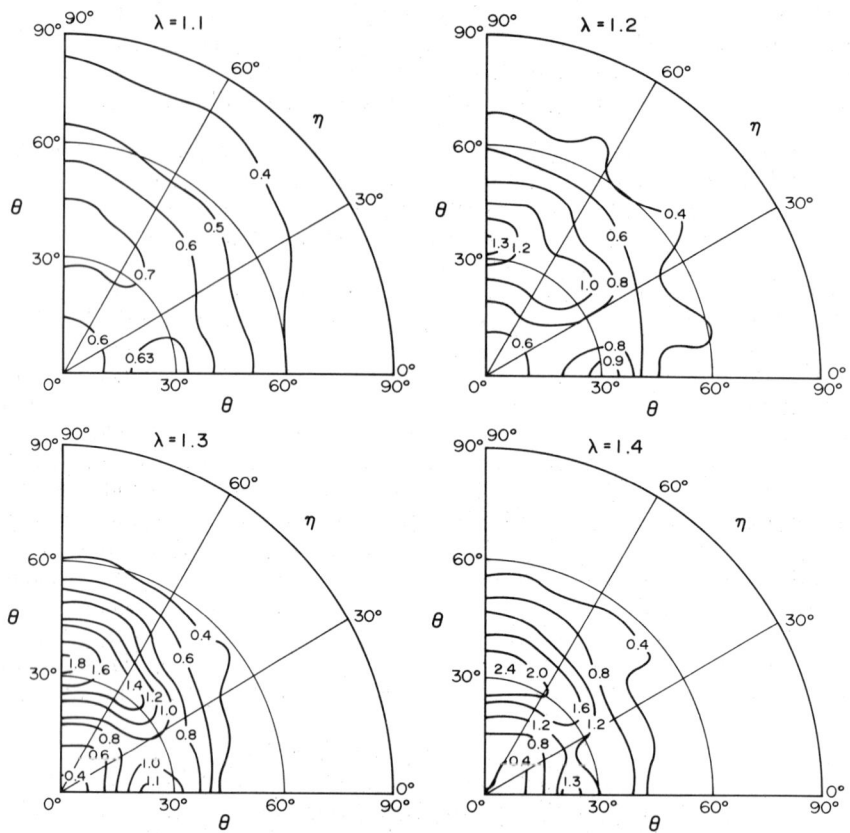

Figure 11 Change in uniaxial orientation distribution function of crystallites, $w(\xi, 0, \eta)$, with the extension ratio of the bulk specimen λ

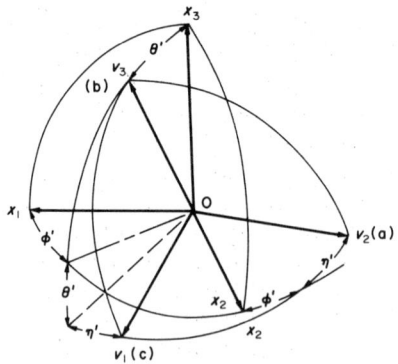

Figure 12 Euler angles, ϕ', θ', and η', specifying the orientation of Cartesian coordinates $0 - v_1 v_2 v_3$ fixed within a segment of crystal lamella with respect to the Cartesian coordinates $0 - x_1 x_2 x_3$ fixed within the bulk specimen

(1) Type C_a: rotation of the crystal grains about their own a-axes. The orientation distribution function of the crystal grains may be formulated as equation (36).

$$w_{C_a}(\theta, \eta) = (W_{C_a}/8\pi^2)[\cos^2\theta(2\cos^2\theta_{0,C_a} - \cos^2\theta)/\cos^4\theta_{0,C_a}]^{2(J_a-1)}(\cos\eta)^{2(I_a-1)} \tag{36}$$

(2) Type C_b: rotation of the crystal grains about their own b axes. The orientation distribution function of the crystal grains may be formulated as equation (37).

$$w_{C_b}(\theta, \eta) = (W_{C_b}/8\pi^2)[\cos^2\theta(2\cos^2\theta_{0,C_b} - \cos^2\theta)/\cos^4\theta_{0,C_b}]^{2(J_b-1)}(\cos\eta)^{2(I_b-1)} \tag{37}$$

(3) Type B: orientation of the crystal grains without any rotation within the lamella. The orientation distribution function of the grains may be given by equation (38).

$$w_B(\theta', \eta') = (W_B/8\pi^2)\lambda^3/\{\lambda^3 - (\lambda^3 - 1)\cos^2\theta'\}^{3/2} \tag{38}$$

(4) Type R: random orientation of the crystal grains representing the degree of imperfection in the spherulitic crystalline texture. The orientation distribution function is given by equation (39).

$$w_R(\theta, \eta) = 1/8\pi^2 \tag{39}$$

In equations (36)–(39), W_i ($i = C_a$, C_b and B) are normalization constants, $\theta_{0,i}$ ($i = C_a$ and C_b) are the angles at which the respective orientation distribution functions $w_{C_a}(\theta, \eta)$ and $w_{C_b}(\theta, \eta)$ give maximum intensities, J_i ($i = a$ and b) are the parameters representing the sharpness of the orientation distribution of the angle θ in the vicinity of $\theta_{0,i}$ for the $w_{C_a}(\theta, \eta)$ and $w_{C_b}(\theta, \eta)$ respectively, and I_i ($i = a$ and b) are the parameters representing the sharpness of distribution of the rotational angle η of crystal grains about their own a and b axes for type C_a and type C_b respectively.

In the undeformed state, the spherulite must be composed of crystal grains of type B and type R orientations. In the deformed state, however, the crystal grains of type B and type R orientations must be transformed to type C_a and type C_b orientation. As discussed previously, two types of crystal reorientation mechanisms resulting in type C_a and type C_b orientations must be localized at particular positions in the uniaxially deformed spherulites in such a way that type C_a and type C_b are accentuated in the polar and equatorial zones respectively. Type C_a orientation must also be accentuated in the particular zone of $\theta' = 45°$, because it is in the maximum shear stress region, and it can be designated as type C_{a45} in contrast to type C_{a0} in the polar zone. Figure 13 is a schematic diagram demonstrating the respective types of orientations of the crystal grains within a spherulite.

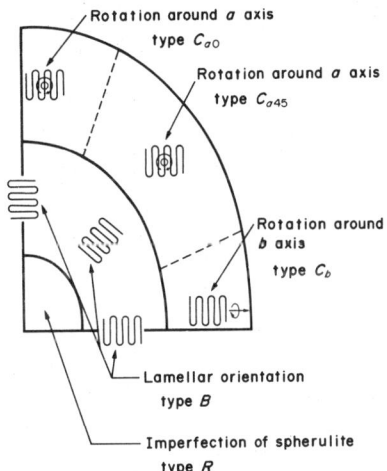

Figure 13 Schematic diagram demonstrating the four types of orientations of the crystal grains, type C_a, type C_b, type B and type R orientations, within a uniaxially deformed spherulite

Fractional concentrations of respective orientations of crystal grains, $X_i(\theta')$ ($i = C_{a0}$, C_{a45}, C_b, B and R), may be formulated as equations (40)–(44), where σ is a parameter representing the ease of transformation of the crystal grains from type B and type R orientations to type C_{a0}, type C_{a45} and type C_b orientations.

$$X_{C_{a0}}(\theta') = f_{a0}\,\sigma\cos^2\theta' \tag{40}$$

$$X_{C_{a45}}(\theta') = f_{a45}\,\sigma\,4\cos^2\theta'\sin^2\theta' \tag{41}$$

$$X_{C_b}(\theta') = (1 - f_{a0} - f_{a45})\sigma\sin^2\theta' \tag{42}$$

$$X_B(\theta') = f_b[1 - \{X_{C_{a0}}(\theta') + X_{C_{a45}}(\theta') + X_{C_b}(\theta')\}] \tag{43}$$

$$X_R(\theta') = (1 - f_b)[1 - \{X_{C_{a0}}(\theta') + X_{C_{a45}}(\theta') + X_{C_b}(\theta')\}] \tag{44}$$

By combining the orientation distribution function $w_i(\theta, \eta)$ with its fractional concentration $X_i(\theta')$, the expansion coefficients W^i_{l0n} for the respective orientation distribution functions of the crystal grains are given by equations (45) and (46), where $i = C_{a0}$, C_{a45} and C_b, and $\xi' = \cos \theta'$. The expansion coefficient W_{10n}^i for the orientation distribution function of the entire crystal grains can be obtained by a sum of W_{10n}^i for respective types of orientation. Using W_{10n} thus obtained, the orientation distribution function of crystallites, $w(\xi, 0, \eta)$, can be determined from equation (4) with a degree of approximation depending on the order of l. Similarly, the expansion coefficient Q_{l0}^j, and furthermore the orientation distribution function of the reciprocal lattice vector of the jth crystal plane $q_j(\zeta_j, 0)$, can be determined from equations (11) and (5) respectively.

$$W_{10n}^i = \frac{1}{2\pi} \int_0^{2\pi} \int_{-1}^1 w_i(0, \eta) \Pi_l^n(\xi) \cos n\eta \, d\xi \, d\eta \int_0^\pi \{N(\theta')/N_0\} X_i(\theta') \sin \theta' d\theta' \tag{45}$$

$$W_{10n}^B = \frac{1}{2\pi} \int_0^{2\pi} \int_{-1}^1 w_B(\theta', \eta') X_B(\theta') \Pi_l^n(\xi') \cos n\eta' \, d\xi' \, d\eta' \tag{46}$$

Figures 14 and 15 show the orientation distribution functions, $w(\xi, 0, \eta)$ and $q_j(\zeta_j, 0)$, calculated as above for the extension ratios of the spherulite from 1.1 up to 1.4 by choosing the parameters so that the calculated results give the best fit to the observed orientation distribution functions in Figures 9 and 11.[78] As can be seen by a comparison of Figure 14 with Figure 11, the development of the two populations in the orientation distribution function is well reproduced by theory. The orientation distribution functions, $q_j(\zeta_j, 0)$, calculated for the principal crystallographic axes shown in Figure 15 also agree fairly well with the observed functions shown in Figure 9. The uniaxial deformation mechanism of spherulites can now be discussed in terms of the changes of the model parameters.[78, 79]

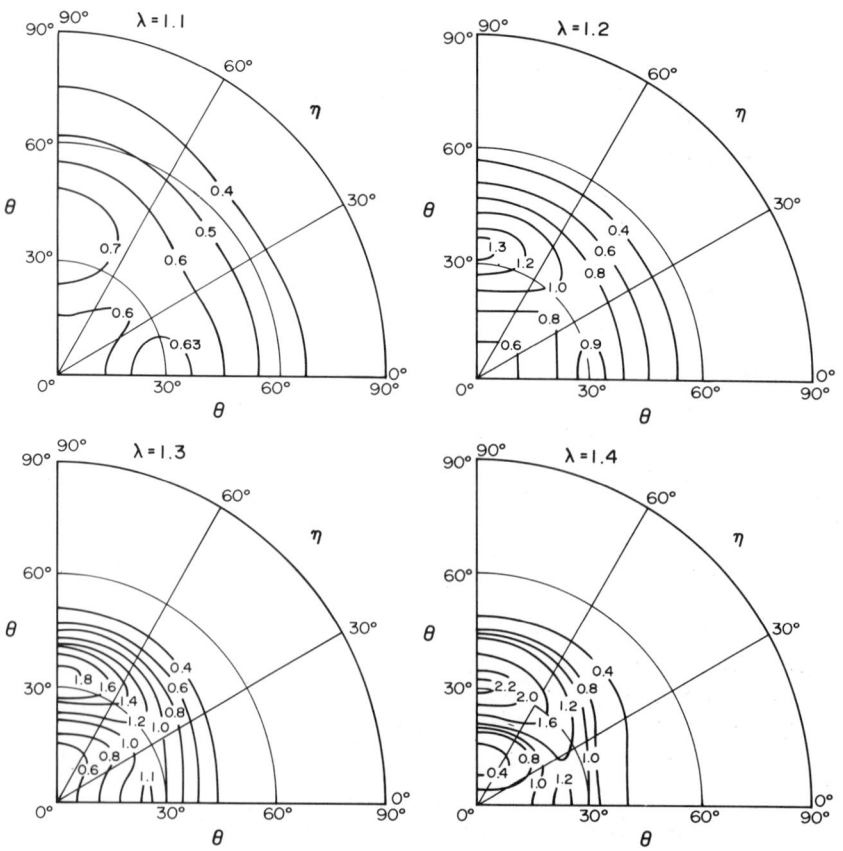

Figure 14   Change in uniaxial orientation distribution function of the crystal grains, $w(\xi, 0, \eta)$, with extension ratio of the spherulite, calculated from the spherulitic deformation model

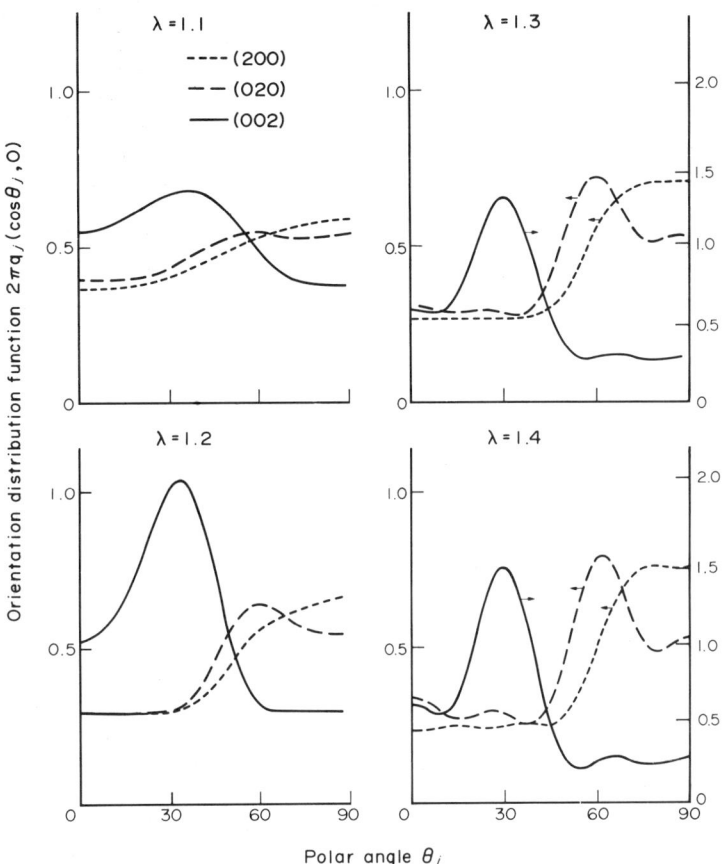

Figure 15 Changes in orientation distribution functions of the principal crystallographic axes of a polyethylene crystal, $q_j(\zeta_j, 0)$, with extension ratio of the spherulite, calculated from the spherulite deformation model

13.4 OPTICAL ANISOTROPY AND ORIENTATION OF STRUCTURAL UNITS

In oriented polymeric materials, the bulk state is usually optically anisotropic. This anisotropy must be attributed to the orientation of certain structural units which are also optically anisotropic. Mathematically, the optical anisotropy in bulk materials can be given in terms of both the average degrees of the orientation distribution function of the structural units and their intrinsic optical constants. The structural units for semicrystalline polymers may be taken as crystal unit cells and noncrystalline chain segments, on the basis of a two-phase hypothesis of crystalline and noncrystalline regions. Therefore, the average degrees of the orientation distribution functions of the units are important parameters not only for characterizing the orientation of polymer molecules in bulk polymers, but also for predicting the anisotropy of the physical properties, such as optical and mechanical properties, of the processed materials.

In this section, some indices of the optical anisotropy in bulk polymers, such as birefringence and absorption and/or emission dichroisms, will be discussed in a general manner based on the optical dichroic ratios which may be related to the generalized orientation factor defined in Section 13.2. Quantitative discussion on visible absorption dichroism can be extended to UV and IR dichroisms without any major change of principle, but with a change in the structural unit interacting with the absorbed ray, as fully discussed by Nomura *et al.*[72] Quantitative descriptions of IR dichroism have been given by Fraser[105] and Beer[106] in terms of the molecular orientation and the direction of the transition moment within the molecule. IR dichroism in biaxially symmetric systems has been discussed by Stein[107] and by Koenig *et al.*[108] UV and visible dichroisms for conjugated double bonds were also discussed by Okajima *et al.*[109] and by Shindo and Stein.[110,111] The theory of absorption/emission dichroisms, such as fluorescence polarization and Raman polarization, as well as NMR, have also been greatly developed. The fluorescence polarization technique was pioneered by Nishijima and co-workers[112-114] and has been discussed by several authors.[73,115-118] Raman polarization and the emission dichroism of fluorescence polarization can be discussed in terms of a common concept.[117,119,120] Differences between them are, as is the case for absorption dichroisms, in the kind of electromagnetic wave employed and in the kind of structural unit concerned with the absorption and emission of the wave. In principle, the NMR technique can give the higher orders of the orientation factors up to the eighth order,[116,121-126] but in practice considerable difficulties are encountered.

13.4.1 General Description of Absorption and Emission Dichroism

Fluorescence polarization has been discussed by a number of authors[73,112-118] in somewhat different ways, which will be generalized here. The fluorescence intensity observed under an analyzer with electric vector Q from polymeric materials excited by polarized light with electric vector P is given by the product of two types of the second rank tensors of optical anisotropies, absorption and emission tensors in equation (47), where K is an instrumental constant, P_i and Q_p are direction cosines of the polarization directions of the polarizer and analyzer, and A_{ij} and E_{pq} are the second rank tensors for the anisotropic absorption and emission of the bulk materials, with respect to the specimen coordinates $0-x_1 x_2 x_3$. (In equation (47) and hereafter, an asterisk * indicates the summation convention.)

$$I(P, Q) = K P_i A_{ij} P_j Q_p E_{pq} Q_q \tag{47*}$$

For simplicity, let us consider the absorption and emission anisotropies of the chromophoric group as having rotational-cylindrical symmetry around an identical vector r_0 fixed within the structural unit by the polar and azimuthal angles, Θ_0 and Φ_0 respectively. The rotational-cylindrical symmetry of absorption and emission anisotropies of the chromophoric group can also be represented in terms of the second rank tensors as equations (48) and (49), where (a^0_{kl}) and (e^0_{rs}) can be related to (A_{ij}) and (E_{pq}) respectively, with the transformation of the second rank tensors giving equations (50) and (51). In equations (50) and (51), for example, t_{ik} is the direction cosine between the x_i axis fixed within the bulk specimen and the v_k axis fixed at the chromophoric group, where the v_3 axis is taken along the vector r_0 and the v_1 and v_2 axes are perpendicular to r_0. Then the fluorescence intensity from the bulk material can be written by averaging over all orientations of the chromophoric groups as in equations (52) and (53).

$$(a^0_{kl}) = \begin{pmatrix} a^0_{11} & 0 & 0 \\ 0 & a^0_{22} & 0 \\ 0 & 0 & a^0_{33} \end{pmatrix} \tag{48}$$

$$(e^0_{rs}) = \begin{pmatrix} e^0_{11} & 0 & 0 \\ 0 & e^0_{22} & 0 \\ 0 & 0 & e^0_{33} \end{pmatrix} \tag{49}$$

$$A_{ij} = t_{ik} t_{jk} a^0_{kk} \tag{50*}$$

$$E_{pr} = t_{pr} t_{qr} e^0_{rr} \tag{51*}$$

$$I(P, Q) = K P_i P_j Q_p Q_q \langle t_{ik} t_{jk} t_{pr} t_{qr} \rangle a^0_{kk} e^0_{rr} \tag{52*}$$

$$= K P_i P_j Q_p Q_q \langle F_{ijpq} \rangle \tag{53*}$$

With the orthogonal-biaxial symmetry of the orientation distribution function of the structural units, $\langle F_{ijpq} \rangle$ in equation (53) may be written as equations (54)-(56),[73] where a_0 and e_0 are given by equations (57) and (58).

$$L_{ij} = \langle F_{iijj} \rangle / (a_0 e_0) = \langle t^2_{i3} t^2_{j3} \rangle \{(a^0_{33} - a^0_{11})/a_0\} \{(e^0_{33} - e^0_{11})/e_0\} + \langle t^2_{i3} \rangle \{(a^0_{33} - a^0_{11})/a_0\} e^0_{11}/e_0$$
$$+ \langle t^2_{j3} \rangle \{(e^0_{33} - e^0_{11})/e_0\} a^0_{11}/a_0 + (a^0_{11}/a_0)(e^0_{11}/e_0) \tag{54}$$

$$L'_{ij} = \langle F_{ijij} \rangle / (a_0 e_0) = \langle t^2_{i3} t^2_{j3} \rangle \{(a^0_{33} - a^0_{11})/a_0\} \{(e^0_{33} - e^0_{11})/e_0\} \tag{55}$$

$$\langle F_{ijkl} \rangle = 0 \tag{56}$$

$$a_0 = a^0_{33} + 2 a^0_{11} \tag{57}$$

$$e_0 = e^0_{33} + 2 e^0_{11} \tag{58}$$

For a given geometry of the optical system with respect to the bulk specimen, one can define the dichroic ratio given by equation (59), where the instrumental constant is cancelled out and the error is minimized. For the incident radiation perpendicular or inclined at $\pi/4$ to the surface of the bulk specimen (the $x_2 x_3$ plane), the dichroic ratios can be related to simple combinations of L_{ij}, which should be first determined as given by equations (60a)–(60d).[127] Equations (60a) and (60b) are the results for the perpendicular radiation, while equations (60c) and (60d) are those for the inclined radiation for which refraction of the incident ray at the surface of the bulk specimen must be avoided by use of immersion[115] or the hemisphere[128] technique, or corrected for appropriately. Using a combination of the dichroic ratios given by equations (61), the component L_{ij} can be obtained by means of equations (62).

$$D_k(P, Q) = I(P, Q)/I(x_3, x_3) \tag{59}$$

$$D_1(x_2, x_2) = L_{22}/L_{33} \quad D_2(x_3, x_2) = L_{32}/L_{33} \quad D_3(x_2, x_3) = L_{23}/L_{33} \tag{60a}$$

$$
\begin{aligned}
D_4(x_3, x_1 \cos \pi/4 + x_2 \sin \pi/4) &= (L_{31} + L_{32})/(2L_{33}) \\
D_5(x_2, x_1 \cos \pi/4 + x_2 \sin \pi/4) &= (L_{21} + L_{22})/(2L_{33})
\end{aligned} \tag{60b}
$$

$$
\begin{aligned}
D_6(-x_1 \cos \pi/4 + x_2 \sin \pi/4, x_3) &= (L_{13} + L_{23})/(2L_{33}) \\
D_7(-x_1 \cos \pi/4 + x_2 \sin \pi/4, x_2) &= (L_{12} + L_{22})/(2L_{33})
\end{aligned} \tag{60c}
$$

$$
\begin{aligned}
D_8(-x_1 \cos \pi/4 + x_2 \sin \pi/4, x_1 \cos \pi/4 + x_2 \sin \pi/4) &= (L_{11} + L_{12} + L_{21} + L_{22} - 4L'_{12})/(4L_{33}) \\
D_9(x_1 \cos \pi/4 + x_2 \sin \pi/4, x_1 \cos \pi/4 + x_2 \sin \pi/4) &= (L_{11} + L_{12} + L_{21} + L_{22} + 4L'_{12})/(4L_{33})
\end{aligned} \tag{60d}
$$

$$D_0 = 2(D_4 + D_6 + D_8 + D_9) + 1 = \sum_{i=1}^{3} \sum_{j=1}^{3} L_{ij}/L_{33} = 1/L_{33} \tag{61}$$

$$
\begin{aligned}
L_{11} &= (2D_8 + 2D_9 - 2D_7 - 2D_5 + D_1)/D_0 \\
L_{12} &= (2D_7 - D_1)/D_0 \\
L_{13} &= (2D_6 - D_3)/D_0 \\
L_{21} &= (2D_5 - D_1)/D_0 \\
L_{22} &= D_1/D_0 \\
L_{23} &= D_3/D_0 \\
L_{31} &= (2D_4 - D_2)/D_0 \\
L_{32} &= D_2/D_0 \\
L_{33} &= 1/D_0
\end{aligned} \tag{62}
$$

In the fluorescence polarization technique, the intrinsic optical constants $(a_{33}^0 - a_{11}^0)/a_0$ and $(e_{33}^0 - e_{11}^0)/e_0$ can be determined. With random orientation, one can obtain α_2 using equation (63), where the value of α_2 varies from zero to unity depending on the absorption and emission anisotropies of the chromophoric group. Alternatively, one can obtain the material constant α_1 independently of the degree of orientation given by equation (64). Then the intrinsic optical constants of absorption and emission anisotropies can be determined by equations (65). Finally, one can calculate the moments of $\langle t_{i3}^2 t_{j3}^2 \rangle$ from the linear combination of L_{ij} and the intrinsic optical constants. As a result, the fluorescence polarization has an additional advantage over ordinary absorption or emission dichroisms, where the values of the intrinsic optical and material constants cannot be obtained.

$$\alpha_2 = (5/2)\{1 - (D_2 + D_3)/2\}/\{1 + (D_2 + D_3)\} = (5/2)\{L_{33} - (L_{23} + L_{32})/2\}/\{L_{33} + L_{23} + L_{32}\} = \{(a_{33}^0 - a_{11}^0)/a_0\}\{(e_{33}^0 - e_{11}^0)/e_0\} \tag{63}$$

$$\alpha_1 = 1 + 3(D_5 - D_7)/(D_8 + D_9 + D_6 - D_3 - 2D_5) = 1 + 3(L_{21} + L_{12}) \bigg/ \sum_{j=1}^{3}(L_{1j} + L_{2j}) = \{(e_{33}^0 - e_{11}^0)/e_0\}/\{(a_{33}^0 - a_{11}^0)/a_0\} \tag{64}$$

$$(a_{33}^0 - a_{11}^0)/a_0 = (\alpha_2/\alpha_1)^{1/2} \qquad (e_{33}^0 - e_{11}^0)/e_0 = (\alpha_2 \alpha_1)^{1/2} \tag{65}$$

For fluorescence polarization, on the other hand, simple relations of the measurable dichroic ratios and/or the components of L_{ij} to the second- and fourth-order orientation factors cannot be obtained. Thus, the moments $M_{ij} = \langle t_{i3}^2 t_{j3}^2 \rangle$, which are derived from L_{ij} and also from the dichroic ratios D_k, may be related to the second- and fourth-order orientation factors through equations (66) and (67).

$$F_{20}^0 = \left(3\sum_{i=1}^{3} M_{i3} - 1\right)\bigg/2 = \left(3\sum_{j=1}^{3} M_{3j} - 1\right)\bigg/2 \tag{66a}$$

$$F_{22}^0 = 3\sum_{i=1}^{3}(M_{i1} - M_{i2}) = 3\sum_{j=1}^{3}(M_{1j} - M_{2j}) \tag{66b}$$

$$F_{40}^0 = \left(35M_{33} - 30 \sum_{i=1}^{3} M_{i3} + 3 \right) \Big/ 8 \tag{67a}$$

$$F_{42}^0 = 15 \left\{ 7(M_{31} - M_{32}) - \sum_{i=1}^{3} (M_{i1} - M_{i2}) \right\} \Big/ 2 \tag{67b}$$

$$F_{44}^0 = 105(M_{11} + M_{22} - 6M_{12}) \tag{67c}$$

In a similar way to the derivation of equation (14), the orientation factors F_{lmn} of the structural unit, possibly of a noncrystalline unit involving the fluorescence, may be related to the above orientation factors using equation (68), where $l(\leq 4)$, $m(\leq l)$ and $n(\leq l)$ are even numbers. For most of the noncrystalline orientation, it may be assumed that the noncrystalline chain segments are randomly oriented around their own u_3 axes, *i.e.* random orientation with respect to the rotational angle η. Thus equation (68) can be rewritten as equation (69), where $l = 2$ or 4, and $m = 0$, 2 or 4.

$$F_{lm}^0 = F_{lm0} \, P_l(\cos \Theta_0) + 2 \sum_{n=2}^{l} \{(l-n)!/(l+n)!\} \, F_{lmn} \, P_l^n(\cos \Theta_0) \cos n\Phi_0 \tag{68}$$

$$F_{lm}^0 = F_{lm0} \, P_l(\cos \Theta_0) \tag{69}$$

13.4.2 Absorption and Emission Dichroism

The absorption or emission dichroisms may be considered as a special case of the fluorescence polarization. With polarized radiation P, the intensity absorbed by the bulk material may be written as equation (70), obtained from equations (47)–(53). Similarly, if an unpolarized incident ray perpendicular to the x_2x_3 plane is used, the observed emission intensity through the analyzer Q is given by equation (71). However, the electric vector of the incident ray is parallel to the x_2x_3 plane and a simple relation cannot be obtained. For the orthogonal-biaxial orientation, the components $\langle A_{ii} \rangle$ in equation (70) may be rewritten as equation (72); that is, the absorption dichroism is expressed as a special case of the fluorescence polarization.

$$A(P) = K P_i P_j \sum_{k=1}^{3} \langle F_{ijkk} \rangle / e_0 = K P_i P_j \langle A_{ij} \rangle = K P_i P_j \langle t_{ik} t_{jk} \rangle a_{kk}^0 \tag{70}*$$

$$E(Q) = K Q_p Q_q \sum_{i=2}^{3} \langle F_{iipq} \rangle \tag{71}*$$

$$\langle A_{ii} \rangle = \langle t_{i3}^2 \rangle (a_{33}^0 - a_{11}^0) + a_{11}^0 \tag{72}$$

The dichroic ratios for the absorption dichroism are also defined by equation (73). Then the dichroic ratios for the perpendicular and inclined radiations can be rewritten as equations (74) and (75) respectively. The dichroic orientation factors may be defined as equations (76) and (77). Substituting the results of equations (74) and (75) into equations (76) and (77), equations (78) and (79) can be obtained.

$$D_k(P) = A(P)/A(x_3) \tag{73}$$

$$D_1(x_2) = \frac{\langle t_{23}^2 \rangle (a_{33}^0 - a_{11}^0) + a_{11}^0}{\langle t_{33}^2 \rangle (a_{33}^0 - a_{11}^0) + a_{11}^0} \tag{74}$$

$$D_2(-x_1 \cos \pi/4 + x_2 \sin \pi/4) = \frac{(\langle t_{13}^2 \rangle + \langle t_{23}^2 \rangle)(a_{33}^0 - a_{11}^0)/2 + a_{11}^0}{\langle t_{33}^2 \rangle (a_{33}^0 - a_{11}^0) + a_{11}^0} \tag{75}$$

$$F_{D,20} = (1 - D_2)/(1 + 2D_2) \tag{76}$$

$$F_{D,22} = (D_2 - D_1)/(1 + 2D_2) \tag{77}$$

$$F_{D,20} = \{(a_{33}^0 - a_{11}^0)/a_0\}\{(3\langle t_{33}^2 \rangle - 1)/2\} = \{(a_{33}^0 - a_{11}^0)/a_0\}F_{20}^0 \tag{78}$$

$$F_{D,22} = \{(a_{33}^0 - a_{11}^0)/a_0\}\{(\langle t_{13}^2 \rangle - \langle t_{23}^2 \rangle)/2\} = \{(a_{33}^0 - a_{11}^0)/a_0\}F_{22}^0/6 \tag{79}$$

As can be seen from equations (78) and (79), the dichroic orientation factors are represented by the product of the orientation of the chromophoric group and its intrinsic optical constants. The constants, however, cannot be obtained from the absorption dichroism alone and must be determined from other sources. When the orientation factors F_{20}^0 and F_{22}^0 are obtained from equations

(78) and (79), equations (68) and (69) also hold for $l=2$, and the orientation factors of the structural units F_{2mn} may be determined, providing that the polar and azimuthal angles Θ_0 and Φ_0 are also known from other sources.

13.5 MECHANICAL ANISOTROPY OF REGENERATED CELLULOSE FILMS

Mechanical anisotropy of polymeric materials has been studied by Raumann *et al.*[47-49] and Ward *et al.*[50-55] as mentioned previously. Raumann's intention was to describe the anisotropy in bulk as completely as possible, whereas Ward interpreted the anisotropy of fibrous materials, such as textile fibers, by using a kind of aggregation model for the orientation of structural units.[50, 55] However, the orientation behavior of structural units in semicrystalline polymers is usually more complicated than that predicted by Ward. Therefore, it must suffice to determine the orientation of the structural units from experimental sources, such as X-ray diffraction and optical dichroisms.

In this section, the mechanical anisotropy of some commercial cellophanes will be discussed quantitatively;[60] firstly, on the basis of the theory of infinitesimal deformations of an orthotropic elastic body in bulk, and secondly, in terms of the biaxial orientation of the crystalline and noncrystalline structural units and their mechanical anisotropy. The simple volume additivity concept of the contribution from both phases to the bulk properties will be the primary problem to be addressed for such bulk phenomena as mechanical properties, in contrast to the treatment of Ward *et al.*, who discussed a monophase system with a rather undefined structural unit.

13.5.1 Mechanical Anisotropy of Bulk Specimens

According to the theory for infinitesimal deformation of an anisotropic elastic body, the mechanical properties can be represented by 36 independent elastic compliances s_{ij} or by the same number of independent elastic stiffnesses c_{ij}. The number of constants can be reduced to nine on the basis of the orthogonal anisotropy of the body with respect to Cartesian coordinates $0-x_1 x_2 x_3$ in the bulk specimen, and can be further reduced to five on the basis of the transverse isotropy about the x_3 axis.

Let us rotate the Cartesian coordinates $0-x_1 x_2 x_3$ about the x_1 axis by an angle δ from x_3 toward the x_2 direction and designate a new set of coordinates as $0-x_1' x_2' x_3'$. Then the elastic compliance or Young's modulus along the x_3' axis for the body which is orthogonal with respect to the $0-x_1 x_2 x_3$ system is given by equation (80), where G_{23} is the shear modulus in the $x_2 x_3$ plane and v_{32} is the Poisson ratio of the strain along the x_2 axis to that along the x_3 axis.

$$s_{33}'(\equiv 1/E_\delta)=\cos^4\delta\, s_{33}+\sin^4\delta\, s_{22}+\cos^2\delta\sin^2\delta(2s_{32}+s_{44})=\cos^4\delta/E_3+\sin^4\delta/E_2+\cos^2\delta\sin^2\delta(1/G_{23}+2v_{32}/E_3) \quad (80)$$

When δ is taken as $0°$, $45°$ and $90°$, equation (80) reduces to equations (81)–(83) respectively.

$$1/E_0 = 1/E_3 \quad (81)$$

$$1/E_{45} = (1/E_2 + 1/E_3)/4 + (1/G_{23} - 2v_{32}/E_3)/4 \quad (82)$$

$$1/E_{90} = 1/E_2 \quad (83)$$

It is possible to measure E_0, E_{45}, E_{90} and v_{32}, and hence one can evaluate G_{23} from equations (81)–(83). The ratios v_{32} were measured for the commercial cellophane test specimens in the standard dry state of $20\,°C$ and 65% RH, $C\text{-}1\text{-}D_{st}$, $C\text{-}2\text{-}D_{st}$ and $C\text{-}3\text{-}D_{st}$, in which the biaxial orientations of crystalline and noncrystalline phases are shown in Figure 16 in terms of equilateral triangular coordinates, and the moduli G_{23} were calculated by using the tensile moduli with δ equal to $0°$, $45°$ and $90°$. These results are included in Figure 17, together with the measured v_{23}. Furthermore, the validity of the assumption of orthogonal mechanical anisotropy was checked for the test specimens by comparing the angular dependence of E_δ calculated from equation (80) using the values of E_0, E_{90}, G_{23} and v_{32} with those observed. The agreement between the calculated and observed results is fairly satisfactory for all of the specimens, as shown in Figure 17, and indicates the validity of the assumption of orthogonal anisotropy with respect to $0-x_1 x_2 x_3$, *i.e.* the thickness, transverse and machine directions.

Raumann *et al.*[47-49] have shown that the shear modulus G_{23} has little dependence upon molecular orientation for uniaxially stretched PE and PET films; Shinohara and Tanzawa[129] have reported similar findings for uniaxially stretched regenerated cellulose films. The latter authors have

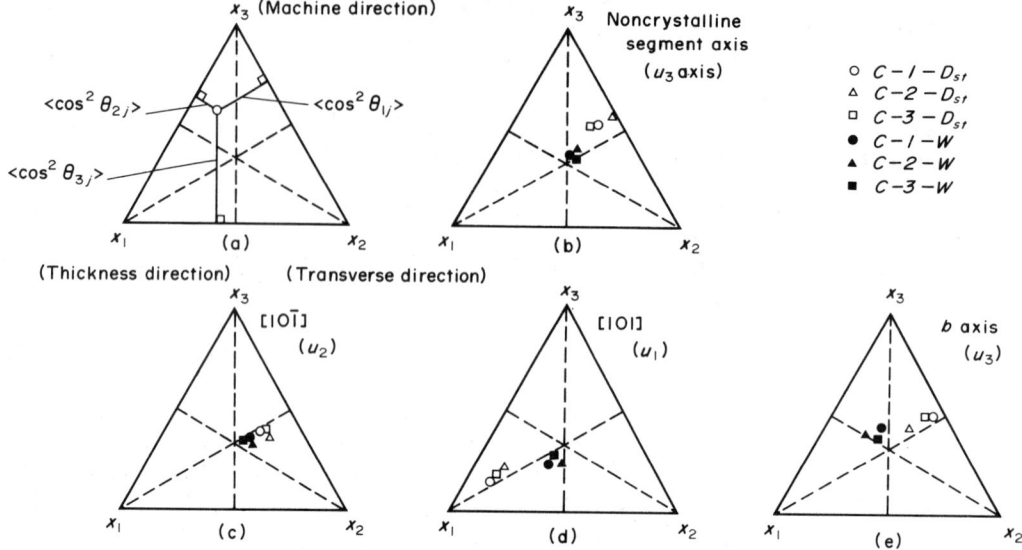

Figure 16 Equilateral triangular coordinates representing the degrees of biaxial orientation of crystallites and noncrystalline chain segments in terms of the three quantities $\langle \cos^2 \theta_{ij} \rangle$, for the three kinds of commercial cellophanes

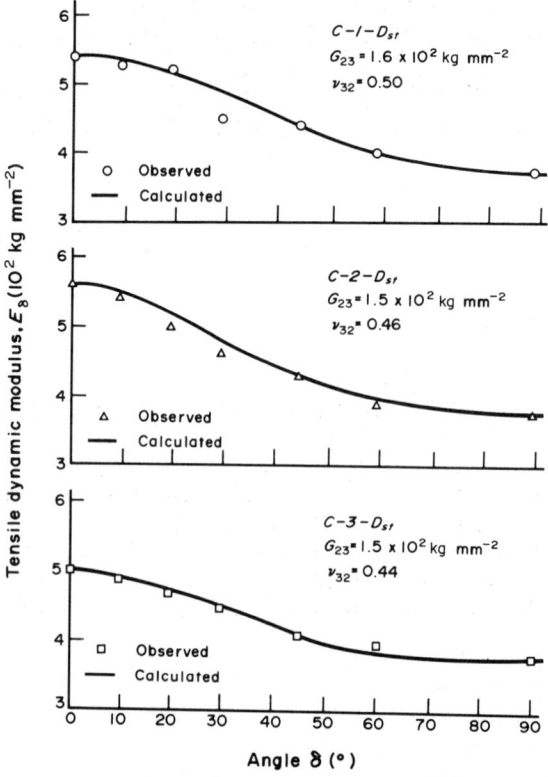

Figure 17 Comparison of mechanical anisotropy of the bulk specimen observed with that calculated from the theory of infinitesimal deformation of an orthogonally anisotropic elastic body

obtained G_{23} as 1.8 GPa for regenerated cellulose films, which is very close to the values obtained by Nomura *et al.*[60]

13.5.2 Mechanical Anisotropy in Relation to Orientation of Crystalline and Noncrystalline Structural Units

The elastic properties of polycrystalline aggregates have been studied by several authors from the standpoint of composite crystals, since the first opposing proposals by Voigt[132] and Reuss[131] for a random aggregate. Voigt took the overall average of the elastic stiffnesses c_{ij}^{c0} of randomly oriented composite crystals to obtain the bulk stiffnesses c_{ij}^c of the aggregate under the homogeneous strain hypothesis, while Reuss derived the elastic compliances s_{ij}^c of the aggregate from the homogeneous stress hypothesis. For oriented aggregates, the most extensive studies have been made by Ward and co-workers[50-55] who used both hypotheses for uniaxially symmetric systems.

From the homogeneous stress hypothesis, the elastic compliance tensor s_{ijkl}^c of the aggregate can be expressed by equation (84). Similarly, from the hypothesis of homogeneous strain, the elastic stiffness tensor c_{ijkl}^c of the aggregate can be expressed by equation (85), where t_{ip}, for example, is the direction cosine between the x_i axis of the aggregate and the u_p axis of the crystal unit cell; s_{pqrs}^{c0} and c_{pqrs}^{c0} are the elastic compliances and stiffnesses of the unit respectively. These expressions for averaging are, of course, two extreme approximations, which neglect the interaction between the units and do not correspond to the real situation of polycrystalline materials, as has been pointed out by several authors.[55, 132-135]

$$s_{ijkl}^c = \langle t_{ip}\, t_{jq}\, t_{kr}\, t_{ls} \rangle s_{pqrs}^{c0} \tag{84}*$$

$$c_{ijkl}^c = \langle t_{ip}\, t_{jq}\, t_{kr}\, t_{ls} \rangle c_{pqrs}^{c0} \tag{85}*$$

These expressions may be expanded to include polyphase systems. For semicrystalline polymers, such as the regenerated cellulose films, the elastic compliances and stiffnesses of the system are given by equations (86) and (87). Here X_c is the degree of crystallinity in terms of volume fraction, and s_{ijkl}^a and c_{ijkl}^a are elastic compliance and stiffness tensors of the noncrystalline phase and are described, much like equations (84) and (85), in terms of the compliances and stiffnesses of the noncrystalline chain segment s_{pqrs}^{a0} and c_{pqrs}^{a0} respectively. Equation (86) is a simple representation of series coupling and equation (87) of parallel coupling of the two phases.

$$s_{ijkl} = X_c\, s_{ijkl}^c + (1 - X_c) s_{ijkl}^a \tag{86}$$

$$c_{ijkl} = X_c\, c_{ijkl}^c + (1 - X_c) c_{ijkl}^a \tag{87}$$

Compliances of the crystalline phase, such as s_{ii}^c, s_{ij}^c and s_{pp}^c, where $i = 1, 2, 3$; $i, j = 1, 2, 3$; and $p = 4, 5, 6$ respectively, can be written as equations (88)–(90), where the mechanical anisotropy of the cellulose crystal is assumed to be transversely isotropic with respect to the crystal b axis (the molecular chain direction). Similarly, stiffnesses of the crystalline phase, such as c_{ii}^c, c_{ij}^c and c_{pp}^c are given by equations (91)–(93). The compliances and stiffnesses of the noncrystalline phase can be described similarly in terms of the degrees of biaxial orientation of the noncrystalline chain segments and stiffnesses.

$$s_{ii}^c = \langle \sin^4 \theta_{i3} \rangle s_{11}^{c0} + \langle \cos^4 \theta_{i3} \rangle s_{33}^{c0} + \langle \cos^2 \theta_{i3} \sin^2 \theta_{i3} \rangle (2 s_{13}^{c0} + s_{44}^{c0}) \tag{88}$$

$$s_{ij}^c = \langle \cos^2 \theta_{i3} \cos^2 \theta_{j3} \rangle (s_{11}^{c0} + s_{33}^{c0} - s_{44}^{c0}) + \langle (1 - \cos^2 \theta_{i3} - \cos^2 \theta_{j3}) \rangle s_{12}^{c0} + \langle (\cos^2 \theta_{i3} + \cos^2 \theta_{j3} - 2\cos^2 \theta_{i3} \cos^2 \theta_{j3}) \rangle s_{13}^{c0} \tag{89}$$

$$s_{pp}^c = \langle \cos^2 \theta_{i3} \cos^2 \theta_{j3} \rangle (4 s_{11}^{c0} - 4 s_{33}^{c0} - 8 s_{13}^{c0} - 4 s_{44}^{c0}) + 2 \langle (1 - \cos^2 \theta_{i3} - \cos^2 \theta_{j3}) \rangle (s_{11}^{c0} - s_{12}^{c0}) + \langle (\cos^2 \theta_{i3} + \cos^2 \theta_{j3}) \rangle s_{44}^{c0} \tag{90}$$

$$c_{ii}^c = \langle \sin^4 \theta_{i3} \rangle c_{11}^{c0} + \langle \cos^4 \theta_{i3} \rangle c_{33}^{c0} + 2 \langle \cos^2 \theta_{i3} \sin^2 \theta_{i3} \rangle (c_{13}^{c0} + 2 c_{44}^{c0}) \tag{91}$$

$$c_{ij}^c = \langle \cos^2 \theta_{i3} \cos^2 \theta_{j3} \rangle (c_{11}^{c0} + c_{33}^{c0} - 4 c_{44}^{c0}) + \langle (1 - \cos^2 \theta_{i3} - \cos^2 \theta_{j3}) \rangle c_{12}^{c0} + \langle (\cos^2 \theta_{i3} + \cos^2 \theta_{j3} - 2\cos^2 \theta_{i3} \cos^2 \theta_{j3}) \rangle c_{13}^{c0} \tag{92}$$

$$c_{pp}^c = \langle \cos^2 \theta_{i3} \cos^2 \theta_{j3} \rangle (c_{11}^{c0} + c_{33}^{c0} - 2 c_{13}^{c0} - 4 c_{44}^{c0}) + (1/2) \langle (1 - \cos^2 \theta_{i3} - \cos^2 \theta_{j3}) \rangle (c_{11}^{c0} - c_{12}^{c0}) + \langle (\cos^2 \theta_{i3} + \cos^2 \theta_{j3}) \rangle c_{44}^{c0} \tag{93}$$

The mechanical constants of cellulose II crystals, such as E_i^{c0}, G_{ij}^{c0} and v_{ij}^{c0}, have not been reported, except for experimental and theoretical investigations of E_3^{c0} by several authors.[1, 136-138] The correct value of E_3^{c0} seems to be around 130 GPa. The ratio of E_2^{c0} (or E_1^{c0}) to E_3^{c0} may be taken as about 1/30, *i.e.* $E_2^{c0} = 4.3$ GPa, by analogy to experimental results on poly(vinyl alcohol) crystals by Sakurada *et al.*[1] However, the values of G_{23}^{c0}, v_{31}^{c0}, and v_{12}^{c0} are quite uncertain, and these are at first assumed to be one-third of E_2^{c0}, *i.e.* 1.4 GPa, 0.3 and 0.5 respectively. From these values, the compliances s_{ii}^{c0}, s_{ij}^{c0} and s_{pp}^{c0} can be determined by $s_{ii}^{c0} = 1/E_i^{c0}$, $s_{ij}^{c0} = s_{ji}^{c0} = -v_{ij}^{c0}/E_j^{c0}$ and $s_{pp}^{c0} = 1/G_{ij}^{c0}$ respectively. The stiffnesses c_{ii}^{c0}, c_{ij}^{c0} and c_{pp}^{c0} can be determined from the inverse matrix of $(s_{ij}^{c0})^{-1}$.

The mechanical constants of the noncrystalline chain segment, such as E_i^{a0}, G_{ij}^{a0} and v_{ij}^{a0}, are also quite uncertain. However, as an approximation we can assume that the relation between the potential energy P(r) and the atomic or molecular distance r holds for a noncrystalline chain and is given by equation (94), where $m = 9$–12, and $n = 1$ or 6 for an ionic or molecular crystal respectively.[139] Then, neglecting the repulsion term in equation (94) for r greater than the equilibrium distance, we obtain the mechanical constants E_2^{a0} from equations (95) or (96), where v_c is the specific volume of the crystalline phase and v_a is that of the noncrystalline phase including sorbed water and plasticizer. The swelling of the noncrystalline phase is assumed to occur only along the lateral direction of the polymer chain, not lengthwise. The mechanical constant E_3^{a0} may also be approximated by equation (97) and the other mechanical constants G_{23}^{a0}, v_{31}^{a0} and v_{12}^{a0} may be assumed to be one-third of E_2^{a0}, 0.5 and 0.5 respectively.

$$P(r) = -c/r^n + d/r^m \tag{94}$$

$$E_2^{a0} = E_2^{c0}/(r_a/r_c)^3 = E_2^{c0}/(v_a/v_c)^{3/2} \tag{95}$$

$$E_2^{a0} = E_2^{c0}/(r_a/r_c)^8 = E_2^{c0}/(v_a/v_c)^{8/2} \tag{96}$$

$$E_3^{a0} = E_3^{c0}/(v_a/v_c) \tag{97}$$

The elastic compliances or stiffnesses of the bulk specimens s_{ii} or c_{ij} can be determined from the mechanical constants of the structural units, their degrees of biaxial orientation and the apparent

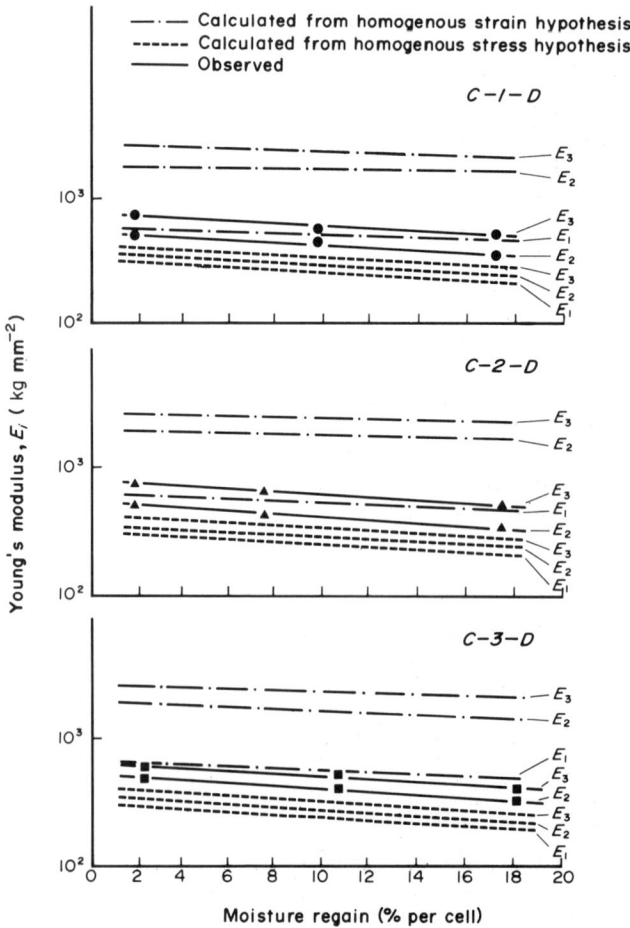

Figure 18 Comparison of the mechanical anisotropy of the bulk specimen measured in the three different dry states with that calculated from the degree of biaxial orientation of crystallites and noncrystalline chain segments and their mechanical anisotropies by using equation (86) or (87) in combination with equation (96)

degree of crystallinity, through equations (86)–(93). The bulk compliances or stiffnesses can be further converted to the Young's moduli E_i by equations (98) and (99), where $i \neq j \neq k$.

$$E_i = 1/s_{ii} \tag{98}$$

$$E_i = \begin{vmatrix} c_{11} & c_{12} & c_{13} \\ c_{21} & c_{22} & c_{23} \\ c_{31} & c_{32} & c_{33} \end{vmatrix} (c_{jj}c_{kk} - c_{jk}^2)^{-1} \tag{99}$$

The results thus calculated for E_3, E_2 and E_1 by using mechanical constants of the noncrystalline chain segment from equation (96) are shown in Figure 18 as a function of moisture regain and compared with the experimental results. As can be seen in Figure 18, the predicted mechanical anisotropy of the bulk specimen and the moisture dependence of the Young's moduli are realized, at least qualitatively, for every specimen. When the calculated and observed results are compared, the moduli based on the averaging under the homogeneous strain hypothesis are too high, and in contrast the moduli under the homogeneous stress hypothesis are rather low. This is because, as has been frequently pointed out, these averages give upper and lower bounds respectively. Both assumptions are made for reasons of mathematical convenience. They neglect the interaction between the structural units and have no physically well-defined correlation with the morphology of the real materials. The fact that the latter average is rather close to the observed results suggests its validity for materials of low degrees of crystallinity, such as regenerated cellulose films.

13.6 REFERENCES

1. I. Sakurada, T. Ito and K. Nakamae, *J. Polym. Sci., Part C*, 1966, **15**, 75.
2. K. Furuhata, Y. Yokokawa and K. Miyasaka, *J. Polym. Sci., Polym. Phys. Ed.*, 1984, **22**, 133.
3. A. Frey, *Naturwissenschaften*, 1925, **13**, 403.
4. A. Frey, *Kolloidchem. Beih.*, 1925, **20**, 209.
5. H. Neubert, *Kolloidchem. Beih.*, 1925, **20**, 244.
6. O. Wiener, *Sachs. Ges. Wiss. Leipzig Math.-Phys. Kl. Abh.*, 1912, **32**, 509.
7. O. Wiener, *Kolloidchem. Beih.*, 1927, **23**, 189.
8. A. Möhring, *Kolloidchem. Beih.*, 1927, **23**, 126.
9. J. M. Preston, *J. Soc. Dyers Colour*, 1931, **47**, 312.
10. J. M. Preston and P. C. Tsien, *J. Soc. Dyers Colour*, 1946, **62**, 242.
11. J. M. Preston and P. C. Tsien, *J. Soc. Dyers Colour*, 1946, **62**, 368.
12. J. M. Preston and Y. F. Su, *J. Soc. Dyers Colour*, 1950, **66**, 357.
13. J. M. Preston and P. C. Tsien, *J. Soc. Dyers Colour*, 1950, **66**, 361.
14. D. R. Morey, *Text. Res. J.*, 1933, **3**, 325.
15. D. R. Morey, *Text. Res. J.*, 1934, **4**, 491.
16. D. R. Morey, *Text. Res. J.*, 1935, **5**, 105, 483, 538.
17. O. Kratky, *Kolloid-Z.*, 1933, **64**, 213.
18. O. Kratky, *Kolloid-Z.*, 1934, **68**, 347.
19. O. Kratky, *Kolloid-Z.*, 1935, **70**, 14.
20. F. Breuer, O. Kratky and G. Saito, *Kolloid-Z.*, 1937, **80**, 139.
21. O. Kratky, *Kolloid-Z.*, 1938, **84**, 149.
22. O. Kratky and P. Platzek, *Kolloid-Z.*, 1938, **84**, 268.
23. P. H. Hermanns, O. Kratky and P. Platzek, *Kolloid-Z.*, 1939, **86**, 245.
24. O. Kratky and P. Platzek, *Kolloid-Z.*, 1939, **88**, 78.
25. O. Kratky, *Angew. Chem.*, 1940, **53**, 153.
26. B. Baule, O. Kratky and R. Treer, *Z. Phys. Chem., Abt. B*, 1941, **50**, 255.
27. B. Baule and O. Kratky, *Z. Phys. Chem., Abt. B*, 1942, **52**, 142.
28. P. H. Hermanns, *Kolloid-Z.*, 1937, **81**, 143.
29. P. H. Hermanns and A. J. Leeuw, *Kolloid-Z.*, 1937, **81**, 300.
30. P. H. Hermanns and A. J. Leeuw, *Kolloid-Z.*, 1938, **82**, 58.
31. P. H. Hermanns, *Kolloid-Z.*, 1938, **83**, 71.
32. P. H. Hermanns, *Kolloid-Z.*, 1939, **86**, 107.
33. P. H. Hermanns, *Recl. Trav. Chim. Pays-Bas*, 1939, **58**, 63.
34. P. H. Hermanns and P. Platzek, *Kolloid-Z.*, 1939, **87**, 296.
35. P. H. Hermanns and P. Platzek, *Kolloid-Z.*, 1939, **88**, 68.
36. P. H. Hermanns and J. Booys, *Kolloid-Z.*, 1939, **88**, 73.
37. P. H. Hermanns, *Kolloid-Z.*, 1939, **88**, 172.
38. P. H. Hermanns, *Proc. K. Ned. Akad. Wet.*, 1939, **42**, 798.
39. P. H. Hermanns, *Kolloid-Z.*, 1939, **89**, 344.
40. P. H. Hermanns and P. Platzek, *Kolloid-Z.*, 1939, **89**, 349.
41. P. H. Hermanns and P. Platzek, *Z. Phys. Chem., Abt. A*, 1939, **185**, 260.
42. J. Booys, H. L. Bredee and P. H. Hermanns, *Recl. Trav. Chim. Pays-Bas*, 1940, **59**, 73.

43. P. H Hermanns, *Naturwissenschaften.*, 1940, **28**, 223.
44. F. H. Müller, *Kolloid-Z.*, 1941, **95**, 138, 306.
45. W. Kuhn and F. Grün, *Kolloid-Z.*, 1942, **101**, 248.
46. L. R. G. Treloar, *Trans. Faraday Soc.*, 1947, **43**, 277, 284; 1954, **50**, 881; L. R. G. Treloar, 'The Physics of Rubber Elasticity', 2nd edn., Oxford University Press, London, 1958.
47. G. Raumann and D. W. Saunders, *Proc. Phys. Soc. London*, 1961, **77**, 1028.
48. G. Raumann, *Proc. Phys. Soc. London*, 1962, **79**, 1221.
49. G. Raumann, *Br. J. Appl. Phys.*, 1963, **14**, 795.
50. I. M. Ward, *Proc. Phys. Soc. London*, 1962, **80**, 1176.
51. D. W. Hadley, I. M. Ward and J. Ward, *Proc. R. Soc. London, Ser. A*, 1965, **285**, 275.
52. P. R. Pinnock, I. M. Ward and J. M. Wolfe, *Proc. R. Soc. London, Ser. A*, 1966, **291**, 267.
53. I. M. Ward, *Appl. Mater. Res.*, 1966, **5**, 224, 228.
54. N. H. Ladizesky and I. M. Ward, *J. Macromol. Sci., Phys.*, 1971, **5**, 661, 745.
55. V. B. Gupta and I. M. Ward, *J. Macromol. Sci., Phys.*, 1967, **1**, 373.
56. Z. H. Stachurski and I. M. Ward, *J. Polym. Sci., Part A-2*, 1968, **6**, 1083, 1817; *J. Macromol. Sci., Phys.*, 1969, **3**, 427, 445.
57. V. B. Gupta and I. M. Ward, *J. Macromol. Sci., Phys.*, 1968, **2**, 89.
58. V. B. Gupta, A. Keller and I. M. Ward, *J. Macromol. Sci., Phys.*, 1968, **2**, 139.
59. H. Takahara, S. Nomura, H. Kawai, Y. Yamaguchi, K. Okazaki and A. Fukushima, *J. Polym. Sci., Part A-2*, 1968, **6**, 197.
60. S. Nomura, S. Kawabata, H. Kawai, Y. Yamaguchi, A. Fukushima and H. Takahara, *J. Polym. Sci., Part A-2*, 1969, **7**, 325.
61. M. Maeda, S. Hibi, F. Itoh, S. Nomura, T. Kawaguchi and H. Kawai, *J. Polym. Sci., Part A-2*, 1970, **8**, 1303.
62. R. J. Samuels, in 'The Science and Technology of Polymer Films', ed. O. J. Sweeting, Wiley, New York, 1968, p. 255; R. J. Samuels, 'Structured Polymer Properties', Wiley, New York, 1974.
63. G. L. Wilkes, *Adv. Polym. Sci.*, 1971, **8**, 91.
64. I. M. Ward, 'Structure and Properties of Oriented Polymers', Applied Science, London, 1975.
65. H. Kawai and S. Nomura, in 'Developments in Polymer Characterization — 4', ed. J. V. Dawkins, Applied Science, London, 1983, p. 211.
66. R. J. Roe and W. R. Krigbaum, *J. Chem. Phys.*, 1964, **40**, 2608.
67. R. J. Roe, *J. Appl. Phys.*, 1965, **36**, 2024.
68. S. Nomura, H. Kawai, I. Kimura and M. Kagiyama, *J. Polym. Sci., Part A-2*, 1970, **8**, 383.
69. R. S. Stein and F. H. Norris, *J. Polym. Sci.*, 1956, **21**, 381.
70. R. S. Stein, *J. Polym. Sci.*, 1958, **31**, 327.
71. R. S. Stein, *J. Polym. Sci.*, 1958, **31**, 335.
72. S. Nomura, H. Kawai, I. Kimura and M. Kagiyama, *J. Polym. Sci., Part A-2*, 1967, **5**, 479.
73. I. Kimura, M. Kagiyama, S. Nomura and H. Kawai, *J. Polym. Sci., Part A-2*, 1969, **7**, 709.
74. R. S. Stein and B. E. Read, *Appl. Polym. Symp.*, 1969, **8**, 255.
75. C. R. Desper and R. S. Stein, *J. Appl. Phys.*, 1966, **37**, 3990.
76. S. Nomura, A. Asanuma, S. Suehiro and H. Kawai, *J. Polym. Sci., Part A-2*, 1971, **9**, 1991.
77. S. Nomura, M. Matsuo and H. Kawai, *J. Polym. Sci., Polym. Phys. Ed.*, 1972, **10**, 2489.
78. K. Fujita, S. Suehiro, S. Nomura and H. Kawai, *Polym. J.*, 1982, **14**, 545.
79. K. Fujita, H. Niwa, S. Nomura and H. Kawai, *J. Polym. Sci., Polym. Phys. Ed.*, 1983, **21**, 1713.
80. H. D. Keith and F. J. Padden, *J. Polym. Sci.*, 1959, **41**, 525.
81. P. H. Geil, *J. Polym. Sci., Part A*, 1964, **2**, 3813.
82. Z. W. Wilchinsky, *Polymer*, 1964, **5**, 271.
83. K. Sasaguri, S. Hoshino and R. S. Stein, *J. Appl. Phys.*, 1964, **35**, 47.
84. K. Sasaguri, R. Yamada and R. S. Stein, *J. Appl. Phys.*, 1964, **35**, 3188.
85. I. L. Hay and A. Keller, *Kolloid Z.Z. Polym.*, 1965, **204**, 43.
86. T. Oda, S. Nomura and H. Kawai, *J. Polym. Sci., Part A*, 1965, **3**, 1993.
87. T. Oda, N. Sakaguchi and H. Kawai, *J. Polym. Sci., Part C*, 1966, **15**, 223.
88. K. Kobayashi and T. Nagasawa, *J. Polym. Sci., Part C*, 1966, **15**, 163.
89. R. S. Moore, *J. Polym. Sci., Part A-2*, 1967, **5**, 711.
90. A. Keller and M. J. Machin, *J. Macromol. Sci., Phys.*, 1967, **1**, 41.
91. R. G. Crystal and D. Hansen, *J. Polym. Sci., Part A-2*, 1968, **6**, 981.
92. T. Oda, N. Sakaguchi and H. Kawai, *Kobunshi Kagaku*, 1968, **25**, 588.
93. T. Oda, M. Motegi, M. Moritani and H. Kawai, *Kobunshi Kagaku*, 1968, **25**, 639.
94. C. A. Garber and E. S. Clark, *J. Macromol. Sci., Phys.*, 1970, **4**, 499; *Int. J. Polym. Mater.*, 1971, **1**, 31.
95. A. Peterlin, *J. Mater. Sci.*, 1971, **6**, 490.
96. E. S. Clark, in 'Structure and Properties of Polymer Films', ed. R. W. Lenz and R. S. Stein, Plenum Press, New York, 1973, p. 267.
97. B. S. Sprague, *J. Macromol. Sci., Phys.*, 1973, **8**, 157.
98. T. Hashimoto, K. Nagatoshi, A. Todo and H. Kawai, *Polymer*, 1976, **17**, 1063, 1075.
99. M. Matsuo, K. Hirota, K. Fujita and H. Kawai, *Macromolecules*, 1978, **11**, 1000.
100. K. Shimamura, S. Murakami, M. Tsuji and K. Katayama. *Nippon Reoroji Gakkaishi*, 1979, **7**, 42.
101. T. Tagawa and K. Ogura, *J. Polym. Sci., Polym. Phys. Ed.*, 1980, **18**, 971.
102. W. R. Krigbaum and R. J. Roe, *J. Chem. Phys.*, 1964, **41**, 737.
103. S. Suehiro, T. Yamada, H. Inagaki, T. Kyu, S. Nomura and H. Kawai, *J. Polym. Sci., Polym. Phys. Ed.*, 1979, **17**, 763.
104. R. J. Samuels, *J. Polym. Sci., Part C*, 1967, **20**, 253; *J. Polym. Sci., Part A-2*, 1968, **6**, 1101.
105. R. D. B. Fraser, *J. Chem. Phys.*, 1953, **21**, 1511; 1958, **28**, 1113; 1958, **29**, 1428.
106. M. Beer, *Proc. R. Soc. London, Ser. A*, 1956, **236**, 136.
107. R. S. Stein, *J. Appl. Phys.*, 1961, **32**, 1280.
108. J. L. Koenig, S. W. Cornell and D. E. Witenhafer, *J. Polym. Sci., Part A-2*, 1967, **5**, 301.
109. S. Okajima, Y. Kobayashi and K. Kawada, *Kogyo Kagaku Zasshi*, 1956, **59**, 1213.

110. Y. Shindo and R. S. Stein, *J. Polym. Sci., Part B*, 1967, **5**, 737.
111. Y. Shindo, B. E. Read and R. S. Stein, *Macromol. Chem.*, 1968, **118**, 272.
112. Y. Nishijima, Y. Onogi and T. Asai, *Rep. Prog. Polym. Phys. Jpn.*, 1965, **8**, 131.
113. Y. Nishijima, Y. Onogi and T. Asai, *J. Polym. Sci., Part C*, 1966, **15**, 237.
114. Y. Onogi and Y. Nishijima, *Rep. Prog. Polym. Phys. Jpn.*, 1976, **19**, 411, 415, 419.
115. C. R. Desper and I. Kimura, *J. Appl. Phys.*, 1967, **38**, 4225.
116. R. J. Roe, *J. Polym. Sci., Part A-2*, 1970, **8**, 1187.
117. D. I. Bower, *J. Polym. Sci., Polym. Phys. Ed.*, 1972, **10**, 2135.
118. S. Hibi, M. Maeda, H. Kubota and T. Miura, *Polymer*, 1977, **18**, 143.
119. S. W. Cornell and J. L. Koenig, *J. Appl. Phys.*, 1968, **39**, 4883.
120. R. G. Snyder, *J. Mol. Spectrosc*, 1971, **37**, 353.
121. J. H. Van Vleck, *Phys. Rev.*, 1948, **74**, 1168.
122. K. Yamagata and S. Hirota, *Oyo Butsuri*, 1961, **30**, 261.
123. V. J. McBriety and I. M. Ward, *Br. J. Appl. Phys.*, 1968, **1**, 1529.
124. V. J. McBriety, I. R. McDonald, and I. M. Ward, *J. Phys. D*, 1971, **4**, 88.
125. V. J. McBriety, *Polymer*, 1974, **15**, 503.
126. V. J. McBriety and I. R. McDonald, *Polymer*, 1975, **16**, 125.
127. S. Nomura, Ph.D. Thesis, Kyoto University, 1970.
128. R. Zbinden, 'Infrared Spectroscopy of High Polymers', Academic Press, New York, 1964, chap. 5.
129. Y. Shinohara and H. Tanzawa, *Kobunshi Kagaku*, 1957, **14**, 488, 495.
130. W. Voigt, 'Lehrbuch der Kristallphysik', Teubner, Leipzig, 1910.
131. A. Reuss, *Z. Angew. Math. Mech.*, 1929, **9**, 49.
132. E. Kröner, *Z. Phys.*, 1958, **151**, 504.
133. J. Bishop and R. Hill, *Philos. Mag.*, 1951, **42**, 414, 1298.
134. Z. Hashin and S. Shtrikman, *J. Mech. Phys. Solids*, 1962, **10**, 335.
135. H. H. Kausch-Blecken von Schmeling, *J. Appl. Phys.*, 1967, **38**, 4213.
136. K. H. Meyer and W. Lotmar, *Helv. Chim. Acta*, 1936, **19**, 68.
137. W. J. Lyons, *J. Appl. Phys.*, 1958, **29**, 1429; 1959, **30**, 796.
138. L. R. G. Treloar, *Polymer*, 1960, **1**, 95, 279, 290.
139. J. E. Lennard-Jones, *Proc. R. Soc. London, Ser. A*, 1924, **106**, 463.

14

Textile Fibres

J. GEORGE TOMKA
University of Leeds, UK

14.1 INTRODUCTION

The first encounter with fibre science and technology is often frustrated by many unfamiliar terms, concepts and units. Use of a suitable 'dictionary' can help with many problems.[1]

The term 'fibre' is defined[1] as: 'A unit of matter characterized by flexibility, fineness and high ratio of length to thickness'. Whilst this definition is undoubtedly comprehensive, it should be added that a 'useful fibre' should be sufficiently uniform and its balance of properties and cost should be in accordance with the intended end-use.

The traditional fibre applications are in clothing and furnishing but the range of fibre end-uses is rapidly expanding, particularly in the industrial sector (*e.g.* geotextiles, composites, insulation and filtration). Even in the traditional areas there is now a growing demand for fibres with enhanced properties for end-uses such as special protective clothing and aerospace furnishing. Such a diversity of end-uses obviously requires fibres with a wide range of properties.

Until the end of the last century, fibres for all end-uses had to be selected from the available natural fibres such as cotton, wool and silk. Today these fibres are still important, accounting for more than 55% of the total fibre production.[2] The first man-made fibres were based on a natural polymer — cellulose. The production share of cellulosic man-made fibres is presently about 9%.[2] Both natural and cellulosic man-made fibres are outside the scope of this chapter which is concerned with organic synthetic polymers.

Originally, the synthetic fibres were viewed as cheaper substitutes for natural fibres, but this image is gradually changing. While even with synthetic fibres the 'selection approach' still prevails, the industry now offers a range of fibres designed specifically for particular end-uses.

The subject of synthetic fibres is very broad and hence only the outlines can be presented here. As far as the references are concerned, the emphasis has been placed on balanced review articles and monographs.[3-19]

14.2 FIBRE PROPERTIES

Both selection and design of a fibre for a given end-use must take into account a wide range of properties.[3] However, satisfactory mechanical performance is always essential. Various aspects of the mechanical properties of polymers are discussed in detail in other chapters of this work and elsewhere.[4-6] For fibres with cylindrical symmetry, the fundamental approach requires determination of five independent viscoelastic functions,[4,5] which is prohibitively time-consuming. Besides, in practice, the fibres assembled in an end-product are often deformed in a complex way. The magnitudes and directions of the applied forces vary with time in a manner which precludes an exact mathematical description. At the same time, the fibres may be exposed to a changing environment. Consequently, it is often necessary to use various end-use oriented tests[7] for a practical evaluation of a fibre.

However, the evaluation of the mechanical properties invariably starts with the testing of tensile properties. Whilst for some fibre applications the tensile properties are directly relevant, there are many cases where it is difficult to relate the fibre performance in the end-product to the fibre tensile properties. Indeed, in some applications 'better' tensile properties could result in a poorer end-product performance (*e.g.* due to fibrillation resulting in poor abrasion resistance). The result of a tensile test is a stress/strain curve. Since fibres are viscoelastic, the relationship between the tensile stress and tensile strain is influenced by the test conditions (temperature and relative humidity of the environment, stress and/or strain time-dependence, *etc.*[3]).

In science and engineering, the tensile properties, such as Young's modulus and tensile strength, are expressed in terms of stress σ, *i.e.* the appropriate load (F) divided by the cross-sectional area of the specimen under zero load (A_0). Accurate determination of the fibre cross-sectional area is difficult and, for fibres of complex cross-sectional shapes (see Section 14.7), it is practically impossible.

It is therefore convenient to use instead of the cross-sectional area a related quantity, linear density, which is defined as mass per unit length; its SI unit is $kg\,m^{-1}$. From the definition of linear density it follows that it is equal to the product of the average cross-sectional area A_0 (m^2) and the density ρ ($kg\,m^{-3}$). Since the linear densities of fibres are usually around 10^{-7}–$10^{-6}\,kg\,m^{-1}$, it is much more convenient to use a tex unit[1] ($1\,tex = 10^{-6}\,kg\,m^{-1} = 1\,g\,km^{-1}$), or its 0.1 submultiple, dtex. Information about other units and other systems used for designation of fibre fineness can be found elsewhere.[1,3,20]

The use of linear density leads to a replacement of stress by a related quantity, namely specific stress s, which is defined as the ratio of stress and the density; thus

$$s \;=\; \sigma/\rho \;=\; F/(A_0\rho) \tag{1}$$

where $A_0\rho$ is the fibre linear density. Hence the specific stress is also defined as the ratio of load and the linear density of the undeformed specimen. The SI unit of specific stress is $N\,m\,kg^{-1}$ but, since the tex unit is more convenient for expressing the linear density of fibres, a suitable unit for expressing the specific stress is $N\,tex^{-1}$ ($1\,N\,tex^{-1} = 10^6\,N\,m\,kg^{-1}$). Submultiples such as $cN\,tex^{-1}$ or $mN\,tex^{-1}$ and traditional non-SI units still persist in practice.[20] The initial modulus and various secant moduli bear units of specific stress and the same applies to the yield stress. Tenacity is defined as the specific stress at the break.

A useful measure of fibre toughness is the specific work of rupture, obtained by integration of the area under the specific stress/strain curve; this quantity is also expressed in units of specific stress ($1\,MJ\,kg^{-1} = 1\,N\,tex^{-1}$). Since the integration is somewhat tedious, various 'tensile factors' are sometimes used as rather crude measures of the fibre toughness. The most popular one is obtained by multiplying the tenacity by the square root of the breaking extension, but other powers of the breaking extension have also been used.

A better understanding of fibre behaviour can sometimes be gained by using 'true specific stress' which takes into consideration the reduction of fibre linear density resulting from the extension. The true specific stress s^*, which is defined as the ratio of load and the actual linear density of the extended fibre (strain e), is related to the specific stress as follows

$$s^* \;=\; s\,(1 \;+\; e) \tag{2}$$

Although the corresponding quantities, expressed in terms of either stress or specific stress, characterize similar properties of the material, it must be emphasized that a change from one to the other is not strictly a unit conversion since it involves another physical quantity, density; it is in fact a change of physical concept. Table 1 gives tensile properties expressed in terms of both stress and

Table 1 Tensile Properties

Material	Density (kg m^{-3})	Young's modulus (GPa)	Specific modulus (N tex^{-1})	Tensile strength (GPa)	Tenacity (N tex^{-1})
Polyester industrial yarn	1380	9.8	7.1	1.04	0.75
Dyneema SK 60 fibre[a]	970	97	100	2.9	3.0
Kevlar 29 yarn[b]	1450	83	57	2.8	1.9
Kevlar 49 yarn[b]	1450	125	86	2.8	1.9
PRD 149 yarn[b]	1480	165	111	2.4	1.6
Steel wire	7850	210	27	2.5	0.3

[a] Polyethylene fibre produced by DSM/Toyobo by gel spinning of high molecular weight polymer; see R. Kirschbaum, H. Yasuda and E. H. M. van Gorp, *Chemiefasern/Text.-Ind.*, 1986, **36/88**, T 134 (E 123).
[b] Aramid fibres produced by Du Pont; see J. E. Van Trump and E. Lahijani, in 'Proceedings of 32nd International SAMPE Symposium, 6 to 9 April, 1987', p. 917 (*Chem. Abstr.*, 1987, **107**, 60 507).

specific stress for several synthetic fibres and for steel. It shows that the transformation affects the relative rating of these materials. The use of specific stress in fibre technology started because of its convenience; it is now recognized that for applications where high stiffness and/or high strength are required together with low mass (and hence low weight) the use of the quantities expressed in terms of specific stress gives the true measure of material performance.

The concern about satisfactory mechanical performance must not obscure the importance of other fibre properties. As far as the thermal properties are concerned, the main emphasis is placed on transition temperatures, in particular the glass transition and the melting temperature. They determine the temperature range for fibre utilization and for care and maintenance treatments of the end-products (*e.g.* washing, drying and ironing conditions of clothing). Contrary to some claims, thermal properties such as specific heat capacity and thermal conductivity are relatively unimportant, since the insulating properties of garments are determined by the amount of trapped air, *i.e.* by the fabric construction.

Amongst electrical properties,[3] attention should be paid to the propensity to static charging[21] and, in conjunction with it, to the electrical resistance, since its lowering results in charge dissipation.

The subject of surface properties falls into two categories. Fibre friction[3] is of the utmost importance for textile processing of fibres and it also affects substantially the mechanical behaviour of fibre assemblies such as yarns and fabrics. The surface energy of fibres[14] affects their soiling, wetting and the transport of liquids through fibre assemblies. The surface properties of fibres can be modified by application of temporary, semipermanent or permanent coatings. Indeed, a major part of textile finishing technology[21,22] is concerned with modifications of surface properties.

In addition to the above-mentioned physical properties, the fibre evaluation must take into account interaction with water.[3,14] The property of absorbing moisture is believed to be an important feature of clothing materials. In addition, other fibre properties are significantly influenced by absorbed water.

The suitability of a fibre for numerous textile end-uses is strongly influenced by its dyeing behaviour (range of suitable dyestuffs, rate of dye uptake and equilibrium dye-uptake, dye fastness[15]) and by its optical properties (refractive index, light reflection and absorption, lustre[3]) determining its colour.[16]

With the growing importance of fibres in biomedical applications,[23] increasing attention is paid to their physiological properties (*e.g.* assimilation in living systems).

Many fibre properties change during the use and maintenance of the end-products. Hence the stability of fibres often plays a decisive role in the assessment of their suitability. Obviously, the emphasis varies with the intended end-use, but the most important aspects of fibre stability include the thermal and thermo-oxidative stability, photostability, hydrolytic stability and solvent resistance.[24] In practice, the stability is usually assessed by measuring the change in physical (mainly mechanical) properties of fibres after an exposure specified in terms of the composition of the environment, temperature, light intensity and time.

Finally, we should mention the flame resistance. Its assessment is a complex task.[25,26] Apart from the ignition temperature and the limiting oxygen index (LOI) it is important to consider features such as char formation, smoke evolution and toxicity of the combustion products.

This brief survey of fibre properties is by no means complete. Its main objective is to draw attention to the range of properties which need to be considered and to point out some sources of more detailed information.

14.3 FACTORS DETERMINING FIBRE PROPERTIES

In common with other 'shaped articles', the properties of fibres are determined by their composition and by their physical structure and geometry (*i.e.* shape and dimensions). The objectives of fibre science and technology are: (1) to establish quantitative relationships between these factors and fibre properties and (2) to find out how to control these factors.

The chemical composition of the majority of synthetic fibres is essentially the same as that of the starting material and is therefore determined by its choice. Although the main component of a synthetic fibre is a polymer, it is important to note that fibres invariably contain various additives (delustrant, antioxidants, other stabilizers, *etc.*). Optimization of additives plays an important role in the successful development of synthetic fibres. The specification of fibre composition should therefore include not only the description of the chemical structure of the polymer (Section 14.4) but also the composition and concentration of all the additives. For additives which are insoluble in the polymer (*e.g.* delustrant and pigments), their particle size is also important.

In spite of all the care taken during the polymer and fibre manufacture, small quantities of impurities (*e.g.* catalyst residues and products of side reactions) are also present, and their effects on fibre properties can be significant. Whilst it is generally desirable to minimize the chemical changes taking place during the fibre formation, it should be noted that certain specialty fibres are made by a chemical transformation of precursor fibres (Section 14.4).

The understanding of the physical structure of synthetic fibres (Section 14.5) and its effect on fibre properties (Section 14.6) is still incomplete. The physical structure is controlled by the fibre-formation conditions (Section 14.8) but the downstream textile processing (*e.g.* dyeing and finishing) may also affect it.

The third factor, fibre geometry, includes features such as fibre dimensions and shape (Section 14.7). The importance of fibre geometry is fully recognized in technological practice. Indeed, the classification of product types manufactured by the industry is based on geometry: (i) mono-filaments (single continuous filaments of diameter ranging from about 0.04 to 2 mm); (ii) continuous filament yarns, which may be either 'flat' (*i.e.* consisting of straight filaments) or textured (consisting of filaments which are coiled or otherwise distorted); the linear densities of the constituent filaments are between 0.1 and 0.4 tex, which, for fibres of circular cross-section, corresponds to diameters between about 10 and 20 μm; (iii) staple, *i.e.* fibres (usually crimped) obtained by cutting or breaking filaments to the required length; the linear densities are between 0.1 and 2.0 tex.

Each type includes a large number of products of different linear densities; cross-sectional variants are also available.

The fibre geometry is determined by the production conditions. The linear density and fibre length can be controlled with sufficient accuracy; the know-how for controlling fibre cross-sectional shape and fibre distortion (curvature) is well established.

14.4 CHEMICAL STRUCTURE

The main features of the chemical structure of a fibre-forming polymer or copolymer are the constitution, configuration, concentration and sequence of all the repeating units, and also the constitution and concentration of the end-groups. They have a dominant effect on the transition temperatures, moisture absorption, dyeing behaviour, stability and flame resistance of the resulting fibre, but they also affect to a certain extent the other fibre properties considered in Section 14.2. The polymer molecular weight and molecular weight distribution (MWD) have a strong effect on mechanical (in particular on tensile) properties of the resulting fibre.

The polymer chemical structure also plays a decisive role in the choice of fibre production process (Section 14.8). A sufficiently broad temperature interval between the melting temperature and the thermal stability limit, and an appropriate melt viscosity (determined by the molecular weight and MWD) are essential for successful melt processing. For polymers which do not have an adequate 'melt processing window', the availability of suitable solvents, sufficiently high solubility and appropriate rheological behaviour of the resulting solutions are essential features dominated by the polymer chemical structure.

In spite of the limitations imposed by the chemical structure upon the choice of fibre-forming polymers, the number of potentially useful polymers is extremely large. Yet, in reality, the majority of commodity fibres are produced from a few well-established polymers. In commerce, fibres are classified by generic names[1,27] which sometimes employ chemical names in a more restricted sense than that which is common in polymer science. The generic names of the most important commodity

Table 2 Main Commodity Synthetic Fibres

Generic name[a]	Chemical name	Abbreviation	Melting temperature (°C)	Spinning method
Polyester	Poly(ethylene terephthalate)	PET	260	Melt
Nylon or	Poly(hexamethyleneadipamide)	Nylon 6,6	260	Melt
polyamide	Poly(pentamethylenecarbonamide)	Nylon 6	220	Melt
Polyolefin	Isotactic polypropylene	PP	170	Melt
Acrylic	Polyacrylonitrile copolymers[b]	—	c	Wet or dry
Modacrylic	Polyacrylonitrile copolymers[b]	—	c	Wet or dry

[a] According to the International Organization for Standardization.
[b] See text.
[c] Decomposes.

fibres, together with the chemical structures of polymers employed in their production, are given in Table 2.

The majority of polyester fibres[28] are based on poly(ethylene terephthalate) (PET), currently accounting for about 50% of all synthetic fibres produced.[2] Modification of the dyeing behaviour of PET fibres can be achieved by using small quantities of comonomers.[28] For example, use of 2 mol% of the sodium salt of 5-sulfoisophthalic acid gives fibres which are dyeable with cationic dyes.

Nylons[29] derived from amino acids are distinguished from each other by the number of carbon atoms in the repeating unit; thus, the repeating unit of nylon z is $-[-NH-(CH_2)_{z-1}-CO-]-$. For nylons produced from diamines and dicarboxylic acids, the numbers of carbon atoms in both reactants are given, starting with the diamine component; thus the repeating unit of nylon u,v is $-[-NH-(CH_2)_u-NHCO-(CH_2)_{v-2}-CO-]-$. Nylon fibres[29] are produced mostly from nylon 6,6 and nylon 6. The dyeing behaviour of nylon fibres can be modified by adjusting the balance of carboxyl and amine end-groups.

Polymers for the manufacture of polyester[28,30] and nylon[29] fibres are produced by melt polycondensation; their MWD is therefore rather narrow ($\bar{M}_w/\bar{M}_n = 2$) and their molecular weights are rather low ($\bar{M}_w$ usually less than 50 000). If required, the molecular weight can be raised by a solid phase post-polycondensation. Polymers used for the production of commodity polyester and nylon fibres, consisting of linear flexible chains with regular chemical structure, are crystallizable. Their melting temperatures (see Table 2) are sufficiently far below their thermal stability limit (approximately 300 °C) so that melt processing is possible. Nevertheless, excessive dwell times in the melt may cause difficulties. In the case of PET, side reactions result in polymer discolouration. In the case of nylon 6,6, the side reactions lead to crosslinking, whilst in nylon 6 the formation of caprolactam due to the ring–chain equilibration may cause problems.

The majority of polyolefin (polyalkene) fibres are produced from isotactic polypropylene[31,32] (PP) although polyethylene fibres are now gaining importance. A wide range of fibre-forming PP grades is now available. Their molecular weights $\bar{M}_w$ range from about 200 000 to 500 000; the MWD can vary widely[32] giving $\bar{M}_w/\bar{M}_n$ between 2 and 12. PP is a crystallizable polymer. Its melting temperature is around 170 °C but unstabilized polymer will degrade when melted. However, stabilizers which ensure sufficient melt stability have been developed and melt processing can be carried out without undue difficulty. Recently developed grades of narrow MWD, so-called 'controlled rheology resins', give improved processing performance.

Both acrylic and modacrylic fibres are based on atactic polyacrylonitrile.[33,34] The generic name 'acrylic fibre' refers to fibres made from linear copolymers that consist of not less than 85 wt% acrylonitrile units.[1] The majority of commercial acrylic fibres contain between 5 and 8% of 'neutral' comonomers, namely vinyl acetate, methyl acrylate or methyl methacrylate. In addition, smaller quantities of various ionic comonomers (*e.g.* sodium styrenesulfonate) are used to provide, together with the ionic end-groups formed from sulfonate and sulfate initiators, the dye sites in the fibres.

The content of acrylonitrile units in modacrylic fibres is, by definition,[1] between 35 and 85 wt%. These fibres usually have a high concentration of halogen-containing comonomers (such as vinylidene dichloride, vinyl bromide and vinyl chloride) which improve their flame resistance. Related to modacrylic fibres, but with a higher content of halogen-containing monomers, are chlorofibres, produced only in small quantities from poly(vinyl chloride), poly(vinylidene dichloride) and from their copolymers, *e.g.* with vinyl acetate.[35,36]

Neither polyacrylonitrile nor its copolymers are melt processable without using additional processing aids in conjunction with special processing techniques. Commercial fibres are therefore spun from solutions (Section 14.8). The 'neutral' comonomers increase the polymer solubility and also result in increased rate of dye uptake in fibres. In order to achieve an acceptable combination of spinning solution concentration and viscosity, the molecular weights $\bar{M}_w$ of fibre-forming polymers are usually kept between 90 000 and 140 000, with $\bar{M}_w/\bar{M}_n$ ranging from 1.5 to 3.

The main reason for the success of the commodity fibres is their versatility. Polymer grades with different molecular weights, with different levels of TiO_2 delustrant and with different stabilizers and other additives (*e.g.* flame retardants and antistatic agents) have been developed. The dyeing behaviour of fibres can be modified by adjusting the polymer end-groups balance (in nylons) or by copolymerization (in acrylics and, less often, in polyesters). In addition to the above 'composition variants', each polymer grade can be processed into fibres which differ in their physical structure and geometry. Downstream textile processes further enhance the usefulness of commodity synthetic fibres, *e.g.* by blending them with cellulosic man-made fibres or with natural fibres.

Polymers used for the manufacture of commodity fibres are also used in the production of multiconstituent fibres, sometimes in conjunction with a modified commodity polymer or a specialty polymer. These fibres consist of two or more polymers forming separate phases. There are essentially two types of such fibres. In the first type, each polymer occupies a discrete region of the fibre cross-section (*e.g.* sheath/core-type fibres, where the sheath is a lower melting homopolymer or copolymer; such fibres are used for thermal bonding). In the second type, the components are dispersed in a random way throughout the fibre, typically forming a matrix/fibril arrangement (*e.g.* incorporation of dyeable polymer fibrils into polypropylene which forms the matrix). Detailed description of these fibres can be found elsewhere.[37]

Although the commodity fibres are suitable for a majority of end-uses, there is a growing demand for high-performance specialty fibres belonging to the following main categories: (1) fibres with enhanced chemical stability and combustion resistance and (2) high strength/high modulus fibres.

A large number of polymers have been considered for production of fibres belonging to the first category.[38] Usually these polymers contain aromatic structures and/or heterocycles joined by stable linkages. However, high chemical stability and combustion resistance of a polymer does not ensure that the fibres are suitable for the manufacture of clothing which gives good protection in a fire. For this application, the fibre should not melt below the decomposition temperature. Thus, melt-processable poly(1,4-phenylene sulfide) forms fibres[39] which are extremely resistant to chemical attack and have a low flammability (LOI = 0.34), but their relatively low melting temperature (285 °C) makes them much less suitable for fire protection. Similar consideration applies to other melt-processable aromatic polymers such as PEEK[40] [poly(oxy-1,4-phenyleneoxy-1,4-phenylene-carbonyl-1,4-phenylene); melting temperature 335 °C] and PEK[41] [poly(oxy-1,4-phenylene-carbonyl-1,4-phenylene); melting temperature 367 °C]. Fibres made from poly(*m*-phenylene iso-phthalamide), which do not melt below their decomposition temperature, are more suitable for these applications in spite of their lower chemical stability.[38] Other commercial flame-resistant fibres are based on an aromatic poly(amide imide)[38] (**1**) and on an aromatic copolyimide.[42] Extremely good flame-resistance combined with an outstanding dimensional stability at high temperatures has been achieved with sulfonated polybenzimidazole (**2**) fibres[43] developed for aerospace applications.

(**1**) R = O or CH$_2$ (**2**)

Although the majority of flame-resistant fibres are produced from fully aromatic and/or hetero-cyclic polymers, an alternative approach, based on utilization of crosslinked polymers, is also feasible. Phenol–formaldehyde fibres[38] were produced by a process involving spinning of a fusible novolac resin followed by crosslinking of the precursor fibre. A recently developed nonflammable fibre with a low smoke-evolution is based on a copolymer of acrylic acid and acrylamide; the carboxylic groups are crosslinked by zinc ions.[44]

We now turn to the second main category of high-performance fibres, *i.e* to high strength/high modulus (HSHM) fibres. Two types of approach, classified respectively as physical and chemical, are utilized in the production of such fibres. The physical approaches are directed to realizing fully the modulus and strength of linear flexible-chain polymers predicted by theoretical calculation for fibres consisting of fully extended linear chains aligned perfectly along the fibre axis (Section 14.6). Conventional methods employed in fibre production from linear flexible-chain polymers result in fibres with a limited chain extension arising from the presence of entanglements and chain folding. Special fabrication techniques (Section 14.8) resulting in HSHM fibres have only been developed commercially for high density polyethylene,[45,46] but a considerable effort is being directed towards extending the physical approach to other linear flexible-chain polymers.

The chemical approaches utilize linear rigid-chain polymers; the chain rigidity prevents the formation of entanglements and chain folds. Nematic solutions of a rigid-chain aromatic polyamide, poly(*p*-phenyleneterephthalamide), in concentrated sulfuric acid are used for the production of commercial HSHM aramid fibres.[47-49] Poly(*p*-phenylenebenzobisthiazole) (3), forming nematic solutions in poly(phosphoric acid) or in a mixture of methanesulfonic acid and chlorosulfonic acid, gives fibres of exceptionally high modulus (up to 265 GPa, corresponding to 180 N tex^{-1}).[47,49]

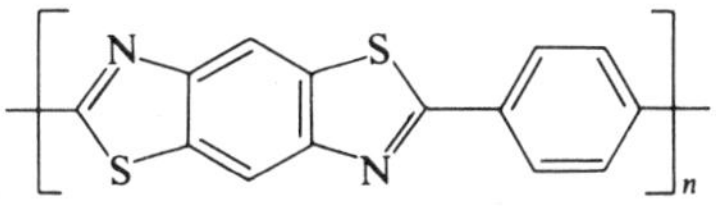

(3)

Aromatic copolyesters which are sufficiently rigid to form a nematic mesophase upon heating[47,49,50] (*e.g.* 4 and 5) also give HSHM fibres, but major industrial exploitation awaits improved competitiveness with established fibres. Exceptionally high strength values have been reported for fibres made from thermotropic nematogenic poly(azomethine)s[51] (*e.g.* 6) but these materials seem to suffer from a poor hydrolytic stability.[50]

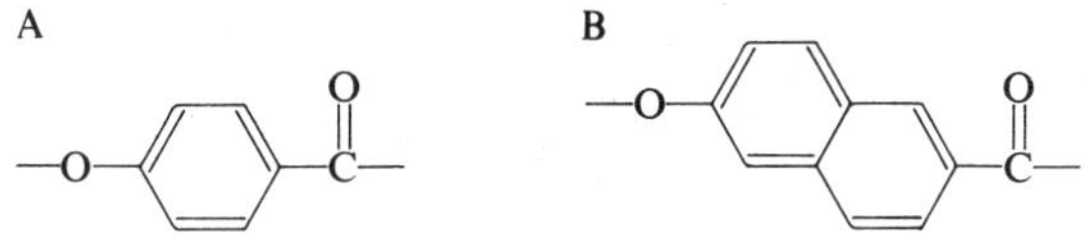

(4) [A] = 70 mol%, [B] = 30 mol%

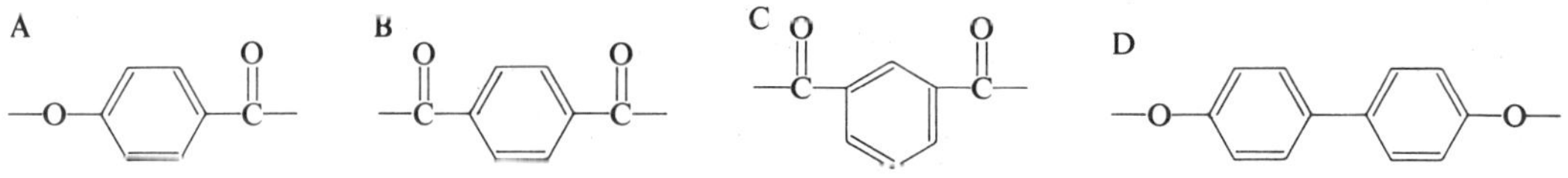

(5) [A] = 60 mol%, [B] = 15 mol%, [C] = 5 mol%, [D] = 20 mol%

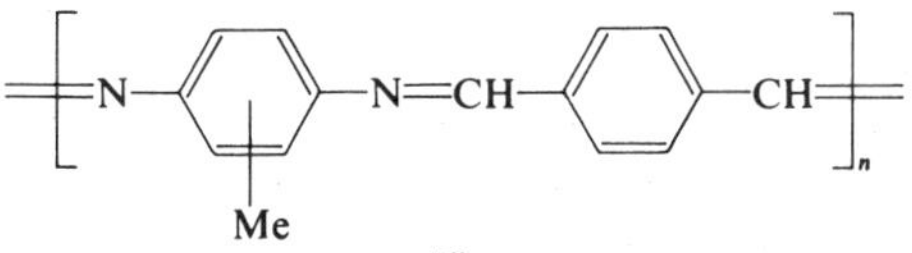

(6)

An aromatic copolyamide (7) which does not form nematic solutions due to its reduced chain rigidity is also used for the production of HSHM fibres.[52] It appears that the chain rigidity is sufficient for the formation of a nematic mesophase in the absence of solvent, and this probably occurs during the high temperature drawing of the fibres, which are spun from conventional solutions.

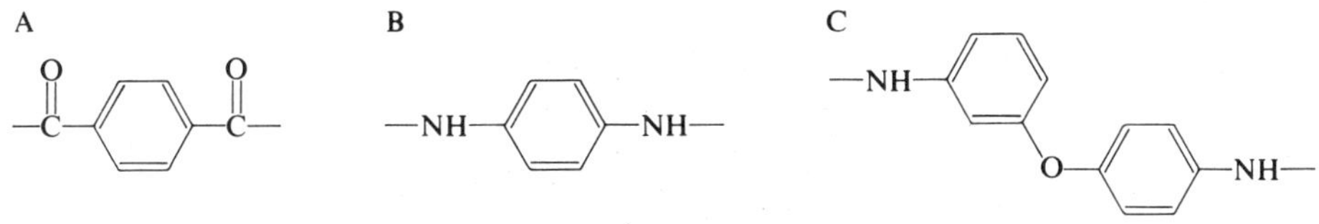

(7) [A] = 50 mol%, [B] = 25 mol%, [C] = 25 mol%

Another kind of chemical approach leading to HSHM fibres is based on formation of the load-bearing chemical bonds by a chemical reaction in the precursor fibre. This occurs in the production of certain types of carbon fibres[53] which are obtained from an acrylic precursor fibre by oxidation followed by carbonization and, optionally, by graphitization.

Apart from the two main categories of specialty fibres, a range of other specialty fibres has been developed. These include, for example, fibres for biomedical applications,[23] water soluble poly(vinyl alcohol) fibres[36] and elastomeric fibres based on segmented polyurethanes.[54, 55] These fibres cannot be included here due to lack of space.

14.5 PHYSICAL STRUCTURE

The main features of the physical structure of a fibre are: (i) the degree of order of the spatial arrangement of the constituent molecules and its distribution within the fibre and (ii) the degree of chain extension with respect to the fibre axis.

Considerations of the intra- and inter-molecular arrangements lead to a spectrum of order/disorder. At one extreme of this spectrum we have a perfect crystalline structure where both the conformation of individual chain molecules and their mutual packing are exactly defined. The corresponding atoms of the repeating units form a perfect three-dimensional lattice. Needless to say, only polymer chains with a regular chemical structure (with respect to both constitution and configuration) can form a true three-dimensional lattice; such polymers are crystallizable.

At the other extreme of the order/disorder spectrum we have an ideal amorphous structure. This is viewed as an assembly of interpenetrating random coils. Except for a liquid-type nearest-neighbour order there is no correlation in the mutual arrangement of individual chains. Such structure is completely disordered at both intra- and inter-molecular levels.

Having defined both extremes of the order/disorder spectrum we now turn to the intermediate states, starting from the ideal amorphous structure. A change from a random conformation to a more extended one, together with a change to a parallel arrangement of chain segments, leads to an 'amorphous structure with correlation';[56] naturally, in the case of linear flexible-chain polymers, such an arrangement can only exist below the glass transition temperature. The continuation of this 'ordering' would lead to a structure with essentially regular chain conformations where all chains are mutually parallel but still displaced along their axes; the lateral packing of chains is still irregular. This structure is equivalent to a nematic glass.[57] Elimination of the axial displacement and improved lateral packing lead to paracrystalline structures. Alternatively, regular packing of chains maintaining a certain degree of conformational disorder will result in condis crystals.[57] We can envisage a further improvement in the degree of order resulting in a crystalline structure with defects (*e.g.* dislocations) and, finally, in a perfect crystalline structure. Naturally, it is possible to generate the same spectrum starting with the perfect crystalline structure and introducing defects of increasing severity in both chain packing and conformation.[58] Although, in principle, we can have a continuous spectrum of order/disorder it is convenient to separate the structures into two broad categories. In 'crystalline structures' there is an identifiable three-dimensional lattice which may be, of course, imperfect. All other structures are designated here as 'noncrystalline'.

Fibres containing only noncrystalline structures usually represent the transient or intermediate stages in the manufacturing sequence. Although even in these fibres there may be localized differences in the degree of order, it is convenient to view them as single-phase structures.

After the completion of the manufacturing sequence, fibres usually contain structures designated as crystalline. Fibres produced from rigid-chain homopolymers are best considered as single-phase structures, *i.e.* as imperfect crystals with defects distributed more or less uniformly within the fibre. In fibres made from linear flexible-chain polymers there is a segregation of regions with different degrees of order. These fibres, referred to as semicrystalline or partially crystalline, are mostly viewed as 'two-phase' structures consisting of crystalline and noncrystalline regions; models consisting of three phases have also been proposed.[59−61] It should be noted that in this context the term 'phase' is not used in its thermodynamic sense (see ref. 58, p. 435).

The second main feature of the physical structure is the degree of chain extension. It must be emphasized that the concept of chain extension[45, 62] (or molecular extent[63]) is qualitatively different from the well-established concept of orientation. Yet, recent developments showed[45, 46] that it is the chain extension rather than the orientation which has a dominant effect on mechanical properties of fibres (*cf.* Section 14.6).

A rigorous treatment of the concept of orientation is rather complex; details can be found in Volume 2 Chapter 13 and elsewhere.[4, 5] Essentially, each chain is viewed as a sequence of segments represented by a set of vectors. The angle between a given vector and the fibre axis is θ_z. The average orientation of the polymer with respect to the fibre axis is expressed by means of the orientation factor

$$f = \frac{3\langle\cos^2\theta_z\rangle - 1}{2} \tag{3}$$

where the averaging includes all the segments of all the chains. More detailed description of the material requires knowledge of the orientation distribution function.

Experimental methods for determination of the orientation factor of fibres are well established. It is most often evaluated from the optical birefringence[4,5,64] or from the sonic modulus.[65-67]

Let us now consider schematic two-dimensional diagrams of two chains, each consisting of 100 segments (Figure 1). Their orientation factors are identical ($f = 0.85$), and so are their orientation distribution functions. Yet, their degrees of chain extension are clearly different; this illustrates the difference between these two concepts. The concept of chain extension is rather new and at this stage it is difficult to suggest an exact definition.[63] The development of this concept will depend on the studies concerned with chain macroconformations in the solid state. Small-angle neutron scattering (SANS) of blends of deuterated and nondeuterated polymer has already produced some interesting results.[68]

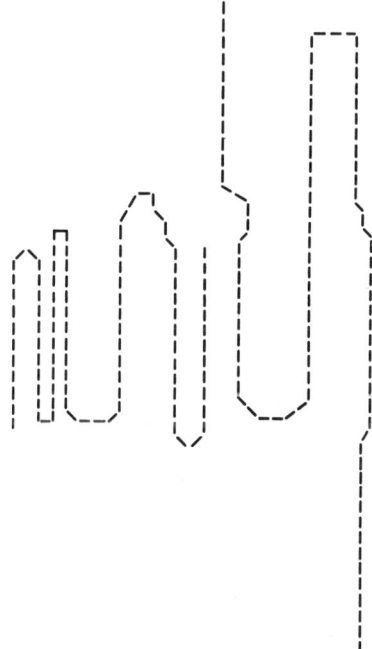

Figure 1   Schematic diagram illustrating the concept of chain extension; the orientation factors of both chains (each consisting of 100 segments) are identical ($f = 0.85$)

We now turn to a description of the structure of semicrystalline fibres produced from linear flexible-chain polymers with a regular chemical structure (PET, nylon 6,6, nylon 6, PP, *etc.*). It is usually viewed as an aggregate consisting of crystalline and noncrystalline regions. This is consistent with their wide-angle X-ray scattering (WAXS) which shows discrete reflections assigned to the crystalline material as well as diffuse features assigned to the noncrystalline material.

There is evidence that the structure of crystalline regions (crystallites) is far from perfect. The chain conformation and packing often differ from those obtained for specimens crystallized under optimum conditions.[69] In some fibres the situation is further complicated by polymorphism; moreover, stable and metastable crystal modifications may occur simultaneously.[70-72] Similarly, the noncrystalline regions are far from being completely disordered; indeed, there is a qualitative difference between their structure and that of an amorphous polymer.

X-Ray crystallography yields, in the first instance, information about the unit cell and the density of the crystalline material. Ultimately, it aims at the determination of both chain conformation and packing in the crystal. Crystallographic data for many fibre-forming polymers can be found in appropriate reviews (polyesters,[28] nylons,[29,73] PP,[32] aramids[47]).

The first stage in a quantitative characterization of semicrystalline fibres is the determination of crystallinity (*i.e.* the weight or volume fraction of crystallites). Since different methods employ different criteria for dividing the structure into crystalline and noncrystalline fractions, the crystallinity values obtained by various methods may differ.

The analysis of WAXS represents the most fundamental approach to the determination of polymer crystallinity. The main problem is separation of the corrected and normalized diffraction trace into contributions from crystalline and noncrystalline regions.[74] This task is sometimes further complicated by polymorphism.[70-72] After separation, the crystallinity is obtained from the ratio of the integrated intensity of coherently diffracted X-rays arising from crystalline regions and the total integrated intensity of coherently diffracted X-rays. Ruland's method,[75] which yields not only crystallinity but also an isotropic disorder parameter of the crystallites, can also be used for fibres provided that they are isotropized; this is accomplished, for example, by cutting the fibres into short lengths (60–80 μm) and pelleting the powder obtained.[76,77]

The density method[58] is undoubtedly employed most frequently for the determination of fibre crystallinity. Needless to say, the fibre must be free of voids and a correction for the presence of TiO_2 delustrant must be used. As already mentioned, the unit cell dimensions (and hence the density of the crystalline regions) are often influenced by the crystallization conditions. Similarly, the density of the noncrystalline regions may be influenced by their orientation. Consequently, realistic values of crystallinity can be obtained only when average densities of crystalline and noncrystalline portions are determined independently for each specimen.[69,78]

Evaluation of crystallinity from the heat of fusion, usually obtained by differential scanning calorimetry (DSC), often suffers from complications due to recrystallization which may take place during heating of the specimen.[79] It should also be noted that the surface energy of the crystallites causes a decrease in the measured heat of fusion; this effect should also be taken into consideration.[80]

Although any generalization is difficult, it would appear that the crystallinity values for conventional drawn fibres obtained from linear flexible-chain polymers vary between approximately 0.3 and 0.6. Even after extensive heat treatment the crystallinity does not usually increase above 0.7.

The determination of crystallinity is followed by a more detailed characterization of the structural features of the crystalline and noncrystalline regions, and of their mutual arrangement (sometimes referred to as 'macrolattice') leading to the formulation of a structural model.

The characterization of crystallites is concerned with their size and orientation. Both these features are evaluated from WAXS. The broadening of the diffraction peaks arises from both the finite size and the lattice disorder of the crystallites. Usually these two effects cannot be separated and the Scherrer equation yields only the 'average apparent size' of crystallites in the direction normal to the appropriate lattice planes. Obviously, the actual size must always exceed the apparent size. Typically,[61,74,78] the apparent lateral dimensions of crystallites are between 3 and 9 nm. The crystallite size in the chain direction is of the same magnitude; it should be emphasized that the meridional reflections obtained by small-angle X-ray scattering (SAXS) are due to axial periodicities entailing both crystalline and noncrystalline regions. Thus, the long period (typically between 10 and 20 nm) corresponds to the mean distance between the centres of axially adjacent crystallites.[81]

The orientation factor f_c of the chain segments constituting the crystallites is evaluated from the azimuthal diffraction traces.[82-84] For commercial drawn fibres the values of f_c are usually above 0.9. It is also possible to evaluate the orientation factors expressing the orientation of other crystallographic directions of the unit cell with respect to the fibre axis.[85]

Whilst X-ray methods yield reasonably accurate information about the crystalline regions, there is no direct method which yields structural information about the noncrystalline regions. Instead, it is necessary to rely on information deduced from several methods. As already mentioned, the noncrystalline regions in oriented semicrystalline fibres differ qualitatively from the isotropic amorphous polymer. Firstly, the material constituting the noncrystalline regions is oriented. The orientation factor of the noncrystalline regions f_n is usually evaluated from optical birefringence Δn using the following relationship

$$\Delta n = X f_c \Delta n_c^* + (1 - X) f_n \Delta n_n^* + \Delta n_f \qquad (4)$$

where X is the crystallinity, f_c is the orientation factor of the crystalline portion obtained from WAXS and Δn_c^* and Δn_n^* are the values of the intrinsic birefringence of the crystalline and noncrystalline material, respectively.[87] The form birefringence (Δn_f) is invariably disregarded. When we consider the problems in accurate evaluation of crystallinity and the difficulties in obtaining reliable intrinsic birefringences[87] it is not surprising that the values of f_n obtained are rather uncertain. An alterntive method based on measurement of the sonic modulus suffers from similar problems.[66,67] In spite of these reservations it is clear that, as expected, f_n is always significantly smaller than f_c.

The second, even more important, difference between the noncrystalline regions and amorphous polymer follows from the fact that the size of the crystallites along the chain direction is much smaller than the chain length. It is therefore obvious that the noncrystalline regions consist mostly of chain segments whose end or ends are anchored in crystallites. This restricts the chain mobility and results in an increased glass transition temperature, an effect which is of great practical importance, particularly for PET fibres.[88]

In order to get a clearer picture of the noncrystalline regions let us consider the fate of the chains emerging from a crystallite (*i.e.* their macroconformation). There are three possibilities:

(i) The chain enters a different crystallite, thus forming a 'tie molecule'. These can be more or less coiled and, in the extreme case, they can approach a fully extended conformation.

(ii) The chain re-enters the same crystallite either in the adjacent position or at some distance (nonadjacent re-entry). The resulting chain folds can be of different lengths.

(iii) The chain terminates in the noncrystalline region thus forming the so-called 'cilium'. These are often disregarded, but it should be noted that the fraction of cilia in a typical fibre-forming polycondensation polymer is far from being negligible, as illustrated by the following example. Let us consider a semicrystalline poly(ethylene terephthalate) with a crystallinity of 0.5 and crystallite thickness of 7 nm. A chain with a molecular weight of 25 000 contains about 130 repeating units; on average, about 65 of them will be incorporated into crystallites and this corresponds to about 10 passages through crystallites. Thus, in this reasonably typical example, the fraction of cilia is roughly 0.1.

The available information about the structure of noncrystalline regions has been deduced mainly from IR[64,89,90] and NMR studies[91-93] and from ESR measurements using stressed fibres.[94] The ratio of the fractions of folded chains and tie molecules is obviously closely related to the concept of chain extension. In spite of considerable progress made recently, there are still many unresolved questions concerning the structure of noncrystalline regions, in particular about the fraction of tie molecules, their length distribution and their location.

The final stage of description of the fibre structure is the formulation of a structural model. The diversity of models described in the literature reflects, to a certain extent, the lack of precise knowledge about fibre structure. Models of fibre structure were reviewed by Hearle[63] who also pointed out their fundamental limitations.

The structure of semicrystalline fibres is usually described in terms of essentially three types of model. The first type views the structure as a continuous noncrystalline matrix in which there are embedded small imperfect crystallites. This model may be appropriate for fibres of low crystallinity which do not show any discrete SAXS reflections.

The second type includes various lamellar models which describe the fibre structure in terms of alternating layers of crystalline and noncrystalline material. Considerations of shape and intensity of SAXS patterns[95] and crystallinity rule out any regular lamellar structure. According to Fischer and Fakirov,[81] the crystalline lamellae consist of mosaic blocks whose size and mutual packing depend on crystallization conditions.

The third type of model is based on a microfibrillar concept. According to Prevorsek and co-workers,[59-61] the fibre consists of three distinct domains. The crystalline domains and the 'intra-fibrillar disordered domains' are joined in series and form the microfibrils. The 'interfibrillar domains' are noncrystalline and consist of extended molecules. A different kind of microfibrillar model has been proposed, for example by Peterlin.[96,97]

The microfibrillar concept seems to be most suitable for explaining structural changes taking place during the tensile deformation of fibres.[61] However, when we reflect upon various models, there is hardly any actual difference between an irregular microfibrillar structure, an imperfect lamellar structure containing mechanical weaknesses due to the boundaries between the mosaic blocks and a structure consisting of crystallites distributed in a noncrystalline matrix and joined by tie molecules. Thus, the differences between various models of semicrystalline fibres are essentially a matter of semantics. Whilst models may have their use in communication. the arguments about them tend to obscure the main issue, which is the chain macroconformation.

Until now, we were concerned with fibres made from crystallizable linear flexible-chain polymers. Acrylic and modacrylic fibres are produced from copolymers derived from atactic polyacrylonitrile, which are noncrystallizable. Yet these fibres are in many aspects similar to the semicrystalline fibres.

In particular, some acrylic fibres, after suitable thermomechanical treatment, show SAXS patterns indicative of a long period[98] usually ascribed to a two-phase structure. The structural investigations reviewed recently by Frushhour and Knorr[33] showed that this paradox is resolved by recognizing the existence of 'ordered regions' where the chains with an irregular helical conformation form a two-dimensional hexagonal lattice. Thus, these fibres can be viewed as two-phase structures where the crystallites are replaced by the two-dimensionally ordered regions.

As already stated, the two-phase concept used for semicrystalline fibres from flexible-chain polymers is unsuitable for fibres produced from rigid-chain homopolymers. Such fibres are highly crystalline. The awkward problem of the chain macroconformation does not arise, since the chain rigidity rules out any chain folding; the extended chains are essentially parallel to the fibre axis. The structure of poly(p-phenyleneterephthalamide) fibres was investigated in detail and the results are summarized in several reviews.[47,48,99] It is worth noting that successful imaging of crystal lattice fringes offered a unique insight into the structure of these fibres.[100] An interesting longer-range feature of these fibres is a regular pleated sheet structure with a periodicity of approximately 500 nm.[101,102]

Investigation of the structure of fibres from rigid and semirigid thermotropic nematogenic copolymers is still in its early stages.[47,49,50,103] Chain folding is again ruled out by the chain rigidity. The highly irregular chemical structure of these copolymers has a strong limiting effect upon the development of crystallinity. It appears that the fibres consist of a paracrystalline array of chains parallel to the fibre axis, with a mosaic of thin crystalline regions formed by the matching chemical sequences.[104]

In structural studies, fibres are generally considered to be transversely isotropic. However, it is sometimes found that commercial fibres have a distinct radial differentiation of structure. This is most pronounced in fibres spun from solutions, which show a distinct skin–core effect.[33] A radial nonuniformity has been also found in some melt-spun fibres, *e.g.* in polyester fibres produced at very high spinning speeds.[105] It should also be noted that in some fibres (*e.g.* aramids[47,99]) there is a preferential radial orientation of certain crystal planes, in contrast to a random radial orientation normally exhibited by conventional commercial fibres.

Finally, it should be mentioned that fibres frequently contain microvoids; their presence is indicated, for example, by strong diffuse SAXS scattering on the equator. Larger voids occurring in fibres spun from solutions can be detected by optical microscopy; these are usually considered to be geometrical rather than structural features.

14.6   EFFECT OF PHYSICAL STRUCTURE ON FIBRE PROPERTIES

When we consider the fibre properties surveyed in Section 14.2 we realize that they are, without exception, influenced by the physical structure. The role of the physical structure is not always dominant, and often it is far from being completely understood. Due to lack of space, discussion of the structure–properties relationships is limited here to three subjects, selected for their importance: (1) mechanical properties, specifically Young's modulus and tensile strength; (2) thermal properties, specifically the phenomena associated with the glass transition; and (3) interaction with low molecular weight compounds.

The majority of research concerned with the relationship between physical structure and mechanical properties of fibres is concerned with the Young's modulus. Theoretical calculations[106–108] showed that at absolute zero the modulus of polyethylene crystals in the chain direction is between 260 and 320 GPa; the value obtained by X-ray measurements[109] assuming homogeneous stress was 240 GPa. The moduli in other directions obtained both by calculation[106,107] and by the X-ray method[109] are considerably lower. The moduli in the chain direction for other polymers, with essentially extended-chain conformations, are somewhat lower (*e.g.* 110 GPa for PET[110]). For polymer crystals with helical chain conformations, the chain modulus is further reduced (*e.g.* 42 GPa for isotactic polypropylene[109]). The average transverse moduli obtained by various experimental methods are typically between 2 and 4 GPa.[66,67,111]

By the mid-1960s, it became well established that the moduli of the then-available fully drawn semicrystalline fibres were less than about one tenth of the corresponding crystal moduli in the chain direction. This was ascribed to the physical structure, in particular to a low fraction of load-bearing extended tie molecules. The discovery of fibres produced from rigid-chain aromatic polyamides[47–49,99] gave the most direct proof that this explanation is correct. The modulus of a recently introduced improved type of poly(p-phenyleneterephthalamide) fibre,[112] PRD 149, is 165 GPa, *i.e.* nearly 70% of the higher estimate of the chain modulus.[113] New processes, developed in particular for polyethylene, resulted in fibres with moduli approaching the theoretical value.[45,46,114]

Fibres from rigid-chain aromatic polyamides (*para*-aramids), which are viewed as single-phase structures, are obviously more amenable to a detailed analysis than fibres from linear flexible-chain polymers with segregated crystalline and noncrystalline regions. The absence of chain folding in *para*-aramids also simplifies the situation. Compilation of early results[115] showed that the chain orientation with respect to the fibre axis has a strong effect on fibre modulus; this was later confirmed by more detailed studies[116,117] which resulted[113] in the formulation of a relationship between the fibre modulus E and the orientation, expressed as $\langle \sin^2 \theta \rangle$:

$$1/E = (1/E_c) + (\langle \sin^2\theta \rangle/2G) \tag{5}$$

where E_c is the modulus in the chain direction (240 GPa) and G is the shear modulus for the longitudinal displacement of the chains (2 GPa). After recasting this relationship in terms of orientation factor f we get

$$E = E_c[1 + (E_c/3G)(1 - f)]^{-1} \tag{6}$$

Obviously, the strong effect of orientation is due to the high E_c/G ratio. It should be noted that the fibre modulus also correlates with the degree of order related to the width of both meridional and equatorial reflections.[118] It may be speculated that the degree of order affects the shear modulus G. It has been pointed out that in commercial Kevlar 49 fibres the increase in modulus was related to both improved orientation and an increase in the apparent lateral size of the crystallite size which occurred simultaneously.[99] Further complications in formulating quantitative relationships between the fibre structure and its modulus are due to the previously mentioned pleated structure and radial nonuniformity of the fibres.

In comparison with the 'simple case' of *para*-aramid fibres, formulation of the structure–modulus relationships for semicrystalline fibres is much more complex. This is due to the segregation of crystalline and noncrystalline regions and to chain folding. The usual approach is based on recasting the structural models into mechanical models proposed by Takayanagi *et al.*,[119] as shown in Figure 2. If it is assumed that the component B is coupled in parallel with an 'in-series' arrangement of components A and C, then the fibre modulus E is expressed as

$$E = \lambda \left[\frac{\phi}{E_C} + \frac{(1 - \phi)}{E_A} \right]^{-1} + (1 - \lambda)E_B \tag{7}$$

where parameters ϕ and λ are related to the volume fractions of the components (see Figure 2), and E_A, E_B and E_C are the moduli of the components.

For conventional fibres, where the crystalline component is considered to be discontinuous, the component C is taken as the crystalline portion of the fibre and, hence, the product $\phi\lambda$ is equal to the crystallinity. In terms of the microfibrillar model proposed by Prevorsek and co-workers,[61] the component A represents the intrafibrillar disordered domains, whilst the component B [volume fraction $(1 - \lambda)$] represents the interfibrillar domains. A similar approach has been used by

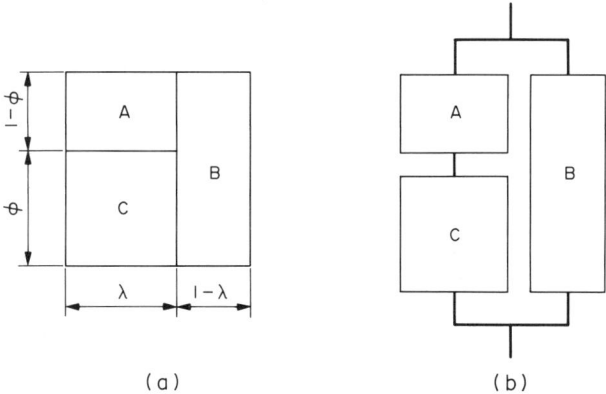

Figure 2 Schematic representation of (a) the components of fibre structure and (b) their assumed mechanical coupling

PS 2—Q

others;[73,110,120,121] a very detailed modelling of oriented nylon 6 has been carried out by Lewis and Ward.[122,123] Mishra and Deopura[124] identified the component B with the load-bearing extended tie molecules and used mechanical modelling for estimating the fraction of these molecules in drawn nylon 6 fibres.

The structure of high modulus polyethylene fibres obtained by optimized drawing of linear polyethylene[114] is viewed as crystalline lamellae linked by intercrystalline bridges.[125,126] Accordingly, the component B is then viewed as crystalline, and its content $(1 - \lambda)$ corresponds to the volume fraction of the material incorporated in the crystalline bridges. A more complex model consisting of four components has been proposed for these fibres by Grubb.[127]

The modelling approaches have been successful in the interpretation of the mechanical behaviour of fibres at low strain, such as the temperature dependence of shear and tensile relaxation of high modulus polyethylene[126] and the effect of moisture on the properties of nylon 6.[122,123] However, the predictive value of these approaches is rather limited. In essence, they only confirm the conclusion that, in order to maximize the modulus, it is essential to increase the number of extended tie molecules.

The task of correlating fibre strength with its structure is extremely complex. In addition to the problem encountered in the analysis of the factors affecting the modulus, it is necessary to separate the effect of flaws from that of the structure itself. In addition, the molecular weight of the polymer is known to have a very significant influence on fibre strength.

Most of the effort concerned with investigation of the strength of fibres produced from linear flexible-chain polymers has concentrated on polyethylene.[45,46,114] Theoretical calculations based on the C—C bond energy indicate that the strength of an individual polyethylene chain should be[108,128] between 15 and 25 GPa. A recent theoretical study concerned with ideal fibres consisting of a perfectly ordered array of fully extended chains of finite length has taken into account both primary and secondary bond failure.[129] This approach showed that the strength of such an ideal fibre is expected to decrease significantly with decreasing molecular weight. Thus, the strength computed for $M = 330\,000$ is around 10 GPa, decreasing to around 3 GPa for $M = 22\,000$; however, the modulus remains constant. These results were evaluated for the strain rate of 100% min^{-1}; later, the effects of strain rate and temperature[130] as well as MWD[131] were also investigated and the predicted trends were in good agreement with experimental results.

The maximum tensile strength achieved so far[45,46,132] is around 3 GPa (corresponding to a tenacity of 3 N tex^{-1}); these fibres, produced by gel spinning (Section 14.8) of high molecular weight polyethylene (e.g. $\bar{M}_w = 1.5 \times 10^6$, $\bar{M}_n = 2 \times 10^5$), were drawn *ca.* 30 times at 120 °C. In spite of some reservations, the polymer molecular weight ($\bar{M}_n$) appears to be the predominant factor controlling the achievable tensile strength of both melt- and solution-spun drawn fibres.[114] Comparison of the tensile strengths and moduli of fibres of a constant molecular weight produced at different draw ratios[46,132] suggests that structural features controlling both these properties are similar.

The theoretical strengths of individual chains of other linear polymers without bulky substituents are similar to the value calculated for polyethylene.[133] The computations for an ordered array of chains of finite length have so far only been carried out for poly(*p*-phenyleneterephthalamide),[134] again predicting a significant effect of molecular weight; however, in this case, the computation predicts that the modulus should also decrease with decreasing molecular weight. As in the case of polyethylene, the achieved strengths are still significantly lower than the calculated values. It is interesting to note that the strength of a recently developed PRD 149 yarn[112] with an enhanced modulus is actually somewhat lower than that of a lower modulus Kevlar 49 yarn (Table 1). This indicates that features such as skin–core differentiation[99] may play an overriding role in determining the strength of *para*-aramid fibres.

The tenacities of commercially available commodity fibres are considerably lower than those of specialty HSHM fibres. It should be emphasized that for many textile end-uses the strength is rather unimportant, provided that it exceeds the level required for successful textile processing. Typically,[135] the tenacities of fibres intended for apparel end-uses are between 0.3 and 0.5 N tex^{-1}; the extension at break (ranging from about 15 to 40%) decreases with increasing tenacity. For industrial commodity fibres, where the strength is considered to be important, some improvements can be achieved by increasing the polymer molecular weight[136] and by optimizing the drawing conditions (Section 14.9). However, the strength achieved (Table 1) is still less than one tenth of the theoretical value for an individual chain.

We now turn to a consideration of thermal properties. The discussion will be limited to the glass transition and to the phenomena arising from devitrification (*i.e.* the transition from a glassy to a rubber-like state) of the noncrystalline component of fibres. Obviously, fibre crystallinity affects the extent of the changes taking place at the glass transition, but the continuity of the crystalline component is also important. Thus, for fibres viewed as single-phase crystalline structures with defects (*para*-aramids), the glass transition is clearly of little importance.

Although the effect of chemical structure on the glass transition temperature is dominant, the shift arising from the physical structure cannot be disregarded.[137] The rise of T_g in semicrystalline oriented PET above the value of 70 °C usually quoted for noncrystalline isotropic material is well documented.[88]

The most important consequence of devitrification of the noncrystalline material in oriented fibres is the irreversible change of fibre length, *i.e.* fibre shrinkage. It is essentially due to re-coiling of extended chains (*i.e.* an entropic effect). In the absence of crystallites the shrinkage of PET fibres may be as high as 70%, depending on fibre orientation and on test conditions; in particular, stress and temperature.[138] The presence of crystallites results in a substantial reduction of fibre shrinkage.[69] In semicrystalline fibres, the shrinkage is strongly influenced by the orientation factor of the noncrystalline regions. For conventional commercial fibres, the boiling water shrinkage (BWS) is typically between 7 and 15%, but fibres with BWS as low as 0.5%, or as high as 60% (in the case of polyester) are also available.

Devitrification of the noncrystalline regions has a significant effect on the mechanical properties of semicrystalline fibres. The changes in modulus are usually monitored using dynamic mechanical measurements.[73,88] As already mentioned, mechanical models of the type shown in Figure 2 can be used for interpretation of the results. However, attention must be paid to the possibility of structural changes taking place during the measurements.

Finally, we shall consider briefly the interaction of fibres with low molecular weight compounds; this relates to properties such as moisture regain, dyeing behaviour and to certain aspects of chemical stability, in particular to hydrolytic and thermo-oxidative stability. Although the effects of chemical structure are dominant, the contribution from the physical structure is also significant. The effect of crystallinity on the equilibrium uptake of the low molecular weight compounds is well understood, since only the noncrystalline regions are accessible. The question of the diffusion rate is more complex, but since it is related to control of the rate of dyeing of textile fibres it has received much attention. The rate of diffusion of dye molecules is determined, not only by the crystallinity, but also by the nature of the macrolattice (*i.e.* the size, shape and mutual arrangement of the crystallites) which affects the diffusion path length.[61,78,88,139] The role of molecular mobility of the chain segments constituting the noncrystalline regions is also highly significant[61,88,139] and it explains the effect of temperature and the orientation factor f_n. Unfortunately, the methods for structural characterization of fibres are still not sufficiently accurate. Consequently, so-called diagnostic rate of dyeing tests carried out under suitable conditions,[15] can detect small differences in fibre structure which elude the fundamental characterization techniques.

14.7 FIBRE GEOMETRY

As already stated in Section 14.3, recognition of the importance of fibre geometry is reflected in the basic product-type classification. The number of new products which are based on variations in geometry is growing rapidly. The main features of fibre geometry include: (a) cross-sectional area, which is related to linear density (see Section 14.2); (b) cross-sectional shape (circular, triangular, trilobal, tetralobal, *etc.*); (c) axial distortion (usually helical or zigzag); (d) fibre length; (e) surface topology (smooth, striated, *etc.*); and (f) size, shape and distribution of deliberately introduced voids.

The effect of geometry on fibre properties will be illustrated by a few examples, starting with mechanical behaviour, namely flexural rigidity. Although the fabric construction has a dominant effect on its stiffness, the flexural rigidity of fibres is also important. For fibres which are identical with respect to chemical and physical structure, the flexural rigidity is proportional to the square of linear density.[3] Hence, with increasing emphasis on comfort, there is an increasing demand for fine fibres with linear densities less than 0.1 tex. On the other hand, the stiffness of carpet pile can be increased by increasing the fibre linear density.

The thermal insulation of a fabric depends on the amount of air trapped in it, and is again determined mainly by the fabric construction. However, the use of textured yarns is invariably beneficial. The thermal insulation of fibre assemblies, *e.g.* in duvets, is improved by using hollow fibres.

The capacity of hydrophobic fibres to retain water can be enhanced substantially by modifying the fibre geometry. Thus, an acrylic fibre with a porous core covered by a thin channelled sheath can retain as much as 30% water.[140] The use of fine profiled fibres affects the area and shape of the channels between the assembled fibres, and this improves the moisture transport through a fabric due to better wicking. Both these modifications result in improved wear comfort of garments which are in contact with the skin.

The use of fibres with modified cross-sectional shape (*e.g.* prismatic or trilobal) affects the appearance of a fabric or carpet. The use of voided fibres or fibres containing continuous channels results in improved soil-hiding.

It should be pointed out that, in order to achieve the desired effect, several geometrical modifications may be used simultaneously (*e.g.* use of fine textured filaments with modified cross-sectional shape). There is no doubt that the approaches based on modifications of fibre geometry are extremely successful, not least because new variants can be produced on an existing plant with only small modifications.

14.8   FIBRE MANUFACTURE

Details of the fibre production routes differ according to the choice of starting material and according to product type and required properties.[27] We shall first outline the production processes for the manufacture of conventional fibres from linear flexible-chain polymers. The first stage in the production sequence is always the spinning (also called 'extrusion'). In this process, the starting material is converted into continuous filaments. The resulting 'as-spun' filaments usually do not have adequate properties. In order to achieve the required strength, stiffness and dimensional stability, the as-spun fibres are drawn above the glass transition temperature. In the production of continuous filament yarns, the drawing is often combined with heat setting and twisting. Drawn yarns can be textured either by false-twist[141] or by other methods.[142] In the manufacture of staple fibres, the drawing and heat setting is usually followed by crimping and cutting.[27] Modern processes often combine several or all stages of the production sequence.

Only a brief discussion of the principles of spinning and drawing is presented here; details and an extensive bibliography are available in reviews concerned with polyester,[28] polyamide,[29] polypropylene[31,32] and acrylic[33] fibres and elsewhere.[27,143] An excellent analysis of the fundamental aspects of fibre formation can be found in Ziabicki's monograph.[18]

The spinning consists of: (1) preparation of the spinning liquid; (2) shaping of the spinning liquid by extrusion through an orifice; (3) solidification of the liquid jet; (4) collection of the as-spun fibres. There are three basic processes used in the manufacture of synthetic fibres, namely melt, dry and wet spinning. In melt spinning, the liquid is obtained by heating the starting material above its melting temperature; the solidification (either vitrification or crystallization) is accomplished by cooling the extrudate. Dry and wet spinning involve dissolution of the starting material in a suitable solvent; in the former process the solidification is achieved by evaporation of the solvent, whilst for the latter it is achieved by coagulation in a nonsolvent, which must be miscible with the solvent employed in the preparation of the spinning solution.

Spinning can be carried out successfully only when the liquid is spinnable, *i.e.* capable of being substantially extended prior to solidification.[18,144] The lack of spinnability of solutions may be due to capillary break-up arising from a high ratio of the surface tension and viscosity; for melts, this ratio is typically less than 10^{-4} m s^{-1} and this mechanism may be disregarded. The other reason for the lack of spinnability may be either ductile or brittle failure of the liquid jet.[144] In practice, the spinnability is often limited by polymer contamination rather than by fundamental material properties.

Melt spinning is the simplest and most economical method of fibre formation.[19,145] A schematic diagram of a typical melt-spinning process for production of a continuous filament yarn is shown in Figure 3. The polymer is melted in an extruder and passed to melt pumps *via* a manifold. Each pump accurately meters the melt through a filter and a spinneret containing the required number of orifices. The molten polymer emerging from the spinneret is cooled in air, and simultaneously subjected to tensile deformation; a cross-flow quench is used to facilitate the cooling. Spin finish is applied to solidified filaments after convergence. The yarn is collected by means of a winder. On existing plants, the winding speeds are limited to about 6 km min^{-1}, but a winder operating at speeds of up to 10 km min^{-1} is already available.[143]

For a given composition of the starting material (Section 14.4), the outcome of the spinning is determined by the processing conditions. Careful control of the melting conditions is essential since the thermomechanical history of the polymer may affect its rheological properties (partly due to possible changes of molecular weight and MWD) as well as its crystallization behaviour.

The orifice design affects the cross-sectional shape of the filaments, whilst their linear density is determined exactly by the ratio of the mass throughput per orifice w and the winding speed v_f.

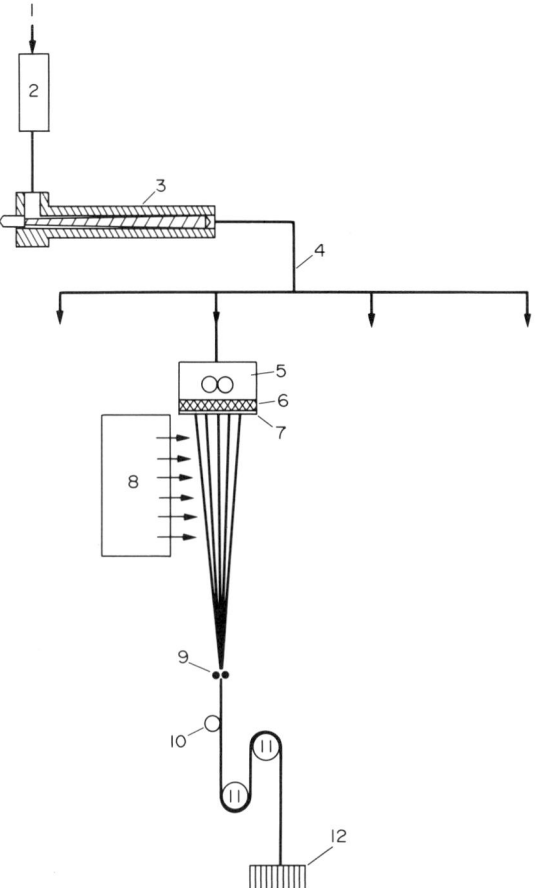

Figure 3 Schematic diagram of a melt-spinning process for production of a continuous filament yarn: 1, polymer granules; 2, dryer; 3, screw extruder; 4, manifold; 5, metering pump; 6, filter; 7, spinneret; 8, quench air source; 9, convergence guide; 10, spin finish applicator; 11, godets; 12, winder

The physical structure of the as-spun filaments is controlled by the solidification conditions, in particular by the rate of cooling and by the spinline stress.[146] Rapid cooling suppresses the crystallization, whilst high spinline stress results in increased orientation. However, since the rate of crystallization increases very substantially with increasing orientation,[147,148] high spinline stress favours the formation of oriented semicrystalline fibres. Neither the rate of cooling nor the spinline stress can be 'selected'; they are determined by the process variables (namely extrusion temperature, throughput, winding speed and the air temperature and velocity) and by the polymer properties (namely molecular weight and MWD, which have a dominant effect on the rheological behaviour of the extrudate). Investigation of the effect of these variables on the fibre structure and properties has been the subject of many theoretical and experimental studies,[18,19] in particular in the case of PET. This is due both to the practical importance of this polymer and to its convenient properties: its rate of crystallization in the absence of orientation is sufficiently low, and its glass transition temperature (*ca.* 70 °C) is sufficiently high to ensure the structural stability of the as-spun fibres at ambient temperature.

For a set of fibres of a constant linear density (*i.e.* w/v_f = constant) produced from a given polymer, the winding speed v_f has a dominant effect on fibre structure.[69,105,149] This is due to the increase of spinline stress with increasing v_f, whilst the changes in the cooling rate are relatively small. Accordingly, an investigation of PET fibres (linear density 0.56 tex), produced at v_f ranging from 2 to 6 km min^{-1}, showed that between 2 and 3.5 km min^{-1} the fibre orientation increases, but the fibres are noncrystalline. The density measurements indicate an increase in the degree of order at speeds above 3.5 km min^{-1}, but WAXS detects crystallinity only above 4.5 km min^{-1}.[69,149] The fibre properties (tenacity, extension at break and shrinkage) changed in accordance with these structural changes.[149]

It should be pointed out that, by modifying the spinline temperature by using a heated tube below the spinneret, it is possible to obtain semicrystalline PET fibres even at v_f less than 1 km min^{-1}.[150] More importantly, it has been shown that, by controlling the air flow below the spinneret, the crystallization can be suppressed even at v_f as high as 10 km min^{-1}.[151]

The spinline regimes which do not result in polymer crystallization are now well understood. The differential equations describing the kinetics of cooling and the forces acting on the spinline have been formulated.[18,152] Although they cannot be solved analytically, computer programs for numerical solution have been developed and successful computer simulation of melt spinning has been carried out.[153,154] Attempts were also made to extend this approach to regimes resulting in crystallization.[155-157] Here, the main problem is a satisfactory formulation of the relationship expressing the effect of both temperature and orientation on the rate of crystallization, which must not be too complex to prevent numerical solution of the spinline equations. The understanding of these regimes is further complicated by the occurrence of a rapid increase in the spinline velocity associated with the onset of crystallization[146] (so-called 'necking'); so far there is no commonly accepted explanation of this phenomenon.

In comparison with melt spinning, solution spinning processes are much more complex from both practical[27] and theoretical[18] points of view. Amongst the commodity fibres, acrylics are the most important generic type produced by these processes. In dry spinning, the solidification is achieved by evaporation of a volatile solvent in a vertical cell 5–10 m high, through which a hot gas (usually air) is passed. Even when the spinneret orifices are circular, the cross-sectional shape of the filaments may be noncircular; this is due to the formation of a relatively rigid skin during the initial rapid solvent removal. The winding speed does not exceed 1 km min^{-1} due to the limitation arising from the rate of solvent diffusion. The as-spun fibres, containing between 10 and 50% solvent, are of low orientation; their drawing, washing and drying is carried out as a separate process.

The fundamental analysis of dry spinning is much less advanced than that of melt spinning.[18] In the formulation of the spinline equations, it is necessary to take into consideration both heat and mass transfer as well as the force balance and the variation of the elongational viscosity with both temperature and solvent concentration.[158]

In wet spinning, the spinneret is normally immersed in the coagulation bath, where the solid but highly swollen 'gel filaments' are formed. The processes taking place in the coagulation bath are extremely complex, involving outward diffusion of the solvent and inward diffusion of the non-solvent, as well as the phase transition of the polymer/solvent/nonsolvent ternary system.[18] Consequently, the fundamental analysis of wet spinning has so far only been carried out with rather severe simplifying assumptions.[159]

In the production of acrylic staple,[33,34] the speed of the highly swollen fibres at the first godet located after the coagulation bath is less than 10 m min^{-1}. These gel fibres are washed and stretched at about 65 °C. The stretching (by 4–10 times) is followed by spin-finish application, drying, crimping and relaxation; alternatively, the crimping may be carried out before the drying. The final speed is normally less than 100 m min^{-1}.

Experimental studies concerned with wet-spun acrylic fibres[33] showed that the spinning conditions (spinning solution composition, composition and temperature of the coagulation bath and spinline stress — determined by the jetting speed and take-up speed) as well as the drawing and drying conditions affect both geometrical and structural features of the resulting fibres. Depending upon the fibre formation conditions, the cross-sectional shape can vary between approximately circular and dog-bone, and there are also considerable differences in fibre porosity. The prominent feature of the physical structure, influenced strongly by the fibre-formation conditions, is the sheath/core differentiation. From the published information it appears that it is still impossible to quantify the contributions due to individual process variables.

We now turn to the drawing of conventional as-spun fibres produced from linear flexible-chain polymers. The majority of published studies on fibre drawing are concerned with melt-spun fibres. In the course of drawing, the fibres are subjected to a large irreversible elongation (between 30 and 600%, corresponding to draw ratios between 1.3 and 7). The response of an as-spun fibre to tensile deformation depends on its chemical and physical structure and on the deformation conditions, namely on temperature and strain rate.[18] At sufficiently low temperatures, all fibres fail at a low strain in a brittle manner, or at somewhat higher temperatures, in a ductile manner. A further increase in temperature results in neck drawing; it is also referred to as 'cold drawing', which is somewhat misleading since it is not necessarily confined to low temperatures. In the course of neck drawing, the fibre shows a yield point followed by neck formation and neck propagation. The draw ratio achieved when the neck propagates over the entire sample is called the natural draw ratio λ_N. After completion of the neck propagation, the fibre is extended uniformly until it breaks. The natural draw ratio of noncrystalline polymers decreases with increasing temperature and with decreasing strain rate; close to the glass transition temperature $\lambda_N \to 1$ and the neck-drawing process is transformed into homogeneous drawing.[4,5,160] The natural draw ratio of noncrystalline PET fibres is known to decrease substantially with increasing preorientation.[161] Semicrystalline fibres of a low orientation may also exhibit neck drawing,[18] but the transition to homogeneous drawing then takes place at temperatures well above T_g. Generally, the factors which result in enhanced molecular mobility (*i.e.* increased temperature, presence of a plasticizer, reduced crystallinity) tend to favour the transition from brittle behaviour to neck drawing and, ultimately, to homogeneous drawing. An increased strain rate is equivalent to a decrease of temperature.

In practice, drawing is achieved by stretching the as-spun fibres between the feed roll or a set of feed rolls (velocity v_0) and a draw roll or a set of draw rolls (velocity v_1); the draw ratio is determined by the ratio v_1/v_0. The drawing is carried out under high molecular mobility conditions favouring homogeneous drawing; hence the drawing temperature is always above T_g and a low crystallinity of the as-spun fibre is desirable. If, for some reason, the structure of the feedstock fibre and the drawing conditions are such that neck drawing is unavoidable, the draw ratio must obviously be higher than the natural draw ratio to avoid the presence of undrawn regions in the drawn fibre. In the production of continuous filament yarns, where uniformity is extremely important, it is essential to stabilize the location of the draw point. This may be achieved, for example, by passing the yarn around a stationary snubbing pin. The drawing of noncrystalline fibres (*e.g.* PET produced at low spinning speeds) is often combined with a heat treatment on a hot plate in order to increase the crystallinity to the required level. In some processes, the drawing may be carried out in two or more stages, or it may be combined with a relaxation stage where a controlled shrinkage takes place. Details of arrangements employed in industrial practice can be found elsewhere.[27]

The effect of draw ratio on molecular orientation resulting from both neck drawing and uniform drawing has been studied in considerable detail, but at this stage very little is known about changes in the degree of chain extension.[68] The increase of optical and mechanical anisotropy caused by neck drawing can be interpreted by means of a phenomenological aggregate model.[4,5] The polymer is viewed as an aggregate of transversely isotropic units. During the tensile deformation, the symmetry axes of these units rotate to become aligned in the draw direction but the optical and mechanical properties of the units remain unchanged. This type of behaviour has been described as a 'pseudoaffine deformation'; the relationship between the orientation and the draw ratio for this type of deformation is shown in Figure 4. Numerous experimental data for 'cold drawing' of noncrystalline PET are in good agreement with this model.[162]

In contrast, changes in optical and mechanical anisotropy resulting from homogeneous drawing of noncrystalline polymers can be interpreted in terms of the theory of rubber elasticity. The majority of experimental evidence is based again on investigation of PET.[64,162-164] Quantitative measurements of birefringence, shrinkage and shrinkage stress led to a suggestion that noncrystalline PET above T_g can be regarded as a rubber-like material with molecular entanglements acting as the network junction points. The behaviour of such a network is distinctly time-dependent as shown, for example, by the decay of shrinkage stress and by the strain rate effects.[91] There is also a further complication arising from crystallization.[163] It is therefore obvious that interpretation of the behaviour of PET in terms of the theory of rubber elasticity should be viewed only as an approximation. In spite of these reservations, it has been found that the increase of orientation of

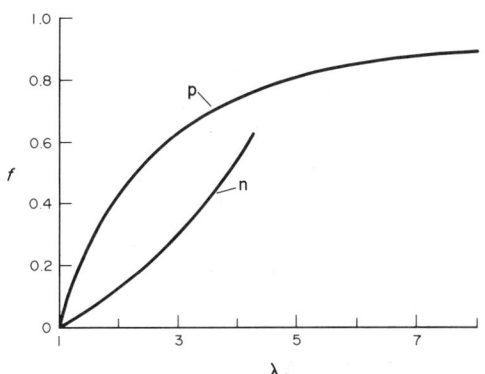

Figure 4 Effect of draw ratio λ on the orientation factor f: curve p = pseudoaffine deformation; curve n = network deformation (the number of random links per network chain is $N = 5.7$)

PET with the draw ratio λ can be expressed by a relationship based on the Gaussian form of the theory of rubber elasticity

$$f = (1/5N)\left(\lambda^2 - \frac{1}{\lambda}\right) \tag{8}$$

where N is the number of random links per network chain. The plot of f *vs.* λ for $N = 5.7$ found by Perena *et al.*[163] is shown in Figure 4 to illustrate the contrast between the network deformation and the pseudoaffine deformation schemes.

The concept of a transient network of limited extensibility was found to be useful in gaining a unified picture of polymer extension taking place during the spinning, drawing and tensile testing. The network extension resulting from the spinning can be expressed in terms of the equivalent draw ratios equal to the network draw ratio required for bringing the isotropic polymer to the same network extension as that existing in the as-spun fibre. It must be emphasized that λ_S is much smaller than the actual drawdown in the spinline. For noncrystalline as-spun fibres, λ_S can be evaluated from fibre shrinkage S measured under such conditions that the network reverts to an isotropic state; then[4, 161, 164]

$$\lambda_S = 1/(1 - S) \tag{9}$$

where shrinkage S is expressed as the length contraction related to the initial length.

During tensile testing the as-spun fibre is extended to its break point; the draw ratio applied during the testing is

$$\lambda_{T,B} = (1 + e_B) \tag{10}$$

where e_B is the strain at break. Thus the total draw ratio at the break point is

$$\lambda_B^* = \lambda_S \lambda_{T,B} \tag{11}$$

The λ_B^* evaluated from the published data for PET fibres[164] ($\bar{M}_n = 27\,000$) is approximately 7.4; these data also indicate that λ_B^* decreases with increasing molecular weight.

Alternatively, the as-spun fibre can be drawn (draw ratio λ_D) and then extended to the break point during the tensile test; hence

$$\lambda_B^* = \lambda_S \lambda_D \lambda_{T,B} \tag{12}$$

If the drawing is carried out under 'moderate conditions' (*i.e.* the drawing temperature is well below the polymer melting temperature, and the draw ratio is such that the strain at break of the drawn fibres is not less than approximately 0.2), then it is often found that λ_B^* for a given polymer remains approximately constant, irrespective of the split between the network extensions taking place during the spinning, drawing and testing. The true specific stress (Section 14.2) at break s_B^* also remains approximately constant.

Instead of relating the total deformation to the isotropic state, it is often preferred to relate it to a reference fibre, usually the least-extended fibre in the set. This approach has been used, for example, by Perez,[165] who showed that suitably transformed stress/strain curves of PET fibres can be superimposed on to the curve of the reference fibre. The superposition was successful in spite of significant differences in the crystallinities of the fibres included in the set. A more elaborate approach based on the network concept has been used by Vasilatos *et al.*[105]

It is worth noting that the values of s_B^* for various fibres produced under moderate drawing conditions from linear flexible-chain polymers such as PET,[105, 165] nylon 6,6[166, 167] and polypropylene[32] are all around 0.6 N tex^{-1}. This suggests that neither differences in the constitution of these fibre-forming polymers nor differences in the physical structure of the resulting fibres substantially affect the network breakdown resulting in fibre break during the testing at ambient temperature. However, by increasing the polymer molecular weight, it is possible to achieve a small increase in s_B^* in spite of the reduction in the network extensibility.

In order to increase the fibre strength, it is obviously necessary to increase the degree of chain extension beyond that determined by the network constraint. Some increase in s_B^* can be achieved by increasing the drawing temperature and the draw ratio. Thus, for a high speed spun nylon 6,6 fibre drawn at 220 °C at the maximum draw ratio, the s_B^* increased to 0.9 N tex^{-1} compared with the value of 0.65 N tex^{-1} for fibres drawn at the ambient temperature.[167] By a high temperature stretching[168] of a drawn nylon 6,6 fibre (249 °C at 92% of the breaking tension) it was possible to increase its s_B^* to about 1.2 N tex^{-1}. This is still considerably lower than the theoretical strength

(Section 14.6) and hence the attempts to overcome the network constraints in nylon 6,6 fibres cannot be rated as successful; the same applies to nylon 6 and PET. In contrast, the superdrawing of melt-spun polyethylene and polypropylene fibres is well established.[114] In the case of polyethylene it was possible, by optimization of polymer molecular weight and spinning as well as drawing conditions, to obtain a fibre with a tenacity of $1.75\,N\,tex^{-1}$ and extension at break of 3.9% (*i.e* s_B^* over $1.8\,N\,tex^{-1}$).[169] It appears that, due to weaker intermolecular interactions in polyalkenes compared with aliphatic polyamides and PET, it is much easier to overcome the network constraints.

The most convincing evidence for the limiting role of the polymer network is the success of gel spinning of high molecular weight polyethylene.[45,46,114] In this process, the polymer is dissolved in a high temperature solvent (*e.g.* decalin) to form a low concentration solution (1–2%). The hot solution is extruded into a cooling bath and this results in the formation of gel filaments with a very low level of chain entanglements.[170] These filaments can therefore be drawn, even after solvent extraction, to very high draw ratios, yielding fibres with exceptionally high modulus and strength (Table 1). The success achieved with polyethylene stimulated attempts to apply gel spinning to other linear flexible-chain polymers.[45]

We now turn to fibre formation from rigid-chain homopolymers such as poly(*p*-phenylene-terephthalamide) and poly(*p*-phenylenebenzobisthiazole) (3). These polymers do not melt below their decomposition temperature, and hence the fibres are obtained by solution spinning.[47–49] In suitable solvents these polymers form nematic solutions, provided that the concentration exceeds a critical value determined by polymer/solvent interactions, temperature and polymer molecular weight. The nematic solution is usually extruded through an air gap (0.5–2 cm) into the coagulation bath; this process is called dry-jet wet spinning. In the production of poly(*p*-phenyleneterephthalamide) fibres, the polymer is dissolved in concentrated sulfuric acid and the coagulant is usually dilute sulfuric acid. The coagulation conditions play a decisive role in the radial differentiation of the fibre structure. The elongational flow taking place during the spinning ensures that the extended chains present in the nematic solution are well aligned along the fibre axis. Hence, no drawing of the as-spun fibres is necessary. However, a short thermal treatment (0.5–5 s) carried out under tension at high temperatures improves the orientation and the degree of order and this results in improved tensile properties.

Thermotropic nematogenic copolyesters[47,49,50] (*e.g.* **4** and **5**) or poly(azomethine)s[51] (*e.g.* **6**) with a sufficiently broad temperature interval between the crystal–nematic transition and the onset of the decomposition can be melt spun. The as-spun fibres are highly oriented[104] even at winding speeds as low as $250\,m\,min^{-1}$. However, the tenacities of these fibres are similar to those of conventional drawn fibres (*i.e.* less than $1\,N\,tex^{-1}$), and their extensions at break are very low (1–2%). Extensive heat treatment (usually several hours or more) at high temperatures can produce significant increases of both tenacity (up to $2.7\,N\,tex^{-1}$) and extension at break;[47,49,50] for some compositions[171] (*e.g.* **5**) there is also a significant increase of modulus (up to $125\,N\,tex^{-1}$). Whilst the increased modulus is due to structural changes occurring during the heat treatment, the increase in tenacity is ascribed to a rise in molecular weight due to a solid phase post-polycondensation. Although for some copolyesters the tensile properties at ambient temperature are similar to those of *para*-aramid fibres, it should be noted that there is a significant drop in modulus around 120–140 °C, which is caused by a mechanical loss process related to the glass transition. This feature, together with the need for a lengthy heat treatment, is obviously hindering the commercial development of these fibres.

14.9 CONCLUDING REMARKS

The unquestionable success of the commodity fibres produced in large quantities from a few well-established polymers is due to their cost effectiveness combined with versatility, which is based on a skilful utilization of additives and relatively minor modifications of the chemical structure of the fibre-forming polymers, on the development of a large number of geometrical variants and on the capability of varying properties such as modulus, tenacity and shrinkage within broad limits. Whilst innovation along these lines will undoubtedly continue, successful development of a commodity fibre based on a new polymer is highly unlikely. In contrast, there is still scope for development of specialty fibres based on new or novel polymers. However, any such fibre must offer either a desirable unique combination of properties or a significant cost advantage in comparison with the specialty fibres already available.

Even from the generalized (and often oversimplified) picture presented here, it should be clear that understanding of the relationship between the fibre-formation conditions, structure and properties is still incomplete. As far as fibres from linear flexible-chain polymers are concerned, it is the concept of the degree of chain extension that needs to be developed and quantified, particularly in relation to the limiting role of the transient network. For fibres from rigid-chain polymers it is particularly the formation of defects during the fibre manufacture and their effect on fibre properties that need to be understood.

14.10   REFERENCES

1. S. R. Beech, C. A. Farnfield, P. Whorton and J. A. Wilkins (eds.), 'Textile Terms and Definitions', 8th edn., The Textile Institute, Manchester, 1986.
2. *Textile Organon*, 1987, **58** (6), 105.
3. W. E. Morton and J. W. S. Hearle, 'Physical Properties of Textile Fibres', 2nd edn., The Textile Institute/ Heinemann, London, 1975.
4. I. M. Ward, 'Mechanical Properties of Solid Polymers', 2nd edn., Wiley, New York, 1983.
5. I. M. Ward (ed.), 'Structure and Properties of Oriented Polymers', Applied Science, London, 1975.
6. W. Brostow and R. D. Corneliussen (eds.), 'Failure of Plastics', Hanser, Munich, 1986.
7. J. E. Booth, 'Principles of Textile Testing', 3rd edn., Butterworth, London, 1983.
8. I. M. Ward (ed.), 'Developments in Oriented Polymers', Elsevier, London, 1982, vol. 1.
9. I. M. Ward (ed.), 'Developments in Oriented Polymers', Elsevier, London, 1987, vol. 2.
10. M. Lewin and S. B. Sello (eds.), 'Handbook of Fiber Science and Technology', Dekker, New York, 1984, vol. II, part B.
11. M. Lewin and J. Preston (eds.), 'Handbook of Fiber Science and Technology', Dekker, New York, 1985, vol. III, part A.
12. M. Lewin and E. M. Pearce (eds.), 'Handbook of Fiber Science and Technology', Dekker, New York, 1985, vol. IV.
13. F. Happey (ed.), 'Applied Fibre Science', Academic Press, London, 1977–79, vols. 1–3.
14. P. K. Chatterjee (ed.), 'Absorbency', Elsevier, Amsterdam, 1985.
15. D. M. Nunn (ed.), 'Dyeing of Synthetic-Polymer and Acetate Fibres', Dyers Company Publications Trust, London, 1979.
16. K. McLaren, 'The Colour Science of Dyes and Pigments', Hilger, Bristol, 1983.
17. A. J. Hughes *et al.*, *Text. Prog.*, 1976, **8** (1).
18. A. Ziabicki, 'Fundamentals of Fibre Formation', Wiley, London, 1976.
19. A. Ziabicki and H. Kawai (eds.), 'High-Speed Fiber Spinning', Wiley, New York, 1985.
20. J. E. Ford, *Textiles*, 1985, **14** (1), 17; *Textiles*, **14** (2), 41.
21. S. B. Sello and C. V. Stevens, in ref. 10, p. 291.
22. E. Kissa, in ref. 10, pp. 143 and 211.
23. S. W. Shalaby, in ref. 11, p. 87.
24. R. H. Peters and R. H. Still, in ref. 13, vol. 2, p. 321.
25. M. Lewin, in ref. 10, p. 1.
26. L. Rebenfeld, B. Miller and J. R. Martin, in ref. 13, vol. 2, p. 463.
27. J. E. McIntyre and M. J. Denton, in 'Encyclopedia of Polymer Science and Engineering', ed. H. F. Mark *et al.*, Wiley, New York, 1987, vol. 6, p. 802.
28. J. E. McIntyre, in ref. 12, p. 1.
29. H. Reimschuessel, in ref. 12, p. 73.
30. J. E. McIntyre, in 'Toluene, the Xylenes and Their Industrial Derivatives', ed. E. G. Hancock, Elsevier, Amsterdam, 1982, p. 400.
31. M. Ahmed, 'Polypropylene Fibers — Science and Technology', Elsevier, Amsterdam, 1982.
32. M. Wishman and G. E. Hagler, in ref. 12, p. 371.
33. B. G. Frushhour and R. S. Knorr, in ref. 12, p. 171.
34. J. C. Masson and A. L. McPeters, *Fiber Prod.*, 1984, **12** (June), 34.
35. J. E. McIntyre, in ref. 17, p. 84.
36. I. Sakurada, in ref. 12, p. 503.
37. S. P. Hersh, in ref. 11, p. 1.
38. A. A. Mohajer and W. J. Ferguson, in ref. 17, p. 97.
39. J. G. Scruggs and J. O. Reed, in ref. 11, p. 335.
40. T. A. Attwood, P. C. Dawson, J. L. Freeman, L. R. J. Hoy, J. B. Rose and P. A. Staniland, *Polymer*, 1981, **22**, 1096.
41. J. B. Rose, *Polym. Prepr., Am. Chem. Soc., Div. Polym. Chem.*, 1986, **27** (1), 480.
42. K. Weinrotter and H. Griesser, *Chemiefasern/Text.-Ind.*, 1985, **35/87**, 409 (E50).
43. A. B. Conciatori, A. Buckley and D. E. Stuetz, in ref. 11, p. 221.
44. P. J. Akers and R. A. Chapman, *Lenziger Ber.*, 1987, **63**, 44.
45. P. J. Lemstra, R. Kirschbaum, T. Ohta and H. Yasuda, in ref. 9, p. 39.
46. P. J. Lemstra, N. A. J. M. van Aerle and C. W. M. Bastiaansen, *Polym. J.*, 1987, **19**, 85.
47. M. G. Dobb and J. E. McIntyre, *Adv. Polym. Sci.*, 1984, **60/61**, 61.
48. M. Jaffe and R. S. Jones, in ref. 11, p. 349.
49. S. L. Kwolek, P. W. Morgan and J. R. Schaefgen, in 'Encyclopedia of Polymer Science and Engineering', ed. H. F. Mark *et al.*, Wiley, New York., 1987, vol. 9, p. 1.
50. G. Calundann and M. Jaffe, in 'Proceedings of The Robert A. Welch Conferences on Chemical Research XXVI. Synthetic Polymers, Houston, Texas, 15 to 17 November, 1982', R. A. Welch Foundation, Houston, 1983, p. 247.
51. P. W. Morgan, S. L. Kwolek and T. C. Pletcher, *Macromolecules*, 1987, **20**, 729.
52. S. Ozawa, *Polym. J.*, 1987, **19**, 119.
53. P. Ehrburger and J.-B. Donnet, in ref. 11, p. 169.

54. A. J. Ultee, in 'Encyclopedia of Polymer Science and Engineering', ed. H. F. Mark *et al.*, Wiley, New York, vol. 6, p. 733.
55. M. Couper, in ref. 11, p. 51.
56. J. W. S. Hearle and R. Greer, *J. Text. Inst.*, 1970, **61**, 240.
57. B. Wunderlich and J. Grebowicz, *Adv. Polym. Sci.*, 1984, **60/61**, 1.
58. B. Wunderlich, 'Macromolecular Physics', Academic Press, New York, 1973, vol. 1.
59. D. C. Prevorsek, P. J. Harget, R. K. Sharma and A. C. Reimschuessel, *J. Macromol. Sci., Phys.*, 1973, **B8**, 127.
60. D. C. Prevorsek, G. A. Tirpak, P. J. Harget and A. C. Reimschuessel, *J. Macromol. Sci., Phys.*, 1974, **B9**, 733.
61. D. C. Prevorsek, R. H. Butler, Y. D. Kwon, G. E. R. Lamb and R. K. Sharma, *Text. Res. J.*, 1977, **47**, 107.
62. P. J. Barham and A. Keller, *J. Mater. Sci.*, 1985, **20**, 2281.
63. J. W. S. Hearle, 'Polymers and Their Properties', Horwood, Chichester, 1982, vol. 1.
64. A. Cunningham, I. M. Ward, H. A. Willis and V. Zichy, *Polymer*, 1974, **15**, 749.
65. H. M. Morgan, *Text. Res. J.*, 1962, **32**, 866.
66. J. H. Dumbleton, *J. Polym. Sci., Part A-2*, 1968, **6**, 795.
67. R. J. Samuels, 'Structured Polymer Properties', Wiley, New York, 1974.
68. J. W. Gilmer, D. Wiswe, H. G. Zachmann, J. Kugler and E. W. Fischer, *Polymer*, 1986, **27**, 1391.
69. H. M. Heuvel and R. Huisman, *J. Appl. Polym. Sci.*, 1978, **22**, 2229.
70. H. M. Heuvel, R. Huisman and K. C. J. B. Lind, *J. Polym. Sci., Polym. Phys. Ed.*, 1976, **14**, 921.
71. H. M. Heuvel and R. Huisman, *J. Polym. Sci., Polym. Phys. Ed.*, 1976, **14**, 941.
72. H. M. Heuvel and R. Huisman, *J. Polym. Sci., Polym. Phys. Ed.*, 1981, **19**, 121.
73. A. J. Owen, in ref. 9, p. 237.
74. A. M. Hindeleh and D. J. Johnson, *Polymer*, 1978, **19**, 27.
75. W. Ruland, *Acta Crystallogr.*, 1961, **14**, 1180.
76. M. Sotton, A.-M. Arniaud and C. Rabourdin, *J. Appl. Polym. Sci.*, 1978, **22**, 2585.
77. M. Sotton, *ACS Symp. Ser.*, 1983, **141**, 193.
78. R. Huisman and H. M. Heuvel, *J. Appl. Polym. Sci.*, 1978, **22**, 943.
79. B. Wunderlich, 'Macromolecular Physics', Academic Press, New York, 1980, vol 3.
80. A. Wlochowicz and W. Przygocki, *J. Appl. Polym. Sci.*, 1973, **17**, 1197.
81. E. W. Fischer and S. Fakirov, *J. Mater. Sci.*, 1976, **11**, 1041.
82. Z. W. Wilchinsky, *J. Appl. Phys.*, 1959, **30**, 792.
83. Z. W. Wilchinsky, *J. Appl. Phys.*, 1960, **31**, 1969.
84. M. D. Danford, J. E. Spruiell and J. L. White, *J. Appl. Polym. Sci.*, 1978, **22**, 3351.
85. H. P. Nadella, H. M. Henson, J. E. Spruiell and J. L. White, *J. Appl. Polym. Sci.*, 1977, **21**, 3003.
86. R. S. Stein and F. H. Norris, *J. Polym. Sci.*, 1956, **21**, 381.
87. A. Wlochowicz, S. Rabiej and J. Janicki, *Acta Polym.*, 1982, **33**, 117.
88. G. Valk, G. Jellinek and U. Schröder, *Text. Res. J.*, 1980, **50**, 46.
89. K. J. Friedland, V. A. Marikhin, L. P. Myasnikova and V. I. Vettegren, *J. Polym. Sci., Polym. Symp.*, 1977, **58**, 185.
90. H. M. Heuvel and R. Huisman, *J. Appl. Polym. Sci.*, 1985, **30**, 3069.
91. H. J. Biangardi and H. G. Zachmann, *J. Polym. Sci., Polym. Symp.*, 1977, **58**, 169.
92. H. W. Spiess, *Adv. Polym. Sci.*, 1985, **66**, 23.
93. H. W. Spiess, in ref. 8, p. 47.
94. H.-H. Kausch, in ref. 6, p. 84.
95. R. Bonart and R. Hosemann, *Makromol. Chem.*, 1960, **39**, 105.
96. A. Peterlin, *Text. Res. J.*, 1972, **42**, 20.
97. A. Peterlin, *Appl. Polym. Symp.*, 1973, **20**, 269.
98. G. Hinrichsen, *J. Polym. Sci., Part C*, 1972, **38**, 303.
99. M. G. Dobb and D. J. Johnson, in ref. 9, p. 115.
100. M. G. Dobb, D. J. Johnson and B. P. Saville, *Polymer*, 1979, **20**, 1284.
101. M. G. Dobb, D. J. Johnson and B. P. Saville, *J. Polym. Sci., Polym. Phys. Ed.*, 1977, **15**, 2201.
102. E. J. Roche, M. S. Wolfe, A. Suna and P. Avakian, *J. Macromol. Sci., Phys.*, 1986, **B24**, 141.
103. J. Blackwell and A. Biswas, in ref. 9, p. 153.
104. A. B. Erdemir, D. J. Johnson, I. Karacan and J. G. Tomka, *Polymer*, 1988, **29**, 597.
105. G. Vasilatos, B. H. Knox and H. R. E. Frankfort, in ref. 19, p. 383.
106. A. Odajima and T. Maeda, *J. Polym. Sci., Part C*, 1966, **15**, 55.
107. G. Wobser and S. Blasenbrey, *Kolloid-Z. Z. Polym.*, 1970, **241**, 985.
108. D. S. Boudreaux, *J. Polym. Sci., Polym. Phys. Ed.*, 1973, **11**, 1285.
109. I. Sakurada, T. Ito and K. Nakamae, *J. Polym. Sci., Part C*, 1966, **15**, 75.
110. T. Thistlethwaite, R. Jakeways and I. M. Ward, *Polymer*, 1988, **29**, 61.
111. L. Holliday, in ref. 5, p. 242.
112. J. E. Van Trump and J. Lahijani, in 'Proceedings of 32nd International SAMPE Symposium, 6 to 9 April, 1987', p. 917 (*Chem. Abstr.*, 1987, **107**, 60 507).
113. M. G. Northolt and R. v. d. Hout, *Polymer*, 1985, **26**, 310.
114. I. M. Ward, *Adv. Polym. Sci.*, 1985, **70**, 1.
115. G. B. Carter and V. T. J. Schenk, in ref. 5, p. 454.
116. M. G. Northolt and J. J. van Aartsen, *J. Polym. Sci., Polym. Symp.*, 1977, **58**, 283.
117. M. G. Northolt, *Polymer*, 1980, **21**, 1199.
118. R. Barton, Jr., *J. Macromol. Sci., Phys.*, 1986, **B24**, 119.
119. M. Takayanagi, K. Imada and T. Kajiyama, *J. Polym. Sci., Part C*, 1966, **15**, 263.
120. A. J. Owen, *J. Mater. Sci.*, 1981, **16**, 2324.
121. J. W. S. Hearle, R. Prakash and M. A. Wilding, *Polymer*, 1987, **28**, 441.
122. E. L. V. Lewis and I. M. Ward, *J. Macromol. Sci., Phys.*, 1980, **B18**, 1.
123. E. L. V. Lewis and I. M. Ward, *J. Macromol. Sci., Phys.*, 1981, **B19**, 75.
124. S. P. Mishra and B. L. Deopura, *J. Appl. Polym. Sci.*, 1982, **27**, 3211.
125. A. G. Gibson, G. R. Davies and I. M. Ward, *Polymer*, 1978, **19**, 683.

126. A. G. Gibson, S. A. Jawad, G. R. Davies and I. M. Ward, *Polymer*, 1982, **23**, 349.
127. D. T. Grubb, *J. Polym. Sci., Polym. Phys. Ed.*, 1983, **21**, 165.
128. H. F. Mark, in 'Polymer Science and Materials', ed. H. F. Mark and A. V. Tobolsky, Wiley-Interscience, New York, 1971, p. 231.
129. Y. Termonia, P. Meakin and P. Smith, *Macromolecules*, 1985, **18**, 2246.
130. Y. Termonia, P. Meakin and P. Smith, *Macromolecules*, 1986, **19**, 154.
131. Y. Termonia, W. Greene, Jr. and P. Smith, *Polym. Commun.*, 1986, **27**, 295.
132. P. Smith, P. J. Lemstra and J. P. L. Pijpers, *J. Polym. Sci., Polym. Phys. Ed.*, 1982, **20**, 2229.
133. K. E. Perepelkin, *Faserforsch. Textiltech.*, 1974, **25**, 251.
134. Y. Termonia and P. Smith, *Polymer*, 1986, **27**, 1845.
135. J. G. Cook, 'Handbook of Textile Fibres: II Man-Made Fibres', 5th edn., Merrow, Durham, 1984.
136. J. R. Schaefgen, in 'The Strength and Stiffness of Polymers', ed. A. E. Zachariades and R. S. Porter, Dekker, New York, 1983, p. 327.
137. R. F. Boyer, *J. Polym. Sci., Polym. Symp.*, 1975, **50**, 189.
138. J. O. Warwicker and B. Vevers, *J. Appl. Polym. Sci.*, 1980, **25**, 977.
139. D. C. Prevorsek, Y. D. Kwon and R. K. Sharma, *J. Mater. Sci.*, 1977, **12**, 2310.
140. E. Körner, G. Blankenstein, P. Dorsch and U. Reinehr, *Chemiefasern/Text.-Ind.*, 1979, **29/81**, 452 (E 62).
141. D. K. Wilson, *Text. Prog.*, 1978, **10** (3).
142. D. K. Wilson and T. Kollu, *Text. Prog.*, 1987, **16** (3).
143. M. J. Denton, *Text. Asia*, 1988, **29** (1), 30; *Text. Asia*, **29** (2), 23.
144. Y. Ide and J. L. White, *J. Appl. Polym. Sci.*, 1976, **20**, 2511.
145. L. Riehl, *Chemiefasern/Text.-Ind.*, 1987, **37/89**, 32 (E3).
146. A. Ziabicki, in ref. 19, p. 21.
147. F. S. Smith and R. D. Steward, *Polymer*, 1974, **15**, 283.
148. G. C. Alfonso, M. P. Verdona and A. Wasiak, *Polymer*, 1978, **19**, 711.
149. H. M. Heuvel and R. Huisman, in ref. 19, p. 295.
150. R. K, Gupta and K. F. Anyeung, *J. Appl. Polym. Sci.*, 1987, **34**, 2469.
151. G. Vassilatos (Du Pont), *US Pat.* 4 687 610 (1987).
152. S. Kase, in ref. 19, p. 67.
153. D. K. Gagon and M. H. Denn, *Polym. Eng. Sci.*, 1981, **21**, 844.
154. H. H. George, *Polym. Eng. Sci.*, 1982, **22**, 292.
155. K. Katayama and M. G. Yoon, in ref. 19, p. 207.
156. A. Ziabicki and L. Jarecki, in ref. 19, p. 225.
157. H. H. George, in ref. 19, p. 271.
158. I. Brezinski, A. G. Williams and H. L. LaNieve, *Polym. Eng. Sci.*, 1975, **15**, 834.
159. J. L. White and T. A. Hancock, *J. Appl. Polym. Sci.*, 1981, **26**, 3157.
160. I. Marshall and A. B. Thompson, *Proc. R. Soc. London, Ser. A*, 1954, **221**, 541.
161. S. W. Allison, P. R. Pinnock and I. M. Ward, *Polymer*, 1966, **7**, 66.
162. F. Rietsch, R. A. Duckett and I. M. Ward, *Polymer*, 1979, **20**, 1133.
163. J. M. Perena, R. A. Duckett and I. M. Ward, *J. Appl. Polym. Sci.*, 1980, **25**, 1381.
164. J. H. Dumbleton, *Text. Res. J.*, 1970, **40**, 1035.
165. G. Perez, in ref. 19, p. 333.
166. C. Lecluse, in 'Nylon: Retrospect and Prospect, Leeds, July 1983', University of Leeds, Leeds, 1983, lecture G.
167. J. G. Tomka, in 'Nylon: Retrospect and Prospect, Leeds, July 1983', University of Leeds, Leeds, 1983, lecture H.
168. R. L. Keefe, Jr. and W. O. Stratton (Du Pont), *Br. Pat.* 1 261 579 (1972).
169. W.-L. Wu and W. B. Black, *Polym. Eng. Sci.*, 1979, **19**, 1163.
170. P. Smith, P. J. Lemstra and H. C. Booij, *J. Polym. Sci., Polym. Phys. Ed.*, 1981, **19**, 877.
171. K. Ueno, H. Sugimoto and K. Hayatsu (Sumitomo Ltd.), *Eur. Pat.* 92 843 (1984).

15

Strength and Toughness

ROBERT J. YOUNG

UMIST, Manchester, UK

15.1 INTRODUCTION

Historically, polymers have not been exploited for mechanical properties such as high strength or toughness. These properties have normally been neglected as long as the mechanical behaviour of the polymer has been adequate for its application. Over recent years, however, this approach has changed dramatically and serious attempts are being made to exploit the full potential of polymer molecules. For example, it has now been realized that some polymer fibres are among the stiffest and strongest materials known.[1] These properties are particularly impressive when the low densities of solid polymers are taken into account, and high-performance polymer fibres are now being used in critical applications such as aircraft structures to replace metallic components, especially when weight saving is at a premium.

A major problem with producing strong and tough materials is that it is normally difficult to have both properties for the same material. High strength is a reflection of the ability of a material to sustain high loads whereas toughness depends upon high energy absorption during fracture. These two attributes are not necessarily (and in fact rarely) found for one material and polymers are no exception.

High toughness involves producing a material for which a large amount of energy is absorbed during fracture. The polymers that were produced originally, such as polystyrene, were relatively brittle but improvements have taken place over recent years particularly through copolymerization and blending whereby extremely tough polymers can be obtained whilst other properties such as stiffness and strength are adequately maintained.

This chapter will describe the current state of knowledge concerning the relationship between polymer structure and strength and toughness. Particular emphasis will be given to recent attempts to produce high strength fibres and to methods of toughening brittle polymers. The production of strong, tough polymers is widely recognized as a critical step in polymers becoming more widely accepted as engineering materials.

15.1.1   Definitions and Testing Methods[2]

The two principal parameters used to describe the mechanical behaviour of polymers are stress (σ) and strain (e). Stress is simply force divided by cross-sectional area and strain is the change in length divided by the original length. Although these two definitions suffice for simple tensile or compression testing both parameters can be represented in three-dimensional space using tensor notation. However, under most circumstances, such as uniaxial deformation, they can be assumed to be simple scalar quantities.

For many materials, the relationship between stress and strain can be expressed at least at low strains by Hooke's Law, which says that stress is proportional to strain and also helps to define the Young's modulus E, which is given by

$$E \;=\; \text{stress/strain} \;=\; \sigma/e \tag{1}$$

Many polymers obey Hooke's law at low strains but for complex systems of stresses it must be generalized in a three-dimensional form.

The general mechanical behaviour of typical polymers is shown schematically in Figure 1. In the case of the amorphous glassy polymer, deformation is linear elastic up to the failure point which occurs at less than about 5% strain. The Young's modulus of such a polymer is generally of the order of 3 GPa. In contrast, the semicrystalline polymer (above its glass transition temperature T_g) has a lower modulus but is capable of undergoing yield and extensive plastic deformation. The specimens generally undergo localized deformation and a stable 'neck' extends along the length of a tensile specimen. The mechanical behaviour of elastomers is very different from that of other polymers as can also be seen in Figure 1. Elastomeric materials have relatively low values of Young's modulus (~ 1 MPa) and are capable of extending reversibly up to over 500% strain before eventual failure. The behaviour has been likened to that of an 'entropy spring'.

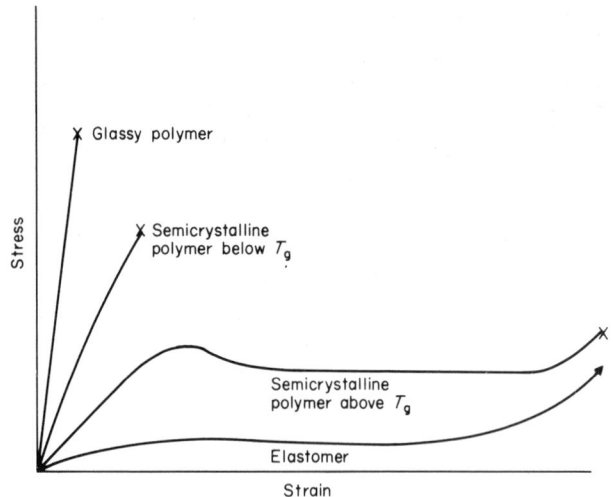

Figure 1   Schematic stress/strain curves for different types of solid polymer

The mechanical behaviour of polymers is strongly dependent upon the temperature and rate of testing. In general they display aspects of both elasticity and viscosity and so have been termed 'viscoelastic'. For example, the Young's modulus of a polymer depends upon the length of time for which the material is loaded. This is shown in Figure 2 in the case of constant stress deformation (creep) and constant strain deformation (stress relaxation). Figure 2a shows the effect of subjecting a polymer to constant load. In this case the strain increases with time until it approaches a constant value at a sufficiently long period of time (which may be several years). The inverse of this behaviour is shown in Figure 2b where a polymer is subjected to constant strain. The stress in the material decreases steadily and then eventually levels off. Such behaviour makes it extremely difficult for engineers to design structures from polymers since it is not possible to assign a unique Young's modulus to the material (unlike metals or ceramics). Because of this, theories of viscoelasticity have been developed to enable more accurate predictions of polymer mechanical behaviour over long periods of time.

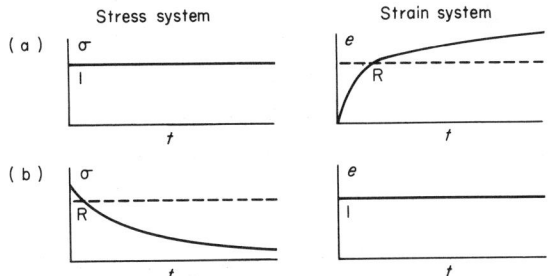

Figure 2 Time-dependent deformation of viscoelastic polymers showing the response (R) to different inputs (I) for (a) creep under constant stress; and (b) stress relaxation under constant strain

A simple tensile test will enable the fracture strength (σ_f) of a polymer to be determined. This is defined as the stress at failure. However, it does not give a full picture of the fracture behaviour of the material since σ_f reflects both the initiation and propagation of fracture. Because of this, fracture mechanics[3] has been developed and this is described in Section 15.3.1.

15.2 STRENGTH

15.2.1 Theoretical Estimates of Strength

Fracture involves the creation of new surfaces in a body, and in a polymer occurs by the breaking of primary (covalent) and/or secondary (*e.g.* van der Waals or hydrogen) bonds, depending upon the structure of the polymer. An estimate of the theoretical strength of a solid polymer can be made using the relatively simple analysis of Kelly and MacMillan[1] developed originally for any type of solid. The material is assumed to be a perfectly elastic isotropic solid. A specimen of such a material under stress is shown schematically in Figure 3a. When a tensile stress σ is applied to the body there is an increase in the interatomic separation x. The variation of interatomic potential energy with the separation of the atoms is shown schematically in Figure 3b and the application of stress clearly

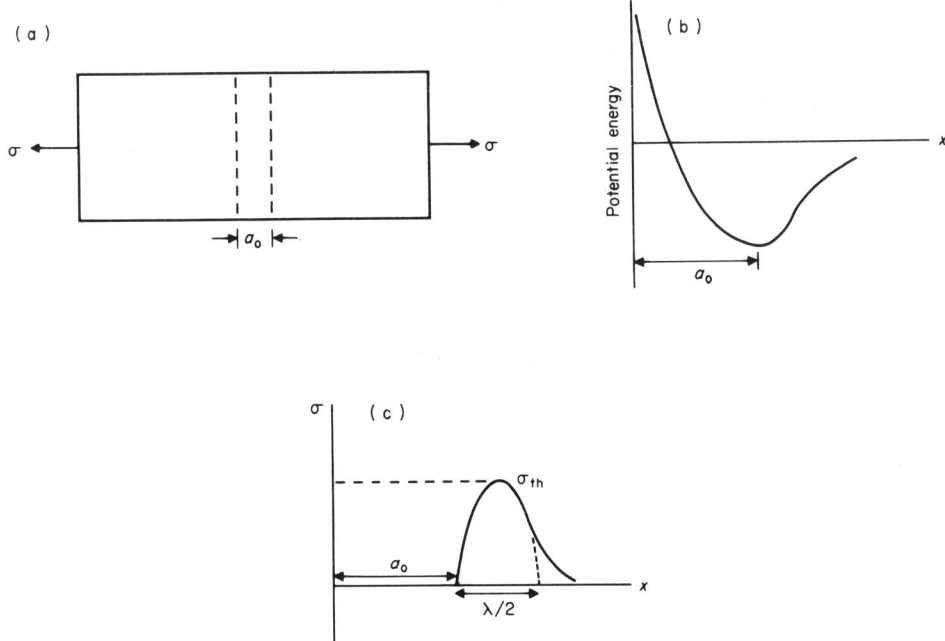

Figure 3 Deformation of an elastically isotropic solid:[3] (a) body under an applied stress σ with atomic planes of equilibrium separation a_0; (b) interatomic potential energy as a function of separation x; (c) dependence of σ upon x

causes an increase in the energy of the solid as x increases from its equilibrium value a_0. The force producing the increase in bond separation is the first derivative of the potential energy with respect to separation and it will increase as the bonds are stretched until it reaches a maximum at the point of inflexion of the curve in Figure 3b. The analysis can be simplified if it is assumed that the dependence of the force (or stress σ for a specimen of uniform cross-sectional area) upon atomic displacement x can be approximated to a sine function of wavelength λ to give a relationship of the form

$$\sigma = \sigma_{th} \sin(2\pi x/\lambda) \tag{2}$$

where σ_{th} is the maximum tensile strength. This function is plotted in Figure 3c and at low strains it approximates to

$$\sigma = \sigma_{th} 2\pi x/\lambda \tag{3}$$

If a_0 is the equilibrium separation of the interatomic planes perpendicular to the tensile axis then Hooke's law gives

$$\sigma = Ex/a_0 \tag{4}$$

where E is the Young's modulus of the material. If the energy required to create new surfaces G_0 is taken into account then it can be shown[1-3] that the maximum tensile strength is given approximately by

$$\sigma_{th} = \sqrt{\frac{EG_0}{2a_0}} \tag{5}$$

When different values of E, G_0 and a_0 are put into this equation it is found for most solids that

$$\sigma_{th} \approx E/10 \tag{6}$$

The term σ_{th} is often called the 'theoretical strength' of the material and is of interest since it gives the upper limit of strength expected for the material. However, in practice the measured values of tensile strength σ_f are invariably much less than $E/10$ for most materials because of the presence of flaws.[2-6] The modulus of polyethylene deformed parallel to the chain direction is of the order of 200–300 GPa[7] and so the theoretical strength for a polyethylene crystal deformed in this direction is of the order of 20–30 GPa.

Obviously the analysis outlined above is very approximate and does not take into account the chemical nature of the bonds in the material. The potential energy between two atoms bound by a covalent bond can be described approximately by a Morse function.[8] Kelly and MacMillan[1] have shown that if such a function is employed for polymers with aliphatic carbon–carbon chain bonds, then an estimate of the strength of the bond can be obtained. They calculated a C–C bond strength of 6.1×10^{-9} N and an atomic separation at failure equivalent to a tensile elongation of 21%. Kausch[9] estimated a bond strength of the order of 3×10^{-9} N using similar considerations. The strength of a particular polymer can then be determined from the density of packing of the covalent bonds within the structure. For example, the cross-sectional area occupied by each chain in a polyethylene crystal is about 0.18 nm². The analyses outlined above therefore imply a theoretical strength of between 16 and 33 GPa for polyethylene crystals deformed parallel to the chain direction. This range is similar to that determined from equation (6) for polyethylene using an entirely different method.

It is well established that, taking the analogy of diamond, a polymer in which the covalently bonded molecular chains are perfectly aligned would be expected to be extremely stiff and strong.[1,10] Hence it is expected that materials such as polymer crystals would have extremely high levels of stiffness and strength when deformed parallel to the chain direction. The levels of strength calculated above support this, although in practice the presence of imperfections and defects tends to greatly reduce the strengths achieved in real materials.[1,2] In order to make an accurate prediction of the strength, a good estimate of the modulus is required. Holliday[11] has summarized the various attempts to calculate crystal moduli for various polymers. One of the best examples is the calculation by Treloar[12] of the modulus of linear polyethylene by consideration of the forces required to cause bending and stretching of the individual bonds using spectroscopically determined force constants. He determined a value of about 180 GPa. A much higher value of 340 GPa was calculated by Shimanouchi and co-workers,[13] who took into account interchain interactions. There have been many other estimates of the chain direction modulus of polyethylene but most tend to lie between these two values. Examples of calculated values of stiffness and strength for polymers are listed in Table 1.[1]

Table 1 Theoretical Tensile Strength and Young's Modulus of Linear Polymers[1]

Polymer	Repeat unit	Predicted modulus (GPa)	Reference	Theoretical tensile strength (GPa)	Reference
Polyethylene	$-[CH_2]-$	182–340	a,b,c	19–36	d,e,f,g
Nylon-6,6	$-[CH_2]_6NHCO-$	196	a		
Poly(ethylene terephthalate)	$-CO\langle\bigcirc\rangle CO[CH_2]_2O-$	122	a		
Poly(tetrafluoroethylene)	$-[CF_2]-$	160	b	15–16	d,e
Cellulose parallel to chain	—	56.5	a	12–19	d,e

ª L. R. G. Treloar, *Polymer*, 1960, **1**, 95, 279, 290.
ᵇ T. Shimanouchi, M. Asahina and S. Enomoto, *J. Polym. Sci.*, 1962, **59**, 93.
ᶜ A. Odejima and T. Maeda, *J. Polym. Sci., Part C*, 1966, **15**, 55.
ᵈ K. E. Perepelkin, *Mekh. Polim.*, 1966, **2**, 845.
ᵉ K. E. Perepelkin, *Sov. Mater. Sci. (Engl. Transl.)*, 1970, **6**, 213.
ᶠ H. F. Mark, in 'Polymer Science and Materials', ed. A. V. Tobolsky and H. F. Mark, Wiley, New York, 1971.
ᵍ N. I. Zhirnov, Ye. G. Koryak-Doronenko and G. M. Bartenev, *Polym. Sci. USSR (Engl. Transl.)*, 1969, **11**, 1396.

Increases in the power of modern computers have enabled the complete theoretical stress/strain curves for polymers, such as polyethylene deformed parallel to the chain direction, to be calculated from first principles using quantum mechanical methods. Boudreaux[14] used the Hartree–Fock self-consistent field method and calculated a Young's modulus of 300 GPa and a strength of 19 GPa at an elongation of 33%. The full stress/strain curve is shown in Figure 4. The calculation was later revised and corrected by Crist and co-workers[15] who predicted a modulus of 405 GPa and a strength of 66 GPa at a strain of 43%. Their stress/strain curve is also shown in Figure 4. Such levels of stiffness and strength are thought to be rather high even though the value of $\sigma_{th} = 66$ GPa represents the stress required to fragment a crystal at absolute zero. The stress will naturally be reduced by the presence of defects at finite temperatures.

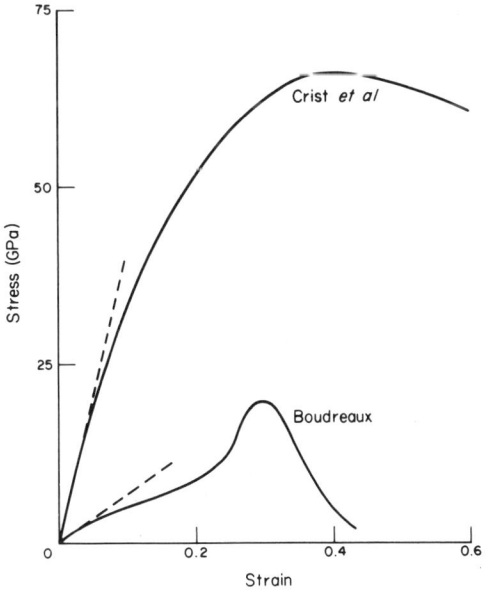

Figure 4 Theoretically calculated stress/strain curves for a polyethylene crystal deformed in tension parallel to the chain direction (after Boudreaux[14] and Crist et al.[15])

It is apparent from the considerations outlined above that there is, as yet, a considerable discrepancy in calculations and a wide range of values for the theoretical strength of even such a simple polymer as polyethylene. However, as will be shown next, the calculated values of σ_{th} ranging from 16 GPa to 66 GPa are well in excess of any measured for real polymers. The reasons for this are discussed in detail.

15.2.2 Strong Polymers

Large single crystal samples of polymers such as polyethylene are not yet available to test the theoretical predictions of strength but macroscopic single crystals of certain substituted poly(diacetylene)s can be prepared by topochemical solid state polymerization reactions.[16-18] It has been possible to obtain full stress/strain curves for poly(diacetylene) single crystal fibres using conventional mechanical testing methods[19-21] and to examine the factors which control the strength of a polymer crystal. Figure 5 shows a stress/strain curve for a poly(diacetylene) single crystal fibre and it can be seen that the deformation is elastic but slightly nonlinear above 2% strain. Fracture takes place at a strain of about 3% and is well below the levels of stress predicted for polyethylene. This is thought to be due to the presence of defects in the poly(diacetylene) single crystals.[19-21] The effect of defects can be demonstrated by following the dependence of the fracture strength of the crystals σ_f upon crystal diameter d as shown in Figure 6a. There is a marked reduction in fibre strength with increasing diameter. It is thought that this behaviour is due to the presence of crystal defects such as surface steps.[21,22] Using a stress–concentration factor analysis it can be shown that for such steps the dependence of σ_f upon d will be given by an equation of the form

$$\frac{1}{\sigma_f} = K_1 + \frac{K_2 d^{\frac{1}{2}}}{\sigma_{th}} \qquad (7)$$

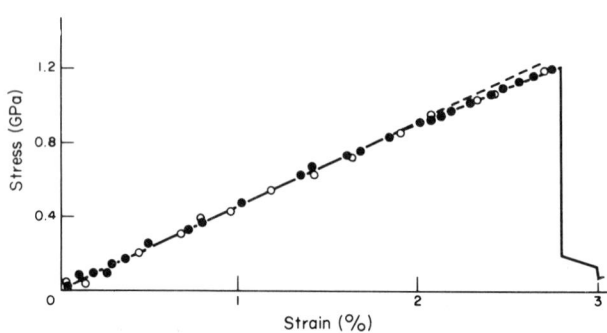

Figure 5 Typical stress/strain curve for a poly(diacetylene) single crystal fibre.[21] The closed circles are for loading and the open ones for unloading

where σ_{th} is the strength of a step free crystal (*i.e.* the theoretical strength) and K_1 and K_2 are constants which depend upon the step geometry and polymer type. The value of σ_{th} can be obtained from the slope of a plot of $1/\sigma_f$ *vs.* $d^{\frac{1}{2}}$ as shown in Figure 6b and is found to be of the order of 3 GPa for the particular poly(diacetylene) fibre used. The full crystal structure of the poly(diacetylene) fibre is known[23] and it is found that each molecule supports a cross-sectional area of about 1 nm². The theoretical strength of 3 GPa therefore corresponds to a force to break a molecule of about 3×10^{-9} N and a fracture strain of the order of 7–10%. Poly(diacetylene)s have a carbon–carbon backbone[16-18] and this value of 3×10^{-9} N is within the range of 3–6×10^{-9} N predicted theoretically[1,9] for the breaking force for a carbon–carbon bond, as described in Section 15.2.1. A molecular strength of this magnitude corresponds to a fracture stress of the order of 16 GPa for a polyethylene crystal which is at the bottom of the range of estimates for this polymer. Hence it is apparent that these investigations into the mechanical properties of poly(diacetylene) single crystal fibres have allowed a critical examination of the theoretical calculations outlined in Section 15.2.1. Although it appears that the theoretical estimates are on the high side, it is quite clear that aligned polymer molecules have the potential of producing structures with high degrees of stiffness and strength.[10]

It is not yet possible to make large single crystals of conventional polymers such as polyethylene, since the solid state polymerization reaction of substituted diacetylenes[16-18] appears to be a special case, specific to this type of monomer. Also, good single crystal samples of poly(diacetylene)s are only generally obtained when relatively large side groups are employed. This means that the reinforcing effect of the polymer backbones is reduced and, although it is found that the backbone is very stiff and strong, the levels of stiffness and strength achieved for the crystals are not so impressive.

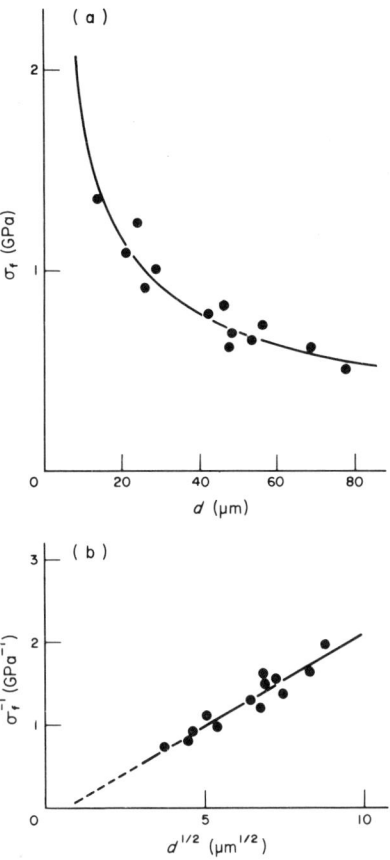

Figure 6 (a) Dependence of fracture stress σ_f upon fibre diameter d for poly(diacetylene) single crystal fibres.[21] (b) Variations of $1/\sigma_f$ with $d^{\frac{1}{2}}$ for poly(diacetylene) single crystal fibres[21]

There has been considerable interest over recent years in producing polymer fibres with high strength and stiffness from flexible molecules such as polyethylene, which have few or no side groups, in which molecular backbones are highly oriented. The techniques used include tensile drawing,[24-28] extrusion,[29-31] flow-controlled crystallization[32-34] and gel spinning.[35,36] The levels of strength and stiffness attained so far as listed in Table 2. Most of the investigations have concentrated upon polyethylene, which has the highest chain modulus (Table 1) and so potentially the best mechanical properties. The best levels of strength reported for this polymer have been for fibres produced by gel spinning[35] or flow-controlled crystallization[32] with values of σ_f up to 4 GPa being reported. This is, of course, well below the strength of 16–66 GPa predicted theoretically for polyethylene crystals (Section 15.2.1). This is because all these techniques produce polymer fibres in which there is a high degree of chain orientation but which are not single crystals. They are microfibrillar and polycrystalline and contain many defects which lead to this reduction in strength. A size dependence of the strength is found for fibres produced by these methods and the dependence of σ_f upon fibre diameter can be analyzed[37] using a relationship of the form of equation (7) which was developed originally for the single crystal fibres.[20,21]

A rather different way of producing polymer fibres with high stiffness and strength is by the solvent spinning of polymer molecules with rigid backbones. The solutions of such molecules tend to be liquid crystalline which assists the orientation process. Also the molecules, which invariably have aromatic groups in the backbone, tend to be inherently stiffer than flexible polymer molecules such as polyethylene. The best known examples are aromatic polyamides such as poly(p-phenyl-eneterephthalamide)[38,39] (**1**) which was developed by DuPont as the fibre Kevlar®. High molecular weight aromatic polyamides can be spun from liquid crystalline solutions using aggressive solvents such as sulfuric acid, to give fibres with high modulus and strength (Table 2). The properties can be improved by heat treatments under tension,[38,39] which improves orientation and removes defects. It can be seen from Table 2 that the properties of aromatic polyamides are significantly better than those of aliphatic polyamides such as nylon-6,6. The problem of aggressive acids can be avoided by

Table 2 Maximum Stiffnesses and Strengths of Oriented Polymer Fibres Produced by a Variety of Techniques[3]

Polymer	Technique	Modulus (GPa)	Strength (GPa)	Reference
Polyethylene	Tensile drawing	68	0.35	a
	Extrusion	70	0.4	b
	Flow crystallization	100	4	c
	Hot-drawing with solvent	120	3.7	d
Polypropylene	Tensile drawing	22	0.93	e
Acetal Polymer	Tensile drawing	35	1.7	f
Aromatic polyamide	Spun from HF	26.6	1.2	g
	(+ heat treatment)	140	1.8	g
Aromatic copolyester	Melt spun	40.2	0.47	h
	(+ heat treatment)	46.7	1.5	h
PBT[i]	Spun from acids	50–110	0.6–1.1	j
	(+heat treatment)	260–300	2.5–3.0	k
Substituted poly(diacetylene)	Solid state polymerized	62	1.7	l
Nylon-6,6	Industrial grade	3.8	0.77	h

[a] G. Capaccio and I. M. Ward, *Polym. Eng. Sci.*, 1975, **15**, 219.

[b] W. T. Mead, C. R. Desper and R. S. Porter, *J. Polym. Sci., Polym. Phys. Ed*, 1979, **17**, 859.

[c] A. J. Pennings and K. E. Meihuizen, in 'Ultra-High Modulus Polymers', ed. A. Ciferri and I. M. Ward, Applied Science, London, 1979, p. 117.

[d] B. Kalb and A. J. Pennings, *J. Mater. Sci.*, 1980, **15**, 2584.

[e] W. N. Taylor and E. S. Clark, *Polym. Prepr., Am. Chem. Soc., Div. Polym. Chem.*, 1977, **18**, 332.

[f] E. S. Clark and L. S. Scott, *Polym. Eng. Sci.*, 1974, **14**, 682.

[g] J. R. Schaefgen, T. I. Bair, J. W. Ballou, S. L. Kwolek, P. W. Morgan, M. Panar and J. Zimmerman in 'Ultra-High Modulus Polymers', ed. A. Ciferri and I. M. Ward, Applied Science, London, 1979, p. 173.

[h] W. C. Wooten, F. E. McFarlane, T. F. Gray and W. Jackson, in 'Ultra-High Modulus Polymers', ed. A. Ciferri and I. M. Ward, Applied Science, London, 1979, p. 227.

[i] PBT = poly(*p*-phenylenebenzobisthiazole).

[j] S. R. Allen, A. G. Filippov, R. J. Farris and E. L. Thomas, *J. Appl. Polym. Sci.*, 1981, **26**, 291.

[k] S. R. Allen, A. G. Filippov, R. J. Farris and E. L. Thomas, paper presented at the conference 'Deformation, Yield and Fracture of Polymers, Cambridge, 1982', PRI, London.

[l] C. Galiotis and R. J. Young, *Polymer*, 1983, **24**, 1023.

using aromatic copolyesters which can be spun from the melt into fibres.[40] However, the properties for these materials do not yet reach those of the aromatic polyamides (Table 2).

The ordered polymer programme of the US Air Force has produced some very exciting developments in the area of strong and stiff polymers with good high temperature resistance.[41-44] The two most interesting polymers are poly(*p*-phenylenebenzobisthiazole), (PBT, **2**) and poly(*p*-phenylenebenzobisoxazole), (PBO, **3**).

(**1**)　　　　(**2**)

(**3**)

As with aromatic polyamides, (**2**) and (**3**) are produced by spinning from concentrated acids and their properties are improved by heat treatment. Fibres of PBT and PBO have been found to have very high levels of modulus, in excess of 300 GPa and strengths of up to 3 GPa[43] as can be seen in Table 2. These types of molecules behave as rigid rods giving rise to liquid crystalline solutions and this aids orientation during spinning. The molecules are inherently rigid and deform principally by bond stretching rather than bending because of the presence of aromatic rings. They also have no side groups which gives rise to good chain packing. It is these factors which lead to the high levels of stiffness and strength. However, the perfection that is present in single crystals is lacking in the oriented polymer fibres which means that they do not have strengths which are significantly better than those of polymer single crystal fibres[20] where there are larger side groups and the molecules are inherently less rigid.

The value of 3–4 GPa appears to be an upper limit currently attainable for the strength of a polymer fibre even though the values of theoretical strength are well above this figure. The shortfall appears to be due to the presence of defects and imperfections. In practice, however, it is not always the absolute values of stiffness and strength that are important but rather the specific strength and stiffness, *i.e.* the strength and stiffness divided by the density. This is especially the case in engineering applications where the strength/weight ratio is important. Because they have relatively low densities, typically 900–1400 kg m⁻³, high modulus/high strength polymer fibres generally compare well in terms of specific mechanical properties with other high strength materials, such as steel, boron fibres and carbon fibres. This is illustrated in Figure 7[45] which shows a plot of specific strength *vs.* specific stiffness for a series of strong materials. Since there is a close relationship between stiffness and strength, most of the materials are grouped along the diagonal. However, the polymer fibres tend to lie at the top of the plot, indicating a relatively high level of strength for their stiffness, whereas carbon fibres tend to lie towards the bottom of the plot, indicating that they have high moduli but relatively low strengths. Current research is aimed at producing materials with strength/modulus properties corresponding to the top right hand corner of the diagram, *i.e.* materials with extremely high levels of stiffness and strength. It can be seen that, at the moment, the greatest success is being achieved with polymer fibres.

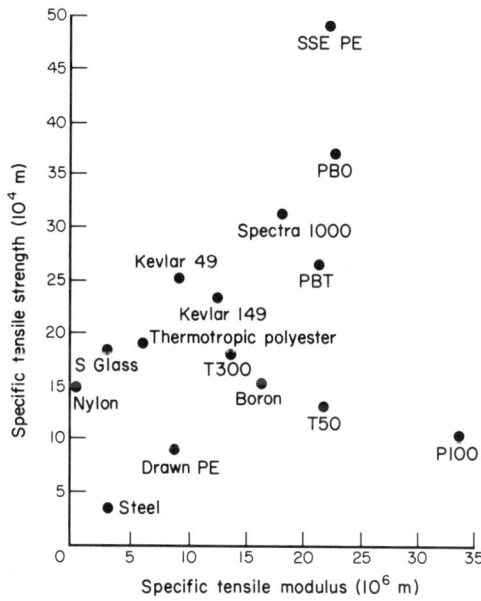

Figure 7 Variation of specific tensile strength with specific tensile modulus for high performance fibres.[45] T300, T50 and P100 are carbon fibres; spectra 1000 is a gel-spun polyethylene; and SSE PE is a solid state extruded polyethylene

15.2.3 Strength of Conventional Polymers

The previous section was concerned with the behaviour of strong polymers. These are invariably highly oriented polymer fibres in which deformation and fracture take place through the stretching and breakage of primary backbone bonds. The behaviour is more complex in the case of isotropic polymers where fracture can occur by breaking either primary covalent bonds or secondary bonds through molecular 'pull-out'.[9,46-48] It is thought that both types of process can take place, but that the extent to which each occurs depends upon the type of polymer undergoing fracture and the testing conditions. Clearly, glassy network polymers such as thermosetting resins will act like diamond and covalent bonds must be severed. However, in thermoplastic glasses it may be possible for fracture to take place by molecules sliding past each other with no molecular fracture. In high molecular weight thermoplastic glasses chain entanglements may act as cross-links preventing molecular pull-out, and the crystals in a semicrystalline polymer may act as similar anchorage points for the molecules.

It is well established that the tensile strength of a glassy thermoplastic depends on the molecular weight of the polymer. The effect of molecular weight and molecular weight distribution upon the mechanical behaviour of polymers has been reviewed by Martin, Johnson and Cooper.[49] The effect of molecular weight upon the tensile strength of poly(methyl methacrylate)[50] (PMMA) is shown in Figure 8. Measurements upon several glassy thermoplastics[49-52] have shown that low molecular weight polymers have very low tensile strengths but that the strength increases as the molecular weight is increased. It is generally found that above a critical molecular weight the polymer reaches a constant level of strength for which any further molecular weight increase causes no more increase in strength.

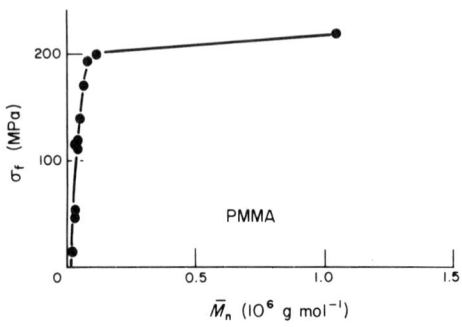

Figure 8 Variation of the strength of PMMA with the number average molecular weight[50] $\bar{M}_n$

It was suggested by Flory[53] many years ago that some of the physical properties of polymers could be related to the number average molecular weight $\bar{M}_n$ by an empirical equation of the form

$$P = A - B/\bar{M}_n \tag{8}$$

where P is the property and A and B are constants. The data in Figure 8 approximately follow a relationship of this form,[50] although the question of whether it is best to correlate mechanical behaviour with $\bar{M}_w$ or $\bar{M}_n$ has not yet been settled.[49]

It is found[54] that the zero strength molecular weight $\bar{M}_0$ (*i.e.* the value of molecular weight for the curve in Figure 8 when extrapolated to $\sigma_f = 0$) can be correlated with the molecular weight at which entanglements start to form in a polymer melt $\bar{M}_e$ as is shown in Table 3. This behaviour has been examined in more detail by Turner,[55] who demonstrated that if it were assumed that the entangled chains in a polymer glass form a network similar to that of a cross-linked rubber, but for which the cross-links are physical rather than chemical, then a relationship between fracture strength and molecular weight $\bar{M}$ of the form

$$\sigma_f = \sigma_{f\infty} - C/\bar{M} \tag{9}$$

Table 3 Comparison of the Zero Strength Molecular Weight $\bar{M}_0$ Measured in Flexure with the Critical Molecular Weight at which Entanglements are Obtained $\bar{M}_e$, Measured using Melt Viscosity[3]

Polymer	$\bar{M}_0$ (g mol^{-1})	$\bar{M}_e$ (g mol^{-1})	Reference
PMMA	22 000	11 000	a
	21 000	27 500	b
Polystyrene	50 000	30 000–35 000	a
	72 000	—	b
Polyethylene	11 000	4000	a
	12 000	—	b

[a] A. N. Gent and A. G. Thomas, *J. Polym. Sci., Part A-2*, 1972, **10**, 571.
[b] D. J. Turner, *Polymer*, 1982, **23**, 626.

is predicted. The parameters $\sigma_{f\infty}$ and C are the strength of a polymer of infinite molecular weight and a material constant respectively. There is now accumulated evidence that entanglements play a vital role in controlling the fracture behaviour of glass thermoplastics and that, in particular, they have an important effect on controlling strength and toughness.[3]

The chemical factors which govern the strength of thermoplastics are complex and still not fully understood. For example it is not possible to say with any certainty if a polymer will be tough or brittle solely from the knowledge of its chemical structure. Vincent[47] attempted to correlate the strengths of a wide variety of thermoplastics with their chemical structures. He was able to show that the strength at the ductile/brittle transition was proportional to the number of chains per unit area.[3] In general, he found that the strongest polymers tend to have linear chains and the weakest ones have large side groups which reduce the density of main chain bonds.[47] In addition it is estimated that only a small proportion (of the order of 1%) of the main chain bonds are fully deformed during fracture of isotropic polymers.[3]

The measured strength of isotropic polymers is affected by a wide variety of testing variables[3] which include: (i) temperature and rate of testing;[5,56] (ii) orientation;[5,57] (iii) environment;[57,58] (iv) pressure;[59,60] (v) presence of notches or defects;[4,5] and (vi) cyclic deformation and fatigue.[2,3]

These factors have been reviewed widely in the references cited above. It is generally found that the strength of polymers increases with decreasing temperature and increasing strain rate,[5,56] although such changes can sometimes induce a ductile/brittle transition and a consequent drop in strength.[5] The effect of temperature upon the deformation of PMMA is shown in Figure 9 where it can be seen that an increase in temperature leads to an increase in elongation but a reduction in strength.

As discussed earlier, oriented polymers tend to be stronger when deformed parallel to the direction of orientation but weaker in the transverse direction.[5,57] This is due to the orientation leading to molecular alignment whereby the relatively strong covalently bonded molecules lie in the direction of orientation with only Van der Waals bonding in the transverse direction.

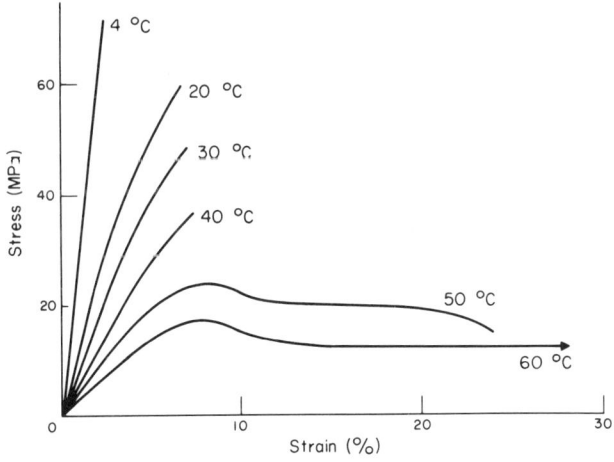

Figure 9 Effect of testing temperature upon stress/strain behaviour of PMMA[3]

Polymers used for engineering purposes tend to be relatively resistant to chemical attack, but the strength of certain polymers can fall dramatically in the presence of particular environments through the process of environmental stress cracking.[5,57,58] The problem is very specific to certain polymer environment combinations such as PMMA with alcohols and polyalkenes in detergents and can lead to premature failure at very low stresses through either plasticization or a reduction in surface energy.

Finally, the presence of sharp notches or cracks can give rise to a reduction in strength of a polymer compared with that of an uncracked specimen[4,5] as shown in Figure 10 for PMMA. The cracks and notches give rise to stress concentrations and failure at relatively low stresses, especially for brittle polymers. Recent improvements in the mechanical properties of polymers have led to tougher materials with improved resistance to brittle fracture. These are discribed next.

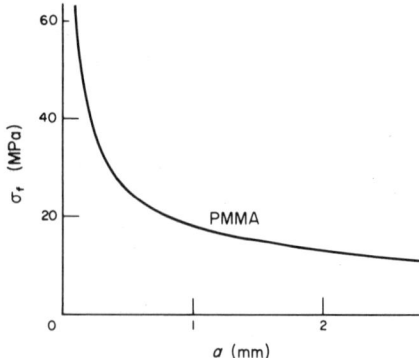

Figure 10   Dependence of the strength of PMMA tensile specimens upon the length of sharp cracks a[5]

15.3   TOUGHNESS

15.3.1   Fracture Mechanics

It was shown in Section 15.2.1 that the theoretical strength of a brittle solid is of the order of $E/10$, *i.e.* one-tenth of Young's modulus. The modulus of a brittle polymer is typically 3 GPa and so the theoretical strength of such a material ought to be 300 MPa. However, the measured strengths of such materials are typically 10–100 MPa, well below the theoretical value.[2] The shortfall in strength was recognized many years ago by Griffith[61] who showed that the relatively low strength of a brittle solid could be explained by the presence of flaws acting as stress concentrators. This hypothesis has been developed and extended by many investigators and is the basis of 'fracture mechanics' which is used to interpret the fracture of many solids, including polymers.

In the early 1960s Berry[5] applied the energy balance of Griffith to explain the dependence of the strength of PMMA upon the size of cracks and notches, shown in Figure 10. He found from experiments in which he carefully controlled the length and quality of cracks in specimens that the dependence of σ_f upon a followed the Griffith[61] equation

$$\sigma_f = \sqrt{\frac{2E\gamma}{\pi a}} \qquad (10)$$

where E is the Young's modulus and γ is the 'surface energy'. However, in practice γ is very much higher than the surface energy of the material due to energy dissipation in the vicinity of the crack tip during crack growth.[3] Also, equation 10 predicts that $\sigma_f \to \infty$ when $a \to 0$, which is clearly not the case for an unnotched specimen. Polymer samples always fail at finite levels of stress due to the presence of flaws which form during loading.[3] It is thought that the 'flaws' in a polymer such as PMMA are crazes which nucleate under stress and then break down into cracks. The Griffith equation is not now used to describe the failure of brittle polymers since the assumption that γ is the surface energy is incorrect. Instead, fracture mechanics, which was developed originally to characterize the fracture of metals and ceramics is normally employed to analyze the fracture behaviour of polymers.

Two main approaches are used in fracture mechanics although they are closely related.[3,62] The first is an energy criterion developed by Orowan[63] which supposes that fracture takes place when sufficient energy is released (from the stress field) during crack growth to supply the energy requirements of the new fracture surfaces created. The fracture of a material is thereby characterized by the material property G_c known as the 'strain energy release rate' or 'fracture energy'.[3] The second approach was developed by Irwin[64] who showed that the stress field around a sharp crack in an elastic material could be uniquely defined by a parameter known as the 'stress intensity factor' K. Fracture then occurs when K reaches a critical value K_c, which is a material property often called the 'fracture toughness'. The parameters G_c and K_c are related for a linear elastic material through the approximate relationship

$$G_c \approx K_c^2/E \qquad (11)$$

The fracture of many glassy polymers can be analyzed using linear elastic fracture mechanics,[3,62] where the mechanical behaviour of the polymer is approximately linear elastic even though small-scale inelastic deformation or yielding or bulk nonlinear elastic behaviour may take place such as with tough polymers or elastomers, when different approaches need to be used.[3,62,64,65]

15.3.2 Fracture Mechanics Testing

The aim of fracture mechanics testing is to obtain a material parameter which is capable of characterizing crack propagation in the material and is independent of test geometry. A wide variety of different geometries have been developed in order to validate this requirement.

Some of the types of specimen which have been used to study the fracture of rigid, brittle polymers are shown in Figure 11.[3] It is possible to measure both K_c and G_c with all of these specimens and in the examples shown the parameters determined are for the tensile-opening mode (Mode I).[3] The value of K is a function of applied stress or load, crack length and a nondimensional function of the specimen dimensions: expressions for the different geometries shown in Figure 11 are listed in Table 4. K_c is determined from the critical stress or load at fracture. However, care should be taken in using K and K_c, since they are often only strictly valid for certain specimen geometries and dimensions and the original references should be consulted to ensure that actual specimens comply with these restrictions. Also the expressions are only valid for materials in which there is small-scale yielding and which obey linear elastic fracture mechanics.[3] It is possible to determine G_c from equation (11) as long as the modulus of the polymer is also known.

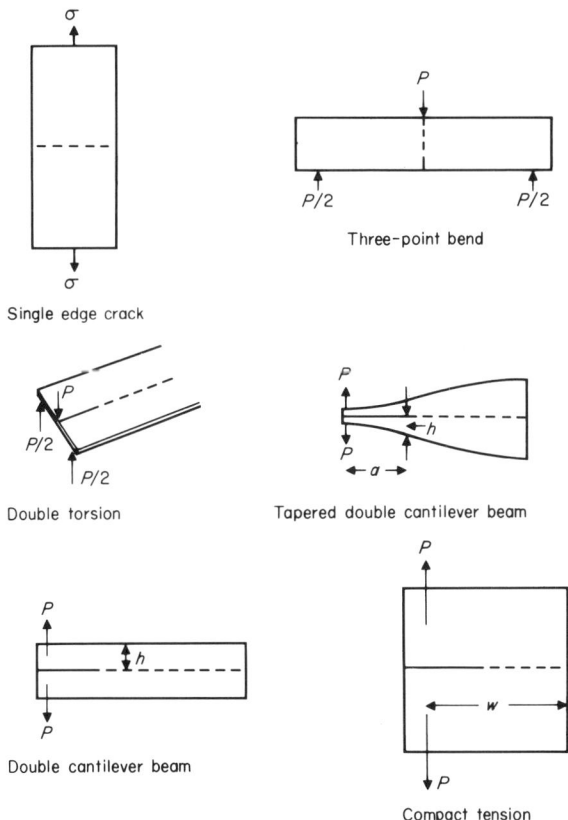

Single edge crack

Three-point bend

Double torsion

Tapered double cantilever beam

Double cantilever beam

Compact tension

Figure 11 Schematic diagrams of various specimens used for fracture mechanics testing of rigid polymers.[3] P = applied load

Although elastomers show high degrees of extension it is possible to analyze their fracture behaviour using fracture mechanics as long as their deformation is elastic (but nonlinear).[2,64] The different geometries used to follow crack propagation in elastomers are shown schematically in Figure 12.[3] Formulae have been developed to determine G_c for these specimens and examples are

Table 4 Expressions for the Stress Intensity Factor K for Fracture Mechanics Specimens Used to Study Crack Growth in Rigid Polymers[3]

Geometry (see Figure 11)	Expression for K	Comments	References
Single edge crack	$K = Q\sigma\, a^{\frac{1}{2}}$ $$Q = \left[1.99 - 0.41\left(\frac{a}{w}\right) + 18.70\left(\frac{a}{w}\right)^2 - 38.48\left(\frac{a}{w}\right)^3 + 53.85\left(\frac{a}{w}\right)^4 \right]$$	w = width of specimen	a
Three point bend	$$K = \frac{Ps}{bw^{\frac{3}{2}}}\left[2.9\left(\frac{a}{w}\right)^{\frac{1}{2}} - 4.6\left(\frac{a}{w}\right)^{\frac{3}{2}} + 21.8\left(\frac{a}{w}\right)^{\frac{5}{2}} - 37.6\left(\frac{a}{w}\right)^{\frac{7}{2}} + 38.7\left(\frac{a}{w}\right)^{\frac{9}{2}} \right]$$	P = applied load s = span between supports $= 4w$ b = sheet thickness	b
Double torsion	$$K = Pl_{\mathrm{m}}\left[\frac{3(1 + v)}{lb^3 b_{\mathrm{n}}} \right]^{\frac{1}{2}}$$ (Plane stress) $$K - Pl_{\mathrm{m}}\left[\frac{3}{lb^3 b_{\mathrm{n}}(1 - v)} \right]^{\frac{1}{2}} \quad (l/2 > 6)$$ (Plane strain)	l_{m} = length of moment arm b_{n} = sheet thickness in plane of crack, *i.e.* for grooved specimen v = Poisson's ratio	c
Tapered double cantilever beam	$$K = 2P\left(\frac{m_1}{bb_{\mathrm{n}}}\right)^{\frac{1}{2}}$$	$$m_1 = \frac{3a^2}{h^3} + \frac{1}{h}$$ specimen is contoured such that m_1 is constant h = height of specimen edge from fracture plane	d,e
Double cantilever beam	$$K = \frac{2P}{(bb_{\mathrm{n}})^{\frac{1}{2}}}\left[\frac{3a^2}{h^3} + \frac{1}{h} \right]^{\frac{1}{2}}$$		d,e
Compact tension	$$K = \frac{P}{bw^{\frac{1}{2}}}\left[29.6\left(\frac{a}{w}\right)^{\frac{1}{2}} - 185.5\left(\frac{a}{w}\right)^{\frac{3}{2}} + 655.7\left(\frac{a}{w}\right)^{\frac{5}{2}} - 1017\left(\frac{a}{w}\right)^{\frac{7}{2}} + 638.9\left(\frac{a}{w}\right)^{\frac{9}{2}} \right]$$		b

[a] W. F. Brown and J. E. Srawley, 'Plane-Strain Crack Toughness Testing of High Strength Metallic Materials', American Society for Testing and Materials, STP 410, 1966.
[b] 'Plane-Strain Fracture Toughness of Metallic Materials', American Society for Testing and Materials, E 399-78.
[c] R. J. Young and P. W. R. Beaumont, *J. Mater. Sci.*, 1976, **11**, 776.
[d] S. Mostovoy, P. B. Crosley and E. J. Ripling, *J. Mater.*, 1967, **2**, 661.
[e] P. B. Crosley, S. Mostovoy and E. J. Ripling, *Eng. Fract. Mech.*, 1971, **3**, 421.

given in Table 5. Again the parameter defining the fracture process G_{c} is generally a function of applied load, crack length and a geometrical factor.

Considerable work has been done over the years in measuring the fracture toughness K_{c} and the fracture energy G_{c} for polymers and other materials. Typical values of these parameters are listed in Table 6 for both polymers and other engineering materials (high values imply a tough material). It should be noted that the values for polymers are particularly sensitive to the rate and temperature of testing and so are not particularly reliable. However, Table 6 does give the reader the opportunity to see the approximate values of these parameters in relationship to those of other materials.

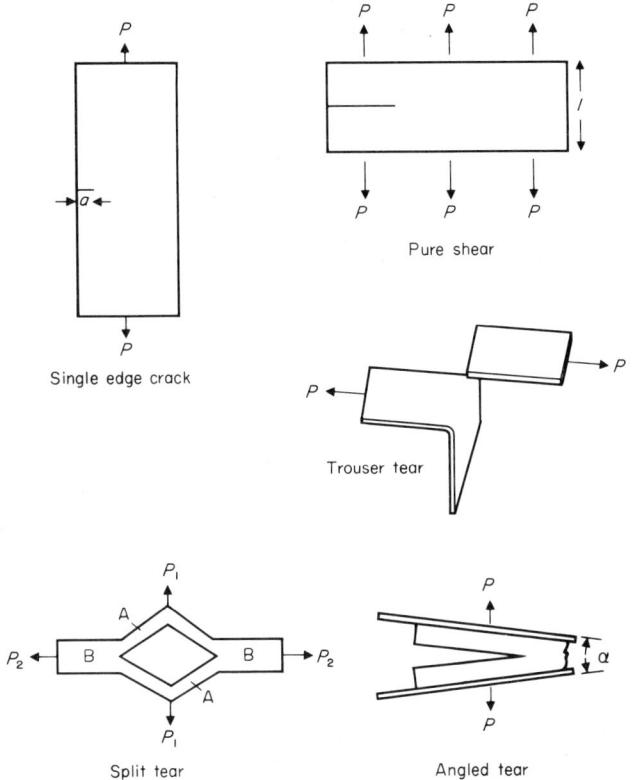

Figure 12 Schematic diagrams of various specimens used for fracture mechanics testing of elastomers.[3] P = applied load

Table 5 Expressions for the Fracture Energy G_c for Fracture Mechanics Specimens Used to Study Crack Growth in Flexible Polymers[3]

Geometry (see Figure 12)	*Expression for G_c*	*Comments*	*References*
Single-edge crack	$G_c = 2k_1 a W_c \qquad k_1 \simeq \pi/\lambda_c^{\frac{1}{2}}$	λ_c = extension ratio at onset of crack growth W_c = critical stored elastic strain energy density	a,b,c
Pure shear	$G_c = l W_c$	l = initial length	a
Trouser tear	$G_c = \dfrac{2P_c \lambda_c}{b} - 2w W_c$ if $\lambda_c \approx 1$, $G_c = \dfrac{2P_c}{b}$	P_c = load at onset of crack growth w = width of specimen arms λ_c = critical extension ratio in arms W_c = strain energy density in arms b = specimen thickness	a
Split tear	$G_c = \dfrac{(\lambda_c^A + \lambda_c^B)}{2b}$ $\times (\sqrt{P_{1c}^2 + P_{2c}^2} - P_{2c})$	λ_c^A and λ_c^B are critical extension ratios in regions A and B respectively	d
Angled tear	$G_c = \dfrac{2P_c}{b}\sin \alpha/2$		e

[a] R. S. Rivlin and A. G. Thomas, *J. Polym. Sci.*, 1953, **10**, 291.
[b] H. W. Greensmith, *J. Appl. Polym. Sci.*, 1963, **7**, 993.
[c] G. J. Lake, 'Conference Proceedings of the Physical Institute, The Yield Deformation and Fracture of Polymers, Cambridge, 1970'.
[d] G. J. Lake, P. B. Lindley and A. G. Thomas, in 'Fracture' ed. L. Averbach, Chapman and Hall, London, 1979.
[e] A. G. Thomas, *J. Appl. Polym. Sci.*, 1960, **3**, 168.

Table 6 Typical Values of the Fracture Energy G_c and the Fracture Toughness K_c for Various Materials[3]

Material	Young's modulus, E (GPa)	G_c (kJ m^{-2})	K_c (MN m$^{-\frac{3}{2}}$)
Rubber	0.001	13	—
Polyethylene	0.15	20	—
Polystyrene	3	0.4	1.1
High impact polystyrene	2.1	15.8	—
PMMA	2.5	0.5	1.1
Epoxy	2.8	0.1	0.5
Rubber-toughened epoxy	2.4	2	2.2
Glass-reinforced thermoset	7	7	7
Glass	70	0.007	0.7
Wood	2.1	0.12	0.5
Aluminum alloy	69	20	37
Steel — mild	210	12	50
Steel — alloy	210	107	150

15.3.3 Toughening Mechanisms

An increase in the toughness of a polymer is reflected by an improvement in the values of K_c and G_c. Original approaches to the analysis of the fracture of solids assumed that fracture involved only the creation of new surfaces.[3,61] However, it was soon realized that since measured values of G_c were well in excess of the surface energy of the material, significant amounts of energy were also dissipated through other processes such as localized plastic deformation in the vicinity of the crack.[3] In general, two mechanisms are responsible for this plastic deformation in rigid polymers; namely 'crazing'[3,66] and 'shear yielding'.[3] Crazing can take place in many brittle thermoplastics in a process whereby small crack-like entities form in a direction perpendicular to the tensile axis, and closer examination shows them to be cracks bridged by small fibrils[66] as shown in Figure 13. Shear yield

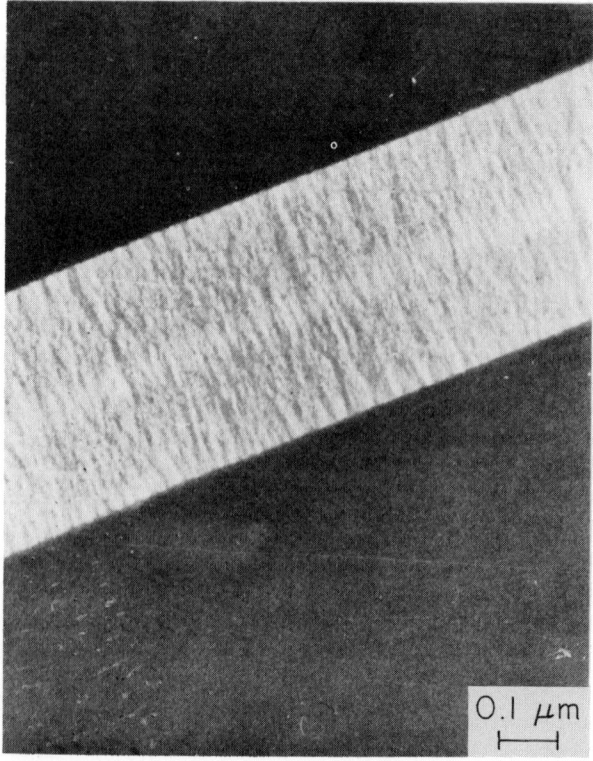

Figure 13 Transmission electron micrograph of the region near the centre of a craze in polystyrene showing craze fibrils (courtesy of Prof. E. J. Kramer)

occurs in tougher thermoplastics and also in rigid cross-linked thermosetting polymers.[3] It is characterized by regions of sheared polymer oriented approximately at 45° to the tensile or compression axis as shown in Figure 14.

Both crazing and shear yielding involve the absorption of energy and most methods of toughening polymers involve modifying the polymer such that more crazing and shear yielding take place. One of the best examples is 'rubber toughening'[67] whereby small (typically 1 μm diameter) rubber particles are dispersed in a brittle polymer. Figure 15 shows an electron micrograph of the dispersion of particles in a rubber-toughened sample of PMMA. The polymer is transparent since the refractive indices of the rubber and matrix are matched and the particles sizes are below the wavelength of light (0.3 μm). The effect of the addition of rubber upon the toughness of the material is shown in Figure 16: the elongation at break of the material increases from about 3% to over 40% with the addition of rubber. The Young's modulus of the rubber particles is several orders of magnitude less than that of the matrix. This leads to a stress concentration around the particles and shear yielding

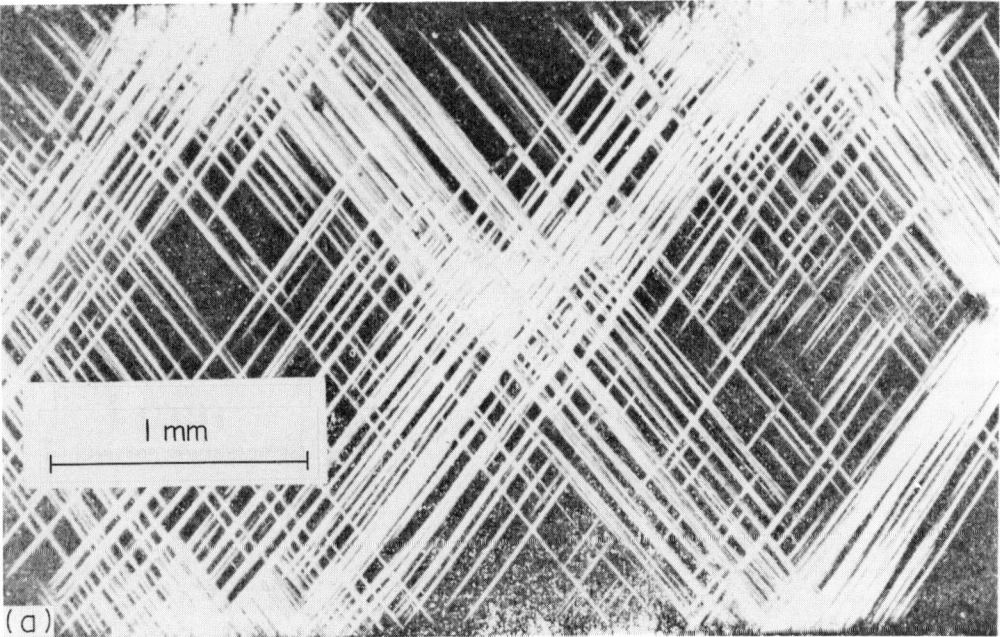

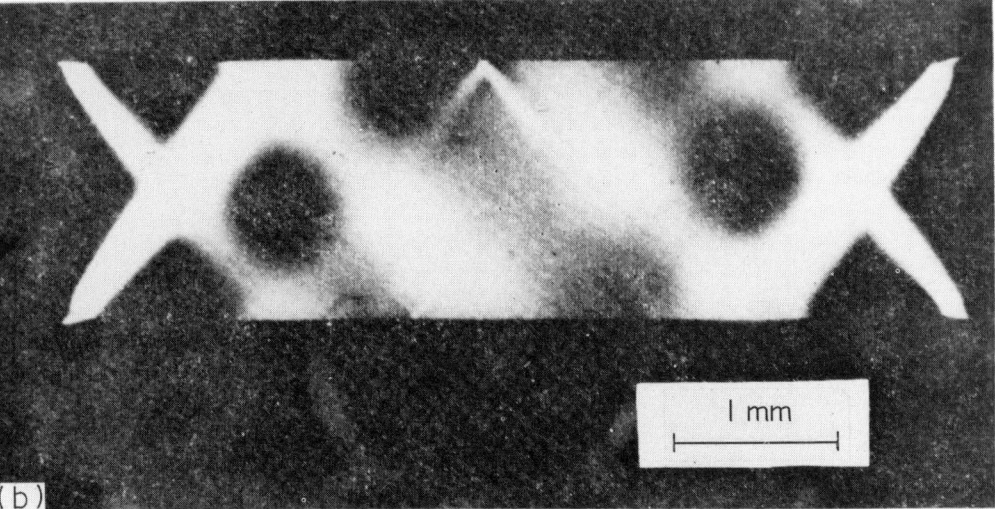

Figure 14 Shear yielding in glassy polymers. The micrographs are from polished sections of polymers deformed just past the yield point and viewed in polarized light: (a) polystyrene showing microshear bands; (b) PMMA showing diffuse shear zones[72]

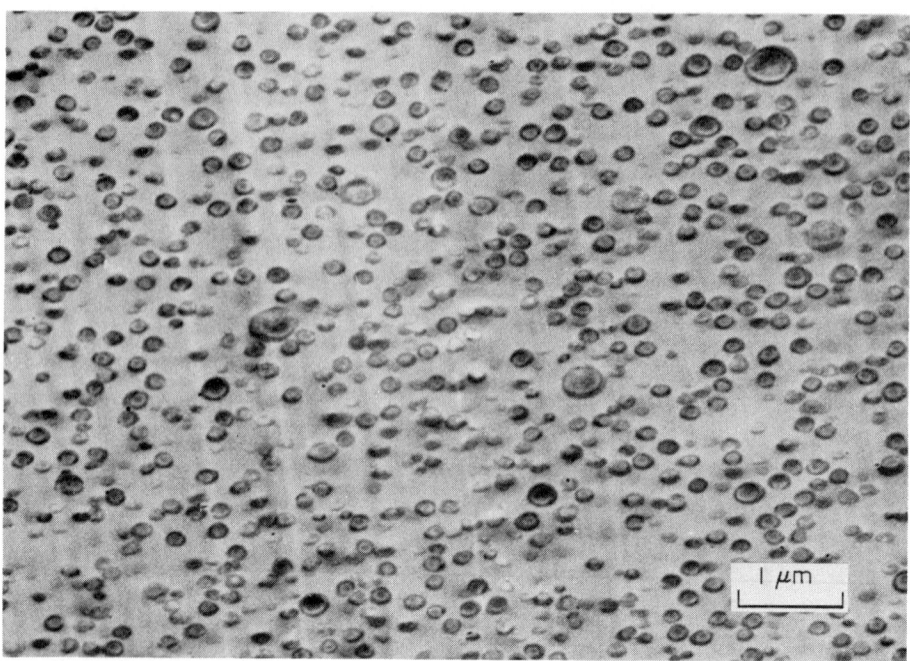

Figure 15 Transmission electron micrograph of a microtomed section of rubber-toughened PMMA showing multistage
rubber particles with core/shell structures (courtesy of D. J. Saunders); Magnification ∼15 000 ×

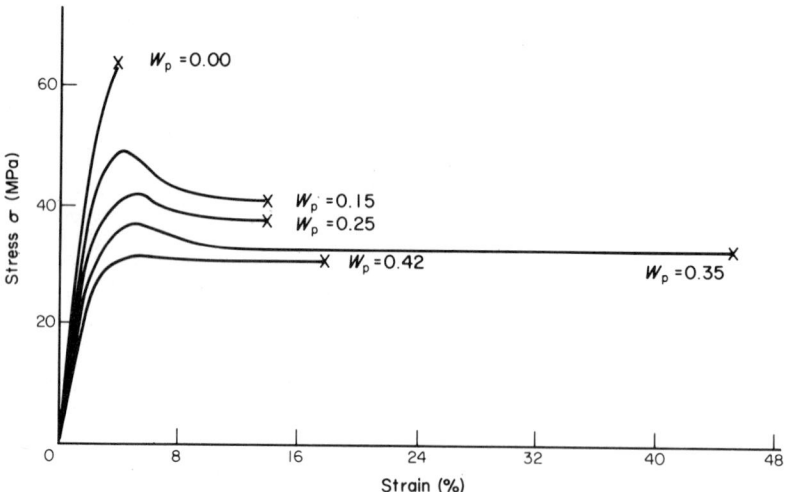

Figure 16 Stress/strain curves for a series of tensile specimens of rubber-toughened PMMA showing the effect of the weight
fraction of particles w_p upon deformation behaviour

or crazing throughout a large volume of the material rather than just at the crack tip. Hence the
polymer absorbs more energy during deformation and is toughened. Figure 17 shows an electron
micrograph of a section obtained from a deformed sample of high impact polystyrene. It can be
clearly seen that crazes have formed at the equators of the rubber particles and that they have
propagated from particle to particle. The best levels of toughening are obtained when there is a
strong bond between the particles and the matrix.[3,67] Many polymers are now toughened by
blending or copolymerizing with a rubber[3,67] and this method of toughening is now well established
for many thermoplastics, thermosets and even adhesives.

15.3.4 Factors Controlling Toughness

The toughness of polymers depends upon material variables, such as polymer structure and
orientation, and external variables, such as environment rate and temperature.[3] The dependence is

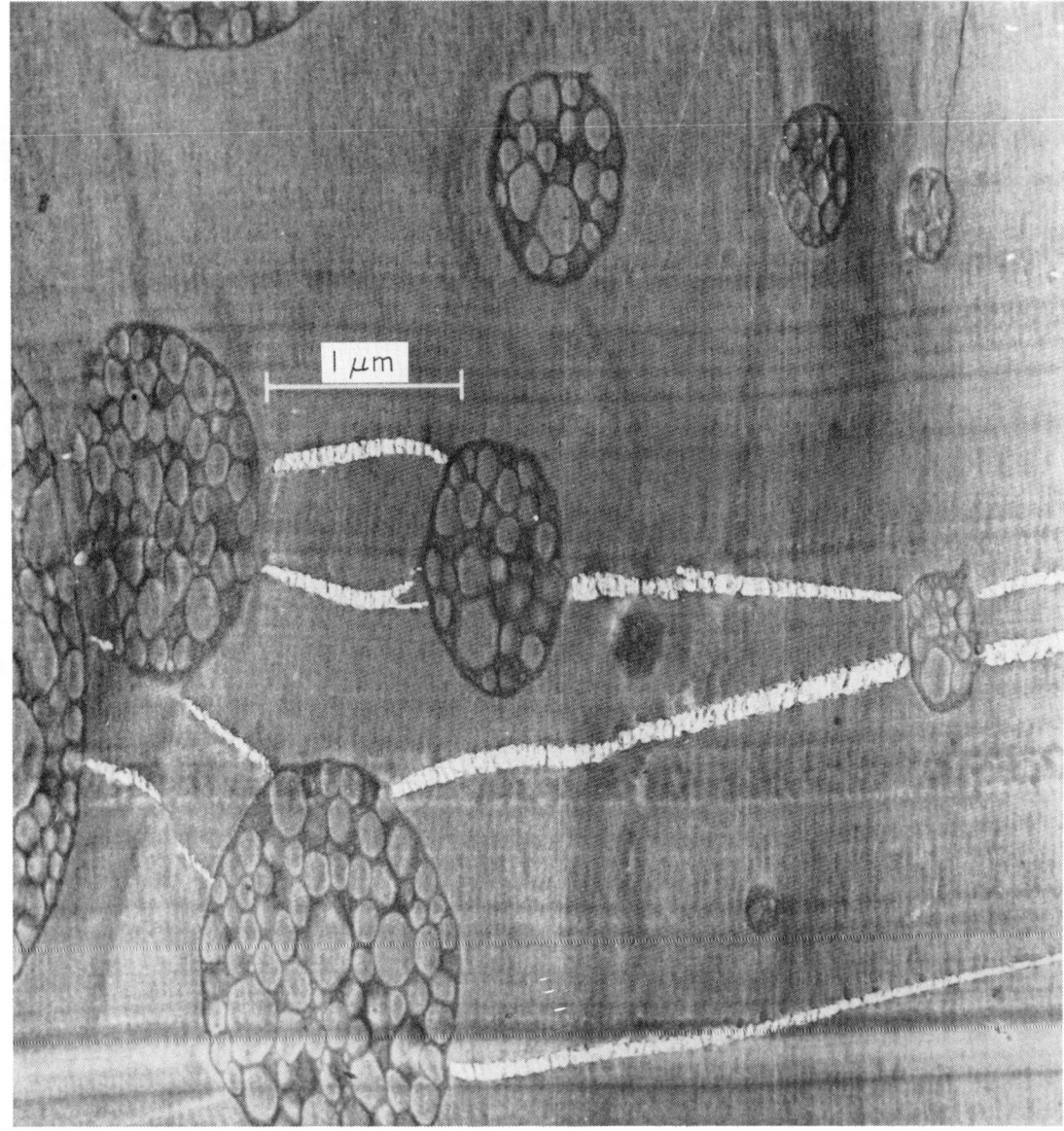

Figure 17 Transmission electron micrograph of a microtomed section of osmium tetroxide-stained high impact polystyrene[73] showing inclusions in the rubber particles and crazing

similar to that of strength described previously in Section 15.2.3. For example, the fracture energy of an elastomer, such as styrene–butadiene rubber, is found to depend upon both testing rate and temperature as shown in Figure 18. This three-dimensional plot shows that the fracture energy G_c increases with both increasing rate and decreasing temperaure.[3] In rubber-toughened systems the toughness of the polymer generally increases with an increase in volume fraction of rubber particles, as shown in Figure 19, although when the particles have a complex structure (*cf.* Figure 15) the morphology of the particles can also have an effect upon the toughening efficiency.[3]

Polymers are prone to premature failure during cyclic deformation or fatigue loading whereby cracks propagate at lower stresses (or values of K_c and G_c) than during monotonic loading.[68] The process of fatigue, which is found for many other materials, takes place by the accumulation of damage at the crack tip. Extensive studies have taken place to characterize the fatigue behaviour of many types of polymers.[3,68]

One other area where problems are often encountered with polymer toughness is in high strain rate deformation or impact.[3,68–70] Materials which are apparently tough at low rates of testing become quite brittle when subjected to impact loading. Significant advances have been made in

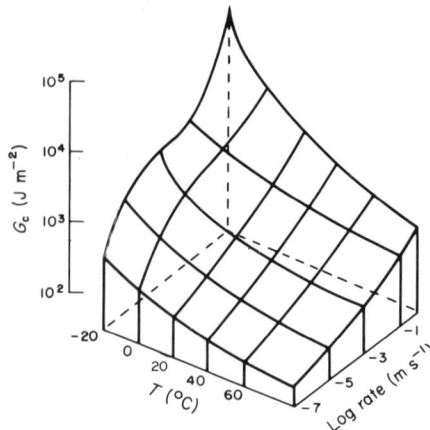

Figure 18 Variation of fracture energy G_c with testing rate and temperature for styrene–butadiene rubber[74]

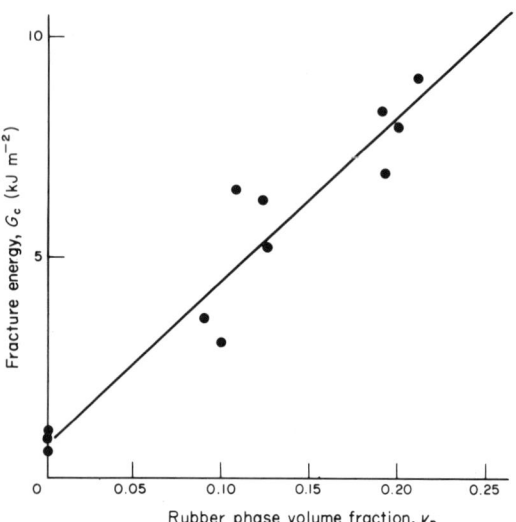

Figure 19 Relationship between the fracture energy G_c and rubber phase volume fraction in a rubber-toughened epoxy resin[75]

recent years both in producing impact-resistant polymers[67] and in understanding what is measured during an impact test.[68–70]

15.4 CONCLUSIONS

It is clear that our understanding of the factors controlling the strength and toughness of solid polymers has led to improvements in material properties and so to polymers being accepted as structural engineering materials in their own right. Polymer fibres can now be produced with extremely high levels of stiffness and strength which rank them among the strongest materials known.

The lack of toughness shown by many polymers can often be overcome by the incorporation of rubber particles through blending or copolymerization, leading to materials with relatively high levels of toughness even during impact loading. Even so, there is still significant scope for polymer chemists, materials scientists and engineers to collaborate further to produce polymers with even more impressive mechanical properties.

15.5 REFERENCES

1. A. Kelly and N. H. MacMillan, 'Strong Solids', 3rd edn., Clarendon Press, Oxford, 1986.
2. R. J. Young, 'Introduction to Polymers', Chapman and Hall, London, 1981.
3. A. J. Kinloch and R. J. Young, 'Fracture Behaviour of Polymers', Applied Science, London, 1983.
4. J. P. Berry, *J. Polym. Sci.*, 1961, **50**, 107.
5. J. P. Berry, in 'Fracture VII', ed. H. Liebowitz, Academic Press, New York, 1971.
6. A. Kelly and N. H. MacMillan, 'Strong Solids', 3rd edn., Oxford Science Publications, 1986.
7. A. Odajima and T. Maeda, *J. Polym. Sci., Part C*, 1966, **15**, 55.
8. J. H. de Boer, *Trans. Faraday Soc.*, 1936, **32**, 10.
9. H. H. Kausch, 'Polymer Fracture', Springer-Verlag, Berlin, Heidelberg, 1978.
10. F. C. Frank, *Proc. R. Soc. London, Ser. A*, 1970, **319**, 127.
11. L. Holliday and J. W. White, *Pure Appl. Chem.*, 1971, **26**, 545.
12. L. R. G. Treloar, *Polymer*, 1960, **1**, 290.
13. T. Shimanouchi, M. Asahina and S. Enomoto, *J. Polym. Sci.*, 1962, **59**, 93.
14. D. S. Boudreaux, *J. Polym. Sci., Polym. Phys. Ed.*, 1973, **11**, 1285.
15. B. Crist, M. A. Ratner, A. L. Brower and J. R. Sabin, *J. Appl. Phys.*, 1979, **50**, 6047.
16. R. J. Young, in 'Developments in Oriented Polymers — 2', ed. I. M. Ward, Applied Science, London, 1987.
17. G. Wegner, *Pure Appl. Chem.*, 1977, **49**, 443.
18. D. Bloor and R. R. Chance (eds.), 'Polydiacetylenes', NATO ASI Series E-102, Martinus Northold, Dordrecht, 1985.
19. R. H. Baughman, H. Gleiter and N. Sendfeld, *J. Polym. Sci., Polym. Phys. Ed.*, 1975, **13**, 1871.
20. C. Galiotis and R. J. Young, *Polymer*, 1983, **24**, 1023.
21. C. Galiotis, R. J. Young and D. N. Batchelder, *J. Polym. Sci., Polym. Phys. Ed.*, 1983, **21**, 2483.
22. D. M. Marsh, in 'Fracture in Solids', ed. D. D. Crucker and J. J. Gilman, Interscience, New York, 1963.
23. P. A. Apgar and K. C. Yee, *Acta Crystallogr., Sect. B*, 1978, **34**, 957.
24. J. M. Andrews and I. M. Ward, *J. Mater. Sci.*, 1970, **5**, 411.
25. G. Capaccio and I. M. Ward, *Polym. Eng. Sci.*, 1975, **15**, 219.
26. G. Capaccio, G. Gibson and I. M. Ward, in 'Ultra-High Modulus Polymers', ed. A. Ciferri and I. M. Ward, Applied Science, London, 1979.
27. T. Kanamoto, A. Tsumita, K. Tanaka, M. Takeda and R. S. Porter, *Polym. J.*, 1983, **15**, 327.
28. A. E. Zachariades and T. Kanamoto, *J. Appl. Polym. Sci.*, 1988, **35**, 1265.
29. W. T. Mead, C. R. Desper and R. S. Porter, *J. Polym. Sci., Polym. Phys. Ed.*, 1979, **17**, 859.
30. A. E. Zachariades, W. T. Mead and R. S. Porter, in 'Ultra-High Modulus Polymers', ed. A. Ciferri and I. M. Ward, Applied Science, London, 1979.
31. R. S. Porter, M. Daniels, M. P. C. Watts, J. R. C. Pereira, S. J. DeTeresa and A. E. Zachariades, *J. Mater. Sci.*, 1980, **16**, 1134.
32. B. Kalb and A. J. Pennings, *J. Mater. Sci.*, 1980, **15**, 2584.
33. A. Zwijnenburg and A. J. Pennings, *Colloid Polym. Sci.*, 1975, **253**, 452.
34. P. J. Barham and A. Keller, *J. Mater. Sci.*, 1985, **20**, 2281.
35. P. J. Lemstra and R. Kirschbaum, in 'Developments in Oriented Polymers — 2', ed. I. M. Ward, Applied Science, London, 1987, p. 39.
36. P. J. Lemstra, N. A. J. M. van Aerle and C. W. M. Bastiaansen, *Polym. J.*, 1987, **19**, 85.
37. J. Smook, W. Hamersma and A. J. Pennings, *J. Mater. Sci.*, 1984, **19**, 1359.
38. M. G. Dobb and D. J. Johnson, in 'Developments in Oriented Polymers — 2', ed. I. M. Ward, Applied Science, London, 1987.
39. J. R. Schaefgen, in 'The Strength and Stiffness of Polymers', ed., A. E. Zachariades and R. S. Porter, Dekker, New York, 1983.
40. J. Blackwell and A. Biswas, in 'Developments in Oriented Polymers — 2', ed. I. M. Ward, Applied Science, London, 1987.
41. S. R. Allen, A. G. Filippov, R. J. Farris and E. L. Thomas, *J. Appl. Polym. Sci.*, 1981, **26**, 291.
42. J. A. Odell, A. Keller, E. D. T. Atkins and M. J. Miles, *J. Mater. Sci.*, 1981, **16**, 3309.
43. S. R. Allen, A. G. Filippov, R. J. Farris and E. L. Thomas, in 'The Strength and Stiffness of Polymers', ed. A. E. Zachariades and R. S. Porter, Dekker, New York, 1983.
44. R. J. Day, I. M. Robinson, M. Zakikhani and R. J. Young, *Polymer*, 1987, **28**, 1833.
45. S. J. Krause, T. B. Haddock, P. G. Lenhert, W.-F. Hwang, G. E. Price, T. E. Helminiak, J. F. O'Brien and W. W. Adams, *Polymer*, 1988, **29**, 1354.
46. E. H. Andrews and P. E. Reed, *Adv. Polym. Sci.*, 1978, **27**, 1.
47. P. I. Vincent, *Polymer*, 1972, **13**, 558.
48. J. G. Williams, *Metal Sci.*, 1980, **14**, 344.
49. J. R. Martin, J. F. Johnson and A. R. Cooper, *J. Macromol. Sci., Rev. Macromol. Chem.*, 1972, **8**, 57.
50. P. I. Vincent, *Polymer*, 1960, **1**, 425.
51. H. W. McCormick, F. M. Brower and L. Kin, *J. Polym. Sci.*, 1959, **39**, 87.
52. J. H. Golden, B. L. Hammant and E. A. Hazell, *J. Polym. Sci., Part A*, 1964, **2**, 4787.
53. P. J. Flory, *J. Am. Chem. Soc.*, 1945, **67**, 2048.
54. A. N. Gent and A. G. Thomas, *J. Polym. Sci., Part A-2*, 1972, **10**, 571.
55. D. J. Turner, *Polymer*, 1982, **23**, 626.
56. P. Beardmore and T. L. Johnston, *Philos. Mag.*, 1971, **23**, 1119.
57. R. P. Kambour, *J. Polym. Sci., Macromol Rev.*, 1973, **7**, 1.
58. E. J. Kramer, in 'Developments in Polymer Fracture — 1', ed. E. H. Andrews, Applied Science, London, 1979.
59. K. D. Pae and S. K. Bhateja, *J. Macromol. Sci., Rev. Macromol. Chem.*, 1975, **13**, 1.
60. R. A. Duckett, *J. Mater. Sci.*, 1980, **15**, 2471.
61. A. A. Griffith, *Philos. Trans. R. Soc. London, Ser. A*, 1920, **221**, 163.
62. J. G. Williams, 'Fracture Mechanics of Polymers', Ellis Horwood, Chichester, 1984.
63. E. Orowan, *Rep. Prog. Phys.*, 1948, **12**, 185.

64. R. S. Rivlin and A. G. Thomas, *J. Polym. Sci.*, 1953, **10**, 291.
65. E. H. Andrews, *J. Mater. Sci.*, 1974, **9**, 887.
66. H. H. Kausch, *Adv. Polym. Sci.*, 1983, **52/53**.
67. C. B. Bucknall, 'Toughened Plastics', Applied Science, London, 1977.
68. R. W. Hertzberg and J. A. Manson, 'Fatigue of Engineering Plastics', Academic Press, New York, 1980.
69. P. E. Reed, in 'Developments in Polymer Fracture — 1', ed. E. H. Andrews, Applied Science, London, 1979.
70. H. R. Brown, *J. Mater. Sci.*, 1973, **8**, 941.
71. E. Plati and J. G. Williams, *Polym. Eng. Sci.*, 1975, **15**, 470.
72. P. B. Bowdon, *Philos. Mag.*, 1970, **22**, 455.
73. R. P. Kambour and D. R. Russell, *Polymer*, 1971, **12**, 237.
74. H. W. Greensmith, L. Mullins and A. G. Thomas, *Trans. Soc. Rheol.*, 1960, **4**, 179.
75. C. Bucknall and T. Yoshii, *Br. Polym. J.*, 1978, **10**, 53.

16

Dynamic-mechanical Properties

PER GRADIN, PAUL G. HOWGATE and RAGNAR SELDÉN
PGI Sundsvall, Sweden
and
RICHARD A. BROWN
University of Manchester, UK

16.1 INTRODUCTION

Knowledge of dynamic-mechanical properties is essential for the design of engineering components, such as antivibration mounts and sound-deadening devices. In addition, studies of these properties have added greatly to the understanding of molecular mobility in polymers.

In this section we introduce the concepts and quantities used in the description of elastic and viscoelastic materials (both isotropic and anisotropic) and their dynamic-mechanical properties. Experimental methods are briefly described in Section 16.2, and the dynamic-mechanical properties of glassy, crystalline and rubbery polymeric materials are described in Sections 16.3 to 16.5.

16.1.1 Elastic Isotropic Materials

For many polymeric materials, their mechanical properties are isotropic; in other words, their behaviour is independent of the direction of measurement. Rubbers and glasses having chains in random orientation fall into this category. Crystalline polymers composed of a random array of crystalline units (*i.e.* polycrystalline polymers) are also isotropic, as are short-fibre reinforced plastics in which the fibres are randomly distributed within the material.[1,2]

In the case of a linearly elastic material, the components of strain are proportional to the applied stress. For a particular type of deformation the ratio of stress to strain defines an elastic modulus (which is a measure of stiffness) and the reciprocal of the modulus is the compliance.

The state of stress or strain within a test piece may be defined by considering an infinitesimally small cubic volume element with edges parallel to orthogonal axes x, y and z. The components of stress (defined as force per unit area) acting normally on planes perpendicular to the directions x, y and z are customarily represented by σ_{xx}, σ_{yy} and σ_{zz} respectively. The stresses acting on planes perpendicular to x and in the direction of z are the shear stress components $\sigma_{yz} = \sigma_{zy}$ and $\sigma_{xy} = \sigma_{yx}$. The stress-induced displacement of a point (x,y,z) within a material can be resolved into components u, v and w parallel to the axes x, y and z. The normal strains are defined by

$$\varepsilon_{xx} = \partial u/\partial x; \quad \varepsilon_{yy} = \partial v/\partial y; \quad \varepsilon_{zz} = \partial w/\partial z \tag{1}$$

The so-called 'engineering' shear strains are

$$\gamma_{yz} = \gamma_{zy} = (\partial v/\partial z) + (\partial w/\partial y) \tag{2a}$$

$$\gamma_{xz} = \gamma_{zx} = (\partial u/\partial z) + (\partial w/\partial x) \tag{2b}$$

$$\gamma_{xy} = \gamma_{yx} = (\partial u/\partial y) + (\partial v/\partial x) \tag{2c}$$

The orthogonal axes x, y, and z are usually taken to be parallel to length, width and thickness for test strips with rectangular cross-sections.

The elastic properties of isotropic materials may be described in terms of just two elastic moduli. This can best be explained by consideration of some simple types of deformation. Figure 1(a) shows the simple shear deformation in which equal tangential forces are applied in opposite directions to the parallel faces of a test piece (of constant volume). The shear stress σ_{yx} is the force per unit area and, from Hooke's law, the shear modulus G is given by

$$G = \sigma_{yz}/\gamma_{yx} \tag{3}$$

where the shear strain $\gamma_{yx} = \tan \alpha$. For small strains

$$G = \sigma_{yx}/\alpha \tag{4}$$

where α is in radians.

The shear modulus can be readily obtained from a torsional deformation of a test piece of constant volume.[3] Figure 1(b) shows a cylindrical rod of length L and circular cross-section radius a under the action of equal and opposite torque Γ applied to its ends. The tangential shear stress σ_t acting on the planes of the cross-section at radius r is given by

$$\sigma_t = \frac{G\phi r}{L} \tag{5}$$

where ϕ is the angle of twist. At equilibrium the applied torque is given by

$$\Gamma = \int_0^a 2\pi r^2 \sigma_t \, dr \tag{6}$$

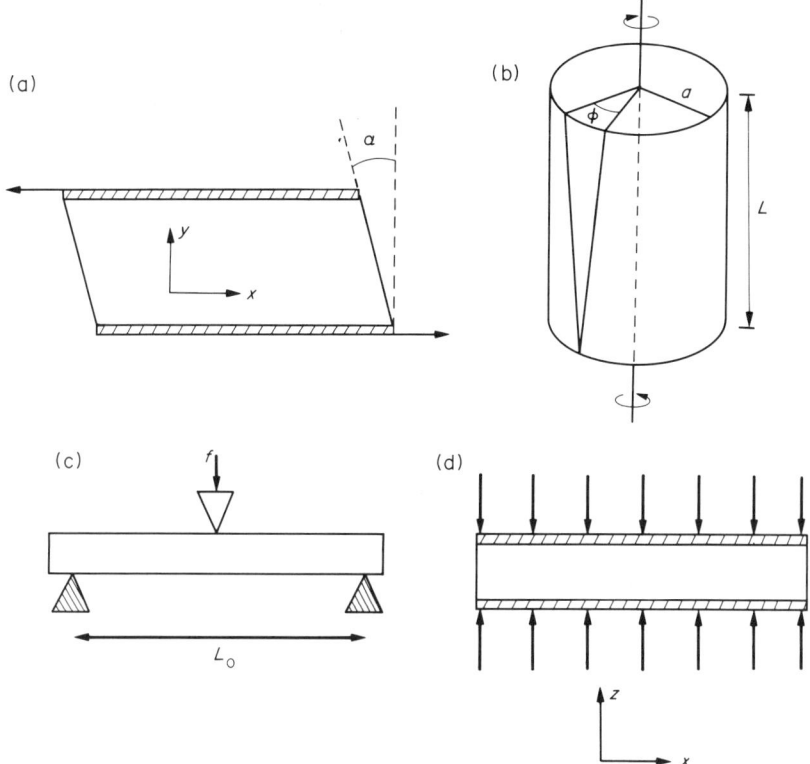

Figure 1 Schematic representations of some simple deformations: (a) shear, (b) torsion, (c) bending, (d) compression

On substitution for σ_t using equation (5)

$$\Gamma \ - \ G(\pi\phi a^4/2L) \tag{7}$$

from which G may be determined, having measured ϕ and knowing the dimensions of the specimen and the applied torque.

In contrast to the shear modulus, the bulk modulus K is best obtained from a deformation in which the shape of the test piece remains the same but the volume is varied. Consider a cubic test piece in which the normal stresses acting on each face are equal to the applied hydrostatic pressure (p) and there are no shear stresses. If the test piece of initial volume V_0 undergoes a volume change ΔV, the volume strain is

$$\Delta V/V_0 \ = \ \varepsilon_{xx} \ + \ \varepsilon_{yy} \ + \ \varepsilon_{zz} \tag{8}$$

and

$$K \ = \ p(\Delta V/V_0)^{-1} \tag{9}$$

One of the simplest deformations to study experimentally is simple extension which involves the longitudinal extension of a sample (whose length is usually much greater than its width or thickness). This type of deformation, however, induces both shear and volume deformations on the specimen. The tensile modulus E is obtained from

$$E \ = \ \sigma_{xx}/\varepsilon_{xx} \tag{10}$$

where σ_{xx} is the applied longitudinal force per unit cross-sectional area, and the longitudinal strain ε_{xx} is $\Delta l/l$ where l and Δl are the original length and change in length respectively (where the specimen is assumed to be aligned in the x direction).

Poisson's ratio v can also be defined in such an experiment, it being the ratio of the lateral strain (in the y or z direction) to the longitudinal strain

$$v \ = \ -\varepsilon_{yy}/\varepsilon_{xx} \ = \ -\varepsilon_{zz}/\varepsilon_{xx} \tag{11}$$

Longitudinal tensile (and compression) strains are also induced when a thin long beam is forced into flexure, and this may therefore also be used to determine E. In one such test a beam of rectangular cross-section, which is supported at two points separated by a distance L_0, is subjected to a force f at its mid-point (see Figure 1c). The resulting deflection γ at the centre is measured, from which E can be calculated

$$E = fL_0^3/4xy^3\gamma \tag{12}$$

where x and y are the width and thickness of the specimen respectively.

For linearly elastic materials

$$E = 9KG/(3K + G) \tag{13}$$

and

$$E = 2G(1+v) \tag{14}$$

From these relationships and equivalent relationships it follows that any two of the constants G, K, E and v are, in principle, sufficient to characterize the elasticity. At low levels of strain, glassy, crystalline and crosslinked elastomers will exhibit the necessary linear behaviour for these relationships to hold provided the conditions are such that the materials are purely elastic.

When forces are applied normally to a thin sheet with large lateral dimensions, as illustrated in Figure 1(d), the lateral strains will be negligible. The ratio of stress to strain ($\sigma_{zz}/\varepsilon_{zz}$) is the longitudinal modulus L. The relationship to the moduli K and G is

$$L = K + 4/(3G) \tag{15}$$

The modulus L governs the velocity of longitudinal ultrasonic waves in an elastic material.

Instead of G, K, E and v, the elastic properties of isotropic materials may be specified in terms of the various compliances such as: shear compliance, $J = G^{-1}$; tensile compliance, $D = E^{-1}$; and compressibility, $B = K^{-1}$.

16.1.2 Elastic Anisotropic Materials

The elastic properties of an anisotropic solid may be defined in terms of the generalized form of Hooke's law

$$\sigma_{ij} = \sum_k \sum_l C_{ijkl}\varepsilon_{kl} \tag{16}$$

or

$$\varepsilon_{ij} = \sum_k \sum_l S_{ijkl}\sigma_{kl} \tag{17}$$

where the subscripts, i, j, k and l each take on values of 1, 2 or 3 corresponding to the cartesian set of axes x_1, x_2 and x_3 used to define the location of the material. It is usual to choose axes to match the shape symmetry of the specimen. The terms σ_{ij} and ε_{kl} are the tensor components of the stress and strain respectively, and C_{ijkl} and S_{ijkl} are the corresponding tensor components of the elastic modulus and compliance.

Equations (16) and (17) are more conventionally presented in the contracted (engineering) forms

$$\sigma_p = \sum_q C_{pq}\varepsilon_q \tag{18}$$

and

$$\varepsilon_p = \sum_q S_{pq}\sigma_q \tag{19}$$

where p and q take values 1 to 6, σ_1, σ_2 and σ_3 are the normal stresses, σ_4, σ_5 and σ_6 the shear stresses, ε_1, ε_2 and ε_3 the extensional strains and ε_4, ε_5 and ε_6 the (engineering) shear strains. Relationships (18) and (19) represent six equations, one for each σ_p and one for each ε_p. The six equations can be written in matrix form and the moduli and compliances can be related by the matrix inversion $[C_{pq}] = [S_{pq}]^{-1}$.

When the test piece is in static equilibrium $C_{pq} = C_{qp}$ and $S_{pq} = S_{qp}$. These relationships reduce the number of independent moduli or compliances from 36 to 21 for the most general case. Further reductions result because of material symmetry. For materials having orthorhombic symmetry the independent constants reduce to nine. In matrix form we then have

$$
\begin{bmatrix}
C_{11} & C_{12} & C_{13} & 0 & 0 & 0 \\
C_{12} & C_{22} & C_{23} & 0 & 0 & 0 \\
C_{13} & C_{23} & C_{33} & 0 & 0 & 0 \\
0 & 0 & 0 & C_{44} & 0 & 0 \\
0 & 0 & 0 & 0 & C_{55} & 0 \\
0 & 0 & 0 & 0 & 0 & C_{66}
\end{bmatrix}
\tag{20}
$$

where the subscripts 1, 2 and 3 indicate the three two-fold symmetry axes; an equivalent matrix can be written for compliance. For unidirectionally reinforced or unidirectionally drawn specimens which are isotropic in the transverse direction, the number of independent elastic constants reduces to five. If we take direction 1 to correspond to the main (longitudinal) symmetry axis of the test piece, the stiffness and compliance matrices become

$$
\begin{bmatrix}
C_{11} & C_{12} & C_{13} & 0 & 0 & 0 \\
C_{12} & C_{22} & C_{23} & 0 & 0 & 0 \\
C_{12} & C_{23} & C_{22} & 0 & 0 & 0 \\
0 & 0 & 0 & \tfrac{1}{2}(C_{22} - C_{23}) & 0 & 0 \\
0 & 0 & 0 & 0 & C_{66} & 0 \\
0 & 0 & 0 & 0 & 0 & C_{66}
\end{bmatrix}
\tag{21}
$$

and

$$
\begin{bmatrix}
S_{11} & S_{12} & S_{12} & 0 & 0 & 0 \\
S_{12} & S_{22} & S_{23} & 0 & 0 & 0 \\
S_{12} & S_{23} & S_{22} & 0 & 0 & 0 \\
0 & 0 & 0 & 2(S_{22} - S_{23}) & 0 & 0 \\
0 & 0 & 0 & 0 & S_{66} & 0 \\
0 & 0 & 0 & 0 & 0 & S_{66}
\end{bmatrix}
\tag{22}
$$

By expanding the summations in equation (18) and equating all the strain components except one to zero, the physical significance of the stiffness components C_{pq} can be explored. By an equivalent procedure the physical significance of the compliance components S_{pq} can be investigated. It is found that C_{pq} represents the ratio of stress component σ_p to a given strain component ε_q when all other strain components are zero. Similarly it is found that S_{pq} represents the ratio of strain component ε_p to a given stress component σ_q when all other stress components are zero. It follows that for transversely isotropic materials

$$
S_{11} = 1/E_1, \quad S_{22} = 1/E_2, \quad S_{12} = -v_{12}/E_1, \quad S_{23} = -v_{23}/E_2 \tag{23a}
$$
$$
2(S_{22} - S_{23})^{-1} = \tfrac{1}{2}E_2(1 + v_{23})^{-1} = G_{23} = \tfrac{1}{2}(C_{22} - C_{23}) \tag{23b}
$$
$$
(S_{66})^{-1} = G_{12} = C_{66}, \quad C_{11} = L_1, \quad C_{22} = L_2 \tag{23c}
$$

In the above equations, E_1 is the tensile modulus measured in direction 1, E_2 is the transverse tensile modulus, G_{12} and G_{23} are shear moduli corresponding to shear stresses in directions 2 and 3 on planes perpendicular to 1 and 2 respectively, and $v_{12} = -\varepsilon_2/\varepsilon_1$ and $v_{23} = -\varepsilon_3/\varepsilon_2$ are Poisson's ratios for successive loading in directions 1 and 2. L_1 and L_2 are longitudinal moduli in directions 1 and 2 respectively. They may be obtained from the velocity of longitudinal ultrasonic waves in an infinite elastic medium when the lateral strains are zero. Alternatively, they may be obtained from the type of test outlined in Figure 1(c). The constants C_{12} and C_{23} are the ratios of lateral stress to longitudinal strain.

For isotropic materials only two independent compliances (S_{11} and S_{12}) or stiffnesses (C_{11} and C_{22}) are required, from equation (18) and (19); *i.e.*

$$S_{11} = E^{-1}, \quad S_{12} = -vE^{-1} \tag{24a}$$

$$\tfrac{1}{2}(S_{11} - S_{12})^{-1} = \tfrac{1}{2}(C_{11} - C_{12}) = G, \quad C_{11} = L \tag{24b}$$

$$\tfrac{1}{3}(S_{11} + 2S_{12})^{-1} = \tfrac{1}{3}(C_{11} + 2C_{12}) = K \tag{24c}$$

Equations (13) to (15) follow directly from the above relationships.

16.1.3 Viscoelastic Materials

Polymers are viscoelastic materials; in other words, their mechanical properties have both elastic and viscous components. Their response to an applied stress depends on the time-scale of the experiment. The phenomenon of creep, which is the increase in strain with time at constant applied stress, and stress relaxation, which is the decrease in stress at constant strain, are examples of time-dependent properties.

If an ideal elastic body is subjected to a shear stress σ_0 for a period of time and then the stress is removed, it will follow the strain pattern shown in Figure 2(a); in this case the shear strain $\varepsilon_0 = J\sigma_0$, where J is the compliance. If, on the other hand, the material is viscoelastic the strain will in general be an increasing function of time until the stress is removed, as shown in Figure 2(b). A creep compliance can be defined as

$$J(t) = \varepsilon(t)/\sigma_0 \tag{25}$$

Figure 3 shows a plot of creep compliance as a function of time for a non-crystalline viscoelastic material. For very short times $J(t)$ is in the region of 10^{-9} Pa^{-1} and the material has the properties of a glassy solid. At somewhat longer times the compliance is very time-dependent (*i.e.* the longer the time-scale of the experiment the greater the compliance). On moving to still longer times $J(t)$ reaches a plateau where it remains approximately constant and has a value in the region of 10^{-5} Pa^{-1}, and the material behaves as a rubbery solid. For very long times (provided the material is not crosslinked) $J(t)$ again starts to increase due to liquid-like flow behaviour.

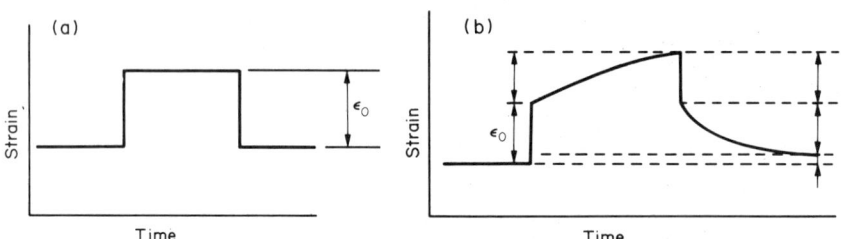

Figure 2 Response to the deformation of (a) an elastic material and (b) a linear viscoelastic material

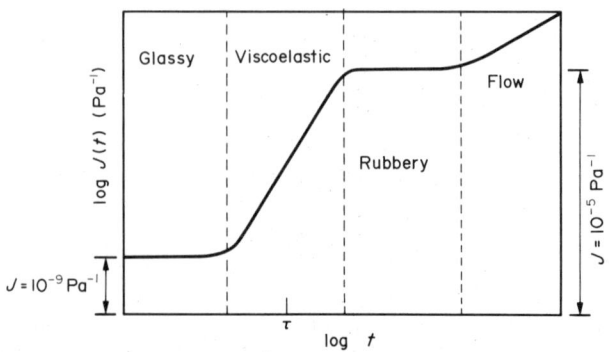

Figure 3 A plot of the creep compliance J(t) as a function of time; the retardation time τ is shown

Whether a material behaves as a glassy solid, a viscoelastic material, a rubber or a fluid depends on the time-scale of the experiment relative to a characteristic time constant (or more correctly a spectrum of time constants) of the material. For creep, the time constant(s) is called the retardation time, τ; its location is roughly indicated in Figure 3.

The retardation time is governed by the rate of molecular motion of the polymer chains, which in turn depends on molecular structure and the temperature and pressure of the experiment. For a material behaving as a typical rubber at ambient temperature, τ is very much smaller than 1 s and therefore the material appears soft. For a glass on the other hand, τ is very much longer than 1 s and the material appears hard. Lowering the temperature or increasing the pressure increases τ and for a particular time-scale can change a material from a rubber to a glass.

In an analogous fashion, the response of a viscoelastic material subjected to a constant shear strain can be described in terms of a stress relaxation modulus

$$G(t) = \sigma(t)/\varepsilon \tag{26}$$

Figure 4 shows a schematic plot of $G(t)$ as a function of time for a non-crystalline viscoelastic polymer.

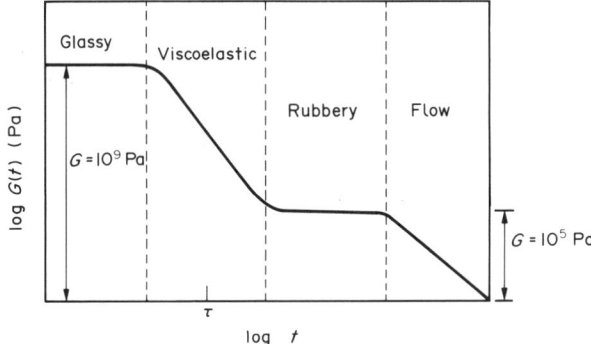

Figure 4 A plot of the stress relaxation modulus $G(t)$ as a function of time; the retardation time τ is shown

16.1.4 Linear Viscoelasticity

Consider first an elastic solid obeying Hooke's law subjected to a series of shear stresses $\sigma_0, \sigma_1, \sigma_2, \ldots, \sigma_{n-1}, \sigma_n$ at times $t = 0, t_1, t_2, \ldots, t_{n-1}, t_n$. The shear strain at the end of the loading history is given by

$$\varepsilon = J\sigma_0 + J[\sigma_1 - \sigma_0] + J[\sigma_2 - \sigma_1] + \ldots + J[\sigma_n - \sigma_{n-1}] = J\sigma_n \tag{27}$$

For viscoelastic materials, however, the strain produced by each of these successive stresses will show a time dependence. Materials are described as linear viscoelastic if under such circumstances they obey the following equation due to Boltzmann

$$\varepsilon(t) = J(t)[\sigma_0] + J(t - t_1)[\sigma_1 - \sigma_0] + J(t - t_2)[\sigma_2 - \sigma_1] + \ldots J(t - t_n)[\sigma_n - \sigma_{n-1}] \tag{28}$$

where $J(t)$ is the creep compliance function. This is a result of the Boltzmann superposition principle which states that the response of a material to a given load is independent of the response of the material to any load which is already on the material. The Boltzmann principle is expressed in integral form for the case of a continuum strain history

$$\varepsilon(t) = \int_0^t J(t - \theta) \, d\theta \tag{29}$$

where θ is time during the strain history (and is taken to start at zero), and the present time t is considered a constant in the integration.

Equations analogous to those for $\varepsilon(t)$ above, can be written for the time-dependent stress relaxation modulus, $G_r(t)$, of a linearly viscoelastic body subjected to successive shear strains,

$\varepsilon_0, \varepsilon_1, \varepsilon_2, \ldots, \varepsilon_{n-1}, \varepsilon_n$

$$\sigma(t) = G_r(t)[\varepsilon_0] + G_r(t - t_1)[\varepsilon_1 - \varepsilon_0] + G_r(t - t_2)[\varepsilon_2 - \varepsilon_1] + G_r(t - t_n)[\varepsilon_n - \varepsilon_{n-1}]$$

(30)

For continuous loading

$$\sigma(t) = \int_0^t G_r(t - \theta)(d\varepsilon(\theta)/d\theta)d\theta$$

(31)

If a material exhibits linear viscoelastic behaviour as defined by equations (28) to (31) its behaviour can be described by sets of linear differential equations. The developments in linear viscoelasticity prior to 1943, from the standpoint of the integral equations of Boltzmann[4] and the various differential formulations, have been comprehensively reviewed by Leaderman.[5] Further insight into the development of the subject is provided by the treatise of Alfrey.[6] The generalized mathematical structure of linear viscoelasticity has been developed by Gross.[7]

Boltzmann and his students showed that the superposition principle was valid for inorganic glasses. Leaderman carried out experiments to check its validity for oriented fibres such as rayon, silk and nylon 6,6. He found marked deviations from the Boltzmann principle under many conditions. He correctly attributed this behaviour to the fact that these crystalline substances when subjected to a load often undergo further orientation and crystallization and at the end of the loading experiment they are structurally different from the starting material. The Boltzmann principle can only be expected to apply to such materials at extremely small loads, and temperatures at which further crystallization does not occur. On the other hand, it has been found that many polymeric materials, particularly non-crystalline systems, do show linear viscoelastic behaviour at small strains.

Ward[3] has discussed attempts to extend linear viscoelastic theory to take into account non-linear effects at small strains and to deal with the situation of large strains.

16.1.5 Dynamic Properties

In defining functions to specify the dynamic viscoelastic properties of polymers we base our treatment on the case of a specimen subjected to small-amplitude harmonic shear stresses and strains. Furthermore, the inertial forces arising from movements of the mass elements of this specimen will be assumed to be negligible: an assumption which becomes invalid for relatively high test frequencies.

16.1.5.1 *Isotropic materials*

If a specimen is subject to a sinusoidal shear strain, and if steady-state conditions have been established, then the shear stress will also vary sinusoidally, but it will lag behind the stress by some phase angle δ. Thus, if the variation of shear stress with time is written as

$$\sigma = \sigma_0 \sin(\omega t + \delta)$$

(32)

one has for the shear strain

$$\gamma = \gamma_0 \sin \omega t$$

(33)

where t is time and ω is the angular frequency (omitting the directional subscripts on σ and γ). Equation (32) can be written

$$\sigma = \sigma_0 \sin \omega t \cos \delta + \sigma_0 \cos \omega t \sin \delta$$

(34)

showing that the shear stress can be considered to consist of two parts: one which is in-phase with the strain and has an amplitude $\sigma_0 \cos \delta$, and the other which is 90° out-of-phase with the strain and has an amplitude $\sigma_0 \sin \delta$. The shear-stress/shear-strain relationship can thus be specified by a modulus G' in-phase with the shear strain and a modulus G'' 90° out-of-phase with the strain, *i.e.*

$$\sigma = \gamma_0 G' \sin \omega t + \gamma_0 G'' \cos \omega t$$

(35)

where

$$G' = (\sigma_0/\gamma_0)\cos\delta, \qquad G'' = (\sigma_0/\gamma_0)\sin\delta \tag{36}$$

The mathematical treatment of cyclic loading can be much simplified by using a complex representation of stress and strain. Writing applied shear stress as

$$\sigma^* = \sigma_0 e^{i(\omega t + \delta)} \tag{37}$$

then the shear strain will vary according to

$$\gamma^* = \gamma_0 e^{i\omega t} \tag{38}$$

where $i = (-1)^{1/2}$. The complex shear modulus G^* is given by

$$G^* = \sigma^*/\gamma^* = (\sigma_0/\gamma_0)e^{i\delta} = (\sigma_0/\gamma_0)(\cos\delta + i\sin\delta) \tag{39}$$

Thus

$$G^* = G' + iG'' \tag{40}$$

where G' is the ratio of the amplitude of the in-phase component of the stress to the strain amplitude, and G'' is the ratio of the amplitude of the out-of-phase stress to the strain amplitude. The modulus G' is called the shear storage modulus (or simply the dynamic shear modulus) since it is directly proportional to the peak energy stored per cycle (as discussed below). The modulus G'' is called the shear loss modulus since it is directly proportional to the energy dissipated as heat per cycle.

The energy dissipated per quarter cycle, ΔW, is given by

$$\Delta W = \int_0^{\pi/2\omega} \sigma \left(\frac{d\gamma}{dt}\right) \tag{41}$$

Substituting for $\sigma[=\sigma_0 \sin(\omega t + \delta)]$ and for $\gamma(=\gamma_0 \sin\omega t)$ and integrating yields

$$\Delta W = \tfrac{1}{2}\sigma_0\gamma_0\cos\delta + \tfrac{1}{4}\pi\sigma_0\gamma_0\sin\delta \tag{42}$$

The first term in equation (42) is the maximum amount of energy, ΔW_m which is stored during the test. Over a complete cycle the stored energy is zero. The second term in the equation is the mechanical energy dissipated per quarter cycle. For the full cycle the energy dissipated, ΔW_0, is $\pi\sigma_0\gamma_0\sin\delta = \pi\gamma_0^2 G''$. The quantity $\Delta W_0/\Delta W_m$ is a measure of energy loss or damping.

$$\Delta W_0/\Delta W_m = 2\pi\tan\delta \tag{43}$$

In many cases G'' is small compared with G', and therefore $|G^*| \simeq G'$. It has become customary to define the dynamic-mechanical behaviour of polymeric materials in terms of G' and $\tan\delta = G''/G'$. The latter ratio is often called the loss factor and is a measure of the amount of energy dissipated compared to the amount of stored energy. Typical values of the loss factor for engineering polymers are between 0.01 to 0.1. Instead of $\tan\delta$, the logarithmic decrement, $\Lambda = \pi\tan\delta$, is sometimes used. Figure 5 shows schematically the variation of G', G'' and $\tan\delta$ with frequency for a non-crystalline polymer.

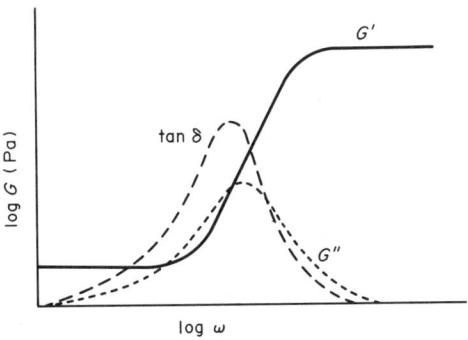

Figure 5 The components of the complex shear modulus, G' and G'', and $\tan\delta$, as functions of frequency ω

In the case of linear viscoelastic materials, the complex shear compliance J^* could equally be used to describe shear

$$\gamma^* = J^* \sigma^* \tag{44}$$

where

$$J^* = (\varepsilon_0/\sigma_0)e^{-i\delta} = (\varepsilon_0/\sigma_0)\cos\delta - (\varepsilon_0/\sigma_0)i\sin\delta = J' - iJ'' \tag{45}$$

and

$$J'/J'' = \tan\delta \tag{46}$$

Figure 6 shows schematically the variation of J', J'' and $\tan\delta$ with frequency for a non-crystalline polymer.

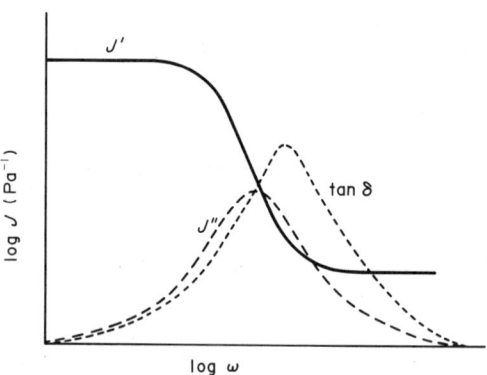

Figure 6   The components of the complex compliance, J' and J'', and $\tan\delta$, as functions of frequency ω

The relationships between the different shear compliances and moduli are

$$J^* = 1/G^*; \quad J' = \frac{G'}{(G')^2 + (G'')^2}; \quad J'' = \frac{G''}{(G')^2 + (G'')^2} \tag{47}$$

The viscoelastic properties of materials in other modes of deformation may also be described in terms of a complex modulus or a compliance. In the case of dynamic tensile tests, the complex modulus E^* is given by

$$E^* = E' + iE'' = E'(1 + i\tan\delta_E) \tag{48}$$

where

$$\tan\delta_E = E''/E' \tag{49}$$

whilst for dynamic compression the complex bulk modulus K^* is given by

$$K^* = K' + iK'' = K'(1 + i\tan\delta_K) \tag{50}$$

where

$$\tan\delta_K = K''/K' \tag{51}$$

Similarly, the complex longitudinal modulus L^* can be obtained from the velocity and attenuation of longitudinal ultrasonic waves

$$L^* = L' + iL'' = L'(1 + i\tan\delta_L) \tag{52}$$

where

$$\tan\delta_L = L''/L' \tag{53}$$

As is the case with the purely elastic deformation functions, the dynamic functions are not independent, and for isotropic materials measurements of the dynamic modulus and loss modulus as a function of frequency for two types of deformation mode are sufficient to characterize the dynamic properties at a given temperature and pressure. When all the stress and strain components have a sinusoidal time-dependence of the same frequency, the relationships between the dynamic functions can be obtained from those applying to the purely elastic functions simply by replacing the real moduli and real compliances with complex moduli and complex compliances; this is known as the correspondence principle. On this basis, from equations (13) and (14)

$$E^* = \frac{9K^*G^*}{3K^* + G^*} \tag{54}$$

and

$$E^* = 2G^*(1 + v^*) \tag{55}$$

16.1.5.2 Anisotropic materials

The dynamic viscoelastic functions for anisotropic materials may be obtained simply be expressing the various components in equations (18) and (19) in complex form, *i.e.*

$$\sigma_p^* = \sum_q C_{pq}^* E_q^* \quad (p, q = 1, 2, \dots 6) \tag{56}$$

and

$$E_p^* = \sum_q S_{pq}^* \sigma_q^* \quad (p, q = 1, 2, \dots 6) \tag{57}$$

For the special case of a transversely isotropic material discussed earlier, it follows from equations (23) applying the correspondence principle that

$$S_{11}^* = 1/E_1^*, \quad S_{22}^* = 1/E_2^*, \quad S_{12}^* = -v_{12}^*/E_1^*, \quad S_{23}^* = -v_{23}^*/E_2^* \tag{58a}$$

$$2(S_{22}^* - S_{23}^*)^{-1} = \frac{1}{2}(1 + v_{23}^*)^{-1}E_2^* = G_{23}^* = \frac{1}{2}(C_{22}^* - C_{23}^*) \tag{58b}$$

$$1/S_{66}^* = G_{12}^* = C_{66}^*, \quad C_{11}^* = L_1^*, \quad C_{22}^* = L_2^* \tag{58c}$$

16.1.6 Mechanical Models

A variety of models have been employed to explain the viscoelastic behaviour of polymeric materials. The Maxwell unit, consisting of a spring and dashpot in series, and the Kelvin (or Voigt) unit, consisting of a spring and dashpot in parallel, are the simplest of these models (see Figures 7 and 8). The Maxwell and Kelvin models lead to analogous equations and have similar limitations. Here we consider only the Maxwell model.

The stress–strain behaviour for the spring in the Maxwell model (assuming only small deformations are involved) is described as linearly elastic

$$\sigma_1 = E_M \varepsilon_1 \quad \text{(Hooke's law)} \tag{59}$$

whilst the response of the dashpot is taken to be that of a linear fluid

$$\sigma_2 = \eta_M \frac{d\varepsilon_2}{dt} \quad \text{(Newton's law)} \tag{60}$$

Figure 7 The Maxwell unit; consisting of a spring and dashpot in series

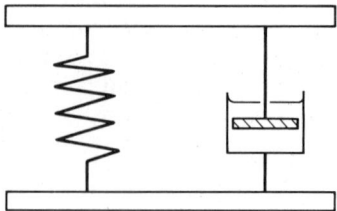

Figure 8   The Voigt unit; consisting of a spring and dashpot in parallel

where η_M is the viscosity coefficient (or simply viscosity) of the fluid. Since the stress is the same for both the spring and dashpot, we have $\sigma_1 = \sigma_2 = \sigma$, whilst the total strain on the system $\varepsilon = \varepsilon_1 + \varepsilon_2$. From equation (59)

$$\frac{d\sigma}{dt} = E_M \frac{d\varepsilon_1}{dt} \tag{61}$$

Combining equations (60) and (61)

$$\frac{d\sigma}{dt} + \sigma = (E_M + \eta_M)\frac{d(\varepsilon_1 + \varepsilon_2)}{dt} \tag{62}$$

and

$$\frac{d\varepsilon}{dt} = \frac{1}{E_M}\frac{d\sigma}{dt} + \frac{\sigma}{\eta_M} \tag{63}$$

One of the applications of the Maxwell model has been in the treatment of stress relaxation. Setting $d\varepsilon/dt = 0$ in equation (63), and also setting $\sigma = \sigma_0$ at $t = 0$, gives

$$\sigma = \sigma_0 \exp(-t/\tau) \tag{64}$$

where $\tau = \eta_M/E_M$ is called the relaxation time. Not surprisingly, this simple model has serious deficiencies. For the case of a constant stress experiment, when $d\sigma/dt = 0$, it predicts Newtonian flow (*i.e.* $\sigma = \eta_M(d\varepsilon/dt)$), which is not generally observed, the experimental behaviour being considerably more complex. Secondly, stress relaxation is not normally described by a simple exponential. Thirdly, the model is not able to adequately describe the dynamic-mechanical response of a polymer subjected to a sinusoidally applied stress. Setting

$$\sigma = \sigma_0 e^{i\omega t} = (G' + iG'')\varepsilon \tag{65}$$

and

$$\eta_M/E_M = \tau$$

in equation (65) yields

$$G' + iG'' = E_M i\omega\tau/(1 + i\omega\tau) \tag{66}$$

Thus

$$G' + E_M\omega^2\tau^2/(1 + \omega^2\tau^2) \tag{67}$$

$$G'' = E_M\omega\tau/(1 + \omega^2\tau^2) \tag{68}$$

and

$$\tan\delta = 1/(\omega\tau) \tag{69}$$

The predictions of equations (67) to (69) are shown in Figure 9. Comparison with Figure 5 shows that the Maxwell model predicts the wrong qualitative features for $\tan\delta$. A more detailed analysis shows that, whilst the G' and G'' curves have approximately the correct shape, equations (67) and (68) do not adequately describe the behaviour of a viscoelastic polymer. A better representation generally requires invoking the concept of a relaxation-time spectrum $H(\tau)$. An infinite set of Maxwell models are arranged in parallel and the single spring constant E_M is replaced by a weighting function, $H(\tau)d(\ln\tau)$ which defines the contribution of the elements with relaxation times lying

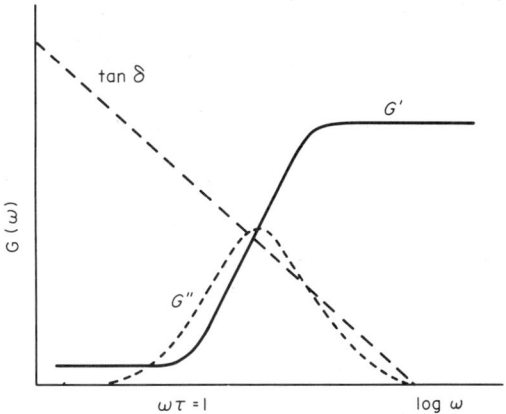

Figure 9 The components of the complex shear modulus, G' and G'', and $\tan \delta$, as predicted by the Maxwell model as a function of frequency ω

between $\ln \tau$ and $\ln \tau + \mathrm{d}(\ln \tau)$. The general stress relaxation modulus is then given by

$$G(t) = \frac{\sigma(t)}{\varepsilon(t)} = G_r + \int_{-\infty}^{\infty} H(\tau)\exp(-t/\tau)\mathrm{d}(\ln \tau) \tag{70}$$

where G_r is the equilibrium modulus. For alternating strain

$$G'(\omega) = G_r + \int_{-\infty}^{\infty} \frac{H(\tau)\omega^2\tau^2}{1 + \omega^2\tau^2}\,\mathrm{d}(\ln \tau) \tag{71}$$

and

$$G''(\omega) = \int_{-\infty}^{\infty} \frac{H(\tau)\omega\tau}{1 + \omega^2\tau^2}\,\mathrm{d}(\ln \tau) \tag{72}$$

From the theory of linear viscoelasticity as developed by Gross,[7] it can be shown that $G(t)$, $G'(\omega)$ and $G''(\omega)$ are related by

$$G(t) = \frac{2}{\pi}\int_0^{\infty} \frac{G'(\omega)}{\omega}\sin \omega t\, \mathrm{d}\omega \tag{73}$$

and

$$G(t) = \frac{2}{\pi}\int_0^{\infty} \frac{G''(\omega)}{\omega}\cos \omega t\, \mathrm{d}\omega \tag{74}$$

16.1.7 Frequency–Temperature Relationship

Measurements to characterize the stiffness and damping behaviour of a polymeric material can be made, in principle, over a wide frequency range and/or a wide temperature range. From a theoretical standpoint it is easiest to interpret results obtained over a range of frequencies at constant temperature. However, the frequency range available experimentally is often limited.

Barrier theories constitute the simplest approach to understanding the effect of temperature on molecular motion.[8] In such theories, motion involves a transition over a potential barrier separating equilibrium states. The relaxation time for the process is given by

$$\tau = (Bh/kT)\exp(\Delta G^{\ddagger}/RT) \tag{75}$$

where B is a measure of the probability of an activated state decomposing, and $\Delta G^{\ddagger}$ is the molar free energy increment between the activated and initial states. Expressed in terms of the enthalpy and entropy of activation, equation (75) becomes

$$\tau = (Bh/kT)\exp(-\Delta S^{\ddagger}/R)\exp(\Delta H^{\ddagger}/RT) \tag{76}$$

or

$$\ln \tau \;=\; \ln B \;+\; \ln(h/kT) \;-\; \Delta S^{\ddagger}/R \;+\; \Delta H^{\ddagger}/RT \tag{77}$$

If B is assumed to be temperature independent, and neglecting the small temperature dependence of the second term, then

$$\frac{\partial \ln \tau}{\partial(1/T)} \;=\; -\frac{1}{R}\frac{\partial \Delta S^{\ddagger}}{\partial(1/T)} \;+\; \frac{1}{RT}\frac{\partial \Delta H^{\ddagger}}{\partial(1/T)} \;+\; \frac{\Delta H^{\ddagger}}{R} \tag{78}$$

Making the assumption that the sum of the first two terms on the right-hand side of equation (78) is negligible compared with the third yields

$$\partial \ln \tau / \partial(1/T) \;=\; \Delta H^{\ddagger}/R \tag{79}$$

Equation (79) has been used to describe secondary relaxation in the glassy region. Such relaxations arise from local motions of polymer chain segments or from motions of side groups. For such a process, the average relaxation time at a given temperature may be obtained from the frequency, $\omega = 1/\tau$, of the peak in $\tan \delta$ *vs.* frequency. Then from equation (79)

$$\partial \ln \omega / \partial(1/T) \;=\; -\Delta H^{\ddagger}/R \tag{80}$$

A plot of $\log f$ *vs.* $1/T$ (where f is the frequency in Hz, *i.e.* $f = \omega/2\pi$) is shown in Figure 10 for the β process (*i.e.* the first relaxation below T_g) in poly(methyl methyacrylate). From the approximately linear plot $\Delta H^{\ddagger} = 82 \text{ kJ mol}^{-1}$.

This approach is not satisfactory for interpretation of the glass-to-rubber relaxation (α process), as shown in Figure 10. The plot of $\log f$ *vs.* $1/t$ is distinctly curved, consistent with $\Delta H^{\ddagger}$ increasing as T_g is approached. This behaviour may be satisfactorily described by an empirical equation proposed by Williams, Landel and Ferry.[9,10] The WLF equation has been derived from free volume theory[11] and from configurational entropy theory.[12]

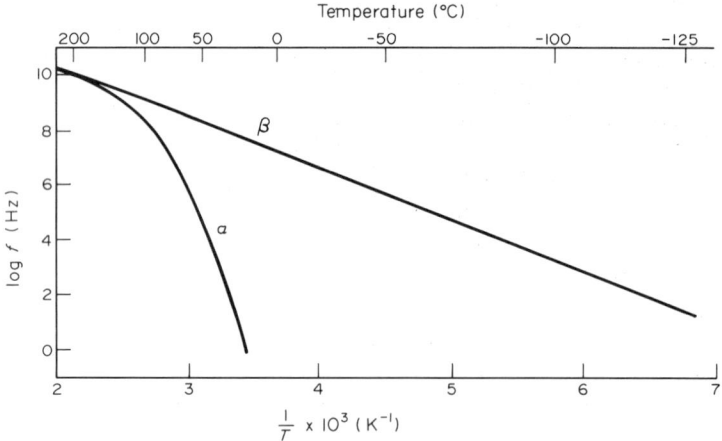

Figure 10 Resonant frequency $f_{\max}$ as a function of reciprocal temperature for α_a and β_a processes in poly(methyl methacrylate). In general the α_a process gives rise to a curved plot while the β_a gives a linear plot. At very high frequencies the α_a and β_a processes merge to give $(\alpha\beta)_a$ for the polymer

For unfilled non-crystalline polymers it is generally found that changing the temperature causes a shift along the frequency axis of all the relaxation processes, without any appreciable change in their magnitude. This effect is illustrated schematically in Figure 11, in which it is shown that if temperature is increased $(T_0 < T_1 < T_2)$ then the relaxation shifts to higher frequencies $(\omega_0 < \omega_1 < \omega_2)$. The WLF equation predicts this variation of relaxation frequency with temperature. For an increase in temperature from T_0 to T_1 the equation takes the form

$$\log a_T \;=\; \log(\tau_1/\tau_0) \;=\; \log(\omega_0/\omega_1) \;=\; -C_1(T_1 - T_0)/[C_2 + (T_1 - T_0)] \tag{81}$$

where C_1 and C_2 are constants for a given polymer. For many polymers, if T_0 is set equal to T_g then $C_1 = 17.4$ and $C_2 = 51.6$.

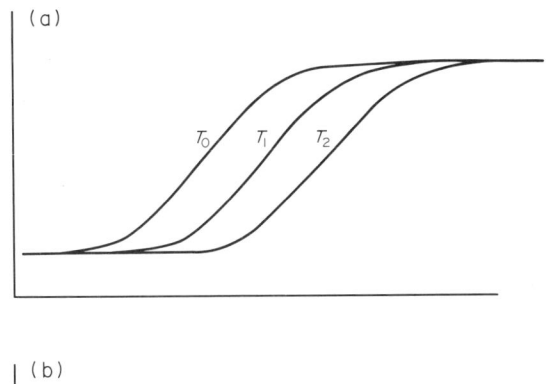

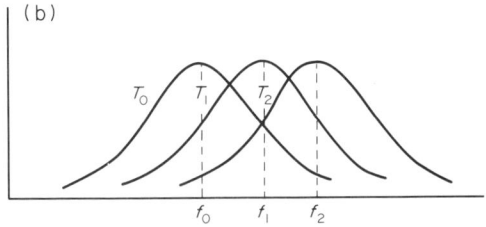

Figure 11 An idealized representation of the effect of changing temperature on the viscoelastic responses. It is assumed that there is no change in the shape of the relaxation spectrum

16.2 EXPERIMENTAL METHODS

The measurement of the dynamic-mechanical properties of polymers requires the separation of the response of the material to cyclic or transient loading into two components: an elastic response and an inelastic response. Frequently, these two are best resolved experimentally as a complex response and a damping. Whilst the complex or total reponse is familiar to most experimentalists, damping frequently poses a conceptual problem. However, the effect of damping is manifested in many ways. For example: (i) the phase lag between stress and strain (*i.e.* load and displacement); (ii) the decrease with time of the amplitude of stress and strain in a freely vibrating system; and (iii) the limited amplitude of a system excited at resonance.

Three types of measurement are generally used. At low frequencies a small specimen acts as a massless spring, and it is sufficient to determine the complex spring constant. At higher frequencies, where this approach breaks down, beam specimens are used and the mechanical properties are derived from the behaviour of longitudinal, torsional or bending waves. At very high frequencies it is possible to treat the specimen as a semi-infinite body. The various methods available have been reviewed.[1-3, 13]

16.2.1 Forced Vibration

Perhaps the most widely used method for engineering polymers is the direct forced-vibration method. The method has the great advantage that the frequency of the excitation can be changed at will without changing the specimen. Hence it is readily possible to determine the frequency dependence of complex modulus and loss factor over a wide range.

16.2.1.1 Tensile method

The tensile method is illustrated in Figure 12. From the amplitude of the load F_a and the resulting displacement amplitude d_a, the components of the complex modulus can be calculated from the relations

$$E' = (F_a/d_a)(L/A)\cos\delta, \quad E'' = E'\tan\delta \tag{82}$$

where A and L are the area of cross-section and the length of the specimen. Note that this treatment is based on the assumption that the specimen can be regarded as a massless spring.

Dynamic-mechanical Properties

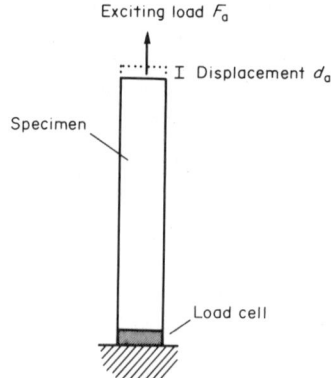

Figure 12   Direct forced-vibration method

The main problems with this kind of experiment are the measurement of the phase angle with sufficient accuracy and the need to achieve a rigid attachment of the specimen. To measure loss factors of the order of 10^{-2} accurately, the phase angle must be measured with an accuracy of about $0.2°$. As this is difficult to achieve, the method is best suited to materials with relatively large loss factors. Furthermore, the equipment must have a resonance frequency which is well above the frequency at which the specimen is being excited. The attachment of the specimen to the test rig is a problem if grips are used, because it is difficult to define the specimen length, as those parts of the specimen inside the grips contribute to the deformation. By using slender beams it is possible to decrease this effect.

16.2.1.2   Cantilever method

For flexural deformations the specimen is clamped at its end (single cantilever) or at both ends (dual cantilever), see Figure 13. In the case of the dual cantilever, the components of E^* are calculated from the relations

$$E' = (F_a/d_a)(L^3/2bh^3)\cos\delta, \qquad E'' = E'\tan\delta \tag{83}$$

where b and h are the width and thickness of the beam respectively and L is the effective distance between the centre and the clamp. Note that a correction must be applied if shear deformation is significant.

16.2.2   Decay of Free Vibration

In the forced-vibration methods described above it is necessary to monitor both load and displacement. If the measurements are performed at resonance, using the decay of free vibration, it is sufficient to measure one variable, load or displacement. However, this method has the drawback that data are obtained at one frequency only, the resonance frequency, which is a function of the dimensions of the specimen and its material properties.

16.2.2.1   Torsion pendulum

Figure 14 is a sketch of a torsional pendulum as used for determination of modulus and damping. After the initial excitation, the natural frequency can be directly determined from the time difference

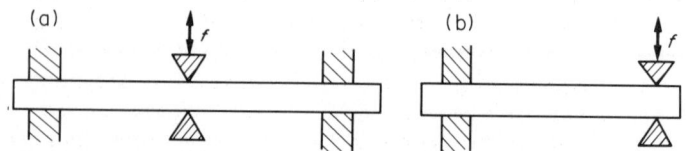

Figure 13   Forced vibration; bending in the (a) dual- and (b) single-cantilever modes

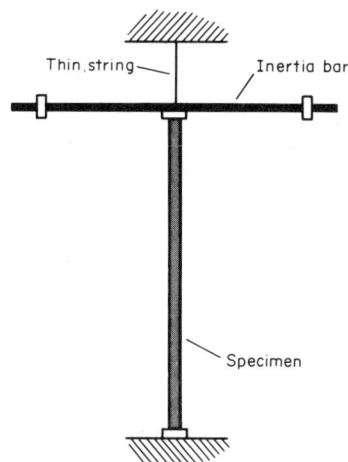

Figure 14 Torsion pendulum method

between the maxima in the curve of torsional deformation *vs.* time. The damping is usually determined from the logarithmic decrement Λ, defined as the logarithm of the ratio between two consecutive torsion maxima. If A_n is the amplitude of the nth torsion maximum and A_{n+1} is the amplitude one period later, then Λ can be calculated according to

$$\Lambda = \ln(A_n/A_{n+1}) \tag{84}$$

For the case where damping is small

$$\tan\delta = \Lambda/\pi \tag{85}$$

For large damping, equation (77) can be corrected to give

$$\tan\delta = (\Lambda/\pi)/[1 - (\Lambda/2\pi)^2]^{1/2} \tag{86}$$

Since it is possible to determine Λ quite accurately over several periods, this is a simple and accurate method for determination of the loss factor. Also, since the decay proceeds most slowly when the damping is low, the method can be successfully used for materials with very small loss factors, *i.e.* too low for direct measurement. This is best carried out with the apparatus enclosed in a vacuum chamber.

The main advantage of the torsion pendulum is its simplicity of operation. The main disadvantage, common to all free decay methods, is that, for a given specimen and given inertia, loss factors and moduli are obtained at one frequency only. Results at different frequencies can be obtained by changing the specimen dimensions or the inertia. This is possible but rather labour-intensive and introduces an additional source of experimental error. Frequencies down to 0.1 Hz can be obtained quite easily, but useful results are rarely obtainable for frequencies higher than 500 Hz.

16.2.2.2 *Suspended beam*

Depending on the system used, the excitation of a beam can result in quasi-longitudinal, torsional and bending waves. Surface waves are obtained only at high frequencies. In practice, only bending waves are used for the determination of moduli and loss factors, because of the relative simplicity with which this wave type can be achieved.

Resonant bending waves are most easily measured for freely suspended beams that are excited at one end and the response measured at the other. The principle is shown in Figure 15: see also refs. 14 and 15. In this arrangement excitation and sensor devices are placed at the antinodes, which eliminates the problem of frequency-dependent standing waves. Any excitation system that can be loosely coupled to the specimen can be used, commonly electromagnetic exciters are favoured. It is important that the coupling is loose in order not to introduce extraneous damping. The same applies to the sensor. A 'non-contact' sensor is preferable for light specimens, but for heavier specimens the sensors can be attached to the specimen provided that its mass is less than $M/30n$, where M is

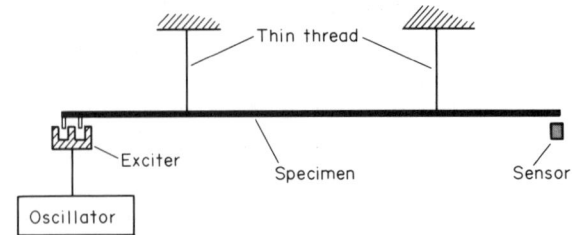

Figure 15   Vibration of a freely suspended beam: excitation at the end

the specimen mass and n is the number of nodes in the beam at the frequency considered. The suspension of the specimen also has to be such that it does not introduce any extraneous damping. If the loss factor of the specimen is greater than 10^{-3}, acceptable conditions can be achieved by suspending the specimen on strings. For lower loss factors, one has to suspend the specimen at its nodal points in order to minimize the energy that is transmitted from the specimen to the suspension. In this case the suspension points have to be moved when specimens with different resonance frequencies are studied. Radiation of acoustic energy to the surroundings can contribute to the damping and can affect the results if small loss factors ($\tan \delta < 10^{-4}$) are to be measured. This effect can be reduced by performing the test *in vacuo*.

The real part of the elastic modulus can be determined from the relation

$$E' = (48\pi^2 M L^3 f_n^2)/(bh^3 \alpha_n^4) \tag{87}$$

where M is the mass of the beam of length L, width b and thickness h, and f_n is the frequency of the natural resonance with n nodes. Values of the parameter α_n can be found elsewhere.[1] Corrections are applied for added mass, shear and rotary inertia.[1]

The loss factor is obtained from

$$\tan \delta = \Delta f / f_n \tag{88}$$

where Δf is the width of the resonance peak at the half-power point (*i.e.* half-height). Equation (88) is not exact since it does not take into account the influence of damping. However, it gives results of acceptable accuracy when $\tan \delta < 0.1$.

A slightly different method can be used[16] to excite the beam. The principle is shown in Figure 16. The excitation consists of a small sinusoidal displacement applied at the mid-point P, the displacement of which is measured.

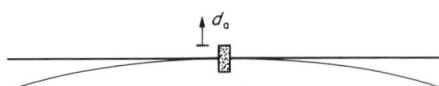

Figure 16   Vibration of a freely suspended beam: excitation at the centre

For very small loss factors, it is often possible to obtain more accurate results by measuring the decay time instead of the band width. The same arrangement as that illustrated in Figure 15 can be used, but instead of measuring the frequency at resonance, the amplitude decay after removal of the excitation is studied as a function of time. If the material is linearly elastic, the energy will decay exponentially, according to $\exp(-\omega t \tan \delta)$. By measuring the decay time t' (*i.e.* the time at which the energy has the value of, for example, 10^{-6} times its initial value) the loss factor can be determined from

$$\tan \delta = \ln(10^6/\omega t') \tag{89}$$

Of course, any parameter that describes the decay of the vibration can be used for the calculation of $\tan \delta$. Note also that if an excitation with a frequency not equal to a resonance frequency is turned off then the amplitude decay will not follow the relation $\exp(-\omega t \tan \delta)$ satisfactorily.

If one is interested only in the loss factor, the decay time from an impulsive excitation such as a hammer blow can be used.[17,18] In this case the frequency dependence of the loss factor can be determined provided that the signal is filtered. However, such a test may not give a very accurate result, since two or more resonance frequencies with different time factors may lie within one filter band width.

16.2.3   Ultrasonic Waves

The methods described so far are most suited to isotropic materials, their application to anisotropic materials being more difficult and generally less accurate. A full characterization of anisotropic materials requires test pieces cut at different angles to the material system axes. The difficulties encountered and the loss in accuracy are particularly severe for samples with low symmetry. In contrast, ultrasonic wave techniques[1,3,13,19] can be applied to anisotropic materials without the need for elaborate sectioning.

With ultrasonic wave propagation techniques, the phase velocity and attenuation of acoustic waves travelling along directions different to the symmetry axes of the material are measured. The storage and loss components of the complex moduli are determined from the measurements. The method is particularly applicable to the measurement of the storage moduli of low loss ($\Lambda < 0.2$) single crystals, fibre-reinforced plastics and oriented thermoplastics, as well as to the study of storage and loss moduli of isotropic thermoplastics. It is also widely used in the study of the viscoelastic behaviour of liquids. Ultrasonic wave test methods extend over the frequency range from 100 kHz to over 100 MHz and therefore also complement the other methods by extending the frequency range.[20-22]

The magnitude of the velocity of an ultrasonic wave travelling in a low loss test piece is related to one or more of the stiffness components $C_{pq'}$ and the density. All the stiffness components can be determined from a single test piece if its dimensions allow experimental measurements to be performed along a sufficient number of directions. The analysis of the results is greatly simplified if wave propagation is carried out along the symmetry axes of the specimen. In this case

$$C_{pp'} = \rho v^2 \tag{90}$$

where $p = 1$ to 6, v is the velocity of the wave and ρ the density. The symmetry of fibre-reinforced plastics and oriented thermoplastics may be assumed to be hexagonal, in which case there are only five independent stiffness components. Measurement of the velocity of longitudinal and transverse waves propagating along and normal to the unique axis (1 axis) of such a test piece allows the longitudinal and transverse tensile, $C_{11'}$, $C_{22'}$, and shear, $C_{66'}$, $C_{44'}$, stiffness components of the test piece to be determined. The fifth stiffness component may be determined from an additional velocity measurement at a known angle to the unique axis.

The most direct method for determination of the velocity of sound waves through materials and their attenuation involves time-of-flight and reduction in amplitude measurements. The method is suitable for studies in the frequency range 0.1 MHz to 10 MHz. At frequencies above this range the attenuation is generally so high that only thin specimens can be used, which greatly reduces the accuracy of the measurements. Other methods are then more suitable,[20,21] enabling measurements to be made over a wider frequency range.

Figure 17 shows schematically one type of apparatus which has been used for time-of-flight and attenuation measurements.[19] Two coaxial ultrasonic transducers, one acting as transmitter and the

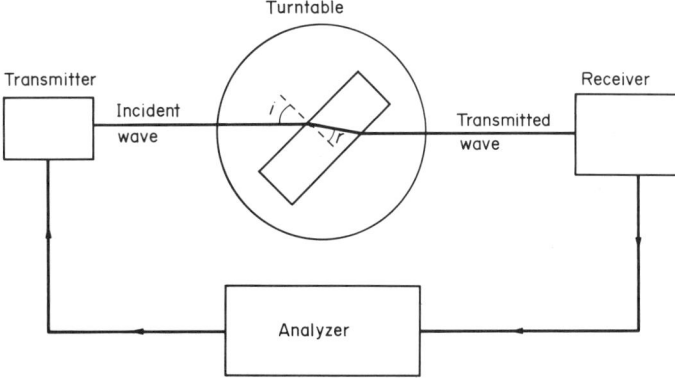

Figure 17   Schematic diagram of an ultrasonic pulse apparatus

other as receiver, are immersed in a thermostatted water bath. For anisotropic materials, test pieces need to be in the form of a rectangular parallelpiped, whilst for isotropic materials, test pieces with only one set of parallel faces are required. A series of high-voltage short duration pulses (*e.g.* 300 V for 10 ns) power the transmitting transducer. The receiving transducer detects the transmitted pulses, which are then amplified by a circuit with an accurate attenuator.

16.2.4 Energy Dissipation

If a perfectly elastic specimen is subject to a dynamic strain then the work done will create strain energy which will be recovered during unloading. If the specimen is viscoelastic part of the energy will be transformed to heat which is not recoverable. During one full cycle the dissipated energy per unit volume (see Section 16.1.5.1) is

$$\Delta W_0 \;=\; \pi E'' \varepsilon_0^2 \tag{91}$$

The energy dissipation per unit time can be written

$$\Delta W_0 \;=\; \pi f E'' \varepsilon_0^2 \tag{92}$$

where $f(=\omega/2\pi)$ is the frequency. If no energy is transferred to the surrounding medium (usually air) or conducted away to material in the immediate neighbourhood, the dissipated energy will cause a temperature increase in the specimen

$$\Delta T \;=\; \pi f E'' \varepsilon_0^2 / \rho C_p \tag{93}$$

where ρ is the density and C_p is the specific heat of the polymer. Usually some part of the energy will be conducted away to neighbouring material, so that the energy available for temperature increase and heat transfer to surrounding air is[23]

$$\Delta W_1 \;=\; K^2 \Delta W_0 \tag{94}$$

If $K = 1$, there is no conduction and all the dissipated energy is for temperature increase and loss to surrounding air. If $K < 1$, part of the energy is conducted away to material in the neighbourhood. The heat transfer to the surrounding air is proportional to the temperature difference between the specimen and the surrounding air. If the specimen surface area is A, its volume V, and if the heat transfer coefficient is K_t, then the rate of energy loss to the surrounding air will be $AK_t(T_s - T_a)$, where T_s is the surface temperature of the specimen and T_a the temperature of the surrounding air. Energy loss per unit time and unit volume is then[23]

$$\Delta W_2 \;=\; (A/V)K_t(T_s \;-\; T_a) \;=\; \beta K_t(T_s \;-\; T_a) \tag{95}$$

where β is the surface to volume ratio. Therefore temperature increase per unit time is

$$\Delta T \;=\; dT/dt \;=\; (\Delta W_1 \;-\; \Delta W_2)/\rho C_p \tag{96}$$

Inserting expressions for ΔW_1 and ΔW_2 into equation (96) and solving the differential equation gives the temperature increase as a function of time

$$T_s(t) \;=\; T_a \;+\; [1 \;-\; \exp(-t\beta K_t/\rho C_p)]\,[K^2 \pi f E'' \varepsilon_0^2]/\beta K_t \tag{97}$$

At steady-state conditions the surface temperature of the specimen becomes

$$T_s(\infty) \;=\; T_a \;+\; K^2 \pi f E'' \varepsilon_0^2 / \beta K_t \tag{98}$$

It is to be noted that in the above derivation it has been assumed that the loss modulus E'' is independent of temperature. This is generally not true and more accurate calculations take this into account. Experimental measurements of surface temperatures for a glass-fibre/polyester laminate during flexural loading are shown in Figure 18, together with theoretical calculations according to equation (97). The value of the heat transfer coefficient K_t was obtained from the steady-state value of the temperature by use of equation (98). Best agreement between experimental and theoretical results was obtained with $K = 0.5$–0.7, meaning that between 50 and 75% of the dissipated energy was conducted away to surrounding materials.[23]

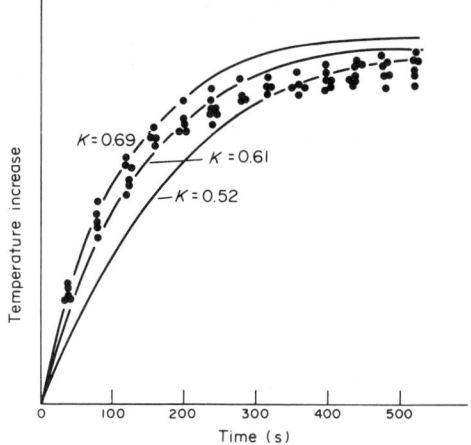

Figure 18 Theoretical (——) and experimental (●) temperature increase for a glass-fibre/polyester laminate

16.2.5 Dynamic-mechanical Thermal Analysis: DMTA

Instruments for determination of dynamic-mechanical properties are available from several manufacturers. For the reasons discussed earlier (see Section 16.1.7) measurements are most often made over a limited range of frequencies and a wide range of temperature: the technique is then called dynamic-mechanical thermal analysis (DMTA). The great majority of exploratory and product-control measurements are made on highly automated DMTA instruments, which give a rapid broad summary of material characteristics of a sample.

16.2.5.1 Commercial instruments

In this section we consider briefly some commercial instruments and provide the names and addresses of a number of manufacturers.

(i) Forced vibration (DMTA)

Generally measurements can be made in tensile, flexure and shear modes. Some instruments allow measurements to be made in compression or torsion modes. Typically, frequency can be varied from 0.01 to 100 Hz; temperature from −150 to +500 °C at heating rates in the range 0.01 to 50 K min^{-1}. Manufacturers include: Du Pont Co. Instrument Systems, Wilmington, DE, USA; I Mass Inc., Hingham, MA, USA; Metravib Instruments, Dardilly, France; Polymer Laboratories Instrument Division, Loughborough, UK; Rheometrics Inc., Piscataway, NJ, USA; and Toyo Baldwin Co. Ltd., Tokyo, Japan.

(ii) Torsion pendulum

Typically, temperature can be varied from −180 to +400 °C at heating rates in the range 0.05 to 5 K min^{-1}. Manufacturers include: Brabender OHC, Duisberg, West Germany; Myrenne GmbH, Roetgen, West Germany; and Zwick GmbH, Ulm, West Germany.

(iii) Torsion braid

In torsional braid analysis (TBA) the sample is impregnated into a multistranded glass braid which acts as support. The method allows analysis of mechanically weak materials, and also materials which are available in very small amounts. Since the specimen is a composite the properties determined are only relative. The main use of TBA is in locating the temperatures of transitions and relaxations. The temperature range available is similar to that given above for conventional torsion pendulums. Manufacturers include: Plastics Analysis Instruments Inc., Princeton, NJ, USA.

(iv) Resonant vibration

Most resonance instruments employ flexural vibrations, although instrumentation is available for longitudinal and torsional modes. Many instruments are purpose-built in individual laboratories. Typically, the temperature can be varied from -150 to $+250\,°C$. Manufacturers include: Bruel and Kjaer, Naerum, Denmark; and Nene Instruments, UK.

(v) Ultrasonic wave

Instruments are often assembled from independently purchased ultrasonic probes, drive units, amplifiers and oscilloscopes. A system is available from Matec Co.

16.2.5.2 Polymer Laboratories DMTA

The practice of DMTA is illustrated here by reference to the Polymer Laboratories PL-DMTA. The same considerations apply to the other commercially available DMTA instruments.[24]

The PL-DMTA is shown schematically in Figure 19. The stress is proportional to the level of AC current fed to the drive coil from the analyzer module. The frequency of oscillation can be selected from 0.01 to 200 Hz. Strain is proportional to the displacement of the drive clamp and is monitored by a non-contracting eddy-current transducer. The stress and strain signals are compared in the analyzer unit, where counting circuits resolve the strain into its in-phase and out-of-phase components. Input of the sample geometry constant allows computation of $\log E'$ and $\tan \delta$. The sample temperature is controlled in the range -150 to $+500\,°C$ by a temperature programmer, and can be ramped up or down at controlled rates (up to 15 K min^{-1}) or be maintained constant for isothermal measurements. Use of a microcomputer and the IEEC interface allows frequency to be multiplexed during a slow thermal scan and all data to be stored for subsequent manipulation.

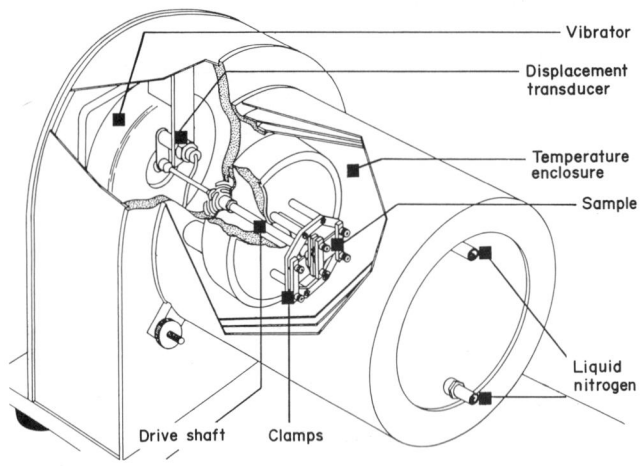

Vibrator
Displacement transducer
Temperature enclosure
Sample
Liquid nitrogen
Drive shaft   Clamps

Figure 19 A schematic diagram of the PL-DMTA head; the dual-cantilever bending mode is illustrated in this case

The normal mode of deformation is bending small bars as dual or single cantilevers, as shown in Figure 20(a). In the single-cantilever mode, considerable thermal expansion, such as occurs at the melting point, can be accommodated because of the lateral compliance of the drive. Alternative geometries are available: *e.g.* shear-sandwich geometry, for measuring the shear modulus of rubbers and soft adhesives (Figure 20b), and tensile geometry suitable for thin films and fibres (Figure 20c). In the tensile mode, single fibres as small as 40 μm in diameter have been studied. Tensile measurements on films and fibres require the application of a static force which must be separately controlled. In the PL-DMTA two options are open: either a force is applied initially and maintained constant by servo-control, or the initial force is automatically reduced in sympathy with the sample stiffness, which avoids excessive creep at high temperatures.

Torsion measurements can be used for melts, with either parallel-plate or cone-and-plate geometry. The latter has the advantage of a uniform ratio of shear rate to strain across the sample.

(a) 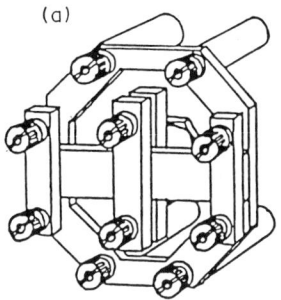 (b) 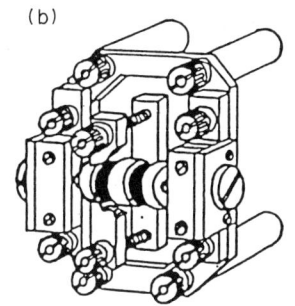 (c)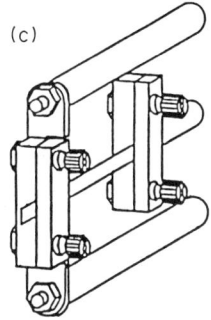

Figure 20 Details of the clamping arrangements available for the PL-DMTA: (a) bending, (b) shear, (c) tensile

Generally, absolute modulus cannot be measured accurately over the whole temperature range, because the arrangements for clamping and measuring cannot be optimized over the range of material properties (glass, rubber) encountered in the full frequency/temperature range. The loss factor is more reliably obtained, provided that it is derived from a direct measurement of phase angle, or is calculated in such a way that the geometry factor cancels out. End corrections are the major source of error. These are determined, and accurate moduli calculated, by measuring samples of different length. For example, in the case of bending, the absolute modulus E can be calculated from the measured moduli E_m obtained for samples of different length l_m from

$$l_m/E_m^{1/3} \quad = \quad (l_m \quad + \quad \Delta l)E^{1/3} \tag{99}$$

where Δl is the end correction. A plot of $l/E_m^{1/3}$ *vs.* l_m is linear with slope $1/E^{1/3}$ and intercept Δl. Once Δl is determined for a particular material and geometry it can be used in subsequent measurements. The effect of Δl can be minimized by using long (or thin) samples, but in practice this is not usually possible.

16.3 GLASSY POLYMERS AND THE GLASS TRANSITION

In the glassy state the amorphous chains are largely frozen into a rigid disordered structure, giving a high modulus and a low loss factor. Some limited mobility is possible and this gives rise to one or more transitions of low magnitude. Usually the relaxation transitions are labelled in alphabetical order, α, β, γ and δ, with decreasing temperature, with the highest transition, the α relaxation, usually being the glass transition. An example is poly(methyl methacrylate) (PMMA) which shows three relaxation transitions in the glassy state,[25] two of which are shown in Figure 21. The highest temperature transition, the α relaxation, is the glass transition, which marks the limit of long-range motions of chain segments. The highest temperature transition in the glassy state is the β relaxation, which is of much lower magnitude. In PMMA this transition has been shown to be associated with side-chain motions of the ester group. The other two transitions in the glassy state, the γ and δ relaxations, are associated with motions of methyl groups attached to the main chain and to the side chains, respectively.

The behaviour of PMMA forms a basis for discussing transitions in many other amorphous polymers. However, the assignment of relaxation processes to molecular processes is not always so straightforward.

16.3.1 The Glass-to-rubber Transition

The glass-to-rubber transition (the α relaxation) changes the properties of amorphous plastics from those of a rigid essentially elastic glass to those of a flexible high loss rubber. The temperature at which this transition occurs will depend on a number of factors, such as chain flexibility, molecular weight, plasticizers, *etc.*

16.3.1.1 Chain flexibility

As examples, it is noted that chains containing mainly methylene sequences, or methylene sequences plus ether linkages, have high chain flexibility and a low glass transition temperature.

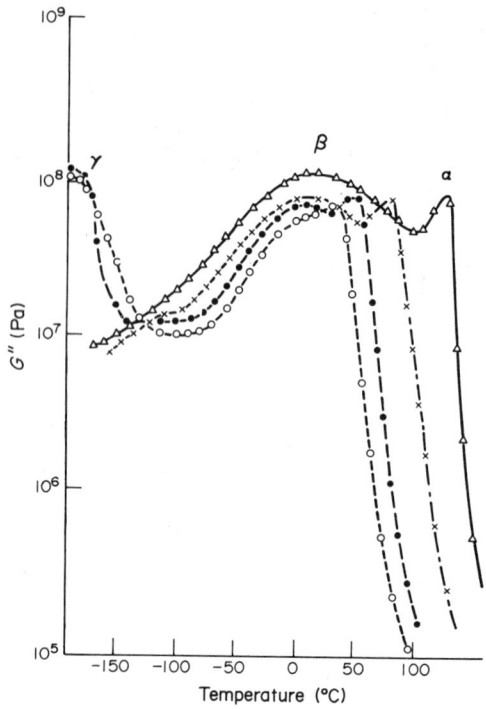

Figure 21   Temperature dependence of shear loss modulus G'' for poly(methyl methacrylate) ($\triangle$, PMMA), poly(ethyl methacrylate) ($\times$, PEMA), poly(n-propyl methacrylate) ($\bullet$, P-n-PMA) and poly(n-butyl methacrylate) ($\bigcirc$, P-n-BMA)

Polymer chains containing stiff phenylene groups, such as poly(phenylene oxide), have a low chain flexibility and a high glass transition temperature.

Side groups also influence chain flexibility. For example, values of the shear modulus and logarithmic decrement for three polymers, polyisobutylene, polystyrene and poly(vinylcarbazole), are shown in Figure 22.[26] As the size and rigidity of the side groups increases, the chain flexibility is lowered and the glass transition temperature is increased. In contrast to this effect of large rigid side-groups, the introduction of lengthy flexible non-polar side-groups decreases T_g.[26]

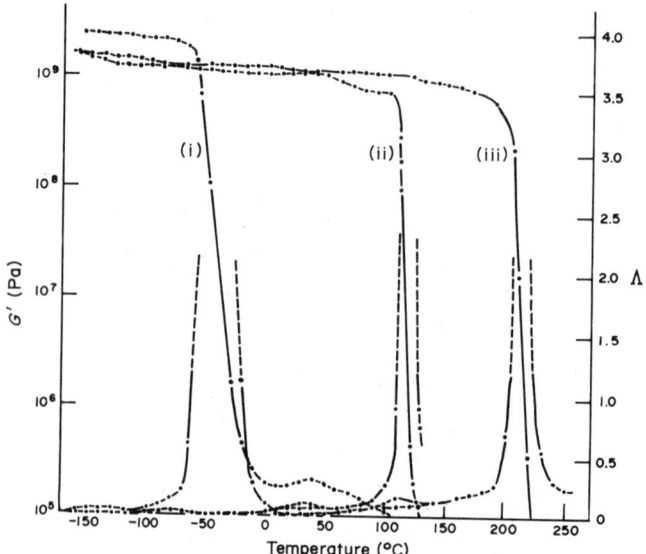

Figure 22   Shear storage modulus G' and damping, $\Lambda = \pi \tan \delta$, in polyisobutylene (i), polystyrene (ii) and poly-(vinylcarbazole) (iii) as a function of temperature at 1 Hz

16.3.1.2 *Molecular weight*

In low molecular weight polymers, the glass transition increases with increasing molecular weight. The relation between molecular weight and T_g can be written

$$T_g = T_{g\infty} - KM_n^{-1} \tag{100}$$

where $T_{g\infty}$ signifies T_g when M_n goes to infinity, M_n is the number-average molecular weight and K is a constant. The value of constant K depends on the polymer type. Equation (100) can be derived by considering the contribution of extra free volume provided by the chain ends.[27, 28] The extra free volume increases the mobility of the chain molecules and decreases T_g.

For amorphous polymers, molecular weight also has a large effect on the dynamic mechanical behaviour above the glass transition region. High molecular weight polymer chains are long enough to temporarily become entangled with each other, thereby forming physical crosslinks. These crosslinks prevent molecular flow and lead to rubber-like behaviour (see Figure 23). On a longer timescale, or at higher temperatures, the crosslinks slowly disentangle leading to viscous flow. The stability of the physical crosslinks increases with increasing molecular weight, thus increasing the length of the rubbery plateau.

16.3.2 Crosslinking

The effect of chemical crosslinking on the dynamic behaviour of phenol–formaldehyde (PF) resin is shown in Figure 23.[29] By increasing the amount of hardener (hexamethylenetetramine) the degree of crosslinking is increased. At a hardener concentration of wt 2%, the PF resin is mainly thermoplastic with a glass transition temperature of 120 °C. At wt 4% hardener concentration, a low degree of crosslinking is obtained and T_g moves to a higher temperature. At the same time the transition becomes broader. At 10 wt % hardener, the plastic is fully crosslinked and a glass transition cannot be detected. The behaviour shown in Figure 23 is typical of many thermosets and crosslinked thermoplastics.[2]

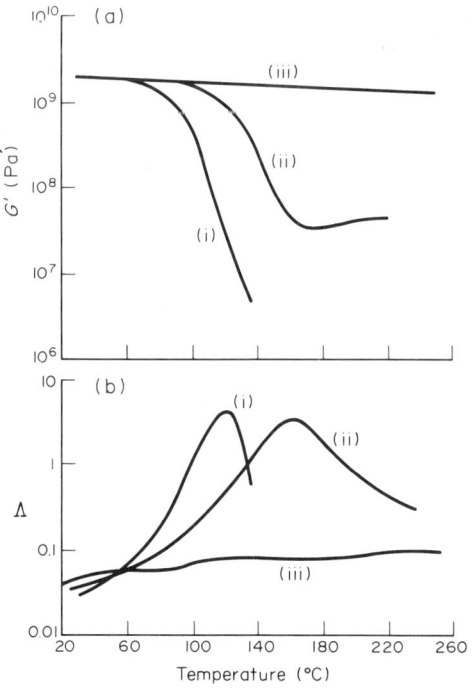

Figure 23 (a) Shear storage modulus G' and (b) damping, $\Lambda = \pi \tan \delta$, for phenol–formaldehyde materials with different degrees of crosslinking using the curing agent hexamethylenetetramine; torsion pendulum measurements. Curing agent concentration: (i) 2%, (ii) 4%, (iii) 10% by weight

16.3.3 Plasticizers

Plasticizers are liquids that are added to plastics to soften them. The softening is brought about by the plasticizer dissolving in the polymer and lowering its glass transition temperature. Figure 24 shows the logarithmic decrement associated with the glass transition of poly(vinyl chloride) plasticized with various amounts of ethyl hexyl phthalate.[30]

Torsion braid analysis is particularly useful in the evaluation of plasticizers. Soft materials each containing different plasticizers or different amounts of a plasticizer can be compared. For example, the film-forming polymer (hydroxypropyl)methylcellulose plasticized with poly(ethylene glycol) 200

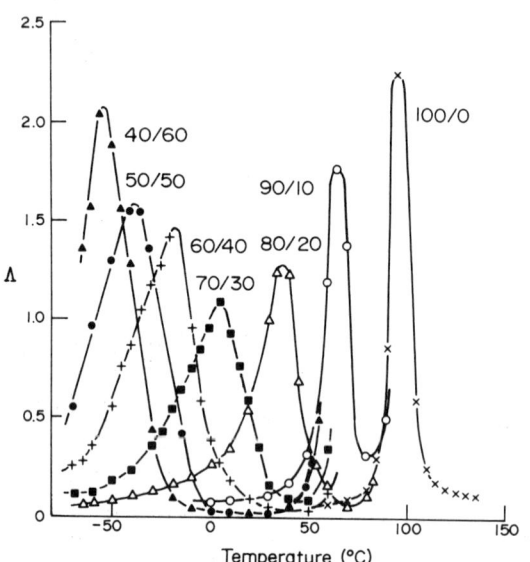

Figure 24   Logarithmic decrement of poly(vinyl chloride) plasticized with various amounts of ethyl hexyl phthalate. (The ratios of poly(vinyl chloride)/ethyl hexyl phthalate by weight are shown)

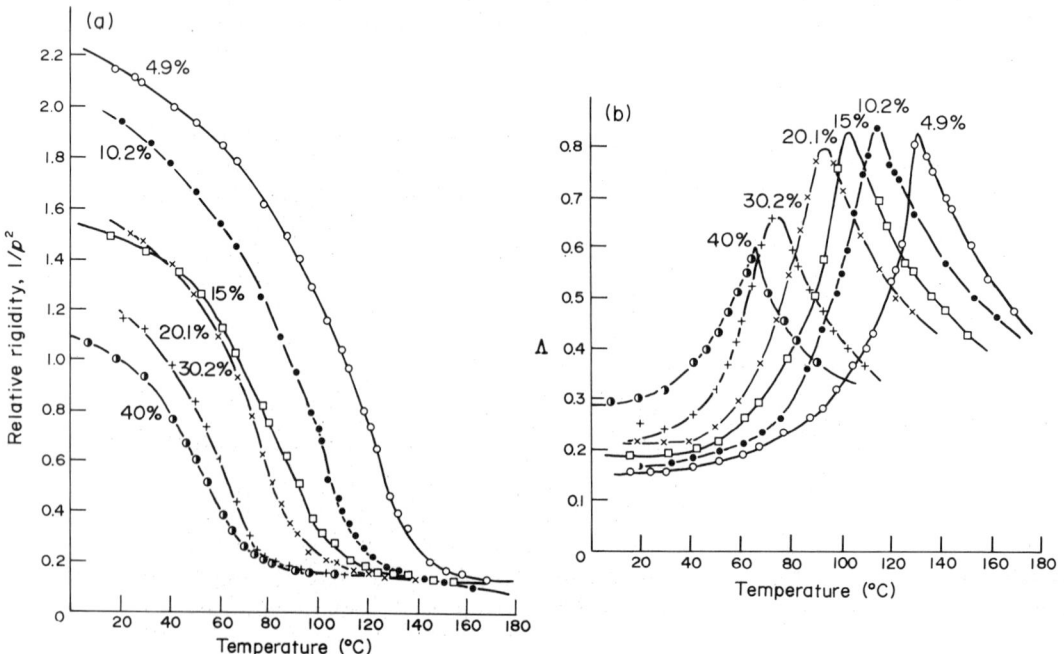

Figure 25   Results of the TBA analysis of (hydroxypropyl)methylcellulose plasticized with poly(ethylene glycol) 200, showing the (a) relative rigidity and (b) logarithmic decrement as a function of temperature

has been examined by this technique.[31] Plots of relative rigidity and logarithmic decrement against temperature are shown in Figure 25; the systematic shift in T_g is apparent.

16.3.4 Fillers

Fillers are frequently added to polymers to lower their price. Usually such fillers are finely ground inorganic materials, such as chalk, silica and clay. As a result of the presence of filler, the mechanical properties of the material are changed. Impact and tensile strengths are usually decreased while hardness and stiffness are increased. The effect of an inert filler (NaCl) on the shear modulus (G') of a polyurethane rubber is shown in Figure 26(a). The value of G' is increased both above and below the glass transition, but the position of the glass transition on the temperature axis is not altered. This behaviour is typical of inert fillers, since the properties of the polymer matrix are not changed.

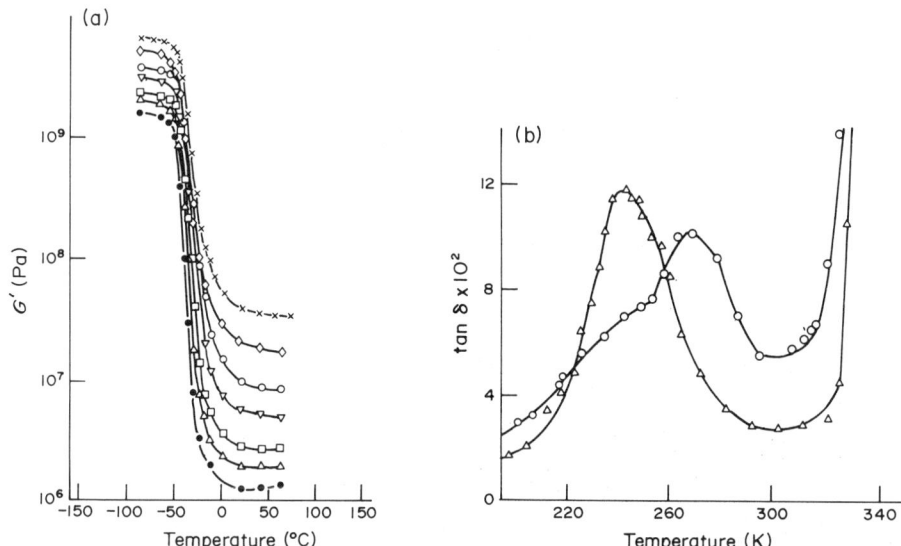

Figure 26 (a) Shear storage modulus G' at 1 Hz *vs.* temperature for a series of urethane rubbers filled with increasing amounts of NaCl. Vol % NaCl: (●) 0; (△) 15.4, (□) 26.0, (▽) 36.2, (○) 46.6 (125–150 μm); (◇) 59.8 (210–300 μm); (×) 69.9 (33–40 μm). (b) Loss factor for poly(oxyethylene) filled with unmodified silica. v_2 is the volume fraction of silica· (△) $v_2 = 0$; (○) $v_2 = 0.31$

If the polymer molecules are adsorbed on the filler surface their mobility is restricted. As a result the glass transition of the adsorbed polymer will increase. Figure 26(b) shows the loss factors as a function of temperature for unfilled and silica-filled poly(oxyethylene). In the filled polymer two transitions can be seen, corresponding to the glass transition of unadsorbed (shoulder) and adsorbed polymer. The large increase in T_g obtained for the adsorbed polymer has been related[32] to the strong polymer–filler interaction.

16.3.5 Copolymers and Blends

For a random copolymer the glass transition varies monotonically between the glass transitions of the homopolymers of the two constituents. The following simple formula holds for the glass transition temperature of a random copolymer

$$T_g = \phi_A T_{gA} + \phi_B T_{gB} \tag{101}$$

where ϕ_A and ϕ_B are the volume fraction of A and B units in the copolymer, and T_{gA} and T_{gB} are the glass transition temperatures of the relevant homopolymers. Alternatively

$$1/T_g = w_A/T_{gA} + w_B/T_{gB} \tag{102}$$

where w_A and w_B are weight fractions.

Similar arguments to those for copolymers hold for blends, block copolymers or graft polymers provided that the polymers are completely miscible with each other. If the two polymers are immiscible they exist in separate phases (or microphases) and two glass transitions are observed. An example of the dynamic-mechanical properties of a graft polymer is shown in Figure 27.[33] Tensile modulus (E') and loss factor at two frequencies, 100 and 1000 Hz, are shown for polystyrene and for a graft copolymer containing polystyrene and polybutadiene. In the graft polymer the polybutadiene and polystyrene components are in separate phases, with the polybutadiene dispersed as small particles (1–10 μm in diameter) in a glassy polystyrene matrix. The dynamic-mechanical measurements show two main transitions, one at 120 °C corresponding to T_g of the polystyrene matrix at these frequencies, the other at -60 °C corresponding to T_g of the polybutadiene.

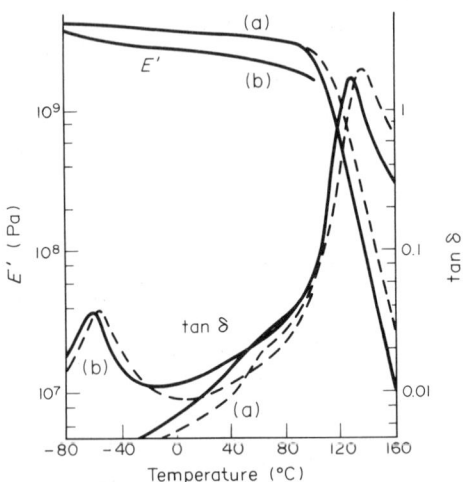

Figure 27   Storage modulus E' and loss factor, tan δ, of polystyrene and a graft copolymer of polystyrene and polybutadiene at (——) 100 Hz and (– – – –) 1000 Hz; (a) polystyrene, (b) graft copolymer

16.4   CRYSTALLINE POLYMERS

The assignment of relaxation transitions to molecular or structural processes in crystalline polymers is not as well established as for amorphous plastics.

The following types of relaxation can be identified: (i) Relaxations in the amorphous phase, *e.g.* the glass transition. (ii) Relaxations within the crystalline phase. These could be movements of defects (such as dislocations) or cooperative movement of chains. (iii) Relaxations which occur in both the crystalline and the amorphous phases. Small differences in the relaxation process may be found depending on which phase is involved. (iv) Relaxations involving large-scale features of the crystalline morphology. Relaxations of this type might involve shear deformations between lamellae or fibrils.

A suitable polymer for illustrating relaxation transitions in crystalline polymers is polyethylene, as much work has been devoted to its study. Tan δ as a function of temperature is shown in Figure 28 for low-density (LDPE) and high-density (HDPE) polyethylene.[3] Three transitions can be seen in LDPE; in order of decreasing temperature they are labelled the α, β and γ relaxations. High-density polyethylene shows a similar behaviour, but the β relaxation is absent and the α relaxation appears to be a composite process of two transitions, labelled α and α'. By systematic variation of the number of chain branching points (*e.g.* number of Me side groups per 1000 chain atoms) it has been shown that the β relaxation is associated with the relaxation of side groups (short branches) in amorphous regions.[34] Figure 29 shows shear modulus, loss modulus and tan δ for linear and highly branched (30 Me groups per 1000 chain atoms) polyethylene.[34] For the linear polyethylene the β relaxation is weak. The α relaxation can be assigned to a crystalline relaxation. This was shown by varying the crystallinity of polyethylene by chlorination.[35] The α loss peak disappeared when evidence of crystallinity disappeared. Annealing experiments on HDPE further showed that the intensity of the α relaxation was inversely proportional to lamellar thickness, suggesting that it is associated with motion of chain folds at the surfaces of the lamellae.[36] The α' relaxation has been shown to be sensitive to the orientation of lamellae with respect to the direction of the applied stress. This transition has been assigned to slip at lamellar boundaries.[37] The γ relaxation increases in

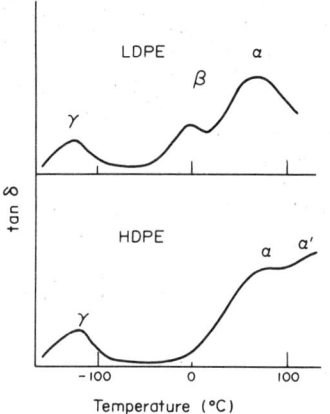

Figure 28 Schematic diagram showing α, α', β and γ relaxation processes in low-density polyethylene (LDPE) and high-density polyethylene (HDPE)

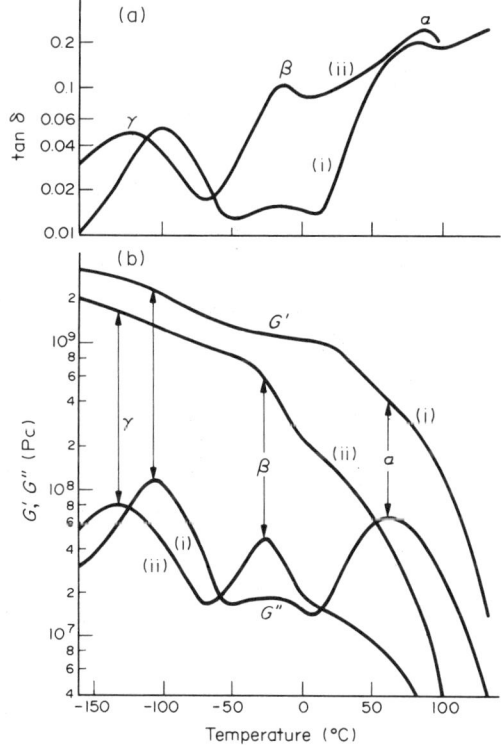

Figure 29 (a) Loss factor, tan δ, and (b) storage modulus G' and loss modulus G'' for linear and branched (30 Me per 1000 chain atoms) polyethylene (PE) as a function of temperature at a frequency between 0.2 and 15 Hz: (i) linear PE, (ii) branched PE

magnitude as the crystallinity decreases, so this relaxation has been assigned to the amorphous regions. Results on solution-crystallized mats and bulk-crystallized samples[37] have indicated that the γ relaxation is also associated with defects, *i.e.* dislocations and point defects within the crystalline lamellae. Therefore the γ relaxation is a composite relaxation.

Figure 30 shows the shear modulus (G') and the logarithmic decrement as a function of temperature for poly(tetrafluoroethylene) (PTFE).[38] The effect of changed crystallinity on the dynamic behaviour is shown. The crystalline structure is different from that of polyethylene, and quenched PTFE also shows recrystallization at $+19\,°C$ and $+30\,°C$. Three main transitions can be identified in PTFE, labelled the α, β and γ relaxations. The α relaxation decreases with increasing crystallinity and is therefore attributed to the amorphous phase; it has been assigned to the glass

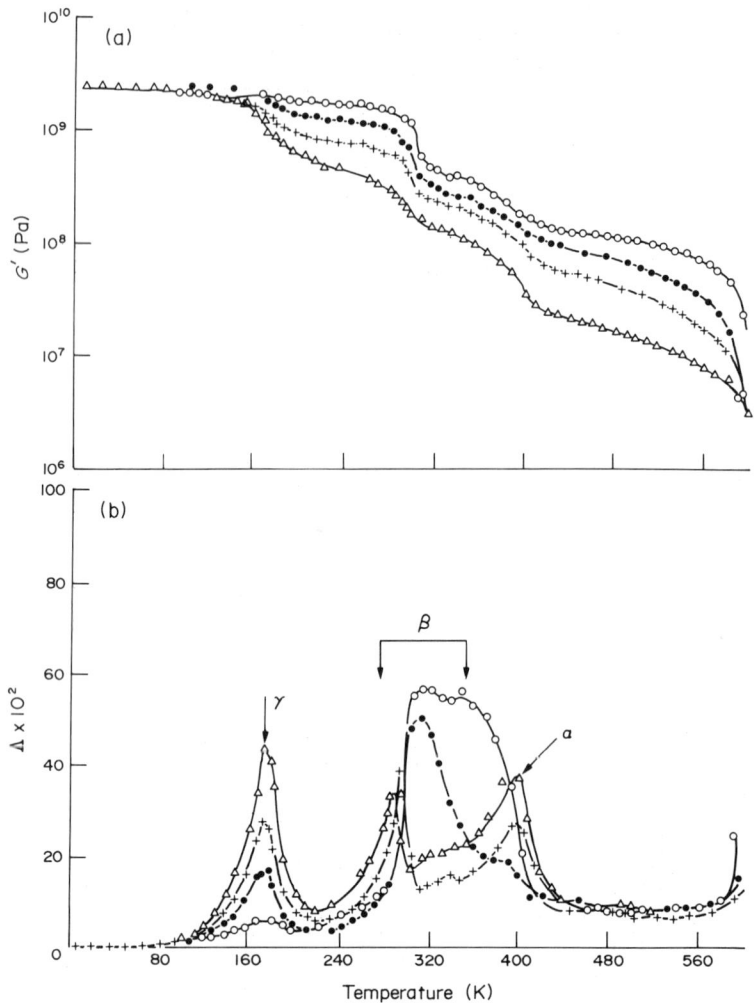

Figure 30 (a) Shear storage modulus G' and (b) damping, $\Lambda = \pi \tan \delta$, for poly(tetrafluoroethylene) of varying crystallinity; torsion pendulum measurements. Crystallinity: ($\bigcirc$) 92%, ($\bullet$) 76%, ($+$) 64%, ($\triangle$) 48%

transition. The β relaxation increases in magnitude and broadens as the crystallinity increases and is therefore attributed to the crystalline phase. The γ relaxation increases in magnitude with decreasing crystallinity and is therefore attributed to the amorphous phase.

Poly(ethylene terephthalate) (PETP), see Figure 31, has two transitions, the α and β relaxations.[39] The α relaxation has been assigned to the amorphous regions, since an increase in crystallinity lowers the height of the peak and broadens the transition. The effect of crystallinity of the β relaxation is very small so it must occurs in both the crystalline and amorphous phases.

Figure 32 shows shear modulus and logarithmic decrement for nylon 6 and nylon 6,6 measured with a torsion pendulum.[40] Strong intermolecular forces existing within the crystalline regions lead to high melting temperatures (*e.g.* 225 °C for nylon 6 and 265 °C for nylon 6,6). At these temperatures there is a steep drop in modulus. The transition occurring around $+50$ °C is the α relaxation and corresponds to the glass transition of the amorphous phase.[41] The transition below -100 °C is the γ relaxation which corresponds to the γ relaxation in polyethylene. The transition between the α and the γ relaxations corresponds to the β relaxation which has been related to movements involving carbonyl groups which have formed hydrogen bonds with absorbed water.[41]

The effect of absorbed moisture on shear modulus and damping in nylon 6,12 is seen in Figure 33.[42] With increasing water content the α relaxation shifts to lower temperatures, whereas the effect on the γ relaxation is rather small. In dry nylon 6,12, the β relaxation is almost absent, but it increases markedly in magnitude with an increase in water content, supporting the assignment of this transition to CO groups with appended H_2O.

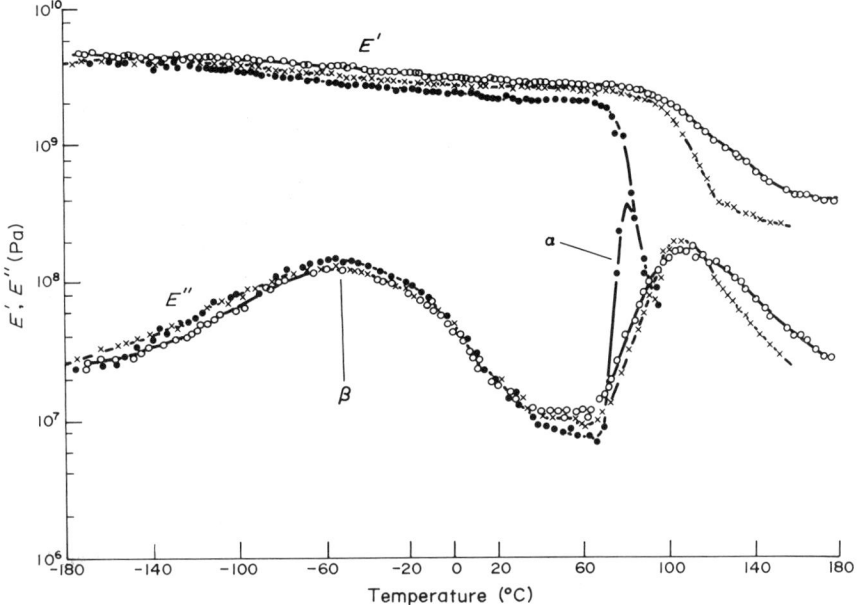

Figure 31   Storage modulus E' and loss modulus E'' for **PETP** with different degrees of crystallinity: (●) 5%, (×) 34%, (○) 50%

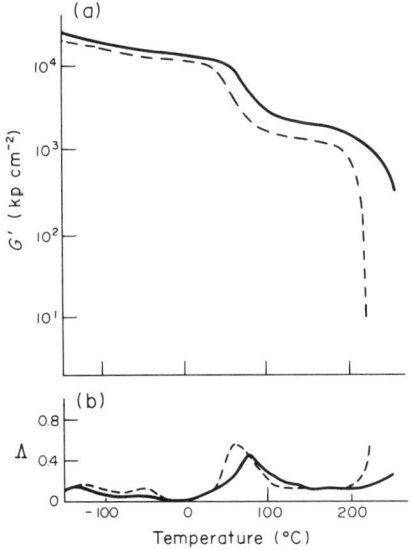

Figure 32   (a) Shear storage modulus G' and (b) damping, $\Lambda = \pi \tan \delta$, in dry nylon 6 (---) and dry nylon 6,6 (——)

16.5   ELASTOMERS

The dynamic-mechanical properties of elastomers have been studied extensively by rubber physicists and technologists for about 50 years. The principal objective in much of this work has been to relate the experimental observations to the known composition and structure of the materials. At first sight it appears that elastomers exhibit extremely complex behaviour, having time-, temperature- and strain-history-dependent hyperelastic properties. This is because elastomers are compounded for practical use and are mixtures of a hyperelastic material (the polymer) with materials exhibiting only short-range elasticity (the filler).

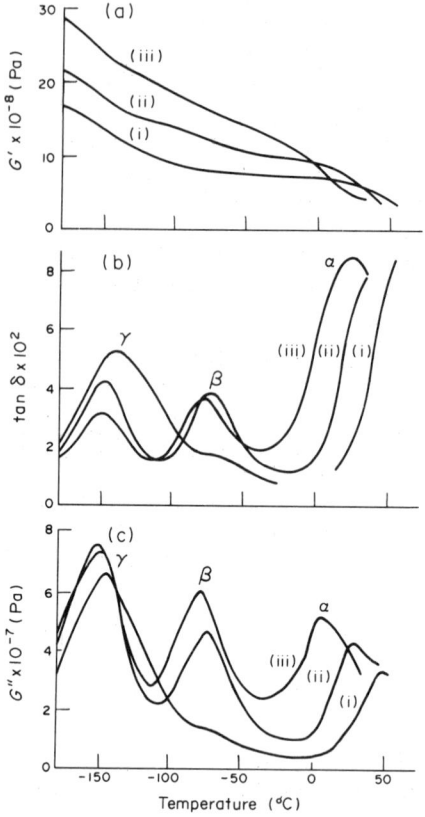

Figure 33   (a) Shear storage modulus, (b) loss factor and (c) loss modulus for nylon 6,12 with different amounts of absorbed water; torsion pendulum measurements. Water content in mol %, (i) 0%, (ii) 14%, (iii) 40%

16.5.1   Influence of T_g

Figure 34 shows log E' and tan δ *vs.* temperature at low strain amplitude in clamped bending for a scragged natural-rubber compound containing 40 parts by weight of HAF carbon black per 100 parts of rubber prepared by means of simple cure system. This compound can be considered to exemplify the behaviour of many engineering elastomers in use today. The two curves can be divided into three distinct regions. The first region, at temperatures well below the glass transition temperature, exhibits an extremely high modulus and a low damping. For commercial elastomers, this region is generally of no practical significance, except as a limit on the low temperature performance of the base polymer. The second region is characterized by a rapid and large drop in modulus with temperature and a high peak in tan δ. Once again this region is only of practical

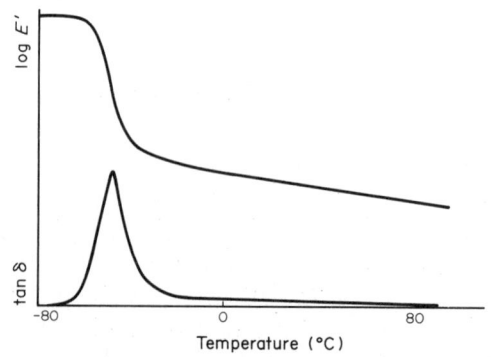

Figure 34   Logarithm of the storage modulus and tan δ against temperature for HAF (High Abrasion Furnace) black filled natural rubber (40 parts black per 100 of rubber) bending at 1 Hz

importance when high damping of small vibrational amplitudes is required. The high temperature sensitivity of both modulus and damping, and the heat build-up inherent in any reasonable amplitude-damping application, make materials with properties in this region difficult to apply, except over a narrow temperature band. The third region, at high temperature, which encompasses the majority of the curve in Figure 34, is referred to as the rubbery region and is characterized by a low-to-medium damping and a low modulus. Materials with these rubbery properties are of main industrial importance.

The transition region (second region) has always been important for those interested in practical damping applications. This region can be moved in temperature, more or less at will, by proper choice of the base polymer and by use of modern compounding techniques. Figure 35 shows results for a 50 wt % acrylonitrile-content rubber (*i.e.* high-nitrile rubber) which exhibits peak damping at room temperature at normal mechanical frequencies. This compound is of little practical use since it offers useful damping only over a narrow temperature range. The problem is often tackled by blending two incompatible elastomers with different transition temperatures in such a way that their transition peaks merge to give a broader damping band. Results for a simple example, a blend of the high-nitrile rubber and a bromobutyl rubber, is shown in Figure 36. The height of the individual damping peaks has been reduced by the blending, but the useful range of damping has been greatly extended. Compared with the original copolymers, the shapes and temperatures of the two peaks are somewhat changed; in particular, the damping transition of the nitrile rubber is broader.

The influence of time and temperature on dynamic properties may be accounted for by the Method of Reduced Variables.[10] The results of this method are shown for the high-nitrile rubber in Figure 37, which gives plots of modulus (E') and loss factor against reduced frequency. Reduced frequency is a single 'universal' frequency axis based upon a particular reference temperature T_s. The reduced curve is produced by shifting the curves of modulus and tan δ *vs.* frequency at various temperatures along the frequency axis, so that they align to produce one 'master curve'. The required

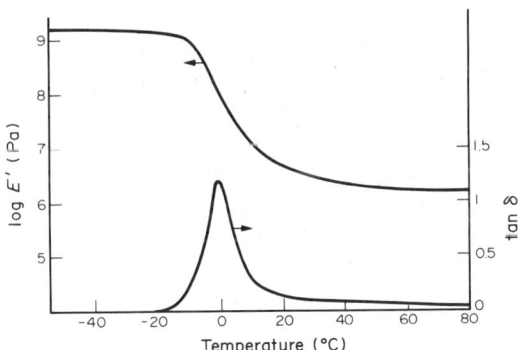

Figure 35 Logarithm of the storage modulus and tan δ against temperature for high-nitrile rubber

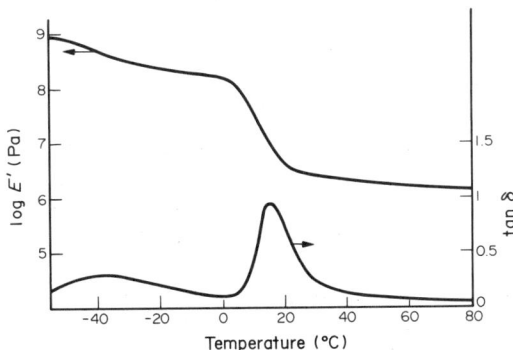

Figure 36 Logarithm of the storage modulus and tan δ against temperature for a blend of high-nitrile rubber and bromobutyl rubber at a frequency of 38 Hz

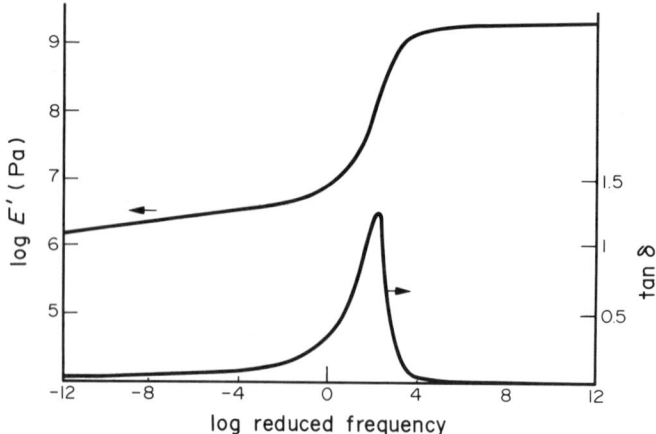

Figure 37 Master curves of $\log E'$ and $\tan \delta$ for high-nitrile rubber (data reduced to 296 K)

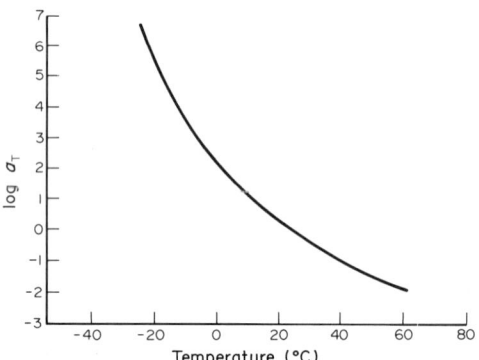

Figure 38 Shift factor against temperature for a high-nitrile rubber

shifts are described by the empirical relationship

$$\log a_T = -8.86(T - T_s)/(101.6 + T - T_s) \tag{103}$$

where a_T is the ratio of the frequencies at T and T_s at which the property considered has the same value; T is the temperature of test for the data being shifted; and T_s is the reference temperature for which the master curve is obtained. Figure 38 shows the curve of shift factor ($\log a_T$) *vs.* temperature used in preparing Figure 37. The important feature to note is that it is non-linear.

The method is remarkably accurate for short-range extrapolation of the frequency range and the plot has practical use, since it is a convenient indicator of the sensitivity of the material properties to temperature. In fact the shape of the curve of $\log a_T$ *vs.* T for any elastomers is the same as that of the high-nitrile compound (see Figure 38). Generally, as the temperature increases away from the glass transition temperature the material properties in the frequency domain become less temperature sensitive.

If such a shifting process were attempted for a mixture of polymers, the situation changes and the simple shifting law is no longer obeyed. Fesko and Tschoegl[43] have developed an extension of the method of reduced variables to account for two-phase systems, such as the bromobutyl-rubber/ nitrile-rubber blend, which takes into account its composite structure. This work allows prediction of the broadening of the damping peak and flattening of the $\log a_T$ *vs.* temperature curve at high temperature, as shown in Figure 39.

It transpires that there is a practical limit to the broadening of the damping peak that can be achieved by simple blending. This is due mainly to limitations of the control of the morphology of the phases in a hyperelastic medium. Other methods have to be sought to force the two interesting environments to remain together. One method is to employ an interpenetrating network (IPN); *i.e.* two separate and physically incompatible polymers produced in a physically entangled state by

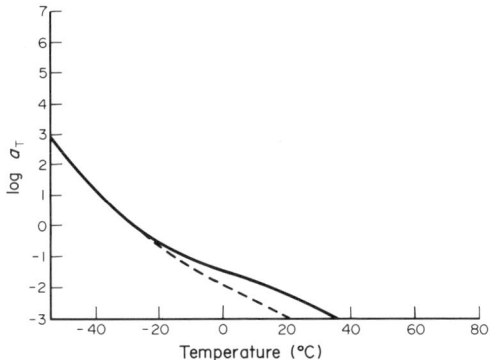

Figure 39 Shift factor against temperature for a blend of high-nitrile rubber and bromobutyl rubber

polymerizing or crosslinking them in separate non-interacting chemical reactions.[44] Another method is to modify one or both of the original polymers by incorporating side chains (by grafting or other means): thus changing the mutual interaction of the polymers in the blend.[45]

16.5.2 Fillers

The principal effects of carbon black on the dynamic properties of elastomers were established as early as 1942. These can be summarized as an increase in both the elastic modulus and damping, compared with the unfilled material, and a pronounced strain-amplitude dependency, which is not found to a significant level in unfilled rubbers. It is true that any filler normally used in elastomers at least increases the elastic modulus, but the reason for the importance of carbon black is that no filler does it better. Not only does carbon black increase the modulus, it also improves the general strength and fatigue properties of elastomeric materials to a level which changes them from a curious hyperelastic novelty into a useful engineering material.

The mechanism of reinforcement of elastomers by carbon black is best understood by reference to the curve of shear modulus *vs.* strain amplitude shown in Figure 40. The filler imparts a strain-amplitude softening to the material. Carbon black forms aggregates within the elastomer which act as rigid bridges, hence restricting the freedom of motion of the chain molecules. This reduction in flexibility increases the modulus at low strain amplitudes. When the amplitude is increased, these aggregates, which have no long-range elasticity, are broken by the stress imposed through the polymer network, thus reducing the modulus (at large strain amplitudes) to the lower limiting value G_∞. It is this breakdown of structure that toughens these materials and protects the elastomeric chain. The breakdown is also manifested as a general increase in damping with strain amplitude. Payne[46] studied this behaviour in detail in the early 1960s, and showed that curves of tan δ *vs.* strain

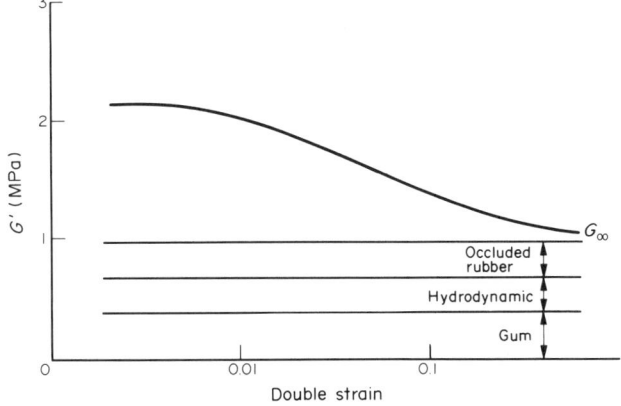

Figure 40 Shear storage modulus against double strain amplitude showing components of G_∞ for natural rubber filled with HAF (High Abrasion Furnace) black (40 parts of by weight black to 100 of rubber) at 21 °C and 0.3 Hz

amplitude go through a small peak in the region where filler breakdown is most pronounced. However, even today, there is no completely satisfactory explanation for the presence of this peak.

The stress on the carbon black aggregate is applied through a distribution of lengths of polymer chains. This results in the smooth curve of modulus *vs.* strain amplitude seen in Figure 40. The figure has been subdivided into four zones as suggested by Medalia[47] after an original idea by Payne.[46] The bottom zone represents the modulus of the gum, *i.e.* the unfilled elastomer. The modulus is raised by the presence of the carbon black through the zone labelled 'hydrodynamic'. This increase in modulus is thought to be analogous to the increase in viscosity of a fluid when 'filled' with an inert material and has been described mathematically on that basis.[48] This treatment, however, falls short in its ability to predict the large strain-amplitude modulus (G_∞), *i.e.* the modulus when all effects due to the filler aggregates are removed. The extra contribution has been ascribed to 'occlusion' within the filler particle of part of a rubber molecule making it less flexible and forming particle-to-particle links. The 'occluded' structure is not broken down by large strain amplitudes in the way that physical aggregates are, and is a permanent addition to the level of G_∞. The lower three zones can thus be considered as direct results of the properties of the polymer/filler system. The phrase 'strain amplification' is frequently used with advantage here, since it implies that the filler increases the modulus by making the elastomer play a more active role in the deformation cycle. The upper zone cannot be considered in this light. The behaviour here is dominated by the progressive breakdown of the filler aggregates, which cannot keep pace with the hyperelastic behaviour of the physically adsorbed elastomer. What is remarkable is the ability, particularly in the case of carbon black, of the agglomerates to slowly but reversibly reform after breakdown, so that the reinforcement reappears after resting. This recovery process, at a slow rate, leads to some additional complexity to the dynamic properties. The strain history effects and the behavioural separability of the components of modulus below and above G_∞, lead to an inversion of the slope of the modulus *vs.* temperature curve at a level of prestrain dependent upon the loading of carbon black.[49] For low levels of prestrain curves, the materials shows a decrease in modulus with temperature, whilst for high prestrain the curves follow the proportionality of modulus to absolute temperature predicted by the Gaussian theory of rubber elasticity.[50]

16.5.3 Oils

As we have seen, the nature of the filler as well as the polymer has a large influence on the dynamic-mechanical properties of elastomeric materials. The effects of other ingredients in the compound can usually be ignored, except in the final stage of property tuning. Two components, however, cannot be treated in this way: plasticizers (oils) and crosslinking agents.

Plasticizers and oils, which can be incorporated in extremely large quantities into certain elastomer compounds, are commercially important for two main reasons. They depress the glass transition temperature and hence improve the working temperature range of an elastomer selected on the basis of other properties. Secondly, they allow the incorporation of more diluent filler than the surface adsorption properties of the polymer would normally allow, hence reducing the volumetric cost of the compound.

Apart from the depression of T_g and the accompanying dispersion of the tan δ peak, the main effect of oil on the dynamic properties is an overall reduction in the elastic modulus, when measured at constant filler loading and crosslink density.[51] This is accompanied initially by a slight rise in the magnitude of the tan δ peak followed, at higher oil levels, by a large fall. This latter fall is associated with the separation of the agglomerates by the interposed oil, which reduces the hysteresis effect.

Many contradictory reports on the effects of oils on elastomers have been published. The problem is the variable effect of the oil on the dispersion of the fillers and curatives during mixing: poor dispersion can produce significantly greater changes in properties than the presence of the oil itself.

16.5.4 Crosslinking

Low crosslink density produces a compound with high hysteresis due to the inadequate three-dimensionality of the structure. At low levels of crosslink density, the resultant compound is of little practical importance due to the low values of important mechanical properties, particularly strength. Payne[52] showed that the value of G'_∞ increased rapidly with increase in degree of cure, whereas the difference between the low strain-amplitude modulus and G'_∞ was essentially unchanged by increase in degree of cure. This difference is the component of modulus which is due to foreshortening of chains by adsorption of polymer on the filler particles. This physical adsorption should have no interaction with the chemically crosslinked network, except at impractically high crosslink densities, and so can be considered separately. Payne also found that $\tan \delta$ followed the general trend of decreasing with increasing crosslink density, but showed no simple relationship with it.

ACKNOWLEDGEMENT

The authors would like to thank Professor R. E. Wetton, Dr. B. E. Read and Dr. E. F. T. White for their help and advice in the preparation of this review.

16.6 REFERENCES

1. B. E. Read and G. D. Dean, 'The Determination of Dynamic Properties of Polymers and Composites', Hilger, Bristol, 1978.
2. L. E. Nielson, 'Mechanical Properties of Polymers and Composites', Dekker, New York, 1974.
3. I. M. Ward, 'Mechanical Properties of Solid Polymers', 2nd edn., Wiley, New York, 1983.
4. L. Boltzmann, *Pogg. Ann. Phys.*, 1876, **7**, 624; *Wied. Ann.*, 1878, **5**, 430.
5. H. Leaderman, 'Elastic and Creep Properties of Filamentous Materials and Other High Polymers', The Textile Foundation, Washington, DC, 1943.
6. T. Alfrey, 'Mechanical Behaviour of High Polymers', Interscience, New York, 1948.
7. B. Gross, 'Mathematical Structure of the Theories of Viscoelasticity', Hermann, Paris, 1953.
8. S. Glasstone, K. J. Laidler and H. Erying, 'The Theory of Rate Processes', McGraw-Hill, New York, 1941.
9. M. L. Williams, R. F. Landel and J. D. Ferry, *J. Am. Chem. Soc.*, 1955, **77**, 3701.
10. J. D. Ferry, 'Viscoelastic Properties of Polymers', 2nd edn., Wiley, New York, 1970.
11. M. H. Cohen and D. J. Turnbull, *J. Chem. Phys.*, 1959, **31**, 1164.
12. G. Adam and J. H. Gibbs, *J. Chem. Phys.*, 1965, **43**, 139.
13. L. Cremer, M. Heckl and E. E. Ungar, 'Structure Born Sound', Springer-Verlag, Berlin, 1979.
14. D. E. Kline, *J. Polym. Sci.*, 1956, **22**, 449.
15. J. W. Pendered and R. E. D. Bishop, *J. Mech. Eng. Sci.*, 1963, **5**, 4.
16. R. F. Gibson and R. J. Plunkett, *J. Compos. Mater.*, 1976, **10**, 325.
17. D. X. Lin and R. D. Adams, *J. Phys. (Orsay, Fr.)*, 1984, **44**, 69.
18. S. A. Suarez, R. F. Gibson and L. R. Deobald, 'Experimental Techniques 8', 1984, vol. 10, p. 19.
19. B. E. Read, G. D. Dean and J. C. Duncan, in 'Physical Methods of Chemistry', ed. B. W. Rossiter, J. F. Hamilton and R. C. Baetzold, Interscience, New York, to be published.
20. A. J. Barlow, G. Harrison, J. Richter, H. Seguin and J. Lamb, *Lab. Pract.*, 1961, **10**, 786.
21. A. M. North, R. A. Pethrick and D. W. Phillips, *Macromolecules*, 1977, **10**, 992.
22. G. L. Petersen, B. Chick and W. Junker, *Ultrason. Symp. Proc.*, 1975, 650.
23. S. V. Hoa and Q. B. Nguyen, *Polym. Comp.*, 1983, 4 (2), 85.
24. R. E. Wetton, in 'Developments in Polymer Characterisation — 5', ed. J. V. Dawkins, Elsevier Applied Science, London, 1986, chap. 5.
25. J. Heijboer, in 'Physics of Non-Crystalline Solids', J. A. Prins, Holland, Amsterdam, 1965, p. 231.
26. P. I. Vincent, in 'Physics of Plastics', ed. P. D. Ritchie, Iliffe Books, London, 1965.
27. T. G. Fox and P. J. Flory, *J. Appl. Phys.*, 1950, **21**, 581.
28. T. J. Fox and P. J. Flory, *J. Polym. Sci.*, 1954, **14**, 315.
29. M. F. Drumm, C. W. H. Dodge and L. E. Nielsen, *Ind. Eng. Chem.*, 1956, **48**, 76.
30. K. Wolf, *Kunststoffe*, 1951, **41**, 89.
31. P. Sakellariou, R. C. Rowe and E. F. T. White, *Int. J. Pharm.*, 1986, **31**, 55.
32. A. Yim, R. S. Chahal and L. E. St. Pierre, *J. Colloid Interface Sci.*, 1973, **43**, 583.
33. H. Oberst, *Ber. Bunsenges. Phys. Chem.*, 1969, **70**, 375.
34. K. A. Wolf, *Z. Electrochem.*, 1961, **65**, 604.
35. K. Schmieder and K. Wolf, *Kolloid Z. Z. Polym.*, 1953, **134**, 149.
36. K. M. Sinnot, *J. Appl. Phys.*, 1966, **37**, 3385.
37. N. G. McCrum and E. L. Morris, *Proc. R. Soc. London, Ser. A*, 1966, **292**, 506.
38. N. G. McCrum, *J. Polym. Sci.*, 1959, **34**, 355.
39. M. Takanyagi, *Mem. Fac. Eng. Kyushi Univ.*, 1963, **23**, No. 1, 1, 41.
40. H. Dreier, *Motortech. Z.*, 1963, **9**, 24.
41. G. Schreyer, 'Konstuieren mit Konststoffen', Hanser, Munchen, 1972, p. 446.
42. K. H. Illers, *Makromol. Chem.*, 1968, **38**, 168.
43. D. G. Fesko and N. W. Tschoegl, *J. Polym. Sci., Part C*, 1971, **35**, 51.
44. L. H. Sperling, A. F. George and V. Huelck, *J. Appl. Polym. Sci.*, 1970, **14**, 2815.
45. P. G. Howgate, 'The Use of Multiple Transition Polymerics in shock and Vibration Damping', paper presented at the 55th Shock and Vibration Symposium, Oct. 1984.
46. A. R. Payne, *Rubber Plast. Age*, 1961, 963.
47. A. I. Medalia, *Rubber Chem. Technol.*, 1978, **51**, 473.
48. E. Guth and O. Gold, *Phys. Rev.*, 1938, **53**, 322.
49. P. G. Howgate, *Proc. Scand. Rubber Conf. 7th* 1983, 284.
50. L. R. G. Treloar, 'The Physics of Rubber Elasticity', 3rd edn., Oxford University Press, Oxford, 1975.
51. M. Gajewski, *Polimery (Warsaw)*, 1974, **19**, 244 (*Engl. Transl. Int. Polym. Sci. Technol.*, 1975, **2** (2), T/15).
52. A. R. Payne, R. E. Whittaker and J. F. Smith, *J. Appl. Polym. Sci.*, 1972, **16**, 1191.

17

Acoustical Properties

RICHARD A. PETHRICK
University of Strathclyde, Glasgow, UK

17.1 INTRODUCTION

Ultrasonic measurements have been performed on polymer solids and solutions for well over 40 years; however, the method is one of the least understood or applied of those presented in these volumes. The two main reasons for the low level of research activity in this area are: (i) the highly labour intensive nature of the experiments; and (ii) a lack of apparent interest by major instrument manufacturers in the production of the ultrasonic techniques. The use of ultrasonic measurements to study the dynamic properties of polymers has been the subject of several reviews[1,4] and book chapters.[5,6] In this chapter, the principles behind the ultrasonic method will be initially described. The review will then consider the type of data which can be obtained for polymer solutions, melts and solids and also identify its potential as a non-destructive testing method for the characterization of polymeric materials.

17.2 PRINCIPLES OF ULTRASONIC PROPAGATION IN MOLECULAR SYSTEMS

Ultrasonic waves, usually generated by the application of a radio frequency electrical signal to a piezo-electric transducer, can propagate through a molecular system in the form of a perturbation of the momentum distributions in a liquid translational temperature or the phonon density of a solid.[3,4] Changes in the translation energy distribution can lead to a perturbation of the populations of other energy states coupled to the momentum distribution *via* inelastic collision. The exchange of energy between various states requires a finite period of time and can hence lead to the possibility of relaxational phenomena. In a simple fluid, such as carbon tetrachloride, inelastic collisions can produce an increase in the population of the various vibrational and rotational states possessed by the molecule. At low frequency or long times, the sound wave can efficiently perturb the population of states feeding energy into vibrationally excited states. Increasing the frequency allows a point to be reached where the rate of perturbation is faster than the rate of energy exchange and hence energy is *not* fed into the excited states (Figure 1). Hence the energy absorbed from the propagating ultrasonic wave decreases as the frequency of observation is increased from below to above the natural exchange rate for the process. In a molecule such as 1,2-dibromo-1,1-difluoroethane, pertubation of the translational temperature can, through vibrational–rotational coupling, induce torsional excitation and eventually rotational isomerization to a more highly populated *cis* isomeric form.[7] The loss of sound energy is a measure of both the energy difference between the isomeric states and the rate of interconversion between the *cis* and *trans* forms.[5] At low frequency, the sound attenuation is large because the efficiency of promoting molecules into the high energy state is high. Increasing the frequency of perturbation allows a condition to be achieved where the rate of inelastic collision is so fast that little or no excitation to the *cis* state occurs. As a consequence energy is retained in the sound wave and the loss of energy decreases markedly. The frequency at which the changes in the absorption occur are directly related to the rate constant for the *cis* to *trans* isomerization process and the energy lost from the sound wave per cycle exhibits a maximum at the frequency of the isomeric conversion. Data may be presented either as a function of frequency at constant temperature or as a function of temperature for a fixed frequency. The theory of ultrasonic propagation has been described elsewhere.[5] The continuum approach to sound propagation considers that the attenuation is described by the Navier–Stokes equation. For a non-relaxing fluid the hydrodynamic

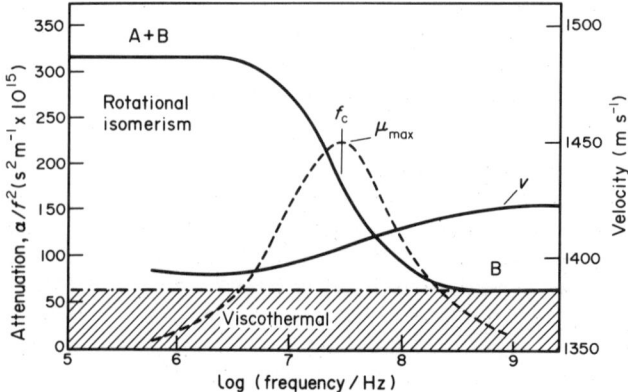

Figure 1   Typical ultrasonic relaxation for a molecular system undergoing a simple rotational isomeric process

equations yield a square law dependence of the absorption on frequency given by

$$\frac{\alpha}{f^2} = \frac{2\pi^2}{\rho c^3}\left(\frac{4}{3}\eta_v + \eta_s\right) + \frac{(\gamma-1)}{C_p}\kappa \tag{1}$$

where ρ is the density, c is the speed of the sound wave, η_s and η_v are respectively the shear and volume viscosities, γ is the ratio of the specific heat at constant pressure C_p to that at constant volume C_v, κ is the thermal conductivity and f is the frequency ($\omega = 2\pi f$) of the observation. At frequencies below 1.0 GHz, the absorption coefficient divided by the frequency squared is found to be independent of frequency for most non-associated liquids. Observations above 10 GHz show that deviations from this simple law occur as a result of inadequacies in the continuum approach when the time constant approaches that of a molecular collision.

In a molecular system with internal degrees of freedom, dynamic storage of energy, which leads to a frequency dependent absorption of sound, occurs, through inelastic collisions. The dispersion associated with a simple rotational isomeric process may be described by the equation

$$\frac{\alpha}{f^2} = \frac{A}{1+(f/f_c)^2} + B \tag{2}$$

17.3 ROTATIONAL ISOMERISM IN SIMPLE MOLECULES

Flory in his book on the statistical mechanics of conformational change in polymer molecules[8] has identified the importance of understanding the factors influencing the intramolecular potential. Relaxation data and associated thermodynamic parameters for a wide range of substituted ethanes have been presented elsewhere.[5,6] A good correlation is observed (Figure 2) between the sum of the Van der Waals radii of the atoms attached to the bond about which rotation occurs and the activation energy for isomerization. Rotation about a carbon–carbon bond involves the atoms on opposite ends of the bond being brought into an eclipsed state and the energy for this process is dominated by repulsive forces between closed electron shell structures. It appears that the dipolar contribution associated with individual bonds influences the energy difference between isomeric states, but makes a relatively small contribution to the energy determining the barrier to internal rotation.

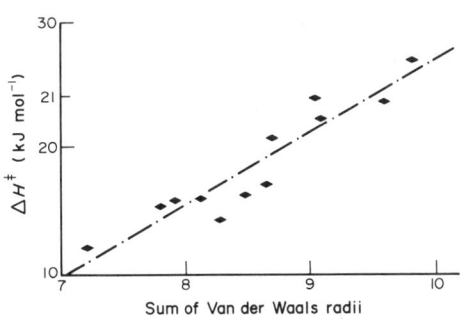

Figure 2 Correlation of barrier heights and Van der Waals radii for simple ethane-like molecules

In partially double-bonded structures, such as the α,β-unsaturated aldehydes,[9] a good correlation is observed between the activation energy for rotational isomerism and the magnitude of the ^{13}C–H proton NMR coupling constant. The coupling constant is a direct measure of the s orbital electron density of the starred carbon atom involved in the isomeric process and hence also a direct measure of the π electron density. The correlation observed in Figure 3 is consistent with the idea that for rotational isomerism to occur, partial localization of electron density on to the carbon atoms forming the α,β linkage must occur.

The above are just two examples of the types of system which have been studied ultrasonically. Two extensive reviews of the data on small molecules have been published elsewhere.[5,6]

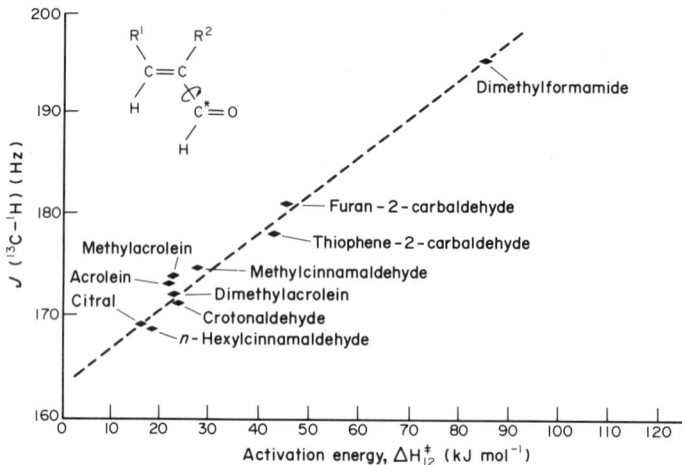

Figure 3 Correlation of barrier height *vs.* the ^{13}C–H shifts for rotational isomerism in α,β-unsaturated aldehydes

17.3.1 Rotational Isomerism in Hydrocarbons

In order to understand the dynamic properties of polymers it is necessary to understand the way in which the intermolecular potentials are influenced by the number of bonds involved in rotational isomeric processes.[10] Butane is the direct analogue of the simple ethane molecules described above. The barrier height restricting internal rotation is the result of non-bonding interaction between terminal methyl groups.[8] Pentane, however, represents a more complex system in that we have to consider the effects on the barrier height of coupling the rotations about the C(2)–C(3) and C(3)–C(4) rather than independent motion of the component elements. Detailed calculations of the non-bonding interactions as a consequence of changes in the rotational angles[8,11] lead to the idea of a potential energy surface (Figure 4). The rotational isomerism involves concomitant rotation of both the C(2)–C(3) and C(3)–C(4) bonds rather than motion about the C(2)–C(3) followed by rotation about the C(3)–C(4) bond. The calculations once more use non-bonded interactions and the profile indicates that the isomeric process proceeds *via* compensation of interactions leading to the formation of the eclipsed state through correlated motions of the terminal methyl groups, the so called 'cog-wheel effect'. Increasing the number of bonds leads to the possibility of even more complex three and higher dimensional representations. It is observed that as the length of the chain increases, so the magnitude of the activation energy falls below a value of twice the single-bond rotational energy expected if the rotational isomeric process were to involve separate motion about two separate bonds. The relaxation data clearly indicate that as the length of the chain increases so the process becomes increasingly more cooperative and involves collective excitation of a number of bonds. A variety of theoretical models are available to describe these processes, one of the most successful being that due to Monnerie.[12] The extent of coupling between the rotational isomeric

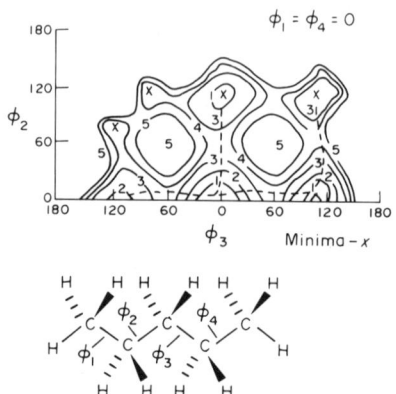

Figure 4 Potential surface for the isomerization of pentane

processes associated with neighbouring bonds is introduced into the theory as an adjustable parameter. An extensive review of models describing the dynamic processes have been published elsewhere.[13]

The validity of the use of the simple additive non-bonding approximation in the calculation of rotational isomeric potentials has been demonstrated in a recent study of various substituted pentanes and hexanes.[14,15] It was observed that not only does the theory predict accurately the energy difference between states but also provides a reasonable estimate of the barrier restricting free rotation. In the case of molecules in which more than one structure can be ascribed to the conformational isomeric isomers involved in the relaxation process, the calculations are able to assist in assigning the states involved. Hydrocarbons with *n* less than 10–14 will tend to generate a single conformational defect, whereas longer chain molecules can form a double conformation defect and hence adopt a hair pin structure.[16] Since most of the longer chains are solids, discussion of their relaxation behaviour will be deferred to later in this chapter.

A close analogue of the hydrocarbons are the fluorocarbons. The fluorine atoms are able to induce quadrupolar interactions leading to an isoenergetic double potential minimum displaced approximately 10° from the equatorial configuration. These minima are of considerable significance as they force the molecule to adopt a helical conformation and this structure is retained in the solid phase. Ultrasonic measurements[17] combined with ^{19}F NMR studies[18] of short chain fluorinated hydrocarbons confirm these theoretical predictions and indicate that facile exchange between *trans* minimum energy states occurs.

17.4 ROTATIONAL ISOMERISM IN POLYMERS

17.4.1 Polystyrene

Studies of the effect of increasing chain length on the ultrasonic relaxation of narrow molecular weight distribution samples of polystyrene[19–22] indicate that both the amplitude and the frequency vary with molecular mass (Figure 5). Studies of dimers, trimers and heptamers indicate that as the number of groups in the chain increases, so the frequency of the relaxation process decreases and the amplitude increases. Analysis of the temperature dependence of the ultrasonic relaxation indicates that both the energy difference between isomeric states and the height of the barrier to rotational isomerism increase with chain length. The trend apparently continues up to a molecular weight of 10 000 and gives rise to the concept of 'end group' effects. In the dimer,[23] the eclipsed state will involve bringing a hydrogen atom and a phenyl ring into close proximity. The trimer[24] will similarly involve these same groups being brought into close proximity, but the strain in the bonds can be reduced by a cooperative realignment of the phenyl group about its pendant linkage. As the length of the chain is increased, so the effects of 1–3, 1–5 and higher order interactions become important. In the case of polystyrene,[19] these long range interactions lead to an increase in the energy difference

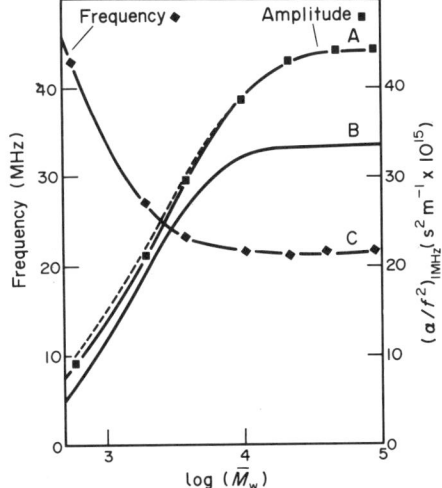

Figure 5 Variation of relaxation parameters with molecular mass for polystyrene solutions in toluene

and also the activation energy for rotational isomerism. Conformational changes of the chain ends are less restricted than those associated with motion of elements in the centre of the polymer, as reflected in the changes in the energy difference and barrier height. Above a molecular weight of 10 000, the number of groups at the end of the polymer relative to those in the main backbone become so small as to have a negligible effect on the observed relaxation. A detailed analysis of the ultrasonic relaxation data indicates that part of the absorption observed in the MHz frequency range arises from 'collective' normal mode processes of the polymer chain.[25]

A recent study[26] of polystyrene over an extended temperature range has indicated that a helix–coil transition can be observed in certain high molecular weight samples at approximately 333 K. It has also been reported that crystals of polystyrene grown above 333 K have a different structure, as revealed by X-ray analysis, compared with those formed below this temperature.

17.4.2   Collective Normal Mode Relaxation

In a wide range of polymer systems it has been observed that in the frequency range 10 kHz–10 MHz, a contribution to the ultrasonic relaxation can be identified due to the dynamic shear viscosity contribution to equation (1). The dynamic shear viscosity arises from cooperative normal mode distortion of the whole polymer molecule (Figure 6). Studies of the ultrasonic relaxation in the frequency range 10 kHz–10 MHz for dilute polystyrene solutions in toluene exemplifies these processes (Figure 7).[25] It is apparent, as predicted by theory,[27,28] that as the molecular weight of the polymer increases so the frequency of the first normal mode relaxation decreases. Similiar effects to those outlined above have been observed in poly(methyl methacrylate),[29]

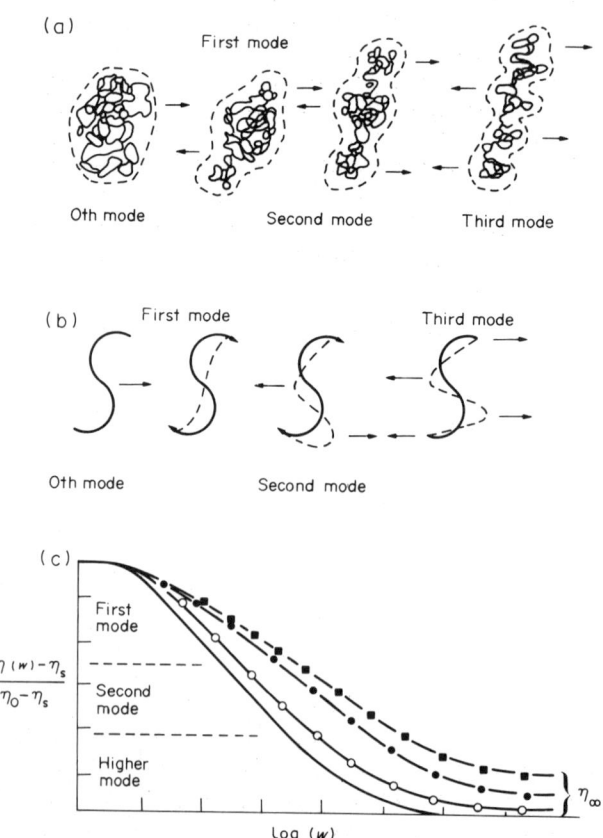

Figure 6   Normal mode relaxation processes in linear polymers: (a) motion of a polymer molecule when subjected to an oscillating shear field; (b) diagrammatic representation of the motion of an ideal flexible chain when in an oscillating shear field; (c) viscoelastic relaxation of a polymer in solution. η_∞ is the high frequency limiting value, η_0 is the zero frequency and η_s is the dynamic shear viscosity. Curves are for the ideally flexible free-draining coil (——), the ideally flexible non-free-draining coil ($\bigcirc$), partially flexible free-draining coil ($\bullet$), and partially flexible non-free-draining coil ($\blacksquare$)

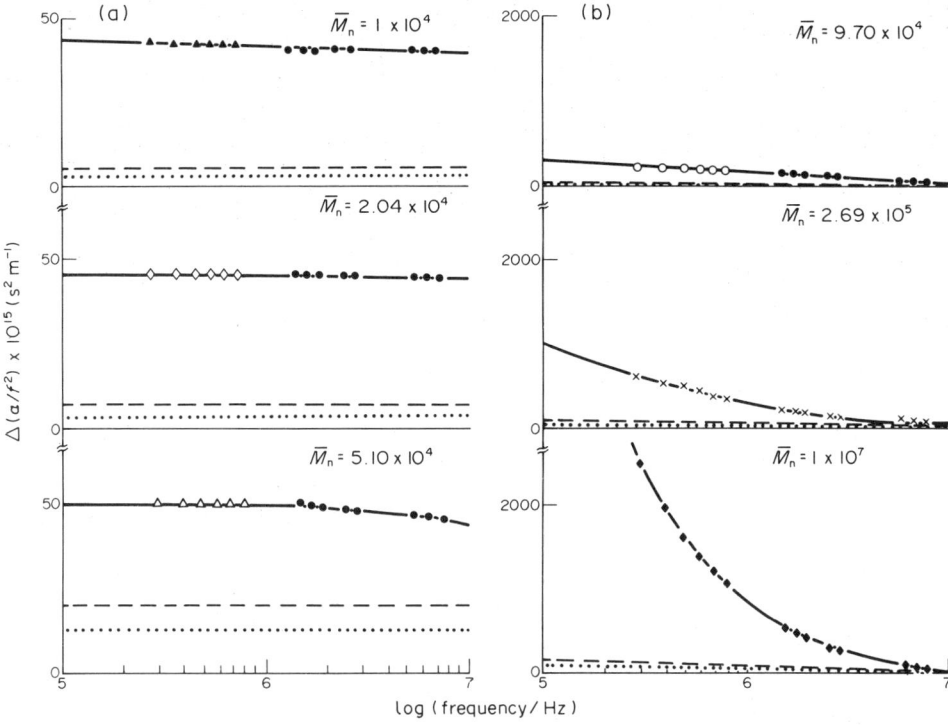

Figure 7 Summary of relaxation behaviour for polystyrene in the frequency range 10 kHz to 10 MHz

poly(α-methylstyrene)[30] poly(vinylpyrrolidone)[31] and polydioxanes.[2] It is clear that in both hydrophobic and hydrophilic polymers both the end group and normal mode effects are clearly observed and are not unique to polystyrene.

17.4.3 Concentrated Polymer Solutions and Melts

The key to understanding the ultrasonic relaxation in concentrated solution and melts is to appreciate that increasing the concentration of the polymer also increases the degree of interaction between individual polymers and leads to chain–chain overlap. Studies of polystyrene[32] and poly(dimethylsiloxane) (PDMS)[33, 34] indicate that it is only when the coils of separate polymers overlap that an additional contribution is observed to the high frequency relaxation processes. As the concentration of the polymer molecules in solution is increased, entanglement will occur and this leads to non-linear changes in the ultrasonic relaxation amplitude variation with concentration as well as modification of the frequency dependence. A comparison of the viscoelastic relaxation for linear high molecular weight PDMS with the ultrasonic relaxation indicates that there are differences in the processes associated with the effects of entanglements modifying the normal mode spectrum and also contributions from unconstrained motions of chain ends.[35] A detailed discussion of the ultrasonic relaxation in concentrated polymer solutions has been recently published elsewhere.[32]

Poly(dimethylsiloxane) also illustrates the complexity of processes which occur within a polymer molecule when undergoing conformational isomerism.[33] Segmental relaxation in PDMS can be studied by ^{1}H NMR, dielectric and nuclear magnetic resonance relaxation.[36, 40] Extrapolation of the low frequency–low temperature relaxation data (Figure 8) to high temperatures indicates that the rate of relaxation would achieve a value in excess of the natural collison frequency for molecular species, *i.e. ca.* 10^{15}–10^{17} s. Such a situation is physically impossible and we find that the actually observed value of 10^{10} s is apparently independent of temperature. This effect has been discussed by Eyring in terms of his 'starvation kinetics' model.[41] At low temperatures, the rate constant is proportional to the probability of a particular segment becoming thermally activated and a simple exponential type of relationship is observed. At higher temperatures the segments have energies in excess of the barrier height and the conformational process is *not* controlled by the probability that

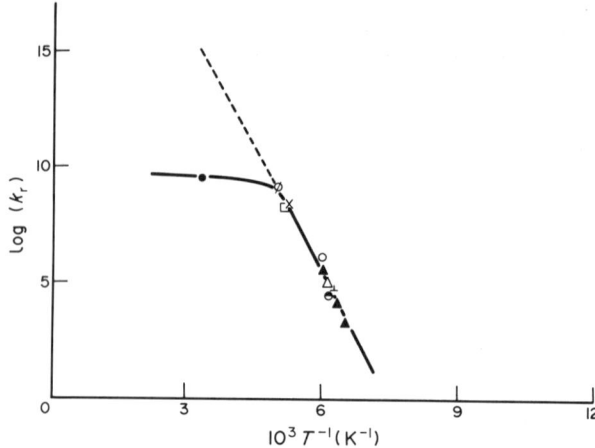

Figure 8   Activation energy plot for rotational isomerism in linear poly(dimethylsiloxane). NMR data: Ø,[37] ○,[38] ×,[39] □,[36] △;[36] dielectric data: ▲,[40] ◓,[40] ⊥;[40] ultrasonic data: ●.[34] Extrapolated---, experimental---

an element can be thermally excited by internal energy redistribution but is dependent upon the rate at which energy is fed into the segment from translational energy. Energy for the conformational excitation comes from inelastic collisions and the high temperature region implies that conformation change is limited by the rate at which energy is fed into the torsional modes by polymer–polymer interactions rather than by the probability that a particular segment achieves an energy in excess of the barrier height. In most polymer systems, thermal activation is the rate-controlling step and simple exponential behaviour is invariably observed. However, a situation can be achieved where the rate of conversion of energy from momentum to vibrational energy can, in systems such as PDMS, lead them to conform to a 'starvation kinetics' model.

17.4.4   Solvent Effects

Considerable speculation has arisen over the years concerning the role of solvent molecules on the local segment relaxation of a polymer molecule. It is often observed that changing the solvent leads to changes in the apparent activation energy for internal rotation (Table 1). In fact it has been found that in certain cases subtraction of the friction coefficient, in the form of the activation energy for viscous flow, yields an almost constant segmental activation energy.[42] Solvent interactions can be envisaged to influence the conformational process in several ways. Inelastic collisions between the solvent and polymer are the primary source of energy for the excitation of the torsional states involved in the conformational process.[43] Changes in the solvent type will alter the probability of energy being transferred from the translational energy to the torsional excitation, but this also alters the pre-exponential factor for the rotational isomeric process. Solvent molecules which undergo specific interactions through polymer solvation can influence conformational change by stabilizing the solvated state and loss of the solvation sheath is required prior to isomerization. In the introduction it was pointed out that the barrier to internal rotation is a composite of non-bonding interactions between closed-shell structures and electrostatic interactions, and alteration of the dielectric constant of the media by change of solvent can significantly influence the magnitude of the

Table 1   Effect of Solvent on the Relaxation of Polystyrene

Solvent	Viscosity (cP)	Energy difference $\Delta H°$(kJ mol^{-1})	Barrier height $\Delta H^{\ddagger}_{21}$(kJ mol^{-1})
Methyl ethyl ketone	0.44	7.1	1.2
Toluene	0.52	1.7	10.7
Decalin	2.4	4.5	10.7
Dibutyl phthalate	20.7	7.0	27.6
Benzene	0.56	3.7	27.6
Carbon tetrachloride	0.84	3.7	27.6

electrostatic contribution to the potential. Evidence for all these factors can be found in the literature[43,44] on rotational isomerism both in small molecules and polymers.

17.4.5 Block Copolymers

A number of studies have been reported on the ultrasonic relaxation of copolymers.[45-48] Two systems will be described here which illustrate important facts of ultrasonic studies.

17.4.5.1 Copolymers with phenylethylene and alkylene units

Strictly alternating copolymers have been prepared and characterized. The polymers have the constitutional repeating unit (**1**) with values of *n* between 0 and 10. The ultrasonic relaxation[47] of the polymers with small values of *n* are very similar to those observed with polystyrene and approximate to a single relaxation process. As the value of *n* is increased above 5, there is evidence of a separate fast relaxation associated with the hydrocarbon moiety. A parallel study of the ^{13}C NMR relaxation indicates that for low values of *n* it is not possible to separate the relaxation[49] rates of the styrene and alkane moieties. Once *n* has a value of greater than 5, separate styrene and alkane moiety relaxation processes can be resolved. Increase in the value of *n* also leads to a lowering of the activation energy for the isomerization process, consistent with the idea that the alkane block reduces long range interactions between styrene groups and covers the rotational isomeric barrier (Figure 9). A similiar trend was also observed in the related series of copolymers based on 1,1-phenylmethylethylene and alkylene constitutional units.[49] These studies indicate that the critical size of group involved in segmental motion is probably between six and eight monomer units and reflects the distance over which significant interactions between chemical entities influence the barrier height in polymer systems.

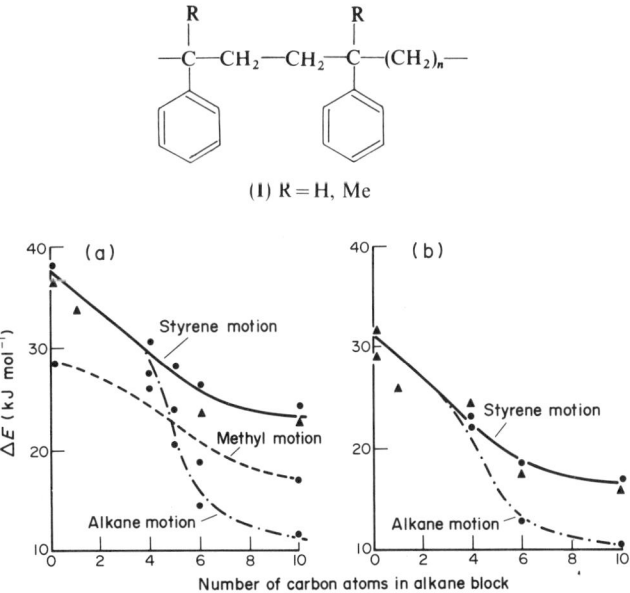

(**1**) R = H, Me

Figure 9 Variation of the activation energy for segmental relaxation with block length and moiety moving: (a) α-methylstyrene–alkane; (b) styrene–alkane. ▲, acoustic; ●, ^{13}C NMR

17.4.5.2 Block-copoly(styrene/butadiene/styrene)

The ultrasonic relaxation of an SBS triblock copolymer in dilute solution in toluene[48] corresponds to a superposition in appropriate proportions of the spectra of polystyrene and polybutadiene. The relaxation in the MHz frequency range is entirely predictable on the basis of the models described above except in one respect. The general relaxation spectrum in the MHz range is a composite of segmental and normal mode motions and the observed behaviour is calculated from the behaviour of polybutadiene and polystyrene combined in proportions corresponding to the weight fraction and a contribution associated with normal mode behaviour of the whole polymer. However, cooling dilute solutions of SBS can lead to the formation of micelles and other phase-separated structures[50] and associated changes are observed in the ultrasonic attenuation. Although these particles are significantly smaller than the wavelength of the sound wave, they are able to increase the high frequency limiting viscosity of the solution. The result of this is that the high frequency limiting attenuation increases markedly as the solution is cooled below its critical micelle-forming temperature (Figure 10). The ultrasonic method therefore provides an ideal method for the study of phase separation and, unlike optical methods, it does not suffer from problems when the refractive index of solvent and cluster are almost identical.

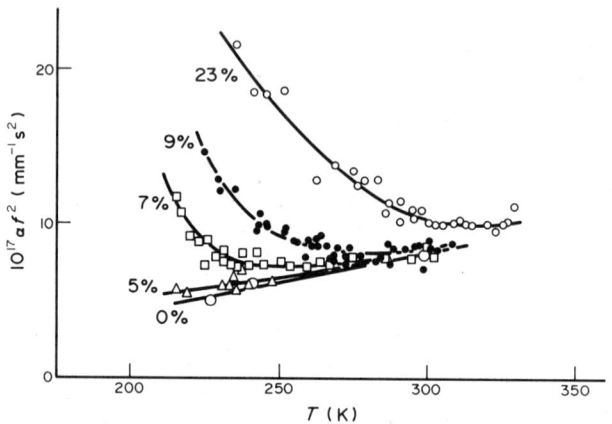

Figure 10   Variation of the high frequency acoustic absorption (750 MHz) with temperature for SBS triblock polymers

17.5   ADIABATIC COMPRESSIBILITY OF POLYMER SOLUTIONS AND MELTS

The preceding discussion has concentrated on the data obtained from the analysis of attenuation data obtained as a function of frequency. A less extensively investigated, yet important, property of acoustic propagation is the sound velocity. The speed of sound propagation in solution depends directly on the adiabatic (isoentropic) compressibility κ [$= 1/(\rho v^2)$, where v is the sound velocity and ρ is the density]. The compressibility is itself frequency dependent and will be influenced by the relaxation processes which give rise to dispersion in the sound attenuation. At low frequencies (Figure 11) all possible motions can contribute to the adiabatic compressibility. Increasing the frequency of observation leads to the progressive freezing out of various contributions to the total compressibility at a particular frequency. The first to be lost are the long range (collective normal mode) relaxation processes; increasing the frequency leads successively to loss of contributions due to local conformational change, polymer–solvent interactions and finally of the solvent relaxation itself.

It is observed that an increase in the concentration of the polymer leads to a linear decrease in the velocity of sound propagation. Corrections to the data can be made for the effects of conformational

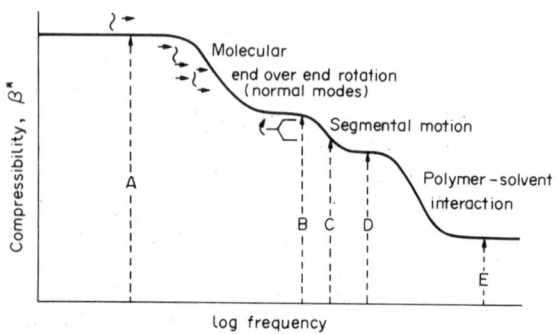

Figure 11   Variation of the adiabitic compressibility of polymer solutions with frequency

change. However, in general the decrement in the velocity can be attributed to effects of polymer–solvent interaction. By use of certain assumptions with regard to the contribution to the total compressibility associated with the polymer chain motion, which is normally negligible in dilute solution in comparison to other effects, it is possible to estimate 'solvation numbers'. Typical values for a series of polymer systems are presented in Table 2. The values of the solvation numbers quoted should *not* be considered as indicative of the number of solvent molecules attached to a monomer, but rather as the average number of molecules resident in a solvent sheath around the polymer during the time scale of the experiment. The values presented in Table 2 are intuitively correct and reflect both the varying strengths of interaction and specificity with change in solvent. The sound velocity measurements have two advantages over conventional thermodynamic methods for the study of polymer solutions. Firstly, it is fairly easy to achieve a precision of measurement of better than 0.1% which compares with 1% achieved in the best conventional studies, and secondly by variation of the frequency of observation it is possible to determine the magnitude of an increment due to a specific molecular process, or to freeze it out from the total measured value (Figure 11).

Table 2 Solvation Numbers for some Polymer–Solvent Systems as Determined by Adiabatic Compressibility Measurements

Polymer	Solvent	Number of solvent molecules interacting per monomer unit
Poly(ethylene oxide)	Toluene	0.20
	Cyclohexane	0.12
	Water	1.74
	Carbon tetrachloride	0.22
Poly(*N*-vinylcarbazole)	Toluene	1.1
Polystyrene	Toluene	0.45
Poly(2,6-dimethyl-1,4-	Toluene	0.65
phenylene oxide)	1-Butanol	0.62
	2-Butanol	0.63
	2-Methyl-1-propanol	0.63
Poly(methyl methacrylate)		
isotactic	Toluene	0.60
syndiotactic	Toluene	0.69
heterotactic	Toluene	0.59
Polybutadiene	Hexane	0.51
	Cyclohexane	0.21
	Ethylbenzene	0.10

The latter principle is demonstrated in a recent study of mixtures of linear and branched chain hydrocarbons.[52] Mixing of two hydrocarbons can lead to a combination of liquid structure and isomeric changes. It is not possible using conventional equilibrium methods to separate the relative contributions of these processes to the excess thermodynamic properties for the mixture. Studies of the rotational isomeric process can be carried out at frequencies below 500 MHz, leaving changes in the contribution due to liquid structure to influence measurements above 1 GHz. The frequency dependent data can thus be separated and excess contributions due to rotational isomeric and liquid structure effects. In fact, it is observed that the main contribution to the excess thermodynamic functions in these mixtures arises from liquid structural contributions as proposed by Patterson[53] rather than from rotational isomeric effects as suggested by Flory.[54]

Adiabatic compressibility measurements have been performed on a number of polymer–solvent systems and have allowed indentification of specific interactions between polymer and solvent as in poly(ethylene oxide)–water[55] or alcohol mixtures,[56] effects of branching as in polyisobutene in hydrocarbon mixtures[57] and helix–coil phenomena in polystyrene[25] and polyether.[58]

The effects of molecular weight have been specifically studied for PDMS (Figure 12).[33] Low molecular weight PDMS behaves like a liquid and exhibits a high compressibility. Increasing the length of the chain leads to the generation of an apparently molecular weight independent value of the compressibility which is lower than that of the low molecular weight fluid. Further increase of the molecular weight above the point at which entanglement occurs leads to a further decrease in compressibility as a consequence of restrictions generated due to polymer–polymer entanglement. The sound velocity once more provides a useful and distinct way of monitoring the generation of various types of interaction in polymer solutions and melts.

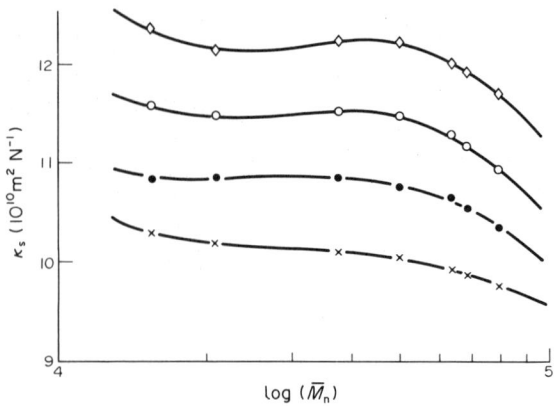

Figure 12 Effect of molecular weight on the adiabitic compressibility for linear poly(dimethylsiloxane): ×: 303; ●, 313; ○, 323; and ◇, 333 K

The above review of ultrasonic relaxation is not intended to be comprehensive, but rather to illustrate the wide range of problems which have been studied using the method. Many of the topics mentioned have been discussed more fully elsewhere.[1-6]

17.6   STUDIES OF REACTING SYSTEMS

In the above discussion, it has been assumed that the entities involved are stable molecules and no significant chemical modification occurs during the experiment. It was, however, shown over 20 years ago,[59] that high intensity low frequency ultrasonic waves can induce degradation of polymers. The mechanism whereby this occurs is not totally understood; however, it is clear that as a result of excitation of the lowest normal mode, either by energy transferred from the excitation of microscopic gas bubbles or purely by selective excitation, that controlled degradation can occur. A polymer of molecular weight 100 000 subjected to high intensity radiation will degrade to a polymer of approximately half its molecular mass and without appreciable change in the distribution coefficient ($\bar{M}_w/\bar{M}_n$). It is clear from these observations that the degradation is a selective process and must involve the generation of an energy bottle-neck within isolated polymer chains.

A more recent application of ultrasonics has been the characterization of the extent of polymerization in a condensation or radical process.[60, 64] The first observations were made by Sokolov,[60] and subsequent reports of measurements on polystyrene,[61] poly(vinyl chloride)[62, 63] and poly(vinyl acetate)[64, 65] have confirmed the utility of the method. It is clear that this type of study is still in its infancy; however, certain facts emerge which demonstrate the importance of this method. The compressibility of a solution containing monomer and polymer is directly related to the proportion of each component present. It is therefore possible to quantitatively estimate the extent of conversion from the observed velocity of sound. In a suspension polymerization, the glass transition of the polymer forming the bead is itself a function of the extent to which unreacted monomer is retained in the system. In this case, observation of the attenuation can indicate the extent to which polymerization has occurred in the system. Unfortunately the data are not sufficiently extensive to estimate the general validity of the method for the monitoring of polymerization in reactors, although the potential has been clearly demonstrated.

In the condensation reaction between a trifunctional alcohol and a difunctional isocyanate, gel formation is marked by significant changes in the velocity and attenuation in the MHz region.[66] The relaxation in the isolated monomers can be ascribed to a combination of intra- and inter-molecular processes. Intermolecular relaxation occurs in the hydrogen-bonded structure formed as a result of hydroxyl interactions. Reaction between the isocyanate and alcohol leads to the possibility of normal mode contributions to the relaxation spectrum and consequent increase in the ultrasonic attenuation. Detailed analysis of the data is difficult because of the complex topography generated by the reaction of di- and tri-functional monomers. This study does however once more illustrate the possibility of using ultrasonic techniques to monitor polymerization processes.

A similar application of ultrasonics is found in the case of phenolformaldehyde resins. The mixture is initially liquid and has a low velocity, high compressibility and low loss. As cure proceeds, the chains become more extended and firstly a gel and then a rigid solid develop. As the medium becomes more viscous, so the velocity of propagation increases. The extended structure generates molecular entities with isomeric relaxations close to the observation frequency and the sound attenuation is observed to rise. As the rigidity of the matrix increases so these latter motions become inhibited and the attenuation once more decreases. The increased rigidity of the media leads to a marked increase in the velocity of sound, the magnitude of which is proportional to the modulus of the matrix formed. It is thus possible from a combination of the attenuation and velocity to define unambigiously the state of cure in this system.

17.7 ULTRASONIC STUDIES OF SOLID POLYMERS

Temperature dependent studies of solid polymers using the ultrasonic method reveal relaxation features which can be directly related to dielectric and other methods of measurement of segmental motion.[24,68]

17.7.1 Theory of Sound Propagation in Solid Polymers

For small amplitude deformations Hooke's law is obeyed and the strain E is a linear function of the stress

$$\dot{\sigma} = E\dot{\varepsilon} \tag{8}$$

where E is the Young's modulus of the system. In the case of a viscoelastic medium, the modulus is a mathematically complex quantity $E^* = E' + iE''$, where E' is the real and E'' the imaginary part of the modulus. The bulk modulus K_s^* has the form

$$K_s^* = K_s' + iK_s'' = \frac{E^*}{3(1 - 2\mu^*)} \tag{9}$$

where μ^* is the complex Poisson ratio. Equation (9) is the equation for starting the discussion of acoustic propagation in solids. At high frequency, the wavelength of the sound wave is much less than the lateral dimensions of the sample, hence the velocity is that of a purely longitudinal wave (v_L) which has the form

$$v_L = \frac{\left(K' + \frac{4}{3}G'\right)^{1/2}}{\rho} = \frac{[E'(1 - \mu')]^{1/2}}{\rho(1 + \mu')(1 - 2\mu')} \tag{10}$$

where ρ is the density of the medium and G^* ($= G' + iG''$) is the corresponding shear modulus. It is convenient to define an equivalent dynamic longitudinal modulus (L^*). Then

$$v_L = \left(\frac{L'}{\rho}\right)^{1/2} \tag{11}$$

where $L^* = L' + iL''$, which is also equal to $K_s^* + \frac{4}{3}G^*$. When the wave propagation is constrained to thin strips ($\lambda \leq b$, the lateral dimensions of a body), equation (11) becomes

$$v_L = \left(\frac{G''}{\rho}\right)^{1/2} \tag{12}$$

The above equations apply only if the attenuation (α) in the solid is small ($\alpha\lambda/2\pi < 1$).

In polymers, it is necessary to allow for the effects of dispersion in the calculation of the equivalent elasticity modulus $M^* = M' + iM''$

$$M' = \frac{\rho v^2[1 - (\alpha\lambda/2\pi)]^2}{[1 + (\alpha\lambda/2\pi)^2]^2} \tag{13}$$

and

$$M'' = \frac{2\rho v^2(\alpha\lambda/2\pi)}{[1 + (\alpha\lambda/2\pi)^2]^2} \tag{14}$$

If v is replaced by v_L, the elastic modulus $M*$ becomes the longitudinal modulus $L*$. Alternatively if v_s is used then the shear modulus $G*$ is obtained.

17.7.2   Molecular Interpretation of the Acoustic Parameters

As with low frequency mechanical data[69,70] at low temperatures where molecular motion is frozen, the modulus will have a value of 10^{10} dyn cm^{-2} (1 dyn $= 10^{-5}$ N). Increasing the temperature will lead to the formation of a rubbery phase which will have a corresponding lower value of the modulus 10^6–10^7 dyn cm^{-2}. A decrease in modulus is associated with the onset of molecular motion in the solid state. The relative ease of rotation is reflected by the position on the time–temperature correlation diagram of the relaxation process. The loss peaks are usually labelled according to their appearance on cooling from the melt. The highest temperature is the α process, followed in order by the β, γ and δ processes. The molecular origins of the relaxation processes are as follows: (i) α process — associated with the glass transition temperature T_g in amorphous polymers and involving relatively large scale motion of between 6–8 monomer units with an activation energy of 80–800 kJ mol^{-1}; (ii) β process — associated with the rotational isomerism of side chain elements with an activation energy of 60–160 kJ mol^{-1}; (iii) γ process — also associated with rotation of groups attached to the main backbone or of side chains with an activation energy of 28–80 kJ mol^{-1}; and (iv) δ process — associated with rotational motion of groups attached to side chains with an activation energy in the range 16–40 kJ mol^{-1}.

The statements above summarize the behaviour found in many amorphous polymers. If the polymer is capable of forming a partially crystalline solid, a multiplicity of relaxations will be observed corresponding to motions in the crystalline and amorphous phases.

Reviews of ultrasonic relaxations in solid polymers have been published elsewhere.[2,4] The interpretation of the data has to a large extent neglected the morphology of the solid. Recent advances in the methods available for the characterization of solid polymers have allowed much of the earlier data to be reinterpreted. The influence of morphology on the form of the ultrasonic relaxation has been described elsewhere.[4] Typical plots of the attenuation against temperature for a wide range of polymers are shown in Figure 13. The relaxation features of many polymers are very broad and are associated with the β processes. Analysis of the temperature dependence of these loss peaks allows determination of the activation energy for local mode motions.

It has been observed in a variety of polymers that the velocity of sound becomes essentially independent of temperature, corresponding to a plateau modulus condition. Analysis of the data using the kinetic theory of rubber elasticity yields the relation[71]

$$E_0 = 3\rho RT/M_e \tag{15}$$

where ρ is the density and M_e is the molecular weight of the polymer chain between entaglement points. Using the relation $v^2 = E_0/\rho$, equation (15) becomes

$$v^2 = 3RT/M_e \tag{16}$$

Analysis of data for polystyrene above 423 K yields a value of M_e equal to 4580, an average of 44 monomer units between each entanglement. For poly(methyl methacrylate) M_e is equal to 1985, yielding 20 monomer units between each entanglement. The corresponding data for poly(vinyl acetate) yield M_e equal to 4680 and hence an average of 55 monomer units between entanglements. Polycarbonates give low values of M_e between 1680 and 2300, which correspond to six to nine units.

17.8   MOLECULAR RELAXATION IN POLYMER SOLIDS

17.8.1   Polystyrene

Dielectric, NMR and low frequency dynamic mechanical studies indicate that the transition detected acoustically at 193 K can be ascribed to the onset of oscillatory motion of the phenyl group.[76,77] Transitions observed at 383 K and 373 K are attributed to the glass transition and this assignment has been confirmed by dilatometry measurements. The observation of two transitions leads to the suggestion that clustering occurs in the solid phase. The lower temperature transition would therefore correspond to the onset of collective motions of polymer chains in the amorphous regions. The low temperature dependence of the modulus between 373 K and 383 K is believed to be

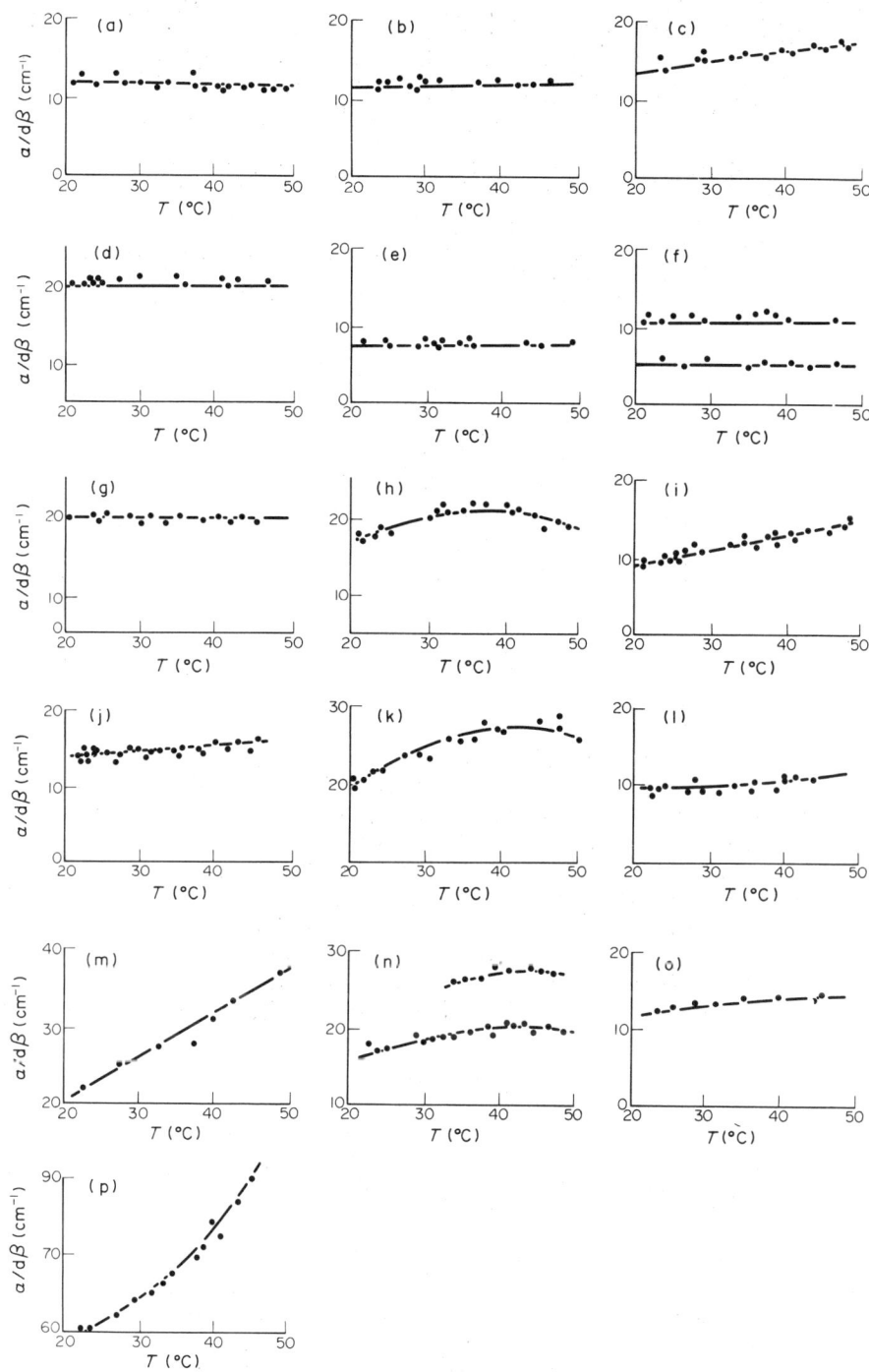

Figure 13 Attenuation plots for a variety of polymer systems as a function of temperature: (a) poly(phenylene sulfide); (b) poly(ether sulfone); (c) polyester DMC; (d) copoly(acrylonitrile/styrene/acrylate); (e) poly(acrylonitrile/butadiene/styrene); (f) polystyrene (toughened); (g) polyester resin; (h) melamine formaldehyde; (i) polyimide; (j) ekonol; (k) nylon 6; (l) poly-(phenylene oxide); (m) polypropylene; (n) phenol formaldehyde; (o) poly(vinyl chloride) (rigid); (p) poly(vinyl chloride) (flexible)

a consequence of the reinforcing action of the clustered region. Recent ultrasonic studies of polymer solutions have indicated that at 340 K a marked increase in the relaxation amplitude occurs associated with a helix–coil type of transition.[78] Conformation transitions associated with certain stereoregular sequences do not become active until high temperatures and in the presence of the stronger intermolecular forces found in the solid may be expected to be shifted to higher temperatures. It is therefore possible that part of the reason for multiple relaxation in the region of T_g may be related to the distribution of stereoregular forms found in the 'atactic' polymer.

17.8.2   Polycarbonate, Poly(ether sulfone) and Polysulfone

These polymers possess a *para*-linked chain-extended diphenylmethane structure (Figure 14). They are the so-called 'engineering polymers' having high melting points and impact strengths. The origin of the high impact strength is still not fully understood but probably is a consequence of the molecular backbone motion occurring well below T_g.[71-81] A comparison of the temperature dependence of the dielectric and low frequency mechanical measurements of polycarbonate have shown two major relaxation processes:[82,84] the α process with activation energy 450 kJ mol^{-1} associated with large scale motion of the polymer chain; the β process with activation energy of 45 kJ mol^{-1} ascribed to more local reorientational motions of the phenyl ring structure. An additional β process has been detected in certain samples and is attributed to either changes in morphology or relaxation of strained regions within the sample. The activation energies for the β process in poly(ether sulfone) and polysulfone are respectively 13.1 and 16.6 kJ mol^{-1}.[85]

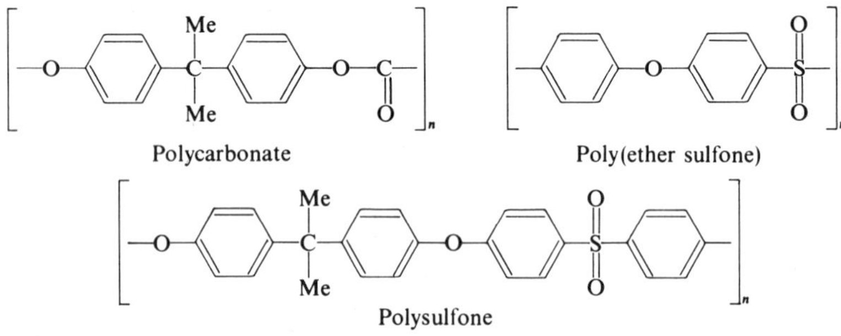

Figure 14   Structures of engineering polymers

The temperature dependence of the ultrasonic velocity (Figure 15), normalized to the glass transition temperature, indicates that both polycarbonate and poly(ether sulfone) possess similiar molecular relaxation characteristics. In contrast, polysulfone either has a much higher sound velocity (modulus) in the rubbery state or (as is more likely) undergoes another relaxation (with associated decrement in the sound velocity) between 363 K and the glass transition. This is probably a consequence of partially crystalline structure existing in the solid at room temperature.

The high impact strength of these materials is ascribed to the β relaxation. It is assumed that the local stress concentration generated by the impact may be dissipated by the facile molecular motion avoiding the rupture of the polymer backbone bonds. Polycarbonate is exceptional in that it possesses a free volume which is 1.5 times greater than that observed for the majority of polymers. As

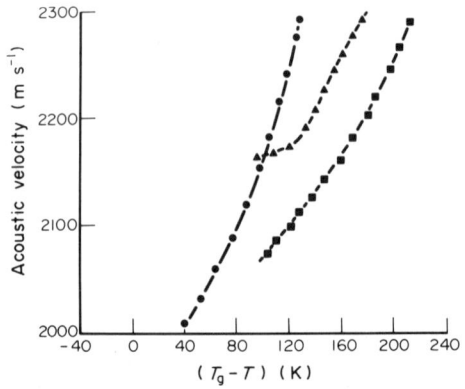

Figure 15   Temperature dependence of the velocity of sound close to T_g: ●, polycarbonate; ▲, polysulfone; ■, poly(ether sulfone)

a consequence the possibility of motion of one chain relative to another must be included in the discussion of the mechanism of the response to impact.

17.8.3 Polytetrafluoroethylene

Commercial polytetrafluoroethylene (PTFE) is usually the product of sintering and has its crystallinity reduced by the introduction of branched chains. PTFE (degree of crystallinity, 93–98%) has a melting point of 620 K and exhibits a helical structure which undergoes unit cell transitions at 292 and 303 K.[86,88] Below 292 K the conformation can be described by a helix containing 13 CF_2 units per 180° twist. The unit cell is triclinic, and essentially perfect three-dimensional order is observed in the polymer. Above 292 K the helix expands to 15 CF_2 groups per twist of 180°, and the unit cell becomes hexagonal. Between 292 and 303 K the packing of the molecules on the hexagonal lattice is disordered due to small displacements arising from angular oscillations of segments about their long axis. Above 303 K the preferred crystallographic direction is lost and segmental oscillations lead to random angular orientation on the lattice. The crystal disorder transitions at 292 and 303 K have been studied by X-ray diffraction,[86] densitometry[89] and specific-heat methods.[88]

Acoustic studies of crystalline PTFE indicate discontinuities in the absorption–temperature plots at 292 and 303 K and broader transitions at 173 and 393 K (Figure 16). The very sharp peaks clearly correspond to the crystalline order–disorder transitions discussed above.[89,90] The peak at 393 K is sensitive to the degree of crystallinity of the sample (increasing in amplitude with increasing degree of crystallinity) and has been assigned to torsional oscillation of chain segments around the chain axis. Studies of the ultrasonic relaxation in solution[91] have indicated that the conformational relaxation associated with an exchange between the non-degenerate *trans* structures should occur in the region observed for the conformational transition in the solid and explain the above behaviour.

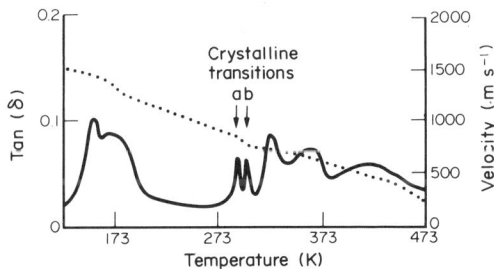

Figure 16 Variation of the acoustic attenuation with temperature for PTFE: (a) 13 CF_2, (b) 15 CF_2

17.8.4 Partially Fluorinated Hydrocarbons

All these polymers exhibit relaxation spectra characteristic of paracrystalline materials. Substitution of hydrogen by fluorine leads to modification of the intermolecular potential and results in a net reduction of the interactions in comparison with polyethylene. However, the quadrupolar interactions lead to an intramolecular potential which has two shallow minima in many of the *trans* forms and hence induces helical forms similar to those found in PTFE.[92] No attempt appears to have been made to quantify the relative order of magnitude of the changes occurring; however, extrapolation of the ultrasonic velocity data to high temperature does indicate that the intramolecular potentials are significantly different.[93] Combination of velocity data with thermal expansion coefficient measurements allow an effective three-dimensional intermolecular force field to be defined.[94–98] Using the model of Tarasov, an analysis may be performed on the data to yield the low temperature specific heat of the polymer. A significant difference between the observed and theoretical values based on contributions from the vibrational modes of the polymer confirms the existence of low frequency 'vibrational' modes which are associated with the time dependence of thermal diffusivity of the polymer due to the existence of crystalline domains.

17.8.5 Poly(vinyl chloride) $[(-CHCHCl-)_n]$

A rigid and comparitively simple polymer, poly(vinyl chloride) (PVC) exhibits a surprisingly rich relaxation spectrum.[99-110] A close examination of the velocity data's dependence on temperature in the region of I_g indicates the existence of a double transition, attributable to the occurrence of supermolecular organization in the solid state. It has been proposed that the effects are due to specific interactions which lead to the formation of regions of high density. Crystallinity is inhibited in this polymer by a lack of regular stereochemistry of the main backbone and the possibility of branched chain structures occurring at irregular intervals along the polymer backbone. The postulate of supermolecular order and specific interactions is supported by the observation that annealing leads to an increase in the density associated with growth of crystalline regions. The relaxation spectrum of PVC is a composite of molecular motions with disordered and paracrystalline regions.

17.8.6 Poly(alkyl methacrylate)s

Poly(methyl methacrylate) and the related alkyl methacrylates have been the subject of extensive investigation.[111,112] The glass transition process in PMMA in the atactic polymer occurs as a loss in the acoustic spectrum at 390 K. The ester group motion occurs at approximately 291 K and the relaxation processes associated with the methoxy and methyl group have been assigned respectively to transitions at 233 and 166 K. Neutron-scattering studies[113] support the above assignment. Partial deuterium substitution of the backbone and side chain allow unambigious identification of the various molecular motions.

In alkyl-substituted methacrylate polymers, an additional transition is detected ultrasonically and has also been observed using dynamic mechanical and dielectric measurements.[114-117] This relaxation is associated with conformational changes of the alkane side chain (Figure 17). The motion of

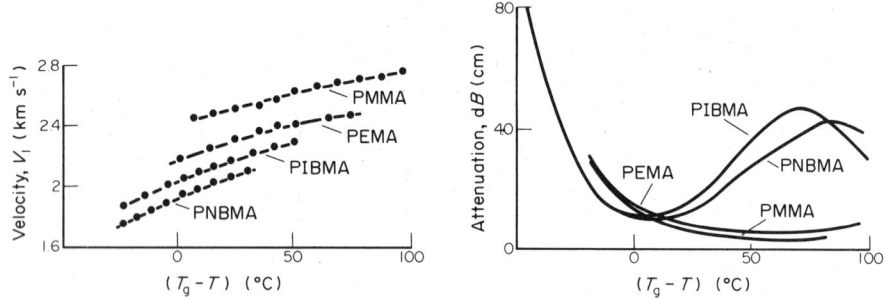

Figure 17 Acoustic relaxation in poly(alkyl methacrylate)s, $\{CH_2C(Me)(CO_2R)\}_n$: PMMA, R = Me; PEMA, R = Et; PNBMA, R = Bu^n; PIBMA, R = Bu^i

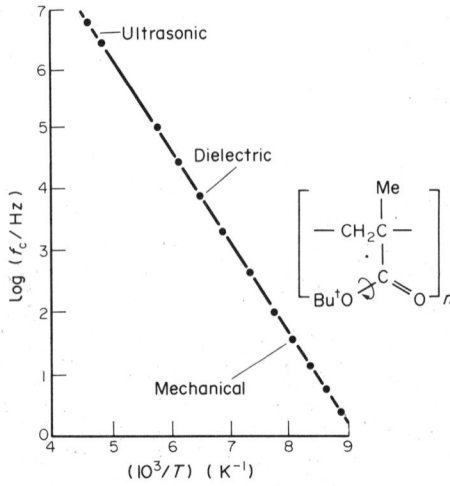

Figure 18 Activation energy for rotational isomerism in poly(alkyl methacrylates)

the side chain in the case of the butyl group exhibits a simple thermal activated behaviour (Figure 18), in contrast with the relaxation of the backbone which is free-volume-controlled. The assignment of the transition at 291 K in PMMA is a little more difficult. It has been ascribed to 'retarded rotation of the methoxyl carbonyl group of the side chain'. It is not, however, clear that we should apparently have two relaxations assigned to the same motion, although similar features have been observed in liquid crystals where correlation theory predicts two features associated with components of the motion in different frames of reference. An alternative explanation for the observation of the features may be based on the fact that PMMA will contain short sequences of syndio-, hetero- and iso-tactic structures. The T_g values of these sequences are respectively 388 (syndio-), approximately 380 (hetero-) and 318 (iso-tactic). It is possible that the incorporation of the isotactic sequence into the atactic polymer will increase its mobility and lower the effective value of T_g. It is therefore possible that a relaxation can be observed associated with the coupled motion of the isotactic sequences and the motion of the side chain groups. In a complex molecular system such as PMMA it is not possible to unambigiously analyze the relaxation spectrum.

17.8.7 Poly(vinyl acetate)

In contrast to PMMA, four relaxations are observed in the acoustic spectrum of poly(vinyl acetate) (PVAC), and occur respectively at 153, 195, 288 and 306 K.[118-122] The lowest temperature transition may be ascribed to methyl group rotation and this is in agreement with neutron-scattering observations. Dilatometric studies indicate the existence at 306 K and 313 K of two transitions connected with the onset of molecular motion and the backbone in amorphous and clustered regions. This is in agreement with dielectric and NMR observations. The transition at 193 K is assigned to the rotational isomerism of the side chain groups.

17.8.8 Polyamides — Nylon $[(-NH(CH_2)_x NHCO(CH_2)_y CO-)_n]$

Nylon is the condensation product of a diamine with a dicarboxylic acid and as such contains a number of potentially hydrogen-bonding sites. For nylon 66, $x = 6$ and $y = 6$. The properties are dominated by the formation of the intermolecular N–H $\cdots$ O=C bonds, which leads to crystallization from solution in the form of lamellar structures of thickness between 500 and 1000 nm. The bulk polymer usually forms spherulites orientated so that the hydrogen bonds are along the radius. In normal bulk polymer, the degree of crystallinity can vary between 20 and 55%.

Acoustic relaxation peaks observed at room temperature[123] are sensitive to water content and have been ascribed to relaxation of the amide group and its solvation sheath.[126-128] A similar feature is observed in the dynamic mechanical spectrum at 238 K and leads to an activation energy of 10 kJ mol^{-1} for this process.

The longitudinal velocity exhibits a change in slope at approximately 348 K and this is associated with the onset of motion in the crystalline regions of the polymer.[125,126] The shear velocity shows a similar trend, the change in slope being less dramatic than in the longitudinal case.

17.8.9 Polyalkene Rubbers

Some of the earliest ultrasonic investigations of solid polymers were performed on natural and synthetic rubbers.[129] Natural or hevea rubber is obtained from *Hevea brasiliensis* and has the chemical structure of poly(*cis*-1,4-isoprene) (Figure 19). The presence of the C=C bond imparts conformational rigidity to the backbone of the polymer. Gutta percha is also naturally occurring, however it has the *trans* rather than the *cis* structure. This simple change in the backbone structure leads to very different physical properties: hevea is amorphous and rubbery at room temperature, flows at 333 K and crystallizes on cooling below 273 K or on stretching. In contrast, gutta percha is in a highly crystalline state and has a unit cell containing two right-handed and two left-handed *cis*-1,4-isoprene units.

Synthetic polymers are obtained by use of Zeigler–Natta catalysis, which leads to a poly(*cis*-1,4-isoprene) structure. These materials are usually strengthened by vulcanization with either sulfur or a peroxide treatment. This process involves the generation of crosslinkages which are either sulfur or oxygen bridges.

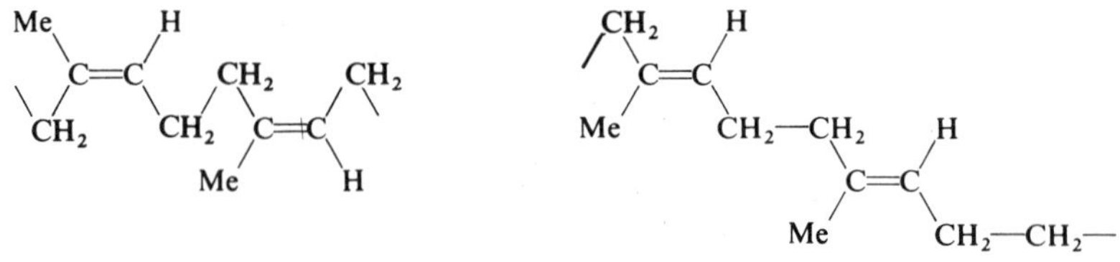

Poly(*cis*-1,4-isoprene)   Poly(*trans*-1,4-isoprene)

Figure 19   Structures of polyisoprene polymers

Acoustic studies indicate that rubbers produce a large attenuation of the sound wave in the frequency range 40 kHz to 10 MHz. The magnitude, rate of change with frequency and position of the loss are all sensitive to the blend of conformational structure, being highest in the *cis* loss and gutta percha being the lowest. Commercial rubbers lie between these extremes. Fillers such as carbon black, iron oxide, china clay, magnesium and lithage have significant effects on the loss and velocity of sound propagation.[135] The fillers appear to slow down the motion of the polymer chains and increase the modulus.

17.8.10   Polyethylene [$(-CH_2CH_2-)_n$]

Polyethylene has one of the simplest structures possible in a polymer yet it has one of the most complex relaxation behaviours observed in any macromolecule.[136–140] It is one of the few polymers to have been investigated by ultrasonics in the region of its melt temperature, T_m. The precise value of T_m depends on the crystallinity but usually lies in the range 389 to 403 K. The denser, more crystalline samples have the higher melt temperatures. The melt temperature is marked acoustically by a sharp decrease in both the longitudinal and transverse velocities. Observations at low frequency (50 kHz) have indicated the existence of the premelt increase in the attenuation which has been ascribed to the onset of molecular motion in areas of the crystalline phase.[141,142] A marked drop in the acoustic velocity in the temperature range 213 to 223 K is associated with the glass transition of the amorphous phase. Increase in the branched chain content has a significant effect on the loss in the room temperature range and only a minor effect on the lower temperature process (Figure 20). Clearly the higher temperature process is associated with the motion of the pendant groups; however, the insensitivity of the lower temperature process to the effects of change in crystallinity indicates that it is not simply the result of motion of chains between the crystalline blocks. A correlation plot of the relaxations, with tentative assignments of processes, is presented in Figure 21. Spin probe dielectric and low frequency dynamic mechanical studies on linear and γ-irradiated material have recently helped to elucidate the nature of the relaxation processes in this polymer and these data are included in the correlation plot.

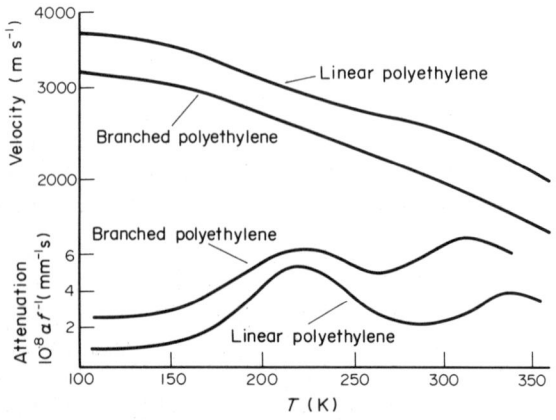

Figure 20   Acoustic relaxation in polyethylene polymers

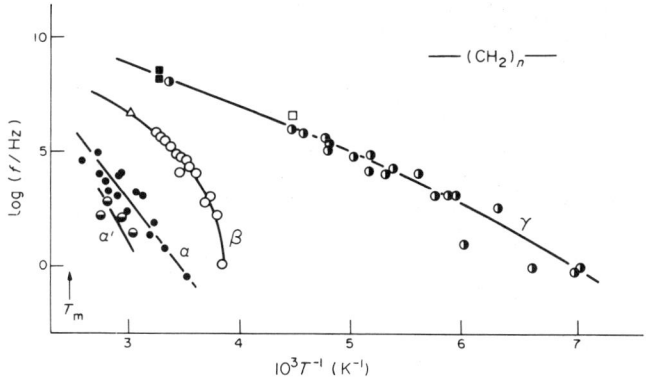

Figure 21 Activation energy plots for various relaxation processes in polyethylene

17.8.11 Polypropylene $[(—CH_2CHMe—)_n]$

Polypropylene is paracrystalline in the bulk state, and as in the case with polyethylene, its relaxation spectrum reflects the complexity of the sample morphology. A well-defined loss process has been observed using ultrasonics in the frequency range 10 to 300 kHz at room temperature and has been assigned to large scale motion of the amorphous regions of the polymer.[143,144] Neutron-scattering data on partially deuterated polymers have provided an unambiguous assignment of the rotational relaxation of the methyl group. As in the case of polyethylene, the relaxation spectrum is sensitive to the effects of thermal annealing.

It is clear from this review that assignment of relaxation features to specific motion of polymer molecules is often very difficult and it is usually essential for the data to be correlated with observations using other methods. In paracrystalline polymers the additional constraints of the ordered regions on the relaxation can yield new relaxation features and therefore electron micro-scopy, X-ray and neutron diffraction can greatly assist in the assignment of the origins of the effects observed.

17.9 WAVELENGTH AND VECTORIAL PROPERTIES OF SOUND PROPAGATION

17.9.1 Drawn-anisotropic Polymers

Polyethylene can form crystalline ordered regions in their solid state.[145] Mechanical work in terms of drawing extension or compression can achieve alignment of the axis of these crystalline regions with a consequent increase in the modulus of the material in the direction of draw and also changes in the relaxation spectrum. The morphology of the crystalline state of polyethylene has been extensively investigated. Using X-ray diffraction techniques it is possible to indicate the alignment of a polymer in terms of its pole diagram.[146,147] Similar data on the orientation of the crystalline phase in the polymer can be obtained from acoustic experiments. In the typical immersion experiment the velocity is determined perpendicular to the axis of rotation of the sample holder. If the direction of alignment is varied, it is possible to explore the various elastic components associated with the anisotropic nature of the crystalline phase (Figure 22). Moseley[148] has proposed a semi-empirical treatment of the propagation equations to describe such an experiment. He considered two different cases: (i) a series addition of force constants; and (ii) a parallel addition of force constants. Only the first case appears to lead to valid results and can be described by an equation of the form

$$\frac{1}{v^2} = \frac{1 - \langle \cos^2 \Theta \rangle}{v_\perp^2} + \frac{\langle \cos^2 \Theta \rangle}{v_\parallel^2} = \frac{\langle \sin^2 \Theta \rangle}{v_\perp^2} + \frac{\langle \cos^2 \Theta \rangle}{v_\parallel^2} \tag{17}$$

where v, $v_\perp$ and $v_\parallel$ are respectively the ultrasonic velocities for samples having an average orientation angle Θ. (For perpendicular alignment and parallel alignment Θ is 90° and 0° respectively). Equation (17) assumes that only the second moment ($\langle \cos^2 \Theta \rangle$) of the orientation distribution is influential in determining the velocity. It has been pointed out by Ward[149] that starting with the more realistic model of the solid being composed of aggregates of structural units, where elastic

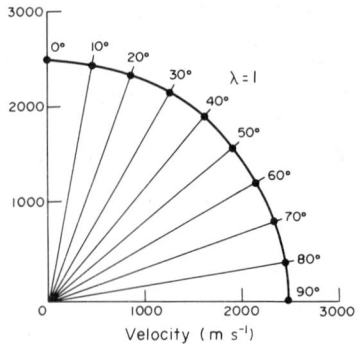

Figure 22   Acoustic pole diagram

constants are identical to those of the highly orientated material, one obtains

$$\frac{1}{E} = \frac{\langle \sin^4 \Theta \rangle}{E_\perp} + \frac{\langle \cos^4 \Theta \rangle}{E_\parallel} + \langle \sin^2 \Theta \cos^2 \Theta \rangle \left[\frac{1}{G} - \frac{2\gamma}{E} \right] \tag{18}$$

where G refers to the torsional modulus of the perfectly orientated material and γ is its Poisson's ratio. Equation (17) can be compared with equation (18) using the relation $E = \rho v^2$. Clearly E is not only a function of the second moment of the orientation distribution, but is also dependent on the fourth moment as well. Since γ is of the order of 0.5, $\gamma/E_\parallel$ will be small as will $1/E_\parallel^2$ relative to $1/G$ or $1/E_\perp$. In this case equation (18) reduces to

$$\frac{1}{E} = \frac{\langle \sin^4 \Theta \rangle}{E_\perp} + \frac{\langle \sin^2 \Theta \cos^2 \Theta \rangle}{G} \tag{19}$$

If the assumption is made that $E_\perp \approx G$, which is not unreasonable for moderate orientation, then equation (19) becomes

$$\frac{1}{E} = \frac{1 - \langle \cos^2 \Theta \rangle}{v^2} \tag{20}$$

This is identical with the Moseley equation (17), if the second term is negligible. The principle limitation of this method is that it measures the average orientation of the total sample. It has, however, been found that a good correlation exists between the orientation data derived from ultrasonic measurements and those obtained from birefringence measurements.[150]

Table 3   Elastic Stiffness Constants for Drawn Polymers

Polymer	State	Stiffness constants (GPa)		
		C_{11}	C_{22}	C_{33}
Polyethylene	Undrawn	3.9	3.9	3.9
	Drawn × 2	4.3	4.3	4.8
	Drawn × 3.5	4.1	4.2	5.1
	Drawn × 8	3.9	3.7	17.9
	Drawn × 11	4.1	4.2	36
	Drawn × 18	4.3	4.4	53
Polypropylene	Undrawn	5.75	6.75	5.75
	Drawn × 1.7	4.7	4.7	6.28
	Drawn × 3.3	3.19	3.4	8.98
	Drawn × 5.5	3.2	3.2	11.3
	Drawn × 10.5	4.16	4.3	15.0
Poly(ethylene terephthalate)	Undrawn	5.8	5.8	5.8
	Drawn × 2	5.7	5.7	7.5
	Drawn × 3	5.7	5.7	10
	Drawn × 4	5.7	5.7	12.5

Data obtained for polyethylene and polypropylene[151,152] are summarized in Table 3. The high value of the modulus observed in the draw direction is consistent with sound waves sensing distortion of the polymer chain (intramolecular bond stretching), while the perpendicular propagation senses the modulus transverse to the fibre axis. The latter is determined by intermolecular bonds and is primarily determined by van der Waals interactions between the fibres. The temperature dependence of the velocity both in polyethylene[153] and polypropylene (Figure 23)[154] reflects the nature of the molecular motions possible in these samples. A change in slope is indicative of the onset of molecular motion in the sample. In the case of polypropylene, there appears to be a shift of the relaxation spectrum on drawing as indicated by an elbow in the high draw ratio. The results for polyethylene indicate that there is a significant component at 45° to the draw direction. It has been reported from X-ray measurements that in the initial stages of draw the crystalline regions can become orientated at 45° to the draw direction; the ultrasonic observations are in agreement with X-ray data.

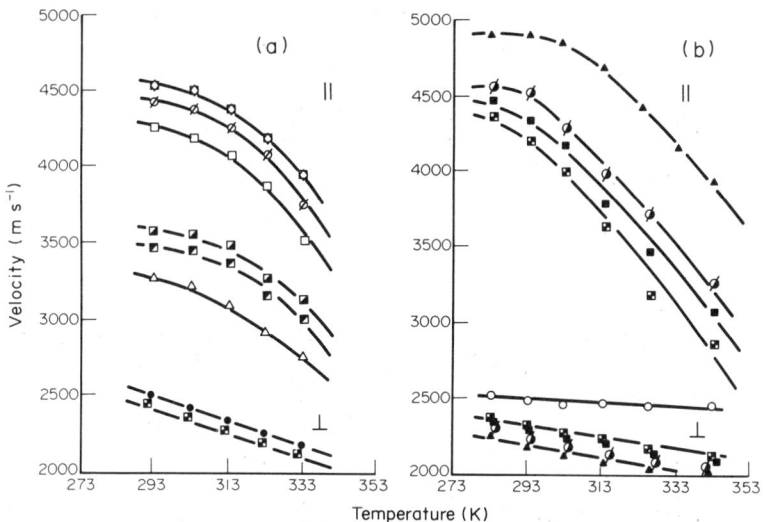

Figure 23 Variation of the acoustic velocity with temperature for various drawn polymers: (a) polyethylene, ●, undrawn polymer; △, $\lambda = 3$; ◪, $\lambda = 4$; ◩, $\lambda = 5$; ▲, $\lambda = 7$; ⊘, $\lambda = 10$; and ▧, $\lambda = 12$; (b) polypropylene, ○, undrawn polymer; ◪, $\lambda = 2.6$; ■, $\lambda = 6$; ◩, $\lambda = 9$; and □, $\lambda = 12$. The symbols ∥ and ⊥ indicate that the respective curves correspond to the components parallel and perpendicular to the draw direction. The curves are sensitive to the molecular weight, molecular weight distribution and drawing temperature used in the sample preparation

17.9.2 Studies of Phase-separated Polymers

Samuels[155] has extended the Moseley treatment to the analysis of two-phase systems, specifically semicrystalline polymers. This theory leads to an equation of the form

$$\frac{3}{2}(\Delta E)^{-1} = \frac{(1-\beta)f_1}{E_1} + \frac{\beta f_2}{E_2} \tag{21}$$

where β is the volume fraction of component 2 while f_1 and f_2 are, respectively, the Herman's orientation functions for components 1 and 2, E_1 and E_2 are the moduli values for the perfectly orientated components and

$$(\Delta E)^{-1} = \left(\frac{1}{E_u} - \frac{1}{E}\right) \tag{22}$$

where E_u and E are respectively the modulus for the unoriented sample and the modulus for the sample being measured. If we consider the case of a randomly orientated (isotropic) sample, *i.e.* $\langle \cos^2 \Theta \rangle = 1/3$, then equation (17) becomes

$$\frac{1}{v_u^2} = \frac{2}{3v^2} + \frac{1}{3v^2} \tag{23}$$

where v_u is the velocity of an unorientated sample. Rearranging, equation (23) becomes

$$v^2 = 2v_u v/3v^2 - v_u^2 \qquad (24)$$

Since typical values of v and v_u are of the order of 14 km s^{-1} and 1.5 km s^{-1} respectively, we can assume that $3v \gg v_u^2$ and thus

$$v^2 = 2v_u/3 \qquad (25)$$

Equation (25) indicates that the value of the observed velocity is directly proportional to the perpendicular component;[156, 157] using equation (21) we see that there will be a linear relation with the volume fraction of the crystalline phase.[158]

A plot of the velocity of various polyethylene samples with linear and branched chain structures (Figure 24) indicates that this prediction is correct and establishes the potential of the ultrasonic method as a direct non-destructive method for the determination of crystallinity in 'isotropic' samples. Thermal annealing of isotropic samples can lead to an increase in the velocity of propagation consistent with an increase in the degree of crystallinity. Investigation of the morphology reveals that this process is usually accompanied by a thickening of the lamellae from which the crystallites are composed. Thermal annealing can also change the size distribution of the crystallites and this can lead to changes in the component observed when the acoustic attenuation is measured over the frequency range 1 to 1000 MHz.[158]

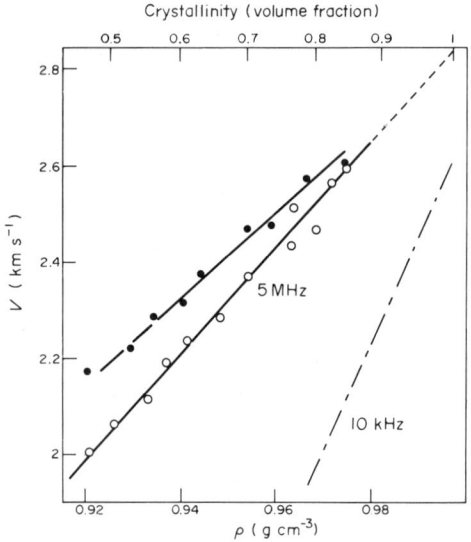

Figure 24 Variation of the ultrasonic velocity with degree of crystallinity for polyethylene: ●, linear; ○, branched chain

By slow cooling and quenching of polyethylene samples it is possible to generate spherulite structures with dimensions in the range 5 to 50 μm. The adiabatic passage of a sound wave through such a medium will lead to the generation of local differences in temperature (Figure 25). Coupling between the chains forming the spherulites will be rather weak, prohibiting the rapid thermal equilibration of temperature between domains which are raised to different temperatures. However, within the lamellae, which are typically 5 to 50 nm thick, there is rapid equilibration of the energy between intermolecular and intramolecular vibrations. The process by which equilibration occurs may be *via* transfer from the crystallites to amorphous regions which can then interact with the surrounding crystallite structure. Calculations of the relative rates of these processes indicate that the equilibration between crystallites and the contiguous amorphous zones will occur with times in the range 10^{-7} to 10^{-9} s, whereas the gross transfer of heat between neighbouring crystallites takes approximately 10^{-4} s. The occurrence of a restriction in the rapid thermal equilibration of energy between spherulites may in part contribute to the low temperature thermal anomalies observed in the low temperature specific heat measurements. Cooling the samples will alter the wavelength and also the number of waves (thermal photons) available. Energy migration within a sample usually occurs *via* phonon–phonon interactions and hence the occurrence of a lower than statistical density

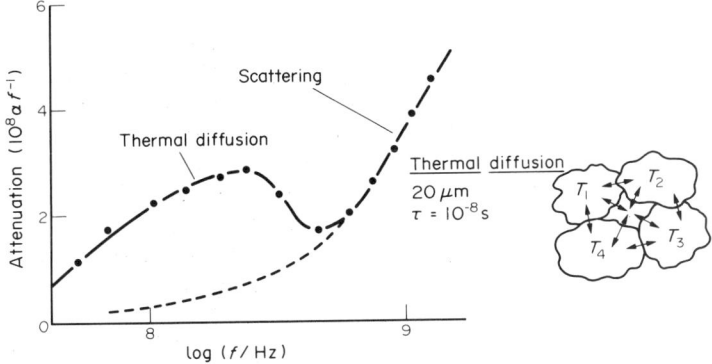

Figure 25 Effects of local thermal differences on the acoustic propagation of sound

of a particular type of wave can lead to a 'bottle-neck' situation and hence an anomaly in the specific heat and also acoustic attenuation.

17.9.3 Drawn Polymers — Scattering Effects

Morphologically related phenomena other than those associated with the anisotropy of the modulus can be observed when a sample is drawn.[154] Polypropylene has a spherullite structure which when subjected to the process of drawing will elongate and may lead to void formation. This hypothesis is supported by the observation of a minimum in the density *vs.* draw ratio (Figure 26). It is also observed that the attenuation of the drawn sample is considerably higher than that of the undrawn sample (Figure 26). Assuming that the increase in the attenuation arises principally from the scattering from air voids, and referring their experimental characters to theory, indicates that the scattering occurs from $0.1 \times 1 \mu m$ voids.

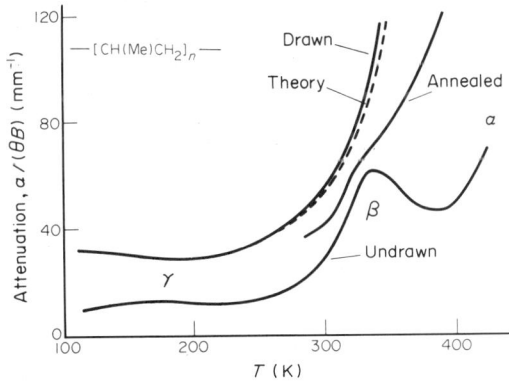

Figure 26 Effects of scattering from microcrystallites on the ultrasonic propagation

17.9.4 Stress Relaxation

Ultrasonics relaxation can be used to monitor stress relaxation in polymer samples.[159] It is observed that the attenuation in samples of drawn polypropylene is higher the lower the temperature used in preparation of the sample. This observation is consistent with the observation that those drawn at the lower temperature are opaque, whereas those drawn at high temperature are clear, transparent and with low acoustic attenuation. A decrease in the attenuation occurs on annealing of the sample at high temperature consistent with a decrease in void content. The overall length of the sample decreases on annealing and this is consistent with the reorganization of the fibrillar structure as a consequence of internal stress. The analysis of the acoustic data indicates that the activation energy is similar to that observed for the β process, whereas the changes in the overall length indicate a larger temperature dependence. It is probable that reorganization of the void structures occurs *via* a relatively local motion of the polymers, whereas the motion of fibrillar structures must involve a pseudo reptation type of motion. The ultrasonic method has considerable potential for this type of study since it does not destroy or significantly modify the sample being investigated.

17.10 CROSSLINKING REACTIONS

Cure of phenolic resins[160,161] and poly(phenylquinoxaline)[162] materials have been followed using ultrasonic techniques. In the case of the phenol resin it is observed that cure of the resin leads initially to a marked increase in the acoustic attenuation followed by a decrease once the gel point has been reached. The acoustic velocity exhibits an initial decrease as the density of the curing resin decreases before it develops a marked increase as the modulus of the materials rises. Systematic investigation of phenolic polymers has indicated that specimens with densities varying between 1.217 and 1.229 $g cm^{-3}$ exhibit relaxations which show little correlation with the temperature at which the cure was performed. The longitudinal velocity was observed to vary with density and not directly with the temperature at which cure was performed or the degree of crosslinking achieved. As in the case of polypropylene, it appears that the acoustic properties are once more correlating with the density of void rather than with a property of the matrix. The sound absorption, however, was found to be a function of the cure temperature, having the lowest value in the highest temperature cure; this is consistent with the lowest void content in this material. Comparison of the absorption data for phenolics and PPQ indicates that relaxation can occur in the former but not in the latter. In the case of the phenolics the magnitude of the acoustic loss was observed to vary with water content implying that the molecular relaxation process is associated with motion of the solvated ether bridge of hydrated residual phenol groups.

17.11 ACOUSTIC MICROSCOPY

In the last five or six years we have seen the appearance of acoustic microscopy. This technique is the direct analogue of light microscopy except that a sound source is used rather than illumination with light. In the earlier microscopes the sample was moved. However, more recently, scanning lasers and segmented detectors have allowed the use of static samples. The principle limitation of the method is its resolution. Using the highest frequencies currently available, approximately 1 GHz, the ultimate resolution is barely 1 μm. An example of an acoustic microscopic picture is shown for a sample illuminated at 100 MHz (Figure 27). It can be seen that the acoustic method allows visualization of small defects and in suitable systems it is possible to detect internal voids within the

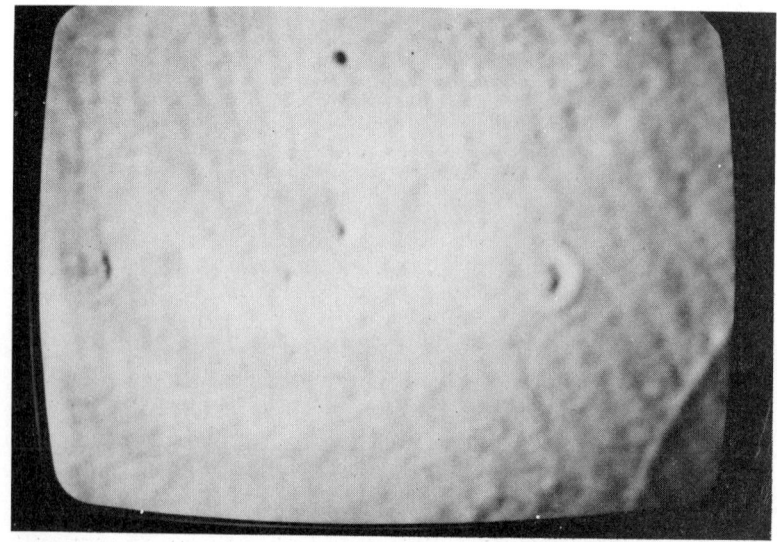

Figure 27 Ultrasonic micrograph of a polymer sample

sample. Acoustic microscopy has found use in the imaging of electronic circuits and also biological systems. The images reflect the acoustic impedance mismatches and hence are able to reveal internal surfaces which can not be observed using optical or electron microscopy. The technique will probably have a limited appeal but is clearly a useful addition to the methods available for the study of structure in polymer solids. The above discussions of scattering, thermal conduction relaxation and anisotropy of propagation explain the origins of the velocity changes which lead to the observation of structure in polymer solids.

17.12 REFERENCES

1. A. M. North and R. A. Pethrick, in 'Developments in Polymer Characterisation — 2', ed. J. V. Dawkins, Applied Science, London, 1980, p. 183.
2. R. A. Pethrick, 'Developments in Polymer Characterisation — 4', ed. J. V. Dawkins, Applied Science, London, 1983, p. 177.
3. R. A. Pethrick, *Sci. Prog.*, 1980, **66**, 571.
4. D. W. Philips and R. A. Pethrick, *J. Macromol. Sci., Rev. Macromol. Chem.*, 1977, **C16**, 79.
5. E. Wyn–Jones and R. A. Pethrick, *Top. Stereochem.*, 1972, **5**, 205.
6. A. M. North and R. A. Pethrick, in 'International Reviews of Science, Physical Chemistry, Series I', ed. A. D. Buckingham and G. Allen, Butterworths, London, 1972, p. 159.
7. R. A. Pethrick and E. Wyn–Jones, *J. Chem. Phys.*, 1968, **49**, 5349.
8. P. J. Flory, 'Statistical Mechanics of Chain Molecules', Wiley Interscience, New York, 1976.
9. E. Wyn–Jones and R. A. Pethrick, *Trans. Faraday Soc.*, 1970, **66**, 2483.
10. M. A. Cochran, A. M. North and R. A. Pethrick, *J. Chem. Soc., Faraday Trans. 2*, 1974, **70**, 1274.
11. R. A. Scott and H. A. Scheraga, *J. Chem. Phys.*, 1966, **44**, 3054.
12. B. Valeur and L. Monnerie, *J. Polym. Sci., Polym. Phys. Ed.*, 1976, **14**, 29.
13. R. T. Bailey, A. M. North and R. A. Pethrick, 'Molecular Motion in High Polymers', Oxford University Press, Oxford, 1981, p. 76.
14. A. M. Awwad, A. M. North and R. A. Pethrick, *J. Chem. Soc., Faraday Trans. 2*, 1982, **78**, 1687.
15. A. M. Awaad, A. M. North and R. A. Pethrick, *J. Chem. Soc., Faraday Trans. 2*, 1983, **79**, 731.
16. D. V. S. Jain, A. M. North and R. A. Pethrick, *J. Chem. Soc., Faraday Trans. 1*, 1974, **70**, 1292.
17. B. T. Poh and R. A. Pethrick, *Mol. Phys.*, 1980, **40**, 539.
18. J. R. Lyerla, Jr. and D. L. Van der Hart, *J. Am. Chem. Soc.*, 1976, **98**, 1697.
19. M. A. Cochran, J. M. Dunbar, A. M. North and R. A. Pethrick, *J. Chem. Soc., Faraday Trans. 2*, 1974, **70**, 215.
20. H. Hassler and H.-J. Bauer, *Kolloid Z.*, 1969, **230**, 194.
21. W. Ludlow, E. Wyn–Jones and J. Rassing, *J. Chem. Phys. Lett.*, 1972, **13**, 477.
22. B. Froehich, C. Noël and L. Monnerie, *Polymer*, 1979, **20**, 529.
23. B. Froehich, C. Noël, B. Jasse and L. Monnerie, *Chem. Phys. Lett.*, 1976, **44**, 159.
24. B. Froehich, B. Jasse, C. Noël and L. Monnerie, *J. Chem. Soc., Faraday Trans. 2*, 1978, **74**, 445.
25. A. M. North, R. A. Pethrick and B. T. Poh, *Polymer*, 1980, **21**, 769.
26. J. H. Dunbar, A. M. North, R. A. Pethrick and B. T. Poh, *Polymer*, 1980, **21**, 764.
27. F. W. Wang and B. H. Zimm, *J. Polym. Sci., Polym. Phys. Ed.*, 1974, **12**, 1619.
28. P. E. Rouse, Jr., *J. Chem. Phys.*, 1953, **21**, 1272.
29. J. H. Dunbar, A. M. North, R. A. Pethrick and D. B. Steinhauer, *J. Chem. Soc., Faraday Trans. 2*, 1975, **70**, 1478.
30. A. M. North, R. A. Pethrick and I. Rhoney, *J. Chem. Soc., Faraday Trans. 2*, 1974, **70**, 223.
31. E. Wyn–Jones and K. R. Crook, *J. Phys. Chem.*, 1969, **30**, 3450.
32. J. H. Dunbar, A. M. North, R. A. Pethrick and D. B. Steinhauer, *J. Polym. Sci., Poly. Phys. Ed.*, 1977, **15**, 263.
33. W. Bell, A. M. North, R. A. Pethrick and B. T. Poh, *J. Chem. Soc., Farada Trans. 2*, 1979, **75**, 1115.
34. W. Bell, J. Daly, A. M. North, R. A. Pethrick and B. T. Poh, *J. Chem. Soc., Faraday Trans. 2*, 1979, **75**, 1452.
35. A. M. Abubaker and R. A. Pethrick, *Polym. Commun.*, 1986, **27**, 194.
36. J. G. Powles, A. Hartland and J. A. E. Kail, *J. Polym. Sci.*, 1961, **55**, 361.
37. L. J. Burnett, C. L. Rottler and D. J. Langhen, *J. Polym. Sci., Polym. Phys. Ed.*, 1978, **16**, 341.
38. C. M. Huggins, L. E. St. Pierre and A. M. Bueche, *J. Phys. Chem.*, 1960, **64**, 1304.
39. H. Kusumoto, I. J. Lawrenson and H. S. Gutowsky, *J. Chem. Phys.*, 1960, **32**, 724.
40. M. E. Baird and D. R. Sengupta, *Polymer*, 1971, **12**, 802.
41. H. Eyring, *Science*, 1978, **199**, 740.
42. A. T. Bullock, G. G. Cameron and P. M. Smith, *J. Chem. Soc., Faraday Trans. 2*, 1974, **70**, 1202.
43. D. V. S. Jain, A. M. North, M. M. Omar and R. A. Pethrick, *J. Chem. Soc., Faraday Trans. 2*, 1982, **78**, 2829.
44. R. J. Abraham and E. Bretschnicider, in 'Internal Rotation in Molecules', ed. W. J. Orrille Thomas, Wiley, New York, 1974, p. 481.
45. K. R. Crook and E. Wyn–Jones, *J. Chem. Phys.*, 1969, **50**, 3445.
46. P. N. Shankar Iyer, A. M. North, R. A. Pethrick and D. B. Steinhauer, *Polymer*, 1975, **16**, 797.
47. A. M. North, R. A. Pethrick and I. Rhoney, *J. Chem. Soc., Faraday Trans. 2*, 1974, **70**, 223.
48. K. Adachi, A. M. North, R. A. Pethrick, G. Harrison and J. Lamb, *Polymer*, 1982, **23**, 1451.
49. A. V. Cunliffe and R. A. Pethrick, *Polymer*, 1980, **21**, 1025.
50. D. J. Meier, in 'Block and Graft Copolymers', ed. J. J. Burke and V. Weiss, Syracuse University Press, Syracuse, NY, 1973, p. 43.
51. W. Bell and R. A. Pethrick, *Eur. Polym. J.*, 1972, **8**, 927.
52. A. M. Awwad, A. M. North and R. A. Pethrick, *J. Chem. Soc., Faraday Trans. 2*, 1983, **79**, 2333.
53. P. De St. Romain, H. T. Van, D. Patterson, V. L. Lam, P. Picker and D. Trancrede, *J. Chem. Soc., Faraday Trans. 1*, 1979, **75**, 1700.

54. R. A. Orwoll and P. J. Flory, *J. Am. Chem. Soc.*, 1967, **89**, 6822.
55. W. Bell and R. A. Pethrick, *Eur. Polym. J.*, 1975, **11**, 293.
56. A. M. North, B. T. Poh and R. A. Pethrick, *Adv. Mol. Relaxation Processes*, 1981, **19**, 209.
57. W. Bell and R. A. Pethrick, *Polymer*, 1982, 369.
58. R. Milne and R. A. Pethrick, *Eur. Polym. J.*, 1974, **10**, 921.
59. G. Gooberman and J. Lamb, *J. Polym. Sci.*, 1960, **42**, 35.
60. S. I. Sokolov, *Zh. Tekh. Fiz.*, 1946, **16**, 283.
61. E. M. Zacharais, *Instrum. Control Syst.*, 1970, **43**, 112.
62. P. Sladky, I. Pelant and L. Parman, *Ultrasonics*, 1979, **16**, 32.
63. P. Sladky, L. Parma and J. Zdrazil, *Polym. Bull. (Berlin)*, 1982, **7**, 401.
64. P. Hauptmann, F. Dinger and R. Sauberlich, *Polymer*, 1985, **26**, 1741.
65. P. Hauptmann, in 'Physical, Optics of Dynamic Phenomena and Processes in Macromolecular Systems', ed. B. Sedlacek, de Gruyter, Berlin, 1985, p. 329.
66. J. R. Emery, D. Durand, M. Tabellont and R. A. Pethrick, unpublished data.
67. G. A. Sofer, A. G. H. Dietz and E. A. Hauser, *Ind. Eng. Chem.*, 1953, **45**, 2743.
68. K. Adachi, G. Harrison, J. Lamb, A. M. North and R. A. Pethrick, *Polymer*, 1981, **22**, 1032.
69. I. M. Ward, 'Mechanical Properties of Solid Polymers', Wiley, New York, 1976.
70. M. G. McCrum, B. Reid and G. Williams, 'Inelastic and Dielectric Properties of Polymers', Wiley, New York, 1967.
71. M. W. Volkenstein, 'Configurational Statistics of Polymer Chains', Mir, Moscow, 1959.
72. R. Buchdahl and L. E. Nielsen, *J. Appl. Phys.*, 1950, **21**, 482.
73. J. A. Sauer and D. E. Kline, *J. Polym. Sci.*, 1955, **18**, 491.
74. B. Maxwell, *J. Polym. Sci.*, 1956, **20**, 551.
75. K. H. Illers and E. Jenckel, *Kolloid Z.*, 1959, **165**, 73.
76. O. Yana and Y. Wada, *Dep. Prog. Polym. Phys. Jpn.*, 1969, **12**, 297.
77. A. Ya. Malkin, E. A. Dzyura and C. V. Vinagradov, *DAN SSSR, Ser. Khim.*, 1969, **188**, 1328.
78. A. M. North, R. A. Pethrick and B. T. Poh, *Polymer*, 1980, **21**, 772.
79. T. Hara, *Jpn. J. Appl. Phys.*, 1967, **6**, 147.
80. B. Hartman and J. Jarzynski, *J. Acoust. Soc. Am.*, 1974, **56**, 1469.
81. K. Marcinčin and A. Romanov, *Polymer*, 1975, **16**, 173.
82. G. Allen, J. McAinsh and G. M. Jeffs, *Polymer*, 1971, **12**, 85.
83. E. Ito and T. Hatakeyama, *J. Polym. Sci., Polym. Phys. Ed.*, 1975, **13**, 2313.
84. I. I. Perepechko and O. V. Startsev, *Sov. Phys. Acoust. (Engl. Transl.)*, 1974, **20**, 456.
85. D. W. Phillips, A. M. North and R. A. Pethrick, *J. Appl. Polym. Sci.*, 1977, **21**, 1859.
86. C. W. Bunn and E. R. Howells, *Nature (London)*, 1954, **174**, 549.
87. H. A. Rigby and C. W. Bunn, *Nature (London)*, 1949, **164**, 583.
88. G. T. Furukawa, R. E. McCoskey and G. J. King, *J. Res. Natl. Bur. Stand.*, 1952, **49**, 273.
89. I. I. Perepechko and V. E. Sorokin, *Sov. Phys. Acoust. (Engl. Transl.)*, 1972, **18**, 485.
90. I. I. Perepechko, *Sov. Phys. Acoust. (Engl. Transl.)*, 1972, **18**, 343.
91. R. A. Pethrick and B. T. Poh, *Mol. Phys.*, 1980, **40**, 539.
92. G. P. Koo, in 'Fluoropolymers', ed. L. A. Wall, Wiley, New York, 1972, p. 507.
93. N. V. Karyakin, I. B. Rabinovich and V. A. Ulyanov, *Polym. Sci. USSR (Engl Transl.)*, 1969, **11**, 3159.
94. S. P. Kabin, *Sov. Phys. Acoust. (Engl. Transl.)*, 1956, **1**, 2542.
95. R. Kono, *J. Phys. Soc. Jpn.*, 1961, **16**, 1580.
96. R. K. Eby and K. M. Sinnott, *J. Appl. Phys.*, 1961, **32**, 1765.
97. R. K. Eby and F. C. Wilson, *J. Appl. Phys.*, 1962, **33**, 2951.
98. G. A. Samara and I. J. Fritz, *J. Polym. Sci., Polym. Lett. Ed.*, 1975, **13**, 93.
99. K. Schmieder and K. Wolf, *Kolloid Z.*, 1952, **127**, 65.
100. W. Sommer, *Kolloid Z.*, 1959, **167**, 97.
101. G. W. Becker, *Kolloid Z.*, 1955, **140**, 1.
102. I. I. Perepechko, L. I. Trepelkova, L. A. Bodrova and L. A. Bunina, *Vysokomol. Soedin., Ser. B*, 1968, **10**, 507.
103. J. A. Sauer and A. E. Woodward, *Rev. Mod. Phys.*, 1960, **32**, 88.
104. D. A. Jackson, H. T. A. Pentecost and J. G. Powles, *Mol. Phys.*, 1972, **23**, 425.
105. K. Neki and P. H. Geil, *J. Macromol. Sci., Phys.*, 1973, **8**, 295.
106. J. A. Manson, S. A. Iobst and R. Acosta, *J. Macromol. Sci., Phys.*, 1974, **9**, 301.
107. Y. Maeda, *J. Polym. Sci.*, 1955, **18**, 87.
108. G. Pezzin, G. Ajroldi, T. Casiraghi, C. Garbughlio and G. Vittadini, *J. Appl. Polym. Sci.*, 1972, **16**, 1839.
109. R. Buchdahl and L. E. Nielsen, *J. Appl. Phys.*, 1950, **21**, 482.
110. J. A. Sauer and D. E. Kline, *J. Polym. Sci*, 1955, **18**, 491.
111. E. A. Woodward, *Pure Appl Chem.*, 1966, **12**, 34.
112. R. F. Boyer, *Polym. Eng. Sci.*, 1968, **8**, 161.
113. J. S. Higgins, G. Allen and P. N. Brier, *Polymer*, 1972, **13**, 157.
114. A. M. North, R. A. Pethrick and D. W. Phillips, *Polymer*, 1977, **18**, 324.
115. K. Shimizu, O. Yano, Y. Wada and Y. Kawamura, *J. Polym. Sci., Polym. Phys. Ed.*, 1975, **13**, 1959.
116. K. Shimizu, O. Yano, Y. Wada and Y. Kawamura, *J. Polym. Sci., Polym. Phys. Ed.*, 1973, **11**, 1641.
117. B. Hartmann and J. Jarzynski, *J. Appl. Phys.*, 1972, **43**, 4304.
118. S. Saito and T. Nakajima, *J. Appl. Phys.*, 1959, **30**, 93.
119. H. Thurn and W. Wolf, *Kolloid Z.*, 1956, **148**, 16.
120. I. I. Perepechko, L. A. Kvachieva, L. A. Ushakov, A. Ya. Svetov and V. A. Grechishkin, *Plast. Massy.*, 1970, **8**, 43.
121. Y. Wada, H. A. Hirose, T. Sano and S. Furutomic, *J. Phys. Soc. Jpn.*, 1959, **14**, 1064.
122. G. P. Mikhailov and M. P. Eidelkant, *Vysokomol. Soedin.*, 1960, **2**, 287.
123. H. J. McSkimin, *J. Acoust. Soc. Am.*, 1951, **23**, 429.
124. A. Levene, W. J. Pullen and J. Roberts, *J. Polym. Sci., Part A*, 1965, **3**, 697.
125. J. R. Asay and A. H. Guenther, *J. Appl. Polym. Sci.*, 1966, **11**, 1087.

126. Y. Wada and K. Yamamoto, *J. Phys. Soc. Jpn.*, 1956, **11**, 887.
127. M. P. Mason and H. J. McSkimin, *Bell Syst. Tech. J.*, 1952, 122.
128. G. Wardel, V. J. McBrierty and D. C. Douglas, *J. Appl. Phys.*, 1974, **45**, 3441.
129. S. C. Mowry and A. W. Nolle, *J. Acoust. Soc. Am.*, 1948, **20**, 432.
130. M. Shen, V. A. Kaniskin, K. Biliyar and R. H. Boyd, *J. Polym. Sci., Polym. Phys. Ed.*, 1973, **11**, 2261.
131. H. J. Naake and K. Tamm, *Acustica*, 1958, **8**, 65.
132. W. P. Mason and H. J. McSkimin, *Bell Syst. Tech. J.*, 1952, 122.
133. J. R. Cunningham and D. G. Ivey, *J. Appl. Phys.*, 1956, **27**, 967.
134. Y. Maeda, *J. Polym. Sci.*, 1955, **18**, 87.
135. P. Hatfield, *Br. J. Appl. Phys.*, 1950, **1**, 252.
136. R. K. Eby, *J. Acoust. Soc. Am.*, 1964, **36**, 1485.
137. B. Hartman and J. Jarzynski, *J. Acoust. Soc. Am.*, 1968, **44**, 387.
138. H. A. Waterman, *Kolloid Z.*, 1963, **192**, 1.
139. N. A. Bordelius and V. K. Semenchenko, *Sov. Phys. Acoust. (Engl. Transl.)*, 1971, **16**, 519.
140. B. Hartmann and J. Jarzynski, *J. Appl. Phys.*, 1972, **43**, 4304.
141. H. A. Flocke, *Kolloid Z.*, 1962, **180**, 118.
142. R. F. Boyer, *J. Polym. Sci., Polym. Symp.*, 1975, **50**, 189.
143. Y. Wada, H. Hirose, T. Asano and S. Fukutomi, *J. Phys. Soc. Jpn.*, 1959, **14**, 1064.
144. M. Takayanagi, K. Imada and T. Kajiama, *J. Polym. Sci.*, 1966, **15**, 263.
145. I. M. Ward and A. Ciferri, 'Ultra-High Modulus Polymers', Applied Science, London, 1979.
146. R. A. Pethrick, unpublished data.
147. K. Thomas, D. E. Meyer, E. C. Fleet and M. Abrahams, *J. Phys. D*, 1973, **6**, 1336.
148. W. W. Moseley, *J. Appl. Polym. Sci.*, 1960, **3**, 266.
149. I. M. Ward, *Text. Res. J.*, 1964, **34**, 806.
150. H. M. Morgan, *Text. Res. J.*, 1962, **32**, 866.
151. J. G. Rider and K. M. Watkinson, *Polymer*, 1978, **19**, 645.
152. O. K. Chan, F. C. Chen, C. L. Choy and I. M. Ward, *J. Phys. D*, 1978, **11**, 617.
153. R. A. Pethrick and D. Crofton, unpublished data.
154. P. K. Datta and R. A. Pethrick, *Polymer*, 1978, **19**, 145.
155. R. J. Samuels, *J. Polym. Sci., Part A-2*, 1965, **3**, 1741.
156. R. J. Urick, *J. Appl. Phys.*, 1947, **18**, 983.
157. H. A. Waterman, *Kolloid Z.*, 1963, **192**, 9.
158. K. Adachi, G. Harrison, J. Lamb, A. M. North and R. A. Pethrick, *Polymer*, 1981, **22**, 1032.
159. K. Adachi, G. Harrison, J. Lamb, A. M. North and R. A. Pethrick, *Polymer*, 1982, **23**, 1451.
160. R. W. Martin, 'The Chemistry of Phenolic Resins', Wiley, New York, 1956.
161. G. A. Sofer, G. H. Dietz and E. A. Hauser, *Ind. Eng. Chem.*, 1953, **45**, 2743.
162. B. Hartman, *J. Appl. Polym. Sci.*, 1975, **19**, 3241.

18

Dielectric Properties

GRAHAM WILLIAMS

University College of Wales, Aberystwyth, UK

18.1 INTRODUCTION

Of the many physical methods which can be used to study polymers in their bulk or solution states those involving the mechanical and electrical properties have been prominent. The need for data on the mechanical properties is evident since most applications of synthetic polymers require the materials to be strong, tough and mechanically stable. The electrical properties are required when polymers are used for cable insulation, for capacitors, for insulation or packaging of electronic devices or as integral parts of electronic devices. Polymers used in these connexions include low and high density polyethylenes, poly(vinyl chloride), the nylons, polystyrene, poly(tetrafluoroethylene) and the acrylate and methacrylate polymers.

The term 'electrical properties' requires clarification. As we shall see, the dielectric permittivity of a material is, in general, a complex quantity when measured in the frequency domain. Its real part is usually called the 'permittivity' and decreases with increasing frequency. Its imaginary part is usually called the 'dielectric loss factor' and may increase or decrease with increasing frequency. A true dc conduction in a material contributes only to the imaginary part of the complex permittivity and leads to a loss factor which is inversely proportional to frequency, while other dielectric processes such as dipole relaxation or space charge formation contribute to both the permittivity and loss behaviour. Measurement of the dielectric properties over a wide range of frequency characterizes the electrical properties of a material in terms of (i) the dispersion of the permittivity and (ii) the loss factor–frequency behaviour — which may be further analyzed to give information on the dielectric relaxation behaviour and the dc conduction behaviour. Often the measurement of the ac conduction of a polymer at a fixed frequency is used to imply the dc conduction in the material. Since most polymers are high impedance materials, such an identification of the ac property with the dc property is often erroneous and leads to much confusion and incorrect interpretation of experimental data for polymers. The presence of a dc conduction process can only be established from an analysis of the frequency dependence of either the dielectric loss factor or the ac conductivity. Also the electrical properties (complex dielectric permittivity, dc conductivity) will, in general, depend on the amplitude of the applied electric field. Phenomena such as dielectric breakdown, carrier injection and electrical 'treeing' will appear at high fields. Such phenomena, which certainly are to be included as part of the dielectric properties of a polymer material (and have great importance for the applications of polymers as dielectric insulators), may not be readily related to the chemical or physical structure of the material. In this account we shall be only concerned with the linear (field-independent) dielectric properties of polymers and shall concentrate on the phenomenological and molecular aspects of dielectric relaxation. Aspects of the electrical conductivity of polymers will be described in Volume 2, Chapter 22.

The early dielectric studies of polymers date back to World War II when the need arose to characterize the permittivity and loss factor for materials which were to be used in electrical cables and in capacitors at frequencies ranging from power frequencies (10 to 10^3 Hz) through audio and radio frequencies to the UHF range (10^5 to 10^8 Hz) and as electrical insulation in radar assemblies (in the microwave range 10^8 to 10^{11} Hz). Extensive studies were carried out at MIT by von Hippel and his co-workers and at London University by Willis Jackson and his co-workers. Methods were developed for the accurate determination of the permittivity and loss of polymers and other solid dielectrics over the entire frequency range. The methods used and the results obtained for a wide range of materials were subsequently published.[1] In parallel with these studies Fuoss and Kirkwood[2,3] carried out detailed experimental studies of the dielectric properties of solid polymers, notably plasticized materials, and analyzed their experimental results using both a phenomenological approach and a molecular theory which showed that the static and complex permittivities could be related to the dipole moments of chain segments, to the angular correlations between dipoles (*via* the Kirkwood '*g* factor') and to the orientational motions of the chain segments.[2,3] Their work laid the basis of the molecular theory of dielectric relaxation in polymers and established structure–property relations for those materials. Following World War II many complementary studies were made of the dielectric and dynamic mechanical properties of amorphous and crystalline polymers. Multiple relaxation regions were observed and mechanisms for the individual processes were sought. Classic dielectric studies were reported by Reddish[4] for poly(ethylene terephthalate) and by Scott *et al.*[5] for poly(chlorotrifluoroethylene). Both studies covered wide ranges of frequency and temperature. The work of Reddish included a contour map representation of data while that of Scott *et al.* varied the degree of crystallinity which enabled them to assign individual relaxations to the amorphous and crystalline phases. The late 1950's and early 1960's saw a great expansion in the documentation of dielectric and dynamic mechanical behaviour for polymers. A remarkable contribution was made by Ishida and his co-workers who gave definitive data for many of the conventional amorphous and crystalline polymers.[6]

The extensive literature on the dielectric and dynamic mechanical properties of solid polymers which had accumulated up to the mid 1960s was reviewed and discussed in the book by McCrum, Read and Williams.[7] By that time most of the 'volume' polymers used today had been made and had found commercial application. It was possible to describe and compare their dielectric and dynamic mechanical properties in that text. In addition, interpretations were given, in phenomenological and molecular terms, of the multiple relaxation regions which are usually observed for solid polymers. The effect on the properties resulting from the variation of chemical structure, tacticity, plasticizer content, nature and extent of crystallinity and sample orientation on the dielectric and mechanical properties were explored systematically and structure–property relations were established. Since the publication of that text in 1967 there have been few new volume polymers but many speciality polymers to study. The experimental methods for low frequency dielectric measurements have improved markedly, especially in the range 10^{-3} to 10^5 Hz, taking full advantage of solid state electronics and computer control of measurement. In addition, dielectric studies have been made over a range of applied pressures (1 to 3000 atm, where 1 atm = 101 825 Pa) which have aided our understanding of multiple relaxation processes. The theory of dielectric relaxation has been advanced during the past twenty years in many ways, particularly by the introduction of the time-correlation function approach to single and multiple relaxations. Thus, at this time, dielectric relaxation offers an attractive complementary method to those of nuclear magnetic resonance, electron spin resonance, quasielastic laser light scattering and neutron scattering for the study of molecular dynamics and relaxation phenomena in polymers. In addition it has the advantages that the molecular probe is well defined (molecular dipole moment) and the frequency range, which can be readily scanned, is unrivalled (10^{-4} to 10^{7} Hz fairly routinely and 10^{7} to 10^{11} Hz with some difficulty).

In the present account we shall not seek to duplicate the detailed accounts of the dielectric properties of polymers contained in the texts of McCrum, Read and Williams,[7] Hedvig[8] and Jonscher[9] or in the reviews by Williams.[10,11] Instead we shall give a structured account which incorporates the recent developments in experimental techniques and theoretical interpretations and gives illustrations of representative dielectric behaviour for selected amorphous, crystalline and liquid crystalline polymers, and for polymers in solution.

18.2 PHENOMENOLOGICAL ASPECTS

18.2.1 The Dielectric Permittivity

Before we consider the behaviour of polymers, it is helpful to define the basic quantities which occur in dielectric and electrical studies. The relative dielectric permittivity, which is also known as the dielectric constant, or specific inductive capacity, of a material occurs naturally in Maxwell's equations for the propagation of electromagnetic radiation through a medium. However it is perhaps simpler to define it in relation to low frequency ('lumped circuit') measurements as the factor of proportionality between the dielectric displacement D and the applied electric field E when the field is applied to a capacitor filled with the dielectric.

$$D = \varepsilon E \tag{1}$$

For an ideal insulating dielectric, ε measured in the frequency domain is a real quantity and is equal to the ratio of the capacitances measured with sample 'in' and sample 'out' of the active electrode region; *i.e.*

$$\varepsilon = C/C_0 \tag{2}$$

where C_0 is the electrode capacitance for sample out. Measurements of permittivity may be made in the time or frequency domains. If an electric field is applied to the dielectric-filled capacitor as a step at $t = 0$ then for an ideal insulating dielectric the induced charge on the capacitor plates would be established instantly so that ε is a constant, independent of time. If the field is removed as a step and the capacitor is simultaneously short circuited then the ideal capacitor would discharge instantly. Measurements made on this ideal system in the frequency domain by application of a steady state ac electric voltage of the form $V = V_0 \exp(i\omega t)$ would show that the capacitor current $I(\omega)$ is a quantity which is $\pi/2$ out of phase with the applied voltage. $I(\omega)$ is given by

$$I(\omega) = i\omega\varepsilon C_0 V(\omega) \tag{3}$$

where $\omega = 2\pi\nu$ Hz. Such behaviour does not obtain for any real dielectric material owing to dielectric relaxation and dc conduction. This leads to a time dependence of the permittivity as measured in the time domain and to the permittivity in the frequency domain becoming a complex quantity, the real and imaginary parts of which exhibit a dependence on frequency. If a step voltage is applied to the real dielectric the charge on the capacitor plates takes time to build up to its steady value so we may define a time dependent permittivity.

$$\varepsilon(t) = C(t)/C_0 \tag{4}$$

In addition the material would, in general, have a dc conduction so the capacitor would pass a time independent leakage current $I(t)$ equal to ΛV where Λ is the conductance (reciprocal of resistance) of the sample-filled capacitor. Λ is proportional to the specific conductivity (in units of Ω^{-1} cm^{-1}) of the dielectric. Note that the time-dependent charge, which arises from relaxation processes or space-charge development, has a derivative which is a current but this is *not* to be associated with the steady leakage current of a dc conductivity.

In the frequency domain the capacitor passes a steady state current which is now a complex quantity, composed of an out-of-phase capacitor current $I_c(\omega)$ and an in-phase loss current $I_L(\omega)$. Defining a complex dielectric permittivity as

$$\varepsilon(\omega) = \varepsilon'(\omega) - i\varepsilon''(\omega) \tag{5}$$

it is readily shown that

$$I_c(\omega) = i\omega\varepsilon' C_0 V(\omega); \quad I_L(\omega) = \omega\varepsilon'' C_0 V(\omega) \tag{6}$$

ε' is termed the permittivity and ε'' is termed the dielectric loss factor. Since $I_L(\omega)$ is in phase with the applied voltage, it follows that energy is dissipated by the dielectric, and the energy loss per cycle is proportional to the loss factor. This is the o.igin of dielectric heating, so well known in the context of microwave cooking. In the early dielectric studies of capacitors and cable insulators the energy dissipation factor for a dielectric was often expressed in terms of the loss tangent tan δ which is defined as

$$\tan \delta = \varepsilon''/\varepsilon' = \text{loss current/capacitor charging current} \tag{7}$$

The dielectric loss factor is made up of two components; one due to relaxation or space-charge conduction processes and the other due to the dc conduction of the material. We may write

$$\varepsilon''(\omega) = \varepsilon''_r(\omega) + \varepsilon''_\sigma(\omega) \tag{8}$$

where

$$\varepsilon''_\sigma(\omega) = 1.8 \times 10^{12}(\sigma/v) \tag{9}$$

and σ is the specific dc conductivity of the material (in units of Ω^{-1} cm^{-1}). In practice the relaxation component $\varepsilon_r(\omega)$ usually takes the form of a bell-shaped curve when ε'' is plotted against log v. The loss due to dc conduction increases with decreasing frequency (equation 9) so in our dielectric studies we expect to see a rising loss at the lowest frequencies which may obscure the loss peak due to relaxation and space-charge processes. Since the specific dc conductivity of a polymer dielectric depends exponentially upon temperature, it follows that studies of dielectric relaxation in such materials at low frequencies and at temperatures greatly exceeding room temperature are often made extremely difficult due to the dominance of the dc conductivity losses.

18.2.2 Frequency–Time Interrelations

The frequency-dependent complex permittivity $\varepsilon_r(\omega)$ and the time-dependent permittivity $\varepsilon(t)$ are related by a Fourier transform according to[7]

$$\frac{\varepsilon_r(\omega) - \varepsilon_\infty}{\varepsilon_0 - \varepsilon_\infty} = 1 - i\omega \int_0^\infty \Phi(t) \cdot \exp(-i\omega t) dt \tag{10}$$

where ε_0 and ε_∞ are the limiting low and high frequency permittivities with respect to all relaxation processes and $\Phi(t)$ is the normalized relaxation function

$$\Phi(t) = \frac{\varepsilon(t) - \varepsilon_\infty}{\varepsilon_0 - \varepsilon_\infty} \tag{11}$$

where $\varepsilon(t)$ describes the evolution in time of the apparent permittivity when a steady dc field is removed as a step at $t = 0$ from a polarized material. $\Phi(t)$ decays from 1 to 0. Note that the dc conductivity does not contribute to $\varepsilon_r(\omega)$ or $\varepsilon(t)$.

This description is phenomenological and is obeyed by dielectrics irrespective of the origins of the relaxation and conduction processes. The components of $\varepsilon_r(\omega)$ may be expressed in terms of integrals involving $\Phi(t)$ according to

$$\frac{\varepsilon'_r(\omega) - \varepsilon_\infty}{\varepsilon_0 - \varepsilon_\infty} = 1 - \omega \int_0^\infty \Phi(t) \sin \omega t \, dt \tag{12a}$$

$$\frac{\varepsilon''_r(\omega)}{\varepsilon_0 - \varepsilon_\infty} = \omega \int_0^\infty \Phi(t) \cos \omega t \, dt \tag{12b}$$

Fourier inversion of these equations give

$$\Phi(t) = \frac{2}{\pi} \int_0^\infty \left[\frac{\varepsilon_0 - \varepsilon_r(\omega)}{\varepsilon_0 - \varepsilon_\infty} \right] \cdot \sin \omega t \, \frac{d\omega}{\omega} \tag{13a}$$

$$\Phi(t) = \frac{2}{\pi} \int_0^\infty \left[\frac{\varepsilon_r''(\omega)}{\varepsilon_0 - \varepsilon_\infty} \right] \cdot \cos \omega t \, \frac{d\omega}{\omega} \tag{13b}$$

According to equations (12), if $\Phi(t)$ is known over the entire relaxation range then the reduced permittivity and loss factor may be calculated at all frequencies in the relaxation region. This connexion is important since it allows dielectric measurements to be made in the time domain at long times (10^{-2} to 10^3 s) and be converted into loss data in the frequency domain at ultra low frequencies, 10 to 10^{-4} Hz, to supplement conventional ac dielectric data obtained by bridge methods in the range 10^2 to 10^6 Hz. Conversely, equations (13) show that if either the reduced permittivity or the reduced loss factor are known over the entire relaxation region then $\Phi(t)$ may be calculated at all times in the relaxation range. As well shall see, the time dependent function $\Phi(t)$ is closely related to the dipole moment time correlation function which describes the randomization of the dipole moment vector in time due to molecular motion.

18.2.3 Relaxation Functions

The simplest form for $\Phi(t)$ is the single exponential decay characterized by a relaxation time τ

$$\Phi(t) = \exp(-t/\tau) \tag{14}$$

Insertion into equations (12) and integrating gives

$$\frac{\varepsilon_r'(\omega) - \varepsilon_\infty}{\varepsilon_0 - \varepsilon_\infty} = \frac{1}{1 + \omega^2 \tau^2} \tag{15a}$$

$$\frac{\varepsilon_r''}{\varepsilon_0 - \varepsilon_\infty} = \frac{\omega \tau}{1 + \omega^2 \tau^2} \tag{15b}$$

These are termed the single relaxation time (SRT) functions and are shown in Figure 1. In practice solid polymers exhibit several processes each of which are broader in the frequency domain than that for a single relaxation time process.[7] It has been customary both for dielectric and dynamic mechanical relaxation data for polymers to simply use empirical modifications of the single

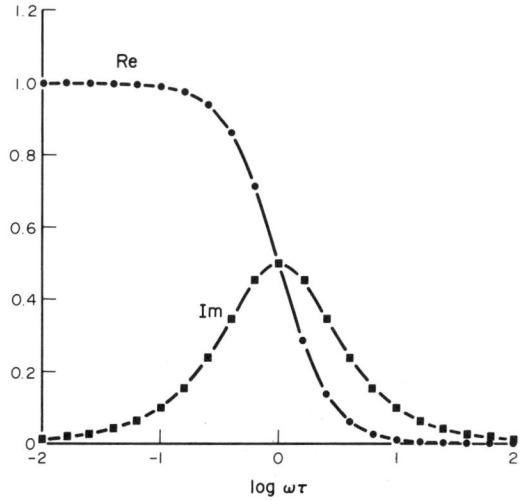

Figure 1 The real part (Re) and the imaginary part (Im) of the normalized permittivity as a function of log $\omega\tau$ for a single relaxation time function

relaxation time expression. For the SRT case insertion of equation (14) into equation (10) gives

$$\frac{\varepsilon_r(\omega)-\varepsilon_\infty}{\varepsilon_0-\varepsilon_\infty}=\frac{1}{1+i\omega\tau} \tag{16}$$

which gave equations (15) on separating real and imaginary parts. Four empirical modifications of equation (16) have been used extensively for polymer relaxations:

(a) Cole and Cole[12]

$$\frac{\varepsilon_r'(\omega)-\varepsilon_\infty}{\varepsilon_0-\varepsilon_\infty}=\frac{1}{1+(i\omega\tau)^p}; \quad 0<p\le 1 \tag{17}$$

(b) Davidson and Cole[13,14]

$$\frac{\varepsilon_r'(\omega)-\varepsilon_\infty}{\varepsilon_0-\varepsilon_\infty}=\frac{1}{[1+(i\omega\tau)]^q}; \quad 0<q\le 1 \tag{18}$$

(c) Havriliak and Negami[15]

$$\frac{\varepsilon_r(\omega)-\varepsilon_\infty}{\varepsilon_0-\varepsilon_\infty}=\frac{1}{(1+(i\omega\tau)^p)^q} \tag{19}$$

$$0<p\le 1; \quad 0<q\le 1$$

(d) Fuoss and Kirkwood[2,3]

$$\frac{\varepsilon_r''(\omega)}{\varepsilon_0-\varepsilon_\infty}=\frac{s\cdot(\omega\tau)^{2s}}{1+(\omega\tau)^{2s}} \tag{20}$$

$$0<s\le 1$$

The Cole–Cole and Fuoss–Kirkwood equations give loss curves which are broadened symmetrically with respect to the frequency of maximum loss (see Figure 1) while the Davidson–Cole loss curve is skewed, being broader on its high frequency side. The Havriliak–Negami function gives curves which are intermediate between those for the two other functions [equations (17) and (18)] and, being a two-parameter fitting function, is often found to give a very good representation of asymmetric loss curves observed for polymers. Other empirical modifications of the SRT equation (16) due to Jonscher, Hill and Dissado have been discussed in some detail (see ref. 9, p. 100).

Since the frequency-dependent permittivity is related to a relaxation function in the time domain (see equation 10), another way of introducing an empirical modification to the SRT function is to modify the relaxation function itself. A general procedure used for many years in relaxation theory[7,16] is to express $\Phi(t)$ as a discrete set of exponential decay functions or as an integral over a continuous distribution of such decay functions so that

$$\Phi(t)=\sum_i g_i \exp(-t/\tau_i) \tag{21a}$$

$$\Phi(t)=\int \phi(\tau) \exp(-t/\tau)d\tau \tag{21b}$$

where g_i is the probability of obtaining a process having $\tau=\tau_i$ and $\phi(\tau)d\tau$ is the probability of obtaining processes in the range τ to $\tau+d\tau$. These are subject to the normalization conditions

$$\Sigma g_i = 1; \quad \int\phi(\tau)d\tau = 1 \tag{22a,b}$$

Insertion of equations (21) into equation (10) gives expressions for the complex permittivity which correspond to discrete or continuous systems of single relaxation time processes

$$\frac{\varepsilon_r(\omega)-\varepsilon_\infty}{\varepsilon_0-\varepsilon_\infty}=\sum_i \frac{g_i}{1+i\omega\tau_i} \tag{23}$$

$$\frac{\varepsilon_r(\omega)-\varepsilon_\infty}{\varepsilon_0-\varepsilon_\infty}=\int \frac{\phi(\tau)d\tau}{1+i\omega\tau} \tag{24}$$

Equations (21a) and (21b) are empirical, as introduced, therefore the derived relations (23) and (24) are also empirical. For molecules moving in a barrier system by thermal activation[7] or a polymer chain undergoing normal modes of motion,[7,16] $\Phi(t)$ takes the form of equation (21a) where the coefficients g_i are functions of the parameters of the model. Clearly, experimental dielectric relaxation data may always be fitted using a suitably chosen distribution of relaxation times. Indeed, it is possible to determine $\phi(\tau)$ from the experimental data using methods developed by Ferry and co-workers[16] [see also ref. 17 for an analysis of dielectric data for poly(methyl acrylate)]. We note also that the empirical relations equations (17)–(20) may also be expressed as equation (24) and the form of the distribution function (discrete or continuous) provides a vehicle for the conversion of frequency-dependent data into the time domain and *vice versa*. A further possibility for modifying the SRT function is to assume a particular empirical form for $\Phi(t)$. This was done by Williams and Watts[18a] who considered the empirical function

$$\Phi(t) = \exp[-(t/\tau)^\beta] \tag{25}$$

$$0 < \beta \leq 1$$

This function, originally introduced by Kohlrausch in 1854 to describe creep in silk and glass threads used as supports in magnetometers, is a very 'slow' function of time and hence gives broad dispersion and loss curves when used in conjunction with equation (10). The integration cannot be expressed generally in closed form. For the special case $\beta = 0.5$ Williams and Watts showed that equations (10) and (25) give

$$\frac{\varepsilon_r(\omega) - \varepsilon_\infty}{\varepsilon_0 - \varepsilon_\infty} = -\frac{1}{2}\left(\frac{\pi}{\tau}\right)^{\frac{1}{2}}\frac{1}{(i\omega)^{\frac{1}{2}}}\exp\left[\frac{k^2}{i\omega}\right]\operatorname{erfc}\left[\frac{k}{(i\omega)^{\frac{1}{2}}}\right] \tag{26}$$

where $2k = \tau^{-\frac{1}{2}}$. The loss curve calculated from equation (26) is broad and asymmetrical in the Davidson–Cole sense but is not as skewed. Subsequently Williams, Watts, Dev and North[18b] calculated dielectric dispersion and loss curves for a range of values of β using series expansions of the Fourier integral and showed that experimental data in the frequency domain for several amorphous polymers were well fitted by this function with β values in the range 0.4 to 0.6. Subsequently it was shown that this empirical representation was widely applicable to relaxation data for glass-forming systems. The data for electric field relaxation in ionic melts, volume, enthalpy and specific heat relaxation in ionic, molecular and polymeric glass forming systems, dielectric and NMR relaxations in molecular and polymeric glass forming systems may be represented quantitatively by the empirical relation, equation (25), which is commonly known as the Kohlrausch–Williams–Watts (KWW) function, or 'stretched exponential' function. These applications have been the subject of reviews[19-22] (see ref. 22(a) for a summary which includes references to the use of the KWW function for physical aging in polymeric glasses). Note that if equations (21b) and (25) are forced to be equivalent then the distribution function can be determined. Lindsey and Patterson[22b] have shown that $\phi(\tau)$ can be expressed analytically and that $\tau\phi(\tau)[\equiv \phi(\ln\tau)]$ can be calculated over the entire range of τ. They showed that the double logarithmic plot of $\tau\phi(\tau)$ against τ was skewed to small values of τ, which is consistent with the sense of asymmetry of the plots of ε_r'' against $\log\omega$ (skewed to high frequencies).

As introduced above in equation (25) the KWW function is empirical. However its wide application to relaxations in glass-forming systems has led to a search for a possible physical basis for such a stretched exponential. Diffusion models,[23] hierarchically constrained models,[24] ultrametric space models,[25] direct-transfer models,[26] fractal representations[27] and dynamic Ising models[28] have been shown to lead to stretched exponential relaxation functions of the KWW type. These different models for the dynamics of a glass-forming system have been compared and contrasted.[19,22,29] We note that quite different models for dynamics may lead to relaxation functions of the KWW type (equation 25) and also that the function appears to result from the cooperative nature of the model in each case. It is important that we consider the molecular theory for the dielectric properties of polymers, and this will be done in the next section. Before we proceed, we note that the theoretical work carried out recently[22-29] and in progress relating to the stretched exponential relaxation function, equation (25), appears to be one of the main thrust areas in condensed matter physics due to its application to charge transport and conduction and relaxation in glass-forming materials of very different chemical compositions.

8.2.4 Temperature and Pressure Dependences of the Average Relaxation Time

The frequency of maximum loss v_m for a single relaxation time process is obtained from equation 15b) as

$$\omega_m\tau = 2\pi v_m\tau = 1 \tag{27}$$

For loss curves broader than a single relaxation time process we may always define an average relaxation time $\langle \tau \rangle$ by the relation

$$\langle \tau \rangle = (2\pi \nu_m)^{-1} \tag{28}$$

$\langle \tau \rangle$ may be related analytically to the different effective relaxation times which occur in the empirical relations (17) to (20) above, and numerically to the effective relaxation time which occurs in the KWW function (equation 25) (see ref. 18). As is well known,[7,16] $\langle \tau \rangle$ obtained from the different loss peaks which occur in solid polymers usually follow either the Arrhenius relation

$$\langle \tau \rangle = A \, \exp(Q/RT) \tag{29}$$

or the Vogel–Fulcher equation

$$\langle \tau \rangle = A' \, \exp[B/(T - T_0)] \tag{30}$$

where Q is an apparent activation energy, and A, A', B and T_0 are constants for a material. The Vogel equation leads to the Williams–Landel–Ferry equation[16] if we write from equation (30)

$$\ln[\langle \tau \rangle_1 / \langle \tau \rangle_2] = \frac{C_1[T_2 - T_1]}{C_2 + T_2 - T_1} \tag{31}$$

where $C_1 = B/(T_1 - T_0)$ and $C_2 = T_1 - T_0$.

There is an extensive literature which relates equation (31) [or its root, equation (30)], to models for free volume or configurational entropy.[7,16,30-32] The plot of $\ln\langle \tau \rangle$ against $(T/K)^{-1}$ is linear for equation (29) giving the apparent activation energy $Q_{app} = d \ln\langle \tau \rangle/d(1/T) = Q$, but the corresponding plot for equation (30) is curved with Q_{app} increasing with decreasing temperature in accord with

$$Q_{app} = R B/(1 - T_0/T)^2; \quad T \geq T_0 \tag{32}$$

Equation (29) is usually found to apply to the broad secondary (β) relaxations in amorphous polymers[6,7,10] and to the relaxation originating in the crystalline phase of a partially crystalline polymer.[6,7] Glass-forming polymers[6,7,16] and molecular[11] and ionic glass-forming liquids[21] all give a primary (α) relaxation, due to the microbrownian motions of chain segments or molecules or ions, which appears to follow the Vogel equation. In many cases T_0 is 30–50 K below the experimentally determined apparent glass transition temperature T_g.

The dielectric properties of polymers have been studied over wide ranges of applied hydrostatic pressure[33-43] (for a review see ref. 10). In general $\langle \tau \rangle$ is a function of (T, P, V) and we may write quite generally

$$\left(\frac{\partial \ln \langle \tau \rangle}{\partial T}\right)_V = \left(\frac{\partial \ln \langle \tau \rangle}{\partial T}\right)_P + \left(\frac{\partial \ln \langle \tau \rangle}{\partial P}\right)_T \left(\frac{\partial P}{\partial T}\right)_V \tag{33}$$

where $(\partial P/\partial T)_V$ is the thermal pressure coefficient defined as the ratio of the isothermal expansion coefficient and the isobaric compressibility coefficient.[33,34] Defining a constant pressure apparent activation energy and a constant volume apparent activation energy and an apparent activation volume by the relations

$$Q_P(T, P) = -RT^2\left(\frac{\partial \ln \langle \tau \rangle}{\partial T}\right)_P; \quad Q_V(T, V) = -RT^2\left(\frac{\partial \ln \langle \tau \rangle}{\partial T}\right)_V; \quad \Delta V(T, P) = RT\left(\frac{\partial \ln \langle \tau \rangle}{\partial P}\right)_T \tag{34a, b, c}$$

then we may write from equations (33) and (34)

$$Q_V(T, V) = Q_P(T, P) - T\left(\frac{\partial P}{\partial T}\right)_V \Delta V(T, P) \tag{35}$$

but

$$T\left(\frac{\partial P}{\partial T}\right)_V = P + \Pi(T, V) \tag{36}$$

where $\Pi(T, V)$ is known as the 'internal pressure' of a material.[34] So equation (35) may be rewritten as

$$Q_V(T, V) = Q_P(T, P) - [P + \Pi(T, V)] \Delta V(T, P) \tag{37}$$

In practice Π may be several kilobars and so at small applied pressures equation (37) may be approximated as

$$Q_V(T,V) \simeq Q_P(T,P) - \Pi(T,V)\Delta V(T,P) \tag{38}$$

Experimentally it is found that ΔV is positive for relaxations in polymers so equation (37) shows that $Q_V < Q_P$, i.e. if the temperature of a material is raised at constant pressure, part of the decrease in $\langle \tau \rangle$ is due to the increase in volume accompanying the increase in temperature.[34] The effect of pressure can be very large for the primary (α) relaxation in amorphous polymers with $(\partial \ln \langle \tau \rangle / \partial P)_T$ lying in the range $(1-5) \times 10^{-3}$ atom^{-1}, or can be extremely small, as found for the secondary (β) relaxation in amorphous polymers [e.g. for poly(ethylene terephthalate)[37] and poly(vinyl chloride)[40]]. Use of $\langle \tau(T,P,V) \rangle$ data provides a test of theories for relaxation in the glass transition region (α process).[44]

18.3 MOLECULAR ASPECTS

18.3.1 Introduction

Solid polymers occur as amorphous materials [e.g. atactic poly(methyl methylacrylate)], partially crystalline materials of low degree of crystallinity [e.g. melt-crystallized poly(ethylene terephthalate)] and high degree of crystallinity (e.g. melt-crystallized high density polyethylene) or as liquid crystalline materials (e.g. polymers having mesogenic groups in the main chain[45] or side chain[46,47]). Motion of dipolar groups in these materials gives rise to multiple dielectric relaxations which may be studied over wide ranges of frequency (or time) and temperature. The dielectric data obtained for different systems has been described in detail in the earlier texts[7-9,48] and specialized reviews.[6,10,11,49] The mechanisms for motion differ from system to system, e.g. motions of chain segments in the glassy state differ from those in the crystalline phase. In order to make clear the connexion between the dielectric properties observed experimentally and the underlying molecular processes we shall briefly outline the theory as it applies to particular cases. We note that the greater part of the dielectrics literature for polymers is concerned with solid state properties. We shall therefore concentrate this part of our account in that area.

18.3.2 Bulk Amorphous Polymers and Polymers in Solution

The static dielectric permittivity ε_0 for an ensemble of dipolar chain molecules is given by the generalized form[7,10,50] of the Kirkwood–Fröhlich equation

$$(\varepsilon - \varepsilon_\infty) = \frac{4\pi}{3kT} \cdot \left[\frac{3\varepsilon_0(2\varepsilon_0 + \varepsilon_\infty)}{(2\varepsilon_0 + 1)^2} \right] \cdot \frac{\langle M(0) \cdot M(0) \rangle}{V} \tag{39}$$

where $M(0)$ is the instantaneous dipole moment of a macroscopic sphere embedded in the medium of volume V and $\langle M(0).M(0) \rangle$ is the mean square dipole moment deduced in the absence of an applied electric field. If the volume V contains N polymer chains and the instantaneous dipole moment of a reference chain i is $P_i(0)$, where $P_i(0)$ is the vector sum of the (equivalent) elementary dipole moments $\mu_{ji}(0)$ in the chain then in the absence of orientation correlations between different chains we may write for flexible chains of high molecular weight.[50]

$$\frac{\langle M(0).M(0) \rangle}{V} = C_r \left[\mu^2 + \sum_{\substack{k' \\ k' \neq k}} \langle \mu_k(0) . \mu_{k'}(0) \rangle \right] \tag{40}$$

where C_r is the number of dipole units per unit volume, and $\mu_{k'}$ is the instantaneous dipole moment of dipole unit k' contained in the same chain as the reference dipole unit k'. The dipole moments in equation (40) are 'liquid' dipole moments which are related to the vacuum dipole moments by

$$\mu = \mu^{(v)} \cdot \frac{(\varepsilon_\infty + 2)(2\varepsilon_0 + 1)}{3(2\varepsilon_0 + \varepsilon_\infty)} \tag{41}$$

so equations (39) and (41) give

$$(\varepsilon_0 - \varepsilon_\infty) = \frac{4\pi C_r}{3kT} \cdot \frac{3\varepsilon_0}{2\varepsilon_0 + \varepsilon_\infty} \cdot \frac{(\varepsilon_\infty + 2)^2}{3} \cdot g\mu^2 \tag{42}$$

where g is the Kirkwood factor which expresses the angular correlations between the dipole groups along the chain

$$g = \left[\mu^2 + \sum_{k'} \langle \boldsymbol{\mu}_k(0) \cdot \boldsymbol{\mu}_{k'}(0) \rangle \right] \Big/ \mu^2 \tag{43}$$

and $\langle \boldsymbol{\mu}_k(0) \cdot \boldsymbol{\mu}_{k'}(0) \rangle / \mu^2 = \langle \cos \theta_{kk'} \rangle$ where $\theta_{kk'}$ is the angle between the dipole moment vectors for dipoles k and k'. Equation (43) is true for flexible chains of high molecular weight where each chain contains equivalent units and there are no orientation correlations between dipoles on one chain and those on other chains. Equation (39) may be used to obtain expressions for $(\varepsilon_0 - \varepsilon_\infty)$ in more complicated cases, e.g. for dipolar copolymers, for polymers of arbitrary molecular weight and for cases where interchain dipole moment correlations are present.

The total relaxation magnitude for the motions of chains in amorphous solid polymers and for polymers in solution is given by equation (42) and since the term $3\varepsilon_0/(2\varepsilon_0 + \varepsilon_\infty)$ is a slow function of ε_0, varying from 1 to 1.5 as ε_0 is raised from 2 to ∞, it follows that $(\varepsilon_0 - \varepsilon_\infty) \propto g\mu^2$. The Kirkwood g factor can be > 1 (correlations leading to enhancement of the dipole moment of the chain) or < 1 (correlations leading to a net cancellation of the dipole moments within the chain). The role of chemical structure, chain conformation and chain configuration (tacticity) on $g\mu^2$ has been discussed by Volkenstein[51] and Flory.[52] The effects of chain conformation on the g factor for linear polyethers, which serve as an illustration of the principles involved, have been described by Cook, Watts and Williams[50,53] who used the model theory of Read[54] for polyethers with the constitutive repeat unit $-\!\!\left[(CH_2)_{p-1} O\right]\!\!-_n$ where $p = 2, 3, 4$ and 5. In this case $g \simeq 1$ if *trans* and *gauche* conformations occur with equal probability, but if either conformation is strongly preferred then g departs markedly from unity (see refs. 50, 54 and 10 for details).

The complex permittivity is obtained as a generalization of the equilibrium theory (equations 39–43) to the dynamic situation as has been outlined by Cook, Watts and Williams.[50] The essential relations are as follows[10,50]

$$\left[\frac{\varepsilon(\omega) - \varepsilon_\infty}{\varepsilon_0 - \varepsilon_\infty} \right] \cdot p(\omega) = 1 - i\omega \int_0^\infty \Lambda(t) \cdot \exp(-i\omega t)\, dt \tag{44}$$

$$\Lambda(t) = \frac{\langle \boldsymbol{M}(0) \cdot \boldsymbol{M}(t) \rangle}{\langle \boldsymbol{M}(0) \cdot \boldsymbol{M}(0) \rangle} \tag{45}$$

$P(\omega)$ is an internal field factor and $\Lambda(t)$ is a time-correlation function which represents the fluctuations of the macroscopic dipole moment of the volume V in time in the absence of an applied electric field. Equations (44) and (45) are a consequence of applying linear-response theory (Kubo–Callen–Green) to the case of dielectric relaxation, as was first described by Glarum[55] in connexion with dipolar liquids. For the special case of flexible polymer chains of high molecular weight having intramolecular correlations between dipoles but no intermolecular correlations between dipoles of different chains we may write

$$\Lambda(t) = \frac{\langle \boldsymbol{\mu}_k(0) \cdot \boldsymbol{\mu}_k(t) \rangle + \sum_{k \neq k'} \langle \boldsymbol{\mu}_k(0) \cdot \boldsymbol{\mu}_{k'}(t) \rangle}{\langle \boldsymbol{\mu}_k(0) \cdot \boldsymbol{\mu}_k(0) \rangle + \sum_{k \neq k} \langle \boldsymbol{\mu}_k(0) \cdot \boldsymbol{\mu}_{k'}(0) \rangle} \tag{46}$$

The denominator in equation (46) is simply $g\mu^2$. The numerator contains the autocorrelation function $\langle \boldsymbol{\mu}_k(0) \cdot \boldsymbol{\mu}_k(t) \rangle = \mu^2 \langle \cos \theta_{kk}(t) \rangle$ for the motion of a reference dipole unit k, and the cross-correlation functions $\langle \boldsymbol{\mu}_k(0) \cdot \boldsymbol{\mu}_{k'}(t) \rangle = \langle \boldsymbol{\mu}_k(0) \cdot \boldsymbol{\mu}_{k'}(0) \rangle \lambda_{kk'}(t)$ where $\lambda_{kk'}(t)$ is a normalized decay function for the motion of dipoles k and k' contained in the same chain. Since $\langle \boldsymbol{\mu}_k(0) \cdot \boldsymbol{\mu}_{k'}(0) \rangle$ decreases rapidly as k and k' are separated along a chain, it follows that the dielectric relaxation behaviour of polymers in the bulk amorphous state or in solution is determined largely by the autocorrelation term $\langle \boldsymbol{\mu}_k(0) \cdot \boldsymbol{\mu}_{k'}(0) \rangle$ and a few near-neighbour cross-correlation terms $\langle \boldsymbol{\mu}_k(0) \cdot \boldsymbol{\mu}_{k'}(t) \rangle$. On the basis of experimental dielectric data for amorphous polymers and copolymers Williams, Cook and Hains[56] have reasoned that the different $\lambda_{kk'}(t)$ are very similar to $\langle \cos \theta_{kk}(t) \rangle$, and that an understanding of multiple dielectric relaxation regions is to be found in the behaviour of the autocorrelation term itself, the cross-correlation terms contributing to the magnitude but not in a dominating fashion to the overall shape of individual relaxation processes.

For materials of low permittivity, $\varepsilon_0 \simeq \varepsilon_\infty$ so $p(\omega) \simeq 1$ in equation (44). For this case comparison with equation (10) shows that the relaxation function $\Phi(t)$ is approximately the same as the molecular time-correlation function $\Lambda(t)$. Multiple relaxations correspond to multiple decay regions for $\Phi(t)$ or $\Lambda(t)$. We shall consider the origins of multiple relaxations in amorphous polymers below but emphasize here that the complex permittivity has a well-defined molecular basis as expressed by equations (44)–(46) irrespective of the details of molecular structure and dynamics. In contrast to NMR and ESR methods, dielectric studies can be made over a wide range of frequency or time, then allowing multiple relaxations to be documented[6-9] and the forms of $\Phi(t)$ and $\Lambda(t)$ to be determined.

As a simple illustration of the use of a dipole moment correlation function, consider the model case where the reference dipole undergoes small-step rotational diffusion governed by the field-free diffusion equation[57]

$$\nabla^2 f(\Omega, t) = \frac{1}{\sin^2\theta} \left[\sin\theta \frac{\partial}{\partial\theta} \left(\sin\theta \frac{\partial}{\partial\theta} f(\Omega, t) \right) + \frac{\partial^2}{\partial\phi^2} f(\Omega, t) \right] = \frac{1}{D_r} \frac{\partial}{\partial t} f(\Omega, t) \tag{47}$$

where $f(\Omega, t)$ is the time-dependent distribution function for the reorientation of the dipole moment and D_r is the rotational diffusion coefficient. For a dipole moment having its initial orientation along the $+Z$ axis, we obtain the time correlation function as[57]

$$\langle \cos\theta(t) \rangle = \int\int f(\Omega, t) \cos\theta \sin\theta \, d\theta \, d\phi \tag{48}$$

where $f(\Omega, t)$ is determined from equation (47) using a Legendre polynominal expansion and the boundary condition $f(\Omega, 0) = \delta(\cos\theta - 1)$ as

$$f(\Omega, t) = \frac{1}{4\pi} \sum_m (2m + 1) \cdot P_m(\cos\theta) \exp[-m(m+1)D_r t] \tag{49}$$

Equations (48) and (49) give

$$\frac{\langle \boldsymbol{\mu}(0) \cdot \boldsymbol{\mu}(t) \rangle}{\mu^2} = \langle \cos\theta(t) \rangle = \exp[-2D_r t] \tag{50}$$

Thus small-step diffusion gives a single relaxation time process with relaxation time $\tau = (2D_r)^{-1}$.

The feature of the correlation function approach is that it provides a field-free method for deducing the dielectric behaviour [using equations (44)–(46)] if the model for motion is specified. Williams[57] has shown how this may be done for a variety of rotational models including free rotation, and rotations in different barrier systems.

Many models for motion in amorphous polymers have been proposed.[6-11] Amorphous polymers usually exhibit two broad relaxation regions called α (associated with the dynamic glass transition and being due to the microbrownian motions of chain segments) and β (due to local chain motions or side group motions and usually observed in the glassy state). We shall discuss these below in relation to specific cases.

18.3.3 Crystalline Polymers

Several dielectric relaxation regions are usually observed for partially crystalline polymers. They arise from motions within the crystals, on their surfaces or in the amorphous regions of the material. Particularly interesting materials, which have been widely studied, are oxidized low density and high density polyethylenes, polyoxymethylene, nylons and poly(vinylidene difluoride) (for reviews see refs. 6, 7, 10, 11, 44, 49, 58). It is useful to consider the theory for the reorientational motion of a chain in a crystal as was first described by Fröhlich[59] and developed by Tuijnman,[60] Booij,[61] Williams, Lauritzen and Hoffman[62,66] and by Boyd and co-workers[63-65] and which has been applied to the α_c relaxation in partially crystalline polyethylene (see ref. 63 for a review).

In the original paper of Fröhlich[59] it was considered that a planar zigzag chain in the crystalline state may reorientate between two equivalent sites 180° apart. The chain moves as a rigid rod if it is short but as its molecular length is increased, the internal twisting of the chain makes a contribution by allowing a lowering of the effective barrier to reorientation. Figure 2 shows a schematic illustration of a hydrocarbon chain in its untwisted and twisted states. Fröhlich specialized the model by assuming that the intermolecular energy had the parabolic form

$$\Delta Q(\phi) = c_1 \phi^2 \qquad 0 \le \phi \le \pi/2 \tag{51a}$$

$$\Delta Q(\phi) = c_1(\pi - \phi)^2 \qquad \pi/2 \le \phi \le \pi \tag{51b}$$

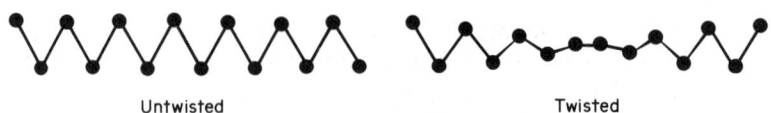

Untwisted   Twisted

Figure 2 Schematic illustration of a hydrocarbon chain in its untwisted and twisted states in the crystal

He assumed that the chain was flexible with an intramolecular (twisting) energy given by $c_2(d\phi dz)^2$ where c_2 is a force constant and z is the coordinate of the chain axis. The configuration of the chain at the barrier condition corresponds to $\phi = \pi/2$ at $z = l/2$ with $(d\phi/dz) = 0$ at $z = 0$ and $z = l$ and with $\phi = 0$ at $z = 0$ and $\phi = \pi$ at $z = l$. The minimum energy required to reach this configuration (which is a saddle point) is obtained as

$$\Delta W(n) = 2\bar{n}\Delta Q_{max} \cdot \tanh(n/2\bar{n}) \tag{52}$$

where
$$\bar{n} = (c_1/c_2)^{\frac{1}{2}} \text{ and } \Delta Q(\phi = \pi/2) = \Delta Q_{max}$$

Equation (52) predicts that chain twisting occurs for all values of n, where n is the number of chain units. Williams, Lauritzen and Hoffman[62] replaced the cusp potential of equation (51) by the more realistic cosine potential equation (53) and by a matched parabola potential, equations (54)

$$\Delta Q(\phi) = \Delta Q_{max}(1 - \cos 2\phi)/2 \tag{53}$$

$$\Delta Q(\phi) = a_1\phi^2 \qquad 0 \le \phi \le \phi_i \tag{54a}$$

$$\Delta Q(\phi) = \Delta Q_{max} - a_2(\pi/2 - \phi)^2 \quad \phi_i \le \phi \le \pi/2 \tag{54b}$$

and showed that the minimum barrier for reorientation $\Delta W(n)$ was strictly proportional to n for $n/\bar{n}$ less than a critical value, and that for larger values of n, chain twisting dominated the reorientation process leading to a rapid fall-off in $\Delta W(n)$ to a constant value, as shown in Figure 3 for the matched parabola model.[62] The plot for $x = 1.0$ corresponds to the Fröhlich result.

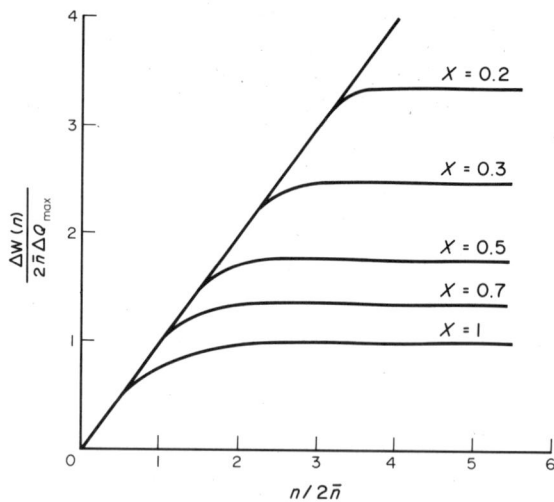

Figure 3 $\Delta W(n)/(2\bar{n}\Delta Q_{max})$ against $n/(2\bar{n})$ for different values of the parameter x for the parabola–antiparabola model. $x = 1$ is the Fröhlich result. The straight line corresponds to rigid molecule rotation in the crystal

Hoffman, Williams and Passaglia[66] applied the results for the cosine and double parabola models to the data for the α_c relaxation in polyethylene, an important feature being that for chain-folded high molecular weight material the experimental and theoretical apparent activation energies become essentially independent of molecular weight, or lamellar thickness at high molecular weights. Mansfield and Boyd[64] considered further the mechanism for the α_c relaxation in poly-ethylene and have discussed chain reorientation models involving twisting and point-defect motions (*e.g.* by a Reneker defect).[67] Mansfield and Boyd emphasize that only the energetics of rotation and twisting are included in the earlier models[59-62] and that the energetics associated with the necessary translation by one CH_2 unit are not included. They carried out conformational-energy modelling for the motion of a reference chain in an orthorhombic array of $C_{22}H_{46}$ alkane molecules. For rigid rotation of the chain by 180°, accompanied by a translation by one CH_2 unit in the lattice they found there were four non-equal stable rotational positions corresponding to the chain being approximately along the *a* and *b* axes. The chain rotates and translates in this barrier system and the barrier to rotation would increase linearly with *n*. Introducing twist into the rotation–translation process gave the result that in any particular case twisting became economical quickly as *n* increased. The twist comprised a uniform succession of bond torsions at large chain lengths and the reorientation of the chain was achieved by the smooth propagation of a twisted region through the crystal leaving a translational lattice mismatch as it proceeds. The calculations yielded the plot shown in Figure 4 of the theoretical dependence of the constant volume activation energy for chain rotation on chain length (alkanes) and lamellar thickness (polyethylenes). Also shown are the experimental dielectric data and they are seen to compare favourably. Thus it appears that chain reorientation in polyethylene crystals occurs by the propagation of a uniformly twisted region (involving ~ 10 CH_2 units). In view of the extensive experimental data available for the α_c process in polyethylenes[7,49,63,65,66] and the detailed theoretical analyses which have been made[59-66] the mechanism for this relaxation process is perhaps the best understood of all those studied for solid polymers.

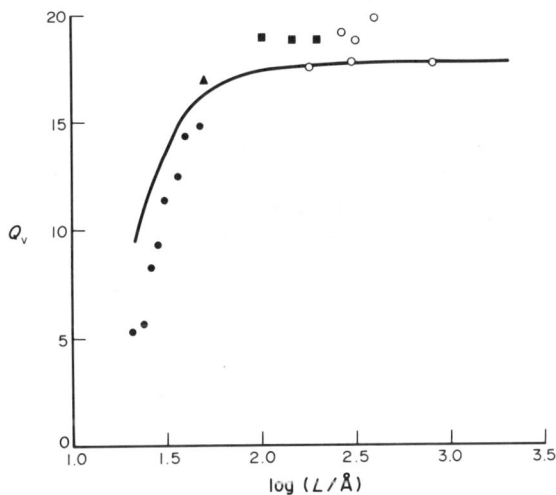

Figure 4 The constant volume apparent activation energy as a function of chain length (for alkanes) or lamellar thickness (for polyethylenes). Continuous line is that from the theory for chain rotation accompanied by twisting (after Mansfield and Boyd)[64]

In addition to the crystalline (α_c) relaxation, other (broader) relaxations are observed in partially crystalline polymers arising from the 'amorphous' regions of the material. These relaxations are related to similar broad absorptions found in amorphous polymers except they are 'perturbed' due to the proximity of amorphous and crystalline regions (see ref. 10 for the example of crystalline poly(ethylene terephthalate) and further discussion below).

18.3.4 Liquid Crystalline Polymers

Thermotropic liquid crystalline polymers having mesogenic groups either in the main chain, side chain or in both main chain and side chain have been prepared in recent years and are of considerable interest both as a new class of polymer and as a new class of liquid crystal (with all the interesting physical properties of anisotropic materials and the responses to moderate electric and magnetic fields). Reviews of their synthesis, characterization and of their physical properties and applications have been given recently[45-47,68-72] and a further account is given in this volume in Chapter 26. The dielectric properties of liquid crystalline polymers are of special interest since they determine the alignment behaviour in applied electric fields. In contrast to bulk amorphous polymers or polymers in solution, extensive angular correlations exists between mesogenic groups in liquid crystals giving rise to local ordering and to anisotropic dielectric properties. We briefly summarize the theory of the dielectric behaviour of a nematic liquid crystalline polymer having mesogenic groups in the main chain or side chain.

The dielectric theory for such a material has been given by Attard[73] and is a development of the theories of Nordio *et al.*[74] and Maier and Meier[75] for the dielectric permittivity of a monodomain of LC material to include the case of partially aligned material. This theory involves the Wigner rotation matrix method for coordinate transformations. A simpler treatment has been given recently by Attard, Araki and Williams[76] and uses a cartesian rotation matrix method for coordinate transformations. Maier and Meier[75] showed that the principal static permittivities for a *monodomain* of a uniaxial liquid crystal are given by

$$\varepsilon_\parallel^0 = \varepsilon_\parallel^\infty + \frac{G}{3kT}\cdot\mu^2[1-(1-3\cos^2\beta)S] \tag{55a}$$

$$\varepsilon_\perp^0 = \varepsilon_\perp^\infty + \frac{G}{3kT}\mu^2[1+(1-3\cos^2\beta)S/2] \tag{55b}$$

$\varepsilon_\parallel^0$ and $\varepsilon_\perp^0$ are the static permittivities measured parallel to and perpendicular to the nematic director *n*, respectively. $\varepsilon_\parallel^\infty$ and $\varepsilon_\perp^\infty$ are the corresponding limiting high frequency permittivities. G is a factor involving the concentration of mesogenic groups (which each have a dipole moment μ) and β is the angle between μ and *n*. S is the second-rank reorientational order parameter of the liquid crystal phase defined as

$$S = \langle 3\cos^2\theta - 1\rangle/2 \tag{56}$$

where θ is the angle between the z axis of the mesogenic group and the director *n*. S is typically 0.5–0.8 for low molar mass liquid crystals, reflecting the fact that the mesogenic group moves in a P_2 potential. The total dipole moment may be expressed in terms of its longitudinal component $\mu_l = \mu\cdot\cos\beta$ and its transverse component $\mu_t = \mu\cdot\sin\beta$ with respect to *n* so equations (55) become

$$\varepsilon_\parallel^0 = \varepsilon_\parallel^\infty + \frac{G}{3kT}\cdot[\mu_l^2(1+2S)+\mu_t^2(1-S)] \tag{56a}$$

$$\varepsilon_\perp^0 = \varepsilon_\perp^\infty + \frac{G}{3kT}[\mu_l^2(1-S)+\mu_t^2(1+S/2)] \tag{56b}$$

Thus the strengths of the relaxations measured parallel to *n* and perpendicular to *n*, i.e. $\Delta\varepsilon_\parallel$ and $\Delta\varepsilon_\perp$, respectively, are weighted sums of terms involving μ_l^2, μ_t^2 and S. Generalization to the case of the complex permittivity in the frequency domain gives

$$\varepsilon_\parallel(\omega) = \varepsilon_\parallel^\infty + \frac{G}{3kT}\cdot[\mu_l^2(1+2S)F_\parallel^l(\omega)+\mu_t^2(1-S)F_\parallel^t(\omega)] \tag{57a}$$

$$\varepsilon_\perp(\omega) = \varepsilon_\perp^\infty + \frac{G}{3kT}\cdot[\mu_l^2(1-S)F_\perp^l(\omega)+\mu_t^2(1+S/2)F_\perp^t(\omega)] \tag{57b}$$

where

$$F_j^i(\omega) = 1 - i\omega\mathscr{F}[F_j^i(t)] \tag{58}$$

where $\mathscr{F}$ indicates a one-sided Fourier transform (*cf.* equations 10 and 44) and $F_j^i(t)$ is a normalized decay (or relaxation) function. $i=1$ or t; $j=\parallel$ or $\perp$. Equations (57) show that four relaxation modes are active in the dielectric spectrum of a LC monodomain. These modes have been represented pictorially by Attard.[73]

If a liquid crystalline material (polymeric or of low molar mass) is partially aligned (uniaxially) by the action of an aligning field (electric or magnetic) then a 'director order parameter' S_d may be defined as

$$S_d = \langle 3\cos^2\theta' - 1\rangle/2 \tag{59}$$

where θ' is the angle between the local director *n* and the symmetry axis (Z) of the partially aligned material. Expressions have been derived[73,76] for the permittivities measured parallel to the symmetry axis of the material (i.e. $\varepsilon_z(\omega)$) and in the XY plane [i.e. $\varepsilon_X(\omega) = \varepsilon_Y(\omega)$]. Experimentally $\varepsilon_z(\omega)$

is most conveniently measured and is given approximately by[73,76]

$$\varepsilon_z(\omega) = \varepsilon_z^\infty + \frac{G}{3kT} \left[\frac{(1+2S_d)}{3} \cdot \{\mu_t^2(1+2S)F_\parallel^1(\omega) + \mu_t^2(1-S)F_\parallel^1(\omega)\} + \frac{2}{3}(1-S_d)\{\mu_t^2(1-S)F_\perp^1(\omega) + \mu_t^2(1+S/2)F_\perp^1(\omega)\} \right] \quad (60)$$

where

$$\varepsilon_z^\infty = \bar{\varepsilon}^\infty + 2S_d(\varepsilon_\parallel^\infty - \varepsilon_\perp^\infty)/3 \quad (61a)$$

$$\bar{\varepsilon}^\infty = (\varepsilon_\parallel^\infty + 2\varepsilon_\perp^\infty)/3 \quad (61b)$$

Thus the dielectric permittivity measured along the Z axis is a weighted sum of four relaxation regions, where the weighting factors involve S, S_d and the dipole moment components μ_t and μ_t. Equation (60) becomes equation (57a) for the case of a homeotropically aligned material (*i.e.* $S_d = 1$) and also reduces to simple forms for the cases of unaligned material ($S_d = 0$) and planarly (or homogeneously) aligned material ($S_d = -\frac{1}{2}$). It is important to stress that as S_d is varied, the variation of the dielectric properties reflect different weightings of the four relaxation processes $F_j^i(\omega)$. The applications of the theory to experimental data for LC polymers will be described below. Note that the theory does not include the Kirkwood g factor for correlations of dipoles on different mesogenic groups along a polymer chain.

18.4 EXPERIMENTAL METHODS

18.4.1 Historical

In contrast with most methods used for studying the molecular dynamics and physical properties of polymers, dielectric studies cover an unusually wide frequency (or time) range. Measurements can be made in the range 10^{-3} to 10^{11} Hz using several methods, each with its own limited frequency range. These have been described earlier[7,8,48] and will not be reproduced here. The advent of modern solid state electronics and on-line computing facilities have brought dielectric relaxation spectroscopy into the forefront of spectroscopic/relaxation methods for the study of polymers, and some of these will be described here. Figure 5 gives a schematic diagram in which the techniques used in particular ranges are indicated.

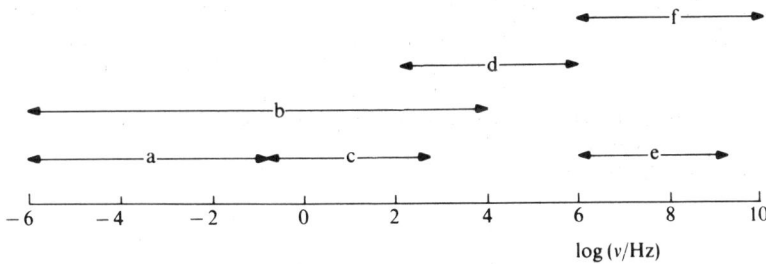

Figure 5 Ranges of some dielectric relaxation spectroscopy techniques: (a) conventional transient methods;[7] (b) newer transient methods;[79] (c) low frequency impedance bridge;[7] (d) conventional impedance bridges and analyzers; (e) high frequency impedance analyzers and time domain spectroscopy; (f) microwave cavities, transmission lines[7]

In practice today the range 10^2 to 10^5 Hz is the most widely used since commercial equipment is readily available and it is the range in which glass-forming polymers exhibit multiple relaxations just above the glass transition temperature. Measurements can be made conveniently at higher and lower frequencies using adaptations of commercially available equipment.

18.4.2 10^{-3} to 10^4 Hz

Since the time taken to balance a low frequency bridge[5,7] can become excessive at frequencies lower than 1 Hz, it is convenient to make transient charge or current measurements and transform these data into the frequency domain by Fourier transformation. The charging current is given by $I(t) = C_0 V[d\Phi(t)/dt]$, and insertion into equation (10) provides the method of transforming charging current data into permittivity and loss data in the frequency domain.[7] In the past the Hamon approximation, which converts a point ($I(t)$, t) into a point ($\varepsilon''(\omega)$, ω) has been widely used for polymer relaxations and this approximation and its improvement by Brather[77] has provided a simple and direct method for studying low frequency behaviour in solid polymers.[78] The transient method has been developed by Mopsik[79] who has described a precision time domain dielectric spectrometer operating automatically in the range 10^{-4} to 10^4 Hz with high accuracy (0.1%) and high sensitivity (tan $\delta = 10^{-5}$). It is shown that direct measurement of the transient charge following application of a 100 V step to a dielectric material, followed by a numerical evaluation of the Fourier transform (equation 10), yields data of exceptional quality for solid polymers such as poly(vinylidene difluoride) (medium loss) and polypropylene (low loss). As one example Figure 6 shows[79] the time-dependence of the capacitance of an unoriented sample of poly(vinylidene difluoride) in its α phase at 40 °C. Figure 7 shows the transformed data as the real and imaginary parts of the complex capacitance determined from Figure 6 and we note that data are obtained over the range 10^{-4} to 10^4 Hz in this example.

Alternatively, low frequency measurements may be made directly in the frequency domain using bridge techniques.[7] A modern unit capable of making automatic measurements in the range 10^{-5} to

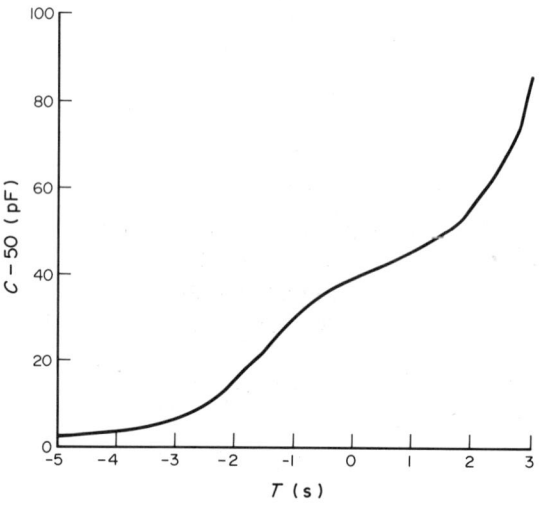

Figure 6   The transient capacitance for an unoriented sample of poly(vinylidene difluoride) at 40 °C (after Mopsik)[79]

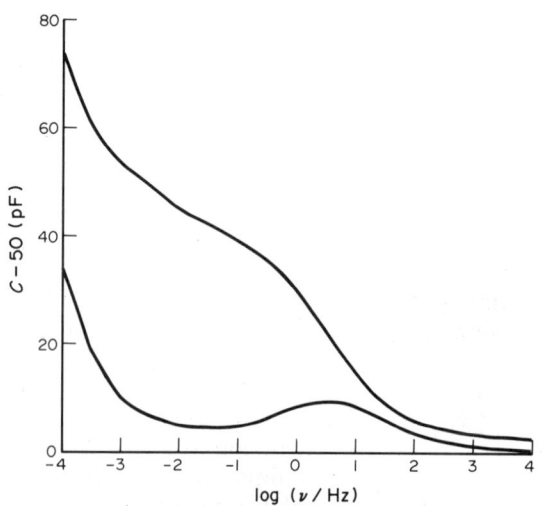

Figure 7   The real and imaginary components of the complex capacitance determined by transformation of the data of Figure 4 for poly(vinylidene difluoride) at 40 °C (after Mopsik)[79]

65 kHz is the Solartron 1250 Frequency Response Analyzer. Although designed originally for fairly low impedance measurements, the addition of a buffer amplifier at the sample input allows measurements to be made on medium and low loss polymers.

18.4.3 10 to 10^5 Hz

A number of instruments are now available in this range and are capable of measuring very large ranges of sample impedance with great accuracy at a large number of frequencies. These include the Gen Rad 1650 Digibridge, the Wayne Kerr B221 and the Hewlett Packard 4272 LCR Meter. Such instruments can be linked to a microcomputer and to peripherals such as disc drives, a plotter and a printer. With proper provision of software this gives an automatic precision dielectric measuring assembly, a commercial example of which is the DETA instrument sold by Polymer Laboratories Ltd., UK. The current availability of instruments of such high quality and precision in this frequency range has allowed dielectric relaxation spectroscopy to take its place along side other modern spectroscopic methods (UV, IR, NMR) for the study of polymer structure and dynamics.

18.4.4 10^6 to 10^{10} Hz

This frequency range has always been the most difficult since it lies between the low frequency lumped circuit range and the microwave range.[7] The Hewlett Packard RF Impedance Analyser, Model 4191A, provides accurate measurements for the range 10^6 to 10^9 Hz. An alternative method is that of Time Domain Spectroscopy (TDS), whose principles have been reviewed,[80,81] which may be used conveniently in the range 10^7 to 10^9 Hz. Cole and co-workers have shown that the TDS method may be applied to conducting solutions,[82] medium loss liquids[83] and to polymer solutions.[84] The method is currently under development and although the essential items of equipment are available commercially (sampling oscilloscope, pulse generator), the experimental set up and its implementation presently require great skill and expertise, so the TDS method is not readily available for use with polymer systems.

18.5 EXPERIMENTAL DATA AND INTERPRETATIONS

18.5.1 Amorphous Polymers

Solid amorphous polymers normally exhibit two relaxation processes which coalesce to form one process at high temperatures. The α process is observed above the glass transition temperature T_g and is due to the microbrownian motions of the chains. The β process, which is normally observed below T_g but can also be observed in a limited range above T_g, is due to limited motions of the main chain or, where present, the motions of dipolar side chains. The frequency–temperature location of these processes is indicated schematically in Figure 8 and the coalescence of α and β processes to form the $\alpha\beta$ process is seen to occur above T_g. The loci of these processes vary greatly with chemical structure, tacticity, plasticizer content and sample orientation. Transition maps have been given[6-9,85] for many amorphous polymers, based on dielectric, NMR and mechanical relaxation results. The loci obtained from the different methods are found to be similar for a given polymer.

As one example of an α process, Figure 9 shows part of the data of Mashimo *et al.*[86] for amorphous poly(vinyl acetate) obtained over the frequency range 10^{-6} to 10^6 Hz and the temperature range 301 to 366 K. The loss curves are asymmetrical in the Davidson–Cole sense and may be fitted approximately using the relations of Davidson–Cole (equation 18), Havriliak–Negami (equation 19) or Kohlrausch–Williams–Watts (equations 10 and 25). Mashimo *et al.*[86] used the Havriliak–Negami equation to fit these data and found that q increased from 0.8 to 0.85 and p increased from 0.4 to 0.8 as temperature was increased from 301 to 366 K. Thus the loss curves narrow and become more symmetrical as temperature is increased to well above T_g. According to these authors the Kirkwood g factor (equation 43) increased from 0.53 at 366 K to 0.73 at 301 K indicating (i) that cross-correlation terms make a substantial contribution to the relaxation behaviour (see equations 44–46) and (ii) that there is a net cancellation of dipole moments between groups along the chain which is decreased when temperature is decreased. Despite the important contribution made by the cross-correlation terms, the α relaxation in this polymer is similar to the α process observed in a variety of amorphous polymers of very different chemical structure, as has been emphasized.[10] The

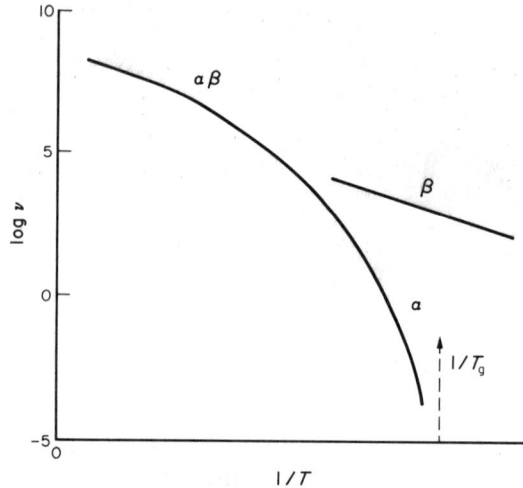

Figure 8 Schematic illustration of the variation of the frequency location of the α, β and $\alpha\beta$ processes with temperature for an amorphous polymer

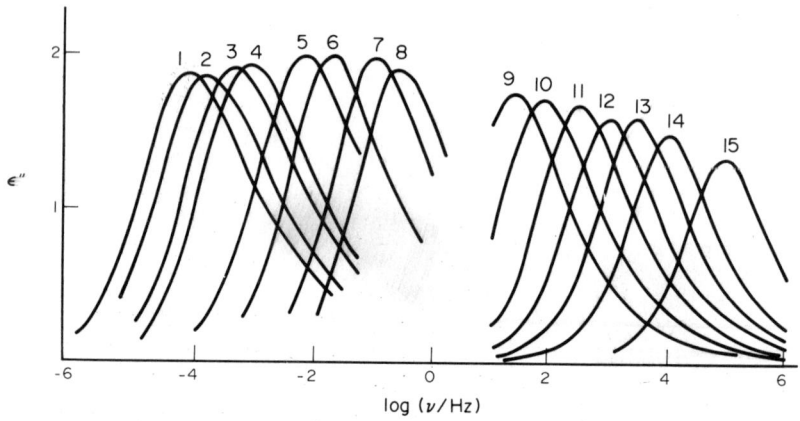

Figure 9 ε'' plotted against $\log_{10}(\nu/\text{Hz})$ for amorphous poly(vinyl acetate). Curves 1 to 15 correspond to 301.2, 301.7, 302.9, 303.6, 305.8, 309.2, 311.1, 315.6, 326.2, 330.5, 335.0, 339.3, 343.7, 349.8 and 365.9 K (after Mashimo *et al.*)[86]

similarity extends to the α process observed in small-molecule glass-forming liquids (*e.g.* di-*n*-butyl phthalate, dipolar solutes in *o*-terphenyl)[20] and it has been established that the shapes of the ε'' *vs.* $\log \nu$ curves are similar for the α process in small-molecule and polymeric glass-forming systems[20, 22] and also that WLF behaviour maintains (equation (31) above). We have discussed the application of the KWW function to such data above (see Section 19.2.3). Below T_g the β process is observed and has a half-width typically 4–6 units of $\log f$. For poly(vinyl acetate) the β process appears much smaller in loss height, $\varepsilon''_{\text{max}}$, than the α process[7] indicating that only a small part of μ^2 is relaxed by the β process.

The frequency–temperature transition map (Figure 8) and the interrelations of α, β and $\alpha\beta$ processes have been discussed extensively[7, 10, 11, 49] and Williams and Watts[39] have proposed a simple phenomenological model which accounts for the coalescence of individual processes, the relative intensity of processes and of the effects of temperature and pressure on the locations of the processes. They consider the motion of a reference group in a chain occurring by (i) partial reorientation in a local environment and (ii) complete reorientation by cooperative motions of the group with its environment. (i) and (ii) rise to the β and α processes respectively and at high temperatures ($T > T_g$) the β process is subsumed into the α process giving the $\alpha\beta$ process. The correlation function for the motion of the reference dipolar group is given by

$$\frac{\langle \boldsymbol{\mu}(0) \cdot \boldsymbol{\mu}(t) \rangle}{\mu^2} = \phi_\alpha(t) \left[\sum_r {}^\circ p_r q_{\alpha r} + \sum_r {}^\circ p_r q_{\beta r} \phi_{\beta r}(t) \right] \tag{62}$$

where $\phi_\alpha(t)$ is the relaxation function for the α process. $\phi_{\beta r}(t)$ is the relaxation function for the partial motion of the dipole in environment r and $^\circ p_r$ is the probability of obtaining environment r. $q_{\alpha r} = [\langle\boldsymbol{\mu}\rangle_r]^2/\mu^2$, $q_{\alpha r} + q_{\beta r} = 1$ and $[\langle\boldsymbol{\mu}\rangle]_r$ is the mean dipole moment residing in environment r following the β_r process. Equation (62) shows that the strength of the dipole relaxation is shared between the α and β processes. If the strength of the β process is suppressed [*e.g.* by increase of pressure as in the case of amorphous poly(ethyl methacrylate)[38]] then the α process shows a corresponding increase in strength, as is confirmed experimentally.[38] At high temperatures when $\phi_\alpha(t)$ decays more rapidly than the $q_{\beta r}(t)$, equation (62) predicts that the β process is subsumed into the α process, as is observed experimentally for poly(ethyl methacrylate)[38] and for other polymers.[10] It is also predicted that the β process will be very broad, and usually small in magnitude, arising as it does from partial reorientations of dipoles in a variety of local environments.

Pressure has a marked effect on the frequency location of the α relaxation, but little on the β relaxation. In a series of papers Williams[34-40] and Saito *et al.*[41-42] studied poly(methyl acrylate), poly(ethyl acrylate), poly(ethyl methacrylate), poly(*n*-butyl methacrylate), poly(propylene oxide), poly(ethylene terephthalate), poly(vinyl chloride) and poly(vinyl acetate). It was shown that $(\partial\log\langle\tau\rangle/\partial P)_T$ was typically in the range $1\to 4\times 10^{-3}$ atm^{-1} for α processes and $0.1\to 1\times 10^{-3}$ atm^{-1} for β processes. Activation volumes (see equation 34c) were in the range 50 to 150 cm^3 mol^{-1} for α processes and 5 to 20 cm^3 mol^{-1} for β processes. Although the effect of pressure on $\langle\tau\rangle_\alpha$ was large experimentally, the constant volume apparent activation energy for α processes was ~ 0.75 of the constant pressure apparent activation energy (see equations 34) showing that temperature (thermal energy) and not volume plays the determining role in the microbrownian motions of chains. Such data also raise questions[66] regarding free volume theories for relaxation which would predict that $Q_V \ll Q_P$ or that Q_V is negative, in complete contradiction with the experimental data.[34-42] Further studies of the effect of pressure on the α and β relaxations have been reported by Naoki *et al.* for a chlorinated polyethylene vulcanizate[87-89] and by Pae for a polyurethane elastomer.[90]

In recent years the structure and local composition of polymer blends has attracted much attention due to the commercial importance of these materials. The question of compatibility of the components is always paramount and Wetton *et al.*[91] showed that dielectric studies provided information which was difficult to obtain by any other method. Random copolymers of styrene and 4-chlorostyrene were blended with poly(2,6-dimethyl-1,4-phenylene oxide) (PPO). It was shown that 'compatible' blends containing $\sim 60\%$ (w/w) of copolymer gave an α relaxation which was far broader than those observed for the parent polymer and copolymer, being indicative of microphase separation in the blend where the domains are smaller than the wavelength of light. This loss-peak-broadening is of the kind observed for mixtures of dibutylphthalate and *o*-terphenyl[92] where microseparation of the components was inferred, and shows that dielectric studies can demonstrate such incompatibility on the microscale in materials which appear to be compatible when judged by optical or DSC measurements.

The effect of solvent on molecular motions and the glass transition temperature may be studied using dielectric relaxation spectroscopy. Adachi *et al.*[93-95] studied polystyrene/toluene,[93] poly(vinyl chloride)/tetrahydrofuran[94] and poly(vinyl acetate)/toluene[95] mixtures. Complicated behaviour was found in the plots of ε'' *vs.* temperature at given frequencies, and up to four relaxations were observed. The main relaxations of the polymer chain were affected by addition of solvent to the bulk polymer. The α relaxation moved rapidly to lower temperatures (at a fixed measuring frequency), consistent with the plasticization of the chains leading to enhanced mobility. Interestingly, the shape of the loss curve for the α relaxation of poly(vinyl acetate)/toluene solutions changed only slightly as the polymer concentration was reduced from 100 to 10% (w/w).

While most of the studies of the dielectric properties of amorphous polymers have been concerned with conventional polymers which exhibit α and β relaxations, whose frequency–temperature locations are essentially independent of molecular weight at the molecular weights normally encountered, certain polymers exhibit a low frequency relaxation which appears to be strongly dependent on molecular weight. The first observation was that by Stockmayer and Baur for poly(propylene oxide), who interpreted the small low frequency molecular-weight-dependent loss peak in terms of the normal mode motions of the cumulative· dipole moment of the entire chain.[96-98] Very recently Adachi and Kotaka[99,100] studied *cis*-1,4-polyisoprene in its undiluted state and observed two relaxation regions, one of which was very dependent on molecular weight. Because of the lack of symmetry in *cis*-1,4-polyisoprene, the polymer has components of the dipole moment both parallel and perpendicular to the chain contour. The perpendicular component gives rise to a loss peak at low temperatures which is independent of polymer molecular weight while the parallel component gives a loss peak at high temperatures whose magnitude and location are strongly dependent upon molecular weight [as is the case for poly(propylene oxide)]. For a sample having $M_w = 1.02 \times 10^5$ the frequencies of maximum loss for the two processes differed by a factor of 10^8. Application of the Rouse theory was successful for $M_w < M_c$, where $M_c (\simeq 10^4)$ is the critical molecular weight for entanglements. In this regime $\langle \tau \rangle \propto M_w^2$. Above M_C it was found that $\langle \tau \rangle \propto M_w^{3.7}$ and this dependence of average dielectric relaxation time on molecular weight was interpreted using the reptation model, as developed by de Gennes, Edwards and Doi, for the motion of a chain in a virtual tube prescribed by the neighbouring polymer chains in the bulk polymer.

18.5.2 Polymers in Solution

The main difficulty with studying the dielectric properties of polymers in solution is that the loss maxima frequently occur above 10^6 Hz and so are not readily studied using conventional equipment. As we have seen above for polymer/diluent systems studied by Adachi and co-workers[93-95] measurements may be made at lower frequencies by going to low temperatures in solvents which do not crystallize out. A recent example is poly(propylene oxide) in toluene reported by Johari and Monnerie.[101] A sample having a molecular weight of 4000 was studied in the ranges 50 Hz to 200 kHz, -200 to $-40\,°C$. For a 47% PPO solution (w/w) the system was a glass below $-117\,°C$. Plots of tan δ *vs.* temperature revealed (i) two relaxation regions, α and α', above T_g and (ii) one broad relaxation, β, below T_g. The α relaxation is due to the segmental chain motions and the α' relaxation is due to the motions of the cumulative dipole moment as described previously for poly(propylene oxide)[96-98] and for *cis*-polyisoprene.[99,100] The β process was considered to be due to the localized motions of the ether groups determined by the local packing in the amorphous glassy state, as envisaged in the phenomenological theory, equation (62) above.

Studies at low temperatures of the segmental motions of polymers in solution have been reported for poly(methyl methacrylate)[102] and various poly(propylene oxide)s.[103] More recently, high frequency dielectric studies up to 3×10^9 Hz have been reported for copoly(methyl methacrylate/methyl acrylate)s in benzene,[104] poly(methyl vinyl ketone) in dioxane[105] and for poly(propylene oxide)s (M_w 440–4000) in benzene.[106] In all cases two relaxation regions were clearly observed. For copoly(MMA/MA)s and PMVK the processes arise from side group motion and main chain segmental motions, while for PPO the high frequency process is due to segmental motion and the low frequency process is due to the motions of the cumulative dipole moment of the chain as described above.[96-100] As one example Figure 10 shows the plot of permittivity and loss factor for poly(methyl vinyl ketone) (4.2%) in dioxane as solvent.[105] The relaxations at 60 MHz and 1 GHz are due to segmental motions of the main chain and the internal rotation of the side acetyl group respectively. This is a good illustration of the information which may be obtained from measurements using modern instrumentation. Such data may be analyzed quantitatively. It is possible to relate the magnitudes of the individual processes to the components of molecular group dipole moments, and also the relaxation frequencies to models for main chain and side chain motions.[105] Such information is not readily obtained from NMR and light scattering studies of solutions and is

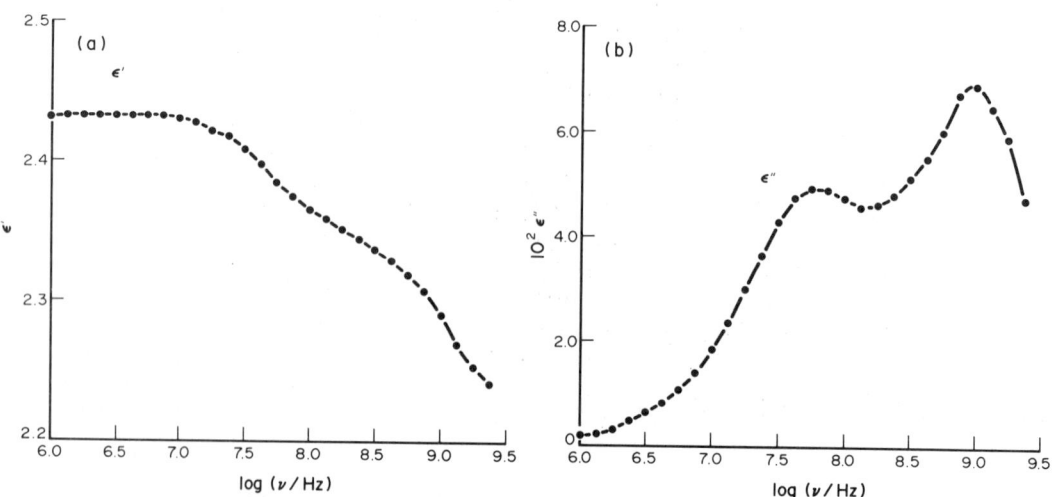

Figure 10 (a) ε' and (b) ε'' plotted against log (ν/Hz) for a 4.2% solution of poly(methyl vinyl ketone) in dioxane at $20\,°C$ (after Mashimo *et al.*)[105]

complementary to the information provided by such studies. There is little doubt that as high frequency measurement methods become more available to polymer scientists the experimental database for chain dynamics of polymers in solution will improve rapidly, and that is urgently needed in view of the many theoretical studies being made currently.

Rod-like polymers [*e.g.* polypeptides, poly(alkyl isocyanates)] and proteins may be studied in solution using dielectric relaxation spectroscopy.[107-116] For dilute solutions the dielectric permittivity gives information on the dipole moment of the isolated molecule and this may be extremely large due to the cumulative effect of individual monomer or repeat dipole contributions to the moment of the chain. Since $\Delta\varepsilon \propto \mu^2$ it follows that large dielectric increments are obtained at low polymer concentrations.[107,109,110] Wada,[107] Bur *et al.*[109,110] have studied the dipole moments and chain rotational dynamics for poly(γ-benzyl-L-glutamate) (PBLG) and poly(alkyl isocyanates) in solution as a function of concentration and molecular weight. At high concentrations such rod-like polymers may form lyotropic liquid crystalline media. The phase behaviour of rods in solution, including the isotropic, biphasic and lyotropic-nematic ranges, was first explained by Flory.[117] A review of the phase behaviour of systems in which the effects of specific interactions between chains and molecular weight distributions of rods are considered in detail has been given recently by Moscicki.[118] Few studies of the dielectric behaviour of concentrated solutions of rod-like polymers have been reported; PBLG was studied in dichloromethane and dioxane up to $\sim$90% polymer (w/w)[115] and poly(n-hexyl isocyanate)[113,114] and a copolymer of butyl and nonyl isocyanates were studied in toluene.[112] Moscicki and Williams[113] showed that the dielectric increment $\Delta\varepsilon$ increased linearly with polymer concentration C in the isotropic phase for the poly(n-hexyl isocyanate)/toluene system; and continued to increase with C well into the biphasic range, then decreased to a small value at high concentrations in the biphasic range and in the lyotropic liquid crystal range. This behaviour parallels the viscosity behaviour of rod-like molecules in solution as concentration is varied over the isotropic, biphasic and liquid crystalline ranges. These authors analyzed the behaviour quantitatively[113,114] with the aid of the theory for the phase behaviour of rods in solution developed by Moscicki and Williams.[114,119-120] In addition, the average relaxation time was observed to increase with increasing polymer concentration in the isotropic range, reached a maximum value in the biphasic range and then showed a sharp decrease as the liquid crystalline range was approached. Using the theories for rod reorientation in solution and the theory of Warchol and Vaughan[121] and Wang and Pecora[122] for rotational diffusion limited to a cone (which approximates to the situation of a rod in the lyotropic-nematic phase), it was found possible to analyze the dielectric relaxation data for isotropic, biphasic and lyotropic-nematic phases quantitatively.[113,114] This work is possibly the most comprehensive study to date of the molecular dynamics and phase behaviour of rod-like molecules in solution.

18.5.3 Crystalline Polymers

Of the many studies of partially crystalline polymers, those for polyethylene, poly(ethylene terephthalate) and poly(vinylidene difluoride) come to mind in view of their commercial interest as insulating dielectrics (polyethylene), as a plastic base for magnetic tapes [poly(ethylene terephthalate)] and as an electroactive polymer [poly(vinylidene difluoride)]. These polymers also exhibit multiple dielectric relaxations which have been studied in detail, and serve as model cases for the dielectric behaviour of partially crystalline polymers.

18.5.3.1 *Polyethylene*

Polyethylene chains are essentially non-dipolar, so in its pure state this polymer rivals poly(tetra-fluoroethylene) as an excellent insulating dielectric, suitable for component insulation and cable insulation (*e.g.* ε'' for submarine telephone cable insulation is typically 10^{-5}). By introducing carbonyl groups (*via* oxidation) or C–Cl groups (*via* chlorination) it is possible to probe the intrinsic motions which occur in the different regions of the partially crystalline polymer. The literature relating to dielectric studies of such polyethylenes is enormous, but has been summarized in key reviews.[7,10,44,63,66] Three relaxation regions are apparent in lightly oxidized or lightly chorinated polyethylenes (high and low density). The highest temperature process (α_c process) is associated with chain rotation in the crystalline regions, and the mechanism for this process is established as being the rotation–translation process, assisted by chain twisting, as described in detail in Section 18.3.3 above. The β process is associated with the motions of chain segments in the 'amorphous' regions of the material and, with its large apparent activation energy of 220 kJ mol^{-1}, is to be regarded as a dynamic glass transition process having similarities to that of the α process in amorphous polymers. This process is prominent in the oxidized low density (branched) polyethylene but is of small magnitude in oxidized high density (linear) polyethylene, confirming its origin as being in the amorphous regions of the material. The γ process may contain contributions from both the amorphous and crystalline regions but Ashcraft and Boyd[63] suggest that it originates in the amorphous regions alone. In a recent study Kakizaki *et al.*[123] have made a comparative study of dielectric, mechanical and NMR relaxations in linear polyethylene and have deduced that the overall γ process arises from motions in an interlamellar amorphous region and at the lamellar surface. The fact that the γ relaxation is extremely broad in the frequency domain precludes a simple model for motion. Ashcraft and Boyd[63] rationalized the broad relaxation in terms of a local conformational reorganization largely dependent on internal chain energetics but partially dependent on environmental packing effects. The variety of local packings produce a broad spectrum of processes [*cf.* β relaxation discussed above for amorphous polymers and equation (62)]. They suggest that the motions involving bond rotation are localized and may be regarded as motion in a site-model situation.

Sayre, Swanson and Boyd[65] have studied the effect of pressure on the α and γ relaxations in oxidized linear polyethylene. $|(\partial \log f_m/\partial P)_T|$ was found to be $\sim 0.75 \times 10^{-3}$ atm^{-1} for the α process and $\sim 1 \times 10^{-3}$ atm^{-1} for the γ process giving apparent activation volumes of $\Delta V_\alpha \simeq 17$ cm^3 mol^{-1} and $\Delta V_\gamma \simeq 14$ cm^3 mol^{-1}. Q_V/Q_P is $\simeq 0.78$ for the α process and is $\simeq 0.91$ for the γ process, showing that the change of thermal energy when sample temperature is changed largely determines the change in relaxation time and that volume effects, although readily measured, are of far lesser importance.

The dielectric loss of cable insulation at cryogenic temperatures ($T \leq 20$ K) is of importance and the careful studies of Yano *et al.*[124] and Gilchrist[125] are of interest. Very small loss curves ($\varepsilon'' \sim 10^{-4}$) were observed in the ranges 1.5 to 4.2 K, 10 Hz to 10 kHz for linear polyethylene, branched polyethylene and copoly(ethylene/vinyl alcohol)s. For linear polyethylene the loss curves corresponded nearly to a single relaxation time process and were assigned to phonon-assisted tunnelling of the protons of OH groups present in the polymer. The loss curves for the other materials were much broader. Gilchrist showed that dielectric relaxation studies at these low temperatures permit detection of three species of oxidation products and that there are substantial hydrogen isotopic effects on the frequency location of these processes involving OOH or OOD groups.

18.5.3.2 *Poly(ethylene terephthalate)*

Condensation polymers, as typified by the aliphatic and aromatic polyesters and the aliphatic polyamides (nylons) have broad distributions of molecular weights and fairly low average molecular weights. Despite these differences from vinyl polymers, it is found that for condensation polymers in the amorphous state (*e.g.* poly(ethylene terephthalate) quenched from the melt) the pattern of α, β and $\alpha\beta$ relaxations and the shapes of the individual relaxation curves resemble closely those found for addition polymers such as poly(vinyl acetate).[7, 10] Such condensation polymers, due to the relative simplicity of the chain backbone, may be crystallized from the melt or from the glassy amorphous state, giving partially crystalline polymers having a degree of crystallinity in the range 0 to $\sim 50\%$. Thus in contrast to linear polyethylene, say, where the disordered material is less than 10% of the bulk, these partially crystalline polymers contain a large volume fraction of 'amorphous' material.

The dielectric properties of poly(ethylene terephthalate)s of different degrees of crystallinity have been reviewed by several workers[6, 7, 10, 11, 126] and will not be detailed here. The following conclusions apply. The α process observed in the amorphous polymer is diminished on isothermal crystallization of the amorphous polymer at 106 °C (see ref. 10, p. 81 for details, also ref. 126), and is replaced by a smaller, but broader, peak which occurs $\sim 10^2$ Hz lower in frequency than that in the normal amorphous material at this temperature. This demonstrates that in the partially crystalline polymer, the amorphous regions are strongly perturbed by the crystallites since these amorphous regions are actually contained within the spherulites, leading to a broader, slower process overall. Thus dielectric studies provide direct evidence for the occurrence of an anomalous 'amorphous' phase, and give information on the chain dynamics in that phase. In contrast, the β process in the partially crystalline polymer, although smaller in magnitude than that in the amorphous polymer, retains the characteristic broad contour and the same frequency–temperature location as that observed in the amorphous polymer. This shows that although the concentration of mobile groups is decreased through crystallization, the local conformational motions which give rise to the β relaxation are unchanged in the abnormal amorphous phase within the spherulites. In a recent paper[127] Coburn and Boyd studied nine samples varying from 4 to 62% crystalline in the ranges -200 to $180\,°C$, $1\,Hz$ to $100\,kHz$ and carried out a complete determination of the relaxation strength, shape, and frequency–temperature location for the α and β processes for all samples. One conclusion of this work is that the strengths of α and β processes extrapolate to zero for a wholly crystalline sample, showing that these processes originate in the amorphous regions of partially crystalline specimens. Further, since the crystallites are intrinsically anisotropic ($\varepsilon_\parallel \neq \varepsilon_\perp$) and a partially crystalline sample may be regarded as an aggregate of amorphous and crystalline regions, it follows that great care is required to relate the measured permittivity to the dielectric properties of the crystalline and amorphous fractions. This is discussed in relation to the determination of the Kirkwood g factor and using the theory developed earlier by Boyd[128] for application to oriented samples of oxidized polyethylene.

The anisotropy of the α and β relaxations for drawn poly(ethylene terephthalate) has been studied by several workers.[129–131] Davies and Ward[129] showed that the α process exhibited little anisotropy but the β process a pronounced anisotropy when measurements were made on sections cut from an oriented rod (extrusion ratio $= 3.3:1$). In a study of the β relaxation in oriented poly(ethylene terephthalate), Hsu *et al.*[130] attempted to modify the Kirkwood equation (equation (42) above) for application to systems of oriented dipoles. They found that the change in the dielectric properties on orientation of a sample could not be explained in terms of the alignment of chains alone, but the effect of conformational changes on dielectric anisotropy needed to be considered. In a further paper[131] these authors showed that satisfactory agreement was obtained between the observed dielectric anisotropy and a theory in which the director order parameter (*cf.* Section 18.3.4) and the angle between the dipole moment vector of a repeat unit and the chain axis were adjustable parameters.

Related polymers such as the aliphatic polyesters[132] and polyamides[7, 133, 134] have been studied by dielectric relaxation spectroscopy. Yemni and Boyd[133] found that the strength of the α process in oriented nylon 610 measured parallel and perpendicular to the draw direction were both lower than in the unoriented material, suggesting that either the amorphous fraction diminishes on alignment or that the amorphous phase chain conformation is changed by orientation. In the extended chain of nylon 610 the dipole moment vectors alternate (and hence cancel) along the chain and this would reduce the relaxation strength on stretching the amorphous chain. However, in a later paper these authors showed[134] that a reduction of the relaxation strength of the α process also occurred for the odd-numbered nylons 7–7 and 11. Since dipole correlations are enhanced by extension of these chains, the reduction of relaxation strength in all cases could not be explained by an anisotropy of the amorphous phase induced by orientation of the partially crystalline material. Also it was not necessary to invoke the disappearance of the amorphous fraction by orientation. Rather, it was shown that the observed reduction in relaxation strength was a direct consequence of the composite nature of these materials (amorphous plus anisotropic crystals) so that the measured permittivity is an average quantity related to the amorphous fraction, the orientation distribution of the crystallites and $\varepsilon_\parallel$ and $\varepsilon_\perp$, as analyzed by Boyd.[128, 136, 137]

18.5.3.3 *Poly(vinylidene difluoride)*

This polymer is of special interest due to its commercial importance as an electroactive material (*e.g.* as piezo- or pyro-electric films). The molecule is highly polar and the material may exist in five crystal forms; in three of which the molecular dipoles are parallel.[135] It is evident from a study of the early literature (see ref. 10 for a review) that the dielectric behaviour of this polymer is the most complex of all the linear polymer systems. Studies have been made for partially crystalline samples having different crystal forms, different degrees of orientation and crystallinity, and electrical and thermal histories. Unoriented materials may have degrees of crystallinity up to $\sim 50\%$ so they are composites of crystalline and amorphous regions with the attendant complications of relating the measured permittivities to the volume fractions of the phases, the permittivities of the phases, including $\varepsilon_\parallel$ and $\varepsilon_\perp$ for the crystals, and the orientation distribution function of the crystallites, as has been discussed by Boyd[128, 136, 137] for different morphologies. In addition the amorphous regions contain extrinsic ionic species which are mobile at the high temperatures at which materials are 'poled' (*i.e.* polarized by a dc electric field) and these ions may obscure the high temperature relaxation behaviour and also affect the poling process. It is not possible to cover adequately all the studies of poly(vinylidene difluoride) in relation to its dielectric, piezoelectric, pyroelectric and ferroelectric behaviour. The literature is extensive and sometimes confusing, but good accounts of the early studies are given in refs. 10, 135, 138–145 and references therein.

The polymer is obtained in the α crystal form (form II) by melt crystallization and in the β crystal form (form I) by drawing specimens. The α and β forms have helical and planar zigzag chain conformations respectively and since both have appreciable dipole moments, the permittivities may be high (in the range 10 to 20). We shall not repeat the descriptions given in detail earlier[10] for the experimental dielectric data of different materials. We summarize here the essential conclusions regarding the four main relaxation regions labelled α_i, α_c, α_a and β in decreasing order of temperature at a fixed measuring frequency.

α_i *process.* At the lowest frequencies and high temperatures there is an increasing low frequency loss and the permittivities reach very large values[141–144] (up to 10^5). The permittivity $\varepsilon'(v)$ and loss factor $\varepsilon''(v)$ follow the $v^{-3/2}$ and v^{-1} laws respectively and such behaviour is interpreted in terms of the diffusion of ionic species.[140–145] This process is very dependent upon the ionic purity of the material.

α_c *process.* This is sometimes obscured by the α_i process but the latter can be partly removed by pretreating specimens with a steady electric field (up to $25\,kV\,cm^{-1}$) at $160\,°C$, resulting in ion migration to the electrodes.[145] The resulting α_c process gives a symmetrical plot of ε'' vs. $\log v$ with a half-width near two decades of frequency and an apparent activation energy of $\sim 100\,kJ\,mol^{-1}$ and all the evidence suggests that this process is due to chain reorientations in the crystals (*cf.* polyethylene). The dielectric increment is substantial, being in the range 2–10, dependent on sample preparation, and increasing markedly with decreasing temperature.[141–146]

α_a *process.* It is generally agreed that this process arises in the amorphous regions of the material[10,140,141] *e.g.* ε''_m decreases with increasing degree of crystallinity. Also the frequency–temperature locus for the α_a process is strongly curved (WLF behaviour), so it seems reasonable to associate this process with the microbrownian motions of chains in the amorphous phase, *i.e.* a dynamic T_g process to be correlated with the glass transition temperature ($\sim 40\,°C$) for the polymer. Figure 11 shows the plots of $\varepsilon'(v)$ and ε'' vs. log frequency for PVDF showing the α_a absorption

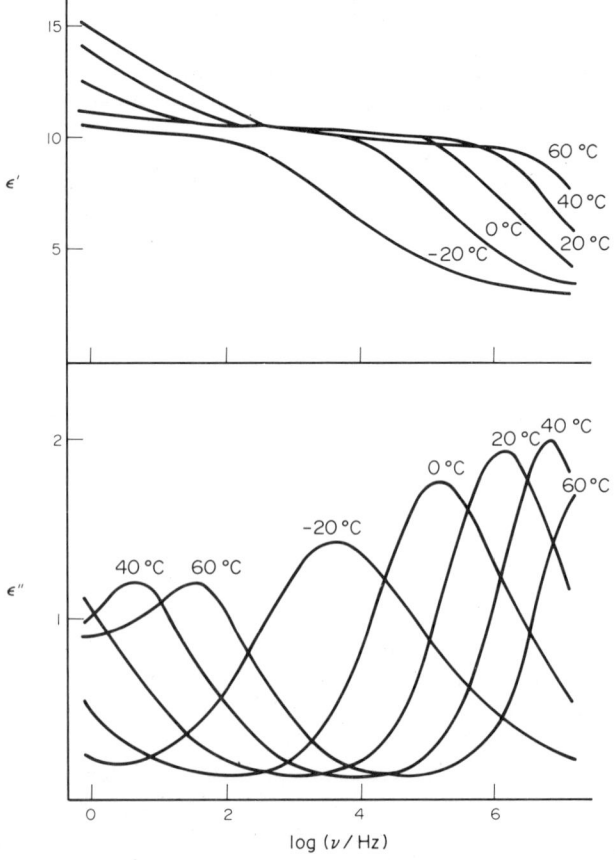

Figure 11 ε' and ε'' plotted against $\log_{10}(v/\mathrm{Hz})$ for poly(vinylidene difluoride) at different temperatures (after Furakawa *et al.*)[146]

(higher frequencies) and the α_c absorption (lower frequencies) at fixed temperature.[146] The striking feature of the α_a absorption is that in contrast to the α absorption in amorphous polymers (see data for poly(vinyl acetate), Figure 9 above) the half-width increases markedly with decreasing temperature, changing from 2 to 6 units of $\log_{10} v$ on going from -1.2 to $-35.4\,°C$.[141] It is likely that this is an example where the constraints imposed by the crystallites on motions in the amorphous regions (*cf.* poly(ethylene terephthalate) discussed above) increase markedly with decreasing temperature.

β *process.* This occurs as a small high frequency shoulder to the α_a process and as a small ($\varepsilon''_{m\beta}/\varepsilon''_{m\alpha} \simeq 0.1$) and broad process. It has an apparent activation energy of about $8\,kJ\,mol^{-1}$ and the process is assigned to be a local mode relaxation in the amorphous regions.[10,140]

Cebe and Grubb[147] have recently studied the high temperature relaxations of PVDF in the α phase and the solution-crystallized γ phase. They have also given a valuable review of the earlier dielectric data, dealing with the low frequency high temperature region where space charge and conductivity effects are of great importance and complicate the experiments and analyses. In particular they show that the permittivity in the range 60–140 °C is very dependent on ionic content. Using the ion-sweeping pretreatment described above, it was found that the permittivity at high temperatures ($>90\,°C$) and low frequencies (200 Hz) for samples after ion sweeping at 140 °C was lowered by up to 50% compared with untreated specimen. As a result it was found possible to observe a marked *decrease* in permittivity for cleaned specimens for $T > 100\,°C$, being indicative of melting of the crystalline material.

The anisotropy of the α_c relaxation in form II PVDF was studied by Miyamoto *et al.*[148] They conclude that molecular orientation cannot be explained by whole-chain rotation in the crystalline phase, and propose a model in which dipole reorientation in the crystals takes place by conformational changes assisted by defect motions.

A significant advance in materials for application has been the preparation of copolymers of vinylidene difluoride and trifluoroethylene (see ref. 146 and refs. therein). It is shown that certain copolymers exhibit Curie points in the permittivity *vs.* temperature plots[146] which are absent in the parent homopolymers. The ferroelectric behaviour of the copolymers is discussed and related to conformational changes which occur in the crystals when an electric field is applied.

18.5.4 Thermotropic Liquid Crystalline Polymers

Thermotropic liquid crystals composed of small molecules are finding increasing application as display materials, information storage materials, optical couplers and optical waveguides. Such materials exist in nematic, smectic and cholesteric states and their microstructures are well known.[149-153] It is not generally appreciated that the alignment behaviour in response to moderate electric or magnetic fields, which forms the basis of most display applications, has its origins in the anisotropy of the dielectric or magnetic relaxation processes of the materials. The action of the applied directing field is to align the local director $\mathbf{n}$ in a mesophase either parallel or perpendicular to the field direction depending on the sign of the anisotropy of the susceptibility[151] (*i.e.* $\Delta\varepsilon = \varepsilon_\parallel - \varepsilon_\perp$ for the case of applied electric fields). The anisotropy of the susceptibility may change sign on increasing the frequency of the applied electric field so that at low frequencies of the directing electric field $\Delta\varepsilon$ is positive leading to $\mathbf{n}\parallel E$ (homeotropic alignment) while at high frequencies $\Delta\varepsilon$ is negative leading to $\mathbf{n}\perp E$ (planar or homogeneous alignment). This forms the basis of the two-frequency addressing of LC displays.[154] The origin of the dielectric anisotropy lies in the molecular factors described in Section 18.3.4 above and in the values of the relaxation frequencies associated with the dispersions of the real part of the component relaxation functions given in equations (57) above. It follows that studies of the dielectric relaxation behaviour of liquid crystals provide essential information for the applications of these anisotropic dielectrics.[155]

Liquid crystalline polymers are a new class of materials that share with polymers their good mechanical properties and with normal liquid crystals their electroactive properties. The main chain thermotropics are generally aromatic polyesters or aromatic polyamides.[45] The dielectric properties of a smectic main chain liquid crystalline polyester have been reported.[156] Data were obtained in the glassy LC state and revealed a small broad dielectric loss due to local segmental motions (β process). The side chain thermotropic LC polymers have more interesting dielectric properties. These materials may be aligned by directing electric or magnetic fields and retain that alignment in the LC state — like low molar mass smectics but unlike low molar mass nematics. The chemical and physical aspects of the LC side chain polymers have been reviewed recently.[46,47,68-72] Extensive dielectric studies of unaligned LC side chain polymers have been made by Kresse and co-workers[157,158] and by Zentel and co-workers[159] for acrylate chains having the mesogenic (liquid crystal forming) group in the side chain. The mesogenic group was an alkyl cyanobiphenyl[157,158] or an alkyl aromatic ester.[159] Multiple relaxations were observed in the latter case and have been assigned to specific group or chain motions. More recently Attard and Williams and co-workers[73,76,160-166] have studied siloxane polymers in different states of alignment including homeotropic, partially homeotropic, random, partially planar and fully planar states, while Haase and co-workers[167] studied acrylate polymers in their homeotropic and homogeneous (planar) states. In the studies of Attard, Williams and co-workers the polymers had the structure (1) where n is

$$\text{Me}_3\text{SiO} \left[\begin{array}{c} \text{Me} \\ | \\ \text{Si} - \text{O} \\ | \\ \text{R} \end{array} \right]_n \text{SiMe}_3 \qquad \text{where R} = -(\text{CH}_2)_m\text{O}\langle\!\!\!\!\!\!\!\!\bigcirc\!\!\!\!\!\!\!\!\rangle\text{CO}_2\langle\!\!\!\!\!\!\!\!\bigcirc\!\!\!\!\!\!\!\!\rangle\text{CN}$$

(1) (2)

approximately 35 and $m = 5$, 6 or 8. It is seen that the methylene spacer is sufficiently long to allow the mesogenic head group to have freedom from the siloxane main chain. Attard and Williams[163] studied a smectic polymer having $m = 8$ and their results for the loss data for the unaligned material are shown in Figure 12. Here $G/\omega = \varepsilon''C_0$ where C_0 is the interelectrode capacitance. The loss curves show a pronounced shoulder at all temperatures indicative of different modes of relaxation for the head group (the polymer backbone has only a small dipole moment). On heating the sample into the isotropic state ($T_{\text{clearing}} \simeq 90\,°\text{C}$) and cooling in the presence of a directing ac electric field of frequency 3 kHz and amplitude 300 V, a fully homeotropic sample is obtained. The data shown in Figure 13 were obtained. The loss curves are now larger in amplitude and the shoulder observed in the aligned material is absent. These data can be understood quantitatively in terms of the theory described in Section 19.3.4 above. For the homeotropically aligned material $S_d = 1$ so the imaginary part of equation (60) gives

$$\varepsilon_Z''(\omega) = \frac{G}{3kT}\cdot[\mu_l^2(1+2S)F_\parallel^1(\omega)+\mu_t^2(1-S)F_\parallel^1(\omega)] \qquad (63)$$

For the unaligned material $S_d = 0$ so equation (60) gives

$$\varepsilon_Z''(\omega) = \frac{G}{3kT}\cdot\left[\left\{\mu_l^2\frac{(1+2S)}{3}F_\parallel^1(\omega)+\mu_t^2\frac{(1-S)}{3}F_\parallel^1(\omega)\right\}+\frac{2}{3}\left\{\mu_l^2\frac{(1-S)}{3}F_\perp^1(\omega)+\mu_t^2\frac{(1+S/2)}{3}F_\perp^1(\omega)\right\}\right] \qquad (64)$$

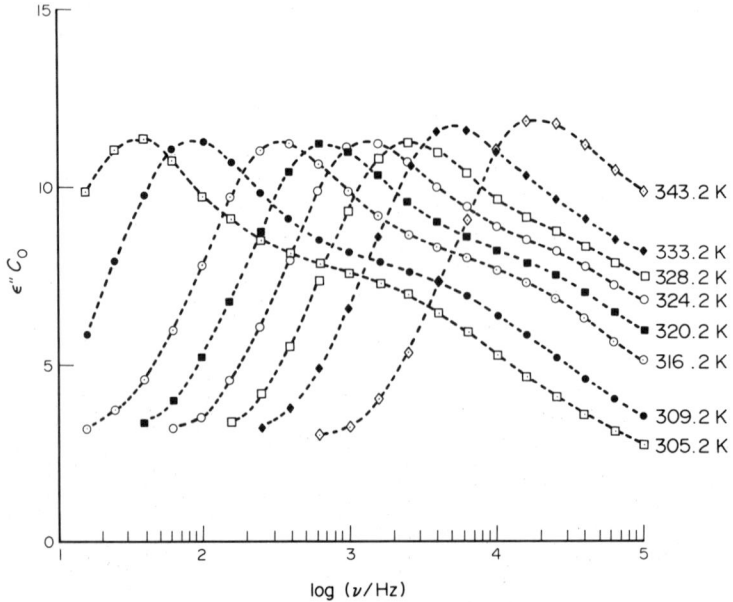

Figure 12 $\varepsilon''C_0$ plotted against $\log_{10}(v/\text{Hz})$ for unaligned smectic liquid crystalline polymer at different temperatures (after Attard and Williams)[163]

where $F_j^i(\omega)$ is the imaginary part of $F_j^i(\omega)$. Thus two modes of absorption are present for the homeotropic material and four modes are present for the unaligned material. Attard and Williams[163] assumed that the overall loss curves shown in Figures 12 and 13 could be resolved into two components, δ and α in increasing order of frequency. They further assumed that the δ process arose solely from the $F_\parallel^1(\omega)$ term and that all remaining terms contributed to the overall α process. Using the experimental data together with equations (63) and (64) they were able to rationalize their data and determined the mesophase order parameter as $S \simeq 0.68$ to 0.78, and the ratio of the longitudinal and transverse dipole moment components as $(\mu_l/\mu_t) \simeq 1.47$ to 1.53. Thus the anisotropic dielectric properties of a LC side chain polymer were analyzed in terms of basic molecular quantities involving the anisotropic motions of the mesogenic head group in the liquid crystalline potential.

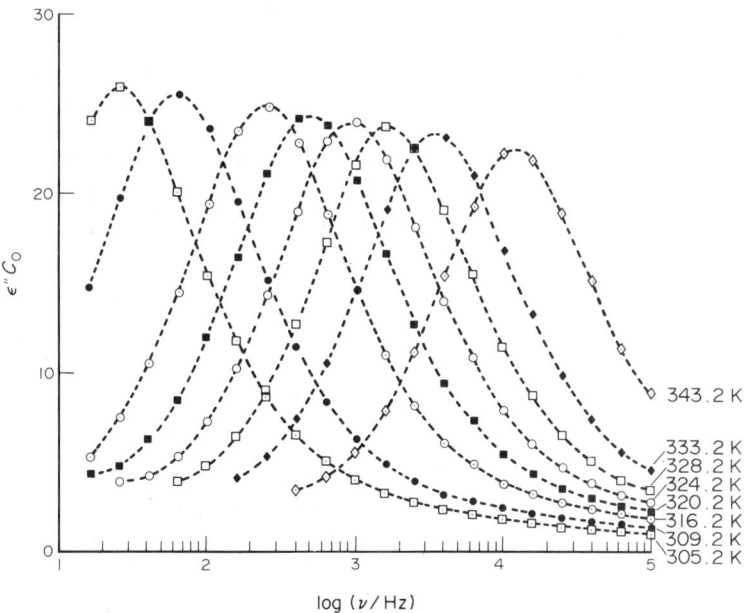

Figure 13 $\varepsilon'' C_0$ plotted against $\log_{10}(v/\mathrm{Hz})$ for homeotropically aligned smectic liquid crystalline polymer at different temperatures (after Attard and Williams)[163]

Figure 14 shows the dielectric loss spectra obtained by Araki and Attard[168] for a nematic LC polymer ($m = 6$ in structure (**2**) above) in its homeotropic, non-aligned and planarly aligned states. The homeotropic and planar states were obtained by cooling the polymer from the melt with saturating ac electric fields of 100 V at 600 Hz and 10 kHz respectively. The different alignments form in accord with the two-frequency addressing principle.[154] The difference in loss spectra for the different states of alignment is apparent, and an isosbestic point is clearly seen seen at ~ 600 Hz. In this polymer the structure present in the loss of curve for the unaligned smectic siloxane polymer (Figure 12) is absent so deconvolution of the curves is not possible. However Attard, Araki and Williams[76] have presented an alternative theory to that given above, which simply assumes that the material is an aggregate of anisotropic regions giving

$$\varepsilon_Z''(\omega) = \varepsilon_\parallel''(1 + 2S_d)/3 + \varepsilon_\perp'' 2(1 - S_d)/3 \tag{65}$$

Equation (65) takes on particularly simple forms for $S_d = 1$, 0 or $-\frac{1}{2}$ and it is found that the data for the three curves of Figure 15 satisfy equation (65) at all frequencies. Equation (65) also provides a simple means of determining the director order parameter S_d for a sample of arbitrary alignment if data for two reference materials are available (*e.g.* any two from homeotropic, random and planar).

The frequency–temperature locations for the resolved δ and α processes have been determined for the smectic siloxane polymer. Figure 15 shows the data of Attard, Moura-Ramos and Williams[169] for both unaligned and aligned polymer. The loci for δ and α processes both follow a WLF equation. Also the apparent glass transition temperature obtained from DSC measurements (0 °C) may be correlated with the premonitory behaviour shown in Figure 15. Thus the anisotropic motions of the mesogenic head groups are cooperative with the main chain motions and exhibit a critical 'slowing

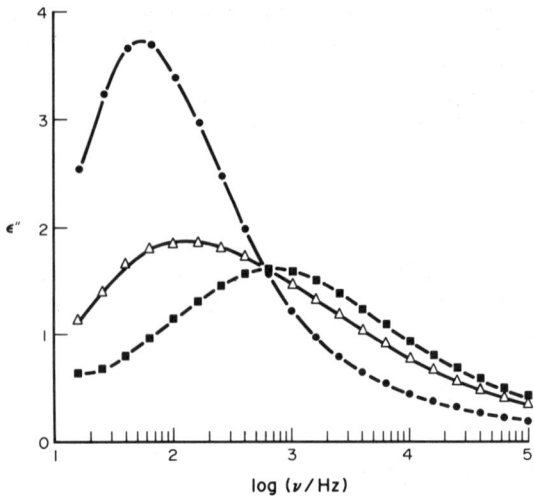

Figure 14 ε'' plotted against $\log_{10}(\nu/\text{Hz})$ for homeotropic ($\bullet$), planar ($\blacksquare$) and unaligned ($\triangle$) nematic liquid crystalline polymer at 307.2 K (after Araki and Attard)[168]

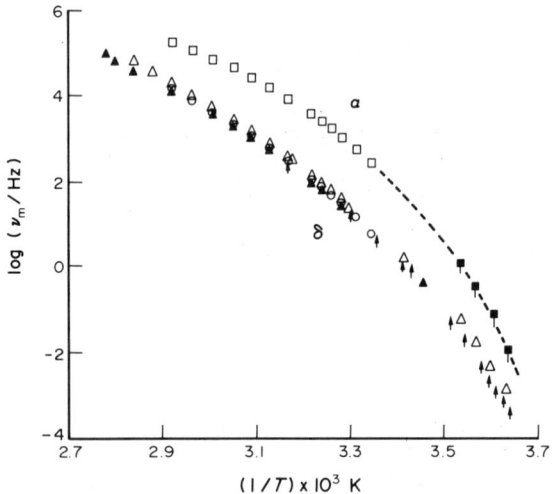

Figure 15 $\log_{10} \nu_m$ against $(T/\text{K})^{-1}$ for a smectic liquid crystalline polymer showing the δ and α processes (after Attard, Moura-Ramos and Williams)[169]

down' as the temperature is decreased into the glass transition range. In the earlier study of a nematic siloxane polymer Attard and co-workers[160] studied the relaxation behaviour through the transition range from liquid crystal to isotropic material and found, as was observed earlier,[157,158] that the relaxation peaks narrowed markedly and the $\log \nu_m$ plot showed a distinct change as the transition region was traversed.

It is evident from these studies[157-169] that dielectric relaxation spectroscopy provides a direct and unambiguous method for studying the molecular dynamics and alignment properties of liquid crystalline side chain polymers. It has the distinct advantage over the NMR, ESR and DSC methods that a wide frequency range (10^{-3} to 10^6 Hz) can be covered at each temperature in the liquid crystal, biphasic and isotropic ranges of these anisotropic materials.

18.6 CONCLUDING REMARKS

In this review the material has been selected to illustrate the main features of the dielectric properties of solid polymers in their amorphous, partially crystalline and liquid crystalline states and of polymers in solution. It should be apparent that the dielectric method provides a powerful and direct means of probing polymer chain dynamics and structure and is to be regarded as being complementary to other methods used such as mechanical relaxation, NMR and ESR relaxation, DSC and quasi-elastic light scattering. We have not been able to discuss several important measuring techniques or certain materials of current interest. For example the technique of thermally stimulated currents[170,171] which is used extensively to detect relaxation processes in solid polymers has not been described. Also polymer electrolytes composed of mixtures of poly(ethylene oxide) or poly(propylene oxide) and metal salts, which are of interest for solid state batteries, have not been described. We note that the dielectric and conduction properties of these electrolytes are complicated[172-177] and that polarization (giving relaxation) and conduction processes appear to be intimately connected. Also the apparent conductivity decreases from $10^{-4}\,\Omega^{-1}\,cm^{-1}$ to $\sim 10^{-11}\,\Omega^{-1}\,cm^{-1}$ as temperature is decreased so the combined effects of changes in ionic mobility and changes in the extent of aggregation of the ions with temperature need to be examined carefully in order to elucidate the molecular and ionic processes responsible for polarization and conduction. The subject of polymer electrolytes is clearly of great interest to electrochemistry and polymer science alike.

With regard to current activities and future work, it appears that much interest will surround studies of the dielectric and electrical properties of thin polymer films which are of interest in electronics, optics and electro-optics. A wide variety of organic polymers are now being synthesized, characterized and investigated. Nematic, smectic and chiral-smectic LC polymers (and their mixtures with low molar mass speciality materials), semiconducting polymers, electrochromic, photochromic and chemically photosensitive polymers are all being investigated with the prospect of their application as media for displays and information storage and for use in photonics and electronics. Also the polymer electrolytes may form an essential element of new solid state batteries. The dielectric properties of these materials will be investigated, with emphasis being placed on non-linear behaviour, anisotropic properties, the interrelation between conduction and polarization phenomena, and on photonic and electronic processes occurring for materials subjected to directing electric fields. These new aspects of the dielectric properties of polymers, taken together with the advances being made in measurement techniques and the molecular theory of relaxation/conduction processes, suggest that dielectric/electrical studies will provide valuable insights into the properties and behaviour of new materials used in the new high technology of thin films and their devices.·

18.7 REFERENCES

1. A. R. Von Hippel, 'Dielectric Materials and Applications', Wiley, New York, 1954.
2. J. G. Kirkwood and R. M. Fuoss, *J. Chem. Phys.*, 1941, **9**, 329.
3. R. M. Fuoss and J. G. Kirkwood, *J. Am. Chem. Soc.*, 1941, **63**, 385.
4. W. Reddish, *Trans Faraday Soc.*, 1950, **46**, 459.
5. A. H. Scott, D. J. Scheiber, A. J. Curtis, J. I. Lauritzen and J. D. Hoffman, *J. Res. Nat. Bur. Stand., Sect. A*, 1962, **66**, 269.
6. Y. Ishida, *J. Polym. Sci.*, 1969, **7**, 1835.
7. N. G. McCrum, B. E. Read and G. Williams, 'Anelastic and Dielectric Effects in Polymeric Solids', Wiley, New York, 1967.
8. P. Hedvig, 'Dielectric Spectroscopy of Polymers', Adam Hilger, Bristol, 1977.
9. A. K. Jonscher, 'Dielectric Relaxation in Solids', Chelsea Dielectrics Press, London, 1983.
10. G. Williams, *Adv. Polym. Sci.*, 1979, **33**, 60.
11. G. Williams, in 'Dynamic Properties of Solid Polymers', ed. R. Pethrick and R. W. Richards, NATO ASI, Reidel, Dordrecht, 1982.
12. K. S. Cole and R. H. Cole, *J. Chem. Phys.*, 1941, **9**, 341.
13. D. W. Davidson and R. H. Cole, *J. Chem. Phys.*, 1950, **18**, 1417.
14. D. W. Davidson, *Can. J. Chem.*, 1961, **39**, 571.
15. S. Havriliak and S. Negami, *J. Polym. Sci., Part C*, 1966, **14**, 99.
16. J. D. Ferry, 'Viscoelastic Properties of Polymers', Wiley, New York, 1972.
17. G. Williams, *Trans. Faraday Soc.*, 1964, **60**, 1556.
18. (a) G. Williams and D. C. Watts, *Trans. Faraday Soc.*, 1970, **66**, 80; (b) G. Williams, D. C. Watts, S. B. Dev and A. M. North, *Trans. Faraday Soc.*, 1971, **67**, 1323.
19. E. W. Montroll and J. T. Bendler, *J. Stat. Phys.*, 1984, **34**, 129.
20. G. Williams, in 'Dielectric and Related Molecular Processes', ed. M. Davies, Specialist Periodic Reports, Chemical Society, London, 1975, vol. 2, p. 151.
21. C. T. Moynihan, *et al.*, *Ann. N.Y. Acad. Sci.*, 1976, **279**, 15.
22. (a) G. Williams, *IEEE Trans. Electr. Insul.*, 1985, **EI-20**, 843; (b) C. P. Lindsey and G. D. Patterson, *J. Chem. Phys.*, 1980, **73**, 3348.
23. M. F. Shlesinger and E. W. Montroll, *Proc. Natl. Acad. Sci. USA*, 1984, **81**, 1280.
24. R. G. Palmer, D. L. Stein, E. Abrahams and P. W. Anderson, *Phys. Rev. Lett.*, 1984, **53**, 958.
25. A. Blumen, J. Klafter and G. Zumoten, *J. Phys. A: Math. Gen.*, 1986, **19**, L27.
26. A. Blumen, *Nuovo Cimento Soc. Ital. Fis. B.*, 1981, **63**, 50.

27. M. F. Shlesinger, *J. Stat. Phys.*, 1984, **36**, 639.
28. G. H. Fredrickson and H. C. Anderson, *Phys. Rev. Lett.*, 1984, **53**, 1244.
29. J. Klafter and M. F. Shlesinger, *Proc. Natl. Acad. Sci. USA*, 1986, **83**, 848.
30. A. J. Kovacs, J. J. Aklonis, J. M. Hutchinson and A. R. Ramos, *J. Polym. Sci., Polym. Phys. Ed.*, 1979, **17**, 1097.
31. J. M. Hutchinson and A. J. Kovacs, *Polym. Eng. Sci.*, 1984, **24**, 1087.
32. S. Matsuoka, G. Williams, G. E. Johnson, E. W. Anderson and T. Furukawa, *Macromolecules*, 1985, **18**, 2652.
33. J. M. O'Reilly, *J. Polym. Sci.*, 1962, **47**, 429.
34. G. Williams, *Trans. Faraday Soc.*, 1964, **60**, 1548, 1556.
35. G. Williams, *Trans. Faraday Soc.*, 1965, **61**, 1564.
36. G. Williams and D. A. Edwards, *Trans. Faraday Soc.*, 1966, **62**, 1329.
37. G. Williams, *Trans. Faraday Soc.*, 1966, **62**, 1321.
38. G. Williams, *Trans. Faraday Soc.*, 1966, **62**, 2091.
39. G. Williams and D. C. Watts, in 'NMR, Basic Principles and Progress', Springer Verlag, Heidelberg, 1971, vol. 4, p. 271.
40. G. Williams and D. C. Watts, *Trans. Faraday Soc.*, 1971, **67**, 1971, 2793.
41. S. Saito, H. Sasabe, T. Nakajima and K. Yada, *J. Polym. Sci., Part A-2*, 1968, **6**, 1297.
42. H. Sasabe and S. Saito, *J. Polym. Sci., Polym. Phys. Ed.*, 1968, **6**, 1401.
43. J. A. Sayre, S. R. Swanson and R. H. Boyd, *J. Polym. Sci., Polym. Phys. Ed.*, 1978, **16**, 1739.
44. G. Williams, *IEEE Trans. Electr. Insul.*, 1982, **EI-17**, 469.
45. B. Wunderlich and J. Grebowicz, *Adv. Polym. Sci.*, 1984, **60/61**, 1.
46. J. H. Wendorff, H. Finkelmann and H. Ringsdorf, *J. Polym. Sci., Polym. Symp.*, 1978, **63**, 245.
47. H. Finkelmann, H. Ringsdorf, W. Siol and J. H. Wendorff, *ACS Symp. Ser.*, 1978, **74**, 22.
48. N. Hill, W. Vaughan, A. H. Price and M. Davies, 'Dielectric Properties and Molecular Behaviour', Van Nostrand, London, 1969.
49. Y. Wada in 'Dielectric and Related Molecular Processes', ed. M. Davies, Specialist Periodic Reports, Chemical Society, London, 1976, **15**, 2041.
50. M. Cook, D. C. Watts and G. Williams, *Trans. Faraday Soc.*, 1970, **66**, 2503.
51. M. V. Volkenstein, 'Configurational Statistics of Polymeric Chains', Wiley–Interscience, New York, 1963.
52. P. J. Flory, 'Statistical Mechanics of Chain Molecules', Wiley–Interscience, New York, 1969.
53. G. Williams, *Trans. Faraday Soc.*, 1963, **59**, 1397.
54. B. E. Read, *Trans. Faraday Soc.*, 1965, **61**, 2140.
55. S. H. Glarum, *J. Chem. Phys.*, 1960, **33**, 1371.
56. G. Williams, M. Cook and P. J. Hains, *J. Chem. Soc., Faraday Trans. 2*, 1972, **68**, 1045.
57. G. Williams, *Chem. Soc. Rev.*, 1978, **7**, 89.
58. A. Tanaka, S. Uemura and Y. Ishida, *J. Polym. Sci., Polym. Phys. Ed.*, 1972, **10**, 2093.
59. H. Fröhlich, *Proc. Phys. Soc., London*, 1942, **54**, 422.
60. C. A. F. Tuijnman, *Polymer*, 1963, **4**, 315.
61. H. C. Booij, *J. Polym. Sci., Part C*, 1967, **16**, 1761.
62. G. Williams, J. I. Lauritzen and J. D. Hoffman, *J. Appl. Phys.*, 1967, **38**, 4203.
63. C. R. Ashcraft and R. H. Boyd, *J. Polym. Sci., Polym. Phys. Ed.*, 1976, **14**, 2153.
64. M. Mansfield and R. H. Boyd, *J. Polym. Sci., Polym. Phys. Ed.*, 1978, **16**, 1227.
65. J. A. Sayre, S. R. Swanson and R. H. Boyd, *J. Polym. Sci., Polym. Phys. Ed.*, 1978, **16**, 1739.
66. J. D. Hoffman, G. Williams and E. Passaglia, *J. Polym. Sci., Part C*, 1966, **14**, 173.
67. D. H. Reneker, *J. Polym. Sci.*, 1962, **59**, 539.
68. A. Ciferri, W. R. Krigbaum and R. B. Meyer, (eds.) 'Polymer Liquid Crystals', Academic Press, New York, 1982.
69. M. G. Dobb and J. E. McIntire, *Adv. Polym. Sci.*, 1984, **60/61**, 61.
70. H. Finkelmann and G. Rehage, *Adv. Polym. Sci.*, 1984, **60/61**, 99.
71. V. P. Shibaev and N. A. Platé, *Adv. Polym. Sci.*, 1984, **60/61**, 173.
72. G. S. Attard and G. Williams, *Chem. Br.*, 1986, **22**, 919.
73. G. S. Attard, *Mol. Phys.*, 1986, **58**, 1087.
74. P. L. Nordio, G. Rigatti and U. Segre, *Mol. Phys.*, 1973, **25**, 129.
75. W. Maier and G. Meier, *Z. Naturforsch.*, 1961, **162**, 262.
76. G. S. Attard, K. Araki and G. Williams, *Br. Polym. J.*, 1988, in press.
77. A. Brather, *Colloid Polym. Sci.*, 1979, **257**, 785.
78. J. L. G. Ribelles and R. D. Calleja, *J. Polym. Sci., Polym. Phys. Ed.*, 1985, **23**, 1505.
79. F. I. Mopsik, *Rev. Sci. Instrum.*, 1984, **55**, 79.
80. T. S. Clarkson, L. Glasser, R. W. Tuxworth and G. Williams, *Adv. Mol. Relax. Interact. Proc.*, 1977, **10**, 173.
81. R. H. Cole, *Annu. Rev. Phys. Chem.*, 1977, **28**, 283.
82. R. H. Cole and D. G. Hall, *J. Phys. Chem.*, 1981, **85**, 1065.
83. R. H. Cole, J. P. Perl, D. T. Wasan and P. Winsor, *J. Mol. Liq.*, 1983, **28**, 103.
84. R. H. Cole, S. Mashimo, P. Winsor, K. Matsuo and W. H. Stockmayer, *Macromolecules*, 1983, **16**, 965.
85. D. W. McCall, in 'Molecular Dynamics and Structure of Solids', ed. R. J. Carter and J. J. Rush, *NBS Spec. Publ. (US)*, 1969, 301.
86. S. Mashimo, R. Nozaki, S. Yagihara and S. Takeishi, *J. Chem. Phys.*, 1982, **77**, 6259.
87. M. Naoki, M. Motomura, T. Nose and T. Hata, *J. Polym. Sci., Polym. Phys. Ed.*, 1975, **13**, 1737.
88. M. Naoki and T. Nose, *J. Polym. Sci., Polym. Phys. Ed.*, 1975, **13**, 1747.
89. M. Naoki and T. Nose, *J. Polym. Sci., Polym. Phys. Ed.*, 1975, **13**, 1893.
90. K. D. Pae, *J. Polym. Sci.*, 1983, **21**, 1195.
91. R. E. Wetton, W. J. Macknight, J. L. Fried and F. E. Karasz, *Macromolecules*, 1977, **11**, 158.
92. M. F. Shears and G. Williams, *J. Chem. Soc., Faraday Trans. 2*, 1973, **69**, 608.
93. K. Adachi, I. Fujihara and Y. Ishida, *J. Polym. Sci., Polym. Phys. Ed.*, 1975, **13**, 2155.
94. K. Adachi and Y. Ishida, *J. Polym. Sci., Polym. Phys. Ed.*, 1976, **14**, 2219.
95. K. Adachi, M. Hattori and Y. Ishida, *J. Polym. Sci., Polym. Phys. Ed.*, 1977, **15**, 693.
96. W. H. Stockmayer, *Pure Appl. Chem.*, 1967, **15**, 539.

97. W. H. Stockmayer and M. E. Baur, *J. Am. Chem. Soc.*, 1964, **86**, 3485.
98. M. E. Baur and W. H. Stockmayer, *J. Chem. Phys.*, 1965, **43**, 4319.
99. K. Adachi and T. Kotaka, *Macromolecules*, 1984, **17**, 120.
100. K. Adachi and T. Kotaka, *Macromolecules*, 1985, **18**, 466.
101. G. P. Johari and L. Monnerie, *J. Polym. Sci., Polym. Phys. Ed.*, 1986, **24**, 2049.
102. P. J. Phillips and G. Singh, *J. Polym. Sci., Polym. Phys. Ed.*, 1975, **13**, 1377.
103. S. Yano, R. R. Rahalkar, S. P. Hunter, C. H. Wang and R. H. Boyd, *J. Polym. Sci., Polym. Phys. Ed.*, 1976, **14**, 1877.
104. S. Mashimo, H. Nakamura and A. Chiba, *J. Chem. Phys.*, 1982, **76**, 6342.
105. S. Mashimo, P. Winsor, R. H. Cole, K. Matsuo and W. H. Stockmayer, *Macromolecules*, 1983, **16**, 965.
106. S. Mashimo, S. Yagihara and A. Chika, *Macromolecules*, 1984, **17**, 630.
107. A. Wada, *Adv. Biophys.*, 1976, **9**, 1.
108. E. H. Grant, R. J. Sheppard and G. P. South, 'Dielectric Behaviour of Biological Molecules in Solution', Clarendon Press, Oxford, 1978.
109. H. Yu, A. J. Bur and L. J. Fetters, *J. Chem. Phys.*, 1966, **4**, 2568.
110. A. J. Bur and D. E. Roberts, *J. Chem. Phys.*, 1968, **51**, 406.
111. M. S. Beevers, D. C. Garrington and G. Williams, *Polymer*, 1977, **18**, 540.
112. J. K. Moscicki, S. M. Aharoni and G. Williams, *Polymer*, 1981, **22**, 1361.
113. J. K. Moscicki, G. Williams and S. M. Aharoni, *Macromolecules*, 1982, **15**, 642.
114. J. K. Moscicki and G. Williams, *J. Polym. Sci., Polym. Phys. Ed.*, 1983, **21**, 197, 213.
115. K. Adachi, Y. Hirose and Y. Ishida, *J. Polym. Sci., Polym. Phys. Ed.*, 1975, **13**, 737.
116. H. Nakamura, S. Mashimo and A. Wada, *Macromolecules*, 1981, **14**, 1698.
117. P. J. Flory, *Proc. R. Soc. London, Ser. A*, 1956, **234**, 73.
118. J. K. Moscicki, in *Adv. Chem. Phys.*, 1985, **63**, 631.
119. J. K. Moscicki and G. Williams, *Polymer*, 1981, **22**, 1451.
120. J. K. Moscicki and G. Williams, *Polymer*, 1982, **23**, 558.
121. M. P. Warchol and W. Vaughan, *Adv. Mol. Relax. Interact. Proc.*, 1978, **13**, 317.
122. C. C. Wang and R. Pecora, *J. Chem. Phys.*, 1980, **72**, 5333.
123. M. Kakizaki, T. Kakudate and T. Hideshima, *J. Polym. Sci., Polym. Phys. Ed.*, 1985, **23**, 787, 809.
124. O. Yano, K. Saiki, S. Tarucha and Y. Wada, *J. Polym. Sci., Polym. Phys. Ed.*, 1977, **15**, 43.
125. J. Le G. Gilchrist, *J. Polym. Sci., Polym. Phys. Ed.*, 1978, **16**, 1773.
126. K. Sawada and Y. Ishida, *J. Polym. Sci., Polym. Phys. Ed.*, 1975, **13**, 2247.
127. J. C. Coburn and R. H. Boyd, *Macromolecules*, 1986, **19**, 2238.
128. R. H. Boyd, *J. Polym. Sci., Polym. Phys. Ed.*, 1983, **21**, 505.
129. G. R. Davies and I. M. Ward., *J. Polym. Sci.*, 1972, **10**, 1153.
130. B. S. Hsu, S. H. Kwan and L. W. Wong, *J. Polym. Sci., Polym. Phys. Ed.*, 1975, **13**, 2079.
131. B. S. Hsu and S. H. Kwan, *J. Polym. Sci., Polym. Phys. Ed.*, 1976, **14**, 1591.
132. M. Ito, S. Nakatani, A. Gokan and K. Tanaka, *J. Polym. Sci., Polym. Phys. Ed.*, 1977, **15**, 605.
133. T. Yemni and R. H. Boyd, *J. Polym. Sci., Polym. Phys. Ed.*, 1976, **14**, 499.
134. T. Yemni and R. H. Boyd, *J. Polym. Sci., Polym. Phys. Ed.*, 1979, **17**, 741.
135. A..J. Lovinger, in 'Developments in Crystalline Polymers', ed. D. C. Bassett, Applied Science, London, 1982.
136. R. H. Boyd, *J. Polym. Sci., Polym. Phys. Ed.*, 1983, **21**, 493.
137. R. H. Boyd, *Macromolecules*, 1984, **17**, 217.
138. R. Hayakawa and Y. Wada, *Adv. Polym. Sci.*, 1973, **11**, 1.
139. S. Osaki and Y. Ishida, *J. Polym. Sci., Polym. Phys. Ed.*, 1975, **13**, 1071.
140. H. Sasake, S. Saito, M. Asahina and H. Kakutani, *J. Polym. Sci., Part A-2*, 1969, **7**, 1405.
141. K. Nakagawa and Y. Ishida, *J. Polym. Sci., Polym. Phys. Ed.*, 1973, **11**, 1503.
142. S. Uemura, *J. Polym. Sci., Polym. Phys. Ed.*, 1972, **10**, 2155.
143. S. Uemura, *J. Polym. Sci., Polym. Phys. Ed.*, 1974, **12**, 1177.
144. S. Yano, K. Tadano, K. Aoki and N. Koizumi, *J. Polym. Sci., Polym. Phys. Ed.*, 1975, **12**, 1875.
145. S. Osaki, S. Uemura and Y. Ishida, *J. Polym. Sci., Part A-2*, 1971, **9**, 585.
146. T. Furukawa, M. Ohuchi, A. Chiba and M. Date, *Macromolecules*, 1984, **17**, 1384.
147. P. Cebe and D. T. Grubb, *Macromolecules*, 1984, **17**, 1374.
148. Y. Miyamoto, H. Miyaji and K. Asai, *J. Polym. Sci., Polym. Phys. Ed.*, 1980, **18**, 597.
149. G. W. Gray, 'Molecular Structure and Properties of Liquid Crystals', Academic Press, London, 1962.
150. G. W. Gray, K. J. Harrison, J. A. Nash, J. Constant, D. S. Hulme, J. Kirton and E. P. Raynes, in 'Liquid Crystals and Ordered Fluids', ed. R. S. Porter and J. S. Johnson, Plenum Press, New York, 1973, vol. 2, p. 617.
151. P. G. de Gennes, 'The Physics of Liquid Crystals', Oxford University Press, London, 1974.
152. S. Chandresekhar, 'Liquid Crystals', Cambridge University Press, London, 1977.
153. G. W. Gray and P. A. Winsor, (eds.), 'Liquid Crystals and Plastic Crystals', Ellis Harwood, Chichester, 1974, vols. 1 and 2.
154. M. Schadt, *Mol. Cryst. Liq. Cryst.*, 1982, **89**, 77.
155. M. J. Clark, *Mol. Cryst. Liq. Cryst.*, 1985, **127**, 1.
156. F. Laupretre, C. Noël, W. N. Jenkins and G. Williams, *Faraday Discuss. Chem. Soc.*, 1985, **79**, 191.
157. H. Kresse and R. V. Talroze, *Makromol. Chem.*, 1981, **2**, 869.
158. H. Kresse, S. Kostromin and V. P. Shibaev, *Makromol. Chem., Rapid Commun.*, 1982, **3**, 509.
159. R. Zentel, G. R. Strobl and H. Ringsdorf, *Macromolecules*, 1985, **18**, 960.
160. G. S. Attard, G. Williams, G. W. Gray, D. Lacey and P. A. Gemmel, *Polymer*, 1986, **27**, 185.
161. G. S. Attard and G. Williams, *Polymer*, 1986, **27**, 2.
162. G. S. Attard and G. Williams, *Polymer*, 1986, **27**, 66.
163. G. S. Attard and G. Williams, *Liq. Cryst.*, 1986, **1**, 253.
164. K. Araki and G. S. Attard, *Liq. Cryst. Lett.*, 1986, **1**, 301.
165. G. S. Attard, K. Araki and G. Williams, *J. Mol. Electron.*, 1987, **3**, 1.
166. G. S. Attard and G. Williams, *J. Mol. Electron.*, 1986, **2**, 107.

167. W. Haase, H. Pranato and F. J. Bormuth, *Ber. Bunsenges. Phys. Chem.*, 1985, **89**, 1229.
168. K. Araki and G.S. Attard, personal communication.
169. G. S. Attard, J. J. Moura-Ramos and G. Williams, *J. Polym. Sci., Polym. Phys. Ed.*, 1987, **25**, 1099.
170. J. van Turnhout, 'Thermally Stimulated Discharge of Polymer Electrets', Elsevier, Amsterdam, 1975.
171. J. van Turnhout, in 'Electrets', ed. G. M. Sessler, Springer, Berlin, 1978.
172. J. J. Fontanella, M. C. Wintersgill, J. P. Calame and G. C. Andeen, *Solid State Ionics*, 1983, **8**, 333.
173. M. C. Wintersgill, J. J. Fontanella, J. P. Calame and G. C. Andeen, *Solid State Ionics*, 1983, **11**, 151.
174. J. J. Fontanella, M. C. Wintersgill, J. P. Calame, F. P. Pursel, D. R. Figueron and G. C. Andeen, *Solid State Ionics*, 1983, **9**, 1139.
175. C. G. Andeen, J. J. Fontanella, M. C. Wintersgill, P. J. Welcher, R. J. Kimble and G. E. Mathews, *J. Phys. C*, 1981, **14**, 3557.
176. M. B. Armand, J. M. Chabagno and M. J. Duclot, in 'Fast Ion Transport', ed. J. N. Mundy and G. K. Shenoy, Pergamon Press, New York, 1979, p. 131.
177. C. Berthier, W. Gorecki, M. Minier, M. B. Armand, J. M. Chabagno and P. Rigand, *Solid State Ionics*, 1983, **11**, 91.

19

Rheo-optical Properties

ROGER I. TANNER

University of Sydney, Australia

19.1 INTRODUCTION

The term 'rheo-optics' seems to have been coined in the 1960s.[1] It designates the use of optical methods to study flow and deformation in materials. The methods used include birefringence, X-ray diffraction, dichroism and light scattering; all these techniques are older than the collective terminology. Thus, in all cases to be considered the light (or X-ray) is affected by the flow; we shall not consider the reverse effect, where, for example, intense light might be used to affect the flow.

We shall not discuss X-ray diffraction (see Volume 1, Chapters 5 and 31) or light scattering (see Volume 1, Chapter 5). Dichroism in materials means that an optically anisotropic state exists in the material due to some permanent orientation of the structure. For example, in a polymer, *e.g.* poly(vinyl alcohol), heating and stretching will align molecules parallel to the stretching. The sheet may be dyed with a solution of iodine, for example, and the iodine will, in turn, line up along the stretched poly(vinyl alcohol) molecules. Conduction electrons associated with the iodine can then circulate up and down the stretched molecule. The result is a linear polarizer. Further details are given in the work of Hecht and Zajac.[2]

The main part of this chapter will deal with birefringence caused by flow (streaming birefringence). Other possibilities (*i.e.* magnetically and electrically induced birefringence) are also possible,[2] but will not be discussed.

19.2 THEORETICAL ASPECTS OF STREAMING BIREFRINGENCE

Flow (or streaming) birefringence is a non-destructive optical method, which may be used to indicate stress patterns in a flowing fluid or to probe the structure of the fluid. In certain classes of flows (*e.g.* simple shearing) values of stress may be accurately determined by this method; for more complex flows the matter is not quite as simple, as we shall discuss below. Stresses in deforming solids can also be measured by an analogous technique.[3]

Birefringence (or optical double refraction) is due to a difference between the principal refractive indices within a material; it may be observed, for example, by bending or twisting a piece of solid poly(methyl methacrylate) between crossed polarizers. Bands of colour appear, indicating splitting of the light due to differences in the principal refractive indices. A similar phenomenon occurs in some flowing fluids. In both cases we can infer the stresses present in the body if a relation between the stress level and the changes in optical properties is known (the stress–optic relation). Flow birefringence was observed by Maxwell[4] (1866) and, independently, by Mach[5] in 1873. In general, it appears to be due to local alignment of elongated particles or macromolecules in the fluid; this, in turn, causes local optical anisotropy.

19.2.1 Optical Relations

To explain the techniques of rheo-optics, we need to recall some simple ideas on the polarization of light in a birefringent medium.

An optically isotropic material is one in which the index of refraction (or the phase velocity of a wave) is the same in all directions. This is true for cubic crystals, such as NaCl, as well as for non-crystalline substances, such as unstressed glass and plastic, water and air, and fluids at rest. In general, however, materials are optically anisotropic; the refractive indices are different in different directions. When monochromatic linearly polarized light enters such an optically 'active' medium it is resolved into waves, called the ordinary and extraordinary, which move with different speeds. The speed of a wave v is linked to the refractive index n by the relation $v = c/n$, where c is the speed of light *in vacuo*. In general, one is looking at the propagation of a light wave in an ideal anisotropic dielectric. Let the electric displacement vector be D (with components D_i) and the electric field be E. For an anisotropic material[6]

$$D_i = \sum_{j=1}^{3} \varepsilon_0 K_{ij} E_j \quad (i = 1, 2, 3)$$ (1)

where ε_0 is the permittivity of free space and K_{ij} is the (relative) dielectric tensor. The dielectric tensor is symmetric[6] and hence it has three orthogonal principal directions. Suppose the z (or x_3) axis is one principal axis of the dielectric tensor. Then the electromagnetic-field equations show that a plane wave propagating along the z-axis is divided into two linearly polarized waves. The direction of vibration of each wave coincides with one of the remaining principal axes of the dielectric tensor. The refractive index tensor is equal to the square root of the dielectric tensor.[7] Let N_{ij} be the refractive index tensor, so that

$$N_{ij} = (K_{ij})^{\frac{1}{2}}$$ (2)

The birefringence Δ of a wave propagated along the z axis is then, as above

$$\Delta = n_1 - n_2$$ (3)

where n_1 and n_2 are principal values of N_{ij}. Note that birefringence is a dimensionless quantity.

The relative phase difference δ between two rays is

$$\delta = \frac{2\pi L}{\lambda_l} [n_1 - n_2]$$ (4)

where L is the length of the path of propagation of the light, λ_l is the light wavelength and the subscripts 1 and 2 stand for principal (optical) directions.

Light travelling within the medium is elliptically polarized, and the major axis of the light or optic ellipse makes an angle χ_0 with the polarization axis. We can study this in some more detail for a fluid medium. Consider a volume of fluid bounded by the planes $z = 0$ and $z = h$, and suppose that the optical properties are independent of z (this will not generally be true, as end regions of the flow may differ slightly in kinematics and stress from the main portions, thereby affecting overall optical properties). Suppose a linearly polarized plane electromagnetic wave, propagating along the z axis, is incident on the surface $z = 0$. The z axis is supposed to be a principal axis of K_{ij} (equation 2) and hence of N_{ij}. Every such wave can be decomposed into monochromatic waves, and these can be factored into components along the two other principal directions of N_{ij}, as mentioned above. Each component will be partly reflected and refracted at each end of the sample, and the resultant change in the wave amplitude due to these effects can be computed. If the light leaving the sample is passed through another polarizer, crossed with respect to the first (*i.e.* with 90° between the axes of the two polarizers), it may be shown[1] that the light intensity I is

$$I = TI_0 = I_0 \sin^2 2\beta \sin^2(\delta/2) \tag{5}$$

where β is the angle between one of the principal axes of the birefringent medium and the (initial) plane of polarization, δ is the retardation given in equation (4), I_0 is the incident intensity and T is the relative transmittance. Inspection of equation (5) shows that there will be zero intensity when $\beta = 0, \pi/2$.

In a non-homogeneous field of stress, there will appear lines of zero light intensity joining the points where $\beta = \frac{1}{2}m\pi(m = 0, 1, 2 \ldots)$. These are the *isoclinics*; if we define the optical angle χ_0 between one principal optical axis and the incoming wave plane of polarization, then χ_0 is always less than 45°, and, given this convention

$$\chi_0 = \beta. \tag{6}$$

Also, from equation (5), the relative transmittance T vanishes when

$$\delta = 2p\pi \quad (p = 0, 1, 2 \ldots) \tag{7}$$

Bands of equal δ are black in monochromatic light, and they are termed *isochromatics*. From suitable measurements (see below), we are therefore able to find δ and hence, from equation (4), the birefringence $\Delta(\equiv n_1 - n_2)$. We now consider the relation of Δ to the stress field in the medium.

19.2.2 Stress–Optic Relations

In an incompressible medium, the rheological state at a point is, as far as the stress is concerned, completely described by the shear stresses (three in a general flow) and in the differences of the normal or direct stresses.[8] We shall denote components of the stress tensor by $\sigma_{ij}(i, j = 1, 2, 3)$ and suppose the stress tensor is symmetric, so that $\sigma_{ij} = \sigma_{ji}$. (If $i \neq j$, σ_{ij} is a shear stress; if $i = j$, σ_{ij} is a direct or normal stress.) We can change axes by rotation to reduce σ_{ij} to principal form — in these (mutually orthogonal) principal axes the principal stresses are σ_1, σ_2 and σ_3 and all shear stresses vanish. The simplest stress–optical relation is to suppose that the dielectric tensor (K_{ij}) and the stress tensor are coaxial, *i.e.* have the same principal axes, and that the differences in principal stresses are proportional to the corresponding differences in (principal) refractive indices. Hence if $z(x_3)$ is a principal axis, and σ_1 and σ_2 lie in the xy plane, we have the simplest stress–optic relation as

$$\Delta \equiv n_1 - n_2 = C(\sigma_1 - \sigma_2) \tag{8}$$

and

$$\chi_0 = \chi_M \tag{9}$$

with σ representing the stress tensor, χ_M the inclination angle of the mechanical or stress ellipse to the original plane of polarization and C the stress–optical coefficient, which is a function of the material and temperature.

From a theoretical viewpoint, the stress–optical relation is found to be true for flexible macromolecules at low extensions[9] and for rigid particles where Brownian motion dominates. Leal and Hinch[10] have shown that the stress and optical ellipses are not coaxial for rigid particles undergoing weak Brownian motion. We shall return to this discussion of the range of validity of equation (8) later. If we accept this relation, then the principal stress differences can be found if Δ is given everywhere.

As an example, a frequently encountered stress field will be discussed to show behaviour of χ_M and hence χ_0. Consider the shearing motion in which the stress tensor is, by symmetry:[8]

$$[\sigma_{ij}] = \begin{bmatrix} \sigma_{xx} & \sigma_{xy} & 0 \\ \sigma_{xy} & \sigma_{yy} & 0 \\ 0 & 0 & \sigma_{zz} \end{bmatrix} \tag{10}$$

Here the usual rheological convention is used, with x being the flow direction, y the shear and z the neutral directions; this simple shear flow is the prototypical viscometric motion.[8]

In this stress field, the principal values σ_1, σ_2, σ_3 are

$$\sigma_3 = \sigma_{zz} \tag{11}$$

and

$$\sigma_{1,2} = \tfrac{1}{2}(\sigma_{xx} + \sigma_{yy}) \pm (\tfrac{1}{4}(\sigma_{xx} - \sigma_{yy})^2 + \sigma_{xy}^2)^{\frac{1}{2}} \tag{12}$$

The rheologically significant quantities in (incompressible) fluid flow are the differences $\sigma_1 - \sigma_2$, $\sigma_2 - \sigma_3$ and $\sigma_3 - \sigma_1$. Alternatively, in shear flows rheologists normally use the shear stress $\sigma_{xy}(\tau)$, $\sigma_{xx} - \sigma_{yy}$ (N_1, the first normal stress difference) and $\sigma_{yy} - \sigma_{zz}$ (N_2, the second normal stress difference). The principal directions corresponding to σ_1, σ_2 and σ_3 are orthogonal, and the first makes an angle $\chi_M^{(1)}$ with the x-axis

$$\tan 2\chi_M^{(1)} = \frac{2\tau}{N_1} \tag{13}$$

The third principal direction is coincident with the z axis and the second is orthogonal to the first and third.

For steady Poiseuille flow of a Newtonian fluid between plane walls, N_1 is zero and $\chi_M^{(1)}$ is always 45°. $\chi_M^{(1)}$ can deviate from 45° for fluids in which N_1 is non-zero. In particular, near the wall, where the shear rate is greatest, the largest deviation is expected to occur. The simple stress–optical law would then indicate that the directions of the principal refractive indices coincide with the principal stresses; one principal refractive index direction lies along the z direction and the other two lie in the xy plane.

The shear flow above is, in some respects, special, and, as an example of a different flow field, we cite elongational flow. Here there are no shear stresses and the principal stress directions are easily identified in terms of the motion.[8]

19.2.3 Validity of the Simple Stress–Optic Law

The question of when, for a given fluid at a given temperature, the proportionality parameter C (equation 8) is a constant, must be discussed. Apart from the intrinsic interest in this question, it is clearly most important to anyone wishing to use streaming birefringence as a tool for finding stresses. Janeschitz-Kriegl[9] has surveyed evidence from polymer melts, concentrating on two-dimensional flows. (It will be appreciated that in a real flow there will be 'end' regions which are not two-dimensional flows, and the measured birefringence will be an average over the light path.) Janeschitz-Kriegl[9] discusses the flow of polymer melts in viscometric (shear-type flows) and in elongational flows. In the latter[9] (see p. 62 of ref. 9) he considers that C is constant (at constant temperature) up to a tensile stress of possibly in the order of 10^6 Pa. It appears that, in some cases, the tests to confirm the constancy of C have not been pursued above 10^4 Pa. For shearing flows of polymer melts, it is stated[9] that no deviation from the coaxiality of the stress and optical tensors has been observed, and also that C is constant in these steady flows. The question of unsteady flows has also been considered. Read[11] has measured C in cyclic sinusoidal deformation and shown that, for poly(methyl methacrylate), C does vary with the frequency of oscillation, that the stress and optical behaviours are not in phase and that a time–temperature superposition curve for C can be constructed. Thus variation of C with temperature is expected, as is also some change with rate of stressing, so that we only expect C to be strictly constant in steady isothermal flows at small shear rates. Retting[12] confirmed this with polystyrene samples in tensile tests. Nevertheless, in transient shear flows (at small shear rate), stresses deduced from the optical rule may be just as reliable as mechanical measurements, because of the difficulties of the latter in highly viscous flows.[9] In dilute solutions there is no reason to expect C to be constant; when a molecular structure is fully extended, the birefringence will saturate, while the stresses continue to increase.[13] In solutions one needs to note that the birefringence is composed of two contributions — intrinsic and form birefringence; dichroism may also appear.[14]

In summary, we see that it is not possible to assume that the simple rheo-optical law, equation (8), is universally applicable. It will generally be expected to apply only when the molecular structure of the material is disturbed by a small amount from its equilibrium configuration; this will be in shearing flows at low stresses. For 'strong' flows of the elongational type, and especially in dilute solutions, where it is easy to deform the structure greatly, one may expect the simple concept to fail.

Similarly, when one is well above the glass transition temperature, there is greater safety in regarding C as constant than when one is near the glass transition temperature. The limitations of the law can often be understood in terms of simple molecular models,[13] and further work is proceeding on this front. A survey of the rheo-optics literature is given by Mackay;[15] in particular, previous theoretical derivations of the stress–optic law are collected; see also Janeschitz-Kriegl,[9] Peterlin,[3] Harrington[16] and Erman and Flory.[17]

19.3 MEASUREMENT TECHNIQUES

A variety of techniques have been used to determine the extinction angle and the birefringence of flows. They all depend on transmitting light through the sample and most depend on intensity measurements. Generally the optical systems used involve a source of light (usually a laser), a polarizer, the sample itself, an analyzer (often with its axis at right angles to the polarizer) and a detection system. Unwanted or parasitic birefringence due to windows in the system will also be present and needs to be considered.[9] The assumption that the light traverses a homogeneous isothermal body of fluid needs to be checked. Janeschitz-Kriegl[9] and Harrington[16] give some mechanical arrangements and discuss some of these problems. Sometimes quarter-wave plates (giving a phase-shift of 45°) are used. Filters, half-shadow plates and lenses may also be present[3,9,16] in the optical system. Methods for detecting outputs include photographic techniques[18] and the use of various compensators (de Sénarmont,[19] Babinet,[16] Brace[20] and Ehringhaus[9]) has been recommended to improve the accuracy of angle determination. Amplitude modulation[16,9] has also been used to improve resolution.

For the rheologist using highly birefringent materials, a plane or circular polariscope consisting of polarizers and quarter-wave plates is adequate (Δ should be greater than 10^{-5} to 10^{-4}). More sensitive methods, using compensators such as the Ehringhaus[9] or de Sénarmont,[19] are capable of measuring birefringences of the order of 10^{-7} and greater. Much birefringence research in flow fields of rheological interest deals with highly birefringent polymer melts and solutions, and hence we will develop the equations for the polariscope, which is the most commonly used arrangement.

A plane polariscope consists of the light source with two polarizers on either side of the test section and a camera; the polarizer nearest the camera is termed the analyzer. A dark-field polariscope has the polarization axis of the analyzer orthogonal to that of the polarizer while a light-field instrument has the axes parallel. Bands representing various values of retardation would result from either method, and are best analyzed by the Walker[9,21] matrix method. Walker developed matrices based on Stokes parameters, which specify the state of polarization due to a set of optical components that are mounted in series. The overall system performance is determined by matrix multiplication. The column matrix for the Stokes parameters when leaving a polarizer (or 'ingoing' to the polariscope), whose axis lies along the arbitrarily chosen x axis, is:

$$\mathbf{I}_i = E_x^2 \begin{vmatrix} 1 \\ 1 \\ 0 \\ 0 \end{vmatrix} \tag{14}$$

where E_x is the x component of the electric vector. The first component is of interest as it specifies the intensity. Using the nomenclature of Walker[21] the 'outgoing' column matrix for a plane polariscope is given by

$$\mathbf{I}_0 = P(\alpha)R(\beta, \delta)\mathbf{I}_i \tag{15}$$

where $R(\beta, \delta)$ represents a matrix for an optical component (in this case the test section) with a relative retardation δ at an angle β to the x axis, and $P(\alpha)$ is the matrix for a polarizer whose axis is at angle α to the x axis.

The relative transmittance ($T = I_{0,1}/I_{i,1}$; where 1 represents the first optical component in equation 14) is found to be

$$T = \tfrac{1}{2} + \tfrac{1}{2}\cos 2\alpha[1 - 2\sin^2 \beta \sin^2 \tfrac{1}{2}\delta] \tag{16}$$

For the dark-field case ($\alpha = \pi/2$), the result has already been given in equation (5), and the relative transmittance is zero when $\delta = 2p\pi$ and $\beta = \tfrac{1}{2}m\pi$, where p, m are integers (0, 1, 2 . . .) as described above.

The birefringence is determined by combining equations (4) and (7)

$$\Delta = p\frac{\lambda_l}{L} \quad (p=0, 1, 2 \ldots) \tag{17}$$

Bands of equal δ will be black in monochromatic light. They are coloured in white light, with the exception of the zero-order band, which is black. These are the isochromatics and are related to the magnitude of the principal refractive indices. Bands of equal β are always black and 'move' if the polarizer and analyzer are rotated in the crossed position. These bands are related to the direction of the principal refractive indices and are the isoclinics.

The properties of the birefringence and isoclinic angles make their determination simple. The polarizer and analyzer are rotated together using white light, and the black zero-order isochromatic can be identified since it appears stationary. The isochromatics are more distinct in monochromatic light and they are determined by 'counting out' from the zero-order isochromatic with the filter in place. The isoclinic angle is found by rotating the polarizer and analyzer to known angles and noting the dark band positions.

In a circular polariscope (*i.e.* one using circularly polarized light[2]), we have

$$\mathbf{I}_0 = P\left(\frac{1}{2}\pi\right) R\left(\frac{3}{4}\pi, \frac{1}{2}\pi\right) R(\beta, \delta) R\left(\frac{1}{4}\pi, \frac{1}{2}\pi\right) \mathbf{I}_i \tag{18}$$

where $R\left(\frac{1}{4}\pi, \frac{1}{2}\pi\right)$ and $R\left(\frac{3}{4}\pi, \frac{1}{2}\pi\right)$ represent quarter-wave plates whose 'fast axes' are at the given angles to the incident radiation. From this, we obtain

$$T = \sin^2 \tfrac{1}{2}\delta \tag{19}$$

Thus, a circular polariscope eliminates the isoclinic angle and the isochromatics are immediately determined.

The birefringence and isoclinic angles may be determined using the above methods. However, the rheologist usually desires the stresses. To do this, the stress–optic relation must be applied, as discussed in Section 20.2. The above discussion deals with steady flow situations. Transient flows (except those with very slow changes) need other methods and recently a number of new ideas have emerged.

In transient flows we usually need to determine the extinction angle χ and the birefringence Δ simultaneously, as they both generally contribute to the magnitude of the output, see equation (5). In elongational flows we only need Δ, but generally there will be a rotational component of the flow. The manual extraction of both quantities is exceedingly tedious. Gortemaker et al.[22] did produce some results manually, following the start of a shear flow, but multiple runs were needed to produce results. Osaki et al.,[23] by altering the incident polarization angle, needed only two runs per point. In order to eliminate errors in duplicate runs, Chow and Fuller[24] have used 'two-colour birefringence' in which two beams [one of 488 nm (blue) and one of 514.5 nm (green)] emitted from the same laser traverse the sample simultaneously. Only a single transient run was needed, and changes in Δ of 2×10^{-7} were detected. Use of phase modulation techniques has been made by Koeman and Janeschitz-Kriegl[25] and Δ values of about 10^{-8} have been resolved. Frattini and Fuller[26] have suggested a phase modulation technique suitable for transient flows. This group has also developed techniques for decoupling flow-induced dichroism and birefringence[27,28] found in suspensions of particles.

19.4 POLYMER PROPERTIES

The principal rheo-optical property of polymers is the stress–optical coefficient, C. Our previous remarks above on the limitations of situations in which C is truly constant should be borne in mind, and only in known situations should the values of C given below be used. Even then, the values given below may be up to 10–20% in error and direct measurement is preferred, if available. In Table 1 the values of C are given in Pa^{-1} ($m^2\ kg^{-1}$); the older cgs unit, the Brewster, is $10^{-13}\ cm\ s^2\ g^{-1}$. There are some variations with the source of the commercial materials, and details of the relevant molecular averages are given by Janeschitz-Kriegl;[9] the range of C for a given polymer is quite small as is the variation with temperature. The wavelength of the light used is an important variable. Note

Table 1 Some Values of the Stress–Optical Coefficient C^a

Polymer	$\bar{M}_w$ range	$T(°C)$	$10^9 C (Pa^{-1})$
Polystyrene	$0.5 \times 10^5 - 8.6 \times 10^5$	153–215	-4 to -5
Polyethylene (low density)	$10^5 - 5.4 \times 10^5$	150	$+2.1$
Polyethylene (high density)	1.6×10^5	150	$+2.4$
		190	$+1.8$
Polypropylene	$4.3 - 4.8 \times 10^5$	210	$+0.9$
Poly(vinyl chloride)	1.4×10^5	210	-0.5
Atactic polystyrene (1% by volume in bromoform)	9×10^6	25	-6.9
Polyisobutylene (Vistanex B-100, 3% by weight in decalin)		25	$+1.7$

^{a} All values for polymer melts are for light of wavelength $\lambda = 633$ nm; the values of λ for the solutions are not known. $\bar{M}_w$ is the weight-average molecular weight.

that C may be either positive or negative, and that the two solutions were prepared with solvents of matching refractive index.[1]

19.5 EXPERIMENTAL STUDIES IN VARIOUS GEOMETRIES

The two types of flow which have been most studied are shearing, a 'weak' flow,[8] and elongation, a 'strong' flow.[8] The former may be generalized to include Poiseuille and Couette flows.

19.5.1 Couette or Concentric Cylinder Flow

The concentric cylinder geometry was used in the first studies of flow birefringence.[4,5] The shear field is inhomogeneous within the gap,[16] and as narrow a gap as possible is needed to reduce this effect. A very small gap gives alignment difficulties and reduces the ability to view through the gap for birefringence measurements.

A check of the stress–optic relation in this geometry was carried out by Philippoff[1] for a polyisobutylene solution. He viewed along the cylinder axis while rotating the inner cylinder. The isoclinic angle and birefringence were measured using de Sénarmont compensation at each shear rate. Using a similar apparatus the shear stress was measured whilst the first normal stress difference was determined in cone and plant geometry.[1,8]

The interpretation of these data was done by noting that (see equations 12 and 13)

$$\sigma_{12} = \frac{1}{2}(\sigma_1 - \sigma_2)\sin 2\chi_m^{(1)} \tag{20}$$

or using the stress–optic relation

$$\Delta \sin 2\chi_0^{(1)} = 2C\sigma_{12} \tag{21}$$

Peterlin[3] plotted his optic and shearing stress data from the two apparatuses according to equation (21) and obtained a linear relation indicating a constant stress–optic coefficient. He also used the normal stress data to calculate $\chi_M^{(1)}$ and compared this to the measured $\chi_0^{(1)}$, and found the two equal at equivalent shear rates. Therefore, the stress–optic relation was confirmed, at least for this system and level of principal stresses ($\sigma_1 - \sigma_2 \approx 2 \times 10^3$ Pa).

Besides being used as a calibration device, concentric cylinders can be used to find the first normal stress difference ($\sigma_{\theta\theta} - \sigma_{rr}$) and to investigate orientation effects in shearing and oscillatory flow. The measurements provide shear orientation information for both the rigid and flexible types of microstructure.

19.5.2 Cone and Plate Flow

To overcome the problem of shear field inhomogeneity a small angle (truncated) cone and plate geometry has been used by Janeschitz-Kriegl and co-workers.[9] The stress–optic relation was found to be true for polymer melts in both transient and steady shear conditions up to shear stresses of the order of 10^4–10^5 Pa (some deviation occurred in the transient measurement and was attributed to mechanical delay of the apparatus).

The design of the apparatus is very involved, certainly more so than for the concentric cylinders. There were also difficulties with 'parasitic' birefringences in the apparatus.[9]

19.5.3 Capillary and Slit Flow

For pressure-driven or Poiseuille flow in these geometries the shear field is inhomogeneous. The shear rate is zero at the symmetry axis and increases to a maximum at the wall for the capillary, whilst the (theoretically infinitely) wide slit has a zero shear rate along its symmetry plane and a maximum at the wall. It is a straightforward process to verify that, for fully developed flow,[8] the shear stress varies linearly across the section, being zero on the centre line. By measuring the pressure drop, the shear stress field is easily calculated.

Although the shear stress is linear with shear direction for both geometries, there is a fundamental difference between the two regarding birefringence measurements. One can view parallel to the z or r direction in the capillary. When viewed along the z axis one obtains information concerning $n_2 - n_3$. This geometry has produced the best estimates of the second normal stress difference N_2 for various polymers.[8,29] However, the interpretation of the data in the r axis experiment is difficult, as the stresses vary in the r direction. The wide slit overcomes this problem, as the stresses change only in the y direction and can be viewed along the z direction (there are some wall effects even here). Thus, one can determine $n_1 - n_2$ by placing the slit in the polariscope with the neutral axis parallel to the wave-normal direction. Wales[9] has recommended a slit width-to-depth ratio of at least 10:1 to minimize the side wall influences. His recommendation comes from birefringence measurements viewed along the y axis (*i.e.* $n_1 - n_2$). Mackay[15] notes a number of difficulties in interpreting fringe patterns in this geometry.

The slit is a useful characterization device, as C found by Mackay[15] agrees within 5–10% of the values given by Wales[9] obtained in a truncated cone and plate for similar melts.

19.5.4 Elongational Flow

Elongational flows have been produced using several apparatuses — fibre spinning, tensile experiments[9] and the four-roll mill.[13] What is immediately obvious is that the stress–optic relation appears to be invalid in a number of instances,[11,13] but there are some cases where the simple stress–optic relation holds true in elongational flow; these are mainly when stresses are small enough.

19.5.5 Further Applications

Finally, we mention some other applications of rheo-optics. It has been used to study liquid crystals,[30] has utilized IR spectroscopy,[31] and heat flow birefringence has been observed.[32] Three-dimensional flow birefringence is discussed by Funatsu *et al.*[33] Multicomponent–multiphase materials have been studied.[34] There has also been considerable theoretical activity.[35]

19.6 REFERENCES

1. R. S. Stein (ed.), *J. Polym. Sci., Part C*, 1964, **5**.
2. E. Hecht and A. Zajac, 'Optics', Addison–Wesley, Reading, MA, 1979.
3. A. Peterlin, in 'Rheology', ed. F. R. Eirich, Academic Press, New York, 1956, vol. 1, p. 643.
4. J. C. Maxwell, *Proc. R. Soc. London*, 1873, **22**, 46.
5. E. Mach, 'Optische-Akustische Versuche', Calve, Prague, 1873.
6. L. D. Landau and E. M. Lifschitz, 'Electrodynamics of Continuous Media', Addition–Wesley, Reading, MA, 1960.
7. E. H. Dill, in 'Proceedings of the 4th International Congress on Rheology, Providence, R. I., 1963', ed. E. H. Lee, Interscience, New York, 1965, part 2, p. 51.

8. R. I. Tanner, 'Engineering Rheology', Oxford University Press, Oxford, 1985.
9. H. Janeschitz-Kriegl, 'Polymer Melt Rheology and Flow Birefringence', Springer, Berlin, 1983.
10. L. G. Leal and E. J. Hinch, *Rheol. Acta*, 1972, **11**, 190.
11. B. E. Read, *Polymer*, 1964, **5**, 1.
12. W. Retting, *Colloid Polym. Sci.*, 1979, **257**, 689.
13. L. G. Leal, in 'Proceedings of the 9th International Congress on Rheology, Mexico, 1984', ed. B. Mena, A. Garcia-Rejon and C. Rangel-Nafaile, Elsevier, Mexico, 1984, vol. 1, p. 191.
14. A. Onuki and M. Doi, *J. Chem. Phys.*, 1986, **85**, 1190.
15. M. E. Mackay, Ph.D Thesis, University of Illinois, 1985.
16. R. E. Harrington, in 'Encyclopedia of Polymer Science and Technology', ed. H. F. Mark, Wiley, New York, 1967, vol. 7, p. 100.
17. B. Erman and P. J. Flory, *Macromolecules*, 1983, **16**, 1601, 1607.
18. H. Wayland, *J. Polym. Sci., Part C*, 1964, **5**, 11.
19. J. T. Edsall, A. Rich and M. Goldstein, *Rev. Sci. Instrum.*, 1952, **23**, 846.
20. O. Snellman and Y. Björnstahl, *Kolloid-Beih.*, 1941, **52**, 403.
21. M. J. Walker, *Am. J. Phys.*, 1954, **22**, 170.
22. F. H. Gortemaker, M. B. Hansen, B. de Cindio, H. M. Laun and H. Janeschitz–Kriegl, *Rheol. Acta*, 1976, **15**, 242, 256.
23. K. Osaki, N. Bessho, T. Kojimoto and M. Kurata, *J. Rheol (N.Y.)*, 1979, **23**, 457.
24. A. W. Chow and G. G. Fuller, *J. Rheol. (N.Y.)*, 1984, **28**, 23.
25. B. Koeman and H. Janeschitz-Kriegl, *J. Phys. E*, 1979, **12**, 625.
26. P. L. Frattini and G. G. Fuller, *J. Rheol. (N.Y.)*, 1984, **28**, 61.
27. P. L. Frattini and G. G. Fuller, *J. Fluid Mech.*, 1986, **168**, 119.
28. S. J. Johnson and G. G. Fuller, *Rheol. Acta*, 1986, **25**, 405.
29. J. L. S. Wales and W. Philippoff, *Rheol. Acta*, 1973, **12**, 25.
30. S. Onogi, in 'Rheology: Proceedings of the 8th International Congress on Rheology, Naples, 1980', ed. G. Astarita, G. Marruci and L. Nicolais, Plenum Press, New York, 1980, vol. 1, p. 127.
31. R. S. Stein, *Polym. J.*, 1985, **17**, 289.
32. D. R. Baalss and S. Hess, *Z. Naturforsch., Teil A*, 1985, **40**, 3.
33. K. Funatsu, Y. Ogata, H. Hayashida and H. Shinokara, *Kagaku Kogaku Ronbunshu*, 1976, **2**, 485.
34. G. L. Wilkes and R. S. Stein, *J. Polym. Sci., Polym. Symp.*, 1977, **60**, 121.
35. E. Kroener and R. Takserman-Krozer, 'in Physics of Optical Dynamic Phenomenological Processes in Macromolecular Systems, Proceedings of Microsymposium on Macromolecules, 27th', ed. B. Sedlacek, de Gruyter, Berlin, 1984, p. 131.

20

Permeation Properties

TIM deV. NAYLOR

The British Petroleum Company plc, Sunbury-on-Thames, UK

20.1 INTRODUCTION

The wide range of polymer permeation properties were first observed and modelled over a century ago.[1,2] However, little further practical or theoretical work was undertaken until after the turn of the century when the need for improved pneumatic tyres developed. This led to the replacement of natural rubber with butyl rubber due to its lower permeability.[3] Subsequently, the need for materials with ever-improved permeation properties has grown unabated. For instance, polymers which are relatively impermeable are used in food packaging and encapsulation of electronic circuits, whilst polymers which are very selectively permeable are also used in food packaging as well as in gas and liquid separation processes. Applications also exist for polymers which can release drugs or pesticides at a controlled rate.

None of these developments could have taken place without an enhanced understanding of the factors which influence the permeation properties of the polymers involved. Very significant advances have been made in the last 30 years, with the rate of advancement increasing dramatically in the last 15 years.

The aims of this chapter are to outline the basic principles of the process of permeation through polymers and to demonstrate how the permeation properties can vary with both the nature of the polymer and those of the permeant. Indications are given of those areas where current research interest lies, along with brief outlines of the main areas of current commercial applications.

20.2 EXPERIMENTAL TECHNIQUES

The process of permeation through non-porous polymers is generally explained in terms of the solution diffusion model. This model postulates that the permeation of a gas through a polymer film occurs in three stages: (1) sorption of the gas on to the polymer, (2) diffusion through the polymer and (3) desorption from the opposite face. Thus it can be seen that the permeability P is given by a combination of the diffusivity of the gas D dissolved in the polymer and its concentration gradient, which in turn is proportional to the gas solubility S in the polymer.

A great variety of techniques have been developed to measure P, D and S. For example, the vacuum microbalance is commonly used, as well as volumetric, optical, radiotracer weighing cup and NMR imaging techniques. Many excellent reviews exist.[4-7]

20.2.1 Steady-state Diffusion

The simplest technique for determining the diffusion coefficient is based on Fick's first law of diffusion. Consider a membrane of thickness x, whose surfaces at $x = 0$ and $x = 1$ are maintained at constant concentrations C_1 and C_2 respectively. At the beginning of the experiment, both the rate of flow and the concentration of permeant at any point in the membrane vary with time. Once equilibrium is established, however, the concentration changes linearly through the membrane from C_1 to C_2. From Fick's law,[5] the flux J of material diffusing through unit area of membrane with time is given by

$$J = D\left(\frac{C_1 - C_2}{x}\right) \tag{1}$$

Generally, C_2, is kept at zero by use of a high vacuum and so, provided the membrane thickness and the surface concentration C_1 are known, the diffusion coefficient or diffusivity D can be obtained from a single measurement of the flux J. It should be noted that D in this instance is dependent upon the sorbed concentration C which itself varies through the polymer. This technique is particularly useful for studying the diffusion of gases in membranes where the activity and concentration in the membrane surface is controlled by the pressure of gas contacting the membrane surface.

In many practical systems, however, the surface concentration is not always known with accuracy and it is conventional to express the flux per unit area, J, in terms of the concentrations of the external phases C^{ext} (or, in the case of gas diffusion, pressure) by

$$J = P\left(\frac{C_1^{\text{ext}} - C_2^{\text{ext}}}{x}\right) \tag{2}$$

where the factor P is termed the permeability, permeability coefficient or permeability constant.

If there is a linear relationship between the concentration of the external phase and the corresponding surface concentration, *i.e.* if Henry's law is obeyed, then

$$C = SC^{\text{ext}} \tag{3}$$

where S is the solubility coefficient, and it follows from equations (1), (2) and (3) that

$$P = DS \tag{4}$$

20.2.2 Time-lag Method

Determination of both the diffusion and solubility coefficients, and hence the permeability coefficient, without prior knowledge of the surface concentration is possible by use of the time-lag experiment.

Measurement of the rate of permeation as a function of time as the system approaches equilibrium enables Fick's second law of diffusion to be solved for the boundary conditions of the experiment. Provided that the diffusion coefficient is independent of the concentration of diffusing species in the membrane, and as $t \to \infty$, the result is[8]

$$Q = \frac{DC_1}{x}\left(t - \frac{x^2}{6D}\right) \tag{5}$$

This linear relationship may be extrapolated to cross the Q axis at time t', as shown in Figure 1. This time lag(t') is given by

$$t' = \frac{x^2}{6D} \tag{6}$$

The time lag is independent of the solubility coefficient and so equation (6) enables D to be obtained directly from permeation experiments. At the steady state, C_1 and hence S may be determined using equation (1).

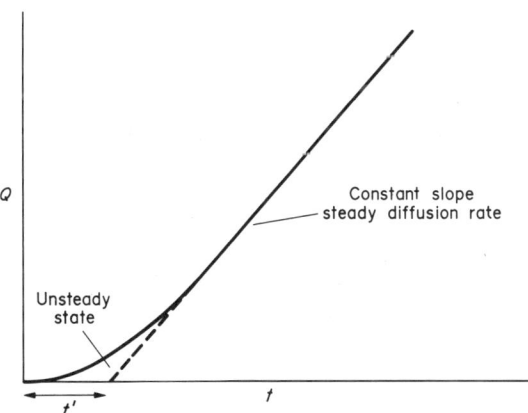

Figure 1 Measurement of time lag t' by extrapolation of plot of amount of material permeating Q *vs.* time t

Time-lag studies are best suited for studying rubbery polymers and tend to be unreliable when applied to glassy polymers or solutions due to the boundary layer effects.[9] More advanced methods have been developed which allow D and S to be determined from step changes in permeant concentration[10,11] and for mixed diffusants.[12]

20.2.3 Concentration and Time Dependence

In many instances where swelling of the sample occurs, such as with organic vapours and the larger gas molecules, both D and S become concentration and time dependent. These effects cannot be disentangled by combining time-lag and steady-state data. Instead it is necessary to study the sorption behaviour as a function of time. Typical experimental procedures utilize either thin films of polymer coated directly on to or suspended from quartz springs[13,14] or 10–100 μm diameter microspheres of polymer[15] where small ($D < 10^{-16}$ m^2 s^{-1}) values of D are expected.

The polymer sample is suspended from a balance under a uniform concentration of vapour at a known temperature and pressure. By studying the equilibrium weight gain or loss at different vapour concentrations, a sorption isotherm may be determined. A typical isotherm observed for Fickian diffusion is shown in Figure 2.

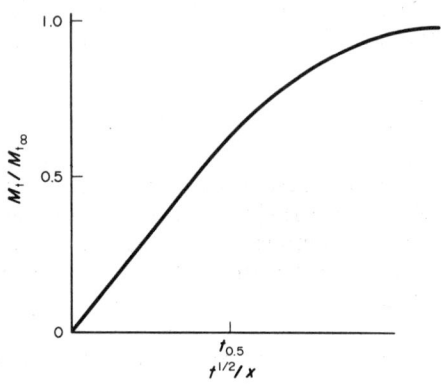

Figure 2 A typical isotherm observed for Fickian diffusion

Very generally, this relationship[16] is expressed by

$$\frac{M_t}{M_{t_\infty}} = kt^n \tag{7}$$

where M_t is the amount of vapour sorbed, M_{t_∞} is the equilibrium quantity sorbed, k is a constant and n can vary depending on the mode of diffusion. In this instance of Fickian diffusion, $n = 1/2$ and it is possible to determine D from $t_{0.5}$ when the amount of vapour taken up is half the final amount taken up at equilibrium,[17] *i.e.*

$$D = \frac{0.04939 x^2}{t_{0.5}} \tag{8}$$

for a film of thickness x.

A significant difference between the sorption of organic vapours and that of permanent gases in polymer membranes is that vapour sorption leads to much larger concentrations of permeant in the membrane. Errors will arise if the resulting increase in membrane thickness is not taken into account when calculating the diffusion coefficient.[18]

The concentration dependence of D can be determined in several ways. For instance, by performing vapour sorption experiments at several different vapour pressures, a range of equilibrium permeant concentrations C_{eq} can be investigated and so the D *vs.* C_{eq} relationship can be obtained, as well as equilibrium solubility data.[19] In the case of polymeric glasses, the previous swelling history of the sample can be very important as it can influence the subsequent results due to hysteresis effects.[20] An average value of the diffusion coefficient $\bar{D}$, may be approximated by the relationship

$$\bar{D} = \frac{1}{C_{eq}} \int_0^{C_{eq}} D(C)\,dC \tag{9}$$

SI units are used throughout this chapter. Thus, for P (the permeation coefficient) the units are mol m^{-1} s^{-1} Pa^{-1}, for D (the diffusion coefficient) the units are m^2 s^{-1}, whilst for S (the sorption coefficient) the units are mol m^{-3} Pa^{-1}. Tabulations of values for P, D and S for some typical polymers exist.[21] Many other units have been used and tables of conversion factors are available.[22,23]

20.2.4 Water in Polymers

Measurement of water-vapour diffusion, solubility and permeability requires special techniques and care because of the unique nature of water. A number of techniques have been developed for the routine measurement of the water-vapour permeability of polymer membranes.[24] More accurate measurements may be obtained by using the techniques described for gases; however, care must be taken to avoid the possibility of water sorbing or desorbing on to or from glass surfaces. With due caution, steady-state permeation rates may be measured in this fashion, but permeation lag-times are unreliable.[25]

Vapour sorption techniques are sometimes better at low water-vapour pressures for hydrophilic polymers, or for all hydrophobic polymers where in water sorption levels tend to be low. However, when large amounts of water are sorbed or desorbed significant temperature changes can occur.

20.3 SORPTION

The term 'sorption' is generally used to describe the initial penetration and dispersal of permeant molecules into the polymer matrix. The term includes adsorption, absorption, incorporation into microvoids and cluster formation. The permeant may undergo several modes of sorption simultaneously in the same polymer. In addition, the distribution of permeant between the different sorption modes may change with concentration, temperature and swelling of the matrix as well as with time.

The extent to which permeant molecules are sorbed and their mode of sorption in a polymer depend upon the enthalpy and entropy of permeant/polymer mixing, *i.e.* upon the activity of the permeant within the polymer at equilibrium. Much information on polymer-solution thermodynamics is given elsewhere in this volume. Here, a brief description is given of the most common types of sorption isotherms which have been utilized to describe permeation in polymers.

Sorption behaviour has been classified on the basis of the relative strengths of the interactions between the permeant molecules and the polymer or between the permeant molecules themselves within the polymer.

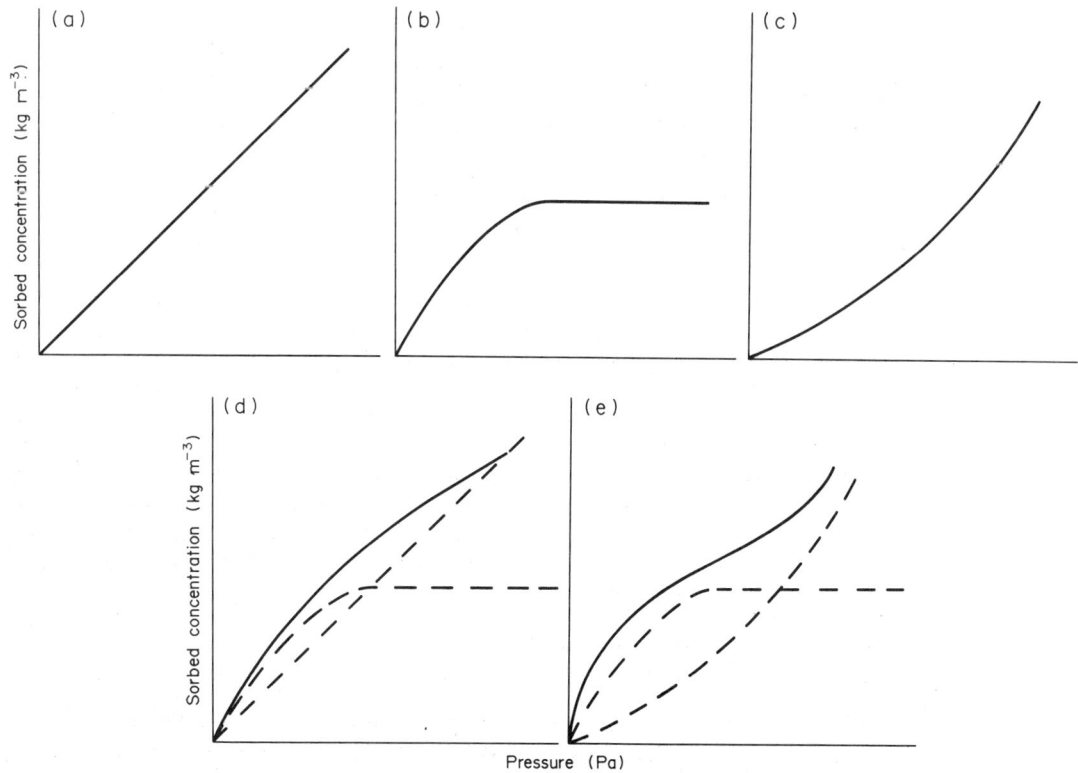

Figure 3 Types of sorption behaviour: (a) type I, Henry's law; (b) type II, Langmuir; (c) type III, Flory–Huggins; (d) dual mode; (e) type IV, BET

Three major types of sorption behaviour have been identified and are shown together in Figure 3 where sorbed concentration is plotted against applied permeant pressure. Two further types of sorption isotherm, which may be considered to be combinations of the above isotherms, are also shown.

20.3.1 Henry's Law Sorption

The simplest type of sorption, shown in Figure 3 as type I, arises when both polymer/permeant and permeant/permeant interactions are weak relative to polymer/polymer interactions, *i.e.* when ideally dilute solution behaviour occurs and Henry's law is obeyed. The solubility coefficient S is a constant independent of sorbed concentration at a given temperature. Consequently, the sorption isotherm shows a linear dependence of concentration *vs.* pressure p (or vapour activity)

$$C = Sp \tag{10}$$

Typically, this type of behaviour is observed when permanent gases are sorbed by rubbery polymers at low ($< 10^5$ Pa) pressure and arises from the very low solubility ($< 0.2\%$) of the permanent gases in the polymer.

20.3.2 Langmuir-type Sorption

Type II behaviour, often known as Langmuir sorption, indicates a tendancy for significant levels of sorption to occur at relatively low pressures. Physically, this represents initial sorption in some kind of specific site or immobilization of permeant molecules in microvoids in the polymer. When nearly all of the sites are occupied, a very small amount of permeant randomly dissolves in the polymer. The concentration C_H of permeant sorbed by these holes or sites is represented by the Langmuir equation

$$C_H = \frac{C'_H bp}{1 + bp} \tag{11}$$

where C'_H is a hole or site saturation constant and b is a hole or site 'affinity' constant, representing the ratio of rate constants for adsorption and desorption. Typically, this type of behaviour occurs when dyes are absorbed by ionic polymers or polymers containing polar groups. Other examples include the sorption of certain gases or vapours by polymers containing dispersed porous particles of high surface area, *e.g.* inorganic fillers such as carbon black or silica gel.

20.3.3 Dual-mode Sorption

A combination of Henry's law sorption with that of the Langmuir type has been used to explain the sorption isotherm commonly observed for the sorption of gases in glassy polymers. The dual-mode theory views sorption occurring in glassy polymers in two distinct modes.[26]

Originally, it was though that some of the sorbed gas molecules are present as a mobile population dissolved in the bulk of the polymer according to Henry's law, and are free to diffuse down a concentration gradient, whilst the remaining gas molecules are bound at a fixed number of adsorption sites or holes occurring within the polymer. It was assumed that the bound and mobile molecules are in equilibrium. The total concentration C_{eq} of sorbed molecules at sufficiently low concentrations (where Henry's law is obeyed) is then obtained from equations (10) and (11)

$$C_{eq} = Sp + \frac{C'_H bp}{1 + bp} \tag{12}$$

Therefore, as the pressure increases, the total sorbed concentration tends towards the limit of Henry's law. Consequently, the mean permeability P is given by the product of the Henry's law constant S and the diffusivity D_D of the mobile fraction of the sorbed molecules

$$P = SD_D \tag{13}$$

However, this theory was subsequently modified to allow for some limited mobility of the 'immobilized' permeant molecules,[27] when the permeability is given by

$$P = SD_D\left(1 + \frac{KR'}{1 + bp}\right) \qquad (14)$$

where $K = C'_H b/S$ and is a ratio characteristic of the relative amounts of gas sorbed in the two modes, and R' is a measure of the degree of immobilization. For total immobilization, $R' = 0$ and equation (14) reduces to equation (13), whereas, for complete mobility, $R' = 1$ and the permeability becomes very pressure-dependent.

The dual-mode sorption model has proved to be very useful in representing the permeation of such gases as CO_2 and C_1–C_5 hydrocarbon vapours through a number of glassy polymers,[26, 28] as well as permeation of vapours through rubbery polymers containing adsorptive fillers.[29] For the sake of simplicity, it is assumed that the diffusion coefficient D_D is independent of the permeant concentration. However, when high permeant concentrations arise, this assumption is no longer valid.[30]

An alternative model known as the 'gas polymer matrix theory' has been proposed, which assumes that there is a continuous variation in the state of the dissolved gas molecules rather than the two distinct states as proposed in the dual-mode theory.[31]

20.3.4 Flory–Huggins Sorption

This type of sorption isotherm arises when the interactions between permeant molecules are strong relative to permeant/polymer interactions, and the solubility coefficient increases continuously with increasing pressure. This sorption can be described for non-polar systems by the Flory–Huggins relationship

$$\ln\frac{P}{P^\circ} = \ln\phi_1 + (1 - \phi_1) + \chi(1 - \phi_1)^2 \qquad (15)$$

where P° is the saturation vapour pressure, ϕ_1 is the volume fraction of permeant in the polymer and χ is the Flory–Huggins interaction parameter. In the simplest formulations of the Flory–Huggins theory, χ is a measure of the heat of mixing of the sorbed molecules with the polymer and, for non-polar systems, can be estimated[32] from solubility parameter theory by

$$\chi = \frac{V_1}{RT}(\delta_1 - \delta_2)^2 \qquad (16)$$

where V_1 is the molar volume of the permeant and δ_1 and δ_2 are the solubility parameters of the permeant and polymer respectively. When $\delta_1 = \delta_2$ the permeant has maximum solubility in the polymer. Otherwise, χ is a positive quantity and, consequently, the extent of sorption decreases as χ becomes larger.

A second system in which permeant/permeant interactions are inherently stronger than polymer/permeant interactions is that of water in hydrophobic polymers.[33] Water is strongly hydrogen-bonded in the liquid state and consequently this leads to association or clustering of water molecules sorbed into the polymer. Once formed, these clusters of water, which may be quite stable, will probably be relatively less mobile than individual molecules. Consequently, if the proportion of clusters increases with increasing sorbed concentration C, as implied by this type of isotherm, then the diffusion coefficient D will decrease with increasing C.[25] This behaviour contrasts sharply with that described above for highly swollen non-polar systems where D increases with C.

20.3.5 Type IV Sorption

It can be seen from Figure 3 that the type IV or Brunnauer, Emmett and Teller (BET) sorption isotherm can be considered to be the sum of type II and type III sorption. It can be regarded as a variant of the BET sorption isotherm. Such behaviour is typical of the sorption of water by highly hydrophilic polymers, *e.g.* poly(vinyl alcohol), poly(acrylic acid) or cellulosic materials.[34] Strong sorption of water molecules on to specific sites, *i.e.* polar groups, occurs initially, whilst at higher pressures either dissolution or clustering phenomena predominate.

20.3.6 Zimm and Lundberg Cluster Theory

The phenomenon of association or clustering of molecules has already been mentioned in connection with water sorption. This idea of clustering has been developed to enable sorption characteristics in general to be interpreted in terms of the degree of clustering of the sorbed molecules.[35, 36] A function has been defined, in terms of the partial volume fraction, the partial molar volume and the thermodynamic activity of the permeant, which gives an indication of the overall tendency for clustering to occur due to the formation of permeant/permeant pairs.

Whilst the cluster theory gives little information regarding the mechanism of sorption and cannot be used to predict sorption isotherms, sorption data can be interpreted for a given system in terms of mean molecular clustering tendencies. For instance, values of the clustering-tendency function determined from water-vapour-sorption isotherms for the hydrophilic polymer collagen, lead to the interpretation that hydrogen bonding occurs between the water and the polar groups on the polymer at low water volume-fractions and that water/water clustering only begins when most of these sorption sites have been covered.

20.3.7 Factors Affecting Sorption

20.3.7.1 Temperature

As the solubility coefficient S is the equilibrium constant relating the distribution of the permeant molecules between the fluid phase and the polymer phase, *i.e.*

$$S = C/p \tag{17}$$

its temperature dependence over small temperature ranges follows an Arrhenius-type relationship

$$S = S_0 \exp(-\Delta H_s/RT) \tag{18}$$

where ΔH_s, the partial molar heat of solution, may be expressed as

$$\Delta H_s = \Delta H_{cond} + \overline{\Delta H}_{mix} \tag{19}$$

where ΔH_{cond} is the molar heat of condensation of the permeant and $\overline{\Delta H}_{mix}$ is the partial molar heat of mixing.[37] ΔH_s is often negative and generally falls in the range 0 ± 10 kJ mol^{-1}. A value of $\overline{\Delta H}_{mix}$ can be estimated from the solubility parameters of the permeant and polymer from the Hildebrand equation[38]

$$\overline{\Delta H}_{mix} \approx E = V_1(\delta_1 - \delta_2)^2 \phi_2^2 \tag{20}$$

where V_1 is the partial molar volume of the permeant and ϕ_2 is the volume fraction of the polymer in the mixture.

For gases such as H_2, N_2 and O_2, which are well above their critical temperature at room temperature, ΔH_{cond} is very small and so ΔH_s is governed by $\overline{\Delta H}_{mix}$. Because interactions between these gases and polymers tend to be weak ΔH_{mix} is positive and hence solubility increases with temperature. In contrast, for the more condensable gases such as CO_2, SO_2, NH_3 and hydrocarbons, ΔH_s is negative due to the large contribution from ΔH_{cond} and hence solubility decreases with increasing temperature. The solubility of liquid permeants in polymers generally increases with temperature, but may decrease depending upon the nature of the polymer/permeant interactions.

20.3.7.2 Permeant size

Because the interactions of simple gas molecules with a polymer matrix are weak, the main factor controlling sorption is the ease of condensation of the gas. Numerous linear relations have been proposed between $\ln S$ and either the critical temperature of the gas T_c or the boiling point T_b or the

Lennard–Jones 6–12 potential force constant, ξ/k.[3,37] Generally, all these parameters increase with molecular size and (provided there are no specific gas/polymer interactions occurring) solubilities tend to increase with permeant size. For instance, a doubling of the molecular volume in a series of aromatic hydrocarbons leads to an increase in the solubility by a factor of about 200.[39]

20.3.8 Sorption Kinetics

The transport of small permeants into glassy polymers has been classified in terms of the relative rates of mobility of the permeant and polymer segments.[40] Three main types of behaviour have been identified and may be distinguished by the shape of their sorption–time curves which may be represented by the general relation given earlier (equation 7).

$$\frac{M_t}{M_{t_\infty}} = kt^n \tag{21}$$

According to this classification, three basic modes of transport may be distinguished; these are described in the following sections.

20.3.8.1 Case I sorption (n = 1/2)

This arises in polymer/permeant systems which obey Fickian diffusion (described later in Section 20.4). The rate of diffusion of permeant molecules is much less than the polymer segment mobility. Under these conditions, the rate at which sorption equilibrium is approached is independent of swelling kinetics. As shown in Figure 4, the sorption curve (M_t/M_{t_∞} vs. $t^{1/2}$) is linear in the initial stage. Above the linear section, both absorption and desorption curves coincide over the entire range of t.

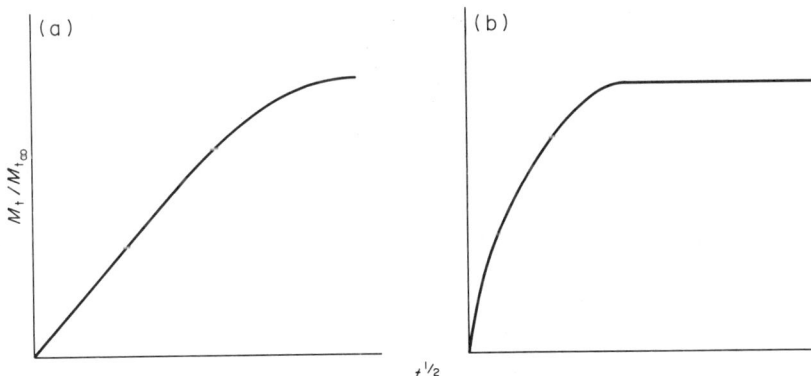

Figure 4 Modes of transport: (a) case I sorption, (b) case II sorption

20.3.8.2 Case II sorption (n = 1)

This may be regarded as the other extreme in which permeant diffusion rates are much faster than polymer relaxation processes. Such a deviation from Fickian diffusion (see Figure 4) typically occurs when organic vapours diffuse into glassy polymers. Consequently, sorption behaviour is strongly dependent on swelling kinetics and sharp step-like permeant boundaries develop which move into the polymer at a constant velocity. For sheet specimens of sufficiently large aspect ratio, the weight gain is linear with time. Typical examples of the occurrence of case II sorption include methanol/poly(methyl methacrylate), acetone/cellulose acetate and acetone/polystyrene.[41,42,43] A system displaying case II sorption behaviour will tend towards case I behaviour with increasing temperature.

Two other features which have noted for case II sorption are the existence of an induction period before the development of the sharp permeant boundary and the acceleration of permeant boundaries approaching one another from opposite directions as they meet. This latter behaviour, which has been observed in systems such as *n*-hexane/polystyrene and *n*-hexane/poly(phenylene oxide),[44,45] has been termed 'super case II diffusion'.

Two theories[46,47] have been developed to explain case II behaviour. One[47] views the permeation of the sharp permeant front as follows. The swelling which occurs in the zone ahead of the front plasticizes the polymer, thus allowing more permeant molecules to enter more rapidly. This 'autocatalytic' response, which arises from the very strongly concentration-dependent viscosity of the polymer, continues until the permeant concentration in the space element reaches an equilibrium value. Consequently, most of the build-up of permeant in an element occurs in a length of time very much less than that necessary for filling the element by another polymer segment under normal viscous flow.

20.3.8.3   *Anomalous or non-Fickian sorption (1/2 < n < 1)*

This occurs when the permeant mobility and polymer segment relaxation rates are similar. A number of models have been proposed to explain the occurrence of these anomalous sorption modes. However none provide a comprehensive explanation and they only describe certain features of a particular type of behaviour.[48,49,50]

20.4   DIFFUSION

The mathematical theory of diffusion in isotropic substances is based on the hypothesis that the rate of transfer of a diffusing substance through the unit area of a section is proportional to the concentration gradient measured normal to the section. This hypothesis is expressed by the empirical relation known as Fick's first law

$$J = -D\frac{\partial C}{\partial x} \tag{22}$$

where J is the flux, or rate of transfer per unit area of section, C is the concentration of diffusing substance, x is the section thickness and D is called the diffusion coefficient or diffusivity. For the unsteady-state condition where the concentration gradient of permeant across the membrane varies with time, the rate of change of permeant concentration at any point (assuming unidirectional flow and that D is a constant independent of x, t and C) is given[51] by Fick's second law.

$$\frac{\partial C}{\partial t} = D\frac{\partial^2 C}{\partial x^2} \tag{23}$$

Depending on the boundary conditions, there are many solutions[18] to equations (22) and (23).

20.4.1   Factors Affecting Diffusivity

For simple gases, where interactions with polymers are weak, the diffusivity D is independent of permeant gas concentration. However, in instances where the permeant, *e.g.* an organic vapour, interacts strongly with the polymer, D becomes dependent on permeant concentration and on other factors such as permeant size and shape, time and temperature.

20.4.1.1   *Concentration*

The concentration dependence of the diffusivity D arises from the presence of permeant molecules within the polymer which weakens the interactions between adjacent polymer chains, which in turn leads to the commonly observed effects of plasticization. Hence, the rate of diffusion increases. Consequently, equation (23) becomes

$$\frac{\partial C}{\partial t} = \frac{\partial D(C)}{\partial x}\frac{\partial C}{\partial x} \tag{24}$$

Unfortunately, there are few rigorous solutions of the diffusion equation for a concentration-dependent diffusivity. For systems in which Henry's law is obeyed and the concentration of sorbed gas is small, as occurs with permeant gases, the dependence of D on concentration has been found to follow the empirical relationship[52]

$$\bar{D} = D(C) = D_0 \exp(\gamma C) \tag{25}$$

where $D(C) = \bar{D}$ is the average diffusion coefficient measured over a small equilibrium concentration range, γ is a plasticizing parameter that can be related to the Flory–Huggins interaction parameter and D_0 is D in the limit as $C \to 0$.

For systems where permeant/polymer interactions are stronger and Henry's law no longer applies, the dependence of the diffusivity on concentration becomes more complex. Empirically, this dependence may be represented by

$$D(C) = D_0 \exp(AB) \tag{26}$$

where A is a temperature-dependent parameter and B may be permeant activity, volume fraction or concentration.

The importance of the concentration dependence of D is shown by the magnitude of the change in D for even small increases in permeant concentration. The exponential dependence found for the benzene/poly(vinyl acetate) system is a consequence of an almost thousandfold increase in D for an increase in permeant volume fraction from 0 to 0.1. However, in the benzene/poly(dimethylsiloxane) system, D is virtually independent of concentration[52] whilst for the benzene/natural rubber system there is only about a threefold increase over the same concentration range.[53]

Clearly the variation of permeant diffusivity with concentration is very dependent on the polymer system studied. However, very generally, it appears that this variation is greater for glassy polymers than for rubbery polymers.

20.4.1.2 *Permeant size and shape*

In glassy polymers, D_0 has been shown to decrease enormously (by up to ten orders of magnitude) with increasing permeant size expressed as van der Waals volume[15] as shown in Figure 5 for poly(vinyl chloride) (PVC). Variations in permeant shape can also have a noticeable effect on D. Flattened or elongated molecules diffuse up to 1000 times faster than spherical molecules of

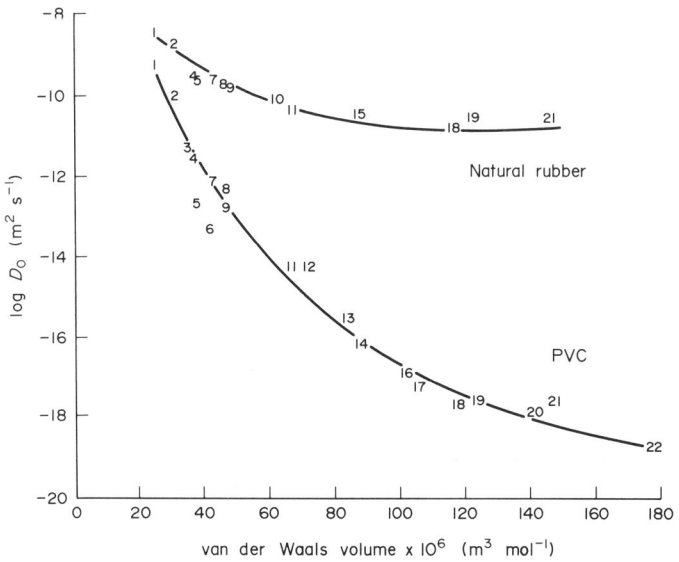

Figure 5 Decrease in D_0 with increasing permeant size for poly(vinyl chloride) (PVC) and natural rubber. Permeants: 1, helium; 2, hydrogen; 3, water; 4, oxygen; 5, argon; 6, krypton; 7, nitrogen; 8, carbon dioxide; 9, methane; 10, ethylene; 11, ethane; 12, methanol; 13, vinyl chloride; 14, ethanol; 15, propane; 16, acetone; 17, *n*-propanol; 18, benzene; 19, *n*-butane; 20, *n*-butanol; 21, *n*-pentane; 22, *n*-hexane (reproduced by permission of W. J. Koros and D. R. Paul, presented at the Symposium on Synthetic Membranes, Midland, MI, 1984)

equivalent molecular volume in glassy polymers.[3, 15, 54] This arises from the increased disturbance to the polymer chains necessary to allow diffusion of the spherical molecules, implying that the anisotropic molecules diffuse along their narrow axes.

In contrast, the effect of permeant size is much less marked in rubbery polymers, as shown in Figure 5 for natural rubber.

20.4.1.3 *Temperature*

Diffusion is an activated process and so, as for reaction rates, the activation energy E_D may be determined from the Arrhenius relationship over small temperature ranges.

$$D = D_0 \exp(-E_D/RT) \tag{27}$$

E_D consists of two terms, the apparent activation energy for diffusion $E_D(C \to 0)$, which is the activation energy of diffusion under conditions in which polymer chain mobility is unaffected by the presence of permeant, and a term which accounts for the reduction in activation energy due to the sorbed permeant,[55] *i.e.*

$$E_D = E_{D(C \to 0)} - yRT \tag{28}$$

For instance, Figure 6 shows how increasing the permeant volume fraction, ϕ, of dichloromethane in poly(vinyl formate) reduces the activation energy for diffusion.

The temperature dependence of D_0 is also affected by permeant size. For instance, Figure 7 shows the increase in E_D with increasing size of permeant in poly(vinyl acetate) at 40 °C. Whilst E_D initially increases with permeant size it tends to level off to a constant value for large permeants. This value

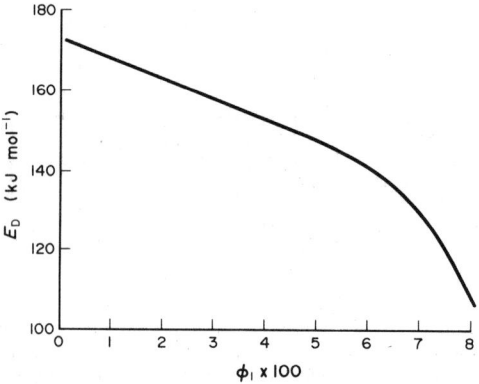

Figure 6 Effect of an increase in permeant volume fraction ϕ of dichloromethane in poly(vinyl formate) on the activation energy for diffusion E_D

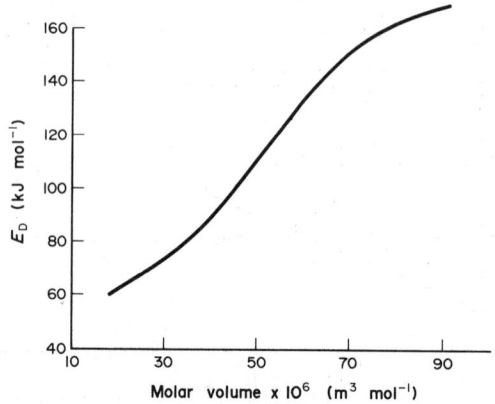

Figure 7 Increase in E_D with increasing size of permeant in poly(vinyl acetate) at 40 °C

has been identified as being equivalent to the activation energy for viscous flow of the polymer.[3,56] This implies that the same molecular motion is involved in the diffusion of large molecules as in the viscous flow of the polymer.

20.4.1.4 *Polymer glass transition temperature*

The glass transition temperature T_g of the polymer has a very marked influence on the diffusion coefficient. This is most clearly shown by the increase in D_0, the diffusion coefficient extrapolated to zero permeant concentration, with decreasing T_g as shown in Figure 8. The large increase in D_0 with decreasing value of T_g also leads to a significant decrease in the concentration dependence of the diffusion coefficient. For instance, the diffusion coefficient of benzene in natural rubber, poly(ethyl acrylate) and poly(methyl acrylate), with T_g values of 200 K, 250 K and 280 K respectively, increases by 2.9, 20 and 340 times on increasing the volume fraction of benzene from 0 to 0.1.

For a given polymer, a change of slope in the plot of D against T often occurs in the vicinity of T_g, as shown in Figure 9.[57] However, this is not always observed for small gas molecules.[58]

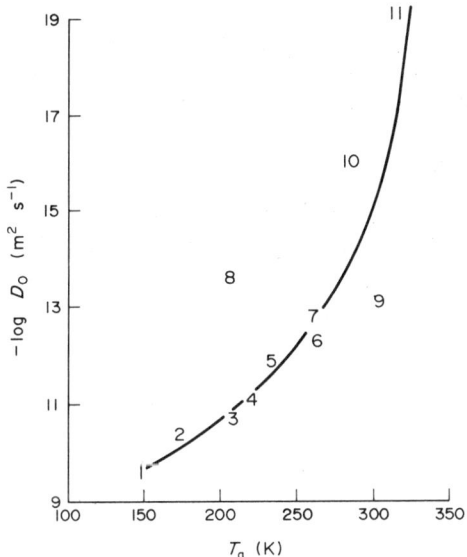

Figure 8 Increase in D_0 with decreasing T_g: 1, silicone rubber; 2, *cis*-polybutadiene; 3, natural rubber; 4, ethylene/propylene rubber; 5, poly(propyl acrylate); 6, polypropylene; 7, poly(ethyl acrylate); 8, polyisobutylene; 9, poly(butyl methacrylate); 10, poly(methyl acrylate); 11, poly(vinyl alcohol) (reproduced by permission of the Federation of Societies for Coating Technologies from *J. Paint Technol.*, 1970, **42**, 16)

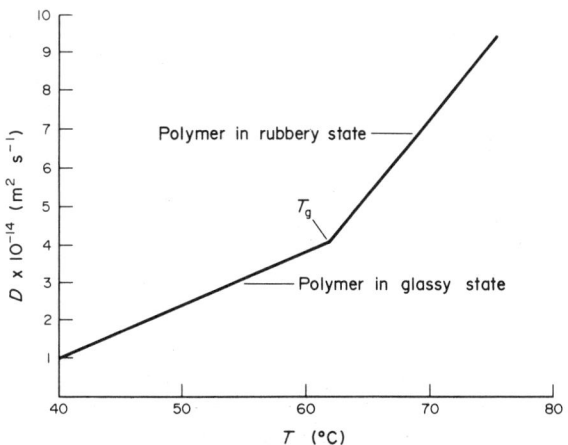

Figure 9 Change of slope at T_g in plot of D *vs.* T for CO_2 in a poly(acrylonitrile-*co*-methyl acrylate copolymer)

20.4.2 Models for Diffusion

Log D_0 may be related to E_D for a wide range of rubbery and glassy polymers and for a variety of permeants by a pair of Arrhenius equations which are valid for values of D_0 over a range of more than nine orders of magnitude.

$$\text{For rubbery polymers} \qquad \log D_0 \;=\; \frac{(E_D \times 10^{-3})}{R} \;-\; 8.0 \tag{29}$$

$$\text{For glassy polymers} \qquad \log D_0 \;=\; \frac{(E_D \times 10^{-3})}{R} \;-\; 9.0 \tag{30}$$

Where D_0 is in $m^2\,s^{-1}$ and E_D/R is in K.[59] The implication is that the mechanism of diffusion is the same for both glassy and rubbery polymers and that the only differences are in the magnitude or frequency of the motion of the polymer segments.

A number of theories have been developed over the years in order to model the variation of diffusion coefficients with concentration and temperature. These may be classified as either 'molecular' or 'free volume' theories.

20.4.2.1 Molecular theories

These are based on specific relative motions of permeant molecules and polymer chains and introduce relevant structural, energy, volume and pressure parameters.[60-63] The energy for diffusion, E_D, is postulated to arise from the need to separate the polymer matrix sufficiently to allow the permeant molecule to make a unit diffusional jump. Whilst the resulting equations describe the variation of E_D with temperature and permeant size, a number of adjustable parameters with no closely defined physical meaning are necessary. Further adjustable parameters are called for, in order to extend the temperature range of the models through T_g, and the calculations become increasingly complex.

20.4.2.2 Free volume theory

The variation of the diffusion coefficient with concentration, temperature, T_g and permeant size can be readily explained in terms of the free volume of the system. In particular, a model proposed by Fujita[64] has been found to describe satisfactorily the diffusion of a number of organic vapours and liquids in polymers above their T_g.

The theory is based on the assumption that a diffusing molecule can only move from one place to another when the local free volume around that molecule exceeds a certain critical value. The probability of finding such sufficient free volume is proportional to $\exp(-B_d/\phi_1 V_f)$, where B_d is a parameter describing the amount of free volume needed and is proportional to σ, the Lennard-Jones size parameter, ϕ is the volume fraction of permeant and V_f is the fractional free volume of the system. The thermodynamic diffusion coefficient D_T may be related to this probability by

$$D_T \;=\; RTA_d \exp\!\left(-\frac{B_d}{V_f}\right) \tag{31}$$

where A_d is related to the size and shape of the permeant by

$$A_d \;=\; \frac{\sigma^2}{M^{1/2}} \tag{32}$$

where M is the molecular weight of the permeant.

The fractional free volume V_f of any system increases with increasing temperature. In polymer systems, V_f decreases with increasing glass transition temperature. Furthermore, since permeants have more free volume than polymer segments, V_f will increase with increasing permeant concentration. Consequently, from equation (31), it follows that D_T increases with increasing permeant concentration and temperature, In addition, D_T increases with decreasing T_g and permeant size.

By relating the fractional free volume of a polymer at temperature T to the fractional free volume at T_g, and to α, the volume expansion coefficient, it is possible to predict the inverse relationship of D to temperature of measurement and to T_g.

This model has given very good correlations with experimental data.[65] However, discrepancies arise at high permeant concentrations and also when the size of the permeant is very different from that of a polymer segment. A more comprehensive theory[66,67] has been developed which can accommodate a wider range of conditions. However, it is more complex and requires further parameters to be determined in separate experiments.

20.5 PERMEATION

As described in the previous sections, the diffusivity D is a kinetic parameter and is related to polymer-segment mobility, whilst the solubility coefficient S is a thermodynamic parameter which is dependent upon the strength of the interactions in the polymer/permeant mixture. Hence D and S are affected in different ways by variables such as permeant concentration and type. However, since the permeation behaviour depends on both D and S, it is clear that the permeation coefficient P will vary in a more complex fashion.

Generally, variations in D can be very large, up to ten orders of magnitude, whilst those for S tend to be much smaller, up to three orders of magnitude. Consequently, variations in D tend to dominate the permeability, but as D is greatly affected by S it is wrong to underestimate the importance of S.

This importance of the sorption contribution to overall permeability may be seen in the temperature dependence of P, which over small temperature changes may be represented by an Arrhenius-type relationship

$$P = P_0 \exp\left(-\frac{E_P}{RT}\right) \tag{33}$$

where E_P is the activation energy for permeation. It can be shown that E_P depends upon contributions from sorption and diffusion of the permeant, *i.e.*

$$E_P = \overline{\Delta H_{mix}} + \Delta H_{cond} + E_D - yRT \tag{34}$$

as indicated by equations (18), (19), (27) and (28). It is found that the sorption terms are dominant.

Since diffusion requires conformational rearrangement of segments within a polymer chain, the behaviour is similar to that which affects the rheological and mechanical properties of the solid polymer in the presence of a permeant. Whilst viscoelastic motions require considerable cooperative chain motions throughout the polymer, permeation behaviour only requires relatively local coordination of segmented motions. Consequently, the time frame for the two processes is quite different.

The following sections indicate how the effects of factors such as polymer functional groups and chain structure can influence chain mobility and hence permeability. In addition, the effects of different types of permeant on polymer permeability are briefly noted, since it is important to recognize that the nature of the permeant can give rise to behaviour which deviates strongly from that of simple gases.

20.5.1 Factors Affecting Permeation

20.5.1.1 *Permeant size and shape*

An increase in size in a series of chemically similar permeants generally leads to an increase in their solubility coefficients due to their increased boiling points, but will also lead to a decrease in their diffusion coefficients due to the increased activation energy needed for diffusion.[5,15,59,68] The overall effect of these opposing trends is that the permeability generally decreases with increasing permeant size, since for many polymer/permeant pairs the sorption coeffcient will only increase by perhaps a factor of ten whilst the diffusion coefficient can vary by ten orders of magnitude, as previously described.

Permeant shape has a noticeable effect on permeability. For instance, flattened or elongated molecules have higher (by up to a factor of 10^3 greater) diffusion coefficients than spherical molecules of equal molecular volume.[15,54] A similar correlation for the dependence of solubility coefficient on shape has been found.[69] Generally, permeant size and shape effects are much more marked in glassy than in rubbery polymers, as mentioned earlier. This arises from the differences in the permeant/polymer mixing processes. In rubbery polymers, energy is required to generate sites for the permeant molecules to occupy but, since increasing permeant size tends to increase the heat of sorption, it follows that larger permeant molecules will be readily sorbed leading to enhanced plasticization of the polymer chains. Consequently, whilst smaller permeants will have a greater diffusion coefficient, the polymer will be less plasticized, whereas the lower diffusion coefficient of the larger permeants will be compensated for by the higher degree of sorption. The overall effect is to minimize the difference in the permeation coefficient for large and small permeants.

In glassy polymers, however, the permeation behaviour is governed by the availability of pre-existing sites or 'holes' as determined by the excess free volume of the system. It has been suggested that these 'holes' have a size distribution and that, depending upon the conditions of formation of the glassy polymer, there are fewer sites available for the larger permeant molecules than for the smaller ones.[70]

20.5.1.2 Permeant phase

The very weak interactions of permanent gases with polymers lead to minimal sorption and hence to little swelling of the polymer. Consequently, the solubility, diffusion and permeability coefficients are constants independent of pressure at a given temperature. The permeation of condensable vapours and liquids is generally much faster than that of permanent gases. Permeants which are good solvents for the polymer swell and plasticize the polymer structure, which gives rise to increased mobility of the polymer chain segments and leads to enhanced permeation rates. For instance, the permeability in a poly(butadiene-*co*-acrylonitrile) rubber increases by over seven orders of magnitude on varying the permeant from gaseous nitrogen to liquid methyl ethyl ketone.[71]

Water is a unique molecule because of its ability to form extensive hydrogen-bonded networks with itself and to hydrogen-bond strongly to other polar materials. These strong interactions result in sorption and diffusion behaviour not exhibited by other permeants. Although alcohols are capable of forming hydrogen bonds, the bond energy is much lower, *e.g.* methanol has a cohesive energy density of only $810 \, \text{kJ dm}^{-3}$ compared with $2120 \, \text{kJ dm}^{-3}$ for water. Consequently, the permeation behaviour of water varies greatly from polymer to polymer, depending upon the strength and type of interaction possible. Hydrophilic polymers such as poly(vinyl alcohol), poly(ethylene oxide) and cellulosics have high water vapour permeabilities, whereas hydrophobic polymers such as polybutadiene or poly(tetrafluoroethylene) (PTFE) have much lower permeabilities, as shown in Table 1.

Table 1   Permeability of Polymer Films to Water Vapour[a]

Polymer	$P \times 10^{-15} (\text{mol m}^{-1} \text{s}^{-1} \text{Pa}^{-1})$[b]
Poly(vinylidene chloride) (Saran)	0.5–1.5
Polytetrafluoroethylene	1.0
Butyl rubber	3.0
Polyethylene (density 0.960)	0.76–1.0
Polyethylene (density 0.922)	2.5–3.75
Polypropylene (density 0.907)	0.5–1.25
Polystyrene	17.5
Polybutadiene	157.7
Cellulose acetate	1839.0
Poly(vinyl alcohol) (pressure = 307 Pa)	14 045.0

[a] C. E. Rogers, in 'Polymer Permeability', ed. J. Comyn, Elsevier Applied Science, 1985, chap. 2.
[b] At 25 °C, 100% relative humidity.

20.5.1.3 Polymer molecular weight

As polymer molecular weight increases, the number of chain ends decreases. The chain ends represent a discontinuity and may form sites for permeant molecules to be sorbed into glassy polymers. For example, for a series of polystyrene samples, the diffusivity of a range of organic vapours (acetone, benzene, *etc.*) decreased by a factor of almost ten as molecular weight increased from 10 000 to 300 000.[15] However, in other systems, molecular weight has been found to have no influence on the transport of liquid permeants.[72, 73]

20.5.1.4 Functional groups

The permeability of permeants which interact weakly with functional groups present in a polymer can be expected to decrease as the cohesive energy of the polymer increases. For example, by increasing the polarity of the substituent group on a vinyl polymer backbone, oxygen permeability is reduced by almost 50 000 times, as shown in Table 2. Functional groups which have specific interactions with a permeant act to increase its solubility in the polymer. This leads to plasticization and hence enhanced permeability.[74] For instance, the very low permeability of poly(vinyl alcohol) to oxygen only applies when the polymer is perfectly dry. Sorption of water vapour plasticizes the polymer by breaking up the strong hydrogen-bonding between the polymer chains and results in a very much higher permeability. Similarly, removal of a functional group which strongly interacts with a permeant from a polymer will reduce its permeability to that permeant. For example, an increase in the acetyl content of cellulose acetate film from 34 to 43 wt % decreased its water permeability by more than a factor of ten.[71]

Table 2 Effect of Functional Groups on Oxygen Permeability of Vinyl Polymers $(CH_2CHX)_n$,[a]

Functional group	$P \times 10^{-17}$ $(mol\,m^{-1}\,s^{-1}\,Pa^{-1})$[b]
H	1.3
Ph	1.1
Me	3.97×10^{-1}
F	3.96×10^{-2}
Cl	2.1×10^{-2}
CN	1.08×10^{-4}
OH	2.64×10^{-5}

[a] S. Steingiser, S. P. Nemphos and M. Salame, in 'Encyclopedia of Chemical Technology', 3rd edn., ed. H. F. Mark *et al.*, Wiley-Interscience, New York, 1978, vol. 3, p. 481.
[b] At 23 °C, 0% relative humidity.

20.5.1.5 Density and polymer structure

Density may be regarded as a guide to the amount of free volume within a polymer. Generally, a reduction in density in a series of polymers results in an increase in permeability.[75] Figure 10 shows a generally good correlation between polymer density and helium permeability. However, there are three polymers which do not fit well. Whilst it could be argued that the small differences might be due to experimental error, much more serious difficulties arise when one includes the appropriate data for the permeability of helium through butyl rubber (density $0.93\,kg\,m^{-3}$ and $P = 2.8 \times 10^{-15}\,mol\,m^{-1}\,s^{-1}\,Pa^{-1}$). Butyl rubber is less permeable to helium than poly(phenylene oxide) (PPO), but it is well above its T_g ($-76\,°C$), whereas the T_g of PPO is 220 °C. Since the solubility of helium in both polymers is low, swelling effects cannot be invoked to explain this apparent anomaly. In general terms the low permeability of butyl rubber is due to the sluggish segmental motion of the polymer chains caused by the steric hindrance of the two pendant methyl groups on every other main chain carbon atom. Poly(phenylene oxide), on the other hand, consists of chains of rigid aromatic groups which, whilst packed quite closely together (accounting for the higher density), are unable to move relative to one another. Consequently, permeation can occur in a relatively unhindered fashion through the microvoids which will exist due to the polymer being below its T_g.

Data shown in Table 3 demonstrate that for polymers with the same substitution pattern it is flexibility of the backbone that dominates the permeation properties. The Si—O backbone allows rapid chain-segment motion to occur in the silicone rubber (polymer C) and substitution of the Si—O linkage by Si—CH$_2$ (polymer B) dramatically reduces the permeability to a value even less than that of butyl rubber (polymer A). Insertion of $(CH_2)_n$ sequences into a siloxane backbone (polymer D) also leads to a decrease in permeability. Similarly, the Si—O backbone substitution of methyl by more bulky substituents decreases the permeability (polymer E).

These decreases in permeability are mainly due to a decrease in permeant diffusivity, caused by the increasing rigidity of the polymer backbone and the decreasing free volume available for the diffusion of permeant molecules. The substitution of bulky functional groups in the side chains

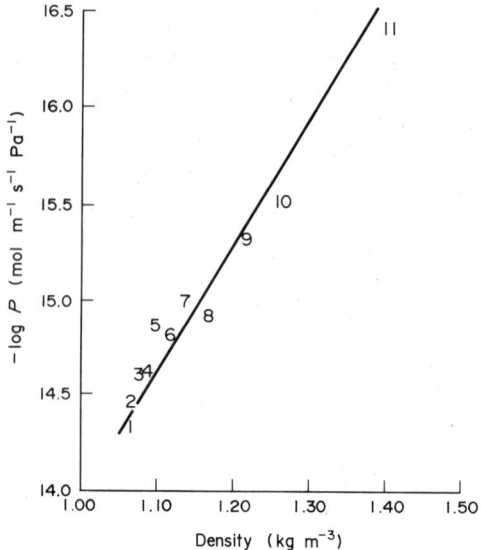

Figure 10 Correlation between polymer density and helium permeability P through polyethylene at 25 °C: 1, poly(tetramethyl-bis-L terephthalate) (**1**); 2, poly(tetraisopropyl-bis-A sulfone) (**2**); 3, poly(2,6-dimethyl-1,4-phenylene oxide) (**3**); 4, poly(tetramethylcyclobutanediol carbonate) (**4**); 5, poly(tetramethyl-bis-A carbonate) (**5**); 6, poly(tetramethyl-bis-L sulfone) (**6**); 7, poly(tetramethyl-bis-A terephthalate) (**7**); 8, poly(tetramethyl-bis-A sulfone) (**8**); 9, poly(bis-A carbonate) (**9**); 10, poly(bis-A sulfone) (**10**); 11, poly(ethylene terephthalate) (**11**)

Table 3 The Effect of Side-chain and Main-chain Substitution on Oxygen Permeability[a]

	Polymer	T_g(°C)	$P \times 10^{-17}$(mol m^{-1} s^{-1} Pa^{-1})[b]
A	—CH$_2$CMe$_2$—	−76	0.84
B	—CH$_2$SiMe$_2$—	−92	0.44
C	—OSiMe$_2$—	−123	437.0

D	Me Me
	  | |
	——Si(CH$_2$)$_8$SiO—
	  | |
	Me Me
		−88	1.0

E	Ph | ——SiO— | Me	−28	0.14

[a] S. A. Stern, V. M. Shah and B. J. Hardy, *J. Polym. Sci., Polym. Phys. Ed.*, 1987, **25**, 1293.
[b] At 35 °C.

appears to have a greater influence on decreasing the diffusivity than substitution of these groups in the polymer backbone.[74, 76, 77] Generally, the diffusivities of small permeant molecules are reduced less relative to larger permeant molecules, implying that the polymer matrix can act rather like a 'molecular sieve'.

Similar results have been found for rigid-backbone polymers where substitution can lead to polymers being able to assume only one non-linear conformation. With these polymers, however, an increase in the rigidity of the structures can lead to increases in permeability, and to high values of size separation efficiency,[75, 78] as shown in Table 4. There are severe hindrances to chain packing in the two polyimides shown in Table 4 due to kinks in the backbone which arise from the sterically hindered methylene carbons and the several non-linear bonds distributed along the chains. Strong hindrances to rotation arise from the rigid imide links, whilst perfluorination of the methyl groups transforms the bisphenol residue into a rigid segment, as shown by use of space-filling models.[79]

The interchain separations in these rigid bulky polymers appears to be large enough to permit relatively free movement of permeants below a certain size. However, for slightly larger molecules

Table 4 The effect of Rigid Polymer Chain Structure on Hydrogen and Methane Permeability[a, b]

Polymer	$P_{H_2} \times 10^{-17}$(mol m^{-1} s^{-1} Pa^{-1})	Ratio P_{H_2}/P_{CH_4}
	14.1	427
	33.4	112
	5.3	78

[a] D. G. Pye, H. H. Hoehn and M. Panar, *J. Appl. Polym. Sci.*, 1976, **20**, 287.
[b] At 35 °C.

this restricted chain spacing combined with the limited local segment motions means that permeation may be selective with respect to size and shape.[80]

20.5.1.6 Crosslinking, orientation and crystallinity

In non-crystalline polymers, diffusion coefficients decrease approximately linearly with crosslink density at low to moderate levels of crosslinking.[5, 18, 81] For instance, the diffusion coefficient of nitrogen in natural rubber is reduced tenfold on crosslinking the rubber with 11% sulfur.[82] Generally, the solubility coefficient is relatively unaffected, except at high degrees of crosslinking or when the permeant swells the polymer significantly.[83] However, crosslinking reduces the mobility of the polymer segments and tends to make the diffusivity more dependent on the size and shape of the permeant molecules and on the permeant concentration.

In crystalline polymers, the crystalline areas act as impermeable barriers to permeating molecules and have the same effect as inert fillers, i.e. they force the permeant molecules to diffuse along longer path lengths.[68, 84] Permeant solubility is proportional to the product of the amorphous volume fraction ϕ_A and the solubility S of the permeant in the amorphous phase.[85] Figure 11 shows the effect of increasing crystallinity, and hence density, on nitrogen permeability.[86] The thermal history of a crystallizable polymer can profoundly affect the permeation properties, since this can affect the number and size of crystallites present.

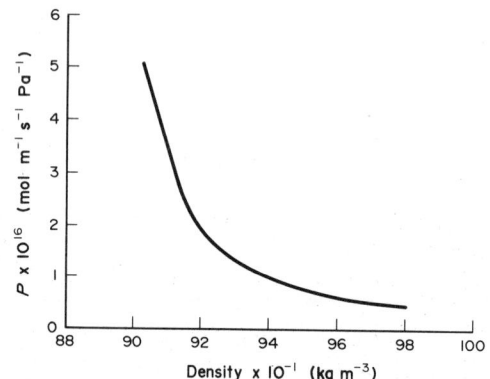

Figure 11 Effect of increasing crystallinity (density) on nitrogen permeability P (reproduced by permission of Wiley from *J. Polym. Sci.*, 1962, **57**, 925)

Orientation of the polymer may also influence the permeation properties. However, the overall effect is highly dependent upon crystallinity. For example, deformation of elastomers has little effect on permeability until crystallization effects occur.[87] Orientation of amorphous polymers can result in a reduction in permeability of around 10–15%, whereas in crystalline polymers, e.g. poly(ethylene terephthalate), reductions of over 50% have been observed.[88] At high degrees of orientation, time-dependent effects on permeability occur in both glassy and semi-crystalline polymers. These effects have been related to the relaxation recovery of strain-induced areas of free volume generated during orientation.[89, 90]

20.5.2 Heterogeneous Systems

Heterogeneous media include filled systems where the filler may or may not be inert to the permeant, block copolymers (where phase separation often occurs) and polymer blends. Multiphase systems occur frequently in industrial practice but, unfortunately, satisfactory theoretical descriptions of their permeation behaviour are not available.

20.5.2.1 Filled polymers

The addition of inert fillers may either increase or decrease permeability depending upon their degree of adhesion and compatibility with the polymer. An inert filler which is compatible with the

Table 5 The Effect of Fillers on Oxygen Permeability[a, b]

Vol % filler	Type	$P \times 10^{-17} (\text{mol m}^{-1}\text{s}^{-1}\text{Pa}^{-1})$
0	Calcium carbonate	1.3
15	Calcium carbonate	2.64
25	Calcium carbonate	5.3
25	Calcium carbonate (surface modified)	0.4

[a] S. Steingiser, S. P. Nemphos and M. Salame, in 'Encyclopedia of Chemical Technology', 3rd edn., ed. H. F. Mark *et al.*, Wiley-Interscience, New York, 1978, vol. 3, p. 482.
[b] At 23 °C, 0% relative humidity

polymer matrix will be impermeable to the permeant molecules and so permeation will be slower due to the reduced area for transport and the increased tortuosity of the permeation path.[5, 91] The degree of tortuosity is dependent on the volume fraction of the filler and the shape and orientation of the particles. When the filler is incompatible with the polymer, voids tend to occur at the interface which leads to an increase in free volume of the system and, consequently, an increase in permeability. The effect of a calcium carbonate filler and of a surface-modified version, which has enhanced polymer compatibility, on the permeability of oxygen in low density polyethylene is shown in Table 5.

20.5.2.2 Polymer blends

The effect on permeability of blending polymers is complex. The blends may be heterogeneous or homogeneous. In the latter case, the permeability is affected by the interaction between the component polymers,[91, 92, 93] whilst for heterogeneous blends interfacial phenomena and the rubbery or glassy nature of the phases are important.[94]

For miscible blends, measured gas permeabilities are often lower than those calculated assuming simple additivity. For example, for blends of polystyrene and poly(phenylene oxide), the measured permeability is considerably lower than that predicted.[95] This has been attributed to a strong polymer/polymer interaction which leads to an increase in density of the blend. Recently, it has been found that for blends of poly(methyl methacrylate) with poly(styrene-*co*-acrylonitrile) permeabilities are up to 30% higher than those predicted.[96] In this case, there is no volume change in the system on mixing. This behaviour is predicted by free volume theory.[92, 97]

Most polymer blends, however, are heterogeneous and consist of a polymeric matrix in which the second polymer is embedded. Consequently, the effects on permeability are very dependent on the degree of heterogeneity of the system and therefore on the method of formation.[94, 98] The nature and type of the polymers used (*i.e.* rubbery, glassy, hydrophilic or hydrophobic) is most important.

20.6 APPLICATIONS

Our increasing comprehension of the ways in which the permeation properties of polymers may be tailored has led to the development of a wide range of applications. This section briefly indicates some of these areas.

20.6.1 Food and Drinks Packaging

Polymers have widely supplemented metals and glass as containers in the food and drinks industry because of their low cost, light weight and controllable permeation properties. However, in order to achieve this, polymers which have the desired permeation properties in combination with strength, toughness, clarity and easy processability have had to be developed. Frequently, in order to attain the necessary combination of properties desired, multiple layers of different polymers are used.

Protection of foods against oxygen ingress is necessary in order to reduce oxidation which can cause flavour or colour changes. For example, in the packaging of cured meats such as ham, antioxidants are permitted. Consequently, in order to prevent the fat from going rancid due to oxidation, the meat is vacuum packed and the packaging must be highly impermeable to oxygen and moisture. Typically, such a packing consists of a nylon 6 resin laminated to a low permeability coated-polyethylene film.

The retention of CO_2 in carbonated drinks [by, for example, poly(ethylene terephthalate)] is important, since loss of 10% or more CO_2 can easily be detected by both taste and observation of the lack of sparkle in the drink. For beer, prevention of oxygen ingress is necessary, since its flavour can be affected by less than 2 p.p.m. of oxygen. Low water-vapour permeability is also important.

Methods have recently been developed which enable the permeation properties of these so-called barrier polymers to be calculated simply from their molecular structure.[99] General references on this topic are available.[88, 100]

20.6.2 Additive Migration

A wide variety of low molecular weight additives, such as plasticizers, antioxidants, UV stabilizers, pigments and antistatic agents are frequently incorporated into polymers to improve a property. Often these additives are simply dissolved in the polymer matrix and, consequently, they are liable to be extracted at a later time. This is especially the case if they are volatile or are soluble in a solvent which may later contact the polymer. One of the most common polymers in which additive migration is encountered is poly(vinyl chloride).[101]

Whilst migration of additives from a polymer matrix with time may be acceptable from the point of view of the polymer properties, contamination of materials, *e.g.* foods, in contact with the polymer is unacceptable. National standards have been laid down for acceptable limits of additive migration in food packaging materials.[102]

20.6.3 Coatings and Encapsulants

Polymeric coatings find a very wide range of applications, from paints for bridges and ships to encapsulants for electronic circuits. In all these applications one of the fundamental aims of the polymer is to provide protection of the substrate from the effects of water, oxygen or other harmful fluids.

For instance, for a polymer coating to provide a metal surface with good resistance to corrosion it must fulfil several roles. Since corrosion occurs at the metal surface only in the presence of water, oxygen and ionic impurities, the coating must be highly impermeable to all of these agents. Consequently, the polymers used are highly crosslinked and contain a high concentration of filler particles to reduce oxygen and water permeabilities. Additionally, the polymers should contain few polar or ionic groups which could interact with water and subsequently lead to swelling of the polymer and so increase its permeability. Furthermore, low permeability to ions, such as those of sodium and chlorine, is also achieved by the polymer having few easily polarizable groups, thus ensuring a low dielectric constant. Whilst there still remains some uncertainty regarding the relative importance of water, oxygen or ionic permeability through the coating, it is generally acknowledged that the adhesion between the coating and the metal surface is often of decisive importance. Much work still remains to be done to better understand the relationship between polymer permeability and surface adhesion.

The wide use of polymers in the electrical industry arises from their excellent electrical insulation and dielectric isolation properties. However, these are dependent upon the polymer permeation properties. For instance, the sheathing of cables and wires is sensitive to the presence of water in the polymer. It has been shown that the most common form of insulation breakdown arises from electrochemical treeing. This arises from the simultaneous effects of an electric field and moisture present in the shielding, either initially present or by permeation.[103]

Polymeric encapsulation has largely displaced techniques of hermetically sealing electronic devices, due to the low cost and ease of production. However, since all polymers are permeable to some small extent to water vapour, problems of adequate protection do still occur. Interestingly, it is not the inherent permeation of water into the polymer which alone causes these problems. Rather, they arise from the presence of trace amounts of catalyst residues in the polymer, which are transported to the surface of the device by the permeating water and which there result in the formation of an aggressive electrolyte solution. This problem is of particular concern with epoxide resins and of lesser concern with silicone encapsulants.[104]

20.6.4 Composites

The use of fibre-reinforced composites has grown dramatically in the last few years. In many applications these materials may come into contact with water (vapour or liquid) as well as organic solvents. Consequently, it is very important to understand how the permeation properties of the polymers used can affect the mechanical properties of the composites.

Water permeation, for instance, can lead to problems of dimensional change due to swelling of the polymer matrix as well as to chemical alternations within the composite. For instance, whilst epoxy resins are generally stable to hydrolysis, the ester groups present in polyesters are prone to hydrolysis.[105,106]

20.6.5 Controlled Release

The periodic administration of chemicals such as drugs, pesticides, fertilizers and herbicides results in a cyclic variation of their concentration with time. Consequently, the concentration levels may vary from dangerously high to ineffectively low levels. Much research carried out over the last 15 years has been aimed at developing techniques for controllably releasing these chemicals.

Many techniques and devices now exist, and the way in which polymers are used varies widely. However, the permeation properties of the polymers are central to the function and design of the devices. In addition to having the desired permeation characteristics, the polymers must be environmentally stable and compatible and possess the required mechanical properties. Very high levels of quality control are demanded.[107,108]

The simplest and most economical types of device are the monolithic devices. In these, the chemical of interest is uniformly dispersed within a polymer matrix, either by dissolution or melting. Subsequently they are cast, moulded or extruded into the required shape. A number of release mechanisms are possible. The simplest occurs when the chemical is dissolved in the polymer, when release occurs *via* diffusion. However, the rate of release is not constant, and the total amount to be released is often limited by the solubility of the chemical in the polymer.

Dispersion, rather than dissolution, of the chemical within the matrix can result in complex release kinetics. However, the rate of release is still determined by the permeation properties of the polymer. Complex shapes have been devised which can give virtually constant release rates.

Reservoir devices form an important class of controlled release vehicles. Here the chemical to be released is surrounded by an appropriate polymer membrane. In many instances, the polymer is relatively impermeable to water and release occurs *via* diffusion of the chemical through the membrane. If the chemical is present as an aqueous solution then the release rate declines with time as its activity within the membrane declines. However, if the chemical exists as a suspension within the reservoir, its activity will remain constant so long as saturation is maintained, thus ensuring a constant release rate. Several commercial devices based on this principle now exist.[109]

20.6.6 Membrane Separations

Membrane separation techniques for fluid mixtures are becoming increasingly important industrially. They offer the benefits of low capital and operating costs in comparison with conventional separation processes such as distillation. A number of different membrane separation technologies have been developed and are utilized in applications such as gas purification, salt removal or recovery and the breaking of azeotropes.[110,111,112] Whilst these techniques utilize different driving forces to achieve their respective separations, they are all dependent upon polymer membranes which are selectively permeable to a desired component in the feed.

20.6.6.1 Reverse osmosis

This technique may be simply described as the application of pressure on a solution in excess of the osmotic pressure of the solvent, which is usually water. The mechanism of separation is governed by the permeation properties of the polymer. Typically, polymers used in reverse osmosis membranes absorb about 5–15 wt % of water at equilibrium.

The driving force for the permeation of water through the membrane arises from the net pressure difference $\Delta P - \Delta \pi$, where ΔP is the hydrostatic-pressure gradient and $\Delta \pi$ is the osmotic-pressure gradient which acts in the opposite direction. The rate of flow of water (the solvent flux) is directly proportional to the membrane thickness. In contrast, the solute flux is independent of both pressure and membrane thickness. Consequently, the development of asymmetric membranes, which consist of a thin (0.1 μm) dense layer of polymer supported by a porous sub-layer, was necessary before this process became economically viable.[113,114]

Many polymers have been screened for use as reverse osmosis membranes, but only a few have proved to be suitable. Examples of current interest include cellulose acetate, aromatic polyamides and certain thin-film composite membranes of polyamides.

Reverse osmosis is largely applied to the desalination of brackish and sea waters and is in use throughout the world. Typically, for an applied pressure of 20–50×10^5 Pa, water fluxes of $1.0 \, \text{m}^3 \, \text{m}^{-2} \, \text{day}^{-1}$ are achievable.

20.6.6.2 Gas separations

Separations of gases (such as hydrogen from purge gas streams, carbon dioxide from hydrocarbon gases and oxygen or nitrogen from air) by polymer membranes are now industrial processes. The basis for these separations lies in the difference in the permeation rates of the gases through the membranes. The driving force for the separation arises from their respective pressure differentials through the membrane. Whilst many polymers exhibit significant differences in their permeability to gases such as hydrogen and methane, it was not until the polymers could be fabricated into asymmetric membranes that sufficiently high fluxes could be attained to make the process economically attractive. A further advance necessary in order to achieve commercial use of these systems was the development of techniques for spinning hollow fibres (200 μm outside diameter) having walls with this asymmetric structure. This enabled the fabrication of hollow fibre modules having very high surface area to volume ratios, typically $4000 \, \text{m}^2 \, \text{m}^{-3}$.[115,116]

Currently, the two most-commonly used polymers for gas separation membranes are polysulfone and cellulose acetate. Polysulfone, for instance, has a separation coefficient α (where α is the ratio of the permeability coefficient of the two gases) of 35 for the hydrogen/methane system and has a hydrogen permeability of $4.3 \, \text{mol} \, \text{m}^{-1} \, \text{s}^{-1} \, \text{Pa}^{-1}$. Much research is currently being undertaken to develop other polymers which will exhibit higher selectivities and fluxes.

20.6.6.3 Electrodialysis

Ionic polymers containing fixed charges may be used in membrane form to separate ions from aqueous solutions. By preventing the permeation of ions having the same charge as that on the membrane, only ions of the opposite charge permeate the membrane under an applied electrical potential gradient. Consequently, by passing an ionic solution, such as sea water, between two membranes of opposite charge the positive and negative ions may be removed. This process, known as electrodialysis has found large-scale applications in desalination, in the production of chlorine and caustic soda, and in the demineralization of cheese whey and fruit juices.[117]

The requirements for these membrane polymers include high ion selectivity and low electrical resistance combined with good chemical and mechanical stability. Typically, the polymers used are based on cationic or anionic derivatives of crosslinked polystyrene, phenol/formaldehyde resin or, more recently, fluorinated polymers.

20.6.6.4 Pervaporation

This technique, which has come into industrial use very recently, enables azeotropic mixtures such as ethanol/water to be separated more simply and cost effectively than by conventional techniques. The polymer membrane is designed to have a high sorption coefficient for just one of the feed components, typically water. Once sorbed, the water diffuses through the membrane and is removed as vapour by a vacuum or stream of inert carrier gas. By correct tailoring of the membrane polymer, permeates containing 99.9 wt % water can be obtained from feed solutions containing only 4 wt % water. Current polymers utilized or proposed include crosslinked poly(vinyl alcohol) and salts of carboxymethyl cellulose.[118,119,120]

ACKNOWLEDGEMENT

The author wishes to thank The British Petroleum Company plc for permission to publish this chapter and would like to acknowledge the kind assitance of friends and colleagues at the Research Centre in its final preparation.

20.7 REFERENCES

1. J. K. Mitchell, *Philadelphia J. Med. Sci.*, 1831, **13**, 36.
2. T. Graham, *Philos. Mag.*, 1866, **32**, 401.
3. G. J. van Amerongen, *Rubber Chem. Technol.*, 1964, **37**, 1065.
4. C. E. Rogers, in 'Engineering Design for Plastics', ed. E. Baer, Reinhold, New York, 1964, chap. 9.
5. J. Crank and G. S. Park, 'Diffusion in Polymers', Academic Press, New York, 1968.
6. H. J. Bixler and O. J. Sweeting, in 'Science and Technology of Polymer Films', ed. O. J. Sweeting, Wiley-Interscience, New York, 1971, vol. II, chap. 1.
7. R. M. Felder and G. S. Huvard, 'Methods of Experimental Physics', Academic Press, New York, 1980, vol. 16C, p. 135.
8. H. A. Daynes, *Proc. R. Soc. London, Ser. A*, 1920, **97**, 286.
9. P. Meares, in 'Diffusion in Polymers', J. Crank and G. S. Park, Academic Press, London, 1968, chap. 10.
10. R. Patterson and P. Doran, *J. Membr. Sci.*, 1986, **27**, 105.
11. S. G. Amarantos and J. H. Petropoulos, *J. Membr. Sci.*, 1979, **5**, 375.
12. K. C. O'Brien, W. J. Koros and T. A. Barbari, *J. Membr. Sci.*, 1986, **29**, 229.
13. M. Bischoff and P. Eyerer, *J. Membr. Sci.*, 1984, **21**, 333.
14. M. Laatikanan and M. Lindstrom, *J. Membr. Sci.*, 1986, **29**, 217.
15. A. R. Berens and H. B. Hopfenburg, *J. Membr. Sci.*, 1982, **10**, 283.
16. T. Alfrey, E. F. Gurnee and W. G. Lloyd, *J. Polym. Sci., Part C*, 1966, **12**, 249.
17. J. Crank, 'The Mathematics of Diffusion', Clarendon Press, Oxford, 1956.
18. J. Crank, 'The Mathematics of Diffusion', 2nd edn., Clarendon Press, Oxford, 1975.
19. H. K. Frensdorff, *J. Polym. Sci., Part A1*, 1964, **2**, 341.
20. M. E. Stewart, H. B. Hopfenburg, W. J. Koros and N. R. McCoy, *J. Appl. Polym. Sci.*, 1987, **34**, 721.
21. H. Yasuda and V. T. Stannett, in 'Polymer Handbook', 2nd edn., ed. J. J. Brandrup and E. H. Immergut, Interscience, New York, 1975, p. III-229.
22. M. B. Huglin and M. B. Zakaria, *Angew. Makromol. Chem.*, 1983, **117**, 1.
23. O. J. Sweeting, in 'Science and Technology of Polymer Films', ed. O. J. Sweeting, Wiley-Interscience, New York, 1971, vol. 1, chap. 1.
24. J. Walker, *Paint Technol.*, 1967, **31**, 15, 22.
25. J. A. Barrie, A. Newman and A. Sheer, in 'Permeability of Plastic Films and Coatings', ed. H. B. Hopfenburg, Plenum Press, New York, 1974, p. 167.
26. W. R. Vieth, J. M. Howell and J. H. Hsieh, *J. Membr. Sci.*, 1976, **1**, 177.
27. J. H. Petropoulos, *J. Polym. Sci., Part A-2*, 1970, **8**, 1797.
28. W. J. Koros, A. H. Chan and D. R. Paul, *J. Membr. Sci.*, 1978, **3**, 117.
29. D. R. Paul and D. R. Kemp, *J. Polym. Sci., Polym. Symp.*, 1973, **41**, 79.
30. S. A. Stern and V. Saxena, *J. Membr. Sci.*, 1980, **7**, 47.
31. D. Raucher and M. D. Sefcik, *ACS Symp. Ser.*, 1983, **223**, chaps. 5 and 6.
32. J. H. Hildebrand and R. L. Scott, 'The Solubility of Non-electrolytes', 3rd edn., Reinhold, New York, 1950.
33. J. A. Barrie and D. Machin, *Trans. Faraday Soc.*, 1971, **67**, 244.
34. C. E. Rogers and D. Machin, in 'Encyclopedia of Polymer Science and Technology', ed. H. F. Mark and N. Gaylord, Interscience, New York, 1970, vol. 12, p. 684.
35. B. H. Zimm and J. L. Lundberg, *J. Phys. Chem.*, 1956, **60**, 425.
36. J. L. Lundberg, *Pure Appl. Chem.*, 1972, **31**, 261.
37. G. Gee, *Q. Rev., Chem. Soc.*, 1947, **1**, 265.
38. J. H. Hildebrand and R. L. Scott, 'The Solubility of Non-electrolytes', 3rd edn., Reinhold, New York, 1950.
39. C. E. Rogers, in 'Encyclopedia of Polymer Science and Technology', ed. H. F. Mark and N. Gaylord, Interscience, New York, 1970, vol. 12, p. 692.
40. T. Alfrey, E. F. Gurnee and W. G. Lloyd, *J. Polym. Sci., Part C*, 1966, **12**, 249.
41. E. H. Andrews, G. M. Levy and J. Willis, *J. Mater. Sci.*, 1973, **8**, 1000.
42. C. Robinson, *Trans. Faraday Soc.*, 1946, **42B**, 12.
43. N. L. Thomas and A. H. Windle, *Polymer*, 1978, **19**, 255.
44. C. H. M. Jacques and H. B. Hopfenburg, in 'Permeability of Plastic Films and Coatings', ed. H. B. Hopfenburg, Plenum Press, New York, 1974.
45. A. H. Windle, in 'Polymer Permeability', ed. J. Comyn, Elsevier Applied Science, London, 1986, p. 115.
46. J. H. Petropoulos and P. P. Roussis, *J. Polym. Sci., Part C*, 1969, **22**, 917.
47. N. L. Thomas and A. H. Windle, *Polymer*, 1982, **23**, 529.
48. E. Bagley and F. A. Long, *J. Am. Chem. Soc.*, 1955, **77**, 2172.
49. A. R. Berens and H. B. Hopfenberg, *Polymer*, 1978, **19**, 489.
50. A. S. Michaels, W. R. Vieth and J. A. Barrie, *J. Appl. Phys.*, 1963, **34**, 1, 13.
51. R. M. Barrer, 'Diffusion In and Through Solids', Cambridge University Press, Cambridge, MA, 1941.
52. A. C. Newns and G. S. Park, *J. Polym. Sci., Part C*, 1969, **22**, 977.
53. M. J. Hayes and G. S. Park, *Trans. Faraday Soc.*, 1955, **51**, 1134.
54. N. Yi-Yan, R. M. Felder and W. J. Koros, *J. Appl. Polym. Sci.*, 1980, **25**, 1755.
55. C. E. Rogers, in 'Polymer Permeability', ed. J. Comyn, Elsevier Applied Science, London, 1986, p. 35.
56. R. J. Kokes and F. A. Long, *J. Am. Chem. Soc.*, 1953, **75**, 6142.

57. D. R. Paul, *ACS Symp. Ser.*, 1985, **285**, 261.
58. H. B. Hopfenburg and V. Stannett, in 'The Physics of Glassy Polymers', ed. R. N. Haward, Applied Science, London, 1973, p. 504.
59. D. W. Van Krevelen, 'Properties of Polymers', 2nd edn., Elsevier, New York, 1976, chap. 18.
60. P. Meares, *J. Am. Chem. Soc.*, 1954, **76**, 3415.
61. W. W. Brandt, *J. Phys. Chem.*, 1959, **63**, 1080.
62. A. T. DiBenedetto and D. R. Paul, *J. Polym. Sci., Part C*, 1965, **10**, 17.
63. R. J. Pace and A. Datyner, *J. Polym. Sci., Polym. Phys. Ed.*, 1979, **17**, 437, 453, 465, 1675, 1693.
64. H. Fujita, *Adv. Polym. Sci.*, 1961, **3**, 1.
65. H. Fujita, A. Kishimoto and K. Matsumoto, *Trans. Faraday Soc.*, 1960, **56**, 424.
66. J. S. Vrentas and J. L. Duda, *J. Polym. Sci., Polym. Phys. Ed.*, 1977, **15**, 403, 417, 441.
67. J. L. Duda, J. S. Vrentas, S. T. Ju and H. T. Liu, *AIChEJ.*, 1982, **28**, 279.
68. C. E. Rogers, in 'Physics and Chemistry of the Organic Solid State', ed. D. Fox, M. M. Labes and A. Weissberger, Interscience, New York, 1965, vol. 2, chap. 6.
69. C. E. Rogers, V. T. Stannett and M. Szwarc, *J. Phys. Chem.*, 1959, **63**, 1406.
70. A. Sfirakis and C. E. Rogers, *Polym. Eng. Sci.*, 1981, **21**, 542.
71. C. E. Rogers, in 'Polymer Permeability', ed. J. Comyn, Elsevier Applied Science, London, 1986, 62.
72. F. Asmussen and K. Uberreiter, *J. Polym. Sci.*, 1962, **57**, 199.
73. F. Asmussen and K. Uberreiter, *Kolloid-Z. Z. Polym.*, 1962, **185**, 1.
74. S. A. Stern, V. M. Shah and B. J. Hardy, *J. Polym. Sci., Polym. Phys. Ed.*, 1987, **25**, 1293.
75. L. Pilato, L. M. Litz, B. Hargitay, R. C. Osbourne, A. G. Farnham, J. H. Kawakarmi, P. E. Fritze and J. E. McGrath, *Polym. Prepr., Am. Chem. Soc., Div. Polym Chem.*, 1975, **16**, 42.
76. T. Nakagawa, Y. Fujiwara and N. Minoura, *J. Membr. Sci.*, 1984, **18**, 111.
77. C. L. Lee, C. H. Lee, H. L. Chapman, M. E. Cifuentes, K. M. Lee, L. O. Merill, K. C. Ulman and K. Venkataraman, in 'Proceedings of 4th BOC Priestley Conference, Leeds, 16–18 September, 1986', The Royal Society of Chemistry, London, 1986, p. 364
78. D. G. Pye, H. H. Hoehn and M. Panar, *J. Appl. Polym. Sci.*, 1976, **20**, 287.
79. H. H. Hoehn, *US Pat.* 3 822 202 (1974) (*Chem. Abstr.*, 1974, **81**, 78 937).
80. R. T. Chern, W. J. Koros, H. B. Hopfenburg and V. T. Stannett, *ACS Symp. Ser.*, 1985, **269**, chap. 2.
81. P. Meares, 'Polymers: Structure and Bulk Properties', Van Nostrand, London, 1965, chap. 12.
82. R. M. Barrer and G. Skirrow, *J. Polym. Sci.*, 1948, **3**, 549.
83. O. T. Turner and A. K. Abell, *Polymer*, 1987, **28**, 297.
84. I. Sobolev, J. A. Meyer, V. T. Stannett, and M. Sware, *Ind. Eng. Chem.*, 1957, **49**, 441.
85. A. S. Michaels and R. B. Parker, *J. Polym. Sci.*, 1959, **41**, 53.
86. H. Alter, *J. Polym. Sci.*, 1962, **57**, 925.
87. J. A. Barrie and B. Platt, *J. Polym. Sci.*, 1961, **54**, 261.
88. R. J. Ashley, in 'Polymer Permeability', ed. J. Comyn, Elsevier Applied Science, London, 1986, p. 286.
89. A. Peterlin, *J. Macromol. Sci., Phys.*, 1975, **B11**, 57.
90. T. L. Smith and R. E. Adam, *Polymer*, 1981, **22**, 299.
91. H. B. Hopfenberg and D. R. Paul, in 'Polymer Blends', ed. D. R. Paul and S. Newman, Academic Press, New York, 1978, vol. 1, chap. 10.
92. D. R. Paul, *J. Membr. Sci.*, 1984, **18**, 75.
93. J. S. Chiou and D. R. Paul, *J. Appl. Polym. Sci.*, 1986, **32**, 2897, 4793.
94. H. Odani, M. Uchikura, Y. Ogino and M. Kurata, *J. Membr. Sci.*, 1983, **15**, 193.
95. G. Morel and D. R. Paul, *J. Membr. Sci.*, 1982, **10**, 273.
96. J. S. Chiou and D. R. Paul, *J. Appl. Polym. Sci.*, 1987, **34**, 1037.
97. W. M. Lee, *Polym. Eng. Sci.*, 1980, **20**, 65.
98. J. Y. Lai, G. J. Wu and S. Shyu, *J. Appl. Polym. Sci.*, 1987, **34**, 559.
99. M. Salame, *Polym. Eng. Sci.*, 1986, **26**, 22, 1543.
100. S. Steingiser, S. P. Nephos and M. Salame, in 'Encyclopedia of Chemical Technology', 3rd edn., ed. H. F. Mark, D. F. Othmer, C. G. Overberger and G. T. Seaborg, Interscience, New York, vol. 3, p. 480.
101. A. M. Husson, F. C. Husson, G. Merle and J. Gole, *J. Macromol. Sci., Phys.*, 1977, **B14**, 553.
102. F. A. Paine and H. Y. Paine, 'A Handbook of Food Packaging', Blackie, Glasgow, 1983.
103. J. C. Bobo, E. Papirer, M. Prigent and J. Schulz, *IEE Conf. Publ.*, 1977, **177**, 235.
104. M. T. Gosey, in 'Polymer Permeability', ed. J. Comyn, Elsevier Applied Science, London, 1986, p. 332.
105. D. H. Kaelble and P. J. Dynes, *J. Adhes.*, 1977, **8**, 195.
106. H. V. Boenig, 'Unsaturated Polyesters: Structure and Properties', Elsevier, Amsterdam, 1964.
107. H. Leeper and H. Benson, *Polym. Eng. Sci.*, 1977, **17**, 42.
108. R. W. Baker and L. Saunders, *NATO Adv. Study Inst. Ser., Ser. C*, 1986, **181**, 581.
109. A. S. Michaels, *Polym. Sci. Technol.*, 1974, **6**, 409.
110. A. K. Fritzsche and R. S. Narayan, *CEER, Chem. Econ. Eng. Rev.*, 1987, **19** (205), 19.
111. T. Nakagawa, *CEER, Chem. Econ. Eng. Rev.*, 1987, **19** (205), 32.
112. R. Rautenbach and R. Albrecht, *Int. Chem. Eng.*, 1983, **27**, 10.
113. S. Loeb and S. Sourirajan, *Adv. Chem. Ser.*, 1962, **38**, 117.
114. R. E. Kesting, 'Synthetic Polymeric Membranes', Wiley, New York, 1985.
115. J. M. S. Henis and M. K. Tripodi, *US Pat.* 4 230 436 (1980).
116. J. M. S. Henis and M. K. Tripodi, *J. Membr. Sci.*, 1981, **8**, 233.
117. J. R. Wilson, 'Demineralisation by Electrodialysis', Butterworths, London, 1960; S. B. Turner, 'Diffusion and Membrane Technology', Reinhold, New York, 1962.
118. I. Casbasso and Z. Z. Liu, *J. Membr. Sci.*, 1985, **24**, 10.
119. P. Aptel, N. Challard, J. Cuny and J. Neel, *J. Membr. Sci.*, 1976, **1** (3), 271.
120. U. Sander and P. Soukup, *J. Membr. Sci.*, 1988, **36**, 463.

21

Ionic Conductivity

JOHN OWEN

University of Salford, UK

21.1   INTRODUCTION

Electrical conductivity is the movement of charge in response to an electric field. In the familiar solid conductors, *i.e.* metals and semiconductors, the charge movement, or current, is due to electrons or holes and is described as *electronic conductivity*.

Ionic conductivity is electrical conductivity due to the motion of ionic charge. Elementary science introduces this phenomenon as a property of liquid electrolyte solutions. In the solid state, ionic conductivity has recently been somewhat overshadowed by electronic, but nevertheless was recognized by Faraday, who observed electrical conductivity in solid lead fluoride at high temperature. The conductivity in this case was due to the motion of fluoride anions within the structure. This type of conductivity in solids has long been of fundamental interest as well as being applied in the interpretation of corrosion.[1] More recently, applications have been found in energy conversion devices[2] and chemical sensors.[3]

Ionic conductivity, like electronic conductivity, can be broken down into a product of three terms: the carrier charge (q), the concentration (number of particles per unit volume, n) and the mobility (the average velocity of a carrier due to an applied electric field of unit strength, b). Thus we may write an expression for the specific conductivity

$$\sigma = qnb \tag{1}$$

In the case of ionic conductivity, many different carriers may be encountered, but ionic mobilities are generally much smaller than the values typically found for electrons. The ionic mobility is conceptually identical to the electrochemists' concept of equivalent conductivity, λ_i, which may be expressed in terms of the specific conductivity, the molar concentration (c_i), and the charge number (z_i)

$$\lambda_i = \frac{\sigma_i}{|z_i|c_i} = Fb_i \tag{2}$$

where F is Faraday's constant. The total ionic conductivity is obtained as the sum of the contributions due to all free ions in the sample

$$\sigma = \sum q_i n_i b_i = \sum |z_i| c_i \lambda_i \tag{3}$$

The selection of conductivities shown in Table 1 illustrates the fact that ions are, on the whole, considerably less mobile in solids than in liquids. For this reason, electrochemistry has been mainly concerned with liquid electrolytes.

Table 1 Conductivities of Various Materials at 25 °C

Substance	Phase	Log_{10} (conductivity/δ cm^{-1})
$H_2SO_4 \cdot 2H_2O$	Liquid	2
$LiCF_3SO_3$/THF	Liquid	-2
Na β-alumina	Crystal	-3
Li_3N	Crystal	-3
$LiI/Li_2S/P_2S_5$	Glass	-3
PEO_8LiClO_4	Semicrystalline	-7
$MEEP \cdot LiClO_4$	Elastomer	-4

Ionic conduction in polymers was, until recent times, regarded as an undesirable phenomenon which was responsible for poor insulation properties. Therefore we find, in the older literature,[4] studies of the intrinsic conductivity: the residual conductivity after removing the last traces of any ions or ionizable impurities. Nowadays there is strong interest in the opposite goal of maximizing the ionic conductivity for special applications such as batteries or fuel cells with elastomeric electrolytes. This is mainly due to the realization[5] that the figure of merit for most applications is not the conductivity itself, but the conductivity divided by the square of the electrolyte thickness. Given a modest conductivity, the excellent film-forming capabilities of elastomers make them prime candidates as new materials in electrochemical devices.

Most polymers are poor ionic conductors because they are primarily organic and contain ions only adventitiously in very small concentrations. The prerequisite for good ionic conductivity is a deliberate addition of ions into an environment which allows a high mobility. The following examples show how ions may be introduced into polymers.

(a) A polymer swollen with a liquid electrolyte solution. An example of this would be microporous poly(vinylidene fluoride) swollen with a solution of lithium perchlorate ($LiClO_4$) in propylene carbonate (Figure 1a).[6] This route can give usefully high conductivity values, but of course the polymer itself is simply a support for the liquid electrolyte.

(b) A polymeric salt, *i.e.* a polymer chain containing bound cationic or anionic moieties and mobile counter-ions. This type of polymer should have the advantage of a transport number of unity for the counter-ion. However, unless the counter-ions are solvated, this type of polymer has been found to have an extremely low conductivity due to the aggregation of ions into clusters. For example, Nafion[7] (Figure 1b) was found to have a conductivity of less than 10^{-13} S cm^{-1} but became a far better conductor ($\sigma < 10^{-8}$ S cm^{-1}) when wetted with (acetyl end-capped) liquid poly(ethylene glycol). However, useful conductivities (10^{-6} S cm^{-1}) were obtained only if $LiClO_4$ was also added to form a supported liquid electrolyte of type (a) above.

Another example of a polymeric salt is poly(diallyldimethylammonium chloride),[8] which is a poorly conducting brittle solid. However, upon plasticization with poly(ethylene glycol) it becomes a good conductor of chloride ions with a conductivity of 10^{-5} S cm^{-1}.

(c) A solution of a salt in an ion-solvating polymer. This is the type of ion-containing polymer which has received most attention as a quasi-solid polymer electrolyte. It offers at least the following

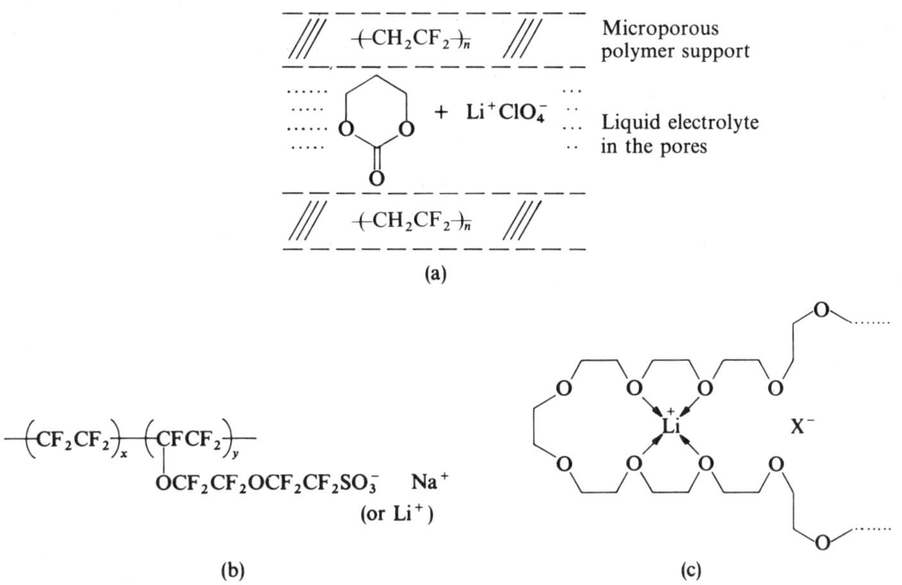

Figure 1 Examples of ions introduced into polymers: microporous poly(vinylidene fluoride) swollen with a solution of lithium perchlorate in propylene carbonate; (b) Nafion; (c) poly(ethylene oxide) plus lithium salt

advantages over materials containing liquid components: (i) negligible vapour pressure; (ii) no transport of solvent along with the ion; (iii) no long-range mobile components except for the ions; and (iv) suitable support for a thin film electrode.

The best known ion-solvating polymer is poly(ethylene oxide) (PEO) (Figure 1c), which can combine with a multitude of inorganic, organic and even polymeric[9] salts to form ion-conducting solutions in which the solvent is an elastomer.

This last type of ion-containing polymer system will be the main subject of the remainder of this chapter. The next section will discuss the nature of the conduction process. This is an area of great interest as the conduction mechanism in an elastomer appears to be distinct from that understood to operate in liquids, solids or glasses. The following sections will outline the chemistry of polymer solvents and the solutions and complexes formed upon adding salt. It will be seen that amorphous materials are required for high conductivity, and the next section will then summarize some recent results on those systems. A section is included on the experimental methods used to determine the conductivity and further characterize transport processes. Finally, the last section will outline the projected applications which have been the main stimulus for research in this area.

The literature on ionic-conducting polymer systems has expanded profusely in the last decade, and detailed reviews and collections of papers may be found in a recent book[10] and conference proceedings.[11]

21.2 THE CONDUCTION MECHANISM

21.2.1 Ionic Mobility in Solids

The theoretical basis of Ohm's law in ionic conductors is by no means obvious; charged particles should accelerate in an electric field rather than reach the constant velocity implied by a constant current. The following analysis shows that Ohmic conductivity should be regarded as a perturbation of diffusion rather than a motion imparted directly by the field.

In a solid structure at absolute zero temperature, ions are trapped in their preferred sites by potential wells formed by the surrounding atoms. Although an electric field can slightly perturb the position of an ion in the well, further motion is prohibited unless the field is very large, and then avalanche breakdown occurs. The ionic conductivity below this breakdown field is therefore zero.

At higher temperatures self-diffusion occurs. This is the random jumping motion of the more energetic ions over the potential barriers between sites. Self-diffusion is difficult to observe directly, but its rate can be inferred from observations of the more familiar Fick's-law diffusion. In the latter case a concentration (or, strictly, activity) gradient introduces a bias into the self-diffusion process so that a net flux is obtained in the direction down the gradient.

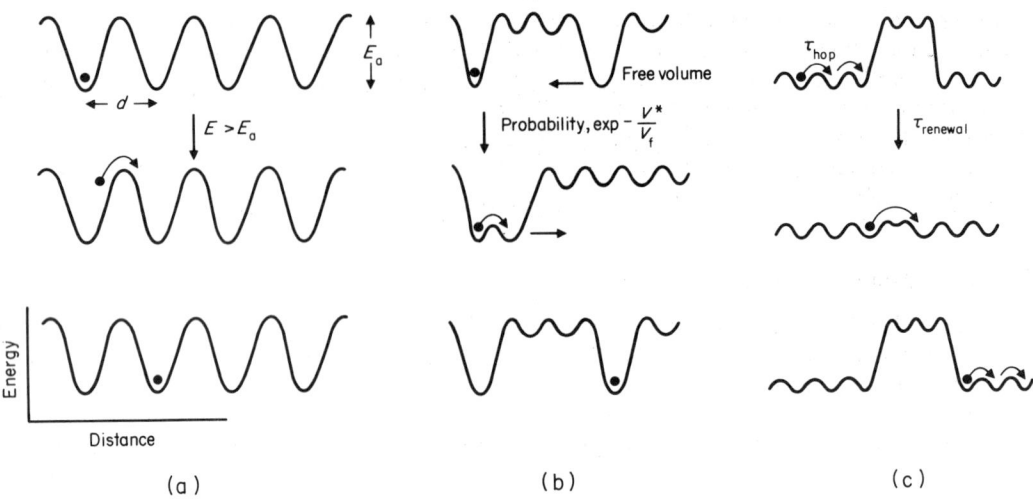

Figure 2 Schematic illustrations of (a) solid-state; (b) free-volume; and (c) DBP mechanisms of ion transport

Ionic conductivity results from an electrical biasing of the same self-diffusion process, such that down-field jumps occur more frequently than up-field jumps. This results in a net down-field drift of current density proportional to the field. Figure 2(a) shows a schematic illustration of this mechanism.

The Nernst–Einstein equation[12] relates either the ionic conductivity to the self-diffusion coefficient

$$\sigma = \frac{nq^2 D}{kT} \tag{4}$$

or the mobility to the diffusion coefficient

$$b/q = D/kT \tag{5}$$

The derivations of expressions for the self-diffusion coefficient and conductivity are simple in the case of a single crystal which contains a small concentration of diffusant dispersed in a regular lattice of mostly vacant sites separated by identical potential barriers. The frequency of jumps per unit volume (f_v) is given by the following expression involving the concentration, the barrier height (E_a) and the thermal energy kT

$$f_v = f_0 n \exp(-E_a/kT) \tag{6}$$

where f_0 is the frequency with which each ion attempts to surmount the barrier, which may be deduced from a knowledge of the vibrational motion of the ion in the well.

The current density i is given by the product of jump distance (d), ionic concentration, ionic charge and the difference in frequency of down-field and up-field jumps

$$i = f_0 dnq\{\exp(-E_a + 1/2qV)/kT - \exp(-E_a - 1/2qV)/kT\} \tag{7}$$

where V is the potential difference between sites due to the field. Rearranging this equation gives

$$i = f_0 dnq \exp(-E_a/kT)\{\exp(1/2qV/kT) - \exp(-1/2qV/kT)\} \tag{8}$$

Ohm's law holds at small values of qV/kT through the approximation

$$i = f_0 dnq^2 V/kT \exp(-E_a/kT) \tag{9}$$

Replacing V by the product of field and jump distance, we obtain the conductivity as the current density divided by the field

$$\sigma = f_0 d^2 nq^2/kT \exp(-E_a/kT) \tag{10}$$

However, real solids are more complex. In the common case where the concentration of the diffusant is much higher than that of vacant sites, one must consider vacancies as the diffusing entities, in a similar fashion to the hole formalism used in semiconductors. More generally, the concentration to be used is that of the diffusing lattice defects, *e.g.* vacancies, interstitial ions, *etc.* The result is still of the Arrhenius form, however, and plots of $\ln(\sigma T)$ *vs.* $1/T$ are linear with a slope of $-E_a/kT$. Consequently, the requirement for high mobility is simply a low activation energy, as expected for an open structure.

21.2.2 Ionic Mobility in Liquids

The Nernst–Einstein expression can also be applied to ionic conduction in liquids. However, the derivation of an expression for the mobility from first principles is much more difficult, because there is no regular lattice. Furthermore, in a liquid any conceivable, though irregular, site network is itself continuously rearranging *via* self-diffusion.

In liquids, however, the mobility is simply related to the viscosity η *via* the Stokes equation[13]

$$\frac{b}{q} = \frac{1}{6\pi r\eta} \tag{11}$$

where r is the effective radius of the ion. The Stokes equation gives good results for dilute electrolyte solutions, and the effective radius of the mobile ion as calculated from conductivity and viscosity measurements can be used to indicate the presence and measure the size of a solvation shell. Alternatively, conductivities can be estimated from known viscosities. In this case the simple requirement for high mobility is a low viscosity.

The Arrhenius equation is not always exactly observed so that plots of $\ln(\sigma T)$ *vs.* $1/T$ often appear to be curved over a large temperature range. This is because the solvent mobility is not only controlled by a thermal-activation process but, more importantly, by the free volume. This aspect will be discussed below, as it can also be applied to elastomers.

21.2.3 Ionic Conductivity in Elastomers

Physically, an elastomer is more liquid-like than solid-like, and therefore the mobility theory for a liquid seems the appropriate choice for ion-conducting polymer systems. The macroscopic viscosity of an elastomer cannot, however, be used in the Stokes equation. This is because the macroscopic viscosity is greatly enhanced by chain entanglement, which does not directly resist ion motion. However, the mobility in a liquid or elastomer can be derived according to the free-volume theory of Cohen and Turnbull,[14] outlined below. Because it involves the glass transition temperature (T_g) as the major parameter, it is particularly applicable to amorphous polymers for which T_g is easily measured.

The disorder inherent in most flexible covalently bonded chains prevents an elastomer from forming a close-packed structure. The resulting vacancies are randomly arranged and are static at temperatures below T_g. Although vacancies are quite abundant, forming about 2.5% of the total volume[15] at T_g, they do not amalgamate into continuous tunnels allowing diffusion of foreign molecules. However, at temperatures above T_g, barriers to rotation about chain bonds and inter-chain interactions are comparable in magnitude to the thermal energy, kT, so that a chain segment is capable of localized motion and can move into a neighbouring vacancy. This process gives a mechanism for a continuous and random redistribution of the free volume and so allows long-range diffusion of small molecules through the structure (Figure 2b).

The ratio of the volume of vacancies to the total volume is called the free-volume fraction, f. This has been shown[15] to be a linear function of temperature in most elastomers, at least in the range $T_g < T < (T + 100)$

$$f = f_{T_g} + \alpha(T - T_g) \tag{12}$$

or $f = \alpha(T - T_0)$. T_0 is the ideal transition temperature, *i.e.* a parameter indicating the temperature at which f extrapolates to zero (about 50 degrees below T_g) and α is the thermal expansion coefficient of the free volume fraction; this has been found to be about 0.0005 K^{-1} for most polymers within the linear range.[15] f_{T_g} is the value of f at T_g and below, where the chains are immobile.

Displacement of a chain segment or other diffusing unit depends on the availability of a sufficiently large vacant space, v^* (the activation volume) in the immediate vicinity. The actual vacant space adjacent to the unit varies around a mean value, which is given by the product of the free-volume fraction, f, the effective volume, v, of the chain segment, and a geometrical correction for overlap, g, which is on the order of unity. The mean free-volume per diffusing unit is much smaller than v^*, but occasionally a free volume of v^* or greater appears according to probability[14] p

$$p = \exp(-v^*/gvf) = \exp[-v^*/gv\alpha(T - T_0)] = \exp[-B/(T - T_0)] \tag{13}$$

where $B = v^*/g\alpha v$. Applied to the motion of the polymer, this probability is a common factor in all forms of transport; viscous, diffusive and electrical. Thus the fluidity, diffusivity and mobility may all be written in the Vogel, Tamman and Fulcher (VTF)[16] form

$$W_T = W_0 \exp[-B/(T - T_0)] \tag{14}$$

where W_0 is a pre-exponential factor, depending on the type of transport and W_T is the value of the property at temperature T. The parameters B and T_0 are related to the WLF parameters, C_1, C_2, and a reference temperature T_r. According to the WLF equation

$$\ln\frac{W_T}{W_{T_r}} = \frac{C_1(T - T_r)}{C_2 + (T - T_r)} \tag{15}$$

while the VTF equation gives

$$\ln\frac{W_T}{W_{T_r}} = -B\left(\frac{1}{T - T_0} - \frac{1}{T_r - T_0}\right) \tag{16}$$

Therefore $T_0 = T_r - C_2$ and $B = C_1 C_2$. The equation for mobility can be written

$$b = A\exp\{-B/(T - T_0)\} \tag{17}$$

or

$$b = A\exp\{-E_a/R(T - T_0)\} \tag{18}$$

where $B = E_a/R$. The latter equation takes the place of the Arrhenius formula used for diffusivity and conductivity in crystalline solids, and the analogous form used for the VTF plot is

$$\log b = A/2.3 - B/2.3(T - T_0) \tag{19}$$

However, E_a is only an apparent activation energy since it is not connected with a model involving a barrier height of that value. A small temperature dependence $(T^{-1/2})$ of A is also predicted, but this has little effect on the linearity of the VTF plot and therefore has not been verified experimentally.

The procedure of fitting data to the VTF equation is more difficult and arbitrary than the Arrhenius plot for solids because of the extra parameter, T_0. One method of extracting parameters is to plot a series of $\log(\sigma T^{1/2})$ vs. $1/(T - T_0)$ with various trial T_0 values, and choosing the T_0 value which gives the best straight line. This method is invalidated if A changes excessively with temperature, as it does when the degree of crystallinity or dissociation depends on temperature. Unfortunately, this has been the case with semicrystalline PEO, on which most conductivity data have been gathered. However, the value of T_0, may be obtained by independent methods, *e.g.* from the temperature dependence of mechanical properties which yields T_0 as $(T_r - C_2)$ from the WLF analysis.[17]

Theoretically, the free-volume model for ionic conduction suggests that the parameter B, unlike T_0, should have different values from those used for polymer segmental motion; B is expected to reflect the relative sizes of the diffusant and the chain segments and should also vary with the type of ion. The VTF equation should therefore only apply to a partial conductivity due to a single carrier, and it is surprising that the total conductivity normally measured should exhibit a VTF form of temperature dependence.

The above treatment is one of the simplest available. The configurational entropy model[19] is more rigorous, but conveniently gives the same form as free-volume theory for the temperature dependence of conductivity. However, the meaning of the parameters is different. T_0 is the temperature at which the conformational entropy ('configurational entropy' is actually a misnomer in a polymer context) becomes zero and

$$B = \frac{\Delta E s_c^* T_0}{k B_1} \tag{20}$$

where ΔE = activation energy for chain segmental motion

$$B_1 = \frac{c_p(\text{liquid}) - c_p(\text{glass})}{T} \tag{21}$$

which is almost constant over a range of temperatures and s_c^* is a constant. In this case there is no direct connection between the value of B and the nature of the diffusing entity, although the polymer structure, and therefore T_0 and c_p are, of course, modified by the presence of ions in large concentrations. One may also note that the formula implies a proportionality between B and T_0 if other factors are equal.

Both models can be modified by adding an activation energy for the escape of an ion from its preferred site before migrating[19]

$$\sigma = A \exp\{-B/(T - T_0) - E_a/kT\} \tag{22}$$

Free-volume theory ignores thermally activated ion-hopping between preferred sites, which dominates the mobility in a solid as seen above. The dynamic bond percolation theory[19](DBP) assumes, on the other hand, a *solid-state*-type jumping of ions along short strings of sites which occasionally connect together according to a *liquid-state*-type self-diffusion mechanism of the sites themselves. (In the context of ionic conduction, the word 'percolation' has been used to describe the accumulation of ion sites into continuous pathways.) A schematic illustration of this process is shown in Figure 2(c).

The DBP theory can be generalized to cover a multitude of situations. It reduces to the Arrhenius equation if the fraction of sites which percolate is large. If the fraction of sites which percolate is small, and the host polymer motion is according to free-volume theory, the result of DBP theory is simply the VTF equation. In this case, the parameter B applies either to the polymer host or to the ions according to the relative rates of ion hopping and string renewal: if the renewal rate is the faster, ion parameters should be used; if ion hopping is the faster, the parameters used are those of the polymer. The latter case is believed to apply in most cases of high ionic conductivity in a polymer host, so that the value of B should be dependent only on the host. The former case may be expected to apply to ions which exhibit very slow desolvation kinetics, *e.g.* Ni^{2+}. (The recent discovery,[20] in the case of $NiBr_2$ dissolved in PEO, that the conductivity of Ni^{2+} is negligible compared with that of the counterion Br^- correlates well with the extremely slow exchange kinetics of that cation with its crown ether complexes.[21]) The DBP theory can also take into account the effect of the density of coordination or other solvating sites; a low site concentration should sharply decrease the fraction of sites which are connected at a given time as well as the hopping rate.[19]

The above theories can considerably enhance the understanding of the ion transport process in a particular polymer and highlight the importance of a low T_g value of the polymer in obtaining a highly conducting electrolyte. However, the models and the evidence to date are contradictory as to whether or not B is an adjustable parameter, and whether or not a single value of B can be used to describe the motion of different species in the same polymer. If we regard B as a constant, the main parameters to be considered are the T_0 value of the polymer and a pre-exponential factor dependent primarily on the ion concentration and hopping rate, as determined by the solvation site concentration.

As will be seen in the next section, the value of T_0 depends on both the nature of the polymer and the amount of added salt, and good correlations may be obtained between conductivity and T_0 according to the VTF equation. However, a rigorous analysis of the variation of B and the pre-exponential factor requires data regarding the partial conductivities, which are generally unavailable, particularly in concentrated solutions as used in the better-conducting polymer electrolytes. This is because the salt is an equilibrium mixture of free ions, ion pairs, ion triplets and higher aggregates.

21.3 THE CHEMISTRY OF POLYMER–SALT COMPLEXES AND SOLUTIONS

21.3.1 The Solvation of Salts by Polymer Solvents

The thermodynamic analysis of the dissolution process in a polymer is identical to that for all solvents. The salt can simply dissolve as a non-conducting ion pair, according to equation (23), or as separate solvated ions as in equation (24). In concentrated solution, multiple ion agglomerates can form (according to equation (25), for example).

$$MX(cryst) \longrightarrow (MX)(solv) \tag{23}$$

$$MX(cryst) \longrightarrow M^+(solv) + X^-(solv) \tag{24}$$

$$2MX(cryst) \longrightarrow M_2X^+(solv) + X^-(solv) \tag{25}$$

The free energy of solution is expressed in the usual way

$$\Delta G = \Delta H - T\Delta S \tag{26}$$

In the simplest analysis, an entropy increase favours dissolution, due to the dissociation of the crystal into the maximum number of independent particles. The main differences in solubility in various solvent/salt combinations reflect differences in the lattice enthalpy of the salt, cohesive enthalpy of the solvent and the solvation enthalpy. Conditions which favour dissolution are: (i) a low lattice energy of the salt; (ii) a low cohesive energy of the polymer; and (iii) a high solvation energy.

The relative stabilities of ion pairs and charged entities in solution also depend on the nature of both salt and solvent. Ion pairs tend to form according to the Coulombic attractive force, especially when the charge/radius ratio is high

$$F = \frac{q_1 q_2}{4\pi\varepsilon r^2} \tag{27}$$

where q_1 and q_2 are the charges on the ions, r is the ion–ion separation and ε is the permittivity.

Polar solvents favour the separation of ions by providing a high permittivity, and by increasing the effective radii of small ions by specific solvation or complexation: cations are Lewis acids and are solvated by electron donors; anions may be regarded as Lewis bases requiring electron acceptors. The extent of specific ion solvation by liquid solvents has been quantified approximately according to the Gutmann donor and acceptor scales of Table 2. Therefore conditions which favour ion dissociation may be summarized as: (iv) a high permittivity of solvent; (v) a high donor number for solvation of a small or multivalent cation; and (vi) a high acceptor number for solvation of a small or multivalent anion.

Table 2 Gutmann Donor and Acceptor Numbers of Liquid Solvents[a]

Solvent	DN	AN	ε_r
Nitrobenzene	4.4	14.8	35.9
Ethylene carbonate	16.4		89.1
Water	18.0	54.8	81.0
Diethyl ether	19.2	3.9	4.3
Dimethyl sulfoxide	29.8	19.3	45.0
Hexamethylphosphoramide	38.8		30.0

[a] Data taken from K. Burger, 'Solvation, Ionic and Complex Formation Reactions in Non-Aqueous Solvents', Elsevier, Amsterdam, 1983.

PEO is, of course, a long-chain extension of the series of α-methyl-ω-methoxyoligo(oxyethylene)s [previously called 'glyme' (glycol methyl ether), 'diglyme', 'triglyme', *etc.*] which are good solvents for metal ions because of the Lewis-base nature of the lone pairs on the oxygen atom. Although the donor number of diethyl ether does not indicate outstanding cation-solvating properties for the oxygen atom, macrocycles based on the oxyethylene (EO) unit, the crown ethers, are among the strongest known complexing agents for cations. The EO unit features strongly in a review of macrocyclic multidentate ligands by Izatt *et al.*,[21] which tabulates equilibrium and rate constants for the complexation of a very large number of combinations of cations and ligands. The reason for this is probably geometrical, *i.e.* the rings formed by EO units tend to place oxygen atoms in favourable positions for the formation of coordination bonds with metal cations. The solvating properties of PEO have been shown to be superior to the closely related poly(oxymethylene) and poly(oxytri-methylene).[22] Therefore, PEO can be considered to be a very cheap universal complexing agent. A large number of salts have been shown to mix with PEO to form solutions and complexes

Table 3 Salts Shown to Form (+) or not Form (−) Adducts with PEO[a]

	Li$^+$	Na$^+$	K$^+$	Mg^{2+}	Ca^{2+}	Zn^{2+}	Cu^{2+}
F$^-$	−	−	−	−	−	−	−
Cl$^-$	+	−	−	+	−	+	+
Br$^-$	+	+	−	+	+	+	+
I$^-$	+	+	+	+	+	+	
NO$_3^-$	+	−	−	+			
SCN$^-$	+	+	+	+	+	+	+
ClO$_4^-$	+	+	+	+	+	+	+
CF$_3$SO$_3^-$	+	+	+	+	+	+	+
BPh$_4^-$	+	+	+	+			

[a] Data from M. B. Armand in ref. 10.

according to Table 3. Indeed, there seems to be no restriction on which cations produce solutions provided they are combined with large monovalent anions. Salts of small, or multivalent, anions have a high lattice energy and unless the anion can be solvated (*e.g.* by a hydrogen-bonding moiety, which is not present in PEO) the solubility is negligibly small.

PEO and its derivatives have dominated the field of ion-conducting polymers. This not only reflects its good solvating power, but also, because of projected applications in lithium batteries, chemical stability to the highly reducing lithium-metal negative electrode and highly oxidizing positive electrodes is required. However, moieties other than ethers have also been found to be effective in cation solvation, as shown in Table 4.

Table 4 Moieties other than Ethers Capable of Ion Solvation within Polymeric Structures

Solvating group	Polymer	Cation solvated
Ester	Poly(ethylene succinate)[a]	Li$^+$
Amino	Poly(ethyleneimine)[b]	Li$^+$
Sulfide	Poly(ethylene sulfide)[c]	Ag$^+$

[a] R. Dupon, B. L. Papke, M. A. Ratner and D. F. Shriver, *J. Electrochem. Soc.*, 1984, **131**, 586. [b] C. K. Chiang, G. T. Davis, C. A. Harding and T. Takahashi, *Macromolecules*, 1985, **18**, 825. [c] S. Clancy, D. F. Shriver and L. A. Ochrymowycz, *Macromolecules*, 1986, **19**, 825.

21.3.2 Effects of Salt Addition on Crystallizable Polymers

The addition of a salt to a liquid complexing-solvent results in a solution at low concentration and a crystalline complex at high concentration. Correspondingly, in semicrystalline PEO, the phase diagram is like the example shown in Figure 3. The diagram shows a liquid-like solution above the eutectic point, the separation of semicrystalline PEO at low salt concentrations below the melting point, and the separation of an almost stoichiometric complex at high salt concentrations. Below the eutectic point crystalline polymer and the stoichiometric complex coexist. An additional feature of the polymer system is the existence of an amorphous phase below the eutectic point. (As for the pure polymer, which exists as a mixture of a crystalline and an amorphous phase below the melting point, we have an apparent violation of the Gibbs phase rule; with two components, *i.e.* polymer and salt, two phases should be able to coexist only at a specified temperature (the melting point) when the pressure is constant. The reason for this is, of course, that the formation of the phases is kinetically controlled and that, even if equilibrium could be achieved, chain ends and other chain defects would create a *de facto* multicomponent system.)

Since the amorphous and crystalline fractions of a high polymer cannot be separated mechanically, *e.g.* by filtration as for the separation of liquid solutions from crystalline complexes, the phase behaviour of polymer–salt systems is more difficult to chart than that of liquid solvents. However, many techniques, *e.g.* polarized light microscopy,[23] differential scanning calorimetry,[24] X-ray diffraction[25] and NMR,[26] have been used to study the partition of the salt and the temperature dependence of phase behaviour.

Crystallinity is generally undesirable in ion-conducting polymers because the crystalline phase is invariably much less conductive than the corresponding amorphous phase. The temperature

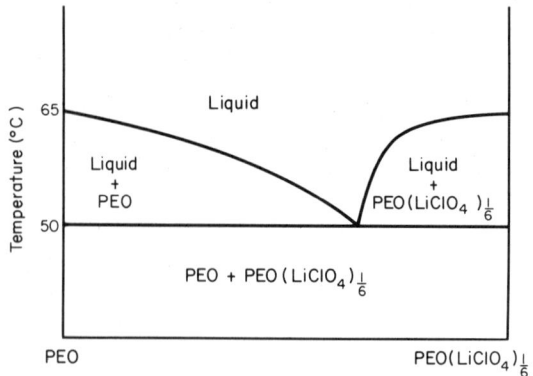

Figure 3 Phase diagram of the PEO/LiClO$_4$ system[61]

dependence of the conductivity in PEO is dominated by the amount and connectivity of the amorphous phase. Acceptable conductivities are only obtained above the eutectic point. The mechanical properties are also affected, because crystals are effective as crosslinks. The effective crosslink density is therefore highly temperature-dependent so that films which are strong at room temperature can become liquid above the melting point. The optimum balance between conductivity and mechanical strength can only exist over a small temperature range unless additional cross-linking agents or fillers[27] are added. The temperature-dependent crystallinity also impedes the study of the temperature dependence of conduction in the amorphous phase.

21.4 IONIC CONDUCTIVITY IN AMORPHOUS SOLVATING POLYMERS

21.4.1 Amorphous Solvating Polymers

If the polymer host is structurally constrained from crystallizing, as is the case in atactic poly(propylene oxide) (PPO), the solutions of salts are also amorphous. Quite large concentrations of salt remain dissolved at low temperatures without crystallization occurring. Many conductivity studies have been made on solutions of salts in this polymer, but the conductivities obtained are lower than those obtained for amorphous varieties of PEO, and the range of salts dissolved is more restricted because of the steric hindrance to solvation.[28] Also, steric hindrance to the backbone C—C bond rotation increases T_g compared with the value for PEO.

Most of the recent studies on polymer electrolytes have been directed at incorporating the EO units into polymers which do not crystallize near room temperature.

The first approaches were urethane crosslinked networks of poly(ethylene glycol)s,[29] which solvated lithium salts in a similar way to PEO but without causing crystallinity unless the number-average molecular weight of the glycol was above about 1500. In this case, however, the necessarily high concentration of crosslinks caused an increase in T_g according to the expression[30]

$$1/T_g \;=\; 1/T_{g0} \;-\; 0.76c \tag{28}$$

where T_{g0} = extrapolated T_g value without crosslinks and c = molar concentration of crosslinks.

The increase in T_g due to the crosslinks was accompanied by a decrease in the ion mobility, as explained in the previous section, and the conductivities obtained were rather low. Also the networks could not be dissolved or liquefied and, therefore, processing problems were anticipated.

The glass transition temperatures of these polymers also increased with salt content, because of a stiffening of the structure, which can be ascribed to the formation of transient crosslinks between solvating moieties through the metal ions. The increase in viscosity of liquid poly(propylene glycol) on salt addition was, in fact, one of the first experimental observations symptomatic of a polymer–salt interaction.[31]

In the case of the urethane-linked poly(ethylene glycol) networks[30] the addition of LiClO$_4$ was found to cause an additional increase in T_g according to

$$1/T_g \;=\; 1/T_{g0} \;-\; 0.267c \tag{29}$$

where T_{g0} = T_g value without salt and c = molar salt concentration. This result was interpreted as the effect of ionic crosslinks.

High molecular weight polymers have elastomeric properties by virtue of their chain entanglements, and there is no need to introduce chemical crosslinks. Recent approaches have used short sequences of EO units as side chains[32] on a linear polymer backbone and, less frequently, in the main chain.

Poly[bis(methoxyethoxyethoxy)phosphazene] (1; MEEP) and its analogues were among the first reported examples of comb-branched amorphous polymers incorporating EO units in side chains.[33] These are still among the best materials from the point of view of ionic conductivity, although little information is available regarding their mechanical strengths. Their precursor, poly(dichlorophosphazene), has a low T_g due to the flexibility of the phosphazene $(P—N)_x$ chain, which is isoelectronic with the equally flexible $(Si—O)_x$ chain of the siloxane polymers. Correspondingly, MEEP has a glass transition temperature of about 190 K which is consistent with its excellent conductivity. The transition temperature increases on salt addition, *e.g.* with 0.5 Ag^+ per EO unit T_g is 280 K.

(1) MEEP

Other comb-branch polymers have been made with methacrylate[34,35] and itaconate backbones.[36] In the latter cases, the T_g values of the precursor and short-chain substituted polymers were rather high, again increasing upon salt addition, and showed lower conductivities than MEEP. However, conductivity values comparable with MEEP were obtained in the methacrylates with 22 units of EO per side chain (although the samples slowly crystallized).

Comb-branch polymers with siloxane backbones have very low T_g values and high conductivities,[37,38,39] but the samples prepared were liquids at room temperature because of the low molecular weight. Solid products were obtained either by crosslinking or blending with PEO.

The polymerization of ethylene oxide macromers prepared by the reaction of monomethylethylene glycol with acetylene has produced comb-shaped polymers, coded $PVMEO_3$, in which the backbone is sterically unhindered, with T_g around 200 K. Addition of salt gave electrolytes with a conductivity behaviour similar to MEEP.[18]

Poly(ethylene glycol) has been chain extended to a high molecular weight amorphous rubbery linear polymer using an oxymethylene link.[40] This polymer, when doped with lithium perchlorate, shows similar conductivity to MEEP and because of its chemical similarity to PEO provides an interesting comparison to show the effect of crystallinity.[41]

In all the above electrolytes, the conductivity changes in a complex way with temperature and salt level, as in the examples of Figure 4, so that electrolytes made from different polymer hosts cannot easily be ranked in order of performance.

21.4.2 Rationalization of Conductivity Data

A study on a large number of network polymers containing salts indicated[17] that the WLF parameters with T_g as the reference temperature were invariant, and roughly equal to the generally accepted values of $C_1 = 10$ and $C_2 = 50$ K over a large number of crosslinked polyethers with a variety of dopant ions. According to these data, B has the value of 500 K ($\pm$ 15%) and $(T_g - T_0)$ is 50 K ($\pm$ 10%) for all the systems studied in that work. Therefore, the two parameters A and T_g were sufficient to determine the conductivity at any temperature. However, other workers have reported a far greater range for B, implying that B is a sample-dependent variable like A and T_g or T_0, and a greater range for $(T_g - T_0)$. For example, the study[18] of $PVMEO_3/LiClO_4$ mentioned above reported B values from 584 to 1440 across the composition range and an explanation was offered in terms of an increase in the activation energy for side-chain rearrangement with salt content. The values for B and T_0 were, in this case, obtained *via* non-linear least-squares analysis of conductivity data. According to T_g values determined by DSC, $(T_g - T_0)$ apparently varied from 28.5 to 48 K. However, those authors accept[60] that the procedure of extracting parameters from the best VTF fit is prone to large experimental errors, and may not be valid at all if the pre-exponential factor has a temperature dependence which has not been accounted for. This point is illustrated in Figures 5(a)

Ionic Conductivity

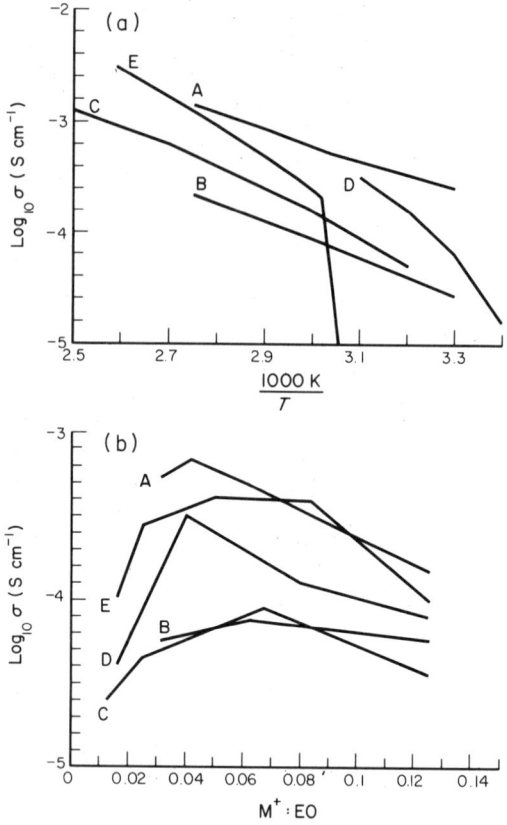

Figure 4 Conductivity *vs.* (a) reciprocal temperature and (b) concentration in a number of polymer electrolytes

	Sample	$EO:M^+$	$T(^\circ C)$
A	MEEP/AgCF$_3$SO$_3$[33]	32	55
B	MEEP/LiCF$_3$SO$_3$[33]	16	55
C	PVMEO$_3$/LiClO$_4$[18]	15	50
D	Chain-extended PEG/LiClO$_4$[41]	24	50
E	PEO/LiClO$_4$[61]	12	65

and 5(b) where the same data are plotted using T_0 values derived from the measured T_g values using two alternative, arbitrary values of $(T_g - T_0)$. At all but the highest salt concentrations (where the degree of salt dissociation may depend significantly on temperature, and the salt concentration is so high that the nature of the conduction process may well be different to the dilute case) reasonably straight lines can be obtained for more than one value of $(T_g - T_0)$ and the resulting B values are independent of salt concentration, but strongly dependent on the choice of $(T_g - T_0)$. The similarity of the B values at constant $(T_g - T_0)$ suggests that B and $(T_g - T_0)$ are constants for the system and that there are only two variables — A and T_g or T_0. A range of values for the constants can be used to give reasonable fits, but the two are interdependent, allowing only one degree of freedom. A unique pair of constants undoubtedly exists, but cannot be established with great accuracy without data at temperatures closer to T_g. The constants which apply to these samples cannot both be compatible with the standard WLF constants; *e.g.* the choice of 50 K for $(T_g - T_0)$ (C_2) requires that $B = 1000$ K, and $C_1 = 20$. An alternative choice, which gives a slightly better fit, is $C_2 = 30$ K, $B = 750$ K and $C_1 = 24$.

In all cases of high conductivity, the polymer–salt interaction causes an increase in T_g (which is, in this case, proportional to the square root of concentration), *e.g.* Figure 5(c), and therefore a decrease in mobility according to the VTF expression. This explains why the addition of salt causes an approximately linear increase in conductivity at very low concentration (where the increase in carrier concentration is dominant) followed by a decrease at high concentration due to the decrease

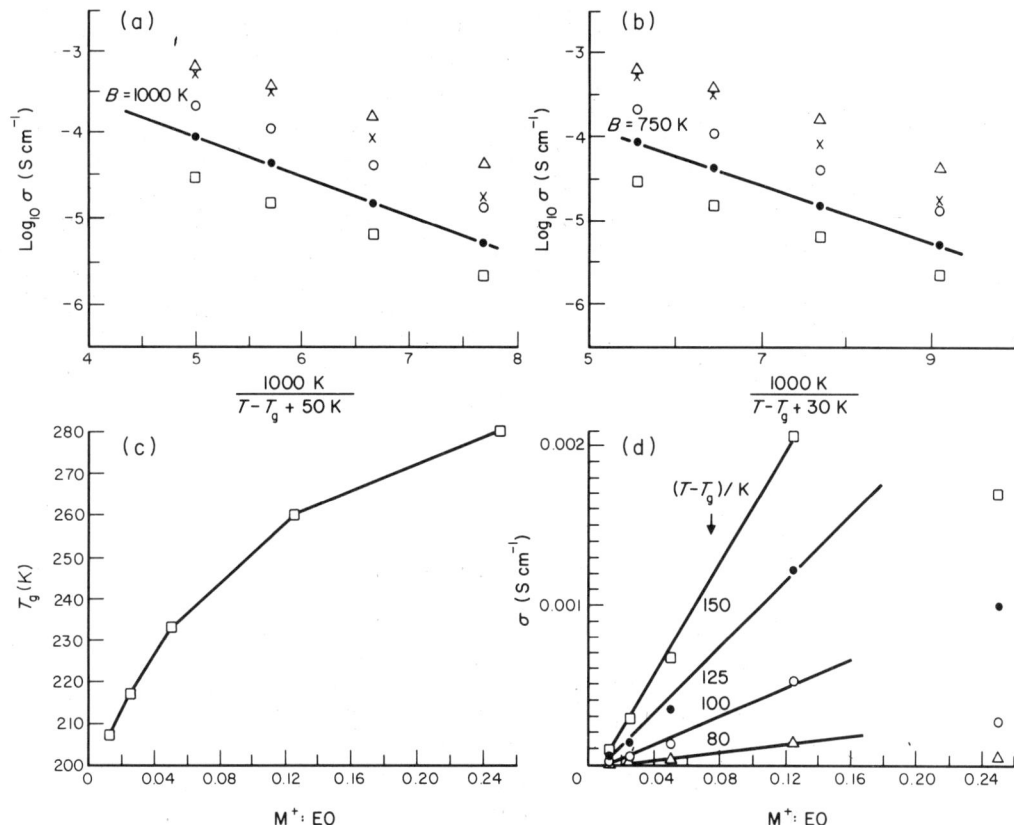

Figure 5 (a) and (b) VTF plots with $T_g - T_0 = 50$ and 30 K; (c) dependence of T_g on salt concentration; (d) plots of conductivity *vs.* concentration at constant $T - T_g$. $M^+:EO = \square$, 0.0125; +, 0.0250; $\diamond$, 0.050; ×, 0.125; $\triangle$, 0.250 (data for PVMEO$_3$/LiClO$_4$ from ref. 18)

in mobility (Figure 4b). Data may be rationalized by plotting conductivity *vs.* concentration using the reduced temperature as the parameter, *e.g.* Figure 5(d). In this example, the effective carrier concentration is roughly proportional to the actual concentration of added salt up to 0.1 cations per EO unit. Therefore the variable A can be replaced by a constant A' multiplied by the concentration, *e.g.*

$$\sigma = A'c\exp\{-B/(T - T_g(c) + 50)\} \tag{30}$$

A' can vary considerably, and can cause large differences in conductivity, between different materials with similar T_g.

Considering the improbability of further reductions in T_g values, the outlook for future discoveries of much higher conductivities would seem to depend on increasing A', and the rather poor properties of the EO unit as viewed in Table 2 suggest that there should be room for improvement. However, the groups which give good solvation properties do not occur in conjunction with low T_g values. There is also a possibility of the discovery of improved salts, although optimization studies have been in progress for many decades in liquid electrolyte studies. Taking these factors into account, the author does not expect to discover room-temperature conductivities much higher than 10^{-4} S cm^{-1} unless static site percolation can be achieved as in the case of solid-state ionic conductors.

21.5 THE MEASUREMENT OF CONDUCTIVITY AND OTHER TRANSPORT PROPERTIES

21.5.1 Ionic Conductivity

The (zero frequency) conductivity is defined as the steady-state current which flows in unit field. The measured quantity is usually the conductance (G), which is the ratio of steady-state current to applied voltage. The conductivity is obtained from the conductance and ratio of length to cross-sectional area (l/A) of a rectangular sample according to

$$\sigma = Gl/A \tag{31}$$

A cell for measuring the conductance of an electrolyte sample must: (a) transform the electronic current from the external circuit into an ionic current in the sample; and (b) transmit the potential difference across the sample to the voltage-measuring device without introducing any additional potentials.

The transformation of current occurs at the electrode/electrolyte interface where either a Faradaic or a displacement current flows. A Faradaic current occurs when an ion present in the electrolyte can be plated on to or removed from the electrode surface in response to the electronic current, *e.g.* when a lithium-based electrolyte is placed in between two lithium-metal electrodes.

The standard method of conductivity measurement has a displacement current at the interface, where ionic charge accumulates at one side to exactly balance the accumulation or depletion of electrons. The charge accumulation creates an interfacial potential which would rise rapidly on application of a direct current. Therefore an alternating signal must be applied. This creates a constant interfacial potential which is determined by interfacial capacitance, current and frequency

$$V_{int} = \frac{I}{2\pi\omega C} \tag{32}$$

The interfacial potential can be minimized by increasing the frequency of the applied signal. For most cases a frequency of 1 kHz, or higher, ensures that any interfacial potentials are negligible.

At high frequencies, however, the current is not the steady-state value. Internal displacement, or capacitive, currents due to ions vibrating in potential wells without jumping the barriers are added to the current due to true conductivity. The displacement current is out of phase with the potential, whereas the conductive current is in phase, and therefore a distinction can be made by examining the complex impedance

$$Z^* = V^*/I^* \tag{33}$$

where out-of-phase components are represented by the imaginary parts.

Standard instruments are available for this measurement. For example, a low-frequency impedance analyzer measures Z^* directly as a function of a scanned frequency. Alternatively, a frequency response analyzer can be used to apply V^* and measure a voltage proportional to I^* after passing the current through a suitable current-to-voltage converter.

The complex impedance of a sample between inert electrodes is generally measured over a wide frequency range, *e.g.* 10^6 Hz to 10 Hz, and plotted as an Argand diagram with frequency as a parameter, *e.g.* Figure 6.

In the complex-plane representation, the parallel combination of conductive and displacement currents in the sample appears as a high-frequency semicircle and the series combination of

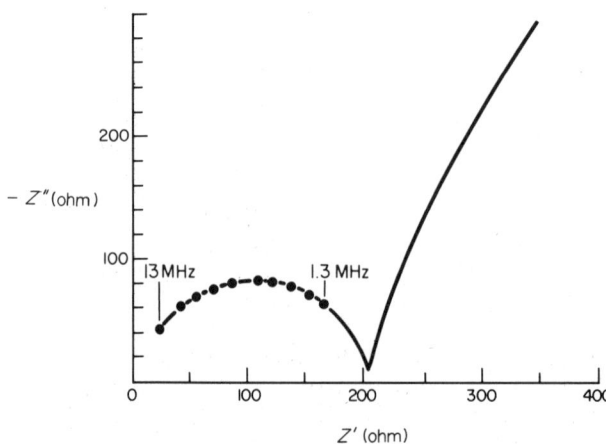

Figure 6   Complex impedance plot for chain-extended PEG/LiClO$_4$:[41] Z' and Z'' are the real and imaginary parts

interfacial displacement current and sample conductive current appears as a low-frequency spur. Ideally, the two should extrapolate to the same point on the real axis, which indicates the resistance of the sample. Many complex impedance data have been discussed in reviews from the electrochemical[42] and the solid-state physical[43] points of view.

The cell design required to achieve a reliable measurement depends on the sample shape and conductivity,[44] although for most polymer electrolytes of interest a path length of about 1 mm and a cross-section of about 10 mm² will suffice. Alternatively, provided stray capacitance and inductance are eliminated, a very short path length and a large area of cross-section can be used. This gives a measurement of the electrical relaxation time constant, which is the ratio of permittivity to conductivity, without a knowledge of the sample dimensions.[44]

The conductivities of the two components of a composite structure can also, in principle, be analyzed by the complex impedance method. Ceramic samples often show two semicircles, one due to a bulk phase and another due to a less conductive grain boundary phase — the two components are effectively in series.[45] In a semicrystalline polymer, however, the 'grain boundary phase' is the more conductive, so that the continuous (parallel) conduction path is more important. The analysis is more complex in this case, but it has been shown that partial crystallinity leads to depressed and skewed semicircles compared with the impedance diagram of corresponding amorphous material.[41]

21.5.2 Transport Number

The transport, or transference, number of an ion is defined as the proportion of the total current carried by that ion. The above method yields the total conductivity due to all ions present, *i.e.* cation and anion in the case of a simple dissolved salt. If the partial conductivity due to a single ion is required, a Faradaic electrode process must be devised, *e.g.* by using electrodes of the same metal as the cation. On application of a constant potential the current decays with time until a steady state is reached corresponding to the condition

$$i_- = 0; \quad i = i_+ = \frac{\sigma_+ \, \mathrm{d}\tilde{\mu}_+}{zF \, \mathrm{d}x} = \sigma_+ \frac{VA}{l} \tag{34}$$

where μ_+ = electrochemical potential of the cation. (The Nernst–Planck equation, rather than the special case of Ohm's law, is used here to account for the concentration gradient that arises during the measurement, which produces complications due to diffusion).

The transference or transport number of the ion is then given as the ratio of the partial to the total conductivity. An equivalent method to this is one based on very low-frequency measurements of the complex impedance,[46] which again requires special electrodes.

The above equation ignores interfacial impedances due to the electrode processes.[47] The consequent error can be estimated by repeating the measurements with different path lengths or, more elegantly, by the four-point method,[48] which measures an internal electrochemical potential difference between two reference electrodes placed inside the electrolyte.

The Tubandt method is a direct approach to the measurement of transport number according to Faraday's law. It is particularly applicable to network polymers,[49] which can be arranged as a stack of discs placed in electrical series between two electrodes made of the same metal as the cation. For each Faraday of charge passed, the disc next to the positive electrode gains t_-/xm moles of salt $M_x^{m+} X_m^{x-}$, which can be determined by weighing (or some other analysis) provided the disc can be separated from the stack at the end of the experiment.

Finally, transport numbers have been determined from the 'liquid junction' potential in a cell containing two electrolytes in series;[50] *e.g.* for a 1:1 salt we have

$$E_j = \frac{RT}{F} \int_1^2 (t_+ - t_-) \, \mathrm{d}\ln a \sim \frac{RT}{F} (t_+ - t_-) \ln \frac{[MX]_2}{[MX]_1} \tag{35}$$

21.5.3 Mobility

A fuller understanding of the conduction process requires measurement of the mobility. This is best obtained *via* non-electrical methods which directly probe self-diffusion.

The most straightforward method of obtaining the mobility is the tracer diffusion method,[51] whereby a small quantity of a radioactive isotope of the relevant ion is introduced, and its diffusion measured by subsequent sectioning and counting. (The measured quantity, the tracer diffusion coefficient, is not quite the same as the self-diffusion coefficient, although a statistical analysis can be used to relate the two.)

NMR can also be used to estimate the mobility. Three methods are in common use.[51] The first is based on the distribution of resonance frequencies corresponding to different ion environments in the sample. Translational motion of the ion causes averaging of the local magnetic fields and a consequent narrowing of the distribution. Thus, under certain conditions, the linewidth is proportional to the mobility. The more sophisticated pulsed methods evaluate a correlation time, *i.e.* the mean residence time of an ion on a site, which is inversely proportional to the mobility. Both of these methods can be used to find the temperature dependence of the mobility, but the calculation of the mobility itself requires a jump distance, which is not normally known. A third approach,[52] the pulsed field gradient method, can evaluate the self-diffusion coefficient directly. However, the lower limit of diffusion coefficient which can be measured at present is about 10^{-9} cm^2 s^{-1}, which is, unfortunately, at the top of the range of diffusion coefficients observed for polymer electrolytes. Nevertheless, this method is of increasing importance in the understanding of highly conducting polymer electrolytes.

21.6 APPLICATIONS OF ION-CONDUCTING POLYMERS

21.6.1 Battery Electrolytes

At the time of writing, the solid-state alkali metal battery[53] is by far the most important projected application of a polymer electrolyte. This is based on a thin film laminated structure containing a lithium-metal negative electrode, a polymer film electrolyte, and a positive electrode made of an oxidizing agent capable of inserting alkali ions into its structure (Figure 7).

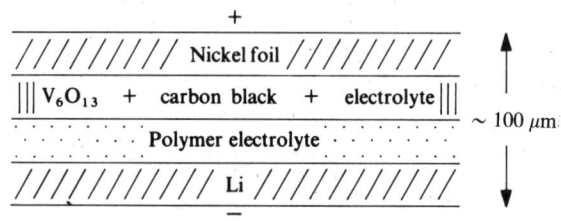

Figure 7   Schematic construction of a lithium battery laminate

A number of polymer electrolytes now have conductivities within the target range of 10^{-5} to 10^{-4} S cm^{-1}. However, a number of problems remain to be solved before expectations can be fulfilled. One is the transport number of the electrode-reversible ion, with which the commonly quoted total conductivity must be multiplied to obtain the operational value. Another, more severe, problem is that of stability to extremes of temperature and electrochemical environment. Some encouraging long-term test results have appeared,[54] but as yet no manufacturer has ventured to market the product.

A particular advantage of the elastomeric electrolyte is in the fabrication of on-chip power sources, in which thin film structures similar to those described above can be deposited directly as components on an integrated circuit. This technology has been demonstrated in principle[55] but encapsulation problems have yet to be overcome due to the extreme reactivity of alkali metals.

21.6.2 Additives to Battery Electrodes

A less demanding role for the polymer electrolyte is as an aid to ion transport within an electrode structure.[56] In this case mechanical stability is not necessarily required, as the electrode particles themselves act as a filler. The stability, however, remains a critical issue.

21.6.3 Electrochromic Displays

Another interesting device is the electrochromic display or window, *i.e.* a device which changes its visual appearance in response to an electrical signal. Such devices can be made by laminating two electrode-coated transparent plates with the polymer electrolyte.[57] Here, at least one of the electrodes should be an electrochromic material — *i.e.* one whose optical absorbance changes with the insertion of ions from the electrolyte and electrons from the external circuit. The electrochemical stability of the polymer may not be such a demanding requirement as the cell potential need not be high. However, a high conductivity is required to minimize the switching time.

21.6.4 Chemical Sensors

A number of sensor designs based on electrochemical cells can benefit from miniaturization and mass production by the use of polymer electrolytes. In this application a high conductivity is not required but, depending on the type of sensor, selectivity of the conductivity to one ion may or may not be required. For example, a hydrogen sensor based on the cell

$$H_2, Pt \mid electrolyte \mid H_xWO_3$$

would not need to be selective to protons. However, a potassium ion sensor

$$K^+(sample) \mid electrolyte \mid K^+(ref)$$

would need to be conductive to potassium ions alone, as in the widely used potassium-sensitive electrode[62] based on PVC plasticized with a valinomycin-doped ester. Here the valinomycin provides a much higher conductivity to potassium over other cations due to selective complexation.

21.6.5 Fuel Cells

The cell

$$H_2(Pt) \mid electrolyte \mid O_2(Pt)$$

is the basis of many commercial fuel cells.[58] Liquid electrolyte fuel cells operating below 100 °C are limited in power output by slow electrode kinetics. The use of a proton-conducting polymer electrolyte with a high thermal stability allows a higher operating temperature. Nafion-based membranes are currently in use, although PEO can also be made into a proton conductor by the addition of phosphoric acid.[59]

21.7 REFERENCES

1. C. Wagner, *Z. Phys. Chem., Abt. B*, 1933, **22**, 181.
2. B. C. H. Steele, *Solid State Ionics*, 1984, **12**, 391.
3. M. Kleitz, G. F. Million-Brody and P. Fabry, *Solid State Ionics*, 1987, **22**, 295.
4. A. E. Binks and A. Sharples, *J. Polym. Sci., Part A-2*, 1968, **6**, 407.
5. M. B. Armand, J. M. Chabagno and M. Duclot, extended abstract in '2nd International Meeting on Solid Electrolytes, St. Andrews, Scotland, Sept. 1978'.
6. K. Tsunemi, H. Ohno and E. Tsuchida, *Electrochim. Acta*, 1983, **28**, 833.
7. K. Shigehara, N. Kobayshi and E. Tsuchida, *Solid State Ionics*, 1984, **14**, 85.
8. L. C. Hardy and D. F. Shriver, *J. Am. Chem. Soc.*, 1985, **107**, 3823.
9. D. J. Bannister, G. R. Davies, I. M. Ward and J. E. McIntyre, *Polymer*, 1984, **25**, 1291.
10. J. R. MacCallum and C. A. Vincent (eds.), 'Polymer Electrolyte Reviews 1', Elsevier Applied Science, London, 1987.
11. *Br. Polym. J.*, 1988, **20**.
12. N. F. Mott and R. W. Gurney, 'Electronic Processes in Ionic Crystals', Clarendon Press, Oxford, 1948.
13. J. O'M. Bockris and A. K. N. Reddy, 'Modern Electrochemistry', Macdonald, London, 1970.
14. M. H. Cohen and D. Turnbull, *J. Chem. Phys.*, 1959, **31**, 1164.
15. M. L. Williams, R. F. Landel and J. D. Ferry, *J. Am. Chem. Soc.*, 1955, **77**, 3701.
16. G. S. Fulcher, *J. Am. Ceram. Soc.*, 1925, **8**, 339.
17. J. F. LeNest, A. Gandini and H. Cheradame, *Br. Polym. J.*, 1988, **20**, 253.
18. J. M. G. Cowie, A. C. S. Martin and A. M. Firth, *Br. Polym. J.*, 1988, **20**, 247.
19. M. A. Ratner, in 'Polymer Electrolyte Reviews 1', ed. J. R. MacCallum and C. A. Vincent, Elsevier Applied Science, London, 1987, chap. 7.
20. R. Huq and G. C. Farrington, *J. Electrochem. Soc.*, 1988, **135**, 524.
21. R. M. Izatt, J. S. Bradshaw, S. A. Nielsen, J. D. Lamb and J. J. Christensen, *Chem. Rev.*, 1985, **85**, 271.
22. A. I. Shotenstein, E. S. Petrov and E. A. Yokovlevla, *J. Polym. Sci., Part C*, 1967, **16**, 1799.
23. D. A. Helmsley, in 'Comprehensive Polymer Science', ed. G. Allen and J. C. Bevington, Pergamon Press, Oxford, 1988, vol. 1, chap. 33.

24. M. J. Richardson, in 'Comprehensive Polymer Science', ed. G. Allen and J. C. Bevington, Pergamon Press, Oxford, 1988, vol. 1, chap. 36.
25. G. R. Mitchell, in 'Comprehensive Polymer Science', ed. G. Allen and J. C. Bevington, Pergamon Press, Oxford, 1988, vol. 1, chap. 31.
26. C. Berthier, W. Gorecki, M. Minier, M. B. Armand, J. M. Chabagno and P. Riqaud, *Solid State Ionics*, 1983, **11**, 91.
27. J. E. Weston and B. C. H. Steele, *Solid State Ionics*, 1984, **7**, 75.
28. M. B. Armand, J. M. Chabagno and M. J. Duclot, in 'Fast Ion Transport in Solids', ed. P. Vashista, J. N. Mundy and G. K. Shenoy, North Holland, Amsterdam, 1979, p. 131.
29. A. Killis, J. F. LeNest and H. Cheradame, *Makromol. Chem., Rapid Commun.*, 1980, **1**, 595.
30. H. Cheradame and J. F. LeNest, in 'Polymer Electrolyte Reviews 1', ed. J. R. MacCallum and C. A. Vincent, Elsevier Applied Science, London, 1987, p. 109.
31. J. Moacanin and E. F. Cuddihy, *J. Polym. Sci., Part C*, 1966, **14**, 313.
32. J. M. G. Cowie, in 'Polymer Electrolyte Reviews 1', ed. J. R. MacCallum and C. A. Vincent, Elsevier Applied Science, London, 1987, chap. 4.
33. P. M. Blonski, D. F. Shriver, P. Austin and H. R. Allcock, *J. Am. Chem. Soc.*, 1984, **106**, 6854.
34. D. J. Bannister, G. R. Davies, I. M. Ward and J. E. McIntyre, *Polymer*, 1984, **25**, 1600.
35. D. W. Xia, D. Soltz and J. Smid, *Solid State Ionics*, 1984, **14**, 221.
36. J. M. G. Cowie and A. C. S. Martin, *Polym. Commun.*, 1985, **26**, 298.
37. K. Nagaoka, H. Naruse, I. Shinohara and M. Watanabe, *J. Polym. Sci., Polym. Lett. Ed.*, 1984, **22**, 659.
38. A. Bouridah, F. Dalard, D. Deroo, H. Cheradame and J. F. LeNest, *Solid State Ionics*, 1985, **15**, 233.
39. D. Fish, I. M. Khan, E. Wu and J. Smid, *Br. Polym. J.*, 1988, **20**, 281.
40. C. V. Nicholas, D. J. Wilson, C. Booth and J. R. M. Giles, *Br. Polym. J.*, 1988, **20**, 289.
41. E. Linden and J. R. Owen, *Solid State Ionics*, 1988, in press.
42. W. I. Archer and R. D. Armstrong, in 'Electrochemistry', The Royal Society of Chemistry, London, 1980, vol. 7, p. 157.
43. A. K. Jonscher, *Phys. Status Solidi A*, 1975, **32**, 665.
44. E. Linden and J. R. Owen, *Br. Polym. J.*, 1988, **20**, 237.
45. J. E. Bauerle, *J. Phys. Chem. Solids*, 1969, **30**, 2657.
46. P. R. Sorensen and T. Jacobsen, *Electrochim. Acta*, 1982, **27**, 1671.
47. D. Fauteux and M. Gauthier, *Proc. Electrochem. Soc.*, 1987, **87-1**, 514.
48. P. G. Bruce, J. Evans and C. A. Vincent, *Solid State Ionics*, 1987, **25**, 255.
49. M. Leveque, J. F. LeNest, A. Gandini and H. Cheradame, *Makromol. Chem., Rapid Commun.*, 1983, **4**, 497.
50. A. Bouridah, F. Dalard, D. Deroo and M. B. Armand, *J. Appl. Electrochem.*, 1987, **17**, 625.
51. A. V. Chadwick and M. R. Worboys, in 'Polymer Electrolyte Reviews 1', ed. J. R. MacCallum and C. A. Vincent, Elsevier Applied Science, London, 1987, chap. 9.
52. S. Bhattacharja, S. W. Smoot and D. H. Whitmore, *Solid State Ionics*, 1986, **18/19**, 306.
53. A. Hooper, *Chem. Ind. (London)*, 1986, 198.
54. G. Vassort, M. Gauthier, P. E. Harvey, F. Brochu and M. B. Armand, 'Proceedings of Symposium on Primary and Secondary Ambient Temperature Lithium Batteries', The Electrochemical Society, Philadelphia, PV88-6, 1988, p. 780.
55. J. R. Upton, J. R. Owen, R. Rudkin, B. C. H. Steele, J. Benjamin and P. Tufton, *Solid State Ionics*, 1988 (proceedings of the conference at Garmisch-PartenKirchen, 1987).
56. J. R. Owen, J. Drennan, G. E. Lagos, P. C. Spurdens and B. C. H. Steele, *Solid State Ionics*, 1981, **5**, 343.
57. S. Pantaloni, S. Passerini and B. Scrosati, *J. Electrochem. Soc.*, 1987, **134**, 753.
58. G. J. Kleywegt and W. L. Driessen, *Chem. Br.*, 1988, **24**, 447.
59. F. Defendini, M. B. Armand, W. Gorecki and C. Berthier, *Electrochem. Soc. Extended Abstr.*, 1986, **86** (2), 893.
60. J. M. G. Cowie, private communication.
61. C. D. Robataille and D. Fauteux, *J. Electrochem. Soc.*, 1986, **133**, 315.
62. W. E. Morf and W. Simon, in 'Ion Selective Electrodes in Analytical Chemistry', ed. H. Freiser, Plenum Press, 1978, p. 212.

22

Electrical Conductivity

DAVID BLOOR

Queen Mary College, London, UK

22.1 INTRODUCTION

The electrical properties of materials are described by four broad categories: insulators, semiconductors, semimetals and metals. The boundaries between categories are somewhat arbitrary and are defined using either the resistivity (Ω cm) or the conductivity (Ω^{-1} cm^{-1} or S cm^{-1}) of the material. Often semimetals and metals are not distinguished. Thus, insulators have conductivities below 10^{-7} S cm^{-1}, semiconductors span 10^{-7} to 10^2 S cm^{-1} and metals have conductivities above 10^2 S cm^{-1}. The extremes of conductivity are of the order of 10^{-18} S cm^{-1} for quartz and 10^6 S cm^{-1} for silver and copper.

The conductivity of a material depends on its electronic energy-level structure.[1] When atoms or molecules are assembled to make a crystalline solid, the degenerate energy-levels produced spread to form non-degenerate energy-bands. The widths of these bands reflect the strength of the interaction between the individual atoms or molecules. The wave functions describing electrons in these band states extend throughout the solid. For a current to flow, an applied electrical-field must impart kinetic energy to electrons by promoting them to higher-energy band-states. For reasonable field-strengths this requires a small gap between filled and empty states, *e.g.* a partially filled band. The level of conductivity is dependent on the density of available filled and empty states.

In insulators there is an energy gap of several electron volts (eV) between the highest filled-electronic-state in the valence band and the lowest empty-state in the conduction band (1 eV is equivalent to 8065.7 cm^{-1}, 1.602×10^{-19} J or 11 604 K). Thus, the empty states are inaccessible by either electric-field or thermal excitation. Semiconductors have energy gaps of less than about 2 eV, at which point thermal excitation across the gap is possible. The number of carriers excited into the conduction band is still small, *e.g.* pure silicon has a conductivity of 10^{-5} S cm^{-1}. This value increases rapidly as the energy gap decreases, because of the exponential form of the Boltzmann distribution, and when dopants are used. Dopants produce states in the energy gap that lie close to either the conduction or the valence bands. Conduction is due to either electrons (n-type semiconductors) or holes (p-type) with conductivities of more than 10^{-1} S cm^{-1}. True metallic behaviour results when the energy gap between filled and empty states disappears. Applied fields can then cause current flow even at very low temperatures. Thus, the conductivity of semiconductors falls at low temperatures, as thermal excitation is quenched, while that of metals increases, as the thermal motion of the lattice and scattering processes are reduced. In semimetals the density of electronic states in the vicinity of the highest filled-level, the Fermi level (E_F), is very low and so is the conductivity. For good conductors, such as copper, the density of states at E_F is high.

The conductivities of polymers span values from insulating to metallic regimes. Applying the simple band-model to polymers with saturated chemical structures, all the valence electrons form strongly localized chemical-bonds and the energy gap is large, *e.g.* polyethylene (PE) has a gap of 8 eV. For polymers with unsaturated (conjugated) backbone structures, the simplest picture is of a chain with equal bond-lengths and one unbonded π-electron per carbon atom, *e.g.* polyacetylene (PA), Figure 1(a). Because two electrons of opposite spin are needed on each site to fill the available states, this chain has a half-filled energy-band. This metallic chain is unstable with respect to dimerization as the π electrons form additional bonds. This costs elastic-deformation energy but gains from the lowering of electronic energy by localizing the electrons in the multiple bonds and opening an energy gap (see Figure 1b). Thus, bond alternation, or the Peierls transition,[2] produces a semiconducting material. However, the properties of conjugated polymers differ distinctly from those of inorganic semiconductors. The three-dimensional bonding in the latter produces a rigid structure that is only slightly deformed by the presence of an added-charge carrier. Hence, the energy-band structure is essentially unchanged by added charges. In contrast, the relaxation of the polymer backbone structure in the vicinity of added charges is large. Thus, new states are produced which lie in the energy gap.[3] These are discussed in more detail in Section 22.4.

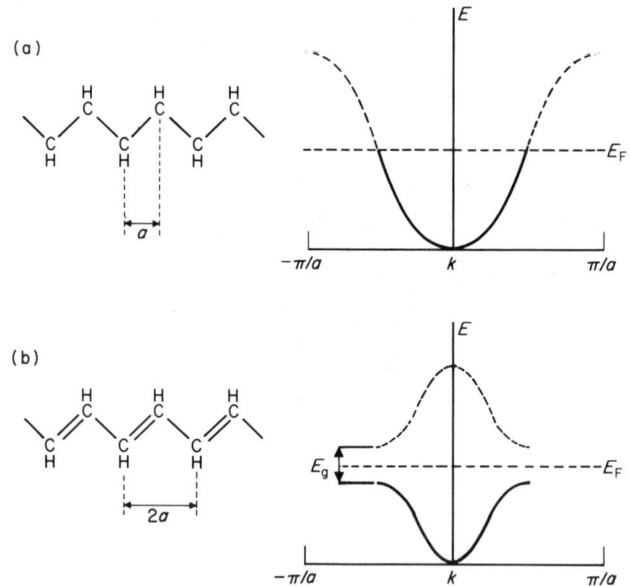

Figure 1 Chemical and electronic band-structure of non-alternated (a) and alternated (b) PA; E_F indicates the Fermi level and E_g the energy gap; $k = 2\pi/\lambda$; filled-band states are shown by the full curve and empty-band states by the broken curve

These ideal models are modified when the real morphology of polymers is taken into account. Chain ends, chain folds, amorphous regions, impurities and other defects will smear out the band states, resulting in whole or partial filling of the gap. Disorder will also tend to localize states. The energy gap which separates valence and conduction band-states in crystalline solids is replaced by a mobility gap in disordered solids.[4] Carrier motion is accomplished by hopping between localized states below this gap and is band-like for states above the gap. Thus, PE exhibits a small defect-induced conductivity, about 10^{-15} S cm^{-1} with very low carrier-mobility. In addition, in an ideal one-dimensional polymer-chain, disorder will localize all states at 0 K.[5] Such Anderson localization will render polymeric semiconductors insulating at low temperature. The coulombic interaction of electrons and holes can also produce localized states (excitons) giving a low-temperature excitonic insulator.[4] Thermal excitation at room temperature can be sufficient to overcome these localization effects. In contrast, polymers will decompose before thermal excitation destroys bond alternation.

Metallic polymers can be obtained by: (a) pyrolysis of insulating or semiconducting polymers; (b) incorporation of metallic particles; (c) action of electron donors and acceptors on conjugated polymers; and (d) producing half-filled band structures. (a) and (b) are the principal routes to commercial products (see Sections 22.5.1 and 22.5.2). The discovery of metallic conductivity in 'doped' conjugated polymers is relatively recent,[6,7] but has been subject to intense activity (see Section 22.4). So far, a linear carbon-backbone polymer with intrinsic metallic behaviour has not been reported. This stems from the fact that some structural deformation can apparently always produce a semiconducting state of lower total energy.

Polythiazyl $(SN)_x$ is an intrinsically metallic polymer.[8] Fibrous highly crystalline samples are obtained by the solid-state polymerization of S_2N_2. The SN repeat unit has one excess electron and the polymer has a half-filled band. The density of states at E_F is low and $(SN)_x$ is an anisotropic three-dimensional semimetal. Below 0.4 K, $(SN)_x$ becomes superconducting; both the transition-temperature and the room-temperature conductivity (2×10^3 S cm^{-1}) are sample dependent. The conductivity of fibrous $(SN)_x$ is enhanced to 2–4×10^4 S cm^{-1} by exposure to halogens, which are assumed to form a layer on the surface of the fibres. Although efforts to prepare polymers related to $(SN)_x$ have proved unsuccessful, new synthetic routes to $(SN)_x$ have yielded a better material.[9] Despite the demonstration of efficient carrier-injection into semiconductors, the problems of preparation and manipulation have hindered the commercial use of $(SN)_x$.

Carbon-chain polymers exhibit a great variety of backbone and side-group structures and electrical properties. In this brief review it will not be possible to mention all these materials, but specific examples will be presented that are typical of the many polymers that have appeared in the literature.

22.2 INSULATING POLYMERS

One of the principal applications of polymers is as electrical insulators. Polymers with saturated structures and those with unsaturated groups, in either the backbone or side-groups, which are well separated by saturated groups, will have intrinsically large energy-gaps and be insulators. The mobility of carriers in these polymers is controlled by the quasi-continuous set of impurity and defect-induced energy-levels within the band gap. Typical conductivities and mobilities are 10^{-15} S cm^{-1} and 10^{-14} m^2 V^{-1} s^{-1} for PE, 10^{-16} S cm^{-1} and 10^{-13} m^2 V^{-1} s^{-1} for poly(tetra-fluoroethylene) (PTFE) and 10^{-15} S cm^{-1} and 5×10^{-11} m^2 V^{-1} s^{-1} for polystyrene (PS). Thus, in addition to a knowledge of the band structure, it is necessary to investigate the motion of carriers and its dependence on sample morphology.

A wide range of quantum chemical methods have been used to calculate the band structures and densities of electronic states for insulating polymers.[10-11] These have been compared with experimental valence-band densities of states determined by X-ray photoelectron spectroscopy (XPS) and UV photoelectron spectroscopy.[10-13] Typical results for PE are shown in Figure 2; the agreement is good and is improved if the calculation includes non-stereoregular chain-segments. Photoconduction, UV absorption and electron-energy loss spectra (EELS) of PE show an absorption onset at 7.6 eV but a photoconduction edge at 8.8 eV.[10,14,15] Thus, the first excited-state is identified as an exciton. Photoemission results place the conduction-band edge just above the vacuum level. The agreement between these results and theoretical calculations is good, though none is able to

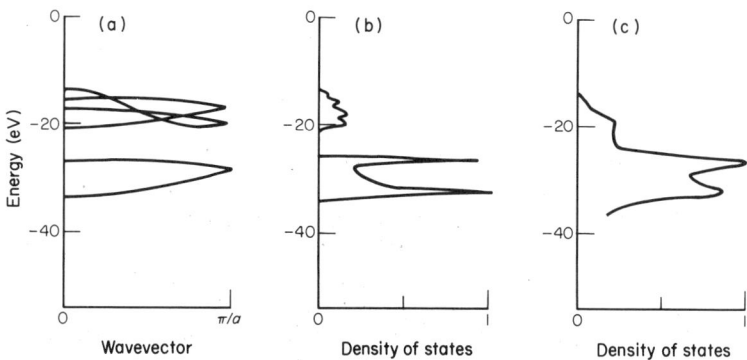

Figure 2 Calculated valence-band structure (a) and density of electronic states (b) compared with the experimental density of states (c) for PE (after J. Delhalle, S. Delhalle and J. M. André, *Chem. Phys. Lett*, 1975, **34**, 430)

reproduce all the experimental data exactly.[10,15] XPS has also been used to determine the valence-band structure of polymers with pendant groups;[10,12,13] here, comparison with molecular spectra has aided interpretation. The development of reliable theoretical models has allowed XPS to be used for structural determination.

The measurement of the electrical properties of highly resistive polymers presents a number of problems. The surface conductivity is often greater than the bulk value and is increased by absorbed impurities. The latter is also affected by impurities. The presence of localized-defect states in the energy gap means that contact with metals is accompanied by charge transfer to the polymer in the absence of an applied field.[16,17] The triboelectric charge appears to be relatively insensitive to sample purity or preparation and even to polymer structure. This indicates that the important defects are structural, with a density that is almost sample independent. The work function of the metal affects the charge transfer, but large experimental uncertainty renders precise functional fits unreliable.

When charge is injected into polymers by an electric field, the low carrier-mobility leads to a space charge near the electrodes and non-linear current–voltage characteristics.[18] These phenomena are of considerable interest in studies of the breakdown of polymeric insulators subjected to high voltages. In practice, this occurs at defects where fields are enhanced and tree-like structures occur, which propagate through partial discharges.[19] Guarded needle electrodes have been used for accurate measurement of the injected charge,[20] showing that injection of energetic carriers is a fast process while subsequent decay in zero-field is slow. Space charge in the parallel electrode geometry has been studied using laser-generated shockwaves, which displace the space charge producing a pulse of induced current. In thin films of poly(ethylene terephthalate), the space charge was localized near the electrodes and decayed by dispersive transport.[21] This is a common feature of carrier transport in polymers[22] and is due to carriers becoming progressively less mobile with time as they encounter deeper traps. The distribution of trap levels can be probed by thermally stimulated currents. After polarization of a sample at low temperatures, currents flow whenever thermal excitation leads to detrapping or, for polar polymers, dipole relaxation. For example, in irradiated samples of PTFE fluorine-deficient sites have been identified as traps.[23]

The determination of the very low carrier-mobilities encountered in insulating polymers, and the results obtained for PE, have been reviewed.[24] Hopping between defect states in PE gives electron and hole mobilities from 10^{-10} to 10^{-15} m^2 V^{-1} s^{-1}, which are much smaller than those estimated from band-structure calculations, 0.2–0.6 m^2 V^{-1} s^{-1} for electrons and 10^{-2}–10^{-4} m^2 V^{-1} s^{-1} for holes. Experimental differentiation of various types of hopping conductivity is difficult, *e.g.* variable-range and multiphonon-assisted local hopping both display an $\exp(T/T_0)^{1/4}$ temperature dependence.[25]

Though most insulating polymers are not intrinsically ferroelectric, polar materials can be produced by charge injection into insulating polymers. Such electrets have been known, studied and utilized for many years.[26] Notable exceptions are poly(vinylidene difluoride) (PVDF) and its copolymers which are ferroelectric.[27] There have been numerous studies of orientation, poling, structure and phase transitions of these polymers. The large piezoelectric effect of PVDF when in the poled state has led to its application in acoustic and ultrasonic transducers.

Examination of the results of AC conductivity measurements shows that the imaginary part of the dielectric function, the dielectric loss-factor, is similar over the range 10^{-4} to 10^5 Hz for many amorphous and partially crystalline polymers. Indeed, it has been argued that there is a 'universal' dielectric response function that is valid for a wide range of solids.[28] In all cases, the dielectric response is due to hopping of electrons (or ions) or reorientation of dipoles. These are fast discontinuous processes which provoke a much slower many-body response which screens the charge or dipole. This combination of processes leads, through the Kramers–Kronig relation, to a 'universal' frequency dependence.

Strong electron-donors or acceptors can effect charge transfer with insulating polymers and produce some increase in conductivity, *e.g.* from 10^{-15} S cm^{-1} to 10^{-12}–10^{-13} S cm^{-1} for PS films exposed to iodine. This can be attributed to carrier hopping between localized charge-transfer sites, *e.g.* the pendant groups. The effect of organic and ionic impurities is particularly important in polymer films used as insulation and protective layers in semiconductor devices.[29] This has been studied for various poly(ester imide)s by measurement of the residual current in a field-effect transistor (FET) coated with impurity-doped polymer.[30]

An important application of both insulating and semiconducting polymers is the xerographic process.[31] In this, a photoconductor is charged uniformly by a corona discharge and, on illumination, is selectively discharged to produce the image. A toner, which consists of metallic particles and a polymer, is charged triboelectrically and used to render the image visible prior to its transfer to paper. An effective toner requires optimized triboelectric properties. This has been achieved by the incorporation of conjugated pendant groups, *e.g.* PS and modified PS. The pendant moieties provide molecular orbitals which determine the sign of charge transfer by their position relative to the E_F of the metal. Surface modification can significantly affect triboelectrification, *e.g.* sulfonated PS films show a sharp rise in charge transfer due to the ease of charging the sulfonate groups. Incorporation of counter-ions, *e.g.* methylene blue, enables further control to be exercised over the degree and sign of charge transfer.

Further discussion of topics touched on here can be found in the literature.[32, 33]

22.3 SEMICONDUCTING POLYMERS

22.3.1 Introduction

The boundary between insulators and semiconducting polymers is somewhat arbitrary. However, the latter are distinguished by the photoconductivity they display when exposed to visible wavelengths. Certain categories of polymers, *e.g.* pyrolyzed and filled polymers, span the semiconductor/metal boundary and are considered separately. This leaves two distinct classes of semiconducting polymers, those with active pendant groups and those with fully conjugated backbones. Work on these two classes has been largely directed towards different goals and has progressed almost independently. Therefore, the former are discussed below and the latter in Section 22.4.

22.3.2 Conjugated-side-chain Polymers

Xerography has driven much of the research into conjugated-side-chain semiconducting polymers.[31] The most studied photoconductive polymer is poly(*N*-vinylcarbazole) (**1**; PNVC).[34] PNVC in its pure form is a hole conductor with a dark conductivity of the order of 10^{-14} S cm^{-1}. The photoresponse is limited to the absorption profile of the carbazole moiety. Both properties are modified when molecular species are added, so that impurities can mask the intrinsic properties of PNVC, *e.g.* purification leads to lower quantum efficiency.

(**1**) Poly(*N*-vinylcarbazole) (PNVC)

Dye molecules sensitize photoconduction by electron transfer between the excited state of the dye molecule and PNVC. For weak acceptors which give a neutral ground-state, *e.g.* 2,4,7-trinitro-9-fluorenone (TNF), photosensitization occurs by direct excitation of electron transer from the PNVC carbazole group to TNF. A charge-transfer band appears in absorption and photocurrent action spectra and the quantum efficiency is high (0.23). Strong acceptors form complexes with ground-state charge-transfer and enhanced conductivity, *e.g.* 2×10^{-6} S cm^{-1} at 77% w/w of I_2. The use of both acceptors and dyes is beneficial, because the action spectrum contains contributions from dye and charge-transfer bands, and degradation due to photochemical reaction is reduced. Coulombic attraction results in the geminate recombination of many electrons and holes. This process was first discussed by Onsager[35] and shows a strong voltage-dependence, quantum efficiency increasing with applied field.[36]

Addition of TNF to PNVC gives an increased electron-mobility. Identical behaviour is found for a polyester matrix, indicating electron hopping between TNF molecules. The hole mobility falls by 10^3 for 50% TNF incorporation, as separation of the carbazole units is increased and charge localization occurs. The mobilities are thermally activated and current-pulse transits exhibit dispersive transport, both effects being symptomatic of hopping conduction. Apparently long lifetimes for bound electron–hole pairs in PNVC have been explained as an artefact of hopping between sites with Gaussian energy disorder.[37] A detailed discussion of conduction and photoconduction in PNVC can be found in the literature.[38]

PNVC has been modified chemically to incorporate substituents on the carbazole groups, (**2**). Methylene blue sensitizer can be bonded to sulfonated PNVC. Bonded nitro- and cyano-groups act as acceptors giving rise to enhanced photoconductivity. Antimony pentachloride produces crosslinking of the carbazole groups to form dimeric carbazylium radicals (**3**).[39] The conductivity of up to $10^{-5}\,S\,cm^{-1}$ has been attributed to electron hopping between the dimer sites. PNVC copolymers with acrylates, methacrylates and styrene, when sensitized with TNF, are found to give similar photoconductivity to PNVC but with better film-properties. Copolymers with mixed carbazole and acceptor pendant groups have been synthesized in order to extend spectral sensitivity, but overall they are less useful than PNVC–acceptor composites. Poly(2-vinylpyridine)s (**4**) and poly(4-vinylpyridine)s (**5**) have been used as antistatic coatings, as constituents of xerographic toners, with PNVC and phthalocyanine dyes as xerographic semiconductors and complexed with iodine for photovoltaic cells and as cathodes for solid-state batteries. Other examples of this class of polymers are poly(N-ethyl-2-vinylcarbazole) and related polymers, polyacenapthylene (**6**), poly(vinylpyrene) (**7**), polyindene, poly(vinylnaphthalene), poly(vinylanthracene), poly(2-vinylquinoline), *etc.* Further details of the properties of these and related materials can be found in the literature.[34,40,41]

(**2**) Chemically modified PNVC   (**3**) Crosslinked PNVC   (**4**) Poly(2-vinylpyridine)

(**5**) Poly(4-vinylpyridine)   (**6**) Polyacenaphthylene   (**7**) Poly(vinylpyrene)

22.4 SEMICONDUCTING AND METALLIC CONJUGATED POLYMERS

22.4.1 Introduction

The interest created by the discovery of metallic conductivity in PA[7] stimulated increased activity in the study of polymers with conjugated backbones. The production of a highly conducting polymer depends on either removal or addition of electrons by accepting or donating moieties. These processes of oxidation and reduction produce polymeric ions, cations or anions, and a counter-ion, either the reduced form of an oxidizing agent or *vice versa*. The overall process has been termed 'doping' in a rather poor analogy with the doping of inorganic semiconductors (see Section 22.4.2). Hence, the processes are referred to as p-type and n-type doping for oxidation and reduction, respectively. The most interesting polymers are those which can be easily oxidized or reduced. Conjugated polymers fall into this general category because the extended π-electron systems are characterized by relatively small ionization-potentials and/or large electron-affinities. Oxidation and reduction are accompanied by either removal or addition of π electrons. This process is aided by chain relaxation, which produces energy levels within the energy gap,[3] and can be achieved without breaking the underlying σ-bonds, which maintain the integrity of the polymer chain. Typical examples of conjugated polymers are shown in structures (**8**)–(**14**).

(**8**) *Trans*-polyacetylene   (**9**) *Cis*-polyacetylene (*cis*-transoid)   (**10**) Poly(1,4-phenylene)

(11) Poly(thio-1,4-phenylene)

(12) Poly(diacetylene)

(13) Heat-treated polyacrylonitrile

(14) Poly(*peri*-naphthalene)

22.4.2 Theoretical Models

Conjugated polymers have been studied theoretically since 1937.[42] Quantum chemical calculations applied to conjugated chains include: (i) Hückel methods, for the rapid assessment of band structures;[43,44] (ii) the Paiser–Parr–Pople configuration interaction method;[45] (iii) complete neglect of differential overlap (CNDO) self-consistent field methods;[46] (iv) *ab initio* Hartree–Fock calculations, which have given very close agreement with experimental results;[47-49] and (v) the valence effective Hamiltonian (VEH) method, which offers a considerable reduction in computational effort.[50]

Although VEH uses a set of parameterized one-electron orbitals, its accuracy approaches that of true *ab initio* methods. VEH gives ground-state properties, ionization potentials and energy gaps typically within 0.1 eV of the experimental values. The evolution of the band structure of poly(1,4-phenylene) (**10**; poly(*p*-phenylene), PPP) during inclusion of sodium is shown in Figure 3. The relaxation of the polymer chain in the vicinity of added electrons produces states in the band gap. These eventually fuse with the valence and conduction bands as Na^+ content increases. These states and their relevance to the properties of real systems has been the focus of much excitement and debate.[51,52]

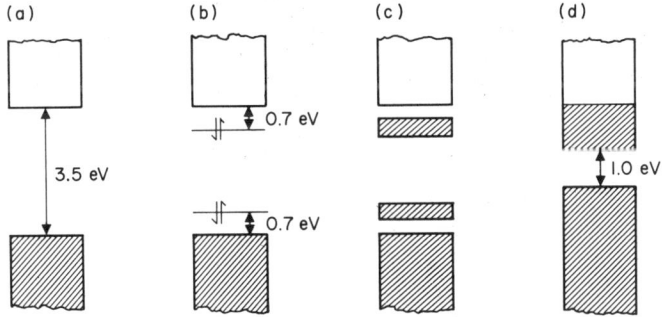

Figure 3 Evolution of band structure of PPP during sodium 'doping': (a) undoped, (b) slight doping, (c) 50 mol % Na and (d) 100 mol % Na (after J. L. Brédas, B. Thémans, J. G. Fripiat, J. M. André and R. R. Chance, *Phys. Rev. B: Condens. Matter*, 1984, **29**, 6721)

Bond-alternation defects in PA, first mooted in 1962,[53] were revived in modified form in 1979.[54] The analysis employs a simple Hamiltonian incorporating electron–phonon interactions and a defect in which the change in bond alternation is extended over 5 to 9 repeat units (see Figure 4). The energy to create this defect (soliton*) and that required to move it are small. Thus, such solitons should be created thermally and be mobile. Single solitons are stable because the two alternative structures of PA have the same ground-state energies, and produce an energy level at the centre of the energy gap. Optical absorption occurs for transitions across the band gap and to the mid-gap state. The charge and spin of the defect will depend on the occupancy of the mid-gap state. It is singly occupied for a neutral soliton (S^0) which has spin 1/2. If the electron is removed, *i.e.* zero occupancy,

---

* The language in this section was introduced by physicists. In chemical terms, neutral solitons in PA are radicals, charged solitons are ions, either cations or anions, polarons are radical ions and bipolarons are di-ions.[3,55,59]

a positively charged soliton (S^+) of zero spin results. Double occupancy gives a negative soliton (S^-) with zero spin. Soliton (S)–antisoliton ($\bar{S}$) pairs occur when a short segment of a PA chain has reversed bond-alternation.

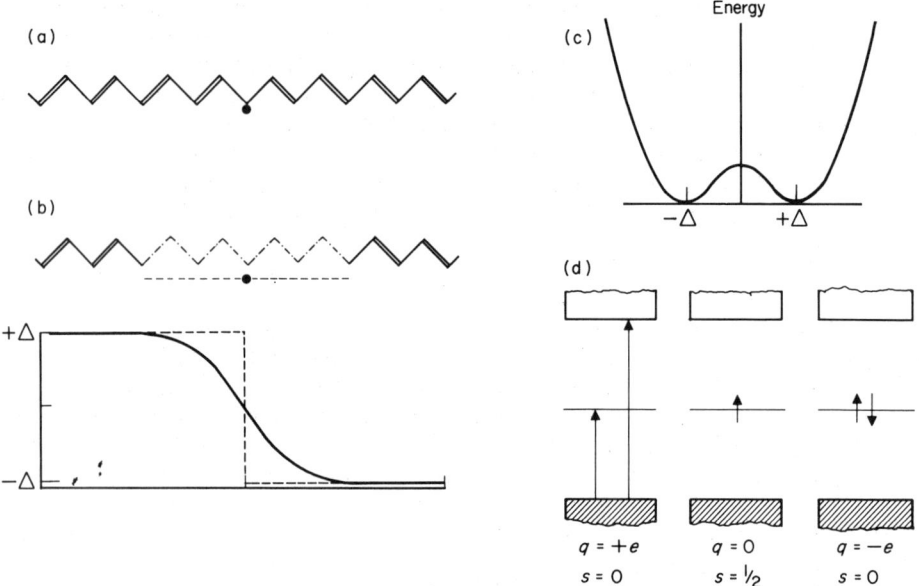

Figure 4 Structure of (a) a Pople–Walmsley and (b) a Su–Schrieffer–Heeger bond-alternation defect (soliton) in PA; the changes in bond orders (a, dotted and b, full line) are shown underneath. The energy of PA as a function of bond order is shown in (c); the spin–charge (s–q) relations for different occupancy of the soliton energy-level are shown in (d)

Isolated solitons are unstable in polymers with structures that do not have equal ground-state energies, *e.g. cis*-polyacetylene (**9**; *cis*-PA, *cis*-polyvinylene) and PPP (**10**). For such polymers, charge exchange will lead to formation of S^0–S^+(S^-) pairs, which will be strongly localized to form a polaron,[55] (Figure 5) well known in inorganic semiconductors.[4] At the polaron site, the polymer structure locally deviates towards the higher-energy structure but never achieves it. A polaron gives rise to two levels in the band gap and several optical-absorption bands occur. If the gain in lattice distortional energy outweighs the cost of bringing like charges close together, two polarons will collapse to form a bipolaron (Figure 5). Bipolarons can have zero spin yet carry charge, a property they share with solitons.

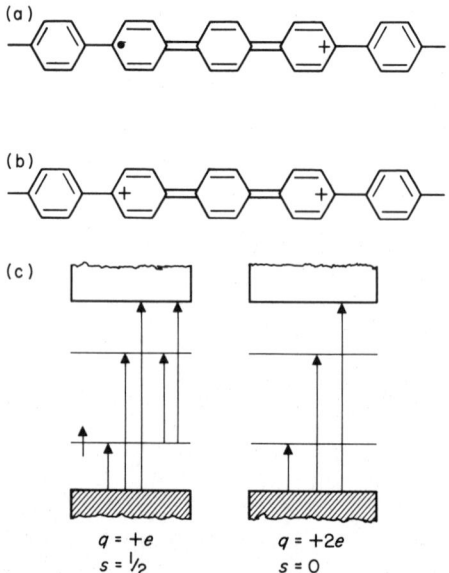

Figure 5 Structure of (a) a polaron and (b) a bipolaron in PPP. The energy-level schemes and allowed optical-transitions are shown in (c) for the polaron (left) and bipolaron (right).

The basic premise of these models is that the band gap in conjugated polymers is due to bond alternation. This has been questioned because the inclusion of coulombic interactions (electron correlation) in the undimerized PA chain can lead to the appearance of a gap.[42] Both bond alternation and electron correlation can contribute to gap formation but it is not clear *a priori* which effect will be larger. Though experimental evidence favours bond alternation, correlation cannot be ignored because it produces excitonic rather than band-like states.[49,56,57] Inclusion of electron correlation for solitons, polarons and bipolarons has been found necessary to improve agreement with experiments, *e.g.* by shifting energy levels and altering transition intensities.[57] The soliton model predicts that photoexcited electron–hole pairs in PA transform into S^+–S^- pairs within 10^{-13} s. Inclusion of electron correlation suggests that geminate recombination to an exciton will intervene and that free carriers result from exciton dissociation above about 120 K,[58] in agreement with experiment.

Addition of either acceptors or donors results in charge transfer with the lowest empty or the highest filled states, which will be the defect states, *i.e.* solitons or bipolarons. This will lead to: (a) formation of localized spinless charge-carriers, with hopping conductivity; (b) screening of coulombic interactions, leading to mobile carriers and merging of localized and band states; and (c) formation of a disordered three-dimensional metal. As noted above, this process has been termed 'doping'. However, the analogy between polymeric and inorganic semiconductors is not a good one, because of the inherent differences between the two types of materials. Furthermore, the levels of doping are very different; in silicon, *etc.*, the dopant levels are fractions of a per cent, while in polymers they are up to tens of a per cent. Thus, it has been argued that the defect model ignores the chemical nature of 'doping'.[59]

The ideal chain-structures employed in theoretical models bear little resemblance to real materials. Disorder has been modelled by either a distribution of conjugation lengths or electron-lattice interactions.[60] Bond length and substitutional disorder has been shown to lead to the disappearance of bond alternation and closure of the band gap, though with a low density of states at the Fermi energy.[61] A simple model has been developed for the influence of defects on the electrical conductivity of doped samples.[62] Pinning of solitons by defects has been discussed, but it is clear that the role of disorder requires further study.[63]

22.4.3 Directly Synthesized Polymers

Conjugated polymers have been prepared by conventional polymerization routes (direct synthesis), electropolymerization and by conversion of unsaturated precursor polymers. Usually, for particular polymers, one of these has become the method of choice. Hence, polymers will be classified by route rather than structure. Exceptions with more than one route in current use will be noted.

PA was the first conjugated polymer to attract attention and is the most thoroughly studied.[64-67] The structure is simple and the emergence of new concepts and physical phenomena has stimulated debate.[52,63,68] Continuing improvements in material properties and potential for use in batteries, photovoltaic cells, electromagnetic screening, *etc.*, have maintained interest. Electrochemical preparation of PA is difficult[69] and the majority of workers have used Ziegler–Natta catalysts. Usually an intractable powder is produced. However, if acetylene gas is brought into contact with the surface of a highly concentrated catalyst solution at 197 K, a film with metallic lustre is produced.[64,70,71] Reaction conditions determine the ratio of *cis*- to *trans*-PA structures, defect density and crystallinity of the product. Such Shirakawa PA has a fibrous sponge-like structure and a very low density. The fibres can be aligned by stretching the films but complete alignment of the PA chains cannot be achieved. Attempts to obtain a greater degree of orientation have included polymerization in a shear flow, synthesis on the surface of biphenyl crystals and polymerization in a liquid-crystal solvent.[72-74] None of these methods has proved as successful as the orientation of precursor polymer (see Section 22.4.5).

cis-PA (**9**) prepared at low temperatures is converted into the *trans*-PA form (**8**; *trans*-polyvinylene) on heating. *cis*-PA has an orthorhombic crystal structure while both orthorhombic and monoclinic forms have been reported for *trans*-PA for samples prepared by different routes.[59,64] Studies of *cis*–*trans* isomerization indicate that the process is sensitive to the initial state of the polymer, *i.e.* to polymerization conditions. Single- and multiple-phase behaviour, with and without changes in degree of crystallinity, have been reported.[75] Block poly(acetylene-*co*-styrene)s have been reported to have a hexagonal structure, which has been attributed to helical PA. However, IR spectra and theoretical calculations do not support this suggestion.[76,77]

The effects of a large number of accepting and donating species on the conductivity of PA have been reported, *e.g.* alkali metals, perchlorates, boron tetrafluoride, transition metal halides, *etc.* Doping can be achieved either by exposure to dopant vapour or solution, *e.g.* potassium naphthalide solution, or electrochemically.[64, 78] In some cases the doping reaction is relatively straightforward, *e.g.*, Scheme 1 for potassium naphthalide solution. In others it is complex, *e.g.* iodine produces I_3^- and I_5^- ions, while, for AsF_5, the ions that have been postulated are AsF_5^-, $As_2F_{10}^{2-}$ and AsF_6^- with by-products such as AsF_3, $HAsF_6$ and $HAsF_5OH$.

$$(CH)_x \ + \ xy\,(nphth^-) \longrightarrow [(CH)^{y-}]_x \ + \ xy(nphth)$$

$$[(CH)^{y-}]_x \ + \ xy(Na^+) \longrightarrow [yNa^+(CH)^{y-}]_x$$

$$nphth^- = naphthalide\ anion$$

Scheme 1

In a cell with Li and PA electrodes containing $LiClO_4$, on external connection of the electrodes, current flows as a result of the electrochemical reactions shown in Scheme 2. Li^+ liberated at the anode is incorporated into the PA as a counter-ion. These processes have been rationalized in terms of the reduction potentials of PA, in the pristine, oxidized and reduced states, and those of the redox couples relevant for the doping process under consideration.[79]

$$\text{Anode} \quad xy\,Li \ = \ xy\,Li^+ \ + \ xy\,e^-$$

$$\text{Cathode} \quad (CH)_x \ + \ xy\,e^- \ = \ [(CH)^{y-}]_x$$

Scheme 2

On doping, the conductivity of PA rises dramatically (see Figure 6). The Curie susceptibility falls and cannot be detected for dopant levels of the order of 1%. The Pauli susceptibility rises from an initially undetectable level at *ca.* 6% dopant concentration. In terms of the soliton model, the Curie susceptibility derives from S^0 defects created during *cis–trans* isomerization. Dopant ions create pinned charged (S^+ or S^-) solitons at the S^0 sites, and conduction occurs by hopping between soliton sites. At high doping levels, screening leads to freeing of the chain carriers, and heavily doped PA behaves as a highly imperfect metal. An alternative picture is that 'doped' PA is really a well-defined compound comparable to the well-known conducting charge-transfer salt crystals.[59] This new metallic phase is produced locally and the conductivity rises at the percolation threshold when

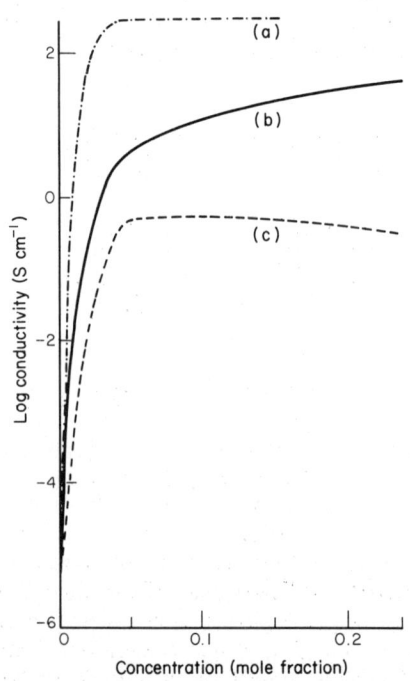

Figure 6 Increase in conductivity of PA on 'doping' with (a) AsF_5, (b) iodine and (c) bromine.

conducting paths are formed through the sample. Below this point, the Pauli susceptibility is small but not necessarily zero. The difference in conductivity and susceptibility is often exaggerated by comparison of conductivity on a logarithmic scale with susceptibility on a linear scale. Evidence in favour of this model is the observation of distinct phase transitions and counter-ion ordering during electrochemical doping of PA with alkali metals and PF_6^- ions.[80]

Many of the apparently discordant experimental results in the early literature probably resulted from insufficient attention being paid to sample quality. Subsequently, both purification and degradation of PA have received attention.[64, 71, 81] Exposure to oxygen initially produces an increase in conductivity, but this 'doping' reaction is soon overtaken by degradation, which leads to the formation of carbonyl and hydroperoxide groups and a decrease in conductivity. Incorporation of stabilizers reduces degradation but does not eliminate it.

Though undoped *trans*-PA has been extensively studied, the evidence for the soliton picture remains contradictory. NMR Overhauser experiments have been interpreted as indicating highly mobile solitons in undoped PA.[82] Electron-spin-echo-detected multiple-quantum NMR, however, indicates localization of the paramagnetic spins (solitons) in both *trans*-and *cis*-PA.[83] Other NMR measurements have included ^{13}C magic-angle spinning, wide line and nutation NMR.[82] Nutation NMR has given clear evidence for bond alternation with bond lengths of 0.136 and 0.144 nm in *trans*-PA, but no evidence of the rapid exchange expected if mobile solitons were present. The interpretation of electron–nuclear double resonance (ENDOR)[84] and electron spin resonance (ESR)[85] results has been debated, because the spectra can be interpreted in solitonic and non-solitonic spin-density distributions.[86] Careful measurement of IR absorption shows no evidence for the discrete mid-gap state associated with neutral solitons. Studies of photocarrier generation have been interpreted in terms of an excitonic intermediate state.[87] The exciton is also seen in EELS spectra.[88] Though direct photoexcitation of $S^+–S^-$ pairs does not seem to occur, charge separation and photoinduced absorptions are best described in terms of soliton states.[87] AC electrical conductivity has been used to study conductivity mechanisms.[89] At all levels of doping, carrier hopping appears to be the dominant conduction-mechanism and has been interpreted with several different models, *e.g.* intersoliton hopping, energy-dependent hopping, *etc.*

Resonant Raman-scattering provides information on the backbone structure and its relation to electronic excited-state energies.[90] Shirakawa *trans*-PA has Raman spectra with an excitation-wavelength independent component and one which shifts to lower frequencies and broadens as the wavelength is decreased. This has been interpreted either as a double-peaked distribution of chain conjugated lengths[91] or as a distribution of electron–phonon, electron–electron and electron–disorder interactions.[60] New IR bands which appear on doping have been interpreted in terms of chain deformation and coulombic interactions, which pin the induced charge to the dopant ion at low doping levels.[92]

The physics of PA–metal and PA–semiconductor interfaces and the applications of PA for photovoltaic devices has been explored.[64, 93] The photoefficiencies observed are low,[65] in the range of 10^{-3} to 10^{-4}, reflecting the high defect-density in PA. The application of PA to batteries has been extensively investigated.[64, 94] Most attention has been focussed on Li/PA electrodes with an electrolyte of $LiClO_4$ or $LiBF_4$ in a non-aqueous solvent. These systems are susceptible to irreversible degradation, and greater stability has been claimed for tetraphenyl-, tetrabutyl- and tetraethyl-borates as reducing agents.[95] A number of all-plastic batteries using ion-conductive membranes have been produced and shown to exhibit good life and performance characteristics.[96] The main limitation on PA batteries is that background oxidation of the electrolyte leads to self-discharge and serious loss of coulombic efficiency.[97] Such problems of stability have limited progress towards application.

A notable advance has been the modification of the synthesis to provide PA samples with fewer conjugation-breaking defects.[98] When stretch aligned and doped with iodine, these have room-temperature conductivity of up to $10^5\ S\,cm^{-1}$. The conductivity is reduced at low temperatures indicating that it is limited by inter- rather than intra-molecular transport.

In addition to PA, related polymers with substituent groups, *e.g.* poly(phenyl acetylene), poly-(1,6-heptadiyne) and copolymers have been investigated.[71, 99] In general, the conductivity achieved on doping is less than that for PA.

Another class of polymer that has received attention is based on PPP, *e.g.* poly(1,3-phenylene), poly(thio-1,4-phenylene) (**11**; poly(*p*-phenylene sulfide), PPS), *etc.*[100] A major advantage of PPP and PPS is that the undoped polymers are solution- and melt-processable. PPS is also available commercially. The physical properties of these polymers, undoped and doped, have been interpreted in terms of the bipolaron model.[85, 88] The validity of this approach has been questioned by chemical analysis, which has shown that these materials are complex mixtures of dissimilar oligomers rather than distinct polymers.[101] Furthermore, doping is accompanied by side reactions leading to crosslinking. PPS was the first polymer to be shown to be soluble in its conducting form. The nature of the solvent, AsF_3, and the rapid development of other soluble conducting-polymers rendered this of academic interest. PPP has been used successfully as a battery electrode.[95] PPP–alkali-metal-alloy composite electrodes have greater cyclability and discharge rate than bare-metal electrodes.[102] Extensive reviews of these and a wide range of related polymers have appeared.[103,104]

Poly(diacetylene)s (**12**; PDA) are uniquely obtainable, by solid-state polymerization, as highly perfect single crystals containing fully extended conjugated chains.[105] They are wide-band-gap materials with low intrinsic conductivity. However, photoconductivity and electric-field-induced spectra reveal that the carriers have extremely high mobility. Attempts to render PDA metallic by doping have been unsuccessful[106] because of side reactions with the pendant groups, essential for solid-state polymerization, and the backbone.[107]

The impetus to find new conductive polymers has led to many reports of conjugated polymers, which remain to be fully characterized. Examples (with maximum conductivities) are polyquinoline (Na doped, 11 $S cm^{-1}$),[108] poly(metal benzodithiolene) (Fe polymer undoped, 0.2 $S cm^{-1}$),[109] poly-(3,4-diisopropylidenecyclobutene) (iodine doped, $10^{-3} S cm^{-1}$)[110] and poly{(E,E)-[6.2]paracyclo-phane-1,5-diene} (iodine doped, $10^{-3} S cm^{-1}$).[111]

Non-carbon-backbone polymers which display good stability and interesting electrical properties are the bridged macrocyclic structures based on phthalocyanine (Pc). Polymers of the types $[(MPc)O]_x$ and $\{[(MPc)O]X\}_x$, where M = Si or Ge and X is a halogen, and $[(MPc)L]_x$, where M is a transition metal and L a conjugated organic ligand, have been studied.[112,113] The halogen-containing polymers have conductivities up to 1 $S cm^{-1}$.

Additional information on PA and other conjugated polymers exists in the numerous reviews and conference proceedings that have appeared over recent years.[59,64-67]

22.4.4 Electrochemically Synthesized Polymers

The production of conductive films by the electropolymerization of pyrole[6] went largely unnoticed until interest was awakened by the work on PA. Electropolymerization has, however, proved to be of great utility in the production of conductive aromatic polymers.[114,115] Polypyrrole [**15**; PPy, poly(2,5-pyrrolediyl)],[116] polythiophene [**16**; PTh, poly(2,5-thiophenediyl)],[117] poly-aniline [**17**; PANi, poly(imino-1,4-phenylene)], polyazulene [**18**; poly(1,3-azulenylene)], polycarbazole, polyindole, polypyrene, polytriphenylene, polyisothianaphthalene, *etc.* have been prepared. The general reaction pathway involves successive additions of monomer radical-cations to oligomer radical-cations, deprotonation to form a neutral $(n+1)$ oligomer and reoxidation back to a radical cation. The overall reaction involves the removal of 2.25 to 2.5 electrons per repeat unit, giving an oxidized polymer with a cationic centre for every 2–4 repeat units, and counter-ions included from the electrolyte to achieve charge neutrality. Thus, the deposited layer is conductive and polymerization can proceed even when the anode is covered with polymer. The oxidized polymer can be reduced to the neutral state with loss of the counter-ion. On reoxidation either the same or different counter-ions can be included. Thus, these polymers display both electronic and ionic conduction effects.[118]

(15) Polypyrrole

(16) Polythiophene

(17) Polyaniline

(18) Polyazulene

The electronic and optical properties of neutral and oxidized PPy have been analyzed in terms of polarons and bipolarons. At low doping levels, polarons (radical cations) are produced, but above 1% doping these combine to form bipolarons (dications). Electrical conductivities for fully oxidized PPy fall in the range $1-300\,S\,cm^{-1}$ depending on preparative conditions. A maximum value of $10^3\,S\,cm^{-1}$ has been reported for stretched samples.[119] Electrically conductive composites with good mechanical strength have been prepared by electropolymerizing pyrrole through porous membranes. However, with careful preparation, thick mechanically strong films can be produced directly. PPy films exhibit excellent stability in inert gases and dry air at elevated temperatures, but degrade much more rapidly in moist air. Reversible absorption of water has been observed, but this can be reduced and film stability improved if appropriate counter-ions are used.[120] Chemical bonding of the counter-ion to PPy has also been attempted.[121]

With n-Si or TiO_2 electrodes, PPy deposition requires photoexcitation to liberate the electrons participating in the polymerization. This has been studied for pattern definition and to mediate photoelectrochemistry at the semiconductor surface.[122,123] Use of PPy as an electrochromic material has been investigated, but this application is limited by side reactions occurring in reduced films. Considerable effort has been expended on developing PPy electrodes for batteries.[124] Self-discharge is a problem, and side effects reduce cycling efficiency and lead to failure of the Li electrode of Li/PPy cells after about 200 cycles. Electrochemical FET devices have been made by bridging gold microelectrodes with PPy and other electropolymerized materials.[125] These have been used to detect low levels of oxidizing species. A promising use of PPy is the production of chemically modified electrodes. Redox groups can be substituted at the pyrrole nitrogen without affecting polymerization behaviour. Efficient electrocatalytic groups can be incorporated into the conductive polymer.[126]

PPy can be prepared by acid and transition-metal halide catalyzed polymerization from solution and gas phases. These routes have been used for the production of composite materials for sensor, battery and other applications.[127]

The facile electrochemical synthesis and outstanding stability of both oxidized and neutral forms against oxidation and hydrolysis are notable advantages of PTh.[117] Polymerization commences in solution in front of the electrode to produce oligomers which precipitate on to the electrode and initiate film growth.[128] Soluble conductive materials have been made by the incorporation of methoxy, alkyl and ether- or amide-containing side-chains at the thiophene 3-position.[129] Spectroscopic studies of PTh and poly(3-methylthiophene) [P3MT, poly(3-methylthiophene-2,5-diyl)] films and solutions have been interpreted in terms of polaron and bipolaron states.[130] P3MT/GaAs photovoltaic devices exhibit good quantum-efficiency and n-GaAs electrodes coated with P3MT have dramatically extended lifetimes when used in photoelectrochemical cells.[131] P3MT microelectrode FET electrochemical transistors have an intrinsic device gain of 10^3 at 10 Hz and can detect $<10^{-15}$ mol of oxidants.[132] Radiation-induced polymerization of thiophene vapour has been studied as a UHV-compatible route to PTh.[133]

The electrochemical production of PANi was first noted over 135 years ago and has been studied intermittently since then; however, interest in this material has expanded rapidly since 1984. It has good stability and displays great structural variety since, in addition to normal oxidation and reduction, it can be protonated and deprotonated.[134,135] In consequence, the conductivity and the spectra of PANi on protonation and electrochemical doping exhibit complex behaviour.[136,137] The doping-induced insulator to 'metal' transition, giving a maximum conductivity of $5\,S\,cm^{-1}$, has been explained in terms of protonation followed by an internal redox reaction producing bipolarons, which are unstable against break-up into polarons.[138,139] This leaves a material with a half-filled polaron-band, *i.e.* a polaronic metal. Whether this sophisticated model is correct for such a structurally complex material is open to question, as transport measurements suggest that PANi has segregated metallic and insulating regions. PANi has been used to protect reactive semiconducting electrodes,[140] to fabricate microelectrode electrochemical transistors[141] and to make electrochromic displays[142] and batteries.[143]

22.4.5 Precursor-route Polymers

The use of soluble precursor-polymers, which can subsequently be converted to a conjugated form, offers a route to pure materials with controlled morphology.[71] It has been known for many years that poly(vinyl chloride) (PVC) can be thermally, chemically and photothermally dehydrochlorinated to produce polyene (PA) chain sequences.[144] Initiation occurs at random, and continuous conjugated-sequences are not obtained. As a result the maximum conductivity for iodine-doped dehydrochlorinated PVC is *ca.* $10^{-3}\,S\,cm^{-1}$.

A better product can be obtained if the precursor polymer contains some conjugated bonds which restrict the final structure and ensure extended conjugated-sequences. The most prominant example. is the Durham precursor-route to PA[71] (equation 1). The precursor polymer is soluble in acetone, ethyl acetate, *etc.*, and can be purified and cast to make films.[145] The thermal elimination reaction produces *cis*-PA as the predominant form at 60 °C. Use of higher temperatures leads to higher proportions of *trans*-PA due to *cis–trans* isomerization. If the precursor film is unoriented the PA produced is amorphous.[146] Stretching the precursor film during elimination leads to a highly oriented non-fibrous form of PA. *trans*-PA produced in this way exhibits a paracrystalline structure.

$$\qquad (1)$$

PA prepared by the Durham route has close to the theoretial density. Thus doping with iodine, *etc.*, is less efficient than with Shirakawa PA because of the low diffusion coefficients. Electrochemical doping shows a much larger band-tailing in unoriented materials and the transition to ordered alkali-metal dopant structures is smeared out.[146] Both are obvious consequences of disorder. However, the predominant bends and twists of the PA chains do not break bond-alternation. Oriented Durham PA has been studied by a variety of spectroscopic methods and shows the expected anisotropy in EELS, optical and Raman spectra.[147] More interesting are the results of photoconduction and photoinduced absorption experiments, which show that the charges which contribute to photocurrents and long-lived photoexcitations result from interchain excitation.[148] Thus, Durham PA has been, and will be, important in gaining a proper understanding of fundamental processes and the role of defects in PA.

The precursor route has also been used effectively for poly(1,4-phenylenevinylene) (poly(*p*-phenylenevinylene), PPPV). This is obtained *via* a sulfonium salt in which the phenyl rings pin the conjugated structure of the final polymer[149] (equation 2). Stretching during thermal elimination gives highly oriented films. Conductivities of $100\,S\,cm^{-1}$ and $2780\,S\,cm^{-1}$ have been reported for AsF_6-doped unoriented and oriented samples. Optical and IR spectra have been reported.[150, 151]

$$\qquad + \; SMe_2 \; + \; HCl \qquad (2)$$

Similar routes have been employed for poly(2,5-thiophenediylvinylene) and poly(2,5-furandiyl-vinylene) (equation 3).[152–154] Conductivities of doped films of these polymers are in the range 20–$60\,S\,cm^{-1}$. A novel route to PPP is *via* the bacterial oxidation of benzene to a substituted dehydrodiol. This is then polymerized using radical initiation to a polymer which, on heating to 140–$240\,°C$, converts to PPP[155] (equation 4).

$$\qquad + \; HCl \; + \qquad (3)$$

$$\qquad + \; 2RCO_2H \qquad (4)$$

22.5 OTHER HIGHLY CONDUCTIVE POLYMER SYSTEMS

22.5.1 Pyrolysis and Beam Modification

The production of carbon fibres and other forms of polymeric carbon by the high-temperature treatment of polymers is a well-established technology.[156] Most of this work has been directed towards the production of high-strength fibres and biocompatible materials for medical applications. A wide range of polymers have been used as precursors including polyesters, polyamides,

PVC, PPP, phenolic resins, *etc.* Pyrolysis is usually conducted at temperatures in excess of 1000 °C with a short final treatment in excess of 2000 °C. The product is graphitic carbon, but the intermediate structures, which are thought to be planar aromatic polymers, are less well-defined.

Polyacrylonitrile (PAN) has attracted attention because an initial oxidation is necessary at 200–300 °C, which is thought to result in a conjugated ladder polymer (13).[157] Doping of thermally oxidized PAN with bromine and iodine gives conductivities up to 10 S cm⁻¹. Heat treatment of spin-coated PAN films in the range 600–1200 °C gives, without doping, materials with conductivities from 0.1 to 800 S cm⁻¹.[158] An alternative approach to a polymer with a planar conjugated structure is the pyrolysis of 3,4,9,10-perylenetetracarboxylicdianhydride. Heat treatment between 700 and 900 °C yields material with a conductivity of up to 250 S cm⁻¹. This reaction is assumed to lead to poly(*peri*-naphthalene) (14).[159] Calculated band-structures indicate a small but finite band-gap of 0.3 eV. This is consistent with conductivities of 10^{-2} S cm⁻¹ for materials pyrolyzed near 500 °C. Conductivity increases from 10^{-5} S cm⁻¹ to 10^{2} S cm⁻¹ have been observed for Kapton (a polyamide formed from 1*H*,3*H*-benzo[1,2-*c*: 4,5-*c'*]difuran-1,3,5,7-tetrone (pyromellitic dianhydride) and 4,4'-diaminophenyl ether) heat treated in the temperature range 550–750 °C.[160] This is well below the range for graphitization. A number of other less well-characterized conductive polymers have been reported for pyrolyses in the range 700–900 °C.[161,162]

Ion implantation is used as a selective doping-mechanism for inorganic semiconductors. Ion bombardment of polymers leads to conductive material. There is, however, evidence that this is due to a carbonized layer on the polymer surface, produced by degradation which arises from uniform heavy damage along the paths of ions in organic solids.[163,164] Thus, ion implantation appears to be an expensive way to produce carbon coatings. There is, however, evidence that the final structures are dependent on that of the initial polymer.[165] In particular, the majority carriers can be either electrons or holes depending on the nature of the polymer used. Electron-beam irradiation has received less attention, though the degradation of organic materials in the electron microscope is a troublesome phenomenon.[166] In the case of PVC, electron-beam irradiation has been shown to result in the production of conjugated sequences.[167]

22.5.2 Conductor-filled Polymers

Fillers such as carbon and metal powders have been used to enhance both the electrical and the thermal conductivity of otherwise insulating polymers.[168] At low levels of loading, the conductivity of the composite differs only slightly from that of the host polymer. Then, at the percolation threshold, when conductive paths are formed, the conductivity rises sharply. Above the percolation threshold the conductivity rises more slowly. The conduction process, involving interparticle tunnelling, has been extensively investigated.[169] The range of conductivities available has led to a variety of practical applications, *e.g.* antistatic components, resistors, semiconducting screens for power cables, electromagnetic-interference screening, *etc.*

Because of its low cost and availability, carbon black has been extensively used. Despite this, much work is still being done to relate particle size, structure, aggregation, surface contamination, loading and interaction of filler and polymer matrix to composite conductivity.[170] Some examples of the different behaviours are shown in Figure 7. Aggregation has significant effects on conductivity, particularly when it produces elongated particles, since the percolation threshold for needles occurs at a considerably lower loading than for spherical particles. Aggregation is affected by the compounding process and period, and the great variety of structure of carbon blacks means that process parameters must be optimized in each case. Melt viscosity and surface tension strongly affect conductivity, as do polymer crystallinity and degree of crosslinking. More troublesome is the tendency of carbon to adsorb moisture and other volatile contaminants, and this has resulted in the use of other fillers.

Metal particles, principally Ni, Cu, Ag, Al and Fe, have been used.[171] These, particularly in the form of metal flakes, offer a higher conductivity (up to 10^{3} S cm⁻¹) than carbon black. Even lower percolation-thresholds can be achieved with conductive fibres. These can be either metal whiskers or metal-coated graphite or glass fibres. The use of lower loading-levels leads to improved mechanical properties, which are influenced by both filler loading and shape.[172]

Filled polymers are sensitive to deformation. The effect can be large in carbon-filled rubbers, particularly for compositions near the percolation threshold. This effect has been used to make self-regulating heaters, since, as the temperature increases, the filled polymer expands and the conductivity falls sharply,[173] resulting in a drastic reduction in current and heat output. A novel application of filled polymers follows from the fact that the percolation threshold depends on sample thickness.

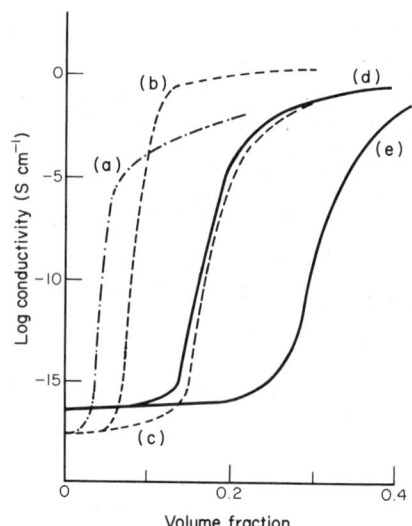

Figure 7 Conductivity *vs.* carbon-black loading for: (a) low-density PE with high melt-index; (b), (c) poly(methyl methacrylate) of high and low melt-index, respectively; and (d), (e) poly(vinyl chloride-*co*-vinyl acetate) with low and high degree of polymerization respectively (after M. Sumita, H. Abe, H. Kayaki and K. Miyasaka, *J. Macromol. Sci., Phys.*, 1986, **B25**, 171)

Thus, a filled polymer can be used to bond semiconductor chips to a substrate and provide geometrically defined interconnections.[174]

An interesting class of conductive composites, which can display good conductivity at low loadings, has been produced by the cocrystallization of an insulating polymer and a conducting organic charge-transfer (CT) compound.[175] The percolation threshold was observed to occur at 0.5% by weight of the CT compound in the system 2,2-bis(4-hydroxyphenyl)propane (bisphenol *A*) polycarbonate/tetrathiotetracene/tetracyanoquinodimethane. This low threshold can be related to the formation of a network of dendritic CT compound crystals in the polymer matrix, *i.e.* so-called 'reticulate doping'. These networks can be formed for a wide variety of polymer matrices with percolation thresholds as low as 0.039% by weight of the CT compound. In addition to enhanced conductivity, the films also have large dielectric constants, particularly for CT compounds with conductivities in excess of 10^{-1} S cm^{-1}.[176] The cast polymer films are flexible and show little dependence of conductivity on deformation. This, and the good long-term stability of the films, has led to their commercial production, despite the need to cast the films from chlorobenzene.

Another interesting composite is that obtained by cospinning of metal phthalocyanines (MPc) or metallo–Pc polymers, *e.g.* [Si(Pc)O]$_n$, with the polyaramid Kevlar.[177] This process produces fibres with conductivities of up to 1 S cm^{-1} with good long-term stability and excellent mechanical properties.[178]

22.6 REFERENCES

1. C. Kittel, 'Introduction to Solid State Physics', 6th edn., Wiley, New York, 1986, chap. 7.
2. R. E. Peierls, 'Quantum Theory of Solids', Clarendon Press, Oxford, 1964.
3. J. L. Brédas and G. B. Street, *Acc. Chem. Res.*, 1985, **18**, 309.
4. N. F. Mott and E. A. Davies, 'Electronic Processes in Non-Crystalline Materials', Clarendon Press, Oxford, 1971.
5. P. W. Anderson, *Phys. Rev.*, 1958, **109**, 1492.
6. A. Dall'Olio, G. Dascola, V. Varacca and V. Bocchi, *C.R. Hebd. Seances Acad. Sci., Ser. C*, 1968, **267**, 433.
7. H. Shirakawa, E. J. Louis, A. G. MacDiarmid, C. K. Chiang and A. J. Heeger, *J. Chem. Soc., Chem. Commun.*, 1977, 578.
8. G. B. Street and W. D. Gill, in 'Molecular Metals', ed. W. E. Hatfield, Plenum Press, New York, 1979, p. 301.
9. A. J. Banister, Z. V. Hauptmann and A. G. Kendrick, *J. Chem. Soc., Chem. Commun.*, 1983, 1016.
10. T. J. Fabish, *CRC Crit. Rev. Solid State Mater. Sci.*, 1979, **8**, 383.
11. J.-M. André, J. Delhalle and J. J. Pireaux, in 'Photo Electron and Ion Probes of Polymer Structure and Properties', ed. D. W. Dwight, T. J. Fabish and H. R. Thomas, American Chemical Society, Washington, DC, 1981, p. 1561.
12. J. J. Pireaux, J. Riga, R. Caudono and J. Verbist, in 'Photo Electron and Ion Probes of Polymer Structures and Properties', ed. D. W. Dwight, T. J. Fabish and H. R. Thomas, American Chemical Society, Washington, DC, 1981, p. 169.
13. W. R. Salanek, in 'Photo Electron and Ion Probes of Polymer Structures and Properties', ed. D. W. Dwight, T. J. Fabish and H. R. Thomas, American Chemical Society, Washington, DC, 1981, p. 121.
14. J. J. Ritsko, in 'Photo Electron and Ion Probes of Polymer Structures and Properties', ed. D. W. Dwight, T. J. Fabish and H. R. Thomas, American Chemical Society, Washington, DC, 1981, p. 35.

15. D. Bloor, *Chem. Phys. Lett.*, 1976, **40**, 323.
16. D. K. Davies, *J. Phys. D*, 1969, **2**, 1533.
17. A. R. Akande and J. Lowell, *J. Phys. D*, 1987, **20**, 565.
18. M. A. Lampert and P. Mark, 'Current Injection in Solids', Academic Press, New York, 1970.
19. R. M. Eichhorn, in 'Engineering Dielectrics', ed. R. Bartnikas and R. M. Eichhorn, American Society for Testing and Materials, Philadelphia, 1983, vol. IIA, p. 355.
20. T. Hibma and H. R. Zeller, *J. Appl. Phys.*, 1986, **59**, 1614.
21. S. R. Kurtz and R. A. Anderson, *J. Appl. Phys.*, 1986, **60**, 681.
22. H. Scher and E. W. Montroll, *Phys. Rev. B: Solid State*, 1975, **12**, 2455.
23. M. M. Perlman and S. Unger, *J. Phys. D*, 1972, **5**, 2115.
24. L. Brehmer, *Acta Polym.*, 1981, **32**, 415.
25. D. Emin, in 'Handbook of Conducting Polymers', ed. T. A. Skotheim, Dekker, New York, 1986, p. 915.
26. G. M. Sessler (ed.), *Top. Appl. Phys.*, 1987, **33**.
27. A. J. Lovinger, in 'Developments in Crystalline Polymers-1', ed. D. C. Bassett, Applied Science, London, 1982, p. 195.
28. A. K. Jonscher, *Nature (London)*, 1977, **267**, 673.
29. G. A. Brown, in 'Polymeric Materials for Electronic Applications', ed. E. D. Feit and C. W. Wilkins Jr., American Chemical Society, Washington, DC, 1982, p. 139.
30. T. Yokoyama, N. Kinjo and Y. Wakashima, *J. Electrochem. Soc.*, 1987, **134**, 975.
31. H. W. Gibson, *Polymer*, 1984, **25**, 3.
32. A. R. Blythe, 'Electrical Properties of Polymers', Cambridge University Press, Cambridge, 1980.
33. J. Mort and G. Pfister (eds.), 'Electronic Properties of Polymers', Wiley, New York, 1982.
34. M. Baijal (ed.), 'Dielectric and Photoconductive Properties of Polymers', Wiley, New York, 1981.
35. L. Onsager, *Phys. Rev.*, 1938, **54**, 554.
36. H. Scher and S. Rackovsky, *J. Chem. Phys.*, 1984, **81**, 1994.
37. B. Ries and H. Bässler, *J. Mol. Electron.*, 1987, **3**, 15.
38. M. Pope and C. E. Swenberg, 'Electronic Processes in Organic Crystals', Clarendon Press, Oxford, 1982, p. 701.
39. R. H. Partridge, *Polymer*, 1983, **24**, 739.
40. E. P. Goodings, *Chem. Soc. Rev.*, 1976, **5**, 95.
41. J. Ulanski, J. K. Jeszka and M. Krysyewski, *Polym.-Plast. Technol. Eng.*, 1981, **17**, 139.
42. A. A. Ovchinnikov, I. I. Ukrainskii and G. V. Kventsel, *Sov. Phys. — Usp. (Engl. Transl.)*, 1973, **15**, 575.
43. M.-H. Whangbo, R. Hoffman and R. B. Woodward, *Proc. R. Soc. London, Ser. A*, 1979, **366**, 23.
44. O. Wennerstrom, *Macromolecules*, 1985, **18**, 1977.
45. J. L. Brédas, M. Dory and J. M. André, *J. Chem. Phys.*, 1985, **83**, 5242.
46. W. K. Ford and C. B. Duke, *Mol. Cryst. Liq. Cryst.*, 1983, **93**, 327.
47. M. Kertesz, *Adv. Quantum Chem.*, 1982, **18**, 161.
48. A. Karpfen, *J. Phys. C*, 1980, **13**, 5673.
49. S. Suhai, *Phys. Rev. B: Condens. Matter*, 1983, **27**, 3506; *Phys. Rev. B: Condens. Matter*, 1984, **29**, 4570.
50. J. L. Brédas, in 'Handbook of Conducting Polymers', ed. T. A. Skotheim, Dekker, New York, 1986, p. 859.
51. W. P. Su, in 'Handbook of Conducting Polymers', ed. T. A. Skotheim, Dekker, New York, 1986, p. 757.
52. A. J. Heeger, in 'Handbook of Conducting Polymers', ed. T. A. Skotheim, Dekker, New York, 1986, p. 729.
53. J. A. Pople and S. H. Walmsley, *Mol. Phys.*, 1962, **5**, 15.
54. W. P. Su, J. R. Schrieffer and A. J. Heeger, *Phys. Rev. Lett.*, 1979, **42**, 1698.
55. R. R. Chance, D. S. Boudreaux, J. L. Brédas and R. Silbey, in 'Handbook of Conducting Polymers', ed. T. A. Skotheim, Dekker, New York, 1986, p. 825.
56. I. I. Ukrainskii, *Phys. Status Solidi B*, 1981, **106**, 55.
57. K. Fesser, A. R. Bishop and D. K. Campbell, *Phys. Rev. B: Condens. Matter*, 1983, **27**, 4804.
58. M. Grabowski, D. Hone and J. R. Schrieffer, *Phys. Rev. B: Condens. Matter*, 1985, **31**, 7850.
59. G. Wegner, *Angew. Chem., Int. Ed. Engl.*, 1981, **20**, 361.
60. B. Horowitz, Z. Vardeny, E. Ehrenfreund and O. Brafman, *J. Phys. C*, 1986, **19**, 7291.
61. B.-C. Xu and S. E. Trullinger, *Phys. Rev. Lett.*, 1986, **57**, 3113.
62. R. H. Baughman and L. W. Shacklette, *Synth. Met.*, 1987, **17**, 173.
63. D. Baeriswyl, *Helv. Phys. Acta*, 1983, **56**, 639.
64. J. C. W. Chien, 'Polyacetylene', Academic Press, Orlando, FL, 1984.
65. J. Simon and J.-J. André, 'Molecular Semiconductors', Springer, Berlin, 1985, chap. 4.
66. H. Kuzmany, M. Mehring and S. Roth (eds.), *Springer Ser. Solid-State Sci.*, 1987, **76**.
67. T. A. Skotheim (ed.), 'Handbook of Conducting Polymers', Dekker, New York, 1986.
68. S. Roth, K. Ehinger, K. Menke, M. Peo and R. J. Schweitzer, *J. Phys. Colloq. (Orsay, Fr.)*, 1983, **C3**, 69.
69. S. A. Chen and H. J. Shy, *J. Polym. Sci., Polym. Chem. Ed.*, 1985, **23**, 2441.
70. H. Shirakawa and S. Ikeda, *Polym. J.*, 1971, **2**, 231.
71. W. J. Feast, in 'Handbook of Conducting Polymers', ed. T. A. Skotheim, Dekker, New York, 1986, p. 1.
72. W. H. Meyer, *Synth. Met.*, 1981, **4**, 81.
73. T. Woerner, A. G. MacDiarmid and A. J. Heeger, *J. Polym. Sci., Polym. Lett. Ed.*, 1982, **20**, 305.
74. K. Araya, A. Mukoh, T. Narahara and H. Shirakawa, *Chem. Lett.*, 1984, 1141.
75. L. C. Dickinson, J. A. Hirsch, F. E. Karasz and J. C. W. Chien, *Macromolecules*, 1985, **18**, 2374.
76. F. S. Bates and G. L. Baker, *Macromolecules*, 1983, **16**, 1013.
77. M. L. Elert and C. T. White, *Macromolecules*, 1987, **20**, 1411.
78. S. Pekker and A. Janossy, in 'Handbook of Conducting Polymers', ed. T. A. Skotheim, Dekker, New York, 1986, p. 45.
79. A. G. MacDiarmid, R. J. Mannmore, R. B. Kaner and S. J. Porter, *Philos. Trans. R. Soc. London, Ser. A*, 1985, **314**, 3.
80. L. W. Shacklette and J. E. Toth, *Phys. Rev. B: Condens. Matters*, 1985, **32**, 5892.
81. J. M. Pochan, in 'Handbook of Conducting Polymers', ed. T. A. Skotheim, Dekker, New York, 1986, p. 1297.
82. T. S. Clarke and J. C. Scott, in 'Handbook of Conducting Polymers', ed. T. A. Skotheim, Dekker, New York, 1986, p. 1127.
83. H. Thomann, H. Jin and G. L. Baker, *Phys. Rev. Lett.*, 1987, **59**, 509.

84. H. Thomann and L. R. Dalton, in 'Handbook of Conducting Polymers', ed. T. A. Skotheim, Dekker, New York, 1986, p. 1157.
85. P. Bernier, in 'Handbook of Conducting Polymers', ed. T. A. Skotheim, Dekker, New York, 1986, p. 1099.
86. M. Mehring and P. K. Kahol, *Springer Ser. Solid-State Sci.*, 1985, **63**, 264.
87. J. Orenstein, in 'Handbook of Conducting Polymers', ed. T. A. Skotheim, Dekker, New York, 1986, p. 1297.
88. G. Crelius, in 'Handbook of Conducting Polymers', ed. T. A. Skotheim, Dekker, New York, 1986, p. 1233.
89. A. J. Epstein, in 'Handbook of Conducting Polymers', ed. T. A. Skotheim, Dekker, New York, 1986, p. 1041.
90. D. N. Batchelder and D. Bloor, in 'Advances in Infrared and Raman Spectroscopy', ed. R. J. H. Clark and R. E. Hester, Wiley, Chichester, 1984, vol. 11, p. 133.
91. R. Tiziani, G. P. Brivio and E. Mulazzi, *Phys. Rev. B: Condens. Matter* 1985, **31**, 4015.
92. E. J. Mele, in 'Handbook of Conducting Polymers', ed. T. A. Skotheim, Dekker, New York, 1986, p. 795.
93. K. Kanicki, in 'Handbook of Conducting Polymers', ed. T. A. Skotheim, Dekker, New York, 1986, p. 543.
94. A. G. MacDiarmid and R. B. Kaner, in 'Handbook of Conducting Polymers', ed. T. A. Skotheim, Dekker, New York, 1986, p. 689.
95. L. W. Shacklette, J. E. Toth, N. S. Murthy and R. H. Baughman, *J. Electrochem. Soc.*, 1985, **132**, 1529.
96. T. Nagatomo, C. Ichikawa and O. Omoto, *J. Electrochem. Soc.*, 1987, **134**, 305.
97. J. B. Schlenoff and J. C. W. Chein, *J. Electrochem. Soc.*, 1987, **134**, 2179.
98. H. Naarmann and N. Theophilou, *Synth. Met.*, 1987, **22**, 1.
99. H. W. Gibson, in 'Handbook of Conducting Polymers', ed. T. A. Skotheim, Dekker, New York, 1986, p. 405.
100. R. L. Elsenbaumer and L. W. Shacklette, in 'Handbook of Conducting Polymers', ed. T. A. Skotheim, Dekker, New York, 1986, p. 213.
101. C. E. Brown, P. Kovacic, C. A. Wilkie, R. B. Cody, R. E. Heim and J. A. Kisinger, *Synth. Met.*, 1986, **15**, 265.
102. T. R. Jow, L. W. Shacklette, M. Maxfield and D. Vernick, *J. Electrochem. Soc.*, 1987, **134**, 1730.
103. V. A. Sergeyev, V. I. Nedel'kin and S. A. Arnautov, *Polym. Sci. USSR (Engl. Transl.)*, 1985, **27**, 1003.
104. G. A. Kossmehl, in 'Handbook of Conducting Polymers', ed. T. A. Skotheim, Dekker, New York, 1986, p. 351.
105. D. Bloor and R. R. Chance (eds.), 'Polydiacetylenes', Nijhof, Dordrecht, 1985.
106. D. Bloor, in 'Quantum Chemistry of Polymers — Solid State Aspects', ed., J. Ladik, J. M. Andre and M. Seel, Reidel, Dordrecht, 1984, p. 191.
107. D. J. Sandman, B. S. Elman, G. P. Hamil, J. Hefter and C. S. Velazquez, *ACS Symp. Ser.*, 1987, **337**, 118.
108. S. E. Tunney, J. Suenaga and J. K. Stille, *Macromolecules*, 1983, **16**, 1398.
109. C. W. Dirk, M. Bousseau, P. H. Barrett, F. Moraes, F. Wudl and A. J. Heeger, *Macromolecules*, 1986, **19**, 266.
110. T. M. Swager and R. H. Grubbs, *J. Am. Chem. Soc.*, 1987, **109**, 894.
111. D. T. Glatzhofer and D. T. Longone, *J. Polym. Sci., Polym. Chem. Ed.*, 1986, **24**, 947.
112. C. W. Dirk, T. Inabe, J. W. Lyding, K. F. Schoch Jr, C. R. Kannewurf and T. J. Marks, *J. Polym. Sci., Polym. Symp.*, 1983, **70**, 1.
113. M. Hanack and A. Leverenz, *Springer Ser. Solid-State Sci.*, 1987, **76**, 318.
114. R. J. Waltman and J. Bargon, *Can. J. Chem.*, 1986, **64**, 76.
115. A. F. Diaz and J. Bargon, in 'Handbook of Conducting Polymers', ed. T. A. Skotheim, Dekker, New York, 1986, p. 81.
116. G. B. Street, in 'Handbook of Conducting Polymers', ed. T. A. Skotheim, Dekker, New York, 1986, p. 265.
117. G. Tourillon, in 'Handbook of Conducting Polymers', ed. T. A. Skotheim, Dekker, New York, 1986, p. 293.
118. P. Burgmayer and R. W. Murray, in 'Handbook of Conducting Polymers', ed. T. A. Skotheim, Dekker, New York, 1986, p. 507.
119. T. Hagiwara and K. Iwata, *Synth. Met.*, 1986, **14**, 61.
120. H. Naarman, *Makromol. Chem., Macromol. Symp.*, 1987, **8**, 1.
121. N. S. Sundaresan, S. Basak, M. Pomerantz and J. R. Reynolds, *J. Chem. Soc., Chem. Commun.*, 1987, 621.
122. M. Okano, K. Itoh, A. Fujishima and K. Honda, *J. Electrochem. Soc.*, 1987, **134**, 837.
123. O. Inganäs, T. A. Skotheim and I. Lundström, in 'Handbook of Conducting Polymers', ed. T. A. Skotheim, Dekker, New York, 1986, p. 525.
124. R. Bittihn, G. Ely, F. Woeffler, H. Münstedt, H. Naarman and D. Naegele, *Makromol. Chem., Macromol. Symp.*, 1987, **8**, 51.
125. M. S. Wrighton, *Science (Washington, D.C.)*, 1986, **231**, 32.
126. G. Bidan and D. Limosin, *Ann. Phys. (Paris)*, 1986, **11** (Suppl. 1), 5.
127. P. Novák, O. Inganäs and R. Bjorklund, *J. Electrochem. Soc.*, 1987, **134**, 1341.
128. P. Lang, F. Char, M. Costa and F. Garnier, *Polymer*, 1987, **28**, 608.
129. M. R. Bryce, A. Chissel, P. Kathirgamanathan, D. Parker and N. R. M. Smith. *J. Chem. Soc., Chem. Commun.*, 1987, 466.
130. A. C. Cheng and L. L. Miller, *Synth. Met.*, 1987, **22**, 71.
131. F. Garnier and G. Horowitz, *Makromol. Chem., Macromol. Symp.*, 1987, **8**, 159.
132. J. W. Thackeray, H. S. White and M. S. Wrighton, *J. Phys. Chem.*, 1985, **89**, 5133.
133. W. R. Salaneck, J. L. Brédas, C. R. Wu and J. O. Nilsson, *Synth. Met.*, 1987, **21**, 57.
134. J. C. Chiang and A. G. Macdiarmid, *Synth. Met.*, 1986, **13**, 193.
135. J. C. LaCroix and A. F. Diaz, *Makromol. Chem., Macromol. Symp.*, 1987, **8**, 17.
136. R. J. Cushman, P. M. McManus and S. C. Yang, *Makromol. Chem., Rapid Commun.*, 1987, **8**, 69.
137. A. P. Monkman, D. Bloor, G. C. Stevens and J. C. H. Stevens, *J. Phys. D*, 1987, **20**, 1337.
138. A. J. Epstein, J. M. Ginder, F. Zuo, H. S. Woo, D. B. Tanner, A. F. Richter, M. Angllopoulos, W. S. Huang and A. G. MacDiarmid, *Synth. Met.*, 1987, **21**, 63.
139. S. Stafstrom, J. L. Brédas, A. J. Epstein, H. S. Woo, D. B. Tanner, W. S. Huang and A. G. MacDiarmid, *Phys. Rev. Lett.*, 1987, **59**, 1464.
140. R. Noufi, A. J. Nozik, J. White and L. F. Warren, *J. Electrochem. Soc.*, 1982, **129**, 2261.
141. E. W. Paul, A. J. Ricco and M. S. Wrighton, *J. Phys. Chem.*, 1985, **89**, 1441.
142. T. Kobayshi, H. Yoneyama and H. Tamura, *J. Electroanal. Chem.*, 1984, **161**, 419.
143. A. Kitani, M. Kaya and K. Sasaki, *J. Electrochem. Soc.*, 1986, **133**, 1069.
144. H. J. Bowley, D. L. Gerrard, W. F. Maddams and M. R. Paton, *Makromol. Chem.*, 1985, **186**, 695.
145. W. J. Feast, M. J. Taylor and J. N. Winter, *Polymer*, 1987, **28**, 593.

146. D. C. Bott, C. S. Braun, J. N. Winter and J. Barker, *Polymer*, 1987, **28**, 601.
147. J. Fink, N. Nücker, B. Scheerer, A. von Felde and G. Leising, *Springer Ser. Solid-State Sci.*, 1987, **76**, 84.
148. P. D. Townsend and R. H. Friend, *J. Phys. C*, 1987, **20**, 4221.
149. D. R. Gagnon, J. D. Capistran, F. E. Karasz, R. W. Lenz and S. Antoun, *Polymer*, 1987, **28**, 567.
150. J. Obrzut and F. E. Karasz, *J. Chem. Phys.*, 1987, **87**, 2349.
151. D. D. C. Bradley, R. H. Friend, H. Lindenberger and S. Roth, *Polymer*, 1986, **27**, 1709.
152. K.-Y. Jen, M. Maxfield, L. W. Shacklette and R. L. Elsenbaumer, *J. Chem. Soc., Chem. Commun.*, 1987, 309.
153. S. Yamada, S. Tokito, T. Tsutsui and S. Saito, *J. Chem. Soc., Chem. Commun.*, 1987, 1448.
154. K.-Y. Jen, T. R. Jow and R. L. Elsenbaumer, *J. Chem. Soc., Chem. Commun.*, 1987, 1113.
155. D. G. H. Ballard, A. Courtis, I. M. Shirley and S. C. Taylor, *J. Chem. Soc., Chem. Commun.*, 1983, 954.
156. E. Fitzer, *Angew. Chem., Int. Ed. Engl.*, 1980, **19**, 375.
157. N. R. Lerner, *Polymer*, 1983, **24**, 800.
158. C. L. Renschler and A. P. Sylwester, *Appl. Phys. Lett.*, 1987, **50**, 1420.
159. J. L. Brédas and R. H. Baughman, *J. Chem. Phys.*, 1985, **83**, 1316.
160. C. Z. Hu and J. D. Andrade, *J. Appl. Polym. Sci.*, 1985, **30**, 4409.
161. T. M. Keller, *J. Polym. Sci., Polym. Lett. Ed.*, 1986, **24**, 211.
162. L. Y. Chiang, *J. Chem. Soc., Chem. Commun.*, 1987, 304.
163. D. Fink, J. P. Biersack, J. T. Chen, M. Städele, K. Tjan, M. Behar, C. O. Olivieri and F. C. Zawislak, *J. Appl. Phys.*, 1985, **58**, 668.
164. L. Calcagno and G. Foti, *Appl. Phys. Lett.*, 1985, **47**, 363.
165. G. E. Wnek, B. Wasserman, M. S. Dresselhaus, S. E. Tuney and J. K. Stille, *J. Polym. Sci., Polym. Lett. Ed.*, 1985, **23**, 609.
166. L. Reimer, *Ultramicroscopy*, 1984, **14**, 291.
167. D. Vesely and D. S. Finch, *Ultramicroscopy*, 1987, **23**, 329.
168. R. A. Crossman, *Polym. Eng. Sci.*, 1985, **25**, 507.
169. R. D. Sherman, L. M. Middleman and S. M. Jacobs, *Polym. Eng. Sci.*, 1983, **23**, 36.
170. E. K. Sichel (ed.), 'Carbon Black–Polymer Composites', Dekker, New York, 1982.
171. S. K. Bhattacharya and A. C. D. Chaklader, *Polym.-Plast. Technol. Eng.*, 1982, **19**, 21.
172. T. S. Chow, *J. Mater. Sci.*, 1980, **15**, 1873.
173. J. Kost, M. Narkis and A. Foux, *Polym. Eng. Sci.*, 1983, **23**, 567.
174. S. Etemad, X. Quan and N. A. Sanders, *Appl. Phys. Lett.*, 1986, **48**, 607.
175. A. Tracz, J. K. Jeszka, E. E. Shafee, J. Olanski and M. Kryszewski, *J. Phys. D*, 1986, **19**, 1047.
176. H. Kita, K. Okamoto and S. Mukai, *J. Appl. Polym. Sci.*, 1986, **31**, 1383.
177. T. Inabe, J. W. Lyding and T. J. Marks, *J. Chem. Soc., Chem. Commun.*, 1983, 1084.
178. K. J. Wynne, A. E. Zacharides, T. Inabe and T. J. Marks, *Polym. Commun.*, 1985, **26**, 162.

23

Surface Properties

DAVID BRIGGS AND DEREK G. RANCE
ICI Wilton, Cleveland, UK
and
BRIAN J. BRISCOE
Imperial College, London, UK

23.1 INTRODUCTION

In the interaction of polymers with other materials, surface properties are clearly critical. In the case of interactions with liquids the important phenomena are wetting and spreading and these, in turn, affect the adhesion of materials applied in the liquid state. In the case of interactions with solids the important phenomena are contact adhesion, hardness and scratching, friction, and wear. Our understanding of these properties, in particular of the way in which they relate to polymer structure, is very far from complete. This chapter attempts to provide a brief account of the measurement techniques used to assess these properties (with the appropriate theory), the limits of application and the fundamental data which can be derived from the results.

23.2 THE INTERACTION OF POLYMER SURFACES WITH LIQUIDS: WETTING, SPREADING AND ADHESION

23.2.1 Thermodynamics of Wetting

When a liquid is brought into contact with a solid, a solid/liquid interface is formed at the expense of both a solid-free and a liquid-free surface. The reversible free energy change G_{SL} per unit area of new interface formed is

$$G_{SL} = \gamma_{SL} - \gamma_S - \gamma_{LV} = -W_A \tag{1}$$

where γ_S is the surface free energy of the solid in a vacuum, γ_{LV} the surface tension of the liquid in equilibrium with its vapour, γ_{SL} the interfacial tension and W_A is the thermodynamic work of adhesion.

If a liquid makes contact with a solid surface, that surface is said to be wetted. The extent of wetting is described by the so-called contact angle θ which a liquid droplet makes with the surface at the three-phase contact line. If the solid adsorbs liquid vapour, then the solid surface tension becomes reduced to γ_{SV} where $\gamma_S = \gamma_{SV} + \pi_e$ and where π_e is known as the spreading pressure. Values of π_e have been found to be negligible for liquids which have a nonzero contact angle on low surface energy substrates such as polymers.[1]

When a liquid wets a solid to the extent that the contact angle becomes zero, the liquid is said to spread such that for each unit area of solid/vapour interface which disappears equivalent areas of solid/liquid and liquid/vapour interfaces are formed. The equilibrium spreading coefficient S is defined as

$$S = \gamma_{SV} - \gamma_{SL} - \gamma_{LV} \tag{2}$$

A liquid will spread spontaneously on a solid surface when $S \geq 0$.

It has been argued[2] that for good adhesion to a substrate the contact angle of an adhesive or coating must be zero so that the liquid will spread. The argument was that only under this condition will an adhesive maximize interfacial contact and hence optimize the extent of molecular attraction across the interface. If this condition were necessary, the number of adhesives which could be used for many substrates would be severely limited. In practice, real surfaces are seldom smooth; a schematic representation of a real surface is shown in Figure 1.

Huntsberger[3,4] has shown that in this case it is more appropriate to consider the free energy change to obtain complete wetting of this surface

$$G_{SL} = -\gamma_{LV}[1 + (\Omega_S/A_S)\cos\theta] \tag{3}$$

where (Ω_S/A_S) is the ratio of the actual to the projected surface area. Equation (3) suggests that a decrease in free energy accompanies wetting for all cases except where $\cos\theta < 0$ and the value of $(\Omega_S/A_S)\cos\theta > 1$. This implies that the complete wetting of any surface at equilibrium is favourable

(a)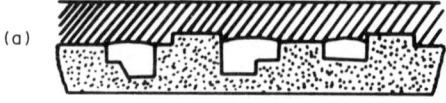

(b)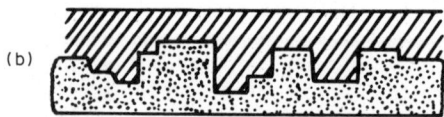

Figure 1 Schematic of the wettting of a solid by a liquid: (a) substrate incompletely wetted; and (b) substrate completely wetted (reproduced from J. R. Huntsberger, *Adv. Chem. Ser.*, 1964, **43**, 180 by permission of the American Chemical Society)

for any adhesive which makes a contact angle $< 90°$ on that surface (and this includes most adhesive situations which are likely to be encountered). Huntsberger suggests therefore that the zero contact angle criterion for good adhesion is invalid. He qualifies this view by adding that the poor interfacial adhesion obtained in some cases is more the result of the slow rate of wetting of adhesives rather than unfavourable thermodynamics. However, in many cases the maximum rate of wetting may occur at zero contact angle.[3] In practice a high rate of wetting must be achieved to allow often highly viscous adhesives to penetrate all the surface irregularities in the short time the adhesive remains in a liquid state after contact. The kinetic aspects of wetting are discussed by Cherry.[5]

23.2.2 Practical Assessment of Wetting

23.2.2.1 *Preparation of clean surfaces for contact angle measurement*

The intrinsic surface properties of a material can only be investigated if the surface is clean at a molecular level. It is likely that a large body of information on surfaces which has been obtained and reported in the literature is misleading because the measurements have been made on surface contaminants. Extreme care must be taken in preparing surfaces for contact angle measurement. For the calculation of thermodynamic parameters it is the equilibrium contact angle of a liquid on a surface which is of interest. In order to obtain a meaningful result on a planar surface the surface must be as smooth as possible and free from contamination.

The tendency for materials of low surface energy to adsorb on polymers is low compared with metals although many surfaces are readily contaminated by airborne hydrocarbon and silicone greases. Polymer surfaces which are intrinsically the cleanest are those which can be drawn as a film in the solid state such as polyethylene, polypropylene and poly(ethylene terephthalate) (PET). In the manufacturing process of these films new surface is generated which only comes into contact with similar polymer surface when it is wound up on a reel. For the polyalkenes, however, migratory additives present in the polymer to improve polymer stability and processability still present a problem in trying to produce a clean surface. Briggs and Kendall[6] have shown that the migratory additives or low molecular weight polymer can be readily removed by extracting the surface of polyalkenes with diethyl ether at ambient temperature.

Many workers are tempted to form smooth polymer surfaces by melting the polymer in contact with an optically flat surface, *e.g.* chrome plate, aluminum or glass. However, contamination such as traces of oxide in the case of metals or silicate in the case of glass invariably transfers to the polymer on separating the materials. The most satisfactory methods for preparing polymer surfaces for contact angle measurements are (a) solution-casting a thin film onto a clean substrate followed by solvent removal; or (b) melt-casting discs of polymer against a carefully cleaned flat surface, followed by removal of the surface layers by polishing. In the latter approach, pregrinding with a series of progressively finer carborundum papers may be necessary in order to obtain a flat surface. The final surface polish can be obtained using the technique described by Ottewill and co-workers[7] for silver iodide surfaces. By this technique the surfaces are mechanically polished on a rotating plate covered with a soft wet cloth which is impregnated with fine alumina powder. Polishing with two alumina powders is recommended: firstly alumina with a particle size of 5 μm followed by material with a particle size 0.05 μm. The surface is regarded as sufficiently smooth for contact angle measurement when the surface roughness could no longer be detected using a $\times 5$ magnifying lens. Residual alumina can be removed from the polished surface by immersing it in a container of distilled water in an ultrasonic bath. The surface is then ready for measurement after final rinsing and drying.

Surface analytical techniques such as X-ray photoelectron spectroscopy (XPS) and secondary ion mass spectroscopy (SIMS) have allowed a much-improved assessment of the extent of surface contamination (see Volume 1, Chapter 24). The development of more rigorous surface-cleaning regimes has been made possible by an iterative process of employing a given cleaning technique followed by analysis of what has been achieved.

23.2.2.2 *Methods of contact angle measurement*

(i) *Planar surfaces*

Equilibrium contact angles can be measured very simply from the profiles of liquid drops (Figure 2a) or bubbles (Figure 2b) resting on a plane surface. These methods are known as the sessile drop and captive bubble methods respectively. The contact angle may be measured indirectly by

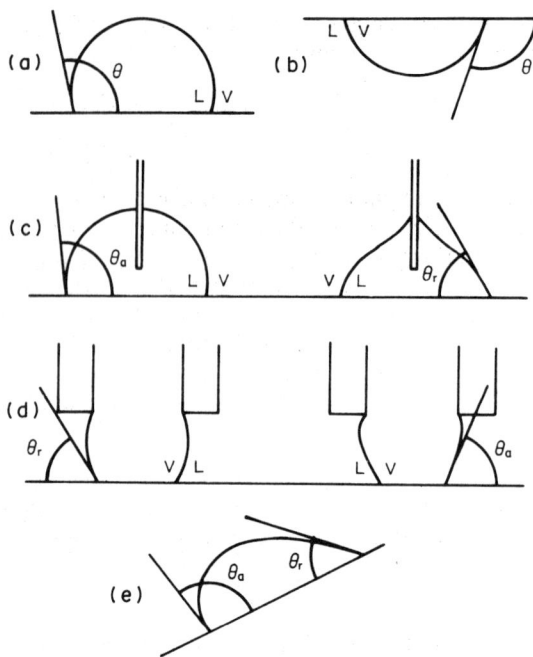

Figure 2 Drop and bubble configurations for measurement of equilibrium, advancing and receding contact angles: (a) equilibrium sessile drop; (b) equilibrium pendant bubble; (c) advancing and receding drop; (d) advancing and receding captive bubble; and (e) drop on tilted plate

drawing a tangent to the profile at the point of three-phase contact after the profile has been enlarged by image projection or photography. The contact angle can be measured directly using a microscope fitted with a goniometer eyepiece. All the equipment necessary to obtain accurate measurements can be readily assembled in the laboratory although commercial instruments are also available.[8,9]

The measurement of drop dimensions is another useful indirect method for obtaining contact angle data. Provided the drops are sufficiently small that gravitational distortion of the drop can be considered negligible, the contact angle may be calculated from the height h and base diameter d of the drop, which is considered to be a segment of a sphere so that $\tan(\theta/2) = 2(h/d)$. This method applies[10] to all drops which have a volume $< 10^{-10}\,\mathrm{m}^3$ and can be extended[11] to drops where $d < 0.5$ mm and $\theta < 90°$. For larger drops, the dependence of the profile shape on contact angle and drop size is considered by Padday.[12] Other indirect methods include a reflected light technique[13] (only valid for $\theta < 90°$) and the interference microscopy method.[14,15] The latter method is particularly useful for measuring contact angles $< 10°$ which are difficult to assess accurately by the tangent method and has been successfully applied to the measurement of the contact angle between two immiscible liquids and a solid.[15] The accuracy with which contact angles can be measured is generally between $\pm 1°$ and $\pm 2°$.

Surfaces are seldom ideally smooth or chemically homogeneous. As a result different contact angles may be obtained depending on whether a liquid drop is advanced or withdrawn across the surface. The contact angles which the three-phase contact line makes with the surface when brought to rest after being slowly advanced or withdrawn across the surface define the advancing and receding contact angles respectively. The most common methods of measurement are shown in Figure 2(c)–2(e). In Figure 2(c) liquid is advanced or withdrawn across the surface by increasing or decreasing the volume of liquid which is dispensed by the needle from a syringe. The needle remains in the drop during measurements to avoid unnecessary vibration or distortion of the drop and does not affect the contact angle. The equivalent captive bubble technique is shown in Figure 2(d). In Figure 2(e) the advancing and receding angles can be measured from a single drop in the limit when the drop begins to move on increasing the angle of tilt of the surface.

The Wilhelmy plate or tensiometric method is widely used for measuring equilibrium, advancing and receding contact angles. This is described fully in the next section as it is one of the few techniques which are generally applicable to materials of cylindrical cross-section such as fibres.

These and other techniques for measuring contact angle, such as capillary rise at a vertical plane plate, are included in the review by Neumann and Good.[16]

(ii) Curved surfaces

The curved surface which has a regular geometry and is therefore amenable to contact angle measurement is a cylinder. Polymeric fibres (and carbon fibres derived from them) find increasingly important applications in high performance polymer composites. Complete wetting of fibres by the matrix is a prerequisite for obtaining the good fibre–matrix adhesion which is believed to be necessary in order to achieve optimum composite mechanical properties.[17,18]

The most appropriate method for determining the wettability of fibres is to use the tensiometric or Wilhelmy plate method. If a thin fibre is suspended below the pan of an electromicrobalance the force F_0 acting on the fibre when it just touches the surface of a liquid (Figure 3a) is given by

$$F_0 = p\gamma_{LV}\cos\theta_a \tag{4}$$

where p is the perimeter of the fibre and θ_a is the advancing contact angle. On further immersion to depth d (Figure 3b) the force F is modified by a buoyancy correction such that

$$F_0 = p\gamma_{LV}\cos\theta_a - \rho g A d \tag{5}$$

where A is the cross-sectional area of the fibre, ρ the density of the liquid and g the acceleration due to gravity. A plot of F against d extrapolated to $d = 0$ allows calculation of θ_a. Similarly, the receding contact angle may be calculated from force measurement on the fibre as it is slowly raised through the liquid. Measurements of this type have been made to characterize high-modulus carbon fibres[19] and Kevlar fibres[18] using a number of standard liquids (see Section 23.3.4). Polymer surfaces have also been characterized in this way after the careful solution coating of polymers on to metal wire.[20]

An alternative method for measuring the contact angle on fibres is described by Carroll.[21] Essentially, he has applied drops of liquid to a fibre (Figure 3c) and derived analytical expressions relating dimensionless forms of drop length, drop radius and fibre radius to contact angle. The method is not as widely used as the tensiometric method; droplets are difficult to place on the thinnest fibres. However, contact angle measurements of molten polyethylene on glass fibre have been reported[22] using this technique.

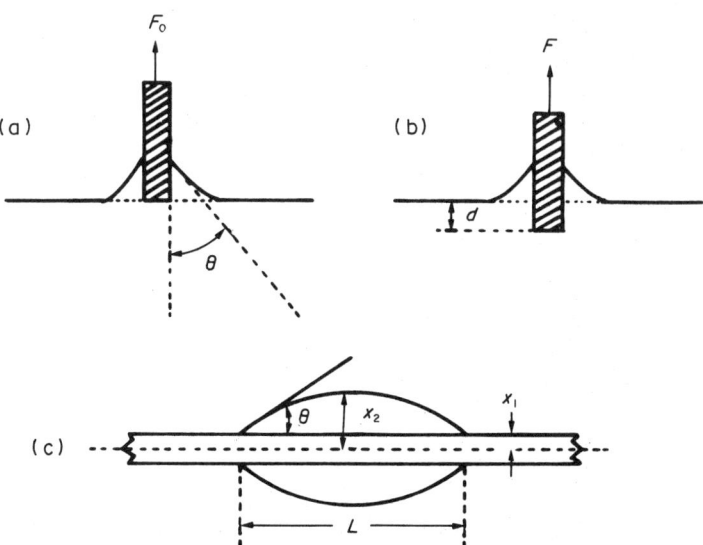

Figure 3 Techniques for measuring contact angles on fibres: (a) and (b) tensiometric or Wilhelmy plate technique for measurement of advancing and receding angles; (a) fibre touching liquid surface; (b) fibre after immersion in liquid to depth d; and (c) drop on fibre technique for measurement of equilibrium contact angle

23.2.2.3 *Sensitivity of contact angles to surface morphology and chemical composition*

The difference between the advancing contact angle and receding contact angle as defined in the previous section is known as the contact angle hysteresis. From a thermodynamic standpoint contact angle hysteresis should not exist if the surface being studied is completely smooth, chemically homogeneous and provided that the advancing and receding angles are measured under conditions where the three-phase contact line moves sufficiently slowly for all the interfaces to remain as close as possible to equilibrium.[23] However, in most practical situations few of these conditions pertain so that measurement of hysteresis in addition to the equilibrium contact angles can provide further information about the nature of surfaces. For example, surfaces which are rough or microporous have provided the focus for detailed study both practically and theoretically, especially by Johnson and Dettre.[24] The roughness factor r of a surface is essentially the ratio of the true area of surface including the ridges and troughs to its apparent area considered as a planar projection (Ω_S/A_S in equation 3). For a rough surface, a drop which wets the entire area of contact without forming voids will, at equilibrium, adopt a configuration which represents a minimum in surface free energy. Under these conditions the roughness ratio is given by Wenzel's equation[25]

$$r = \cos\phi/\cos\theta \qquad (6)$$

where ϕ is the apparent contact angle of the drop to the horizontal surface and θ is the intrinsic contact angle (that which would be obtained on an ideally smooth surface). Wenzel's equation shows that provided the intrinsic contact angle is $<90°$ then the apparent contact angle is reduced on a rough surface. This provides the basis for improving wetting and adhesion of adhesives to the surfaces of plastics.[26] Surface roughness of these materials may be introduced by surface abrasion techniques such as sand blasting or chemical etching. Rough surfaces which cannot be completely penetrated by liquids produce so-called composite surfaces in which air becomes trapped in the surface troughs; this can lead to localized stress concentrations. In this case the relationship between ϕ and θ differs from equation (6); full details of this and other aspects of rough surfaces are given by Johnson and Dettre.[24]

Other surface morphological changes which are amenable to study by contact angles include the transitions from one crystalline phase to another. Crystalline polymers or metals have a higher molecular or atomic density and hence a different surface free energy than their amorphous equivalents. Therefore crystalline phase transitions should be detected by contact angle measurement. This has been demonstrated by some elegant measurements by Neumann and co-workers[27,28] who studied the capillary rise of liquids at vertical surfaces of PET and poly(tetra-fluoroethylene) (PTFE). The precision which this technique offers allows the two crystalline transitions in PTFE around ambient temperature and the T_g transition in PET at 75 °C to be readily identified from the sharp change in slope of the liquid height *vs.* temperature curves which accompanies these transitions.

Chemical heterogeneity can also have a pronounced effect on contact angle and hysteresis. For example, chemical heterogeneity can be deliberately introduced in the surface of a block copolymer where the two blocks have a markedly different surface free energy. Even homopolymers such as PTFE, poly(vinyl chloride) (PVC) or poly(methyl methacrylate) (PMMA) have heterogenous surfaces where the low energy component is the polymer and the high energy component may be provided by carboxylic acid, sulfonate or sulfate groups resulting from the incorporation of initiator fragments at the ends of polymer chains during the polymerization process. A model for heterogenous surfaces which is analogous to that for rough surfaces was proposed by Johnson and Dettre.[29] Readers are referred to their work for details.

Examples of other morphological or chemical differences at surfaces which can be detected by contact angle techniques include the effect of polymer orientation,[30] the time-dependent processes of small molecule migration from and into the surface of polymers, and polymer molecular re-orientation.[31–33]

23.2.2.4 *Selection of liquids for wetting measurements*

The liquids used for contact angle measurements must be carefully selected depending on the type of information which is required. For the measurement of Zisman's critical surface tension of wetting for polymers (see Section 23.2.3.2) a series of pure liquids of similar polarity are required. Liquids may be classified into three series: (a) nonpolar liquids such as a homologous series of *n*-alkanes or di-*n*-alkyl ethers; (b) polar liquids such as halogenated hydrocarbons and esters; and (c) polar liquids which in addition have hydrogen-bonding capability such as water, glycerol and formamide.

The only other criterion for choice of liquid is that they must produce a finite contact angle on the surface; while a series of *n*-alkanes can be used to characterize very low energy surfaces such as fluorinated polymers, this is not a viable approach for polymers of higher surface energy such as polystyrene, PVC or PET.

Calculations of the surface free energies of polymers using the Kaelble[34] and Owens–Wendt[35] approach (see Section 23.2.3.3) are possible from contact angle measurements of well-characterized liquids such as those listed in Table 1. Some of these liquids are also useful for assessing the specific acid–base type of interfacial interactions[36] described in Section 23.2.4.1.

Table 1 Surface Free Energy of Various Liquids at 20 °C

Liquid	γ_{LV} (mJ m^{-2})	γ_{LV}^d (mJ m^{-2})	$\gamma_{LV} - \gamma_L^d$ (mJ m^{-2})
Water	72.8	21.8	51.0
Glycerol	63.4	37.0	26.4
Formamide	58.2	39.5	18.7
Methylene iodide	50.8	49.5	1.3
Ethylene glycol	48.3	29.3	19.0
Dimethyl sulfoxide	43.54	34.86	8.68
Tricresyl phosphate	40.70	36.24	4.46
Pyridine	38.00	37.16	0.84
Dimethyl formamide	37.30	32.42	4.88
2-Ethoxyethanol	28.6	23.6	5.0
n-Hexadecane	27.6	(27.6)	—

A general observation that should be emphasized is that pure liquids must be used to obtain contact angle data if the calculated thermodynamic quantities are to be meaningful. Mixtures of liquids of differing surface tension appear at first sight to offer a wide range of liquids with different surface tension. For example, Dann[37] measured the contact angles of ethylene glycol/2-ethoxy-ethanol and ethanol/water mixtures on nonpolar surfaces, but using this approach he found widely variable results. A drawback to this approach is that the adsorption of each component of a liquid mixture at the three interfaces (SL, LV, SV) will be different. As a result the preferred adsorption of one component at any single interface is sufficient to influence the surface tension forces which act at the three-phase contact line, thus effecting the contact angle. Liquid mixtures would only provide useful thermodynamic data if the liquid were an ideal solution and the liquid layers at both the LV and SL interfaces were the same composition as the bulk liquid. These conditions are seldom met for most liquid pairs.

23.2.3 Estimation of Thermodynamic Parameters from Contact Angle Measurements

23.2.3.1 *Equations of Young and Dupre*

The equilibrium contact angle of a liquid on a solid surface represents the balance of the attraction between molecules in the first few molecular layers of both phases and the extent of interaction between solid and liquid molecules in their free surfaces. It was shown by Young[38] that the different tensions acting in the interfaces at the three-phase contact line may be vectorially resolved in a direction parallel to the surface

$$\gamma_{LV} \cos \theta = \gamma_{SV} - \gamma_{SL} \tag{7}$$

Hence, from equation (1), the Young–Dupre equation is obtained by substitution

$$W_A = \gamma_{LV}(1 + \cos \theta) + \pi_e \tag{8}$$

23.2.3.2 *Zisman's critical surface tension of wetting*

In the Young equation (equation 7) only γ_{LV} and $\cos \theta$ are directly measurable. However, γ_{SV} and γ_{SL} have been found to be useful parameters for predicting the adhesion between materials. Fox and

Zisman[39] obtained an estimate of the surface free energy of a solid by plotting $\cos\theta$ *vs.* γ_{LV} for a homologous series of liquids on the surface. They found a linear relationship such that

$$\cos\theta = 1 + b(\gamma_{LV} - \gamma_c) \tag{9}$$

where γ_c is defined as the critical surface tension of wetting for the solid and b is a constant. The critical surface tension of wetting is therefore the value of γ_{LV} at which the liquid just wets a surface with zero contact angle. The choice of liquids for this work is important. Zisman[2,40] noticed that it was possible to obtain different values of γ_c depending on whether he used a series of nonpolar, polar or hydrogen-bonding liquids. Katazaki and Hata[41] have analyzed these results to determine which value of γ_c was most appropriate in terms of its approximation to γ_S. They concluded that the highest value of γ_c should be selected because under these conditions γ_{SL} is a minimum. This argument follows from an examination of equation (7) where $\gamma_{LV} = \gamma_c$ and $\cos\theta = 1$. Values of γ_c for some common polymeric solids are shown in Table 2. This table shows the correlation between γ_c for the polymers and the molecular structure of the constitutive units.

23.2.3.3   *Estimation of the surface free energy of solids*

Fowkes[42] suggested that the total surface free energy of a solid or liquid could be described as the sum of components which arise from different intermolecular forces *e.g.* dispersion (d), polar (p) and hydrogen bonding (h)

$$\gamma = \gamma^d + \gamma^p + \gamma^h \tag{10}$$

Also, where two interacting phases possess intermolecular forces of the same type, the interfacial interaction will contribute a term to the thermodynamic work of adhesion

$$W_A = W_A^d + W_A^p + W_A^h \tag{11}$$

Using a geometric mean approximation to describe the dispersion force interaction between phases 1 and 2, Fowkes also showed that for dispersion force interactions alone, the interfacial free energy is given by

$$\gamma_{12} = \gamma_1 + \gamma_2 - 2(\gamma_1^d \gamma_2^d)^{1/2} \tag{12}$$

Kaelble and Uy[43] together with Owens and Wendt[35] suggested that the effects of all polar interactions, including hydrogen-bonding interactions, could be considered as one term in equations (10) and (11); the work of adhesion then simplifies to two terms, W_A^d and W_A^p. By assuming that polar interactions between two phases could be approximated by a geometric mean expression, they obtained from equation (12)

$$\gamma_{SL} = \gamma_S + \gamma_{LV} - 2(\gamma_S^d \gamma_{LV}^d)^{1/2} - 2(\gamma_S^p \gamma_{LV}^p)^{1/2} \tag{13}$$

Combining equation (13) with the Young–Dupre equation (equation 8) and neglecting the π_e term gives

$$\gamma_{LV}(1 + \cos\theta) = 2(\gamma_S^d \gamma_{LV}^d)^{1/2} + 2(\gamma_S^p \gamma_{LV}^p)^{1/2} \tag{14}$$

Values of γ_{LV}^p for different liquids were obtained from $\gamma_{LV}^p = \gamma_{LV} - \gamma_{LV}^d$, as indicated in Table 1. From measurements of the contact angle of two such liquids (water/methylene iodide or water/formamide are commonly used) on a solid surface, γ_S^p and γ_S^d can be obtained from two simultaneous equations constructed using equation (14). The sum of these values then provides an estimate of the surface free energy of the solid. Kaelble[34] recommends the use of as many different liquid pairs as possible and suggests that the solutions of each pair of simultaneous equations should be computer fitted to provide the best values for the surface free energy components.

Values of γ_S^p, γ_S^d and γ_S obtained in this way for a range of polymers are shown in Table 2. A very detailed analysis of different methods of estimating γ_S is included in the book by Wu.[44]

23.2.4   Factors Effecting Adhesion

The Young–Dupre equation (equation 8) (for polymers π_e can be taken as zero) shows that W_A can only vary by a factor of two. This cannot account for the very large variations in adhesion actually observed within systems (*e.g.* by variation of substrate for a single adhesive, or by variation of the

Table 2 Critical Surface Tension of Wetting and Surface Free Energy for Polymers

Polymer	Chemical structure of unit compared with ethylene	γ_c at 20°C (mN m^{-1})	γ_s^d (mJ m^{-2})	γ_s^p (mJ m^{-2})	γ_s (mJ m^{-2})
Polytetrafluoroethylene	4H replaced by F	18.5	18.6	0.5	19.1
Polytrifluoroethylene	3H replaced by F	22	19.9	4.0	23.9
Poly(vinylidene fluoride)	2H replaced by F	25	23.2	7.1	30.3
Poly(vinyl fluoride)	1H replaced by F	28	31.3	5.4	36.7
Low density polyethylene	—	31	33.2	—	33.2
Polypropylene	1H replaced by Me	31	30.2		30.2
Poly(methyl methacrylate)	1H replaced by ester	39	35.9	4.3	40.2
Poly(vinyl chloride)	1H replaced by Cl	39	40.0	1.5	41.5
Poly(vinylidene chloride)	2H replaced by Cl	40	42.0	3.0	45.0
Polystyrene	1H replaced by benzene ring	43	41.4	0.6	42.0
Poly(ethylene terephthalate)	Polar monomers inserted	43	43.2	4.1	47.3
Poly(hexamethylene adipamide)	(ester, amide) in hydrocarbon chain	46	35.9	4.3	40.2

adhesive for a common substrate) which can be of two orders of magnitude. Practical adhesion is a highly complex effect involving, besides interfacial thermodynamics, interfacial rheology and fracture mechanics. These are outside the scope of this chapter, but the book by Wu[44] offers a sound guide. The fracture process in practical systems is *irreversible* and hence consideration only of W_A (the thermodynamic *reversible* work of adhesion) is inappropriate. However, achieving optimal wetting of the substrate is clearly vital to maximize interfacial interactions. The form of these interactions is obviously important; in particular the role of acid–base interactions can be highlighted since these can be assessed by contact angle measurements.

23.2.4.1 Acid–base interactions

Calculations of surface free energies from contact angle data in Section 23.2.3.3 were simplified by assuming that all polar interactions at interfaces could be taken into one term and treated mathematically in the same way as dispersion force interactions. However, Fowkes[45,46] has suggested that hydrogen-bonding interactions at interfaces are highly specific and must therefore be considered separately. The method which Fowkes developed is based on the earlier work of Drago and co-workers,[47,48] who regarded hydrogen bonding as a subset of Lewis acid–base interactions. Drago's results suggested that the total enthalpy of interaction of an acid–base pair in solution could be described by the sum of dispersion force interactions and acid–base interactions; contributions due to dipole–dipole interactions were considered negligible. Drago measured the enthalpy of acid–base interaction ΔH^{ab} for different Lewis acids and bases in a neutral solvent such as carbon tetrachloride, characterizing each acid A and base B by two constants C and E such that

$$-\Delta H^{ab} = C_A C_B + E_A E_B \tag{15}$$

Liquids and polymer surfaces can include three types of hydrogen-bonding molecules: (a) proton acceptors (electron donors or bases) such as esters, ketones, ethers or aromatics including polymers such as PMMA, polystyrene, copoly(ethylene vinyl acetate), polycarbonate; (b) proton donors (electron acceptors or acids) such as partially halogenated molecules, including polymers such as PVC, chlorinated polyethylenes or polypropylenes, poly(vinylidene fluoride) and copoly(ethylene acrylic acid); and (c) both proton acceptors and proton donors such as amides, amines and alcohols, including polyamides, polyimides and poly(vinyl alcohol).

Fowkes and Maruchi[36] tested Drago's hypothesis applied to the interaction of liquids with a solid surface by measuring the contact angles of various acidic and basic liquids (which had been characterized by Drago in terms of their E and C parameters) on different acidic and basic surfaces. Fowkes calculated W_A and W_A^d for the liquid and solid and examined the difference between these two quantities, which according to the preceding equations is given by

$$W_A^p + W_A^{ab} - \pi_e = \gamma_{LV}(1 + \cos\theta) - 2(\gamma_S^d \gamma_{LV}^d)^{1/2} \tag{16}$$

where the work of adhesion due to acid–base interactions W_A^{ab} is a more general term which also includes hydrogen-bonding interactions. Fowkes discovered that for either acidic liquids on acidic surfaces or for basic liquids on basic surfaces the left-hand side of equation (21) is always zero. This implies that liquids and solids which may be polar but have the same electron-accepting (or donating) capability interact by dispersion forces alone. For a basic surface, only acidic liquids were found to produce a positive contribution to W_A^{ab}.

These observations led Fowkes to conclude that the contribution to interfacial interactions by purely dipolar forces W_A^p is very small and may be neglected. Later work by Fowkes[49] confirmed that polarity as measured by dipole moments played a negligible role in the intermolecular interactions between liquids and solids. Moreover, since the π_e term in equation (21) has also been shown[10] to be approximately zero for high energy liquids (such as water and methylene iodide) on low energy surfaces (such as polyethylene) the right-hand side of equation (21) provides an expression for W_A^{ab}. Every parameter in this equation can be either measured experimentally or calculated without making any mathematically convenient assumptions. Values of γ_{LV}^d for any liquid can be calculated from the contact angle of that liquid on untreated polyethylene, as polyethylene interacts with other materials solely through dispersion force interactions. Similarly γ_S^d for any low energy surface may be calculated from the contact angle of methylene iodide on that surface. Fowkes states that methylene iodide can interact with other materials by dispersion forces alone and since in this situation $\gamma_{LV}^d = \gamma_{LV}$, γ_S^d from equations (7) and (12) simply becomes $\gamma_S^d = \gamma_{LV}(1 + \cos\theta)^2/4$.

Fowkes and Mostafa[50] relate W_A^{ab} to the enthalpy of acid–base interaction ΔH^{ab} from equation (15) by

$$W_A^{ab} = -f(C_A C_B + E_A E_B)n^{AB} \tag{17}$$

where n^{AB} is the number of acid–base pairs per unit area of surface and $f \simeq 1$ converts the enthalpy per unit area to surface free energy. Hence calculation of W_A^{ab} from contact angle data using equation (17) not only provides a measure of the strength of the acid–base bond between solid and liquid but is also a function of the number density of acid–base pairs which interact at the SL interface. For monofunctional surfaces this method therefore provides a technique for estimating the number density of interacting groups in the solid surface, assuming that E and C parameters for the liquids and the surface functionality are known. The surface site density of basic acetate groups in copoly(ethylene/vinyl acetate) has been determined in this way[50] from the contact angles of acidic phenol/tricresyl phosphate mixtures. The practical bond strength will clearly be improved if W_A^{ab} can be maximized. Pretreatment of surfaces for the bonding of adhesives or coatings often results in the generation of highly polar surface groups which also have an electron-accepting or electron-donating capacity. Good adhesion will be obtained with those coatings which contain a chemical group that can produce a high enthalpy of acid–base interaction.

23.2.4.2 *The role of surface treatments*

Although strictly outside the scope of this chapter it is important to note the importance of surface modification processes for adhesion enhancement of many polymers, particularly for polyalkenes and PTFE. A great deal of controversy has been generated over the years as to the reasons for the poor initial adhesion of these polymers, the effects of the modification processes ('pretreatments') and the reasons for the enhanced adhesion performance. The surface characterization methods discussed in Volume 1, Chapter 24 have been crucial in helping to answer these questions and the interested reader is referred to refs. 51–55 for further information.

23.3 THE INTERACTION OF SOLID POLYMERS WITH OTHER SOLIDS

23.3.1 Contact Adhesion: Viscoelastic Solids

23.3.1.1 *Thermodynamics of solid–solid interactions: the adhesion of smooth elastic solids*

The previous section dealt with solid–liquid interactions and described how these studies provide a means of characterizing polymeric surfaces. The main properties sensed are the various surface free energy components and the surface topography. In addition, indirect measurements of polymer–vapour interactions *via* the estimation of the equilibrium-spreading pressure may also be made with some difficulty.[56, 57] The study of certain solid–polymer contacts provides similar types of information although the appropriate analysis is very different. The contacting solid, unlike the analogous liquid, is capable of storing mechanical strain and thus the treatment of solid–solid contacts invariably requires a consideration of stored and dissipated mechanical work.

If two smooth elastic bodies are brought into contact, say a deformable rubber and a rigid plane, the surface forces create an adhesive function and work must be done to separate the bodies.[58-68] Rather than summing the surface forces across the interface to account for the adhesion, the net surface energy change G_{ss} on creating the interface may be computed as

$$G_{ss} = \gamma_{s1} + \gamma_{s2} - \gamma_{s12} \tag{18}$$

where γ_{s1} and γ_{s2} are the surface free energies of the solids and γ_{s12} the interfacial free energy. The surface free energies may be evaluated from wetting studies as described in Section 23.2. The two solids 'wet' one another but spreading is constrained by the elastic strains in the contact. This problem has been analyzed in detail by many authors. A very useful approach is to balance the change in elastic energy and surface free energy as the elastic body deforms.[58, 61] One analysis[58] then yields a result for the pull-off force P as

$$P = \frac{3}{2}\pi R G_{ss} \tag{19}$$

for an elastic sphere of radius R in contact with a plane. The relationship has been shown under rather special circumstances to give values of G_{ss} close to those obtained from liquid-wetting studies for large rubber contacts[58] and the contacts formed between fine PET monofilaments.[69, 70] This type of analysis is inappropriate when the contacts are neither smooth nor perfectly elastic.

23.3.1.2   The adhesion of contacting rough elastic bodies

When two rough elastic bodies are brought into contact, the force required to separate them is generally less than if they are smooth.[61, 71−74] Unlike some low viscosity liquids the solid will not ingress into all of the voids produced by the roughness and the contact area will be reduced. However, this is not the main reason why the adhesion is reduced. The Wenzel ratio (equation 6) described in Section 23.2.2.3 is not a useful parameter. Now there are three types of asperity interactions. For simplicity we will consider a smooth deformable elastomer and a rough rigid plane. Some asperity contacts 'pull' the surfaces together such that a force P_a is needed to separate them. The value of P_a is approximately $3/2\pi R_a G_{ss}$ where R_a is the asperity radius (equation 19); $R_a \ll R$. There will be a distribution of asperity heights and for many solids the distribution is normal with a variance of, say, σ. A second group of asperities, the highest ones, will attempt to 'push' the surfaces apart with a force which approximates to

$$\frac{E\sigma^{3/2}R_a^{1/2}}{(1-v)}$$

where E is the Young's modulus of the elastomer and v the Poisson's ratio. The third group of asperities will not be in contact and will have no influence on the processes. There is now an energy balance with some asperities pulling the surfaces together and some pushing the surfaces apart. A nondimensional parameter called the adhesion index,[72] α_a, describes the relative contributions

$$\alpha_a = (4\sigma/3)(4E/3\pi R_a^{1/2}G_{ss})^{2/3} \tag{20}$$

The adhesion is now a function of the modulus E of the deformable solid; for smooth contacts this is not the case. Where α_a is small, *i.e.* for low modulus and smooth surfaces, the adhesion is greatest and *vice versa*. The form of α_a has been derived and Figure 4 shows a comparison of theory and experiment. The enhanced adhesion at small values of α_a is notable but the mechanism which produces this effect is not properly resolved.[73, 74]

23.3.1.3   Inelastic effects and contact adhesion

Section 23.2 dealt mainly with the *equilibrium* wetting of liquids on polymers and it was not appropriate to consider in detail dynamic effects as these phenomena are very largely governed by

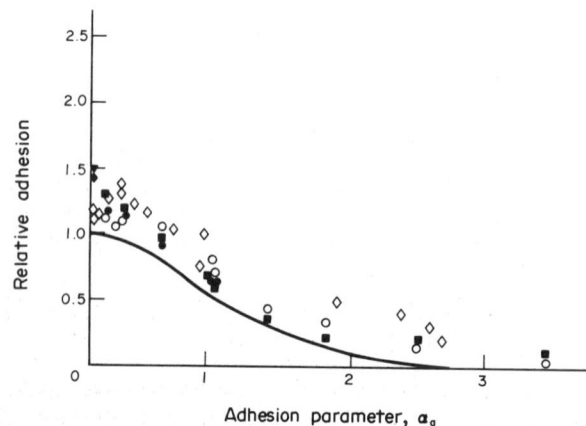

Figure 4 Relative adhesion as a function of the adhesion parameter α_a (equation 20) for various silicone elastomers contacting rough PMMA. The relative adhesion is the ratio of the rough value divided by the smooth ($\alpha_a = 0$) value. The normalization is quite effective if the peeling velocities are comparable[72, 73, 74]

the rheology of the liquid.[5,44] The analogy which may be offered with solid–solid contacts is that the magnitude of the adhesion is strongly controlled by the rheology of the deforming solid. The simplest approach is to recognize that whilst surface forces, as introduced by G_{ss}, are the origin of the adhesion, the magnitude is scaled by the volume work, $G(T,v)$, dissipated in the interface zones near the propagating crack.[59,60,63-65,75] T is the temperature and v the crack velocity. Fracture mechanics may be usefully applied but here it is sufficient to quote an empirical result which accounts for the total adhesive work, G_T, as[64]

$$G_T(T,v) = G_{ss}G_v(T,v) \qquad (21)$$

or[59,60]

$$G_T(T,v) - G_{ss} = G'_v(T,v)G_{ss} \qquad (22)$$

The quantities $G_v(T,v)$ or $G'_v(T,v)$ have all of the time–temperature superimposition properties associated with viscoelastic deformations of polymers and the WLF transform may effectively be applied to normalize $G(T, v)$ data.[60,64,76] Environmental changes may influence G_{ss} and thus G_{ss} is a direct scalar.[64,65] However, generally $G_v(T,v) \gg G_{ss}$ and factors of 10 to 10^3 are typical.

23.3.1.4 Irreversible effects and contact adhesion

If the elastic limit is exceeded during the loading or unloading of a contact the treatments described in Section 23.3.1.3 are not applicable. Permanent deformation during loading is discussed under the heading of *hardness* in the next section. Plastic or viscoelastic plastic deformation during unloading falls within the province of fracture mechanics and may involve junction failure at a plane away from the original interface.

A second type of irreversible phenomena commonly occurs at the interface itself. They may be irreversible chemical reactions of the sort deliberately produced in certain structural adhesive joints. In addition, in certain polymeric contacts, diffusion processes may produce either the segregation of low molecular weight species at the interface or diffusion bonding by the intermixing of the polymeric chains. The former will reduce adhesion and the latter will increase it. In principle, for both processes the time dependent value of the adhesion may be used to monitor the extent of the diffusion processes[44,77-79] and hence provide an estimate of the diffusion coefficient. For these and other reasons it is often found that the adhesion is a strong function of the temperature and contact time.

23.3.1.5 Experimental techniques for measuring contact adhesion

The simplest experiment is the peel test, where a strip of polymer is peeled from a rigid plane: Figure 5 shows this configuration and examples of other experimental arrangements which have been used successfully by various authors.[59] In some cases, for example the blister test, the necessary analysis is often only approximate.

23.3.1.6 Contact adhesion: summary

The measured contact adhesion between two solid polymers and between solid polymers and other bodies is a function of many variables. The important ones are the contact geometry, the surface topography, the surface mechanical and transport properties, the surface forces, the loading and unloading times and the level of surface contamination. Carefully constructed experiments provide a means of quantifying these properties, albeit in some cases only approximately.

23.3.2 Contact Deformation: Hardness

The classical quotation which accurately conveys the level of our understanding of hardness compares the subject to our appreciation of the 'storminess of the seas'.[80,81] Both are readily appreciated but less easily quantified. Hardness measurements and their interpretation are fraught with difficulties and ambiguities. This is particularly so for organic polymers. The potential value of these measurements is, however, enormous and well proven in metallurgy. Hardness studies have

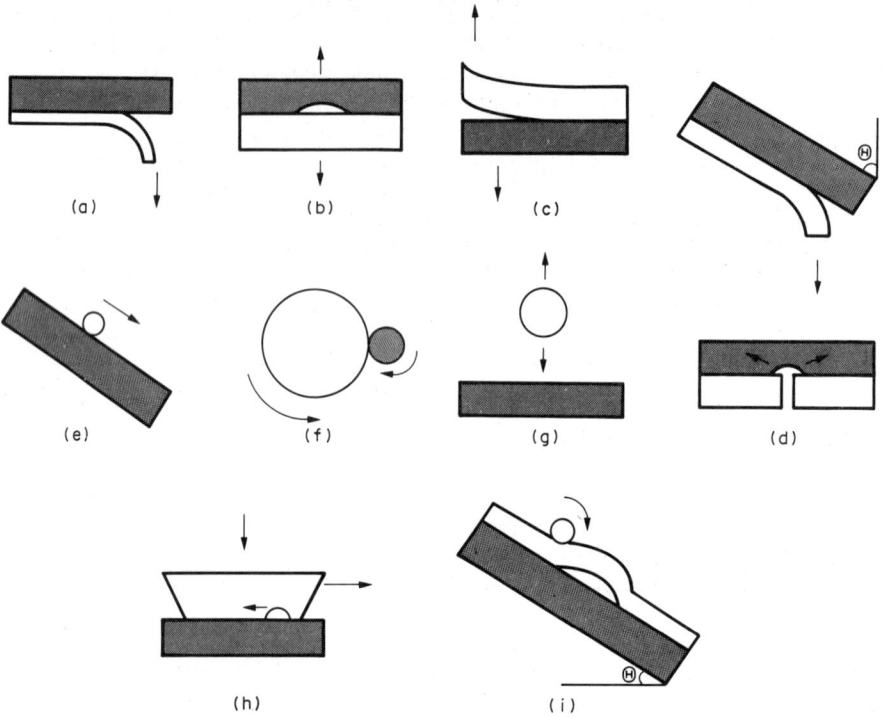

Figure 5 Various forms of the adhesion experiment suitable for studying elastomers in contact with rigid substrates: (a) peel test; (b) and (c) growth of a crack in tension; (d) blister test; (e) rolling; (f) the 'tackometer'; (g) rebound; (h) Schallamach waves; and (i) propagation of a macro dislocation in ironing. The rigid substrate is shaded (adapted from ref. 59)

not been as widely used in polymeric systems for a number of reasons which will be introduced in this section. The numerical value of the hardness is simply the normal load divided by the projected contact area in the plane which is orthogonal to the load axis.

First we should note three general points. There are two types of hardness measurement; normal indentation hardness (NIH) and scratch hardness (SH). The latter technique has been used since ancient times and is the basis of the Mohs' hardness scale.[82] NIH is a relatively new technique and is used to construct the Brinell,[83] Vickers[84] and Shore[85] scales. In both experiments a rigid indentor is used to deform a softer body; for our purposes we will identify two types of indentor: cones and pyramids with included angles 2θ, and spheres with radii R. The conical indentor produces deformations where the strain is independent of the depth of penetration and proportional to $\tan(90° - \theta)$ or $\tan\theta'$; $\theta' = (90° - \theta)$. For spheres the strain is proportional to r/R where r is the radius of the contact region. Of more immediate consequence, however, is the fact that cones impose much higher strains than spheres. The third and final general point is that the hardness can be measured during or after loading. This distinction is not necessary in metallic systems where elastic or viscoelastic recovery effects are negligible; in polymeric systems elastic and viscoelastic effects may predominate.

23.3.2.1 *Elastic and viscoelastic deformations under normal load and dynamic loading*

For a cone or wedge of slope angle θ' (semiangle θ) the hardness, H, is $W/(\pi r^2)$, where W is the load and r the contact radius in the plane of the surface. A first order treatment gives

$$H = \frac{E \tan\theta'}{2(1 - v^2)} \tag{23}$$

For a spherical indentor of radius R

$$H = \frac{4Er}{\pi 3R(1 - v^2)} \tag{24}$$

Both cases are for deformation *under* load and rather careful experimentation is required in order to compute the contact radius r. The reduced modulus, $E^* = E/(1 - v^2)$ is then obtained. In principle $E^*(t)$ could be measured as a function of time t by the measurement of $r(t)$. It is, however, simpler to use a dynamic test such as impact or rolling. The Shore hardness technique[81,85] actually measures resilience in this context: an indentor falls from a height h_i and rebounds to a height h_r. The quantity $h_i - h_r$ is a measure of the energy dissipated either by plastic flow or viscoelastic losses.[86,87] The experimental configuration may be reversed, for example as in a deformable ball impacting a plane. The contact time is relatively independent of h_i and is typically 10 μs. The same general type of deformation process may be produced by rolling a sphere or cylinder over a flat surface of the polymer at a velocity V. The polymer is repeatedly loaded and unloaded with a mean contact time of r/V. The work dissipated is sensed *via* the rolling friction, F_d[88,89] where

$$F_d = \phi\alpha \tag{25}$$

where ϕ is the work done per unit distance rolled; for a sphere of radius R

$$\phi = 2.5 \times 0.17 W^{4/3} R^{-2/3} (1 - v^2)^{1/3} E^{-1/3} \tag{26}$$

and α is the fraction of this work which is dissipated through viscoelastic losses; $\alpha \cong \pi \tan \delta$, where δ is the loss angle at the deformation frequency. The contact deformation frequency is approximately V/r, where V is the rolling velocity. Thus $F_d(V)$ senses $E^{*-1/3} \tan \delta$. The same result holds for sliding a well-lubricated sphere; the deformations generated during rolling are nominally identical to those produced in lubricated sliding. The work done for a lubricated cone is to a first order given by[89]

$$\phi = \frac{W}{\pi} \cot \theta \tag{27}$$

This result has a consequence in the materials selection of compounds for automobile tyres. High traction is required during braking on wet (lubricated) roadways and hence a high value of $\tan \delta$ is required. Unfortunately, high values of $\tan \delta$ mean greater heat generation in dry conditions.[90]

23.3.2.2 *Elastic–plastic and viscoelastic–plastic deformations under normal loads*

In the context of metals a hardness value is synonymous to the yield stress; plastic deformation effects dominate. The classical approach is to say that[81]

$$H = cY \tag{28}$$

where Y is a yield stress; for a ductile metal $c \cong 3$. This assumes that two thirds of the contact stress is stored as a hydrostatic stress beneath the contact. Slip-line field analyses have been developed to predict the effects of varying θ and the friction.[61] Polymers are not rigid plastic solids but viscoelastic–viscoplastic materials and some of the load is supported by time-dependent elastic deformation and longer time viscoplastic relaxations. The result is that for these materials $c < 3$, say 1.5. Advanced theories[61,91–93] have c as a function of the ratio E/Y. These theories are based, not on slip-line field analysis, but on the expansion of a plastically deforming cavity, which creates a hydrostatic stress in an infinite elastic body.[61]

The relationships generated express P/Y as a function of $(E/Y)\tan \theta'$ for cones and $(E/Y)(r/R)$ for spheres. P is the mean normal contact stress which is nearly proportional to H. Tan θ' and (r/R) define the strain and (Y/E) the elastic strain capacity of the solid. Low values of the ratio favour elastic response (Section 23.3.2.1) whilst for high values a rigid–plastic response occurs where $c \cong 3$ and slip-line field analysis is appropriate.[61] Figure 6 shows the computed hardness values of PMMA and PTFE at a 20 s loading time under load; as θ increases the hardness decreases. Similar results have been obtained for copper and in that case the trend has been attributed to interface friction effects. An example of the influence of interface lubrication on NIH is shown in Figure 7 for PMMA; lubrication reduces the hardness and this reduction is not a strong function of θ. Thus c in equation (28) is $f(\tan \theta', \tau, t, E/Y)$ where τ is the interface shear stress which is described later in this chapter. As $E/Y \tan \theta'$ increases, so does c. The larger the ratio E/Y, the less pronounced is the variation in the hardness with $\tan \theta'$ as elastic effects are less important.[91] The variation of the normal hardness with θ or $\tan \theta'$ is thought to be largely due to subtle changes in the mode of deformation in the subsurface of the polymer[61] as the strain is varied.

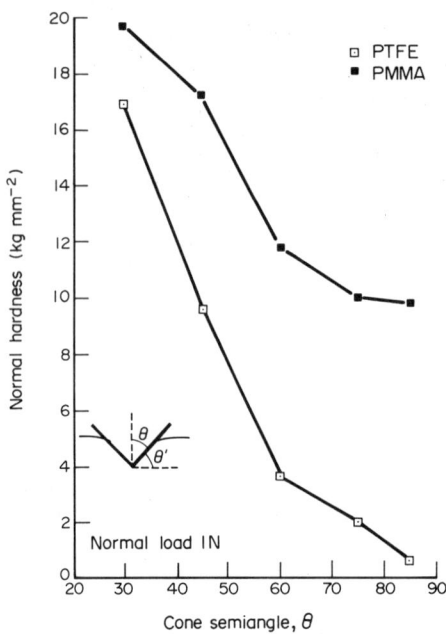

Figure 6 The 20 s normal indentation in hardness (NIH) of PTFE (□) and PMMA (■) at 20 °C for cones of semiangle θ. The normal load is 1 N; the surfaces are unlubricated. The NIH is a strongly decreasing function of θ

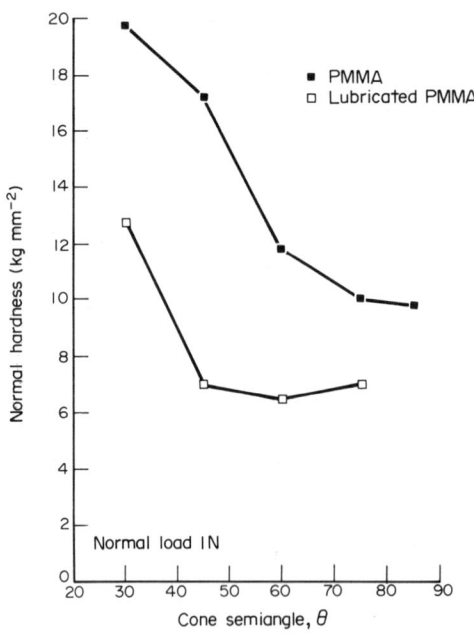

Figure 7 The NIH of PMMA as a function of the cone semiangle θ and the state of lubrication; unlubricated (■) and lubricated (□) with a silicone fluid. Other conditions as for Figure 6. The lubricated value is reduced but the trend with θ is retained

The fact that the hardness value senses each of E, Y and the interfacial traction is potentially useful, but it does create experimental problems. On unloading a deformed contact it will partially and perhaps completely recover through time-dependent viscoelastic relaxations. Estimating hardness using the approach of sizing a residual deformation, which is adopted in metals, leads to error. In the case of the deformation by cones most of the initial recovery is in the depth of the indentation and relatively little occurs in the diameter.[61,92] The hardness can be measured providing that the edge of the deformed cone can be identified. Thin reflective coatings or even ink coatings may be used to improve the contrast. The elastic recovery on unloading can be regarded as the elastic part of the load support.[61,92] If the depth under load is h_1 and on recovery is h_u, then $h_1/(h_1 - h_u)$ is clearly a function of Y/E. Baltá Calleja[91] has used this approach to estimate Y/E using an analysis developed by Lawn.[93,94]

The elastic–plastic solutions have not been developed in detail to include time effects. Time dependence is a major feature of polymer hardness. Figure 8 shows the extent of the penetration of a cone under a constant load W into series of polymer plates of different creep compliance; the hardness may be approximated from the penetration depth h as $W/(h \tan \theta)^2 \pi$. The initial penetration is very rapid followed by long-term creep. It is generally sufficient to write

$$H(t) = K_0 t^{-\beta} \tag{28}$$

where K_0 and β are motion constants. The values of K_0 and β computed from hardness data may be compared with data obtained from creep measurements where an analogous expression is applicable. In several cases the agreement is very reasonable and hence hardness measurements may be used to sense creep processes.[81,91,92,95] The recent review by Baltá Calleja[91] details how a hardness measurement may be used to monitor a variety of processing variables and indicates how the hardness value is influenced by morphological characteristics for certain polymers.

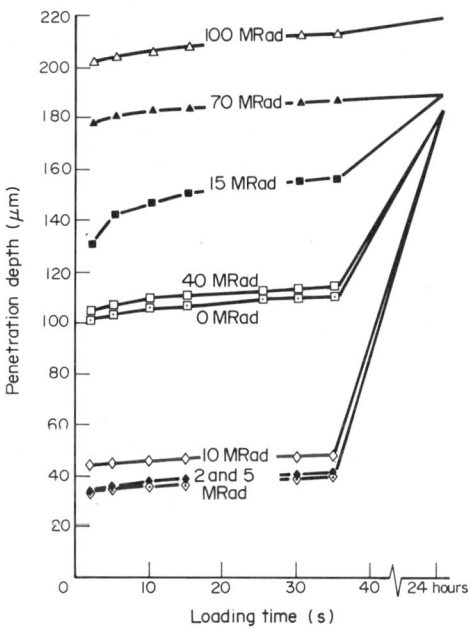

Figure 8 Penetration depth as a function of loading time for samples of gamma-irradiated PTFE. Conditions are load 1 N; cone semiangle $\theta = 30\,°C$; 20 °C; contact unlubricated

Before leaving the topic of normal hardness it is appropriate to mention further the role of lubrication. The effect of lubrication is to produce a reduction in the apparent hardness; see Figure 7. Lubrication influences the stress fields and facilitates both elastic relaxation and plastic flow. First order theoretical treatments are available.[61]

23.3.2.3 Scratching

Scratching was introduced earlier in the context of viscoelastic losses in rolling and lubricated sliding; all the frictional work is dissipated as heat and no damage to the polymer occurs; *i.e.* no track is left behind. If the strain level is increased, a rubber will tear and a ductile polymer will flow. Additional work must be done to enable this damage to occur.[89,96] In elastomers various types of tears are produced at and beyond a certain value of ϕ and discontinuous sliding motion occurs;[89,97] see Figure 9 and the next section. The area shown in Figure 9 is proportional to the toughness of the

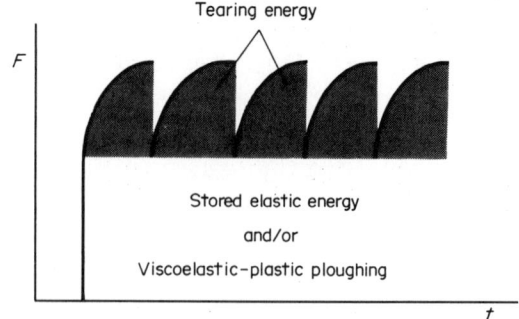

Figure 9 Typical trace of frictional force (F) against sliding distance (t) where intermittent motion occurs. This case is for a sharp cone scratching a rubber and the shaded areas indicate the work done in tearing the elastomer

rubber.[97,98] A certain amount of cracking or tearing is also apparent with thermoplastics but it is also accompanied by extensive plastic deformation. The plastic deformation may take several forms depending upon how the deformed material is displaced. With cones a critical value of θ' is reached as 2θ, the cone angle, is reduced where chips of polymers are machined from the surface.[98] The result of the plastic deformation is that a semipermanent groove or scratch is produced of width, say, $2r$. A number of simple models have been developed to describe the origin of the friction and the observed width of the groove.

Figure 10 shows the ideal rigid plastic case for a cone of included angle 2θ. The area A_1 supports the load W and the polymer is sheared over an area A_2 where

$$A_1 = \frac{1}{2}\pi r^2 \quad \text{and} \quad A_2 = r^2 \cot\theta = r^2 \tan\theta' \tag{30}$$

The friction is then

$$F = \frac{2}{\pi}\cot\theta W \tag{31}$$

or

$$\mu = \frac{F}{W} = \frac{2}{\pi}\tan\theta' \tag{32}$$

Figure 11 is an example of the quality of this model. The value of r may be used to compute a scratch hardness value H_s. If the normal load is supported over the area A_1 and $W = H_s A_1$, then

$$H_s = 2W/\pi r^2 \tag{33}$$

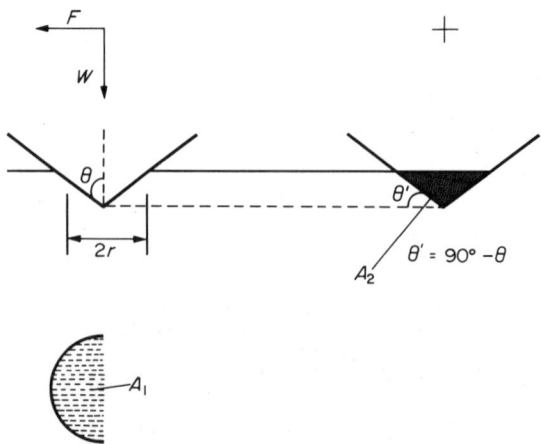

Figure 10 Sections of the plastic deformation produced by a rigid cone during sliding. The load W is supposed to be supported by the area A_1 and the frictional force must displace an area A_2

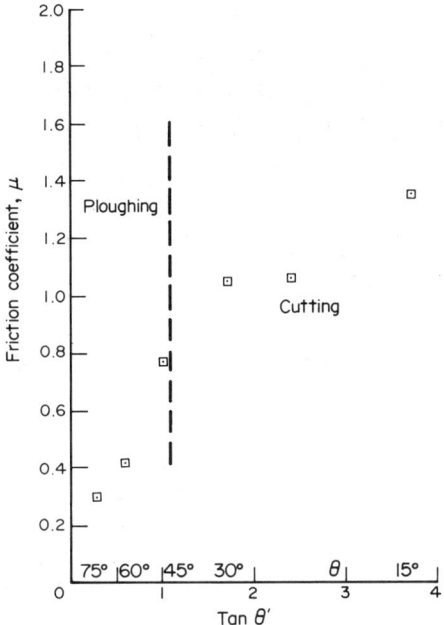

Figure 11 Plot of friction coefficient $\mu = F/W$ against $\tan \theta'$ (see Figure 10) for cones scratching PMMA. Various deformation modes occur and the regimes of two of these are shown. This curve is typical of polymeric response; μ is given by equations (25) and (27) for small $\tan \theta'$ when viscoelastic losses dominate. Where plastic ploughing occurs $\mu = (2/\pi)\tan \theta'$ (equation 32). Once cutting occurs the effective θ' is defined by an internal shear angle in the polymer and not explicitly by the actual value of θ'

Figure 12 shows H_s as a function of θ for PTFE and PMMA. The trend is similar to that seen in the normal hardness (see Figure 6) for PTFE but this is not the case for PMMA. Also shown in Figure 12 is the influence of lubrication on H_s for PMMA. Unlike the normal hardness behaviour, lubrication *increases* the apparent hardness, although for the PMMA system there are several

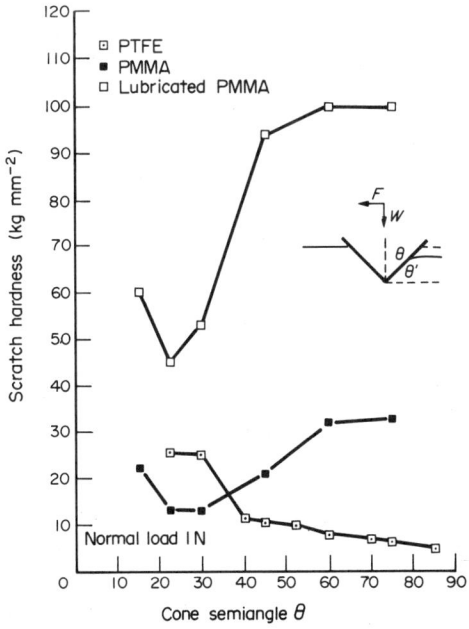

Figure 12 Scratch hardness of lubricated and unlubricated PMMA and of PTFE as a function of the semicone angle θ for a cone scratching the polymers at a velocity of *ca.* 1 mm s^{-1} and a load of 1 N. Note that lubrication has the effect of increasing the hardness. This set of data may be compared with those illustrated in Figures 6 and 7. The values of the SH and NIH are similar but their variations with θ are not generally comparable

peculiar trends in the hardness which are produced by subtle changes in the scratching mechanisms.[99] The correlation between normal and scratch hardness values is discussed briefly in the next section.

23.3.2.4   Polymer hardness: summary

The hardness measurement provides a means of measuring certain surface mechanical properties such as E and Y. The analytical procedures required to produce unique values of E and Y as functions of time are not fully developed. The best approach is to recognize that observed hardness is generally a function of a parameter $(E/Y)(\tan \theta)$ or $(E/Y)(r/R)$ and that a consideration of contact time and the state of lubrication is essential. Bowman and Bevis[100] have discussed the relative importance of E and Y for polymers. The experimental techniques are rather straightforward in principle but may be difficult to apply in certain cases. The technique has the virtue of sampling small regions of surface and hence may be used to study anisotropic mechanical properties.[91] The load dependence of the NIH may also be used to depth profile and study the extent of plasticized surfaces[101] and transcrystalline layers.[91,102] Hardness studies may also be used to investigate chemically modified surfaces.[91,103]

Scratch hardness is useful for sampling large areas of surface and studying the time dependence of the deformation. For various reasons, some of which have been introduced, the correlation between the numerical values obtained in NIH and SH studies is often poor.

23.3.3   Contact Deformation: Friction

23.3.3.1   General time-dependent aspects of friction

If two solid bodies, for example, one rigid and the other a polymer, are slid over one another, the work done is just FV where V is the imposed relative velocity and F the frictional force. Both F and V may fluctuate with time and under these conditions the motion is discontinuous. In the limit 'stick-slip' motion occurs. There are periods when the relative velocity is zero (the stick) followed by periods of relative motion (the slip). The response observed is a function of the properties of the contact and the viscoelastic characteristics of the machine; see Figure 9. If the structure of the machine is subcritically damped, continuous motion will be observed, otherwise the time-dependent characteristics of the contact generate discontinuous motion and hence a fluctuation in $F(t)$. This will occur if $F(V=0) > F(V \neq 0)$ and/or $\mathrm{d}F/\mathrm{d}V < 0$. $F(V \neq 0)$ is termed the dynamic friction F_d, and $F(V=0)$ the static friction F_s. It is common for F_s to exhibit stochastic properties. Figure 9 shows the magnitudes of F_s for the friction generated by a sharp indentor on a rubber. Similar effects are seen for many contacts. The values of F_s have been shown in one case (fine crossed PET monofilaments) to follow a gamma distribution, whose variance is a function of the applied load.[104] The prediction of friction and the modelling of its origins may thus have two parts: the prediction of the most probable value and its distribution function. This chapter will only discuss the former.

23.3.3.2   Models to account for the origins of friction

Section 23.3.2 described the origins of rolling friction and scratching friction in terms of the volumes of material deformed and the nature of the deformation, in particular the energy dissipation processes involved. These processes are commonly described as the *ploughing components* of friction and it will be realized that first order models are available to describe the friction in terms of the contact mechanics of the deformation and the rheology of the *bulk* solid.[88,89,97] The contact configurations involve rigid indentors or perhaps rigid asperities contacting a polymer body; see Figure 13(a). If the contact geometry is reversed and the polymer, say a sphere, is slid over a smooth rigid plane as in Figure 13(b), the dissipation processes invoked previously are inappropriate. The stress is now considered to be transmitted *via* adhesive forces; see Section 23.3.1. Adhesive force in conjunction with applied load creates junctions of area A which are sheared when a frictional force F is applied in a direction parallel to the contact plane. The quantity F/A is termed the interfacial shear stress, τ, such that

$$F = \tau A$$

This equation is the basis of the adhesion model of friction and accounts for the *adhesion component*. The idea that frictional work could be usefully described by two components, the ploughing and adhesion contributions, was initially developed for metals and has been extended to polymers.[88,89,105,106] Extensive studies have shown that it is often sufficient to consider that the two components are noninteracting and that the friction is just the simple sum of the predictions of the two models. This approach is described as the 'two term, noninteracting' model and should be regarded as a useful first order approximation only.

The deformation models have been discussed in Section 23.3.2; the next subsection describes some of the adhesion models.

23.3.3.3 Adhesion models of friction

Unlike the deformation models where the friction scales with the apparent contact area, the adhesion models assume that the friction is proportional to the *real* area of contact. The direct measurement of the real area of contact is generally impracticable and it is normal to resort to analytical models or computer simulations.[107] Surfaces are invariably rough and the type and scale of the surface topography as well as the deformation characteristics of the solids control the real contact area.[106] Some of the topographical factors were introduced in Section 23.3.1.2 for elastic bodies. At the simplest level the contact area for elastic contacts can be considered as arising from the contact of many spherical asperities of radius R_a with a plane. For one asperity the contact area A is given by[61]

$$A = \pi (D W_a R_a)^{2/3} \tag{34}$$

where D is $\frac{3}{4}\{(1 - v_1^2)(E_1 + (1 - v_2^2)/E_2\}$ and W_a is the load on the asperity. The subscripts refer to the two bodies. This is the Hertz equation and generally R_a may be written as a mutual radius of curvature where

$$\frac{1}{R_a} = \frac{1}{R_{a1}} - \frac{1}{R_{a2}} \tag{35}$$

R_{a1} and R_{a2} are the radii of curvature in a particular plane. If the solid bodies are molecularly smooth or highly deformable (a soft rubber), equation (34) may be used to calculate A directly. Generally this is not the case and tribologists estimate A by modelling. If the contact load is sufficient to induce plastic flow at the asperities a good approximation is[107]

$$A = \frac{W}{P_0} \tag{36}$$

where P_0 is close to the flow stress or NIH; the limiting normal stress that the asperities can endure. It now only remains to consider the properties of τ, the interface shear stress.

The properties of τ have been studied quite extensively.[89,97,105,108] It is an interface rheological property which manifests many of the characteristics of the bulk shear yield stress although the numerical values of τ are significantly less than the bulk analogue. Thus to a first order

$$\tau = \tau_0 + \alpha P \quad \text{(constant } T, V)$$
$$\tau = \tau_{0'} \exp(\theta/RT) \quad \text{(constant } P, V) \tag{37}$$
$$\tau = \tau_{0''} \ln V/V_0 \quad \text{(constant } T, P)$$

where τ_0, $\tau_{0'}$, $\tau_{0''}$, α, θ and V_0 are characteristic material constants. For example, the parameter α is of a similar magnitude to the bulk coefficient but the value of τ_0 for an interface is typically one tenth of the bulk equivalent.[109] The study of the properties of τ, which usually involves the use of thin solvent-deposited films, provides a means of investigating the rheology of polymers under abnormal conditions. Typically, contact pressures, which are primarily hydrostatic, may exceed 10^9 Pa and nominal rates of shear may be greater than 10^6 s^{-1}. The temperature dependence of τ may also be used to investigate the pressure and strain rate dependence of thermally induced molecular relaxations such as the glass transition temperature.[109,110] There are, however, some uncertainties which may arise from transient loading effects; typically the contact region may only be under quasihydrostatic stress for a few milliseconds. In these short contact times it appears that the polymer may not reach its equilibrium state and this is reflected in the shear stress. This type of behaviour is termed viscoelastic retardation in compression.[111]

Combining equations

$$\mu = \frac{\tau A}{W} = (\tau_0 + \alpha P_0)\frac{W}{P_0}$$

$$\mu = \frac{\tau_0}{P_0} + \alpha \tag{38}$$

at high loads. The friction coefficient under these conditions is the sum of two terms; τ_0/P_0 and α. For many systems $\alpha > \tau_0/P_0$ and thus to a reasonable approximation $\mu \cong \alpha$. The value of α for organic polymers ranges from 0.05 to 0.8.[108] Although there have been attempts to interpret the significance of α and τ_0 in microscopic terms using Eyring theory these parameters are best regarded as empirical constants.[112] The energy dissipation processes associated with τ are not understood in detail and may correspond to interface fracture or cohesive flow in regions away from the original interface. The similarity between contact adhesion (mode I crack propagation) and shear-induced interface rupture or friction-induced fracture (mode II crack propagation) has been identified but not widely and generally explored. The exception is one facet of the friction of elastomers which involves the propagation of macroscopic interfacial dislocation; the so-called Schallamach waves. A Schallamach wave is shown in Figures 5 and 13. Relative motion is accommodated not by sliding at the interface but by the propagation of stored length which is associated with a buckle or 'ruck' in the interface. The propagation of the ruck through the interface corresponds to the peeling and reforming of the interface. In essence, the process is quite similar to a contact adhesion experiment and may be analyzed as such. If the contact area is A and the wave frequency is f then the work done

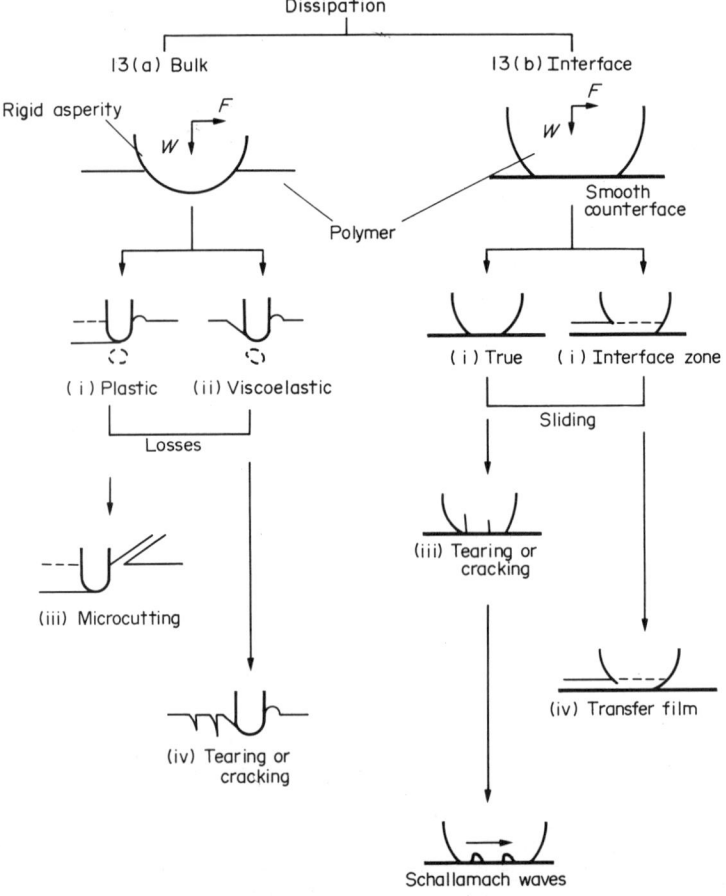

Figure 13 The various damage modes which are precursors to wear grouped according to the two term model of friction; bulk and interface dissipation processes or adhesion and deformation respectively. Bulk modes involve viscoelastic and plastic losses microcutting and cracking or tearing. Interface modes may involve 'true' interface sliding or cohesive (interface zone) shear in the polymer and the creation of transfer films. True sliding may involve the propagation of Schallamach waves as well as cracking or tearing. In addition, thermally induced softening and chemical erosion may occur

is $G_T(v)Af$ neglecting the work recovered on reforming the contact which is small. The mean wave velocity is v. The work input is FV and if all the work is involved in propagation then

$$FV = G_T(v)Af \qquad (39)$$

The values of $G_T(v)$ as measured from Schallamach wave propagation are similar to those obtained by more conventional means;[113,114] see Figure 5.

Two final points are worth making in conclusion to this section on the adhesion model of friction. First, the energy dissipated per unit volume is very high and high surface temperatures may be produced at sliding interfaces. Polymers are poor thermal conductors and their properties are also often very sensitive to temperature variation. Unusual trends are often therefore observed when friction is studied as a function of sliding velocity. Many of these trends may be quantitatively explained by considering $A(T)\tau(T)$. The value of τ usually decreases as T increases but A increases as T increases. The net effect on the friction is that it may decrease initially with velocity (or temperature) but rapidly increase at high velocities (or temperatures). Another consequence of these high temperatures is either surface melting or extensive thermal degradation of the surface.[115] The surface-melting process invariably produces what are called transfer films on the counterface; see Figure 13(b)(iv). This leads to what is called transfer wear (see next section). The frictional work in these cases may be correlated with the melt rheology of the polymer.[116]

There is also another form of transfer which is thought to occur under purely isothermal conditions; this is sometimes called 'cold' transfer: Figure 13(b)(ii). PTFE is the best example and its transfer mechanisms have been studied extensively.[89,97,115,117,118] During sliding the adhesion-induced interface tractions first produce extensive interfacial reorientation of the polymer's molecular chains of morphology. The polymer's oriented interface is then drawn out onto the counterface as a coherent, thin and highly oriented layer. This interfacial reorientation is the major reason why the friction of PTFE is small. Many polymers produce transfer films but not all have the same pronounced orientation as the PTFE films.

23.3.3.4 Friction: summary

Friction is not difficult to measure but it is difficult to explain in a comprehensive way. The question as to how the energy is fed into the contact regions is currently best explained using the two-term model in spite of its intrinsic weaknesses. The current ideas on energy dissipation processes largely derive from the understanding of energy dissipation in polymers *per se* with a belief that the deformation conditions such as strain rate and pressure correspond to circumstances not encountered in conventional mechanical testing. The study of friction does provide a means of investigating the rheology of polymers but most often it is a knowledge of the rheology which enables a rationalization of friction behaviour to be made.

23.3.4 Wear of Polymers

The previous sections have identified two major friction mechanisms; the ploughing and the adhesion components. Wear processes which arise from frictional energy dissipation may be classed similarly (see Figure 13). Alternatively, a number of generic classes may be adopted based upon what is now classical tribological terminology. This would include abrasive wear, fatigue wear, transfer wear, chemical wear and delamination wear to name a few.[119] Transfer wear was introduced in the previous section. The adhesive bonding at the interface is sufficiently strong to 'pull out' polymer particles from the matrix. A transfer film of polymer then forms on the counterface. The film is then displaced and wear results. The overall process is not well understood but the subject has been reviewed extensively,[89,115,119–122] particularly recently in the context of polymeric composites.[123,124] As with the rationalization of friction behaviour a knowledge of the mechanical properties of the polymer is crucial to the interpretation of wear. The primary damage modes and their location must be identified: Figure 13 is a first order pictorial account. Similarly, certain types of wear measurement can provide limited information on the mechanical properties of polymer surfaces although there are few examples of this approach in the literature. The most notable is the so-called Ratner–Lancaster correlation[119,121,125] and we will restrict ourselves to a brief discussion of this empirically based correlation. It is applied to abrasive wear when polymers are slid over very rough counterfaces such as abrasive papers and abraded metals.

Wear is a result of frictional work μWV. To remove polymer the strain must be large and hence is proportional to, say, $\tan \theta'$, where θ' is the mean asperity slope angle. The efficiency of the work done per removal operation is inversely proportional to the toughness of the polymer which may be approximated as $\sigma_y \varepsilon_y$, where σ_y and ε_y are the stress and strains at rupture in a macroscopic test. The volume of material deformed at the interface will be inversely proportional to the hardness H. If the rate of wear is expressed as W/V then

$$W/V = \frac{K\mu W \tan \theta}{H\sigma_y \varepsilon_y} \tag{40}$$

where K is a parameter which includes such inaccessible parameters as the number of asperity engagements required to detach a wear particle.

There is dispute regarding the value of this relationship and in particular as to whether a fatigue parameter rather than a toughness should be used. In addition, wear debris rapidly modifies the surface topography. Nevertheless, the relationship is found to hold in practice for a wide range of polymers sliding over a variety of substrates. Abrasion studies therefore afford a convenient means of ranking the toughness or fatigue life of polymers, particularly when only small samples are available.

Additional wear regimes are introduced in Figure 13 and the reader is referred to refs. 89, 115, 119–123 for details. Their discussion is outside the scope of this review.

23.3.4 Wear: summary

Wear is a complex process. It involves the dissipation of frictional work and most of this work ultimately appears as heat. A part of this work produces irreversible damage either involving mechanical rupture or chemical degradation. The type and rate of wear are not predictable but correlations with bulk mechanical properties are invaluale. Certain wear experiments thus provide a means of indirectly evaluating surface mechanical and rheological properties such as toughness or fatigue life.

23.4 REFERENCES

1. R. J. Good, *J. Colloid Interface Sci.*, 1975, **52**, 308.
2. W. A. Zisman, *Adv. Chem. Ser.*, 1964, **43**, 1.
3. J. R. Huntsberger, *J. Paint Technol.*, 1967, **39**, 199.
4. J. R. Huntsberger, *Adhes. Age*, 1978, Dec. issue, 23.
5. B. W. Cherry (ed.), 'Polymer Surfaces', Cambridge University Press, Cambridge, 1981.
6. D. Briggs and C. R. Kendall, *Int. J. Adhes. Adhes.*, 1982, **2**, 13.
7. R. H. Ottewill, D. F. Billett, G. Gonzalez, D. B. Hough and V. M. Lovell, in 'Wetting, Spreading and Adhesion', ed. J. F. Padday, Academic Press, London, 1978, p. 183.
8. Rame–Hart Inc., 43 Broomfield Ave, Mountain Lakes, NJ 07046, USA.
9. Kernco Instruments Co., 19 Walt Whitman Road, Huntingdon Station, New York 11746, USA.
10. G. L. Mack, *J. Phys. Chem.*, 1936, **40**, 159.
11. F. E. Bartell and H. H. Zuidema, *J. Am. Chem. Soc.*, 1936, **58**, 1449.
12. J. F. Padday, *Proc. R. Soc. London, Ser. A*, 1972, **330**, 561.
13. T. Fort and H. T. Patterson, *J. Colloid Sci.*, 1963, **18**, 217.
14. G. C. Sawicki, in 'Wetting, Spreading and Adhesion', ed. J. F. Padday, Academic Press, London, 1978, p. 361.
15. I. C. Callaghan, D. H. Everett and A. J. P. Fletcher, *J. Chem. Soc., Faraday Trans. 1*, 1983, **79**, 2723.
16. A. W. Neumann and R. J. Good, *Surf. Colloid Sci.*, 1979, **11**, 31.
17. M. Yamamoto, S. Yamada, Y. Sakatani, M. Taguchi and Y. Yamaguchi, *Proceedings of the International Conference on Carbon Fibres, Their Composition and Applications*, Plastics Institute, London, 1971, p. 179.
18. L. S. Penn, F. A. Bystry and H. J. Marchionni, *Polym. Compos.*, 1983, **4**, 26.
19. L. T. Drzal, J. A. Mescher and D. L. Hall, *Carbon*, 1979, **17**, 375.
20. F. A. Bystry and L. S. Penn, *Surf. Interface Anal.*, 1983, **5**, 98.
21. B. J. Carroll, *J. Colloid Interface Sci.*, 1976, **57**, 488.
22. M. S. Akutin, M. L. Kerber, I. O. Stal'nova, *Zavod. Lab.*, 1973, **39**, 716.
23. W. A. Zisman, *J. Paint Technol.*, 1972, **44**, 42.
24. R. E. Johnson and R. H. Dettre, *Adv. Chem. Ser.*, 1964, **43**, 112, 136.
25. R. N. Wenzel, *Ind. Eng. Chem.*, 1936, **28**, 988.
26. D. M. Brewis (ed.), 'Surface Analysis and Pretreatment of Plastics and Metals', Applied Science, London, 1982.
27. A. W. Neumann and W. J. Tanner, *J. Colloid Interface Sci.*, 1970, **34**, 1.
28. A. W. Neumann, Y. Harnoy, D. Stanga and A. V. Rapacchietta, in 'Colloid and Interface Science', ed. M. Kerker, Academic Press, London, 1976, vol. 3, p. 301.
29. R. E. Johnson, Jr. and R. H. Dettre, *J. Phys. Chem.*, 1964, **68**, 1744.

30. R. J. Good, J. A. Kvikstad and W. O. Bailey, *J. Colloid Interface Sci.*, 1971, **35**, 314.
31. D. Briggs, D. G. Rance, C. R. Kendall and A. R. Blythe, *Polymer*, 1980, **21**, 895.
32. A. Baszkin and L. Ter-Minassian-Saraga, *Polymer*, 1978, **19**, 1083.
33. J. D. Andrade, D. E. Gregonis and L. M. Smith, in 'Surface and Interfacial Aspects of Biomedical Polymers: I', ed. J. D. Andrade, Plenum, New York, 1985, p. 15.
34. D. H. Kaelble, 'Physical Chemistry of Adhesion', Wiley Interscience, New York, 1971.
35. D. K. Owens and R. C. Wendt, *J. Appl. Polym. Sci.*, 1969, **13**, 1741.
36. F. M. Fowkes and S. Maruchi, *Coat. Plast. Prepr. Pap. Meet. Am. Chem. Soc., Div. Org. Coat. Plast. Chem.*, 1977, **37**, 605.
37. J. R. Dann, *J. Colloid Interface Sci.*, 1970, **32**, 302, 321.
38. T. Young, *Philos. Trans. R. Soc. London*, 1805, **95**, 65.
39. H. W. Fox and W. A. Zisman, *J. Colloid Sci.*, 1950, **5**, 514.
40. W. A. Zisman, in 'Polymer Science and Technology', vol. 9A, 'Adhesion Science and Technology', ed. L.-H. Lee, Plenum, New York, 1975, p. 55.
41. Y. Kitazaki and T. Hata, *J. Adhes.*, 1972, **4**, 123.
42. F. M. Fowkes, *Ind. Eng. Chem.*, 1964, **56**(12), 40.
43. D. H. Kaelble and K. C. Uy, *J. Adhes.*, 1970, **2**, 50.
44. S. Wu, 'Polymer Interface and Adhesion', Dekker, New York, 1982.
45. F. M. Fowkes *J. Adhes.*, 1972, **4**, 155.
46. F. M. Fowkes, in 'Polymer Science and Technology', vol. 12A, 'Adhesion and Adsorption of Polymers', ed. L.-H. Lee, Plenum, New York, 1980, p. 43.
47. R. S. Drago, G. C. Vogel and T. E. Needham, *J. Am. Chem. Soc.*, 1971, **93**, 6014.
48. R. S. Drago, L. B. Parr and C. S. Chamberlain, *J. Am. Chem. Soc.*, 1977, **99**, 3203.
49. F. M. Fowkes, in 'Physicochemical Aspects of Polymer Surfaces', ed. K. L. Mittal, Plenum, New York, 1983, vol. 2, p. 583.
50. F. M. Fowkes and M. A. Mostafa, *Ind. Eng. Chem., Prod. Res. Dev.*, 1978, **17**, 3.
51. D. M. Brewis (ed.), 'Surface Analysis and Pretreatments of Plastics and Metals', Applied Science, London, 1982, especially chaps. 9 and 10.
52. D. M. Brewis and D. Briggs (eds.), 'Industrial Adhesion Problems', Orbital Press, Oxford, 1985.
53. D. M. Brewis, *Prog. Rubber Plast. Technol.*, 1985, **1**(4), 1.
54. D. M. Brewis and D. Briggs, *Polymer*, 1981, **22**, 7.
55. A. Chew, R. H. Dahm, D. M. Brewis, D. Briggs and D. G. Rance, *J. Colloid Interface Sci.*, 1986, **110**, 88.
56. F. M. Fowkes, D. C. McCarthy and M. A. Mostafa, *J. Colloid Interface Sci.*, 1980, **78**, 200.
57. P. Hu and A. W. Adamson, *J. Colloid Interface Sci.*, 1977, **59**, 605.
58. K. L. Johnson, K. Kendall and A. D. Roberts, *Proc. R. Soc. London, Ser. A*, 1971, **324**, 301.
59. B. J. Briscoe, in 'Polymer Surfaces', ed. D. Clarke and J. Feast, Wiley, New York, 1973, p. 25.
60. D. Maugis, in 'Microscopic Aspects of Adhesion and Lubrication', ed. J. M. Georges, Elsevier, Amsterdam, 1982, p. 221.
61. K. L. Johnson, 'Contact Mechanics', Cambridge University Press, London, 1986.
62. D. Tabor, *J. Colloid Interface Sci.*, 1977, 58, 2.
63. K. Kendall, *J. Phys. D*, 1971, **4**, 1186.
64. A. N. Gent and R. P. Petrich, *Proc. R. Soc. London, Ser. A*, 1969, **310**, 433.
65. E. H. Andrews and A. J. Kinloch, *Proc. R. Soc. London, Ser. A*, 1972, **332**, 385.
66. A. D. Roberts, *Rubber Chem. Technol.*, 1979, **52**, 23.
67. K. Kendall, *J. Phys. D*, 1973, **6**, 1782.
68. J. A. Greenwood and K. L. Johnson, *Philos. Mag., Part A*, 1981, **43**, 697.
69. B. J. Briscoe and S. L. Kremnitzer, *J. Phys. D*, 1979, **12**, 505.
70. M. J. Adams, B. J. Briscoe and S. L. Kremnitzer, in 'Physico-chemical Aspects of Polymer Surface', ed. K. Mittal, Plenum, New York, 1982.
71. K. L. Johnson, in 'The Mechanics of the Contact Between Deformable Bodies', ed. A. D. de Pato and J. J. Kalker, Delft University Press, 1975, p. 26.
72. K. N. G. Fuller and D. Tabor, *Proc. R. Soc. London, Ser. A*, 1975, **345**, 327.
73. G. A. D. Briggs and B. J. Briscoe, *J. Phys. D*, 1977, **10**, 2453.
74. K. N. G. Fuller and A. D. Roberts, *J. Phys. D*, 1981, **14**, 221.
75. D. Maugis and M. Barquin, in 'Adhesion and Adsorption of Polymers Part A', ed. L. H. Lee, Plenum, New York, 1980, p. 302.
76. J. A. Greenwood, *Philos. Mag., Part A*, 1981, **43**, 281.
77. R. M. Christensen, *Int. J. Fract.*, 1979, **15**, 3.
78. R. P. Campion, *J. Adhes.*, 1975, **7**, 1.
79. S. N. Nau and A. G. Thomas, *J. Polym. Sci., Polym. Phys. Ed.*, 1980, **18**, 511.
80. H. O'Neille, 'The Hardness of Metals and its Measurement', Chapman and Hall, London, 1934.
81. D. Tabor, 'The Hardness of Metals', Oxford University Press, Oxford, 1951.
82. F. Mohs, 'Grundriss der Mineralogie', Dresden Press, 1822.
83. A. Wahlberg, *J. Iron Steel Inst. London*, 1901, **59**, 243.
84. R. Smith and G. Sandland, *Proc. Inst. Mech. Eng.*, 1922, **1**, 623.
85. A. F. Shore, *J. Iron Steel Inst*, 1918, **2**, 59.
86. E. Southern and A. G. Thomas, *J. Appl. Polym. Sci.*, 1972, **16**, 1641.
87. A. D. Roberts and A. G. Thomas, *Wear*, 1975, **33**, 45.
88. F. P. Bowden and D. Tabor, 'Friction and Lubrication of Solids', Clarendon Press, Oxford, 1954 and 1964, parts 1 and 2.
89. B. J. Briscoe and D. Tabor, in 'Fundamental of Tribology', ed. N. P. Suh and N. Saka, MIT Press, Boston, 1980, p. 733.
90. R. Bond, *Proc. R. Soc. London, Ser. A*, 1985, **399**, 1.
91. J. Baltá Calleja, *Adv. Polym. Sci.*, 1985, **66**, 117.
92. D. Tabor, 'Indentation Hardness and its Measurement: Some Cautionary Comments, Special Technical Publication 889', American Society for Testing and Materials, Philadelphia, 1986, p. 129.

93. B. R. Lawn and V. R. Howes, *J. Mater. Sci.*, 1981, **16**, 2745.
94. B. J. Briscoe, P. D. Evans and J. K. Lancaster, *J. Phys. D*, 1987, **20**, 346.
95. A. G. Atkins, A. Silvero and D. Tabor, *J. Inst. Met.*, 1966, **94**, 369.
96. A. Schallamach, *J. Polym. Sci.*, 1952, **9**, 385.
97. B. J. Briscoe, in 'Friction and Traction', ed. D. Downson and C. Taylor, Westbury House, IPC Press, Guildford, 1981, p. 81.
98. B. J. Briscoe, P. D. Evans and J. K. Lancaster, *J. Phys. D*, (in press).
99. B. J. Briscoe, P. D. Evans and J. K. Lancaster, in 'Mechanisms and Surface Distress', ed. D. Dowson and C. M. Taylor, Butterworths, London, 1986.
100. J. Bowman and M. Bevis, *Colloid Polym. Sci.*, 1977, **255**, 954.
101. B. J. Briscoe, G. L. Davies and T. A. Stolarski, *Tribol. Int.*, 1984, **17**, 129.
102. E. J. Baltá Calleja, *Colloid Polym. Sci.*, 1976, **254**, 258.
103. J. Martinez-Salazar, D. R. Rueda, M. E. Cagiao, E. Lopez-Cabarcos and E. J. Baltá-Calleja, *Polym. Bull.*, 1983, **10**, 553.
104. B. J. Briscoe, A. B. Winkler and M. J. Adams, *J. Phys. D*, 1985, **18**, 2143.
105. B. J. Briscoe, *Philos. Mag., Part A*, 1981, **43**, 511.
106. D. Tabor, 'Friction', in Institute of Mechanical Engineers Bicentenial Conference, June 1987.
107. R. Sayles, in 'Tribology in Powder Processing and Conveying', ed. B. J. Briscoe and M. J. Adams, Hilger, London, 1988.
108. B. J. Briscoe and A. C. Smith, *Rev. Deform. Behav. Mater. III*, 1980, no. 3, 151.
109. B. J. Briscoe and A. C. Smith, *Polymer*, 1981, **22**, 1587.
110. B. J. Briscoe and A. C. Smith, *J. Appl. Polym. Sci.*, 1983, **28**, 3827.
111. B. J. Briscoe and A. C. Smith, *J. Phys. D*, 1982, **15**, 579.
112. D. Tabor, in 'Microscopic Aspects of Adhesion and Lubrication', ed. J. M. Georges, Elsevier, Amsterdam, 1982.
113. A. D. Roberts and A. G. Thomas, *Wear*, 1975, **33**, 45.
114. G. A. D. Briggs and B. J. Briscoe, *Wear*, 1975, **35**, 357.
115. B. J. Briscoe, *Tribol. Int.*, 1981, **14**, 231.
116. K. Tanaka and Y. Uchiyama, in *Polym. Sci. Technol.*, 1974, **A5**, 499.
117. C. M. Pooley and D. Tabor, *Proc. R. Soc. London, Ser. A*, 1972, **329**, 251.
118. H. Chizchos, 'Systems Approach to Tribology', Elsevier, Amsterdam, 1979.
119. J. K. Lancaster, in 'Polymer Science; a Material Science Handbook', ed. A. D. Jenkins, North Holland, London, 1972, p. 959.
120. L. H. Lee, *ACS Symp. Ser.*, 1985, **287**.
121. B. J. Briscoe, in 'Physical and Chemical Aspects of Polymer Surfaces', ed. K. L. Mittal, Plenum, New York, 1981, p. 387.
122. K. Tanaka, N. Eiss *et al.*, in 'New Directions in Lubrication, Materials, Wear and Surface Interactions: Tribology in the 80's', ed. W. R. Loomis, Noyes Publication, New York, 1985.
123. K. Freidrich (ed.), 'Friction and Wear of Polymer Composites', Elsevier, Holland, 1986.
124. B. J. Briscoe and P. J. Tweedale, ICCM Conference Publication, Elsevier, Holland, 1987.
125. S. E. Ratner, L. I. Farberova, D. V. Radyukeuich and E. G. Lure, *Sov. Plast., Engl. Transl.*, 1964, **7**, 37.

24

Adsorption

IAN D. ROBB

Unilever Research, Port Sunlight, UK

24.1 INTRODUCTION

The behaviour of macromolecules at interfaces is of major importance in biological systems and in controlling the stability of colloidal dispersions. Adsorbed polymers are generally studied in association with an overwhelming mass of inert solid, and a complete description of an adsorbed polymer's segment-density distribution has only recently[1] been obtained by neutron scattering. Initial ideas on the behaviour of adsorbed polymers came from isotherm determinations, though, as with many detailed models based only on thermodynamics, they were often misleading. Spectroscopic measurements have given estimates of the fraction of polymer segments attached to the solid surface, and hydrodynamic techniques have been used to estimate their extension into solution. For

Table 1   Publications Dealing with the Adsorption of Polymers[a]

Substrate	Polymer	Solvent	Ref.
Glass/SiO$_2$	Polystyrene	Various organic solvents	8, 9, 32, 53, 73, 85, 268–270
	Poly(methyl methacrylate)	Various organic solvents	18, 48, 53, 59, 75, 190, 212
	Poly(alkyl methacrylate)	Decalin	109
	Poly(vinyl acetate)	CHCl$_3$, CCl$_4$, C$_7$H$_8$	47, 191
	Poly(vinyl alcohol)	Water	62, 192
	Polysiloxanes	Various solvents	16, 48, 56, 58
	Poly(oxyethylene glycol)	Various solvents	50, 149, 193–5
	Polyesters	Various solvents	14, 49, 69
	Poly(acrylic acid)	Water	149
	Poly(vinylpyrrolidone)	Water	149, 196
		Various organic solvents	197
	Dextran	Water	5
	Ethylcellulose	Various solvents	198
	Gelatin	Water	156
	Poly(ethyleneimine)	Water	199
	Poly(1,2-dimethyl-5-vinylpyridinium methyl sulfate)	Water	157
	Poly(styrenesulfonate)	Water	13
	Poly(dialkyldimethylammonium chloride)	Water	200
	Poly(ethylene-co-vinyl acetate)	CCl$_4$, CHCl$_3$, C$_6$H$_6$	201
		C$_2$HCl$_3$, C$_6$H$_{12}$	61, 63
	Poly(styrenesulfonate-co-2-vinylpyridine)	Water	202
	Poly(styrene-co-2-vinylpyridine)	Various organic solvents	76, 202
	Poly(styrene-co-acrylonitrile)	C$_2$HCl$_3$	203
	Proteins	Water	204
Carbon	Polyisobutylene	Various organic solvents	70, 208, 209
	Polystyrene	Various organic solvents	20, 46
		C$_7$H$_8$	210, 211
		CDCl$_3$	44
	Poly(oxyethylene glycol)	Water	185, 213, 214, 215
	Poly(oxyethylene glycol)	Various organic solvents	185
	GRS rubber	Various solvents	216
	Poly(methyl methacrylate)	C$_7$H$_8$, 2-butanone	20, 46
	Poly(vinyl acetate)	MeOH, C$_7$H$_8$, 2-butanone	46, 217
	Polybutadiene	C$_7$H$_8$	20
	Poly(vinyl β-cyanoethyl ether)	Acetone	57
	Poly(sulfonic acid)s	Water	218
	Poly(styrene-co-butadiene)	C$_7$H$_8$, decalin, 2-pentanone	20

Substrate	Polymer	Solvent	Ref.
Inorganic crystals			
TiO_2	Polystyrene	C_6H_{12}	205
	Poly(methacrylic acid)	Water	206
	Poly(vinyl acetate-*co*-crotonic acid)	Water	79
Al_2O_3	Poly(styrene-*co*-vinylferrocene)	CCl_4, $CHCl_3$	77
	Poly(vinyl acetate)	CCl_4, $C_2H_4Cl_2$	60
$CaCO_3$	Poly(acrylate ester)s	C_7H_8	207
	Polyacrylamide	Water	220
	Lignosulfonates	Water	165
	Poly(sulfonic acid)s	Water	218
	Poly(vinyl chloride)	PhCl, THF	6, 7, 14
	Polycaprolactone	PhCl	221
	Poly(methyl methacrylate)	Various solvents	190
Dolomite	Poly(acrylic acid)	Water	222
Hydroxyapatite	Poly(acrylic acid)	Water	169
$BaSO_4$	Poly(4-vinylalkylpyridinium fluoride)	Water	223
	Sodium carboxymethylcellulose	Water	19
	Poly(acrylic acid)	Water	19
AgI	Poly(oxyethylene)	Water	224
	Poly(vinylpyrrolidone)	Water	225
Metals			
Chrome	Polystyrene	C_6H_{12}, C_6H_6	4, 26
Mercury	Polystyrene	C_6H_6, C_6H_{12}	26
	Poly(vinylpyridine)	Water	161–3
Steel	Gelatin	Water	156
Iron	Poly(methyl methacrylate)	CH_3CN, C_6H_6, C_7H_8	59
	Poly(vinyl acetate)	CCl_4, $C_2H_4Cl_2$, C_6H_6	60
	Poly(dimethylsiloxane)	CCl_4, C_7H_{16}, C_6H_6	58
	Poly(lauryl methacrylate)	C_7H_8, $C_{10}H_{22}$	67
Aluminum	Poly(acrylic acid)	Water	117
Tin	Poly(vinyl acetate)	CCl_4, $C_2H_4Cl_2$, C_6H_6	60
Clays			
Kaolin	Poly(1,2-dimethyl-5-vinylpyridinium methyl sulfate)	Water	226
	Poly(acrylic acid)	Water	172
	Polyacrylamide	Water	171, 227
	Poly(*N*-alkylpiperidinium chloride)	Water	228
	Poly(oxyethylene)	Water	229
	Poly(vinyl alcohol)	Water	230
Bentonite	Poly(SO_2-*co*-dimethyldialkylammonium chloride)	Water	173
Tungstic acid sols	Polyacrylamide	Water	72
Montmorillonites	Poly(oxyethylene)	Water	231
	Cellulose ethers	Water	232

Table 1 (*continued*)

Substrate	Polymer	Solvent	Ref.
Polymers			
Cellulose	Sodium carboxymethylcellulose	Water	150
	Poly(dimethylpiperidinium dialkylammonium chloride)	Water	158
	Poly(dimethyldialkylammonium chloride)	Water	233, 234
	Poly(ethyleneimine)	Water	31, 235
	Polyacrylamide	Water	10, 236, 237
	Poly(cationic ionenes)	Water	238
	Poly(vinyl acetate)	Various solvents	29, 86, 239
	Poly(methyl methacrylate)	C_6H_6, binary solvents	86
Nylon	Poly(acrylic acid)	Water	159, 240
	Poly(oxyethylene)	Water, CCl_4, C_6H_6	30
	Sodium carboxymethylcellulose	Water	160
Cellulose ester	Poly(vinyl acetate)	$CHCl_3$	241
	Poly(ethyl vinyl ether-*co*-maleic anhydride)	Water	93
Polystyrene latex	Polylysine	Water	219
	Poly(acrylic acid-*co*-N-1-naphthyl acrylamide)	Water	153
	Poly(methyl methacrylate-*co*-acrylic acid)	Water	152
	Poly(acrylic acid-*co*-lauryl methacrylate)	Water	242
Paraffin wax	Poly(acrylic acid)	Water	154
Polyethylene	Poly(2-sulfoethyl methacrylate)	Water	151

[a] These publications contain little or no information about adsorbed polymer conformations (publications dealing with conformations are listed in Table 2).

easy reference, Table 1 lists some of the publications available that deal with the adsorption of polymers.

Theoretical descriptions of adsorbed polymers, discussed in detail elsewhere,[2] have been based either on the random walk approach or on statistical mechanics[3] (Monte Carlo calculations), both approaches giving essentially similar results. Comparison between theory and experiment is mainly on the basis of trends observed on variation of certain parameters because: (a) many of the parameters used in the theories, *e.g.* segment–surface free energies or polymer chain flexibility, are not known quantitatively for real systems; (b) the relative spacing between polymer adsorbing groups and adsorption sites on the surface depends on the systems studied; and (c) the nature of solid surfaces of practical interest is often difficult to treat theoretically. Nevertheless, the data emerging from the considerable array of techniques show general agreement with currently accepted theories.

Perhaps one of the more neglected areas is the detailed study of the kinetics of adsorption, from which information on the diffusion of polymers on or near the interface may be obtained.

24.2 ADSORPTION ISOTHERMS

24.2.1 Determination

The first, though not necessarily the most important, piece of information needed for understanding the adsorption of polymers is the adsorption isotherm. This is almost always determined by difference in polymer concentration before and after adsorption, though Stromberg[4] has directly measured adsorption to flat metal surfaces of radiolabelled polystyrene. Problems in isotherm determination can come from fractionation of the polymer, degradation by shear or biodegradation of water-soluble polymers.[5]

Adsorption isotherms are normally determined by keeping a fixed solid/liquid ratio and increasing the polymer concentration. Provided there is no fractionation taking place in the adsorption process, the isotherm will be independent of the solid/liquid ratio up to about 10% phase volume of solids. However, fractionation often occurs, and may be the result of using (i) polydisperse polymers on a non-porous substrate where the high molecular weight fraction is sometimes preferentially adsorbed[6-8] or (ii) polydisperse polymers on porous substrates where low molecular weight materials can be preferentially adsorbed.[5,9,10] When fractionation occurs, there is clearly the possibility that the isotherm will depend on the solid/liquid ratio.

The order of addition of polymers, additives and substrate also needs to be carefully controlled. The amount of polymer adsorbed in the presence of other strongly competing molecules may not be the same as if the polymer were allowed to adsorb first, followed by addition of the same concentration of competing molecules. At true thermodynamic equilibrium, the adsorbed amounts would, of course, be independent of the method of addition, though the time taken to reach this equilibrium may be very long. The order of addition can influence not only the amounts adsorbed, but also subsequent flocculation behaviour, as demonstrated by Fleer[11] with poly(vinyl alcohol) adsorbed on to AgI.

24.2.2 Reversibility and Fractionation

Isotherms of high molecular weight polymers are normally high-affinity types, a result of the low probability of breaking many polymer–surface bonds simultaneously. Often the concentration of polymer that desorbs into pure solvent (from which it had adsorbed) is below the limit of detection experimentally, leading (wrongly) to the assumption that polymer adsorption is irreversible. Desorption of weakly bonded[4,12,13] or low molecular weight polymers[14-16] has been observed. Isotherms of strongly binding polymers should essentially consist of two lines, an initial sharply rising part followed by a plateau, almost independent of free-polymer concentration. Experimentally observed isotherms often have a rounded shoulder at the junction, which Cohen Stuart *et al.*[17] have shown is due to the polydispersity of polymers. High molecular weight polymers normally displace adsorbed low molecular weight ones, and it is this fractionation which produces the roundness of the shoulder. The theory[17] predicts that the roundness of the isotherms should be greater in poor solvents, where the molecular weight dependence of plateau adsorption is greater, than in good solvents, though Hamori *et al.*[18] did not observe this experimentally.

Fractionation on adsorption can result from differences in chemical composition as well as molecular weight. This has been observed[19] with the adsorption of sodium carboxymethylcellulose on $BaSO_4$, where the lower molecular weight (and more highly charged) molecules were initially adsorbed preferentially.

24.2.3 Kinetics of Adsorption

Though the kinetics of adsorption is an area deserving more attention, a number of factors that influence it have been identified experimentally. These factors include the molecular weight of the polymer, the thermodynamic quality of the solvent, the polymer concentration and the characteristics of the substrate, such as porosity. In the simplest model, where the rate of uptake of the polymer is controlled by its diffusion to the solid surface, it would be expected that adsorption should be slower for higher molecular weight polymers. On flat surfaces, this has been found by several workers. Stromberg et al.[4] showed that, for narrow molecular weight distribution polystyrenes, plateau adsorption on to a metal surface was attained at shorter times by polymers of molecular weight 38 100 than those of molecular weight 76 000.

These and other data[20-23] also show that, at lower polymer concentrations, longer times are needed to reach the (lower) plateau adsorption. For higher molecular weight polymers, rearrangements of the polymers on the surfaces appears to delay attainment of equilibrium. Using ellipsometry, Stromberg et al.[24] found that the root-mean-square thickness of polystyrene (molecular weight 5.37×10^5) adsorbing on to a flat chrome surface, took about 100 min to reach its equilibrium value, whereas for another polystyrene (molecular weight 1.9×10^6), equilibrium thickness had not been reached after 300 min. In addition, fractionation by molecular weight, involving exchange of larger molecules for smaller ones on the surface, will delay attainment of equilibrium.[21] Botham and Thies,[25] using infrared (IR) techniques to study adsorption of poly(isopropyl acrylate), showed that polymers of average molecular weight $>2 \times 10^6$ took much longer to equilibrate (~ 100 h) than shorter chain poly(methyl methacrylate). Thus entanglement, surface diffusion and exchange appear to play a significant part in determining the kinetics of adsorption. Grant et al.,[26] studying the adsorption of polystyrene on to chrome and mercury surfaces from good (benzene) and poor (cyclohexane) solvents, showed that adsorption was faster on to mercury and faster (for both surfaces) from the good solvent. In good solvents, polymers have fewer segments in contact with the surface than in poor ones, so that diffusion across the surface may be faster, allowing equilibrium to be reached faster.

The time for rearrangement of polymers was studied with poly(vinylpyrrolidone) (PVP) on silica.[27] PVP molecules (molecular weight 20 000), which initially adsorbed in a flat conformation, adopted the equilibrium more-extended conformation in less than 2 min from the addition of a saturating amount of extra PVP. Thus, for polymers of molecular weight $<50 000$, simple conformational changes are rapid by comparison with the slow stage of adsorption.[26] An additional indication of the importance of surface diffusion comes from the influence of polyelectrolytes on the growth of crystals[28] such as $CaSO_4$. Polyelectrolytes can prevent crystal growth, even though they contain adsorbing groups (e.g. carboxyl) that would individually (e.g. in surfactants) have no influence on crystal growth. Thus the multisegment nature of polymers can lead to their slow diffusion across surfaces, as illustrated further below.

Where the solid substrate is not flat, but porous or absorbent, the time taken to reach equilibrium can be much longer.[9,29,30] For poly(vinyl acetate) adsorbing on to cellulose filter pulp,[29] initial adsorption is very rapid, with slow but measurable adsorption continuing for as long as 10 days. The initial adsorption on to cellulose by poly(ethyleneimine) was reported[31] to be a second-order kinetics process, though no unequivocal explanation exists for this. Where the substrates consist of colloidal particles, the rate of adsorption has been shown to increase with intensity of shaking.[20]

Desorption of polymer molecules from a surface can be either rapid, i.e. completely displaced within 1–2 h,[32] or, as is more often found, no *measurable* desorption occurs at all.

Measurement of the kinetics of adsorption, continuously and in real time, is difficult, but has been done recently using total internal reflection fluorescence (TIRF). With this technique, fluorescent-labelled macromolecules adsorb at a glass–water interface, and the label is excited by light entering the glass from the side opposite the adsorbing macromolecules. If the light is totally internally reflecting, only those macromolecules close to the surface are excited, and the fluorescence intensity of the evanescent wave is proportional to the amount of adsorbed macromolecule. The initial studies on the application of TIRF were with proteins and other biopolymers, with data on adsorption, desorption and exchange being obtained.[33-43] In the systems chosen, adsorption was often complete after about 30 min and was controlled by diffusion through the bulk to the surface. The technique used allowed the rate of flow of the macromolecular solution over the glass surface to be altered, so that diffusion could be studied. Even more interesting, the diffusion of macromolecules across the surface can be studied uniquely by this technique. After adsorption has reached equilibrium, an initial high intensity pulse of light is used to saturate (bleach) the fluorescent label, so that further fluorescence, for a short interval, comes only from other labelled molecules diffusing into the area of study. The surface-diffusion coefficient, calculated from the recovery of the fluorescent intensity, was found to be 5×10^{-9} cm^2 s^{-1} for bovine serum albumin on glass.[35]

The relation between the kinetics of adsorption and time for equilibrium conformation to be reached is uncertain. Nuclear magnetic resonance (NMR)[44] experiments have shown that equilibrium conformation is reached much slower than adsorption from solution, and direct comparative experiments on adsorption kinetics and conformation are needed.

24.2.4 Influence of Solution Conditions

Usually,[45-50] more polymer is adsorbed from poorer solvents. Variation in solvent–substrate interactions can complicate this simple finding, which was predicted theoretically by Scheutjens and Fleer.[51,52] In good solvents, repulsion between segments (*e.g.* in tails) of adsorbed polymers will be greater than in poor solvents, leading to lower adsorption in the former. The lower adsorption in good solvents has been wrongly attributed[46] to less-extended conformations.

Increases in temperature normally lead to better solvent quality with organic solvents, so that less polymer should adsorb at higher temperatures. This has been observed for some systems,[49] though the reverse has also been observed,[14,18] as well as independence from temperature.[49,53] The effect of temperature on the free energy of adsorption per segment is difficult to predict. The heat of adsorption per segment may stay constant, though entropy effects in changing solvent quality could be important. Dickinson and Lal[54] claimed that the influence of excluded volume, *i.e.* from the increasing quality of solvent, has only a small effect on the critical energy of adsorption (the minimum energy per segment/substrate bond required for adsorption), so that temperature effects should be governed by interaction between the adsorbed segments. Theoretical studies[3,51] show this critical energy to be about 0.4–0.5 kT ($k = 1.380\,622 \times 10^{-23}$ J K^{-1}). Isosteric heats of adsorption have been measured by a few workers[15,55-57] though only once have these quantities been compared to heats of adsorption measured calorimetrically.[55] In this case, for poly(oxyethylene) adsorbing on to silica, the isosteric heat was endothermic (polymer molecular weight 6130) but endothermic for larger molecules (molecular weight 40 000). The calorimetric heat for both was exothermic. No obvious explanation exists for these results, and further work in this area is needed.

The importance of a second component in the solvent has been demonstrated in many reports. Where these second components are much more strongly adsorbed on to the solid surface than the main solvent molecules, low amounts (<1%) can have a marked effect on both the amount of polymer adsorbed and its configuration. Perkel and Ullman[58] noticed this with the influence of trace amounts of water on the adsorption of poly(dimethylsiloxane)s on to iron and glass.

Reduction in the amount of polymer adsorbed on to hydroxylic surfaces have been observed[48,59,60] with increasing amounts of the second solvent, acetonitrile, a strongly adsorbing solvent. The increased extension of chains into solution in the presence of minor fractions of strongly adsorbing solvents has been clearly demonstrated by Thies,[61] using IR measurements. In addition, where the minor solvent component interacts strongly with the polymer, the amount of polymer adsorbed can increase or decrease, depending on whether the polymer/solvent quality decreases or increases.[27,62,63]

24.2.5 Competitive Adsorption Between Two Polymers

With two polymers of different chemical type, kinetic studies have shown that equilibrium adsorption and desorption are reached more rapidly than normally associated with polymer adsorption. Botham and Thies[64] found that poly(vinyl acetate) (molecular weight 28 000) could displace polystyrene (molecular weight 105 000) within one hour. This displacement study confirms the expected greater energy of the carbonyl group's interaction with silica compared with that of styrene. It is also demonstrated by the results of the competitive adsorption of polystyrene and poly(methyl methacrylate)[65,66] on to silica. Here it was shown that the latter polymer was adsorbed in strong preference to the former; indeed, no polystyrene was adsorbed until the poly(methyl methacrylate) had been completely removed from the solution.

Ethyl cellulose displaces poly(vinyl acetate) from silica[64] though it only partially displaces poly(methyl methacrylate). Similar results were found for alkyl-substituted methacrylates.[67] Competition for adsorption sites leads,[23] as expected, to polymers adopting more extended conformations on the surface. The relative affinity of polymers for surfaces naturally depends on substrate type. Although poly(vinyl acetate) displaces polystyrene from silica, this does not apply to displacement from graphitized carbon,[46] where weaker dispersion forces are involved.

The above examples are ample evidence of the exchange that occurs between uncharged macromolecules in the adsorbed and free state. Unlike these, polyelectrolytes at low ionic strengths do not exchange, even when the more weakly adsorbing polymer is on the surface and a more strongly adsorbing one is in solution. This is discussed further in the section on polyelectrolyte adsorption.

24.2.6 The Effect of Polymer Characteristics

The amount of polymer adsorbed on most substrates depends on the molecular weight, though this dependence is different for non-porous and porous surfaces. Below a molecular weight of about 5000–10 000, adsorption increases with molecular weight. Oligomers having between about three and ten segments show[16,68-70] low-affinity isotherms,[16,68] and their adsorption can be reversible[69] even though their conformations[15,68,69] may be quite flat. For larger molecules, the plateau adsorption level, A_s, generally increases with increasing molecular weight, though sometimes A_s can be independent of it,[71] depending on the solvent used in the system.[50,266]

With very high molecular weight polymers, *i.e.* those in the range 10^6–10^7, there appears to be little regular dependence on molecular weight in good solvents,[72] though for polystyrene adsorbed on to metals from cyclohexane, higher molecular weights have given more adsorption.[24]

This molecular weight dependence of adsorption leads to fractionation, when the polymer is polydisperse,[8,73,74] and exchange can take place freely. Hamori[18] showed that there were significant differences in the adsorption behaviour of atactic and isotactic poly(methyl methacrylate) (PMMA) on silica from acetonitrile. Isotactic PMMA was adsorbed whereas atactic PMMA was not. Miyamoto *et al.*[75] used this difference to separate isotactic and syndiotactic polymers. Botham and Thies[25] found that isotactic poly(isopropyl acrylate) had a higher fraction of segments in direct contact with the surface than its syndiotactic or atactic equivalents. These experiments show the importance, even if second order, of the orientation and accessibility of the polymers' adsorbing groups to the solid surface.

Experiments with carefully designed copolymers have given some insight into the relation between segment adsorption energy and amount of polymer adsorbed. Quite small quantities of strongly adsorbing groups incorporated into an otherwise weakly bound polymer can enhance the level of adsorption significantly. This has been demonstrated by several studies of the adsorption of copolymers having a wide range of compositions.[32,76,77] With the adsorption of poly(methyl methacrylate) on iron powder,[59] the introduction of only 1% acid groups into the polymer chain nearly trebled the amount adsorbed. For the adsorption of poly(styrene-co-methyl methacrylate) on to silica (styrene being the weakly adsorbing group),[71] the amount of copolymer adsorbed increased sharply with small amounts of the ester group, and remained almost constant above about 5% ester content.

Similar results have been found by other workers.[65,78] Where one of the chain units has a repulsive[79] interaction with the solid surface, *e.g.* carboxylate groups with an anatase (titanium oxide) surface, less polymer is adsorbed with increasing content of that chain unit. For the intermediate case of the co-units having the same adsorbing group, *e.g.* oxyethylene and oxypropylene adsorbing on to charcoal,[71] the amount adsorbed changed almost linearly with copolymer composition, probably a result of the changing solvent quality.

The effects of copolymers having block or statistical distributions of monomers on the colloid stability of silica have been studied.[76,80] Adsorbed copolymers of styrene and methyl methacrylate with a block distribution of units produced colloid stability over the entire composition range, whereas statistically distributed ones produced stability only over part of that range.

24.2.7 Substrate Type and Preparation

The behaviour of adsorbed macromolecules would be expected to depend on the nature of the surface, as demonstrated with silica. Flame silicas contain hydroxyl groups (silanols) in two main forms, an A and a B form.[81] The former ($\sim$ 1.4 hydroxyl groups per 100 Å^2) exist as single hydroxyls, whereas the latter ($\sim$ 3.2 hydroxyl groups per 100 Å^2) occur in pairs. Heating to about 500°C results in dehydroxylation and the removal of water, leaving only the single isolated A form. This is reflected in the IR spectra of silica.

In organic solvents, the spectra of silica are different from those in water, although they do depend on the treatment of the silica prior to immersion in the liquid. Thies[82] showed that the adsorption and configuration of poly(methyl methacrylate) on to silica depended on the pretreatment of the silica. Several other authors have found similar results.[16,50,83,84]

On flat surfaces, it is generally observed that plateau adsorption (A_s) either increases or remains constant with increasing molecular weight. However, on porous surfaces the reverse is often observed, i.e. A_s decreases with increasing molecular weight. This is simply due to the size of the pores being too small to allow access to larger molecules, thus effectively presenting smaller polymers with more surface area on to which they can adsorb.

Kotera et al.[85] found that, for polystyrene adsorbing on to porous glass, A_s passed through a maximum at a molecular weight of 20 000. Polymers of this size had a radius of gyration approximately the same as the pore radius. Day et al.,[11] using non-adsorbing polymers, showed that penetration of polymers into pores was dependent only on the relative size of the polymer and pore. Cellulose is an interesting porous substrate, as it adsorbs relatively large amounts of solvents in small pores, giving rise to negative adsorption of polymers.[86] The rate of adsorption on to porous substrates may well be slower than on to non-porous ones. Bogacheva et al.[9] found that polystyrene adsorbing on to porous silica did not reach equilibrium for 30 or more days, whereas others have reported[10,71,85,86] equilibrium on porous substrates after about 20 to 30 h.

24.3 CONFORMATION OF ADSORBED POLYMERS

The best description of the conformation of adsorbed polymers comes from the segment-density distribution, determined using neutron scattering. As this technique is relatively inaccessible to many research workers, other techniques need to be used in a complementary manner to obtain a reasonable idea of the adsorbed polymer's conformation. These techniques tend to measure either the fraction of segments attached to the surface or the extension of the polymer into solution. The advantages and limitations of these techniques are discussed below, dealing first with those giving larger adsorbed thicknesses. Table 2 lists some of the publications available that deal with the conformations of adsorbed polymers.

24.3.1 Techniques for Estimating the Extension of Adsorbed Polymers

24.3.1.1 Ellipsometry

The technique of ellipsometry provides an independent measurement of the thickness of an adsorbed film and the amount of material in that film. It relies on the fact that the state of polarization of light changes after reflection from a surface. When elliptically polarized light is reflected from a bare surface, there is a change in both the amplitude ratio and the phase difference of the components. This can be expressed mathematically in the following way

$$\tan \Psi = \frac{(A_p^r / A_n^r)}{(A_p^i / A_n^i)} \tag{1}$$

where A_p / A_n is the amplitude ratio of the light polarized parallel (p) and normal (n) to the plane of incidence for the reflected (r) and incident (i) beams. The phase difference $(\delta_p - \delta_n)$ of the two components changes on reflection, so that τ (the film thickness) is given by

$$\tau = (\delta_p^r - \delta_n^r) - (\delta_p^i - \delta_n^i) \tag{2}$$

Detailed descriptions of the background theory and apparatus have been published.[21,87-89] Interpretation of the data requires an assumption concerning the segment-density distribution, and careful cleaning of the substrate (usually a metal) is required.[84]

Table 2　Publications Dealing with Conformations of Adsorbed Polymers

Substrate	Polymer	Solvent	Measured Quantity	Technique	Ref.
Glass/SiO$_2$	Poly(methyl methacrylate)	Benzene	p	EPR	243
		Benzene	p	IR	71
		C$_2$HCl$_3$	p	IR	64, 66, 71, 82
		CCl$_4$	p	EPR	243–6
		CHCl$_3$	p	IR	111
		CCl$_4$	p	EPR	243–6
		CHCl$_3$	p	IR	111
		CHCl$_3$	p	EPR	132, 246
	Poly(vinylpyrrolidone)	Water	p	EPR	27, 131
		Water	p	IR	113
		D$_2$O	p	NMR	119
		CHCl$_3$	p	EPR	27, 132
		CHCl$_3$	p	IR	113
	Polystyrene	C$_2$HCl$_3$	p	IR	71, 111
		CCl$_4$	p	IR	114, 247, 248
		C$_6$H$_{12}$	p	IR	247
		Decalin	p	IR	111
		C$_7$H$_8$	τ	Viscosity	12
		C$_2$HCl$_3$	p	IR	66
		Benzene	τ	Porous flow	249
	Poly(vinyl acetate)	CHCl$_3$	p	IR	250
		CHCl$_3$	p	EPR	130
		C$_2$HCl$_3$	p	IR	61, 64
		CCl$_4$	p	IR	116, 250
		C$_6$H$_{12}$	p	IR	61
	Polyesters	CHCl$_3$	p	IR	112, 115
		CCl$_4$	p	IR	114
	Poly(alkyl methacrylate)	C$_2$H$_4$Cl$_2$	p	IR	251
		Decalin	p	IR	109
		CCl$_4$	p	IR	116
		C$_2$HCl$_3$	p	IR	25
	Polycarbonate	C$_2$H$_4$Cl$_2$	p	IR	251
	Poly(oxyethylene)	CCl$_4$	p	IR	110
		CCl$_4$	p	Calorimetry	68
		Water	p	Calorimetry	55

Substrate	Polymer	Solvent		Method	References
	Polybutadiene	n-C$_7$H$_{16}$	τ	Isotherms	83
	Poly(dimethylsiloxane)	C$_6$H$_{14}$, CCl$_4$	p	IR	15
	Poly(4-vinylpyridine)	CHCl$_3$	p	IR	111
	Poly(acrylonitrile-co-butadiene)	Benzene, C$_2$HCl$_3$	p	IR	78, 80
	Poly(styrene-co-methyl methacrylate)	C$_2$HCl$_3$	p	IR	65
	Poly(ethylene-co-vinyl acetate)	C$_2$HCl$_3$	p	IR	64
Alumina	Poly(methyl methacrylate)	C$_2$HCl$_3$	p	IR	82
	Poly(oxyethylene)	CCl$_4$	p	IR	23
Chrome	Polystyrene	THF/MeOH	τ	Ellipsometry	148
		C$_6$H$_{12}$	τ	Ellipsometry	21, 24, 84, 89, 143, 147, 148, 258–260
	Poly(methyl methacrylate)	CH$_3$CN, C$_4$H$_9$Cl	τ	Ellipsometry	147, 148
	Poly(oxyethylene)	MeOH	τ	Ellipsometry	84, 148
		Water	τ	Ellipsometry	22, 84, 148
	Poly(vinylpyrrolidone)	Water	τ	Ellipsometry	22, 84
	Poly(ethylene-o-phthalate)	Ethyl acetate	τ	Ellipsometry	112
Platinum	Polystyrene	C$_6$H$_{12}$	τ	Ellipsometry	147, 148
	Sodium polyacrylate	Water	τ	Ellipsometry	144
Mercury	Polystyrene	C$_6$H$_{12}$	τ	Ellipsometry	261
Gold	Polystyrene	THF/MeOH	τ	Ellipsometry	147
	Poly(ethylene-o-phthalate)	Ethyl acetate	τ	Ellipsometry	112
AgI	Poly(vinyl alcohol)	Water	τ	Electrophoresis	97, 107, 262
	Poly(oxyethylene)	Water	τ	Viscosity	97
BaSO$_4$	Poly(oxyethylene)	Water	τ	Electrophoresis	263
	Sodium polyacrylate	Water	p	EPR	129, 168
	Sodium carboxymethylcellulose	Water	p	EPR	129, 164, 168
Gibbsite	Poly(vinyl alcohol)	Water	τ	Electrophoresis	265
Cellulose ester	Poly(vinyl acetate)	CHCl$_3$	τ	Porous flow	241
	Polyacids	Water	τ	Porous flow	264
	Poly(maleic anhydride-co-ethyl vinyl ether)	Water	τ	Porous flow	93
Polystyrene latex	Poly(vinyl alcohol)	Water	τ	PCS	98, 102, 252, 253
		Water	τ	Electrophoresis	102
		Water	τ	Sedimentation	101, 102, 253, 254
	Poly(oxyethylene)	Water	τ, p	Neutron scattering	255, 256
	Poly(styrenesulfonate)	Water	τ, p	Neutron scattering	1
	Poly(oxyethylene-co-oxypropylene)	Water	τ	PCS and electrophoresis	257

24.3.1.2   Viscometry

An alternative approach to estimating the thickness of an adsorbed layer measures the apparent change in dimensions of an orifice, *i.e.* a pore or a capillary, caused by the adsorbed layer. This approach has been thoroughly studied by Priel and Silberberg,[12,90,91] who measured the viscosities of polystyrene dissolved in toluene. If the radius of the capillary R is reduced to $R-\tau$ by the adsorbed polymers, having thickness τ, then the flow time (t) for a given polymer solution in the polymer-coated viscometer would be related to t_0, the flow time for an equal volume of solvent in an uncoated viscometer, by

$$t/t_0 = (\eta/\eta_0)[R/(R-\tau)]^4 \tag{3}$$

where η is the viscosity of the polymer solution and η_0 the viscosity of the solvent.

To achieve reproducible results, careful cleaning of the apparatus is necessary, flow times of about 1000 s must be measured to $\pm 10^{-3}$ s and a temperature stability of $\pm 1.5 \times 10^{-4}$ K should be maintained.[90,91] Varoqui *et al.*[92,93] have also studied the thickness of adsorbed polymer layers, using porous membranes in addition to viscometers.

24.3.1.3   Photon correlation spectroscopy

Of the hydrodynamic methods available for determining the adsorbed-layer thickness, photon correlation spectroscopy is potentially one of the more reliable and accurate. If monochromatic light is passed through a dispersion of colloidal particles, the scattered light will have its frequency changed by the motion of the particles, *i.e.* there will be a Doppler frequency shift between the incident and scattered beams. The combined scattered and incident beams fluctuate in intensity and these fluctuations have an autocorrelation[94-96] function $|g'(k, \tau)|$, where $k[=(4\pi/\lambda)\sin(\theta/2)]$ is the scattering vector, λ is the wavelength of the light in the dispersion medium, θ is the scattering angle and τ is the correlation time delay.

For monodisperse non-interacting particles, having a diffusion coefficient D_0

$$|g'(k, \tau)| \propto \exp(-D_0 k^2 \tau) \tag{4}$$

For spherical particles, D is related to the hydrodynamic radius r by the Stokes–Einstein equation

$$D = kT/6\pi\eta r \tag{5}$$

where η is the viscosity of the solution and k is Boltzmann's constant. Thus, if D for a particle is determined before and after polymer adsorption, the change can be attributed to an increase, Δr, in the hydrodynamic radius, caused by the polymer. It is most important that the state of aggregation of the particles (if any) does not change following addition of the polymer. Added polymer often causes some flocculation, particularly at low coverage,[97] so that care must be taken with polymer and particle concentrations as well as order of addition. It is advisable to use low angle light-scattering to determine whether[98] slight changes in the state of aggregation have occurred. Dispersions that can be studied need not be monodisperse, though the polydispersity should not change on addition of the polymer.

24.3.1.4   Centrifugation

The centrifugation of particles under a force field is well known and has been used extensively to determine the molecular weights of biological macromolecules such as proteins.[99,100] This molecular weight is derived from the measurement of the radius of the sedimenting particles. The increase in radius of a colloidal particle caused by an adsorbed layer can thus be measured, as demonstrated by Garvey *et al.*[101] with poly(vinyl alcohol) adsorbed on to polystyrene latices. Alternatively, slow-speed centrifugation of colloidally stable particles leads to a hexagonally close-packed sediment, giving a measure of the particle radius and thus a thickness of the adsorbed layer.[102]

24.3.1.5   Electrophoresis

In an electrophoresis experiment, charged colloidal particles move under the influence of an external electrical field in a manner largely determined by the magnitude of the charge or potential on the particle surface and the distance of the slipping plane from the surface.[103,104] One approach to determining the adsorbed-layer thickness is to assume that the adsorbed layer does not change the surface charge (or potential) or distribution of ions away from the surface. Then the thickness of the adsorbed polymer can be calculated from the measured surface potential of the particle, with and without the adsorbed polymer.[97,99,102,105,106] An alternative approach[107] uses the fact that the dependence of the particle mobility on the concentration of a potential-determining ion depends on the distance of the slipping plane from the surface of the ion. Koopal and Lyklema[107] used this technique to study poly(vinyl alcohol) adsorbing on to AgI particles, finding thicknesses increasing with surface coverage, and of a magnitude expected from the solution dimensions of the polymer.

24.3.2 Techniques for Estimating *p*, the Fraction of Segments Attached to the Surface

24.3.2.1 *Infrared spectroscopy*

This is the most commonly reported method for estimating *p*. The method relies on a shift occurring in the IR adsorption band of a polymer segment, upon adsorption at a surface. A review of these shifts has been given by Rochester.[108] Commonly, carbonyl groups of polymers have been studied in systems of silica in organic solvents.[64,109–112] In these systems, the shift in frequency is large enough to be able to separate the spectra of the bound and free carbonyl groups, whereas, in aqueous systems, this shift is small.[113]

It is also possible to obtain information on the adsorbed polymer by studying the IR shift of the binding group on the solid surface in addition to that of the binding group of the polymer. With silica, the spectra of the surface hydroxyl groups can be used to determine the fraction and, hence, number of sites occupied.[114–116] After comparing the number of these sites with the number of bound groups on the polymer, Dietz[115] concluded that some of the carbonyls of adsorbed polyesters were bonded to two hydroxyl groups of the silica. Joppien[114] similarly found that carbonyl groups of some polymers were bound to more than one hydroxyl group of silica.

The adsorption of poly(acrylic acid) on to aluminum metal has been studied using external reflection techniques.[117] The spectra showed the presence of carboxylate ions in the polymer adjacent to the metal surface, indicating a redox mechanism (proton abstraction) for the adsorption reaction. No such abstraction occurred on copper, silver or gold surfaces.

24.3.2.2 *Magnetic resonance*

When a macromolecule adsorbs at a solid–liquid interface, the molecular motion of the polymer's backbone becomes slower, and the longer correlation time of the motion is reflected in the relaxation times of protons (^{1}H NMR) or free electrons (electron paramagnetic resonance, EPR) that are attached closely to the backbone. Provided there is slow exchange between segments associated with the surface (trains) and those in loops or tails, the spectrum of the whole molecule will be resolvable into the fraction of the chain in each state. As discussed in more detail later, these fractions obtained by EPR and NMR may well be quite different from, but complementary to, those obtained by IR measurements.

(i) Nuclear magnetic resonance

The slow molecular motion of the segments of adsorbed polymers makes their simple high-resolution spectra too broad to be very useful. Pulsed techniques[118,9] have been the most successful. These techniques employ a train of pulses that have different effects on fast and slowly relaxing nuclei. The pulses are chosen so that the signal from the slowly relaxing nuclei, *i.e.* those in loops or tails, is still observed, whereas that from nuclei in trains is zero, thus allowing the relative number of nuclei in each state to be observed. The results for poly(vinylpyrrolidone) on SiO_2 showed that *p* decreased with increasing surface coverage, as expected from theory. The NMR technique has the advantage that it measures the segmental mobility of the polymer directly, though as protons are normally the nuclei under observation, solvents without protons need to be used.

(ii) Electron paramagnetic resonance

The EPR spectra from nitroxides are reasonably well-established and understood[120–122] and are principally determined by the correlation-time and anisotropy of the motion. Stable nitroxide spin labels may be attached to hydrocarbon backbones in various ways, allowing both surfactant[123] and polymers to be spin labelled. With polymers the spin labels can be attached directly to the backbone[5,124] or to the pendant side chains,[126-8] and most polymers, with the important exception of poly(oxyethylene), can be labelled randomly along the backbone. The last point is important if conclusions about polymer conformation are to be drawn. Perhaps the major attraction of the EPR method is that it can be used in aqueous or non-aqueous solvents,[129-131] and for most high-area spin-free substrates, such as colloids or fibrous materials. The main disadvantage of the method is that the added label can influence the adsorption of the whole polymer[132] or the properties of the polymer itself.

Like NMR, EPR spectra from labels on the less mobile segments can be separated from those on more mobile segments, allowing p to be measured. Magnetic resonance measures segments close to or in contact with the solid surface, whereas IR distinguishes only those in direct contact with the surface. Nevertheless, the change in p with surface coverage (θ) measured by EPR and IR is in reasonable agreement.

24.3.2.3 Calorimetry

Killman and co-workers[55,68] measured the heat involved when poly(oxyethylene) is adsorbed on to silica. After subtracting the heat of immersion of the silica, and the heat of dilution of the polymer, the heat evolved was related to the number of polymer/substrate contacts. Killmann and Winter[68] showed that a linear correlation exists between the adsorption enthalpy and the IR extinction of the hydrogen bond between the silanol and the ether group of the polymer.

24.3.2.4 Neutron scattering

Small-angle neutron scattering (SANS) has recently provided the most comprehensive description of adsorbed-polymer conformation, by allowing the complete segment-density distribution to be determined. The scattering of neutrons by a nucleus is characterized by a single parameter b, the scattering length.[133] The scattering lengths of protons and deuterons are quite different, so that mixtures of protonated and deuterated solvents give a wide range of scattering lengths. In studies of polymers adsorbed on to particles, the scattering length of the solvent is adjusted (contrast matched) to that of the particle, so that the spatial distribution of the polymer can be determined. The theory for this was developed in detail by Crowley[134] and a more detailed description of the method has been given by Cohen Stuart et al.[135] Barnett et al.[136] used SANS to measure the segment-density distribution for poly(vinyl alcohol) adsorbed on to polystyrene latex. For the system of poly(oxyethylene) adsorbed on to polystyrene latex[137] the segment-density distribution was approximately exponential, extending 5 nm from the surface, with a large tail extending about a further 4 nm. The hydrodynamic thickness was measured at about 18 nm, so that SANS may not be able to detect the longest tails, predicted by theory.[138]

24.3.3 Thickness of Adsorbed Layers

The thickness of an adsorbed polymer layer, together with the quality of the solvent, are the important parameters determining the magnitude of the steric barriers in colloid stabilization. As the distribution of polymer segments away from a surface is approximately exponential,[51,136,139] the thickness measured may depend on the technique used. The influence of various parameters on adsorbed-layer thickness is discussed below.

24.3.3.1 Molecular weight

The extension of polystyrene from a metal surface in a theta solvent has been measured as a function of molecular weight using ellipsometry. The thickness is less than twice the radius of gyration, $2R_g$, though it varies linearly with the half-power of molecular weight, $M^{1/2}$, as predicted for loop and tail segments.[52,140,141] The thicknesses of adsorbed polystyrene layers measured by viscometry[12] are much larger than those obtained by ellipsometry, even after allowing for the different solvents used. This larger thickness measured by hydrodynamic methods has been predicted by theory[92,138] and is due to the presence of long tails. Garvey *et al.*[101,102] also found that the thickness of adsorbed poly(vinyl alcohol) layers varied linearly with $M^{1/2}$ by both ultracentrifugation and intensity fluctuation spectroscopy. Klein and Luckham,[142] studying polymers adsorbed between mica surfaces, showed that the thickness of adsorbed poly(oxyethylene) varied as $M^{0.4}$, as predicted from scaling arguments.

24.3.3.2 Solvent quality

Adsorbed-layer thickness increases with better solvent quality, as found by Kawaguchi and Takahashi[143] studying polystyrene adsorbed on to chrome from cyclohexane, by ellipsometry. Similarly[144] the thickness of a layer of a polyelectrolyte adsorbed on to platinum increased with decreasing ionic strength. Poly(vinylpyrrolidone) adsorbed on to silica[145] from aqueous solution had more segments in contact with the surface in poorer solvents, indicating thinner layers in the poorer solvents. These data simply highlight the importance of interactions between the segments of adsorbed polymers as an important factor in the adsorption of polymers, especially polyelectrolytes.

The energy of adsorption per segment is another important factor in understanding the adsorption of polymers, though only one determination of its magnitude has been made.[146] Poly(vinylpyrrolidone) was adsorbed on to silica from one solvent, and the displacement of the polymer was studied using other competing solvents. By knowing the critical displacement concentration and the adsorption energy for the second solvent, the adsorption energy for the polymer segments on silica was estimated to be about $4kT(k = 1.380\,622 \times 10^{-23}\,\text{J K}^{-1})$.

24.3.3.3 Surface coverage

Theory[2] predicts that the thickness of adsorbed polymer layers increases with increasing coverage, even where the energy of adsorption per polymer segment is high. Polymers adsorbed in flat conformations gain entropy by adopting more-extended conformations, thus exposing surface sites to additional molecules. The total energy of the system declines as a result. Experimentally, this predicted increase in layer thickness with increasing coverage has been observed,[12,84,147] though thicknesses independent of surface coverage have also been observed.[84,148]

24.4 POLYELECTROLYTE ADSORPTION

Many industrial and bioloical processes involve proteins, polysaccharides, dispersants or flocculants, many of which are charged. Despite this, far fewer studies of the adsorption of polyelectrolytes have been made than for uncharged molecules. With polyelectrolytes, the repulsion between the charged polymer segments can be much greater than with uncharged polymers, sometimes quite dominating the adsorption. An example of this is[149] the zero adsorption of poly(acrylic acid) on silica at high pH compared to its adsorption at low pH. The repulsion can be reduced by adding electrolyte, so that sodium carboxymethylcellulose will not adsorb on to cellulose from pure water,[150] but will adsorb from a moderately concentrated electrolyte solution. In addition, if the surface allows some ion exchange to reduce the net charge at the surface, more polyelectrolyte can adsorb. Thus, polyelectrolytes adsorb on to calcium and barium salts, where desorption of anions from the crystal surface accompanies the polymer adsorption. The literature on polyelectrolyte adsorption is reviewed below according to substrate type.

24.4.1 Hydrophobic Surfaces

There are several studies of the adsorption of polyelectrolytes on to hydrophobic surfaces, showing the effect of both pH and ionic strength on the amount adsorbed. In all instances where data are available,[93,151-3] the amount adsorbed increases with decreasing pH and increasing electrolyte concentration. Buscall's results[152] on the adsorption of graft copolymers (where the backbone was uncharged and the grafted segments were charged) on to polystyrene latex show increasing levels of adsorption with increasing ionic strength, illustrating that repulsion between segments within the adsorbed layer is important in controlling the amount adsorbed. Similarly

Awad and Morawetz[153] have shown that the greater the distance between charged groups, the greater the amount adsorbed.

In one of the few studies of the kinetics of adsorption of polyelectrolytes, Greene[151] showed that the amount adsorbed varied as the square-root of time, and the initial rate of adsorption was greater from higher ionic-strength solutions. Both these results are consistent with the kinetics of adsorption being controlled by diffusion, although the dependence of the rate of adsorption on the polymer dimensions, between 0.6 and 0.12 mol dm^{-3} of added electrolyte, was not as predicted.

The conformation of sodium poly(styrenesulfonate) on polystyrene latices has been studied by Cosgrove *et al.*[1] using neutron scattering. At low ionic strengths and for positively or negatively charged latices, the polyelectrolyte adopted a flat conformation, with loops or tails developing only at high ionic strength. The main reason for the lack of tails or loops at low ionic strength is probably the high repulsion there would be between these segments. Another investigation of polyelectrolyte conformation on hydrophobic surfaces, by van Vliet and Lyklema,[154] showed that poly(methacrylic acid) had similar conformational transitions in the adsorbed and free state. Pefferkorn *et al.*[93] studied the adsorption of a copolymer of maleic acid and ethyl vinyl ether on to cellulose acetate membranes. The data clearly show that increasing the ionic strength leads to lower thicknesses for the adsorbed polyelectrolyte layer, in agreement with ellipsometric data on platinum.[144]

24.4.2 Hydrophilic Surfaces

These surfaces, such as silica or cellulose, are often negatively charged in aqueous systems at neutral pH, and thus provide an additional barrier to adsorption, through segment–surface repulsion, for anionic polyelectrolytes. Thus, Jopien[149] found that poly(acrylic acid) did not adsorb on to silica above about pH 7 and that the adsorption of copolymers of crotonic acid with vinyl acetate on to anatase (TiO_2) decreased with increasing pH above the polymer's precipitation pH in free solution.[79] Evans and Evans[150] also found that sodium carboxymethylcellulose (SCMC) did not adsorb on to cotton from pure H_2O but did in the presence of strong electrolyte. SCMC adsorbed on to microcrystalline cellulose showed few segments in contact with the surface as measured by EPR[155] and the extended conformation produced a stable dispersion of cellulose particles, even in high ionic-strength solutions. At low ionic strengths the dispersions were less stable because of the weak adsorption of the SCMC. Amphoteric polyelectrolytes have a minimum in their segment–segment repulsion at their isoelectric point, and thus would be expected to have a maximum adsorption at that pH. This was observed for the adsorption of gelatin on to several oxide substrates.[156]

Where the charge on the polyelectrolyte is opposite to that of the substrate, a more complex dependence on electrolyte concentration would be expected. At low levels of adsorption electrolyte may have little influence, though once the surface charge had been neutralized, additional electrolyte would be required to reduce the segment–segment repulsion in the adsorbed layer. Something of this behaviour has been demonstrated by Shyluk[157] for the adsorption of poly(1,2-dimethyl-5-vinyl-pyridinium methyl sulfate) on to silica. The initial adsorption from both water and electrolyte is of a high-affinity type, though added electrolyte allows the final level of adsorption to be greater. If the negative charge on the substrate is increased, greater adsorption[158] of cationic polyelectrolyte is observed. Cole and Howard's[159] data on the adsorption of poly(acrylic acid) onto nylon show similar trends to those mentioned above. Studies on either side of the isoelectric point of nylon showed that the cationic charge on the substrate could be reversed by adsorption of the poly-electrolyte.[160]

24.4.3 Metals

Using metals as substrates allows additional electrical and optical techniques to be used to study the adsorbed polymer. Miller and co-workers[161-3] have studied differential capacitance curves at the mercury/water interface, from which some aspects of the polymer conformation have been deduced. Values of p, the fraction of segments in the surface, were estimated from the surface-charge density, and found to increase with increasing energy of the segment–surface bond.

Takahashi *et al.*[144] have studied the adsorption of sodium polyacrylate on to platinum using ellipsometry, though the exact pH of the experiments is not clear. As expected, more polymer adsorbed with increasing ionic strength and, more interestingly, the thickness of the adsorbed layer decreased with increasing ionic strength.

24.4.4 Crystals

The adsorption of polyelectrolytes on to pure crystals is different from that on to most other surfaces, particularly at high pH, where significant amounts of polymer can be adsorbed. This is likely to be a result of two factors, the first of which is that, although crystals may have an overall negative charge, there will be both positive and negative adsorption sites on the surface. The second factor is that desorption of ions from the surface may reduce the charge produced by the adsorbed polyelectrolyte. Several studies have been reported for polyelectrolytes adsorbed on to $BaSO_4$, which can be prepared reproducibly in a pure form, and, as a substrate, has the advantage of (a) relatively high surface areas ($> 10 \, m^2 \, g^{-1}$), (b) low solubility and (c) a chemically stable surface over a wide pH range. It has been shown[129] that the adsorption of sodium polyacrylate on to $BaSO_4$ crystals is accompanied by desorption of SO_4^{2-}, partially reducing the negative charge on the surface conferred by the adsorbed polyelectrolyte. There is, of course, a net increase in negative charge on the particle after adsorption of anionic polyelectrolytes. This has been noticed for sodium carboxymethylcellulose on $BaSO_4$,[164] and lignin sulfonates on $CaCO_3$.[165] In a similar way, calcium and phosphate ions were released from hydroxyapatite following adsorption of poly(carboxylic acid)s.[166]

Adsorption of polyelectrolytes on to insoluble calcium[167] and barium[129,164,168] salts occurs to a significant extent over a wide range of pH, with added electrolyte enhancing the adsorption particularly in the higher pH range.[129] Van Lierde[169] found less sodium polyacrylate adsorbed on to dolomite on increasing the pH above pH 7. Sodium polyacrylate adsorbed on $BaSO_4$ in the absence of added electrolyte adopts a flat conformation at all coverages,[129] as detected by EPR spectroscopy. In the presence of added electrolyte, adsorbed polyelectrolytes have more extended conformations,[129,168] and the amount of polymer adsorbed increases with decreasing polymer charge density.[164] The conformation of adsorbed polyelectrolyte is more extended when the density of surface adsorbing sites is lower.[170]

In contrast to uncharged macromolecules, polyelectrolyte adsorbed on crystals does not exchange freely with non-adsorbed polyelectrolyte in the absence of added electrolyte.[19] This is a result of the electrical repulsion from the adsorbed polymer preventing the close approach of unadsorbed polymers, thus giving a form of kinetic equilibrium rather than thermodynamic equilibrium.

24.4.5 Clays

As clays contain both hydroxyl and charged groups on their surfaces, the adsorption of polyelectrolytes on to them shows some similarity to their adsorption on to silica and crystal surfaces. Michaels and Morelos[171] showed that sodium polyacrylate was not adsorbed on to kaolinite above pH 8 and Mortensen[172] also found a decreasing adsorption for this system with increasing pH. For both cationic[173] and anionic[172] polyelectrolytes adsorbing on to clays, more is adsorbed with increasing ionic strength. Ueda and Harada[173] have shown that, with the adsorption of a cationic polyelectrolyte on to bentonite, the cation-exchange capacity decreases, whereas the anion-exchange capacity increases. This was interpreted as adsorption being accompanied by an ion-exchange process at the surface. This ion exchange was not demonstrated by other experiments, though it would be consistent with data from the polyacrylate/$BaSO_4$ system,[129] where SO_4^{2-} desorption accompanied adsorption by the polyelectrolyte.

24.5 POLYMERS BETWEEN TWO SURFACES

A common use of the adsorption of polymers is to control the stability of colloidal particles. Here the conformation of the adsorbed polymer is of vital importance. The behaviour of polymers between approaching surfaces has been studied extensively in the last few years, particularly since the application of smooth mica surfaces that allow very close separations to be studied.[174,175] The effect of polymers on the force between two surfaces depends on whether the polymers adsorb on the surface or not. It has been calculated[176-9] that, with non-adsorbing polymers, the force between the two approaching surfaces passes through a maximum (repulsion) with decreasing separation until the force becomes attractive at close separations. The maximum in repulsion occurs at a separation slightly less than twice the radius of gyration of the polymer, and arises from the polymer's loss of conformational entropy. At closer separations, polymer chains are forced to migrate from the gap between the surfaces, leaving that region depleted of polymer and the adjacent solution rich in polymer. The attractive force can then be viewed either as coming from the solvent leaving the gap to dilute the adjacent polymer-rich solution, or as due to the pressure of the adjacent polymer on the back of the plate. This attractive force, which gives rise to depletion flocculation, will be greater in good solvents than in poor ones. This flocculation has been observed[180] with polystyrene latices stabilized by anchored poly(oxyethylene) chains, when addition of free poly(oxyethylene) caused a reversible flocculation to occur. Similar flocculation (or phase-separation behaviour) has been reported by other workers.[181-182] The free polymer concentration at which phase separation occurred decreased with increasing molecular weight,[180,181,183] and was lower in good solvents compared to poor solvents. De Gennes,[184] using scaling principles, predicted that the force between particles due to non-adsorbing polymer should be attractive at all separations. This conclusion appears to be incorrect, and may arise from the fact that the total free energy change on moving the surfaces from infinity to touching is negative. However, this does not preclude there being a limited region where the free energy change with distance is positive.

Where the polymers are irreversibly bound to the surfaces, repulsion or attraction between the surfaces depends largely on the solvent quality for the polymer and the density of packing of the polymer on each surface. At low surface coverages, where bridging between the surfaces can occur, both theory[267] and experiment[185,186] show that attraction exists at certain separations. However, if the surfaces are saturated with polymer, and the solvent quality is good, then the force between the surfaces due to the polymer is repulsive at all distances.[185-7] Even if the polymer/solvent system is near to θ conditions,[188] there will still be repulsion arising from compression of the polymer layer, reducing the number of conformations possible. For conditions under which the polymer segments can associate with themselves,[189] an attraction at large separations has been measured.

Initial experiments with polymers between the mica sheets were complicated by surface diffusion and exchange or fractionation of the polymers. With the use of more monodisperse polymers, the technique is now demonstrating good qualitative agreement with theory, and is being used to confirm predictions of the influence of polymers on colloid stability.

24.6 REFERENCES

1. T. Cosgrove, T. M. Obey and B. Vincent, *J. Colloid Interface Sci.*, 1986, **111**, 409.
2. G. J. Fleer and J. Lyklema, in 'Adsorption from Solution at the Solid/Liquid Interface', ed. G. D. Parfitt and C. H. Rochester, Academic Press, London, 1983, p. 153.
3. M. Lal, M. A. Turpin, K. A. Richardson and D. Spencer, *ACS Symp. Ser.*, 1975, **8**, 16.
4. R. R. Stromberg, W. H. Grant and E. Passaglia, *J. Res. Natl. Bur. Stand., Sect. A*, 1964, **68**, 391.
5. J. C. Day, B. Allince and A. A. Robertson, *Can. J. Chem.*, 1978, **56**, 2951.
6. R. E. Felter, *J. Polym. Sci., Part C*, 1971, **34**, 227.
7. R. E. Felter, *J. Polym. Sci., Polym. Lett. Ed.*, 1974, **12**, 7.
8. G. J. Howard and S. J. Woods, *J. Polym. Sci., Part A-2*, 1972, **10**, 1023.
9. Y. E. K. Bogacheva, A. V. Kiselev, Yu S. Nikitin and Yu A. El'tekov, *Vysokomol. Soedin., Ser. A*, 1968, **10**, 574.
10. T. Lindstöm and C. Söremark, *J. Colloid Interface Sci.*, 1976, **55**, 305.
11. G. J. Fleer, Ph.D. Thesis, Agricultural University of Wageningen, Netherlands, 1971.
12. Z. Priel and A. Silberberg, *J. Polym. Sci., Polym. Phys. Ed.*, 1978, **16**, 1917.
13. J. Marra, H. A. van der Schee, G. J. Fleer and J. Lyklema, 'Adsorption from Solution', Academic Press, London, 1982.
14. R. R. Stromberg and W. H. Grant, *J. Res. Natl. Bur. Stand., Sect. A*, 1963, **67**, 601.
15. K. I. Brebner, G. R. Brown, R. S. Chalal and L. E. St. Pierre, *Polymer*, 1981, **22**, 56.
16. K. I. Brebner, R. S. Chalal and L. E. St. Pierre, *Polymer*, 1980, **21**, 533.
17. M. A. Cohen Stuart, J. M. H. M. Scheutjens and G. J. Fleer, *J. Polym. Sci., Polym. Phys. Ed.*, 1980, **18**, 559.
18. E. Hamori, W. C. Forsman and R. E. Hughes, *Macromolecules*, 1971, **4**, 193.
19. D. Bain, M. C. Cafe, I. D. Robb and P. Williams, *J. Colloid Interface Sci.*, 1982, **88**, 467.
20. G. S. Sadakne and J. L. White, *J. Appl. Polym. Sci.*, 1973, **17**, 453.
21. R. R. Stromberg, E. Passaglia and D. J. Tutas, *J. Res. Natl. Bur. Stand., Sect. A*, 1963, **67**, 431.
22. E. Killmann and H.-G. Wiegand, *Makromol. Chem.*, 1970, **132**, 239.
23. G. R. Joppien, *Makromol. Chem.*, 1975, **176**, 1129.
24. R. R. Stromberg, D. J. Tutas and E. Passaglia, *J. Phys. Chem.*, 1965, **69**, 3955.
25. R. Botham and C. Thies, *J. Colloid Interface Sci.*, 1969, **31**, 1.
26. W. H. Grant, L. E. Smith and R. R. Stromberg, *Faraday Discuss. Chem. Soc.*, 1975, **59**, 209.
27. I. D. Robb and R. Smith, *Eur. J. Polym. Sci.*, 1974, **10**, 1005.
28. B. R. Smith and A. E. Alexander, *J. Colloid Interface Sci.*, 1970, **34**, 81.
29. J. E. Luce and A. A. Robertson, *J. Polym. Sci.*, 1961, **51**, 317.
30. G. J. Howard and P. M. McConnell, *J. Phys. Chem.*, 1967, **71**, 2991.
31. P. Nedelcheva and G. V. Stoilkov, *J. Appl. Polym. Sci.*, 1976, **20**, 2131.
32. R. A. Botham and C. Thies, *J. Colloid Interface Sci.*, 1973, **45**, 512.
33. R. A. Watkins and C. R. Robertson, *J. Biomed. Mater. Res.*, 1977, **11**, 915.
34. N. L. Thompson, T. P. Burghardt and D. Axelrod, *Biophys. J.*, 1981, **33**, 435.
35. T. P. Burghardt and D. Axelrod, *Biophys. J.*, 1981, **33**, 455.

36. B. K. Lok, Y.-L. Cheng and C. R. Robertson, *J. Colloid Interface Sci.*, 1983, **91**, 104.
37. T. P. Burghardt and D. Axelrod, *Biochemistry*, 1983, **22**, 979.
38. B. K. Lok, Y.-L. Cheng and C. R. Robertson, *J. Colloid Interface Sci.*, 1983, **91**, 87.
39. H. Bader, R. van Wagenen, J. D. Andrade and H. Ringsdorf, *J. Colloid Interface Sci.*, 1984, **101**, 246.
40. N. L. Thompson and D. Axelord, *Biophys. J.*, 1983, **43**, 103.
41. G. K. Iwamoto, L. C. Winterton, R. S. Stoker, R. A. van Wagenen, J. D. Andrade and D. F. Mosher, *J. Colloid Interface Sci.*, 1985, **106**, 459.
42. N. L. Thompson, *Biophys. J.*, 1982, **38**, 327.
43. S. A. Rockhold, R. D. Quinn and R. A. van Wagenen, *J. Electroanal. Chem. Interfacial Electrochem.*, 1983, **150**, 261.
44. T. Cosgrove and J. Fergie–Woods, *Colloids Surf.*, 1987, **25**, 91.
45. M. J. Shick and E. N. Harvey, *Adv. Chem. Ser.*, 1967, **87**, 63.
46. M. J. Shick and E. N. Harvey, *J. Polym. Sci., Part C*, 1969, **7**, 495.
47. K. Mizuhara, K. Hara and T. Imoto, *Kolloid-Z.Z. Polym.*, 1969, **229**, 17.
48. B. V. Ashmead and M. J. Owen, *J. Polym. Sci., Part A-2*, 1971, **9**, 331.
49. R. R. Stromberg, A. R. Quasius, S. D. Toner and M. S. Parker, *J. Res. Natl. Bur. Stand. (U.S.)*, 1959, **62**, 71.
50. G. J. Howard and P. M. McConnell, *J. Phys. Chem.*, 1967, **71**, 2974.
51. J. M. H. M. Scheutjens and G. J. Fleer, *J. Phys. Chem.*, 1979, **83**, 1619.
52. J. M. H. M. Scheutjens and G. J. Fleer, *J. Phys. Chem.*, 1980, **84**, 178.
53. F. W. Rowland and F. R. Eirich, *J. Polym. Sci., Part A-1*, 1966, **4**, 2401.
54. E. Dickinson and M. Lal, *Adv. Mol. Relaxation Interact. Processes*, 1980, **17**, 1.
55. E. Killmann and R. Eckart, *Makromol. Chem.*, 1971, **144**, 45.
56. R. S. Chahal and L. E. St. Pierre, *Macromolecules*, 1968, **2**, 152.
57. T. M. Polonskii, V. P. Zakordonskii and M. N. Soltys, *Makromol. Granitse Razdela Faz*, 1971, 62.
58. R. Perkel and R. Ullman, *J. Polym. Sci.*, 1961, **54**, 127.
59. S. Ellerstein and R. Ullman, *J. Polym. Sci.*, 1961, **55**, 123.
60. J. Koral, R. Ullman and F. R. Eirich, *J. Phys. Chem.*, 1958, **62**, 541.
61. C. Thies, *Macromolecules*, 1968, **1**, 335.
62. Th. F. Tadros, *J. Colloid Interface Sci.*, 1974, **46**, 528.
63. C. Thies, *J. Colloid Interface Sci.*, 1968, **27**, 734.
64. R. A. Botham and C. Thies, *J. Polym. Sci., Part C*, 1970, **30**, 369.
65. R. A. Botham, C. P. Shank and C. Thies, 'Colloid Morphological Behavior, Block Graft Copolymers, Proceedings of American Chemical Society Symposium', 1970, p. 247.
66. C. Thies, *J. Phys. Chem.*, 1966, **70**, 3783.
67. G. Steinberg, *J. Phys. Chem.*, 1967, **71**, 292.
68. E. Killmann and K. Winter, *Angew. Makromol. Chem.*, 1975, **43**, 53.
69. B. J. Fontana, *J. Phys. Chem.*, 1966, **70**, 1801.
70. E. Davidson and J. J. Kipling, in 'Chemistry, Physics and Application of Surface Active Substances, Proceedings of the International Congress on Surface Active Substances, 4th, Brussels, 1964', ed. F. Asinger, Gordon and Breach, London, 1967, vol. B, p. 1029.
71. J. A. Herd, A. J. Hopkins and G. J. Howard, *J. Polym. Sci., Part A*, 1971, **34**, 211.
72. K. Furusawa, Y. Tezuka and N. Watanabe, *J. Colloid Interface Sci.*, 1980, **73**, 21.
73. C. vander Linden and R. van Leemput, *J. Colloid Interface Sci.*, 1978, **67**, 63.
74. R. E. Felter and L. N. Ray, Jr., *J. Colloid Interface Sci.*, 1970, **32**, 349.
75. T. Miyamoto, S. Tomoshige and H. Inagaki, *Polym. J.*, 1974, **6**, 564.
76. G. J. Howard and M. J. McGrath, *J. Polym. Sci., Polym. Chem. Ed.*, 1977, **15**, 1705.
77. A. D. Barrios and G. J. Howard, *Makromol. Chem.*, 1981, **182**, 1081.
78. P. J. Clark and G. J. Howard, *J. Polym. Sci., Polym. Chem. Ed.*, 1973, **11**, 2305.
79. W. Schmidt and F. R. Eirich, *J. Phys. Chem.*, 1962, **66**, 1907.
80. G. J. Howard, *Appl. Polym. Symp.*, 1974, **25**, 89.
81. C. G. Armistead, A. J. Tyler, F. H. Hambleton, S. A. Mitchell and J. Hockey, *J. Phys. Chem.*, 1969, **73**, 3947.
82. C. Thies, *J. Polym. Sci., Part C*, 1971, **34**, 201.
83. Yu. E. El'tekov and A. V. Kiselev, *J. Polym. Sci., Polym. Symp.*, 1977, **61**, 431.
84. E. Killmann and M. V. Kuzenko, *Angew. Makromol. Chem.*, 1974, **35**, 39.
85. A. Kotera, K. Furusawa and H. Nakanishi, *Rep. Prog. Polym. Phys. Jpn.*, 1973, **16**, 73.
86. F. S. Chan, P. S. Minhas and A. A. Robertson, *J. Colloid Interface Sci.*, 1970, **33**, 586.
87. F. L. McCrackin, E. Passaglia, R. R. Stromberg and H. L. Steinberg, *J. Res. Natl. Bur. Stand., Sect. A*, 1963, **67**, 363.
88. J. A. de Feijter, J. Benjamins and F. A. Veer, *Biopolymers*, 1978, **17**, 1759.
89. R. R. Stromberg, L. E. Smith and F. L. McCrackin, *Symp. Faraday Soc.*, 1970, **4**, 192.
90. Z. Priel, M. Sasson and A. Silberberg, *Rev. Sci. Instrum.*, 1973, **44**, 135.
91. Z. Priel and A. Silberberg, *Isr. J. Chem.*, 1974, **12**, 1023.
92. R. Varoqui and P. Dejardin, *J. Chem. Phys.*, 1977, **66**, 4395.
93. E. Pefferkorn, A. Schmitt and R. Varoqui, *J. Membr. Sci.*, 1978, **4**, 17.
94. J. C. Brown, P. N. Pusey, J. W. Goodwin and R. H. Ottewill, *J. Phys. A: Math. Gen.*, 1975, **8**, 664.
95. H. Z. Cummins and E. R. Pike (eds.), 'Photon Correlation and Light-Beating Spectroscopy', Plenum Press, New York, 1974.
96. B. J. Berne and R. Pecora, 'Dynamic Light Scattering', Wiley, New York, 1976.
97. G. J. Fleer, L. K. Koopal and J. Lyklema, *Kolloid-Z.Z. Polym.*, 1972, **250**, 689.
98. R. S. Duckworth, A. Lips and E. J. Staples, *Faraday Discuss. Chem. Soc.*, 1978, **65**, 288.
99. J. W. Williams, 'Ultracentrifugation of Macromolecules', Academic Press, New York, 1972.
100. C. Tanford, 'Physical Chemistry of Macromolecules', Wiley, 1967, chap. 6.
101. M. J. Garvery, Th. F. Tadros and B. Vincent, *J. Colloid Interface Sci.*, 1974, **49**, 57.
102. M. J. Garvery, Th. F. Tadros and B. Vincent, *J. Colloid Interface Sci.*, 1976, **55**, 440.
103. D. J. Shaw, 'Electrophoresis', Academic Press, London, 1969.

104. R. J. Hunter, 'Zeta Potential in Colloid Science', Academic Press, London, 1981.
105. D. E. Brooks and G. V. F. Seaman, *J. Colloid Interface Sci.*, 1973, **43**, 670.
106. D. E. Brooks, *J. Colloid Interface Sci.*, 1973, **43**, 687.
107. L. K. Koopal and J. Lyklema, *Faraday Discuss. Chem. Soc.*, 1975, **59**, 230.
108. C. H. Rochester, *Powder Technol.*, 1976, **13**, 157.
109. B. J. Fontana and J. R. Thomas, *J. Phys. Chem.*, 1961, **65**, 480.
110. E. Killmann and H.-J. Strasser, *Angew. Makromol. Chem.*, 1973, **31**, 169.
111. C. Theis, P. Peyser and R. Ullman, in 'Chemistry, Physics and Application of Surface Active Substances, Proceedings of the International Congress on Surface Active Substances, 4th, Brussels, 1964', ed. F. Asinger, Gordon and Breach, London, 1967, vol. B, p. 1041.
112. P. Peyser, D. J. Tutas and R. R. Stromberg, *J. Polym. Sci., Part A-1*, 1967, **5**, 651.
113. J. C. Day and I. D. Robb, *Polymer*, 1980, **21**, 408.
114. G. Joppien, *Makromol. Chem.*, 1974, **175**, 1931.
115. E. Dietz, *Makromol. Chem.*, 1976, **177**, 2113.
116. M. Korn and E. Killmann, *J. Colloid Interface Sci.*, 1980, **76**, 19.
117. D. L. Allara, 'Adhesion and Adsorption of Polymers', ed. L.-H. Lee, Plenum Press, New York, 1980, vol. 12B, p. 751.
118. T. Cosgrove and K. G. Barnett, *J. Magn. Reson.*, 1981, **43**, 15.
119. K. G. Barnett, T. Cosgrove, B. Vincent, D. S. Sissons and M. Cohen Stuart, *Macromolecules*, 1981, **14**, 1018.
120. L. J. Berliner (ed.), 'Spin Labelling', Academic Press, New York, 1976.
121. H. M. Schwartz, J. R. Bolton and D. C. Borg (eds.) 'Biological Application of Electron Spin Resonance', Wiley, New York, 1972.
122. R. F. Boyer and S. E. Keinath (eds.), 'Molecular Motion in Polymers by ESR', Hardwood, New York, 1980.
123. J. C. Williams, R. Mehlhorn and A. D. Keith, *Chem. Phys. Lipids*, 1971, **7**, 207.
124. A. T. Bullock, G. G. Cameron and P. Smith, *Eur. Polym. J.*, 1975, **11**, 617.
125. P. Törmälä, K. Silvennoinen and J. J. Lindberg, *Acta. Chem. Scand.*, 1971, **25**, 2659.
126. A. T. Bullock, G. G. Cameron and P. Smith, *Polymer*, 1972, **13**, 89.
127. M. C. Cafe and I. D. Robb, *Polymer*, 1979, **20**, 513.
128. M. C. Cafe, N. G. Pryce and I. D. Robb, *Polymer*, 1976, **17**, 91.
129. M. C. Cafe and I. D. Robb, *J. Colloid Interface Sci.*, 1982, **86**, 411.
130. T. M. Liang, P. N. Dickenson and W. G. Miller, *ACS. Symp. Ser.*, 1980, **142**, 1.
131. K. K. Fox, I. D. Robb and R. S. Smith, *J. Chem. Soc., Faraday Trans. 1*, 1974, **70**, 1186.
132. I. D. Robb and R. Smith, *Polymer*, 1977, **18**, 500.
133. J. W. White, *Proc. R. Soc. London, Ser. A*, 1975, **345**, 119.
134. T. L. Crowley, Ph.D. Thesis, Oxford University, 1984.
135. M. A. Cohen Stuart, T. Cosgrove and B. Vincent, *Adv. Colloid Interface Sci.*, 1986, **24**, 143.
136. K. G. Barnett, T. Cosgrove, T. L. Crowley, Th. F. Tadros and B. Vincent, 'The Effect of Polymers on Dispersion Properties', ed. Th. F. Tadros, Academic Press, 1981, 183.
137. T. Cosgrove, T. L. Crowley, M. A. Cohen Stuart and B. Vincent, *ACS Symp. Ser.*, 1984, **240**, 147.
138. J. M. H. M. Scheutjens, G. J. Fleer and M. A. Cohen Stuart, *Colloids Surf.*, 1986, **21**, 285.
139. A. T. Clark, M. Lal, M. A. Turpin and K. A. Richardson, *Faraday Discuss. Chem. Soc.*, 1975, **59**, 189.
140. C. A. J. Hoeve, *J. Chem. Phys.*, 1966, **44**, 1505.
141. C. A. J. Hoeve, *J. Polym. Sci., Part C*, 1970, **30**, 361.
142. J. Klein and P. F. Luckham, *Macromolecules*, 1986, **19**, 2007.
143. M. Kawaguchi and A. Takahashi, *J. Polym. Sci., Polym. Phys. Ed.*, 1980, **18**, 2069.
144. A. Takahashi, M. Kawaguchi and T. Kato, 'Adhesion and Adsorption of Polymers', ed. L. H. Lee, Plenum Press, New York, 1980, vol. 12B, p. 729.
145. A. T. Clark, I. D. Robb and R. Smith, *J. Chem. Soc., Faraday Trans. 1*, 1976, **72**, 1489.
146. M. A. Cohen Stuart, G. J. Fleer and J. M. H. M. Scheutjens, *J. Colloid Interface Sci.*, 1984, **97**, 526.
147. H. Gebhard and E. Killmann, *Angew. Makromol. Chem.*, 1976, **53**, 171.
148. E. Killmann, J. Eisenlauer and M. Korn, *J. Polym. Sci., Polym. Symp.*, 1977, **61**, 413.
149. G. R. Joppien, *J. Phys. Chem.*, 1978, **82**, 2210.
150. P. G. Evans and W. P. Evans, *J. Appl. Chem.*, 1967, **17**, 276.
151. B. W. Greene, *J. Colloid Interface Sci.*, 1971, **37**, 144.
152. R. Buscall, *J. Chem. Soc. Faraday Trans. 1*, 1981, **77**, 909.
153. N. M. Awad. and H. Morawetz, *J. Polym. Sci., Polym. Phys. Ed.*, 1981, **19**, 245.
154. T. van Vliet and J. Lyklema, *J. Colloid Interface Sci.*, 1978, **63**, 97.
155. R. Harrop, G. O. Phillips, I. D. Robb and P. A. Williams, *Prog. Food Nutr. Sci.*, 1982, **6**, 331.
156. F. R. Eirich and A. Kudish, *Pap. Meet.—Am. Chem. Soc., Div. Org. Coat. Plast. Chem.*, 1970, **30**, 501.
157. W. P. Shyluk, *J. Polym. Sci., Part A-2*, 1968, **6**, 2009.
158. F. Onabe, *J. Appl. Polym. Sci.*, 1978, **22**, 3495.
159. D. Cole and G. J. Howard, *J. Polym. Sci., Part A-2*, 1972, **10**, 993.
160. T. Suzawa and H. Shirahama, *Colloid Polym. Sci.*, 1979, **257**, 732.
161. I. R. Miller and D. C. Grahame, *J. Colloid Sci.*, 1961, **16**, 23.
162. I. R. Miller, *J. Phys. Chem.*, 1960, **64**, 1790.
163. I. R. Miller, *Trans. Faraday Soc.*, 1961, **57**, 301.
164. P. A. Williams, R. Harrop, G. O. Phillips, G. Pass and I. D. Robb, *J. Chem. Soc., Faraday Trans. 1*, 1982, **78**, 1733.
165. J. C. Le Bell and P. Stenius, *Paperi Ja Puu*, 1977, **59**, 477.
166. J. C. Voegel, S. Gillmeth and R. M. Frank, *J. Colloid Interface Sci.*, 1981, **84**, 108.
167. C. H. Nestler, *J. Colloid Interface Sci.*, 1968, **26**, 10.
168. P. A. Williams, R. Harrop, G. O. Phillips, I. D. Robb and G. Pass, 'The Effect of Polymers on Dispersion Properties', ed. Th. F. Tadros, Academic Press, New York, 1982, p. 361.
169. A. van Lierde, *Int. J. Miner. Process.*, 1974, **1**, 81.

170. I. D. Robb and M. Sharples, *J. Colloid Interface Sci.*, 1982, **89**, 301.
171. A. S. Michaels and O. Morelos, *Ind. Eng. Chem.*, 1955, **47**, 1801.
172. J. L. Mortensen, *Clays Clay Miner., Proc. Conf.*, 1960, **9**, 530.
173. T. Ueda and S. Harada, *J. Appl. Polym. Sci.*, 1968, **12**, 2395.
174. J. N. Israelachvili and G. E. Adams, *J. Chem. Soc. Faraday Trans. 1*, 1978, **74**, 975.
175. J. N. Israelachvili, R. K. Tandon and L. R. White, *J. Colloid Interface Sci.*, 1980, **78**, 430.
176. A. T. Clark and M. Lal, *J. Chem. Soc., Faraday Trans. 2*, 1981, **77**, 981.
177. R. I. Feigin and D. H. Napper, *J. Colloid Interface Sci.*, 1980, **74**, 567.
178. R. I. Feigin and D. H. Napper, *J. Colloid Interface Sci.*, 1980, **75**, 525.
179. J. M. H. M. Scheutjens and G. J. Fleer, *Adv. Colloid Interface Sci.*, 1982, **16**, 361.
180. C. Cowell, R. Li-In-On and B. Vincent, *J. Chem. Soc., Faraday Trans. 1*, 1978, **74**, 337.
181. H. de Hek and A. Vrij, *J. Colloid Interface Sci.*, 1981, **84**, 409.
182. D. B. Siano and J. Bock, *J. Polym. Sci., Polym. Lett. Ed.*, 1982, **20**, 151.
183. C. Pathmananohoran, H. de Hek and A. Vrij, *Colloid Polym. Sci.*, 1981, **259**, 769.
184. P.-G. de Gennes, *Macromolecules*, 1982, **15**, 492.
185. J. Klein, *Makromol. Chem., Macromol. Symp.*, 1986, **1**, 125.
186. J. Klein and P. F. Luckham, *Nature (London)*, 1984, **308**, 836.
187. P. M. Claesson and C.-G. Gölander, *J. Colloid Interface Sci.*, 1987, **117**, 366.
188. K. Ingersent, J. Klein and P. Pincus, *Macromolecules*, 1986, **19**, 1374.
189. E. Perez and J. E. Proust, *J. Colloid Interface Sci.*, 1987, **118**, 182.
190. F. M. Fowkes and M. A. Mostafa, *Ind. Eng. Chem. Prod. Res. Dev.*, 1978, **17**, 3.
191. K. Mizuhara, K. Hara and T. Imoto, *Kolloid-Z.Z. Polym.*, 1970, **238**, 442.
192. G. Christakos, U. Hofmann, B. Weick and G. Joppien, *Ber. Bunsenges. Phys. Chem.*, 1984, **88**, 1067.
193. H. Hommel, A. P. Legrand, J. Lecourtier and J. Desbarres, *Eur. Polym. J.*, 1979, **15**, 993.
194. J. Eisenlauer, E. Killmann and M. Korn, *J. Colloid Interface Sci.*, 1980, **74**, 120.
195. E. Killmann, N. Gütling and Th. Wild, *Ber. Bunsenges. Phys. Chem.*, 1984, **88**, 1141.
196. L. Gargallo and E. Cid, *Colloid Polym. Sci.*, 1977, **255**, 556.
197. M. A. Cohen Stuart, G. J. Fleer and J. M. H. M. Scheutjens, *J. Colloid Interface Sci.*, 1984, **97**, 526.
198. K. Hara, K. Mizuhara and T. Imoto, *Kolloid-Z.Z. Polym.*, 1970, **238**, 438.
199. R. E. Hostetler and J. W. Swanson, *J. Polym. Sci., Polym. Chem. Ed.*, 1974, **12**, 29.
200. E. Kokufuta and K. Takahashi, *Macromolecules*, 1986, **19**, 351.
201. K. Hara and T. Imoto, *Kolloid-Z.Z. Polym.*, 1970, **237**, 297.
202. E. Pefferkorn, Q. Tran and R. Varoqui, *J. Polym. Sci., Polym. Chem. Ed.*, 1981, **19**, 27.
203. G. J. Howard and M. J. McGrath, *J. Polym. Sci., Polym. Chem. Ed.*, 1977, **15**, 1721.
204. V. Hlady, D. R. Reinecke and J. D. Andrade, *J. Colloid Interface Sci.*, 1986, **111**, 555.
205. A. V. Kiselev, A. S. Nazanskii and Yu. A. El'tekov, *Kolloidn. Zh.*, 1975, **37**, 556.
206. G. Lopatin and F. R. Eirich, 'Proceedings of the International Congress of Surface Activity, 3rd, Köln', University Printers, Mainz, 1960, vol. 2, p. 97.
207. I. I. Maleyer, T. M. Polonskii and M. N. Soltys, *Vysokomol. Soedin., Ser. A*, 1968, **10**, 2122.
208. J. S. Binford, Jr. and A. M. Gessler, *J. Phys. Chem.*, 1959, **63**, 1376.
209. E. R. Gilliland and E. B. Gutoff, *J. Appl. Polym. Sci.*, 1960, **3**, 26.
210. H. L. Frisch, M. Y. Hellman and J. L. Lundberg, *J. Polym. Sci.*, 1959, **38**, 441.
211. V. K. Dunn and R. D. Vold, *J. Colloid Interface Sci.*, 1976, **54**, 22.
212. T. Miyamoto and H.-J. Cantow, *Makromol. Chem.*, 1972, **162**, 43.
213. K. Urano, and Y. Nishimura and Y. Yanaga, *Nippon Kagaku Kaishi*, 1975, **8**, 1444.
214. A. V. Kiselev, N. V. Kovaleva, V. V. Khopina, G. A. Chirkova and Yu. A. El'tekov, *Kolloidn. Zh.*, 1972, **34**, 934.
215. A. V. Kiselev, N. N. Kileveya, V. V. Khopina and Yu. A. El'tekov, *Vysokomol. Soedin., Ser. A*, 1972, **14**, 2343.
216. I. M. Kolthoff, R. G. Gutmacher and A. Kahn, *J. Phys. Chem.*, 1951, **55**, 1240.
217. A. V. Kiselev, N. V. Kovaleva, O. G. Kryukova and V. V. Khopina, *Kolloidn. Zh.*, 1970, **32**, 438.
218. C. R. Earl, Ph.D. Thesis, Polytechnic Institute of Brooklyn, New York, 1970.
219. B. C. Bonekamp and J. Lyklema, *J. Colloid Interface Sci.*, 1986, **113**, 67.
220. A. Monies, *Powder Technol.*, 1980, **27**, 227.
221. R. E. Felter, *J. Polym. Sci., Polym. Lett. Ed.*, 1974, **12**, 147.
222. U. Adam and I. D. Robb, *J. Chem. Soc., Faraday Trans. 1*, 1983, **79**, 2745.
223. T. Bartels and J. Arends, *J. Polym. Sci., Polym. Chem. Ed.*, 1981, **19**, 127.
224. M. Sigiura and E. Fuji, *Bull. Soc. Sci. Photogr. Jpn.*, 1967, **17**, 41.
225. M. Sigiura, *Bull. Chem. Soc. Jpn.*, 1970, **43**, 2604.
226. W. P. Shyluk, *J. Appl. Polym. Sci.*, 1964, **8**, 1063.
227. R. I. S. Gill and T. M. Herrington, *Colloids Surf.*, 1986, **22**, 51.
228. A. P. Black, F. B. Birkner and J. J. Morgan, *J. Colloid Interface Sci.*, 1966, **21**, 626.
229. D. Kronberg, J. Kuortti and Per Stenius, *Colloids Surf.*, 1986, **18**, 411.
230. L. E. Kuznetsova and N. N. Serb–Serbina, *Kolloidn. Zh.*, 1968, **30**, 853.
231. J. E. Glass, H. Ahmed and A. Karunasena, *Colloids Surf.*, 1986, **21**, 335.
232. J. E. Glass, H. Ahmed, S. D. Seneker and G. J. McCarthy, *Colloids Surf.*, 1986, **21**, 323.
233. F. Onabe, *J. Appl. Polym. Sci.*, 1979, **23**, 2909.
234. L. Winter, L. Wågberg, L. Ödberg and T. Lindström, *J. Colloid Interface Sci.*, 1986, **111**, 537.
235. G. G. Allan, K. Akagane, A. N. Neogi, W. M. Reif and T. Mattila, *Nature (London)*, 1970, **225**, 175.
236. R. H. Pelton, *J. Colloid Interface Sci.*, 1986, **111**, 475.
237. E. R. Hendrickson and R. D. Neuman, *J. Colloid Interface Sci.*, 1986, **110**, 243.
238. L. S. Sandell and P. Luner, *J. Appl. Polym. Sci.*, 1974, **18**, 2075.
239. F. S. Chan and A. A. Robertson, *J. Colloid Interface Sci.*, 1970, **33**, 598.
240. D. Cole and G. J. Howard, *J. Polym. Sci., Part A-2*, 1972, **10**, 1013.
241. J. Gramain, *C.R. Hebd. Seances Acad. Sci., Ser. C*, 1974, **278**, 1401.

242. R. Buscall and T. Corner, *Colloids Surf.*, 1986, **17**, 39.
243. H. Sakai, T. Fujimori and Y. Imamura, *Rep. Prog. Polym. Phys. Jpn.*, 1979, **22**, 509.
244. H. Sakai and Y. Imamura, *Bull. Chem. Soc. Jpn.*, 1980, **53**, 1749.
245. H. Sakai and Y. Imamura, *Rep. Prog. Polym. Phys. Jpn.*, 1978, **21**, 447.
246. H. Sakai, T. Fujimori and Y. Imamura, *Bull. Chem. Soc. Jpn.*, 1980, **53**, 3457.
247. C. vander Linden and R. van Leemput, *J. Colloid Interface Sci.*, 1978, **67**, 48.
248. M. Kawaguchi, A. Sakai and A. Takahashi, *Macromolecules*, 1986, **19**, 2952.
249. F. W. Roland and F. R. Eirich, *J. Polym. Sci., Part A-1*, 1966, **4**, 2033.
250. A. V. Kiselev, V. I. Lygin, I. N. Solomonova, D. O. Osmanova and Yu. A. El'tekov, *Kolloidn. Zh.*, 1967, **30**, 386.
251. Yu. S. Lipatov, L. M. Sergeeva, T. T. Todosiichuk and T. S. Khramova, *Kolloidn. Zh.*, 1975, **37**, 280.
252. Th. van den Boomgaard, T. A. King, Th. F. Tadros, H. Tang and B. Vincent, *J. Colloid Interface Sci.*, 1978, **66**, 68.
253. M. C. Barker and M. J. Garvey, *J. Colloid Interface Sci.*, 1980, **74**, 331.
254. Th. F. Tadros and B. Vincent, *J. Colloid Interface Sci.*, 1979, **72**, 505.
255. K. G. Barnett, T. Cosgrove, B. Vincent, A. N. Burgess, T. L. Crowley, T. King, J. D. Turner and Th. F. Tadros, *Polymer*, 1981, **22**, 283.
256. T. Cosgrove, T. L. Crowley, B. Vincent, K. G. Barnett and Th. F. Tadros, *Symp. Faraday Soc.*, 1981, **16**, 101.
257. J. B. Kayes and D. A. Rawlins, *Colloid Polym. Sci.*, 1979, **257**, 622.
258. M. Kawaguchi and A. Takahashi, *J. Polym. Sci., Polym. Phys. Ed.*, 1980, **18**, 943.
259. A. Takahashi, M. Kawaguchi, H. Hirota and T. Kato, *Macromolecules*, 1980, **13**, 884.
260. G. G. Fuller, *AIP Conf. Proc.*, 1985, **137**, 263.
261. R. R. Stromberg and L. E. Smith, *J. Phys. Chem.*, 1967, **71**, 2470.
262. J. Lyklema, *Pure Appl. Chem.*, 1976, **46**, 149.
263. A. A. Baran, I. I. Kochevga and I. M. Solomentseva, *Kolloidn. Zh.*, 1977, **39**, 5.
264. R. Varoqui, P. Dejardin and E. Pefferkorn, *Stud. Surf. Sci. Catal.*, 1982, **10**, 107.
265. B. V. Kavanagh, A. M. Posner and J. P. Quirk, *Faraday Discuss. Chem. Soc.*, 1975, **59**, 242.
266. G. J. Howard and P. M. McConnell, *J. Phys. Chem.*, 1967, **71**, 2981.
267. J. M. H. M. Scheutjens and G. J. Fleer, *Macromolecules*, 1985, **18**, 1882.
268. E. K. Bogacheva, A. V. Kiselev and Yu. A. El'tekov, *Kolloidn. Zh.*, 1969, **31**, 176.
269. Ye. K. Bogacheva and Yu. A. El'tekov, *Polym. Sci. USSR (Engl. Transl.)*, 1974, **16**, 714.
270. K. Furusawa, H. Nakanishi and A. Kotera, in 'Proceedings of the International Congress on Surface Science', 1975, p. 221.

25

Ionomers

C. W. LANTMAN, W. J. MacKNIGHT and R. D. LUNDBERG*
University of Massachusetts, Amherst, MA, USA

25.1 INTRODUCTION AND HISTORICAL BACKGROUND

Perhaps the oldest branch of polymer science is the study of ion-containing polymers, since most naturally occurring polymers contain ionic species. The area of structure–property relationships for synthetic polymer materials containing salt groups has seen expansive growth in recent years. This is evident from the scientific interest manifested in the technical literature and by the appearance of several monographs devoted to the subject.[1-8] Clearly, the range of polymeric materials within the field of ionic polymers is extremely varied, extending from naturally occurring biopolymers through to ceramics and inorganic glasses. The class of materials to be discussed here occupies an intermediate position between purely organic structures on the one hand and purely inorganic structures on the other. The term ionomer was coined by the Du Pont Co. to describe such materials and has come into general use to mean a polymer which is composed of a hydrocarbon backbone containing pendant acid groups which are neutralized partially or completely to form salts. The concentration of the salt groups may vary but the hydrocarbon backbone is always the majority component. In general, unless otherwise specified, we shall consider the term ionomer to mean a polymer containing less than 10 mol% salt groups. This specifically excludes classical polyelectrolytes, which contain salt groups on alternate backbone atoms.

Considerable industrial and academic research effort has been expended on ionomers. A brief historical context for these studies will be presented and then a representative survey of the existing experimental data on ionomeric systems will be discussed. Despite extensive study, several questions in the field of ionomers remain unanswered. Of central importance is the question of structure. The exact distribution of salt groups in the bulk and in solution is not yet known. A brief survey of the existing theories and models will be presented and some of the current areas of ionomer research will be described.

* Now at Exxon Chemical Co., Linden, NJ, USA.

The introduction of ionic groups into hydrocarbon resins dates back to the 1930s when methods for the carboxylation of elastomers appeared in the patent literature. In the 1950s, Goodrich introduced one of the first elastomers based on ionic interactions, a copoly(butadiene/acrylonitrile/acrylic acid). These materials could be neutralized with zinc salts and plasticized to break ionic association at elevated temperature. Such ionic elastomers displayed enhanced tensile properties and improved adhesion compared to conventional copolymers.

A second family of ionic elastomers was introduced in the 1950s by Du Pont under the trade name Hypalon. These materials were based on the sulfonation of chlorinated polyethylene. After suitable curing with various metal oxides, these materials possess a combination of ionic and covalent crosslinks. Upon neutralization, such an elastomer changes from a material with the general properties of low density polyethylene to an extremely tough, flexible and optically clear thermoplastic. This dramatic change in properties was initially explained on the basis of ionic interactions. It seemed natural to assume that these elastomeric properties were due to the presence of ionic crosslinks. Such crosslinks were originally envisioned as a divalent cation bridging two pendant anionic groups. Subsequent work with monovalent species led to the realization that the 'ionic crosslinks' possess a more complicated structure. Factors such as the coordinating tendency of the counterion, the low polarity of the hydrocarbon backbone, and the influence of polar impurities such as water have to be considered. On closer examination, it was found that the early simplified concept of the ionic crosslink was inadequate.

A breakthrough occurred in the mid 1960s when Du Pont introduced copoly(ethylene/methacrylic acid) under the tradename Surlyn; these copolymers were partially neutralized with sodium and zinc cations. These modified polyethylenes possess remarkable clarity and tensile properties superior to those of conventional polyethylene. The development of Surlyn was an important factor in stimulating research in ionomers. The Surlyn systems emphasized the versatility of ionomer structures and the unique properties available from the modification of the polyethylene backbone. Many of the features which are peculiar to ionomers were recognized at this time; notably, the idea that multi-ion clusters would be formed due to the low dielectric constant of the hydrocarbon matrix. It was only with more detailed X-ray diffraction studies and mechanical property measurements that the morphology of these materials was gradually revealed.

During the 1970s and 1980s, several novel ionomer systems have been developed. Of particular interest are the perfluorinated resins and halato-telechelic polymers as well as the various sulfonated materials. These systems will be discussed in greater depth in the following sections.

25.2 PREPARATION OF IONOMERS

In general, the preparation of ionomers is a straightforward procedure. The particular acid group of interest can be introduced onto the hydrocarbon backbone either by direct copolymerization or post-synthesis reaction. The following five important groups of ionomers illustrate the various methods of preparation. These ionomer families are: ethylene-based materials, ionic elastomers, modified polystyrenes, perfluorinated resins and halato-telechelic polymers.

25.2.1 Ethylene-based Materials

The preparation of ethylene-based ionomers is an example of copolymerization. Briefly, copolymers of ethylene and methacrylic or acrylic acid are synthesized by a high pressure, free radically initiated process similar to that used for the production of low density polyethylene. The resulting copolymer is then neutralized by fluxing with the metal-neutralizing agent of choice. Mixing can be achieved in plastics-processing equipment, such as an extruder for large quantities or a two-roll mill for smaller amounts. A controlled amount of base in solution is added to the molten polymer and the solvent is subsequently removed by vacuum extraction or simply flashed off. It is also possible to dissolve the copolymer in a suitable solvent and add the base — often as an alkoxide in methanol. The ionomer may then precipitate out as it forms. The resulting ionomer has the structure (**1**). Typical property values for Surlyn resins[9] are listed in Table 1.

$$\underset{(\mathbf{1})}{+CH_2CH_2)_n(CH_2\underset{\underset{CO_2^-\ M^+}{|}}{\overset{\overset{Me}{|}}{C}})_m}$$

Table 1 Properties of Surlyn Resins

	Na resin	Zn resin
Tensile impact (kJ m^{-2})	1160	925
Notched izod (J m^{-1})	610	no break
Specific gravity (g cm^{-3})	0.94	0.94
Flexural modulus (MPa)	220	130
Tensile strength (MPa)	29	21.4
Yield strength (MPa)	12.4	8.3
Elongation (%)	450	500
Hardness, shore D	60	54
Melt flow index (g/10 min)	1.3	5.5
Density (lb in^{-3})^a	0.034	0.034
Melting point (°C)	94	95
Heat deflection temperature (455 kPa, °C)	40	41
Coefficient of thermal expansion (−20 to 32 °C, 10^{-5} K^{-1})	14	16
Freezing point (°C)	67	84

^{a} lb in^{-3} ≡ 27.68 g cm^{-3}; 0.034 lb in^{-3} ≡ 0.941 g cm^{-3}.

25.2.2 Ionic Elastomers

The preparation of ionic elastomers typifies the second method of ionomer synthesis: post-synthesis modification. Sulfonation of poly(ethylene/propylene/diene monomer) [poly(EPDM)] for example (Scheme 1), permits the preparation of an ionomer whose ionic content is proportional to the amount of sulfonating agent used[10-12]. These reactions are conducted in solution and so direct neutralization of the acid functionality to the desired level can be obtained. The neutralized ionomer is then isolated by conventional techniques such as coagulation in a nonsolvent or by solvent flashing. An alternate approach to the preparation of an ionic elastomer involves reactive milling of the polymer in an extruder.[13] The extruder sulfonation of poly(EPDM) can be achieved with the same sulfonating agents which are typically employed in solution. Such continuous melt sulfonation processes are generally conducted at elevated temperatures and occur quite rapidly. Typical reaction conditions are 10 minutes at 100 °C. Neutralization of the resulting acid form with metal stearates can also be conducted in the extruder.

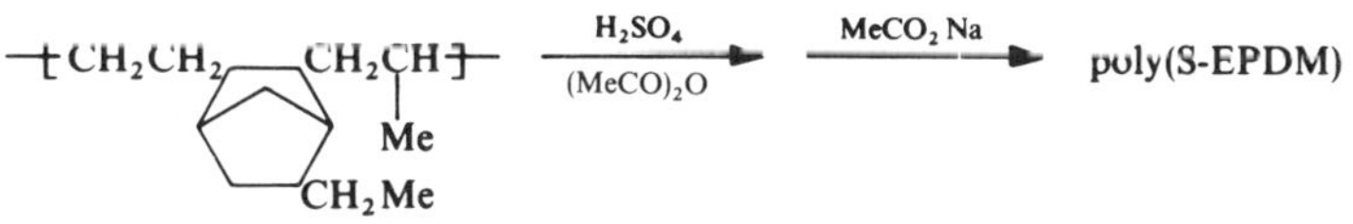

Scheme 1

A novel ionic elastomer was developed during the 1970s based on a polypentenamer backbone.[14-20] Carboxylic, thioglycolic, sulfonic and phosphonic acid salts were prepared as shown in Scheme 2. The starting polypentenamer was obtained *via* ring-opening polymerization of cyclopentane. Hydrogenated derivatives were prepared using toluenesulfonyl hydrazide, presumably without backbone degradation. With these materials, the effects of acid group and of crystallinity on ionomer properties could be studied.

25.2.3 Polystyrene-based Materials

Polystyrene-based ionomers have been extensively studied. These materials can be prepared either by post-reaction modification[21] or by copolymerization.[22] One method of preparation is to copolymerize styrene and methacrylic acid using benzoyl peroxide as the free radical initiator. Following reaction at elevated temperature (*ca.* 80 °C), the product is dissolved in benzene and titrated with sodium hydroxide in methanol to effect neutralization.

Several studies of the post-reaction modification of polystyrene have been undertaken.[23-27] Of particular interest are the sulfonation and carboxylation processes outlined in Scheme 3. Viscosity

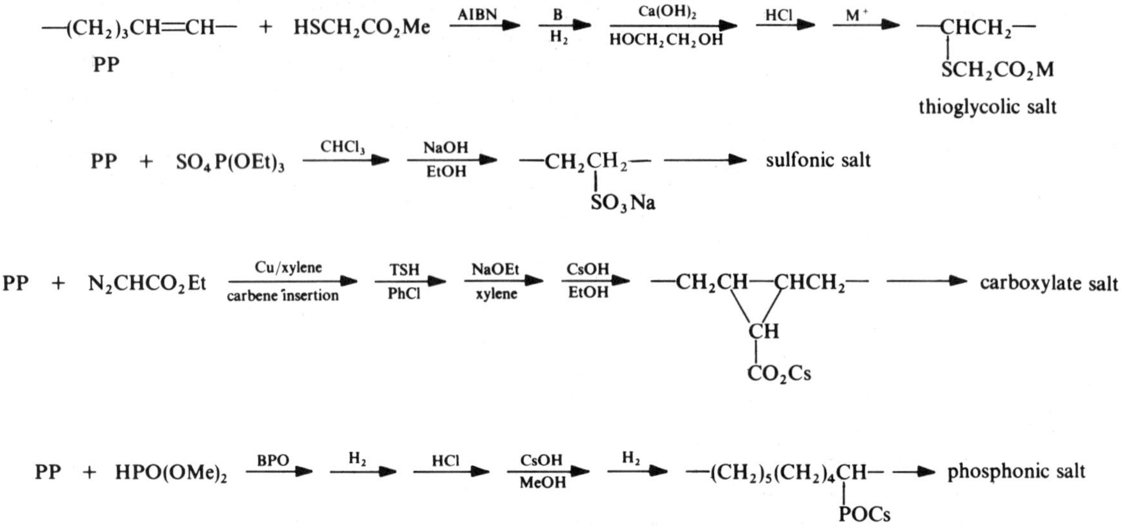

$$-(CH_2)_3CH=CH- \ + \ HSCH_2CO_2Me \ \xrightarrow{AIBN} \ \xrightarrow[H_2]{B} \ \xrightarrow[HOCH_2CH_2OH]{Ca(OH)_2} \ \xrightarrow{HCl} \ \xrightarrow{M^+} \ \begin{array}{c} -CHCH_2- \\ | \\ SCH_2CO_2M \end{array}$$

PP

thioglycolic salt

$$PP \ + \ SO_4P(OEt)_3 \ \xrightarrow[EtOH]{CHCl_3} \ \xrightarrow{NaOH} \ \begin{array}{c} -CH_2CH_2- \\ | \\ SO_3Na \end{array} \ \longrightarrow \ \text{sulfonic salt}$$

$$PP \ + \ N_2CHCO_2Et \ \xrightarrow[\text{carbene insertion}]{Cu/xylene} \ \xrightarrow[PhCl]{TSH} \ \xrightarrow[xylene]{NaOEt} \ \xrightarrow[EtOH]{CsOH} \ \begin{array}{c} -CH_2CH-CHCH_2- \\ \diagdown \ \diagup \\ CH \\ | \\ CO_2Cs \end{array} \ \longrightarrow \ \text{carboxylate salt}$$

$$PP \ + \ HPO(OMe)_2 \ \xrightarrow{BPO} \ \xrightarrow{H_2} \ \xrightarrow{HCl} \ \xrightarrow[MeOH]{CsOH} \ \xrightarrow{H_2} \ \begin{array}{c} -(CH_2)_5(CH_2)_4CH- \\ | \\ POCs \end{array} \ \longrightarrow \ \text{phosphonic salt}$$

BPO = dibenzyl peroxide
TSH = toluenesulfonyl hydrazide

Scheme 2

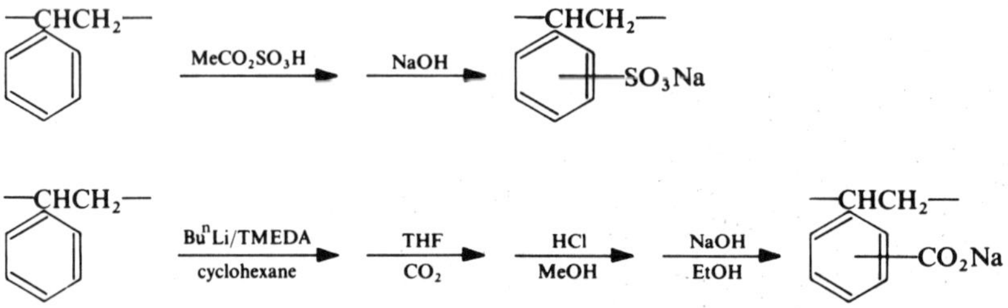

Scheme 3

measurements in mixed solvents have shown that these functionalization reactions do not degrade the backbone. In addition to sodium cations, salts of lithium, magnesium, zinc and barium may also be prepared.

25.2.4 Perfluorinated Resins

A current industrial application of ionomers is their use as permselective membranes for the chloralkali process. The ionomers used in these membranes are based on a poly(tetrafluoroethylene) backbone containing occasional ether linkages with ionic side groups. These are based upon either sulfonate or carboxylate salts.

Ionomers based on perfluorosulfonic acid derivatives were developed by Du Pont and bear the tradename Nafion. They may be prepared as in Scheme 4.[6,28] The sodium counterions can be readily exchanged for other metal ions by soaking in an aqueous electrolyte solution.

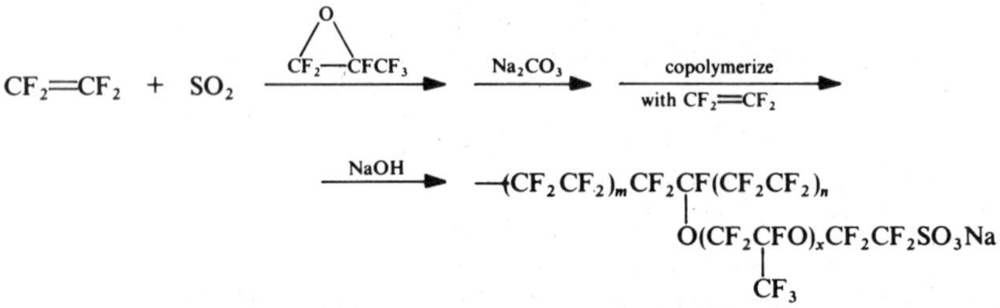

Scheme 4

In addition to the perfluorosulfonic acid derivatives, carboxylic acid forms have been developed. These may be obtained by modification of Nafion films[6] or by direct synthesis.[29] In the case of modification, the ionomer is converted from a thionyl chloride to a carboxylic acid. Typical reducing agents are hydriodic acid, hydrobromic acid and hydrophosphonic acid. The carboxylic acid form can be made directly by copolymerization of tetrafluoroethylene and a carboxylated perfluorovinyl ether. The synthesis of the ether is a difficult task, however, as cyclization can occur. One synthetic route to the ether is shown in Scheme 5. After copolymerization with tetrafluoroethylene, the ester group can be quantitatively hydrolyzed to yield the sodium salt.

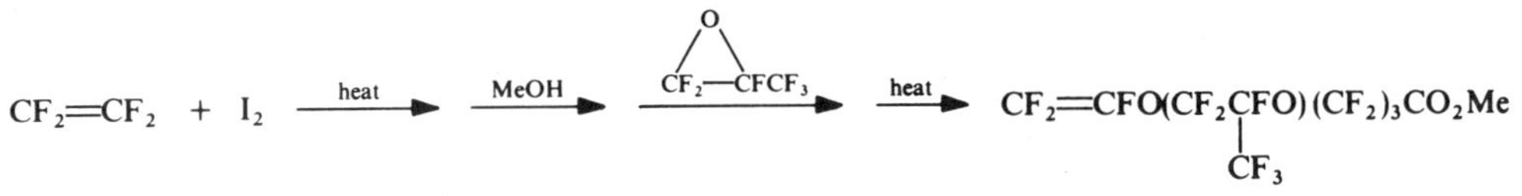

$$CF_2{=}CF_2 \;+\; I_2 \xrightarrow{\text{heat}} \xrightarrow{\text{MeOH}} \xrightarrow{\overset{O}{\overset{\triangle}{CF_2-CFCF_3}}} \xrightarrow{\text{heat}} CF_2{=}CFO(CF_2\underset{\underset{CF_3}{|}}{C}FO)(CF_2)_3CO_2Me$$

Scheme 5

25.2.5 Halato-telechelic Polymers

The synthesis of low molecular weight difunctional carboxyl-terminated butadiene-based polymers *via* either free radical or anionic polymerization is well established. Teyssie and co-workers[30-37] have converted such carboxyl-terminated polymers to salt forms, which they refer to as halato-telechelic polymers, by neutralization with metal alkoxides in appropriate solvents. The quantitative removal of low molecular weight reaction products is necessary to drive the reaction to completion and fully realize the ionomeric properties of these materials.

Telechelic poly(isobutylenesulfonate) ionomers have recently been reported,[38-43] based on either linear telechelic polyisobutylene dialkenes (**2**) or star-shaped trialkenes (**3**). Sulfonation is carried out in hexane solution at room temperature with excess acetyl sulfate generated *in situ* by the addition of sulfuric acid to acetic anhydride. After the reaction, neutralization is accomplished in tetrahydrofuran solution with ethanolic sodium hydroxide. Model compounds and proton NMR measurements have been used to verify the sulfonation process.

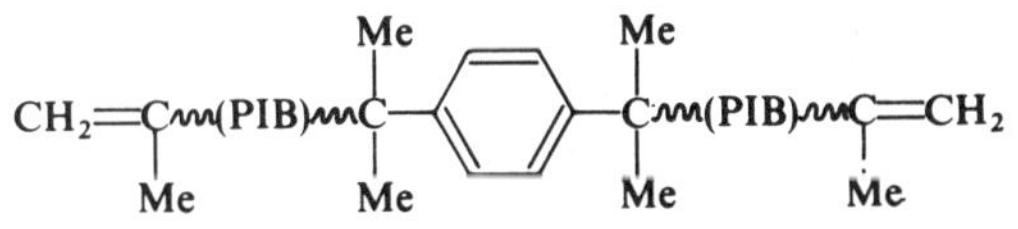

(2) PIB = polyisobutylene

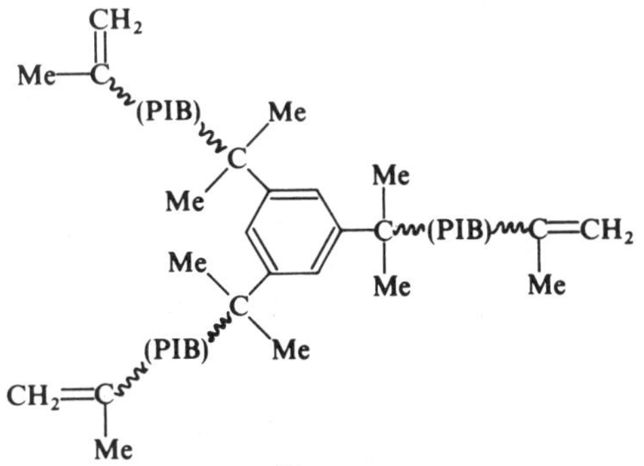

(3)

These five materials are representative of the wide variety of ionomers which have emerged in recent years. A summary of available ionic elastomers or flexible plastics is given in Table 2.

25.3 STRUCTURE OF IONOMERS

25.3.1 Theory

As with any material, knowledge of the underlying structure of ionomers is necessary in order to understand and optimize their properties. In the case of ionomers however, microscopic structure is

Table 2 Commercial Ionomer Systems

Polymer system	Tradename	Comment
Copoly(ethylene/methacrylic acid)	Surlyn (Du Pont)	modified thermoplastic
Copoly(butadiene/acrylic acid)	Hycar (Goodrich)	high green strength elastomer
Perfluorsulfonate ionomer	Nafion (Du Pont)	multiple membrane uses
Perfluorocarboxylate ionomer	Flemion (Asahi Glass)	chloralkali membranes
Telechelic polybutadiene ionomer	Hycar (Goodrich)	specialty uses
Chlorosulfonated polyethylene	Hypalon (Du Pont)	elastomeric sheeting
Sulfonated ethylene/propylene/diene terpolymer	Ionic elastomer (Uniroyal)	thermoplastic elastomer

difficult to assess. Within the same polymer molecule are combined the nonpolar covalent nature of the hydrocarbon backbone and the polar ionic nature of the chemically attached salt groups.

The first attempt to predict the spatial arrangement of the salt groups in ionomers was the theoretical work of Eisenberg.[44] In this work, the fundamental structural unit is assumed to be the contact ion pair where the anion and cation are separated from each other only by their ionic radii. This assumption is shown to be a reasonable one since the work required to separate the ion pair in media of low dielectric constant is nearly two orders of magnitude greater than the available disruptive thermal energy. Associated ion pairs, triplets, quartets, *etc.* in which the charges are as close to one another as possible are defined as multiplets. Three factors which govern the formation of multiplets are considered: the dimensions of polymer chains and of ion pairs, the tension on the chains resulting from ionic aggregation, and the electrostatic energy released upon multiplet formation. The possibility of multiplets forming clusters in which the multiples are separated by nonionic material is also considered. Thus a single ion pair can be regarded as the smallest possible multiplet while the maximum multiplet size is governed by its geometry and the restriction that each ion pair must be attached to a hydrocarbon backbone. Clusters may be formed by the association of multiplets. This association is favored by electrostatic interactions between multiplets and opposed by the retractive elastic forces of the backbone chains.

The calculation of the maximum possible size for the multiplets is performed with the assumption of a spherical geometry where the attached hydrocarbon chains are restricted to the multiplet's surface. It then follows that the multiplet radius r_m is given by

$$r_m = 3v_p/s \tag{1}$$

where v_p is the volume of an ion pair and s is the area of the hydrocarbon chain in contact with the surface. For an ethylene/sodium methacrylate polymer, this leads to $r_m = 3$ Å, which suggests eight ion pairs per multiplet. Hence it is concluded that the multiplets cannot be very large if they are spherical, though such a conclusion does not hold for other geometries. In particular, if the multiplets are lamellar, there is essentially no limit to their size due to steric restrictions alone.

The formation of clusters will be driven by electrostatic interactions between multiplets. The nature of these interactions is governed by cluster structure. It is envisioned that cluster formation proceeds as follows. A multiplet of maximum size, approximately eight ion pairs, is completely coated with a hydrocarbon 'skin'. Therefore it is impossible for another multiplet to approach to a distance less than the thickness of a hydrocarbon chain, even though the multiplet of maximum size is expected to attract other multiplets through electrostatic interactions. Thus the cluster consists of a central core composed of a multiplet of maximum size surrounded at a distance by other multiplets of various sizes from ion pairs upwards. The size of the cluster is limited by the elastic forces arising from the backbone chain, which tend to pull the cluster apart.

The calculation of the electrostatic force is based on the assumption that energy is released when multiplets aggregate to form a cluster, the work depending on the dielectric constant of the medium and the geometry of the cluster. The opposing elastic forces are calculated on the basis of rubber elasticity theory. Since the elastic forces are proportional to temperature and the electrostatic forces vary only slightly with temperature, it is apparent that these two forces must balance at some critical temperature T_c above which clusters become unstable. Furthermore, it is suggested that a critical multiplet concentration for the formation of clusters exists since the elastic forces will be greater than the electrostatic forces for small ionic concentrations.

The result of Eisenberg's calculations of the number of ion pairs n per cluster is

$$n = \rho \frac{N_A}{M} \left[\frac{4l^2}{3kT_c} \left(\frac{\bar{h}^2}{\bar{h}_0^2} \right)^2 \frac{M}{M_0} \frac{k'}{\kappa} \frac{e^2}{4\pi\varepsilon_0 d} + 2 \left(\frac{M_0 M}{\rho N_A} \right)^{2/3} \right]^{3/2} \tag{2}$$

where ρ is the density, N_A is Avagadro's number, M is the average molecular weight of the chain between pendant acid groups, M_0 is the average molecular weight per chain repeat unit, l is the length of a backbone bond, $\bar{h}^2$ is the mean square end-to-end distance for the free chain, $\bar{h}_0^2$ is the mean square end-to-end distance calculated for a freely jointed chain, κ is the dielectric constant, $(1/4\pi\varepsilon_0)^{-1} = 1$ dyne cm^2 statcoulomb^{-1} (1 dyne $= 10^{-5}$ N), e is the electronic charge in an ion pair, d is the distance between the centers of positive and negative charge in an ion pair and k' is a parameter related to the electrostatic energy per ion pair released upon cluster collapse. As mentioned, the magnitude of k' depends on the cluster geometry and so a particular cluster structure must be assumed in order to calculate n.

A similar theory of multiplets and clusters has been developed by Dreyfus.[45] Again starting with the assumption of contact ion pairs, multiplets are postulated to exist with a maximum size determined by the steric hindrance of the surrounding nonionic coating. The three main contributions to the free energy controlling cluster formation are: (1) the residual electrostatic energy between multiplets; (2) the steric repulsion between monomers; and (3) the entropy needed to deform the coils from their natural conformation. On this basis, the radius ρ_c of the cluster and the number of ion pairs per cluster η_c are calculated as

$$\rho_c = l \left(\frac{e^2}{\kappa d k_B T} \right)^{3/(2j+1)} \frac{d^3}{jv} \tag{3}$$

$$\eta_c = \frac{2\pi d^6}{j^2 \sigma^3} \left(\frac{e^2}{\kappa d k_B T} \right)^{6/(2j+1)} \tag{4}$$

where the polymer chains are modelled as freely jointed rods of length l and cross-sectional area σ, v is the volume of one monomer and j is the number of ion pairs in a multiplet. Other symbols have the same meaning as previously defined. In the case of a polyethylene matrix, $j = 2$ is shown to be the most probable value. Using this value of j leads to the following scaling predictions

$$\rho_c \sim d^{2.4} \sigma^{-1} \tag{5}$$

$$\eta_c \sim d^{4.8} \sigma^{-3} \tag{6}$$

Dreyfus has also considered the spatial arrangement of the clusters. In order to deposit their charges inside the clusters, the segments of the chains have to be somewhat extended. This constraint is minimized by choosing a structure with the least distance between first neighbors for a given density of lattice sites. The diamond lattice is one simple structure which meets this criterion. Assuming such a lattice and requiring the conservation of charge leads to the following prediction for the distance D between first neighbor clusters

$$D \doteq 1.6 (\rho_c^2 l)^{1/3} c^{-1/3} \tag{7}$$

where c is the molar concentration of charged monomers.

Although the distance D between clusters is so large that no significant potential can have an influence on their positions, a local order is nonetheless predicted. This surprising prediction is based on the fact that the clusters are linked by many segments whose average length cannot be easily extended to accomodate changes in D. This results in an entropic chain tension which holds the clusters locally on a regular lattice. Perhaps the most interesting aspect of this hypercrystal model is its implications for solvation. In the process of hydration there is a strong tendency for the clusters to collapse into channels which, in the present model, would be along the tetrahedral bonds of the diamond lattice. The resulting three dimensional network is of great significance in understanding the cationic transport properties of ionomers.

The transport properties of perfluorinated ionomers are of particular interest due to their use as membrane separators in chloralkali cells. Gierke and Hsu[6] have developed a 'cluster network' model for these systems which suggests that the ionic clusters are inverted micellar structures. In this model, the absorbed water phase is predicted to separate into approximately spherical domains and the ion-exchange groups are near the interface, probably imbedded in the aqueous phase. Based on water transport measurements, it is suggested that these clusters are connected by short narrow channels. The resulting structure allows the sulfonic acid sites to be hydrated and at the same time minimizes the unfavorable interactions between water and the fluorocarbon matrix. Consistent with this model, ion transport in perfluorosulfonated membranes is controlled by percolation which implies that connectivity of the ionic clusters is critical.

An additional aspect of ionomer structure is the conformational behavior of the hydrocarbon chains. Eisenberg originally made the simplifying assumption that on average the chains would not undergo dimensional changes as a result of clustering of the ionic species. This restriction was later removed by Forsman[46] who showed that chain dimensions must actually increase as a result of ionic association, though the amount of such expansion will depend on the shape and symmetry of the cluster. The results of these calculations are embodied in equation (8):

$$\alpha^2 = \frac{\langle L^2 \rangle}{\langle L_0^2 \rangle} = 1 + \left(\frac{m}{cN_A}\right)^{2/3} \left(\frac{f}{M_0}\right)^{1/3} \left(\frac{1}{2k}\right)(1 + \psi(vf)^{2/3}) \qquad (8)$$

where α is the expansion due to clustering, m is the number of interacting groups in a cluster, c is the concentration, k is the proportionality factor between the unperturbed mean end-to-end distance and the molecular weight, f is the fraction of repeat units participating in cluster formation, v is the volume fraction of polymer and ψ is a dimensionless factor determined by the geometry of the clusters. This analysis assumes equally spaced interacting groups along a Gaussian backbone chain and clusters composed of a constant number of nearest-neighbor groups. These theoretical predictions of chain expansion have been confirmed by experimental measurements[47,48] on sodium salts of lightly sulfonated polystyrene.

The preceding theories comprise an excellent first approach to the problem of ionomer structure by choosing solutions of salts in media of low dielectric constant as model systems. As Eisenberg has pointed out, experimental and theoretical studies of such systems have shown that multiplets definitely exist. However, it may prove to be the case that the use of the contact ion pair as the fundamental structure unit is an oversimplification. More sophisticated theoretical treatment is clearly needed in the area of ionomer structure. It is possible that the coordinating tendency of the metal ions is of overriding importance and that the fundamental structural units consist of metal ions coordinated to an appropriate number of anions. Certainly steric considerations severely limit the size of multiplets if they are of spherical geometry and if they are close packed. A variety of model geometries are discussed in the following section. It is equally important, and a great contribution of Eisenberg's work, to note that the steric limitations are relaxed if intervening hydrocarbon material is allowed in the structure. There can be little doubt that the size of the cluster is determined largely by the balance between electrostatic and elastic forces.

25.3.2 Morphological Studies and Models

One of the most useful techniques for studying microscopic structure is X-ray diffraction. In the case of ionomers, X-ray scattering has proven particularly important in attempts at structure determination. The value of this technique can be seen in Figure 1, where the X-ray scattering from low-density polyethylene and from copoly(ethylene/methacrylic acid) and its sodium salt are compared.[61] The presence of polyethylene-like crystallinity is apparent in all three samples, as is readily seen by the 110 and 200 reflections from the orthorhombic polyethylene unit cell. The crystallinities of the acid copolymer and ionomer are less than that of the parent polyethylene but are quite similar to each other. The ionomer contains a new feature however: a peak centered at approximately $2\theta = 4°$. This peak, which will be referred to as the ionic peak, appears to be a common feature of all ionomers regardless of the presence or absence of backbone crystallinity. In addition, the ionomer peak possesses the following characteristics: (1) the ionomer peak occurs in all ionomers, regardless of the nature of the cation; it is observed with lithium counterions as well as heavy metals, divalent or trivalent cations, and quaternary ammonium ions; (2) both the magnitude and the location of the ionomer peak are dependent on the nature of the cation; thus the peak occurs at lower angles for cesium cations at a given concentration than for corresponding lithium cations; in addition, the intensity of the peak is several times greater for cesium than for lithium; (3) the ionomer peak is relatively insensitive to temperature; it was found that the peak for the material shown in Figure 1 persisted to at least $300°C$; and (4) the ionomer peak is destroyed or moved to lower angles when the ionomer is saturated with water. The scattering in the vicinity of the peak at $2\theta = 20°$ of the water-saturated sample is different from that of the acid copolymer.

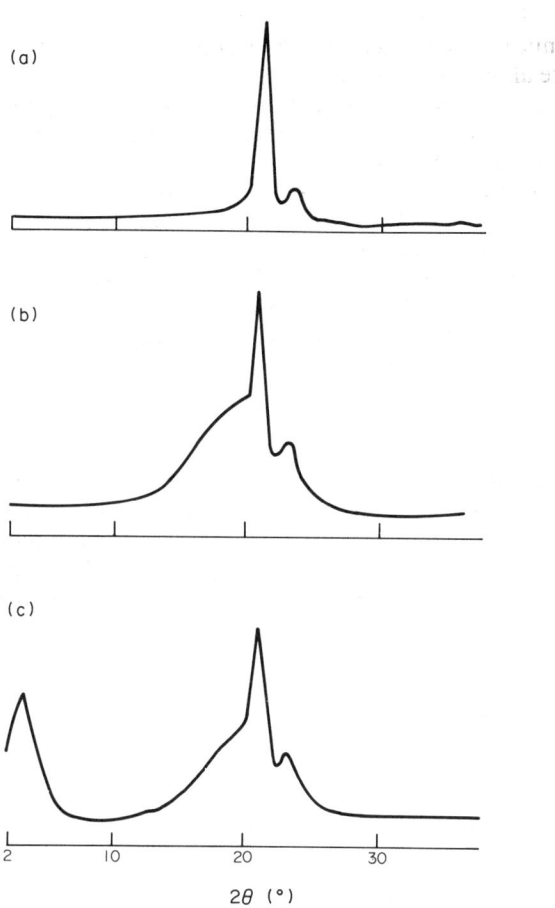

Figure 1 X-ray diffraction scans of (a) low density polyethylene, (b) copoly(ethylene/methacrylic acid) (5.8 mol% acid), and (c) the 100% neutralized sodium salt of (b) (reproduced by permission of the American Chemical Society from F. C. Wilson, R. Longworth and D. J. Vaughan, *Polym. Prepr., Am. Chem. Soc., Div. Polym. Chem.*, 1968, **9**, 505)

It is well known that the interpretation of scattering data is model dependent. The procedure is to assume a reasonable model, fit the experimentally observed data to it and then deduce model parameters from the best fit. Several models for the distribution of salt groups in ionomers have been proposed, based mainly on analysis of the ionomer peak. Those appearing up to 1979 are summarized in ref. 4. The existing models are effectively based on two different approaches: either (1) that the ionomer peak arises from structure within a scattering entity, or (2) that the peak is due to interference effects between scatterers.

Representative of the first approach is the 'shell–core' model[49] originally proposed in 1974 and later elaborated.[50, 51] In essence, the shell–core model postulates that in the dry state a cluster of about 0.1 nm in radius is shielded from surrounding unincorporated matrix ions by a shell of hydrocarbon chains. This is shown schematically in Figure 2. The surrounding matrix ions will be attracted to the cluster by electrostatic forces but cannot approach more closely than the outside of the hydrocarbon shell. This mechanism establishes a preferred distance between the cluster and the matrix ions. This distance is assumed to be of the order of 2 nm and accounts for the spacing of the ionomer peak. As a consequence of swelling and deformation studies, this model has been refined to allow for either ion-depleted zones or lamellar geometries. The assumption of lamellar structures proves particularly interesting as it allows for nonspherical ionic domains which may extend over large regions of the material.

Figure 3 schematically illustrates the interference model where it is assumed that the ionomer peak arises from a preferred interparticle distance. Recently, Yarusso and Cooper[52] have proposed an interpretation of the ionomer peak which is based on the liquid-like scattering from hard spheres originally described by Fournet.[53] The Fournet model is quantitatively capable of fitting the X-ray peak from sulfonated polystyrene ionomers. In the case of zinc-neutralized material, about half of the ionic groups were found to aggregate into well ordered domains ('clusters') with the remainder

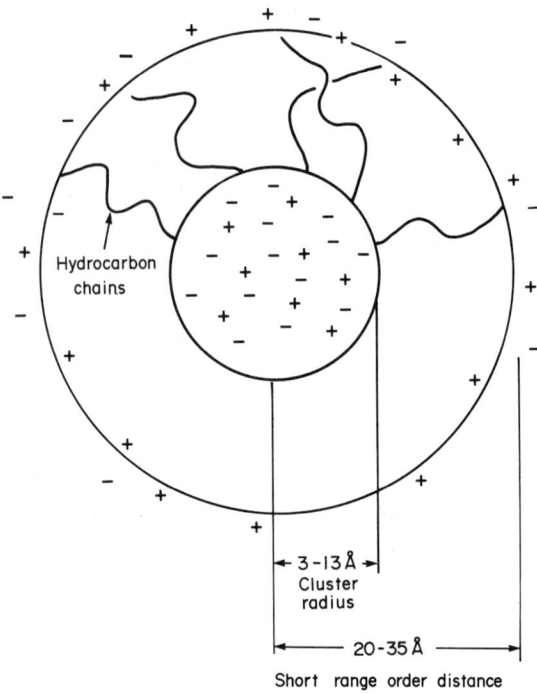

Figure 2 'Shell–core' scattering model for small angle X-ray scattering from ionomers

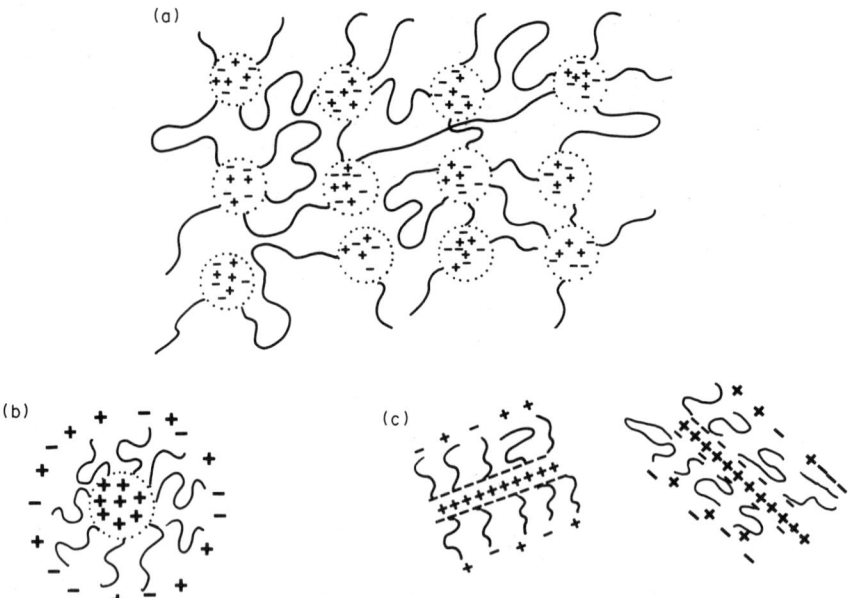

Figure 3 (a) Intercluster (lattice) scattering model for small angle X-ray scattering from ionomers compared with (b) shell–core and (c) lamellar intracluster models

dispersed in the matrix ('multiplets'). The clusters had an estimated diameter of 2.0 nm and their distance of closest approach was found to be 3.4 nm center-to-center. It is clear that this model, although based on quite different physical principles, yields structural parameters which are very similar to whose obtained from the shell–core model and which are consistent with what may be termed the 'standard' model for ionomer structure of multiplets and clusters.

It is enticing to attempt direct imaging of the ionic domains within ionomers by electron microscopy. Many of the electron microscopy studies prior to 1979 have been reviewed.[4] Due to the small size of the clusters and the difficulty encountered in sample preparation, it may be stated that none of these studies make a convincing case for the direct imaging of ionic clusters. The entire problem was reexamined in 1980.[54] Ionomers based on copoly(ethylene/methacrylic acid), sulfonated polystyrenes, sulfonated polypentenamers and sulfonated poly(EPDM) were examined. The transfer theory of imaging was used to interpret the images of both solvent-cast and microtomed thin films. In the case of microtomed sulfonated poly(EPDM), phase-separated regions of size 300 nm were found. Osmium tetroxide staining showed domains averaging less than 3 nm in size within these regions. Unfortunately, the section thickness prohibited an accurate determination of the size distribution of the detailed shape of these domains.

In summary, there is a considerable body of experimental and theoretical evidence that salt groups in ionomers exist in two different environments, termed multiplets and clusters. The multiplets are thought to consist of small numbers of ion dipoles (perhaps up to six or eight) which associate to form higher multipoles. These multiplets are dispersed in the hydrocarbon matrix and are not phase-separated from it. The clusters are thought to be small (<5 nm) microphase-separated regions which are rich in ion pairs but also contain considerable quantities of hydrocarbon. The proportion of salt groups which resides in either of the two environments is determined by the nature of the backbone, the total concentration of salt groups, their distribution along the backbone and their chemical nature. The details of the local structure of clusters is not yet known and neither is the mechanism by which the clusters interact with low-molecular weight impurities such as water. Up to the present time, attempts to image the clusters by electron microscopy have been only partially successful at best.

25.4 PROPERTIES OF IONOMERS

Having described the current understanding of ionomer structure, some general properties of ionomers are now considered. These properties naturally reflect the underlying structural detail. The effects of the inclusion of salt groups will be surveyed in the areas of melt rheology, dynamic mechanical properties and finally dielectric behavior. For clarity, only one family of ionomers will be discussed, materials based on (ethylene/carboxylic acid) copolymers. The effects described are generally typical of other ionomeric materials as well, though the exact details will vary from system to system.

25.4.1 Melt Rheology

Several groups[55-61] have studied the response of (ethylene/carboxylic acid) copolymers to an applied stress or strain. This type of measurement has been the principle method for determining the unusual features of ionomer behavior. The results of such studies provide clear evidence for the presence of ionic clusters.

By utilizing the principles of linear viscoelasticity,[62,63] it is possible to correlate mechanical behavior with molecular structure. The quantity of interest is the complex shear modulus G^* given as

$$G^* = G' + iG'' \tag{9}$$

where G' is the shear storage modulus and G'' is the shear loss modulus. These are related by the loss tangent

$$\tan \delta = G''/G' \tag{10}$$

In the region of a molecular relaxation, the complex shear modulus will drop dramatically, while the loss tangent and shear loss modulus will exhibit maxima. For materials above their melting or flow temperatures, the dynamic viscosity η' is often measured.

These measurements are performed at sufficiently low stress that the Boltzmann superposition principle is valid, that is if one stress σ_0 is applied at time t_0 and a second stress σ is added at time t then the resulting total strain $\gamma(t)$ is

$$\gamma(t) = \frac{\sigma_0}{G'(t_0)} + \frac{\sigma}{G'(t-t_0)} \tag{11}$$

It is also necessary to consider the effects of temperature on the frequency-dependent molecular-relaxation spectrum. Briefly, if the effect of an increase in temperature on the relaxation spectrum is to simply shift the spectrum without change of shape to higher frequencies, then the so-called time–temperature superposition principle applies. This is generally the case when only one type of mechanism contributes to the relaxation behavior.

When ethylene-based copolymers are stressed above the melting temperature of polyethylene, it has generally been found that the viscosity and the activation energy for viscous flow are greater for the acid copolymer than for the analogous polyethylene or ester derivative, that is when compared at identical temperatures, viscosity–shear rate curves for an acid copolymer exhibit an enhanced viscosity and greater shear rate dependency. These results were initially understood in terms of hydrogen bonding. It was suggested that intermolecular hydrogen bonds between carboxylic acid groups could act as temporary quasi-crosslinks and substantially affect the flow properties. Subsequent comparisons of steady and dynamic shear results, as well as the construction of master viscosity curves, have changed this interpretation.

Excellent time–temperature superposition is obtained over a large temperature range for the acid copolymers, as seen in Figure 4. It would be expected for superposition to be impossible if the rheological response were determined by an equilibrium concentration of hydrogen-bonded quasi-crosslinks. This expectation arises from the fact that the equilibrium concentration of hydrogen-bonded dimers changes greatly with temperature and should give rise to differently shaped viscosity–shear rate curves at different temperatures. The excellent superposition of the data at different temperatures indicates that the rheological response of the melt is not dependent on the equilibrium number of hydrogen-bonded dimers. Therefore the association and dissociation of hydrogen bonds occurs more rapidly than the time scale of these rheological measurements.

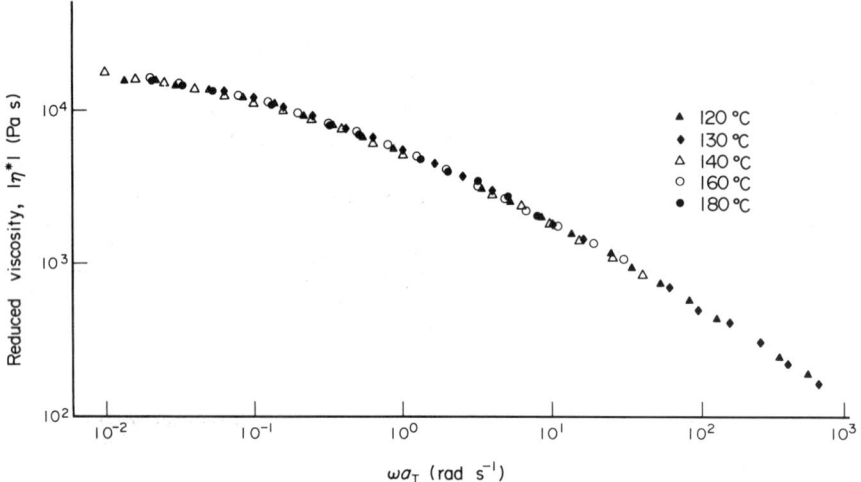

Figure 4 Viscosity master curve [absolute value of the complex viscosity ($|\eta'|$) *vs.* frequency × shift factor (ωa_T, rad s^{-1})] for copoly(ethylene/methacrylic acid (3.5 mol % acid, $\bar{M}_w = 86\,000$, $\bar{M}_n = 9400$) (reproduced by permission of John Wiley & Sons, Inc. from T. R. Earnest, Jr., and W. J. MacKnight, *J. Polym. Sci., Polym. Phys. Ed.*, 1978, **16**, 143)

Differences in the rheological response of the acid and ester copolymers can be successfully interpreted on the basis of differences in their glass transition temperatures. This is done using a WLF (Williams–Landel–Ferry) equation

$$\log(a_T) = \frac{-A(T - T')}{B + (T - T')} \tag{12}$$

where a_T is the shift factor at temperature T, T' is an arbitrary reference temperature and A, B are constants. By choosing reference temperatures for the acid and ester copolymers which differ by the same amount as their glass transition temperatures, the rheological data can be well described. The shapes of the master curves for the acid and ester derivatives at the same reference temperature are identical and can be made to superimpose by simple horizontal shifts. These results indicate that differences in the melt rheology between ester and acid copolymers can be removed by simple temperature or frequency shifts and that differences in viscous flow activation energy are due to differences in glass transition temperatures.

The low shear viscosity of the salts of (ethylene/carboxylic acid) copolymers greatly increases with both degree of neutralization and ionic content. An example of this increase is shown in Figures 5 and 6, where results for an acid copolymer are compared with those for its sodium salt. In general, there is little difference in melt rheology due to cation type; the largest effects are due to neutralization. The activation energy for viscous flow increases with salt content, but is only slightly higher than for the acid copolymer form when calculated at constant shear rate and when shear-thinning effects are considered. Activation energies calculated at constant stress are significantly higher for the salts than for the parent acid copolymers.

For the salt forms, there is no correspondence between the steady state apparent viscosity and the absolute value of the complex viscosity as determined in dynamic shear. As can be seen in Figure 7, the steady shear viscosities are significantly higher than the corresponding dynamic viscosities. It is important to note that this particular material is neutralized to about 70%. In this case, the two

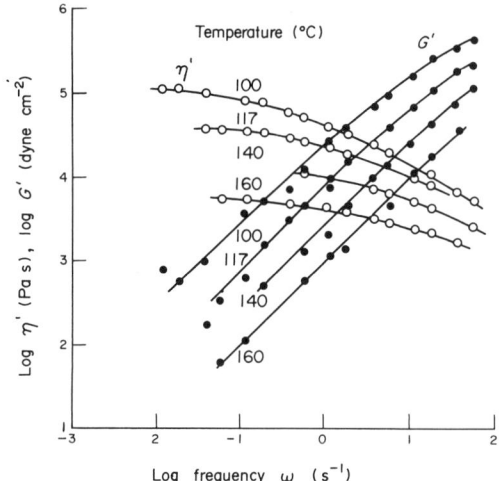

Figure 5 Dynamic viscosity η' and modulus G' of coply(ethylene/methacrylic acid) (95.9–4.1 mol %) as a function of angular frequency (1 dyne $= 10^{-5}$ N)

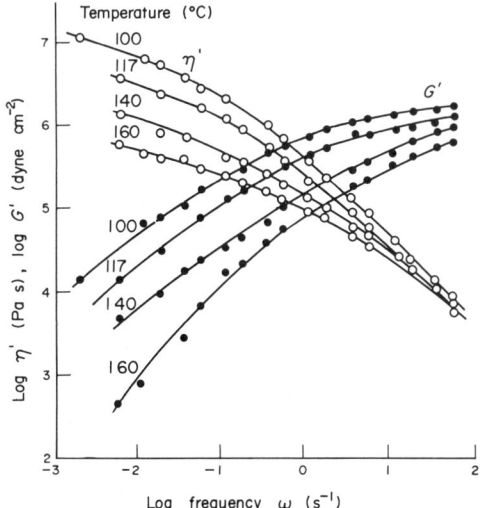

Figure 6 Dynamic viscosity η' and modulus G' of the sodium salt of copoly(ethylene/methacrylic acid) (95.9–4.1 mol %) as a function of angular frequency (1 dyne $= 10^{-5}$ N) (reproduced by permission of John Wiley & Sons, Inc. from K. Sakamoto, W. J. MacKnight and R. S. Porter, *J. Polym. Sci., Part A-2*, 1970, **8**, 277)

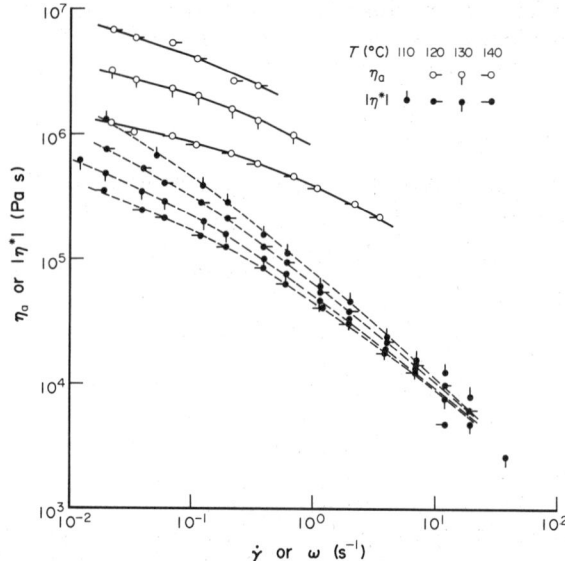

Figure 7 Apparent viscosity *vs.* shear rate and the absolute value of the complex viscosity *vs.* frequency for the 70% neutralized sodium salt of copoly(ethylene/methacrylic acid) (4.1 mol % acid)

experiments are sensitive to different types of segmental motion. Steady shear will involve large scale cooperative segmental motions which allow the chains to slip past one another. Hence ionic associations have to be broken continuously and presumably reformed in order for flow to occur. The dynamic viscosity, on the other hand, reflects both long range and short range segmental motions depending on the frequency of measurement. The results in Figure 7 imply that ionic aggregates influence long range motions more than short range motions. It is clear that any long range segmental motion must be accompanied by dissociation of an ion pair from an aggregate or cluster. This mechanism may be the rate-determining step for the flow of ionomers and certainly plays an important role in determining their melt–rheological properties.

Unlike the ester and acid derivatives, the salts of copoly(ethylene/methacrylic acid) do not yield master curves. Attempts to superimpose the storage modulus/frequency data lead to pseudo-master curves at best, as shown in Figure 8. The low-frequency storage modulus is observed to decrease with increasing temperature. This behavior can be explained in the following way. At short times (high frequencies) the viscoelastic response is dominated by deformation occurring in the hydrocarbon matrix. As the frequency is reduced (longer times), the viscoelastic response of the ionic phase

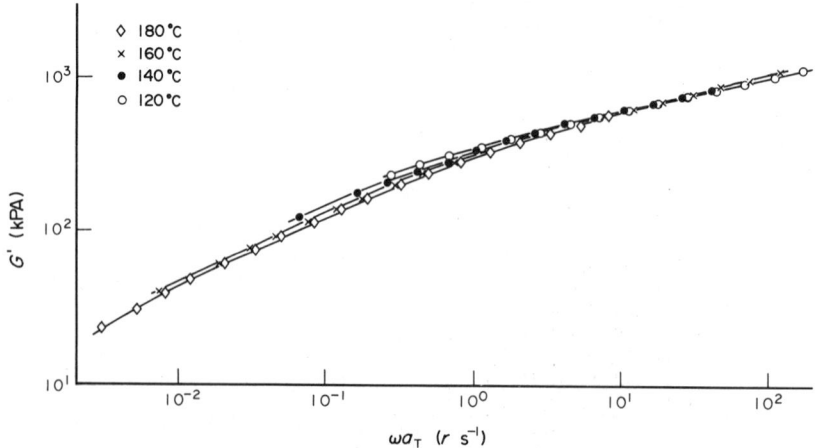

Figure 8 Pseudo-master curve of the storage modulus/frequency data for the 70% neutralized sodium salt of copoly(ethylene/methacrylic acid (3.5 mol % acid) at the temperatures indicated (T. R. Earnest, Jr., Ph.D. Thesis, University of Massachusetts, 1978)

becomes more important. The temperature dependence of the modulus on the structure and viscoelastic properties of the ionic phase therefore decreases as temperature increases.

25.4.2 Dynamic Mechanical Properties

Ionomers may also be examined at lower temperatures where less mobility is present.[64-71] In general the response for both the parent acid copolymers and their salt derivatives resembles that of low density polyethylene. At the lowest temperatures there is a large loss peak, the γ relaxation occurring near $-120\,°C$. This relaxation can be decomposed into two peaks for both the acid copolymer and its salt derivatives. The relaxation strength of the lower temperature peak correlates well with the degree of crystallinity, while the higher temperature peak is proportional in size to the amount of amorphous material. Since the degree of crystallinity in ethylene/carboxylic acid copolymers and ionomers is very low, the γ relaxation is perhaps best described by a crankshaft motion of short hydrocarbon segments in the amorphous phase.

In the acid copolymers, a relaxation occurs at a temperature between 0 and $50\,°C$, increasing in temperature with increasing fraction of acid comonomer. This peak, labelled the β' peak, is assigned to micro-Brownian segmental motion accompanying the glass transition. The onset of this motion occurs at a higher temperature in the acid copolymer than in branched polyethylene. This increase in the β' relaxation temperature can be attributed to the crosslinking effect of dimerized carboxyl groups.

As the acid groups are neutralized, the intensity of the β' peak diminishes and a new relaxation, labelled β, appears at lower temperatures between -20 and $0\,°C$. Figure 9 shows the effect of neutralization to the sodium salt on the β and β' peaks for annealed copoly(ethylene/methacrylic acid). As more and more acid groups are neutralized, the β' peak disappears completely and the

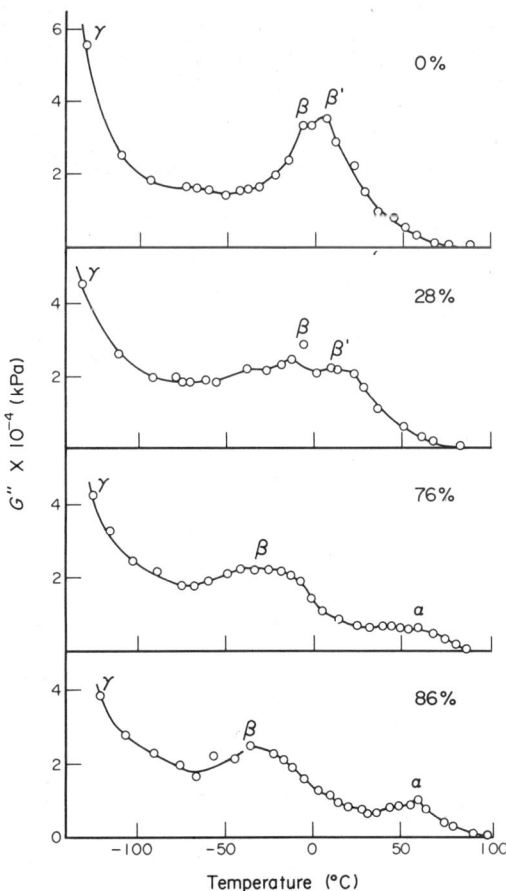

Figure 9 Temperature dependence of G'' for annealed copoly(ethylene/methacrylic acid) (4.1 mol % acid) neutralized to various degrees with acid as indicated (W. J. MacKnight, L. W. McKenna and B. E. Read, *J. Appl. Phys.*, 1967, **38**, 4208)

β relaxation shifts to lower temperatures, approaching those of the β relaxation in branched polyethylene. The β peak is therefore assigned to a relaxation occuring in the amorphous branched polyethylene phase from which most of the ionic material has been excluded. It should be noted that the relaxation strength and temperature of the β peak will be dependent on both the degree of crystallinity and the perfection of phase separation. As a result, the relaxation spectrum is expected to be very sensitive to the material's thermal history. Water saturation, which destroys the previously discussed ionomer X-ray peak, shifts the β relaxation to lower temperatures and greatly enhances its magnitude. These results indicate that water breaks up the ionic phase, thereby creating more amorphous phase.

In acid copolymers containing relatively small amounts of acid comonomer, a high temperature relaxation occurs which is related to the crystalline α peak in polyethylene. At higher concentrations (and therefore lower crystallinities), this relaxation is not observed. As seen in Figure 9, a new relaxation, also labelled α, becomes observable as the acid groups are neutralized. This relaxation occurs near 50 °C and moves to higher temperatures as a function of neutralization and of increasing ionic content. Neutralization with divalent cations causes a further increase in the temperature and magnitude of this relaxation. The α peak has been assigned to motions involving the ionic phase and can be considered as a softening temperature of the ionic domains.

On the basis of these measurements and related physicochemical studies, it is possible to process a structure for ethylene-based ionomers. The un-ionized acid copolymer is thought to consist of two phases: a crystalline polyethylene phase and an amorphous phase consisting of polyethylene crosslinked by hydrogen-bonded carboxylic dimers. The ionized copolymers exhibit three distinct phases: a crystalline polyethylene phase, an amorphous polyethylene phase, and a dispersed ionic phase consisting of ionic domains.

25.4.3   Dielectric Properties

Several studies of the dielectric properties of copoly(ethylene/methacrylic acid) and their salts have been conducted.[72-75] These measurements are analogous to those described in the preceding section except that in this case an AC electrical field is passed through the material rather than a sinusoidal stress field. The accessible range of electrical frequencies is of course different from the mechanical frequencies generally used. As the polar components of the material respond to the applied field, a storage and loss component are recorded and their ratio provides the dielectric loss tangent.

In general, the same features observed mechanically are also found in the dielectric experiments. The β' relaxation of the parent acid copolymer is observed, indicating that motions of both hydrocarbon and dipolar groups are involved in this process. In the ionomers, the dielectric α and β peaks correlate with the α and β mechanical relaxations discussed earlier. The relatively large magnitude for the α loss peak indicates that most of the salt groups are involved in this process. Curvature of the frequency–temperature plot for the α relaxation supports its assignment as a glass transition of the ionic phase. Similarly, the frequency–temperature plot for the β process is curved, suggesting that this relaxation is associated with the glass transition of the hydrocarbon phase, which is rendered dielectrically active by the inclusion of carboxylic acid or extraneous carbonyl groups.

The adsorption of water has a tremendous effect on dielectric properties. In the case of ionomer salts, the α peak shifts to lower temperatures due to a plasticization effect on the ionic phase and an additional peak appears near -43 °C whose magnitude is approximately proportional to the content of water but whose position is only slightly affected by increasing water content. Model calculations of the relaxation strength using the dipole moment of water show that motions of the water molecule itself can account for this low-temperature peak.

The effects of varying degrees of neutralization on the dielectric properties of (ethylene/carboxylic acid) copolymers are shown in Figure 10. The same general trends are observed as in the mechanical studies depicted in Figure 9. As the degree of neutralization increases, the β' relaxation diminishes while the β peak increases in magnitude and moves to lower temperatures. The α relaxation, however, increases in magnitude and moves to higher temperatures as the degree of neutralization increases.

There are several reasons why quantitative analysis of the dielectric–relaxation behavior of ionomers is difficult. As discussed above, there is a significant effect of even the smallest amounts of water on the position and magnitude of the α ionic phase relaxation. The removal of the last traces of water is very difficult due to the possibility of the coordination of water to the ionic species.

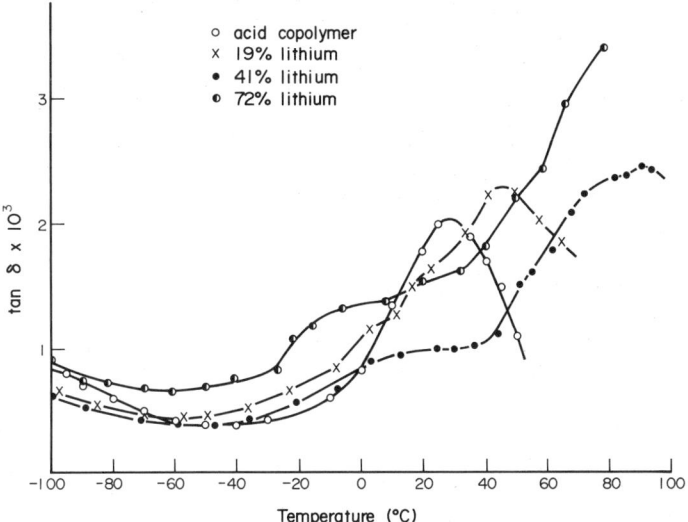

Figure 10 Temperature dependence of dielectric loss factor for various lithium salts of copoly(ethylene/methacrylic acid) (4.1 mol % acid) neutralized to the degrees shown (reproduced by permission of John Wiley and Sons, Inc. from P. J. Phillips and W. J. MacKnight, *J. Polym. Sci., Part A-2*, 1970, **8**, 727)

Another factor contributing to the uncertainty in the interpretation of dielectric measurements is the possibility of space–charge effects. Migration of ionic impurities and of the cations themselves will result in large values of the dielectric constant as well as large values of tan δ. These effects cannot be subtracted directly from the observed dielectric response. In general though, the results of dielectric measurements are consistent with the mechanical results.

25.5 APPLICATIONS AND CURRENT RESEARCH

Most applications of ionomers make use of the ionic associations present in these materials. Changes in physical properties due to ionic aggregation are readily apparent in polymer melts and in elastomeric systems. For example, markedly enhanced elastomeric green strength and increased melt viscosity are characteristic of ionomer-based systems. In the case of polyethylene-based metal carboxylate ionomers, the improved melt properties are utilized in heat sealing for packaging applications. Other properties attributable to ionic aggregation in polyethylene-based ionomers are toughness and outstanding abrasion resistance, oil resistance and adhesion to a wide variety of substrates. These properties lead to a variety of applications such as packaging, molding of high impact objects, projective coatings, foams, adhesives and modifiers.

The interaction of various polar agents with the ionic groups and the ensuing property changes are unique to ionomer systems. This plasticization is important for ionic elastomers such as sulfonated poly(EPDM). The plasticization of ionic clusters by interaction with metal stearates is used to induce softening transitions. This plasticization process is required to achieve processibility of these thermoplastic elastomers. The presence of metal sulfonate groups provides strong crosslinks at ambient temperatures. The addition of a suitable polar additive such as zinc stearate causes these elastomers to become thermoplastic at elevated temperatures and melt processible. Metal sulfonated poly(EPDM) can be compounded with fillers, rubber-processing oils and selected polymers to make a variety of elastomeric materials. These find application in such areas as adhesives, impact modifiers, footwear and general rubber goods.

Perfluorinated ion exchange polymers derived from tetrafluoroethylene perfluoroethers and either sulfonic or carboxylic acids are highly resistant to harsh chemical environments and are used in a large number of applications, including membranes for caustic chlorine cells, dehydration units, electrochemical cells, fuel cells and catalysts for a variety of chemical reactions. The main advantage of a membrane process is the ability to produce a high concentration caustic solution without requiring energy-intensive evaporation steps. Use of perfluorinated membranes results in long separator lifetimes and high product purity with low power consumption.

Ionomers continue to be an active area of ongoing research. Some of the current areas of interest are new ionomer materials, ionomer blends and ionomer solutions. A recent development in the search for new ionomers is the modification of aromatic backbones. The carboxylation of poly(2,6-dimethyl-1,4-phenylene oxide) has recently been reported.[76] Cesium and sodium salts have been prepared as well as ester derivatives over a wide range of ionic contents. These materials show great promise for use in polymer blends. Another recently announced material is sulfonated poly(oxy-1,4-phenylene-carbonyl-1,4-phenylene).[77] The sodium salts of this material exhibit the characteristic ionomer X-ray peak. Due to its elevated glass transition temperature, this ionomer presents several possibilities for high temperature applications.

Pioneering work in the area of ionomer blends has been undertaken by Eisenberg and co-workers,[78] studying ion-pair interactions as a means of obtaining miscibility in blends of urethanes and styrene ionomers. In blends of copoly(styrene/sodium methacrylate) with urethanes containing quaternary ammonium salts in the hard segments, ion-pair/ion-pair interactions induce miscibility. Ion-pair/dipole interactions, by contrast, do not yield miscible blends. The properties of the resulting ionomer blends have also been reported.[79] Such blends are clearly of interest in the development of new polymeric materials.

Solution properties of ionomers are currently of great interest. Initial studies of the solution viscosity of copoly(styrene/sodium sulfonated styrene) in mixed solvents has revealed very unusual behavior.[80] At relatively low solute concentrations these solutions exhibit unexpected thickening behavior and an unusual viscosity–temperature dependence. In effect, the solution viscosity is observed to rise with increasing temperature, a phenomenon not observed with nonionic polymers. These phenomena are explained by a temperature-dependent interaction between the polar co-solvent and the sulfonate group. In media of low dielectric constant[81] the metal sulfonate groups are thought to be largely un-ionized in solution, suggesting that the solution viscosity is dominated by interactions between ion pairs rather than free ions. However, when these ionomers are dissoved in media of high dielectric constant, polyelectrolyte behavior is observed. Similar results have been observed in solutions of other ionomer systems.[82-84] Scattering measurements are being made on ionomer solutions[85-88] in an effort to establish their structure. The variation between associating solutions and polyelectrolyte behavior can be understood as a variation between ion pair and free ion interactions respectively. Analogs of these ionomer solutions are of technological interest for the recovery of petroleum. For example, sulfonate ionomers are very effective viscosifiers for oil-based drilling fluids designed specifically for high temperature applications.[89]

Clearly, ionomers present a promising and challenging area of science, ranging from novel synthetic and physical chemistry to new applications of materials with unique properties.

25.6 REFERENCES

1. L. Holliday (ed.), 'Ionic Polymers', Halstead–Wiley, New York, 1975.
2. A. Eisenberg and M. King (eds.), 'Ion-Containing Polymers', Academic Press, New York, 1977.
3. A. Eisenberg and F. E. Bailey, *ACS Symp. Ser.*, 1986, **302**.
4. W. J. MacKnight and T. R. Earnest, Jr., *J. Polym. Sci., Macromal. Rev.*, 1981, **16**, 41.
5. A. D. Wilson and H. J. Proser (eds.), 'Developments in Ionic Polymers — 1', Applied Science, London, 1983.
6. A. Eisenberg and H. J. Yeager, *ACS Symp. Ser.*, 1982, **180**.
7. E. P. Ottocka, *J. Macromol. Sci., Rev. Macromol. Chem.*, 1971, **5**, 275.
8. W. J. MacKnight and R. D. Lundberg, *Rubber Chem. Tech.*, 1984, **57**, 652.
9. Du Pont Technical Bulletin No. E-68761, 1985 (courtesy of R. Longworth).
10. H. S. Makowski, R. D. Lundberg, L. Westerman and J. Bock, *Polym. Prepr., Am. Chem. Soc., Div. Polym. Chem.*, 1978, **19**, 292.
11. R. Neumann, W. J. MacKnight and R. D. Lundberg, *Polym. Prepr., Am. Chem. Soc., Div. Polym. Chem.*, 1978, **19**, 298.
12. H. S. Makowski, R. D. Lundberg, L. Westerman and J. Bock, *Polym. Prepr., Am. Chem. Soc., Div. Polym. Chem.*, 1978, **19**, 304.
13. B. Siadat, R. D. Lundberg and R. W. Lenz, *Polym. Eng. Sci.*, 1980, **20**, 530.
14. K. Sanui, R. W. Lenz and W. J. MacKnight, *J. Polym. Sci., Polym. Chem. Ed.*, 1974, **12**, 1965.
15. K. Sanui, R. W. Lenz and W. J. MacKnight, *J. Polym. Sci., Polym. Lett. Ed.*, 1974, **11**, 427.
16. K. Sanui, W. J. MacKnight and R. W. Lenz, *Macromolecules*, 1974, **7**, 101.
17. D. Rahrig, C. Azuma and W. J. MacKnight, *J. Polym. Sci., Polym. Phys. Ed.*, 1978, **16**, 59.
18. D. Rahrig, W. J. MacKnight and R. W. Lenz, *Macromolecules*, 1979, **12**, 195.
19. C. Azuma and W. J. MacKnight, *J. Polym. Sci., Polym. Chem. Ed.*, 1978, **16**, 547.
20. H. Tanaka and W. J. MacKnight, *J. Polym. Sci., Polym. Chem. Ed.*, 1978, **16**, 2975.
21. H. S. Makowski and R. D. Lundberg, *Polym. Prepr., Am. Chem. Soc., Div. Polym. Chem.*, 1978, **19**, 287.
22. A. Eisenberg and M. Navaratil, *Macromolecules*, 1973, **6**, 604.
23. H. S. Makowski, R. D. Lundberg, L. Westerman and J. Bock, *ACS Monogr.*, 1980, **187**.
24. R. D. Lundberg, *Polym. Prepr., Am. Chem. Soc., Div. Polym. Chem.*, 1978, **19**, 455.
25. R. D. Lundberg and H. S. Makowski, *J. Polym. Sci., Polym. Phys. Ed.*, 1980, **18**, 1821.
26. R. D. Lundberg and H. S. Makowski, *Polym. Prepr., Am. Chem. Soc., Div. Polym. Chem.*, 1978, **19**, 287.
27. R. D. Lundberg, H. S. Makowski and L. Westerman, *Polym. Prepr., Am. Chem. Soc., Div. Polym. Chem.*, 1978, **19**, 310.

28. R. L. Dotson and K. E. Woodard, ref. 6, p. 311.
29. H. Ukihashi and M. Yamabe, ref. 6, p. 427.
30. G. Broze, R. Jérôme and P. Teyssié, *Macromolecules*, 1981, **14**, 224.
31. G. Broze, R. Jérôme, P. Teyssié and C. Marc, *Macromolecules*, 1982, **15**, 920.
32. G. Broze, R. Jérôme, P. Teyssié and C. Marc, *Macromolecules*, 1982, **15**, 1300.
33. G. Broze, R. Jérôme, P. Teyssié and C. Marc, *Macromolecules*, 1983, **16**, 996.
34. G. Broze, R, Jérôme and P. Teyssié, *J. Polym. Sci., Polym. Phys. Ed.*, 1983, **21**, 2205.
35. G. Broze, R. Jérôme, P. Teyssié and C. Marc, *Macromolecules*, 1983, **16**, 1771.
36. G. Broze, R. Jérôme and P. Teyssié, *J. Polym. Sci., Polym. Lett. Ed.*, 1983, **21**, 237.
37. R. Jérôme, J. Horrion, R. Fayt and P. Teyssié, *Macromolecules*, 1984, **17**, 2447.
38. J. P. Kennedy and R. F. Storey, *Org. Coat. Appl. Polym. Sci. Proc.*, 1982, **46**, 182.
39. Y. Mohajer, D. Tyagi, G. L. Wilkes, R. F. Storey and J. P. Kennedy, *Polym. Bull.*, 1982, **8**, 47.
40. S. Bagrodia, Y. Mohajer, G. L. Wilkes, R. Storey and J. P. Kennedy, *Polym. Bull.*, 1982, **8**, 281.
41. S. Bagrodia, G. L. Wilkes and J. P. Kennedy, *J. Rheol.*, 1983, **28**, 474.
42. Y. Mohajer, S. Bagrodia, G. L. Wilkes, R. F. Storey and J. P. Kennedy, *J. Appl. Polym. Sci.*, 1984, **29**, 1943.
43. G. Tant, G. L. Wilkes and J. P. Kennedy, *Polym. Prepr., Am. Chem. Soc., Div. Polym. Chem.*, 1985, **26**, 32.
44. A. Eisenberg, *Macromolecules*, 1970, **3**, 147.
45. B. Dreyfus, *Macromolecules*, 1985, **18**, 284.
46. W. C. Forsman, *Macromolecules*, 1982, **15**, 1032.
47. T. R. Earnest, Jr., J. S. Higgins, D. L. Handlin and W. J. MacKnight, *Macromolecules*, 1981, **14**, 192.
48. W. C. Forsman, W. J. MacKnight and J. S. Higgins, *Macromolecules*, 1984, **17**, 490.
49. W. J. MacKnight, W. P. Taggart and R. S. Stein, *J. Polym. Sci., Polym. Symp.*, 1974, **45**, 113.
50. E. J. Roche, R. S. Stein, T. P. Russell and W. J. MacKnight, *J. Polym. Sci., Polym. Phys. Ed.*, 1980, **18**, 1497.
51. M. Fujimura, T. Hashimoto and H. Kawai, *Macromolecules*, 1982, **15**, 136.
52. D. J. Yarusso and S. L. Cooper, *Macromolecules*, 1983, **16**, 1871.
53. G. Fournet, *Acta Crystallogr.*, 1951, **4**, 293.
54. D. L. Handlin, W. J. MacKnight and E. L. Thomas, *Macromolecules*, 1980, **14**, 795.
55. L. L. Blyler, Jr., *Rubber Chem. Technol.*, 1969, **42**, 823.
56. L. L. Blyler, Jr., and T. W. Haas, *J. Appl. Polym. Sci.*, 1969, **13**, 2721.
57. T. Limm and T. W. Haas, *Adv. Polym. Sci., Eng. Proc. Symp.*, 1972.
58. K. Sakamoto, W. J. MacKnight and R. S. Porter, *J. Polym. Sci., Part A-2*, 1970, **8**, 277.
59. T. R. Earnest, Jr. and W. J. MacKnight, *J. Polym. Sci., Polym. Phys. Ed.*, 1978, **16**, 143.
60. T. R. Earnest, Jr., Ph.D. thesis, University of Massachusetts, 1978.
61. R. Longworth and D. J. Vaughan, *Nature (London)*, 1968, **218**, 85.
62. J. D. Ferry, 'Viscoelastic Properties of Polymers', 2nd. edn., Wiley, New York, 1961, chap. 11.
63. N. G. McCrumm, B. E. Read and G. Williams, 'Anelastic and Dielectric Effects in Polymeric Solids', Wiley, New York, 1967.
64. R. W. Rees and D. J. Vaughan, *Polym. Prepr., Am. Chem. Soc., Div. Polym. Chem.*, 1965, **6**, 287.
65. E. P. Otocka and T. K. Kwei, *Macromolecules*, 1968, **1**, 401.
66. E. P. Otocka and T. K. Kwei, *Macromolecules*, 1968, **1**, 244.
67. E. P. Otocka and T. K. Kwei, *Macromolecules*, 1969, **2**, 110.
68. R. Longworth and D. J. Vaughan, *Polym. Prepr., Am. Chem. Soc., Div. Polym. Chem.*, 1968, **9**, 525.
69. W. J. MacKnight, L. W. McKenna and B. E. Read, *J. Appl. Phys.*, 1967, **38**, 4208.
70. W. J. MacKnight, T. Kajiyama and L. McKenna, *Polym. Eng. Sci.*, 1968, **8**, 267.
71. L. W. McKenna, T. Kajiyama and W. J. MacKnight, *Macromolecules*, 1969, **2**, 58.
72. W. J. MacKnight and F. A. Emerson, in 'Dielectric Properties of Polymers', ed. F. E. Karasz. Plenum Press, New York, 1971.
73. B. E. Read, E. A. Carter, T. M. Conner and W. J. MacKnight, *Br. Polym. J.*, 1969, **1**, 123.
74. P. J. Phillips and W. J. MacKnight, *J. Polym. Sci., Part A-2*, 1970, **8**, 727.
75. R. M. Neumann and W. J. MacKnight, *J. Polym. Sci., Polym. Symp.*, 1978, **63**, 281.
76. Y. Huang, G. Cong and W. J. MacKnight, *Macromolecules.*, 1986, **19**, 2267.
77. C. Bailly, D. J. Williams, F. E. Karasz and W. J. MacKnight, *Polymer*, 1987, **28**, 1009.
78. M. Rutkowski and A. Eisenberg, *J. Appl. Polym. Sci.*, 1985, **30**, 3317.
79. C. G. Bazuin and A. Eisenberg, *J. Polym. Sci., Polym. Phys. Ed.*, 1986, **24**, 1021.
80. R. D. Lundberg and H. S. Makowski, *J. Polym. Sci., Polym. Phys. Ed.*, 1980, **18**, 1821.
81. R. D. Lundberg and R. R. Phillips, *J. Polym. Sci., Polym. Phys. Ed.*, 1982, **20**, 1143.
82. R. D. Lundberg, *J. Appl. Polym. Sci.*, 1982, **27**, 4623.
83. J. Niezette, J. Vanderschueren and L. Aras, *J. Polym. Sci., Polym. Phys. Ed.*, 1984, **22**, 1845.
84. J. C. Salamone, C. G. Tsai, Q. Anwaruddin, A. P. Olson, M. Arnold, T. Nagabhusharram, S. Sawan and A. C. Watterson, *J. Polym. Sci., Polym. Chem. Ed.*, 1984, **22**, 2005.
85. C. W. Lantman, W. J. MacKnight, R. D. Lundberg, S. K. Sinha and D. G. Peiffer, *Polm. Prepr., Am. Chem. Soc., Div. Polym. Chem.*, 1986, **27**, 327.
86. M. Hara, A. H. Lee and J. Wu, *Polym. Prepr., Am. Chem. Soc., Div. Polym. Chem.*, 1986, **27**, 177.
87. S. B. Clough, D. Cortelek, T. Nagabhusharram, J. C. Salamone and A. C. Watterson, *Polym. Eng. Sci.*, 1984, **24**, 385.
88. C. W. Lantman, W. J. MacKnight, R. D. Lundberg, D. G. Peiffer and S. K. Sinha, *Polym. Prepr., Am. Chem. Soc., Div. Polym. Chem.*, 1986, **27**, 292.
89. R. C. Portnoy, E. R. Werlein, R. D. Lundberg and D. G. Peiffer, *Proc. Am. Chem. Soc., Div. Polym. Mat. Sci. Eng.*, 1986, **55**, 860.

Subject Index